Wheat
Science and Trade

Wheat Science and Trade

Edited by

Brett F. Carver

A John Wiley & Sons, Inc., Publication

Blackwell Publishing was acquired by John Wiley & Sons in February 2007. Blackwell's publishing program has been merged with Wiley's global Scientific, Technical, and Medical business to form Wiley-Blackwell.

Editorial Office
2121 State Avenue, Ames, Iowa 50014-8300, USA

For details of our global editorial offices, for customer services, and for information about how to apply for permission to reuse the copyright material in this book, please see our website at www.wiley.com/wiley-blackwell.

Library of Congress Cataloguing-in-Publication Data

Carver, Brett Frederick, 1958-
Wheat : science and trade / Brett F. Carver. – 1st ed.
p. cm.
Includes bibliographical references and index.
ISBN-13: 978-0-8138-2024-8 (alk. paper)
ISBN-10: 0-8138-2024-3 (alk. paper)
1. Wheat–Genetics. 2. Wheat–Diseases and pests. 3. Wheat–Breeding.
4. Wheat trade. I. Title.
SB191.W5C355 2009
633.1′1–dc22

2008049844

A catalog record for this book is available from the U.S. Library of Congress.

Set in 10.5 on 12 pt Ehrhardt by SNP Best-set Typesetter Ltd., Hong Kong
Printed and Bound in Singapore by Markono Print Media Pte Ltd

1 2009

Dedication

to
Paul Jackson, Jr., and Henry Jo Von Tungeln,
for their lifetime and relentless dedication
to furthering the science and trade of wheat,
so all might prosper.

Contents

Foreword

"No one can be a statesman who is entirely ignorant of the problems of wheat." This statement from Socrates carries as much impact now, as when it was first uttered over two thousand years ago. Indeed just as the system of government of ancient Greece has become globalized, so too has their staple cereal. At around 680 million tonnes from 224 million hectares in 2008–2009, wheat is the world's greatest source of food calories from the biggest crop area; it is the most widely consumed grain across all its diverse products, and the agricultural commodity most traded internationally (124 million tonnes).

When wheat price more than doubled in the year up to March 2008, alarm bells were ringing for politicians; eight months later it has fallen back to where it was, which although an alarming fall for farmers, marks a return to around the smoothed long-term trend price. This price in real terms is as low as wheat has ever been, resulting from the steady decline in the real price over the last 60 years, a huge bonus for consumers and world development, and a reflection of productivity growth based on science and technology. Reasonable wheat prices are vital for poor consumers; only productivity gains can deliver this and at the same time allow wheat farming to remain economically viable. Whether researchers can continue to achieve these gains—while doing so sustainably, including avoiding expansion of the world's wheat area, and delivering better wheat products—is the subject of this book.

While cost reductions through input efficiencies per ton of grain are important, wheat productivity gains over the last 60 years largely reflect yield increase: world wheat yield has risen from 1.0 to 3.0 t ha^{-1} in this period, and continues to advance at about 1% per annum. Wheat occupies a very broad range of agroecologies, from favorable irrigated and high-rainfall environments to unfavorable, semiarid environments, yet generally similar relative progress, beginning in the developed world, has now been seen under almost all these circumstances. Underpinning this is a huge international research effort, spanning all aspects of genetic improvement and crop management.

At the beginning, this was based entirely in the public sector: the last 30 years has seen a significant shift to private-sector research, not just in agricultural chemicals and machinery, but also in variety delivery. For many obvious reasons yield must continue to grow, and to do so at better than the current 1%. While rural extension and yield gap closing are important in achieving this, vital will be wheat research investments at an intensity (research costs per value of product) at least as high as in the past, along with gains in research efficiency. Close communication and sharing of ideas and materials among wheat researchers, as is so evident in the chapters of this book, should engender efficiency gains, since the challenges are daunting for individual research teams and are often common around the world; achieving efficient private–public complementation remains another important goal.

Also evident is the fact that environmental issues and the industrial and nutritional quality of grain have become increasingly important for researchers and farmers alike. In the former area, progress has been made through yield increase itself, and through the gradual adoption of conservation tillage and more efficient input delivery to the crop. It is not helped by misguided notions of low-input farming nor, at the other extreme, input subsidies and excessive price support. Driven by intense market competition, wheat industrial quality in major wheat exporting nations has probably been improved in the last 30 years as significantly as has yield itself. In developing countries, nutritional quality could assume greater importance as consumers have less opportunity to favorably supplement high-wheat diets. Many quality improvements are foreshadowed herein.

It is not clear how much yield could increase with better pest, nematode, disease, and weed control because modern agriculture invests heavily to keep these losses low (probably <10%); but losses could be higher in neglected parts of the developing world. Either way, the whole world needs less costly control methods, especially easy-to-incorporate and more durable host-plant resistance, which is probably the most fruitful field for transgene deployment. Although modern wheat fields are often deplored because of their apparent uniformity, the researcher knows that it is the underlying hidden genetic diversity, in the face of biotic and abiotic stresses, and the diversity of the management and cropping system, which is the diversity that really counts. At the same time, the researcher must remain always vigilant against evolving biotic stress organisms. Over 25% of the book is devoted to biotic stresses of wheat.

Wheat is currently the premier food grain of the world, because of its versatility in production and use, and because of the huge body of research knowledge underpinning modern wheat production and marketing which can be seen here. Wheat is the grain crop *par excellence* of the vast temperate semiarid and subhumid regions of the world, regions which will grow in importance as water and land competition elsewhere intensifies. Because of these advantages, wheat should remain the premier food grain through to peak world population and beyond. The world needs this to be the case, but continuing research across all of the diverse fields affecting wheat productivity and utilization that are described herein will be essential.

R.A. (Tony) Fischer, CSIRO Plant Industry, Canberra, Australia
December 2008

Preface

Wheat is the cosmopolite of edible plants. It can be used in worldly ways and grown in worldly places. Its sphere of influence is global. Scientists marvel at its genetic complexity, but its complexity is perfectly fitting for such a versatile and planetary crop plant.

Much of what the world's agricultural society produces in food derives from wheat and other cereal crops. Hence to no surprise, a critical and nutritious part of the human diet comes from wheat—indeed no other grain crop can produce as many types of food. Calling it a staple may be a bland understatement, when wheat can dominate the ingredient list of appetizer to dessert, with versatility beyond what the mill can generate. Wheat farmers do not simply produce a wheat crop; they produce a food ingredient, and it is this distinction on which *Wheat: Science and Trade* is based.

In a specialized and highly focused era of scientific discovery, our literary base can easily become highly fragmented. This tendency applies no less to the literature for wheat. Whether the subject addresses soil management, epidemiology of a various array of pathogens, or genetic approaches to improving wheat productivity, one can find a significant piece of literature devoted entirely to that subject area within the past 30 years. This book represents a concerted attempt to swing the literary pendulum back to center. This is a book about *wheat* and the products derived from it—not strictly about bread wheat, not about wheat breeding, and not only about wheat quality, but all of that, and more. Consider it a drill-to-mill treatise of the current knowledge base and futuristic visions for wheat to flourish in a global environment and in a world market.

Wheat: Science and Trade was designed as a hub for directing students, practitioners, and scientists into four knowledge centers, or sections. The primary intended audience includes a wide spectrum of agricultural scientists working in the general research areas of crop science and soil science, and more specifically in weed science, plant pathology, entomology, genetics, cultivar development, physiology, taxonomy, cereal chemistry, food processing, and agribusiness. Another principal use of this book should include the classroom for advanced undergraduate students and graduate students studying crop production and utilization. Vocational agricultural teachers and practitioners, and the clientele they serve, will also find this book to be a critical resource. No matter the audience, the reader will likely reach equilibrium with the information provided herein and that which is extensively cited in a multitude of references.

The four sections approach wheat science and trade from the making of a wheat plant, to the making of a wheat crop, to the making of a wheat cultivar, and finally to the making of a wheat industry. The various sections provide the following: (i) fresh perspectives on classic tenets that define the evolutionary and phenological development of wheat (Section I, Chapters 1–3); (ii) a comprehensive view of some of the primary pathogens, pests, and abiotic stress factors that must be overcome to ensure a viable and marketable product, and the practices that can be adopted to maximize profitability (Section II, Chapters 4–12); (iii) the genetic components that define wheat improvement and cultivar development, from dissection of a myriad of traits critical to the total wheat industry to the development of novel genetic resources critical to its continued world-wide production (Section III, Chapters 13–18); and (iv) a contemporary look at the functional properties that allow wheat products to appear from one end of the grocery store to the other, and the forces that drive wheat from family farm to river barge (Section IV, Chapters 19–23).

Chapters 20 to 22 provide a three-dimensional perspective on what determines wheat quality and how wheat quality can be manipulated to benefit humankind, even beyond the natural power of wheat's flour. Each of these chapters begins from a common base to which most might consider the one and only product of wheat—milled flour. Each chapter extends from that base to take the subject of wheat quality to very different levels, whether to describe the techniques used to predict quality from an end-use perspective, or to reveal the basic biochemical components which may lay the foundation for end use, or to consider areas which newly define wheat quality. Finally, we are reminded that certain uses of wheat demand different pricing structures, but more determines wheat price than the wheat itself.

Excellence reflected in this book emanates from the talented panel of authoritative contributors with whom I have had the honor to work. My sincerest appreciation is extended to all of them. I trust you will be equally rewarded by their insight as I.

Brett F. Carver, PhD
Editor, Regents Professor
Department of Plant and Soil Sciences
Oklahoma State University
Stillwater, OK
September 2008

Acknowledgements

I must first give tribute to the honor of simply having the opportunity to study the science and trade of wheat in one of the most intensive areas of wheat production, wheat research, and wheat marketing in the world—the US Great Plains. Twenty-five years of participation in the wheat improvement community of this region laid a firm foundation and provided a fertile environment to edit a text on the very subject of wheat science and trade. Initial formulation of some of its content leaned on the expertise of a few persons to whom I am most grateful: Mark Hodges, John Oades, Carl Griffey, and Brad Seabourn.

The authors and I acknowledge several colleagues who reviewed substantial parts of the manuscript draft: Bob Hunger, Gary Muehlbauer, Phil Bregitzer, Liuling Yan, Brad Seabourn, Perry Gustafson, Elizabeth Ross, and Surjani Uthayakumaran. Perry Gustafson and coauthors of Chapter 1 extend their thanks to Kathleen Ross for her careful editing of the manuscript and the reference list. Greg Rebetzke and coauthors (Chapter 11) acknowledge the use of data and figures kindly provided by Drs. David Bonnett, Marc Ellis, and Xavier Sirault. Stephen Baenziger and Ron DePauw (Chapter 13) wish to extend their special thanks to C. Davidson and F.R. Clarke for their comments and review, to Aidan Beaubier for literature acquisitions and grammar critique, and to D. Schott, S. Inwood, and R. Lamberts for their assistance with figures and plates. Katrien Devos and coauthors of Chapter 15 wish to thank colleagues Jeff Bennetzen, Alicia Massa, Etienne Paux, Pierre Sourdille, Pavla Suchánková, Wolfgang Spielmeyer, Jan Šafář, Hana Šimková, and Xiangyang Xu for sharing unpublished data.

The staff at Wiley-Blackwell could not have been more accommodating and understanding, and for that I extend my deepest appreciation to Shelby Hayes Allen, Erica Judisch, and commissioning editor Justin Jeffryes.

I am grateful to my academic home, Oklahoma State University, for allowing the time and resources to focus intensively on editorial responsibilities. Certain key individuals are recognized for piloting the Oklahoma State University wheat breeding program during my repeated grounding, including a very capable wheat breeder for 25-plus years, Wayne Whitmore, and three dedicated graduate students, Rima Thapa, Jana Morris, and Shuwen Wang.

My thanks go to Jessica Evans, who as an undergraduate student at Oklahoma State University devoted countless hours perusing the reference lists and each text citation for accuracy and synchrony. Travis Collins, another OSU undergraduate student, provided expert support upon Jessica's graduation. Melanie Bayles, a cotton geneticist, supplied much appreciated expertise in assembly of the index. Thanks also go to Debbie Porter and Vickie Brake, who provided computer support in the final organization and production of figures, plates, and tables. I am especially grateful for the extended patience and understanding of my wife, Terri, who eloquently coined a new term with an old word, *LAB*—life after book.

Brett F. Carver

Contributors

Kim B. Anderson, PhD
Professor and Crop Marketing Specialist
Charles A. Breedlove Professor in Agribusiness
Department of Agricultural Economics
Oklahoma State University
Stillwater, OK
USA

Robert Asenstorfer, PhD
Research Fellow
University of Adelaide
School of Agriculture Food and Wine
Waite Campus
Glen Osmond
Australia

David Backhouse, PhD
Senior Lecturer in Plant Pathology
University of New England
Armidale, New South Wales
Australia

P. Stephen Baenziger
Eugene W. Price Distinguished Professor
Department of Agronomy and Horticulture
University of Nebraska
Lincoln, NE
USA

Ian Batey, PhD, MRACI
Honorary Research Fellow
Food Science Australia and Wheat CRC
North Ryde (Sydney)
Australia

Arthur D. Bettge
USDA-ARS
Western Wheat Quality Lab
Pullman, WA
USA

Robert E. Blackshaw, PhD
Weed Scientist
Agriculture and Agri-Food Canada Research Center
Lethbridge, Alberta
Canada

Ann E. Blechl, PhD
Research Geneticist
USDA-ARS
Western Regional Research Center
Albany, CA
USA

B. Wade Brorsen, PhD
Regents Professor and Jean & Patsy Neustadt Chair
Department of Agricultural Economics
Oklahoma State University
Stillwater, OK
USA

Scott C. Chapman, PhD
Crop Adaptation Scientist
CSIRO Plant Industry
Brisbane
Australia

Xianming Chen, PhD
Research Plant Pathologist
USDA-ARS
Wheat Genetics, Quality, Physiology and Disease Research Unit
Pullman, WA
USA

Anthony G. Condon, PhD
Crop Physiologist
CSIRO Plant Industry
Canberra
Australia

Geoffrey Cornish, BSc, Grad Dip Teaching, Assoc Dip Appl Chem, MRACI
Leader Wheat Quality Research
Grain Quality Research Laboratory
South Australian Research and Development Institute
Adelaide
Australia

Li Day, PhD
Senior Research Scientist
Food Science Australia
Werribee
Australia

Ronald M. DePauw, CM, SOM, FAIC, FCSA, BA, Msc, PhD
Senior Principal Wheat Breeder
SemiArid Prairie Agricultural Research Centre
Agriculture and Agri-Food Canada
Swift Current, Saskatchewan
Canada

Katrien M. Devos, PhD
Professor
Department of Crop and Soil Sciences, and Department of Plant Biology
University of Georgia
Athens, GA
USA

Jaroslav Doležel, PhD
Principal Investigator and Research Group Leader
Laboratory of Molecular Cytogenetics and Cytometry
Institute of Experimental Botany
Olomouc
Czech Republic

Jeffrey T. Edwards, PhD
Small Grains Extension Specialist
Oklahoma State University
Stillwater, OK
USA

Catherine Feuillet, PhD
Research Director
Structure, Function and Evolution of the Wheat Genomes Laboratory
INRA—Genetics, Diversity and Ecophysiology of Cereals Unit
Clermont-Ferrand
France

Mark E. Fowler
Director of Technical Services
International Grains Program
Kansas State University
Manhattan, KS
USA

Gurjeet S. Gill, PhD
Associate Professor
School of Agriculture, Food and Wine
University of Adelaide
South Australia
Australia

Robert A. Graybosch, PhD
Research Geneticist
USDA-ARS-GFBU
University of Nebraska
Lincoln, NE
USA

Perry Gustafson, PhD
Research Geneticist
USDA-ARS
University of Missouri
Columbia, MO
USA

Marion O. Harris, PhD
Department of Entomology
North Dakota State University
Fargo, ND
USA

Louis S. Hesler, PhD
Research Entomologist
North Central Agricultural Research Laboratory
USDA-ARS
Brookings, SD
USA

Gavin Humphreys, PhD
Research Scientist
Agriculture and Agri-Food Canada
Cereal Research Centre
Winnipeg
Canada

Yue Jin, PhD
Research Plant Pathologist
USDA-ARS
Cereal Disease Laboratory
St. Paul, MN
USA

Huw D. Jones, PhD, FIBiol
Principal Investigator and Research Group Leader
Cereal Transformation Laboratory
Plant Science Department
Rothamsted Research Harpenden
United Kingdom

James Kolmer, PhD
Research Plant Pathologist
USDA-ARS
Cereal Disease Laboratory
St. Paul, MN
USA

Rui Hai Liu, MD, PhD
Associate Professor
Department of Food Science
Cornell University
Ithaca, NY
USA

Philippe Lucas
Directeur de Recherche
Institut National de la Recherche Agronomique
Le Rheu
France

Drew J. Lyon, PhD
Dryland Cropping Systems Specialist
University of Nebraska-Lincoln
Panhandle Research and Extension Center
Scottsbluff, NE
USA

XueFeng Ma, PhD
Senior Scientist
Ceres, Inc.
Thousand Oaks, CA
USA

Ronald L. Madl, PhD
Director, Bioprocessing and Industrial Value Added Program
Co-director, Center for Sustainable Energy
Grain Science Department, Kansas State University
Manhattan, KS
USA

Daryl Mares, PhD
Senior Research Fellow
University of Adelaide
School of Agriculture Food and Wine
Waite Campus
Glen Osmond
Australia

David Marshall, PhD
Research Leader and Professor
USDA-ARS
Department of Plant Pathology
North Carolina State University
Raleigh, NC
USA

Kendall L. McFall
Senior Vice President and
Chief Operating Officer
Engrain, LLC
Adjunct Instructor
Department of Grain Science
Kansas State University
Manhattan, Kansas
USA

C. Lynne McIntyre, PhD
Research Geneticist
CSIRO Plant Industry
Brisbane
Australia

Gregory S. McMaster, PhD
USDA-ARS, Agricultural Systems Research Unit
Fort Collins, CO
USA

Kolumbina Mrva, PhD
Senior Research Fellow
University of Adelaide
School of Agriculture Food and Wine
Waite Campus
Glen Osmond
Australia

Eviatar Nevo, PhD
Professor of Evolutionary Biology
Director International Graduate Center of Evolution
Institute of Evolution
University of Haifa
Mount Carmel, Haifa
Israel

Julie M. Nicol, PhD
Senior Soil Borne Wheat Pathologist
ICARDA-CIMMYT Wheat Improvement Program
CIMMYT Global Wheat Program
Ankara
Turkey

Ivan Ortiz-Monasterio, PhD
Agronomist and Wheat Harvest Coordinator
CIMMYT Global Wheat Program
Mexico, D.F.
Mexico

Timothy C. Paulitz, PhD
Research Plant Pathologist
USDA–ARS
Root Disease and Biological Control Research Unit
Washington State University
Pullman, WA
USA

David R. Porter, PhD
Professor and Head
Plant and Soil Sciences
Oklahoma State University
Stillwater, OK
USA

Gary J. Puterka, PhD
USDA–ARS
Plant Science Research Laboratory
Stillwater, OK
USA

Olga Raskina, PhD
Senior Scientist
Institute of Evolution
University of Haifa
Mount Carmel, Haifa
Israel

William R. Raun, PhD
Regents Professor
Plant and Soil Sciences
Oklahoma State University
Stillwater, OK
USA

Greg J. Rebetzke, PhD
Research Geneticist
CSIRO Plant Industry
Canberra
Australia

Richard A. Richards, PhD
Research Program Leader
CSIRO Plant Industry
Canberra
Australia

Andrew S. Ross, PhD
Associate Professor
Department of Crop and Soil Science
Oregon State University
Corvallis, OR
USA

Jackie C. Rudd, PhD
Texas AgriLife Research
Texas A&M System
Amarillo, TX
USA

Yong-Cheng Shi, PhD
Professor
Department of Grain Science and Industry
Kansas State University
Manhattan, KS
USA

Richard W. Smiley, PhD
Professor of Plant Pathology
Oregon State University
Pendleton, OR
USA

John B. Solie, PhD
Biosystems and Agricultural Engineering
Oklahoma State University
Stillwater, OK
USA

Daryl J. Somers, PhD
Research Chair
Molecular Breeding and Biotechnology
Vineland Research and Innovation Centre
Vineland Station, Ontario
Canada

Richard M. Trethowan, PhD
Professor of Plant Breeding
University of Sydney
Plant Breeding Institute
Camden, New South Wales
Australia

Maarten van Ginkel, PhD, Ir
Deputy Director General—Research
ICARDA
Aleppo
Syria

Anthony F. van Herwaarden, PhD
Crop Agronomist
CSIRO Plant Industry
Brisbane
Australia

Donghai Wang, PhD
Associate Professor
Department of Biological and Agricultural Engineering
Kansas State University
Manhattan, KS
USA

Michelle Watt, PhD
Research Scientist
CSIRO Plant Industry
Canberra
Australia

Colin Wrigley, MSc, PhD, FRACI
Honorary Research Fellow
Food Science Australia and Wheat CRC
North Ryde (Sydney)
Australia

Xiaorong Wu, PhD
Research Associate
Department of Biological and Agricultural Engineering
Kansas State University
Manhattan, KS
USA

Liuling Yan, PhD
Assistant Professor
Plant and Soil Sciences
Oklahoma State University
Stillwater, OK
USA

Wheat
Science and Trade

Section I
Making of a Wheat Plant

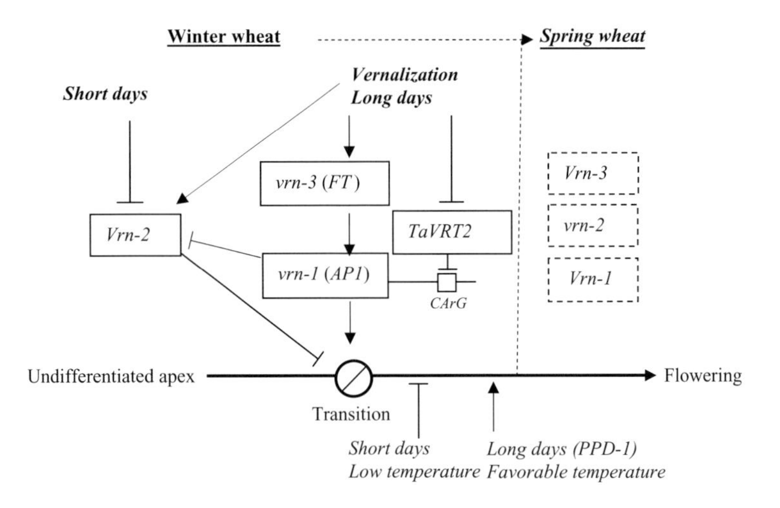

"History . . . celebrates the battlefields whereon we meet our death, but scorns to speak of the plowed fields whereby we thrive; it knows the names of the king's bastards, but cannot tell us the origin of wheat. That is the way of human folly."

Jean-Henri Fabre, 1823–1915

Chapter 1
Wheat Evolution, Domestication, and Improvement

Perry Gustafson, Olga Raskina, XueFeng Ma, and Eviatar Nevo

SUMMARY

(1) Wheat is the world's largest and most important food crop for direct human consumption; therefore, continued wheat improvement is paramount for feeding an ever-increasing human population.

(2) Wheat improvement is tightly associated with the characterization and understanding of wheat evolution and the genetic diversification of various wheat species and relatives. The evolution of the genus *Triticum* mainly resulted from inter- and intraspecific hybridization, polyploidization, and recurrent formation of wheat and its wild relatives.

(3) An understanding of the process of species domestication and genome evolution that has occurred and is still occurring within the primary, secondary, and tertiary gene pools is critical for further exploring the improvement of wheat production.

(4) Also critical is to evaluate the relative importance of the evolutionary processes, such as the pivotal genome concept and cell cycle differences, driving and shaping the structural rearrangements and genomic changes occurring within the genomes of polyploid wheat.

INTRODUCTION

Cereals, including wheat (*Triticum* spp.), rice (*Oryza sativa* L.), and maize (*Zea mays* L.), are the major food crops for all humans and are the principal resources that have led to the emergence of human civilization as we know it today. Domestication of cereals during the past 10 millennia is one of the most dramatic demonstrations involving humans' manipulation of the evolutionary processes of speciation, natural selection, and adaptation. Plant domestication revolutionized human cultural evolution and is primarily responsible for the advances that have occurred in our civilization. A post-Pleistocene global temperature increase following the ice age may have induced the expansion of economically important thermophilous plants, which in turn promoted the change from a world of hunter–gatherer societies to complex foraging and plant-cultivating societies. The shift from foraging to steady production definitely led to the occurrence of an incipient agriculture in many parts of the world and to a decline in genetic diversity of the world's crops, which has been accelerated by domestication and the recent breeding of modern cultivars.

The evolution of domestication should be considered in two main contexts: (i) the evolution of new species of crop plants by humans through strong artificial selection (see Darwin 1859); (ii) the evolution of human civilizations, and the current consequences of population explosion and increasing world hunger in developing countries. Both contexts provide fascinating insight into the evolutionary process.

WHEAT DOMESTICATION AND HUMAN CIVILIZATION

The earliest signs of crop domestication appeared 10,000–12,000 years ago in the Fertile Crescent of the Near East, in Central America, and in southern China, involving different crops and independent cradles of domestication. Cereal domestication was founded, in the Fertile Crescent of the Near East, on crop reliability, yield, and suitability for storage. Recent botanical, genetic, and archaeological evidence has pointed to a small core area within the Fertile Crescent—near the upper reaches of the Tigris and Euphrates rivers, in present-day southeastern Turkey–northern Syria—as the cradle of cereal agriculture (Lev-Yadun et al., 2000). Further evidence is needed to clarify when and where wheat domestication and agriculture, as driving forces of modern civilization, originated. Was it spread in time or space, or both, in the Fertile Crescent?

The genetic changes required for wheat domestication to occur were relatively straightforward and rapid, including selection for nonshattering, free-threshing, nonbrittle rachis and hull-less spike characteristics and for higher yield. In all cereals, the gene complexes for ease of harvest, yield, and suitability for short- and long-term storage have been critical for domestication. The domestication of cereals was essential for human populations' change to an agriculture-based society. Evolution of any crop species from their wild progenitors to full domestication, and the emergence of agricultural ecologies from pre-agricultural ones, clearly established human movement from hunter–gatherer societies to sedentariness, urbanization, culture, and an unprecedented population explosion (Harlan 1975, p. 295).

Considerable progress has been achieved in characterizing the wild ancestry of Old World crops, including cereals. The wild progenitors of most of our cultivated plants have been satisfactorily identified by comparative morphology and genetic analysis. The distribution and ecological ranges of wild relatives have also been established. Furthermore, comparisons between wild types and their cultivated counterparts have revealed many of the evolutionary changes that resulted in domestication. Research has enabled us to assess the relative importance of the evolutionary forces driving wheat evolution—hybridization, migration, drift, and natural selection—interacting in generating the contemporary wheat genotype. Studies suggested that, besides polyploid hybridization, natural selection played a large role and oriented wheat evolution primarily through the mechanisms of diversifying and balancing selection regimes (Nevo et al., 2002).

Wheat has become the world's largest and most important food crop for direct human consumption, with an annual harvest of more than 620 million tonnes produced in over 40 countries for more than 35% of the global population (Williams 1993). The US produces approximately 55 to 60 million tonnes per year and supplies about 40% of the world's exports. Wheat is currently grown from 67°N, in Norway, Finland, and Russia, to 45°S, in Argentina and Chile. The world's main wheat-producing regions are in temperate and southern Russia, the central plains of the US, southern Canada, the Mediterranean Basin, northern China, India, Argentina, and Australia. Wheat makes up 29%–30% of the world's total cereal production and is humans' most important source of protein. As a crop for direct human consumption, only rice comes close to matching wheat production. As a food grain, wheat is the major dietary component throughout the world; in 1996 it served as the source of over 55% of the world's carbohydrates (http://www.fao.org).

Wheat cultivars are superior to most other cereals in their nutritive value (Zohary and Hopf 2000). Besides the grain containing from 60% to 80% starch, it also contains from 7% to 22% storage protein, which in elite wild genotypes can reach as much as 17% to 28% (Avivi 1978, 1979; Avivi et al., 1983; Grama et al., 1983; Nevo et al., 1986; Levy and Feldman 1987). The gluten proteins in the seed endosperm impart unique bread-baking qualities to wheat dough, which has made wheat the staple food in the ancient and modern world for billions of people. Only minor amounts of wheat are occasionally used as animal feed, with the amount being highly dependent on wheat prices compared with other feed grains. A very small portion of the world's wheat and wheat flour is used for industrial purposes such as starch and gluten production (Morrison 1988).

Global efforts to increase wheat production and to keep up with population growth and rising demand have been relatively successful in maintaining a steady increase in wheat yield, representing roughly a threefold increase over production levels of the 1960s. It should be noted that, despite dramatic increases in global wheat production, in 2003 more than 800 million people in the world suffered daily from severe undernourishment and hunger (http://www.fao.org). Protein deficiency is one of the most serious problems and threatens to become a real nutritional disaster in the near future, primarily in Asia and Africa, where about 80% of the human diet is protein supplied by plants. The urgent need to increase high-quality protein sources is exacerbated by major problems affecting cultivated crops, including the cereals, with respect to the reduction in genetic diversity (Plucknett et al., 1983, 1987).

In the future, the nutritional composition of the world wheat supply will become even more critical as world demand for wheat continues to grow and world wheat stocks continue to decrease. World population, which currently stands at well over 6 billion, was projected in 2001 to reach 8.3 billion by the year 2030 and 9.3 billion by 2050 (http://www.fao.org). Income growth and urbanization, which are shifting consumer preference away from rice, coarse grains, and tubers to more wheat-based food products and meat, are also expected to continue to increase in many developed and developing countries. World demand is projected to require approximately a 66% increase in agricultural production by 2040. In addition, our ability to bring more land into wheat cultivation is rapidly diminishing due to population growth, environmental pressures, and the increasingly limited availability of arable land (Young 1999). The need for future improvement in wheat production will clearly coincide with a loss of flexibility and availability of traditional resources.

The success of wheat improvement programs to meet future demands will require complementing the traditional breeding approaches with innovative, nontraditional methodologies that will enhance genetic variation in wheat. One of many approaches to improving wheat production will be the manipulation of secondary and tertiary gene pools for new sources of biotic and abiotic stress tolerance. A key to the successful manipulation of the primary, secondary, and tertiary gene pools is to fully understand the evolution of the cultivated wheat species. Unfortunately, some loss of genetic diversity involving most of the world's crops, including wheat, has accelerated in recent decades. The dynamic conservation of wheat germplasm and wild wheat relatives offers one of the best hopes for sustained wheat improvement (Nevo 1998). However, it is clear that conservation of germplasm is not the only answer. To achieve a more efficient and comprehensive utilization of the gene pools of wheat and wild wheat relatives, it is critical that we learn how to predict, screen, manipulate, maintain, and properly evaluate genetic diversity and resources (Nevo 2001). After all, plant breeding is basically an accelerated manipulation of natural evolution. Once we understand the evolutionary processes involved in the formation and stabilization of wheat, we can better design wheat improvement programs that will enable a more efficient restructuring of gene complexes within and between wheat, wheat-related species, and genera to capitalize on the value-added traits that may be economically important for wheat improvement.

WHEAT CULTIVATION

Until the late 19th century, all cultivated wheat existed as highly heterogeneous landraces. Wheat cultivars were morphologically uniform mixtures of inbred lines and hybrid segregates, the products of low levels of random crossing within a landrace. Any artificial selection was primarily for increased yield, larger seed size, better flour quality, and adaptation to a wider range of climatic and farming regimes (Feldman et al., 1995). Many landraces still exist today, in fields in many regions around the world and in germplasm collections. Over the past century of modern breeding, attempts to produce cultivars that meet the advanced agriculture demands of an ever-increasing population has resulted in the landraces being almost wholly displaced by genetically uniform cultivars. The result of modern agriculture has been a marked narrowing of the genetic base in probably all advanced agricultures (Harlan 1975, 1976, 1992). While wheat yields have kept up with population demands in advanced agricultures (e.g., Avery 1985), genetic homogeneity has also dramatically increased due to modern agricultural practices (Frankel and Bennett 1970; Frankel and Hawkes 1975; Harlan 1975, 1976, 1992; Frankel and Soulé 1981; Nevo et al., 1982; Nevo 1983, 1986, 1989, 1995, 1998, 2001; Plucknett et al., 1983, 1987; Lupton 1987; Nevo and Beiles 1989). Consequently, the genetic base of many cultivated crops, including wheat, has been narrowed and placed under serious risk (Frankel and Soulé 1981; Plucknett et al., 1987). A global network of gene banks has been established to provide plant breeders with the genetic resources for maintaining germplasm collections and for developing more resistant and tolerant crops that will improve production (Lupton 1987; Plucknett et al., 1987; Brown et al., 1989, 1990).

Dynamic *in situ* conservation of landraces and wild relatives, the best hope for improving cultivated plants (Feldman and Sears 1981), is being actively discussed as an optimal conservation strategy (Hawkes 1991; Heyn and Waldman 1992; Valdes et al., 1997; Maxted et al., 1997; Nevo 1998). Just as important as the conservation of diverse germplasm is the achievement of a more efficient and comprehensive utilization of conserved wild gene pools. It is essential to be able to efficiently predict, screen, and evaluate promising genetic diversity and resources, thereby optimizing crop improvement (Nevo 1983, 1989, 1992, 1995, 2001; Peng et al., 2000a,b,c). The analysis of genetic diversity across the geographic range, at both the macro- and microscale, will unravel patterns and forces driving wheat genome evolution and lay open the full potential of its genetic resources for utilization.

ORIGIN, DOMESTICATION, AND EVOLUTION OF WHEAT

Modern wheat cultivars belong primarily to two polyploid species: hexaploid bread wheat [*T. aestivum* ($2n = 6x = 42$ chromosomes)] and tetraploid hard or durum-type wheat [*T. turgidum* L. (Thell.) ($2n = 4x = 28$)] used for macaroni and low-rising bread. The cultivated diploid species *T. monococcum* L. einkorn wheat ($2n = 2x = 14$) is a relic and is only found in some mountainous Mediterranean regions. Wheat is predominantly self-pollinated; hence, genetic diversity is represented in the wild by numerous clones, in vast national and international germplasm collections, and in current cultivation by some 25,000 different cultivars. Wild and primitive wheat forms have hulled grains and brittle ears that disarticulate at maturity into individual spikelets, with each spikelet having a wedge-shaped rachis internode at its base, and an arrowlike device that inserts the seed into the ground (Zohary 1969). By contrast, all cultivated wheat forms have nonbrittle ears that stay intact after maturation, thus depending on humans for reaping, threshing, and sowing. The nonbrittleness and nakedness of cultivated wheat is controlled by the Q locus (Luo et al., 2000), located on chromosome 5 of genome A, which may have arisen from the *q* gene of the hulled varieties by a series of mutations (Feldman et al., 1995).

Polyploidy, a form of plant evolution

The evolution of the genus *Triticum* serves as one of the best models of polyploidy, one of the most common forms of plant evolution (Elder and Turner 1995; Soltis and Soltis 1999). The gradual shift to a steady-production-based agriculture has been the main driving force behind the domestication of wheat. The evolution of domestication can be considered as the evolution of new crop species by natural and artificial selection, and the evolution of human civilization as we know it. Unfortunately, this has resulted in a massive population explosion and greatly increased world hunger in many regions of the world.

From a practical perspective, a large number of simply inherited dominant or recessive genes conferring different types of resistances are still available in wheat germplasm and wild relatives of wheat. A solid knowledge of the mechanisms of polyploidization will help scientists in manipulating gene pools to improve cultivated wheat. Scientists and historians have long been searching for an explanation of the evolution and domestication of the various forms of cultivated wheat (diploid, tetraploid, and hexaploid; *T. monococcum, T. turgidum, and T. aestivum*, respectively). The origin of polyploid wheat is complex because its evolution, since the various grass species diverged, has involved a long-established massive intervention of human and environmental selection pressures. The evolution within the entire Triticeae tribe included early widespread intra- and intergenome hybridization followed by introgression, gene flow, gene fixation, and rapid diversification within and among the ancestral diploid and polyploid species (Kellogg et al., 1996). Unequal rates of evolution, parallel evolution, DNA sequence deletion and/or amplification, and silencing during the evolution of present-day wheat species has been postulated to explain the complexity in phylogenetic relationships (McIntyre 1988; Appels et al., 1989; Feldman 2001).

Evolutionary studies involving various plant taxa have demonstrated that not only wheat, but also many polyploids, evolved from different progenitor populations. Independently formed polyploids most likely came in contact to hybridize with each other, thus resulting in ever-expanding primary and secondary germplasm pools (reviewed by Soltis and Soltis 1999). The formation of many allopolyploids was also accompanied by considerable genomic changes and structural reorganization within all or some of the parental genomes, including rapid nonrandom coded and noncoded sequence elimination, genic silencing, intergenomic colonization by repeats and transposable elements, intergenomic homogenization of divergent DNA sequences, DNA methylation changes, and other genomic modifications (Ozkan et al., 2001; Liu and Wendel 2002; Ma and Gustafson 2005). These genomic changes have been well demonstrated in the polyploids of the Triticeae tribe (Feldman et al., 1997; Kashkush et al., 2002; Han et al., 2003; Ma et al., 2004; Ma and Gustafson 2005, 2006). Such genomic changes, coupled with the likely repeated occurrence of polyploid formation, also contribute to the conflicting determinations of phylogenetic relationships and origins of many species, including wheat.

The origin, evolution, and domestication of cereals were among the major events shaping the development and expansion of human culture and will continue to shape the world in which we live. The domestication of cereals, which occurred approximately 10,000 years ago, was critical in laying the groundwork for the Neolithic revolution that transformed humanity to more centralized, sedentary farming societies (for a complete discussion see Kimber and Feldman 1987; also see especially Feldman 2001). There is no question that the various grass species (approximately 10,000 species), growing in every habitat in the world, and our understanding of the evolution of grasses are critical to developing the potential for grasses to feed the world's ever-increasing population.

Polyploidy has been defined as the presence of more than one genome per cell and is probably the most common mode of speciation in plants (Stebbins 1950; Wendel 2000). Polyploids are classified into autopolyploids, which are formed from intraspecific chromosome doubling, and allopolyploids, which are the result of the interspecific or intergeneric hybridization of two or more genomes from differentiated species (Stebbins 1947). Polyploidy is one of the most important evolutionary events leading to a massive increase of genetic diversity, thus allowing species to adapt to varying environments. The most important and best-characterized group of allopolyploids, from an agricultural point of view, is the wheat genus (Kimber and Sears 1987; Feldman 2001). The evolutionary development of the various cultivated wheat species comprises several converging and diverging polyploid events involving several *Triticum* and *Aegilops* species from the Triticeae tribe.

It has been estimated that the Triticeae tribe began diverging from its progenitor approximately 35 million years ago (MYA) and that the *Triticum* group separated out about 11 MYA. The formation of the various polyploid wheat species within the *Triticum* genus began approximately 10,000 years ago. Since the early 1900s it has been known that the wheat species and the entire Triticeae tribe have a basic chromosome number of $n = 1x = 7$. Cultivated wheat consists of diploid (einkorn; $2n = 2x = 14$, AA), tetraploid (emmer, durum, rivet, Polish, and Persian; $2n = 4x = 28$, BBAA), and hexaploid (spelt, bread, club, and Indian shot; $2n = 6x = 42$, BBAADD) species. The various diploid genomes of the Triticeae tribe appear to be highly conserved in gene order along the seven pairs of chromosomes (Gale and Devos 1998). The chromosomes (1 through 7) in the various diploid genomes (B, A, and D) are considered to be evolutionarily related, that is, homoeologous in nature. When combined in the same nucleus, homoeologues can be induced to pair with each other. The importance of this fact in controlling our ability to make interspecific crosses and manipulate genes from one species to another will be considered elsewhere in this and other chapters.

Origin of the A genome

It is apparent that the key to understanding the evolution of wheat involves an elucidation of the evolution of the tetraploid wheat species. Early cytogenetic studies led to the conclusion that the A genome of the tetraploid species, *T. timopheevi* and *T. turgidum*, was contributed by *T. monococcum* (Sax 1922; Kihara 1924; Lilienfeld and Kihara 1934). However, it became apparent that diploid einkorn wheat actually comprised two biological species, *T. monococcum* and *T. urartu* Tum. Ex Gand. Chapman et al. (1976) determined that the A genome originated from *T. urartu*. Konarev et al. (1979) concluded, from studies of the immunological properties of seed-storage proteins, that the A genome in *T. turgidum* was contributed by *T. urartu*, and the A genome of *T. timopheevi* (Zhuk.) Zhuk. (GGAA) was contributed by *T. monococcum*. However, Nishikawa et al. (1994) suggested that the A genomes in both diploid species were contributed by *T. urartu*.

Clearly, the diploid component of the *Triticum* genus is composed of two defined species, *T. urartu* and *T. monococcum*. *Triticum monococcum* is the only cultivated diploid wheat species and was first found in Greece by Boissier (1884). Both *T. urartu* and *T. monococcum* have been identified in natural habitats ranging from southwestern Iran, northern Iraq, Transcaucasia, eastern Lebanon, southeastern Turkey, western Syria, and beyond into neighboring Mediterranean areas (Kimber and Feldman 1987). The sterility of their hybrids (Johnson and Dhaliwal 1976, 1978) confirms that they are valid biological species. It has been established that *T. urartu* contains approximately 4.93 pg DNA (http://data.kew.org/cvalues/introduction.html) and is the donor of the A genome to all polyploid wheat species. Dvořák et al. (1988) showed that variation in A-genome repeated nucleotide sequences, present in both tetraploid wheat species, was more related to the A genome of *T. urartu* than to the A genome of *T. monococcum*.

Origin of the B genome

Both the B and G genomes of tetraploid wheat have undergone massive changes following ancestor divergence and polyploidization, and they are widely considered to be modified S genomes having evolved from a common ancestor. Gu et al. (2004) indicated that the B genome diverged before the separation of the A and D genomes. There has been considerable controversy over the donor of the S-genome progenitor, but it was correctly identified as an ancestor of *Ae. speltoides* Tausch ($2n = 2x = 14$) in 1956 (Sarkar and Stebbins 1956; Riley et al., 1958; Shands and Kimber 1973; Dvořák and Zhang 1990; Daud and Gustafson 1996) and contains 5.15 pg DNA (http://data.kew.org/cvalues/introduction.html). Cytoplasmic analyses have shown that *Ae. speltoides* was the maternal donor of not only tetraploid but also hexaploid wheat (Wang et al., 1997). It is clear that the B genome has undergone significant intergenomic noncoded and coded DNA changes (in both the diploid and the tetraploid wheats) since the formation of tetraploid wheat. The B-genome component of polyploid wheat is the largest of the wheat genomes and, because of the large degree of change at the DNA level, the true donor of the B genome since polyploid formation has been very difficult to establish. It would be useful to test representatives of A- and B-genome donors from across their geographic ranges to settle some of the past and present controversies over their origins. See the discussions based on genomic *in situ* hybridization (GISH) molecular cytogenetics by Belyayev et al. (2000) and Yen and Baenziger (1996). Regardless, since the cytoplasm donor is the female in the original cross creating the polyploid and is always listed first in any pedigree, the tetraploid genome designations should technically be BBAA or GGAA.

Emmer and durum wheat

Origin of Triticum turgidum

Triticum urartu exists only in its wild form, contains 4.93 pg DNA (http://data.kew.org/cvalues/introduction.html), and supplied the male parent of tetraploid wheat (Feldman and Sears 1981), including several cultivated species. The most important are *T. turgidum* (BBAA), containing 12.28 pg DNA (http://data.kew.org/cvalues/introduction.html), and the sometimes cultivated, non-free-threshing *T. timopheevi*, which contains 11.30 pg DNA (http://data.kew.org/cvalues/introduction.html) and includes wild subspecies *T. timopheevi araraticum* (Jakubz.) Mac Key and cultivated subspecies *timopheevi* = *T. turgidum* ssp. *timopheevi* (Zhuk.). *Triticum turgidum* is further divided into several species, including *T. turgidum* ssp. *dicoccoides* (Korn. Ex Asch. and Graebn.), which is well known as the progenitor of all modern cultivated polyploid wheat species—that includes *T. turgidum* ssp. *durum* (Desf.) Husn., which is widely cultivated and is commonly called durum or macaroni wheat.

Outside the Fertile Crescent area, where *T. dicoccoides* wheat (Color Plate 1) reached the range of *Ae. tauschii*, the two species hybridized (Van Zeist 1976; Van Zeist and Bakker-Heeres 1985) and formed the hexaploid wheat group. This key hybridization event most likely occurred in the Caspian Sea region approximately 10,000 years ago. Notably, *T. dicoccoides* wheat is more adapted to Mediterranean environments, whereas noncultivated hexaploid wheat grows in cooler and more continental parts of Europe and western Asia (Fig. 1.1). Thus, *T. dicoccoides* deserves a particular in-depth study as suggested by Aaronsohn (1910, 1913), since it is the main genetic resource for improving both tetraploid and hexaploid wheat. Here we review theoretical and applied studies on *T. dicoccoides* that are important for all polyploid wheat improvement, including genetic structure across its range and its genetic resources. These topics are critically important to overcome the dangerous process of homogeneity occurring in all cultivated wheat gene pools. In addition, we will discuss the genome organization and evolution of *T. dicoccoides*.

Genetic and morphological evidence clearly indicates that the cultivated tetraploid *turgidum* wheat is closely related to the wild wheat, *T. dicoccoides* (Korn), which is native to the Near East and is traditionally called wild emmer wheat (Zohary 1969; Chapman et al., 1976; Miller 1987, 1992; Harlan 1992; Zohary and Hopf 1993; Feldman et al., 1995). *Triticum dicoccoides* (Fig. 1.1) is a tetraploid containing the A (male) and B (female) genomes and is the female progenitor of all hexaploid wheat species. *Triticum aestivum* is the most important of the hexaploid wheats, followed by several primitive hulled types (spelta wheat) and numerous modern free-threshing forms (Zohary and Hopf 1993).

Origin of Triticum dicoccoides (wild emmer)

Triticum dicoccoides is an annual, is predominantly self-pollinated, and has large and brittle ears with large elongated grains (Nevo et al., 2002), similar to cultivated emmer and durum wheats. It is the only wild ancestor in the genus *Triticum* that is cross-compatible and fully interfertile with cultivated *T. turgidum* wheat. Hybrids between wild *T. dicoccoides* and all members of the *T. turgidum* complex show normal chromosome pairing in meiosis. Natural hybrids do occasionally form between cultivated tetraploid wheat and *T. dicoccoides*, so *T. dicoccoides* is sometimes ranked as the

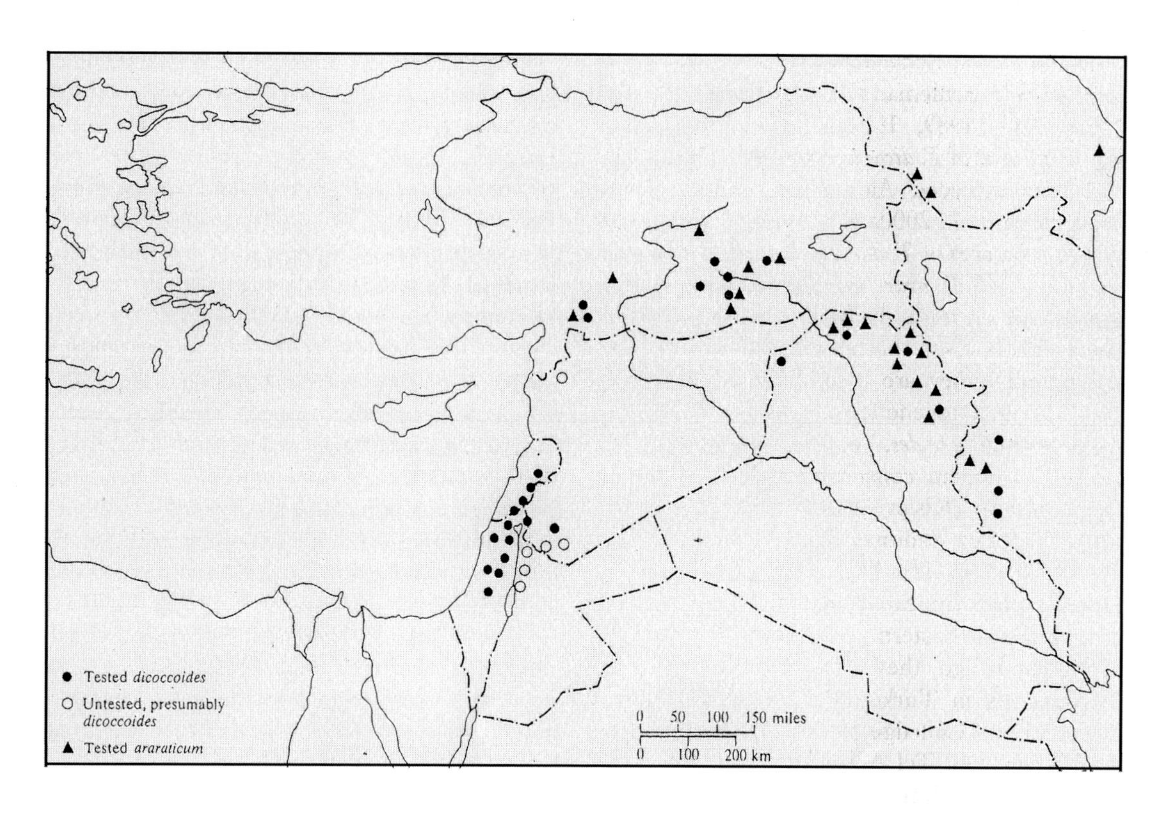

Fig. 1.1 Distribution of wild tetraploid wheat: (●,○) wild emmer wheat, *Triticum turgidum* ssp. *dicoccoides* (*T. dicoccoides*); (▲) wild Timopheev's wheat, *T. timopheevi* ssp. *araraticum* (*T. araraticum*). (●,▲) Collections were tested cytogenetically. Adapted from Zohary and Hopf (1993) and references therein.

wild subspecies of the *T. turgidum* complex. Because of its central place in the evolution of cultivated wheat, wild emmer is among the best sources for obtaining insights into wheat evolution and improvement (Xie and Nevo 2008).

Triticum dicoccoides is a valid biological species (Miller 1992) that has a unique ecological niche in nature, where the seed dispersal mechanism involves "wild type" rachis disarticulation (brittle rachis), and spikelet morphology reflects adaptive-specialized traits that ensure survival in nature (Zohary 1969). Under the human selection system of reaping, threshing, and sowing, the selection and maintenance of the nonbrittle phenotype was highly advantageous and resulted in accelerated domestication (Miller 1992). Wild and domesticated forms also differ in kernel morphology (Van Zeist 1976); in cultivated tetraploid species the grain is wider, thicker, and rounder than in *T. dicoccoides*. Unique chromosomal translocations (Kawahara et al., 1993; Nishikawa et al., 1994; Joppa et al., 1995; Kawahara and Nevo 1996) and genetic polymorphisms (Nevo et al., 1982; Nevo and Beiles 1989; Fahima et al., 1998, 1999; Nevo 1998, 2001) also characterize *T. dicoccoides*. This combined evidence justifies its traditional classification as a separate species, as implied in the name *T. turgidum* ssp. *dicoccoides*, and it is clearly the progenitor of cultivated tetraploid and hexaploid wheat.

Triticum dicoccoides is found in Israel and Syria (which are its centers of distribution based on genetic diversity), Jordan, Lebanon, southeast Turkey, northern Iraq, and western Iran (Nevo and Beiles 1989; Nevo 1998). It was discovered in 1906, in eastern Galilee on the slopes of Mt. Hermon by Aaronsohn, who recognized its potential importance for all wheat improvement

(Aaronsohn and Schweinfurth 1906; Aaronsohn 1910, 1913; Schiemann 1956; Feldman 1977; Nevo 1983, 1989, 1994, 2001). The genetic resource value of *T. dicoccoides* for wheat improvement far exceeded Aaronsohn's vision (Nevo 2001; Peng et al., 2000a,b,c). In the northeastern distribution area of *T. dicoccoides*, where the sympatric area of *T. araraticum* is located, the two species are separated by strong sterility barriers (Maan 1973). Even though they are similar morphologically, they are practically indistinguishable without cytogenetic analysis.

Triticum dicoccoides, like *T. boeoticum*, was collected for human consumption long before its domestication (Kislev et al., 1992; Zohary and Hopf 1993; Lev-Yadun et al., 2000; Nesbitt 2001). Brittle *T. dicoccoides*–like plants with relatively narrow grains appeared in early Neolithic and Natufian Near Eastern settlements. However, 9–10 millennia ago, they also coexisted with non-brittle seeds in Turkey (Jarmo, Iraq, Cayonu) (Hillman and Colledge 1998), in northern Syria (Tel Aswad and Tel Abu Hureira) (Zohary and Hopf 1993; Nesbitt 1998; Nesbitt and Samuel 1998; Lev-Yadun et al., 2000), and in Syria (Tell Mureybet I and II; 9000–8000 BC). *Triticum dicoccoides* was also discovered in Neolithic sites in Syria (Jerf el-Ahmar, Mureybet III, and Djade) and Turkey (Cayonu) (8000–7500 BC) and in sites near pre-Neolithic Turkey (Hallan Cemi Tepesi) and Iraq (Neolithic Qermez Dere and M'lefaat) (Nesbitt 1998; Lev-Yadun et al., 2000). Domesticated forms appeared in core-area Neolithic sites in Syria (Tell Abu Hureira 2A) and Turkey (Cafer Huyuk) about 7500 BC, and soon thereafter in Turkey (Cayonu and Nevali Cori) (Kislev et al., 1992; Nesbitt and Samuel 1998). From the early beginnings of agriculture in the Near East, 10,000 years ago and throughout the Chalcolithic and Bronze times, emmer was the principal wheat of newly established farming settlements; approximately 7000 years ago it spread from there to Egypt, the Indian Subcontinent, and Europe.

Patterns of allozyme diversity in wild *T. dicoccoides* suggest the following: (i) during the evolutionary history of wild *T. dicoccoides*, diversifying and balancing natural selections, through climatic, edaphic, and biotic factors, were major agents of creating genetic structure and maintaining differentiation; (ii) wild *T. dicoccoides* harbors large amounts of genetic diversity that can be utilized to improve both tetraploid and hexaploid wheat.

Wild *T. dicoccoides* grows extensively in the catchment areas of the Upper Jordan Valley (in northern Israel, in the eastern Upper Galilee Mountains, and the Golan Heights). Elsewhere in the Fertile Crescent (Fig. 1.1), populations of wild *T. dicoccoides* are semi-isolated and isolated and display a patchy structure. The highly subdivided, archipelago-type ecological population structure of wild *T. dicoccoides* is matched by its genetic population structure. Substantially more gene differentiation has been found within and between populations that were sometimes geographically very close within Israel, than between wild *T. dicoccoides* populations in Israel and Turkey (Nevo and Beiles 1989), where 40% of the *T. dicoccoides* genetic diversity existed within populations and 60% existed between populations. Only 5% of the genetic diversity was found between the Israel and Turkey metapopulations. This conclusion was reinforced based on edaphic, topographic, and temporal differentiation, on local microclimatic differentiation, on the extreme case of local isozyme differentiation in the Golan Heights (Nevo et al., 1982; Golenberg and Nevo 1987; Nevo et al., 1988a,b), and on recent DNA analyses (Fahima et al., 1999; Li et al., 1999, 2000a,b,c,d). The DNA results suggested that at least part of the noncoding regions were also subjected to natural selection. Genetic diversity was eroding across coding and noncoding regions of the *T. dicoccoides* genomes during and following domestication (Fahima et al., 2001). The *T. dicoccoides* genomes have been molded, in part, by diversifying natural selection from various ecological stresses.

The genetic differentiation within and between populations of *T. dicoccoides* was also reflected by an analysis of allele distribution (Nevo and Beiles 1989), which showed that 70% of all variant alleles were not widespread but revealed a definite localized somewhat sporadic distribution. Likewise, the analysis of genetic distances between populations supported the conclusion that sharp local differentiation over short geographic distances was the rule, and the frequency of some common alleles (>10%) was localized and high. The population genetic structure of wild *T. dicoccoides* is obviously a mosaic and reflects the underlying ecological heterogeneity, which has been derived from local and regional geological, edaphic, climatic, and biotic differentiations. The genetic landscape is definitely not random between loci, populations, and habitats, and it most likely displays adapted patterns predictable on the basis of environmental factors. Could these polymorphisms represent adaptation to fluctuating environments? It has been possible to decide if selection is responsible for the occurrence of many DNA variants across the coding and noncoding regions of a genome, and it is clear that major DNA changes can and do occur within and between *T. dicoccoides* populations over a relatively short time frame, paralleling that of allozymes.

Nevo and Beiles (1989) predicted that neither migration nor genetic drift could have generated the patterns observed between loci and alleles of wild *T. dicoccoides* and that selection remained a vital explanatory model. Environmental selection also partly affected loci differentially, but differently from migration. This was supported by data from Nevo and Beiles (1989) for three reasons: (i) variation was found among loci; (ii) in an autocorrelation analysis, positive correlations were found in different distant *T. dicoccoides* groups, and not necessarily in the first one as would be expected if migration determined the interpopulation genetic structure; and (iii) the predominance of negative correlations in the larger distant groups was found to be due to decreasing ecological similarity often observed with increasing distance.

The maintenance of polymorphisms in wild *T. dicoccoides* may be explicable by both spatial and temporal variation in selection. Theory indicates that selection, acting differentially in space, coupled with limited migration, which is typical of wild *T. dicoccoides*, will maintain a substantial amount of polymorphism (Karlin and McGregor 1972; Hedrick 1986; Nevo et al., 2000). Thus, different polymorphisms will be favored in different climatic and edaphic niches, from regional to local, and at miniscule levels within a locality. Microniche ecological selection (e.g., climatic factors related to temperature, available water, and biotic and abiotic stresses) could be a major cause of genetic differentiation rather than stochastic processes.

Origin of hexaploid wheat

There are two main forms of hexaploid wheat, including *T. zhukovskyi* Men. & Er., which was the result of a recent hybridization involving *T. timopheevi* and *T. monococcum*, the only example of hexaploid wheat to have the $GGAAA^mA^m$ constitution (Upadhya and Swaminathan 1963). The most important hexaploid wheat group comprises *T. aestivum* (BBAADD) and its several subspecies containing 21 pairs of chromosomes with seven pairs belonging to each of the A, B, and D genomes (Sears 1954; Okamoto 1962) and containing 17.33 pg DNA (http://data.kew.org/cvalues/introduction.html).

Triticum aestivum originated approximately 10,000 years ago after the domestication of tetraploid wheat and was derived from the hybridization of a primitive tetraploid (BBAA), as the female, and *T. tauschii* ssp. *strangulata* [*Ae. tauschii* (Coss.) Schmal, also known as *Ae. squarrosa*, DD, $2n = 2x = 14$, 5.10 pg DNA], as the male (Kihara 1944; McFadden and Sears 1944, 1946a,b; Kimber and Sears 1987; Kimber and Feldman 1987; Dvořák et al., 1998). The first primitive hexaploid wheat was probably a hulled-type like *T. aestivum* var. *spelta*, *macha*, or *vavilovii*. The current free-threshing types, *T. aestivum* var. *aestivum*, *sphaerococcum*, or *compactum*, were the result of a mutation at the *Q* gene locus (Muramatsu 1986) followed by selection. All polyploid wheat species are disomic in inheritance due to complete diploidlike chromosome pairing, which is controlled by two main homoeologous pairing genes *Ph1* (Riley and Chapman 1967) and *Ph2* (Mello-Sampayo 1971) and several minor genes (for a complete review, see Sears 1977). As previously stated, since the cytoplasm donor of hexaploid wheat was the female in the original cross creating the polyploid, it should be listed first in any pedigree or genome designation; therefore, hexaploid genome designations should be stated as BBAADD as noted by Feldman (2001).

Research has shown that hexaploid wheat is less variable than its diploid progenitors, suggesting the possibility of a genetic bottleneck caused by a very limited number of initial hybridizations (Appels and Lagudah 1990). However, given the obviously large distribution area of primitive tetraploid wheat and *Ae. tauschii* populations within the cradle of agriculture (for an excellent review of cultivation regions, see Feldman 2001), the natural occurrence of multiple tetraploid wheat and *Ae. tauschii* hybrids could be a more common occurrence than originally thought. The presence of several sets of alleles and microsatellites established that hexaploid wheat resulted from several hybridizations (Dvořák et al., 1998; Talbert et al., 1998; Lelley et al., 2000; Caldwell et al., 2004; Zhang et al., 2008). Zohary and Hopf (1993) suggested that these hybridizations are still occurring today.

Clearly, under certain environmental conditions, some degree of outcrossing in wheat does occur (Griffin 1987; Martin 1990) and hybrids of various ploidy levels can be formed (McFadden and Sears 1947; Ohtsuka 1998), both of which would indicate less of an evolutionary bottleneck in the development of hexaploid wheat than previously suggested. In addition, hexaploid wheat originated and still originates in a region where all of the progenitors reside, thus allowing for a continuous intercrossing and backcrossing with diploid progenitors. Even with the presence of ploidy and genome differences, the various types of primitive wheat species are capable of widespread intercrossing, culminating in intraspecific hybrid swarms which would significantly increase the potential for gene flow over time. This is possible because all polyploid wheat progenitors share at least one common genome, which can serve as a buffer or a pivot around which unpaired homoeologous chromosomes can pair. Any homoeologous chromosome pairing, within the genomes of the Triticeae tribe, can and does allow for the occurrence of additional gene recombination and exchange. Since tetraploid and hexaploid wheat are predominantly self-pollinated, homozygosity for any gene exchanges favored by natural selection would be rapidly achieved and available for artificial selection.

This presence of modified genomes along with unmodified or pivotal genomes, widespread throughout the Triticeae tribe, was originally suggested by Zohary and Feldman (1962) and later in wheat–rye hybrids by Gustafson (1976), and it has been shown to occur more often than expected. The presence of the pivotal (buffering) genomes in primitive polyploid wheat crosses made possible the rapid and very successful expansion of wheat in a very short time. This manner of polyploid speciation allowed for a greater degree of gene flow and genome modification than that which has been observed in any diploid system of speciation. This process of wheat polyploid speciation needs to be kept clearly in mind when attempting to make wide crosses (interspecific and intergeneric) to introduce genes from other species and genera into wheat. The presence of pivotal genomes in polyploid wheat complexes makes it easier for breeders to understand the processes involved in manipulating gene complexes from related grass species into wheat. It also makes very clear the importance of maintaining and expanding all existing diploid and polyploid germplasm collections of wheat and wheat relatives as vast primary, secondary, and tertiary gene pools for future use in wheat improvement.

The polyploid wheat species represent a converging form of evolution where several genomes have been combined. This form of species hybridization coupled with inbreeding has resulted in a very successful polyploid that is highly adaptable to a wide range of environmental growing conditions. The evolution of polyploid wheat and its intimate connection with the transition of human societies from hunting-and-gathering to an agricultural culture occurred over a long time and involved vast mixtures of wild and increasingly domesticated populations, and of hulled and free-threshing forms, ultimately resulting in a diverse and dynamic wheat gene pool.

The genetic composition of polyploid wheat species fully accounts for their successful establishment. The evolutionary development of a genetic system conferring diploidization (*Ph* mutant; for a complete review, see Sears 1977), thus preventing multivalent chromosome formation with deleterious intergenomic exchange, was critical for the stabilization of all polyploid wheat species. Mutations, within the various wheat genomes, also played a major role in allowing wheat to increase in variability, stabilize as a species, and become the major food crop. The two and three genomes present within tetraploid and hexaploid wheat, respectively, and the self-pollinating character of all species, resulted in the accumulation of mutations that became available for selection. This allows for individuals within populations to become a main driving force upon which natural selection operates. This form of gene formation, modification, and stabilization is one of the most powerful processes in plant evolution.

GENOME EVOLUTION AND MODIFICATION

We now have a voluminous amount of information concerning the ancestors and evolutionary processes that created polyploid wheat. To fully understand the genomic evolution of polyploid wheat, it is important to ask why each of the diploid genomes comprising polyploid wheat is so massive relative to other grass species such as rice. The B genome is 5.15 pg DNA (Furuta et al., 1986); the A genome is 4.93 pg DNA (Bennett and Smith 1976); and the D genome is 5.10 pg DNA (Rees and Walters 1965). On the other hand, rice contains only 0.6–1.0 pg DNA in japonica and indica types, respectively (Bennett and Leitch 1997; http://data.kew.org/cvalues/introduction.html). However, the various wheat genomes and the rice genome appear to have similar genetic composition with a good macro-level syntenic relationship (Gale and Devos 1998; Sorrells et al., 2003; Tang et al., 2008). Flavell et al. (1974) and Gu et al. (2004) established that over 80% of the hexaploid wheat genome comprised noncoded highly repeated DNA sequences and highly active and nonactive retrotransposons. The intergenic regions (Bennetzen 2000; Feuillet and Keller 2002; Wicker et al., 2003) and genic regions (Gu et al., 2004) of many species are mainly composed of retrotransposons, and the vast numbers of these retrotransposons are correlated with genome size (Kidwell et al., 2002). Most of the retroelements in the three wheat genomes are not colinear, which suggests that their present location was the result of genome divergence after the individual A, B, and D genome parental species were combined (Gu et al., 2004). When analyzing the glutenin genes of wheat, Gu et al. (2004) found that more genes from the glutenin region of the A genome contained retrotransposons than occurred in orthologous regions of either the B or D genome.

Is all or most of the noncoded DNA present in hexaploid wheat really "junk" DNA? From an evolutionary view, it is highly unlikely that any genome would expend a vast percentage of its energy production maintaining DNA that was of little or no value. The reason behind the presence and function of vast amounts of noncoded DNA in the wheat genomes remains largely unknown. To fully understand and be able to manipulate wheat genome evolution, the function and purpose of this noncoded DNA needs to be investigated.

There is an abundance of data supporting the ability of a genome to increase and/or decrease in DNA amount over time, compared with that observed in its original progenitor. Such genomic changes (deletions and additions, gene conversions, transposon activation and silencing, chromosomal rearrangements, epigenetic events, etc.) are known to occur widely in grass genomes (Feldman et al., 1997; Liu et al., 1998; Ozkan et al., 2001; Shaked et al., 2001; Kashkush et al., 2002; Han et al., 2003; Ma et al., 2004; Ma and Gustafson 2005, 2006) and other polyploid plant genomes, including, for example, *Brassica napus* polyploids (for an excellent article, see Gaeta et al., 2007). The frequency of such events is not uniform across individual chromosomes or within complete genomes. The selection pressures acting on DNA deletion or insertion in either a plant or animal genome can be different, depending on whether or not changes are located in repeated DNA, heterochromatin regions, or gene-rich regions. Diaz-Castillo and Golic (2007) noted in

Drosophila that gene structure and expression were influenced by the location of genes proximal to heterochromatin and were evolving at a rate in response to their chromosomal location.

No fully satisfactory explanation has been suggested for why these major evolutionary genome modification, deletion, and addition processes take place mainly in the noncoded portion of the genome. Wicker et al. (2003) and Gaut et al. (2007) have made a strong case for illegitimate recombination having a major influence on genome evolution. Illegitimate recombination is capable of generating deletions, inversions, gene conversions, and duplications within any chromosome of any genome. However, it is difficult to envision illegitimate recombination as the main cause for such a sizeable DNA deletion, of up to about 10% or more in the genomes of many allopolyploid cereals. It is likely that no single explanation will answer the question of why the cereal genomes vary so much in size. It will most certainly require a number of working hypotheses and a large body of new evidence and knowledge bearing on the problems associated with the evolution of genome size in grasses to resolve this question. See a recent review on synteny and colinearity in plant genomes by Tang et al. (2008).

We can propose one possible cause for many of the observed vast changes in grass genome composition. Clearly every grass genome goes through its cell cycle at a specific rate, which varies with each genome. Van't Hof and Sparrow (1963) first proposed the existence of a relationship between DNA content, nuclear volume, and mitotic cell cycle, and suggested that any mitotic cell cycle is greatly influenced by the amount of DNA present in the genome. They made it clear that the amount of DNA present in a genome does affect cell cycle, and ultimately plant growth, regardless of whether or not it was coded. Recently, Francis et al. (2008) concluded that the speed of DNA replication was identified as the limiting factor in the cell cycle. Therefore, it follows that individual genome cell cycle differences cause problems of maintaining their synchrony when two or more genomes, with different volumes of DNA, are placed together in a cell.

For example, in the wheat–rye hybrid triticale (×*Triticosecale* Wittmack), Bennett and Kaltsikes (1973) showed that the meiotic duration of wheat and rye differed from that observed in the hybrid, and the hybrid had a meiotic cell cycle closer to the wheat parent. Their observations made it clear that if one genome of a hybrid has not completed its cell cycle by the time cell wall formation has initiated, the possibility of breakage-fusion-bridges occurring in the genome with the lagging cell cycle will be greatly increased, most likely resulting in DNA elimination or addition. This is what happens in a wheat–rye hybrid and can be readily seen in the formation of large aberrant nuclei that are readily visible in the early cenocytic stages of endosperm development before cellularization takes place (Fig. 1.2). The formation of cell walls at the first division of the embryo would definitely cause breakage-fusion-bridges to occur immediately and lead to the decrease—or even increase—of DNA present in the genome with the lagging cell cycle.

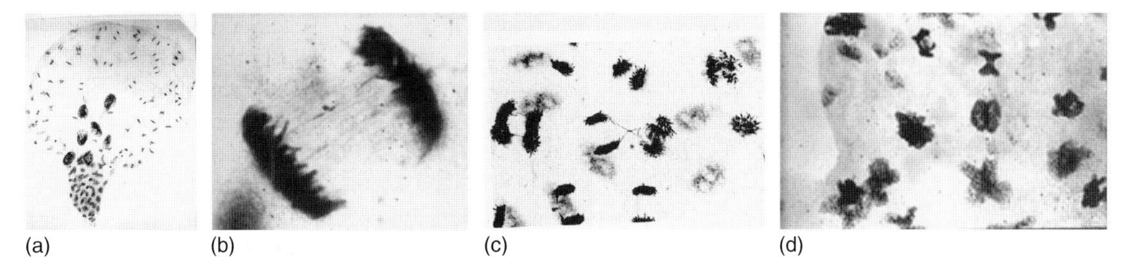

Fig. 1.2 (a) A wheat–rye hybrid (triticale) seed only 48 hours after pollination with a cenocytic endosperm and a cellular embryo (arrow); (b) a nuclear division (24 hours after pollination) showing bridges that have formed during anaphase; (c) nuclear divisions (48 hours after pollination) showing rye telomeres that have formed bridges during anaphase; and (d) nuclear divisions (72 hours after pollination) showing rye telomeres that have formed bridges during anaphase.

As has been observed in the early endosperm development of wheat–rye hybrids, deletions within the rye genome were clearly detected (Gustafson and Bennett 1982; Bennett and Gustafson 1982). Deletions and increases in DNA have even been detected within heterochromatic regions of the genus *Secale* (Gustafson et al., 1983). Variations in DNA content have been observed in polyploids of the Triticeae tribe (Feldman et al., 1997; Ozkan et al., 2001; Kashkush et al., 2002; Liu and Wendel 2002; Han et al., 2003; Ma et al., 2004; Ma and Gustafson 2005, 2006), in rice (Wang et al., 2005), in maize (Messing et al., 2004), in a few *Hordeum* species (Jakob et al., 2004), and in synthetic polyploids of *Brassica* (Song et al., 1995). In some induced dihaploids of *Nicotiana* (Dhillon et al., 1983; Leitch et al., 2008) and *Gossypium* species (Grover et al., 2007), an increase in DNA has actually been detected. For an excellent review of polyploids showing a decrease in DNA over time, see Leitch and Bennett (2004). Any genome cell cycle differences could easily be the major cause of genome variation in DNA content between an allopolyploid and its parental species.

MECHANISMS FOR CHROMOSOME EVOLUTION

As stated previously, the evolution of the genus *Triticum* serves as a good model of polyploidy, one of the most common forms of plant evolution (Elder and Turner 1995; Soltis and Soltis 1999). From a practical perspective, large stores of simply inherited genes that confer different types of resistance are available in wheat and its wild relatives via germplasm collections. Knowledge of the mechanisms of polyploidization will help plant breeders to enrich the gene pool of cultivated wheat. The origin and co-evolution of A and B genomes of tetraploid wheat has long been controversial (Feldman and Sears 1981). Unknown are the details of the co-evolution of A and B genome repetitive sequence arrays in allotetraploid wheat. There is no reason to regard the process of allopolyploidization as a mechanical combination of sequences from two genomes, but still less is known about the interaction between sequences from different arrays in chromatin fractions (Wendel 2000). In this section we draw attention to several critical points of speciation-related chromosomal changes.

Chromosomal rearrangements and repetitive DNA

Major structural chromosome rearrangements including deletions, duplications, translocations, and inversions are often associated with cytogenetically detectable heterochromatic regions composed of repetitive DNA, and they frequently appear in heterochromatin–euchromatin borders (Badaeva et al., 2007). Chiasmata in meiosis appear very close to the terminal and intercalary C-bands and mark the point of exchange (Loidl 1979). Well-studied intraspecific C-banding polymorphisms can be regarded as a manifestation of this interdependence. The diploid–polyploid *Aegilops–Triticum* complex exemplifies abundant C-banding polymorphism based on chromosomal rearrangements (Badaeva et al., 1996, 1998, 2002, 2004, 2007; Friebe and Gill 1996; Rodríguez et al., 2000a,b; Maestra and Naranjo 1999, 2000). A good example of this is where the combination of C-banding techniques and fluorescence *in situ* hybridization (FISH) with ribosomal RNA genes, 5S and 18S-5.8S-26S rDNA (45S rDNA), and with a D-genome-specific repetitive DNA sequence pAs1 revealed species-specific patterns of heterochromatin, rDNA, and pAs1 clusters for six D-genome-containing allopolyploid *Aegilops* species: *Ae. cylindrica*, *Ae. ventricosa*, *Ae. uniaristata*, *Ae. crassa*, *Ae. vavilovii*, and *Ae. juvenalis* (Badaeva et al., 2002). A wide spectrum of chromosomal rearrangements, particularly species-specific, and genome-specific redistribution of repetitive DNA clusters led to hypothesizing the phylogenetic relationships in this group of polyploid *Aegilops* species.

Heterochromatin

An inherent feature of heterochromatin is the complex composition of tandem repeats of various types (Sharma and Raina 2005) and transposable elements (TEs), predominantly retrotransposons (elements of Class I that transpose via RNA intermediates) (Lipman et al., 2004). Three groups of retroelements—Ty1-*copia*, Ty3-*gypsy*, and *LINE*—were found in large quantities in heterochromatin of the diploid B/G-genome progenitor *Ae. speltoides* (Belyayev et al., 2001). Clusters of tandem repeats and TEs form species-specific and chromosome-specific heterochromatin patterns. There is a certain correlation between distribution-clustering of retroelements and chromosome location of tribe-specific and species-specific sequences within cereal genomes (Color Plate 2). Thus, the independently discovered tribe-specific tandem repeat *Spelt 52* (Anamthawat-Jonsson and Heslop-Harrison 1993; Friebe et al., 2000; Giorgi et al., 2003) and *Ae. speltoides*–specific tandem repeat *Spelt 1* (Salina et al., 1997; Pestsova et al., 1998) cluster together with retroelements at the same chromosomal locations corresponding to AT-enriched heterochromatin. Moreover, this complex of distal–terminal chromosomal regions enriched by TEs of different types and tribe- and species-specific tandem repeats could be classified as a faster-evolving part of the genome (Belyayev and Raskina 1998) (Color Plate 2, note the green signal on the distal regions of chromosome 4 after GISH).

Repetitive DNA

The repetitive DNA fraction plays an important role during polyploidization and post-polyploidization changes (Dvořák and Zhang 1990, 1992; Dvořák and Dubcovsky 1996; Feldman and Levy 2005; Ma and Gustafson 2005, 2006). In the genomes of allopolyploid wheat, *T. dicoccoides* (B and A genomes) and *T. aestivum* (B, A, and D genomes), the distribution pattern of highly repetitive DNA clusters and Ty1-*copia* retroelements differs from those of its diploid progenitors, *T. urartu* and *Ae. speltoides* (Color Plate 2a–d) (Raskina et al., 2002). Significant intercalary repositioning and decay of a majority of distal–terminal clusters of AT-positive heterochromatin were observed in the B genome of allopolyploid wheat in contrast to the S genome of *Ae. speltoides* (Color Plate 2c). Similar differences have been observed between populations of *Ae. speltoides* involving a series of distal–terminal chromosomal rearrangements (Raskina et al., 2004a,b). Multiple translocations and deletions occurred, which led to the current heterochromatin pattern (Color Plate 2e) and a majority loss of *Ae. speltoides*–specific tandem repeat *Spelt 1* clusters (A. Belyayev, pers. comm.). Since new allopolyploids continue to occur in the periphery distribution areas of existing species (Grant 1981), we will continue to observe allopolyploidization involving the *Ae. speltoides* genome containing numerous chromosomal rearrangements and mobile elements (Raskina et al., 2004a,b). The present-day B genome of wild and cultivated wheat carries from zero to two *Spelt 1* clusters per haploid genome in contrast to the G genome of existing allopolyploids, which contains up to six *Spelt 1* clusters (Salina et al., 2006). These data are in accordance with the purported independent origin of the B and G genomes of allopolyploid wheat (Jiang and Gill 1994b; Rodríguez et al., 2000a).

Ongoing permanent intragenomic mutagenesis in plant populations is a generator of heterozygosity leading to intraspecific genetic variability and creates the basis for natural selection under changing environments. Significant inter-B/A-genomic interactions, in allotetraploid wild emmer wheat, revealed major substitutions of part of the A-genome heterochromatin clusters by satellite DNA from the B genome (Color Plate 2c) (Belyayev et al., 2000). Enrichment of these clusters with mobile Ty1-*copia* retroelements suggests an important role of TEs in rebuilding and homogenizing the allopolyploid genome, leading to stabilization of *T. dicoccoides* as a new species. Retroelements are known to play a large role in gene and genome evolution (Flavell et al., 1997; Kidwell and Lisch 2001; Bennetzen 2002). In *T. aestivum*, substitution of part of the heterochromatin from the "youngest" D genome by repetitive DNA from the A and B genomes was revealed to a far lesser degree (Color Plate 2d, 14 red D-genome chromosomes marked by asterisk).

Transposable elements can directly change molecular composition and/or DNA amount in the regions of insertions. They also can mediate ectopic chromosomal exchanges when homologous and/or nonhomologous chromosome recombination moves sequences within and between genomes. Furthermore, insertions of TEs may create new crossing-over "hot" spots that provoke transposable element-mediated homologous or nonhomologous chromosome rearrangements. The latter include spontaneous translocations, inversions, and deletions and are potential mechanisms for rapid genome reorganization during speciation and stabilization of any allopolyploid. For example, in the wheat 4AL–7BS translocation (Naranjo 1990), a cluster of Ty1-*copia* retrotransposons was detected (Color Plate 2b), and chromosomes 4A and 7B were also involved in a 4AL–7BL translocation, which was detected in a natural population (Raskina et al., 2002).

The manipulation of repetitive DNA complexes plays an important role in evolutionary genome transformation. Changes in repetitive DNA may cause chromosomal rearrangements and, in turn, chromosomal rearrangements may cause repetitive DNA change through mechanisms of concerted evolution (Elder and Turner 1995). Therefore, these processes are interdependent.

Repatterning of rDNA arrays in the wheat genome

In addition to the direct detection of major chromosomal rearrangements, it is also possible to indirectly estimate the level of microevolutionary genomic change by evaluating the repatterning of well-determined chromosomal markers and by the mobility of rDNA clusters. It is obvious that speciation-related chromosome structure change establishes further increases or decreases in the number of rDNA sites or their repositioning, but the dynamics of rDNA clusters may be regarded as an indicator for significant intragenomic processes (Jiang and Gill 1994a; Dubcovsky and Dvořák 1995; Raskina et al., 2004b).

The location, number, and mobility of rDNA clusters have been described in many plant species and may involve major loci, small numbers of copies of the repeat unit, or fragments of a repeat unit, which are often known not to be transcribed (for review, see Heslop-Harrison 2000). There is evidence that rDNA repeat sites may alter chromosomal location without the involvement of translocations or other chromosomal rearrangements (Dubcovsky and Dvořák 1995). Schubert and Wobus (1985) examined the mobility of nucleolar organizing regions (NORs) in *Allium* and proposed TE activity as one of the possible sources for rDNA movement. Recent studies proposed that transposons (*En/Spm*-like; elements of Class II that move by extinction and reintegration) might be involved in rDNA repatterning (Raskina et al., 2004a). The ability of some classes of transposons (Pack-MULES, Helitrons) to capture entire genes and move them to different parts of the genome has been documented (Jiang et al., 2004; Lai et al., 2005). Therefore, any interaction of ribosomal genes and TEs relative to evolutionarily significant chromosomal repatterning appears to be of tremendous interest yet remains largely unexplored. Certain remodeling of chromosome-specific repetitive DNA patterns may lead to meiotic abnormalities. In extremes, these abnormalities are capable of causing reproductive (postzygotic) isolation (Grant 1981).

A series of *in situ* hybridization (ISH) experiments revealed permanent clustering of different TEs in the NOR (which contains 45S rDNA loci) as well as near or within clusters of 5S rDNA (Belyayev et al., 2001, 2005; Raskina et al., 2004a,b). Therefore, we can suggest that the possible association of TEs and rDNA loci arise due to, first, the insertion preference of the TEs in the rDNA arrays. Indeed, rDNA arrays are common targets for several LINE retrotransposons (Eickbush and Eickbush 2003; Averbeck and Eickbush 2005) and also for some Class II transposons (Penton and Crease 2004). Second, these two components may accumulate preferentially within the same genomic context, perhaps driven over time by selection against insertions elsewhere in the genome (e.g., heterochromatin in the case of retroelements). Third, a possible functional relationship exists between the dispersion of TEs and rDNA genes. Additional molecular-bioinformatic studies may further explain TE–rDNA gene interactions.

Repetitive DNA and mobile elements as perpetual generators of diversity and evolution

Speciation in wild diploid and polyploid wheat is tightly connected with significant repatterning of rDNA sites (Dubcovsky and Dvořák 1995). Considering rDNA in terms of temporary genome changes, we face a certain paradox. On one hand, rDNA is the most conservative fraction in the eukaryotic genome, and ribosomal RNA genes undergo minimal changes over hundreds of millions of years. On the other hand, this conservatism appears to be a source of genome instability. Due to the similarity of rDNAs, any chromosome that carries extended rDNA arrays has the potential for involvement in heterologous synapses and recombination (Raskina et al., 2004b). Thus, any rDNA cluster could consist of several layers of different origins, especially in a polyploid species, with a high level of interhaplome invasion (Belyayev et al., 2000). This is clearly seen in wheat (Color Plate 2f), where differently labeled nontranscribed spacers (NTS) of 5S rRNA genes of different origins (short A1 and long G1) show slightly different positions inside the rDNA cluster on chromosome 1A of *T. dicoccoides* (Baum et al., 2004).

Due to the known capability of mobile elements to provoke ectopic exchanges, the consequences of TE–rDNA interaction make it possible to propose that the collocation of different TEs within recombinogenic hot spots could intensify homologous and heterologous recombination. Moreover, TE-mediated intragenomic transfer of rDNA fragments and the inheritance of such mutations may cause significant evolutionary changes in chromosomal distribution of rDNA clusters (Raskina et al., 2004a,b).

Another possible consequence of the physical association of rDNA–TE within the 45S rDNA region (NOR) could be the loss of chromosome satellites with all their genetic content. McClintock (1946) suggested that TEs could cause chromosomal breakages. High concentrations of TEs around 45S rDNA increase the fragility of this site (Color Plate 2c, chromosomes 1 and 6). In the case of a satellite loss, the remainder of the 45S rDNA block will be in the telomeric position as has been detected in many wheat and *Aegilops* species (Mukai et al., 1990; Dubcovsky and Dvořák 1995; Badaeva et al., 1996; Baum et al., 2004). When chromosomes are broken, the breakpoints become highly unstable and acquire the ability to fuse with other broken ends (McClintock 1941). However, the breakpoints are eventually stabilized, and the reconstructed chromosomes are transmitted to the daughter cells. This phenomenon, known as healing of breakpoints, involves the addition of repetitive telomere sequences at the breakpoints by telomerase, the enzyme that normally synthesizes the telomere sequence at normal chromosome terminals (Tsujimoto et al., 1997). According to Tsujimoto et al. (1999), rDNA sequences provide insight into the properties of telomerase activity at the breakpoints. The telomere sequences initiate two- to four-nucleotide motifs in the original rDNA sequence. These motifs are also found in the repeat unit of telomere sequences. Thus, it has been documented in many plant species that rDNA in terminal positions could stimulate *de novo* rapid synthesis of telomeres.

Therefore, we can emphasize that single Class I or II TEs constantly form distinct clusters in or around regular and irregular rDNA sites, and that the presence of TEs in or around rDNA sites increases the possibility of recombination and satellite loss. Apparently this event is common in plant karyotype evolution, since in many plant species rDNA clusters in terminal positions have been detected.

THE POTENTIAL OF WILD EMMER IN WHEAT IMPROVEMENT

Studies on wild emmer (*T. dicoccoides*), the progenitor of most tetraploid and hexaploid wheats, have revealed rich genetic resources applicable to wheat improvement, given its diverse single- and multilocus adaptations to stressful abiotic and biotic environments (Xie and Nevo 2008). The available resources have been described (Zohary 1970; Feldman 1979; Lange and Jochemsen 1992; Grama et al., 1983; Nevo 1995, 2001; Nevo et al.,

2002) and they include genotypic variation for traits such as: (i) germination, biomass, earliness, nitrogen content, and yield; (ii) amino acid composition; (iii) grain protein content and storage protein genes (HMW glutenins); (iv) disease resistances, including resistance to powdery mildew [*Blumeria graminis* (DC) E.O. Speer f. sp. *tritici*], leaf rust (*Puccinia triticina* Eriks.), stripe or yellow rust (*P. striiformis* Westend. f. sp. *tritici* Eriks.), stem rust (*Puccinia graminis* Pers.:Pers. f. sp. *tritici* Eriks. & E. Henn.), and *Soilborne wheat mosaic virus* (WSBMV); (v) high photosynthetic yield; (vi) salt tolerance; (vii) herbicide resistance; (viii) amylases and α-amylase inhibitors; and (ix) micronutrients such as Zn and Fe. This is only a preliminary list of the vast potential genetic resources existing in wild emmer that remain to be exploited for wheat improvement.

Quantitative trait loci (QTLs) and beneficial cryptic, agronomically important alleles have now been extensively described. The current genetic map of *T. dicoccoides*, with 549 molecular markers and 48 significant QTLs for 11 traits of agronomic importance (Peng et al., 2000b), will permit the unraveling of beneficial alleles of candidate genes that are otherwise hidden. These beneficial alleles could be introduced into cultivated wheat by marker-assisted selection.

The Near East, in general, and Israel, in particular (Nevo 1986), are the centers of origin and diversity of wild emmer, where it developed wide genetic adaptations against multiple pathogens and diverse ecological stresses. Genetic diversity is transferable from the wild to the cultivated gene pool, so genes of wild emmer are directly accessible for future wheat improvement. Consequently, exploration of *in situ* and *ex situ* collections (with optimized sampling strategies) along with utilization programs should maximize the contribution of wild emmer to wheat improvement. Among the potential donors for improving wheat, wild emmer occupies a very important and unique position due to its direct ancestry of bread wheat and its rich and largely adaptive genetic diversity. This was first suggested by Aaronsohn (1913) and later elaborated on by many authors (see Feldman 1977; Nevo 1983, 1995, 2001, 2006; Xie and Nevo 2008).

There are many ongoing programs around the world utilizing genes of wild emmer for wheat improvement, primarily involving genes coding for resistance to powdery mildew and the rusts, for high protein content, and for improved baking quality. Cultivars based on introgression of *T. dicoccoides* genes have appeared and will continue to appear in the near future. With *T. dicoccoides* at least three to four backcrosses with bread wheat are a necessity in breeding programs to minimize linkage drag (Groenewegen and van Silfhout 1988; Reader and Miller 1991). Wheat improvement programs will continue utilizing *T. dicoccoides* and other wheat relatives (Xie and Nevo 2008). Extensive work on transferring genes for high protein content from *T. dicoccoides* to cultivated wheat is currently underway in several laboratories (e.g., Weizmann Institute; US Department of Agriculture, Fargo, North Dakota; and University of California, Davis).

CONCLUDING REMARKS ON THE PROCESS OF WHEAT EVOLUTION

The molecular diversity and divergence of wheat species displays parallel ecological–genetic patterning and demonstrates the following: (i) significant genetic diversity and divergence exists at single- and multilocus structures of allozymes, random amplified polymorphic DNA, simple-sequence repeats, and single-nucleotide polymorphisms over very short distances of several to a few dozen meters; (ii) genetic patterns across coding and largely noncoding genomic regions are correlated with, and predictable by, environmental stress (climatic, edaphic, biotic) and heterogeneity (the niche-width variation hypothesis), displaying significant niche-specific and niche–unique alleles and genotypes; and (iii) genomic organization of wheat, including the noncoding genome, is nonrandom, heavily structured, and at least partly, if not largely, adaptive. The process of wheat evolution defies explanation by genetic drift, neutrality, or near neutrality alone as the primary driving forces. The main viable model to explain wheat genomic organization seems to be natural selection, primarily diversifying and balancing, and cyclical selection over space and time according to the two- or multiple-niche ecological models. Natural selection interacts with mutation, migration, and stochastic factors, but it overrides them in orienting the evolutionary processes of wheat.

FUTURE PERSPECTIVES

What will be the next step in wheat improvement in the current genomic and postgenomic eras? Conceptually, in-depth probing of comparative genome structure and function are the major challenges, including analyses of the intimate relationship between the coding and noncoding regions of the wheat genomes. Such studies will unravel genome evolution and highlight the rich genetic potentials for wheat improvement residing in wheat and various wheat relatives, including *Triticum* and *Aegilops* species as well as other Triticeae.

We believe that the theoretical and applied perspectives for future wheat improvement will encompass the following. First is characterization of the genome structure, function, regulation, and evolution at macro- and microgeographic scales of wheat and wheat-related species. Second is to combine multilocus markers and fitness-related traits to produce direct estimates of adaptive fitness differentiations within and between populations. Third, a critical activity will be to analyze the genetic system determining the enormous genetic flexibility of the various wheat and wheat-related species in diverse ecological contexts, mutation rates in different elements of the genome, recombination properties of the genome with their genetic and ecological control, and genomic distribution and function of structural genes (primarily abiotic and biotic stress genes). Fourth, it would be prudent to characterize the interface between ecological and genomic spatiotemporal dynamics and adaptive systems, to characterize genome evolution and the polyploidization processes, and to conduct colinearity studies between the grasses, including model species with small genome size such as rice and *Brachypodium distachyon*.

REFERENCES

Aaronsohn, A. 1910. Agricultural and botanical explorations in Palestine. USDA, Bureau of Plant Industry Bull. 180.

Aaronsohn, A. 1913. The discovery of wild wheat. City Club Bull. Chicago 6:167–173.

Aaronsohn, A., and G. Schweinfurth. 1906. Die Auffindung des wilden Emmers (*Triticum dicoccum*) in Nordpalästina. p. 213–220. Altneuland Monatsschrift für die Wirtschaftlicher Erschliessung Palästinas, No. 7–8, Berlin.

Anamthawat-Jonsson, K., and J.S. Heslop-Harrison. 1993. Isolation and characterization of genome-specific DNA sequences in Triticeae species. Mol. Gen. Genet. 240:151–158.

Appels, R., and E.S. Lagudah. 1990. Manipulation of chromosomal segments from wild wheat for the improvement of bread wheat. Aust. J. Plant Physiol. 17:253–366.

Appels, R., P. Reddy, C.L. McIntyre, L.B. Moran, O.H. Frankel, and B.C. Clarke. 1989. The molecular–cytogenetic analysis of grasses and its applications to studying relationships among species of the Triticeae. Genome 31:122–133.

Averbeck, K.T., and T.H. Eickbush. 2005. Monitoring the mode and tempo of concerted evolution in the *Drosophila melanogaster* rDNA locus. Genetics 171:1837–1846.

Avery, D. 1985. Farm dilemma: The global bad news is wrong. Science 230:408–412.

Avivi, L. 1978. High grain protein content in wild tetraploid wheat, *Triticum dicoccoides* Korn. p. 372–380. *In* S. Ramanujam (ed.) Proc. Int. Wheat Genetic Symp., 5th., New Delhi, India. 23–28 Feb. 1987. Indian Soc. Genet. and Plant Breeding, Indian Agric. Res. Inst., New Delhi, India.

Avivi, L. 1979. Utilization of *Triticum dicoccoides* for the improvement of grain protein quantity and quality in cultivated wheats. Monogr. Genet. Agrar. 4:27–38.

Avivi, L., A.A. Levy, and M. Feldman. 1983. Studies on high protein durum wheat derived from crosses with the wild tetraploid wheat *Triticum turgidum* var. *dicoccoides*. p. 199–204. *In* S. Sakamoto (ed.) Proc. Int. Wheat Genet. Symp., 6th, Kyoto, Japan. 28 Nov.–3 Dec. 1983. Plant Germ-Plasm Inst., Fac. Agric., Kyoto Univ., Kyoto, Japan.

Badaeva, E.D., A.V. Amosova, O.V. Muravenko, T.E. Samatadze, N.N. Chikida, A.V. Zelenin, B. Friebe, and B.S. Gill. 2002. Genome differentiation in *Aegilops*: 3. Evolution of the D-genome cluster. Plant Syst. Evol. 231:163–190.

Badaeva, E.D., A.V. Amosova, T.E. Samatadze, S.A. Zoshchuk, N.G. Shostak, N.N. Chikida, A.V. Zelenin, W.J. Raupp, B. Friebe, and B.S. Gill. 2004. Genome differentiation in *Aegilops*: 4. Evolution of the U-genome cluster. Plant Syst. Evol. 246:45–76.

Badaeva, E.D., O.S. Dedkova, G. Gay, V.A. Pukhalskyi, A.V. Zelenin, S. Bernard, and M. Bernard. 2007. Chromosomal rearrangements in wheat: Their types and distribution. Genome 50:907–926.

Badaeva, E.D., B. Friebe, and B.S. Gill. 1996. Genome differentiation in *Aegilops*: 2. Physical mapping of 5S and 18S-26S ribosomal RNA gene families in diploid species. Genome 39:1150–1158.

Badaeva, E.D., B. Friebe, S.A. Zoshchuk, A.V. Zelenin, and B.S. Gill. 1998. Molecular cytogenetic analysis of tetraploid and hexaploid *Aegilops crassa*. Chromosome Res. 6:629–637.

Baum, B.R., G.L. Bailey, A. Belyayev, O. Raskina, and E. Nevo. 2004. The utility of the nontranscribed spacer of the 5S rDNA units grouped into unit classes assigned to haplomes—a test on cultivated wheat and wheat progenitors. Genome 47:590–599.

Belyayev, A., and O. Raskina. 1998. Heterochromatin discrimination in *Aegilops speltoides* by simultaneous genomic in situ hybridization. Chromosome Res. 6:559–565.

Belyayev, A., O. Raskina, A. Korol, and E. Nevo. 2000. Coevolution of A and B genomes in allotetraploid *Triticum dicoccoides*. Genome 43:1021–1026.

Belyayev, A., O. Raskina, and E. Nevo. 2001. Chromosomal distribution of reverse transcriptase-containing retroelements in two Triticeae species. Chromosome Res. 9:129–136.

Belyayev, A., O. Raskina, and E. Nevo. 2005. Variability of Ty3-gypsy retrotransposons chromosomal distribution in populations of two wild Triticeae species. Cytogenet. Genome Res. 109:43–50.

Bennett, M.D., and J.P. Gustafson. 1982. The effect of telomeric heterochromatin from *Secale cereale* L. on triticale (X *Triticosecale* Wittmack): II. The presence or absence of blocks of heterochromatin in isogenic backgrounds. Can. J. Genet. Cytol. 24:93–100.

Bennett, M.D., and P.J. Kaltsikes. 1973. The duration of meiosis in a diploid rye, a tetraploid wheat and the hexaploid triticale derived from them. Can. J. Genet. Cytol. 15:671–679.

Bennett, M.D., and I.J. Leitch. 1997. Nuclear DNA amounts in angiosperms—583 new estimates. Ann. Bot. 80:169–196.

Bennett, M.D., and J.B. Smith. 1976. Nuclear DNA amounts in angiosperms. Philos. Trans. R. Soc. Lond. B Biol. Sci. 274:227–274.

Bennetzen, J.L. 2000. Comparative sequence analysis of plant nuclear genomes: Microcolinearity and its many exceptions. Plant Cell 12:1021–1030.

Bennetzen, J.L. 2002. Mechanisms and rates of genome expansion and contraction in flowering plants. Genetica 115:29–36.

Boissier, P.E. 1884. Flora orientalis. p. 673–679. Vol. 5. Soc. Phys. Genev., Soc. Linn. Londin., Geneva, Basel, and Lyon, Switzerland.

Brown, A.H.D., M.T. Clegg, A.L. Kahler, and B.S. Weir (ed.) 1990. Plant population genetics, breeding and genetic resources. Sinauer, Sunderland, MA.

Brown, A.H.D., G.J. Lawrence, M. Jenkin, J. Douglass, and E. Gregory. 1989. Linkage drag in backcross breeding in barley. J. Hered. 80:234–239.

Caldwell, K.S., J. Dvořák, E.S. Lagudah, E. Akhunov, M.C. Luo, P. Wolters, and W. Powell. 2004. Sequence polymorphism in polyploid wheat and their D-genome diploid ancestor. Genetics 167:941–947.

Chapman, V., T.E. Miller, and R. Riley. 1976. Equivalence of the A genome of bread wheat and that of *Triticum urartu*. Genet. Res. 27:69–76.

Darwin, C.D. 1859. On the origin of species by means of natural selection, or preservation of favoured races in the struggle for life. p. 352. John Murray, London, UK.

Daud, H.M., and J.P. Gustafson. 1996. Molecular determination of a B genome progenitor of wheat (*Triticum aestivum*) using a genome-specific repetitive DNA sequence. Genome 39:543–548.

Dhillon, S.S., E.A. Wernsman, and J.P. Miksche. 1983. Evaluation of nuclear DNA content and heterochromatin changes in anther-derived diploids of tobacco (*Nicotina tabacum*) cv. Coker 139. Can. J. Genet. Cytol. 25:169–173.

Diaz-Castillo, C., and K.G. Golic. 2007. Evolution of gene sequence in response to chromosomal location. Genetics 177:359–374.

Dubcovsky, J., and J. Dvořák. 1995. Ribosomal RNA multigene loci: Nomads of the Triticeae genomes. Genetics 140:1367–1377.

Dvořák, J., and J. Dubcovsky. 1996. Genome analysis of polyploid species employing variation in repeated nucleotide sequences. p. 133–145. *In* P.P. Jauhar (ed.) Methods of genome analysis in plants. CRC Press, Boca Raton, FL.

Dvořák, J., M-C. Luo, Z-L. Yang, and H-B. Zhang. 1998. The structure of *Aegilops tauschii* genepool and the evolution of hexaploid wheat. Theor. Appl. Genet. 97:657–670.

Dvořák, J., P.E. McGuire, and B. Cassidy. 1988. Apparent sources of the A genomes of wheats inferred from the polymorphism in abundance and restriction fragment length of repeated nucleotide sequences. Genome 30:680–689.

Dvořák, J., and H.B. Zhang. 1990. Variation in repeated nucleotide sequences sheds light on the origin of the wheat B and G genomes. Proc. Natl. Acad. Sci. USA 87:9640–9644.

Dvořák, J., and H.B. Zhang. 1992. Reconstruction of the phylogeny of the genus *Triticum* from variation in repeated nucleotide sequences. Theor. Appl. Genet. 84:419–429.

Eickbush, D.G., and T.H. Eickbush. 2003. Transcription of endogenous and exogenous R2 elements in the rRNA gene locus of *Drosophila melanogaster*. Mol. Cell Biol. 23:3825–3836.

Elder, J.F., and B.J. Turner. 1995. Concerted evolution of repetitive DNA sequences in eucaryotes. Q. Rev. Biol. 70:297–320.

Fahima, T., J.H. Peng, J.P. Cheng, M.S. Roder, Y.I. Ronin, Y.C. Li, A.B. Korol, and E. Nevo. 2001. Molecular genetic maps in tetraploid wild emmer wheat, *Triticum dicoccoides*, based on microsatellite and AFLP markers. Abstract. Isr. J. Plant Sci. 49:154.

Fahima, T., M.S. Röder, A. Grama, and E. Nevo. 1998. Microsatellite DNA polymorphism divergence in *Triticum dicoccoides* accessions highly resistant to yellow rust. Theor. Appl. Genet. 96:187–195.

Fahima, T., G.L. Sun, A. Beharav, T. Krugman, A. Beiles, and E. Nevo. 1999. RAPD polymorphism of wild emmer wheat population, *Triticum dicoccoides*, in Israel. Theor. Appl. Genet. 98:434–447.

Feldman, M. 1977. Historical aspects and significance of the discovery of wild wheats. Stadler Symp. 9:121–124.
Feldman, M. 1979. Genetic resources of wild wheats and their use in breeding. Monogr. Genet. Agrar. 4:9–26.
Feldman, M. 2001. Origin of cultivated wheat. p. 3–58. *In* A.P. Bonjean and W.J. Angus (ed.) The world wheat book: A history of wheat breeding. Lavoisier Publishing, Paris, France.
Feldman, M., and A.A. Levy. 2005. Allopolyploidy—a shaping force in the evolution of wheat genomes. Cytogenet. Genome Res. 109:250–258.
Feldman, M., B. Liu, G. Segal, S. Abbo, A.A. Levy, and J.M. Vega. 1997. Rapid elimination of low-copy DNA sequences in polyploid wheat: A possible mechanism for differentiation of homoeologous chromosomes. Genetics 147:1381–1387.
Feldman, M., F.G.H. Lupton, and T.E. Miller. 1995. Wheats: *Triticum* spp. (Gramineae–Triticinae). p. 184–192. *In* J. Smartt and N.W. Simmonds (ed.) Evolution of crop plants. 2nd ed. Longman Scientific and Technical Press, London, UK.
Feldman, M., and E.R. Sears. 1981. The wild gene resources of wheat. Sci. Am. 244:102–112.
Feuillet, K., and B. Keller. 2002. Comparative genomics in the grass family: Molecular characterization of grass genome structure and evolution. Ann. Bot. 89:3–10.
Flavell, R.B., M.D. Bennett, J.B. Smith, and D.B. Smith. 1974. Genome size and the proportion of repeated sequence DNA in plants. Biochem. Genet. 12:257–269.
Flavell, A.J., S.R. Pearce, J.S. Heslop-Harrison, and A. Kumar. 1997. The evolution of Ty1-*copia* group retrotransposons in eukaryote genomes. Genetica 100:185–195.
Francis, D., M. Stuart Davies, and P.W. Barlow. 2008. A strong nucleotypic effect on the cell cycle regardless of ploidy level. Ann. Bot. 101:747–757.
Frankel, O.H., and E. Bennett. 1970. Genetic resources in plants—their exploration and conservation. Blackwell Scientific Publications, Oxford, UK.
Frankel, O.H., and J.G. Hawkes. 1975. Crop genetic resources for today and tomorrow. Cambridge University Press, Cambridge, MA.
Frankel, O.H., and M.E. Soulé. 1981. Conservation and evolution. Cambridge University Press, New York.
Friebe, B., and B.S. Gill. 1996. Chromosome banding and genome analysis in diploid and cultivated polyploid wheats. p. 39–60. *In* P.P. Jauhar (ed.) Methods of genome analysis in plants. CRC Press, Boca Raton, FL.
Friebe, B., L.L. Qi, S. Nasuda, P. Zhang, N.A. Tuleen, and B.S. Gill. 2000. Development of a complete set of *Triticum aestivum–Aegilops speltoides* chromosome addition lines. Theor. Appl. Genet. 101:51–58.
Furuta Y., K. Nishikawa, and S. Yamaguchi. 1986. Nuclear DNA content in diploid wheat and its relatives in relation to the phylogeny of tetraploid wheat. Jpn. J. Genet. 61:97–105.
Gaeta, R.T., J.C. Pires, F. Iniguez-Luy, E. Leon, and T.C. Osborn. 2007. Genomic changes in resynthesized *Brassica napus* and their effect on gene expression and phenotype. Plant Cell 19:3403–3417.
Gale, M.D., and K.M. Devos. 1998. Comparative genetics in the grasses. Proc. Natl. Acad. Sci. USA 95:1971–1974.
Gaut, B.S., S.I. Wright, C. Rizzon, J. Dvořák, and L.K. Anderson. 2007. Recombination: An underappreciated factor in the evolution of plant genomes. Nature Rev. 8:77–84.
Giorgi, D., R. D'Ovidio, O.A. Tanzarella, C. Celeoni, and E. Peorceddu. 2003. Isolation and characterization of S genome specific sequences from *Aegilops* sect. Sitopsis species. Genome 46:478–489.
Golenberg, E.M., and E. Nevo. 1987. Multilocus differentiation and population structure in a selfer, wild emmer wheat, *Triticum dicoccoides*. Heredity 58:951–956.
Grama, A., Z.K. Gerechter-Amitai, and A. Blum. 1983. Wild emmer as donor of genes for resistance to stripe rust and for high protein content. p. 187–192. *In* S. Sakamoto (ed.) Proc. Int. Wheat Genet. Symp., 6th, Kyoto, Japan. 28 Nov.–3 Dec. 1983. Plant Germ-Plasm Inst., Fac. Agric., Kyoto Univ., Kyoto, Japan.
Grant, V. 1981. Plant speciation. 2nd ed. Columbia University Press, New York.
Griffin, W.B. 1987. Out-crossing in New Zealand wheats measured by occurrence of purple grain. N.Z. J. Agric. Res. 30:287–290.
Groenewegen, L.J.M., and C.H. van Silfhout. 1988. The use of wild emmer in Dutch practical wheat breeding. p. 184–189. *In* M.L. Jorna and L.A.J. Slootmaker (ed.) Cereal breeding related to integrated cereal production. Pudoc, Wageningen, The Netherlands.
Grover, C.E., H. Kim, R.A. Wing, A.H. Paterson, and J.F. Wendel. 2007. Microcolinearity and genome evolution in the AdhA region of diploid and polyploid cotton (*Gossypium*). Plant J. 50:995–1006.
Gu, Y.Q., C.C. Crossman, X. Kong, M. Luo, F.M. You, D. Coleman Derr, J. Dubcovsky, and O.D. Anderson. 2004. Genomic organization of the complex α-gliadin gene loci in wheat. Theor. Appl. Genet. 109:648–657.
Gustafson, J.P. 1976. The evolutionary development of triticale: The wheat–rye hybrid. p. 107–135. *In* M.K. Hect, W.C. Steere, and B. Wallace (ed.) Evolutionary biology. Vol. 9. Plenum Press, New York.
Gustafson, J.P., and M.D. Bennett. 1982. The effect of telomeric heterochromatin from *Secale cereale* L. on triticale (× *Triticosecale* Wittmack): I. The influence of several blocks of telomeric heterochromatin on early endosperm development and kernel characteristics at maturity. Can. J. Genet. Cytol. 24:83–92.
Gustafson, J.P., A.J. Lukaszewski, and M.D. Bennett. 1983. Somatic deletion and/or redistribution of telomeric heterochromatin in the genus *Secale* and triticale. Chromosoma 88:293–298.
Han, F.P., G. Fedak, T. Ouellet, and B. Liu. 2003. Rapid genomic changes in interspecific and intergenomic hybrids and allopolyploids of Triticeae. Genome 46:716–723.
Harlan, J.R. 1975. Crops and man. ASA, Madison, WI.
Harlan, J.R. 1976. Genetic resources in wild relatives of crops. Crop Sci. 16:329–333.
Harlan, J.R. 1992. Crops and man. 2nd ed. ASA, CSSA, Madison, WI.

Hawkes, J.G. 1991. International work-shop on dynamic in-situ conservation of wild relatives of major cultivated plants: Summary and final discussion and recommendations. Isr. J. Bot. 40:529–536.

Hedrick, P.W. 1986. Genetic polymorphism in heterogeneous environments: A decade later. Annu. Rev. Ecol. Syst. 17:535–566.

Heslop-Harrison, J.S. 2000. RNA, genes, genomes and chromosomes: Repetitive DNA sequences in plants. Chromosomes Today 13:45–56.

Heyn, C.C., and M. Waldman. 1992. In situ conversation of plant with potential economic value. *In* R.P. Adams and J.E. Adams (ed.) Conservation of plant genes: DNA banking and in vitro biotechnology. Academic Press, Inc., San Diego, CA.

Hillman, G., and S. Colledge (ed.) 1998. The transition from foraging to farming in southwest Asia. Proc. Int. Workshop, Groningen, The Netherlands. Sept. 1998.

Jakob, S.S., A. Meister, and F.R. Blattner. 2004. Considerable genome size variation of *Hordeum* species (Poaceae) is linked to phylogeny, life form, ecology, and speciation rates. Mol. Biol. Evol. 21:860–869.

Jiang, N., Z. Bao, X. Zhang, S.R. Eddy, and S.R. Wessler. 2004. Pack-MULE transposable elements mediate gene evolution in plants. Nature 431:569–573.

Jiang, J., and B. Gill. 1994a. New 18S-26S ribosomal gene loci: Chromosomal landmarks for the evolution of polyploid wheats. Chromosoma 103:179–185.

Jiang, J., and B. Gill. 1994b. Different species-specific chromosome translocations in *Triticum timopheevii* and *T. turgidum* support the diphyletic origin of polyploid wheats. Chromosome Res. 2:59–64.

Johnson, B.L., and H.S. Dhaliwal. 1976. Reproductive isolation of *Triticum boeoticum* and *Triticum urartu* and the origin of the tetraploid wheats. Am. J. Bot. 63:1088–1096.

Johnson, B.L., and H.S. Dhaliwal. 1978. *Triticum urartu* and genome evolution in the tetraploid wheats. Am. J. Bot. 65:907–918.

Joppa, L.R., E. Nevo, and A. Beiles. 1995. Chromosome translocations in wild populations of tetraploid emmer wheat in Israel and Turkey. Theor. Appl. Genet. 91:713–719.

Karlin, S., and J.L. McGregor. 1972. Polymorphisms for genetics and ecological systems with weak coupling. Theor. Popul. Biol. 3:210–238.

Kashkush, K., M. Feldman, and A.A. Levy. 2002. Gene loss, silencing and activation in a newly synthesized wheat allotetraploid. Genetics 160:1651–1656.

Kawahara, T., and E. Nevo. 1996. Screening of spontaneous major translocations in Israeli populations of *Triticum dicoccoides* Koern. Wheat Information Serv. 83:28–30.

Kawahara, T., E. Nevo, and A. Beiles. 1993. Frequencies of translocations in Israeli populations of *Triticum dicoccoides* Korn. p. 8020. Proc. Int. Botany Congr., XV, Yokohama, Japan. 28 Aug.–3 Sept. 1993.

Kellogg, E.A., R. Appels, and R.J. Mason-Gamer. 1996. When genes tell different stories: The diploid genera of Triticeae (Gramineae). Syst. Bot. 21:321–347.

Kidwell, M.G., and D.R. Lisch. 2001. Perspective: Transposable elements, parasitic DNA and genome evolution. Evolution 55:1–24.

Kidwell, K., G. Shelton, V. DeMacon, M. McClendon, J. Smith, J. Baley, and R. Higginbotham. 2002. Spring wheat breeding and genetics. p. 24–26. *In* J. Burns and R. Veseth (ed.) Field Day Proc., 2002: Highlights of research progress. Washington State University, Pullman, WA.

Kihara, H. 1924. Cytologische und genetische Studien bei wichtigen Getreidearten mit besonderer Rücksicht auf das Verhalten der Chromosomen und die Sterilität in den bastarden. Mem. Coll. Sci. Kyoto Imp. Univ. Serv. Bull. 1:1–200.

Kihara, H. 1944. Discovery of the DD-analyser, one of the ancestors of *Triticum vulgare* (Japanese). Agric. Hort. (Tokyo) 19:13–14.

Kimber, G., and M. Feldman. 1987. Wild wheats, an introduction. p. 1–142. Special Report 353. College of Agriculture, University of Missouri, Columbia, MO.

Kimber, G., and E.R. Sears. 1987. Evolution in the genus *Triticum* and the origin of cultivated wheat. *In* E.G. Heyne, (ed.) Wheat and wheat improvement. 2nd ed. ASA, CSSA, SSSA, Madison, WI.

Kislev, M.E., D. Nadel, and I. Carmi. 1992. Epi-palaeolithic (19 000 BP) cereal and fruit diet at Ohalo II, Sea of Galilee, Israel. Rev. Palaeobot. Palynol. 3:161–166.

Konarev, V.G., I.P. Gavrilyuk, N.K. Gubareva, and T.I. Peneva. 1979. About nature and origin of wheat genomes on the data of biochemistry and immunochemistry of grain proteins. Cereal Chem. 56:272–278.

Lai, J., Y. Li, J. Messing, and H.K. Dooner. 2005. Gene movement by Helitron transposons contributes to the haplotype variability of maize. Proc. Natl. Acad. Sci. USA 102:9068–9073.

Lange, W., and G. Jochemsen. 1992. Use of the gene pools of *Triticum turgidum* ssp. *dicoccoides* and *Aegilops squarrosa* for the breeding of common wheat (*T. aestivum*), through chromosome-doubled hybrids. Euphytica 59:197–212.

Leitch, I.J., and M.D. Bennett. 2004. Genome downsizing in polyploid plants. Bot. J. Linn. Soc. 82:651–663.

Leitch, I.J., L. Hanson, K.Y. Lim, A. Kovarik, M.W. Chase, J.J. Clarkson, and A.R. Leitch. 2008. The ups and downs of genome size evolution in polyploid species of *Nicotiana* (Solanaceae) Ann. Bot. 101:805–814.

Lelley, T., M. Stachel, H. Grausgruber, and J. Vollmann. 2000. Analysis of relationships between *Aegilops tauschii* and the D genome of wheat utilizing microsatellites. Genome 43:661–668.

Levy, A.A., and M. Feldman. 1987. Increase in grain protein percentage in high-yielding common wheat breeding lines by genes from wild tetraploid wheat. Euphytica 36:353–359.

Lev-Yadun, S., A. Gopher, and S. Abbo. 2000. The cradle of agriculture. Science 288:1602–1603.

Li, Y.C., T. Fahima, A. Beiles, A.B. Korol, and E. Nevo. 1999. Microclimatic stress and adaptive DNA differentiation in wild emmer wheat, *Triticum dicoccoides*. Theor. Appl. Genet. 99:873–883.

Li, Y.C., T. Fahima, A.B. Korol, J.H. Peng, M.S. Röder, V.M. Kirzhner, A. Beiles, and E. Nevo. 2000a. Microsatellite diversity correlated with ecological-edaphic and genetic factors in three microsites of wild emmer wheat in North Israel. Mol. Biol. Evol. 17:851–862.

Li, Y.C., T. Fahima, T. Krugman, A. Beiles, M.S. Röder, A.B. Korol, and E. Nevo. 2000b. Parallel microgeographic patterns of genetic diversity and divergence revealed by allozyme, RAPD, and microsatellites in *Triticum dicoccoides* at Ammiad, Israel. Conserv. Genet. 1:191–207.

Li, Y.C., T. Fahima, J.H. Peng, M.S. Röder, V.M. Kirzhner, A. Beiles, A.B. Korol, and E. Nevo. 2000c. Edaphic microsatellite DNA divergence in wild emmer wheat, *Triticum dicoccoides* at a microsite: Tabigha, Israel. Theor. Appl. Genet. 101:1029–1038.

Li, Y.C., M.S. Röder, T. Fahima, V.M. Kirzhner, A. Beiles, A.B. Korol, and E. Nevo. 2000d. Natural selection causing microsatellite divergence in wild emmer wheat at the ecologically variable microsite at Ammiad, Israel. Theor. Appl. Genet. 100:985–999.

Lilienfeld, F., and H. Kihara. 1934. Genomanalyse bei *Triticum* and *Aegilops*: V. *Triticum timopheevi* Zhuk. Cytologia 6:87–122.

Lipman, Z., A-V. Gendrel, M. Black, M.W. Vaughn, N. Dedhia, W.R. McCombie, K. Lavine, V. Mittal, B. May, K.D. Kasschau, J.C. Carrington, R.W. Doerge, V. Colot, and R. Martienssen. 2004. Role of transposable elements in heterochromatin and epigenetic control. Nature 430:471–476.

Liu, B., J.M. Vega, and M. Feldman. 1998. Rapid genomic changes in newly synthesized amphiploids of *Triticum* and *Aegilops*: II. Changes in low-copy coding DNA sequences. Genome 41:535–542.

Liu, B., and J.F. Wendel. 2002. Epigenetic phenomena and the evolution of plant allopolyploids. Mol. Phylogenet. Evol. 29:365–379.

Loidl, J. 1979. C-band proximity of chiasmata and absence of terminalisation in *Allium flavum* (Liliaceae). Chromosoma 73:45–51.

Luo, M.C., Z.L. Yang, and J. Dvořák. 2000. The Q locus of Iranian and European spelt wheat. Theor. Appl. Genet. 100:602–606.

Lupton, F.G.H. 1987. Wheat breeding: Its scientific basis. Chapman and Hall, London, UK.

Ma, X-F., P. Fang, and J.P. Gustafson. 2004. Polyploidization-induced genome variation in triticale. Genome 47:839–848.

Ma, X-F., and J.P. Gustafson. 2005. Genome evolution of allopolyploids: A process of cytological and genetic diploidization. Cytogenet. Genome Res. 109:236–249.

Ma, X-F., and J.P. Gustafson. 2006. Timing and rate of genome variation in triticale following allopolyploidization. Genome 49:950–958.

Maan, S.S. 1973. Cytoplasmic and cytogenetic relationships among tetraploid *Triticum* species. Euphytica 22:287–300.

Maestra, B., and T. Naranjo. 1999. Structural chromosome differentiation between *Triticum timopheevii* and *T. turgidum* and *T. aestivum*. Theor. Appl. Genet. 98:744–750.

Maestra, B., and T. Naranjo. 2000. Genome evolution in Triticeae. p. 155–167. *In* E. Olmo and C.A. Redi (ed.) Chromosomes today. Vol. 13. Birkhäuser Verlag, Basel, Boston, MA.

Martin, T.J. 1990. Out-crossing in twelve hard red winter wheat cultivars. Crop Sci. 30:59–62.

Maxted, N., B.V. Ford-Lloyd, and J.G. Hawkes. 1997. Plant genetic conservation: The *in situ* approach. Chapman and Hall, London, UK.

McClintock, B. 1941. The stability of broken ends of chromosomes in *Zea mays*. Genetics 26:234–282.

McClintock, B. 1946. Maize genetics. Carnegie Institution of Washington Year Book 45:176–186.

McFadden, E.S., and E.R. Sears. 1944. The artificial synthesis of *Triticum spelta*. Rec. Genet. Soc. Am. 13:26–27.

McFadden, E.S., and E.R. Sears. 1946a. The origin of *Triticum spelta* and its free-threshing hexaploid relatives. J. Hered. 37:81–89.

McFadden, E.S., and E.R. Sears. 1946b. The origin of *Triticum spelta* and its free-threshing hexaploid relatives: Hybrids of synthetic *T. spelta* with cultivated relatives. J. Hered. 37:107–116.

McFadden, E.S., and E.R. Sears. 1947. The genome approach in radical wheat breeding. Agron. J. 39:1011–1026.

McIntyre, C.L. 1988. Variation at isozyme loci in Triticeae. Plant Syst. Evol. 160:123–142.

Mello-Sampayo, T. 1971. Genetic regulation of meiotic chromosome pairing by chromosome 3D of *Triticum aestivum*. Nature New Biol. 230:22–23.

Messing, J., A.K. Bharti, W.M. Karlowski, H. Gundlach, H.R. Kim, Y. Yu, F. Wei, G. Fuks, C.A. Soderlund, K. F. Mayer, and R.A. Wing. 2004. Sequence composition and genome organization of maize. Proc. Natl. Acad. Sci. USA 101:14349–14354.

Miller, T.E. 1987. Systematics and evolution. p. 1–30. *In* F.G.H. Lupton (ed.) Wheat breeding: Its scientific basis. Chapman and Hall, London, UK.

Miller, T.E. 1992. A cautionary note on the use of morphological characters for recognizing taxa of wheat (genus *Triticum*). p. 249–253. Préhistoire de l'agriculture: Nouvelles approches expérimentales et ethnographiques. Monographie du CRA. No. 6. CNRs.

Morrison, W.R. 1988. Lipids. p. 373–439. *In* Y. Pomeranz (ed.) Wheat chemistry and technology. AACC, Washington, DC.

Mukai, Y., T.R. Endo, and B.S. Gill. 1990. Physical mapping of the 5S rRNA multigene family in common wheat. J. Hered. 8:290–295.

Muramatsu, M. 1986. The *vulgare* super gene, *Q*: Its universality in durum wheat and its phenotypic effects in tetraploid and hexaploid wheats. Can. J. Genet. Cytol. 28:30–41.

Naranjo, T. 1990. Chromosome structure of durum wheat. Theor. Appl. Genet. 79:397–400.

Nesbitt, M. 1998. The transition from foraging to farming in Southwest Asia. Proc. Int. Workshop, Groningen, The Netherlands. Sept. 1998.

Nesbitt, M. 2001. Wheat evolution: Integrating archaeological and biological evidence. p. 37–59. *In* P.D.S. Caligari and P.E. Brandham (ed.) Wheat taxonomy: The legacy of John Percival. Vol. 3, Linnean, Special Issue. Linnean Society, London, UK.

Nesbitt, M., and D. Samuel. 1998. Wheat domestication: Archaeobotanical evidence. Science 279:1433.

Nevo, E. 1983. Genetic resources of wild emmer wheat: Structure, evolution and application in breeding. p. 421–431. *In* S. Sakamoto (ed.) Proc. Int. Wheat Genet. Symp., 6th, Kyoto, Japan. 28 Nov.–3 Dec. 1983. Plant Germ-Plasm Inst., Fac. Agric., Kyoto Univ., Kyoto, Japan.

Nevo, E. 1986. Genetic resources of wild cereals and crop improvement: Israel, a natural laboratory. Isr. J. Bot. 35:255–278.

Nevo, E. 1989. Genetic resources of wild emmer wheat revisited: Genetic evolution, conservation and utilization. p. 121–126. *In* T.E. Miller and R.M.D. Koebner (ed.) Proc. Int. Wheat Genet. Symp., 7th, Cambridge, UK. 13–19 July 1988. Inst. Plant Science Res., Cambridge Laboratory, Cambridge.

Nevo, E. 1992. Origin, evolution, population genetics and resources for breeding of wild barley, Hordeum spontaneum, in the Fertile Crescent. p. 19–43. *In* P. Shewry (ed.) Barley: Genetics, molecular biology and biotechnology. CAB Int., Wallingford, UK.

Nevo, E. 1994. Evolutionary significance of genetic diversity in nature: Environmental stress, pattern and theory. p. 267–296. *In* C.L. Markert, J.G. Scandalios, H.A. Lim, and O.L. Serov (ed.) Isozymes: Organization and roles in evolution, genetics and physiology. Int. Congr. on Isozymes, 7th, Novosibirsk, Russia. 6–13 Sept. 1992. World Scientific, Hackensack, NJ.

Nevo, E. 1995. Genetic resources of wild emmer, *Triticum dicoccoides* for wheat improvement: News and views. p. 79–87. *In* Proc. Int. Wheat Genet. Symp., 8th, Beijing, China. 20–25 July, 1993. China Agricultural Scientech Press, Beijing.

Nevo, E. 1998. Genetic diversity in wild cereals: Regional and local studies and their bearing on conservation ex-situ and in-situ. Genet. Resour. Crop Evol. 45:355–370.

Nevo, E. 2001. Genetic resources of wild emmer, *Triticum dicoccoides*, for wheat improvement in the third millennium. Isr. J. Plant Sci. 49:77–91.

Nevo, E. 2006. Genome evolution of wild cereal diversity and prospects for crop improvement. Plant Genet. Resour. 4:36–46.

Nevo, E., and A. Beiles. 1989. Genetic diversity of wild emmer wheat in Israel and Turkey: Structure, evolution and application in breeding. Theor. Appl. Genet. 77:421–455.

Nevo, E., A. Beiles, A.B. Korol, Y.I. Ronin, T. Pavliček, and W.D. Hamilton. 2000. Extraordinary multilocus genetic organization in mole crickets, gryllotalpidae. Evolution 54:586–605.

Nevo, E., A. Beiles, and T. Krugman. 1988a. Natural selection of allozyme polymorphisms: A microgeographic climatic differentiation in wild emmer wheat, *Triticum dicoccoides*. Theor. Appl. Genet. 75:529–538.

Nevo, E., A. Beiles, and T. Krugman. 1988b. Natural selection of allozyme polymorphisms: A microgeographic differentiation by edaphic, topographical and temporal factors in wild emmer wheat *Triticum dicoccoides*. Theor. Appl. Genet. 76:737–752.

Nevo, E., E.M. Golenberg, A. Beiles, A.H.D. Brown, and D. Zohary. 1982. Genetic diversity and environmental associations of wild wheat, *Triticum dicoccoides*, in Israel. Theor. Appl. Genet. 62:241–254.

Nevo, E., A. Grama, A. Beiles, and E.M. Golenberg. 1986. Resources of high-protein genotypes in wild wheat, *Triticum dicoccoides* in Israel: Predictive method by ecology and allozyme markers. Genetica 68:215–227.

Nevo, E., A.B. Korol, A. Beiles, and T. Fahima. 2002. Evolution of wild emmer and wheat improvement. p. 364. *In* E. Nevo, A.B. Korol, A. Beiles, and T. Fahima (ed.) Population genetics, genetic resources, and genome organization of wheat's progenitor, *Triticum dicoccoides*. Springer-Verlag, Berlin.

Nishikawa, K., S. Mizuno, and Y. Furuta. 1994. Identification of chromosomes involved in translocations in wild emmer. Jpn. J. Genet. 69:371–376.

Ohtsuka, I. 1998. Origin of the central European spelt wheat. p. 303–305. *In* A.E. Slinkard (ed.) Proc. Int. Wheat Genet. Symp., 9th, Saskatoon, SK. 2–7 Aug. 1998. University Extension Press, Saskatoon, SK, Canada.

Okamoto, M. 1962. Identification of the chromosomes of common wheat belonging to the A and B genomes. Can. J. Genet. Cytol. 4:31–37.

Ozkan, H., A.A. Levy, and M. Feldman. 2001. Allopolyploidy-induced rapid genome evolution in the wheat (*Aegilops–Triticum*) group. Plant Cell 13:1735–1747.

Peng, J.H., T. Fahima, M.S. Röder, Q.Y. Huang, A. Dahan, Y.C. Li, A. Grama, and E. Nevo. 2000a. High-density molecular map of chromosome region harboring stripe-rust resistance genes YrH52 and Yr15 derived from wild emmer wheat, *Triticum dicoccoides*. Genetica 109:199–210.

Peng, J.H., T. Fahima, M.S. Röder, Y.C. Li, A. Grama, and E. Nevo. 2000b. Microsatellite high-density mapping of the stripe rust resistance gene YrH52 region on chromosome 1B and evaluation of its marker-assisted selection in the F2 generation in wild emmer wheat. New Phytol. 146:141–154.

Peng, J.H., A.B. Korol, T. Fahima, M.S. Röder, Y.I. Ronin, Y.C. Li, and E. Nevo. 2000c. Molecular genetic maps in wild emmer wheat, *Triticum dicoccoides*: Genome-wide coverage, massive negative interference, and putative quasi-linkage. Genome Res. 10:1509–1531.

Penton, E.H., and T.J. Crease. 2004. Evolution of the transposable element Pokey in the ribosomal DNA of species in the subgenus *Daphnia* (Crustacea: Cladocera). Mol. Biol. Evol. 21:1727–1739.

Pestsova, E.G., N.P. Goncharov, and E.A. Salina. 1998. Elimination of a tandem repeat of telomeric heterochromatin during the evolution of wheat. Theor. Appl. Genet. 97:1380–1386.

Plucknett, D.L., N.J.H. Smith, J.T. Wiliams, and N.M. Anishette. 1983. Crop germplasm conservation and developing countries. Science 220:163–169.

Plucknett, D.L., N.J.H. Smith, J.T. Wiliams, and N.M. Anishette. 1987. Gene banks and the world's food. Princeton University Press, Princeton, NJ.

Raskina, O., A. Belyayev, and E. Nevo. 2002. Repetitive DNAs of wild emmer wheat *Triticum dicoccoides* and their relation to S-genome species: Molecular-cytogenetic analysis. Genome 45:391–401.

Raskina, O., A. Belyayev, and E. Nevo. 2004a. Activity of the En/Spm-like transposons in meiosis as a base for chromosome repatterning in a small, isolated, peripheral population of *Aegilops speltoides* Tausch. Chromosome Res. 12:153–161.

Raskina, O., A. Belyayev, and E. Nevo. 2004b. Quantum speciation in *Aegilops*: Molecular cytogenetic evidence from rDNA clusters variability in natural populations. Proc. Natl. Acad. Sci. USA 101:14818–14823.

Reader, S.M., and T.E. Miller. 1991. The introduction into bread wheat of a major gene for resistance to powdery mildew from wild emmer wheat. Euphytica 53:57–60.

Rees H., and M.R. Walters. 1965. Nuclear DNA and the evolution of wheat. Heredity 20:73–82.

Riley, R., and V. Chapman. 1967. Effect of 5BS in suppressing the expression of altered dosage of 5BL on meiotic chromosome pairing in *Triticum aestivum*. Nature 216:60–62.

Riley, R., J. Unrau, and V. Chapman. 1958. Evidence on the origin of the B genome of wheat. J. Hered. 49:91–98.

Rodríguez, S., B. Maestra, E. Perera, M. Díez, and T. Naranjo. 2000a. Pairing affinities of the B- and G-genome chromosomes of polyploid wheats with those of *Aegilops speltoides*. Genome 43:814–819.

Rodríguez, S., E. Perera, B. Maestra, M. Díez, and T. Naranjo. 2000b. Chromosome structure of *Triticum timopheevii* relative to *T. turgidum*. Genome 43:923–930.

Salina, E.A., K.Y. Lim, E.D. Badaeva, A.B. Scherban, I.G. Adonina, A.V. Amosova, T.E. Samatadze, T.Y. Vatolina, S.A. Zoschuk, and A.R. Leitch. 2006. Phylogenetic reconstruction of *Aegilops* section Sitopsis and the evolution of tandem repeats in the diploids and derived wheat polyploids. Genome 49:1023–1035.

Salina, E.A., E.G. Pestsova, and A.V. Vershinin. 1997. Spelt1—new family of cereal tandem repeats. Russ. J. Genet. 33:352–357.

Sarkar, P., and G.L. Stebbins. 1956. Morphological evidence concerning the origin of the B genome in wheat. Am. J. Bot. 43:297–304.

Sax, K. 1922. Sterility in wheat hybrids: II. Chromosome behavior in partially sterile hybrids. Genetics 7:513–552.

Schiemann, E. 1956. Fünfzig Jahre *Triticum dicoccoides*. Ber. Dtsch. Bot. Ges. 69:309–322.

Schubert, I., and U. Wobus. 1985. In situ hybridization confirms jumping nucleolus organizing regions in *Allium*. Chromosoma 92:143–148.

Sears, E.R. 1954. The aneuploids of common wheat. Mo. Agric. Exp. Stn. Res. Bull. 572:1–59.

Sears, E.R. 1977. An induced mutant with homoeologous pairing in common wheat. Can. J. Genet. Cytol. 19:585–593.

Shaked, H., K. Kashkush, H. Ozkan, M. Feldman, and A.A. Levy. 2001. Sequence elimination and cytosine methylation are rapid and reproducible responses of the genome to wide hybridization and allopolyploidy in wheat. Plant J. 13:1749–1759.

Shands, H., and G. Kimber. 1973. Reallocation of the genomes of *Triticum timopheevi* Zhuk. p. 101–108. *In* E.R. Sears and L.M.S. Sears (ed.) Proc. Int. Wheat Genet. Symp., 4th, Columbia, MO. 6–11 Aug. 1973. Mo. Agric. Exp. Stn., Columbia, MO.

Sharma, S., and S.N. Raina. 2005. Organization and evolution of highly repeated satellite DNA sequences in plant chromosomes. Cytogenet. Genome Res. 109:15–26.

Soltis, D.E., and P.S. Soltis. 1999. Polyploidy: Recurrent formation and genome evolution. Trends Ecol. Evol. 14:348–352.

Song, K.M., P. Lu, K.I. Tang, and T.C. Osborn. 1995. Rapid change in synthetic polyploids of *Brassica* and its implications for polyploid evolution. Proc. Natl. Acad. Sci. USA 92:7719–7723.

Sorrells, M.E., C.M. La Rota, C.E. Bermudez, et al. 2003. Comparative DNA sequence analysis of wheat and rice genomes. Genome Res. 13:1818–1827.

Stebbins, G.L. 1947. Types of polyploids: Their classification and significance. Adv. Genet. 1:403–420.

Stebbins, G.L. 1950. Variation and evolution in plants. Columbia University Press, New York.

Talbert, L.E., L.Y. Smith, and N.K. Blake. 1998. More than one origin of hexaploid wheat is indicated by sequence comparison of low-copy DNA. Genome 41:402–407.

Tang, H., J.E. Bowers, X. Wang, R. Ming, M. Alam, and A.H. Paterson. 2008. Synteny and collinearity in plant genomes. Science 320:486–488.

Tsujimoto, H., N. Usami, K. Hasegawa, T. Yamada, K. Nagaki, and T. Sasakuma. 1999. De novo synthesis of telomere sequences at the healed breakpoints of wheat deletion chromosomes. Mol. Gen. Genet. 262:851–856.

Tsujimoto, H., T. Yamada, and T. Sasakuma. 1997. Molecular structure of a wheat chromosome end healed after gametocidal gene-induced breakage. Proc. Natl. Acad. Sci. USA 94:3140–3144.

Upadhya, M.D., and M.S. Swaminathan. 1963. Deoxyribonucleic acid and the ancestry of wheat. Nature 200:713–714.

Valdes, B., V.H. Heywood, F.M. Raimondo, and D. Zohary (ed.) 1997. Proc. Workshop on Conservation of the Wild Relatives of European Cultivated Plants. Herbarium Mediterraneum Panormitanum, Palermo, Italy.

Van't Hof, J., and A.H. Sparrow. 1963. A relationship between DNA content, nuclear volume, and minimum mitotic cycle time. Proc. Natl. Acad. Sci. USA 49:897–902.

Van Zeist, W. 1976. On macroscopic traces of food plants in southwestern Asia (with some references to pollen data). Philos. Trans. R. Soc. Lond. Biol. Sci. 272:27–41.

Van Zeist, W., and J.A.H. Bakker-Heeres. 1985. Archaeological studies in the Levant, 1. Neolithic sites in the Damascus Basin: Aswad, Ghoraife, Ramad. Palaeohistoria 24:165–256.

Wang G., N.T. Miyashita, and K. Tsunewaki. 1997. Plasmon analyses of *Triticum* (wheat) and *Aegilops*: PCR-single-stand conformational polymorphism (PCR-SSCP) analyses of organeller DNAs. Proc. Natl. Acad. Sci. USA 94:14570–14577.

Wang, W.Y., X.L. Shi, B.L. Hao, S. Ge, and J.C. Luo. 2005. Duplication and DNA segmental loss in the rice genome: Implications for diploidization. New Phytol. 165:937–946.
Wendel, J.F. 2000. Genome evolution in polyploids. Plant Mol. Biol. 42:225–249.
Wicker, T., N. Yahiaoui, R. Guyot, E. Schlagenhauf, Z-D. Liu, J. Dubcovsky, and B. Keller. 2003. Rapid genome divergence at orthologous low molecular weight glutenin loci of the A and A^m genomes of wheat. Plant Cell 15:1186–1197.
Williams, P.C. 1993. The world of wheat. p. 557–602. *In* Grains and oilseeds: Handling, marketing, processing. 4th ed. Canadian International Grains Institute, Winnipeg, MB, Canada.
Xie, W., and E. Nevo. 2008. Wild emmer: Genetic resources, gene mapping and potential for wheat improvement. Euphytica 164:603–614.
Yen, Y., and P.S. Baenziger. 1996. Chromosomal locations of genes that control major RNA-degrading activities in common wheat (*Triticum aestivum* L). Theor. Appl. Genet. 93:645–648.
Young, A. 1999. Is there really spare land? A critique of estimates of available cultivable land in developing countries. Environ. Dev. Sustain. 1:3–18.
Zhang Q., Y. Dong, X. An, A. Wang, Y. Zhang, X. Li, L. Gao, X. Xia, Z. He, and Y. Yan. 2008. Characterization of HMW glutenin subunits in common wheat and related species by matrix-assisted laser desorption/ionization time-of-flight mass spectrometry. J. Cereal Sci. 47:252–261.
Zohary, D. 1969. The progenitors of wheat and barley in relation to domestication and agricultural dispersal in the Old World. p. 47–66. *In* P.J. Ucko and G.W. Dimbley (ed.) The domestication and exploitation of plants and animals. Duckworth, London, UK.
Zohary, D. 1970. Wild wheats. p. 239–247. *In* O.H. Frankel and E. Bennett (ed.) Genetic resources in plants—their exploitation and conservation. Blackwell, Oxford, UK.
Zohary, D., and M. Feldman. 1962. Hybridization between amphiploids and the evolution of polyploids: I. The wheat (*Aegilops–Triticum*) group. Evolution 16:44–61.
Zohary, D., and M. Hopf. 1993. Domestication of plants in the old world—the origin and spread of cultivated plants in West Asia, Europe, and the Nile valley. 2nd ed. Oxford University Press, New York.
Zohary, D., and M. Hopf. 2000. Domestication of plants in the old world. 3rd ed. Clarendon Press, Oxford, UK.

Chapter 2
Development of the Wheat Plant

Gregory S. McMaster

SUMMARY

(1) Wheat development is important in creating structures such as leaves and roots needed to capture resources, and also to create the structures ultimately needed to produce viable seed or the desired quality for grain.

(2) Wheat canopy development can be considered at many scales of the plant but often is first viewed at the highest scale of the whole-plant canopy. The canopy can also be considered as the result of the appearance, growth, and abortion or senescence of shoots or tillers. At the lowest scale, each shoot consists of a basic phytomer unit.

(3) A phytomer unit is normally considered to be the leaf, the node plus internode above the node, and an axillary bud. The axillary bud gives rise to new shoots. The root nodal bud should also be considered part of the vegetative phytomer unit. A shoot therefore can be viewed as the appearance, growth, and abortion or senescence of phytomers, or components of the phytomers, that leads to dynamically changing canopies over the growing season and among years.

(4) Regardless of the scale considered, morphological naming schemes have been developed to uniquely identify all parts of the plant, and phenology growth staging scales describe the progress of the tiller or canopy through the life cycle.

(5) Wheat development is orderly and predictable. Genetics provides the orderliness, and environmental factors, mainly temperature, are used to predict development. Thermal time is used as an estimate of the biological clock the wheat plant uses to mark time. Thermal time can be calculated many different ways, but most fundamentally an average temperature is estimated over a time interval (often daily) and used in a temperature-response function to determine the effectiveness of temperature on development rate.

(6) The external phenological progression through the life cycle has also been coordinated with developmental events occurring at the shoot apex, resulting in the complete developmental sequence of the tiller or canopy.

(7) Simulation models of wheat have increasingly incorporated these developmental concepts to varying degrees.

(8) Much of the work describing wheat development is quite empirical, but molecular biology is contributing newfound understanding to underlying genes and mechanisms controlling developmental events. Genetic pathways controlling flowering and plant stature are notable achievements in new understanding.

INTRODUCTION

Development and growth are related, but distinct, processes. As with many terms, precise definitions that satisfy all can be difficult. Therefore, for the purpose of this chapter, growth is defined as the permanent increase in volume, and development is the initiation and differentiation of organs and the progression of stages through which cells, organs, and plants pass during their life cycle. Often, but not always, growth and development occur simultaneously.

J.W. von Goethe recognized the orderly development of plants in the late 18th century. Since that time, extensive research has quantified the orderly and predictable nature of how the wheat plant proceeds from germination through its life cycle to maturity. This orderliness is observed at all scales of the wheat plant, from the whole canopy to the whole shoot, as well as within the shoot. Genes determine the orderly sequence of wheat development, and the rate and timing of development respond to environmental conditions.

Development serves two purposes. First, it creates the structures such as leaves, roots, and stems needed to capture resources (e.g., light, water, nutrients) that are then used to produce a viable seed. Second, development produces the inflorescence structures needed to produce a viable ovule that can be pollinated at anthesis. Therefore, development is essential in producing the yield potential and the ultimate realization of that potential in final yield.

This chapter discusses the development of the wheat plant. Given the vast research on wheat development and extensive breadth of aspects that can be covered, the goal is to capture and challenge current thinking about wheat development as it fits into the overall purposes of this book. This chapter examines, first, different scales of wheat development, and second, how a shoot develops since the canopy is merely a collection of shoots. The underlying building block of a shoot is the phytomer, and this is discussed to better understand shoot or canopy development. Environmental impacts on development are considered, particularly the critical role of temperature. This chapter presents an overview of simulating wheat development and concludes with new information emerging from molecular biology that provides better mechanistic understanding of wheat development.

SCALES OF PLANT DEVELOPMENT

A cursory walk through a wheat field readily exposes the dynamic nature of wheat development, with many different scales to consider (Masle-Meynard and Sebillotte 1981). Variation in seedling emergence or among wheat spikes that are spatially distributed and differing in size and greenness reflect variation in the canopy. This variation likely will be greater in a larger field with varying soils and topography than within small areas within a field. Further variation might be introduced by variation in genes within a plant community (e.g., single cultivar or multiple cultivars) which control development, though breeders attempt to minimize this variation within a given cultivar. A higher level of resolution is represented by the shoots that comprise the canopy. Examination of individual shoots reveals other differences, such as height, leaf number and size, and stage of development, and this is the result of the basic building block of a shoot, the phytomer. Therefore, at least three scales of plant development can be important: the canopy, the shoot, and the phytomer. Selection of the appropriate scale is dependent on the research or production problem and level of understanding desired.

Canopies

A wheat field is the collection of plants that form the canopy. Plants emerge and grow at different rates, and some die prematurely. Differential patterns of seedling emergence cause some of the variation in development observed in the field. The pattern of seedling emergence varies greatly depending on many nongenetic factors that affect germination and seedling emergence such as planting depth and rate, and soil moisture, temperature, and strength. These factors are highly dependent on the climate, soils, and management practice; spatial variation even at the microscale of less than one meter can cause variation in seedling emergence. Genetic factors such as strength of seed dormancy will influence the pattern of seedling emergence, but even expression of this trait is not completely independent of environmental influences.

The importance is that the pattern of seedling emergence is reflected in variation in the development observed within a field throughout the growing season. For instance, seedlings that emerge earliest normally are bigger, reach developmental stages earlier, and are higher yielding than those seedlings that emerge later (Gan and Stobbe 1995). Interestingly, although seedlings that emerge later may reach developmental stages later, their development rate often is faster. However, the increased development rate is insufficient to offset their delayed emergence; thus they may still reach maturity slightly later (Nuttonson 1948; Angus et al., 1981; O'Leary et al., 1985) or, if simultaneously, then at the cost of reduced growth.

Shoots or tillers

The canopy is a collection of shoots or tillers which appear, grow, abort, and ultimately senesce, albeit at variable rates among individuals within the canopy. The cumulative number of shoots that appear is dependent on many factors, including the density of plants, the genotype, the environment, and management (Darwinkel 1978; Masle-Meynard and Sebillote 1981; Fraser et al., 1982; Masle 1985). Seedling emergence is important not only in determining the number of plants, but also in determining the number of shoots that a plant produces that will likely survive to produce a spike. This is because a specific axillary bud that produces a tiller has a window of time during which it can appear (Klepper et al., 1982), and once this window passes, that axillary bud will not further differentiate and grow. In most instances, tillers that appear the earliest and from axillary buds on the main stem will be the last ones to abort. This has implications for final yield prediction, as these tillers and the main shoot (i.e., the first shoot to emerge from the seed) are the primary yield-producing shoots (Power and Alessi 1978; McMaster et al., 1994).

Some propose that because the main shoot is usually the most productive, management should focus on producing stands of uniculm plants (i.e., only main shoots). Producing tillers, particularly those that abort before physiological maturity, merely "waste" resources. This approach seems most viable for high-production environments that are less variable in precipitation and extreme events. In highly variable environments, such as many semiarid wheat production regions, tillering can partially adjust for winter kill and other causes of plant loss, and it can provide an adjustment for planting rates normally used that are less than optimal for favorable years (McMaster et al., 2002). Further, when planting at high densities and reducing tiller spikes, main-stem spikes usually will be less productive, so little gain is obtained by the uniculm approach.

Just as later emerging plants usually require less thermal time (i.e., fewer growing degree-days) to reach a given developmental stage, the same has been observed for shoots that appear later, as well as the variation among the main stem and different tillers in reaching a developmental stage being reduced as the plant approaches physiological maturity (Hay and Kirby 1991). This increasing synchrony among shoots results in less canopy variation over time.

Despite differences in developmental stage among shoots on the plant, the pattern of development is the same for each shoot. Further, the response to environmental factors is similar among shoots, although specific responses of individual shoots will vary because they are in different environments within the canopy and the vascular connections and root systems of each shoot will be different. This will be discussed in greater detail later in the chapter.

Phytomers

The shoot is composed of subunits, or building blocks, called phytomers (Gray 1879; Bateson 1894). The phytomer has generally been defined as the leaf, node, internode above the node, and the axillary bud (Wilhelm and McMaster 1995). Each phytomer component can appear, change over time, and/or abort or senesce. In wheat, prior to the beginning of stem elongation, each phytomer has a minimal internode. The node is the region where the leaf attaches to the shoot and the vascular tissue connects the shoot and leaf; the nodes are closely stacked just below the shoot apex located in the crown of the plant. Once the signal to begin internode elongation occurs, the intercalary meristem of the most recently formed node at the apical region of the node begins cell division and expansion that extends the shoot apex above the crown and ultimately to the top of the canopy. Similarly, the leaf appears, grows, and ultimately senesces. The axillary bud can produce either another shoot, as in wheat, or an inflorescence structure (e.g., either a shoot or the ear of maize, *Zea mays* L.).

For plants such as wheat that can produce nodal roots, the phytomer concept should be expanded to include this component (Forster et al., 2007). Given this definition, generally the phytomer has been viewed as a vegetative building block, yet it is clear that the basic unit is repeated throughout the grass inflorescence (Forster et al., 2007).

The phytomer concept has proven to be a useful botanical abstraction for providing a foundation to understand plant development and architecture. It is the appearance, growth, and senescence of these phytomers that determine the characteristics of individual shoots. This dynamic interplay of phytomers can be considered analogous to a composition of music called a canon (a familiar simple form being a round), where individual phytomers repeat a part against and with other phytomers as do the melodies of a canon (Hargreaves and McMaster 2008). Similarly, the integration of these phytomers within a plant has been outlined in a generic pattern by Rickman and Klepper (1995) for a variety of grasses. This is an excellent demonstration of the consistent and orderly development of the grass shoot apex.

MORPHOLOGICAL NAMING SCHEMES

Naming schemes which uniquely identify each part of the plant can have many advantages. Certainly they aid in communicating precisely what part of the plant is being measured. Naming schemes also can provide insight into how the plant perceives its environment and the efficacy of management practices. Although various nomenclatures have been proposed and modified, most are quite similar.

Leaves

A simple leaf-naming scheme has been proposed by Jewiss (1972) and Klepper et al. (1982, 1983a), in which true leaves (not including the coleoptilar leaves) are numbered acropetally for each culm. The first leaf is designated L1, with successive leaves L2, L3, up to the last leaf formed on the shoot (i.e., the flag leaf). A numerical leaf staging system was proposed by Haun (1973):

$$\text{Haun stage} = (n - 1) + L_n/L_{n-1}, \; (0 < L_n/L_{n-1} \leq 1)$$

in which n is the number of leaves that have appeared on the culm, L_{n-1} is the blade length of the penultimate leaf, and L_n is the blade length of the youngest visible leaf extending from the sheath of the penultimate leaf.

One increasingly important application of knowing leaf number in crop management is that many pesticides are to be applied at certain leaf stages. For instance, the herbicide imazamox is to be applied at the fourth-leaf stage for control of many grass weeds when growing the cultivar Above.

Tillers

The system first proposed by Jewiss (1972) for naming tillers has been modified and extended (Masle-Meynard and Sebillotte 1981; Fraser et al., 1982; Kirby et al., 1985a; Masle 1985). The modified system proposed by Klepper et al. (1982, 1983a) is increasingly being adopted. In this system, the leaf axil and parent culm are used to name the tiller. The first culm to emerge from the seed is the main stem (MS), with subsequent new culms being tillers that can be considered primary, secondary, tertiary, and so on, based on the parent culm. Tillers appearing from axillary buds in the axils of leaves on the MS are considered primary tillers, those from axils of leaves on primary tillers are secondary tillers, and so on. All tillers are designated with a "T" and then numbers. For a primary tiller, the number is a single digit that refers to the leaf number with which the axillary bud is associated. For example, T1 is the tiller emerging from the first leaf (L1) on the MS. Secondary tillers are given a two-digit designation, with the first digit referring to the primary tiller number and the second digit referring to the leaf number. For example, T21 is the tiller emerging from the first leaf (L1) on the primary tiller T2. This system continues for tertiary tillers with a three-digit designation, and so on. The somewhat anomalous coleoptile tiller, which emerges from the axil of the coleoptile leaf, is designated as either T0 (Klepper et al., 1982, 1983a) or TC (Kirby and Eisenberg 1966; Kirby and Appleyard 1984).

Knowledge of the presence or absence of specific tillers has been used to provide information on how the wheat plant perceives its environment. For instance, the proportion of T0 tillers present can indicate the seedbed conditions for seedling emergence. Also, in returning to the window of time that a tiller can appear, the absence of a given tiller indicates conditions were sufficiently stressful to prevent the development and growth of the axillary bud.

Inflorescence parts

The leaf and tiller morphological naming schemes have been extended to the wheat inflorescence. Klepper et al. (1983b) devised a numerical index for the developmental stages of the inflorescence, which extended phenological growth staging scales discussed later in this chapter. However, the morphological naming scheme was not completely developed. Wilhelm and McMaster (1996) proposed a spikelet-naming scheme similar to the leaf-naming scheme: the first spikelet at the base of the spike is designated S1, with subsequent spikelets numbered acropetally until the terminal spikelet. Each floret is also numbered acropetally from the base of the spikelet (e.g., F1, F2, etc.). If referring to the caryopsis, then "C" is used rather than "F" for floret. Each part of the inflorescence can be identified by combining the spikelet and floret/caryopsis designations. For instance, S1F1 refers to the basal floret on the first spikelet of the inflorescence. If the shoot-naming scheme is added, the specific inflorescence on the plant can be identified (e.g., T1S1F1). As with tillers, missing florets or spikelets, or very small caryopsis, indicates a stressful condition leading to abortion or reduced growth of these organs.

Roots

The wheat root system consists of seminal and nodal roots. Seminal roots (usually five to six roots from one seed) are those originating from primordia found in the seed, and nodal roots are those produced from primordia developed after germination (Klepper et al., 1984). The naming system created by Klepper et al. (1984) applies to both seminal and nodal roots. At each node of the shoot, two roots can appear at opposite sides of the node (X and Y zones); two other roots can appear at opposite sides of the node (A and B zones), but rotated 90° from the X and Y zones. Naming schemes for shoots can be applied for nodal roots, with seminal roots originating from the MS. In this system, the timing of root appearance is integrated with leaf appearance (Rickman et al., 1995). The timing, appearance, and growth of tillers and roots as a function of nitrogen fertilizer was examined and found to be strongly impacted by N fertility (Belford et al., 1987).

SHOOT DEVELOPMENT

Morphological naming schemes allow for nondestructive identification of plant parts and provide a context for understanding the developmental processes leading to the appearance, growth, and abortion or senescence of the plant parts. The developmental processes leading to the plant parts are a result of the developmental sequence of the shoot apex and the external developmental stages of the entire shoot (i.e., phenology). Most developmental events occurring at the shoot apex cannot be observed without destructive sampling and often magnification. The external developmental stages have traditionally focused on developmental events that can be observed fairly readily without magnification or destructive sampling.

Phenology

Human cultures have long recognized that plants go through fairly consistent stages of development each year. Common terminology has often referred to these stages as growth stages, but developmental stages is the more appealing term as often little "growth" is involved in the developmental event (e.g., anthesis and physiological maturity). Many so-called growth staging scales have been developed to describe wheat phenology, with many similarities between them (e.g., Frank et al., 1997). Discussions and comparisons among some scales are provided by Bauer et al. (1983), Landes and Porter (1989), and Harrell et al. (1993, 1998).

Of the many growth staging scales, four currently have the greatest usage: Feekes (Large 1954), Zadoks et al. (1974), Haun (1973), and the BBCH Scale (Lancaster et al., 1991). All scales consider some basic developmental events such as germination and emergence, leaf production, tillering, internode elongation, flowering or anthesis, stages of grain ripening, and physiological maturity (Fig. 2.1), with particular scales emphasizing different processes.

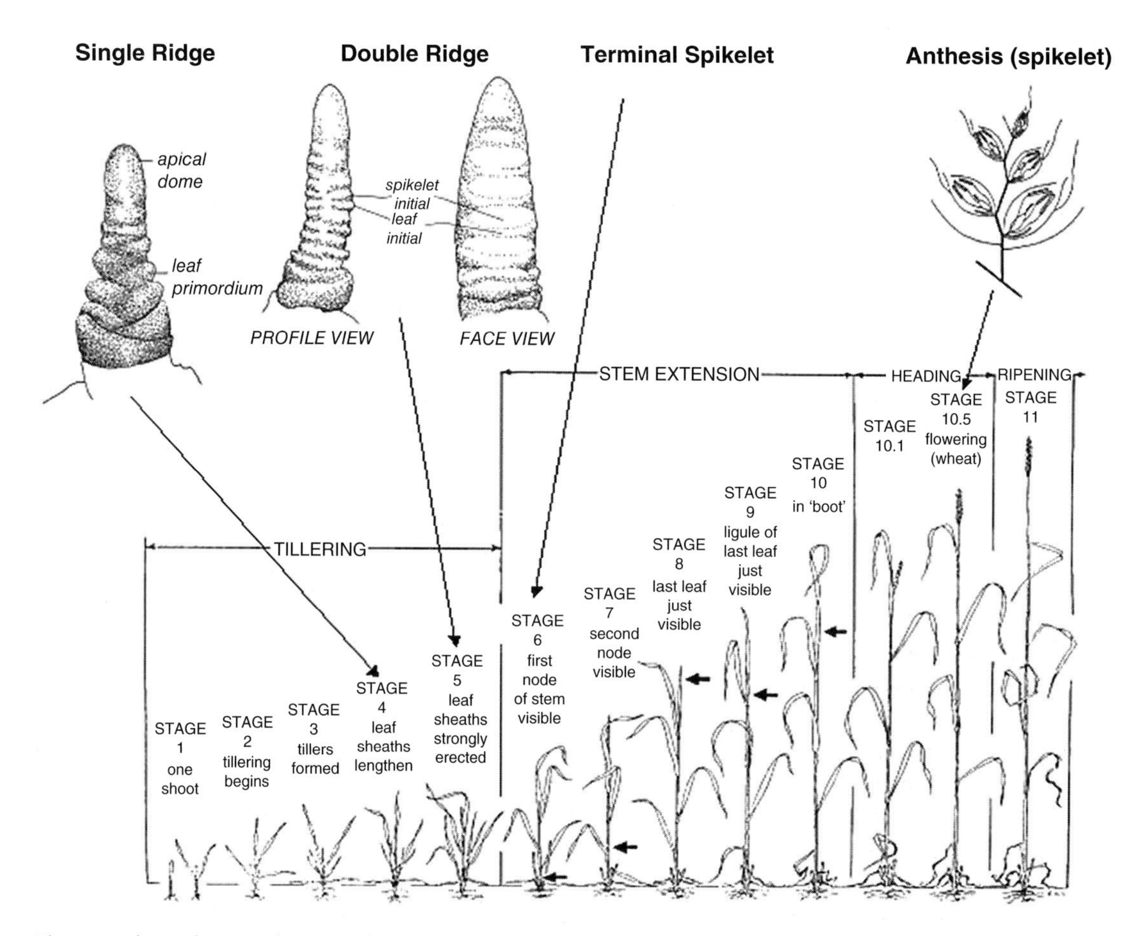

Fig. 2.1 The Feekes growth stage scale (Large 1954) correlated with the approximate timing of certain shoot apex developmental stages. See Table 2.1 for descriptions of developmental stages. (Adapted from McMaster 2005.)

All cultivars and each shoot go through the same developmental event (again, this is the orderliness of plant development), but cultivars and shoots vary on the timing (beginning and ending of the developmental event) and duration. This is a result of both differences in genotype and the cultivars' responses to environmental conditions. The period of grain filling illustrates this well. Cultivar trials clearly demonstrate that within the same environment cultivars vary in the time of internode elongation, anthesis, and physiological maturity, and that this also affects the duration of grain filling. Synchrony among shoots in reaching a developmental stage increases as physiological maturity is approached (Hay and Kirby 1991).

Clear definitions of many developmental stages are not always reported. Descriptions for the Feekes (Large 1954) and Zadoks (Zadoks et al., 1974) scales are provided in Table 2.1. For instance, the start of internode elongation and the developmental stage of jointing are two stages with some ambiguity, particularly for determination in the field. Both scales only discuss stem elongation in terms of when nodes are above the soil surface; therefore the beginning of internode elongation is earlier than this, as often the first node is formed below the soil surface. Jointing in the Feekes scale is defined as when the first node is observed to be one inch (2.54 cm) above the soil surface, although many practitioners merely record when the node is first observed at any distance above the soil surface. The time in which internode elongation begins prior to jointing is difficult to discern in the field without destructive sampling; thus internode elongation is sometimes erroneously assumed to begin at jointing. The concept of first-hollow-stem stage was created to call attention to the developmental difference between initiation of internode elongation and jointing in wheat production systems with grazing cattle (*Bos taurus* L.). Producers depend on staging to know the proper time to remove cattle from wheat pasture just before jointing so that further development and extension of the shoot apex is not impaired (Redmon et al., 1996).

Booting is defined as the stage when the spike can be felt within the whorl of leaf sheaths but is not visible. Given the ambiguity of measuring this stage relative to when booting actually begins and that it is a continuum until heading, a useful definition is to assume booting begins when the flag leaf has formed its ligule and continues until heading. Heading is defined as beginning when the first spikelet of the spike (i.e., head or ear) first appears above the ligule of the flag leaf at the top of the canopy. Normally the first spikelet to appear is the terminal spikelet, and heading is completed when the basal spikelet (= S1 in the morphological naming scheme) appears. Generally awns are ignored in observing the beginning of heading, and occasionally the spike emerges "sideways" from the sheath, and many observers consider this heading.

Physiological maturity is somewhat difficult to quantify and to consistently determine among different observers. Unlike maize that has a black layer in the kernel to indicate when maximum dry weight has been reached (the definition of physiological maturity), wheat has no such discernible trait. The Feekes scale defines harvest maturity as when the kernel is difficult to divide along the crease. Similarly, Zadoks et al. (1974) equate 90% ripeness of rice (*Oryza sativa* L.) to when the seed cannot be dented by the fingernail. Neither scale clearly defines physiological maturity. An association of maturity and maximum dry weight has been suggested when all glumes, paleas, and lemmas of the spike have lost all green color; all leaves will have senesced before this time and internodes will have lost all green color (Hanft and Wych 1982). This is a rather obvious time of maximum dry weight, as all sources of new carbohydrates via photosynthesis are gone and reserves should be allocated to grains or exhausted.

The location of the shoot apex changes with developmental stage and has implications for the environment of the shoot apex and certain management practices such as cattle grazing mentioned previously. Until the time that internode elongation begins, the shoot apex is located in the crown of the plant. The crown is normally located

Table 2.1 Description of principal developmental stages in the context of Feekes (Large, 1954) and Zadoks (Zadoks et al. 1974) developmental scales, with suggested measurement characteristics.

Stage	Description	Measurement Characteristics
Germination	Feekes—no stage Zadoks—01 (start of imbibition), 03 (imbibition complete), 05 (radicle emerged from caryopsis), 07 (coleoptile emerged from caryopsis)	Start of imbibition is when the seed begins to swell
Emergence	Feekes—no stage Zadoks—09 (leaf just at coleoptile tip)	Beginning of emergence is when the first true leaf emerges through the coleoptile and tip is visible above the soil surface
Tillering	Feekes—1.0 (main shoot only), 2.0 (beginning of tillering) Zadoks—20 (main shoot only), 21–29 (main shoot and 1 tiller through 9 tillers)	Beginning of tillering is when the first tiller is visible (likely T0 or T1)
Single ridge	Neither scale describes	Begins when the shoot apex shape changes from dome to more elongated, and leaf primordia begin to form a ridge around apex (Fig. 2.1)
Double ridge	Neither scale describes	Begins when the formation of double ridges (bottom ridge = leaf, top ridge = spikelet) around the apex occurs (Fig. 2.1)
Terminal spikelet	Neither scale describes	When the apical spikelet primordium appears and noted by rotation of 90º from plane of previous spikelets
Internode elongation	Neither stage clearly notes the beginning of internode elongation; rather it is assumed to be in stem elongation when node(s) are visible above the soil surface. Feekes—6 (first node of stem visible at base of shoot above soil surface), 7 (second node visible, next to last leaf just visible) Zadoks—31 (first node detectable) through 36 (sixth node detectable)	Beginning of internode elongation is when the first node is visible (but normally will be below the soil surface)
Jointing	Feekes—implicitly assumed to be when the first node is visible above the soil surface (Stage 6) and when plant growth habit changes from prostrate to upright Zadoks—not described	Occurs when first node is visible above the soil surface
Flag leaf	Feekes—8 (last leaf visible but still rolled, ear beginning to swell), 9 (ligule of last leaf just visible), 10 (sheath of last leaf completely grown out, ear swollen but not yet visible) Zadoks—39 (flag leaf ligule/collar just visible), 41 (flag leaf sheath extending, booting)	Cannot know if it is the flag leaf while it is appearing; flag leaf growth is considered complete when the ligule is visible and no new leaf is emerging
Booting	Feekes—see above for 8–10 Zadoks—41 (see above), 43 (boot just visibly swollen), 45 (boot swollen), 47 (flag leaf sheath opening), 49 (first awns visible)	As booting is a continuum, begins when the flag leaf ligule is visible and ends when heading begins
Heading	Feekes—10.1 (first ears just visible), 10.2 (1/4 of heading process completed), 10.3 (half completed), 10.4 (3/4 completed), 10.5 (all ears out of sheath) Zadoks—50 (first spikelet of inflorescence just visible), 53 (1/4 of inflorescence emerged), 55 (half of inflorescence emerged), 57 (3/4 emerged), 58 (inflorescence completely emerged)	Measure based on one inflorescence (main stem best), and heading begins when first spikelet is visible (awns not considered) and ends when the inflorescence has completely emerged from the flag leaf
Anthesis	Feekes—10.5.1 (beginning of flowering, 10.5.2 (flowering complete to top of ear), 10.5.3 (flowering over at base of ear), 10.5.4 (flowering over, kernel watery-ripe) Zadoks—61 (beginning of anthesis), 65 (anthesis halfway), 69 (anthesis complete)	Measure based on one inflorescence (main stem best); ends when first anther (yellow) is visible on inflorescence, and ends when no more anthers appear on the inflorescence
Physiological maturity	Feekes—11.1 (milky ripe), 11.2 (mealy ripe, contents of kernel soft but dry), 11.3 (kernel hard, difficult to divide by thumbnail), 11.4 (ripe for cutting, straw dead) Zadoks—a series of stages from milk development (77–77), dough development (83–87), and ripening and seed dormancy (91–99)	No scale definitively identifies when physiological maturity is reached (defined as maximum dry weight); Hanft and Wych (1982) indicate when all components of the spike (glumes, paleas, lemmas), internode tissue, and leaves have lost all green color

Developmental stage descriptions derived from Bauer et al. (1983) and the original references. Some developmental stages are shown in Fig. 2.1.

at a depth of about 2 to 3 cm below the soil surface. If planting is deeper than this, as is often the case, then the coleoptilar internode will elongate to form the crown at this depth. As the internodes elongate, the shoot apex moves from the soil environment (i.e., crown) into the aerial environment within the canopy. In many wheat production systems, the danger of cold temperatures (i.e., frost) is present at this time and significant damage to the shoot apex spikelet and floret primordia that are forming and differentiating can occur, resulting in yield loss. At heading, the shoot apex or spike emerges above the canopy and the glumes, paleas, lemmas, and awns (if present) are exposed to full sunlight. Primarily the flag leaf and spike photosynthetic tissues contribute new assimilates for grain filling, with the spike contributing up to about 10% of the flag leaf photosynthetic rate (Weyhrich et al., 1995).

Shoot apex

Developmental events occur at many places within the plant; however, the shoot apex is the site of many of the most important developmental events such as leaf, tiller, and inflorescence primordia production. As previously mentioned, these events are not represented in developmental scales (Fig. 2.1). As with phenology, the shoot apex of all cultivars and shoots has the same developmental sequence (Fig. 2.2), but cultivars and shoots vary in the timing and duration of the developmental event. Furthermore, this developmental sequence is generally shared by most grasses, particularly the annual cereal crops such as barley (*Hordeum vulgare* L.) and rice (Rickman and Klepper 1995). Many reviews of wheat shoot apex development are available that provide considerable detail (e.g., Barnard 1955; Bonnett 1966; Kirby and Appleyard 1984; McMaster 1997).

The developmental sequence of the shoot apex begins with the embryo within the seed. Wheat typically has three to four leaf primordia formed in the embryo (Bonnett 1966; Baker and Gallagher 1983a; Hay and Kirby 1991). Once germination of the seed has occurred, further development and growth of the existing leaf primordia occurs, resulting in leaf appearance of L1, L2, and so on. Coinciding with the appearance of the first leaves is the initiation of more leaf primordia. Leaf primordia are initiated up to the stage of double ridge. At double ridge, the leaf primordium forms a ridge around the apex and this primordium will not further differentiate and grow (Fig. 2.1). The ridge above the leaf primordium is the spikelet primordium. Primordia formed prior to double ridge will continue to differentiate and grow resulting in the continuation of leaf appearance until the flag leaf appears, which is about the second leaf to appear after the developmental stage of jointing.

Developing the leaf area index (LAI) of the canopy is a function of the appearance, growth, and senescence of leaves on each shoot comprising the canopy. Leaf size increases on a shoot up to about the 10th leaf, although the flag leaf tends to be slightly smaller than the penultimate leaf (Gallagher 1979; Hay and Wilson 1982; Rawson et al., 1983; Kirby et al., 1985b). Under unstressed conditions, leaves do not begin senescing until about 6.5 phyllochrons (time interval for appearance of successive leaves) after first appearing, and abiotic stresses such as water deficit and low N availability will enhance the senescence rate (McMaster et al., 1991; Wilhelm et al., 1993). Therefore leaves of increasing size accumulate on a shoot, and leaf senescence begins with the smallest and oldest leaves. Generally peak LAI is reached at the time the flag leaf completes growth.

The rate of leaf primordia initiation (plastochron) and appearance (phyllochron) are critical to many subsequent developmental events. For instance, until the leaf primordium has been initiated and begins to grow, the axillary bud that can form a new tiller is not initiated. This is why the beginning of tiller appearance is delayed from leaf appearance in Fig. 2.2. The rate of leaf appearance therefore "controls" the window of time that the tiller can appear. Leaf primordium differentiation and growth also determines the formation of the node, which is the point where vascular tissue enters the leaf.

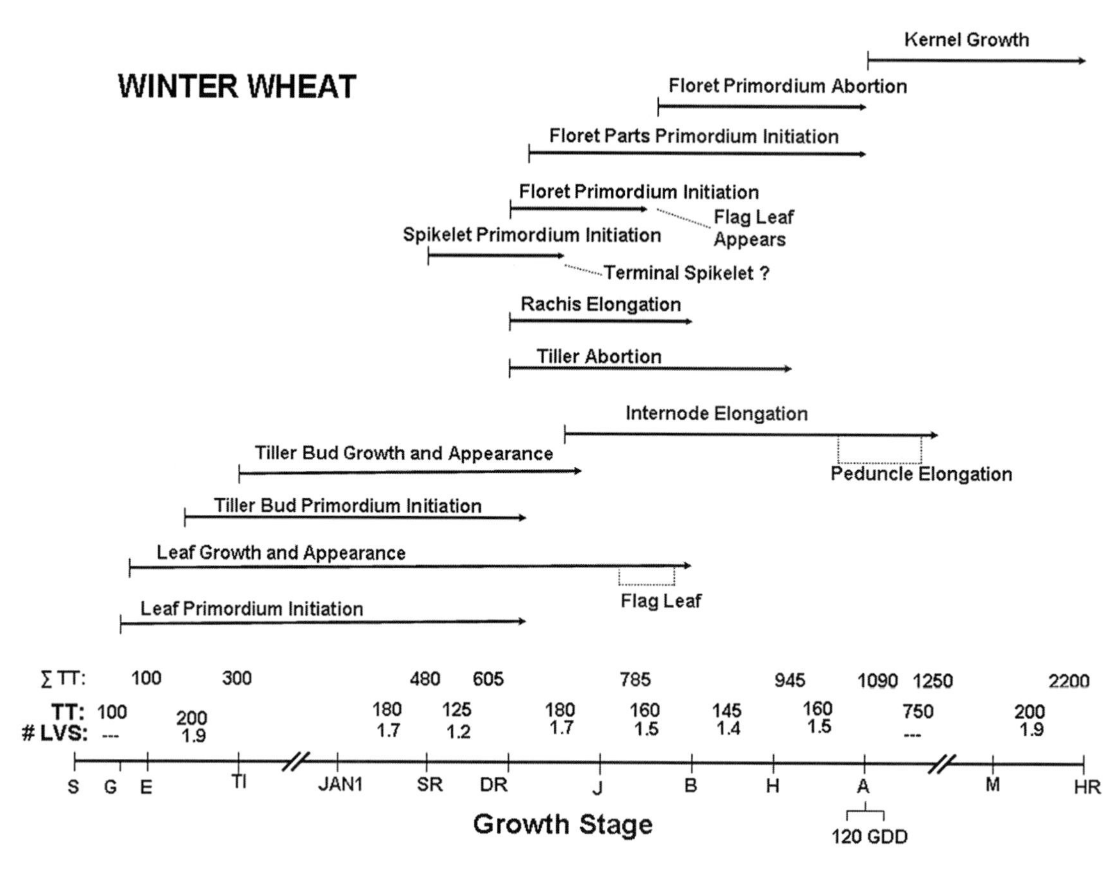

Fig. 2.2 Winter wheat shoot apex developmental sequence correlated with developmental stages (Large 1954) for conditions with no stresses. The timeline is presented as thermal time (TT, in growing degree-days, GDD, using 0ºC base and Method 1 of McMaster and Wilhelm 1997) and number of leaves (#LVS). Question marks indicate areas of uncertainty or significant variation among cultivars. See Table 2.1 for descriptions of developmental stages. (Adapted from McMaster et al., 1992b.)

An important transition of the shoot apex is from leaf to spikelet primordium initiation, and this occurs at the double ridge stage. The general rule is the rate of spikelet primordia initiation is about two to three times greater than leaf primordia initiation (Baker and Gallagher 1983a,b; Hay and Kirby 1991). Extensive physiological research has examined the environmental factors influential in determining when this switch occurs. Of these factors, the vernalization requirement and photoperiod sensitivity of the genotype play a critical role in the switch.

Several vernalization genes (*VRN*) have been identified as further discussed in Chapter 3, and their presence or absence determines the vernalization requirement for the genotype. The physiological conceptual model is that cool temperatures are needed to satisfy the vernalization requirement and that vernalization is a process of accumulating days with cool temperatures until a sufficient period of time with cool temperatures has occurred. Winter wheat cultivars with a high vernalization requirement may require 6 weeks or more of cool temperatures to complete their vernalization requirement. Temperatures within the range of about 2–7 °C seem most effective for vernalizing, with decreasing effectiveness below or above this range (McMaster et al., 2008). High temperatures (>30 °C) can also "undo" some of the vernalization (i.e., devernalize) that had previously occurred.

Photoperiod sensitivity varies considerably among genotypes and is the result of which photoperiod (*PPD*) genes are present (also discussed in Chapter 3). The conceptual model is that photoperiod or day length must reach a certain threshold to produce the signal that induces the primordium switch. Cultivars with low photoperiod sensitivity tend to have a low day length requirement so that it is met under most environments. Some uncertainty has existed on how to combine the vernalization and photoperiod responses in developmental models, with the view often held that vernalization requirements must be at least largely met before the photoperiod response can occur (McMaster et al., 2008).

It is difficult to precisely determine the vernalization response and photoperiod sensitivity of a genotype, but efforts to estimate these from loci present in the genotype appear promising (White et al., 2008). Genes for frost tolerance can also complicate the relationship (Prasil et al., 2004). Until recently, the mechanisms and genetic pathways for the developmental switch were not known. Chapter 3 provides an excellent summary of the genetic pathway as currently understood and explains the variation observed among genotypes.

Once spikelet primordium initiation begins at double ridge, spikelet primordia are produced until near the time that internode elongation begins, when the terminal spikelet is initiated (Fig. 2.2). Shortly after a spikelet primordium is initiated, floret primordia are initiated acropetally within the spikelet. Floret primordium differentiation then occurs to produce the structures of the floret. As many as 10 floret primordia may be initiated within a spikelet, and initiation occurs until about when the flag leaf appears. After this time, floret primordium abortion tends to occur basipetally within a spikelet until the time of anthesis. The end result of the overlapping sequence of spikelet and floret primordia initiation is that, in the mature spike, central spikelets begin differentiation first and thus have the longest period for initiating florets and longest time for those florets to fully develop. This is why central spikelets have the most kernels and kernel number declines acropetally and basipetally from central spikelets.

Yield potential has been determined by the time of flowering, as all florets have been produced. Pollination of fully developed florets is critical in determining grain set and final yield. Environmental conditions determine the success of grain set, with temperature and water stress normally the most important factors. In many semiarid wheat production regions, hot and dry conditions can significantly reduce grain set both through loss of pollen viability and early seed abortion.

Wheat is self-pollinated, and pollination follows the pattern of floret primordia initiation within the spike discussed previously. This has implications for final kernel size. Earlier fertilization events allow greater time for seed development and growth, and potentially larger final kernel size. Seed development and growth proceeds through a clear set of steps, and extensive research has been published on that subject [Herzog (1986) and McMaster (1997) cite many references]. Ontogeny of the seed mainly consists of development of the embryo and the endosperm tissue. Embryo development is completed prior to endosperm maturity, with the loss of seed moisture in the endosperm that is necessary for combine harvesting (usually about 120 g kg^{-1} moisture).

Growth of all kernels is characterized by a sigmoidal pattern that is frequently divided into three phases: lag phase, linear phase, and maturation phase (Herzog 1986). The lag phase primarily consists of cell division, the linear phase is driven by maximum cell expansion growth rates, and the maturation phase is mostly the loss of seed moisture. Abiotic and biotic factors strongly influence final kernel size, consummated both by the duration of the kernel growth (primarily the length of the linear phase) and by the rate of growth (the slope of the linear phase). Both of these growth factors are quite variable among cultivars, but in most instances, the duration of kernel growth is more important than rate of kernel growth in determining final kernel size (Herzog 1986; McMaster 1997). A management consideration in heat-stressed environments is managing when grain filling occurs as a means of avoiding adverse high temperatures.

Figure 2.2 presents a time series of the presence of sources and sinks, and indirectly source and sink activity. Although not shown, root growth is the first main significant plant component early in the life cycle. As leaves and tillers begin to appear, they are initially an equal sink (for growth) as roots. At the time of the initiation of internode elongation, leaf appearance has nearly stopped, and stem tissue is the main sink component for carbohydrates and nutrients. Internode growth is occurring normally at peak canopy LAI and photosynthetic capacity, and this is often a period when carbohydrate reserves are stored in the stem material. When grain filling starts, internode elongation has nearly ceased, and the seeds are the main sink component. Although components overlap somewhat in their primary growth period, a general rule is that only one component is primarily growing at a time.

Integrating phenology, the shoot apex, and phytomers

The orderly development of the wheat plant begins with the developmental events occurring at the shoot apex. By producing the leaf and spikelet primordia on a shoot, the basic building blocks of the phytomers are added to comprise the shoot. The rate of appearance of vegetative phytomers can be determined by rate of leaf primordium initiation (plastochron) or leaf appearance (phyllochron). Similarly, reproductive phytomers in the spike are determined by the rate of spikelet primordium initiation and floret primordium differentiation in the spikelet. While a great number of studies have examined leaf, spikelet, and floret primordium initiation and subsequent growth, equations to accurately predict the rates have not been very successful (McMaster and Wilhelm 1995).

Phytomers (or the subcomponents of a phytomer) appear, grow, and senesce or abort on a shoot (Fig. 2.2). Some phenological stages of the shoot may be defined by the state of a specific phytomer. For instance, the beginning of the boot stage is when the leaf in the last vegetative phytomer (i.e., the flag leaf) has completed growth on the shoot. Often the phenological stage is the result of a collection of phytomers on the shoot. Illustrations of this are: the tillering stage, when axillary buds of different vegetative phytomers differentiate and grow; jointing, which occurs when one or more internodes of vegetative phytomers have elongated so that the first node is elevated above the soil surface; and flowering, which occurs when several reproductive phytomers have produced anthers.

ENVIRONMENTAL FACTORS INFLUENCING SHOOT DEVELOPMENT

The relationship between plant development and the environment was undoubtedly recognized by prehistoric civilizations. While the orderliness of wheat development at various scales outlined above occurs regardless of the environment, abiotic and biotic factors and some management practices change developmental rates and the beginning and ending of the process (and therefore duration). Abiotic factors are used to predict development. The mechanisms of action of management practices and biotic factors are best viewed by how they alter the predictive nature of abiotic factors.

Abiotic factors such as temperature, light, water, fertility, and CO_2 control wheat development, though temperature is viewed as the most important factor (Klepper et al., 1982; Baker and Gallagher 1983a,b; Bauer et al., 1985, 1986; Frank et al., 1987; Porter and Delecolle 1988; Masle et al., 1989; McMaster 2005). Temperature strongly impacts cell cycling time, reaction rates, and progression through the life cycle. Light can strongly influence some developmental events, but the other abiotic factors are considered of secondary importance and usually affect only certain developmental processes or events (Masle et al., 1989). Further, factors such as water and nutrients often seem to have threshold levels before influencing development. Given the importance of temperature in controlling and predicting wheat development, this section focuses primarily on the role of temperature.

Temperature

Reamur (1735) formalized the relationship between phenology and temperature by creating the concept of heat units, which is more commonly known today as thermal time. Thermal time acknowledges that, for most developmental processes, temperature is a better predictor than calendar time (i.e., days, hours, etc.). Many studies (Friend et al., 1962; Cao and Moss 1989; Jame et al., 1998; Yan and Hunt 1999; Streck et al., 2003; Xue et al., 2004) show a curvilinear response of the developmental process to temperature. Thermal time is an attempt to explain the observed differential response to temperature. Many forms of describing and quantifying thermal time have been developed, yet all forms are empirical in approach as the direct mechanisms of temperature effects are not well known (Wang 1960; Shaykewich 1995; Jamieson et al., 2007).

Estimation of thermal time has two elements: (i) average temperature (T_{avg}) over some time interval and (ii) use of T_{avg} in a temperature-response function to estimate the rate of the developmental process or "passage of time" (i.e., the effectiveness of a specific temperature on development rate). The most accurate calculation of T_{avg} is to use the integral of temperatures over the time interval of interest, but in practice the maximum (T_{max}) and minimum (T_{min}) temperature of the interval are used:

$$T_{avg} = (T_{max} + T_{min})/2 \tag{2.1}$$

This simple estimation of T_{avg} is generally accurate, but errors increase as day length deviates from 12 hours and if sudden changes in temperature occur during the time interval. Selection of the time interval depends on availability of data and the problem, but most frequently a daily time interval is used, with day/night and hourly intervals sometimes used. Somewhat problematic is that equation 2.1 has been interpreted in different ways, resulting in over a 20% difference in the estimation of T_{avg} for a growing season using the same weather data, and it often is unclear how the equation is being interpreted (McMaster and Wilhelm 1997).

Once T_{avg} is determined, a key area of divergence in thermal time approaches is the temperature-response function. The most basic version is to estimate thermal time (TT) linearly above the base temperature (T_{base}) at which the development rate is zero (Fig. 2.3a):

$$TT = T_{avg} - T_{base}, (\mathrm{TT} \geq 0) \tag{2.2}$$

Thermal time in this instance is often expressed as growing degree-days (GDD, °C·days), and the value can be either summed or used directly in an algorithm. For example, it may require an average of 105 GDD for a leaf blade to grow completely (Frank and Bauer 1995), or if the final blade length is 105 mm the rate of growth can be 1 mm GDD^{-1}. Normally, predicting when a developmental event begins or ends assumes a summation of thermal time.

The base temperature is important in defining the minimum temperature at which development can occur; T_{base} can be difficult to directly measure, and it likely changes both among cultivars and possibly with progression through the life cycle (Angus et al., 1981; Weir et al., 1984; McMaster and Smika 1988; Slafer and Rawson 1995a,b; Madakadze et al., 2003). Yet often this does not introduce significant error for predicting development in the field, as little development occurs near T_{base}, inaccuracies in calculating T_{avg} mask slight differences in setting T_{base}, and little difference in predictive accuracy is noted by using a range of temperatures for T_{base} (McMaster and Smika 1988). Often T_{base} is set to 0°C (Gallagher 1979; Baker and Gallagher 1983a,b; Slafer and Rawson 1994; McMaster et al., 2008).

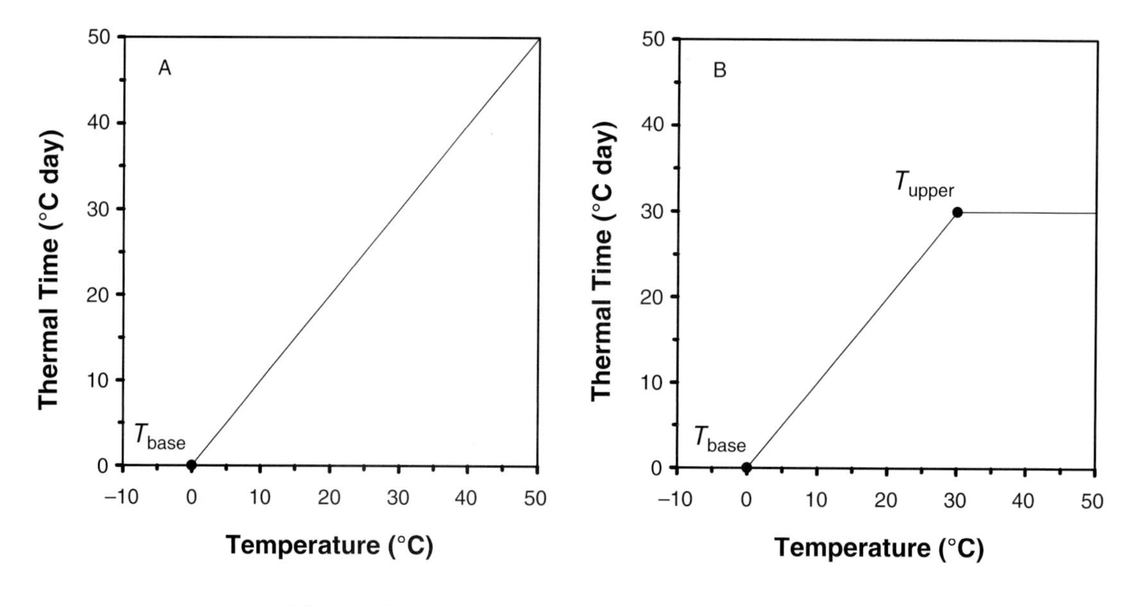

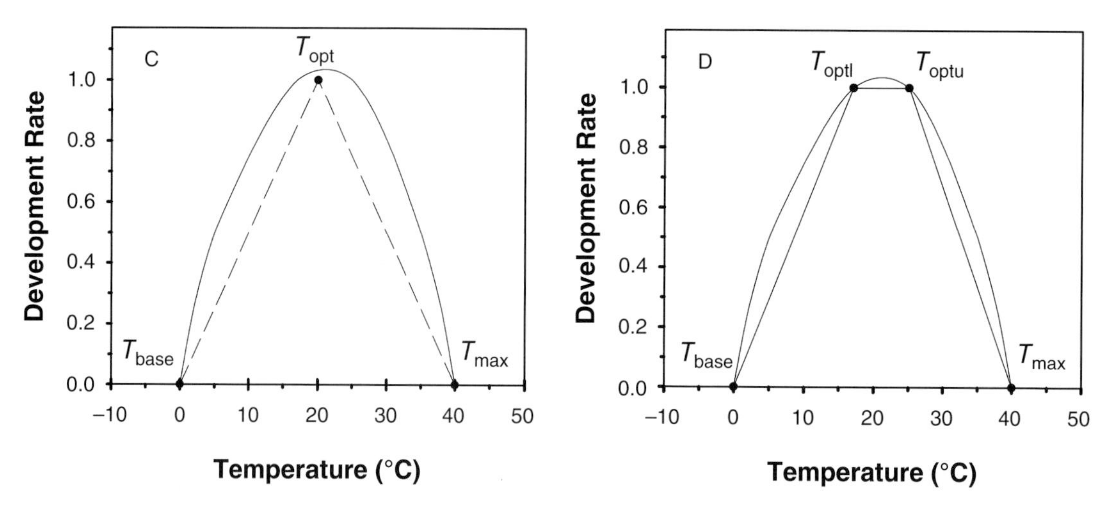

Fig. 2.3 Development rate as a function of temperature. Examples of different approaches for estimating thermal time are shown in each graph. Curvilinear temperature response function shown in parts c and d are based on Friend et al. (1962), Cao and Moss (1989), and Jame et al. (1998). T_{base} is the base temperature (i.e., development rate is zero), T_{opt} is the optimum temperature (i.e., development rate is at the maximum), T_{optl} and T_{optu} are the lower and upper temperatures for the optimum temperature range, and T_{max} is the maximum temperature (i.e., development rate is zero).

Equation 2.2 uses only one cardinal temperature (T_{base}) in calculating thermal time, and as temperature increases so does the development rate. Incorporating a second cardinal temperature for an upper temperature threshold into equation 2.2 recognizes that development rate does not increase indefinitely with temperature. Upper threshold temperatures, if used, are generally set to at least 30 °C when used in equation 2.2 and limit accumulated thermal time to the upper threshold (Fig. 2.3b). Most wheat production systems do not often exceed average daily temperatures greater than 30 °C, and therefore, errors in setting this upper threshold seem relatively minor for field predictions (McMaster et al., 2008).

If another cardinal temperature, related to the optimum temperature (T_{opt}) for development rate, is added along with changing the definition of the upper temperature threshold to be the maximum temperature that development rate is greater than zero (T_{max}), then a better representation of the observed normal temperature-response curve is improved. A two-segmented linear model can be used with thermal time increasing linearly from T_{base} to T_{opt}, then linearly decreasing from T_{opt} to T_{max} (Fig. 2.3c). Dividing T_{opt} into a lower (T_{optl}) and upper (T_{optu}) temperature optimum, between which development rate is maximum, results in a three-segmented linear model that closely approximates the observed temperature response curve (Fig. 2.3d). The final adjustment is to use a curvilinear model such as those proposed by Yan and Hunt (1999) and Streck et al. (2003). If the curvilinear models are parameterized correctly, they should mimic the observed temperature-response function shown in Fig. 2.3c,d.

Largely because air temperature is more readily measured and available, it is used in calculating T_{avg}. The assumption is that the relationship between air temperature and shoot apex temperature (where many developmental events occur) is closely associated. Theory would suggest using either soil temperature at the depth of the shoot apex (approximately 2–3 cm) when the shoot apex is located in the soil or plant canopy temperature when the apex is located in the plant canopy (Peacock 1975). Experimental support for this theory is related to a long history of root–shoot temperature experiments (Hay and Wilson 1982; Jamieson et al., 1995; Stone et al., 1999; Vinocur and Ritchie 2001). However, the possible theoretical gain of using soil and plant canopy temperature was not always realized in the field for wheat (McMaster and Wilhelm 1998; McMaster et al., 2003). This was explained by (i) the shoot apex and other intercalary meristems such as for leaf growth are located over a vertical space where temperatures vary considerably (Skinner and Nelson 1995), (ii) the whole plant senses temperature and that influences signal and resource movement throughout the plant, and (iii) the relationship between air temperature and shoot apex temperature is quite stable in many environments.

All developmental processes show the curvilinear response to temperature shown in Fig. 2.3c,d. However, the duration of grain filling, a critical component of determining final yield, merits special mention here. As temperature increases, both grain filling rates and the accumulation of thermal time are increased. The net effect of these two responses is that kernel weight usually is less under higher temperatures than cooler temperatures because the duration of grain filling is decreased more than the rate of growth is increased (Marcellos and Single 1971; Sofield et al., 1974; Wiegand and Cuellar 1981; Bhullar and Jenner 1983; Herzog 1986; Wardlaw et al., 1989).

Nontemperature environmental factors

Environmental factors in addition to temperature can influence wheat growth and development, but it is difficult to summarize the diversity of responses, or nonresponses, to various environmental factors. Cultivars can vary in their responses to similar treatments, and of course, the ever-present genotype × environment interaction further complicates understanding the responses. Many experiments, particularly those in the field, have not carefully measured the level of the environmental factor, the levels change during the course of the experiment, and an insufficient number of treatments are applied to fully understand and quantify the response surface. This has resulted in a qualitative understanding of response; however, quantitative characterization of the response is often lacking and identification of threshold levels for a factor is unknown.

Regardless of these difficulties, one general rule is that growth of an organ is reduced and abortion or senescence rates are increased when the environmental factor is limiting. For instance, leaf, internode, and kernel growth is reduced and leaf senescence and tiller, spikelet, and floret abortion is increased when water, nutrients, CO_2, or light is limiting or under high salinity conditions (Gifford 1977; Whingwiri and Kemp 1980; McMaster et al., 1987; Maas and Grieve 1990; Jedel and Hunt 1990; Maas et al., 1994).

Although temperature and photoperiod are the most important factors controlling wheat phenology, phenological responses to water availability have been reported (Angus and Moncur 1977; Bauer et al., 1985; Baker et al., 1986; Frank et al., 1987; Davidson and Chevalier 1987, 1992). McMaster and Wilhelm (2003) observed that later developmental stages, particularly anthesis and physiological maturity, were most sensitive to water deficits and required less thermal time to reach the developmental stage under water deficits than well-watered conditions. Most notably, the grain filling duration was significantly reduced by water deficits, as commonly observed in many experiments and for most crops (McMaster et al., 2009). In general, water deficits must be quite substantial before phenological responses are noted. Physiological and genetic mechanisms controlling these phenological responses are not well understood, but the likely indirect effect of water deficits increasing canopy temperature, and therefore thermal time accumulation, might partly explain the general response of reaching a developmental stage earlier under water deficits (McMaster et al., 2009).

Occasional responses of some developmental stages to nutrients have been reported (Frank and Bauer 1984). At low levels of N, developmental stages such as double ridge, terminal spikelet, and anthesis were delayed (Whingwiri and Stern 1982; Longnecker et al., 1993). Nitrogen limitation seems to have the greatest effect on decreasing the duration of grain filling, presumably by increasing leaf senescence. Other factors have a less reported or consistent effect on phenology than nutrients. For instance, if ambient CO_2 levels were doubled (700 ppm), the time required to reach floral induction and anthesis was slightly shortened (Marc and Gifford 1984), and the spikelet, floret, and grain filling development phases were reduced by elevated CO_2 (550 ppm, Li et al., 1997, 2000).

Leaf, spikelet, and floret primordia initiation appears to be minimally reduced by factors other than temperature and light, although leaf appearance and final spikelet and floret number are often reduced by factors other than temperature and light (Whingwiri and Stern 1982; Longnecker et al., 1993; Li et al., 1997). Three possible explanations that are not mutually exclusive could account for the variable responses noted. One explanation has been mentioned before: until a threshold level is reached, little response to an abiotic factor is observed. An example of this would be the response of reduced spikelet primordium initiation and delayed developmental stages observed only at low levels of N (Longnecker et al., 1993). Another explanation is that the duration of the phase is shortened by the abiotic factor that affects the final number of primordia initiated. An illustration of this would be Frank et al. (1987) showing that water stress mainly affected spikelet primordium initiation by decreasing the initiation phase. The third explanation for the reduced response of many developmental processes to environmental factors is that primordium initiation mainly responds to hormonal controls and then is driven by cell cycling time, which is primarily a temperature-response process that requires little resources in terms of water and nutrients. If these resources are limiting to the extent that cell division is impacted, then in annual plants such as wheat, the shoot or shoot apex is in danger of death. However, the differentiation and growth of these primordia (e.g., leaves, tillers) are strongly influenced by many environmental factors

(Frank and Bauer 1982, 1984; Bauer et al., 1986; Maas and Grieve 1990; Cutforth et al., 1992), and this might be hypothesized to be the result of lack of resources required for growth of the organ.

As an illustration of this, leaf primordia initiation showed little response to elevated CO_2 (Li et al., 1997), yet a slight positive relationship between CO_2 and the rate of leaf appearance seems to exist (McMaster et al., 1999). Similarly, differentiation and growth of the axillary bud, resulting in the appearance of a new tiller, is strongly regulated by water, nutrients, and carbohydrate availability, as well as light intensity and quality, although the relationship is sometimes stronger for tillers than the main stem (Fraser et al., 1982; Longnecker et al., 1993; Maas et al., 1994). Further, younger tillers or secondary and tertiary tillers often respond more to resource limitations. One hypothesis is that these tillers are most dependent on the whole plant for resources that every shoot is competing for and are farthest removed from access to the resource in terms of vascular connections (water, nutrients, and so on) or location within the canopy (i.e., light intensity and quality).

DIGITAL TECHNOLOGIES FOR WHEAT DEVELOPMENT

Probably more digital technologies are available for modeling wheat growth and development than for any other crop. Digital technologies include simulation and regression models, decision support systems (DSS), web sites, and specific computer programs. This section will briefly focus on simulation models as related to wheat development.

Many models simulate wheat production and yield (McMaster 1993), and at the risk of omitting many deserving models, several that have been of historical significance or played an influential role in wheat modeling include AFRCWHEAT1/2 (Porter 1984, 1993; Weir et al., 1984), APSIM (Asseng et al., 1998), DSSAT/CERES-Wheat/CropSim (Ritchie and Otter 1985; Ritchie 1991; Hunt and Pararajasingham 1995; Jones et al., 2003; Hoogenboom et al., 2004), Sirius (Jamieson et al., 1995, 1998), SUCROS (van Laar et al., 1992), SWHEAT (van Keulen and Seligman 1987), and WINTER WHEAT (Baker et al., 1985). A diversity of approaches exists in how growth and development are simulated in these wheat process-based models. However, for most simulation models, the earliest approaches that still remain popular today use the energy- or carbon-driven approach, in which sunlight energy is captured by the plant, converted to biomass, and then partitioned within the plant. Developmental detail in these models is usually minimal and consists of calculating canopy LAI and growth of general plant components of leaf, stem, roots, and seeds. Often these models have quite accurate grain yield predictions for variable environments and cultivars.

Phenology modeling has been one of the most successful components in existing wheat simulation models, and the ability to simulate genotype phenology across a broad range of environments is quite reliable. Many alternative approaches are available for use in predicting phenology, and approaches differ in input requirements and number of developmental stages simulated. All models are based on the thermal time approach (with the many variations that exist), with some emphasizing the role of vernalization and photoperiod (e.g., DSSAT, AFRCWheat2, Sirius) more than others. One area of divergence in phenology submodels is whether leaf number or strictly thermal time is used to estimate the time interval between developmental stages. The AFRCWHEAT2, Sirius, MODWht3 (Rickman et al., 1996), and SHOOTGRO (McMaster et al., 1992b; Zalud et al., 2003) models are based primarily on the leaf number approach, while the others use the more common approach of strict thermal time.

Almost all models do not explicitly consider the phenological responses to water deficits (and usually nutrient deficits) for most, or all, developmental stages (McMaster et al., 2009). Exceptions include the SHOOTGRO model and the PhenologyMMS decision support tool (http://arsagsoftware.ars.usda.gov), where the number of leaves or thermal time between developmental events from emergence through maturity is adjusted due to water and N levels.

Beginning in the mid-1980s, some research efforts directed greater attention at incorporating the developmental concepts discussed in this chapter into process-based simulation models (e.g., leaf appearance and tillering) and exploration of canopy architecture. Developmental processes other than phenology have been incorporated to varying degrees in whole-plant simulation models. The AFRCWHEAT1/2 model added two interesting components generally missing in existing wheat models at that time: a population ecology element and greater developmental detail derived from European scientists. For example, detailed tillering and leaf dynamics (e.g., appearance, growth, and senescence–abortion) and the effect on canopy LAI were simulated and then used to estimate biomass.

Simultaneously and independently another effort was underway in the US that resulted in the developmentally driven SHOOTGRO (McMaster et al., 1991, 1992a,b; Wilhelm et al., 1993; Zalud et al., 2003) and MODWht3 (Rickman et al., 1996) models. SHOOTGRO is slightly more developmentally detailed than MODWht3, but less detailed in the root system and simulating biomass production. SHOOTGRO provides the foundation to simulate the development and growth of each morphologically identified shoot (main stem and tillers) on the median plant of up to six age classes, or cohorts, based on time of seedling emergence (http://arsagsoftware.ars.usda.gov). All processes in Fig. 2.2 except for leaf primordia initiation and floret primordia differentiation are simulated. Soil water content determines the thermal time required for germination, and seedling emergence rates are simulated to establish the cohorts. Following germination, sequential developmental events are simulated using the number of leaves produced (e.g., phyllochron) between developmental events up to anthesis, and thermal time after anthesis. SHOOTGRO explicitly includes the effect of water and N availability on all developmental and growth processes. As shoots appear, the appearance, growth or size, and senescence or abortion of each leaf blade and sheath, internode, and spike components on each shoot are simulated. Spike development and growth is simulated by the appearance of spikelets and spikelet differentiation into florets, fertilization of florets, and subsequent growth of each kernel.

The Sirius model has one of the most developed leaf appearance submodels of any wheat simulation model (Jamieson et al., 2007). As with the SHOOTGRO model, the assumption used is that the developmental "clock" from emergence to anthesis is best represented by the rate of leaf appearance and final number of leaves. Based on vernalization requirement and photoperiod sensitivity of the cultivar being simulated and on leaf ontogeny, the final leaf number is determined (Brooking et al., 1995; Brooking 1996; Robertson et al., 1996). This allows for an elegant quantitative description of both spring and winter wheat leaf appearance and integration with developmental events.

Canopy architectural, or functional–structural, modeling for a variety of species has increased dramatically since the mid-1990s, in part a result of faster and low-cost computational resources. Modeling efforts have tended to focus more on the functional aspect of the plant such as simulating the biophysical environment of the canopy and resource allocation, and a model usually requires either "setting" the canopy architecture or relatively simple attempts to create the structure (Norman and Campbell 1983; Grant 2001; Dingkuhn et al., 2005; Evers et al., 2005; Renton et al., 2005). Use of L-systems (Prusinkiewicz 1998) or the phyllochron in many of these models has successfully created the plant architecture. Models such as MODWht3, SHOOTGRO, Sirius, and AFRCWHEAT2 might provide further opportunities for simulating greater canopy architectural detail.

Crop simulation modeling is beginning to benefit from the advent of object-oriented design and programming languages such as C++ and Java. Initial efforts have tended to view the plant as a collection of objects that equate to leaf, stem, root, and seed components (Sequeira et al., 1991, 1997). Recent attempts have begun to incorporate the phytomer approach of building plant canopies into the object-oriented design that can also be scaled up, or aggregated, into lower levels of resolution, such as the seed component of earlier designs (Drouet and Pages 2007; Hargreaves and McMaster 2008).

LINKING MOLECULAR BIOLOGY AND FUNCTIONAL GENOMICS TO DEVELOPMENT

Since sequencing of the model crucifer *Arabidopsis thaliana* genome (Arabidopsis Genome Initiative 2000), and subsequent work with the model cereal rice (Delseny 2007), data are rapidly emerging on genes and genetic pathways related to a number of developmental processes. The hope is that the conservation of gene content and function from Arabidopsis, rice, and other cereals would provide insight for understanding wheat development. Indeed, it appears this hope has been justified in some instances and a generic model has emerged for some processes and traits. Efforts are underway to transform data from molecular biology into information and understanding on the physiological level. Unfortunately, linkage of molecular biology with many developmental processes has not been well established. Considerable challenges remain for understanding and characterizing the function of a gene, identifying the complex genetic pathways for a process, and determining the environmental effects on gene expression before integration with breeding and whole-plant physiology is successful (Edmeades et al., 2004; White et al., 2004a,b).

One area of notable success has been the elucidation of the genetic controls of the flowering pathway (e.g., Hay and Ellis 1998; Laurie et al., 2004 for barley; Beales et al., 2007), and most recent understanding gained for wheat is thoroughly discussed in Chapter 3. Functional orthologues of many of the genes involved have been identified between Arabidopsis, rice, wheat, barley, and other cereal crops. Although the genetic pathway controlling flowering is qualitatively well established, few attempts have been made to predict the time of flowering for plants in the field based on which loci are present and how gene expression responds to the environment. An effort showing promise is using the specific alleles of the vernalization, photoperiod, and earliness *per se* genes to establish the parameters in predicting wheat phenology in the field (White 2006; White et al., 2008). Earlier efforts using neural networks for flowering time of Arabidopsis may also provide alternative approaches for determining wheat flowering time (Welch et al., 2003).

Incorporation of semidwarfing genes was essential for the Green Revolution, and molecular biology is adding greater insight into the variability and functioning of different semidwarfing genes. Functional orthologues of the *GAI* genes of Arabidopsis have been identified for wheat (the *Rht* genes) and other cereal crops. Some efforts have been made to incorporate this knowledge into wheat simulation modeling of plant height (Baenziger et al., 2004).

While other similar examples could be made of new insights emerging from identifying genes and their function and linking the genes to wheat development and modeling (Fowler et al., 1999), most examples have focused on known genes that appear to have more limited genetic pathways controlling the developmental process. With increasing complexity of the genetic pathway controlling a trait, quantitative expression of multiple genes and their response to the environment currently seems nearly intractable given our current understanding. Some overviews of these challenges are discussed elsewhere (White and Hoogenboom 2003; Edmeades et al., 2004; White et al., 2004a,b; Hammer and Jordan 2007; Yin and Struik 2007).

One area where the linkage between molecular biology and wheat development is poorly understood involves the plastochron and phyllochron. Given the variation among cultivars in these traits, and particularly the difference between spring and winter wheat genotypes, this has considerable impact on our understanding and predictions of wheat development and the building of canopies by phytomers.

FUTURE PERSPECTIVES

A rich history of research has elucidated the complex and interactive interplay among development, phenology, and growth of individual organs on different shoots of the wheat plant. From this work has emerged the general pattern of wheat development and how development responds to the environment. Knowledge of wheat development has increasingly been incorporated into improving wheat management and breeding. Several challenges clearly remain in furthering our knowledge of development and applying it to improved wheat production.

Simulation models and DSS provide tools for quantifying, synthesizing, and applying developmental concepts to many diverse problems. Increased emphasis on translating current and newly gained knowledge of wheat development into electronic forms should improve the currently limited availability of tools designed to address specific problems. Adoption of these technologies is currently quite low, and barriers to adoption need to be addressed.

Continued development of simulation models and DSS likely will further clarify gaps in our knowledge of wheat development. In particular, mechanisms controlling developmental processes and addressing the ubiquitous genotype × environment interaction will need much greater emphasis to extend the robustness of digital analysis and modeling. Successful linkage of wheat physiology and simulation models with breeding has been hindered by the lack of developmental detail in the models and of ways to address the genotype interaction with the environment. The challenges of incorporating knowledge of mechanisms controlling wheat development gained from molecular biology into physiology and simulation models is a daunting problem, but one worth pursuing.

REFERENCES

Angus, J.F., D.H. Mackenzie, R. Morton, and C.A. Schafer. 1981. Phasic development in field crops: 2. Thermal and photoperiodic responses of spring wheat. Field Crops Res. 4:269–283.

Angus, J.F., and M.W. Moncur. 1977. Water stress and phenology in wheat. Aust. J. Agric. Res. 28:177–181.

Arabidopsis Genome Initiative. 2000. Analysis of the genome sequence of the flowering plant *Arabidopsis thaliana*. Nature 408:796–815.

Asseng, S., B.A. Keating, I.R.P. Fillery, P.J. Gregory, J.W. Bowden, N.C. Turner, J.A. Palta, and D.G. Abrecht. 1998. Performance of the APSIM-wheat model in western Australia. Field Crops Res. 57:163–179.

Baenziger, P.S., G.S. McMaster, W.W. Wilhelm, A. Weiss, and C.J. Hays. 2004. Putting genes into genetic coefficients. Field Crops Res. 90:133–144.

Baker, C.K., and J.N. Gallagher. 1983a. The development of winter wheat in the field: 1. Relation between apical development and plant morphology within and between seasons. J. Agric. Sci. (Cambridge) 101:327–335.

Baker, C.K., and J.N. Gallagher. 1983b. The development of winter wheat in the field: 2. The control of primordium initiation rate by temperature and photoperiod. J. Agric. Sci. (Cambridge) 101:337–344.

Baker, J.T., P.J. Pinter, Jr., R.J. Reginato, and E.T. Kanemasu. 1986. Effects of temperature on leaf appearance in spring and winter wheat cultivars. Agron. J. 78:605–613.

Baker, D.N., F.D. Whisler, W.J. Parton, E.L. Klepper, C.V. Cole, W.O. Willis, D.E. Smika, A.L. Black, and A. Bauer. 1985. The development of WINTER WHEAT: A physical physiological process model. p. 176–187. *In* W.O. Willis (ed.) Wheat yield improvement. USDA-ARS Publ. 38. National Technical Information Service, Springfield, VA.

Barnard, C. 1955. Histogenesis of the inflorescence and flower of *Triticum aestivum* L. Aust. J. Bot. 3:1–24.

Bateson, W. 1894. Materials for the study of variation treated with especial regard to discontinuity in the origin of species. MacMillan Press, London, UK.

Bauer, A., A.B. Frank, and A.L. Black. 1985. Estimation of spring wheat grain dry matter assimilation from air temperature. Agron. J. 77:743–752.

Bauer, A., A.B. Frank, and A.L. Black. 1986. Estimation of spring wheat spike water concentration and grain maturity from air temperature. Agron. J. 78:445–450.

Bauer, A., D. Smika, and A. Black. 1983. Correlation of five wheat growth stage scales used in the Great Plains. USDA-ARS Publ. ATT-NC-7. USDA-ARS, Peoria, IL.

Beales, J., A. Turner, S. Griffiths, J. Snape, and D. Laurie. 2007. A *Pseudo-Response Regulator* is misexpressed in the photoperiod insensitive *Ppd-D1a* mutant of wheat (*Triticum aestivum* L.). Theor. Appl. Genet. 115:721–733.

Belford, R.K., B. Klepper, and R.W. Rickman. 1987. Studies of intact shoot–root systems of field-grown winter wheat:

II. Root and shoot developmental patterns as related to nitrogen fertilizer. Agron. J. 79:310–319.

Bhullar, S.S., and C.F. Jenner. 1983. Responses to brief periods of elevated temperature in ears and grains of wheat. Aust. J. Plant Physiol. 10:549–560.

Bonnett, O.T. 1966. Inflorescences of maize, wheat, rye, barley, and oats: Their initiation and development. Agric. Exp. Stn. Bull. 721. Urbana-Champaign, IL.

Brooking, I.R. 1996. The temperature response of vernalization in wheat—a developmental analysis. Ann. Bot. 78:507–512.

Brooking, I.R., P.D. Jamieson, and J.R. Porter. 1995. The influence of daylength on the final leaf number in spring wheat. Field Crops Res. 41:155–165.

Cao, W., and D.N. Moss. 1989. Temperature effect on leaf emergence and phyllochron in wheat and barley. Crop Sci. 29:1018–1021.

Cutforth, H.W., Y.W. Jame, and P.G. Jefferson. 1992. Effect of temperature, vernalization and water stress on phyllochron and final main-stem leaf number of NY320 and Neepawa spring wheats. Can. J. Plant Sci. 72:1141–1151.

Darwinkel, A. 1978. Patterns of tillering and grain production of winter wheat at a wide range of plant densities. Neth. J. Agric. Sci. 26:383–398.

Davidson, D.J., and P.M. Chevalier. 1987. Influence of polyethylene glycol–induced water deficits on tiller production in spring wheat. Crop Sci. 27:1185–1187.

Davidson, D.J., and P.M. Chevalier. 1992. Storage and remobilization of water-soluble carbohydrates in stems of spring wheat. Crop Sci. 32:186–190.

Delseny, M. 2007. Rice: A model plant for cereal genomics. p. 397–411. *In* J-F. Morot-Gaudry, P. Lea, and J-F. Briat (ed.) Functional plant genomics. Science Publishers, Enfield, NH.

Dingkuhn, M., D. Luquer, B. Quilot, and P. de Reffye. 2005. Environmental and genetic control of morphogenesis in crops: Towards models simulating phenotypic plasticity. Aust. J. Agric. Res. 56:1289–1302.

Drouet, J-L., and L. Pages. 2007. GRAAL-CN: A model of GRowth, Architecture, and ALlocation for Carbon and Nitrogen dynamics within whole plants formalized at the organ level. Ecol. Model. 206:231–249.

Edmeades, G.O., G.S. McMaster, J.W. White, and H. Campos. 2004. Genomics and the physiologist: Bridging the gap between genes and crop response. Field Crops Res. 90:5–18.

Evers, J.B., J. Vos, C. Fournier, B. Andrieu, M. Chelle, and P.C. Struik. 2005. Towards a generic architectural model of tillering in Gramineae, as exemplified by spring wheat (*Triticum aestivum*). New Phytol. 166:801–812.

Forster, B.P., J.D. Franckowiak, U. Lundqvist, J. Lyon, I. Pitkethly, and W.T.B. Thomas. 2007. The barley phytomer. Ann. Bot. 100:725–733.

Fowler, D.B., A.E. Limin, and J.T. Ritchie. 1999. Low-temperature tolerance in cereals: Model and genetic interpretation. Crop Sci. 39:626–633.

Frank, A.B., and A. Bauer. 1982. Effect of temperature and fertilizer N on apex development in spring wheat. Agron. J. 74:504–509.

Frank, A.B., and A. Bauer. 1984. Cultivar, nitrogen, and soil water effects on apex development in spring wheat. Agron. J. 76:656–660.

Frank, A.B., and A. Bauer. 1995. Phyllochron differences in wheat, barley, and forage grasses. Crop Sci. 35:19–23.

Frank, A.B., A. Bauer, and A.L. Black. 1987. Effects of air temperature and water stress on apex development in spring wheat. Crop Sci. 27:113–116.

Frank, A.B., V.B. Cardwell, A.J. Ciha, and W.W. Wilhelm. 1997. Growth staging in research and crop management. Crop Sci. 37:1039–1040.

Fraser, J., C.T. Dougherty, and R.H.M. Langer. 1982. Dynamics of tiller populations of standard height and semi-dwarf wheats. N. Z. J. Agric. Res. 25:321–328.

Friend, D.J.C., V.A. Helson, and J.E. Fisher. 1962. Leaf growth in Marquis wheat, as regulated by temperature, light intensity, and daylength. Can. J. Bot. 40:1299–1311.

Gallagher, J.N. 1979. Field studies of cereal leaf growth: I. Initiation and expansion in relation to temperature and ontogeny. J. Exp. Bot. 30:625–636.

Gan, Y., and E.H. Stobbe. 1995. Effect of variations in seed size and planting depth on emergence, infertile plants, and grain yield of spring wheat. Can. J. Plant Sci. 75:565–570.

Gifford, R.M. 1977. Growth pattern, carbon dioxide exchange and dry weight distribution in wheat growing under different photosynthetic environments. Aust. J. Plant Physiol. 4:99–110.

Grant, R.F. 2001. A review of the Canadian ecosystem model—*ecosys*. p. 173–263. *In* M.J. Shaffer, L. Ma, and S. Hansen (ed.) Modeling carbon and nitrogen dynamics for soil management. Lewis Publishers, Boca Raton, FL.

Gray, A. 1879. Structural botany. Ivsion, Blakeman, Taylor, and Company, New York.

Hammer, G.L., and D.R. Jordan. 2007. An integrated systems approach to crop improvement. p. 45–61. *In* J.H.J. Spiertz, P.C. Struik, and H.H. van Laar (ed.) Scale and complexity in plant systems research. Springer, Wageningen, The Netherlands.

Hanft, J.M., and R.D. Wych. 1982. Visual indicators of physiological maturity of hard red spring wheat. Crop Sci. 22:584–587.

Hargreaves, J.N.G., and G.S. McMaster. 2008. A canonical composition of phytomers for building plant canopies from the bottom up: Phytomer CANON in D(evelopment). Proc. 2008 Int. Symp. on Crop Modeling and Decision Support, Nanjing, China. 19–22 April 2008 [Online]. Available at http://klia.cn/iscmds/ (verified 11 May 2008).

Harrell, D.M., W.W. Wilhelm, and G.S. McMaster. 1993. SCALES: A computer program to convert among three developmental stage scales for wheat. Agron. J. 85:758–763.

Harrell, D.M., W.W. Wilhelm, and G.S. McMaster. 1998. SCALES 2: Computer program to convert among developmental stage scales for corn and small grains. Agron. J. 90:235–238.

Haun, J.R. 1973. Visual quantification of wheat development. Agron. J. 65:116–119.

Hay, R.K.M., and R.P. Ellis. 1998. The control of flowering in wheat and barley: What recent advances in molecular genetics can reveal. Ann. Bot. 82:541–554.
Hay, R.K.M., and E.J.M. Kirby. 1991. Convergence and synchrony: A review of the coordination of development in wheat. Aust. J. Agric. Res. 42:661–700.
Hay, R.K.M., and G.T. Wilson. 1982. Leaf appearance and extension in field-grown winter wheat plants: The importance of soil temperature during vegetative growth. J. Agric. Sci. (Cambridge) 99:403–410.
Herzog, H. 1986. Source and sink during the reproductive period of wheat: Development and its regulation with special reference to cytokinins. Parey, Berlin, Germany.
Hoogenboom, G., J.W. Jones, P.W. Wilkens, C.H. Porter, W.D. Batchelor, L.A. Hunt, K.J. Boote, U. Singh, U.O. Uryasev, W.T. Bowen, A.J. Gijsman, A. du Toit, J.W. White, and G.Y. Tsuji. 2004. Decision support system for agrotechnology transfer, version 4.0 [CD-ROM]. University of Hawaii, Honolulu, HI.
Hunt, L.A., and S. Pararajasingham. 1995. CROPSIM-WHEAT: A model describing the growth and development of wheat. Can. J. Plant Sci. 75:619–632.
Jame, Y.W., H.W. Cutforth, and J.T. Ritchie. 1998. Interaction of temperature and daylength on leaf appearance rate in wheat and barley. Agric. For. Meteorol. 92:241–249.
Jamieson, P.D., I.R. Brooking, J.R. Porter, and D.R. Wilson. 1995. Prediction of leaf appearance in wheat: A question of temperature. Field Crops Res. 41:35–44.
Jamieson, P.D., I.R. Brooking, M.A. Semenov, G.S. McMaster, J.W. White, and J.R. Porter. 2007. Reconciling alternative models of phenological development in winter wheat. Field Crops Res. 103:36–41.
Jamieson, P.D., M.A. Semenov, I.R. Brooking, and G.S. Francis. 1998. Sirius: A mechanistic model of wheat response to environmental variation. Eur. J. Agron. 8:161–179.
Jedel, P.E., and L.A. Hunt. 1990. Shading and thinning effects on multi- and standard-floret winter wheat. Crop Sci. 30:128–133.
Jewiss, O.R. 1972. Tillering in grasses—its significance and control. J. Br. Grassl. Soc. 27:65–82.
Jones, J.W., G. Hoogenboom, C.H. Porter, K.J. Boote, W. D. Batchelor, L.A. Hunt, P.W. Wilkens, U. Singh, A.J. Gijsman, and J.T. Ritchie. 2003. The DSSAT cropping system model. Eur. J. Agron. 18:235–265.
Kirby, E.J.M., and M. Appleyard. 1984. Cereal development guide. 2nd ed. Arable Unit, National Agricultural Centre, Coventry, UK.
Kirby, E.J.M., M. Appleyard, and G. Fellowes. 1985a. Effect of sowing date and variety on main shoot leaf emergence and number of leaves of barley and wheat. Agronomie 5:117–126.
Kirby, E.J.M., M. Appleyard, and G. Fellowes. 1985b. Leaf emergence and tillering in barley and wheat. Agronomie 5:193–200.
Kirby, E.J.M., and B.E. Eisenberg. 1966. Some effects of photoperiod on barley. J. Exp. Bot. 17:204–213.
Klepper, B., R.K. Belford, and R.W. Rickman. 1984. Root and shoot development in winter wheat. Agron. J. 76:117–122.
Klepper, B., R.W. Rickman, and R.K. Belford. 1983a. Leaf and tiller identification on wheat plants. Crop Sci. 23:1002–1004.
Klepper, B., R.W. Rickman, and C.M. Peterson. 1982. Quantitative characterization of vegetative development in small cereal grains. Agron. J. 74:789–792.
Klepper, B., T.W. Tucker, and B.D. Dunbar. 1983b. A numerical index to assess early inflorescence development in wheat. Crop Sci. 23:206–208.
Lancaster, P.D., H. Bleiholder, T. Van der Boom, P. Langeluddeke, R. Stauss, E. Weber, and A. Witzenberger. 1991. A uniform decimal code for growth stages of crops and weeds. Ann. Appl. Biol. 119:561–601.
Landes, A., and J.R. Porter. 1989. Comparison of scales used for categorizing the development of wheat, barley, rye and oats. Ann. Appl. Biol. 115:343–360.
Large, E.C. 1954. Growth stages in cereals. Plant Pathol. 3:128–129.
Laurie, D.A., S. Griffiths, R.P. Dunford, V. Christodoulou, S.A. Taylor, J. Cockram, J. Beales, and A. Turner. 2004. Comparative genetic approaches to the identification of flowering time genes in temperate cereals. Field Crops Res. 90:87–99.
Li, A-G., Y-S. Hou, G.W. Wall, A. Trent, B.A. Kimball, and P.J. Pinter, Jr. 2000. Free-air CO_2 enrichment and drought stress effects on grain filling rate and duration in spring wheat. Crop Sci. 40:1263–1270.
Li, A-G., A. Trent, G.W. Wall, B.A. Kimball, Y-S. Hou, P.J. Pinter, Jr., R.L. Garcia, D.V Hunsaker, and R.L. Lamorte. 1997. Free-air CO_2 enrichment effects on rate and duration of apical development of spring wheat. Crop Sci. 37:789–796.
Longnecker, N., E.J.M. Kirby, and A. Robson. 1993. Leaf emergence, tiller growth, and apical development of nitrogen-deficient spring wheat. Crop Sci. 33:154–160.
Maas, E.V., and C.M. Grieve. 1990. Spike and leaf development in salt-stressed wheat. Crop Sci. 30:1309–1313.
Maas, E.V., S.M. Lesch, L.E. Francois, and C.M. Grieve. 1994. Tiller development in salt-stressed wheat. Crop Sci. 34:1594–1603.
Madakadze, I.C., K.A. Stewart, R.M. Madakadze, and D.L. Smith. 2003. Base temperatures for seedling growth and their correlation with chilling sensitivity for warm-season grasses. Crop Sci. 434:874–878.
Marc, J., and R.M. Gifford. 1984. Floral initiation in wheat, sunflower, and sorghum under carbon dioxide enrichment. Can. J. Bot. 62:9–14.
Marcellos, H., and W.V. Single. 1971. Quantitative responses of wheat to photoperiod and temperature in the field. Aust. J. Agric. Res. 23:533–540.
Masle, J. 1985. Competition among tillers in winter wheat: Consequences for growth and development of the crop. p. 33–54. *In* W. Day and R.K. Atkin (ed.) Wheat growth and modeling. Plenum Press, New York.
Masle, J., G. Doussinault, G.D. Farquhar, and B. Sun. 1989. Foliar stage in wheat correlates better to photothermal time than to thermal time. Plant Cell Environ. 12:235–247.
Masle-Meynard, J., and M. Sebillotte. 1981. Study on the heterogeneity of a wheat stand: I. Concept of stand structure. Agronomie 1:207–216.

McMaster, G.S. 1993. Another wheat (Triticum spp.) model? Progress and applications of crop modeling. Riv. Agronomia 27:264–272.

McMaster, G.S. 1997. Phenology, development, and growth of the wheat (*Triticum aestivum* L.) shoot apex: A review. Adv. Agron. 59:63–118.

McMaster, G.S. 2005. Phytomers, phyllochrons, phenology and temperate cereal development. J. Agric. Sci. (Cambridge) 143:137–150.

McMaster, G.S., B. Klepper, R.W. Rickman, W.W. Wilhelm, and W.O. Willis. 1991. Simulation of aboveground vegetative development and growth of unstressed winter wheat. Ecol. Model. 53:189–204.

McMaster, G.S., D.R. LeCain, J.A. Morgan, L. Aiguo, and D.L. Hendrix. 1999. Elevated CO_2 increases CER, leaf and tiller development, and shoot and root growth. J. Agron. Crop Sci. 183:119–128.

McMaster, G.S., J.A. Morgan, and W.W. Wilhelm. 1992a. Simulating winter wheat spike development and growth. Agric. For. Meteorol. 60:193–220.

McMaster, G.S., J.A. Morgan, and W.O. Willis. 1987. Effects of shading on winter wheat yield, spike characteristics, and carbohydrate allocation. Crop Sci. 27:967–973.

McMaster, G.S., D.B. Palic, and G.H. Dunn. 2002. Soil management alters seedling emergence and subsequent autumn growth and yield in dryland winter wheat–fallow systems in the Central Great Plains on a clay loam soil. Soil Tillage Res. 65:193–206.

McMaster, G.S., and D.E. Smika. 1988. Estimation and evaluation of winter wheat phenology in the central Great Plains. Agric. For. Meteorol. 43:1–18.

McMaster, G.S., J.W. White, L.A. Hunt, P.D. Jamieson, S.S. Dhillon, and J.I. Ortiz-Monasterio. 2008. Simulating the influence of vernalization, photoperiod, and optimum temperature on wheat developmental rates. Ann. Bot. 102:561–569.

McMaster, G.S., J.W. White, A. Weiss, P.S. Baenziger, W.W. Wilhelm, J.R. Porter, and P.D. Jamieson. 2009. Simulating crop phenological responses to water deficits. pp. 277–300. *In* L.R. Ahuja, V.R. Reddy, S.A. Anapalli, and Q. Yu (ed.) Modeling the response of crops to limited water: Recent advances in understanding and modeling water stress effects on plant growth processes, Vol. 1, Advances in Agricultural Systems Modeling. ASA-SSSA-CSSA, Madison, WI.

McMaster, G.S., and W.W. Wilhelm. 1995. Accuracy of equations predicting the phyllochron of wheat. Crop Sci. 35:30–36.

McMaster, G.S., and W.W. Wilhelm. 1997. Growing degree-days: One equation, two interpretations. Agric. For. Meteorol. 87:289–298.

McMaster, G.S., and W.W. Wilhelm. 1998. Is soil temperature better than air temperature for predicting winter wheat phenology? Agron. J. 90:602–607.

McMaster, G.S., and W.W. Wilhelm. 2003. Phenological responses of wheat and barley to water and temperature: Improving simulation models. J. Agric. Sci. (Cambridge) 141:129–147.

McMaster, G.S., W.W. Wilhelm, and P.N.S. Bartling. 1994. Irrigation and culm contribution to yield and yield components of winter wheat. Agron. J. 86:1123–1127.

McMaster, G.S., W.W. Wilhelm, and J.A. Morgan. 1992b. Simulating winter wheat shoot apex phenology. J. Agric. Sci. (Cambridge) 119:1–12.

McMaster, G.S., W.W. Wilhelm, D.B. Palic, J.R. Porter, and P.D. Jamieson. 2003. Spring wheat leaf appearance and temperature: Extending the paradigm? Ann. Bot. 91:697–705.

Norman, J.M., and G.S. Campbell. 1983. Application of a plant–environment model to problems in irrigation. p. 155–188. *In* D. Hillel (ed.) Advances in irrigation. Vol. 2. Academic Press, New York.

Nuttonson, M.Y. 1948. Some preliminary observations of phenological data as a tool in the study of photoperiodic and thermal requirements of various plant material. p. 29–143. *In* A.E. Murneek and R.O. Whyte (ed.) Vernalization and photoperiodism symposium. Chronica Botanica, Waltham, MA.

O'Leary, G.J., D.J. Connor, and D.H. White. 1985. Effect of sowing time on growth, yield and water-use of rain-fed wheat in the Wimmera, Vic. Aust. J. Agric. Res. 36:187–196.

Peacock, J.M. 1975. Temperature and leaf growth in *Lolium perenne*: II. The site of temperature perception. J. Appl. Ecol. 12:115–123.

Porter, J.R. 1984. A model of canopy development in winter wheat. J. Agric. Sci. (Cambridge) 102:383–392.

Porter, J.R. 1993. AFRCWHEAT2: A model of the growth and development of wheat incorporating responses to water and nitrogen. Eur. J. Agron. 2:64–77.

Porter, J.R., and R. Delecolle. 1988. Interaction of temperature with other environmental factors in controlling the development of plants. p. 133–156. *In* S.P. Long and F.I. Woodward (ed.) Symp. Soc. Exp. Biol. (No. XXXXII): Plants and temperature. Soc. Exp. Biol., Cambridge, MA.

Power, J.F., and J. Alessi. 1978. Tiller development and yield of standard and semidwarf spring wheat varieties as affected by nitrogen fertilizer. J. Agric. Sci. (Cambridge) 90:97–108.

Prasil, I.T., P. Prasilova, and K. Pankova. 2004. Relationships among vernalization, shoot apex development and frost tolerance in wheat. Ann. Bot. 94:413–418.

Prusinkiewicz, P. 1998. Modelling of spatial structure and development of plants: A review. Sci. Hortic. 74:113–149.

Rawson, H.M., J.H. Hindmarsh, R.A. Fischer, and Y.M. Stockman. 1983. Changes in leaf photosynthesis with plant ontogeny and relationships with yield per ear in wheat cultivars and 120 progeny. Aust. J. Plant Physiol. 10:503–514.

Reamur, R.A.F.D. 1735. Observations du thermomètre, fait à Paris pendant l'année 1735, compares avec celles qui ont été faites sous la ligne, à L'Isle de France, à Alger et en quelques-unes de nos isles de l'Amérique. Mémoires de l'Académie des Sciences, Paris, France.

Redmon, L.A., E.G. Krenzer, Jr., D.J. Bernardo, and G.W. Horn. 1996. Effect of wheat morphological stage at grazing termination on economic return. Agron. J. 88:94–97.

Renton, M., J. Hanan, and K. Burrage. 2005. Using the canonical modeling approach to simplify the simulation of function in functional–structural plant models. New Phytol. 166:845–857.

Rickman, R.W., and B. Klepper. 1995. The phyllochron: Where do we go in the future? Crop Sci. 35:44–49.

Rickman, R.W., B. Klepper, and D.A. Ball. 1995. An algorithm for predicting crown root axes of annual grasses. Agron. J. 87:1182–1186.

Rickman, R.W., S.E. Waldman, and B. Klepper. 1996. MODWht3: A development-driven wheat growth simulation. Agron. J. 88:176–185.

Ritchie, J.T. 1991. Wheat phasic development. p. 31–54. *In* J. Hanks and J.T. Ritchie (ed.) Modeling plant and soil systems. ASA-CSSA-SSSA, Madison, WI.

Ritchie, J.T., and S. Otter. 1985. Description and performance of CERES-Wheat: A user-oriented wheat yield model. p. 159–175. *In* W.O. Willis (ed.) ARS wheat yield project. USDA-ARS-38. Natl. Tech. Inf. Serv., Springfield, VA.

Robertson, M.J., I.R. Brooking, and J.T. Ritchie. 1996. The temperature response of vernalization in wheat: Modelling the effect on the final number of mainstem leaves. Ann. Bot. 78:371–381.

Sequeira, R.A., R.L. Olson, and J.M. McKinion. 1997. Implementing generic, object-oriented models in biology. Ecol. Model. 94:17–31.

Sequeira, R.A., P.J.H. Sharpe, N.D. Stone, K.M. El-Zik, and M.E. Makela. 1991. Object-oriented simulation: Plant growth and discrete organ to organ interactions. Ecol. Model. 58:55–89.

Shaykewich, C.F. 1995. An appraisal of cereal crop phenology modeling. Can. J. Plant Sci. 75:329–341.

Skinner, R.H., and C.J. Nelson. 1995. Elongation of the grass leaf and its relationship to the phyllochron. Crop Sci. 35:4–10.

Slafer, G.A., and H.M. Rawson. 1994. Sensitivity of wheat phasic development to major environmental factors: A re-examination of some assumptions made by physiologists and modelers. Aust. J. Plant Physiol. 21:393–426.

Slafer, G.A., and H.M. Rawson. 1995a. Rates and cardinal temperatures for processes of development in wheat: Effects of temperature and thermal amplitude. Aust. J. Plant Physiol. 22:913–926.

Slafer, G.A., and H.M. Rawson. 1995b. Base and optimum temperatures vary with genotype and stage of development in wheat. Plant Cell Environ. 18:671–679.

Sofield, I., L.T. Evans, and I.F. Wardlaw. 1974. The effects of temperature and light on grain filling in wheat. R. Soc. N. Z. Bull. 12:909–915.

Stone, P.J., I.B. Sorensen, and P.D. Jamieson. 1999. Effect of soil temperature on phenology, canopy development and yield of cool-temperate maize. Field Crops Res. 63:169–178.

Streck, N.A., A. Weiss, Q. Xue, and P.S. Baenzinger. 2003. Improving predictions of developmental stages in winter wheat: A modified Wang and Engel model. Agric. For. Meteorol. 115:139–150.

van Keulen, H., and N.G. Seligman. 1987. Simulation of water use, nitrogen nutrition and growth of a spring wheat crop. Simulation Monographs. Pudoc, Wageningen, The Netherlands.

van Laar, H.H., J. Goudriaan, H. van Keulen. 1992. Simulation of crop growth for potential and water limited production situations (as applied to spring wheat). Simulation Reports CABO-TT, 27. CABO-DLO/TPE-WAU, Wageningen, The Netherlands.

Vinocur, M.G., and J.T. Ritchie. 2001. Maize leaf development biases caused by air–apex temperature differences. Agron. J. 93:767–772.

Wang, J.Y. 1960. A critique of the heat unit approach to plant response studies. Ecology 41:785–790.

Wardlaw, I.F., I.A. Dawson, P. Munibi, and R. Fewster. 1989. The tolerance of wheat to high temperatures during reproductive growth: I. Survey procedures and general response patterns. Aust. J. Agric. Res. 40: 1–13.

Weir, A.H., P.L. Bragg, J.R. Porter, and J.H. Rayner. 1984. A winter wheat crop simulation model without water or nutrient limitations. J. Agric. Sci. (Cambridge) 102:371–382.

Welch, S.M., J.L. Roe, and Z. Dong. 2003. A genetic neural network model of flowering time control in *Arabidopsis thaliana*. Agron. J. 95:71–81.

Weyhrich, R.A., B.F. Carver, and B.C. Martin. 1995. Photosynthesis and water-use efficiency of awned and awnletted near-isogenic lines of hard red winter wheat. Crop Sci. 35:172–176.

Whingwiri, E.E., and D.R. Kemp. 1980. Spikelet development and grain yield of the wheat ear in response to applied nitrogen. Aust. J. Agric. Res. 31:637–647.

Whingwiri, E.E., and W.R. Stern. 1982. Floret survival in wheat: Significance of the time of floret initiation relative to terminal spikelet formation. J. Agric. Sci. (Cambridge) 98:257–268.

White, J.W. 2006. From genome to wheat: Emerging opportunities for modeling wheat growth and development. Eur. J. Agron. 25:79–88.

White, J.W., M. Herndl, L.A. Hunt, T.S. Payne, and G. Hoogenboom. 2008. Simulation-based analysis of effects of *Vrn* and *Ppd* loci on flowering in wheat. Crop Sci. 48:678–687.

White, J.W., and G. Hoogenboom. 2003. Gene-based approaches to crop simulation: Past experiences and future opportunities. Agron. J. 95:52–64.

White, J.W., G.S. McMaster, and G.O. Edmeades. 2004a. Physiology, genomics and crop response to global change. Field Crops Res. 90:1–3.

White, J.W., G.S. McMaster, and G.O. Edmeades. 2004b. Genomics, physiology, and global change: What have we learned? Field Crops Res. 90:165–169.

Wiegand, C.L., and J.A. Cuellar. 1981. Duration of grain filling and kernel weight of wheat as affected by temperature. Crop Sci. 21:95–101.

Wilhelm, W.W., and G.S. McMaster. 1995. The importance of the phyllochron in studying the development of grasses. Crop Sci. 35:1–3.

Wilhelm, W.W., and G.S. McMaster. 1996. Spikelet and floret naming scheme for grasses with spike inflorescences. Crop Sci. 36:1071–1073.

Wilhelm, W.W., G.S. McMaster, R.W. Rickman, and B. Klepper. 1993. Above ground vegetative development and growth of winter wheat as influenced by nitrogen and water availability. Ecol. Model. 68:183–203.

Xue, Q., A. Weiss, and P.S. Baenziger. 2004. Predicting leaf appearance in field-grown winter wheat: Evaluating linear and non-linear models. Ecol. Model. 175:261–270.

Yan, W., and L.A. Hunt. 1999. An equation for modelling the temperature response of plants using only the cardinal temperatures. Ann. Bot. 84:607–614.

Yin, X., and P.C. Struik. 2007. Crop systems biology: An approach to connect functional genomics with crop modeling. p. 63–73. *In* J.H.J. Spiertz, P.C. Struik, and H.H. van Laar (ed.) Scale and complexity in plant systems research. Springer, Wageningen, The Netherlands.

Zadoks, J.C., T.T. Chang, and C.F. Konzak. 1974. A decimal code for the growth stages of cereals. Weed Res. 14:415–421.

Zalud, Z., G.S. McMaster, and W.W. Wilhelm. 2003. Parameterizing SHOOTGRO 4.0 to simulate winter wheat phenology and yield in the Czech Republic. Eur. J. Agron. 19:495–507.

Chapter 3
The Flowering Pathway in Wheat

Liuling Yan

SUMMARY

(1) In international and industrial markets, wheat cultivars are categorized into two classes based on their growth habit: winter wheat and spring wheat. Wheat cultivars are also divided into sensitive and insensitive types based on responses of their flowering time to photoperiod.
(2) Flowering time may be regulated by plant development factors independent of vernalization requirement and photoperiod.
(3) Genetic studies have identified internal factors controlling flowering time and have determined their external environmental cues.
(4) Three major vernalization genes have been cloned, based on unambiguous segregation of growth habit by a single gene in each of three mapping populations. Several orthologues of known flowering genes in other plant species have also been characterized in wheat.
(5) New molecular information on wheat flowering genes, combined with comparative studies on flowering pathways in other species, has allowed the formation of a flowering pathway model in wheat.
(6) A better understanding of genetic and molecular mechanisms for control of variation in vernalization requirement duration among winter wheat cultivars, the improved ability to control and design various lengths of developmental phases, and the ability to establish a gene network to regulate flowering constitute key research areas for revealing flowering mechanisms and for breeding novel cultivars of wheat.

OVERVIEW OF FLOWERING INDUCTION IN WHEAT

In international and industrial markets, wheat (*Triticum aestivum* L.) cultivars are classified into two distinct types, spring wheat and winter wheat, based on their growth habits. Understanding of plant growth habit has evolved since it was described by Klippart (1857), although the work of Gassner (1918) is usually cited as the first report on different growth habits in plants (reviewed in Chouard 1960). Wheat sown before winter was originally defined as winter wheat, whereas wheat sown during spring was originally defined as spring wheat (Chouard 1960; Crofts 1989). An original tenet, according to an international survey of wheat breeders and scientists, was that winter wheat did not possess any of the dominant *Vrn* alleles (i.e., the current homoeologous series of dominant vernalization genes, *Vrn-1*) and required extended exposure to low temperature for flowering when grown under nonvernalizing and long-day conditions (Crofts 1989). Contemporary understanding of the term winter wheat is that flowering is accelerated by a period of exposure to low temperature, a process known as vernalization (Law 1987; Amasino 2005). In contrast, the transition from vegetative to reproductive development in spring wheat cannot be accelerated by vernalization. Understanding of vernalization in wheat has rapidly advanced to the molecular level since the beginning of this century, promulgated by the successful cloning of three major vernalization genes in wheat, *VRN-1*, *VRN-2*, and *VRN-3* (Yan et al., 2003, 2004b, 2006).

Response to photoperiod places wheat cultivars into sensitive and insensitive types. Wheat is usually classified as a long-day (LD, >14 hours of light) plant, because it typically flowers earlier when exposed to longer days. Photoperiod insensitivity (or photoperiod neutrality), which arose by mutation from this LD sensitivity, enables wheat to flower earlier without LD treatment (Laurie et al., 1994; Law and Worland 1997; Snape et al., 2001a). Genes which confer photoperiod response in wheat have not been cloned, but the orthologous photoperiod gene *PPD-H1* was cloned in barley (*Hordeum vulgare* L.) that provides intuitive information on the mechanism of the LD response in wheat (Turner et al., 2005; Beales et al., 2007). Wheat, as a member of the winter grass subfamily *Festucoideae*, was initially considered a short-day–long-day (SD–LD) dual-induction plant, that is, SD (<10 hours of light) and/or low temperature treatment primarily induces the developmental transition and LD secondarily induces growth to flower (Heide 1994). It has been rarely noted that a SD period can accelerate the developmental transition of some winter cultivars, thus resulting in earlier flowering without vernalization (Evans 1987). The replacement of vernalization by SD was not found or characterized in numerous modern wheat cultivars, leading to the general classification of wheat as a LD plant (Dubcovsky et al., 2006).

Besides vernalization and photoperiod, plant development provides an additional internal mechanism to regulate flowering time in wheat (Laurie 1995; Snape et al., 2001a). Without treatment with vernalization or SD, winter wheat may still eventually flower due to consequences of plant development or interchangeability between plant development age and vernalization (Wang et al., 1995a,b). A group of earliness *per se* (*EPS*) genes are responsible for fine regulation of flowering time in both spring and winter wheat cultivars (Snape et al., 2001a; Valárik et al., 2006). However, no *EPS* genes have been cloned in wheat.

Vernalization response, photoperiod sensitivity, and other developmental processes provide the necessary adaptive and protective mechanisms to ensure successful reproduction in diverse environments, and ultimately they allow wheat to be the most widely grown crop worldwide. Genetic studies on wheat flowering time have been extensively reviewed (Laurie et al., 1995; Law and Worland 1997; Hay and Ellis 1998; Snape et al., 2001a). Recent advances in molecular understanding of wheat flowering time genes have been reviewed by scientists working directly with cereal crops (Cockram et al., 2007; Trevaskis et al., 2007) and those working with the model plant, Arabidopsis (Henderson et al., 2003; Searle and Coupland 2004; Sung and Amasino 2004; Roux et al., 2006). The following sections explore the current state of our knowledge of wheat flowering time genes, with an emphasis on how these genes were technically cloned from large and complex genomes and how they function in various genetic backgrounds, with the outlook that this knowledge may be applied to wheat improvement.

GENETIC LOCATIONS OF FLOWERING TIME GENES

Genetic loci regulating vernalization response

Vernalization is believed to be the most important adaptive mechanism allowing winter wheat to synchronize plant development with changes in seasonal climate (Flood and Halloran 1986; Rawson et al., 1998; Kirby et al., 1999; Griffiths et al., 2003). Major genetic loci responsible for vernalization effects were located in certain genomic regions using molecular markers to map the growth habit in populations generated from crosses between spring and winter wheat or barley.

VRN-1 on the long arm of homoeologous chromosomes 5

The *VRN-1* gene is the first vernalization gene that was found to be dominant for spring growth habit in hexaploid wheat, *T. aestivum* L. ($2n = 6x = 42$, genome BBAADD). In this context, *Vrn* denotes the dominant allele, *vrn* denotes the recessive allele, *VRN* denotes the locus without allele specificity, and nonitalicized VRN denotes the gene product or protein. This nomenclature applies to symbols for other genes. The *VRN-1* genes were mapped in colinear regions of the long arm of chromosome 5A, followed by original *VRN-2* on 5B, and original *VRN-3* on 5D (reviewed in Dubcovsky et al., 1998). These three genes on group 5 were subsequently found to be homoeoallelic to each other; therefore, they were renamed as *VRN-A1*, *VRN-B1*, and *VRN-D1*, respectively, by McIntosh et al. (1998). The orthologous gene *VRN-H1* in *Hordeum vulgare* (originally *SH2*) (Laurie et al., 1995) and *VRN-R1* in *Secale cereale* L. rye (originally *SP1*) (Plaschke et al., 1993) were also mapped in colinear genomic regions in these species.

The orthologous *VRN-*A^m*1* gene in diploid wheat *T. monococcum* ($2n = 2x = 14$, genome A^mA^m) was mapped in association with growth habit in an F_2 segregating population generated from a cross between spring type G2528 and winter type G1777 (Dubcovsky et al., 1998). This segregating population of diploid wheat hosted subsequent work to successfully clone the *VRN-*A^m*1* gene using the positional cloning approach (Yan et al., 2003).

VRN-A^m*2 on chromosome 5*A^m *in a genomic region translocated from chromosome 4*A^m

A second vernalization gene was found in the diploid wheat *T. monococcum* and designated *VRN-*A^m*2*; it was located in the distal region of chromosome $5A^mL$ within a segment that was translocated from chromosome $4A^m$ (Dubcovsky et al., 1998). This gene resides at a chromosomal location orthologous to the *VRN-H2* locus (originally *SH*) on chromosome 4H in barley (Takahashi and Yasuda 1971). The *VRN-*A^m*2* gene was also mapped in an F_2 segregating population generated from a cross between spring type DV92 and winter type G3116, facilitating its subsequent cloning via the positional cloning approach (Yan et al., 2004b).

Orthologous genes in hexaploid wheat for *VRN-*A^m*2* could be *VRN-A2*, *VRN-B2*, and *VRN-D2*, but these proposed genes have never been genetically detected in hexaploid wheat. One reason is that only when all three homoeologous genes have a recessive allele could spring growth habit caused by this group of genes be detected in this species (Dubcovsky et al., 1998).

VRN-B3 on the short arm of chromosome 7B

The origin and nomenclature of the *VRN-B3* gene in wheat has a more complicated history. The *VRN-H3* locus (also *SH3* or *SGH3*) was first reported on chromosome 1H in barley based on its loose linkage (45 cM) with the *BLP* (black lemma and pericarp) locus on this chromosome (Yasuda 1969). The chromosomal location of *VRN-H3* was determined by mapping a population generated from crossing spring type 'Tammi' with winter type 'Hayakiso 2'. When the Tammi-derived spring barley genetic stock, BGS213, which had the expected allele combination of recessive *vrn-H1* and dominant *Vrn-H2Vrn-H3* (Takahashi and Yasuda 1971), was crossed with winter barley *H. vulgare* ssp. *spontaneum* C. Koch (Thell) expected to have recessive *vrn-H1vrn-H3* and dominant *Vrn-H2*, segregation for growth habit in the F_2 population was controlled by a single locus, *VRN-H3*, as expected (Yan et al., 2006). However, this locus was not mapped to chromosome 1H, the expected location of *VRN-H3*; instead, it was mapped to the short arm of chromosome 7H. This finding of *VRN-H3* on chromosome 7H in barley lead to application of gene symbol *VRN-B3* for an orthologous gene in hexaploid wheat on chromosome 7B.

Chromosome 7B of hexaploid wheat was already reported to carry a flowering time gene designated as *E* in a mapping population generated from a cross between 'Chinese Spring' (CS) and a line with the substituted 'Hope' chromosome 7B in CS [CS(Hope7B)] (Law 1966; Law and Wolfe 1966). The *E* locus was later found to be sensitive to vernalization and was renamed *VRN5* (Law and Worland 1997) but was later renamed again as *VRN-B4*

(McIntosh et al., 1998). Based on the same genetic effects dominant for spring growth habit and similar chromosomal locations between the wheat *VRN-B4* gene and the barley *VRN-H3* gene, the wheat *VRN-B4* gene was recently renamed once again as *VRN-B3* (Yan et al., 2006).

Other vernalization genes in wheat

Near-isogenic lines (NILs) of 'Triple Dirk' were developed to evaluate roles of each *VRN-1* gene in vernalization. Three independent lines possessed only one dominant spring allele, *Vrn-A1* in the Triple Dirk D line, *Vrn-B1* in Triple Dirk B line, or *Vrn-D1* in Triple Dirk E line; each line had a recessive winter *vrn-1* allele on the other two homoeologous chromosomes (Pugsley 1971, 1972). However, an additional line 'Triple Dirk F' (TDF) was found to carry an allele for spring growth habit that differed from any of these three dominant *Vrn-1* alleles. This TDF line was thus believed to carry a dominant *Vrn-4* allele for spring growth habit, compared with the winter Triple Dirk C line. The gene symbol *VRN-4* was proposed to keep as its original name for future use, since the same name for the gene on chromosome 7B had been replaced by *VRN-B3* as previously described. The presence of the dominant *Vrn-4* gene was confirmed by Goncharov (2003), who transferred *Vrn-4* from the near-isogenic TDF line to 'Gabo-2'.

A dominant gene *Vrn-8* for spring growth habit was introgressed from *T. sphaerococcum* into hexaploid wheat by Stelmakh and Avsenin (1996). It was found to be allelic to *Vrn-4* but not to any of the three homoeologous *VRN-1* genes (Goncharov 2003). Goncharov also demonstrated that *VRN-6^{sc}* and *VRN-7^{sc}* that were introgressed from rye into hexaploid wheat (Stelmakh and Avsenin 1996) were not allelic to *Vrn-4* or any *VRN-1* genes. Therefore, in addition to *VRN-1*, *VRN-2*, and *VRN-3* that have been mapped in precise chromosomal locations, at least three more loci (*VRN-4*, *VRN-6^{sc}*, and *VRN-7^{sc}*) may condition vernalization responses in wheat.

Genetic loci regulating photoperiod sensitivity

Flowering is expected to occur either for spring wheat without a vernalization requirement or for winter wheat in which any requirement for vernalization has been satisfied. However, flowering time will be significantly altered when these plants are placed under various environmental stimuli such as photoperiod.

The only example of precise location of *PPD* genes in wheat is a series of homoeoallelic *PPD-1* genes on chromosome group 2. A *PPD-1* locus was first reported by detailed studies of photoperiod insensitivity in near-isogenic lines generated from cultivar Mari and the photoperiod-sensitive cultivar Cappelle-Desprez; this locus was originally named *PPD-D1* on chromosome 2D (Worland and Law 1986). Homoeologous loci were *PPD-2* on chromosome 2B (Scarth and Law 1983; Mohler et al., 2004) and *PPD-3* on chromosome 2A (Law et al., 1978; Scarth and Law *1984*). Following the nomenclature recommended by McIntosh et al. (1998), these homoeoallelic genes were renamed *PPD-A1* for *PPD-3*, *PPD-B1* for *PPD-2*, and *PPD-D1* for *PPD-1*.

In barley the orthologue to the *PPD-1* genes is *PPD-H1*, a major determinant of photoperiod response (Laurie et al., 1995; Decousset et al., 2000). The later-flowering response is controlled by the recessive *ppd-H1* allele, which resulted from a mutation that impairs gene function (Turner et al., 2005). Cloning of *VRN-H1* has facilitated isolation of the wheat *PPD-1* genes (Beales et al., 2007).

Genetic loci regulating plant development processes

The *EPS* genes affect flowering time in a subtle way, which generally results in a difference of only a few days, independently of vernalization and photoperiod (Snape et al., 2001a; Bullrich et al., 2002; Valárik et al., 2006). The minor effects of *EPS* genes on flowering time can be masked in segregating populations generated from crosses between spring and winter lines or between parental lines with different sensitivities to photoperiod. The only example for fine mapping of *EPS* genes is *EPS-A^{m}1* on chromosome 1A^{m} in *T. monococcum*, which has been delimited within a 0.9-cM interval and encompassed within a colinear region between wheat and rice (Valárik et al., 2006).

Quantitative trait loci affecting flowering time

Previous studies with aneuploid and substitution lines of CS indicated the presence of genes affecting flowering time on almost every chromosome of hexaploid wheat (Law et al., 1998). Laurie et al. (1995) predicted 25 loci controlled the duration of the life cycle based on comparative studies of *VRN*, *PPD*, and *EPS* genes between diploid barley and hexaploid wheat. However, approximately 80 genes have been reported to affect flowering time in Arabidopsis (Levy et al., 2002; Tasma and Shoemaker 2003). Due to the presence of three homoeologous genomes, it would not be surprising if more than 200 orthologous genes were found to affect flowering time in hexaploid wheat.

When two parental lines with diverse genetic backgrounds are used to generate a population, flowering time may be mapped by quantitative trait loci (QTLs), and one QTL may appear in one population but not in another. Many QTLs for flowering time have been reported, and the presence of certain QTLs may be affected by vernalization or photoperiod treatment, or both of them, or neither of them. In addition, interactions among genes or QTLs will cause greater complexity of gene effects on flowering time in hexaploid wheat.

Epistatic interactions

Dominance or recessiveness of genes and their epistatic interactions can be genetically determined in a segregating population. Alleles *Vrn-1* and *Vrn-3* are dominant for spring growth habit, whereas *Vrn-2* is dominant for winter growth habit (Takahashi and Yasuda 1971; Dubcovsky et al., 1998). Epistatic interaction between *VRN-A^{m}1* and *VRN-A^{m}2* was detected in a *T. monococcum* population in which only these two genes segregated for growth habit (Tranquilli and Dubcovsky 2000).

Based on current understanding of the three vernalization genes in diploid wheat and barley, only the allele combination of recessive *vrn-1* and *vrn-3* and dominant *Vrn-2* (*vrn-1Vrn-2vrn-3*) confers winter growth habit in diploid wheat (Pugsley 1971; Takahashi and Yasuda 1971; Tranquilli and Dubcovsky 2000; von Zitzewitz et al., 2005; Yan et al., 2006; Szűcs et al., 2007). Due to the presence of three homoeoalleles for a given gene, the analysis of gene action (dominance or recessiveness) and multigenic epistatic interactions will be much more complicated in hexaploid wheat. The recombinant *vrn-A1vrn-B1vrn-D1Vrn-A2vrn-B2vrn-D2vrn-A3vrn-B3vrn-D3* is but one example of 2^{9} gametic possibilities for nine genes determining growth habit in hexaploid wheat. Precise separation of phenotypes becomes impractical with even more genes likely involved in hexaploid wheat. Nevertheless, cloning and characterization of a gene will greatly facilitate analysis of allelic variation and genotypic identification.

POSITIONAL CLONING OF FLOWERING TIME GENES IN WHEAT

VRN-A^{m}1, an orthologue of *AP1*, promotes flowering

Using 6,190 gametes from the G2528 × G1777 F_2 population of diploid wheat *T. monococcum*, *VRN-A^{m}1* was delimited in a 0.03-cM interval containing two MADS-box genes, *AP1* (*APETALA 1*) (Mandel et al., 1992) and *AGLG1* (*AGAMOUS LIKE 2 in GRASSES*), which are orthologues of two Arabidopsis meristem identity genes (Yan et al., 2003). The Arabidopsis *AP1* gene is responsible for the apical transition from vegetative to reproductive phase (Mandel et al., 1992), whereas *AGLG1* belongs to *AGL2*, which is involved in flower development. Only two candidates present in the final version of the wheat physical contigs fit the hypothesis that alternatively either *AP1* or *AGLG1* is *VRN-A^{m}1*.

The *AP1* gene was identified as being *VRN-A^{m}1* based on allelic variation and gene expression profiles (Yan et al., 2003). The presence of a 20-bp deletion adjacent to a putative *CArG*-box for the MADS-box protein binding site in the promoter region of the dominant allele *Vrn-A^{m}1* raised the possibility that this deletion might have impaired the recognition site, so that a repressor cannot bind to the dominant allele but can bind to the recessive allele. Validating this identity between *AP1* and *VRN-A^{m}1* was the discovery that different lengths of deletions existed in a similar region in several independent spring wheat accessions (Yan et al., 2003), with subsequent confirmation of their dominant effects in controlling spring growth habit (Dubcovsky et al., 2006).

In hexaploid wheat, a dominant *Vrn-A1a* allele in most spring accessions has an insertion of a foldback element in the promoter region flanked by 9-bp host-direct duplication; a dominant *Vrn-A1b* allele has two mutations in the same 9-bp host-direct duplication and a 20-bp deletion in the 5′ untranslated region; also, a dominant *Vrn-A1c* allele has no mutation in the promoter region but has a deletion in the first intron, which is similar to the dominant *Vrn-B1* and *Vrn-D1* genes in polyploid wheat (Yan et al., 2004a; Fu et al., 2005). The combined *Vrn-A1*, *Vrn-B1*, and *Vrn-D1* mutations just described explained the spring growth habit of all spring wheat cultivars tested in the previous studies, whereas no mutations were observed in the *vrn-1* genes in the A, B, or D genome of any winter wheat cultivars tested in the same studies. The *Vrn-A1a* allele has the largest contribution to early flowering time. It has been incorporated into spring wheat cultivars in Canadian breeding programs to provide frost avoidance in short-season environments (Iqbal et al., 2007).

Expression of a dominant *Vrn-A^{m}1* allele was observed in nonvernalized plants of spring wheat, but the recessive *vrn-A^{m}1* allele in winter wheat was not expressed until such plants were vernalized. The *vrn-A^{m}1* transcriptional levels were progressively increased during vernalization (Yan et al., 2003).

Regulation of *vrn-1* expression by vernalization was similarly observed in different genotypes and near-isogenic lines of hexaploid wheat. Without vernalization, *Vrn-1* was strongly expressed in spring wheat but not in winter wheat. Vernalization strongly induced *vrn-1* expression and thus accelerated flowering time in winter wheat (Danyluk et al., 2003; Murai et al., 2003; Trevaskis et al., 2003; Loukoianov et al., 2005).

Acceleration of flowering time by *Vrn-1* in spring wheat, and its replacement by vernalization in winter wheat, was validated in subsequent transgenic experiments. A reduction in *VRN-1* transcript levels by RNA interference (RNAi) delayed flowering time for 2 to 3 weeks in transgenic plants of the hexaploid spring wheat cultivar Bobwhite (Loukoianov et al., 2005). Ectopic expression of the wheat *VRN-1* in Arabidopsis not only promoted flowering but also altered development of floral organs, demonstrating pleiotropic effects of *VRN-1* in plants (Adam et al., 2007). An artificial mutant having a deletion of a region including the promoter and MADS box of the *VRN-1* gene prevented the shift from vegetative to reproductive phase (Shitsukawa et al., 2007).

VRN-A^{m}2, a CCT-domain-containing gene, represses flowering

Using 5,698 gametes from the DV92 × G3116 F_2 population in *T. monococcum*, *VRN-A^{m}2* was delimited within a complete physical contig of 438,828 kb sequenced from BAC clones of DV92 (Yan et al., 2004b). Three genes, including two Zinc finger–CCT domain transcription factors *ZCCT1* and *ZCCT2*, were left in the candidate interval. The proteins ZCCT1 and ZCCT2 show similarities to Arabidopsis Constans (CO) and CO-like proteins that regulate flowering time (Putterill et al., 1995), but the similarities are restricted to the conserved CCT (CONSTANS, CONSTANS-LIKE, TOC1) domains, indicating that *ZCCT1* and *ZCCT2* are unique genes in wheat.

The *ZCCT1* gene was identified as *VRN-A*^{m}*2* based on allelic variation (Yan et al., 2004b). In spring wheat DV92 ZCCT1 has a point mutation in the CCT domain that results in loss of its function. This mutation is also present in 22 independent spring wheat accessions but absent in all winter accessions tested. In addition, 17 independent spring accessions were found to have a complete deletion of this gene.

Transcripts of *Vrn-A*^{m}*2* in leaves of winter wheat were down-regulated progressively during vernalization (Yan et al., 2004b). Down-regulation of *VRN-2* expression by vernalization was also confirmed in the hexaploid winter wheat cultivar Jagger (Yan et al., 2004b) and Triple Dirk lines (Loukoianov et al., 2005). The flowering repression by *VRN-2* was confirmed by RNAi in transformed Jagger. Reduction of the RNA level of *VRN-2* by RNAi accelerated the flowering time of transgenic plants by more than one month (Yan et al., 2004b). No allelic variation in *VRN-2* has been reported yet in tetraploid (*T. turgidum* ssp. *durum* L.) or hexaploid forms, but the presence of *VRN-2* gene structure and function in polyploid wheat has been demonstrated by its expression in normal and transgenic plants described previously and by sequencing the orthologous *VRN-2* gene from BAC clones of the tetraploid *T. durum* wheat cultivar Langdon (Dubcovsky and Dvorak 2007).

VRN-B3, an orthologue of *FT*, promotes flowering

Both *VRN-B3* in wheat and *VRN-H3* in barley were cloned in parallel experiments (Yan et al., 2006) The *VRN-3* gene is an orthologue of Arabidopsis *Flowering Locus T* (*FT*) (Samach et al., 2000; Hayama et al., 2003). The product of the *FT* gene was believed to be "florigen," a hypothetical flowering hormone (Chailakhyan 1936) that has been pursued for years. Considerable progress has been made in this research area according to recent studies (reviewed in Corbesier and Coupland 2006 and Zeevaart 2006). The rapid cloning of *VRN-3* in two temperate species greatly benefits from a complete sequence of the rice colinear region, including the heading date gene *Hd3a*, an *FT* orthologue in rice (Kojima et al., 2002).

In spring wheat, the dominant *Vrn-B3* allele was associated with the insertion of a retroelement in its promoter region in the cultivar Hope, whereas in barley, mutations in the dominant *Vrn-H3* allele occurred in the first intron of this gene (Yan et al., 2006). Variation in the noncoding intronic region in *VRN-A3* (= *FT-A*) and *VRN-D3* (= *FT-D*) was also tightly associated with heading date in a large collection of diverse germplasm (Bonnin et al., 2008). The *vrn-3* gene in winter types of wheat and barley was up-regulated by vernalization and long days. Winter wheat plants transformed with the dominant *Vrn-B3* allele carrying the promoter retroelement insertion flowered significantly earlier than non-transgenic plants (Yan et al., 2006).

It is particularly noteworthy that the genetic effect of the dominant *Vrn-H3* allele was detected in two barley populations (BG213 × *H. spontaneum* and BG213 × 'Igri'), which featured a recessive *vrn-H1* (winter) genetic background; the genetic effect of the dominant *Vrn-B3* allele was detected in a wheat CS × CS(Hope7B) population with a dominant spring allele *Vrn-D1* in its background (Pugsley 1971, 1972).

Successes in positional cloning of vernalization genes

Several technical points may be gleaned from the successful cloning of *VRN-1*, *VRN-2*, and *VRN-3* from large and complex genomes of wheat.

1. Cloning of the three vernalization genes greatly benefited from Mendelian segregation of the vernalization requirement according to a single gene. The phenotype was precisely and consistently identified when plants were grown under controlled greenhouse conditions without vernalization.
2. Diploid wheat or barley was used as a model species for cloning of the genes present in hexaploid wheat. Once the target gene was cloned, orthologous genes in hexaploid wheat could be readily isolated by PCR.

3. Large mapping populations were necessarily constructed to decrease the number of candidate genes: two for *VRN-1*, three for *VRN-2*, and only one for *VRN-3*.
4. An orthologue of any flowering time gene (e.g., *VRN-1* vs. *AP1*, *VRN-3* vs. *FT*) or of any genes containing a conserved domain of known flowering time genes (e.g., *VRN-2* vs. *Constans*) in Arabidopsis (or other species) should provide a good candidate for a target gene.
5. Comparative maps with smaller genome size [*Oryza sativa* L., *Sorghum bicolor* (L.) Moench, and currently *Brachypodium distachyon*] provided much help in developing molecular markers in wheat during chromosome walking, although evolutionary events involved in deletion or insertion (in the *VRN-1* region), translocations (in the *VRN-2* region), and duplication and inversion (in the *VRN-3* region) among species were frequently observed.
6. Variation in gene expression profiles could be observed between differing alleles and between plants before and after vernalization treatment.
7. Independent natural mutations present in a large collection of germplasm, and their tight association with the target phenotype, provided convincing evidence for identification of a target gene.
8. Using RNA interference is a powerful technique to validate candidate genes in transgenic wheat.
9. It was fortunate that *VRN-2* was cloned using the BAC library of DV92 that has a recessive *vrn-2* allele. The gene is still present in the library, because DV92 has a point mutation rather than a complete deletion, which is common in the diploid wheat accessions.

Orthologues of other known flowering time genes

The barley *PPD-H1* gene was recently cloned and identified as a member of the pseudoresponse regulator (*PRR*) family (Turner et al., 2005). The PRR proteins are characterized by a pseudoreceiver domain and CCT domain of the protein. Cloning of *PPD-H1* has facilitated isolation of orthologous *PPD-1* genes in wheat. The *PPD-D1a* allele that is insensitive to photoperiod and confers early flowering in SD or LD plants contains a 2-kb deletion upstream from the coding region of the wheat *PRR* gene on chromosome 2D; photoperiod insensitivity caused by *PPD-B1* on chromosome 2B is due to a mutation outside the sequenced region or to a closely linked gene (Beales et al., 2007).

Three orthologues for Arabidopsis *VIN3* (Vernalization-INsensitive 3)—*TmVIL1*, *TmVIL2*, and *TmVIL3*—were isolated and mapped in the centromeric regions of chromosomes 5, 6, and 1, respectively, in *T. monococcum* (Fu et al., 2007). The *VIN3* (Sung and Amasino 2004) and *VIN3*-Like-1 (*VIL1*) (Sung and Amasino 2006) genes were up-regulated by low temperature to repress mitotically stable expression of *FLC* (*Flowering Locus C*), a MADS-box gene repressing flowering time (Michaels and Amasino 1999; Sheldon et al., 1999). The *TmVIL* genes have similar gene structure and transcription regulation to *VIN3*/*VIL*, suggesting that *TmVIL* might have retained similar function in this gene family (Fu et al., 2007).

The *TaVRT2* gene, or *T. aestivum* vegetative-to-reproductive transition gene 2, is a MADS-box gene that is regulated by vernalization and photoperiod (Kane et al., 2005). The *TaVRT2* gene exhibits an inverse pattern of expression relative to *TaVRT1* (i.e., *VRN-1* or *AP1*) in hexaploid wheat.

Concomitant transcriptional profiles of flowering time genes

When *VRN-A^{m}2* was cloned, increased expression of *VRN-A^{m}1* was found to be concomitant with decreased expression of *VRN-A^{m}2* (Yan et al., 2004b; Loukoianov et al., 2005; Dubcovsky et al., 2006). When *VRN-3* was cloned, it was observed to have a transcription profile similar to *VRN-1* but opposite to *VRN-2*. These relationships in gene expression profiles confirmed that *VRN-1*, *VRN-2*, and *VRN-3* act in the same vernalization pathway, consistent with their epistatic interactions observed in previous genetic experiments (Takahashi and Yasuda 1971; Tranquilli and Dubcovsky 2000).

In each of the Triple Dirk spring isogenic lines carrying a single dominant *Vrn-1* allele at one of the *Vrn-A1*, *Vrn-B1*, or *Vrn-D1* loci, plus two recessive *vrn-1* alleles at the other two loci, only the dominant *Vrn-1* allele was transcribed in the seedling stage; however, a few weeks later, transcripts from the recessive *vrn-1* alleles were also detected (Loukoianov et al., 2005). The recessive alleles in these spring-type plants carrying both dominant *Vrn-1* and recessive *vrn-1* alleles were preceded in transcription compared with the recessive *vrn-1* alleles in winter-type plants. This phenomenon also occurred in diploid spring plants with *VRN-1* in the heterozygous condition. These observations lead to a hypothesis that a positive feedback regulatory loop may coordinate transcription of recessive *vrn-1* genes to enhance *VRN-1* transcriptional levels in wheat.

The *VRN-1* gene is up-regulated not only by vernalization but also by LD in NILs either with or without a photoperiod-insensitive gene (Danyluk et al., 2003; Murai et al., 2003). Integration of signals from vernalization and photoperiod to regulate flowering time is consistent with their interactions observed in genetic experiments (Fowler et al., 2001).

The *VRN-A^{m}2* gene is down-regulated not only by vernalization but also by SD in winter lines of *T. monococcum* (Dubcovsky et al., 2006). However, no *VRN-A^{m}1* transcripts were observed in the SD-treated plants until transferred to LD, whereas *VRN-A^{m}1* transcripts were observed during the vernalization process. This comparative study suggested that two different repressors inhibited *VRN-1* expression. A more recent study showed that one of them is TaVRT2 protein, which binds to the *VRN-1* promoter (Kane et al., 2007), supporting the hypothesis that the impaired *CArG*-box located in the *VRN-1* promoter in natural mutants is the most likely regulatory site (Yan et al., 2003; Dubcovsky et al., 2006). Except *TaVRT2*, *VRN-2* is the only gene that has been found to repress flowering in wheat, but no evidence shows how *VRN-2* directly represses *VRN-1*.

COMPARATIVE STUDIES ON FLOWERING PATHWAYS IN PLANTS

Flowering pathways in model species

The flowering pathway in Arabidopsis has been extensively studied, and the rapid progress in this model plant species has been discussed elsewhere (Amasino 2004, 2005; Baurle and Dean 2006; Jaeger et al., 2006; Sung and Amasino 2006). It was indeed fortunate that the vernalization phenomenon also exists in Arabidopsis, in which rapid-flowering accessions do not require vernalization as in spring wheat and vernalization-requiring accessions behave as winter annuals such as winter wheat (Amasino 2005).

A major vernalization gene in Arabidopsis is *FLC* (Michaels and Amasino 1999; Sheldon et al., 1999), which is a central repressor of flowering and is positively regulated by *Frigida* (*FRI*) (Johanson et al., 2000; Gazzani et al., 2003) but negatively regulated by *VIN3* that is induced by vernalization (Sung and Amasino 2004) or by genes in the autonomous pathway (Marquardt et al., 2006). The *FLC* gene delays flowering by repressing expression of *FT* in the leaf and *Suppressor of Overexpression of CONSTANS 1* (*SOC1*) in meristem tissue, preventing up-regulation of a bZIP transcription factor FD, a partner of *FT* in the induction of flowering (Abe et al., 2005; Wigge et al., 2005; Searle et al., 2006). The *SOC1* and *FT* genes activate *AP1* and *Leafy* (*LFY*), a flowering signal integrator, by inducing floral meristem identity. *CO*, a major gene controlling flowering by photoperiod in Arabidopsis (a LD plant like wheat) (Putterill et al., 1995), plays a central role in the regulation of this pathway by inducing transcription of *FT* (reviewed in Thomas 2006).

The flowering pathway in rice, a SD plant, is mainly regulated by photoperiod. Rice has no vernalization requirement, or more precisely, it has not been tested how orthologues of vernalization genes, if any, are regulated by low temperature in rice. Analyses of natural variation showed that

Heading Date 1 (*HD1*) is an ortholog of *CO* (Yano et al., 2000), but *HD1* acts in an opposite way, repressing *FT* and flowering in rice, compared with *CO* in Arabidopsis (Kojima et al., 2002). The interaction of *HD1* with *FT* is altered such that *FT* expression is inhibited under LD conditions (Hayama et al., 2003; Hayama and Coupland 2004).

Among these three species, two related genes *AP1* and *FT* appear to play the same role by promoting flowering. However, unlike wheat, Arabidopsis or rice has not shown allelic variation at *AP1* or *FT*, with no subsequent variation in responses to vernalization. The wheat *VRN-2* gene has no clear orthologues in Arabidopsis or rice, and the Arabidopsis *FLC* has no clear orthologues in rice or wheat, but *FLC* and *VRN-2* have an analogous function to the rice *HD1* that represses flowering under LD. These observations suggest that these three species have relatively conserved genes but with diverse functions (*AP1*, *FT*) or have different genes but with similar functions (*VRN-2*, *FLC*, *HD1*).

Barley shows the same responses to both vernalization and photoperiod as wheat. Its diploid genome, wealthy genetic sources, and conserved gene structure and function have provided transferable information for understanding the flowering pathway in wheat (Danyluk 2003; Trevaskis et al., 2003, 2007; von Zitzewitz et al., 2005; Yan et al., 2005, 2006; Dubcovsky et al., 2006; Szűcs et al., 2007).

A model for the wheat flowering pathway

Various models for the wheat flowering pathway have been proposed (Yan et al., 2003, 2006; Loukoianov et al., 2005; Dubcovsky et al., 2006; Kane et al., 2007; Trevaskis et al., 2007). Current understanding of wheat flowering time genes yields a model for the wheat flowering pathway as updated in Fig. 3.1.

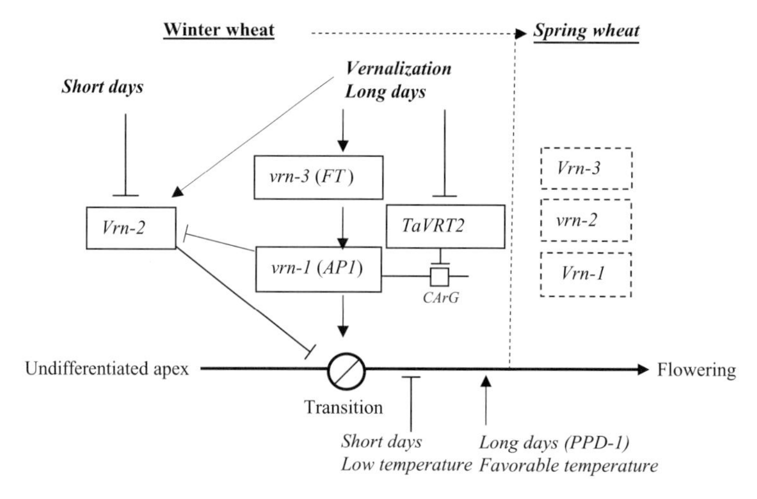

Fig. 3.1 An updated model of the wheat flowering time pathway: Thin arrows indicate promotion and "⊥" indicates repression. Winter wheat has a genotype consisting of dominant *Vrn2* and recessive *vrn1* and *vrn3* alleles (left side); a mutation for any of them will result in a spring wheat genotype (right side). In winter wheat, *vrn1* is up-regulated by vernalization (under long days, LD) by releasing a repressor TaVRT2 that binds to the *vrn1* promoter. This recessive *vrn1* is also activated by *vrn3*, which is up-regulated by vernalization. The *Vrn2* allele is down-regulated by short days (SD) or vernalization, and the repression on *Vrn2* is released by *vrn1* during vernalization. Long days via *PPD-1* and moderately high temperature are required for photoperiod-sensitive cultivars of spring wheat and winter wheat in which any requirement for vernalization has been satisfied to accelerate flowering.

In this model, winter growth habit and photoperiod sensitivity would be considered ancestral traits, whereas spring growth habit and photoperiod insensitivity are mutated traits in domesticated wheat cultivars (Yan et al., 2003; Beales et al., 2007). A mutation in any one of the dominant *Vrn-2* or recessive *vrn-1* or *vrn-3* alleles in winter wheat produces a spring wheat genotype (Takahashi and Yasuda 1971; Pugsley 1972; Dubcovsky et al., 1998; Tranquilli and Dubcovsky 2000; Snape et al., 2001a). Vernalization can promote *vrn-3* to induce *vrn-1* (Yan et al., 2006) or bypass *vrn-3* and directly promote *vrn-1* (Danyluk et al., 2003; Trevaskis et al., 2003; Yan et al., 2003; Loukoianov et al., 2005) by releasing the repressor TaVRT2 on *vrn-1* (Kane et al., 2005, 2007). The *Vrn-2* gene can be repressed by vernalization (Yan et al., 2004b) or through release of *vrn-1* during vernalization (Trevaskis et al., 2007), or it can be repressed by short days alone, which replace the role of vernalization (Dubcovsky et al., 2006). After primary induction of flowering by short days and/or low temperature, secondary induction of growth to flowering occurs via long days and moderately high temperature (Heide 1994). The *PPD-1* gene can promote flowering of wheat cultivars sensitive to LD photoperiod (Beales et al., 2007), but overexposure of winter wheat to low temperature or short days will delay flowering.

FUTURE PERSPECTIVES

Successful cloning of the three vernalization genes greatly benefited from clear segregation of a qualitative trait controlled by singular vernalization genes in three independent mapping populations. However, the genetic basis of quantitative differences in vernalization requirement in winter wheat has not been adequately addressed. Relatively little is known about why various durations of vernalization are required among winter wheat cultivars, which are generally classified into three types: a weak winter type that is stimulated to flower by brief exposure to low temperature, a moderate winter type that requires 2 to 4 weeks' cold exposure to induce flowering, and a strong winter type requiring 4 to 8 weeks' cold exposure (Berry et al., 1980; Crofts 1989; Baloch et al., 2003).

The phenomenon of quantitative vernalization requirement in winter type Arabidopsis can be explained by two alternative hypotheses. One is that gradual expression of *VIN3* induced by vernalization results in a quantitative reduction in *FLC* mRNA levels that negatively correlate with flowering time (Sung and Amasino 2004). The other is that multiple inputs from a network of pathways in response to vernalization, photoperiod, and light quality are integrated, since *FLC* is not the only target of the vernalization process (Reeves and Coupland 2001; Simpson and Dean 2002). Neither of these mechanisms applies to wheat since no *FLC* homologue exists in wheat.

Winter wheat is planted on approximately 17.6 million hectares each year in the US. This amounts to 75% of all wheat grown in the US and is approximately 11% of the world wheat supply and nearly 35% of world wheat exports. Advances in understanding the genetic basis and molecular mechanisms of winter wheat growth and development are critically important to maintaining the productivity of winter wheat wherever it is grown. To reveal the mechanism of how vernalization requirement duration is biologically "measured" to permit flowering would be particularly valuable for improving adaptation of winter wheat. Manipulation of genes for vernalization requirement duration will allow the development of winter wheat cultivars capable of adapting to anticipated global climate shifts. Scientists predict an increase of approximately 3 °C worldwide toward the end of this century (Kerr 2007). Increased temperatures can lead to insufficient vernalization or failed vernalization.

Another key area for future research efforts lies in the regulation of developmental phases to produce wheat for different purposes in world agriculture. As an easily observable trait, flowering time and heading date have been used to represent plant developmental transition in extensive studies on the effects of genetic loci and environmental stimuli. A recent study also showed a positive, linear relationship between stem-elongation stage (or the first-hollow-stem stage) and heading date for a large sample of commercially released winter wheat genotypes in multiple field environments (Edwards et al., 2007). However, the developmental linkage between stem elongation and heading date may not be absolute. No tight association was previously detected between these two phenological events, as many genetic and environmental factors are involved in wheat development (Kirby et al., 1999), indicating that flowering time or heading date might not be the most appropriate phenotype for monitoring the developmental transition.

The life cycle of wheat from sowing to maturity is marked by several critical physiological and morphological stages, including seedling emergence, stem elongation, jointing, heading, flowering, and maturity (Hay and Kirby 1991; Snape et al., 2001b; Gonzalez et al., 2002). At the jointing stage, the plant starts to produce terminal spikelets and should be at Zadoks stage 31, according to scales developed for the Triticeae (Haun 1973; Zadoks et al., 1974; McMaster 2005). At this point, the plant apex has completed the transition from vegetative to reproductive development. The life cycle of the wheat plant can be more simply dissected into three phases: the first phase from seedling emergence to stem elongation (EM–SE), the second phase from stem elongation to heading date (SE–HD), and the third phase from heading date to physiological maturity (HD–PM) (Whitechurch and Slafer 2001).

Each developmental phase plays a key role in the life cycle, ultimate adaptation range, and end use of wheat. The vernalization genes have major effects on the rate of primodia production, whereas the photoperiod genes affect the timing of terminal spikelet production and stem elongation (Snape et al., 2001a). Delayed stem elongation may be selected for an extended vegetative phase to generate more biomass as a forage resource in dual-purpose production systems (Redmon et al., 1996). A longer vegetative phase is also a key characteristic of a cultivar better fit for biomass production as a supplemental biofuel feedstock. In contrast, accelerated stem elongation may be selected to achieve a longer reproductive phase to increase the number of fertile florets (Gonzalez et al., 2003). In addition, proper timing of stem elongation is needed to avoid late-winter freeze and early-spring frost injury (Fowler et al., 2001). An improved genetic understanding of each developmental phase will benefit production of wheat for many different purposes.

Studies in Arabidopsis established the complex network of gene interactions that regulate the transition from vegetative to reproductive development for flowering in plants. In a similar way, such a gene network will be established in wheat and extended to developmental phases before and after flowering. Allelic variation in the flowering time genes allowed the determination of critical regulatory sites, facilitating further studies on their upstream or downstream genes or proteins that interact with cloned genes or their encoded proteins using yeast–hybrid screen systems. Many genes known to affect flowering time in Arabidopsis, such *FCA*, *GI*, *LD*, *SOC1*, have been found to have orthologous expressed sequence tags (ESTs) in wheat, and their functions and respective phenotypes await further investigation. Forward genetics is still a most convincing approach to discover novel flowering time genes unique in wheat by cloning genes responsible for certain traits, such as replacement of vernalization by short days and delay of growth rate by low temperature and short days in winter wheat.

It is expected that molecular markers will be used to create specific gene combinations which elicit a difference in flowering time as small as a few days. Such differences could be used opportunistically to avoid critical heat and drought periods or to produce never-flowering types as a unique and high-quality forage resource. As global climate is never static, this research is particularly important for winter wheat to have the necessary plasticity to grow and thrive in changing environments of the world wheat area.

REFERENCES

Abe, M., Y. Kobayashi, S. Yamamoto, Y. Daimon, A. Yamaguchi, Y. Ikeda, H. Ichinoki, M. Notaguchi, K. Goto, and T. Araki. 2005. FD, a bZIP protein mediating signals from the floral pathway integrator *FT* at the shoot apex. Science 309:1052–1056.

Adam, H., F. Ouellet, N.A. Kane, Z. Agharbaoui, G. Major, Y. Tominaga, and F. Sarhan. 2007. Overexpression of *TaVRN-1* in Arabidopsis promotes early flowering and alters development. Plant Cell Physiol. 48:1192–1206.

Amasino, R. 2004. Vernalization, competence, and the epigenetic memory of winter. Plant Cell 16:2553–2559.

Amasino, R.M. 2005. Vernalization and flowering time. Curr. Opin. Biotechnol. 16:154–158.

Baloch, D.M., R.S. Karow, E. Marx, J.G. Kling, and M.D. Witt. 2003. Vernalization studies with Pacific Northwest wheat. Agron. J. 95:1201–1208.

Baurle, I., and C. Dean. 2006. The timing of developmental transitions in plants. Cell 125:655–664.

Beales, J., A. Turner, S. Griffiths, J. Snape, and D. Laurie. 2007. A pseudo-response regulator is misexpressed in the photoperiod insensitive *Ppd-D1a* mutant of wheat (*Triticum aestivum* L.). Theor. Appl. Genet. 115:721–733.

Berry, G.J., P.A. Salisbury, and G.M. Halloran. 1980. Expression of vernalization genes in near-isogenic wheat lines: Duration of vernalization period. Ann. Bot. 46:235–241.

Bonnin, I., M. Rousset, D. Madur, P. Sourdille, C. Dupuits, D. Brunel, and I. Goldringer. 2008. *FT* genome A and D polymorphisms are associated with the variation of earliness components in hexaploid wheat. Theor. Appl. Genet. 116:383–394.

Bullrich, L., M.L. Appendino, G. Tranquilli, S. Lewis, and J. Dubcovsky. 2002. Mapping of a thermo-sensitive earliness *per se* gene on *Triticum monococcum* chromosome $1A^m$. Theor. Appl. Genet. 105:585–593.

Chailakhyan, M.K. 1936. New facts in support of the hormonal theory of plant development. C. R. Acad. Sci. URSS 13:79–83.

Chouard, P. 1960. Vernalization and its relation to dormancy. Annu. Rev. Plant Physiol. 11:191–238.

Cockram, J., H. Jones, F.J. Leigh, D. O'Sullivan, W. Powell, D.A. Laurie, and A.J. Greenland. 2007. Control of flowering time in temperate cereals: Genes, domestication, and sustainable productivity. J. Exp. Bot. 58:1231–1244.

Corbesier, L., and G. Coupland. 2006. The quest for florigen: A review of recent progress. J. Exp. Bot. 57:3395–3403.

Crofts, H.J. 1989. On defining a winter wheat. Euphytica 44:225–234.

Danyluk, J., N.A. Kane, G. Breton, A.E. Limin, D.B. Fowler, and F. Sarhan. 2003. TaVRT-1, a putative transcription factor associated with vegetative to reproductive transition in cereals. Plant Physiol. 132:1849–1860.

Decousset, L., S. Griffiths, R.P. Dunford, N. Pratchett, and D.A. Laurie. 2000. Development of STS markers closely linked to the *Ppd-H1* photoperiod response gene of barley (*Hordeum vulgare* L.). Theor. Appl. Genet. 101:1202–1206.

Dubcovsky, J., and J. Dvorak. 2007. Genome plasticity a key factor in the success of polyploid wheat under domestication. Science 316:1862–1866.

Dubcovsky, J., D. Lijavetzky, L. Appendino, and G. Tranquilli. 1998. Comparative RFLP mapping of *Triticum monococcum* genes controlling vernalization requirement. Theor. Appl. Genet. 97:968–975.

Dubcovsky, J., A. Loukoianov, D. Fu, M. Valarik, A. Sanchez, and L. Yan. 2006. Effect of photoperiod on the regulation of wheat vernalization genes *VRN-1* and *VRN-2*. Plant Mol. Biol. 60:469–480.

Edwards, J.T., B.F. Carver, and M.E. Payton. 2007. Relationship of first hollow stem and heading in winter wheat. Crop Sci. 47:2074–2077.

Evans, L.T. 1987. Short day induction of inflorescence initiation in some winter wheat varieties. Aust. J. Plant Physiol. 14:277–286.

Flood, R.G., and G.M. Halloran. 1986. Genetics and physiology of vernalization response in wheat. Adv. Agron. 39:87–125.

Fowler, D.B., G. Breton, A.E. Limin, S. Mahfoozi, and F. Sarhan. 2001. Photoperiod and temperature interactions regulate low-temperature-induced gene expression in barley. Plant Physiol. 127:1676–1681.

Fu, D., M. Dunbar, and J. Dubcovsky. 2007. Wheat *VIN3*-like PHD finger genes are up-regulated by vernalization. Mol. Genet. Genomics 277:301–313.

Fu, D.L., P. Szucs, L.L. Yan, M. Helguera, J.S. Skinner, J. von Zitzewitz, P.M. Hayes, and J. Dubcovsky. 2005. Large deletions within the first intron in *VRN-1* are associated with spring growth habit in barley and wheat. Mol. Genet. Genomics 273:54–65.

Gassner, G. 1918. Beiträge zur physiologischen Charakteristik sommer- und winter-anneller Gewächse, insbesondere der Getreidepflanzen. Z. Bot. 10:419–430.

Gazzani, S., A.R. Gendall, C. Lister, and C. Dean. 2003. Analysis of the molecular basis of flowering time variation in Arabidopsis accessions. Plant Physiol. 132:1107–1114.

Goncharov, N.P. 2003. Genetics of growth habit (spring vs. winter) in common wheat: Confirmation of the existence of dominant gene *Vrn4*. Theor. Appl. Genet. 107:768–772.

Gonzalez, F.G., G.A. Slafer, and D.J. Miralles. 2002. Vernalization and photoperiod responses in wheat pre-flowering reproductive phases. Field Crops Res. 74:183–195.

Gonzalez, F.G., G.A. Slafer, and D.J. Miralles. 2003. Grain and floret number in response to photoperiod during stem elongation in fully and slightly vernalized wheats. Field Crops Res. 81:17–27.

Griffiths, S., R.P. Dunford, G. Coupland, and D.A. Laurie. 2003. The evolution of *CONSTANS*-like gene families in barley, rice, and Arabidopsis. Plant Physiol. 131:1855–1867.

Haun, J.R. 1973. Visual quantification of wheat development. Agron. J. 65:116–119.

Hay, R.K.M., and R.P. Ellis. 1998. The control of flowering in wheat and barley: What recent advances in molecular genetics can reveal. Ann. Bot. 82:541–554.

Hay, R.K.M., and E.J.M. Kirby. 1991. Convergence and synchrony: A review of the coordination of development in wheat. Aust. J. Agric. Res. 42:661–700.

Hayama, R., and G. Coupland. 2004. The molecular basis of diversity in the photoperiodic flowering responses of Arabidopsis and rice. Plant Physiol. 135:677–684.

Hayama, R., S. Yokoi, S. Tamaki, M. Yano, and K. Shimamoto. 2003. Adaptation of photoperiodic control pathways produces short-day flowering in rice. Nature 422:719–722.

Heide, O.M. 1994. Control of flowering and reproduction in temperate grasses. New Phytol. 128:347–362.

Henderson, I.R., C. Shindo, and C. Dean. 2003. The need for winter in the switch to flowering. Annu. Rev. Genet. 37:371–392.

Iqbal, M., A. Navabi, R-C. Yang, D.F. Salmon, and D. Spaner. 2007. Molecular characterization of vernalization response genes in Canadian spring wheat. Genome 50:511–516.

Jaeger, K.E., A. Graf, and P.A. Wigge. 2006. The control of flowering in time and space. J. Exp. Bot. 57:3415–3418.

Johanson, U., J. West, C. Lister, S. Michaels, R. Amasino, and C. Dean. 2000. Molecular analysis of *FRIGIDA*, a major determinant of natural variation in Arabidopsis flowering time. Science 290:344–347.

Kane, N.A., Z. Agharbaoui, A.O. Diallo, H. Adam, Y. Tominaga, F. Ouellet, and F. Sarhan. 2007. *TaVRT2* represses transcription of the wheat vernalization gene *TaVRN-1*. Plant J. 51:670–680.

Kane, N.A., J. Danyluk, G. Tardif, F. Ouellet, J-F. Laliberte, A.E. Limin, D.B. Fowler, and F. Sarhan. 2005. *TaVRT-2*, a member of the StMADS-11 clade of flowering repressors, is regulated by vernalization and photoperiod in wheat. Plant Physiol. 138:2354–2363.

Kerr, R.A. 2007. Climate change: Global warming is changing the world. Science 316:188–190.

Kirby, E.J.M., J.H. Spink, D.L. Frost, R. Sylvester-Bradley, R.K. Scott, M.J. Foulkes, R.W. Clare, and E.J. Evans. 1999. A study of wheat development in the field: Analysis by phases. Eur. J. Agron. 11:63–82.

Klippart, J.H. 1857. An essay on the origin, growth, disease, etc. of the wheat plant. Ohio State Bot. Agric. Annu. Rep. 12:562–816.

Kojima, S., Y. Takahashi, Y. Kobayashi, L. Monna, T. Sasaki, T. Araki, and M. Yano. 2002. *Hd3a*, a rice ortholog of the Arabidopsis *FT* Gene, promotes transition to flowering downstream of *Hd1* under short-day conditions. Plant Cell Physiol. 43:1096–1105.

Laurie, D.A., N. Pratchett, J.H. Bezant, and J.W. Snape. 1994. Genetic analysis of a photoperiod response gene in the short arm of chromosome 2(2H) of *Hordeum vulgare* (barley). Heredity 72:619–627.

Laurie, D.A., N. Pratchett, J.H. Bezant, and J.W. Snape. 1995. RFLP mapping of five major genes and eight quantitative trait loci controlling flowering time in a winter × spring barley (*Hordeum vulgare* L.) cross. Genome 38:575–585.

Law, C.N. 1966. The location of genetic factors affecting a quantitative character in wheat. Genetics 53:487–498.

Law, C.N. 1987. The genetic control of day-length response in wheat. p. 225–240. *In* J.G. Atherton (ed.) Manipulation of flowering. Butterworths, London, UK.

Law, C.N., E. Suarez, T.E. Miller, and A.J. Worland. 1998. The influence of the group 1 chromosomes of wheat on ear-emergence times and their involvement with vernalization and day length. Heredity 80:83–91.

Law, C.N., J. Sutka, and A.J. Worland. 1978. A genetic study of day-length response in wheat. Heredity 41:185–191.

Law, C.N., and M.S. Wolfe. 1966. Location of genetic factors for mildew resistance and ear emergence time on chromosome 7B of wheat. Can. J. Genet. Cytol. 8:462–470.

Law, C.N., and A.J. Worland. 1997. Genetic analysis of some flowering time and adaptive traits in wheat. New Phytol. 137:19–28.

Levy, Y.Y., S. Mesnage, J.S. Mylne, A.R. Gendall, and C. Dean. 2002. Multiple roles of Arabidopsis *VRN-1* in vernalization and flowering time control. Science 297:243–246.

Loukoianov, A., L. Yan, A. Blechl, A. Sanchez, and J. Dubcovsky. 2005. Regulation of *VRN-1* vernalization genes in normal and transgenic polyploid wheat. Plant Physiol. 138:2364–2373.

Mandel, M.A., C. Gustafsonbrown, B. Savidge, and M.F. Yanofsky. 1992. Molecular characterization of the Arabidopsis floral homeotic gene *Apetala1*. Nature 360:273–277.

Marquardt, S., P.K. Boss, J. Hadfield, and C. Dean. 2006. Additional targets of the Arabidopsis autonomous pathway members, *FCA* and *FY*. J. Exp. Bot. 57:3379–3386.

McIntosh, R.A., G.E. Hart, K.M. Devos, M.D. Gale, and W.J. Rogers. 1998. Catalogue of gene symbols for wheat. *In* A.E. Slinkard (ed.) Proc. Int. Wheat Genetics Symp., 9th, Vol. 5, Saskatoon, Saskatchewan. 2–7 Aug. 1998. Univ. Ext. Press, Saskatoon, Canada.

McMaster, G.S. 2005. Phytomers, phyllochrons, phenology and temperate cereal development. J. Agric. Sci. 143:137–150.

Michaels, S.D., and R.M. Amasino. 1999. FLOWERING LOCUS C encodes a novel MADS domain protein that acts as a repressor of flowering. Plant Cell 11:949–956.

Mohler, V., R. Lukman, S. Ortiz-Islas, M. William, A. Worland, J. van Beem, and G. Wenzel. 2004. Genetic and physical mapping of photoperiod insensitive gene *Ppd-B1* in common wheat. Euphytica 138:33–40.

Murai, K., M. Miyamae, H. Kato, S. Takumi, and Y. Ogihara. 2003. *WAP1*, a wheat *APETALA1* homolog, plays a central role in the phase transition from vegetative to reproductive growth. Plant Cell Physiol. 44:1255–1265.

Plaschke, J., A. Börner, D.X. Xie, R.M.D. Koebner, R. Schlegel, and M.D. Gale. 1993. RFLP mapping of genes affecting plant height and growth habit in rye. Theor. Appl. Genet. 85:1049–1054.

Pugsley, A.T. 1971. A genetic analysis of the spring–winter habit of growth in wheat. Aust. J. Agric. Res. 22:21–31.

Pugsley, A.T. 1972. Additional genes inhibiting winter habit in wheat. Euphytica 21:547–552.

Putterill, J., F. Robson, K. Lee, R. Simon, and G. Coupland. 1995. The *Constans* gene of *Arabidopsis* promotes flowering and encodes a protein showing similarities to zinc finger transcription factors. Cell 80:847–857.

Rawson, H.M., M. Zajac, and L.D.J. Penrose. 1998. Effect of seedling temperature and its duration on development of wheat cultivars differing in vernalization response. Field Crops Res. 57:289–300.

Redmon, L.A., E.G. Krenzer, D.J. Bernardo, and G.W. Horn. 1996. Effect of wheat morphological stage at grazing termination on economic return. Agron. J. 88:94–97.
Reeves, P.H., and G. Coupland. 2001. Analysis of flowering time control in Arabidopsis by comparison of double and triple mutants. Plant Physiol. 126:1085–1091.
Roux, F., P. Touzet, J. Cuguen, and V. Le Corre. 2006. How to be early flowering: An evolutionary perspective. Trends Plant Sci. 11:375–381.
Samach, A., H. Onouchi, S.E. Gold, G.S. Ditta, Z. Schwarz-Sommer, M.F. Yanofsky, and G. Coupland. 2000. Distinct roles of *CONSTANS* target genes in reproductive development of Arabidopsis. Science 288:1613–1616.
Scarth, R., and C.N. Law. 1983. The location of the photoperiod gene, *Ppd-2*, and an additional genetic factor for ear-emergence time on chromosome 2B of wheat. Heredity 51:607–619.
Scarth, R., and C.N. Law. 1984. The control of the day-length response in wheat by the group 2 chromosomes. Z. Pflanzenzuecht. 92:140–150.
Searle, I., and G. Coupland. 2004. Induction of flowering by seasonal changes in photoperiod. EMBO J. 23:1217–1222.
Searle, I., Y. He, F. Turck, C. Vincent, F. Fornara, S. Krober, R.A. Amasino, and G. Coupland. 2006. The transcription factor FLC confers a flowering response to vernalization by repressing meristem competence and systemic signaling in Arabidopsis. Genes Dev. 20:898–912.
Sheldon, C.C., J.E. Burn, P.P. Perez, J. Metzger, J.A. Edwards, W.J. Peacock, and E.S. Dennis. 1999. The *FLF* MADS box gene: A repressor of flowering in Arabidopsis regulated by vernalization and methylation. Plant Cell 11:445–458.
Shitsukawa, N., C. Ikari, S. Shimada, S. Kitagawa, K. Sakamoto, H. Saito, H. Ryuto, N. Fukunishi, T. Abe, S. Takumi, S. Nasuda, and K. Murai. 2007. The einkorn wheat (*Triticum monococcum*) mutant, maintained vegetative phase, is caused by a deletion in the *VRN-1* gene. Genes Genet. Syst. 82:167–170.
Simpson, G.G., and C. Dean. 2002. Arabidopsis, the Rosetta stone of flowering time? Science 296:285–289.
Snape, J.W., K. Butterworth, E. Whitechurch, and A.J. Worland. 2001a. Waiting for fine times: Genetics of flowering time in wheat. Euphytica 119:185–190.
Snape, J.W., R. Sarma, S.A. Quarrie, L. Fish, G. Galiba, and J. Sutka. 2001b. Mapping genes for flowering time and frost tolerance in cereals using precise genetic stocks. Euphytica 120:309–315.
Stelmakh, A.F., and V.I. Avsenin. 1996. Alien introgression of spring habit dominant genes into bread wheat. Euphytica 89:65–68.
Sung, S., and R.M. Amasino. 2004. Vernalization in *Arabidopsis thaliana* is mediated by the PHD finger protein VIN3. Nature 427:159–164.
Sung, S., and R.M. Amasino. 2006. Molecular genetic studies of the memory of winter. J. Exp. Bot. 57:3369–3377.
Szűcs, P., J. Skinner, I. Karsai, A. Cuesta-Marcos, K. Haggard, A. Corey, T. Chen, and P. Hayes. 2007. Validation of the *VRN-H2/VRN-H1* epistatic model in barley reveals that intron length variation in *VRN-H1* may account for a continuum of vernalization sensitivity. Mol. Genet. Genomics 277:249–261.
Takahashi, R., and S. Yasuda. 1971. Genetics of earliness and growth habit in barley. p. 388–408. *In* R.A. Nilan (ed.) Proc. Int. Barley Genet. Symp., 2nd, Pullman, WA. 6–11 July 1969. Washington State University Press, Pullman, WA.
Tasma, I.M., and R.C. Shoemaker. 2003. Mapping flowering time gene homologs in soybean and their association with maturity (E) loci. Crop Sci. 43:319–328.
Thomas, B. 2006. Light signals and flowering. J. Exp. Bot. 57:3387–3393.
Tranquilli, G.E., and J. Dubcovsky. 2000. Epistatic interactions between vernalization genes *Vrn-A*^{m}*1* and *Vrn-A*^{m}*2* in diploid wheat. J. Hered. 91:304–306.
Trevaskis, B., D.J. Bagnall, M.H. Ellis, W.J. Peacock, and E.S. Dennis. 2003. MADS box genes control vernalization-induced flowering in cereals. Proc. Natl. Acad. Sci. USA 100:13099–13104.
Trevaskis, B., M.N. Hemming, E.S. Dennis, and W.J. Peacock. 2007. The molecular basis of vernalization-induced flowering in cereals. Trends Plant Sci. 12: 352–357.
Turner, A., J. Beales, S. Faure, R.P. Dunford, and D.A. Laurie. 2005. The pseudo-response regulator *Ppd-H1* provides adaptation to photoperiod in barley. Science 310:1031–1034.
Valárik, M., A. Linkiewicz, and J. Dubcovsky. 2006. A microcolinearity study at the earliness per se gene *Eps-A m 1* region reveals an ancient duplication that preceded the wheat–rice divergence. Theor. Appl. Genet. 112:945–957.
von Zitzewitz, J., P. Szűcs, J. Dubcovsky, L. Yan, E. Francia, N. Pecchioni, A. Casas, T.H.H. Chen, P.M. Hayes, and J.S. Skinner. 2005. Molecular and structural characterization of barley vernalization genes. Plant Mol. Biol. 59: 449–467.
Wang, S.Y., R.W. Ward, J.T. Ritchie, R.A. Fischer, and U. Schulthess. 1995a. Vernalization in wheat: I. A model based on the interchangeability of plant-age and vernalization duration. Field Crops Res. 41:91–100.
Wang, S-Y., R.W. Ward, J.T. Ritchie, R.A. Fischer, and U. Schulthess. 1995b. Vernalization in wheat: II. Genetic variability for the interchangeability of plant age and vernalization duration. Field Crops Res. 44:67–72.
Whitechurch, E.M., and G.A. Slafer. 2001. Responses to photoperiod before and after jointing in wheat substitution lines. Euphytica 118:47–51.
Wigge, P.A., M.C. Kim, K.E. Jaeger, W. Busch, M. Schmid, J.U. Lohmann, and D. Weigel. 2005. Integration of spatial and temporal information during floral induction in Arabidopsis. Science 309:1056–1059.
Worland, A.J., and C.N. Law. 1986. Genetic analysis of chromosome 2D of wheat: I. The location of genes affecting height, day length insensitivity, hybrid dwarfism and yellow rust resistance. Z. Pflanzenzuecht. 96:331–345.
Yan, L., D. Fu, C. Li, A. Blechl, G. Tranquilli, M. Bonafede, A. Sanchez, M. Valarik, S. Yasuda, and J. Dubcovsky.

2006. The wheat and barley vernalization gene *VRN-3* is an orthologue of *FT*. Proc. Natl. Acad. Sci. USA 103:19581–19586.

Yan, L., M. Helguera, K. Kato, S. Fukuyama, J. Sherman, and J. Dubcovsky. 2004a. Allelic variation at the *VRN-1* promoter region in polyploid wheat. Theor. Appl. Genet. 109:1677–1686.

Yan, L., A. Loukoianov, G. Tranquilli, A. Blechl, I.A. Khan, W. Ramakrishna, P. San Miguel, J.L. Bennetzen, V. Echenique, D. Lijavetzky, and J. Dubcovsky. 2004b. The wheat *VRN-2* gene is a flowering repressor down-regulated by vernalization. Science 303:1640–1644.

Yan, L., A. Loukoianov, G. Tranquilli, M. Helguera, T. Fahima, and J. Dubcovsky. 2003. Positional cloning of wheat vernalization gene *VRN-1*. Proc. Natl. Acad. Sci. USA 100:6263–6268.

Yan, L., J.V. Zitzewitz, J.S. Skinner, P.M. Hayes, and J. Dubcovsky. 2005. Molecular characterization of the duplicated meristem identity genes *HvAP1a* and *HvAP1b* in barley. Genome 48:905–912.

Yano, M., Y. Katayose, M. Ashikari, U. Yamanouchi, L. Monna, T. Fuse, T. Baba, K. Yamamoto, Y. Umehara, Y. Nagamura, and T. Sasaki. 2000. *Hd1*, a major photoperiod sensitivity quantitative trait locus in rice, is closely related to the Arabidopsis flowering time gene *CONSTANS*. Plant Cell 12:2473–2484.

Yasuda, S. 1969. Linkage and pleiotropic effects on agronomic characters of the genes for spring growth habit. Barley Newsl. 12:57–58.

Zadoks, J., T. Chang, and C. Konzak. 1974. A decimal code for growth stages of cereals. Weed Res. 14:415–421.

Zeevaart, J.A.D. 2006. Florigen coming of age after 70 years. Plant Cell 18:1783–1789.

Section II
Making of a Wheat Crop

"No wheat that has ever yet fallen under my observation exceeds the white which some years ago I cultivated extensively; but which, from inattention during my absence from home of almost nine years has got so mixed or degenerated as scarcely to retain any of its original characteristic properties. But if the march of the Hessian fly, southerly, cannot be arrested; . . . this white wheat must yield the palm to the yellow bearded, which alone, it seems, is able to resist the depredations of that destructive insect."

George Washington,
in correspondence with John Bordley, 1788

Chapter 4
Systems-Based Wheat Management Strategies

Jeffrey T. Edwards

SUMMARY

(1) Systems-based management approaches have increased wheat yield over the past few decades by matching genotype to management style. The intensive wheat management system is an excellent example of how this concept can be utilized to its fullest potential.
(2) Integrated systems, such as the dual-purpose wheat production system, have demonstrated how producers can optimize many parts of a complex system to improve profitability of the system as a whole.
(3) Wheat has lagged other grain crops in adoption of conservation-tillage and no-till management systems. There are several reasons for this, but recent advances in planting technologies and crop rotation are resulting in increased adoption of no-till by wheat producers.
(4) Further genetic improvement is needed to introduce cultivars with improved stress tolerance to reclaim production areas that have been lost due to nonoptimal soil pH or salt accumulation.
(5) The success in systems-based wheat production strategies can serve as models for other grass-based systems, such as grass-based cellulosic biofuel production.

INTRODUCTION

This chapter will focus on management of the wheat crop and how management factors combine to create a production system. It is, therefore, prudent to discuss what is meant by the word *management*. Webster's defines "manage" as *to handle or direct with a degree of skill* (Woolf 1979). Too often management is confused with scouting, which is defined as *to explore an area to obtain information* (Woolf 1979). To properly manage wheat requires foresight and planning and is proactive in nature. In contrast, scouting is primarily reactive in nature and is only one component of a management strategy. This chapter will put forth examples of wheat management systems which incorporate proactive management strategies to successfully address unique challenges faced by producers and will outline a few of the challenges and questions that remain unanswered.

ADVANCES IN WHEAT MANAGEMENT

Yield building versus yield protecting factors

The most recent edition of *Wheat and Wheat Improvement* was published in 1987 (Heyne 1987). This publication remains a valuable resource that outlines and defines many of the basic tenets of wheat production. In addition, numerous extension management guides are available to provide guidance in the basics of wheat production (Alley et al., 1993; Bitzer et al., 1997; Royer and Krenzer 2000; Wiersma and Bennett 2001). The majority of the management operations described by Tucker et al. (1987) have not changed substantially, but significant advances in wheat production have emerged over the past 20 years. Improved cultivars, more efficacious herbicides and fungicides, and better equipment are just a few of the modern advances in wheat technology brought about by public and private research. Much of the improvement in wheat management has resulted from understanding how these other advances can be utilized together in a systems approach to management. That is, the understanding of rudimentary agronomic components of wheat production has not changed, but our understanding of how these components interact to form a profitable and sustainable management system has changed appreciably.

Alley et al. (1993) eloquently showed how the various components of wheat management converge to determine final grain yield. They described yield as a function of kernels per head, weight per kernel, and heads per unit area. Factors such as cultivar selection, plant nutrition, and precision planting are considered to be *yield building factors*. Adjustments in these management factors can actually improve the yield potential of the crop. Weed control, insect control, disease control, lodging control, and harvest management are listed as *yield protecting factors*. These yield protecting factors can only preserve inherent yield potential of the crop; they cannot create new yield potential. Arguably, the greatest improvements in wheat yield at the farm level over the past few decades emanated from better understanding how yield building and yield protecting factors interact in a system.

Intensive wheat management

An example of successful on-farm adoption of a systems approach to wheat management is the implementation of European-style intensive wheat management in the soft red winter wheat production areas of the US. Prior to introduction of this system, wheat in the southeastern US had largely been treated as a second-class crop that did not receive the same attention and level

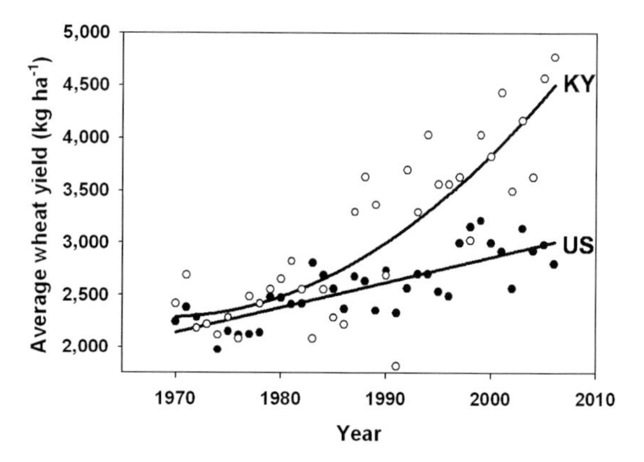

Fig. 4.1 Average wheat yield has increased more rapidly in soft red winter wheat production areas, such as Kentucky, than in the US as a whole. Much of this increase can be attributed to the introduction and implementation of European-style intensive wheat management practices in the 1980s. (Adapted from National Agricultural Statistics Service, 2008.)

of management as summer crops. Introduction of the intensive wheat management system in the mid-1980s changed much of this mindset (Snyder 2000). Producers in this region quickly began to realize that wheat, like other crops, was responsive to increased levels of inputs and management. Since that time, average wheat yields have increased much more rapidly in the US Southeast than in the US as a whole (Fig. 4.1).

Matching cultivar to environment

The basic concepts of intensive wheat management center on matching cultivar with management to create yield potential and protecting that yield potential once it is established. In fact, some have proposed that fine-tuning the matching of management practices with cultivar selection may have as much influence on yield as cultivar improvement. Cooper et al. (2001), for example, reported that much of the differential in grain yield and grain protein that is attributed to genotype × environment interactions in Australia is actually better characterized by what they described as genotype × management interactions. Bell et al. (1995) attributed 28% of wheat yield improvement from 1968 to 1990 to genetic gains; however, they attributed 48% of yield improvement over this same period to improved management, specifically to increased levels of nitrogen fertilizer. Wheat breeding during this time did result in significant gains in nonyield components such as end-use quality, and it is likely that increased nitrogen fertilizer applications would not have been possible without the introgression of semidwarfing genes. These studies reinforce the interdependency of yield building and yield protecting factors and how they must be used collectively to form a wheat management system.

Fertility and pest management

The first step in optimizing and/or intensifying wheat management is selecting the right cultivar and matching nitrogen fertility to the cultivar (Karlen and Gooden 1990). Finding the correct combination of these two elements often dictates the difference between negative and positive cash flow from operations (Karlen and Gooden 1990). The majority of successes in intensive wheat management have hinged on improved nitrogen fertility and finding cultivars that respond to this improved fertility. Austin (1999) attributed much of the advance in wheat yield in the UK to shorter-statured wheat cultivars with increased straw strength, which allowed greater responsiveness to nitrogen fertilizer. Abeledo et al. (2003) made similar observations in winter barley. Moreover, producers who implement intensive wheat management strategies often place greater emphasis on straw strength and tolerance to high nitrogen fertilizer rates than on disease resistance.

In this example, the basic assumption is that a foliar fungicide will be used after flag leaf emergence and that disease resistance can be viewed as subordinate to overall yield potential. In contrast, low-input wheat producers will view disease resistance as crucial to wheat cultivar performance and suitability. These producers are often willing to sacrifice top-end yield potential for consistent performance despite spatial and temporal environmental variation. Similarly, market forces and grain quality standards sometimes trump raw yield potential in the decision-making hierarchy of bread wheat producers.

Quantity is not the only nitrogen-related consideration for intensive wheat management. Increased emphasis on input application timing and accuracy has been a covariate in this success. Timing, split nitrogen applications, and allocations to each split are all integral parts of the overall intensive management scheme. A fall application of nitrogen in intensive wheat management systems is generally limited to 15 to 30 kg ha^{-1} and is often combined with a preplant application of phosphorus and potassium. It is also important to understand that split application of nitrogen in intensive wheat management systems generally refers to a split *spring* application, not just a fall–spring split. These applications typically occur at Feekes growth stage (GS) 3 and 6 (Large 1954). By delaying the bulk of nitrogen application until just before or at jointing in winter wheat, producers have been able to better assess crop nitrogen needs and adjust nitrogen application accordingly. Further, by having a precise knowledge of crop nitrogen requirements, producers are able to increase efficiency and better utilize nitrogen as a management tool. Nitrogen timing, for example, has been shown to be an effective tool in managing tiller numbers in winter wheat (Weisz et al., 2001). Management of tiller numbers is a critical tool in achieving the desired 500–600 wheat heads per square meter at harvest (Alley et al., 1993).

Just as monitoring of crop fertility needs has resulted in an increase in wheat yield over the past couple decades, intensified monitoring of insect infestations and disease incidence and severity have helped protect the yield created through improved cultivars and increased nitrogen rates. Ransom et al. (2006) attribute much of the genotype × management interaction mentioned earlier in this chapter to differing levels of disease control with fungicide. Further, increased nitrogen fertilizer rates associated with increased nitrogen fertility and increased wheat head densities can necessitate the use of foliar fungicides and plant growth regulators. These examples further demonstrate the need to manage wheat by a systems approach rather than in piecemeal steps. The profitability of implementing any one improved management factor such as increased nitrogen rates or foliar fungicides depends heavily on how and when other management factors are implemented.

Timeliness and precision

An old adage states that the difference between a successful farmer and an unsuccessful farmer is about three to five days. One cannot discount the effect of timely and accurate application of yield building factors and yield protecting factors such as nitrogen fertility, weed control, insect control, and foliar disease control. Tramlines have long been used to aid in application precision and timeliness (see Color Plate 3). Tramlines are simply rows in the drill that are sealed off and not planted. At first, creating tramlines meant having a sprayer with a boom width that was a multiple of the drill width and then sealing the appropriate planter rows with duct tape. As drills and spray booms grew larger and as air drills became more popular, the need for electronically controlled tramline systems increased. Most recently, GPS-enabled precision guidance systems and autosteer technologies have allowed producers to create tramlines after the fact. These technologies allow producers to establish virtual tramlines that can be followed year after year. The essential concept, however, remained the same: increase precision and limit traffic. By limiting wheel traffic in fields, producers are able to reduce the area affected by compaction and reduce the area that may require subsoiling to alleviate soil compaction (Sweeney et al., 2006). If it is assumed that normal, unrestricted field traffic results in 90% of the field area being subjected to wheel traffic, implementing a light-bar guidance system would reduce the area subjected to wheel traffic to 30% (Watson and Lowenberg-DeBoer 2004).

In many cases producers have substituted precision-application aids for skilled labor (Watson and Lowenberg-DeBoer 2004). That is, the skill level required to follow a set of tramlines or a differentially corrected GPS signal is much less than that required to follow a foam marker. The wider the implement being used, the greater is the benefit. Outfitting farm equipment with a light-bar guidance system, for example, would allow a 13% increase in field speed (Watson and Lowenberg-DeBoer 2004). Use of an autoguidance system would allow a 20% increase in field speed. These technologies would allow farmers to increase the size of their operation from 728 ha to 1,052 and 1,255 ha by implementing the use of a light-bar or autoguidance system, respectively (Watson and Lowenberg-DeBoer 2004). Whatever the system, the ability to reduce dependence on skilled labor while simultaneously increasing precision and timeliness has been a positive development in wheat production technology over the past 10–15 years.

Previous crop management

Previous crop management is a rudimentary component of intensive wheat production systems. A maize (*Zea mays* L.)–wheat–soybean (*Glycine max* L.) rotation, for example, is the predominant cropping system in many areas of the midwestern and midsouthern US. Selection of the right maize hybrid in this rotation sometimes determines whether or not wheat sowing will even be possible the following fall. Later-maturing hybrids generally perform better than earlier-maturing hybrids, but a gain in maize yield can result in a significant decrease in wheat yield. Likewise choosing a late-maturing wheat cultivar may adversely impact the subsequent soybean crop. To combat this dilemma, some producers have chosen to continue to produce full-season wheat cultivars but elect to utilize technologies such as stripper headers that allow for harvest of higher-moisture grain. This in turn allows the farmer to grow a full-season cultivar but have the same harvest timing as that of a short-season cultivar (see Color Plate 4). Similarly, some maize producers have found they can produce full-season maize hybrids prior to wheat if they are equipped and willing to harvest at higher-than-optimal grain moisture and make use of on-farm drying equipment. Whatever the system used, the key to profitable and sustainable management in modern production systems is to anticipate the effect that a decision today will have 6 months or even 6 years from now. All components of wheat management are intertwined, and one factor is rarely changed without having an intended or unintended influence on another factor of production.

Limitations of the system

Although intensification of wheat management practices has been regarded as a large step forward in areas such as the southeastern US (Snyder 2000), it has not been a universal success in all areas. In areas where intensively managed crops such as vegetables are already the dominant system, few benefits have been observed to further increases in nitrogen fertility or increased inputs (Mohamed et al., 1990). In spring wheat production areas, delaying the bulk of nitrogen applications until wheat jointing can result in insufficient tillering and lower yields. Similarly, in areas with less predictable rainfall patterns, such as the US Great Plains, rainfall may not occur for several weeks after nitrogen application. Thus, delaying nitrogen applications until jointing can sometimes result in excessive nitrogen losses due to volatilization or the failure to move nitrogen into the rooting zone in time to have full impact on wheat yield. Again, the key to a systems approach is to understand which factors are limiting and to determine whether or not those factors are controllable.

DUAL-PURPOSE WHEAT

Description of the system and area of adaptation

The dual-purpose wheat production system is an integrated crop–livestock system that is common in some areas of the US, southern Europe, South America, and Australia. In this system, wheat is sown in early fall with the expectation that wheat growth will be sufficient to allow for grazing by stocker cattle (*Bos taurus* L.) within 8–10 weeks. Cattle remain on wheat pasture until the wheat canopy reaches the first-hollow-stem stage of growth in late winter. First hollow stem occurs just before jointing (approximately Feekes GS 5) and is characterized by 1.5 cm of hollow stem below the growing point (Redmon et al., 1996) (see Color Plate 5). By diversifying the farming enterprise between crop and livestock production on the same acreage, the dual-purpose system spreads risk, increases income, and increases nutrient utilization efficiency (Redmon et al., 1995).

There are several reasons why the dual-purpose wheat production system fits in some areas and not others. First and foremost, many farmers and ranchers in areas where the system is utilized are experienced cattle producers who understand stocker cattle management and marketing. Also critical is that wheat fields seldom have snow cover which would prevent grazing. Wet soil conditions reduce cattle weight gain and increase damage to the developing wheat crop, so it is also beneficial if the winter months are typically the driest months of the year. Finally, many farmers and ranchers that utilize the system are in a monocrop, continuous wheat production system so that rotational summer crops do not delay wheat sowing in the fall.

Dual-purpose wheat production requires several modifications from normal wheat production practices. Grazing restrictions for many of the most popular wheat herbicides, for example, prevent timely herbicide application in the fall when weeds are small and most susceptible to chemical control. Likewise, fall-applied, soil-activated herbicides frequently have reduced efficacy due to soil movement from cattle hooves. For these reasons, dual-purpose wheat producers often find themselves applying herbicides late in the spring to well-established weeds that are hardened from grazing and winter damage. Weed control is just one of many management practices that must be modified for successful dual-purpose wheat production.

Characterizing a suitable dual-purpose cultivar

The requirements for a dual-purpose wheat cultivar are different than for grain-only cultivars. A suitable dual-purpose wheat cultivar needs to have the ability to germinate well under hot soil conditions, a semierect growth habit, prolific tillering ability, and nonprecocious arrival of first-hollow-stem stage. Dual-purpose wheat production requires rapid accumulation of plant biomass in a narrow window of time. Plant biomass accumulation is a function of cumulative light interception by the crop canopy, so wheat producers using the dual-purpose wheat production system place great emphasis on rapid emergence and vigorous growth early in the fall. To aid in rapid canopy closure and establishment of leaf area, seeding densities are generally much higher in dual-purpose than in grain-only systems. In fact, seeding densities of 4–5 times the normal recommended seeding density have been shown profitable in dual-purpose systems (unpublished data).

While important, rapid plant emergence is sometimes difficult to obtain in dual-purpose systems. Soil temperatures during September can often exceed 35°C in geographic areas where dual-purpose production is common. There is an inverse relationship between soil temperature during germination and the coleoptile length of wheat. As the coleoptile of modern semidwarf wheat cultivars is already shortened relative to older, taller cultivars, further shortening of the coleoptile due to hot soil conditions can dictate optimal sowing depth. If the coleoptile does not protrude through the soil surface and the first true leaf emerges below the soil surface, complete stand loss frequently occurs (see Color Plate 6). For this reason, producers sowing wheat early into hot soil conditions often choose to sow wheat at a depth of 2.5 cm or less with anticipation for rainfall, rather than sow deeper to reach sufficient moisture to immediately induce germination.

Another issue associated with hot soil conditions is high-temperature germination sensitivity or postharvest seed dormancy. These two terms are often used interchangeably, and some question remains as to whether a true distinction exists between the two. Whatever the terminology used, it has long been known that some cultivars require a greater period of after-ripening than others and as a result do not germinate well when sown early. The degree of postharvest dormancy is under both genetic and environmental controls. Inhibitory substances in the red seed coat of some cultivars, for example, prolong dormancy (Ching and Foote 1961). Environmental conditions during both seed maturation and germination impact seed dormancy. For example, dormancy is strengthened with decreasing air temperature during grain-fill (Reddy et al., 1985). Likewise soil temperatures greater than 20°C after sowing will strengthen dormancy, lengthen the after-ripening requirement, and delay emergence. While most seed will germinate once soil conditions cool, delayed emergence due to dormancy will result in less cumulative light interception by the crop canopy and reduced total forage production. For this reason, cultivars with strong after-ripening requirements are frequently avoided for dual-purpose production systems.

A semierect growth habit will allow wheat to be easily grazed by cattle without extensive damage to the wheat plant. Cultivars that are too erect generally do not tolerate grazing very well, as much of the green leaf area and photosynthetic capacity of the plant is removed by cattle during the grazing process, which can reduce subsequent grain yield (Arzadun et al., 2003). Conversely, cultivars that are too prostrate frequently have less available forage than more erect cultivars. While these cultivars recover from grazing well, they do not typically provide adequate forage production. So, an intermediate or semierect to semi-prostrate growth habit is desirable for dual-purpose wheat cultivars.

Fertility management

Fertilization practices in dual-purpose wheat are much more reliant on preplant applications of nitrogen fertilizer than grain-only systems. There are two primary reasons for this. First, added nitrogen in the fall encourages early wheat growth and adequate forage production. Second, the removal of cattle in early March leaves little time for top-dress applications of nitrogen, as the plants are quickly approaching the critical time for canopy closure. Still, most modern dual-purpose wheat producers choose to make some type of top-dress nitrogen application in late winter. Top-dress nitrogen amounts vary by producer but extension recommendations indicate a need for 30 kg ha^{-1} of nitrogen for every 1,000 kg ha^{-1} of wheat forage produced. This nitrogen requirement comes despite that 61%–77% of the nitrogen consumed in wheat forage by ruminant animals is excreted in feces (33%) and urine (66%) (Phillips et al., 1995). Nonuniform distribution of urine and feces and losses due to volatilization and runoff necessitate top-dress nitrogen applications.

Grazing termination and impact on grain yield

Timing of cattle removal from wheat pasture has been the subject of investigation since the late 1800s (Redmon et al., 1995). Modern semidwarf cultivars are more sensitive to timing of grazing termination than older, taller cultivars. Increases in cattle weight for grazing past the first-hollow-stem stage have generally not been sufficient to offset reductions in wheat yield; therefore, first-hollow-stem stage is considered the optimal developmental period for removal of cattle from wheat pasture (Redmon et al., 1996). Research by Fieser et al. (2006) indicates that beef gains for grazing past first hollow stem can offset grain yield losses when cattle average-daily-gains are greater than 1.5 kg $steer^{-1}$ day^{-1} and wheat yields are less than 3,000 kg ha^{-1}; however, first hollow stem was still recommended as the optimal timing for grazing termination of wheat pasture.

Even when cattle are removed from wheat pasture in a timely fashion, wheat yield reductions generally occur in a dual-purpose wheat production system relative to a grain-only system. Some of this reduction is due to planting date. Planting date for dual-use wheat is much earlier than for grain-only systems. Hossain et al. (2003) demonstrated that optimal sowing dates for forage production are approximately one month earlier than for grain-only systems (Fig. 4.2). Early-September sowing dates have been shown to increase duration of grazing by as much as 24 days, but the tradeoff was a 22% reduction in grain yield relative to a late-September sowing date (Hossain et al., 2003). Io optimize the forage and grain components of the system, a mid-September planting date was considered optimal.

Actual yield reduction from the act of grazing is debatable. Georgeson et al. (1892) reported a wheat grain yield reduction of 202 kg ha^{-1} when wheat pasture was grazed but then reported no yield reduction in a later study (Georgeson et al., 1896). Similar mixed results followed for the next 100 years (Swanson 1935; Dunphy et al., 1982; Winter and Thompson 1987; Redmon et al., 1996), but most investigators agree that a yield penalty is associated with grazing wheat pasture. The severity of the yield reduction, however, is quite debatable and depends on several factors. Stocking density is probably the largest animal-related influence on dual-purpose wheat yield. Arzadun et al. (2003) reported that heavier

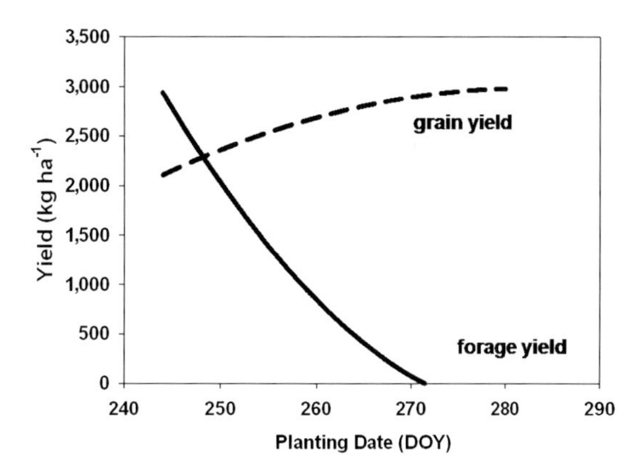

Fig. 4.2 Wheat forage production increases with early sowing, while grain yield benefits from later sowing. Most dual-purpose producers in the US choose September 15 [day of year (DOY) 258] as the optimal compromise between forage and grain yield. (Adapted from Hossain et al., 2003.)

stocking rates decreased wheat grain yield by 29% but simultaneously increased cattle gains by 38%. Similar to Fieser et al. (2006) they concluded that the value of beef cattle gains were enough to offset the reduction in wheat grain yield. In a later study Arzadun et al. (2006) indicated that part of the stocking rate effect was due to the amount of green leaf area left in the field after grazing events. They simulated grazing intensity, and thereby stocking density, by clipping wheat plants to specified heights and found that a 7-cm clipping height had no effect on subsequent grain yield in 2 out of 3 years, but a 3-cm clipping height reduced grain yield every year of their experiment.

Rainfall likely played some role in the amount of yield reduction associated with grazing in these experiments. Cattle traffic on wet or waterlogged soils damages wheat plants and likely increases the yield penalty associated with grazing (see Color Plate 7). Of course, the actual yield penalty is relative to the yield potential of the wheat crop prior to grazing. This is one reason that dual-purpose wheat production has found favor in areas with suboptimal wheat production environments. The cattle component in the systems decreases risk by generating cash flow even if conditions are not favorable for grain yield.

While later first hollow stem is a desirable trait for dual-purpose wheat producers, early maturity is frequently the paramount selection criterion for wheat growers. In areas where double-cropping is prevalent, early maturity will allow for earlier sowing of the following crop; earliness of sowing often translates to increased double-crop yield. Early maturity is preferred in areas such as the US southern Great Plains due to frequent drought. In these areas subsoil moisture is often the only source of water for much of the grain-fill season. Ambient temperatures are generally lower during grain-fill of early-maturing wheat cultivars than for later-maturing cultivars. This translates to lower evaporative demand and frequently to higher yield among cultivars developed primarily in grain-only systems. Dual-purpose wheat producers, however, have had to accept earlier first hollow stem to achieve early heading (Edwards et al., 2007). With cooler ambient temperatures during onset of first hollow stem, however, a 2-week delay in first hollow stem may only translate to a 2- to 3-day delay in heading and wheat maturity.

NO-TILL WHEAT PRODUCTION

Conservation tillage refers to soil management practices that leave at least 30% residue cover on the soil surface. While adoption of conservation tillage practices has remained fairly stable since the mid-1990s, the proportion of all cropland devoted to no-till crop production steadily increased from 14.7% in 1996 to 19.6% in 2002 (Fawcett and Towery 2004).

Why no-till has increased

The most likely factor driving the increase of no-till in the US has been the introduction of herbicide-tolerant, bioengineered crops (Fawcett and Towery 2004). Specifically, the introduction of glyphosate-tolerant soybean, maize, and cotton (*Gossypium hirsutum* L.) has driven the adoption of no-till in the midwestern and southern US. Thus far, genetically modified wheat has not reached commercial production, so adoption of no-till management systems for wheat has been much slower. In the wheat-dominated southern Great Plains, for example, no-till practices account for less than 5% of the acreage and adoption of conservation tillage remains less than 20% (Ali 2002). In environments where monocrop wheat is the primary cropping system, the lack of glyphosate-tolerant wheat cultivars has hampered the adoption of no-till systems. Progress is being made, however. Soft red winter wheat producers in the southeastern and midsouthern US frequently used no-till practices when seeding soybean or maize but then tilled before sowing wheat. These producers found over time, though, that the long-term benefits of no-till systems as a whole offset temporary reductions in wheat yield and now operate under continuous no-till for all crops.

One of the reasons no-till acreage in wheat-dominated production systems has increased over the past 5 years is the decreased labor and fuel requirements of no-till production systems. Previous research indicated that savings from reduced machinery, labor, tractor fuel, and repair costs associated with no-till production practices were negated by increased herbicide costs (Epplin et al., 1983). However, the price of glyphosate (480 g of emulsifiable concentrate per liter) has declined from a US average of $12 per liter in 1999 to $5 per liter in 2005, thus reducing by more than one-half the cost of herbicidal control of summer weeds from harvest in June until planting in September. Coupled with that, record oil prices have resulted in drastic increases in farm diesel costs. The primary labor and fuel savings for a no-till wheat production system is in seedbed preparation, especially in more intensive production systems such as those used in Europe (Fig. 4.3). Many producers view no-till wheat production as a way of trading high-priced, conventional practices burdened with high environmental impact (e.g., tractor-engine emissions, fuel consumption, carbon loss, and erosion from tillage operations) for low-priced, low-environmental-impact production practices.

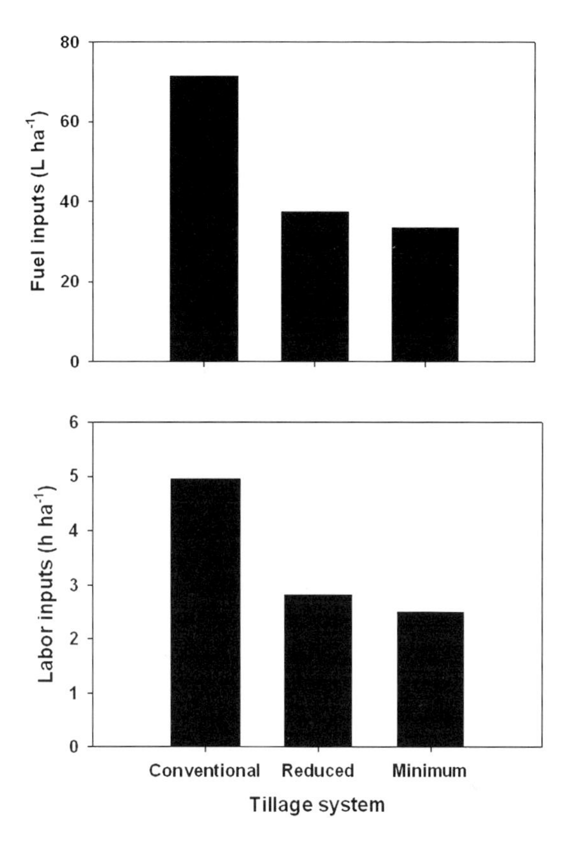

Fig. 4.3 Fuel and labor inputs are reduced by almost 50% when conservation-tillage measures are introduced into an intensive wheat production system. (Adapted from Lithourgidis et al., 2006.)

Advances in equipment technology have also increased adoption of no-till practices. Drift-reducing nozzles have allowed greater use of glyphosate in areas adjacent to susceptible crops (Derksen et al., 1999). Another area that has seen great advancements in no-till wheat production systems has been in no-till drill design and performance. Proper seedbed preparation can dictate the efficiency of all subsequent operations performed throughout the year. Wheat sown into a poorly prepared seedbed or with poorly calibrated equipment has little chance of obtaining optimal yield potential. Stockton et al. (1996) found that actual field emergence of wheat was as low as 30% and averaged 57% of viable seed. They indicated that the primary reason for poor emergence was poor control of seeding depth, causing most seed to be sown too deep. Many conventional wheat drills at the time lacked linkage between the disk openers and press wheels.

Wheat drill design and performance has improved dramatically since the work of Stockton et al. (1996). Most conventional drills now have the press wheel directly connected to the disk opener as a standard design and seeding depth is more precise and consistent. The first no-till drills were essentially heavy conventional drills or conventional drills with a coulter-caddy attachment. The results from these drills in heavy residue such as maize stalks were often undesirable. Common problems included poor soil penetration and inadequate seed-to-soil contact (see Color Plate 8). Modern drills, however, can deliver consistent and even seed distribution in a wide range of seedbed conditions, from true no-till to minimum till to conventional till.

Long-term experiments

As time progresses more critical data is emerging from long-term no-till experiments. In a 41-year study Tarkalson et al. (2006) found that wheat yields were 12% higher in no-till than conventional till systems, and grain sorghum (*Sorghum bicolor* L.) yields were 35% higher. They attributed most of this effect to increased soil moisture retention. Rooting characteristics also play a part in the success of no-till systems. In Switzerland, Qin et al. (2004) found that wheat root density was actually higher in the top 5 cm of soil in no-till systems than in conventional-till systems, but lower from 10 to 30 cm soil depth. Overall the root density was lower in no-till systems, but root diameter was larger and root density in the planted row was greater than in the conventional-till system. Increased root diameter and the resulting root channels after decomposition of old roots likely play a leading role in improved water infiltration and bulk density of soils in long-term no-till systems and the amount of carbon they sequester.

Much of the no-till wheat research over the past 20 years has emphasized crop rotation. Studies have shown that crop rotations generally create synergistic effects on yield among the crops, but occasional antagonistic effects have been observed (Tanaka et al., 2007). Kelley and Sweeney (2007), for example, observed that crop sequence significantly influenced wheat yield. In their 4-year experiment, wheat following soybean or maize produced greater yield than wheat following grain sorghum. Increased nitrogen fertilization, however, decreased the yield differential among the rotational sequences. Their work indicated greater nitrogen efficiency can be obtained in no-till systems by subsurface banding of nitrogen compared to broadcast applications, leading them to conclude that nitrogen immobilization probably had more negative effect on wheat yield than any allelopathic effects associated with grain sorghum.

There is some debate about the influence of no-till production systems on water infiltration and storage. Norwood (2000) found that soil moisture in the top 1.8 m of the soil profile at wheat planting was 9% and 20% less for wheat following soybean or sunflower than for wheat following maize or grain sorghum. Baumhardt et al. (1993), however, found no differences in water infiltration when comparing wheat tillage treatments. Similarly, Unger (1994) found that tillage had no effect on wheat yield or water storage. In areas that frequently experience extreme water-deficit stress, such as the Middle East, no-till has resulted in greater soil water content in the top 30 cm of soil, which resulted in significantly greater seedling survival in this harsh environment (Klein et al., 2002).

FUTURE PERSPECTIVES

Although we now know much more about wheat production systems and how they interact than we did just 20 years ago, considerable research and extension work lies ahead. For example, there is a need to continue to focus on improved wheat cultivars that respond to increased inputs in intensive management systems. Increases in wheat yield through intensive wheat management systems were first made possible through increased harvest index and then through increased total biomass (Shearman et al., 2005). Most recently, improvements in postanthesis radiation-use efficiency have been observed in some cultivars (Shearman et al., 2005). Further advances in breeding for increased radiation-use efficiency and more partitioning of assimilate to grain would likely result in advances in grain yield. Those increases, however, would likely require even more nitrogen fertilizer. For long-term sustainability, advances in crop nitrogen-use efficiency will also be needed to sustain productivity.

Future intensification might arise as the result of intensification of the cropping system as a whole rather than intensification of wheat production alone. Shorter-season wheat cultivars offer opportunity for double-cropping in areas previously not considered to be a viable option. When combined with shorter-season soybean cultivars and sorghum and maize hybrids, these shorter-season wheat cultivars might offer the opportunity to produce as many as four crops in 2 years, especially in areas with relatively long growing seasons and readily available irrigation (Edwards et al., 2005).

Several areas of the dual-purpose wheat production system need attention in future research efforts. Crop simulation models that can effectively and accurately predict fall forage production by winter wheat are needed to improve the decision-making process for cattle stocking density on wheat pastures. If properly modified and calibrated, such models would also prove useful in determining forage regrowth during grazing and if adjustments to cattle stocking density or feed supplementation are prudent. Similarly, a model that effectively and accurately predicts wheat phenology in a dual-purpose production system would be useful in predicting first hollow stem and optimal timing of grazing termination in wheat pastures. Elucidation of the genetic linkage between vernalization requirement in winter wheat and timing of first hollow stem would be an essential component in development of such a model. Finally, development of wheat cultivars with increased daily biomass production through greater radiation-use efficiency and/or increased water-use efficiency would likely reduce the yield penalty associated with grazing wheat.

Many of the current efforts directed toward cellulose-based biofuel feedstock production could perhaps use the dual-use wheat production system as a model for future research needs. Switchgrass (*Panicum virgatum* L.), for example, is a potential cellulose-based biofuel feedstock that might provide opportunity for both livestock and feedstock production opportunities within the same cropping season. Critical factors related to crop establishment, fertilization, fertilization timing, and grazing termination that have been well established for the dual-purpose wheat enterprise could likely offer guidelines for initial dual-use recommendations, and research methodologies and approaches could be used in developing protocols for dual-use biofuel feedstock production research.

Future research in no-till seeding systems that will allow sowing of wheat into heavy maize residue immediately following maize harvest would likely increase yield and acreage in some areas of the Midwest. In areas dominated by monocrop wheat such as the Great Plains, increased infrastructure and improved information on rotational crops such as canola (*Brassica napus* L.), grain sorghum, and sunflower (*Helianthus annuus* L.) will likely result in increased adoption of conservation and no-till programs in this region. Conventional, and perhaps transgenic, breeding efforts that will reduce the incidence and severity of diseases worsened by continuous no-till wheat production, such as those caused by foliar, root, and crown infections, will also likely result in increased adoption of no-till.

There has also been some debate about the suitability of no-till in forage-only or dual-purpose wheat systems. Krenzer et al. (1989) observed as much as a 16% increase in soil bulk density and a 270% increase in soil strength from cattle traffic in a dual-purpose no-till production system. This compaction was, however, restricted to the top 12 cm in silt loam soil and the top 20 cm in sandy soil. More work is needed to evaluate the effect that compaction has on subsequent forage and grain yield. Likewise, more research is needed to investigate the extent of alleviation of compaction through natural methods, such as freezing and thawing, or through use of rotational crops with tap roots such as winter canola.

Finally, advances in stress tolerance of wheat would likely allow reclamation of production in more marginal production areas that have been lost due to abiotic stress factors. Advances in our understanding of mechanisms governing salt tolerance and tolerance of low or high pH would allow for production in areas that were once highly productive but have since been abandoned or diverted to more tolerant crop species. Tolerance to moisture stress, whether it be too much moisture due to waterlogged soils or not enough moisture due to drought, would significantly increase the wheat production potential in many areas of the world.

REFERENCES

Abeledo, L., D. Calderini, and G. Slafer. 2003. Genetic improvement of yield responsiveness to nitrogen fertilization and its physiological determinants in barley. Euphytica 133:291–298.

Ali, M.B. 2002. Characteristics and production costs of U.S. wheat farms. USDA-ERS Stat. Bull. SB974–5. Available at http://www.ers.usda.gov/Publications/SB974-5/ (verified 2 April 2008).

Alley, M.M., D.E. Brann, E.L. Stromberg, E.S. Hagood, A. Herbert, E.C. Jones, and W.K. Griffith. 1993. Intensive soft red winter wheat production: A management guide. Va. Coop. Ext. Serv., Blacksburg, VA.

Arzadun, M.J., J.I. Arroquy, H.E. Laborde, and R.E. Brevedan. 2003. Grazing pressure on beef and grain production of dual-purpose wheat in Argentina. Agron. J. 95:1157–1162.

Arzadun, M.J., J.I. Arroquy, H.E. Laborde, and R.E. Brevedan. 2006. Effect of planting date, clipping height, and cultivar on forage and grain yield of winter wheat in Argentinean Pampas. Agron. J. 98:1274–1279.

Austin, R.B. 1999. Yield of wheat in the United Kingdom: Recent advances and prospects. Crop Sci. 39:1604–1610.

Baumhardt, R.L., J.W. Keeling, and C.W. Wendt. 1993. Tillage and residue effects on infiltration into soils cropped to cotton. Agron. J. 85:379–383.

Bell, M.A., R.A. Fischer, D. Byerlee, and K. Sayre. 1995. Genetic and agronomic contributions to yield gains: A case study for wheat. Field Crops Res. 44:55–65.

Bitzer, M., J. Herbek, J. Green, J. Grove, D. Hershman, D. Johnson, J. Martin, S. McNeill, L. Murdock, D. Overhults, L. Townsend, R. Trimble, and D. van Sanford. 1997. A comprehensive guide to wheat management in Kentucky. ID-125. Univ. Ky. Coop. Ext. Serv., Lexington, KY.

Ching, T.M., and W.H. Foote. 1961. Post-harvest dormancy in wheat varieties. Agron. J. 53:183–186.

Cooper, M., D.R. Woodruff, I.G. Phillips, K.E. Basford, and A.R. Gilmour. 2001. Genotype-by-management interactions for grain yield and grain protein concentration of wheat. Field Crops Res. 69:47–67.

Derksen, R.C., H.E. Ozkan, R.D. Fox, and R.D. Brazee. 1999. Droplet spectra and wind tunnel evaluation of venturi and pre-orifice nozzles. Trans. ASAE 42:1573–1580.

Dunphy, D.J., M.E. McDanial, and E.C. Holt. 1982. Effect of forage utilization on wheat grain yield. Crop Sci. 22:106–109.

Edwards, J.T., B.F. Carver, and M.E. Payton. 2007. Relationship of first hollow stem and heading in winter wheat. Crop Sci. 47:2074–2077.

Edwards, J.T., L.C. Purcell, and D.E. Karcher. 2005. Soybean yield and biomass responses to increasing plant population among diverse maturity groups: II. Light interception and utilization. Crop Sci. 45:1778–1785.

Epplin, F.M., T.F. Tice, S.J. Handke, T.F. Peeper, and E.G. Krenzer. 1983. Economics of conservation tillage systems for winter wheat production in Oklahoma. J. Soil Water Conserv. 38:294–297.

Fawcett, R., and D. Towery. 2004. Conservation tillage and plant biotechnology: How new technologies can improve the environment by reducing the need to plow. Conserv. Tech. Inf. Cent., West Lafayette, IN.

Fieser, B.G., G.W. Horn, J.T. Edwards, and E.G. Krenzer. 2006. Timing of grazing termination in dual-purpose winter wheat enterprises. Prof. Anim. Sci. 22:210–216.

Georgeson, C.C., F.C. Burtis, and W. Shelton. 1892. Experiments with wheat. Bull. 33. Kans. Agric. Exp. Stn., Manhattan.

Georgeson, C.C., F.C. Burtis, and W. Shelton. 1896. Experiments with wheat. Bull. 59. Kans. Agric. Exp. Stn., Manhattan.

Heyne, E.G. (ed.) 1987. Wheat and wheat improvement. Agron. Monogr. 13. 2nd ed. ASA, CSSA, SSSA, Madison, WI.

Hossain, I., F.M. Epplin, and E.G. Krenzer. 2003. Planting date influence on dual-purpose winter wheat forage yield, grain yield, and test weight. Agron. J. 95:1179–1188.

Karlen, D.L., and D.T. Gooden. 1990. Intensive management practices for wheat in the southeastern coastal plains. J. Prod. Agric. 3:558–563.

Kelley, K.W., and D.W. Sweeney. 2007. Placement of preplant liquid nitrogen and phosphorus fertilizer and nitrogen rate affects no-till wheat following different summer crops. Agron. J. 99:1009–1017.

Klein, J.D., I. Mufradi, S. Cohen, Y. Hebbe, S. Asido, B. Dolgin, and D.J. Bonfil. 2002. Establishment of wheat seedlings after early sowing and germination in an arid Mediterranean environment. Agron. J. 94:585–593.

Krenzer, E.G., Jr., C.F. Chee, and J.F. Stone. 1989. Effects of animal traffic on soil compaction in wheat pastures. J. Prod. Agric. 2:246–249.

Large, E.C. 1954. Growth stages in cereals: Illustration of the Feekes scale. Plant Pathol. 3:128–129.

Lithourgidis, A.S., K.V. Dhima, C.A. Damalas, I.B. Vasilakoglou, and I.G. Eleftherohorinos. 2006. Tillage effects on wheat emergence and yield at varying seeding rates, and on labor and fuel consumption. Crop Sci. 46:1187–1192.

Mohamed, M.A., J.J. Steiner, S.D. Wright, M.S. Bhangoo, and D.E. Millhouse. 1990. Intensive crop management practices on wheat yield and quality. Agron. J. 82:701–707.

National Agricultural Statistics Service. 2008. Data available at http://www.nass.usda.gov/ (verified 13 March 2008).

Norwood, C.A. 2000. Dryland winter wheat as affected by previous crops. Agron. J. 92:121–127.

Phillips, W.A., G.W. Horn, and M.E. Smith. 1995. Effect of protein supplementation on forage intake and nitrogen balance of lambs fed freshly harvested wheat forage. J. Anim. Sci. 73:2687–2693.

Qin, R., P. Stamp, and W. Richner. 2004. Impact of tillage on root systems of winter wheat. Agron. J. 96:1523–1530.

Ransom, J.K., G.J. Endres, and B.G. Schatz. 2006. Sustainable improvement of wheat yield potential: The role of crop management. J. Agric. Sci. 145:55–61.

Reddy, L.V., R.J. Metzger, and T.M. Ching. 1985. Effect of temperature on seed dormancy of wheat. Crop Sci. 25:455–458.

Redmon, L.A., G.W. Horn, E.G. Krenzer, Jr., and D.J. Bernardo. 1995. A review of livestock grazing and wheat grain yield: Boom or bust? Agron. J. 87:137–147.

Redmon, L.A., E.G. Krenzer, Jr., D.J. Bernardo, and G.W. Horn. 1996. Effect of wheat morphological stage at grazing termination on economic return. Agron. J. 88:94–97.

Royer, T.A., and E.G. Krenzer (ed.) 2000. Wheat management in Oklahoma: A handbook for Oklahoma's wheat industry. Publ. E-831. Okla. Coop. Ext. Serv., Stillwater, OK.

Shearman, V.J., R. Sylvester-Bradley, R.K. Scott, and M.J. Foulkes. 2005. Physiological processes associated with wheat yield progress in the UK. Crop Sci. 45:175–185.

Snyder, C.S. 2000. Soft red winter wheat—high yields achieved with intensive management. Better Crops Plant Food 84:23–25.

Stockton, R.D., E.G. Krenzer, Jr., J. Solie, and M.E. Payton. 1996. Stand establishment of winter wheat in Oklahoma: A survey. J. Prod. Agric. 9:571–575.

Swanson, A.F. 1935. Pasturing winter wheat in Kansas. Bull. 271. Kans. Agric. Exp. Stn., Kansas State Univ., Manhattan, KS.

Sweeney, D.W., M.B. Kirkham, and J.B. Sisson. 2006. Crop and soil response to wheel-track compaction of a claypan soil. Agron. J. 98:637–643.

Tanaka, D.L., J.M. Krupinsky, S.D. Merrill, M.A. Liebig, and J.D. Hanson. 2007. Dynamic cropping systems for sustainable crop production in the northern Great Plains. Agron. J. 99:904–911.

Tarkalson, D.D., G.W. Hergert, and K.G. Cassman. 2006. Long-term effects of tillage on soil chemical properties and grain yields of a dryland winter wheat-sorghum/corn-fallow rotation in the Great Plains. Agron. J. 98:26–33.

Tucker, B., J.H. Stiegler, P.W. Unger, A.L. Black, R.L. Westerman, A.D. Halvorson, M.M. Alley, L.S. Murphy, G.M. Paulsen, D.L. Carter, and A.P. Appleby. 1987. Management of the wheat crop. p. 232–416. *In* E.G. Heyne (ed.) Wheat and wheat improvement. 2nd ed. ASA, CSSA, SSSA, Madison, WI.

Unger, P.W. 1994. Tillage effects on dryland wheat and sorghum production in the southern Great Plains. Agron. J. 86:310–314.

Watson, M., and J. Lowenberg-DeBoer. 2004. Who will benefit from GPS auto guidance in the Corn belt. Purdue Univ. Agric. Econ. Rep., Feb. 2004. Purdue Univ., West Lafayette, IN.

Weisz, R., C.R. Crozier, and R.W. Heiniger. 2001. Optimizing nitrogen application timing in no-till soft red winter wheat. Agron. J. 93:435–442.

Wiersma, J.J., and J.M. Bennett. 2001. The small grains field guide. Publ. MI-07488-S. Univ. Minn. Coop. Ext. Serv., St. Paul, MN.

Winter, S.R., and E.K. Thompson. 1987. Grazing duration effects on wheat growth and grain yield. Agron. J. 79:110–114.

Woolf, H.B. (ed.) 1979. Webster's new collegiate dictionary. Merriam Company, Springfield, MA.

Chapter 5

Diseases Which Challenge Global Wheat Production—the Wheat Rusts

James Kolmer, Xianming Chen, and Yue Jin

SUMMARY

(1) Wheat rusts have been important throughout the history of wheat cultivation and are currently important diseases that are responsible for regularly occurring yield losses in wheat. New races of wheat rusts have recently emerged worldwide, complicating efforts to develop rust resistant cultivars.

(2) Wheat leaf rust, caused by the rust fungus *Puccinia triticina* Eriks., is the most common and widespread rust of wheat worldwide. Although the alternate host is not present in North America, *P. triticina* is highly diverse for virulence as many races are found annually.

(3) Many leaf rust resistance genes no longer provide effective resistance since virulent leaf rust races have been selected by resistant cultivars. Resistance gene *Lr34* has provided non-race-specific resistance for many years in wheat cultivars grown worldwide. Gene *Lr46* and other characterized sources of adult-plant partial resistance have also provided durable resistance to leaf rust. Wheat cultivars with high levels of resistance can be developed by combining *Lr34*, *Lr46*, and other effective resistance genes.

(4) Wheat stripe rust, caused by *Puccinia striiformis* Westend. f. sp. *tritici* Eriks., is an important disease in areas where wheat grows and matures in cool temperatures. The occurrence of wheat stripe rust has increased in the southern–mid Great Plains and the southeastern states of the US. Although an alternate host has never been found, *P. striiformis* f. sp. *tritici* is highly diverse as many different races are found worldwide.

(5) Race-specific stripe rust resistance genes have been rendered ineffective due to the increase of virulent races. Wheat cultivars with adult-plant resistance genes such as *Yr18* that condition non-race-specific partial resistance or genes that condition high-temperature adult-plant resistance (HTAP) have had long-lasting resistance to stripe rust. Wheat cultivars with durable resistance to stripe rust can be developed by using combinations of these resistance genes.

(6) Wheat stem rust, caused by *Puccinia graminis* Pers.:Pers. f. sp. *tritici* Eriks. & E. Henn., is potentially a highly destructive disease of wheat. The alternate host, *Berberis vulgare*, has largely been eradicated from North America, reducing the number of stem rust races. Major epidemics of wheat stem rust occurred periodically from 1900 to 1954 in North America, causing severe yield losses. The widespread cultivation of stem rust resistant winter wheat and spring wheat since the 1950s has greatly reduced the population size of *P. graminis* f. sp. *tritici*, resulting in many fewer stem rust infections annually.

(7) Wheat cultivars with the adult-plant non-race-specific resistance gene *Sr2* have had durable resistance to stem rust. Stem rust races with virulence to *Sr31* and *Sr38* have recently been found in eastern Africa. These races are virulent to many CIMMYT cultivars with *Sr31* and also have virulence to many US wheat cultivars. Gene *Sr2* and other sources of resistance can be used to develop cultivars with resistance to these stem rust races.

INTRODUCTION

The rusts of wheat are among the most important and common diseases of wheat in the US and worldwide. Rust has afflicted wheat for thousands of years as references to wheat rust can be found in the Bible and the classical literature of ancient Greece and Rome (Chester 1946). In the early 20th century widespread epidemics of wheat rusts provided the impetus for early advances in genetics of disease resistance in plants, epidemiology of plant pathogens, and genetics of host–parasite interactions. Today the rust diseases continue to cause regular yield losses worldwide, threatening the sustainable production of wheat. The continuing evolution of virulent rust races in response to the release of rust resistant wheat cultivars poses a constant challenge to wheat researchers.

WHEAT LEAF RUST

Distribution and epidemiology

Leaf rust, caused by *Puccinia triticina* Eriks., is the most common and widely distributed of the three rust diseases of wheat. In the US, leaf rust is commonly found on soft red winter wheat grown in the southeastern states and Ohio Valley region, on hard red winter wheat from Texas to South Dakota, and on hard red spring wheat in South Dakota, North Dakota, and Minnesota. Leaf rust also occurs on spring wheat that is grown in California, and to a lesser extent on winter and spring wheat in Oregon, Washington, and Idaho. Worldwide, leaf rust is a major continent-wide disease in the western prairies of Canada; the South American region of Argentina, Chile, Brazil, and Uruguay (German et al., 2007); the Central Asia region of northern Kazakhstan and Siberia; southern and central Europe; the Middle East; and parts of the Indian Subcontinent (Roelfs et al., 1992). Leaf rust is less widespread and occurs at a local level in eastern and western Australia, China, eastern Africa, and in South Africa.

Leaf rust is characterized by the uredinial stage of small round, brown to orange pustules that occur on the upper and lower leaf surfaces, and less frequently on the leaf sheaths (Color Plate 9). The pustules remain discrete, without coalescing. The uredinia are capable of producing up to 3,000 urediniospores per day (Roelfs et al., 1992) if the host plant tissue remains healthy. The urediniospores are 20 μm in diameter, echinulate, dikaryotic ($n+n$), wind-disseminated, and deposited in rain events on host plants in the immediate vicinity and potentially also on hosts hundreds of kilometers distant. The urediniospores germinate on wheat plants, producing the specialized infection structures of appressoria, germ tube, and penetration peg (Harder 1984) that allow the fungus to penetrate the host stomata. Further specialized structures, the substomatal vesicle and haustoria, are produced which allow the fungus to obtain nutrients from host mesophyll cells without killing them. Infectious hyphae of the fungus spread throughout the mesophyll layer. Leaf rust infections are initially visible as faint flecks on leaf surfaces 3–4 days after inoculation. Uredinia erupt and break through the epidermal leaf surface 8–10 days after initiation of the infection process. The clonally produced urediniospores can cycle indefinitely on wheat hosts.

Resistance responses to leaf rust are characterized by small uredinia surrounded by necrosis, or by an abundance of hypersensitive flecks produced in response to infection (Color Plate 9). Non-hypersensitive resistance is characterized by fewer and small uredinia compared to a susceptible response. As the uredinia age, teliospores are formed in the uredinia. Teliospores are dark, 16 μm wide, and thick walled and have two dikaryotic cells. The pycniospores and aeciospores, which are spore stages associated with sexual reproduction on alternate hosts, are not produced on the wheat host.

Leaf rust infections occur in the US during September to November on fall-planted winter wheat from Texas to South Dakota, and from the Gulf Coast states to North Carolina. Leaf rust infected volunteer winter wheat plants that survive the summer are the inoculum source for the fall-planted winter wheat crop. The optimal conditions for infection are temperatures near 20 °C with free moisture on the leaf surface for at least 8 hours, and 25 °C is the optimal temperature for growth. Overnight periods of dew formation are optimal for germination of urediniospores. The infection process can occur at temperatures from 2 to 30 °C; however, longer periods of dew are required at the lower temperatures. Temperature and moisture conditions in the fall months throughout much of the winter wheat region of the US allow *P. triticina* to become established over a large geographical area.

Throughout the US winter wheat region leaf rust inoculum sources are a combination of (i) infections that survive the winter either as urediniospores or mycelia and (ii) windblown urediniospores carried in the southerly winds from infected winter wheat to the south. The relative importance of overwintering infections and exogenous sources of inoculum will vary based on conditions in the fall that affect leaf rust infection, temperatures during the winter that allow leaf rust to survive, and the maturity of the crop. Wheat cultivars with later maturity will be more affected by leaf rust infections from exogenous sources. Severe epidemics result from the combination of overwintering leaf rust infections, windblown urediniospores deposited in rain events, regular dew periods, and temperatures greater than 25 °C during the time when winter wheat is breaking dormancy and resuming growth and development.

In south Texas and along the Gulf Coast, winter temperatures often exceed 20 °C, with only infrequent freezes. In this region where *P. triticina* regularly overwinters, leaf rust can reach high severity levels in March and April (Roelfs 1989). From Oklahoma to eastern Virginia, winter temperatures usually do not exceed 20 °C and freezing temperatures are common. In this area leaf rust overwinters in isolated pockets, resulting in foci of leaf rust infections that are apparent in the spring when temperatures are regularly above 20 °C and leaf rust is rapidly increasing. Leaf rust severities in this region usually reach maximum levels in April and mid-May. In the area from Kansas north to South Dakota and east to the Ohio Valley, winter low temperatures range from −10 to −20 °C, and high temperatures rarely exceed 15 °C. Leaf rust can survive in winter wheat in this region as mycelium resulting from infections during the fall months. The overwintering survival of leaf rust in this region is also dependent on adequate snow cover to protect the wheat leaves from freeze damage during periods of extreme low temperatures. Leaf rust severities are at maximum levels in May in Kansas to June in South Dakota and the Ohio Valley states.

In Minnesota, South Dakota, and North Dakota, leaf rust infection on the spring wheat is usually first observed in mid-to-late June, with maximum severity levels in mid-to-late July. Daytime temperatures in the summer in this area are often greater than 25 °C, with frequent rain events and dew periods. All leaf rust infections on spring wheat originate from windblown urediniospores from winter wheat growing in the southern Great Plains or from local fields of winter wheat. Leaf rust epidemics in the spring wheat region are most severe when initial infections occur at the tillering stage, which allows additional generations of urediniospores to infect the crop.

Origin and historical importance

Puccinia triticina was introduced to North America with the first agricultural settlements and wheat cultivation in the early 17th century (Chester 1946). The origin of *P. triticina* is likely the Fertile Crescent region of southwest Asia and the Middle East (Wahl et al., 1984), which is also the origin of diploid, tetraploid, and hexaploid wheat. The most susceptible alternate host of *P. triticina* is *Thalictrum speciosissimum* (= *Thalictrum flavum*), which is native to southern Europe, western Asia, and Turkey. It is likely that the center of origin of *P. triticina* is a region where the alternate and telial hosts overlap, which would be southwest Asia.

In early rust research in the US, the importance of leaf rust was not recognized since all yield losses due to rust in wheat were then attributed to stem rust (caused by *P. graminis* Pers.:Pers. f. sp. *tritici* Eriks. & E. Henn.). A common attitude was that *P. triticina* caused little or no damage to wheat (Chester 1946). This mistaken opinion was most likely because the disease only affected the leaves, did not cause grain shriveling, and was not visible when the grain was harvested. Replicated yield loss studies in field plots with sulfur or fungicide treatments determined that leaf rust was a major cause of yield loss in wheat.

Mains (1930) examined the importance of leaf rust on yield in the 1920s and determined that the commonly grown soft red winter wheat cultivars that were susceptible to leaf rust suffered losses that ranged from 25% to over 90%. Even a resistant cultivar had losses over 10% due to the premature death of heavily infected flag leaves. Caldwell et al. (1934) determined that seven winter wheat cultivars which varied from resistant to susceptible suffered losses from 15% to 28% due to leaf rust infections. In the early hard red winter wheats Johnston (1931) showed that resistant and susceptible cultivars suffered maximum losses of 22% and 55%, respectively, due to leaf rust. In 1938 leaf rust caused a yield loss of 25%–30% statewide in Oklahoma (Chester 1939). In hard red spring wheat in Canada, losses due to leaf rust were over 50% in susceptible cultivars and from 12% to 28% in resistant cultivars (Peturson et al., 1945; Samborski and Peturson 1960). Chester (1946) developed a predictive curve for estimating yield loss due to leaf rust based on the growth stage in which wheat was defoliated, using data from 68 yield studies. Losses ranged from an average of 10% if defoliation occurred in the dough stage of grain development to an average of 95% if defoliation occurred in the jointing stage. In a test with isogenic spring wheat lines that differed by only a single leaf rust resistance gene, Dyck and Lukow (1988) showed a 22% difference in yield between the resistant and susceptible lines. Martin et al. (2003) determined that isolines of winter wheat that had resistance genes *Lr41* and *Lr42* had a 63% and 26% yield increase compared with lines that lacked these genes. Herrera-Foessel et al. (2006) estimated yield losses over 50% due to leaf rust in durum wheat (*T. turigidum* ssp. *durum*) in Mexico.

Leaf rust continues to cause regular losses in present-day wheat cultivars. In 2007 yield loss due to leaf rust in the hard red winter wheat crop in Kansas was estimated to be 14% (Kansas Department of Agriculture, Topeka, Kansas). Khan et al. (1997) developed a yield loss model for southern US soft red winter wheat that predicted a 1% yield loss for every 1% increase in rust severity at the milky-ripe stage of grain development. Leaf rust resistance is a high priority in wheat germplasm developed at the International Maize and Wheat Improvement Center (CIMMYT). Wheat production in Africa, South America, and Asia where CIMMYT wheat germplasm is grown has suffered 1%–20% losses due to leaf rust (Marasas et al., 2004). CIMMYT has estimated a 27 : 1 benefit-to-cost ratio for development of leaf rust resistant wheat cultivars. This same study also determined that breeding for leaf rust resistance in wheat was economically justified even if yield losses in areas with yields of 4 t ha^{-1} were only 0.2%–0.8%.

Effects on grain and flour quality

In the study by Caldwell et al. (1934) 75% of the yield loss to leaf rust was due to reduced number of kernels per head. Susceptible winter wheat cultivars with leaf rust infection had 14%–17% fewer kernels than the same cultivars treated to control rust. In their study the weight of individual kernels was also 6%–7% lower in the leaf-rust-infected treatments, and kernel weight was reduced from 1.5% to 12%, depending on the resistance level of the cultivar. Martin et al. (2003) showed that kernel weight decreased by 9%–14% in susceptible winter wheat lines compared with their resistant isolines. Other studies with susceptible and resistant spring wheat types showed a larger effect, with a 6%–39% reduction in kernel weight (Waldron 1936; Peturson et al., 1945). Leaf rust infections that occur before flowering will result in fewer kernels per head, while infections that occur during grain filling will result in lighter kernels (Chester 1946). Thus different yield components may be affected by leaf rust in winter wheat compared with spring wheat.

Leaf rust infection generally reduces the protein content in harvested grain (Caldwell et al., 1934; Peturson et al., 1945, 1948; Dyck and Lukow 1988), although in some tests protein content was unaffected or increased. In spring wheat heavy leaf rust infection generally resulted in flour with increased loaf volume (Peturson et al., 1945) and farinograph absorption (Dyck and Lukow 1988). Color of harvested grain and flour was also affected by leaf rust. In spring wheat yellow pigmentation of the flour was increased (Peturson et al., 1945); in winter wheat, increased yellow pigmentation of the grain was noted (Caldwell et al., 1934). Everts et al. (2001) determined that leaf rust affected the softness equivalent parameter and may reduce flour yield of soft red winter wheat. Reduced flour yield due to leaf rust was also determined in some tests with spring wheat (Peturson et al., 1945).

Taxonomy, life cycle, and host range

Leaf rust on wheat was originally placed in the highly complex species of *P. rubigo-vera* by Winter (1884). Leaf rusts with telial hosts on grasses and alternate hosts in the Boraginacea were placed into this single species. Eriksson (1899) described the leaf rust on wheat as a single species, *P. triticina*. Jackson and Mains (1921) determined that the most compatible alternate host for leaf rust on wheat was *Thalictrum flavuum* (= *Thalictrum speciosissimum*), which is in the Ranunculaceae. Mains (1932) preferred to group leaf rust of wheat within the complex group of *P. rubigo-vera*. Based on nondiscrete spore morphology and host range, Cummins and Caldwell (1956) also chose to place leaf rust of wheat within a complex species, *P. recondita*, with alternate hosts in Boraginacea and also Ranunculacea. In North America most rust workers referred to leaf rust on wheat as *P. recondita* f. sp. *tritici*, while in Europe leaf rust on wheat was placed in the more narrowly defined species of *P. triticina* Eriks. (Savile 1984) based on small and consistent differences in spore morphology.

D' Oliveira and Samborski (1966) conducted infection experiments of telial grass hosts with aeciospores derived from naturally infected plants of *Thalictrum speciosissimum* and *Anchusa* spp. and other genera in the Boraginacea in Portugal. They showed that the leaf rusts that differed for infection on *Thalictrum speciosissimum*, and hosts in the Boraginacea, also differed for telial hosts. Anikster et al. (1997) showed that the leaf rust from common wheat (*T. aestivum* ssp. *aestivum*), wild emmer wheat (*T. turgidum* ssp. *dicoccoides*), and durum wheat belonged to a distinct group with the alternate host *Thalictrum speciosissimum*. A second distinct group of leaf rusts with alternate hosts in the Boraginacea and telial hosts of wild wheats and rye was also described. The two groups could not be successfully crossed using either *Thalictrum speciosissimum* or alternate hosts in the Boraginacea. Based on these results, it was apparent that the leaf rust on wheat is a distinct species from the other leaf rusts on wild wheats and rye. *Puccinia triticina* Eriks. has since been used by most workers to describe leaf rust on wheat.

Puccinia triticina is a macrocyclic rust (Webster 1980) with five distinct spore stages on two taxonomically distinct hosts. The teliospores on wheat leaf tissue germinate to produce haploid basidiospores that are clear and hyaline. The basidiospores infect the alternate host, producing haploid pycnia that appear as circular yellow pustules on the upper leaf surface. The pycnia produce haploid pycniospores, which are carried by insects or rain to other pycnia. Fertilization in the pycnial structures occurs with the transfer of nuclei from pycniospores to flexuous hyphae in compatible combinations of opposite mating types. After fertilization dikaryotic aecia develop on the underside of the leaf beneath the pycnial infections. Aecial cups are produced in the aecia, from which dikaryotic aeciospores are released and wind-disseminated to infect wheat or other telial hosts. Infections from aeciospores result in production of urediniospores. Sexual reproduction in *P. triticina* has been observed in Portugal (d' Oliveira and Samborski 1966), where one of the alternate hosts, *Thalictrum speciosissimum* is commonly found. However, throughout most of the world the disease is spread by the asexual cycling of dikaryotic urediniospores on wheat, as suitable alternate hosts are usually not present. In North America the native *Thalictrum* spp. are resistant to basidiospore infection (Jackson and Mains 1921; Saari et al., 1968). Infected plants of *Thalictrum* with aeciospores that were pathogenic to wheat have been reported in northern Kazakhstan. *Isopyrum fumarioides* has been noted as an alternate host of *P. triticina* in Siberia (Chester 1946).

Worldwide the primary host of *P. triticina* is common hexaploid wheat. Leaf rust caused by *P. triticina* has also been observed on tetraploid durum and emmer wheats in Europe, South America, Israel, Ethiopia, and Mexico (Ordoñez and Kolmer 2007b), and diploid *Aegilops speltoides* (Yehuda et al., 2004) in Israel. *Puccinia triticina* is also present on wild goatgrass, *Ae. cylindrica*, in the southern Great Plains of the US. For all of these nonhexaploid wheat hosts, only certain races or virulence phenotypes of *P. triticina* were pathogenic to these hosts, indicating a high degree of telial host specificity in *P. triticina*. Infections of *P. triticina* have not been noted in natural stands of wild wheat relatives such as *Ae. sharonensis*, *T. timopheevi*, *Ae. tauschii* (syn. *T. tauschii*), or *T. monococcum*, but infections can be obtained on these species in inoculated greenhouse tests. A different species of leaf rust, designated as *P. tritici-duri* with *Anchusa* spp. as the alternate host, occurs on durum wheat in Morocco (Viennot-Bourgin 1941; Ezzahiri et al., 1992).

Genetic variation in *P. triticina*

Virulence variation

Annual nationwide surveys of leaf rust virulence phenotypes have been conducted in Canada since 1931 (Johnson 1956) and in the US since 1926 (Johnston et al., 1968). The wheat cultivars Malakof with *Lr1*, Webster (*Lr2a*), Carina (*Lr2b*, *LrB*), Loros (*Lr2c*), Brevit (*Lr2c*, *LrB*), Hussar (*Lr11*), Democrat (*Lr3*), and Mediterranean (*Lr3*) were designated as the International Standard set of leaf rust differentials and were used in the early race identification studies. Virulence phenotypes of *P. triticina* are currently identified in the US by testing single-pustule isolates for virulence to near-isogenic lines of 'Thatcher' wheat with genes *Lr1*, *Lr2a*, *Lr2c*, *Lr3*, *Lr9*, *Lr16*, *Lr24*, *Lr26*, *Lr3ka*, *Lr11*, *Lr17*, *Lr30*, *LrB*, *Lr10*, *Lr14a*, *Lr18*, *Lr21*, *Lr28*, and winter wheat lines with *Lr41* and *Lr42* (Long and Kolmer 1989; Kolmer et al., 2007b). *Puccinia triticina* and wheat interact in a gene-for-gene manner (Samborski and Dyck 1968). For each *Lr* gene in wheat there is a corresponding locus in *P. triticina* with alleles that condition avirulent responses in the presence of host resistance genes and alternate alleles that condition virulent responses in the presence or absence of resistance genes (Kolmer and Dyck 1994). The range of seedling infection types in the Thatcher isogenic lines is shown in Color Plate 9b. In the US, up to 70 different virulence phenotypes are identified annually (Kolmer et al., 2007a), with the three most common phenotypes accounting for 25%–30% of isolates. Similar surveys of virulence phenotypes in *P. triticina* are conducted in Canada (McCallum and Seto-Goh 2006), in Australia at the Plant Breeding Institute at Cobbity, and in France (Goyeau et al., 2006).

The high degree of virulence variation in *P. triticina* in North America is directly related to the presence of susceptible hosts and the continual use of race-specific leaf rust resistance genes in the different classes of wheat. In the southern US many winter wheat cultivars that are initially resistant become susceptible to leaf rust due to the emergence and increase of virulent leaf rust races. The susceptible winter wheat cultivars allow a very large population of *P. triticina* to become established over a wide geographical area in the fall and survive during the winter. Mutations to virulence to leaf rust resistance genes are a recurrent event in such a large population. Since isolates of *P. triticina* are highly heterozygous for virulence alleles (Samborski and Dyck 1968; Kolmer 1992), a single mutation in an avirulent isolate would be sufficient to gain virulence to a resistance gene.

The wheat cultivars Renown with gene *Lr14a* (1937) and Pawnee with *Lr3* (1943) were the first cultivars with race-specific leaf rust resistance genes to be released in Canada and in the US, respectively. Previous to the release of these cultivars, leaf rust Race 9 (International Standard race designation), which is avirulent to both genes, was the most common race throughout both Canada (Johnson 1956) and the US (Johnston et al., 1968). Race 9 declined in frequency during the 1940s and is currently only rarely found on common wheat in North America, occurring almost exclusively on *Ae. cylindrica*. Isolates with virulence to *Lr3* and *Lr14a* rapidly increased in Canada and the US. Subsequently the release of winter and spring wheat cultivars with additional race-specific *Lr* genes has resulted in a highly diverse *P. triticina* population.

The cultivar Mediterranean with *Lr3* is a major ancestor to soft red winter wheat. Genes *Lr9*, *Lr10*, *Lr11*, *Lr12*, *Lr18*, *Lr24*, and *Lr26* have been present in soft red winter wheat cultivars that are grown in the southern and eastern US (Kolmer 2003). The presence of these genes over time has selected leaf rust isolates with corresponding virulences in the southern and eastern US. Many of the early hard red winter wheats were derived from crosses with 'Hope' (*Lr14a*), Pawnee (*Lr3*), and Mediterranean (*Lr3*). Currently, hard red winter wheat cultivars in the southern to central Great Plains region have genes *Lr9*, *Lr16*, *Lr17*, *Lr24*, *Lr26*, *Lr41*, and *Lr42*, and possibly *Lr34*. Hard red spring wheat cultivars in the northern Great Plains have genes *Lr1*, *Lr2a*, *Lr10*, *Lr13*, *Lr16*, *Lr23*, and *Lr34* (Oelke and Kolmer 2004). The selection and increase of leaf rust races with virulence to these genes in the different wheat classes has resulted in distinct regional populations of *P. triticina* virulence phenotypes in the US. In 2005 (Kolmer et al., 2007b) the frequency of isolates with virulence to genes *Lr2a* and *Lr16* were highest in the north central spring wheat region; virulence to genes *Lr11*, *Lr18*, and *Lr26* was highest in the southern soft red wheat region, and virulence to *Lr24* and *Lr41* was highest in the hard red winter wheat region.

Molecular variation

Genetic variation in populations of *P. triticina* has also been examined using various types of genetic markers. Molecular markers have the attribute of being neutral and thus not directly selected, as virulence to specific resistance genes is. Molecular markers can be used to assess the underlying genetic variation among isolates within and between populations, providing further insight into the genetic relationships between different populations. In Canada phenotypes of *P. triticina* that were identical or closely related for virulence, had identical or highly related random amplified polymorphic DNA (RAPD) phenotypes (Kolmer et al., 1995). Major groups of *P. triticina* isolates could be determined based on either virulence polymorphism or RAPD polymorphism since there was a significant correlation between the two types of markers. Isolates of *P. triticina* from international collections were also grouped into distinct groups based on continental region and virulence and RAPD polymorphism (Kolmer and Liu 2000). Park et al. (2000) showed that multiple isolates of the same virulence phenotype of *P. triticina* from different countries in western Europe also had identical RAPD phenotypes. The relationship between virulence phenotype and molecular polymorphism is maintained since *P. triticina* reproduces throughout the world almost exclusively by asexual urediniospores. In an experimental population of *P. triticina* derived from aeciospores, the disequilibria between individual virulence and RAPD markers was often eliminated or reduced (Liu and Kolmer 1998b).

In 1996 isolates of *P. triticina* with virulence to *Lr17* began to increase in the Great Plains region of the US (Long et al., 2000) and Canada (Kolmer 1998). These isolates were selected by the winter wheat cultivar Jagger with *Lr17*, which has been widely grown throughout Texas, Oklahoma, Kansas, and Nebraska. The isolates with *Lr17* virulence were unique in that they were also virulent to *Lr3bg* and *LrB*, and avirulent to *Lr28*. By 2001 these isolates had become widespread in almost all wheat growing regions of North

America and were the most common virulence phenotypes in the US. Further analysis with (amplified fragment length polymorphism AFLP) markers (Kolmer 2001a) indicated that the isolates with *Lr17* virulence had very distinct molecular phenotypes compared to all other isolates in North America. This indicated that the isolates with *Lr17* virulence were most likely introduced to the Great Plains region from either Mexico or the Pacific Northwest and were not derived by mutation from the previously existing population. New virulence phenotypes of *P. triticina* were also introduced to Australia in the mid-1980s (Park et al., 1995). In recent years a virulence phenotype of *P. triticina* with virulence to many durum cultivars has been found in France, Spain, Mexico, Argentina, and Chile (Singh et al., 2004; Ordoñez and Kolmer 2007b). This virulence phenotype may have had a single origin and subsequently spread to the other durum producing regions.

Recently locus-specific microsatellite or simple sequence repeat (SSR) markers have been developed for *P. triticina* (Duan et al., 2003; Szabo and Kolmer 2007). These markers can be used to determine molecular genotypes of *P. triticina* since heterozygotes can be distinguished from homozygotes. The SSR markers have been used to differentiate *P. triticina* populations in Central Asia (Kolmer and Ordoñez 2007) and to describe genetic diversity in *P. triticina* populations in France (Goyeau et al., 2007). These locus-specific markers will be extremely valuable for assessing genetic variation in *P. triticina* and patterns of migration between populations in different continental regions.

Leaf rust resistance in wheat

Race-specific resistance

The tremendous amount of genetic variation for virulence in *P. triticina* populations combined with the ability of urediniospores to be wind-disseminated over thousands of kilometers has made breeding for stable leaf rust resistance in wheat a continually challenging task. Time and again wheat cultivars with a single race-specific gene for leaf rust resistance have been quickly rendered susceptible because of the selection and increase of virulent leaf rust races. In the southeastern states since the mid-1970s, *Lr9* derived from *Ae. umbellulata*, *Lr11* derived from Hussar wheat, and *Lr1* derived from various common wheats, have been widely used in soft red winter wheat cultivars and have selected phenotypes of *P. triticina* with virulence to these genes (Fig. 5.1a). Currently *Lr1* and *Lr11* do not provide effective resistance, and cultivars with *Lr9* are moderately resistant, but this resistance would quickly erode if cultivars with *Lr9* were grown over a larger area.

In Texas and Oklahoma isolates with virulence to *Lr1* quickly increased in the late 1970s and early 1980s after the release of cultivars with this gene (Fig. 5.1b). Virulence to *Lr24* appeared shortly after the release of the hard red winter wheat cultivar Agent with *Lr24* in 1971. By the mid-1970s virulence to *Lr24* was common in the winter wheat region of the Great Plains. In the mid-1980s the cultivar Siouxland with *Lr24* and *Lr26* was widely grown from Texas to South Dakota. Isolates with virulence to *Lr24* and *Lr26* increased up to the early 1990s. Starting in 2002 isolates with virulence to *Lr24* increased again due to widespread cultivation of 'Jagalene', with *Lr24*. Although *Lr24* was originally derived from *Ae. elongatum*, and *Lr26* from *Secale cereale*, the nonwheat origin of both genes did little to enhance their durability of resistance. The cultivar Jagger, released in the mid-1990s with *Lr17*, selected isolates with virulence to this gene, as these reached nearly 90% of isolates in Texas and Oklahoma in 2001. Isolates of *P. triticina* with virulence to *Lr41*, derived from *Ae. tauschii*, were found even before winter wheat cultivars with this gene were released in the late 1990s in the southern Great Plains. Isolates with virulence to *Lr41* have increased such that cultivars with this gene ('Thunderbolt', 'Overley', and 'OK Bullet') are now susceptible to leaf rust.

Selection of isolates for virulence to specific resistance has also occurred in the spring wheat region of Minnesota, North Dakota, and South Dakota, even though leaf rust does not frequently overwinter in this area. Cultivars with *Lr1* and

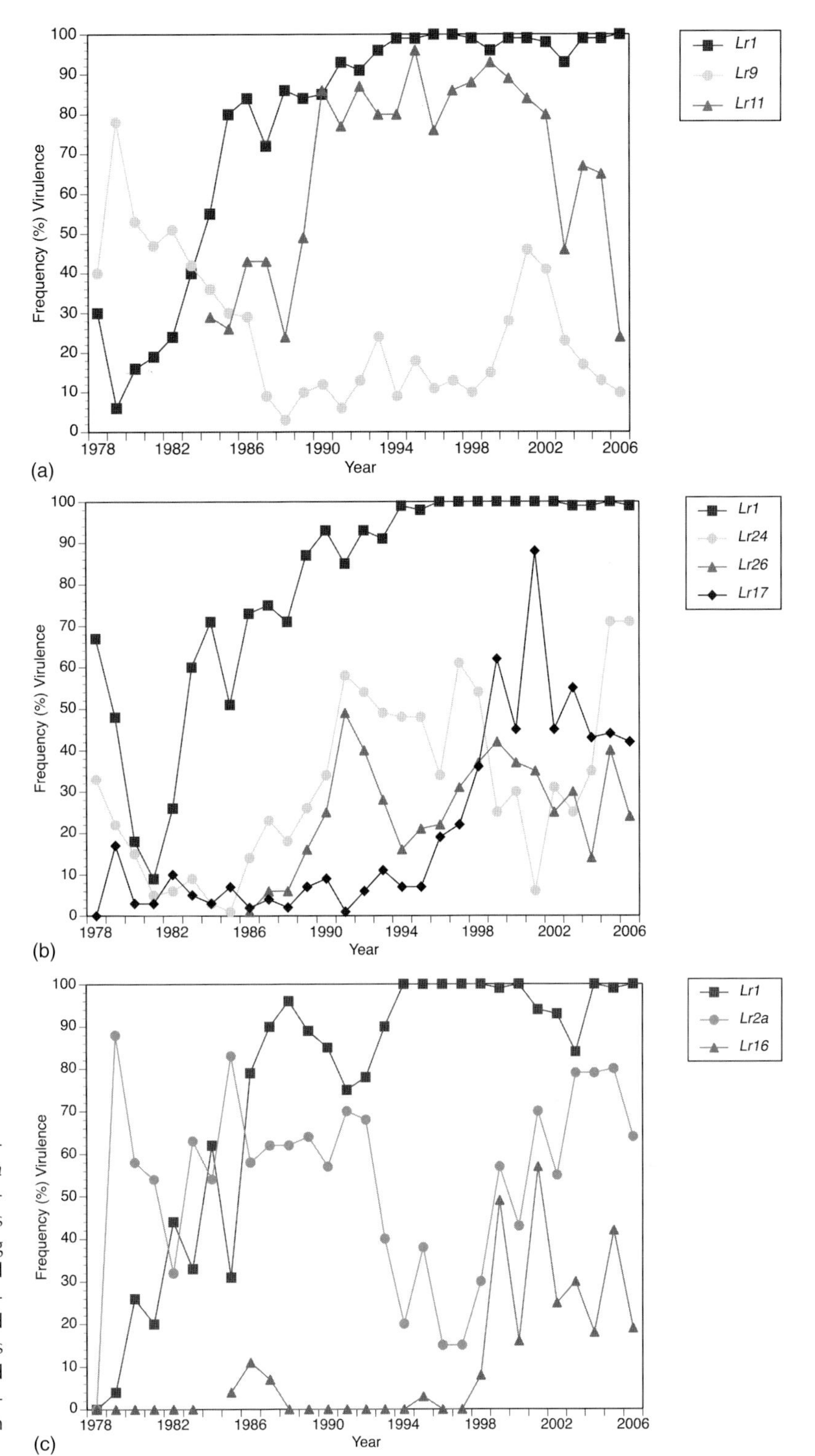

Fig. 5.1 Changes in frequency (%) of *Puccinia triticina* isolates with virulence to specific leaf rust resistance genes in different wheat growing regions of the US. (a) Soft red winter wheat region—southeastern states. (b) Hard red winter wheat region—Texas and Oklahoma. (c) Hard red spring wheat region—Minnesota, North Dakota, and South Dakota.

Lr2a were widely grown in this region starting in the mid-1970s. Isolates with virulence to both genes increased rapidly and were over 65% in the early 1990s (Fig. 5.1c). Virulence to *Lr16* has increased from the mid-1990s since many current spring wheat cultivars have this gene. In Australia cultivars with either combination of genes *Lr13*, *Lr23*, and *Lr34* or genes *Lr1*, *Lr13*, and *Lr23* are currently resistant to leaf rust (Bariana et al., 2007). Cultivars with various combinations of *Lr13*, *Lr24*, *Lr34*, and *Lr37* are considered moderately susceptible to leaf rust.

Leaf rust resistance genes up to *Lr60* have been designated (McIntosh et al., 2007). Genes *Lr1* (Cloutier et al., 2007), *Lr10* from common wheat (Feuillet et al., 2003), and *Lr21* from *Ae. tauschii* (Huang et al. 2003) have been sequenced. The three genes have NBS-LRR regions typical of resistance genes with isolate specificity. Genes *Lr1* and *Lr10* are widely ineffective, and *Lr21* has provided effective resistance in spring wheat cultivars in the US and Canada. Isolates with virulence to *Lr21* would be expected to increase if this gene was used in a winter wheat cultivar in the US. Many of the other *Lr* genes for which virulent isolates of *P. triticina* have not been found have also not been widely used in wheat improvement programs.

Durable leaf rust resistance in wheat

The development of wheat cultivars with high levels of effective durable resistance will depend on genes that confer nonspecific resistance or gene combinations that have proven to be effective over time. The cultivar Frontana released in Brazil in 1946 has been a valuable source of durable nonspecific leaf rust resistance. Dyck et al. (1966) backcrossed leaf rust resistance from Frontana into Thatcher. Backcross lines with the adult-plant gene *Lr13* were characterized, yet none of the lines was as resistant as Frontana because an additional gene was needed to recover the original resistance in Frontana. Dyck and Samborski (1982) characterized gene *LrT2* in a group of wheat cultivars that included 'Terenzio' and Frontana. Later, *LrT2* was determined to be the additional gene in Frontana, designated as *Lr34*, and mapped to chromosome 7DS (Dyck 1987). Singh and Rajaram (1992) determined that Frontana also carries other genes besides *Lr34* that condition adult-plant leaf rust resistance. Frontana was used as a leaf rust resistant parent in spring wheat programs in Minnesota and also at CIMMYT. The Minnesota cultivar Chris was derived from crosses with Frontana and released in 1966 as the first spring wheat in the US to have *Lr34*. The CIMMYT cultivars Penjamo 62, Lerma Rojo, and Nainari 60 also had *Lr34*.

Wheat lines and cultivars with *Lr34* optimally express leaf rust resistance in the adult-plant stage. Isolates of *P. triticina* with complete virulence to lines with *Lr34* have not been found in North America (Kolmer et al., 2003, McCallum and Seto-Goh 2006), despite the presence of wheat cultivars with *Lr34* for over 40 years. In field plots lines with only *Lr34* can have moderate to high levels of leaf rust severity, although these can usually be distinguished from completely susceptible lines if leaf rust readings are made when known susceptible lines are at near-terminal severity (Color Plate 9a). Lines with *Lr34* can also express resistance in seedling plants at cooler temperatures (Singh 1992b; Pretorius et al., 1994). The presence of *Lr34* enhances the response of other effective resistance genes in the same wheat genotype (German and Kolmer 1992). The presence of *Lr34* is also associated with a distinctive leaf-tip necrosis (Singh 1992a) that can vary between genotypes and environments. Wheat cultivars with other *Lr* genes combined with *Lr34* are often more resistant than lines with only *Lr34* or the other genes singly. Spring wheat cultivars with combinations of *Lr13*, *Lr16*, and *Lr34* were highly resistant in Canada (Samborski and Dyck 1982; Liu and Kolmer 1997) and the US (Ezzahiri and Roelfs 1989).

Diagnostic molecular markers closely linked to *Lr34* have been developed (Bossolini et al., 2006; Lagudah et al., 2006) that will greatly simplify selection of breeding materials with *Lr34*. In a survey of wheat classes in the US using the *Lr34* marker *csLV34*, the allele associated with the presence of *Lr34* was completely absent in soft

Table 5.1 US wheat cultivars tested for the presence of *csLV34* alleles and *Lr34*.

Class	Cultivar	*csLV34*[a]	*Lr34*[b]
Soft red winter	Fultz	–	
	Monon	–	
	Knox62	–	
	Arthur	–	
	Caldwell	–	–
	CK 9663	–	–
	Pioneer 26R61	–	–
	Saluda	–	
	McCormick	–	
	NC Neuse	–	
	Pocahontas	–	
	SS550	–	
Hard red winter	Triumph	–	
	Bison	–	
	Scout 66	–	
	Siouxland	–	
	Victory	–	
	Sturdy	+	+
	Ogallala	+	
	Duster	+	
	Santa Fe[c]	+	
	Fuller[c]	+	
	2137	–	
	Overley	–	
	Deliver	+	
	Endurance	–	
	Cutter	–	
	TAM 107	–	

Class	Cultivar	*csLV34*	*Lr34*
Hard red spring	Thatcher	–	–
	Chris	+	
	Era	+	+
	Waldron	–	
	Marshall	+	
	Wheaton	+	+
	Butte 86	-	
	Grandin	+	+
	Norm	+	+
	Russ	-	
	Oxen	+	
	BacUp	–	
	Keene	–	
	HJ98	–	
	Reeder	–	
	Alsen	+	+
	Briggs	+	+
	Steele	–	
	Oklee	+	
	Glenn	–	
Pacific Northwest spring wheat	Alpowa	–	
	Alturas	–	
	Hank	-	
	Hyak	–	
	Jefferson	–	
	Madsen	–	
	Nick	–	
	Scarlet	–	
	Stephens	–	

Source: Adapted from Kolmer et al. (2008).
[a]+ = allele associated with *Lr34*; – = allele associated with lack of *Lr34*.
[b]+ = genetic analysis indicated presence of *Lr34*; – = *Lr34* not present; blank indicates not tested.
[c]Winter wheat cultivars derived from 'Jagger' may lack *Lr34* yet have the *csLV34* allele associated with *Lr34*.

red winter wheat (Table 5.1). The allele associated with *Lr34* was most common in spring wheat cultivars bred for the northern Great Plains, was present at low frequency in older and current hard red winter wheat cultivars, and was not present in cultivars from the Pacific Northwest.

The CIMMYT cultivar Pavon 76 has provided an additional source of nonspecific adult-plant leaf rust resistance. The adult-plant resistance gene *Lr46* in Pavon 76 is on chromosome 1BL (Singh et al., 1998). Gene *Lr46* is likely present in CIMMYT germplasm that has been selected for adult-plant resistance to leaf rust. Cultivars and germplasm with combinations of *Lr34*, *Lr46*, and additional adult-plant resistance can be highly resistant, approaching complete immunity (Singh et al., 2000). Development of diagnostic molecular markers associated with *Lr46* (Rosewarne et al., 2006) will aid in selection of lines with this adult-plant resistance.

Combinations of adult-plant resistance genes with *Lr* genes effective in seedlings can also provide good levels of durable leaf rust resistance. The Minnesota spring wheat cultivar Norm released in 1992 has remained highly resistant to leaf rust. Norm was determined to have genes *Lr1*, *Lr10*, *Lr13*, *Lr16*, *Lr23*, and *Lr34* (Oelke and Kolmer 2005). Genes *Lr1*, *Lr10*, and *Lr13* are now widely ineffective, but isogenic Thatcher lines with *Lr16*, *Lr23*, and *Lr34* have effective resistance in field plots when compared to the completely susceptible Thatcher (Oelke and Kolmer 2004). Gene *Lr23* is highly temperature-dependent in expression of resistance. In ambient greenhouse temperatures of 15–25 °C, lines with this gene expressed variable infection types ranging from moderate to large uredinia to small uredinia surrounded by necrosis (Dyck and Johnson 1983). At 25 °C in growth cabinets, lines with *Lr23* expressed very low hypersensitive infection types to US isolates of *P. triticina*. Although *P. triticina* isolates with *Lr16* have been detected in the spring wheat region of the US, cultivars with this gene still have some resistance in field plots (Oelke and Kolmer 2004). 'Knudson', released by AgriPro-Coker in 2002, has also been very resistant and was determined to have *Lr3*, *Lr10*, *Lr13*, *Lr16*, *Lr23*, and *Lr34* (Kolmer and Oelke 2006). Spring wheat genotypes with combinations of *Lr16*, *Lr23*, and *Lr34* have shown good levels of resistance that has not been significantly eroded by virulence changes in the *P. triticina* population.

Additional genes from wheat germplasm that have shown good levels of durable resistance have also been characterized. Barcellos et al. (2000) determined that the Brazilian cultivar Toropi had two genes that conditioned adult-plant leaf rust resistance which were also associated with leaf-tip necrosis. Mishra et al. (2005) determined that the Indian cultivar C 306 had a single adult-plant resistance gene associated with leaf-tip necrosis that was independent of *Lr34*. The landrace-derived cultivars from Uruguay, 'Americano 25e' and 'Americano 44d', were shown to have unique adult-plant resistance genes that were not *Lr34* (Kolmer et al., 2007c). The Canadian spring wheat cultivar AC Taber was determined to have an effective adult-plant resistance gene other than *Lr13* or *Lr34* (Liu and Kolmer 1997). Navabi et al. (2003) estimated two to four effective adult-plant resistance genes were present in five CIMMYT lines.

Quantitative trait loci (QTLs) that affect resistance in adult plants have also been mapped to chromosome regions. Xu et al. (2005a,b) identified QTLs that affected final rust severity, infection rate, infection duration, and latent period on chromosomes 2B, 7BL, and 2DS in the soft red winter wheat germplasm line CI 13227. The Swiss cultivar Forno was determined to have a major QTL for adult-plant resistance on chromosome 1BS and minor regions for resistance on 2DL, 3DL, 4BS, and 5AL (Schnurbusch et al., 2004). These additional sources of adult-plant resistance can be used in wheat improvement programs to diversify germplasm for effective leaf rust resistance.

Association with other disease resistance genes

An intriguing aspect of adult-plant leaf rust resistance in wheat is an association with resistance to other diseases of wheat. Lines with *Lr34* also have adult-plant, nonspecific resistance to stripe rust (caused by *P. striiformis* Westend. f. sp. *tritici* Eriks.) (McIntosh 1992; Singh 1992c). The stripe rust resistance associated with *Lr34* has been designated as *Yr18*. The *Lr34* locus also exhibits a pleiotropic effect on barley yellow dwarf virus reaction (Singh 1993). Spielmeyer et al. (2005) showed that lines segregating for adult-plant resistance to leaf and stripe rust due to *Lr34/Yr18* also had adult-plant resistance to powdery mildew. The 7DS chromosomal region of *Lr34* may condition a generalized nonspecific response that acts against biotrophic wheat pathogens. The adult-plant resistance gene *Lr46* is also associated with nonspecific stripe rust resistance, which has been designated as *Yr29* (Williams et al., 2002). Navabi et al. (2005) also showed that adult-plant resistance genes other than *Lr34/Yr18* conditioned resistance to both leaf rust and stripe rust. Since adult-plant resistance to leaf and stripe rust was highly associated, selection of germplasm with resistance to both rust diseases could be accomplished by testing for resistance to only one disease.

The Minnesota spring wheat cultivar Thatcher was released in 1935 on the basis of resistance to stem rust and good breadmaking quality characteristics (Hayes et al., 1936). Many subsequent spring wheat cultivars in Canada and the US (Kolmer et al., 1991) have had stem rust resistance derived from Thatcher. Dyck (1987) determined that isogenic lines of Thatcher with *Lr34* had better seedling and adult-plant resistance to stem rust than Thatcher. The Canadian wheat cultivars Roblin (Dyck 1993) and Pasqua (Liu and Kolmer 1998a) have some Thatcher in their pedigrees and also have *Lr34*. In crosses derived from both Roblin and Pasqua, progeny lines with *Lr34* have been associated with higher stem rust resistance. A stem rust resistance suppressor on chromosome 7DL (Kerber and Green 1980) is present in Thatcher and other Thatcher-derived lines. Kerber and Aung (1999) determined that Thatcher lines with *Lr34* had the same stem rust response as did Thatcher lines nullisomic for 7DL that lacked the suppressor. Hence *Lr34* appeared to inactivate the stem rust resistance suppressor.

Gavin-Vanegas et al. (2007) determined that progenies derived from a Thatcher line with *Lr34* crossed with a stem rust susceptible line segregated for two effective stem rust resistance genes in adult plants in the absence of *Lr34*, and segregated for three genes when all progeny lines were fixed for *Lr34*. In this study segregation of resistance to stem rust races that had high infection types to seedlings of Thatcher, but low infection types to Thatcher lines with *Lr34*, was strongly correlated with segregation of stem rust resistance in adult plants to a mixture of stem rust races. The presence of *Lr34* allowed the expression of additional stem rust resistance gene(s) in Thatcher that were most likely derived from 'Iumillo' durum.

Thatcher and the cultivar Chris had low seedling infection types to the stem rust isolate Ug-99, which has appeared in eastern Africa and is highly virulent to many US and CIMMYT wheat cultivars (Jin and Singh 2006). The stem resistance in Chris is most likely due to the presence of stem rust resistance derived from Thatcher that is enhanced by the presence of *Lr34*. Thatcher lines with *Lr34* show good resistance to Ug-99 in field plots in Kenya. Since so few current wheat cultivars have effective resistance to Ug-99, the Thatcher stem rust resistance enhanced by *Lr34* may be an important future source of stem rust resistance.

Leaf rust resistance in durum wheat

Cultivated durum wheat is generally highly resistant to the *P. triticina* isolates found on common wheat (*T. aestivum* L.). Genes *Lr14a* derived from 'Yaroslav' emmer (*T. turgidum* ssp. *dicoccum*) and *Lr23* derived from 'Gaza' durum are present in common wheat. Genes *Lr10* and *Lr33* (Dyck 1994) may also be present in durum wheat. Genetic studies of leaf rust resistance in durum wheat to *P. triticina* isolates from common wheat have often indicated the presence of one to two seedling resistance genes that were expressed in a recessive or dominant manner (Statler 1973; Zhang and Knott 1990, 1993). Phenotypes of *P. triticina* that are virulent to durum cultivars were described in Mexico (Singh 1991). Using isolates collected from durum wheat, Singh et al. (1993) determined that a collection of CIMMYT durum cultivars varied for seedling resistance and adult-plant resistance genes. The durum cultivar Altar C84 and three other durum cultivars had a single gene that conditioned seedling resistance, in addition to two adult-plant resistance genes.

A severe epidemic of leaf rust on durum wheat occurred during 2001–2003 in northwest Mexico (Singh et al., 2004). New phenotypes of *P. triticina* had emerged that were highly virulent to Altar C84. Increased levels of leaf rust infections were also noted in France (Goyeau et al., 2006) and Spain (Martinez et al., 2005). Isolates of *P. triticina* from durum wheat collected during 2002–2004 in France, Spain, Mexico, Argentina, and Chile were highly similar for virulence to the *Lr* genes in the Thatcher isogenic lines, a collection of durum cultivars (Ordoñez and Kolmer 2007b), and for molecular SSR variation (Ordoñez and Kolmer 2007a), suggesting a recent common origin. Singh et al. (2004) identified CIMMYT durum germplasm that was resistant to the new *P. triticina* durum virulent phenotypes in Mexico. The resistant lines were also resistant in the other countries where higher infections of leaf rust on durum wheat had occurred.

Herrera-Foessel et al. (2005) described up to five genes in nine CIMMYT durum lines that conditioned resistance to the *P. triticina* race that was virulent to Altar C84. Herrera-Foessel et al. (2007) determined that one of these genes mapped to the *Lr3* locus on chromosome 6B, and a second resistance gene was closely linked to the *Lr3* locus.

WHEAT STRIPE RUST

Distribution and epidemiology

Stripe rust is an important disease of wheat worldwide. The disease has been reported in more than 60 countries and has caused yield losses in Africa, Asia, Australia, New Zealand, Europe, North America, and South America (Stubbs 1985; Chen 2005). In the US the disease has been most common in states west of the Rocky Mountains since the late 1950s but has become increasingly frequent in the eastern and midwestern states since 2000 (Chen 2005). In recent years significant losses in wheat due to stripe rust have been reported in California, Oregon, Washington, Idaho, Montana, Colorado, Texas, Oklahoma, Kansas, Nebraska, South Dakota, Louisiana, Arkansas, Missouri, Alabama, and Georgia.

Stripe rust infection can occur at any growth stage when green plant tissue is available. The first visible symptom of infection appears as chlorotic spots that resemble viral symptoms. Uredinia of stripe rust are yellow to orange in color; thus the disease is commonly called yellow rust. Stripe rust uredinia, 0.3–0.5 mm by 0.5–1.0 mm, are much smaller than uredinia of stem rust and leaf rust. Uredinia can form on both sides of leaves but are more abundant on the upper surface (Color Plate 10). Uredinia can also form on leaf sheaths, glumes, awns, and on immature green kernels. Uredinia form in patches around infection sites on seedlings and are arranged in stripes between leaf veins on adult plants. Depending on the level of plant resistance, uredinia can be surrounded by chlorosis or necrosis on seedling leaves. On adult-plant leaves of highly susceptible plants, uredinia continue to develop in stripes from the initial infection sites, without necrosis or chlorosis. On resistant cultivars, necrotic stripes develop from the initial infection sites. Responses on resistant wheat genotypes vary from no visible symptom to various sizes or lengths of necrotic patches or stripes with varying amounts of sporulation (Color Plate 10b). Uredinia erupt to release urediniospores. Each uredinium can produce thousands of urediniospores over a period of days. Urediniospores are spherical, 15–20 μm in diameter, and echinulate.

Urediniospores are dispersed mainly by wind but also can be spread by insects, animals, and humans. A minimum of three hours of dew formation on the plant surface is needed for urediniospores to germinate and infect plants (Rapilly 1979). The optimum temperatures for spore germination are 10–12 °C, and the minimum temperature for germination is just above 0 °C (Newton and Johnson 1936). Urediniospores do not germinate well when the temperature is above 20 °C. Germ tubes penetrate into plant tissue through stomata without forming appressoria as in other wheat rusts (Marryat 1907; Allen 1928). After a germ tube passes through a stoma, it forms a substomatal vesicle, from which branched hyphae grow intercellularly. Haustoria are formed from the hyphae and grow into host cells. The uredinia are produced from intercellular hyphae and emerge on the plant surface. Under optimum temperature conditions (13–16 °C), it takes 12–13 days from initial infection to sporulation of new uredinia (Hungerford 1923). Stripe rust will continue to produce new uredinia and urediniospores in stripes further up and down the leaf from the initial site of infection.

Urediniospores are the sole initial inoculum source for the stripe rust pathogen. Initial inoculum can be local and/or from outside a region depending upon climatic conditions that influence survival of stripe rust during the summer or winter. In the Pacific Northwest of the US and the adjacent area of British Columbia and Alberta in Canada, the stripe rust fungus is able to survive during the summer and winter most years. In this region, urediniospores can be found at almost any time of year, especially west of the Cascade Mountains where the mild winters and cool summers are favorable to stripe rust survival. The cool night temperatures and dry conditions during summer in the major wheat growing areas east of the Cascade Mountains allow urediniospores to retain viability for extended periods. Urediniospores that are produced in late summer and early fall from spring wheat fields, volunteer plants, and grasses infect the winter wheat crops. The fungus overwinters in plant tissue as mycelia and often as viable urediniospores on plants (Hungerford 1923). According to Rapilly (1979), stripe rust can survive temperatures as low as – 10 °C. Snow cover provides favorable conditions for both the wheat plants and stripe rust to survive the winter. Heavy local survival of the stripe rust pathogen over the winter will lead to early epidemic development in the spring. Stripe rust forecasting models based on December and January temperatures have been successfully used in the Pacific Northwest to predict stripe rust epidemics (Coakley et al., 1982, 1984; Line 2002; Chen 2005).

Stripe rust can also overwinter in regions with cold winters. Stripe rust can be endemic in the Gallatin Valley and Flathead Lake area in Montana (Sharp and Heln 1963) where the fungus has overwintered. In 2006, stripe rust severity was unusually high in Alberta but light in the US Pacific Northwest, indicating winter survival of the rust in Alberta. In North Dakota and Minnesota stripe rust generally does not overwinter. However, in 2006, stripe rust was observed at St. Paul, Minnesota, on April 26, a month earlier than the first observation of the disease at Pullman, Washington, and also much earlier than reports of the disease in Missouri, Illinois, and Indiana, which indicated that the pathogen survived the 2005–2006 winter in Minnesota. The cold winters and hot summers in the northern Great Plains region usually limits stripe rust development in this region.

In California stripe rust regularly overwinters and survives the summer on wheat crops and grasses at high elevations along the coast (Tollenaar and Houston 1966), and on wheat grown as forage or cover crops. Survival of stripe rust on volunteer wheat plants in irrigated fields resulted in a widespread epidemic in California in 1974 (Line 1976). Northeastern California, where stripe rust oversummers, may be a source of inoculum for southern Oregon, southern Idaho, and northern Utah, as well as for central California. Stripe rust inoculum is exchanged among California, Arizona, New Mexico, and northwestern Mexico.

The stripe rust pathogen generally does not survive the summers in the Great Plains, Mississippi Valley, and the southeastern states due to extended hot and humid conditions and the long period between harvest and planting of wheat crops. The late-planted wheat crops at high elevations in Mexico may provide initial inoculum for infection in the fall in the southern Great Plains (Line 2002). Wheat crops in the Rocky Mountain areas from Colorado to western Texas may contribute stripe rust inoculum to the Great Plains. Stripe rust infections occur soon after emergence of wheat in the late fall and early winter in Texas, Louisiana, Arkansas, and Mississippi. If stripe rust occurs, it usually develops slowly in the winter and at a faster rate in the early spring, providing inoculum for areas further north and east. The scope and severity of epidemics depend on inoculum in the south, on wind directions, and on temperature and moisture conditions in the region east of the Rocky Mountains. Stripe rust has been reported in New York, but has never caused significant damage north of Ohio and Virginia.

Origin and historical importance

Stripe rust of wheat is a long established disease in Asia and Europe (Stubbs 1985). Stripe rust was first recognized in the US in 1915 in Arizona (Carleton 1915). However, examination of herbarium specimens has indicated that the disease was present in the western US before 1892 (Humphrey et al., 1924) and possibly occurred in California in the 1700s (Smith 1961). The Caucasus region is the presumed origin of stripe rust (Stubbs 1985; Line 2002).

From 1957 to 2005, the US experienced four waves of regional epidemics of stripe rust. The first was from 1958 to 1961, and the epidemics were concentrated in the Pacific Northwest and California. Severe epidemics of wheat stripe rust occurred in 1960 and 1961 in Washington (Hendrix 1994) and in 1961 in Oregon (Shaner and Powelson 1971). Stripe rust also was severe in Idaho and Montana (Pope et al., 1963) in those years. In California, the yield losses were estimated at 28%–56% in the Sutter Basin north of Sacramento (Tollenaar and Houston 1966).

The second period of stripe rust epidemics occurred from 1974 to 1978. In 1974, California had an 8% yield loss in wheat production. In 1976, yield losses were 17% in Washington, 13% in Oregon, and 11% in Idaho. The third wave of epidemics occurred from 1980 to 1984. In 1980 and 1981, stripe rust epidemics were widespread in the Pacific Northwest, and yield losses in Washington were estimated to be 13% in 1980 and 11% in 1981. Oregon and Idaho had 5%–9% yield losses during the same period. Yield losses were reduced in 1980–1981 because of the widespread use of fungicides. From the mid-1980s to the late 1990s, yield losses caused by stripe rust were reduced due to widely grown resistant cultivars in the Pacific Northwest and California and the use of fungicides. Yield losses remained below 5% in the western US.

The most recent stripe rust epidemics occurred from 1999 to 2005. In 1999, stripe rust caused a 7% yield loss in California but was not severe in other states. In 2000, stripe rust was widespread and caused severe damage in the south central states with a 7% loss in Arkansas. In 2001 losses in Kansas and Colorado were estimated to be 7% and 8%, respectively. In 2003 states with major yield losses were Kansas (11%), Nebraska (10%), and California (21%). In 2005 stripe rust was very widespread and occurred in more than 30 states, with significant yield losses in Kansas (8%), Texas (15%), Oklahoma (5%), Nebraska (4%), California (5%), Arkansas (5%), and Louisiana (5%).

Taxonomy, life cycle, and host range

Stripe rust on cereal crops and grasses is caused by different *formae speciales* of *P. striiformis*, a basidiomycete rust fungus. The disease was first described by Gadd in 1777 (Eriksson and Henning 1896). Schmidt (1827) named the stripe rust fungus as *Uredo glumarum*. Westendorp (1854) used *P. striaeformis* for stripe rust collected from rye. Fuckel (1860) described stripe rust as *P. straminis*. *Puccinia glumerum*, as described by Eriksson and Henning (1894), was used as the name for the stripe rust fungus until Hylander et al. (1953) revived the name *P. striiformis* Westend.

Since stripe rust pathogens on different cereal crops and grasses are separated into different *formae speciales* (Eriksson 1894), *P. striiformis* Westend. f. sp. *tritici* Eriks. is considered the valid name for the stripe rust pathogen infecting wheat. In addition to wheat stripe rust, Eriksson described stripe rusts on barley as f. sp. *hordei*, on rye as f. sp. *secalis*, on *Elymus* spp. as f. sp. *elymi*, and on *Agropyron* spp. as f. sp. *agropyron*. Later, three more *formae speciales* were proposed: *P. striiformis* f. sp. *poae* on Kentucky bluegrass (*Poa pratensis* L.) (Britton and Cummins 1956; Tollenaar 1967), f. sp. *dactylidis* on orchardgrass (*Dactylis glomerata* L.) (Manners 1960; Tollenaar 1967), and f. sp. *leymi* on *Leymus secalinus* (Georgi) Tzvel (Niu et al., 1991). More recently, Wellings et al. (2004) considered stripe rust on *Hordeum* spp. in Australia to be a new *formae specialis*, different from both *P. striiformis* f. sp. *tritici* and *P. striiformis* f. sp. *hordei*. Not all these *formae speciales* are equally and clearly separated by host specialization. Wheat stripe rust mostly infects wheat but can infect some barley cultivars, while barley stripe rust can infect some wheat cultivars. However, stripe rust of barley does not infect bluegrass, and bluegrass stripe rust does not infect wheat or barley (Chen et al., 1995).

Puccinia striiformis has a hemicyclic life cycle of urediniospores, teliospores, and basidiospores. Teliospores are formed along the sides of uredinia and are the same size and shape as urediniospores, but have black cell walls. Teliospores form more rapidly and more abundantly under hot and humid conditions than under cool and dry conditions. The barley stripe rust fungus (*P. striiformis* f. sp. *hordei*) is more likely to produce telia than *P. striiformis* f. sp. *tritici*. Mature teliospores are two-celled and become diploid with one nucleus in each cell. Teliospores germinate readily to produce haploid single-celled basidiospores that are unable to infect cereals and grasses. Despite early intensive studies, an alternate host for *P. striiformis* was not found (Eriksson and Henning 1894; Mains 1933; Tranzschel 1934; Straib 1937; Hart and Becker 1939). Alternate host plants for *P. striiformis* may not exist (Hassebrauk 1970), or with the short dormancy of teliospores and readily produced basidiospores, the alternate host may escape infection (Wright and Lennard 1978; Rapilly 1979). In the absence of an alternate host, the teliospores and basidiospores are not functional in the *P. striiformis* life cycle.

Puccinia striiformis f. sp. *tritici* is able to infect a broader range of grass species than stem rust or leaf rust. Hassebrauk (1965) listed about 320 grass species of 50 genera in the Gramineae family that were naturally or artificially infected by wheat stripe rust. The most susceptible genera are *Aegilops*, *Agropyron*, *Bromus*, *Elymus*, *Hordeum*, *Secales*, and *Triticum*. Some *Aegilops* spp., especially *Ae. cylindrica* (common goatgrass), are highly susceptible to stripe rust. These grasses near wheat fields can contribute to stripe rust development as sources of early inoculum, but have limited roles in rust survival because they mature earlier than wheat crops. The importance of grasses as hosts in wheat stripe rust epidemics may vary from region to region. Generally, wild grasses play a less important role than wheat crops and volunteer wheat in the initiation, development, spread, and survival of stripe rust. However, grasses can serve as reservoirs in maintaining diversity of stripe rust races.

Genetic variation in *Puccinia striiformis* f. sp. *tritici*

Virulence variation

The wheat stripe rust pathogen is highly variable for virulence to stripe rust resistance genes in wheat. In 2000, 42 races of wheat stripe rust were found in the US (Chen et al., 2002). Early studies of physiological specialization in *P. striiformis* f. sp. *tritici* used several sets of differential cultivars, different inoculation methods, and varying environmental conditions, which compromised continuity from year to year or among investigators. The differential set used in Europe did not identify important races in the US (Line 2002). Line et al. (1970) first developed a uniform system to identify and describe races of the wheat stripe rust pathogen in the US. The current differential set includes 20 wheat cultivars and lines with various combinations of genes *Yr1*, *Yr2*, *Yr3a*, *Yr4a*, *Yr6*, *Yr7*, *Yr8*, *Yr9*, *Yr10*, *Yr17*, *Yr19*, *Yr20*, *Yr21*, and other genes currently undesignated (Chen et al., 2002). A total of 126 races of *P. striiformis* f. sp. *tritici* has been identified since the establishment of this US differential set (Line and Qayoum 1992; Chen et al., 2002, 2007; Chen 2005, 2007).

The emergence of the majority of *P. striiformis* f. sp. *tritici* races in the US can be related to selection by wheat cultivars with race-specific resistance. Line and Qayoum (1992) discussed races selected by widely grown wheat cultivars in the US before 1987. The appearance and rapid development of races with virulence to seedlings of the cultivar Express caused yield losses in California in 1999. The widely grown cultivars RSI 5, Bonus, and Summit in California apparently selected races with virulence to *Yr1*, *Yr9* (on the 1RS.1BL wheat–rye translocation), and the seedling resistance in Express. When these cultivars were released they were highly resistant to the previously detected races but became susceptible within a few years under commercial production. The average length of time of a cultivar with race-specific *Yr* genes retaining effective stripe rust resistance is 3.5 years (Chen 2005).

The soft winter wheat cultivar Stephens has been very popular in the Pacific Northwest since its release in 1978. Wheat stripe rust races with virulence to seedlings of Stephens were first identified in 1977 (Line and Qayoum 1992) and since then have been predominant in this region. Since 2004, races with virulence to Stephens and virulence to *Yr9* have been predominant throughout the US (Chen 2007). The widely grown cultivars Jagger and Jagalene, which likely have stripe rust resistance from Stephens, may have contributed to the widespread occurrence of these races in the

Great Plains region. Because Stephens, Jagger, and Jagalene also have high-temperature adult-plant (HTAP) resistance, these cultivars have not been severely infected. However, races with *Yr9* virulence caused severe epidemics from 2000 to 2005 on susceptible cultivars that lacked any effective resistance.

Other epidemics of stripe rust caused by the introduction of new races are well documented. The best example includes the introduction of the wheat stripe rust pathogen to eastern Australia in the late 1970s (Wellings and McIntosh 1987; Wellings 2007), western Australian in 2002 (Wellings et al., 2003), and South Africa in the mid-1990s (Pretorius et al., 1997). These long-distance introductions of the wheat stripe rust pathogen from one continent to another were thought to be caused by inadvertent human activities. Once present in a new continent, stripe rust spreads quickly to neighboring countries. This was seen in the spread of the barley stripe rust pathogen from Colombia to other South American countries and to Mexico and the US from 1975 to 1991 (Chen et al., 1995). The races with virulence to *Yr9* which have rendered many cultivars with the 1RS.1BL wheat–rye translocation highly susceptible to stripe rust were likely an introduction from outside the US.

Since the group of races with *Yr9* virulence appeared in the US in 2000, numerous new races with additional virulence have subsequently been found. In 2000, the most common races were virulent to the differential lines 'Lemhi', 'Heines VII', 'Lee', 'Fielder', 'Express', 'AVS/6*Yr8', 'AVS/6*Yr9', 'Clement', 'Compair', and 'Produra'. Since then a large number of races with additional virulence to the differential cultivars Tres, Stephens, Yamhill, and Chinese 166 have been detected (Fig.5.2). Races with virulence to Stephens, Yamhill, *Yr8*, and *Yr9* have been the most common races in the US since 2003. In the Pacific Northwest, races with virulence to the cultivars Moro (*Yr10* and *YrMor*) and Paha were detected in 2005. These new races have caused several previously resistant cultivars to become susceptible, or they have reduced the resistance level in cultivars with race-specific resistance and non-race-specific HTAP resistance (Chen 2005, 2007).

Molecular variation

Chen et al. (1993) used RAPD markers to examine molecular variation in *P. striiformis* f. sp. *tritici*. DNA polymorphism was detected among races and among single-spore isolates within races.

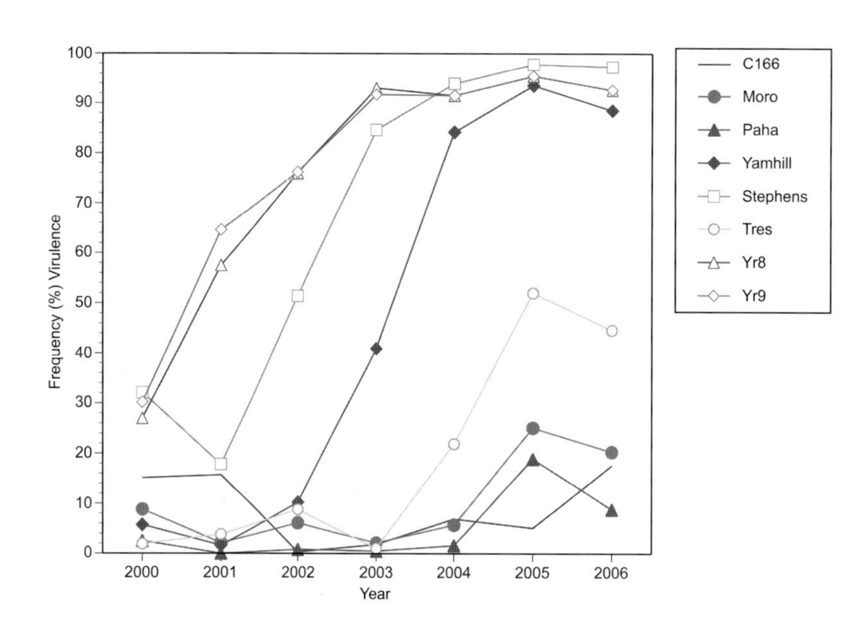

Fig. 5.2 Frequency (%) of isolates of *Puccinia striiformis* f. sp. *tritici* with virulences to selected wheat genotypes in the US from 2000 to 2006. 'Chinese 166' (C166) has resistance gene *Yr1*; 'Moro', *Yr10* and *YrMor*; 'Paha', *YrPa1*, *YrPa2*, and *YrPa3*; 'Yamhill', *Yr2*, *Yr4a*, and *YrYam*; 'Stephens', *Yr3a*, *YrS*, and *YrSte*; and 'Tres', *YrTr1*, and *YrTr2*. The *Yr8* and *Yr9* near-isogenic lines are in the Avocet-susceptible (AVS) background.

Races with virulence to *Yr1* differed for RAPD markers from races avirulent to *Yr1*, indicating different origins of the two groups of races. More recently, Markell et al. (2004) used AFLP markers to study the new wheat stripe rust races in the south central states of the US. The new group of races that was collected after 2000 was virulent to *Yr8* and Yr9 and was clearly different for AFLP genotype compared with races collected before 2000. The results, together with virulence data (Chen et al., 2002; Chen 2005, 2007), indicate that the new group of races was introduced to the US.

In Europe, wheat stripe rust isolates with identical virulences and AFLP phenotypes have been found in the UK and Denmark (Hovmoller et al., 2002). A single population of *P. striiformis* f. sp. *tritici* is present in northwest Europe with regular migration of the rust from the UK to Germany and France. In France two distinct populations of *P. striiformis* f. sp. *tritici* in the north and south of the country were described based on AFLP and virulence variation (Enjalbert et al., 2005).

Stripe rust resistance in wheat

Working with stripe rust of wheat, Biffen (1905) was the first to show that disease resistance in plants was inherited according to Mendel's laws. His research provided the scientific basis for breeding resistant cultivars to control stripe rust and other plant diseases. Much of the early research on resistance and genetics of stripe rust resistance in wheat was conducted in Europe, which was reviewed by Röbbelen and Sharp (1978). Lupton and Macer (1962) determined the genetics of stripe rust resistance in seven wheat genotypes and used the *Yr* symbols to designate resistance genes in wheat.

Breeding for stripe rust resistance in wheat in the US had little incentive from insignificant yield losses during the 40-year period after stripe rust was discovered in 1915 (Line 2002). After the widespread stripe rust epidemics in the early 1960s, wheat breeders and geneticists started work on developing stripe rust resistant cultivars. The first cultivar, Gaines, was released with moderate stripe rust resistance in 1961 by Dr. O. Vogel, a USDA-ARS scientist at Pullman, Washington. This cultivar, and the sister-line cultivar Nugaines, laid the foundation for non-race-specific HTAP resistance that is widely distributed among wheat cultivars grown in the Pacific Northwest and other regions. Since the early 1960s, US scientists have conducted studies to identify, characterize, and map stripe rust resistance genes and to develop cultivars with stripe rust resistance (Röbbelen and Sharp 1978; Line 2002; Chen 2005).

Race-specific resistance

Stripe rust resistance genes up to *Yr40* have been designated (McIntosh et al., 2007) and more than 30 genes with provisional designations have been listed. Races of stripe rust with virulence to most of the race-specific Yr genes have been found (Chen 2005). Currently, only a few race-specific *Yr* genes that are expressed in seedling plants are effective against all identified races in the US. These genes include *Yr5*, *Yr15*, *Yr26*, and *Yr40* (Chen 2005; Kuraparthy et al., 2007).

Although *Yr* genes that are race-specific are generally not durable, the deployment of multiple gene combinations either in homogeneous pure-line cultivars, multiline cultivars, or cultivar mixtures can prolong the effective life span of the *Yr* genes. The most successful example of using a multiline cultivar to control stripe rust is Rely, a club wheat cultivar widely grown in the US Pacific Northwest since its release in 1991 (Allan et al., 1993). Cultivar mixtures combine two or more cultivars with different resistance genes in the same field to reduce disease damage (Finckh and Mundt 1992). Recently more than 30% of the wheat acreage in Washington was planted to mixtures of two or three cultivars. This approach reduces yield losses caused by stripe rust; it also minimizes abiotic stresses such as winter and drought damage. Many wheat breeding programs in the US are currently combining resistance genes, such as *Yr5* and *Yr15*, into elite lines to develop new cultivars with a high level of resistance. In Australia, cultivars with combinations of genes *Yr6*, *Yr7*, *Yr17*, *Yr18*, and *Yr33* are common (Bariana et al., 2007).

High-temperature adult-plant resistance

In contrast to race-specific resistance, HTAP resistance cannot be detected in seedling plants. Cultivars with only HTAP resistance are highly susceptible in the seedling stage. At higher temperatures in older plants resistance is most apparent, with more resistance expressed in flag leaves than in lower leaves. The typical HTAP resistance in the spring wheat cultivar Alpowa is shown in Color Plate 10b. The HTAP resistance was first characterized by Line (1972) and further described in later studies (Qayoum and Line 1985; Milus and Line 1986a,b; Chen and Line 1995a,b; Line and Chen 1995).

The HTAP resistance in Gaines and Nugaines has remained effective for more than 40 years under frequent stripe rust epidemic conditions. Gaines and Nugaines were widely grown in the Pacific Northwest until 1981 (Line 2002), when new cultivars with higher levels of resistance were developed. The cultivar Luke with a much higher level of HTAP resistance than Gaines or Nugaines was released in 1970. Currently, wheat cultivars widely grown in the Pacific Northwest, such as Stephens, 'Madsen', 'Eltan', and 'Rod' soft white winter wheat, 'Bauermeister' hard red winter wheat, and Alpowa, Express, and 'Louise' spring wheat have HTAP resistance.

In the Great Plains, the widely grown cultivar Jagger with HTAP resistance has reduced losses to stripe rust in this region since 2000. Cultivars with HTAP resistance have also reduced yield losses compared with susceptible cultivars in California and the south central states where stripe rust develops in early growth stages (Uauy et al., 2005). Cultivars with HTAP resistance can affect the epidemiology of stripe rust, as stripe rust infections on these cultivars produce fewer urediniospores, thus reducing the overall level of stripe rust inoculum. Increased use of cultivars with HTAP resistance will decrease inoculum levels and thus reduce yield losses over a wide geographic area.

To identify wheat lines with HTAP resistance, adult plants should be tested at defined high (10–35 °C) and low (4–20 °C) temperatures with stripe rust races that are virulent to seedlings of the same genotype. Data from field plots sometimes indicate the presence of HTAP resistance, especially in areas where stripe rust infection occurs in the seedling stage. However, a mixture of avirulent and virulent races to specific *Yr* genes can confound these results. Expression of HTAP resistance can be influenced by temperature, by growth stage at which infections first occur, and by the amount of inoculum. Infection type and disease severity are most commonly used to measure HTAP resistance. Infection types are relatively stable, but can be affected by temperature and plant growth stage. Severity levels of cultivars with HTAP resistance tend to vary across regions and years due to differing levels of rust inoculum and temperatures.

The genetics of HTAP resistance has been studied in numerous wheat cultivars. Segregation for stripe rust resistance in adult plants differed from segregation in seedling plants, which indicated that different genes control resistance in different growth stages (Allan et al., 1966). Milus and Line (1986a) determined that Gaines had one gene and Nugaines and Luke had two genes for HTAP resistance. Gaines and Nugaines had a gene in common, but the HTAP genes in Luke were different. Chen and Line (1995a,b) determined two to three genes or QTLs conferred HTAP resistance in 'Druchamp' and Stephens. The cultivars differed for HTAP resistance genes and for race-specific resistance genes. In these studies, HTAP resistance genes in all cultivars were partially recessive and there was an additive effect when two or more genes were present in a genotype.

Several genes that condition adult-plant or HTAP resistance have been reported (with their known cultivar source): *Yr11* ('Joss Cambier'), *Yr12* ('Frontier'), *Yr13* ('Hustler'), *Yr14* ('Kador'), *Yr16* ('Bersee'), *Yr18* ('Jupateco 73R'), *Yr29* ('Pavon F76'), *Yr30* ('Opata 85'), *Yr34* ('WAWHT2046'), *Yr36* ('Glupro'), and *Yr39* (Alpowa). In addition a large number of QTLs that condition HTAP or adult plant stripe rust resistance have been characterized (Worland and Law 1986; Chen et al., 1998; Boukhatem et al., 2002; Chen 2005; Santra et al., 2006; William et al., 2006; Chen and Lin 2007; Chen and Zhao

2007; Chhuneja et al., 2007; Lin and Chen 2007). The relative abundance of HTAP resistance genes should facilitate incorporation of durable resistance into adapted wheat germplasm.

In 1983, Roy Johnson defined durable resistance as "resistance that remains effective in a cultivar that is widely grown for a long period of time in an environment favorable to the disease" (Johnson 1983). Based on this definition, HTAP resistance has been durable and therefore should be widely used in breeding programs. Although race-specific resistance genes are generally not durable, these genes can be used in combination to prolong their life span. Wheat cultivars with more effective and durable stripe rust resistance can be attained by combining HTAP resistance with effective race-specific resistance.

Slow-rusting resistance

Although there are several definitions given for slow-rusting resistance in the literature, the definition originally given by Caldwell (1968) and followed by Parlevliet (1979) and Singh and Rajaram (1992) is still descriptive. Slow-rusting resistance was characterized as slow disease development in the field despite a susceptible infection type and by one or more resistance components, such as longer latent period, smaller uredinium size, low receptivity, and reduced sporulation. Slow-rusting and HTAP resistances are similar in that both are expressed mostly in adult plants and are characterized by low disease severity in the field. However, cultivars with slow-rusting resistance have a susceptible infection type, while cultivars with HTAP resistance have lower infection types than susceptible cultivars. Slow-rusting resistance usually does not include expression of a hypersensitive response, while HTAP resistance involves some degree of hypersensitivity.

Slow-rusting resistance to stripe rust is present in wheat cultivars. Stripe rust development is much slower on the cultivars Heines VII and Yamhill compared with susceptible cultivars. This resistance is likely due to a longer latent period for stripe rust sporulation. In field plots these two cultivars had lower stripe rust severity compared with susceptible cultivars, even though the predominant races were virulent on the race-specific resistance genes in both cultivars. In field plots at Pullman, Washington, many European wheat cultivars such as 'Cappelle Desprez', 'Vilmorin 23', and 'Hybrid 46' have exhibited slow-rusting resistance. Several spring wheat cultivars, such as 'Eden', 'Macon', and 'Scarlet', have consistently had susceptible infection types but have rust severities lower than highly susceptible cultivars. However, the slow-rusting resistance in these cultivars is not as effective as the HTAP resistance in the cultivars Alpowa, Express, and Louise.

WHEAT STEM RUST

Distribution and epidemiology

Although not the most widespread or common among the wheat rusts, stem rust of wheat can potentially be the most damaging. In the US epidemics have been most frequent and severe in spring wheat and durum wheat in the northern states Minnesota, South Dakota, North Dakota, and to a lesser degree in winter wheat in the southern and central Great Plains states and the Ohio Valley. Stem rust also occurs less frequently in the Pacific Northwest region of Washington and Idaho. Worldwide, stem rust is mostly found in regions with a continental climate where summer temperatures regularly exceed 25 °C. Stem rust has caused losses in wheat in Canada (Kolmer 2001b), the southern Cone of South America (German et al., 2007), continental Europe, the Indian Subcontinent, Australia (Park 2007), eastern Africa (Wanyera et al., 2006), and China (Roelfs et al., 1992).

Stem rust appears as elongated blisterlike pustules, or uredinia, most frequently on the leaf sheaths of a wheat plant, but also on true stem tissues, leaves, glumes, and awns (Color Plate 11). On the leaf sheath and glumes, uredinia rupture the epidermis and give a ragged appearance. Masses of brownish-red urediniospores, up to 10,000 per day, are produced in the uredinia and are easily shaken off plants. The uredinial stage is the most visible and is the disease stage on green plants. Urediniospores are elongated, 10–15 μm long, and echinulate. The urediniospores are disseminated to newly emerged tissues of the same plant or adjacent plants, where these spores are the source of new infections, or the spores can be windblown over long distances. In the case of long-distance dispersal, spore depositions on crops in a new area are often associated with rain showers. Stem rust pustules develop mostly on the underside of leaves, but may penetrate and sporulate on the upper side. In general, leaves of adult plants are not as receptive as stem tissue for stem rust infection. As infected plants mature, uredinia are replaced by telia, changing color from red to dark brown to black; thus the disease is also called black stem rust. Teliospores are firmly attached to plant tissue. The telial stage is not important in the epidemiology of the disease unless the alternate host common barberry (*Berberis vulgaris*) is present.

The infection process and specialized infection structures for *P. graminis* f. sp. *tritici* are the same as previously described for *P. triticina*, except that a 3-hour period of light is required following 6–8 hours of dew to complete the development of appresoria and penetration peg. Spore germination is optimal at 15–24 °C, and can occur up to 30 °C. The optimal temperature for sporulation is 30 °C and can occur up to 40 °C. A dew period of 6 hours is optimal for spore germination and the subsequent infection process (Roelfs et al., 1992).

In the US, stem rust is rarely observed on winter wheat in the fall. Infections on susceptible cultivars are generally not obvious until the spring after the wheat crop has reached heading. The disease is most obvious when a crop is approaching maturity. Infections of stem rust can severely damage crops that are within 2 weeks of harvest. Infections in winter wheat along the Gulf Coast and Texas in early spring are likely due to infections that survived the winter on fall-planted winter wheat or volunteer plants (Roelfs 1989), though field observations of overwintering events are rare. Stem rust infections on wheat are usually first observed in Texas and Louisiana in the last two weeks of April or the first two weeks of May and reach maximum severity by the middle of May.

Stem rust epidemics during and after the 1930s were due to the presence of a large pathogen population on winter wheat in the southern states that was windblown to the spring wheat region in the northern states and Canada. In years following the 1950s epidemics, however, there has been an increase of resistant cultivars in the southern US. This has reduced the opportunity for the pathogen to infect and overwinter in the south, resulting in very small population size (Kolmer et al., 2007a). There is only a small likelihood that stem rust overwinters on fall-planted winter wheat in the central Great Plains and the Midwest. In the northern spring wheat regions, stem rust occurs most frequently on susceptible wheat lines, and the initial inoculum for the region is almost exclusively from infected wheat in the southern and central Great Plains. In Minnesota and North Dakota, stem rust on wheat is usually first observed in mid-to-late June, with maximum severity in the last week of July or the first week of August. In the Pacific Northwest where plants of the alternate host common barberry are present, both overwintering urediniospores and aeciospores can contribute to the initial infections, but stem rust does not normally develop to epidemic status in this region. Comparisons of area of distribution, optimal temperatures for infection and growth, and alternate hosts for *P. triticina*, *P. graminis* f. sp. *tritici* and *P. striiformis* f. sp. *tritici* are summarized in Table 5.2.

Origin and historical importance

Although the origin of *P. graminis* f. sp. *tritici*, one of the specialized forms in the *P. graminis* species complex, is not clear, one or more of the susceptible alternate hosts of *Berberis* spp. was the likely source of the fungus (Leppik 1967). The rust most likely evolved in a region where the aecial and telial hosts overlapped (Wahl et al., 1984). However, it is difficult to restrict the center of origin to a specific region because of the large number and broad distribution of susceptible species of both the pycnial and telial hosts. Likely, *P. graminis* f. sp. *tritici* originated on a close wild

Table 5.2 Optimal temperatures for infection and growth, areas of distribution, and alternate hosts of *Puccinia triticina* (wheat leaf rust), *P. striiformis* f. sp. *tritici* (wheat stripe rust), and *P. graminis* f. sp. *tritici* (wheat stem rust) in the US.

	Infection temperature	Growth temperature	Distribution in US	Alternate hosts
Puccinia triticina	15–25 °C	20–30 °C	Common and widespread in Great Plains, southeastern US, Ohio Valley; Lower incidence in northeast and Pacific Northwest	*Thalictrum speciosissimum*—present in southern Europe—not native to North America
Puccina striiformis f. sp. *tritici*	10–12 °C	13–16 °C	Common and widespread in Pacific Northwest, California, and Gulf Coast; can be common and severe in southern and mid–Great Plains; lower incidence in northern Great Plains and northeastern US	No alternate host found
Puccinia graminis f. sp. *tritici*	15–24 °C	25–35 °C	Overwinters in south Texas and Gulf Coast; infections found on susceptible wheat in Great Plains and southeastern US; historically most destructive in spring wheat area of Great Plains	*Berberis vulgare*—once common and widespread—currently present in low numbers in Great Plains and Pacific Northwest

relative of common wheat, such as emmer wheat or a grass in the tribe Triticeae. It is not known when stem rust was first established in wheat in North America. Both wheat and susceptible *Berberis* spp. were introduced by early settlers, and the pathogen could have been introduced through imported *Berberis* spp. that had stem rust infections, or urediniospores and teliospores on wheat or barley straw, or urediniospores attached to clothing and implements. The capability of stem rust urediniospores to survive in transcontinental air currents indicates that the introduction could also have been a natural event.

In the early 1900s stem rust epidemics were frequent in the north central US and Manitoba and Saskatchewan in Canada, since all bread wheat cultivars were susceptible and millions of barberry plants were present in the Great Plains region. Only durum cultivars had some resistance to stem rust. Severe epidemics occurred in 1904, 1916, 1919, 1923, and 1927. The 1916 epidemic is especially notable since the large yield losses spurred the national barberry eradication program in the US (Campbell and Long 2001). In 1919 losses of 20% in spring wheat occurred in North Dakota and Minnesota, and 10% losses occurred in the winter wheat grown in Nebraska and Kansas. Even after the release of stem rust resistant spring wheat cultivars, severe epidemics with losses over 50% occurred from 1935 to 1937 and 1950 to 1954 due to the emergence of virulent stem rust races (Kolmer 2001b; Leonard 2001). A sustainable wheat industry could be maintained in the Great Plains of North America only if stem rust resistant cultivars were widely grown in both the winter wheat and spring wheat regions. Since the 1950s epidemics, the incidence of stem rust has been greatly reduced, a cumulative result due to the effect of barberry eradication in reducing the number of stem rust races (Leonard 2001) and the widespread planting of stem rust resistant spring wheat and winter wheat cultivars throughout the Great Plains region. Stem rust is virtually nonexistent today in production fields, and is seen almost exclusively in plots of susceptible winter and spring wheats.

The importance of stem rust in causing grain yield losses in wheat is readily apparent since telia of the fungus can be seen on the stems as the crop matures. Stem rust infections rupture the host plant epidermal tissue, causing an increased loss of water. Nutrients and water diverted by the fungus in the production of urediniospores also contribute to added stress of wheat that can contribute to premature death (Roelfs 1985). Stem rust infected plants are more susceptible to winterkill, produce fewer tillers, and have small heads. Lodging of plants caused by broken straw can occur due to severe stem infections. Severe infections in the last few weeks before harvest can greatly reduce grain yield due to the loss of water during the critical period of grain filling. Grain from stem rust infected wheat is often shriveled, which may reduce the market grade.

Taxonomy, life cycle, and host range

Wheat stem rust belongs to one of several *formae speciales* in *P. graminis*. The fungus is heteroecious, alternating between a telial host in the Poaceae and an aecial host in the Berberidaceae, and macrocyclic, with five spore states that are distinct in morphology and function. Illustrations of the life cycle of *P. graminis* f. sp. *tritici* can be readily found in various monographs and textbooks (Roelfs 1985; Agrios 1997).

The dikaryotic (n+n) teliospore is the dormant spore stage and germinates after breaking dormancy, usually after overwintering. The fungus has a brief diploid state when two nuclei fuse in a germinating teliospore. Meiosis follows, producing single-celled, hyaline haploid (n) basidiospores. Basidiospores are windborne to infect the alternate host of susceptible species in *Berberis* spp. and *Mahonia* spp. There are a large number of species in *Berberis* and *Mahonia* listed as susceptible to *P. graminis* (Roelfs 1985), but the common barberry, *B. vulgare*, is considered to be the most important. After infection on the alternate host, flask-shaped pycnia develop on the upper leaf surface, producing single-celled pycniospores (n) and receptive hyphae (n) that serve as gametes. Fertilization occurs when a pycniospore fuses with a receptive hypha of the opposite mating type, resulting in dikaryotic hyphae (n+n) that develop into a cluster of tubular or cuplike aecia on the underside of the leaf surface. Aeciospores (n+n) are produced in chains in aecia and are windborne to infect the telial hosts (grasses), but do not infect the aecial host. After an infection is established on a telial host, blisterlike uredinia develop at the infection site. Urediniospores (n+n) are produced as soon as 7 days after initial infection. Urediniospores are continuously produced in a sporulating uredinium for a sustained period of up to several weeks as long as the host tissue remains viable. The urediniospores are clonally produced for a number of generations and initiate new infections on the telial hosts. Under favorable conditions, urediniospores can reproduce rapidly in wheat crops, generating large quantities of inoculum. Teliospores are produced as the wheat plants approach maturity.

In most areas where wheat stem rust is important, the pathogen survives through the noncrop seasons as mycelia in tissues of dormant crop plants, volunteer plants, or alternative telial hosts (Roelfs 1985). Urediniospores could also be transported from one region to another following the succession of crops within an epidemiological zone. Thus the disease cycle can be completed without the presence of the alternate host. In North America, *P. graminis* f. sp. *tritici* is found primarily on common and durum wheat, barley, foxtail barley (*Hordeum jubatum*), and jointed goatgrass, although artificial inoculation can induce infection on many other grasses.

Genetic variation in *Puccinia graminis* f. sp. *tritici*

Race surveys of *P. graminis* f. sp. *tritici* have been conducted in the US since 1919 (Stakman et al., 1919). The initial set of differentials used to identify wheat stem rust races were the common wheat cultivars Little Club, Marquis, Kanred, and Kota; the durum cultivars Arnautka, Speltz Marz, Mindum, Kubanka, and Acme; the emmers 'White Spring', 'Khapli'; and diploid einkorn wheat (*T. monococcum* ssp. *monococcum*). These various wheats had multiple genes for stem rust resistance and were gradually replaced by differentials with single genes in Chinese Spring or in other backgrounds (McIntosh et al., 1995). The stem rust resistance genes present in the current differential set used in the US and Canada include *Sr5*, *Sr6*, *Sr7b*, *Sr8a*, *Sr9a*, *Sr9b*, *Sr9d*, *Sr9e*, *Sr9g*, *Sr10*, *Sr11*, *Sr17*, *Sr21*, *Sr30*, *Sr36*, and *SrTmp* (Roelfs and Martens 1988). Wheat and *P. graminis* f. sp. *tritici* interact in a gene-for-gene manner (Loegering and Powers 1962; Green 1964); thus the frequency of races with virulence to a resistance gene is related to the frequency of virulence alleles in the rust pathogen.

The alternate host of stem rust, common barberry, was prevalent throughout the north central US in the late 1890s and early 1900s. Pycnial–aecial infections on the barberry plants contributed to the initial inoculum and also to the race diversity of stem rust. The barberry eradication program in the 1920s removed millions of barberry from this region, thus delaying the initial onset of epidemics in the spring wheat region via reduction of initial inoculum and also reducing the race diversity of stem rust (Campbell and Long 2001). From 1919 to the 1950s, 10–38 stem rust races were detected annually in the US, whereas from the 1960s to the present time, fewer than 10 races were usually detected (Groth and Roelfs 1987) (Fig. 5.3). The total amount of wheat stem rust inoculum has been reduced since the 1960s due to the increased use of highly resistant winter and spring wheat cultivars and of winter wheat cultivars with shorter maturity, thus reducing late onsets of stem rust in the southern Great Plains.

In 2006 a single race designated as QFCS accounted for 25 of the 27 total collections, along with single collections of race MCCD (= race 56) and race TTTT (Long et al., 2007). Race QFCS is widely avirulent on many of the *Sr* genes commonly present in spring and winter wheat in the US. The combination of smaller population size and the lack of sexual recombination has stabilized selection of races in the wheat stem rust population and thus has made the resistance of the *Sr* genes much more effective and durable. The use of highly resistant wheat cultivars in Australia (Park 2007) has reduced the number of stem rust races and the overall levels of inoculum.

Roelfs and Groth (1980) determined in 1975 that there were six distinct groups of *P. graminis* f. sp. *tritici* races in North America. Races in each

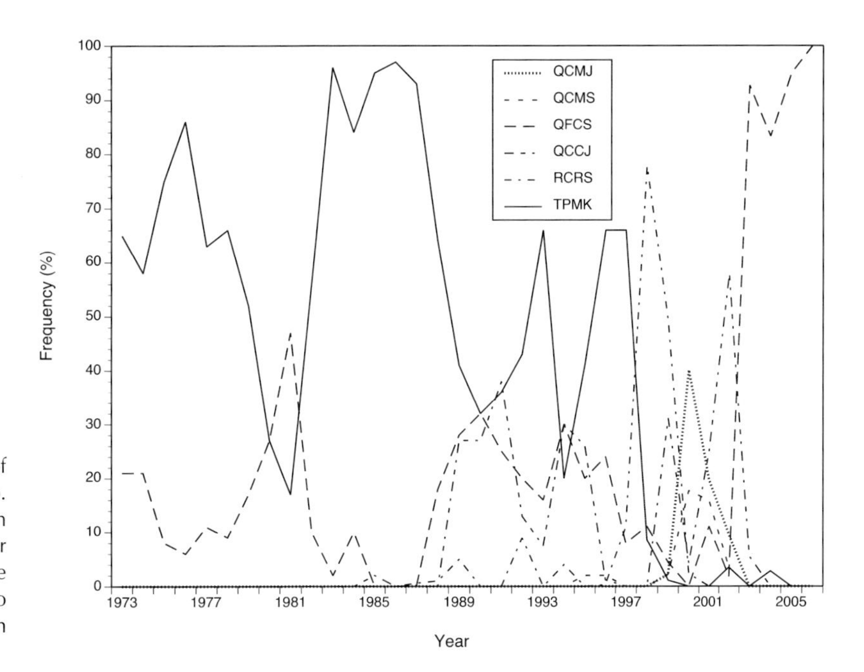

Fig. 5.3 Frequency (%) of races of *Puccinia graminis* f. sp. *tritici* collected from wheat in the US Great Plains. The four letter code race nomenclature is based on infection types to single-gene differentials in Roelfs and Martens (1988).

group were highly related for virulence yet were very distinct from other clusters. Burdon and Roelfs (1985) showed that isozyme variation in wheat stem rust populations was completely correlated with the race groupings. The distinct groups of wheat stem rust races that were described in 1975 in North America were likely ancestral groups of races that existed before barberry eradication. Removal of the sexual cycle would have prevented any further genetic exchange between the different groups of races.

In 1999 a new stem rust race initially designated as Ug-99 was found in Uganda (Pretorius et al., 2000). This race was notable since it was the first to have virulence to gene *Sr31*. Race Ug-99 (designated as TTKS on wheat stem rust differentials) was also virulent to gene *Sr38*. Gene *Sr31* is found in about 30% of CIMMYT germplasm, and *Sr38* is present in European, Australian, and a few CIMMYT wheats (Singh et al., 2006). In addition to these virulences, Ug-99 is virulent to a number of other *Sr* genes present in US spring and winter wheat (Jin and Singh 2006). As of 2007, Ug-99 was found in Uganda, Kenya, Sudan, Ethiopia, Yemen, and Iran. This stem rust race has the potential to spread through the Middle East and then to India and Pakistan, where it could cause devastating grain yield losses in wheat. Since it is widely virulent to many wheat cultivars in this region, it will be imperative to find sources of resistance and incorporate these into current wheat breeding programs.

Stem rust resistance in wheat

In the early part of the 20th century, wheat breeding in the spring wheat region of the northern Great Plains was largely a struggle against stem rust, the most important disease of wheat at the time. Marquis was the most widely grown spring wheat cultivar in the north central US and in Canada. Marquis, developed by the Canada Department of Agriculture in Ottawa and released in 1907, had early maturity and was popular with the milling industry since it was the first high-quality bread wheat developed in North America. However, Marquis and the other bread wheat cultivars were very susceptible to both stem rust and leaf rust. The stem rust epidemics of the early 1900s caused such losses that durum wheat, which was generally resistant to stem rust, had replaced bread wheat in Minnesota. Efforts were thus made to develop bread wheat and durum wheat cultivars that were resistant to stem rust.

The first hard red spring wheat bred for stem rust resistance in North America was 'Ceres', released by the North Dakota Agricultural Experiment Station in 1926 and grown in both the US and Canada. Ceres was developed by crossing the stem rust resistant wheat cultivar Kota with Marquis, and selecting stem rust resistant progeny. However, stem rust race 56, with virulence to the resistance in Kota and Ceres, increased throughout the Great Plains region after the release of Ceres. Widespread epidemics of stem rust on Ceres occurred from 1935 to 1947 due to the increased presence of race 56.

Ceres was largely replaced by Thatcher wheat, developed by the Minnesota Agricultural Experiment Station and released in 1935. Thatcher was developed from a cross between two lines, derived from Kanred/Marquis and from Iumillo durum/Marquis, which combined the good quality characteristics of Marquis with stem rust resistance in Kanred (*Sr5*, *Sr16*) and Iumillo durum (*Sr9g* and *Sr12*) (Kolmer 2001b). The Thatcher resistance of *Sr5*, *Sr9g*, *Sr12*, and *Sr16* was highly effective against race 56 and the other stem rust races. Many cultivars subsequently released in Canada and the US were derived from Thatcher (Kolmer et al., 1991). None of the known race-specific genes in Thatcher could account for the adult-plant field resistance in Thatcher, which became apparent with the increase of stem rust races with virulence to the seedling genes in Thatcher. The adult-plant resistance in Thatcher appeared to be nonspecific and durable (Kolmer et al., 1991). Genetic analyses indicated that stem rust resistance in Thatcher was complex. At least two genes that condition resistance in adult plants were likely derived from Iumillo durum and were independent of the race-specific genes (Nazareno and Roelfs 1981; Gavin-Vanegas et al., 2007).

Other undesignated race-specific seedling resistance genes were also detected in Thatcher (Nazareno and Roelfs 1981; Knott 2000).

Another source of stem rust resistance in spring wheat during that period was the adult-plant resistance gene *Sr2*, derived from Yaroslav emmer (McIntosh et al., 1995). The stem rust resistant cultivar Hope and the breeding line H-44 were developed by selecting progeny from a cross between Marquis and Yaroslav emmer. Breeding line H-44 was subsequently used as a parent in breeding programs in Canada. The resistance conditioned by *Sr2* is characterized by fewer uredinia and lower overall severity of stem rust at maturity. The cultivar Renown released in 1937 was the first stem rust resistant cultivar with *Sr2* to be released in Canada. Cultivars with *Sr2* were resistant to race 56 and the other stem rust races then present. Gene *Sr2* is present in wheat cultivars in Australia and the CIMMYT program (McIntosh et al., 1995) and in Canadian cultivars (Liu and Kolmer 1998a). Gene *Sr2* is present in some hard red winter and soft red winter cultivars in the US (Roelfs 1988) since Hope was used as a parent in these programs.

From 1950 to 1954 race 15B of wheat stem rust was widespread in the Great Plains region. This race caused high stem rust severities on spring wheats with *Sr2* and the Thatcher stem rust resistance. In 1954 yield losses in spring wheat were over 40% in North Dakota and reached 18% in Minnesota. In response to race 15B, the cultivar Selkirk with genes *Sr2*, *Sr6*, *Sr7b*, *Sr9d*, *Sr17*, and *Sr23* was highly resistant to race 15B and was released in 1954 by the Canada Department of Agriculture. Selkirk was widely grown throughout the spring wheat region of the US and Canada through the 1960s. The *Sr6* gene was particularly effective against race 15B and was actively selected in spring and winter wheat breeding programs in the central and northern Great Plains. The gene is also present with high frequency in CIMMYT germplasm. The widespread use of *Sr6* likely contributed to the rapid decline of race 15B in North America in the late 1950s. Wheat cultivars from Kenya also had *Sr6*. The cultivar Chris was released by the Minnesota Agricultural Experiment Station in 1966 and was the first US cultivar to have *Sr6*.

The resistance derived from Hope, Thatcher, and *Sr6* was the main component of stem rust resistance in the spring wheat region from the 1930s through the 1970s. However, the frequency of *Sr2* and of the Thatcher adult plant resistance has apparently declined in current cultivars given the following observations. Thatcher was moderately resistant to race TTKS (Ug-99) (Pretorius et al., 2000) in field tests in Kenya where race TTKS has been predominant. The Thatcher stem rust resistance is also expressed in seedling plants, but seedling tests with race TTKS failed to detect the Thatcher type of resistance in spring wheat cultivars from the northern Great Plains released between 1996 and 2005 (Jin and Singh 2006). Stem rust race TPMK, which became common after the decline of race 15B (Fig. 5.3), produces moderate to high infection types on seedling and adult plants of Thatcher. Selection of stem rust resistant germplasm in breeding programs based on resistance to TPMK could have placed higher selection pressure on the low infection types conditioned by race-specific genes such as *Sr6* and *Sr9b* than on the nonspecific resistance derived from Thatcher. Resistance due to *Sr2* also appeared to be absent in these cultivars. The absence of *Sr2* in spring wheat was likely associated with intense breeding for Fusarium head blight (caused by *Fusarium graminearum*) resistance, for which a major QTL is located on chromosome 3BS (Anderson et al., 2001) in close repulsion linkage with *Sr2* (Spielmeyer et al., 2003).

High levels of stem rust resistance in hard red winter wheat, in combination with early maturity, have had an impact in reducing the *P. graminis* population size in North America. The stem rust resistance gene *SrTmp*, derived from 'Triumph 64', was likely present in some of the initial hard red winter wheat germplasm (Roelfs and McVey 1979). This gene was effective against the majority of stem rust races in North America, but race 56 and race 15B were virulent to *SrTmp*. The resistance conditioned by *SrTmp*, combined with the early maturity of the Triumph background, may have contributed to reducing the amount of stem rust present in winter wheat. The *SrTmp* gene is present in several current cultivars (Jin and Singh 2006) and in nearly 10% of breeding lines in the hard red winter wheat region. Other major components of stem rust resistance in hard red winter wheat are *Sr24* and resistance derived from the 1RS.1AL wheat–rye translocation. Originally derived from *Thinopyrum ponticum*, *Sr24* has been common in hard red winter wheat cultivars since the release of Agent in 1967, and to a lesser degree in soft red winter and spring wheat cultivars. Stem rust races with virulence on *Sr24* have not been detected in North America, and the gene is currently present in nearly 50% of the current hard red winter wheat breeding lines and cultivars. Because *Sr24* is tightly linked with *Lr24* (McIntosh et al., 1995), selection for leaf rust resistance produced lines with stem rust resistance. The 1RS.1AL translocation in 'Amigo', with the rye chromosome introduced from 'Insave F.A.' rye via a triticale with greenbug resistance (Sebesta et al., 1994), gives effective resistance to stem rust races in North America as well as to race TTKS.

Derived from Petkus rye, *Sr31* is on the 1RS.1BL wheat–rye translocation with *Lr26* and *Yr9* (McIntosh et al., 1995). Gene *Sr31* is present in some current hard red winter wheat cultivars and in a few soft red winter wheat cultivars in the US, and is present in CIMMYT-derived wheat cultivars that are grown worldwide. This gene provided a very high level of resistance to all known stem rust races prior to the emergence of TTKS in Uganda in 1999. Derived from *T. timopheevi*, *Sr36* is the most common stem rust resistance gene in soft red winter wheat in the US and has been an important source of resistance in Australia (McIntosh et al., 1995). In addition, *Sr36* conditions resistance to the current stem rust races in the US as well as race TTKS (Jin and Singh 2006).

Genes *Sr9b*, *Sr11*, and *Sr17* are also present in spring and winter wheat in North America. Gene *Sr9b* is closely linked with *Lr13* and is present in several spring wheat lines that were selected for leaf rust resistance from Frontana. Virulence to *Sr9b* is common in stem rust races in North America. Virulence to *Sr11* is present in North America and Australia. The origin of *Sr17* is assumed to be Yaroslav emmer; thus it may also be present in wheats with *Sr2*, though many stem rust races are virulent to this gene.

In Australia wheat cultivars with *Sr24*, *Sr26*, *Sr30*, *Sr36*, and *Sr38* have been released in the past 40 years (Park 2007). Gene *Sr26* is derived from *Th. ponticum*, is present only in Australian cultivars, and stem rust races with virulence to this gene have not been detected. Gene *Sr30 is* present in a number of Australian cultivars and also CIMMYT germplasm (McIntosh et al., 1995). Gene *Sr38*, derived from *T. ventricosa*, is linked with *Lr37* and *Yr17* and was selected in Australian germplasm based on resistance to all three rusts. Currently gene designations up to *Sr46* have been given for stem rust resistance genes in wheat (McIntosh et al., 2007).

FUTURE PERSPECTIVES

The emergence of new races of wheat leaf rust, wheat stripe rust, and wheat stem rust continually challenge wheat breeders and plant pathologists to develop effective sources of durable resistance to these pathogens. The eventual cloning and sequencing of genes such as *Lr34*/*Yr18* and *Sr2* will provide greater insight into how these non-specific resistance genes function, thus offering the potential of designing new resistance genes that will offer greater resistance durability. These genes will likely differ for functional domains and specificity in comparison with the race-specific NBS-LRR resistance genes.

The discovery that the herbicide glyphosate (Anderson and Kolmer 2005) greatly reduces rust infections in glyphosate-resistant wheat is an exciting development that could be utilized immediately to reduce losses in wheat due to rusts, and it may lead to other transgenic strategies to design rust resistance genes or genes for tolerance or resistance to chemical applications in wheat. The characterization of avirulence genes in flax rust (*Melampsora lini*) and the corresponding resistance genes in flax (*Linum usitatissimum*), and an understanding of how their coded functional proteins interact (Dodds et al., 2007), will lead to greater understanding and potential exploitation of gene-for-gene specificity in the wheat rusts. Of course any of these strategies will depend on the eventual governmental, regulatory, and consumer acceptance of transgenic wheat.

In the meantime the frequency and severity of rust epidemics on wheat can be reduced by greater utilization of resources that are already available. Diagnostic markers for genes such as *Lr34/Yr18* should greatly simplify selection of these genes, potentially allowing genes with durable rust resistance to become fixed in wheat germplasm. Increased planting of wheat cultivars with high levels of leaf rust resistance in the US, and removal of susceptible cultivars, will reduce the size of the *P. triticina* population that regularly overwinters, thus reducing the chance of mutation for increased virulence to newly deployed resistance genes. Australia has adopted minimum disease standards for release of wheat cultivars (Wallwork 2007), with the specific goal to reduce the amount of rust inoculum and thus prolong the effective life span of a rust resistance gene by reducing the chances of new virulent races emerging. The effectiveness of this approach can be seen in comparing the effective life span of *Lr24* in the US versus Australia. In the US, virulence to *Lr24* appeared almost immediately after the release of cultivars with this gene, while in Australia races of *P. triticina* with virulence to *Lr24* did not appear until 17 years after cultivars with this gene were first grown (Park et al., 2002). Genes *Lr1* and *Lr13* have also been much more effective in Australia compared with the US due to the greater use of highly resistant cultivars and the reduced size of the *P. triticina* population.

For the stem rust race Ug-99, the challenge will be to incorporate *Sr2* into wheat germplasm in regions where this race poses an immediate threat. This gene was common in older CIMMYT wheat cultivars such as Pavon 76, but more recent germplasm releases appear to lack *Sr2*. Although Ug-99 is virulent to many genes that were originally derived from hexaploid wheat, several genes derived are from wild wheat relatives that condition resistance to this race. It should be feasible to add these genes into wheat germplasm by using tightly linked molecular markers and testing with avirulent stem rust races.

The reproductive capacity, long-distance aerial dispersal, and high degree of genetic variation in the wheat rust fungi almost certainly ensure that new races with virulence to important and widely used resistance genes will continue to arise and pose a threat to wheat production. Ultimately losses due to these pathogens can be avoided only by continued vigilance by wheat breeders and plant pathologists.

REFERENCES

Agrios, G.N. 1997. Plant pathology. 4th ed. Academic Press, Orlando, FL.

Allan, R.E., C.J. Peterson, R.F. Line, G.L. Rubenthaler, and C.F. Morris. 1993. Registration of 'Rely' wheat multiline. Crop Sci. 33:213–214.

Allan, R.E., L.H. Purdy, and O.A. Vogel. 1966. Inheritance of seedling and adult reaction of wheat to stripe rust. Crop Sci. 6:242–245.

Allen, R.E. 1928. A cytological study of *Puccinia glumarum* on *Bromus marginatus* and *Triticum vulgare*. J. Agric. Res. 36:487–513.

Anderson, J.A., and J.A. Kolmer. 2005. Rust control in glyphosate tolerant wheat following application of the herbicide glyphosate. Plant Dis. 89:1136–1142.

Anderson, J.A., R.W. Stack, S. Liu, B.L. Waldron, A.D. Field, C. Coyne, B. Moreno-Sevilla, J.M. Fetch, Q.J. Song, P.B. Cregan, and R.C. Frohberg. 2001. DNA markers for Fusarium head blight resistance QTLs in two wheat populations. Theor. Appl. Genet. 102:1164–1168.

Anikster, Y., W.R. Bushnell, T. Eilam, J. Manisterski, and A.P. Roelfs. 1997. *Puccinia recondita* causing leaf rust on cultivated wheats, wild wheats, and rye. Can. J. Bot. 75:2082–2095.

Barcellos, A.L., A.P. Roelfs, and M.I.B. de Moraes-Fernandes. 2000. Inheritance of adult plant leaf rust resistance in the Brazilian wheat cultivar Toropi. Plant Dis. 84:90–93.

Bariana, H.S., G.N. Brown, U.K. Bansal, H. Miah, G.E. Standen, and M. Lu. 2007. Breeding triple rust resistance for Australia using conventional and marker-assisted selection technologies. Aust. J. Agric. Res. 58:576–587.

Biffen, R.H. 1905. Mendel's law of inheritance of disease resistance. J. Agric. Sci. 4:421–429.

Bossolini, E., S.G. Krattinger, and B. Keller. 2006. Development of simple sequence repeat markers specific for the *Lr34* resistance region of wheat using sequence information from rice and *Aegilops tauschii*. Theor. Appl. Genet. 113:1049–1062.

Boukhatem, N., P.V. Baret, D. Mingeot, and J.M. Jacquemin. 2002. Quantitative trait loci for resistance against yellow rust in two wheat-derived recombinant inbred line populations. Theor. Appl. Genet. 104:111–118.

Britton, M., and G.B. Cummins. 1956. The reaction of species of *Poa* and grasses to *Puccinia striiformis*. Plant Dis. Rep. 40:643–645.

Burdon, J.J., and A.P. Roelfs. 1985. Isozyme and virulence variation in asexually reproducing populations of *Puccinia graminis* and *P. recondita* on wheat. Phytopathology 75:907–913.

Caldwell, R.M. 1968. Breeding for general and/or specific plant disease resistance. p. 263–272. *In* K.W. Finlay and K.W. Shepard (ed.) Proc. Int. Wheat Genet. Symp., 3rd, Canberra, Australia. 5–9 Aug. 1968. Australian Academy of Science, Canberra, Australia.

Caldwell, R.M., H.R. Kraybill, J.T. Sullivan, and L.E. Compton. 1934. Effect of leaf rust (*Puccinia triticina*) on yield, physical characters, and composition of winter wheats. J. Agric. Res. 48:1049–1071.

Campbell, C.L., and D.L. Long. 2001. The campaign to eradicate the common barberry in the United States. p. 16–50. *In* P.D. Peterson (ed.) Stem rust of wheat, from ancient enemy to modern foe. APS Press, St. Paul, MN.

Carleton, M.A. 1915. A serious new wheat rust in this country. Science 42:58–59.

Chen, X.M. 2005. Epidemiology and control of stripe rust on wheat. Can. J. Plant Pathol. 27:314–337.

Chen, X.M. 2007. Challenges and solutions for stripe rust control in the United States. Aust. J. Agric. Res. 58: 648–655.

Chen, X.M., and F. Lin. 2007. Identification and molecular mapping of genes for all-stage and high-temperature adult-plant resistance to stripe rust in 'Express' wheat. Phytopathology 97:S21.

Chen, X.M., and R.F. Line. 1995a. Gene action in wheat cultivars for durable high-temperature adult-plant resistance and interactions with race-specific, seedling resistance to stripe rust caused by *Puccinia striiformis*. Phytopathology 85:567–572.

Chen, X.M., and R.F. Line. 1995b. Gene number and heritability of wheat cultivars with durable, high-temperature, adult-plant resistance and race-specific resistance to *Puccinia striiformis*. Phytopathology 85:573–578.

Chen, X.M., R.F. Line, and H. Leung. 1993. Relationship between virulence variation and DNA polymorphism in *Puccinia striiformis*. Phytopathology 83:1489–1497.

Chen, X.M., R.F. Line, and H. Leung. 1995. Virulence and polymorphic DNA relationships of *Puccinia striiformis* f. sp. *hordei* to other rusts. Phytopathology 85:1335–1342.

Chen, X.M., R.F. Line, Z.X. Shi, and H. Leung. 1998. Genetics of wheat resistance to stripe rust. p. 237–239. *In* R.A. McIntosh, G.E. Hart, K.M. Devos, M.D. Gale, and W.J. Rogers (ed.) Proc. Int. Wheat Genet. Symp., 9th, Vol. 3, Saskatoon, Saskatchewan. 2–7 Aug. 1998. Univ. Saskatchewan, Saskatoon, Canada.

Chen, X.M., M.K. Moore, E.A. Milus, D.L. Long, R.F. Line, D. Marshall, and L. Jackson. 2002. Wheat stripe rust epidemics and races of *Puccinia striiformis* f. sp. *tritici* in the United States in 2000. Plant Dis. 86:39–46.

Chen, X.M., L.N. Penman, and K.L. Richardson. 2007. Races of *Puccinia striiformis*, the stripe rust pathogen in the United States in 2006. Phytopathology 97:S21.

Chen, X.M., and J. Zhao. 2007. Identification of molecular markers for *Yr8* and a gene for high-temperature, adult-plant resistance against stripe rust in the AVS/6*Yr8 wheat line. Phytopathology 97:S21.

Chester, K.S. 1939. The 1938 wheat leaf-rust epiphytotic in Oklahoma. Plant Dis. Rep. 112:1–19.

Chester, K.S. 1946. The nature and prevention of the cereal rusts as exemplified in the leaf rust of wheat. Chronica Botanica, Waltham, MA.

Chhuneja, P., S. Kaur, T. Garg, M. Ghai, S. Kaur, M. Prashar, R.K. Goel, B. Keller, H.S. Dhaliwal, and K. Singh. 2007. Mapping of adult-plant stripe rust resistance genes in diploid A genome wheat species and their transfer to bread wheat. Theor. Appl. Genet. 116:313–324.

Cloutier, S., B.D. McCallum, C. Loutre, T.W. Banks, T. Wicker, C. Feulliet, B. Keller, and M.C. Jordan. 2007. Leaf rust resistance gene *Lr1*, isolated from bread wheat (*Triticum aestivum* L.) is a member of the large psr567 family. Plant Mol. Biol. 65:93–106.

Coakley, S.M., W.S. Boyd, and R.F. Line. 1982. Statistical models for predicting stripe rust on winter wheat in the Pacific Northwest. Phytopathology 72:1539–1542.

Coakley, S.M., W.S. Boyd, and R.F. Line. 1984. Development of regional models that use meteorological variables for predicting stripe rust disease in winter wheat. J. Clim. Appl. Meteorol. 23:1234–1240.

Cummins, G.B., and R.M. Caldwell. 1956. The validity of binomials in the leaf rust complex of cereals and grasses. Phytopathology 46:81–82.

Dodds, P.N., A.M. Catanzariti, G. Lawrence, and J.G. Ellis. 2007. Avirulence proteins of rust fungi: Penetrating the host–haustorium barrier. Aust. J. Agric. Res. 58:521–517.

d'Oliveira, B.D., and D.J. Samborski. 1966. Aecial stage of *Puccinia recondita* on Ranunculaceae and Boraginaceae in Portugal. p. 133–154. *In* R.C.F. Macer and M.S. Wolfe (ed.) Proc. Eur. Brown Rust Conf., 1st, Cereal Rust Conferences, Cambridge, UK. 29 June–4 July, 1964. Plant Breeding Inst., Cambridge, UK.

Duan, X., J. Enjalbert, D. Vautrin, C. Solignac, and T. Giraud. 2003. Isolation of 12 microsatellite loci, using an enrichment protocol, in the phytopathogenic fungus *Puccinia triticina*. Mol. Ecol. Notes 3:65–67.

Dyck, P.L. 1987. The association of a gene for leaf rust resistance with the chromosome 7D suppressor of stem rust resistance in common wheat. Genome 29:467–469.

Dyck, P.L. 1993. Inheritance of leaf rust and stem rust resistance in 'Roblin' wheat. Genome 36:289–293.

Dyck, P.L. 1994. The transfer of leaf rust resistance from *Triticum turgidum* ssp. *dicoccoides* to hexaploid wheat. Can. J. Plant Sci. 74:671–673.

Dyck, P.L., and R. Johnson. 1983. Temperature sensitivity of genes for resistance in wheat to *Puccinia recondita*. Can. J. Plant Pathol. 5:229–234.

Dyck, P.L., and O.M. Lukow. 1988. The genetic analysis of two interspecific sources of leaf rust resistance and their effect on the quality of common wheat. Can. J. Plant Sci. 68:633–639.

Dyck, P.L., and D.J. Samborski. 1982. The inheritance of resistance to *Puccinia recondita* in a group of common wheat cultivars. Can. J. Genet. Cytol. 24:273–283.

Dyck, P.L., D.J. Samborski, and R.G. Anderson. 1966. Inheritance of adult-plant leaf rust resistance derived from the common wheat varieties Exchange and Frontana. Can. J. Genet. Cytol. 8:665–671.

Enjalbert, J., X. Duan, M. Leconte, M.S. Hovmoller, and C. de Vallavielle-Pope. 2005. Genetic evidence of local adaptation of wheat yellow rust (*Puccinia striiformis* f. sp. *tritici*) within France. Mol. Ecol. 14:2065–2073.

Eriksson, J. 1894. Über die Spezialisierung des Parasitismus bei den Getreiderostpilzen. Ber. Dtsch. Bot. Ges. 12:292–331.

Eriksson, J. 1899. Nouvelles études sur la rouille brune des céréales. Ann. Sci. Nat. Bot. Eighth Ser. 9:241–288.

Eriksson, J., and E. Henning. 1894. Die Hauptresultate einer neuen Untersuchung über die Getreiderostpilze. Z. Pflanzenkr. 4:197–203.

Eriksson, J., and E. Henning. 1896. Die Getreideroste. Norstedt & Söner, Stockholm, Sweden.

Everts, K.L., S. Leath, and P.L. Finney. 2001. Impact of powdery mildew and leaf rust on milling and baking quality of soft red winter wheat. Plant Dis. 85:423–429.

Ezzahiri, B., S. Diouri, and A.P. Roelfs. 1992. *Anchusa italica* as an alternate host for wheat leaf rust in Morocco. Plant Dis. 76:1185.

Ezzahiri, B., and A.P. Roelfs. 1989. Inheritance and expression of adult plant resistance to leaf rust in Era wheat. Plant Dis. 73:549–551.

Feuillet, C., S. Travella, N. Stein, L. Albar, L. Nublat, and B. Keller. 2003. Map-based isolation of the leaf rust disease resistance gene *Lr10* from the hexaploid wheat (*Triticum aestivum* L.) genome. Proc. Natl. Acad. Sci. USA 100:15253–15258.

Finckh, M.R., and C.C. Mundt. 1992. Stripe rust, yield, and plant competition in wheat cultivar mixtures. Phytopathology 82:82–92.

Fuckel, L. 1860. Enumeratio fungorum nassoviae. Jahrb. Ver. Naturkd. Herzogthum Nassau 15:9.

Gavin-Vanegas, C.D., D.F. Garvin, and J.A. Kolmer. 2007. Genetics of stem rust resistance in the spring wheat cultivar Thatcher and the enhancement of stem rust resistance by *Lr34*. Euphytica 159:391–401.

German, S.E., A. Barcellos, M. Chaves, M. Kohli, P. Campos, and L. de Viedma. 2007. The situation of common wheat rusts in the Southern Cone of America and perspectives for control. Aust. J. Agric. Res. 58:620–630.

German, S.E., and J.A. Kolmer. 1992. Effect of gene *Lr34* in the enhancement of resistance to leaf rust of wheat. Theor. Appl. Genet. 84:97–105.

Goyeau, H., F. Halkett, M.F. Zapater, J. Carlier, and C. Lannou. 2007. Clonality and host selection in the wheat pathogenic fungus *Puccinia triticina*. Fungal Genet. Biol. 44:474–483.

Goyeau, H., R. Park, B. Schaeffer, and C. Lannou. 2006. Distribution of pathotypes with regard to host cultivars and French wheat leaf rust populations. Phytopathology 96:264–273.

Green, G.J. 1964. A color mutation, its inheritance and the inheritance of pathogenicity in *Puccinia graminis* Pers. Can. J. Bot. 42:1643–1664.

Groth, J.V., and A.P. Roelfs. 1987. The concept and measurement of phenotypic diversity in *Puccinia graminis* on wheat. Phytopathology 77:1395–1399.

Harder, D.E. 1984. Developmental ultrastructure of hyphae and spores. p. 333–374. *In* W.R. Bushnell and A.P. Roelfs (ed.) The cereal rusts. Vol. I. Academic Press, Orlando, FL.

Hart, H., and H. Becker. 1939. Beiträge zur Frage des Zwischenwirtes für *Puccinia glumarum*. Z. Pflanzenkr. (Pflanzenpathol.) Pflanzenschutz 49:559–566.

Hassebrauk, K. 1965. Nomenklatur, geographische Verbreitung und Wirtsbereich des Gelbrostes, *Puccinia striiformis* West. Mitt. Biol. Bundesanst. Land-Forstwirtsch. Berlin-Dahlem 116:1–75.

Hassebrauk, K. 1970. Der Gelbrost *Puccinia striiformis* West. 2 Befallsbid. Morphologie und Biologie der Sporen. Infektion und weitere Entwicklung. Wirkungen auf die Wirtsflanze. Mitt. Biol. Bundesanst. Land-Forstwirtsch., Berlin-Dahlem 139:1–110.

Hayes, H.K., E.R. Ausemus, E.C. Stakman, C.H. Bailey, H.K. Wilson, R.H. Bamberg, M.C. Markley, R.R. Crim, and M.N. Levine. 1936. Thatcher wheat. Bull. 325. Minn. Agric. Res. Stn., St. Paul, MN.

Hendrix, J.W. 1994. Stripe rust, what it is and what to do about it. Wash. State Univ. Stn. Circ. 424:1–6.

Herrera-Foessel, S.A., R.P. Singh, J. Huerta-Espino, J. Crossa, J. Yuen, and A. Djurle. 2006. Effect of leaf rust on grain yield and yield traits of durum wheats with race-specific and slow-rusting resistance to leaf rust. Plant Dis. 90:1065–1072.

Herrera-Foessel, S.A., R.P. Singh, J. Huerta-Espino, J. William, G.M. Rosewarne, A. Djurle, and J. Yuen. 2007. Identification and mapping of *Lr3* and a linked leaf rust resistance gene in durum wheat. Crop Sci. 47:1459–1466.

Herrera-Foessel, S.A., R.P. Singh, J. Huerta-Espino, J. Yuen, and A. Djurle. 2005. New genes for leaf rust resistance in CIMMYT durum wheats. Plant Dis. 89:809–814.

Hovmoller, M.S., A.F. Justesen, and J.K.M. Brown. 2002. Clonality and long-distance migration of *Puccinia striiformis* f. sp. *tritici* in north-west Europe. Plant Pathol. 51:24–32.

Huang, L., S.A. Brooks, W. Li, J.P. Fellers, H.N. Trick, and B.S. Gill. 2003. Map-based cloning of leaf rust resistance gene *Lr21* from the large and polyploid genome of wheat. Genetics 164:655–664.

Humphrey, H.B., C.W. Hungerford, and A.G. Johnson. 1924. Stripe rust (*Puccinia glumarum*) of cereals and grasses in the United States. J. Agric. Res. 29:209–227.

Hungerford, C.W. 1923. Studies on the life history of stripe rust, *Puccinia glumarum*. J. Agric. Res. 24:607–620.

Hylander, N., I. Jørstad, and J.A. Nannfeldt. 1953. Enumeratio uredionearum Scandinavicarum. Opera Bot. 1:1–102.

Jackson, H.S., and E.B. Mains. 1921. Aecial stage of the orange leaf rust of wheat, *Puccinia triticina* Eriks. J. Agric. Res. 22:151–172.

Jin, Y., and R.P. Singh. 2006. Resistance in U.S. wheat to recent eastern African isolates of *Puccinia graminis* f. sp. *tritici* with virulence to resistance gene *Sr31*. Plant Dis. 476–480.

Johnson, T. 1956. Physiologic races of leaf rust of wheat in Canada 1931 to 1955. Can. J. Agric. Sci. 36:323–332.

Johnson, R. 1983. Genetic background of durable resistance. p. 5–26. *In* F. Lamberti, J.M. Waller, and N.A. Van der Graaff (ed.) Durable resistance in crops. Plenum Press, New York.

Johnston, C.O. 1931. Effect of leaf rust infection on yield of certain varieties of wheat. J. Am. Soc. Agron. 23: 1–12.

Johnston, C.O., R.M. Caldwell, L.E. Compton, and L.E. Browder. 1968. Physiologic races of *Puccinia recondita* f. sp. *tritici* in the United States from 1926 through 1960. USDA Tech. Bull. 1393.

Kerber, E.R., and T. Aung. 1999. Leaf rust resistance gene *Lr34* associated with nonsuppression of stem rust in the wheat cultivar Canthatch. Phytopathology 89:518–521.

Kerber, E.R. and G.J. Green. 1980. Suppression of stem rust resistance in the hexaploid wheat cv. Canthatch by chromosome 7DL. Can. J. Bot. 12:1347–1350.

Khan, M.A., L.E. Trevathtan, and J. Robbins. 1997. Quantitative relationship between leaf rust and yield in Mississippi. Plant. Dis. 81:769–772.

Knott, D.R. 2000. The inheritance of stem rust resistance in Thatcher wheat. Can. J. Plant Sci. 80:53–63.

Kolmer, J.A. 1992. Gametic phase disequilibria in two populations of *Puccinia recondita* (wheat leaf rust fungus). Heredity 68:505–513.

Kolmer, J.A. 1998. Physiologic specialization of *Puccinia recondita* f. sp. *tritici* in Canada in 1996. Can. J. Plant Pathol. 20:176–181.

Kolmer, J.A. 2001a. Molecular polymorphism and virulence phenotypes of the wheat leaf rust fungus *Puccinia triticina* in Canada. Can. J. Bot. 79:917–926.

Kolmer, J.A. 2001b. Early research on the genetics of *Puccinia graminis* and stem rust resistance in wheat in Canada and the United States. p. 51–82. *In* P.D. Peterson (ed.) Stem rust of wheat: From ancient enemy to modern foe. APS Press, St. Paul, MN.

Kolmer, J.A. 2003. Postulation of leaf rust resistance genes in selected soft red winter wheats. Crop Sci. 43:1266–1274.

Kolmer, J.A., and P.L. Dyck. 1994. Gene expression in the *Triticum aestivum–Puccinia recondita* f. sp. *tritici* gene-for-gene system. Phytopathology 84:437–440.

Kolmer, J.A., P.L. Dyck, and A.P. Roelfs. 1991. An appraisal of stem and leaf rust resistance in North American hard red spring wheats and the probability of multiple mutations in populations of cereal rust fungi. Phytopathology 81:237–239.

Kolmer, J.A., Y. Jin, and D.L. Long. 2007a. Wheat leaf and stem rust in the United States. Aust. J. Agric. Res. 58:631–638.

Kolmer, J.A., and J.Q. Liu. 2000. Virulence and molecular polymorphism in international collections of the wheat leaf rust fungus *Puccinia triticina*. Phytopathology 90:427–436.

Kolmer, J.A., J.Q. Liu, and M. Sies. 1995. Virulence and molecular polymorphism in *Puccinia recondita* f. sp. *tritici* in Canada. Phytopathology 85:276–285.

Kolmer, J.A., D.L. Long, and M.E. Hughes. 2007b. Physiological specialization of *Puccinia triticina* on wheat in the United States in 2005. Plant Dis. 91:979–984.

Kolmer, J.A., D.L. Long, E. Kosman, and M.E. Hughes. 2003. Physiologic specialization of *Puccinia triticina* on wheat in the United States in 2001. Plant Dis. 87:859–866.

Kolmer, J.A., and L.M. Oelke. 2006. Genetics of leaf rust resistance in the spring wheats 'Ivan' and 'Knudson'. Can. J. Plant Pathol. 28:223–229.

Kolmer, J.A., L.M. Oelke, and J.Q. Liu. 2007c. Genetics of leaf rust resistance in three Americano landrace-derived wheat cultivars from Uruguay. Plant Breed. 126:152–157.

Kolmer, J.A., and M.E. Ordoñez. 2007. Genetic differentiation of *Puccinia triticina* populations in Central Asia and the Caucasus. Phytopathology 97:1141–1149.

Kolmer, J.A., R.P. Singh, D.F. Garvin, L. Viccars, H.M. William, J. Huerta-Espino, F.C. Ogbonnaya, H. Raman, S. Orford, H.S. Bariana, E.S. Lagudah. 2008. Analysis of the *Lr34*/*Yr18* rust resistance region in wheat germplasm. Crop Sci. 48:1841–1852.

Kuraparthy, V., P. Chhuneja, H.S. Dhaliwal, S. Kaur, R.L. Bowden, and B.S. Gill. 2007. Characterization and mapping of cryptic alien introgression from *Aegilops geniculata* with new leaf rust and stripe rust resistance genes *Lr57* and *Yr40* in wheat. Theor. Appl. Genet. 114:1379–1389.

Lagudah, E.S., H. McFadden, R.P. Singh, J. Huerta-Espino, H.S. Bariana, and W. Spielmeyer. 2006. Molecular characterization of the *Lr34*/*Yr18* slow rusting resistance gene region in wheat. Theor. Appl. Genet. 114:21–30.

Leonard, K.J. 2001. Stem rust—future enemy? p. 119–146. *In* P.D. Peterson (ed.) Stem rust of wheat: From ancient enemy to modern foe. APS Press, St. Paul, MN.

Leppik, E. 1967. Some viewpoints on the phylogeny of rust fungi: IV. Biogenic radiation. Mycologia 59:568–579.

Lin, F., and X.M. Chen. 2007. Genetics and molecular mapping of genes for race-specific all-stage resistance and non–race specific high-temperature adult-plant resistance to stripe rust in spring wheat cultivar Alpowa. Theor. Appl. Genet. 114:1277–1287.

Line, R.F. 1972. Recording and processing data on foliar diseases of cereals. p. 175–178. *In* Proc. Eur. Mediterranean Cereal Rusts Conf. Faculty of Science, Charles University, Prague, Czechoslovakia.

Line, R.F. 1976. Factors contributing to an epidemic of stripe rust on wheat in the Sacramento Valley of California in 1974. Plant Dis. Rep. 60:312–316.

Line, R.F. 2002. Stripe rust of wheat and barley in North America: A retrospective historical review. Annu. Rev. Phytopathol. 40:75–118.

Line, R.F., and X.M. Chen. 1995. Successes in breeding for and managing durable resistance to wheat rusts. Plant Dis. 79:1254–1255.

Line, R.F., and A. Qayoum. 1992. Virulence, aggressiveness, evolution, and distribution of races of *Puccinia striiformis* (the cause of stripe rust of wheat) in North America, 1968–87. USDA Tech. Bull. 1788.

Line, R.F., F.L. Sharp, and R.L. Powelson. 1970. A system for differentiating races of *Puccinia striiformis* in the United States. Plant Dis. Rep. 54:992–994.

Liu, J.Q., and J.A. Kolmer. 1997. Genetics of leaf rust resistance in Canadian spring wheats AC Domain and AC Taber. Plant Dis. 81:757–760.

Liu, J.Q., and J.A. Kolmer. 1998a. Genetics of stem rust resistance in wheat cvs. Pasqua and AC Taber. Phytopathology 88:171–176.

Liu, J.Q., and J.A. Kolmer. 1998b. Molecular and virulence diversity and linkage disequilibria in asexual and sexual populations of the wheat leaf rust fungus, *Puccinia recondita*. Genome 41:832–840.

Loegering, W.Q., and H.R. Powers. 1962. Inheritance of pathogenicity in a cross of physiological races 111 and 36 of *Puccinia graminis* f. sp. *tritici*. Phytopathology 52:547–554.

Long, D.L., and J.A. Kolmer. 1989. A North American system of nomenclature for *Puccinia recondita* f. sp. *tritici*. Phytopathology 79:525–529.

Long, D.L., J.A. Kolmer, Y. Jin, M.E. Hughes, and L.A. Wanschura. 2007. Wheat rusts in the United States in 2006. Annu. Wheat Newsl. 53:130–141.

Long, D.L., K.J. Leonard, and M.E. Hughes. 2000. Virulence of *Puccinia triticina* on wheat in the United States from 1996 to 1998. Plant Dis. 84:1334–1341.

Lupton, F.G.H., and R.C.F. Macer. 1962. Inheritance of resistance to yellow rust (*Puccinia glumarum* Erikss. & Henn.) in seven varieties of wheat. Trans. Br. Mycol. Soc. 45:21–45.

Mains, E.B. 1930. Effect of leaf rust (*Puccinia triticina* Eriks.) on yield of wheat. J. Agric. Res. 40:417–446.

Mains, E.B. 1932. Host specialization in the leaf rust of grasses, *Puccinia rubigo-vera*. Pap. Michigan Acad. Sci., Arts and Letters 17:289–393.

Mains, E.B. 1933. Studies concerning heteroecious rusts. Mycologia 25:407–417.

Manners, J.G. 1960. *Puccinia striiformis* Westend. var *dactylidis* var. nov. Trans. Br. Mycol. Soc. 43:65–68.

Marasas, C.N., M. Smale, and R.P. Singh. 2004. The economic impact in developing countries of leaf rust resistance in CIMMYT-related spring wheat. Econ. Program Pap. 04-01. CIMMYT, D.F., Mexico.

Markell, S.G., E.A. Milus, and X.M. Chen. 2004. Genetic diversity of *Puccinia striiformis* f. sp. *tritici* in the United States. p. A2.43. *In* Proc. Int. Cereal Rusts & Powdery Mildews Conf., 11th, John Innes Centre, Norwich, UK. 22–27 Aug. 2004.

Marryat, D.C.E. 1907. Notes on the infection and histology of two wheats immune to attacks of *Puccinia glumarum*, yellow rust. J. Agric. Sci. 2:129–138.

Martin, J.N., B.F. Carver, R.M. Hunger, and T.S. Cox. 2003. Contributions of leaf rust resistance and awns to agronomic and grain quality in winter wheat. Crop Sci. 43:1712–1717.

Martinez, F., J.C. Sillero, and D. Rubiales. 2005. Pathogenic specialization of *Puccinia triticina* in Andalusia from 1998 to 2000. J. Phytopathol. 153:344–349.

McCallum, B.D., and P. Seto-Goh. 2006. Physiological specialization of *Puccinia triticina*, the causal agent of wheat leaf rust, in Canada in 2004. Can. J. Plant Pathol. 28:566–576.

McIntosh, R.A. 1992. Close genetic linkage of genes conferring adult-plant resistance to leaf rust and stripe rust of wheat. Plant Pathol. 41:523–527.

McIntosh, R.A., C.R. Wellings, and R.F. Park. 1995. Wheat rusts: An atlas of resistance genes. CSIRO, Sydney, Australia, and Kluwer, Dordrecht, The Netherlands.

McIntosh, R.A., Y. Yamazaki, K.M. Devos, J. Dubcovsky, J. Rogers, and R. Appels. 2007. Catalogue of gene symbols for wheat. 2007 Supplement [Online]. KOMUGI integrated wheat science database. Available at http://www.shigen.nig.ac.jp/wheat/komugi/genes/symbolClassList.jsp (verified 6 Dec. 2007)

Milus, E.A., and R.F. Line. 1986a. Number of genes controlling high-temperature, adult-plant resistance to stripe rust in wheat. Phytopathology 76:93–96.

Milus, E.A., and R.F. Line. 1986b. Gene action for inheritance of durable, high-temperature, adult-plant resistance to stripe rust in wheat. Phytopathology 76:435–441.

Mishra, A.N., K. Kaushal, S.R. Yadav, G.S. Shirsekar, and H.N. Pandey. 2005. A leaf rust resistance gene, different from *Lr34*, associated with leaf tip necrosis in wheat. Plant Breed. 124:517–519.

Navabi, A., R.P. Singh, J. Huerta-Espino, and J.P. Tewari. 2005. Phenotypic association of adult-plant resistance to leaf and stripe rusts in wheat. Can. J. Plant Pathol. 27:396–403.

Navabi, A., R.P. Singh, J.P. Tewari, and K.G. Briggs. 2003. Genetic analysis of adult-plant resistance to leaf rust in five spring wheat genotypes. Plant Dis. 87:1522–1529.

Nazareno, N.R.X, and A.P. Roelfs. 1981. Adult plant resistance of Thatcher wheat to stem rust. Phytopathology 71:181–185.

Newton, M., and T. Johnson. 1936. Stripe rust, *Puccinia glumerum*, in Canada. Can. J. Res. 14:89–108.

Niu, Y.C., Z.Q. Li, and H.S. Shang. 1991. *Puccinia striiformis* West. f. sp. *leymi* and f. sp. *elymi*, two new formae speciales. Acta Univ. Agric. Boreali Occident. 19:58–62.

Oelke, L.M., and J.A. Kolmer. 2004. Characterization of leaf rust resistance in hard red spring wheat cultivars. Plant Dis. 88:1127–1133.

Oelke, L.M., and J.A. Kolmer. 2005. Genetics of leaf rust resistance in spring wheat cultivars Norm and Alsen. Phytopathology 95:773–778.

Ordoñez, M.E., and J.A. Kolmer. 2007a. Simple sequence repeat diversity of a world-wide collection of *Puccinia triticina* from durum wheat. Phytopathology 97:574–583.

Ordoñez, M.E., and J.A. Kolmer. 2007b. Virulence phenotypes of a world-wide collection of *Puccinia triticina* from durum wheat. Phytopathology 97:344–351.

Parlevliet, J.E. 1979. Components of resistance that reduce the rate of epidemic development. Annu. Rev. Phytopathol. 17:203–222.

Park, R.L. 2007. Stem rust of wheat in Australia. Aust. J. Agric. Res. 58:558–566.

Park, R.F., H.S. Bariana, C.R. Wellings, and H. Wallwork. 2002. Detection and occurrence of a new pathotype of

Puccinia triticina with virulence for *Lr24* in Australia. Aust. J. Agric. Res. 53:1069–1076.

Park, R.F., J.J. Burdon, and R.A. McIntosh. 1995. Studies in the origin, spread, and evolution of an important group of *Puccinia recondita* f. sp. *tritici* pathotypes in Australasia. Eur. J. Plant Pathol. 101:613–622.

Park, R.F., A. Jahoor, and F.G. Felsenstein. 2000. Population structure of *Puccinia recondita* in western Europe during 1995, as assessed by variability in pathogenicity and molecular markers. J. Phytopathol. 148:169–179.

Peturson, B., M. Newton, and A.G.O. Whiteside. 1945. The effect of leaf rust on the yield and quality of wheat. Can. J. Res. 23:105–114.

Peturson, B., M. Newton, and A.G.O. Whiteside. 1948. Further studies on the effect of leaf rust on the yield, grade, and quality of wheat. Can. J. Res. 26:65–70.

Pope, W.K., E.L. Sharp, and H.S. Fenwick. 1963. Stripe rust of wheat in the Pacific Northwest in 1962. Plant Dis. Rep. 47:554–555.

Pretorius, Z.A., W.H.P. Boshoff, and G.H.J. Kema. 1997. First report of *Puccinia striiformis* f. sp. *tritici* on wheat in South Africa. Plant Dis. 81:424.

Pretorius, Z.A., F.J. Kloppers, and S.C. Drijepondt. 1994. Effect of inoculum density and temperature on three components of leaf rust resistance controlled by *Lr34* in wheat. Euphytica 74:91–96.

Pretorius, Z.A., R.P. Singh, W.W. Wagoire, and T. Payne. 2000. Detection of virulence to wheat stem rust resistance gene *Sr31* in *Puccinia graminis* f. sp. *tritici* in Uganda. Plant Dis. 84:203.

Qayoum, A., and R.F. Line. 1985. High temperature, adult-plant resistance to stripe rust of wheat. Phytopathology 75:1121–1125.

Rapilly, F. 1979. Yellow rust epidemiology. Annu. Rev. Phytopathol. 17:59–73.

Röbbelen, G., and E.L. Sharp. 1978. Mode of inheritance, interaction and application of genes conditioning resistance to yellow rust. Verlag Paul Parey, Berlin and Hamburg.

Roelfs, A.P. 1985. Wheat and rye stem rust. p. 4–38. *In* A.P. Roelfs and W.R. Bushnell (ed.) The cereal rusts. Vol. II. Academic Press, Orlando, FL.

Roelfs, A.P. 1988. Resistance to leaf and stem rusts in wheat. p. 10–22. *In* N.W. Simmonds and S. Rajaram (ed.) Breeding strategies for resistance to the rusts of wheat. CIMMYT, D.F., Mexico.

Roelfs, A.P. 1989. Epidemiology of the cereal rusts in North America. Can. J. Plant Pathol. 11:86–90.

Roelfs, A.P., and J.V. Groth. 1980. A comparison of virulence phenotypes in wheat stem rust populations reproducing sexually and asexually. Phytopathology 70:855–862.

Roelfs, A.P., and J.W. Martens. 1988. An international system of nomenclature for *Puccinia graminis* f. sp. *tritici*. Phytopathology 78:526–533.

Roelfs, A.P. and D.V. McVey. 1979. Low infection types produced by *Puccinia graminis* f. sp. *tritici* and wheat lines with designated genes for resistance. Phytopathology 69:722–730.

Roelfs, A.P., R.P. Singh, and E.E. Saari. 1992. Rust diseases of wheat: Concepts and methods of disease management. CIMMYT, D.F., Mexico.

Rosewarne, G.M., R.P. Singh, J. Huerta-Espino, J. William, S. Bouchet, S. Cloutier, H. McFadden, and E.S. Lagudah. 2006. Leaf tip necrosis, molecular markers, and β1-proteasome subunits associated with the slow rusting genes *Lr46*/*Yr29*. Theor. Appl. Genet. 112:500–508.

Saari, E.E., H.C. Young, and M.F. Kernkamp. 1968. Infection of North American *Thalictrum* spp. with *Puccinia recondita* f. sp. *tritici*. Phytopathology 58:939–943.

Samborski, D.J., and P.L. Dyck. 1968. Inheritance of virulence in wheat leaf rust on the standard differential wheat varieties. Can. J. Genet. Cytol. 10:24–32.

Samborski, D.J., and P.L. Dyck. 1982. Enhancement of resistance to *Puccinia recondita* by interactions of resistance genes in wheat. Can. J. Plant Pathol. 4:152–156.

Samborski, D.J., and B. Peturson. 1960. Effect of leaf rust in the yield of resistant wheats. Can. J. Plant Sci. 620–622.

Santra, D.K., M. Santra, C. Uauy, K.G. Campbell, X.M. Chen, J. Dubcovsky, and K.K. Kidwell. 2006. Identifying QTL for high-temperature adult-plant resistance to stripe rust in wheat (*Triticum aestivum* L.). p. 179. *In* Abstracts, Plant and Animal Genome Conf. XIV, San Diego, CA. 14–18 Jan. 2006. Scherago International, New York.

Savile, D.B.O. 1984. Taxonomy of the cereal rust fungi. p. 79–114. *In* W.R. Bushnell and A.P. Roelfs (ed.) The cereal rusts. Vol. I. Academic Press, Orlando, FL.

Schmidt, J.K. 1827. Allgemeine ökonomisch-technische Flora oder Abbildungen und Beschreibungen aller in Bezug auf Ökonomie und Technologie, merkwürdigen Gewächse. Vol. I: 27. Jena, Germany.

Schnurbusch, T., S. Paillard, A. Schori, M. Messmer, G. Schachermayr, H. Winzeler, and B. Keller. 2004. Dissection of quantitative and durable leaf rust resistance in Swiss winter wheat reveals a major resistance QTL in the *Lr34* chromosome region. Theor. Appl. Genet. 108:477–484.

Sebesta, E.E., E.A. Wood, Jr., D.R. Porter, J.A. Webster, and E.L. Smith. 1994. Registration of Gaucho greenbug-resistant triticale germplasm. Crop Sci. 34:1428.

Shaner, G., and R.L. Powelson. 1971. Epidemiology of stripe rust of wheat, 1961–1968. Tech. Bull. 117. Oregon Agric. Exp. Stn., Corvallis, OR.

Sharp, E.L., and E.R. Heln. 1963. Overwintering of stripe rust in winter wheat in Montana. Phytopathology 53:1239–1240.

Singh, R.P. 1991. Pathogenicity variations of *Puccinia recondita* f. sp. *tritici* and *P. graminis* f. sp. *tritici* in wheat-growing areas of Mexico during 1988 and 1989. Plant Dis. 75:790–794.

Singh, R.P. 1992a. Association between gene *Lr34* for leaf rust resistance and leaf tip necrosis in wheat. Crop Sci. 32:874–878.

Singh, R.P. 1992b. Expression of wheat leaf rust resistance gene *Lr34* in seedlings and adult plants. Plant Dis. 76:489–491.

Singh, R.P. 1992c. Genetic association of leaf rust resistance gene *Lr34* with adult plant resistance to stripe rust in bread wheat. Phytopathology 82:835–838.

Singh, R.P. 1993. Genetic association of gene *Bdv1* for tolerance to barley yellow dwarf virus with genes *Lr34* and *Yr18* for adult plant resistance to rusts in bread wheat. Plant Dis. 77:1103–1106.

Singh, R.P., E. Bechere, and O. Abdalla. 1993. Genetic analysis of resistance to leaf rust in nine durum wheats. Plant Dis. 77:460–463.

Singh, R.P., D.P. Hodson, Y. Jin, J. Huerta-Espino, M.G. Kinyua, R. Wanyera, P. Njau, and R.W. Ward. 2006. Current status, likely migration and strategies to mitigate the threat to wheat production from race Ug-99 (TTKS) of stem rust pathogen. *In* CAB reviews: Perspectives in agriculture, veterinary science, nutrition, and natural resources 1, No. 054 [Online]. Available at http://www.cababstractsplus.org/cabreviews/Reviews.asp?action=display&openMenu=relatedItems&ReviewID=25637&Year=2006.

Singh, R.P., J. Huerta-Espino, W. Pfeiffer, and P. Figueroa-Lopez. 2004. Occurrence and impact of a new leaf rust race on durum wheat in northwestern Mexico from 2001 to 2003. Plant Dis. 88:703–708.

Singh, R.P., J. Huerta-Espino, and S. Rajaram. 2000. Achieving near-immunity to leaf rust and stripe rust in wheat by combining slow rusting resistance genes. Acta Phytopathol. Entomol. 35:133–139.

Singh, R.P., A. Mujeebkazi, and J. Huerta-Espino. 1998. *Lr46*—A gene conferring slow rusting resistance to leaf rust in wheat. Phytopathology 88:890–894.

Singh, R.P., and S. Rajaram. 1992. Genetics of adult-plant resistance in Frontana and three CIMMYT wheats. Genome 35:24–31.

Smith, R.E. 1961. Wheat rust as viewed in the early agricultural press in California. Plant Dis. Rep. 45:709–712.

Spielmeyer, W., R.A. McIntosh, J.A. Kolmer, and E.S. Lagudah. 2005. Powdery mildew resistance and *Lr34/Yr18* genes for durable resistance to leaf and stripe rust cosegregate at a locus on the short arm of chromosome 7D of wheat. Theor. Appl. Genet. 111:731–735.

Spielmeyer, W., P.J. Sharp, and E.S. Lagudah. 2003. Identification and validation of markers linked to broad-spectrum stem rust resistance gene *Sr2* in wheat (*Triticum aestivum* L.). Crop Sci. 43:333–336.

Stakman, E.C., M.N. Levine, and J.G. Leach. 1919. New biologic forms of *Puccinia graminis*. J. Agric. Res. 16: 103–105.

Statler, G.D. 1973. Inheritance of leaf rust resistance in Leeds durum wheat. Crop Sci. 13:116–117.

Straib, W. 1937. Über Resistenz bei Gerste gegenüber Zwergrost und Gelbrost. Züchter 9:305–311.

Stubbs, R.W. 1985. Stripe rust. p. 61–102. *In* A.P. Roelfs and W.R. Bushnell (ed.) The cereal rusts. Vol. II. Academic Press, Orlando, FL.

Szabo, L.S., and J.A. Kolmer. 2007. Development of simple sequence repeat markers for the plant pathogenic rust fungus *Puccinia triticina*. Mol. Ecol. Notes 7: 708–710.

Tollenaar, H. 1967. A comparison of *Puccinia striiformis* f. sp. *poae* on bluegrass with *P. striiformis* f. sp. *tritici* and f. sp. *dactylidis*. Phytopathology 57:418–420.

Tollenaar, H., and B.R. Houston. 1966. Effect of temperature during uredospore production and of light on in vitro germination of uredospores from *Puccinia striiformis*. Phytopathology 56:787–789.

Tranzschel, W. 1934. Promežutočnye chozjarva rzavčiny chlebov i ich der USSR. (The alternate hosts of cereal rust fungi and their distribution in the USSR). Bull. Plant Prot. Ser. 2:4–40.

Uauy, C., J.C. Brevis, X.M. Chen, I.A. Khan, L.F. Jackson, O. Chicaiza, A. Distelfeld, T. Fahima, and J. Dubcovsky. 2005. High-temperature adult plant (HTAP) stripe rust resistance gene *Yr36* from *Triticum turgidum* ssp. *dicoccoides* is closely linked to the grain protein content locus *Gpc-B1*. Theor. Appl. Genet. 112:97–105.

Viennot-Bourgin, G. 1941. Diagnose Latine die *Puccinia tritic*-duri nov. sp. Ann. Ed. Natl. Grignon Paris C. Amat Ser. 3. 2:146.

Wahl, I., Y. Anikster, J. Manisterski, and A. Segal. 1984. Evolution at the center of origin. p. 39–78. *In* W.R. Bushnell and A.P. Roelfs (ed.) The cereal rusts. Vol I. Academic Press, Orlando, FL.

Waldron, L.R. 1936. The effect of leaf rust accompanied by heat upon yield, kernel weight, bushel weight, and protein content of hard red spring wheat. J. Agric. Res. 53: 399–414.

Wallwork, H. 2007. The role of minimum disease resistance standards for the control of cereal diseases. Aust. J. Agric. Res. 58:588–592.

Wanyera, R., M.G. Kinyua, Y. Jin, and R.P. Singh. 2006. The spread of stem rust caused by *Puccinia graminis* f. sp. *tritici*, with virulence on *Sr31* in wheat in Eastern Africa. Plant Dis. 90:113.

Webster, J. 1980. Introduction to fungi. Cambridge University Press, Oxford, UK.

Wellings, C.R. 2007. *Puccinia striiformis* in Australia: A review of the incursion, evolution, and adaptation of stripe rust in the period 1979–2006. Aust. J. Agric. Res. 58:567–575.

Wellings, C.R., J.J. Burdon, and F.J. Keiper. 2004. The biology of *Puccinia striiformis* on *Hordeum* spp. in Australia: The case for a new forma specialis. p. A1.50. *In* Proc. Int. Cereal Rusts & Powdery Mildews Conf., 11th, John Innes Centre, Norwich, UK. 22–27 Aug. 2004.

Wellings, C.R., and R.A. McIntosh. 1987. *Puccinia striiformis* f. sp. *tritici* in Eastern Australia—possible means of entry and implications for plant quarantine. Plant Pathol. 36:239–241.

Wellings, C.R., D.G. Wright, F. Keiper, and R. Loughman. 2003. First detection of wheat stripe rust in Western Australia: Evidence for a foreign incursion. Aust. Plant Pathol. 32:321–322.

Westendorp, G.D. 1854. Quatrième notice sur quelques *Cryptogames* récemment découvertes en Belgique. Bull. Acad. R. Sci. Belg. 21:229–246.

William, H.M., R.P. Singh, J. Huerta-Espino, G. Palacios, and K. Suenaga. 2006. Characterization of genetic loci conferring adult plant resistance to leaf rust and stripe rust in spring wheat. Genome 49:977–990.

Williams, M., R.P. Singh, J. Huerta-Espino, S. Ortiz Islas, and D. Hoisington. 2002. Molecular marker mapping of leaf rust resistance gene *Lr46* and its association with stripe rust resistance gene *Yr29* wheat. Phytopathology 93:153–159.

Winter, G. 1884. Repertorium Rabenhorstii fungi europaei et extraeuraopaei. Centuria XXXI et XXXII. Hedwigia 23:164–175.

Worland, A.J., and C.N. Law. 1986. Genetic analysis of chromosome 2D of wheat: I. The location of genes affecting height, day-length insensitivity, hybrid dwarfism and yellow rust resistance. Z. Pflanzenzücht 96:331–345.

Wright, R.G., and J.H. Lennard. 1978. Mitosis in *Puccinia striiformis*: 1. Light microscopy. Trans. Br. Mycol. Soc. 70:91–98.

Xu, X.Y., G.H. Bai, B. Carver, G.E. Shaner, and R.M. Hunger. 2005a. Mapping of QTLs prolonging the latent period of *Puccina triticina* infection in wheat. Theor Appl. Genet. 110:244–251.

Xu, X.Y., G.H. Bai, B. Carver, G.E. Shaner, and R.M. Hunger. 2005b. Molecular characterization of slow leaf-rusting resistance in wheat. Crop Sci. 45:758–765.

Yehuda, P.B., T. Eilam, J. Manisterski, A. Shimoni, and Y. Anikster. 2004. Leaf rust on *Aegilops speltoides* caused by a new forma specialis of *Puccinia triticina*. Phytopathology 94:94–101.

Zhang, H., and D.R. Knott. 1990. Inheritance of leaf rust resistance in durum wheat. Crop Sci. 30:1218–1222.

Zhang, H., and D.R. Knott. 1993. Inheritance of adult plant resistance to leaf rust in six durum wheat cultivars. Crop Sci. 33:694–697.

Chapter 6

Diseases Which Challenge Global Wheat Production—Root, Crown, and Culm Rots

Richard W. Smiley, David Backhouse, Philippe Lucas, and Timothy C. Paulitz

SUMMARY

(1) With few exceptions, root, crown, and culm rots are especially prevalent in cropping systems characterized by high residue retention, reduced tillage, or high frequency of host crops. Most of these diseases are not yet effectively managed by genetic resistance, fungicides, or biological agents. Optimal disease management generally requires changing the soil environment to reduce survival of the pathogens between susceptible host crops, or the virulence of pathogens during the infective stage.

(2) Rotation to nonhosts is an effective management strategy for common root rot, take-all, Cephalosporium stripe, and eyespot but not crown rot, Pythium root rot, or Rhizoctonia root rot.

(3) Other management practices that reduce damage caused by several of these diseases include preventing growth of volunteer cereals and weed hosts during the interval between crops, banding a portion of the fertilizer below the seed, adjusting the planting date, using a seed drill that causes intense soil disturbance in the seed row, and protecting seedlings by applying a fungicide to the seed. Only eyespot can be controlled by applying fungicide to the foliage. Crop management systems to control take-all and eyespot are optimized using models.

(4) Severity of take-all and Rhizoctonia root rot may increase at the beginning of a wheat monoculture and then begin to decline in severity. These disease-decline phenomena are mediated through influences of the soil microbiota.

(5) Wheat cultivars with useful levels of resistance are available to suppress damage by common root rot, crown rot, Cephalosporium stripe, and eyespot. Molecular markers are used to detect resistance genes in seedlings, and DNA-based real-time polymerase chain reaction (PCR) assays are available to identify and quantify these pathogens in soil or plants. These assays plus data interpretation based on disease epidemiology have been used commercially to predict potential grain yield loss from several of these diseases.

INTRODUCTION

Most wheat (*Triticum* spp.) diseases caused by root-, crown- and lower culm-infecting fungi are not yet effectively managed by genetic resistance or by application of a fungicide or biological control agent. The best management strategy for many of these diseases continues to depend upon changing the soil environment in ways that either influence the survival of the pathogen between susceptible host crops or the pathogen's virulence during the infective stage (Cook and Veseth 1991). The soilborne plant-pathogenic fungi that cause the seven root, crown, and culm rots summarized in this chapter (Fig. 6.1) are heavily influenced by soil physical and chemical properties, by interactions with associated microbes and microfauna in soil and on plant surfaces, and by the capacity of plants to serve as hosts for growth and multiplication. The complexity of factors affecting these pathogens before and during pathogenesis is immense.

Updated summaries of these diseases cannot be complete without acknowledging the great contributions that have been made by large numbers of practicing farmers, soil microbiologists, soil ecologists, soil physicists, soil chemists, agricultural engineers, agronomists, plant pathologists, botanists, geneticists, and wheat breeders. An introduction to these founding contributions can be found in Butler (1961), Baker and Snyder (1965), Garrett (1970), Griffin (1972), Bruehl (1975, 1987), Schippers and Gams (1979), Krupa and Dommergues (1979), Cook and Baker (1983), and Parker et al. (1985). Additional reference books are cited in appropriate sections of this chapter.

Many similarities occur among pathogens and diseases of wheat and other grasses. Complementary insights into the biology of pathogens causing four of these wheat diseases are summarized from a different perspective in treatises by Smith et al. (1989), Clarke and Gould (1993), Couch (1995), and Smiley et al. (2005a).

The content of this chapter also would be incomplete without referencing guidelines for studying these pathogens and gaining a greater visual appreciation of disease symptoms and pathogen characteristics. Methods to isolate and study the pathogens are provided by Singleton et al. (1992), among others. Excellent color images are available in Zillinsky (1983), Murray et al. (1998), Wallwork (1992, 2000), Bailey et al. (2003), and Bockus et al. (2009).

COMMON ROOT ROT

Common root rot is a name originally applied to a complex of diseases caused by several species of *Bipolaris* ('*Helminthosporium*') and *Fusarium* (Butler 1961). Current usage restricts this name to root and stem base diseases caused by *Bipolaris sorokiniana*, although this pathogen frequently occurs in association with Fusarium crown rot (Windels and Wiersma 1992; Smiley and Patterson 1996; Fernandez and Chen 2005). Common root rot was considered a very serious disease, especially in Canada and Australia, in the early- to mid-20th century (Butler 1961; Tinline et al., 1991) and can cause grain yield losses as great as 25% (Wildermuth et al., 1992). Common root rot has received less attention in recent years, possibly because of the increasing importance of Fusarium crown rot, and because most research on this pathogen now focuses on the foliar disease, spot blotch (Kumar et al., 2002). The significance of common root rot for modern cultivars in contemporary cropping systems is not well known and is probably underestimated.

Symptoms and epidemiology

All parts of the wheat plant can be infected. Initial symptoms are dark necrotic lesions. The classic symptom in soilborne infections is a dark lesion on the subcrown internode (Color Plate 12a), which can extend up to the crown and, in severe infections, up the lower internodes on the stems. Plants with severe subcrown internode symptoms have reduced root growth, especially of crown roots (Kokko et al., 1995). The roots are not usually the major site of infection, although they may show some browning (Fedel-Moen and Harris 1987; Kokko et al., 1995). Lesions of common root rot (Color Plate 12a) are much darker than those of Fusarium crown rot (Color Plate 12b). On the stems, they appear streaky rather than uniform around the circumference of the stem. Severe common root rot is often associated with water stress (Piccinni et al., 2000). Plants may be stunted, and whiteheads (Color

Disease	Principal pathogen(s)	Primary infection site	Primary symptoms	Conditions leading to highest disease severity	Practices that generally minimize crop damage	Efficacy of seed-applied fungicides	Efficacy of genetic resistance
Common root rot	*Bipolaris sorokiniana*	stem base, crown, subcrown internode	lesion on subcrown internode; stunting; whiteheads	cultivation; planting wheat annually; drought	crop rotation; summer fallow; shallow sowing; delay autumn planting date; plant resistant cultivar	poor to fair	poor to fair
Fusarium crown rot	*Fusarium culmorum, F. pseudograminearum*	stem base, crown, subcrown internode	browning of lower leaf sheath; lesion on subcrown internode; dry rot of crown and roots; whiteheads	no-till or minimum tillage; host crops annually or 2-year rotations; planting winter wheat too early; high N fertility; drought	destroy infested crop residue; crop rotation; judicious application of N; delay autumn planting date; plant resistant cultivar; shallow sowing	poor; more effective for spring than autumn plantings	poor to fair; currently not widely deployed
Pythium root rot	*Pythium ultimum, P. irregulare,* and others	seed, roots	seed rot; damping-off; root rot; stunting; delayed maturity	cool, wet soil; planting host crops annually; planting winter wheat too late or spring wheat too early	plant when soil temperature favors rapid seed germination; band a starter fertilizer below seed; control green bridge; plant fresh seed	poor to fair; more effective for spring than autumn plantings	not available
Rhizoctonia root rot	*Rhizoctonia solani* AG-8, *R. oryzae*	roots	root rot; stunting; delayed maturity	no-till; planting host crops annually; planting winter wheat too late or spring wheat too early	control green-bridge; disruption of seed row at planting; pre-plant tillage; band a starter fertilizer below seed; summer fallow	poor for wheat planted in autumn; poor to fair for spring wheat	not available
Take-all	*Gaeumannomyces graminis* var. *tritici, G. graminis* var. *avenae*	roots	root rot; stunting; blackened stem base in wet or humid sites; whiteheads	planting host crops annually; planting winter wheat too early; major increase in soil pH after lime application	crop rotation; control green-bridge; manage form of N absorbed by roots; delay autumn planting date	poor to fair; more effective for spring than autumn plantings	not available
Cephalosporium stripe	*Cephalosporium gramineum*	roots & crown	long leaf stripes merging into dark veins in leaf sheath; stunting; whiteheads	no-till or minimum tillage; acid soil; host crops annually or 2-year rotations; planting winter wheat too early; freeze/thaw cycles during winter	crop rotation; plant resistant winter wheat; plant spring wheat; destroy infested crop residue; lime acid soils; control green-bridge; delay autumn planting	not effective	fair to good
Eyespot	*Oculimacula yallundae, O. acuformis*	stem base	elliptical lesions on lower culm; plant lodging; whiteheads	host crops annually; planting winter wheat too early	plant resistant winter wheat; delay autumn planting date; plant spring wheat; crop rotation; apply foliar fungicide	fair to good; not necessary for spring plantings	good

Fig. 6.1 Synopsis of primary pathogens, infection sites, symptoms, and effects of crop management practices for diseases addressed in Chapter 6.

Plate 13a) form if plants are water-stressed late in the season.

The main source of inoculum is the conidia, which are large and strongly pigmented and can survive for several years in soil (Wildermuth and McNamara 1991). *Bipolaris sorokiniana* can be seedborne (Couture and Sutton 1980), but seed transmission is unlikely to occur from common root rot–infected plants unless spot blotch is also present.

Incidence of infection is related to the density of conidia in soil (Tinline et al., 1988). Spore density, and disease incidence and severity, are higher in cultivated soils than in no-tillage systems where seed is directly drilled into residue of a previous crop (Reis and Abrao 1983; Mathieson et al., 1990; Wildermuth et al., 1997). This may be related to dispersion of spores within the soil during cultivation. A high proportion of spore production from wheat residues is from crowns (Duczek 1990), and removal of residue by burning has been shown to reduce the severity of common root rot (Wildermuth et al., 1997).

High spore populations and disease severity occur under continuous wheat (Wildermuth and McNamara 1991; Conner et al., 1996). Continuous wheat also increases the average aggressiveness of field populations towards wheat (El-Nashaar and Stack 1989). Wheat and barley (*Hordeum vulgare* L.) are among the most susceptible hosts, with oat (*Avena sativa* L.) being less susceptible while legumes are generally resistant (Wildermuth and McNamara 1987). Spore populations decline under fallow or nonhosts, such as oilseeds or legumes, compared with cereals (Wildermuth and McNamara 1991).

The effect of nutrients on common root rot is equivocal. Most attention has been paid to chloride, which reduces disease severity in some experiments but has no consistent effect in others (Windels et al., 1992; Tinline et al., 1993). Low nitrogen levels possibly reduce disease severity, so that severity in continuous wheat with no added nitrogen is less than in wheat–legume sequences, despite the rotation effect (Dalal et al., 2004; Fernandez and Zentner 2005). This is probably due to an interaction between high nitrogen levels and water use (Dalal et al., 2004).

Incidences of common root rot and Fusarium crown rot are often inversely related when both diseases are present. Severity of common root rot declined as incidence of Fusarium crown rot increased in a long-term trial (Wildermuth et al., 1997). This may represent competition between the pathogens, because *B. sorokiniana* is a poor competitor with *Fusarium* species in plant tissue (Tinline 1977). However, *B. sorokiniana* is strongly antagonized by the crown rot *Fusarium* species in culture, and symptoms of the two diseases are similar. It is therefore possible that *B. sorokiniana* is difficult to detect by isolation or symptoms in the presence of *Fusarium* species. Because of possible interactions, reports of common root rot when it co-occurs with Fusarium crown rot must be interpreted with caution.

Causal organism

Bipolaris sorokiniana (Sacc.) Shoemaker [teleomorph = *Cochliobolus sativus* (S. Ito and Kurib) Drechsler ex Dastur] has a worldwide distribution and a wide host range among small grain cereals and grasses (Kumar et al., 2002). This pathogen is widely reported in older literature as *Helminthosporium sativum* Pammel, C.M. King & Bakke. The sexual state is readily produced in the laboratory (Singleton et al., 1992) but has not been reported in the field.

Some evidence exists for host specialization within the species. Isolates from barley are more virulent to barley roots than to wheat roots, and vice versa (Conner and Atkinson 1989). Wheat isolates vary greatly in their virulence on wheat leaves (Duveiller and Garcia Altamirano 2000) but no clear evidence for races has been found.

Bipolaris sorokiniana produces several sesquiterpenoid toxins, the most important of which is prehelminthosporol (Kumar et al., 2002). Isolates with low prehelminthosporol production in culture tend to have reduced virulence on barley roots (Apoga et al., 2002), but evidence for a major role of toxins in pathogenesis is not available.

Disease management

Rotation to nonhosts is the primary management strategy for common root rot. This must be combined with effective management of grassy weeds. Some benefit may also be gained by rotation among cereal hosts, because this acts against selection for highly aggressive strains (Conner et al., 1996). Zero-tillage (no-till or direct-drill) reduces common root rot severity, and shallow sowing can also reduce contact with inoculum (Tinline and Spurr 1991). Avoidance of stress in the crop reduces severity. In general, management practices directed at more conspicuous diseases like take-all and Fusarium crown rot will also reduce common root rot in most areas.

A moderate level of resistance is available among wheat cultivars (Wildermuth et al., 1992). Most current breeding work is directed against spot blotch, but there is evidence that at least some sources of resistance to spot blotch will also be effective against common root rot (Arabi et al., 2006).

FUSARIUM CROWN ROT

Fusarium crown rot is a generic term for diseases of stem bases caused by several species of *Fusarium*. These diseases are also widely known as foot rot. Crown rot or similar diseases have been reported from all areas where wheat is grown (Nelson et al., 1981; Summerell et al., 2001b). The importance of this disease complex has risen with increasing adoption of residue retention and reduced tillage practices, which favor the buildup of inoculum and greater levels of infection (Windels and Wiersma 1992; Burgess et al., 1993; Smiley et al., 1996a).

The fungi which cause Fusarium crown rot can also incite Fusarium head blight. The focus of most community concern and research in recent years has been on head blight because of the potential for mycotoxin contamination of food and feedstuffs. However, Fusarium crown rot remains a serious problem, especially given the difficulties in managing it within the constraints imposed by modern cropping systems.

Symptoms and epidemiology

Symptoms of crown rot are first seen as necrotic lesions or more general browning on leaf sheaths and stem tissue. Infections from inoculum in the soil may appear first as brown lesions on the subcrown internode (Color Plate 12b) while those from surface residue occur through the crown (Color Plate 13b; Summerell et al., 1990). Infected crown roots exhibit a dry, brown discoloration.

Infection from rain-splashed conidia may be through leaf sheaths above the soil surface, leading to browning first appearing at nodes above the crown (Jenkinson and Parry 1994). For all sites of primary infection, the near-uniform browning may then progress for several internodes up the stem (Color Plate 13c, left side). A pink discoloration under leaf sheaths or in other tissues may also be seen, and orange sporodochia (conidial masses) can form on nodes under high humidity. If infected plants are water-stressed during grain filling, premature ripening may occur, leading to whiteheads (Color Plate 13a). The whitehead symptom generally does not occur when moisture is adequate for optimal plant growth.

Following physiological maturity the fungi aggressively colonize stem tissue and survive as mycelium in the infested residues. *Fusarium culmorum* can also survive as chlamydospores in the soil.

Fusarium crown rot in most environments behaves as a monocyclic disease with incidence being dependent on initial inoculum in the soil or crop residue (Backhouse 2006). Secondary spread appears to be limited in dry environments, but splash dispersal of conidia can be significant in more humid environments (Jenkinson and Parry 1994). It is not known what role ascospores may play in infections of stem bases.

High nitrogen levels favor disease in two ways. The increased leaf area predisposes plants to late-season water stress under dry conditions, increasing severity (Cook 1980). High nitrogen levels also increase incidence of infection (Smiley et al., 1996a), presumably by increasing susceptibility. Zinc deficiency may also increase susceptibility (Grewal et al., 1996).

Causal organisms

A large number of *Fusarium* species are capable of causing stem base disease in wheat (Akinsanmi et al., 2004). Of these, *F. culmorum* (W. G. Smith) Sacc., *F. pseudograminearum* O'Donnell & T. Aoki (teleomorph = *Gibberella coronicola* T. Aoki & O'Donnell), *F. graminearum* Schwabe [teleomorph = *G. zeae* (Schwein.) Petch], and *F. avenaceum* (Fr.) Sacc. (teleomorph = *G. avenacea* R.J. Cook) have been considered the most important species worldwide.

Fusarium culmorum has the most widespread recorded distribution from wheat stem bases among these species, being found on all continents. Parry et al. (1994) suggested that *F. culmorum* was typically found in warmer, drier cereal growing areas. However, surveys in Australia and North America indicate that it is most prevalent in the cooler or higher rainfall parts of the wheat growing areas in these regions (Smiley and Patterson 1996; Backhouse et al., 2004). *Fusarium culmorum* differs from the other species in that chlamydospores play an important part in epidemiology (Sitton and Cook 1981). Infection rates are less affected by surface plant residues compared with other species associated with crown rot (Windels and Wiersma 1992). No teleomorph is known, but evidence for recombination has been found in field populations (Tóth et al., 2004), and it is likely that a sexual state does exist.

Fusarium pseudograminearum was formerly known as *F. graminearum* Group 1 (Aoki and O'Donnell 1999). The teleomorph, *G. coronicola*, is rarely found in the field (Summerell et al., 2001a). This species is the most important cause of crown rot in Australia and South Africa (Burgess et al., 1975; Van Wyk et al., 1987; Backhouse and Burgess 2002; Chakraborty et al., 2006). It is also prevalent in western North America (Smiley and Patterson 1996; Clear et al., 2006), occurs at low frequency in the Mediterranean region and Asia (Bentley et al., 2006; Tunalı et al., 2008), and has not been reported from South America or Europe north of the Alps.

The role of *F. graminearum* as a cause of Fusarium crown rot is unclear because of uncertainty about the identity of fungi reported under this name. In literature prior to the 1980s *F. pseudograminearum* was reported as *F. graminearum*, and even recently the two species have not always been distinguished. O'Donnell et al. (2004) segregated *F. graminearum* from eight cryptic sister species, many of which also occurred on cereals. *Fusarium graminearum* in the strict sense is best known as a head blight pathogen. It is frequently isolated from wheat stem tissue and shows a similar range of aggressiveness to wheat crowns as *F. pseudograminearum* in pathogenicity tests under controlled conditions (Akinsanmi et al., 2004). However, surveys in several countries which have used modern taxonomic concepts have generally failed to report *F. graminearum* as a significant component of the Fusarium crown rot complex compared with other species (Pettitt et al., 2003; Akinsanmi et al., 2004; Backhouse et al., 2004; Smiley et al., 2005b). The teleomorph, *G. zeae*, occurs readily in the field, unlike other species in the complex.

Fusarium avenaceum is typically prevalent in cooler climates such as eastern Canada and northern Europe (Hall and Sutton 1998; Pettitt et al., 2003), although it occurs more widely as a minor component of the disease complex. Pathogenicity to wheat subcrown internodes and crowns is similar to that of *F. culmorum, F. graminearum*, and *F. pseudograminearum* (Fernandez and Chen 2005; Smiley et al., 2005b). The teleomorph, *G. avenacea*, has not been found in the field. *Fusarium avenaceum* appears to have a broader host range than the other species but is more frequently associated with diseases of legumes (Satyaprasad et al., 2000).

Disease management

Chemical control of Fusarium crown rot is generally unsuccessful. Management therefore depends mainly on cultural practices and resistance. Because incidence of disease caused by *F. pseudograminearum*, *F. graminearum*, and *F. avenaceum* is strongly correlated with the quantity of infested residue (Windels and Wiersma 1992; Smiley et al., 1996a; Wildermuth et al., 1997; Backhouse 2006), practices which either remove the residue or allow its decomposition are effective management tools. Stubble burning, either after harvest or immediately before sowing, can maintain the disease at low levels even in continuous wheat production (Burgess et al., 1993, 1996). However, this practice has become unacceptable in many areas. Amounts of surface residue and stubble burning may have less effect where *F. culmorum* is present (Windels and Wiersma 1992; Smiley et al., 1996a), presumably because of the presence of chlamydospores in the soil (Bateman et al., 1998).

Rotation with nonhosts allows time for natural mortality of the pathogen. The rotations that can be used, and their effectiveness, differ between pathogens. *Fusarium pseudograminearum* has the narrowest host range, and almost any noncereal host, including grain legumes and oilseeds as well as sorghum [*Sorghum bicolor* (L.) Moench] can be used (Burgess et al., 1996; Kirkegaard et al., 2004). Maize (*Zea mays* L.) has not been recorded as a host for *F. pseudograminearum*, but would be a poor choice if *F. culmorum* or *F. graminearum*, which do infect maize, are present. Crop rotation may have less effect on *F. culmorum* than on *F. pseudograminearum*, presumably because of the longevity of chlamydospores relative to mycelium in stubble (Sitton and Cook 1981; Summerell and Burgess 1988).

Nitrogen management has a strong impact on disease incidence and severity, particularly in low-rainfall environments where plants mature without the benefit of late-season rainfall (Cook and Veseth 1991; Smiley et al., 1996a). Application of nitrogen at rates greater than required for the expected or attained grain yield typically increases the expression of whiteheads, reduces grain yield and test weight, and increases grain protein content. In low-rainfall regions where low nitrogen inputs are required to produce low-protein soft-white wheat, severity of crown rot is greatly increased when growers apply higher rates of nitrogen to produce high-protein market classes of wheat.

Wheat cultivars differ in susceptibility to crown rot, although the range available among released cultivars is usually small (Wallwork et al., 2004). Moreover, cultivars or lines expressing resistance in some instances are often highly variable in responses over seasons and geographic areas (R.W. Smiley, unpublished data). Durum wheat (*T. turgidum* ssp. *durum*) tends to be more susceptible than bread wheat (*T. aestivum* L.) (Kirkegaard et al., 2004) and should be avoided in high disease-risk situations. Wheat lines with higher levels of resistance have been identified in a wide range of backgrounds, including bread wheat, *T. zhukovskyi* Menabde et Ericzjan, *T. dicoccum* Schrank, and synthetic wheat (Wallwork et al., 2004; Nicol et al., 2007). Resistance appears to be expressed against all pathogen species (Miedaner 1997; Wallwork et al., 2004). Both seedling and adult-plant forms of partial resistance have been identified for *F. pseudograminearum*. For screening purposes, seedling but not adult-plant resistance has been inversely correlated with the genetically determined depth at which crown tissue is formed for each wheat genotype (Wildermuth et al., 2001). Genotypes with seedling resistance form crowns at more shallow depth than susceptible genotypes, possibly enabling them to partially escape or delay infection. Most work on resistance to *Fusarium* in wheat has been done with head blight, but the lack of correlation between resistance for head blight and resistance for crown rot in rye (*Secale cereale* L.) (Miedaner et al., 1997) suggests that head blight resistance will not necessarily be effective against crown rot.

PYTHIUM ROOT ROT

Pythium root rot of wheat is caused by numerous species of *Pythium*. These species have a broad host range including maize, barley, oat, rye, and many broadleaf crops such as pea (*Pisum sativum*

L.), chickpea (*Cicer arietinum* L.), lentil (*Lens culinaris* Medik.), and soybean [*Glycine max* (L.) Merr.].

Pythium species have a worldwide distribution and are found in most agricultural soils (Hendrix and Campbell 1970; Martin and Loper 1999). As many as six species have been isolated from a single soil sample and more than 30 species have been isolated from wheat (Farr et al., 2007), although not all species are equally virulent (Chamswarng and Cook 1985; Ingram and Cook 1990; Higginbotham et al., 2004b). Because of the ubiquitous nature of these pathogens and the chronic nature of the disease, Pythium root rot was termed "the common cold of wheat" (Cook and Veseth 1991).

The impact of *Pythium* on wheat was not realized until field trials were conducted with the fungicide metalaxyl. This oomycete-specific fungicide increased yields as much as 0.8 t ha^{-1} in the Pacific Northwest US (Cook et al., 1980; Smiley et al., 1996b). Treatment of *Pythium*-infested soil with fumigation increased yields 13%–36% (Cook et al., 1987). *Pythium* root rot has also been reported on wheat in Australia (Pankhurst et al., 1995), southeast US (Milus and Rothrock 1997), and Turkey (Tunalı et al., 2008) but can probably be isolated from soils in most wheat growing areas.

Symptoms and epidemiology

Pythium primarily infects juvenile tissues, including embryos, emerging seedlings, root tips, lateral roots, and root hairs. High pathogen populations can reduce emergence of seedlings and stands of wheat (Color Plate 14a), due to death of the seedling from infection either before emergence (preemergence damping-off) or after emergence (postemergence damping-off). Rotted seeds can also be found in soil. However, with cereals such as wheat, these symptoms are rare. Usually, an embryo-infected seedling will emerge successfully and then remain stunted (Fukui et al., 1994).

Diagnosis of *Pythium* is difficult in wheat because of the lack of distinctive aboveground symptoms. In general, wheat will appear stunted, but without disease-free plants for comparison, this generalized mild stunting may not be noticed in the field. However, because the pathogen destroys root tips, feeder roots, and root hairs, the ability of the plant to take up water and nutrients is reduced and symptoms of nutrient deficiency or water stress can become apparent. Maturity can be delayed, plant height is reduced, plants have fewer tillers, and heads are poorly filled. Infected roots may appear yellow-brown in color, but usually the rotted roots quickly disintegrate and are not recovered.

Pythium species survive as thick-walled oospores or sporangia that are produced in infected roots. When roots decay the inoculum is released into the soil. Most inoculum is present in the top 10–15 cm of soil. The pathogen can colonize clean wheat straw, chaff, or green manure as a nutrient source to support mycelium growth and to increase inoculum density (Cook et al., 1990). In the Pacific Northwest, *Pythium* populations averaged 350–400 propagules per gram of soil (Cook et al., 1990), exceeding the threshold of 200 propagules per gram of soil needed to cause growth reductions (Fukui et al., 1994). When a seed is placed in the soil, or a root tip grows near a *Pythium* spore, seed or root exudates stimulate the germination of the spore, resulting in rapid chemotrophic attraction to and infection of the seed or root (Hering et al., 1987; Fukui et al., 1994; Martin and Loper 1999). Spore germination can occur within a few hours and infection within 10–24 hours. Many *Pythium* spp. are also capable of rapid mycelial growth. In wet soils, some species can form motile swimming spores (zoospores) which are chemotactically attracted to root tips and seeds.

Pythium diseases are favored by cool, wet, poorly drained soils with high clay content and low pH (Fukui et al., 1994). These cool wet conditions are often associated with delayed sowing of fall-planted crops (Smiley et al., 1996b) or result from excessive crop residue in no-till systems, which reduce the warming and drying of the soil during the spring (Cook et al., 1990). However, improved water infiltration often associated with long-term no-till may reduce the occurrence of *Pythium* diseases. Maximum infection by *P. ultimum* and *P. irregulare* occurs at 10 °C and 5 °C (Ingram and Cook 1990), but some species can act as snow molds, infecting under snow cover at 0–3 °C.

Causal organisms

Although 30 species of *Pythium* have been associated with wheat, most reports focus upon *P. arrhenomanes* Drechs., *P. graminicola* Subr., *P. ultimum* Trow, *P. aristosporum* Vanterpool, *P. irregulare* Buisman, *P. torulosum* Coker & Patterson, *P. sylvaticum* Campbell & Hendrix, and *P. heterothallicum* Campbell & Hendrix. *Pythium debaryanum* Hesse is frequently cited in older literature but this is not currently recognized as a valid species and many records may be incorrect. Recent surveys using classical and molecular techniques have identified 13 species of *Pythium* on wheat in the Pacific Northwest US, including a new species, *P. abappressorium* (Paulitz and Adams 2003, Paulitz et al., 2003a). Most species are capable of causing significant reductions in root biomass (Higginbotham et al., 2004b), but the most virulent were *P. ultimum*, *P. irregulare* group 1, and *P. irregulare* group IV sensu Matsumoto (identified as *P. debaryanum* in that paper).

Pythium species produce oospores, which result from the fertilization of oogonia by antheridia. Oospores are generally spherical, from 15–40 μm in diameter, and result from sexual recombination. Asexual sporangia can germinate directly to form hyphae or indirectly to form zoospores, and can be spherical, filamentous, or lobed in shape. Identification is based on morphology of sporangia, oospores, oogonia, and antheridia (van der Plaats-Niterink 1981), or on molecular methods based on sequencing of the internal transcribed spacer (ITS) region of the rDNA (Lévesque and de Cock 2004; Schroeder et al., 2006).

Disease management

Although no high-level resistance or tolerance is found in adapted wheat cultivars, minor differences do occur in susceptibility among cultivars (Higginbotham et al., 2004a). Resistance is not a management option at the present time. Crop rotation is not used for management because of the wide host range of *Pythium* species. However, different hosts select for different *Pythium* species (Ingram and Cook 1990) and recent surveys have demonstrated shifts in species composition resulting from different crop rotations or cropping systems (Schroeder et al., 2007). For example, *P. irregulare* Group I is strongly associated with legume rotations (lentil and pea). Because of long survival of *Pythium* oospores in dry soils during the summer, traditional summer fallow may not be effective, although many species decline to low numbers in a fallow system (Schroeder and Paulitz 2006).

Seed treatments reduce early damage to seed and seedlings (Smiley et al., 1996b; Cook et al., 2002b) and can increase grain yield through improved stand establishment (Color Plate 14a), but they will not reduce root rot in mature plants because no commercially registered fungicide is systemically translocated downward into the roots. Effective seed treatments include the oomycete-specific metalaxyl and mefanoxam, and the broad-spectrum thiram. Biological seed treatments with bacteria (*Pseudomonas, Bacillus, Enterobacteria*) have controlled Pythium root rot in the greenhouse and have resulted in some yield increases in the field (Weller and Cook 1986; Kim et al., 1997; Cook et al., 2002b; Kageyama and Nelson 2003).

Some management practices can mitigate the effects of Pythium root rot, such as banding a starter fertilizer directly below the seed to maintain seedling vigor when a portion of the root system becomes rotted (Cook et al., 2000). Because old seed germinates more slowly, giving a greater window of opportunity for *Pythium* to infect, only new seed should be planted. Volunteer crop plants and weeds should be killed with preplant herbicides at least 3 weeks before planting to minimize the "green-bridge" effect (Smiley et al., 1992; Pittaway 1995). Plants that are dying from treatment with the herbicide glyphosate can serve as reservoirs of *Pythium* inoculum, because the necrotrophic pathogen can extensively colonize the root system when the plant defense system is reduced by inhibition of a key enzyme in the shikimate pathway by this herbicide (Lévesque and Rahe 1992).

RHIZOCTONIA ROOT ROT AND BARE PATCH

Rhizoctonia root rot and bare patch of wheat occur throughout the world (MacNish and Neate 1996; Mazzola et al., 1996a). The disease is caused by a complex of *Rhizoctonia* species that infects roots and seeds of wheat, barley, and other cereals, resulting in aboveground stunting, pruning of root tips, reduced tillering, and reduced yield. Some groups of *Rhizoctonia*, including *R. solani* AG-8 and *R. oryzae*, have a wide host range and will also infect broadleaf rotation crops (Cook et al., 2002a; Paulitz 2002; Paulitz et al., 2002a).

Symptoms and epidemiology

The main aboveground symptom of *Rhizoctonia* root rot is plant stunting. Plants can be stunted in patches or individually. When in patches, called bare patch, the classic symptom is expressed as severe stunting of nearly all plants in roughly circular patches that can be small or up to many meters in diameter. Patches are first noticeable about one month after planting spring wheat and as late as early spring for winter wheat (Color Plate 14b). Bare patch is associated with *R. solani* anastomosis group 8 (AG-8) and occurs most commonly in areas with low rainfall and lighter-textured soils such as sandy loam soils with low organic matter. Roots of young seedlings are pruned off, resulting in plants deficient in nutrients such as phosphorus, which can cause purpling of the leaves in phosphorus-deficient soils—hence the name "purple patch" in some regions. *Rhizoctonia* can also cause generalized stunting and uneven plant heights without bare patches, especially in areas of higher precipitation with continuous annual cropping.

Rhizoctonia solani AG-8 causes a characteristic "spear tipping," where the root tips appear reddish-brown and are tapered to a fine point (Color Plate 12c). Brownish lesions 1–3 mm long are present on the root. The pathogen can rot the root cortex, leaving a temporarily intact stele, giving the roots a constricted or pinched appearance at the site of infection (Paulitz et al., 2002a). Seminal and crown root growth is inhibited because of death of root tips. Maturity is delayed and tiller formation is reduced (Smiley and Wilkins 1993). *Rhizoctonia oryzae* can cause preemergence and postemergence damping-off, but this is not common with *R. solani* AG-8.

Rhizoctonia primarily survives from year to year in intact roots and crop debris. *Rhizoctonia oryzae* can also survive as microsclerotia. Hyphae of *Rhizoctonia* species are capable of spreading long distances in soil from a food base (Garrett 1970; Bailey et al., 2000), moving through soil pores and along the surfaces of soil particles. Most of the inoculum is present in the top 10–20 cm of soil (Neate 1987). *Rhizoctonia* is not uniformly or randomly distributed in the field but has an aggregated spatial structure (Paulitz et al., 2003b). Roots are infected by hyphae growing out from crop debris or infected roots (Gill et al., 2002), and infections of seedling roots causes more damage to plants than infections of mature plant roots.

Greenhouse studies have shown that *Rhizoctonia* spreads faster in sandy soil than in finer-textured soil (Gill et al., 2000). *Rhizoctonia* in Australia is associated with light-textured soils, especially the sandy calcareous soils of South Australia, but has also been found in red clays and acid soils (MacNish and Neate 1996). In the Pacific Northwest US, bare patch is more common in sandy loam soils with low soil organic matter than in silt loam soils.

Increased disease is associated with reduced-tillage or no-till (Weller et al., 1986; Pumphrey et al., 1987; Smiley and Wilkins 1993; Roget et al., 1996), at least in the initial years of no-till following conversion from conventional tillage (Schroeder and Paulitz 2006). The mechanism for this increase is unknown. Tillage may disrupt hyphal networks (Gill et al., 2001a) or promote flushes of microbial activity that inhibit *Rhizoctonia*.

In moist soil, root rot caused by *R. solani* AG-8 is most severe at 12–15°C (Smiley and Uddin 1993; Gill et al., 2001c). However, in dry soil the pathogen is equally and highly virulent at temperatures from 10–25°C (Gill et al., 2001c). Pathogen virulence in warm, moist soils is somewhat suppressed by a high level of activity by associated soil microorganisms (Gill et al., 2001b). *Rhizoctonia oryzae* has a higher temperature optimum for disease (Ogoshi et al., 1990; Smiley and Uddin 1993), causing maximum disease at 20–27°C.

Reduction in root mass can result in nutrient deficiency. Increasing nutrient availability with starter fertilizer banded below or beside the seed enables plants to tolerate the disease even where seedling infection rates are amplified by the starter fertilizer (Smiley et al., 1990; Cook et al., 2000). Application of nitrogen has shown variable effects on disease, and application of zinc reduced bare patch in zinc-deficient soils in Australia (MacNish and Neate 1996) but not in zinc-sufficient soils in the US (Cook et al., 2002a).

Infected roots of grassy weeds and crop volunteers enable *Rhizoctonia* to survive and expand inoculum density between crops. When these plants are killed with preplant or in-crop herbicides, this nectrophic pathogen can extensively colonize the dying root system, serving as a bridging reservoir of amplified *Rhizoctonia* inoculum. If a preplant herbicide is applied to infected plants soon before planting, the newly planted wheat seed germinates and seedlings emerge at a time when the *Rhizoctonia* inoculum level is at a maximum level (Smiley et al., 1992). This phenomenon is known as the *green bridge* and is especially evident following application of glyphosate, which curtails plant defenses because of inhibition of a key enzyme in the shikimate pathway (Lévesque and Rahe 1992). Residual levels of certain other herbicides in soil from a previous crop may also increase Rhizoctonia disease (Smiley and Wilkins 1992).

Causal organisms

A complex of species causes root rot on wheat, including *Rhizoctonia solani* Kühn (teleomorph = *Thanatephorus cucumeris* Donk), *R. oryzae* Ryker and Gooch (teleomorph = *Waitea circinata* Warcup and Talbot), and binucleate species with a sexual stage in the genus *Ceratobasidium*. *Rhizoctonia solani* is divided into a number of subgroups called anastomosis groups (AGs), based on fusion of hyphae between isolates (Sneh et al., 1991); AG-8 is associated with stunting and bare patch symptoms on wheat and barley. Other weakly virulent AGs also have been isolated from wheat, including AGs 2, 2-1, 4, 5, 9, and 10 (Ogoshi et al., 1990; T.C. Paulitz, unpublished data).

Rhizoctonia oryzae also is an important cause of root rot in the Pacific Northwest US (Ogoshi et al., 1990; Smiley and Uddin 1993; Mazzola et al., 1996b; Paulitz et al., 2002a,b). The sexual state, *W. circinata*, consists of subgroups that attack rice (*Oryza sativa* L.; *W. circinata* var. *oryzae*) and turfgrasses (*W. circinata* var. *circinata* and *W. circinata* var. *agrostis*). The most pathogenic isolates on wheat in the Pacific Northwest appear to be var. *circinata*, based on DNA sequencing (P.A. Okubara, pers. comm.).

Binucleate *Rhizoctonia* AGs CI, E, H, K, and D have also been isolated from wheat, although their pathogenicity on wheat roots is not well known (Mazzola et al., 1996a; Tunalı et al., 2008). Most of these binucleate isolates appear to be weakly virulent on cereals and more virulent on broadleaf rotation crops (T.C. Paulitz, unpublished data). *Rhizoctonia* AG-D is also known as *Rhizoctonia cerealis* Van der Hoeven, which causes sharp eyespot, a basal culm disease similar in appearance to eyespot caused by *Oculimacula* species and discussed later in this chapter.

Rhizoctonia species can be difficult to identify because they do not produce spores. *Rhizoctonia solani* forms rather thick hyphae (4–15 μm diam.) with characteristic right-angle branching, dolipore septa near each hyphal branch, and a slight constriction at the branching point (Sneh et al.,

1991). Some species form microsclerotia in culture, which are irregularly shaped dark-brown aggregations of thick-walled, monilioid cells. However, these are not common with *R. solani* AG-8. *Rhizoctonia solani* and *R. oryzae* are multinucleate. *Rhizoctonia oryzae* branches at 30°–50° from the main hyphae and forms abundant irregularly shaped orange-pink or salmon- to brown-colored sclerotia in culture, 1–3 mm in diameter. Both species can survive in root pieces as rounded, thick-walled monilioid cells.

Rhizoctonia is difficult to quantify because of low population densities in soil. Rhizoctonia propagules can be extracted by sieving of organic matter, plating of soil pellets, and elutriation or baiting with various plant materials (Singleton et al., 1992). A semiquantitative method using wood toothpicks as bait has been particularly useful (Paulitz and Schroeder 2005) and quantitative DNA-based methods using real-time polymerase chain reaction (PCR) have been developed to identify and quantify several *Rhizoctonia* species from soil and plants (Okubara et al., 2008).

Disease management

The most effective cultural management practice involves controlling the green bridge by killing volunteer crop plants and weeds with a preplant herbicide at least three weeks before planting (Smiley et al., 1992). Keeping the field fallow without a living host also reduces the severity of disease, but only if the fallow period is long enough for inoculum levels of the fungus to be reduced (Roget et al., 1987). Paulitz (unpublished data) showed that chemical fallow over two consecutive years was not enough to reduce inoculum of *R. solani* AG 2-1 or *R. oryzae* in a higher precipitation area of the Pacific Northwest, but could reduce hyphal activity in low-rainfall areas by the end of the fallow season, and that this effect could be carried over to the following crop. Reduced mechanical fallow was more effective than chemical fallow.

Increased disturbance in the seed row can also reduce disease (Roget et al., 1996). Farmers in the US greatly reduced the impact of Rhizoctonia bare patch by using very heavy direct-seed drills that caused extensive and deep disruption of soil in the seed row. However, such drills are now seldom used because they require large tractors and high amounts of energy. A lighter, paired-row direct-seed drill configuration also reduced Rhizoctonia disease, possibly because of a more open canopy and quicker soil warming between pairs of rows and residue removal between paired rows (Cook et al., 2000).

Burning or otherwise removing stubble from no-till fields does not generally reduce disease severity (Smiley et al., 1996a; T.C. Paulitz, unpublished data) or the amount of *Rhizoctonia* inoculum (T.C. Paulitz, unpublished data), possibly because the pathogen mainly survives in the root system. However, stubble or crop residue does affect the soil temperature and moisture during the spring, often resulting in cooler soils which are more conducive for *Rhizoctonia* damage to young seedlings.

Protective seed treatments with chemicals such as tebuconazole, difenoconazole, thiram, and fludioxonil often result in better seedling health, expressed as more tillers, roots, and greater plant height, but in most cases grain yield is not statistically increased (Mazzola et al., 1996a; Paulitz and Scott 2006). Effects of crop rotation have been variable because of the wide host range of *Rhizoctonia* on other rotation crops (MacNish and Neate 1996; Mazzola et al., 1996a; Cook et al., 2002a). No genetic resistance to Rhizoctonia root rot has been detected in adapted cultivars of wheat but resistance appears to reside in wild relatives of wheat that have not yet been exploited (Smith et al., 2003a,b).

TAKE-ALL

Take-all is the most damaging root disease of wheat worldwide and can cause severe grain yield losses when consecutive cereal crops are grown (Asher and Shipton 1981; Hornby et al., 1998). Take-all is the most important limiting factor for winter wheat production in Western Europe.

The pathogen causes root necrosis of wheat and, to a lesser extent, of barley, rye, and some grasses.

Symptoms and epidemiology

The take-all fungus survives during the intercrop period on root and shoot debris of a previously infected crop, or on grasses and volunteer cereals. It infects seminal roots, causing characteristic black necrosis, sometimes after root surface colonization by brown runner hyphae of the fungus. Secondary infections occur mainly from root-to-root contact, extending disease to the neighboring plants, but spreading only a short distance resulting in a patchy distribution of the disease within a crop (Cook 2003; Gosme et al., 2007). The root system can become entirely affected. In moist conditions a black necrosis can develop on the lower stem (Color Plate 12d), but this symptom seldom occurs in low-rainfall regions where wheat matures with little or no summer rainfall or irrigation. Perithecia (sexual stage) may form on the lower stem and discharge ascospores, but the importance of ascospores in the existing crop is likely limited because perithecia are formed late in the growing season. Due to restricted capture of nitrogen and water by roots (Schoeny et al., 2003), infected plants develop poorly and appear as stunted patches in spring. Premature ripening in summer is expressed as whiteheads (Color Plates 13a, 14c) bearing shrivelled grains.

Severe disease that develops early in the season causes dramatic yield losses by reducing all yield components (Schoeny et al., 2001). Even when disease severity is low or moderate, major yield losses are to be expected when a dry period occurs during grain formation and filling. Take-all is severe when the autumn and winter are mild and humid, allowing early infections to greatly increase the level of primary inoculum (Hornby 1978). As the take-all fungus is not a good saprophyte, high levels of inoculum are the result of host plants being infected. Wheat monocultures increase inoculum levels through several seasons, increasing disease severity to a maximum generally between the second and fourth year. In regions where soil organic matter is plentiful and the soil environment is favorable for microbial growth, continued production of wheat then leads to a decline of disease severity until a balance is reached at a severity less than that during years of maximum disease expression. Several biological hypotheses were proposed to explain this phenomenon, named take-all decline (Hornby 1979). Current agreement focuses on the important role of fluorescent pseudomonads as biological antagonists of the take-all pathogen (Weller et al., 2007).

Mathematical models (Brasset and Gilligan 1989) incorporate components for primary and secondary infection, together with reduction of inoculum potential over time. A routine DNA-based assay recently refined as a real-time PCR test is provided as a service to Australian farmers to quantify take-all pathogens in soil samples and to serve as a basis for predicting potential yield loss (Herdina and Roget 2000).

Causal organism

Gaeumannomyces graminis (Sacc.) Arx & Olivier var. *tritici* Walker (*Ggt*) is responsible for take-all of wheat, triticale (*Triticosecale rimpaui* Wittm.), barley, and rye, in decreasing order of susceptibility. These crops can also be attacked by *G. graminis* var. *avenae* (*Gga*), which causes take-all of oat and take-all patch of turfgrass (Smiley et al., 2005a). Varieties *Ggt* and *Gga* can be differentiated by host range and pathogenicity tests or by measuring the mean length of ascospores: 70–105 μm for *Ggt* and 100–130 μm for *Gga* (Freeman and Ward 2004). Two other *G. graminis* varieties are known to colonize roots of wheat but are considered as weakly or nonpathogenic, *G. graminis* var. *graminis* causing dieback of Bermudagrass [*Cynodon dactylon* (L.) Pers.], and *G. graminis* var. *maydis* causing take-all on maize (Freeman and Ward 2004).

Gaeumannomyces species such as *Ggt* and *Gga* are homothallic, meaning individual isolates are able to produce perithecia. When *Gaeumannomyces*-like fungi are isolated from plants with take-all symptoms but fail to produce perithecia in culture, the isolates should be examined for the possibility that they are an anamorphic state (*Phialophora* or *Harpophora* spp.; Gams 2000) of recently described heterothallic species requiring the presence of both mating types to produce the sexual stage. These fungi seldom produce a sexual stage in nature and require mixtures of mating types to produce the sexual stage in culture. They are easily misidentified as *Ggt* or *Gga*, may be underreported, and are likely to occur on cereals in nature. *Gaeumannomyces incrustans* Landschoot & Jackson and *Magnaporthe poae* Landschoot & Jackson were the first-described heterothallic species of *Gaeumannomyces*-like fungi (Clarke and Gould 1993). Anamorphs of *M. poae*, misreported initially as *P. graminicola* and later shown to be individual mating types of *M. poae* (Clarke and Gould 1993), cause summer patch of perennial grasses (Smiley et al., 2005a). Anamorphic states of *G. incrustans*, *M. poae*, and related fungi cause a high-temperature form of take-all on wheat and other cereals in pot tests (Smiley et al., 1986; Elliott 1991).

DNA probes have also been developed to identify isolates of *G. graminis* varieties and related species (Clarke and Gould 1993; Augustin et al., 1999; Herdina and Roget 2000; Rachdawong et al., 2002; Freeman and Ward 2004). Molecular tools have revealed genetic polymorphism within *Ggt* populations (Ward and Gray 1992; Bateman et al., 1997). Characterizations of *Ggt* populations from monoculture wheat crops, using both restriction fragment length polymorphism (RFLP) and random amplified polymorphic DNA (RAPD) markers, have revealed two genetic groups called G_1 and G_2 (Lebreton et al., 2004). Isolates of G_1 were dominant in the first and sixth wheat crops, and G_2 isolates were dominant in the third and fourth wheat crops. Aggressiveness of group G_2 was significantly greater than that of group G_1, which corresponds with observed peaks of disease during a wheat monoculture. A linear relationship between G_1 and G_2 frequencies and disease severity on wheat roots occurred in fields monitored over three consecutive seasons (Lebreton et al., 2007).

Disease management

The most important cultural practice used to control take-all is crop rotation (Asher and Shipton 1981; Hornby et al., 1998; Ennaïfar et al., 2007). Other practices that influence take-all severity following the buildup of inoculum during preceding host crops include tillage (Ennaïfar et al., 2005), sowing date and density, application of fertilizer or lime, and grass weed control. Fungicide seed treatment provides consistent but partial efficacy by reducing primary infections (Schoeny and Lucas 1999; Bailey et al., 2005; Ennaïfar et al., 2005). No resistant cultivar is currently available and sources of resistance are scarce (Cook 2003).

Effects of soil cultivation have been variable. Direct-seeded crops have had a lower disease incidence in Britain (Brooks and Dawson 1968) and either a higher level of disease (Moore and Cook 1984) or no effect (Schroeder and Paulitz 2006) in the Pacific Northwest US.

Late sowing will allow a longer period of inoculum reduction and less favorable temperature conditions at the time of possible infections, thus reducing the frequency of primary infections. Reducing the sowing density reduces the amount of primary infection as well as secondary infections capable of transmitting the disease from plant to plant (Colbach et al., 1997).

Take-all generally becomes more severe immediately following application of lime to acid soils, particularly when applications causing a large change in acidity occur during intervals in which environmental conditions and host frequency are also conducive to disease expression (Asher and Shipton 1981). However, host genotypes, isolates of *Ggt*, and other soil microbes each vary in tolerance to acidity. In some regions the occurrence of severe take-all has been associated more with acid than neutral soils, resulting in less severe take-all following application of lime (Hornby et al., 1998).

Application of the ammonium-ion form of nitrogen fertilizer generally leads to a reduction of take-all severity compared to the nitrate form or a mixture of these ions (Huber et al.,

1968; Lucas et al., 1997). When absorbed by roots, the ammonium ion reduces rhizosphere pH and stimulates antagonistic components of the root-surface microflora, such as fluorescent pseudomonads (Smiley 1978; Sarniguet et al., 1992a,b).

Efficient control of grass weeds and volunteer cereals is important for eliminating additional inoculum buildup (Ennaïfar et al., 2005; Gutteridge et al., 2005) that may occur without contributing to take-all decline, as reported for blackgrass (*Alopecurus myosuroides* Hudsen), barren brome [*Anisantha sterilis* (L.) Nevski], and rye brome (*Bromus secalinus* L.) (Dulout et al., 1997; Gutteridge et al., 2005).

Attempts have been made to develop a biological method to control the disease based on various hypotheses proposed to explain the phenomenon of take-all decline (Lucas and Sarniguet 1998). The most intensive work on this topic has involved fluorescent pseudomonads that produce antibiotic compounds (Weller et al., 2007). Due to often inconsistent performance when these biocontrol agents are applied to soil in an inundative biological control, Cook (2007) stressed the importance of managing the resident rhizobacteria with the cropping system to achieve a conservation biological control.

CEPHALOSPORIUM STRIPE

Cephalosporium stripe is a vascular wilt caused by a pathogen with a host range within the Poaceae (Farr et al., 2007). Spring cereals are susceptible but most economic damage occurs on winter cereals (wheat, barley, oat, rye, and triticale) in cool, temperate regions of North America, Europe, Africa, and Japan.

Symptoms and epidemiology

One or more distinct longitudinal chlorotic stripes appear in leaves during jointing (Color Plate 15a). A dark brown leaf vein (Color Plate 15b) extends from the base of each leaf stripe, through the leaf sheath, and into the culm. Leaf striping often does not occur on all leaves and tillers of affected plants. Affected leaves senesce prematurely, plants are typically stunted, and heads may ripen prematurely to produce whiteheads (Color Plate 13a). As plants mature the culm of infected tillers may darken at and below nodes. When seedlings are heavily infected they may exhibit a mosaic-like yellowing in late winter or early spring and may die before stripes develop.

The pathogen is disseminated in infested seed (Murray 2006), but most disease is caused by infection of roots by soilborne conidia (Mathre and Johnston 1975). The pathogen directly penetrates adventitious (coronal) roots and lower stems and moves into xylem vessels (Stiles and Murray 1996; Douhan and Murray 2001). Conidia produced in the xylem and those entering directly through wounds are carried upward in the transpiration stream and lodge and multiply at stem nodes and in leaf veins. Occlusion of xylem vessels by conidia impedes the transport of water and nutrients (Wiese 1972).

The pathogen produces a chlorosis-inducing toxin Graminin A and an exogenous polysaccharide that are not required for pathogenicity and virulence (Van Welt and Fullbright 1986). They are important to development of disease symptoms (Kobayasi and Ui 1979; Creatura et al., 1981; Rahman et al., 2001) and survival of the pathogen in dead straw (Bruehl 1975; Wiese and Ravenscroft 1978).

Foliar tissue infested during the parasitic phase on living plants is returned to the soil during harvest and tillage. The pathogen does not persist in root tissue. In regions where summer rainfall is common the fungus can survive for up to 3 years in infested residue but is mostly destroyed within one year if the residue is buried by tillage. In regions where most precipitation occurs during the winter period the rate of straw decomposition is slower and survival of the pathogen longer. Acid soils favor saprophytic survival in straw and production and survival of conidia (Murray and Walter 1991).

Reduction in grain yield is correlated with numbers of spores in soil (Martin et al., 1986;

Specht and Murray 1990; Bockus et al., 1994). Cool, wet weather during autumn and winter favors profuse sporulation on infested plant debris at or near the soil surface (Wiese and Ravenscroft 1975, 1978; Mathre and Johnston 1977). Disease incidence is often more prevalent when seed is planted into wet soil (Bruehl 1968; Pool and Sharp 1969; Anderegg and Murray 1988). Numbers of propagules in soil decline rapidly during spring.

Pathogen entry into the plant does not require preexisting tissue damage (Anderegg and Murray 1988) but is strongly amplified by injuries caused by emerging secondary roots and tillers (Douhan and Murray 2001), freezing of roots (Bailey et al., 1982; Martin et al., 1989), root breakage during freeze and thaw cycles (Bruehl 1968), insects (Slope and Bardner 1965), and acid soils (Bockus and Claassen 1985; Anderegg and Murray 1988; Stiles and Murray 1996).

Causal organism

Cephalosporium gramineum Y. Nisik. & Ikata produces small unicellular conidia in a slimy exopolysaccharide matrix (Stasinopoulos and Seviour 1989). The *Cephalosporium* stage occurs in the xylem of living plants and on the surface of dead straw at or near the soil surface. In most but not all regions the fungus also produces a saprophytic sporodochial stage (*Hymenula cerealis* Ellis & Everh.) mostly near nodes of dead straw that was infested while living (Wiese and Ravenscroft 1978). Sporodochia are formed during cool, wet periods from late autumn to early spring. Hyaline conidia are produced in great abundance on moistened sporodochia.

Disease management

Cephalosporium stripe incidence and severity increase with frequency of winter wheat production (Latin et al., 1982; Bockus et al., 1983). Disease is greatest where wheat is grown annually and is much less damaging in two-year rotations where summer rainfall is common and in three-year rotations where precipitation occurs mostly during the winter. Rotations are most effective when volunteer cereals and grass weeds are controlled during the overwinter fallow period to prevent an increase in pathogen inoculum density. Cephalosporium stripe is especially damaging in direct-drill (no-till) planting systems (Latin et al., 1982; Bockus et al., 1983). The amount of pathogen inoculum can be greatly reduced by burning, removing, or deeply burying infested residue (Pool and Sharp 1969; Wiese and Ravenscroft 1975; Bockus et al., 1983; Christian and Miller 1984). Planting as late as possible during the autumn reduces disease incidence and severity (Bruehl 1968; Pool and Sharp 1969; Martin et al., 1989) by limiting the colonization of root and crown tissue by the pathogen (Pool and Sharp 1969; Douhan and Murray 2001), provided planting is not delayed such that yield potential is reduced more than may be caused by Cephalosporium stripe in earlier plantings (Raymond and Bockus 1984). Disease severity on winter wheat grown on acid soils can be reduced by applying lime (Bockus and Claassen 1985; Anderegg and Murray 1988), but the benefit is mostly limited to years when root wounding from frozen soil is minor (Murray et al., 1992).

Winter wheat cultivars with partial resistance are available (Bockus 1995; Murray et al., 2001; Mundt 2002). Susceptible cultivars are consistently susceptible and resistant cultivars vary widely in disease reaction from year to year (Martin et al., 1989). Repeated plantings of moderately resistant cultivars reduce the level of pathogen inoculum in soil and adequately manage the loss of yield over time (Shefelbine and Bockus 1989; Murray et al., 2001). Two mechanisms of resistance have been described. The pathogen may be excluded from entering the plant, resulting in lower levels of disease incidence, or the pathogen may have restricted ability to move through root and crown tissue, resulting in fewer infected tillers and delayed symptom development (Morton and Mathre 1980; Mathre and Johnston 1990; Douhan and Murray 2001). Genes conveying a high level of resistance to *C. gramineum* were derived from a wheat–*Thinopyrum* amphiploid (Mathre et al., 1985), were characterized (Cai et al., 1996, 1998), and are being introgressed into commercial cultivars.

Fungicide and microbial products do not effectively suppress Cephalosporium stripe. Spring cereals are susceptible but mostly escape infection.

EYESPOT

Eyespot is caused by a fungus which produces lesions on the lower culms, just above the soil surface. The disease is also called strawbreaker foot rot because, when severe, it causes the stem to break and the plant to lodge. Eyespot occurs commonly on fall-planted wheat but can also be observed on barley and oat, and occasionally on wheat planted during early spring. Although reported in many countries, eyespot is most important in regions with temperate climates such as in the Pacific Northwest US and Western Europe (Nelson and Sutton 1988).

Symptoms and epidemiology

Initial symptoms appear on seedlings during autumn or early spring as dark lens-shaped lesions with diffuse margin and occasionally a central black "pupil" on the outer leaf sheaths, just above the soil. The infection progresses inward from sheath to sheath and into the stem (Color Plate 13c, right-hand side). Invasion of the stem reduces translocation of water and nutrients, reducing yield mainly through a reduction of kernels per head and kernel weight (Ponchet 1959). Severe eyespot lesions (Color Plate 13d) weaken the stems to the extent that they collapse (Ray et al., 2006), causing plants to lodge (Color Plate 14d).

The main source of inoculum is mycelium of the fungus surviving on infested crop debris remaining from a previous crop. Infection occurs via spores formed on the debris, which are mostly spread over short distances in rain-splash droplets. The optimum temperature for sporulation is 5°C (Fitt et al., 1988). Frequent rains are necessary to assure inoculum dispersal, and high humidity is required for sporulation and infection (Rowe and Powelson 1973).

Growth of the fungus inward through successive leaf sheaths on an infected tiller is a function of accumulated temperature and differences in susceptibility between cultivars (Ponchet 1959; Rapilly et al., 1979). The earliest infections lead to earlier and more severe penetrations of the stems and the greatest reduction in yield. Forecasting models have been developed (Rapilly et al., 1979; Siebrasse and Fehrmann 1987), but most of them do not take into account the effects of cultivar resistance, diversity within the fungus populations, and crop management practices (Colbach and Saur 1998). The models are mainly used at a regional level as an alert system to enable advisors and farmers to optimize fungicide applications. Typical decisions at the field level call for a justification to apply a fungicide if the disease exceeds a threshold of at least two outer leaf sheaths penetrated on more than 20% of the tillers at Zadoks growth stage 30 (Fitt et al., 1988). A model defining crop management systems that reduce the risk of eyespot is also available (Colbach et al., 1999).

Causal organisms

Eyespot is caused by *Oculimacula yallundae* (Wallwork & Spooner) Crous & W. Gams and by *O. acuformis* (Boerema, R. Pieters & Hamers) Crous & W. Gams (Crous et al., 2003). These fungi were previously classified as *Tapesia yallundae* and *Tapesia acuformis*, respectively (Robbertse et al., 1995), and are the teleomorphic stages of *Helgardia herpotrichoides* (Fron) Crous & W. Gams and *H. acuformis* (Niremberg) Crous & W. gams (Crous et al., 2003). Before being considered as two species these fungi were described as *Pseudocercosporella herpotrichoides* var. *herpotrichoides* and *Pseudocercosporella herpotrichoides* var. *acuformis* (Niremberg 1981). These two varieties were initially thought to correlate with two pathotypes known as the wheat-type (W-type) and rye-type (R-type), respectively, in reference to their pathogenicity on wheat only or on wheat and rye (Priestley et al., 1992). They were also distinguished as the N- and L-type in reference to normal or slow mycelium growth (Cavelier et al., 1987), but further examination of more strains found this distinction to be incomplete (Lucas et al., 2000).

Globose, greyish brown apothecia bearing ascospores of *O. yallundae* develop on decaying stems and leaf sheaths late in the season and are mainly found on stubble during the intercrop period (Wallwork and Spooner 1988; Hunter 1989). Although *O. yallundae* has been considered the more important causal agent of eyespot, *O. acuformis* increased significantly to become the dominant species during the 1990s in the Pacific Northwest US (Douhan et al., 2003) and Western Europe (Lucas et al., 2000).

The forms which are more commonly observed in a growing wheat crop are the anamorphs *H. herpotrichoides* and *H. acuformis*. They can be differentiated through examination of conidial and cultural characteristics: *H. herpotrichoides* produces either curved or curved and straight conidia (4 septate, 35–80 μm × 1.5–2.5 μm) on fast growing, even-edged colonies and squirrel-grey or olive-grey mycelia, and *H. acuformis* with only straight conidia (4–6 septa, 43–120 μm × 1.2–2.3 μm) on slow-growing, feathery or uneven-edged colonies and grey to brown-grey mycelia (Nirenberg 1981). The two species can be identified more rapidly and accurately by molecular markers used in a PCR assay combined with restriction enzyme digestion of an amplified ribosomal DNA fragment (Gac et al., 1996a,b). A real-time PCR assay now allows workers to simultaneously identify and quantify *O. yallundae* and *O. acuformis* (Walsh et al., 2005).

Disease management

Crop rotation remains the best preventive method for managing this disease. Intensification of agriculture in Western Europe led to an increase in grain yield losses due to eyespot and this disease became, from the 1970s, the main target of fungicides applied between the tillering and stem-extension stages of wheat growth. The most active fungicides have been the antimicrotubular benzimidazole group, especially carbendazim. In the early 1980s, because of the selection of resistant strains, the benzimidazoles were replaced by C-14 demethylation inhibitors (DMIs) such as prochloraz (imidazole group) or flusilazole (triazole group). In the early 1990s the efficiency of these DMIs was also compromised by fungicide resistance in some regions (Leroux and Gredt 1997). The most-used current fungicide is cyprodinil, but decreased sensitivity from repeated applications has also been observed with this fungicide (Babij et al., 2000).

Moderate resistance to eyespot, provided by the gene *Pch2*, was first incorporated in the French winter wheat cultivar Cappelle-Desprez (Muranty et al., 2002). This gene remained durable despite widespread exploitation and has been transferred into many cultivars. The gene *Pch1* was transferred (Maia 1967) from *Aegilops ventricosa* Tausch. into *Triticum persicum* Vavilum ex Zhuk., and the F_1 hybrid was backcrossed with the bread wheat cultivar Marne, producing the resistant line VPM_1; initials refer to ventricosa, persicum, and Marne. The resistance to eyespot conferred by *Pch1* is higher than that conferred by *Pch2* (Hollins et al., 1988; Jahier et al., 1989), which only acts at the seedling stage (Muranty et al., 2002), and now is being emphasized in wheat breeding programs in Europe and the US (Allan et al., 1993). These eyespot resistance genes remain durable and there is no evidence of differences with respect to *Oculimacula* species.

FUTURE PERSPECTIVES

Knowledge of the etiology and control of root, crown, and culm rots continues to improve with advances in technology. The following examples illustrate promising areas of emerging research.

Development of molecular procedures is greatly expanding the precision of pathogen identification and, consequently, also supporting a constant evolution in pathogen taxonomy and phylogeny. Examples include identification of DNA sequences of the ITS region of *Pythium* (Martin 2000; Lévesque and de Cock 2004; Schroeder et al., 2006), *Rhizoctonia* (González et al., 2001), *Gaeumannomyces* (Freeman and Ward 2004), and *Fusarium* (O'Donnell et al., 2004). Results of these tests suggest that many *Pythium* species reported in the literature may be conspecific with others, and other described species may actually be complexes of cryptic species. Many sequences of *Pythium* species on GenBank may also be misidentified. Likewise, newly recognized polymorphism within *G. graminis* var. *tritici* has been revealed (Lebreton et al., 2004), and it appears that anastomosis groups of *R. solani* may function as phylogenetic and biological species which have evolved separately on different host plants and no longer exchange genetic material (González et al., 2006). PCR techniques were used to show that *Cephalosporium gramineum* can become seedborne in wheat (Vasquez-Siller and Murray 2003), and real-time PCR procedures were developed to identify and quantify several of these pathogens in DNA extracts from soil and plants (Lees et al., 2002; Schroeder et al., 2006; Okubara et al., 2008).

Real-time PCR is also now used to make routine farm management decisions. A commercial soil diagnostic laboratory in South Australia uses a DNA extract from soil to identify and estimate population levels of several nematode species and inoculum levels of fungal pathogens, including *F. culmorum*, *F. pseudograminearum*, *G. graminis* var. *tritici*, *G. graminis* var. *avenae*, and *R. solani* AG-8 (Ophel-Keller et al., 2008). Predictions of potential disease risk are communicated back to farmers through a network of agronomic advisors (Herdina and Roget 2000). This application of modern technology will undoubtedly facilitate more rapid and effective surveys of species distribution and inoculum density.

The growing body of literature on molecular aspects of pathogenicity will continue to refine understanding of pathogenic variation within pathogen populations, as well as to improve development of genetic resistance. Considerable new molecular information is available regarding the pathogenicity and genetic structure of *B. sorokiniana* (Kumar et al., 2002), *Fusarium* species causing crown rot (O'Donnell et al., 2004; Monds et al., 2005; Chakraborty et al., 2006), and *G. graminis* var. *tritici* (Lebreton et al., 2004). All *Fusarium* populations studied thus far appear to be recombining, suggesting an unrecognized role for ascospores in epidemiology. Current studies of the roles of avenacinase, melanin, and laccases in take-all, and the link between *G. graminis* var. *tritici* and *Magnaporthe grisea* (Cook 2003; Freeman and Ward 2004) are likely to provide new options for achieving genetic resistance. Likewise, despite the lack of specific genetic resistance to *Pythium* in wheat, a better understanding is emerging for innate resistance and the role of signaling pathways, especially the jasmonic acid and ethylene pathways (Vijayan et al., 1998; Okubara and Paulitz 2005). Elicitors produced by *Pythium* have also been identified, which may be perceived by the host plant (Veit et al., 2001). Higher levels of eyespot resistance may result from studies of the determinants of pathogenicity by *Oculimacula yallundae* and *O. acuformis*, including analysis of the infection process and importance of tissue susceptibility.

Identification and deployment of genetic resistance has been especially difficult for species of *Pythium*, *Rhizoctonia*, *Fusarium*, and *Gaeumannomyces*. Strong advances are currently being made in mapping QTLs for partial seedling resistance to Fusarium crown rot (Boville et al., 2006; Collard et al., 2006). This work is also being extended to include adult-plant resistance. However, these pursuits need to be balanced by a deeper understanding of the components of resistance, including resistance to penetration, resistance to stem colonization, plant reaction to infection, and sensitivity to toxins to enable a more definitive identification of the QTL.

Identification of loci for resistance to *Fusarium* and other pathogens are complemented by improvements in the precision and speed of assays to detect disease resistance or to link phenotypic reactions to sources of genetic resistance (Cowger and Mundt 1998; Wildermuth et al., 2001; Mitter et al., 2006). Continued development of markers for detecting the presence of resistance genes in seedlings will further improve the efficiency of wheat breeding programs. Plant breeders in France (INRA) are also attempting to clone the *Pch1* gene for resistance to eyespot.

New insights into disease epidemiology are emerging through development of models to predict incidence and severity of take-all (Ennaïfar et al., 2007) and eyespot (Colbach et al., 1999). These models provide a tool to more effectively identify and combine the most efficient methods that individually provide only partial disease control (Ennaïfar et al., 2005). They also provide a tool to more effectively respond to society's demand to reduce the use of pesticides. One example is the current multidisciplinary emphasis on developing crop rotation and management systems to control eyespot, based on the use of multiresistant, hardy winter wheat cultivars in France (Savary et al., 2006).

Extensive research continues to be focused on the pursuit of biological control and enhanced soil suppressiveness for diseases such as take-all, common root rot, Pythium root rot, and Rhizoctonia root rot. The greatest focus has been applied to take-all (Hornby et al., 1998; Weller et al., 2007). Additionally, *B. sorokiniana* is a poor saprophytic competitor, sensitive to suppression in soils (Bailey and Lazarovits 2003), and sensitive to several potential biocontrol agents (Kumar et al., 2002). Likewise, Rhizoctonia root rot often becomes severe during the initial transition from conventional tillage to no-till (Schroeder and Paulitz 2006), but long-term no-till farms and annual wheat experiments in the US show little *Rhizoctonia* disease (Smiley et al., 1996a; T.C. Paulitz, unpublished data). Natural suppression to *R. solani* in cereal crops has been documented (Lucas et al., 1993; Roget 1995; Mazzola et al., 1996a). Wiseman et al. (1996) demonstrated that suppression was dependent upon a microbial component, and disease incidence and severity have been inversely correlated with microbial biomass (Smiley et al., 1996a). Many other instances of *Rhizoctonia*-suppressive soils have been described (Mazzola et al., 1996a; Sneh et al., 1996) but specific mechanisms for suppression are generally unknown.

Complementation and collaboration among regional or national research programs throughout the world have been highly effective for identifying and deploying germplasm with higher levels of resistance to common root rot, eyespot, and Fusarium crown rot. However, these efforts are not effectively funded and coordinated. Greater institutional collaboration and funding linkages are needed to improve the efficiency of coordination between international organizations such as CIMMYT and ICARDA, and public and commercial programs in countries where wheat is an important crop.

REFERENCES

Akinsanmi, O.A., V. Mitter, S. Simpfendorfer, D. Backhouse, and S. Chakraborty. 2004. Identity and pathogenicity of *Fusarium* spp. isolated from wheat fields in Queensland and northern New South Wales. Aust. J. Agric. Res. 55:97–107.

Allan, R.E., G.L. Rubenthaler, C.F. Morris, and R.F. Line. 1993. Registration of three soft white winter wheat germplasm lines resistant or tolerant to strawbreaker footrot. Crop Sci. 33:1111–1112.

Anderegg, J.C., and T.D. Murray. 1988. Influence of soil matric potential and soil pH on Cephalosporium stripe of winter wheat in the greenhouse. Plant Dis. 72: 1011–1016.

Aoki, T., and K. O'Donnell. 1999. Morphological and molecular characterization of *Fusarium pseudograminearum* sp. nov., formerly recognized as the Group 1 population of *F. graminearum*. Mycologia 91:597–609.

Apoga, D., H. Akesson, H.B. Jansson, and G. Odham. 2002. Relationship between production of the phytotoxin prehelminthosporol and virulence in isolates of the plant pathogenic fungus *Bipolaris sorokiniana*. Eur. J. Plant Pathol. 108:519–526.

Arabi, M.I.E., A. Al-Daoude, and M. Jawhar. 2006. Interrelationship between spot blotch and common root rot in barley. Australasian Plant Pathol. 35:477–479.

Asher, M.J.C., and P.J. Shipton (ed.) 1981. Biology and control of take-all. Academic Press, London, UK.

Augustin, C., K. Ulrich, E. Ward, and A. Werner. 1999. RAPD-based inter- and intravarietal classification of fungi of the *Gaeumannomyces–Phialophora* complex. J. Phytopathol. 147:109–117.

Babij, J., Q. Zhu, P. Brain, and D.W. Hollomon. 2000. Resistance risk assessment of cereal eyespot, *Tapesia yallundae* and *Tapesia acuformis*, to the anilinopyrimidine fungicide cyprodinil. Eur. J. Plant Pathol. 106:895–905.

Backhouse, D. 2006. Forecasting the risk of crown rot between successive wheat crops. Aust. J. Exp. Agric. 46:1499–1506.

Backhouse, D., A.A. Abubakar, L.W. Burgess, J.I. Dennis, G.J. Hollaway, G.B. Wildermuth, H. Wallwork, and F.J. Henry. 2004. Survey of *Fusarium* species associated with crown rot of wheat and barley in eastern Australia. Australasian Plant Pathol. 33:255–261.

Backhouse, D., and L.W. Burgess. 2002. Climatic analysis of the distribution of *Fusarium graminearum*, *F. pseudograminearum* and *F. culmorum* on cereals in Australia. Australasian Plant Pathol. 31:321–327.

Bailey, K.L., B.D. Gossen, R.K. Gugel, and R.A.A. Morrall (ed.) 2003. Diseases of field crops in Canada. 3rd ed. Univ. Extension Press, Univ. Saskatchewan, Saskatoon, Canada.

Bailey, K.L., and G. Lazarovits. 2003. Suppressing soil-borne diseases with residue management and organic amendments. Soil Tillage Res. 72:169–180.

Bailey, J.E., J.L. Lockwood, and M.V. Wiese. 1982. Infection of wheat by *Cephalosporium gramineum* as influenced by freezing of roots. Phytopathology 72:1324–1328.

Bailey, D.J., W. Otten, and C.A. Gilligan. 2000. Saprotrophic invasion by the soil-borne fungal plant pathogen *Rhizoctonia solani* and percolation thresholds. New Phytol. 146:535–544.

Bailey, D.J., N. Paveley, C. Pillinger, J. Foulkes, J. Spink, and C.A. Gilligan. 2005. Epidemiology and chemical control of take-all on seminal and adventitious roots of wheat. Phytopathology 95:62–68.

Baker, K.F., and W.C. Snyder (ed.) 1965. Ecology of soil-borne plant pathogens: Prelude to biological control. John Murray, London, UK.

Bateman, G.L., G.M. Murray, R.J. Gutteridge, and H. Coskun. 1998. Effects of method of straw disposal and depth cultivation on populations of *Fusarium* spp. in soil and brown foot rot in continuous winter wheat. Ann. Appl. Biol. 132:35–47.

Bateman, G.L., E. Ward, D. Hornby, and R.J. Gutteridge. 1997. Comparisons of isolates of the take-all fungus, *Gaeumannomyces graminis* var. *tritici*, from different cereal sequences using DNA probes and non-molecular methods. Soil Biol. Biochem. 29:1225–1232.

Bentley, A.R., B. Tunalı, J.M. Nicol, L.W. Burgess, and B.A. Summerell. 2006. A survey of *Fusarium* species associated with wheat and grass stem bases in northern Turkey. Sydowia 58:163–177.

Bockus, W.W. 1995. Reaction of selected winter wheat cultivars to Cephalosporium stripe. p. 112. *In* Biological and cultural tests for control of plant diseases. Vol. 10. APS Press, St. Paul, MN.

Bockus, W.W., R.L. Bowden, R.M. Hunger, W.L. Morrill, T.D. Murray, and R.W. Smiley (ed.) 2009. Compendium of wheat diseases and insects. 3rd ed. APS Press, St. Paul, MN.

Bockus, W.W., and M.M. Claassen. 1985. Effect of lime and sulfur application to low-pH soil on incidence of Cephalosporium stripe in winter wheat. Plant Dis. 69:576–578.

Bockus, W.W., M.A. Davis, and T.C. Todd. 1994. Grain yield responses of winter wheat coinoculated with *Cephalosporium gramineum* and *Gaeumannomyces graminis* var. *tritici*. Plant Dis. 78:11–24.

Bockus, W.W., J.P. O'Connor, and P.J. Raymond. 1983. Effect of residue management method on incidence of Cephalosporium stripe under continuous winter wheat production. Plant Dis. 67:1323–1324.

Boville, W.D., W. Ma, K. Ritter, B.C.Y. Collard, M. Davis, G.B. Wildermuth, and M.W. Sutherland. 2006. Identification of novel QTL for resistance to crown rot in the doubled haploid wheat population 'W21MMT70' × 'Mendos'. Plant Breed. 125:538–543.

Brasset, P.R., and C.A. Gilligan. 1989. Fitting of single models for field disease progress data for the take-all fungus. Plant Pathol. 38:397–407.

Brooks, D.H., and M.G. Dawson. 1968. Influence of direct-drilling of winter wheat on incidence of take-all and eyespot. Ann. Appl. Biol. 61:57–64.

Bruehl, G.W. 1968. Ecology of Cephalosporium stripe disease of winter wheat in Washington. Plant Dis. Rep. 52:590–594.

Bruehl, G.W. (ed.) 1975. Biology and control of soil-borne plant pathogens. Am. Phytopathol. Soc., St. Paul, MN.

Bruehl, G.W. 1987. Soilborne plant pathogens. Macmillan Publishing. Co., New York, NY.

Burgess, L.W., D. Backhouse, B.A. Summerell, A.B. Pattison, T.A. Klein, R.J. Esdaile, and G. Ticehurst. 1993. Long-term effects of stubble management on the incidence of infection of wheat by *Fusarium graminearum* Schw. Group 1. Aust. J. Exp. Agric. 33:451–456.

Burgess, L.W., D. Backhouse, L.J. Swan, and R.J. Esdaile. 1996. Control of Fusarium crown rot of wheat by late stubble burning and rotation with sorghum. Australasian Plant Pathol. 25:229–233.

Burgess, L.W., A.H. Wearing, and T.A. Toussoun. 1975. Surveys of Fusaria associated with crown rot of wheat in eastern Australia. Aust. J. Agric. Res. 26:791–799.

Butler, F.C. 1961. Root and foot rot diseases of wheat. Science Bull. 7. Dep. Agric., New South Wales, Sydney.

Cai, X., S.S. Jones, and T.D. Murray. 1996. Characterization of an *Agropyron elongatum* chromosome conferring resistance to Cephalosporium stripe in common wheat. Genome 39:56–62.

Cai, X., S.S. Jones, and T.D. Murray. 1998. Molecular cytogenetic characterization of *Thinopyrum* and wheat-*Thinopyrum* translocated chromosomes in a wheat *Thinopyrum* amphiploid. Chromosome Res. 6:183–189.

Cavelier, N., D. Rousseau, and D. Lepage. 1987. Variabilité de *Pseudocercosporella herpotrichoides*, agent du piétin-verse des céréales: Comportement in vivo de deux types d'isolats et d'une population en mélange. Z. Pflanzenkrankh. Pflanzenschutz. 94:590–599.

Chakraborty, S., L. Liu, V. Mitter, J.B. Scott, O.A. Akinsanmi, S. Ali, R. Dill-Macky, J. Nicol, D. Backhouse, and S. Simpfendorfer. 2006. Pathogen population structure and epidemiology are keys to wheat crown rot and Fusarium head blight management. Australasian Plant Pathol. 35:643–655.

Chamswarng, C., and R.J. Cook. 1985. Identification and comparative pathogenicity of *Pythium* species from wheat roots and wheat-field soils in the Pacific Northwest. Phytopathology 75:821–827.

Christian, D.G., and D.P. Miller. 1984. Cephalosporium stripe in winter wheat grown after different methods of straw disposal. Plant Pathol. 33:605–606.

Clarke, B.B., and A.B. Gould (ed.) 1993. Turfgrass patch diseases caused by ectotrophic root-infecting fungi. APS Press, St. Paul, MN.

Clear, R.M., S.K. Patrick, D. Gaba, M. Roscoe, T.K. Turkington, T. Demeke, S. Pouleur, L. Couture, T.J. Ward, and K. O'Donnell. 2006. Trichothecene and zearalenone production, in culture, by isolates of *Fusarium pseudograminearum* from western Canada. Can. J. Plant Pathol. 28:131–136.

Colbach, N., P. Lucas, and J.M. Meynard. 1997. Influence of crop management on take-all development and disease cycles on winter wheat. Phytopathology 87:26–32.

Colbach, N., J.M. Meynard, C. Duby, and P. Huet. 1999. A dynamic model of the influence of rotation and crop management of the disease development of eyespot: Proposal of cropping systems with low disease risk. Crop Prot. 18:451–461.

Colbach, N., and L. Saur. 1998. Influence of wheat crop management on eyespot development and infection cycles. Eur. J. Plant Pathol. 104:37–48.

Collard, B.C.Y., R. Jolley, W.D. Bovill, R.A. Grams, G.B. Wildermuth, and M.W. Sutherland. 2006. Confirmation of QTL mapping and marker validation for partial seedling resistance to crown rot in wheat line '2-49'. Aust. J. Agric. Res. 57:967–973.

Conner, R.L., and T.G. Atkinson. 1989. Influence of continuous cropping on severity of common root rot in wheat and barley. Can. J. Plant Pathol. 11:127–132.

Conner, R.L., L.J. Duczek, G.C. Kozub, and A.D. Kuzyk. 1996. Influence of crop rotation on common root rot of wheat and barley. Can. J. Plant Pathol. 18:247–254.

Cook, R.J. 1980. *Fusarium* foot rot of wheat and its control in the Pacific Northwest. Plant Dis. 64:1061–1066.

Cook, R.J. 2003. Take-all of wheat. Physiol. Mol. Plant Pathol. 62:73–86.

Cook, R.J. 2007. Management of resident plant growth-promoting rhizobacteria with the cropping system: A review of experience in the US Pacific Northwest. Eur. J. Plant Pathol. 119:255–264.

Cook, R.J., and K.F. Baker. 1983. The nature and practice of biological control of plant pathogens. Am. Phytopathol. Soc., St. Paul, MN.

Cook, R.J., C. Chamswarng, and W.H. Tang. 1990. Influence of wheat chaff and tillage on *Pythium* populations in soil and *Pythium* damage to wheat. Soil Biol. Biochem. 22:939–947.

Cook, R.J., B.H. Ownley, H. Zhang, and D. Vakoch. 2000. Influence of paired-row spacing and fertilizer placement on yield and root diseases of direct-seeded wheat. Crop Sci. 40:1079–1087.

Cook, R.J., W.F. Schillinger, and N.W. Christensen. 2002a. Rhizoctonia root rot and take-all of wheat in diverse direct-seed spring cropping systems. Can. J. Plant Pathol. 24:349–358.

Cook, R.J., J.W. Sitton, and W.A. Haglund. 1987. Influence of soil treatments on growth and yield of wheat and implications for control of Pythium root rot. Phytopathology 77:1172–1198.

Cook, R.J., J.W. Sitton, and J.T. Waldher. 1980. Evidence for *Pythium* as a pathogen of direct-drilled wheat in the Pacific Northwest. Plant Dis. 64:102–103.

Cook R.J., and R.J. Veseth. 1991. Wheat health management. APS Press, St. Paul, MN.

Cook, R.J., D.M. Weller, A.Y. El-Banna, D. Vakoch, and H. Zhang. 2002b. Yield responses of direct-seed wheat to fungicide and rhizobacteria treatments. Plant Dis. 87:780–784.

Couch, H.B. 1995. Diseases of turfgrasses. 3rd ed. Keiger, Malabar, FL.

Couture, L., and J.C. Sutton. 1980. Effect of dry heat treatments on survival of seed borne *Bipolaris sorokiniana* and germination of barley seeds. Can. Plant Dis. Survey 60:59–61.

Cowger, C., and C.C. Mundt. 1998. A hydroponic seedling assay for resistance to Cephalosporium stripe of wheat. Plant Dis. 82:1126–1131.

Creatura, P.J., G.R. Safir, R.P. Scheffer, and T.D. Sharkey. 1981. Effects on *Cephalosporium gramineum* and a toxic metabolite on stomates and water status of wheat. Physiol. Plant Pathol. 19:313–323.

Crous, P.W., J.Z. Ewald Groenewald, and W. Gams. 2003. Eyespot of cereals revisited: ITS phylogeny reveals new species relationships. Eur. J. Plant Pathol. 109:841–850.

Dalal, R.C., E.J. Weston, W.M. Strong, K.J. Lehane, J.E. Cooper, G.B. Wildermuth, A.J. King, and C.J. Holmes. 2004. Sustaining productivity of a Vertosol at Warra, Queensland, with fertilisers, no-tillage or legumes: 7. Yield, nitrogen and disease-break benefits from lucerne in a two-year lucerne-wheat rotation. Aust. J. Exp. Agric. 44:607–616.

Douhan, G.W., and T.D. Murray. 2001. Infection of winter wheat by a beta-glucuronidase-transformed isolate of *Cephalosporium gramineum*. Phytopathology 91:232–239.

Douhan, G.W., T.D. Murray, and P.S. Dyer. 2003. Population genetic structure of *Tapesia acuformis* in Washington State. Phytopathology 93:650–656.

Duczek, L.J. 1990. Sporulation of *Cochliobolus sativus* on crown and underground parts of spring cereals in relation to weather and host species, cultivar, and phenology. Can. J. Plant Pathol. 12:273–278.

Dulout, A., P. Lucas, A. Sarniguet, and T. Doré. 1997. Effects of wheat volunteers and blackgrass in set-aside following a winter wheat crop on soil infectivity and soil conduciveness to take-all. Plant Soil 197:149–155.

Duveiller, E., and I. Garcia Altamirano. 2000. Pathogenicity of *Bipolaris sorokiniana* isolates from wheat roots, leaves and grains in Mexico. Plant Pathol. 49:235–242.

Elliott, M.L. 1991. Determination of an etiological agent of bermudagrass decline. Phytopathology 81:1380–1384.

El-Nashaar, H.M., and R.W. Stack. 1989. Effect of long-term continuous cropping of spring wheat on aggressiveness of *Cochliobolus sativus*. Can. J. Plant Sci. 69:395–400.

Ennaïfar, S., P. Lucas, J-M. Meynard, and D. Makowsky. 2005. Effects of summer fallow management on take-all of winter wheat caused by *Gaeumannomyces graminis* var. *tritici*. Eur. J. Plant Pathol. 112:167–181.

Ennaïfar, S., D. Makowsky, J.M. Meynard, and P. Lucas. 2007. Evaluation of models to predict take-all incidence in winter wheat as a function of cropping practices, soil, and climate. Eur. J. Plant Pathol. 118:127–143.

Farr, D.F., A.Y. Rossman, M.E. Palm, and E.B. McCray. 2007. Fungal databases [Online]. USDA-ARS, Beltsville, MD. Available at http://nt.ars-grin.gov/fungaldatabases/ (verified 13 June 2008).

Fedel-Moen, R., and J.R. Harris. 1987. Stratified distribution of *Fusarium* and *Bipolaris* on wheat and barley with dryland root rot in South Australia. Plant Pathol. 36:447–454.

Fernandez, M.R., and Y. Chen. 2005. Pathogenicity of Fusarium species on different plant parts of spring wheat under controlled conditions. Plant Dis. 89:164–169.

Fernandez, M.R., and R.P. Zentner. 2005. The impact of crop rotation and N fertilizer on common root rot of spring wheat in the Brown soil zone of western Canada. Can. J. Plant Sci. 85:569–575.

Fitt, B.D.L., A. Goulds, and R.W. Polley. 1988. Eyespot (*Pseudocercosporella herpotrichoides*) epidemiology in relation to prediction of disease severity and yield loss in winter wheat—A review. Plant Pathol. 37:311–328.

Freeman, J., and H. Ward. 2004. *Gaeumannomyces graminis*, the take-all fungus and its relatives. Mol. Plant Pathol. 5:235–252.

Fukui, R., G.S. Campbell, and R.J. Cook. 1994. Factors influencing the incidence of embryo infection by *Pythium* spp. during germination of wheat seeds in soils. Phytopathology 84:695–702.

Gac, M.L., F. Montfort, and N. Cavelier. 1996a. An assay based on the Polymerase Chain Reaction for the detection of N- and L-types of *Pseudocercosporella herpotricoides* in wheat. J. Phytopathol. 144:513–518.

Gac, M.L., F. Montfort, N. Cavelier, and A. Sailland. 1996b. Comparative study of morphological, cultural and molecular markers for the characterization of *Pseudocercosporella herpotrichoides* isolates. Eur. J. Plant Pathol. 102:325–337.

Gams, W. 2000. *Phialophora* and some similar morphologically little-differentiated anamorphs of divergent ascomycetes. Stud. Mycol. 45:187–199.

Garrett, S.D. 1970. Pathogenic root-infecting fungi. Cambridge Univ. Press, Cambridge, UK.

Gill, J.S., K. Sivasithamparam, and K.R.J. Smettem. 2000. Soil types with different texture affects development of Rhizoctonia root rot of wheat seedlings. Plant Soil 221:113–120.

Gill, J.S., K. Sivasithamparam, and K.R.J. Smettem. 2001a. Influence of depth of soil disturbance on root growth dynamics of wheat seedlings associated with *Rhizoctonia solani AG-8* disease severity in sandy and loamy sand soils of Western Australia. Soil Tillage Res. 62:73–83.

Gill, J.S., K. Sivasithamparam, and K.R.J. Smettem. 2001b. Soil moisture affects disease severity and colonisation of wheat roots by *Rhizoctonia solani AG-8*. Soil Biol. Biochem. 33:1363–1370.

Gill, J.S., K. Sivasithamparam, and K.R.J. Smettem. 2001c. Effect of soil moisture at different temperatures on Rhizoctonia root rot of wheat seedlings. Plant Soil 231:91–96.

Gill, J.S., K. Sivasithamparam, and K.R.J. Smettem. 2002. Size of bare-patches in wheat caused by *Rhizoctonia solani AG-8* is determined by the established mycelial network at sowing. Soil Biol. Biochem. 34:889–893.

González, D., D.E. Carling, S. Kuninaga, R. Vilgalys, and M.A. Cubeta. 2001. Ribosomal DNA systematics of *Ceratobasidium* and *Thanatephorus* with *Rhizoctonia* anamorphs. Mycologia 93:1138–1150.

González, D., M.A. Cubeta, and R. Vilgalys. 2006. Phylogenetic utility of indels within ribosomal DNA and beta-tubulin sequences from fungi in the *Rhizoctonia solani* species complex. Mol. Phylogenet. Evol. 40:459–470.

Gosme, M., L. Willocquet, and P. Lucas. 2007. Size, shape and intensity of aggregation of take-all disease during natural epidemics in second wheat crops. Plant Pathol. 56:87–96.

Grewal, H.S., R.D. Graham, and Z. Rengel. 1996. Genotypic variation in zinc efficiency and resistance to crown rot disease (*Fusarium graminearum* Schw. Group 1) in wheat. Plant Soil 186:219–226.

Griffin, D.M. 1972. Ecology of soil fungi. Syracuse Univ. Press, Syracuse, NY.

Gutteridge, R.J., J.P. Zhang, J.F. Jenkyn, and G.L. Bateman. 2005. Survival and multiplication of *Gaeumannomyces graminis* var. *tritici* (the wheat take-all fungus) and related fungi on different wild and cultivated grasses. Appl. Soil Ecol. 29:143–154.

Hall, R., and J.C. Sutton. 1998. Relation of weather, crop, and soil variables to the prevalence, incidence, and severity of basal infections of winter wheat in Ontario. Can. J. Plant Pathol. 20:69–80.

Hendrix, F.F., and W.A. Campbell. 1970. Distribution of *Phytophthora* and *Pythium* species in soils in the continental United States. Can. J. Bot. 48:377–384.

Herdina, and D.K. Roget. 2000. Prediction of take-all disease risk in field soils using a rapid and quantitative DNA soil assay. Plant Soil 227:87–98.

Hering, T.F., R.J. Cook, and W.H. Tang. 1987. Infection of wheat embryos by *Pythium* species during seed germination and the influence of seed age and soil matric potential. Phytopathology 77:1104–1108.

Higginbotham, R.W, K.K. Kidwell, and T.C. Paulitz. 2004a. Evaluation of adapted wheat cultivars for tolerance to Pythium root rot. Plant Dis. 88:1027–1032.

Higginbotham, R.W., T.C. Paulitz, and K.K. Kidwell. 2004b. Virulence of *Pythium* species isolated from wheat fields in eastern Washington. Plant Dis. 88:1021–1026.

Hollins, T.W., K.D. Lockley, J.A. Blackman, P.R. Scott, and J. Bingham. 1988. Field performance of Rendezvous, a wheat cultivar with resistance to eyespot (*Pseudocercosporella herpotrichoides*) derived from *Aegilops ventricosa*. Plant Pathol. 37:251–260.

Hornby, D. 1978. The problems of trying to forecast take-all. p. 151–158. *In* P.R. Scott and A. Bainbridge (ed.) Plant disease epidemiology. Blackwell Scientific Publications, Oxford, UK.

Hornby, D. 1979. Take-all decline: A theorist's paradise. p. 133–156. *In* B. Schippers and W. Gams (ed.) Soil-borne plant pathogens. Academic Press, London, UK.

Hornby, D., G.L. Bateman, R.J. Gutteridge, P. Lucas, A.E. Osbourn, E. Ward, and D.J. Yarham. 1998. Take-all disease of cereals: A regional perspective. CAB Int., Wallingford, UK.

Huber, D.M., C.C. Painter, H.C. McKay, and D.L. Petersen. 1968. Effect of nitrogen fertilization on take-all of winter wheat. Phytopathology 58:1470–1472.

Hunter, T. 1989. Occurrence of *Tapesia yallundae*, teleomorph of *Pseudocercosporella herpotrichoides*, on unharvested wheat culms in England. Plant Pathol. 38:598–603.

Ingram, D.M., and R.J. Cook. 1990. Pathogenicity of four *Pythium* species to wheat, barley, peas and lentils. Plant Pathol. 39:110–117.

Jahier, J., A.M. Tanguy, and G. Doussinault. 1989. Analysis of the level of eyespot resistance due to genes transferred to wheat from *Aegilops ventricosa*. Euphytica 44: 55–59.

Jenkinson, P., and D.W. Parry. 1994. Splash dispersal of conidia of *Fusarium culmorum* and *Fusarium avenaceum*. Mycol. Res. 98:506–510.

Kageyama, K., and E.B. Nelson. 2003. Differential inactivation of seed exudate stimulation of *Pythium ultimum* sporangium germination by *Enterobacter cloacae* influences biological control efficacy on different plant species. Appl. Environ. Microbiol. 69:1114–1120.

Kim, D.S., R.J. Cook, and D.M. Weller. 1997. *Bacillus* sp. L324–92 for biological control of three root diseases of wheat grown with reduced tillage. Phytopathology 87:551–558.

Kirkegaard, J.A., S. Simpfendorfer, J. Holland, R. Bambach, K.J. Moore, and G.J. Rebetzke. 2004. Effect of previous crops on crown rot and yield of durum and bread wheat in northern NSW. Aust. J. Agric. Res. 55:321–334.

Kobayasi, K., and T. Ui. 1979. Phytotoxicity and antimicrobial activity of Graminin A, produced by *Cepha-losporium gramineum*, the causal agent of Cephalosporium stripe disease of wheat. Physiol. Plant Pathol. 14:129–133.

Kokko, E.G., R.L. Conner, G.C. Kozub, and B. Lee. 1995. Effects of common root rot on discoloration and growth of the spring wheat root system. Phytopathology 85:203–208.

Krupa, S.V., and Y.R. Dommergues (ed.) 1979. Ecology of root pathogens. Elsevier Sci. Publ., Amsterdam, The Netherlands.

Kumar, J., P. Schafer, R. Huckelhoven, G. Langen, H. Baltruschat, E. Stein, S. Najarajan, and K.H. Kogel. 2002. *Bipolaris sorokiniana*, a cereal pathogen of global concern: Cytological and molecular approaches towards better control. Mol. Plant Pathol. 3:185–195.

Latin, R.X., R.W. Harder, and M.V. Wiese. 1982. Incidence of Cephalosporium stripe as influenced by winter wheat management practices. Plant Dis. 66:229–230.

Lebreton, L., M. Gosme, P. Lucas, A.Y. Guillerm-Erckelboudt, and A. Sarniguet. 2007. Linear relationship between *Gaeumannomyces graminis* var. *tritici* (Ggt) genotypic frequencies and disease severity on wheat roots in the field. Environ. Microbiol. 9:492–499.

Lebreton, L., P. Lucas, F. Dugas, A.Y. Guillerm, A. Schoeny, and A. Sarniguet. 2004. Changes in population structure of the soilborne fungus *Gaeumannomyces graminis* var. *tritici* during continuous wheat cropping. Environ. Microbiol. 6:1174–1185.

Lees, A.K., D.W. Cullen, L. Sullivan, and M.J. Nicolson. 2002. Development of conventional and quantitative real-time PCR assays for the detection and identification of *Rhizoctonia solani* AG-3 in potato and soil. Plant Pathol. 51:293–302.

Leroux, P., and M. Gredt. 1997. Evolution of fungicide resistance in the cereal eyespot fungi *Tapesia yallundae* and *Tapesia acuformis* in France. Pestic. Sci. 51:321–327.

Lévesque, C.A., and A.W.A.M. de Cock. 2004. Molecular phylogeny and taxonomy of the genus *Pythium*. Mycol. Res. 108:1363–1383.

Lévesque, C.A., and J.E. Rahe. 1992. Herbicide interactions with fungal root pathogens, with special reference to glyphosate. Annu. Rev. Phytopathol. 30:579–602.

Lucas, J.A., P.S. Dyer, and T.D. Murray. 2000. Pathogenicity, host specificity, and population biology of *Tapesia* spp., causal agents of eyespot disease in cereals. Adv. Bot. Res. 33:226–258.

Lucas, P., M.-H. Jeuffroy, A. Schoeny, and A. Sarniguet. 1997. Basis for nitrogen fertilisation management of winter wheat crops infected with take-all. Aspects Appl. Biol. 50:255–262.

Lucas, P., and A. Sarniguet. 1998. Biological control of soilborne pathogens with resident versus introduced antagonists: Should diverging approaches become strategic convergence? p. 351–370. *In* P. Barbosa (ed.) Conservation biological control. Academic Press, New York, NY.

Lucas, P., R.W. Smiley, and H.P. Collins. 1993. Decline of Rhizoctonia root rot on wheat in soils infested with *Rhizoctonia solani* AG-8. Phytopathology 83:260–265.

MacNish, G.C., and S.M. Neate. 1996. Rhizoctonia bare patch of cereals: An Australian perspective. Plant Dis. 80:965–971.

Maia, N. 1967. Obtention de blés tendres résistants au piétin-verse par croisements interspécifiques blés × *Aegilops*. Comptes Rendus Hebdomadaires des Séances de l'Académie d'Agriculture de France 53:149–154.

Martin, F.N. 2000. Phylogenetic relationships among some *Pythium* species inferred from sequence analysis of the mitochondrially encoded cytochrome oxidase II gene. Mycologia 92:711–727.

Martin, F.N., and J.E. Loper. 1999. Soilborne plant diseases caused by *Pythium* spp: Ecology, epidemiology, and prospects for biological control. Crit. Rev. Plant Sci. 18:111–181.

Martin, J.M., R.H. Johnston, and D.E. Mathre. 1989. Factors affecting the severity of Cephalosporium stripe of winter wheat. Can. J. Plant Pathol. 11:361–367.

Martin, J.M., D.E. Mathre, and R.H. Johnston. 1986. Winter wheat genotype responses to *Cephalosporium gramineum* inoculum levels. Plant Dis. 70:421–423.

Mathieson, J.T., C.M. Rush, D. Bordovsky, L.E. Clark, and O.R. Jones. 1990. Effects of tillage on common root rot of wheat in Texas. Plant Dis. 74:1006–1008.

Mathre, D.E., and R.H. Johnston. 1975. Cephalosporium stripe of winter wheat: Infection processes and host response. Phytopathology 65:1244–1249.
Mathre, D.E., and R.H. Johnston. 1977. Physical and chemical factors affecting sporulation of *Hymenula cerealis*. Trans. Brit. Mycol. Soc. 69:213–215.
Mathre, D.E., and R.H. Johnston. 1990. A crown barrier related to Cephalosporium stripe resistance in wheat relatives. Can. J. Bot. 68:1511–1514.
Mathre, D.E., R.H. Johnston, and J.M. Martin. 1985. Sources of resistance to *Cephalosporium gramineum* in *Triticum* and *Agropyron* species. Euphytica 34:419–424.
Mazzola, M., R.W. Smiley, A.D. Rovira, and R.J. Cook. 1996a. Characterization of *Rhizoctonia* isolates, disease occurrence and management in cereals. p. 259–267. *In* B. Sneh, S. Jabaji-Hare, S. Neate, and G. Dijst (ed.) *Rhizoctonia* species: Taxonomy, molecular biology, ecology, pathology and disease control. Kluwer Academic Publishers, Dordrecht, The Netherlands.
Mazzola, M., O.T. Wong, and R.J. Cook. 1996b. Virulence of *Rhizoctonia oryzae* and *R. solani* AG-8 on wheat and detection of *R. oryzae* in plant tissue by PCR. Phytopathology 86:354–360.
Miedaner, T. 1997. Breeding wheat and rye for resistance to *Fusarium* diseases. Plant Breed. 116:201–220.
Miedaner, T., G. Gang, C. Reinbrecht, and H.H. Geiger. 1997. Lack of association between *Fusarium* foot rot and head blight resistance in winter rye. Crop Sci. 37:327–331.
Milus, E.A., and C.S. Rothrock. 1997. Efficacy of bacterial seed treatments for controlling Pythium root rot of winter wheat. Plant Dis. 81:180–184.
Mitter, V., M.C. Zhang, C.J. Liu, R. Ghosh, M. Ghosh, and S. Chakraborty. 2006. A high-throughput glasshouse bioassay to detect crown rot resistance in wheat germplasm. Plant Pathol. 55:433–441.
Monds, R.D., M.G. Cromey, D.R. Lauren, M. di Menna, and J. Marshall. 2005. *Fusarium graminearum*, *F. cortaderiae* and *F. pseudograminearum* in New Zealand: Molecular phylogenetic analysis, mycotoxin chemotypes and coexistence of species. Mycol. Res. 109:410–420.
Moore, K.J., and R.J. Cook. 1984. Increased take-all of wheat with direct drilling in the Pacific Northwest. Phytopathology 74:1044–1049.
Morton, J.B., and D.E. Mathre. 1980. Identification of resistance to Cephalosporium stripe in winter wheat. Phytopathology 70:812–817.
Mundt, C.C. 2002. Performance of wheat cultivars and cultivar mixtures in the presence of Cephalosporium stripe. Crop Prot. 1:93–99.
Muranty, H., J. Jahier, A.J. Tanguy, A.J. Worland, and C. Law. 2002. Inheritance of resistance of wheat to eyespot at the adult stage. Plant Breed. 121:536–538.
Murray, T.D. 2006. Seed transmission of *Cephalosporium gramineum* in winter wheat. Plant Dis. 90:803–806.
Murray, T.D., D.W. Parry, and N.D. Cattlin (ed.) 1998. A color handbook of diseases of small grain cereal crops. Iowa State Univ. Press, Ames, IA.
Murray, T.D., L. Pritchett, S.S. Jones, and S. Lyon. 2001. Reaction of winter wheat cultivars and breeding lines to Cephalosporium stripe. p. S21. *In* Biological and cultural tests for control of plant diseases. APS Press, St. Paul, MN.
Murray, T.D., and C.C. Walter. 1991. Influence of pH and matric potential on sporulation of *Cephalosporium gramineum*. Phytopathology 81:79–84.
Murray, T.D., C.C. Walter, and J.C. Anderegg. 1992. Control of Cephalosporium stripe of winter wheat by liming. Plant Dis. 76:282–286.
Neate, S.M. 1987. Plant debris in soil as a source of inoculum of Rhizoctonia in wheat. Trans. Br. Mycol. Soc. 88:157–162.
Nelson, K.E., and J.C. Sutton. 1988. Epidemiology of eyespot on winter wheat in Ontario. Phytoprotection 69:9–21.
Nelson, P.E., T.A. Toussoun, and R.J. Cook (ed.) 1981. Fusarium: Diseases, biology, and taxonomy. Pennsylvania State Univ. Press, University Park, PA.
Nicol, J.M., N. Bolat, A. Bağcı, R.T. Trethowan, M. William, H. Hekimhan, A.F. Yildirim, E. Şahin, H. Elekçioğlu, H. Toktay, B. Tunalı, A. Hede, S. Taner, H.J. Braun, M. van Ginkel, M. Keser, Z. Arisoy, A. Yorgancılar, A. Tulek, D. Erdurmuş, O. Büyük, and M. Aydogdu. 2007. The international breeding strategy for the incorporation of resistance in bread wheat against the soil borne pathogens (dryland root rot and cyst and lesion nematodes) using conventional and molecular tools. p. 125–137. *In* H.T. Buck, J.E. Nisi, and N. Salomón (ed.) Wheat production in stressed environments. Springer Dordrecht, The Netherlands.
Nirenberg, H.I. 1981. Differentiation of *Pseudocercosporella* strains causing foot rot diseases of cereals: 1. Morphology. Z. Pflanzenkrankh. Pflanzenschutz. 88:241–248.
O'Donnell, K., T.J. Ward, D.M. Geiser, H.C. Kistler, and T. Aoki. 2004. Genealogical concordance between the mating type locus and seven other nuclear genes supports formal recognition of nine phylogenetically distinct species within the *Fusarium graminearum* clade. Fungal Genet. Biol. 41:600–623.
Ogoshi, A., R.J. Cook, and E.N. Bassett. 1990. *Rhizoctonia* species and anastomosis groups causing root rot of wheat and barley in the Pacific Northwest. Phytopathology 80:784–788.
Okubara, P.A., and T.C. Paulitz. 2005. Root defense responses to fungal pathogens: A molecular perspective. Plant Soil 274:215–226.
Okubara, P.A., K.L. Schroeder, and T.C. Paulitz. 2008. Identification and quantification of *Rhizoctonia solani* and *R. oryzae* using real-time PCR. Phytopathology 98:837–847.
Ophel-Keller, K., A. McKay, D. Hartley, Herdina, and J. Curran. 2008. Development of a routine DNA-based testing service for soilborne diseases in Australia. Australasian Plant Pathol. 37:243–253.
Pankhurst, C.E., H.J. McDonald, and B.G. Hawke. 1995. Influence of tillage and crop rotation on the epidemiology of *Pythium* infections of wheat in a red-brown earth of South Australia. Soil Biol. Biochem. 27:1065–1073.
Parker, C.A., A.D. Rovira, K.J. Moore, P.T.W. Wong, and J.F. Kollmorgen (ed.) 1985. Ecology and manage-

ment of soilborne plant pathogens. APS Press, St. Paul, MN.

Parry, D.W., T.R. Pettitt, P. Jenkinson, and A.K. Lees. 1994. The cereal Fusarium complex. p. 301–320. *In* J. P. Blakeman and B. Williamson (ed.) Ecology of plant pathogens. CAB Int., Wallingford, UK.

Paulitz, T.C. 2002. First report of *Rhizoctonia oryzae* on pea. Plant Dis. 86:442.

Paulitz, T.C., and K. Adams. 2003. Composition and distribution of *Pythium* communities from wheat fields in eastern Washington State. Phytopathology 93:867–873.

Paulitz, T.C., K. Adams, and M. Mazzola. 2003a. *Pythium abappressorium*—a new species from eastern Washington. Mycologia 95:80–86.

Paulitz, T.C., and K.L. Schroeder. 2005. A new method for quantification of *Rhizoctonia solani* and *R. oryzae* from soil. Plant Dis. 89:767–772.

Paulitz, T.C., and R.B. Scott. 2006. Effect of seed treatments for control of Rhizoctonia root rot in spring wheat, 2005. Fungic. Nematic. Tests 61:ST014.

Paulitz, T.C., R.W. Smiley, and R.J. Cook. 2002a. Insights into the prevalence and management of soilborne cereal pathogens under direct seeding in the Pacific Northwest, U.S.A. Can. J. Plant Pathol. 24:416–428.

Paulitz, T.C., J. Smith, and K. Kidwell. 2002b. Virulence of *Rhizoctonia oryzae* on wheat and barley cultivars from the Pacific Northwest. Plant Dis. 87:51–55.

Paulitz, T.C., H. Zhang, and R.J. Cook. 2003b. Spatial distribution of *Rhizoctonia oryzae* and rhizoctonia root rot in direct-seeded cereals. Can. J. Plant Pathol. 25:295–303.

Pettitt, T., X.M. Xu, and D. Parry. 2003. Association of *Fusarium* species in the wheat stem rot complex. Eur. J. Plant Pathol. 109:769–774.

Piccinni, G., C.M. Rush, K.M. Vaughn, and M.D. Lazar. 2000. Lack of relationship between susceptibility to common root rot and drought tolerance among several closely related wheat lines. Plant Dis. 84:25–28.

Pittaway, P.A. 1995. Opportunistic association between *Pythium* species and weed residues causing seedling emergence failure in cereals. Aust. J. Agric. Res. 46:655–662.

Ponchet, J. 1959. La maladie du piétin-verse des céréales: *Cercosporella herpotrichoides* Fron. Importance agronomique, biologie, épiphytologie. Annales des Epiphyties 10:45–98.

Pool, R.A.F., and E.L. Sharp. 1969. Some environmental and cultural factors affecting Cephalosporium stripe of winter wheat. Plant Dis. Rep. 53:898–902.

Priestley, R.A., F.M. Dewey, P. Nicholson, and H.N. Rezanoor. 1992. Comparison of isoenzyme and DNA markers for differentiating W-, R- and C-pathotypes of *Pseudocercosporella herpotrichoides*. Plant Pathol. 41:591–599.

Pumphrey, F.V., D.E. Wilkins, D.C. Hane, and R.W. Smiley. 1987. Influence of tillage and nitrogen fertilizer on Rhizoctonia root rot (bare patch) of winter wheat. Plant Dis. 71:125–127.

Rachdawong, S., C.L. Cramer, E.A. Grabau, V.K. Stromberg, G.H. Lacey, and E.L. Stromberg. 2002. *Gaeumannomyces graminis* vars. *avenae*, *graminis*, and *tritici* identified using PCR amplification of avenacinase-like genes. Plant Dis. 86:652–660.

Rahman, M., C.C. Mundt, T.J. Wolpert, and O. Riera-Lizarazu. 2001. Sensitivity of wheat genotypes to a toxic fraction produced by *Cephalosporium gramineum* and correlation with disease susceptibility. Phytopathology 91:702–707.

Rapilly, F., P. Eschenbrenner, E. Choisnes, and F. La Croze. 1979. La prévision du piétin-verse sur blé d'hiver. Perspectives Agricoles 23:30–40.

Ray, V.R., M.J. Crook, P. Jenkinson, and S.G. Edwards. 2006. Effect of eyespot caused by *Oculimacula yallundae* and *O. acuformis*, assessed visually and by competitive PCR, on stem strength associated with lodging resistance and yield of winter wheat. J. Exp. Bot. 57:2249–2257.

Raymond, P.J., and W.W. Bockus. 1984. Effect of seeding date of winter wheat on incidence, severity, and yield loss caused by Cephalosporium stripe in Kansas. Plant Dis. 68:665–667.

Reis, E.M., and J.J.R. Abrao. 1983. Effect of tillage and wheat residue management on the vertical distribution and inoculum density of *Cochliobolus sativus* in soil. Plant Dis. 67:1088–1089.

Robbertse, B., G.F. Campbell, and P.W. Crous. 1995. Revision of *Pseudocercosporella*-like species causing eyespot disease of wheat. S. Afr. J. Bot. 61:43–48.

Roget, D.K. 1995. Decline in root rot (*Rhizoctonia solani* AG-8) in wheat in a tillage and rotation experiment at Avon, South Australia. Aust. J. Exp. Agric. 35:1009–1013.

Roget, D.K., S.M. Neate, and A.D. Rovira. 1996. Effect of sowing point design and tillage practice on the incidence of Rhizoctonia root rot, take-all and cereal cyst nematode in wheat and barley. Aust. J. Exp. Agric. 36:683–693.

Roget, D.K., N.R. Venn, and A.D. Rovira. 1987. Reduction of Rhizoctonia root rot of direct-drilled wheat by short-term chemical fallow. Aust. J. Exp. Agric. 27:425–430.

Rowe, R.C., and R.L. Powelson. 1973. Epidemiology of *Cercosporella* footrot of wheat: Spore production. Phytopathology 63:981–984.

Sarniguet, A., P. Lucas, and M. Lucas. 1992a. Relationships between take-all, soil conduciveness to the disease, populations of fluorescent pseudomonads and nitrogen fertilizers. Plant Soil 145:17–27.

Sarniguet, A., P. Lucas, M. Lucas, and R. Samson. 1992b. Soil conduciveness to take-all of wheat: Influence of the nitrogen fertilizers on the structure of populations of fluorescent pseudomonads. Plant Soil 145:29–36.

Satyaprasad, K., G.L. Bateman, and E. Ward. 2000. Comparisons of isolates of *Fusarium avenaceum* from white lupin and other crops by pathogenicity tests, DNA analyses and vegetative compatibility tests. J. Phytopathol. 148:211–219.

Savary, S., B. Mille, B. Rolland, and P. Lucas. 2006. Patterns and management of crop multiple pathosystems. Eur. J. Plant Pathol. 115:123–138.

Schippers, B., and W. Gams (ed.) 1979. Soil-borne plant pathogens. Academic Press, London, UK.

Schoeny, A., F. Devienne-Barret, M.H. Jeuffroy, and P. Lucas. 2003. Effect of take-all infections on nitrate uptake in winter wheat. Plant Pathol. 52:52–59.

Schoeny, A., M.H. Jeuffroy, and P. Lucas. 2001. Influence of take-all epidemics on winter wheat yield formation and yield loss. Phytopathology 91:694–701.

Schoeny, A., and P. Lucas. 1999. Modeling of take-all epidemics to evaluate the efficacy of a new seed-treatment fungicide on wheat. Phytopathology 89:954–961.

Schroeder, K.L., P.A. Okubara, and T.C. Paulitz. 2007. Geographic distribution of *Rhizoctonia* and *Pythium* species in soils from dryland cereal cropping systems in eastern Washington Phytopathology 97:S105 (abstract).

Schroeder, K.L., P.A. Okubara, J.T. Tambong, C.A. Lévesque, and T.C. Paulitz. 2006. Identification and quantification of pathogenic *Pythium* spp. from soils in eastern Washington using real-time PCR. Phytopathology 96:637–647.

Schroeder, K.L., and T.C. Paulitz. 2006. Root diseases of wheat and barley during the transition from conventional tillage to direct seeding. Plant Dis. 90:1247–1253.

Shefelbine, P.A., and W.W. Bockus. 1989. Decline of Cephalosporium stripe by monoculture of moderately resistant winter wheat cultivars. Phytopathology 79:1127–1131.

Siebrasse, G., and H. Fehrmann. 1987. An enlarged model for the chemical control of eyespot (*Pseudocercosporella herpotrichoides*) in winter wheat. Z. Pflanzenkrankh. Pflanzenschutz. 94:137–149.

Singleton, L.L., J.D. Mihail, and C.M. Rush (ed.) 1992. Methods for research on soilborne phytopathogenic fungi. APS Press, St. Paul, MN.

Sitton, J.W., and R.J. Cook. 1981. Comparative morphology and survival of chlamydospores of *Fusarium roseum* 'Culmorum' and 'Graminearum'. Phytopathology 71:85–90.

Slope, D.B., and R. Bardner. 1965. Cephalosporium stripe of wheat and root damage by insects. Plant Pathol. 14:184–187.

Smiley, R.W. 1978. Antagonists of *Gaeumannomyces graminis* from the rhizoplane of wheat in soils fertilized with ammonium or nitrate nitrogen. Soil Biol. Biochem. 10:169–174.

Smiley, R.W., H.P. Collins, and P.E. Rasmussen. 1996a. Diseases of wheat in long-term agronomic experiments at Pendleton, Oregon. Plant Dis. 80:813–820.

Smiley, R.W., P.H. Dernoeden, and B.B. Clarke. 2005a. Compendium of turfgrass diseases. 3rd ed. APS Press, St. Paul, MN.

Smiley, R.W., M.C. Fowler, and K.L. Reynolds. 1986. Temperature effects on take-all of cereals, caused by *Phialophora graminicola* and *Gaeumannomyces graminis*. Phytopathology 76:923–931.

Smiley, R.W., J.A. Gourlie, S.A. Easley, and L.M. Patterson. 2005b. Pathogenicity of fungi associated with the wheat crown rot complex in Oregon and Washington. Plant Dis. 89:949–957.

Smiley, R.W., A.G. Ogg, and R.J. Cook. 1992. Influence of glyphosate on Rhizoctonia root rot, growth, and yield of barley. Plant Dis. 76:937–942.

Smiley, R.W., and L. Patterson. 1996. Pathogenic fungi associated with Fusarium foot rot of winter wheat in the semiarid Pacific Northwest. Plant Dis. 80:944–949.

Smiley, R.W., L.-M. Patterson, and C.W. Shelton. 1996b. Fungicide seed treatments influence emergence of winter wheat in cold soil. J. Prod. Agric. 9:559–563.

Smiley, R.W., and W. Uddin. 1993. Influence of soil temperature on Rhizoctonia root rot (*R. solani* AG-8 and *R. oryzae*) of winter wheat. Phytopathology 83:777–785.

Smiley, R.W., W. Uddin, S. Ott, and K.E.L. Rhinhart. 1990. Influence of flutolonil and tolclofos-methyl on root and culm diseases of winter wheat. Plant Dis. 74:788–791.

Smiley, R.W., and D.E. Wilkins. 1992. Impact of sulfonylurea herbicides on Rhizoctonia root rot, growth, and yield of winter wheat. Plant Dis. 76:399–404.

Smiley, R.W., and D.E. Wilkins. 1993. Annual spring barley growth, yield, and root rot in high- and low-residue tillage systems. J. Prod. Agric. 6:270–275.

Smith, J.D., N. Jackson, and A.R. Woolhouse. 1989. Fungal diseases of amenity turf grasses. E. & F.N. Spon, London, UK.

Smith, J.D., K.K. Kidwell, M.A. Evans, R.J. Cook, and R.W. Smiley. 2003a. Assessment of spring wheat genotypes for disease reaction to *Rhizoctonia solani* AG 8 in controlled environment and no-till field conditions. Crop Sci. 43:694–700.

Smith, J.D., K.K. Kidwell, M.A. Evans, R.J. Cook, and R.W. Smiley. 2003b. Evaluation of spring cereal grains and wild *Triticum* relatives for resistance to *Rhizoctonia solani* AG 8. Crop Sci. 43:701–709.

Sneh, B., L. Burpee, and A. Ogoshi. 1991. Identification of *Rhizoctonia* species. APS Press, St. Paul, MN.

Sneh, B., S. Jabaji-Hare, S. Neate, and G. Dijst (ed.) 1996. *Rhizoctonia* species: Taxonomy, molecular biology, ecology, pathology and disease control. Kluwer Academic Publishers, Dordrecht, The Netherlands.

Specht, L.P., and T.D. Murray. 1990. Effects of root-wounding and inoculum density on Cephalosporium stripe in winter wheat. Phytopathology 80:1108–1114.

Stasinopoulos, S.J., and R.J. Seviour. 1989. Exopolysaccharide formation by isolates of *Cephalosporium* and *Acremonium*. Mycol. Res. 92:55–60.

Stiles, C.M., and T.D. Murray. 1996. Infection of field-grown winter wheat by *Cephalosporium gramineum* and the effect of soil pH. Phytopathology 86:177–183.

Summerell, B.A., and L.W. Burgess. 1988. Stubble management practices and the survival of *Fusarium graminearum* Group 1 in wheat stubble residues. Australasian Plant Pathol. 17:88–93.

Summerell, B.A., L.W. Burgess, D. Backhouse, S. Bullock, and L.J. Swan. 2001a. Natural occurrence of perithecia of *Gibberella coronicola* on wheat plants with crown rot in Australia. Australasian Plant Pathol. 30:353–356.

Summerell, B.A., L.W. Burgess, T.A. Klein, and A.B. Pattison. 1990. Stubble management and the site of penetration of wheat by *Fusarium graminearum* Group 1. Phytopathology 80:877–879.

Summerell, B.A., J.F. Leslie, D. Backhouse, W.L. Bryden, and L.W. Burgess (ed.) 2001b. Fusarium: Paul E. Nelson memorial symposium. APS Press, St. Paul, MN.

Tinline, R.D. 1977. Multiple infections of subcrown internodes of wheat (*Triticum aestivum*) by common root rot fungi. Can. J. Bot. 55:30–34.

Tinline, R.D., K.L. Bailey, L.J. Duczek, and H. Harding (ed.) 1991. Proc. Int. Workshop on Common Root Rot of Cereals, 1st, Saskatoon, Canada. 11–14 August 1991. Agric. Canada, Saskatoon.

Tinline, R.D., and D.T. Spurr. 1991. Agronomic practices and common root rot in spring wheat: Effect of tillage on disease and inoculum density of *Cochliobolus sativus* in soil. Can. J. Plant Pathol. 13:258–266.

Tinline, R.D., H. Ukrainetz, and D.T. Spurr. 1993. Effect of fertilizers and of liming acid soil on common root rot in wheat, and of chloride on the disease in wheat and barley. Can. J. Plant Pathol. 15:65–73.

Tinline, R.D., G.B. Wildermuth, and D.T. Spurr. 1988. Inoculum density of *Cochliobolus sativus* in soil and common root rot of wheat cultivars in Queensland. Aust. J. Agric. Res. 39:569–577.

Tóth, B., Á. Mesterházy, P. Nicholson, J. Téren, and J. Varga. 2004. Mycotoxin production and molecular variability of European and American isolates of *Fusarium culmorum*. Eur. J. Plant Pathol. 110:587–599.

Tunalı, B., J.M. Nicol, D. Hodson, Z. Uçkun, O. Büyük, D. Erdurmuş, H. Hekimhan, H. Aktaş, M.A. Akbudak, and S.A. Bağcı. 2008. Root and crown rot fungi associated with spring, facultative and winter wheat in Turkey. Plant Dis. 92:1299–1306.

van der Plaats-Niterink, A.J. 1981. Monograph of the genus Pythium: Studies in mycology 21. Centraalbureau voor Schimmelcultures, Baarn, The Netherlands.

Van Welt, S.L., and D.W. Fullbright. 1986. Pathogenicity and virulence of *Cephalosporium gramineum* is independent of in vitro production of extracellular polysaccharides and graminin A. Physiol. Mol. Plant Pathol. 28:299–307.

Van Wyk, P.S., O. Los, G.D.C. Pauer, and W.F.O. Marasas. 1987. Geographic distribution and pathogenicity of *Fusarium* species associated with crown rot of wheat in the Orange Free State, South Africa. Phytophylactica 19:271–274.

Vasquez-Siller, L.M., and T.D. Murray. 2003. Detection of *Cephalosporium gramineum* in wheat seed with PCR. Phytopathology 93:S87 (abstract).

Veit, S., J.M. Worle, T. Nurnberger, W. Koch, and H.U. Seitz. 2001. A novel protein elicitor (PaNie) from *Pythium aphanidermatum* induces multiple defense responses in carrot, *Arabidopsis*, and tobacco. Plant Physiol. 127:832–841.

Vijayan, P., J. Shockey, C.A. Lévesque, R.J. Cook, and J. Browse. 1998. A role for jasmonate in pathogen defense of Arabidopsis. Proc. Natl. Acad. Sci. USA 95:7209–7214.

Wallwork, H. 1992. Cereal leaf and stem diseases. South Aust. Res. Dev. Inst., Adelaide.

Wallwork, H. 2000. Cereal root and crown diseases. South Aust. Res. Dev. Inst., Adelaide.

Wallwork, H., M. Butt, J.P.E. Cheong, and K.J. Williams. 2004. Resistance to crown rot in wheat identified through an improved method for screening adult plants. Australasian Plant Pathol. 33:1–7.

Wallwork, H., and B. Spooner. 1988. *Tapesia yallundae* the teleomorph of *Pseudocercosporella herpotrichoides*. Trans. Brit. Mycol. Soc. 91:703–705.

Walsh, K., J. Korimbocus, N. Boonham, P. Jennings, and M. Hims. 2005. Using real-time PCR to discriminate and quantify the closely related wheat pathogens *Oculimacula yallundae* and *Oculimacula acuformis*. J. Phytopathol. 153:715–721.

Ward, W.E., and R.M. Gray. 1992. Generation of a ribosomal DNA probe by PCR and its use in identification of fungi within the *Gaeumannomyces–Phialophora* complex. Plant Pathol. 41:730–736.

Weller, D.M., and R.J. Cook. 1986. Increased growth of wheat by seed treatments with fluorescent pseudomonads, and implications of *Pythium* control. Can. J. Plant Pathol. 8:328–334.

Weller, D.M., R.J. Cook, G.E. MacNish, N. Bassett, R.L. Powelson, and R.R. Petersen. 1986. Rhizoctonia root rot of small grains favored by reduced tillage in the Pacific Northwest. Plant Dis. 70:70–73.

Weller, D.M., B.B. Landa, O.V. Mavrodi, K.L. Schroeder, L. De La Fuente, S. Blouin Bankhead, R. Allende Molar, R.F. Bonsall, D.V. Mavrodi, and L.S. Thomashow. 2007. Role of 2,4-diacetylphloroglucinol-producing fluorescent *Pseudomonas* spp. in the defense of plant roots. Plant Biol. 9:4–20.

Wiese, M.V. 1972. Colonization of wheat seedlings by *Cephalosporium gramineum* in relation to symptom development. Phytopathology 62:1013–1018.

Wiese, M.V., and A.V. Ravenscroft. 1975. *Cephalosporium gramineum* populations in soil under winter wheat cultivation. Phytopathology 65:1129–1133.

Wiese, M.V., and A.V. Ravenscroft. 1978. Sporodochium development and conidium production in *Cephalosporium gramineum*. Phytopathology 68:395–401.

Wildermuth, G.B., and R.B. McNamara. 1987. Susceptibility of winter and summer crops to root and crown infection by *Bipolaris sorokiniana*. Plant Pathol. 36:481–491.

Wildermuth, G.B., and R.B. McNamara. 1991. Effect of cropping history on soil populations of *Bipolaris sorokiniana* and common root rot of wheat. Aust. J. Agric. Res. 42:779–790.

Wildermuth, G.B., R.B. McNamara, and J.S. Quick. 2001. Crown depth and susceptibility to crown rot in wheat. Euphytica 122:397–405.

Wildermuth, G.B., R.D. Tinline, and R.B. McNamara. 1992. Assessment of yield loss caused by common root rot in wheat cultivars in Queensland. Aust. J. Agric. Res. 43:43–58.

Wildermuth, G.B., G.A. Thomas, B.J. Radford, R.B. McNamara, and A. Kelly. 1997. Crown rot and common root rot in wheat grown under different tillage and stubble treatments in southern Queensland, Australia. Soil Tillage Res. 44:211–224.

Windels, C.E., J.A. Lamb, and T.E. Cymbaluk. 1992. Common root rot and yield responses in spring wheat from chloride application to soil in northwestern Minnesota. Plant Dis. 76:908–911.

Windels, C.E., and J.V. Wiersma. 1992. Incidence of *Bipolaris* and *Fusarium* on subcrown internodes of spring barley and wheat grown in continuous conservation tillage. Phytopathology 82:699–705.

Wiseman, B.M., S.M. Neate, K.O. Keller, and S.E. Smith. 1996. Suppression of *Rhizoctonia solani* anastomosis group 8 in Australia and its biological nature. Soil Biol. Biochem. 28:727–732.

Zillinsky, F.J. 1983. Common diseases of small grain cereals: A guide to identification. CIMMYT, D.F., Mexico.

Chapter 7

Diseases Which Challenge Global Wheat Production—Powdery Mildew and Leaf and Head Blights

David Marshall

SUMMARY

(1) Conservation tillage practices have provided wheat producers with many crop production benefits but have exacerbated the effects of residue-borne pathogens.

(2) Not only does wheat residue increase the quantity of primary inoculum available to infect subsequent wheat crops, it also serves as the primary means for powdery mildew, leaf blight, and head blight pathogens to complete their life cycle, thereby increasing variability in the pathogen populations.

(3) Crop diversity and host-plant resistance when used together are the two primary methods of managing residue-borne diseases of wheat. Emphasis must be placed on methods of identifying and incorporating durable host-plant resistance that does not encourage resistance-breaking variability in the pathogen populations.

INTRODUCTION

Perhaps the most underrated phenomenon in agriculture is the introduction or intensification of new, unsuspected, or underappreciated problems associated with changes in agricultural practices (Hunter and Leake 1933). Taking root in the 1950s, conservation tillage had become a common and widespread practice by the mid-1980s in many parts of the world, particularly in wheat (*Triticum* spp.), maize (*Zea mays* L.), soybean [*Glycine max* (L.) Merr.], and cotton (*Gossypium hirsutum* L.). First used for its control of soil erosion, conservation tillage was also found to improve soil structure, conserve soil moisture, insulate the soil to temperature fluctuation, store soil carbon, and reduce farming costs due to reduced tractor use (Minoshima et al., 2007). Combined with the advent of highly effective herbicides and better field equipment, conservation tillage has replaced plowing and disking as the standard method of cropland management in many areas.

At the heart of conservation tillage is the management of crop residue. Following harvest of the primary economic yield (typically grain or lint), what remains in the field is the crop residue. For simplicity, crop residue is equated here with stover and postharvest aboveground biomass. Conservation tillage or no-till results in crop residue left on the soil surface not incorporated into the soil or removed for other purposes such as energy production. When crop residue remains on the soil surface, it is degraded much more slowly than if incorporated into the soil where microorganisms can use it as a food source. As a result, pathogens that can colonize the crop residue are provided a safe harbor between actively growing crops. Subsequently, if crop diversity is minimized, or if a pathogen can attack multiple crops, then conditions become conducive for pathogen and disease increase over time and space.

In developing countries, one major positive impact of no-till is its advantage, especially in dry years, to conserve soil moisture, thereby reducing the risk of crop failure in areas without irrigation. On-farm labor is reduced with no-till, because land preparation is minimized and labor is not needed to remove weeds. No-till can lead to better stand establishment through greater moisture availability, it reduces turnaround time to plant a second crop, and it provides a larger number of beneficial insects for pest control (Ekboir et al., 2002). No-till practices led to higher yields in Ghana with increased food availability, more time for other activities, and reduced labor and effort. No-till also expands the markets for agrochemicals, especially herbicides. However, no-till technologies may result in threatened sustainability of the system due to increases in new weeds, pests, and diseases. Clearly, by combining increased pathogen survival on crop residue with other factors conducive to disease increase (e.g., susceptible cultivars and favorable climate), previous episodic disease problems can become recurring problems. These factors can account for the reemergence or intensification in wheat of powdery mildew, the fungal leaf spots (septoria leaf blotch, stagonospora leaf blotch, and tan spot), and Fusarium head blight.

POWDERY MILDEW

Taxonomy and life history

The causal fungus of powdery mildew is the heterothallic ascomycete *Blumeria graminis* (DC) E.O. Speer f. sp. *tritici* (syn. *Erysiphye graminis* DC ex Merat f. sp. *tritici* E. Marchal). Because it is an obligate parasite, the powdery mildew fungus is not typically considered a residue-borne pathogen. However, following successive cycles of asexual, conidial infections on green tissue, the fungus produces dark, round-shaped cleistothecia on both green and senescing host tissue. The cleistothecia contain the sexually produced ascospores. Because they are the result of recombination, the ascospores may have new virulence combinations. The cleistothecia remain in crop residue and are much hardier and longer lived than the cottony-appearing mycelia and conidia. Ascospores require a maturation period and can be released from cleistothecia in the fall, winter, or spring to serve as primary inoculum to infect wheat.

During the growing season conidia produced on wheat plants are wind-dispersed. Conidia germinate and infect plants under cool, moist conditions. Infection does not require free water on the plant surfaces, but high relative humidity (near 100%) favors infection. Under optimum conditions, a new crop of conidia are produced every 7 to 10 days.

Identification and symptomology

Powdery mildew on wheat is recognized by small, white pustules of cottony mycelia (masses of fungal threads of hyphae that make up the body of the fungus), conidiophores, and conidia. These occur on the upper and lower surfaces of the leaves, as well as on leaf sheaths and heads. As these patches sporulate and age, they become a dull tan or gray color. Chlorotic patches may later surround the mildew colonies. The disease typically begins in the lower leaves and spreads to the upper leaves and heads.

Optimum development of powdery mildew occurs between 15 and 22 °C air temperatures. The disease is associated with dense plant growth, such as high seeding rates and high nitrogen fertilization, and also with cool, humid weather conditions. Although favored by high humidity, water on the leaves inhibits conidial germination; thus long periods of rain will limit disease development. Heavily diseased leaves turn yellow and die prematurely. Plants are most susceptible during periods of rapid growth, especially from stem elongation through heading.

Distribution and losses

Although distributed worldwide, powdery mildew is economically important in most temperate regions of the world, particularly on winter wheat (Bennett 1981). Because of its common occurrence, powdery mildew can cause extensive plant damage to susceptible cultivars under climatic conditions conducive to disease. Early infection can cause losses of entire leaves. Under conducive conditions, the disease can weaken stems and thereby intensify lodging. In addition to reducing leaf area available for photosynthesis, the fungus also serves as a metabolic sink, drawing photosynthates to itself. Reductions in grain yield, grain weight, and grain quality can occur. Reported yield losses have varied from 5 to 45% (Fried et al., 1981; Leath and Bowen 1989; Griffey et al., 1993; Conner et al., 2003). The disease also reduces kernel softness, flour yield, and protein content in soft wheat (Johnson et al., 1979; Everts et al., 2001).

Pathogen variability

Blumeria graminis tritici has acquired a high degree of specialization on wheat, with races attacking specific cultivars. In Hungary, between 1971 and 1999, small shifts in virulence in the powdery mildew population were found every 3 to 5 years, while major shifts were found about every 5 to 7 years (Szunics et al., 2001). Genetic drift, mutation, and directional selection all contribute to shifts in powdery mildew populations (Limpert et al., 1999; Paillard et al., 2000). The frequency of virulence to specific *Pm* genes changes over time and space (Niewoehner and Leath 1998; Imani et al., 2002; Brown and Hovmoller 2002; Parks et al., 2008).

STAGONOSPORA NODORUM BLOTCH

Taxonomy and life history

The causal fungus of stagonospora nodorum blotch (previously known as septoria nodorum blotch) has the telomorphic stage of *Phaeosphaeria nodorum* (E. Muller) Hedjar. (Shoemaker and Babcock 1989; Cunfer and Ueng 1999). This heterothallic ascomycete produces ascospores in pseudothecia that mature in wheat residue. Ascospores were shown to be produced and deposited on wheat from August to October in the northern hemisphere and from February to April in the southern hemisphere (Arseniuk et al., 1998). Following primary infection of the leaves, the fungus forms asexual pycnidia and pycnidiospores of the anamorphic stage, *Stagonospora nodorum* (Berk.) Castellani and Germano. Changes in fungal nomenclature, coupled with the use of the full Latin name of the anamorphic stage in the common name of the disease, has caused some confusion in what to call the disease (Cunfer and Ueng 1999). Pycnidiospores serve as the repeating stage of the fungus over the growing season and are spread by splashing rainfall (Shah et al., 2001). Epidemics are favored by wet, windy conditions and by air temperatures between 20 and 27 °C. Infected seed can serve as a source of primary inoculum (Bennett et al., 2007).

Identification and symptomology

The disease is also known as glume blotch. On wheat heads, it is first apparent on glumes as small, irregular, gray to brown spots, which can enlarge and take on a chocolate-brown color. These small spots or blotches then coalesce, often producing small black-brown fungal fruiting bodies in the center of the lesions. Severe head infections occur when there is excessive rainfall between flowering and maturity. The fungus can infect the seed, which can serve as a source of primary inoculum for subsequent wheat crops.

The lesions also can occur on leaves, leaf sheaths, and stem nodes. Early leaf symptoms appear as tan-brown spots having an oval or lens shape. As these spots enlarge and merge, the necrotic spots turn a light-gray color and form irregular patterns on the leaf. Pycnidia may be apparent and scattered within the lesions. As the infected plant nears maturity, the glumes and nodes become symptomatic.

Distribution and losses

Stagonospora nodorum blotch is distributed worldwide, particularly in areas prone to warm, humid, and wet conditions during the growing season (Eyal et al., 1987). The disease reduces grain yield, grain weight, and grain quality (Eyal et al., 1987; McKendry et al., 1995). Yield losses up to 50% have been reported (King et al., 1983; Eyal et al., 1987; Bhathal et al., 2003).

Pathogen variability

In the US, Adhikari et al. (2008) showed that sexual reproduction was common in *Phaeosphaeria nodorum* in both spring wheat and winter wheat growing areas. Moreover, they showed that *Phaeosphaeria nodorum* populations could be genetically differentiated from separate geographic locations. Differences in aggressiveness among *Stagonospora nodorum* isolates have been found (Ali and Adhikari 2008). Host-selective toxins have been found to be produced by *S. nodorum* that are important in disease development and determination of virulence (Liu et al., 2004; Friesen et al., 2008). The frequency of occurrence of *Phaeosphaeria nodorum* pseudothecia indicates ample possibility for the fungus to recombine (Cowger and Silva-Rojas 2006).

SEPTORIA TRITICI BLOTCH

Taxonomy and life history

The telomorphic stage of the causal fungus of septoria tritici blotch is *Mycosphaerella graminicola* (Fuckel) Schroeter, a heterothallic ascomycete (Sanderson 1972; Kema et al., 1996b; Cunfer and Ueng 1999). Ascospores are produced in pseudothecia during the growing season on infected leaves and sometimes on glumes (Garcia and Marshall 1992). More commonly, pseudothecia are formed as the crop matures or when infections are heavy and lesions coalesce. Pseudothecia can survive saprophytically on wheat residue and produce ascospores to infect subsequent wheat crops (Scott et al., 1988; Hunter et al., 1999; Hoorne et al., 2002; Eriksen and Munk 2003). Ascospores are primarily windblown and can serve as primary inoculum (Shaw and Royle 1989). The anamorphic stage of the fungus is referred to as *Septoria tritici* Desm. Conidia are produced in and around pycnidia and are spread by splashing rain within the wheat canopy. Conidia serve as the repeating stage of the fungus during the growing season.

Identification and symptomology

Septoria leaf blotch is also known as wheat leaf blotch and speckled leaf blotch. Symptoms often appear in the lower leaves of tillering plants as irregular-to-oval, longitudinal, reddish-brown lesions. The lesions often develop pycnidia, visible as tiny black specks. The center parts of the lesions tend to become a light brown to white color. Under conditions conducive to disease development, lesions often coalesce, causing entire leaves to die prematurely. Under very severe conditions, lesions may be found on tips of glumes. Prolonged, wet-weather conditions, particularly during the jointing stage with air temperatures from 15 to 20 °C, are needed for severe septoria tritici blotch to occur (Hess and Shaner 1987; Chungu et al., 2001; Henze et al., 2007).

Distribution and losses

Septoria tritici blotch is widespread and an economically important disease worldwide (Cook et al., 1991; Scharen 1999). As is typical of leaf-spotting pathogens, when infections occur and become severe on the flag leaf or the leaf immediately below it, losses in grain yield and volume weight will be greatest (Thomas et al., 1989). The disease decreases the radiation-use efficiency of the wheat canopy, and the disease is made more severe with increasing leaf nitrogen concentration (Olesen et al., 2003). Although *S. tritici* is able to infect other grass genera, including *Agropyron, Agrostis, Brachypodium, Bromus, Dactylis, Festuca, Hordeum, Glyceria, Poa*, and *Secale*, these alternate hosts are not believed to play an important role in septoria tritici blotch distribution and epidemiology (Eyal 1999). The importance of seed-borne inoculum of *S. tritici* is somewhat unclear (Eyal 1999). Significant losses to both grain yield and grain weight can be realized.

Pathogen variability

Genetic diversity is very high in *M. graminicola* (Schnieder et al., 2001; Zhan et al., 2003); however, pathogenic variability is more limited. Early research found distinct physiological specialization in *S. tritici* isolates at the genus level in wheat (*T. durum*– and *T. aestivum*–specific isolates) (Eyal et al., 1985; Kema et al., 1996a). However, later studies showed distinct, qualitative differences in *S. tritici* × wheat cultivar interactions (Eyal 1999).

TAN SPOT

Taxonomy and life history

The causal fungus of tan spot is typically referred to by its telomorphic stage, *Pyrenophora tritici-repentis* (Died.) Drechs. [syn. *P. trichostoma* (Fr.) Fuckel], a homothallic ascomycete. Female and male gametes from the same mycelium are formed in an ascocarp called a pseudothecium. The fungus saprophytically colonizes dead and dying leaves, leaf sheaths, and stems to form mature perithecia-containing ascospores (Odvody et al., 1981). Pseudothecia require a cold period to mature and are visible as dark, raised pinpoints on wheat straw. Ascospores produced in pseudothecia can serve as primary inoculum and infect a wheat crop in the fall, winter, or spring depending on crop rotation, environmental conditions, and geographic location (Wright and Sutton 1990). The anamorphic stage of the fungus is *Drechslera tritici-repentis* (Died.) Shoemaker (syn. *Helminthosporium tritici-repentis* Died.). Conidiophores and conidia produced by this asexual stage are typically found in lesions on leaves, sheaths, and stems and make up the repeating cycle of the disease throughout the growing season. Conidia are primarily wind-disseminated within and between wheat crops or from other grasses to wheat (Schilder and Bergstrom 1992). Environmental conditions, host genotype, and pathogen virulence all influence disease onset, spread, and severity (Ciuffetti and Tuori 1999).

Identification and symptomology

The disease is also known as yellow leaf spot. In the field, the disease appears on leaves as oval- to irregular-shaped, yellow, tan, or brown lesions, each containing a small dark spot and surrounded by a yellow border. Individual lesions can coalesce over part of, or an entire, leaf surface. On more resistant cultivars, lesions are typically smaller and darker in color. The disease progresses from the lower leaves to the upper leaves as the plant grows and matures. The disease progresses rapidly under frequent rains with cool, cloudy, and humid weather conditions. The tan spot pathogen can also infect seed and be dispersed by seed. *Pyrenophora tritici-repentis* infection can produce a red smudge or black point symptom on seed that can impact grain grading as well as subsequent seedling infection (Fernandez et al., 2001).

Distribution and losses

Tan spot is an important disease on wheat throughout the world (Ciuffetti and Tuori 1999). The disease tends to be more damaging in continuous wheat areas and where the previous wheat crop residue is not incorporated into the soil or burned off (Bockus and Shroyer 1998; Carignano et al., 2008). Grain losses are most severe when the disease spreads to and damages the flag leaf. The fungus attacks a wide range of grass species that can serve as alternate hosts (Krupinsky 1992; Ali and Francl 2003). Grain yield reductions varying from 3% to 50% and kernel weight reductions up to 13% have been documented (Evans et al., 1999; Bhathal et al., 2003).

Pathogen variability

At the DNA level, genetic variability is extensive in *Pyrenophora tritici-repentis*, but that level of variability does not particularly equate to pathogenic or geographic variability (Friesen et al., 2005). To determine pathogenic variability, visible lesion types (necrosis, chlorosis, or both) were first used to distinguish specific pathotypes and races of *P. tritici-repentis* (Lamari and Bernier 1989a). Isolates were assigned to four pathotypes (which later became races 1 through 4) based on their ability to produce necrotic and chlorotic reactions (race 1), necrosis only (race 2), chlorosis only (race 3), or neither lesion type (race 4) on specific wheat differential lines (Lamari and Bernier 1989b). Additional races were identified as the number of specific *P. tritici-repentis* isolate × wheat differential lines was expanded and characterized, although race 1 has been found to predominate in many wheat growing regions (Lamari et al., 2005; Singh et al., 2007).

The symptomology produced by specific *P. tritici-repentis* isolates on certain wheat genotypes was subsequently found to be based on host-selective toxins produced by the fungus. For races 1 through 8, these toxins produce identifiable symptoms unique to each isolate–wheat genotype combination. For example, *P. tritici-repentis* race 1 isolates produced necrotic lesions (due to production of toxin Ptr ToxA) on the cultivars Glenlea and Katepwa; chlorotic lesions (due to production of Ptr ToxC) on the line 6B365; and a resistant reaction (no discernable toxin production) on 6B662, 'Auburn', and 'Salamouni' (Lamari et al., 1995; Ciuffetti et al., 1998; De Wolf et al., 1998; Strelkov et al., 2002; Strelkov and Lamari 2003). However, further research showed that some *P. tritici-repentis* isolates did not have the complement of host-selective toxins anticipated by the phenotypic race designations (Andrie et al., 2007). Thus, neither disease phenotype nor host-selective toxin genotype alone can be used to distinguish all races of the pathogen.

FUSARIUM HEAD BLIGHT

Taxonomy and life history

The causal fungus of Fusarium head blight has the telomorphic stage, *Gibberella zeae* (Schwein.) Petch, a homothallic ascomycete (Parry et al., 1995; McMullen et al., 1997). The anamorphic stage is *Fusarium graminearum* Schwabe. In North America, *F. graminearum* predominates, whereas other *Fusarium* species, notably *F. culmorum*, *F. poae*, and *F. avenaceum* may be found to a lesser degree in North America or may predominate in other areas (Sutton 1982; Ireta and Gilchrist 1994). The fungus survives between crops principally as mycelia or immature perithecia on wheat residue (Gilbert and Fernando 2004; Guenther and Trail 2005). The fungus may also be found in soil as mycelia, macroconidia, or chlamydospores (Nyvall 1970; Bai and Shaner 1994). Infected seed, whether fallen to the ground during harvest or planted the following season, may also serve as a source of inoculum. Because the fungus also infects maize and rice (*Oryza sativa* L.), residue from these crops may serve as a source of inoculum (Sutton 1982; Bai and Shaner 1994). Ascospores released from perithecia infect wheat anthers during anthesis, as well as infecting the wheat glume, palea, and rachis (Bennett 1931; Pugh et al., 1933; Markell and Francl 2003). Under favorable conditions, the fungus may spread throughout the wheat head. Wet, humid conditions during flowering are associated with severe Fusarium head blight (Gilbert et al., 2008).

Identification and symptomology

The disease is also known as head scab. The fungus can also contribute to seedling blight, and to root and crown rot diseases discussed in Chapter 6. The disease is primarily recognized on the head by premature ripening or bleaching of one or several spikelets any time after flowering. Diseased spikelets are often a light yellow color during the dough stage. As the disease advances, the light pink or salmon color of the fungus may appear at the base or along the edges of infected spikelets. The peduncle below infected heads may show dark coloration. When the wheat matures, developing perithecia may be evident on the heads as tiny purple-black specks. Infected seed is usually shrunken or shriveled, appearing either bleached or pink in color.

Distribution and losses

The disease has worldwide distribution, although outbreaks are considerably influenced by local weather conditions during flowering and by cropping history in a given field. Grain yield and grain weight losses are variable, with average losses of 5%–15%, but losses as high as 70% in epidemic years (Ireta and Gilchrist 1994). Losses also occur from mycotoxin contamination of grain, which can be hazardous to livestock and humans ingesting contaminated grain products. The most prevalent toxins produced by *F. graminearum* are the trichothecenes, in particular deoxynivalenol. The disease attacks many cereal crops, including wheat, barley (*Hordeum vulgare* L.), rice, rye (*Secale cereale* L.), oat (*Avena sativa* L.), and maize. Other grass species can also serve as alternate hosts (Inch and Gilbert 2003; Goswami and Kistler 2004).

Pathogen variability

Understanding the population structure of *F. graminearum* has been somewhat complicated by differences in many factors including species groupings, mating types, and chemotypes (Ireta and Gilchrist 1994; Guo et al., 2008; Miedaner et al., 2008). Depending on the types and origins of isolates evaluated, and the methods used for evaluation, different conclusions have been reached as to the nature of genetic diversity within *F. graminearum* (Xu et al., 2004; Gale et al., 2007; Leslie et al., 2007; Starkey et al., 2007). Pathogenic diversity and selection for more complex chemotypes have been suggested as major factors in the spread of Fusarium head blight in North America (Ward et al., 2008).

MANAGEMENT OF RESIDUE-BORNE DISEASES

Given the many positive aspects of reduced or no-till farming systems, it is clear that the management of residue-borne diseases of wheat must respond to this new paradigm. The management techniques most likely to impact residue-borne diseases are crop diversity and host-plant resistance (Cook 2006; Anderson 2008). Used together, disease resistant crops and diversification of those crops in time and space should help to minimize the effects of residue-borne diseases (Holtzer et al., 1996; Hanson et al., 2007).

Crop diversity

The frequency and specific crop (or fallow) used in a crop diversification scheme will influence occurrence, severity, and duration of residue-borne disease outbreaks (Peairs et al., 2005; Krupinsky et al., 2007). Rotation to a nonhost crop can be effective in reducing disease incidence in subsequent wheat crops. However, the number of rotations depends on how quickly the wheat residue is disturbed and broken down (Fernandez et al., 1998). In western Australia, a 1-year rotation is sufficient, but in eastern Australia, a 2- or 3-year rotation is needed because of the different effects which climatic conditions have on residue degradation (Summerell and Burgess 1989; Bhathal and Loughman 2001).

Host-plant resistance

Powdery mildew

Wheat cultivars vary widely in their reaction to powdery mildew, and new races of the fungus can develop to overcome resistant cultivars (Brown and Hovmoller 2002; Hsam and Zeller 2002). Breeding for powdery mildew resistance in wheat began in the early 1930s with the pioneering host–pathogen interaction work of Mains (Mains 1934). Many of the initial powdery mildew resistance (*Pm*) genes identified were major or seedling genes, which conferred a hypersensitive resistance reaction (Briggle 1966). Later studies, particularly in Europe, showed that widespread deployment of single, major genes typically resulted in the selection of races capable of overcoming the resistance (Svec and Miklovicova 1998).

Researchers have identified about 50 *Pm* genes or alleles, occurring at 33 loci, with 5 of the loci (*Pm1*, *Pm3*, *Pm4*, *Pm5*, and *Pm8*) having multiple alleles (Hsam and Zeller 2002; Miranda et al., 2006). A number of the genes and alleles for powdery mildew resistance probably have not been deployed. As a result, germplasm developed with these *Pm* genes and combinations of these *Pm* genes could have a high level of resistance to populations of *B. g. tritici* (Engle et al., 2006).

In the absence of genetic analyses or reliable molecular markers, it is still possible to postulate the presence of specific *Pm* genes using pathogen isolates with particular virulence–avirulence profiles, because the wheat–powdery mildew pathosystem typically follows a gene-for-gene relationship. By comparing the reactions of well-characterized *B. g. tritici* isolates on a differential host series with their reactions on unknown wheat genotypes, the identity of *Pm* genes in the unknown lines can be deduced. Leath and Heun (1990) used gene postulation to identify known *Pm* genes in 22 soft red winter wheat cultivars.

In addition to major-gene resistance, wheat also has partial, adult-plant, or minor-gene resistance to powdery mildew. This partial resistance may take the form of defeated yet not completely susceptible major genes such as *Pm3c* (Nass et al., 1981), *Pm4b* (Mingeot et al., 2002), and *Pm5* (Keller et al., 1999). In addition, adult-plant resistance has been identified in the cultivars Knox (Shaner 1973), Knox 62, and Massey (Griffey and Das 1994). This resistance delays infection and reduces growth and reproduction of *B. g. tritici* in adult plants (Gustafson and Shaner 1982). Because this adult-plant (slow mildewing) resistance is thought to be race-nonspecific (Elen and Skinnes 1988), the probability of *B. g. tritici* developing races highly adapted to adult-plant resistance is much less than that for seedling, hypersensitive resistance (Hautea et al., 1987). Molecular markers would be useful for quantitative mildew resistance, which is difficult to assess phenotypically. Two major quantitative trait loci (QTLs) and about 18 minor QTLs have been identified in different environments and at different developmental stages (Keller et al., 1999; Mingeot et al., 2002; Bougot et al., 2006; Jakobson et al., 2006; Tucker et al., 2007).

Stagonospora nodorum blotch

Much of the early work on stagonospora nodorum blotch indicated that resistance was under polygenic control (Nelson and Marshall 1990). Major gene resistance has also been found, with some reported differences in the numbers of genes controlling resistance to leaf symptoms (Wilkinson et al., 1990; Ma and Hughes 1995; Kim et al., 2004). Two QTLs have been identified (Liu et al., 2004). Ali et al. (2008) were able to identify wheat genotypes having multiple resistance to *Stagonospora nodorum*, *S. tritici*, and *Pyrenophora tritici-repentis* (Ali et al., 2008). With the recent finding of host-selective toxins involved in the *S. nodorum* × wheat interaction, greater precision and insight may be available on methods to breed for resistance to stagonospora nodorum blotch (Friesen et al., 2008).

Septoria tritici blotch

Even though there are isolate × cultivar interactions, resistance to septoria tritici blotch is assessed quantitatively, typically by determining the percentage of leaf area covered by lesions (Chartrain et al., 2004a). Resistance in wheat to *S. tritici* has been shown to be both monogenically and polygenically controlled (van Ginkel and

Scharen 1988; Somasco et al., 1996; Simon and Cordo 1997; Arraiano et al., 2001; Brading et al., 2002; McCartney et al., 2003). Chartrain et al. (2004b) found that disease severity was not correlated between the seedling and adult stages. Sources of resistance to septoria tritici blotch have been found in primary, secondary, and tertiary gene pools of wheat (Mergoum et al., 2007; Simon et al., 2007; Singh et al., 2006). Twelve genes for resistance to septoria tritici blotch have been named and mapped (Goodwin 2007).

Tan spot

A major component in the management of tan spot is host-plant resistance, which is the most effective, economical, and environmentally friendly way to control tan spot. Race-specific resistance to tan spot is the result of insensitivity to one or more of the host-selective toxins, plus other factors that slow or inhibit pathogen growth and development (Friesen et al., 2001b, 2003). However, in the presence of toxin sensitivity, quantitative resistance has also been identified (Faris and Friesen 2005). Resistance to races 1 and 5 of *Pyrenophora tritici-repentis* was identified in spring wheat seedlings under greenhouse conditions (Ali et al., 2008). Resistance to race 1 was found in a large array of wheat–alien species derivatives and synthetic wheat lines (Xu et al., 2004; Oliver et al., 2008).

In a field evaluation of adult-plant resistance to race 1, Duveiller et al. (2007) showed that resistance in spring wheat was controlled by two gene pairs. Evans et al. (1999) found a significant correlation between lesion length as determined under greenhouse conditions and area under the disease progress curve as determined in the field, thereby indicating that progress in breeding for resistance could be made with either or both techniques. Although some research has been conducted to use Ptr toxins to screen wheat for differences in susceptibility, and conditions for *in vitro* toxin production have been studied, the correlation between toxin production and pathogenicity has been poor (Brown and Hunger 1999).

Fusarium head blight

Resistance to Fusarium head blight has been difficult to characterize and incorporate into adapted cultivars because of the need to have both resistance to infection and resistance to deoxynivalenol accumulation (Gartner et al., 2008). There are also differences in genetic control of resistance to floret infection (Type I), resistance to spread in the spike (Type II), and resistance to deoxynivalenol accumulation (Schroeder and Christensen 1963; Mesterhazy 1995; Somers et al., 2003; Bai and Shaner 2004). Quantitative trait loci (QTLs) for resistance have also been identified (Chen et al., 2006; Yu et al., 2008). Molecular markers for resistance have accelerated the incorporation of effective resistance into wheat germplasm (Anderson 2007). The Chinese cultivar Sumai 3 has a high level of Type II resistance with good combining ability, and it has been used throughout the world to improve Fusarium head blight resistance (Bai and Shaner 2004).

FUTURE PERSPECTIVES

A common thread to each of these diseases is the ability of the causal fungi to survive and undergo sexual recombination on wheat residue. When infested residue is not brought into contact with soil, it can serve as a safe harbor for the pathogen to survive between crops. Moreover, it allows the pathogen to complete its life cycle, thus increasing genetic variability and potentially new virulence combinations. Increased amounts of wheat residue on the soil surface increases the likelihood of pathogens surviving on the residue, thereby providing a source of inoculum that otherwise may have been eliminated or minimized by tillage (Bailey and Duczek 1996).

This does not mean that increased residue will inevitably lead to greater disease. Disease incidence and severity depend on many factors, including a virulent pathogen, a susceptible host, and environmental conditions conducive for infection, growth, reproduction, and spread of the pathogen. In addition, the level of disease on the residue crop, how the residue was handled (if at all), and the nutritional status of the previous crop and subsequent crop all influence pathogen growth and development on the residue (Gilbert and Woods 2001; Krupinsky and Tanaka 2001). Interestingly, combine harvesters may contribute to the dispersal of residue-inhabiting pathogens by moving pathogen spores from ground level up into atmospheric wind currents (Friesen et al., 2001a). Thus, the widespread adoption of reduced and no-till residue management aids in the survival, increase, and genetic variability of potentially damaging residue-inhabiting fungi. This has caused some shifts and could possibly cause even greater changes in global wheat research priorities. Emphasis must be placed on coupling crop diversification strategies with host resistance in order to minimize these important diseases.

REFERENCES

Adhikari, T.B., S. Ali, R.R. Burlakoti, P.K. Singh, M. Mergoum, and S.B. Goodwin. 2008. Genetic structure of *Phaeosphaeria nodorum* populations in the north-central and midwestern United States. Phytopathology 98:101–107.

Ali, S., and T.B. Adhikari. 2008. Variation in aggressiveness of *Stagonospora nodorum* isolates in North Dakota. J. Phytopathol. 156:140–145.

Ali, S., and L.J. Francl. 2003. Population race structure of *Pyrenophora tritici-repentis* prevalent on wheat and noncereal grasses in the Great Plains. Plant Dis. 87:418–422.

Ali, S., P.K. Singh, M.P. McMullen, M. Mergoum, and T.B. Adhikari. 2008. Resistance to multiple leaf spot diseases in wheat. Euphytica 159:167–179.

Anderson, J.A. 2007. Marker-assisted selection for Fusarium head blight resistance in wheat. Int. J. Food Microbiol. 119:51–53.

Anderson, R.L. 2008. Diversity and no-till: Keys for pest management in the U.S. Great Plains. Weed Sci. 56:141–145.

Andrie, R.M., I. Pandelova, and L.M. Ciuffetti. 2007. A combination of phenotypic and genotypic characterization strengthens *Pyrenophora tritici-repentis* race identification. Phytopathology 97:694–701.

Arraiano, L.S., A.J. Worland, C. Ellerbrook, and J.K.M. Brown. 2001. Chromosomal location of a gene for resistance to septoria tritici blotch (*Mycosphaerella graminicola*) in the hexaploid wheat 'Synthetic 6x'. Theor. Appl. Genet. 103:758–764.

Arseniuk, E., T. Goral, and A.L. Scharen. 1998. Seasonal patterns of spore dispersal of *Phaeosphaeria* spp. and *Stagonospora* spp. Plant Dis. 82:187–194.

Bai, G., and G. Shaner. 1994. Scab of wheat: Prospects for control. Plant Dis. 78:760–766.

Bai, G.H., and G.E. Shaner. 2004. Management and resistance in wheat and barley to Fusarium head blight. Annu. Rev. Phytopathol. 42:135–161.

Bailey, K.L., and L.J. Duczek. 1996. Managing cereal diseases under reduced tillage. Can. J. Plant Pathol. 18:159–167.

Bennett, F.T. 1931. *Gibberella saubinetii* (Mont) Sacc. on British cereals: II. Physiological and pathological studies. Ann. Appl. Biol. 18:158–177.

Bennett, F.G.A. 1981. The expressions of resistance to powdery mildew infection in winter wheat cultivars: 1. Seedling resistance. Ann. Appl. Biol. 98:295–303.

Bennett, R.S., M.G. Milgroom, R. Sainudiin, B.M. Cunfer, and G.C. Bergstrom. 2007. Relative contribution of seed-transmitted inoculum to foliar populations of *Phaeosphaeria nodorum*. Phytopathology 97:584–591.

Bhathal, J.S., and R. Loughman. 2001. Ability of retained stubble to carry-over leaf diseases of wheat in rotation crops. Aust. J. Exp. Agric. 41:649–653.

Bhathal, J.S., R. Loughman, and J. Speijers. 2003. Yield reduction in wheat in relation to leaf disease from yellow (tan) spot and septoria nodorum blotch. Eur. J. Plant Pathol. 109:435–443.

Bockus, W.W., and J.P. Shroyer. 1998. The impact of reduced tillage on soilborne plant pathogens. Annu. Rev. Phytopathol. 36:485–500.

Bougot, Y., J. Lemoine, M.T. Pavoine, H. Guyomarch, V. Gautier, H. Muranty, and D. Barloy. 2006. A major QTL effect controlling resistance to powdery mildew in winter wheat at the adult plant stage. Plant Breed. 125:550–556.

Brading, P.A., E.C.P. Verstappen, G.H.J. Kema, and J.K.M. Brown. 2002. A gene-for-gene relationship between wheat and *Mycosphaerella graminicola*, the Septoria tritici blotch pathogen. Phytopathology 92:439–445.

Briggle, L.W. 1966. Three loci in wheat involving resistance to *Erysiphe graminis* f. sp. *tritici*. Crop Sci. 6:461–465.

Brown, J.K.M., and M.S. Hovmoller. 2002. Aerial dispersal of pathogens on the global and continental scales and its impact on plant disease. Science 297:537–541.

Brown, D.A., and R.M. Hunger. 1999. Regulation of *in vitro* Ptr-toxin production by *Pyrenophora tritici-repentis* isolates by environmental parameters and accumulation of Ptr-toxin in culture over time. J. Phytopathol. 147:25–29.

Carignano, M., S.A. Staggenborg, and J.P. Shroyer. 2008. Management practices to minimize tan spot in a continuous wheat rotation. Agron. J. 100:145–153.

Chartrain, L., P.A. Brading, J.C. Makepeace, and J.K.M. Brown. 2004a. Sources of resistance to septoria tritici blotch and implications for wheat breeding. Plant Pathol. 53:454–460.

Chartrain, L., P.A. Brading, J.P. Widdowson, and J.K.M. Brown. 2004b. Partial resistance to Septoria tritici blotch

(*Mycosphaerella graminicola*) in wheat cultivars Arina and Riband. Phytopathology 94:497–504.

Chen, J., C.A. Griffey, M.A. Saghai Maroof, E.L. Stromberg, R.M. Biyashev, W. Zhao, M.R. Chappell, T.H. Pridgen, Y. Dong, and Z. Zeng. 2006. Validation of two major quantitative trait loci for Fusarium head blight resistance in Chinese wheat line W14. Plant Breed. 125:99–101.

Chungu, C., J. Gilbert, and F. Townley-Smith. 2001. Septoria tritici blotch development as affected by temperature, duration of leaf wetness, inoculum concentration, and host. Plant Dis. 85:430–435.

Ciuffetti, L.M., L.J. Francl, G.M. Ballance, W.W. Bockus, L. Lamari, S.W. Meinhardt, and J.B. Rasmussen. 1998. Standardization of toxin nomenclature in the *Pyrenophora tritici-repentis*/wheat interaction. Can. J. Plant Pathol. 20:421–424.

Ciuffetti, L.M., and R.P. Tuori. 1999. Advances in the characterization of the *Pyrenophora tritici-repentis*–wheat interaction. Phytopathology 89:444–449.

Conner, R.L., A.D. Kuzyk, and H. Su. 2003. Impact of powdery mildew on the yield of soft white spring wheat cultivars. Can. J. Plant Sci. 83:725–728.

Cook, R.J. 2006. Toward cropping systems that enhance productivity and sustainability. Proc. Natl. Acad. Sci. USA 103:18389–18394.

Cook, R.J., R.W. Polley, and M.R. Thomas. 1991. Disease-induced losses in winter wheat in England and Wales 1985–1989. Crop Prot. 10:504–508.

Cowger, C., and H.V. Silva-Rojas. 2006. Frequency of *Phaeosphaeria nodorum*, the sexual stage of *Stagonospora nodorum*, on winter wheat in North Carolina. Phytopathology 96:860–866.

Cunfer, B.M., and P.P. Ueng. 1999. Taxonomy and identification of *Septoria* and *Stagonospora* species on small-grain cereals. Annu. Rev. Phytopathol. 37:267–284.

De Wolf, E.D., R.J. Effertz, S. Ali, and L. Francl. 1998. Vistas of tan spot research. Can. J. Plant Pathol. 20:349–444.

Duveiller, E., R.C. Sharma, B. Cukadar, and M. van Ginkel. 2007. Genetic analysis of field resistance to tan spot in spring wheat. Field Crops Res. 101:62–67.

Ekboir, J., K. Boa, and A.A. Dankyi. 2002. Impact of no-till technologies in Ghana. CIMMYT Economics Program Paper, No. 02–01. CIMMYT, D.F., Mexico.

Elen, O.N., and H. Skinnes. 1988. Partial resistance to powdery mildew in wheat seedlings. Nor. J. Agric. Res. 1:61–66.

Engle, J.S., D.S. Marshall, and L. Whitcher. 2006. Proposed major powdery mildew genes in eastern and southern wheat germplasm. Phytopathology 96:S33 (abstract).

Eriksen, L., and L. Munk. 2003. The occurrence of *Mycosphaerella graminicola* and its anamorph *Septoria tritici* in winter wheat during the growing season. Eur. J. Plant Pathol. 109:253–259.

Evans, C.K., R.M. Hunger, and W.C. Siegerist. 1999. Comparison of greenhouse and field testing to identify wheat resistant to tan spot. Plant Dis. 83:269–273.

Everts, K., S. Leath, and P.L. Finney. 2001. Impact of powdery mildew and leaf rust on milling and baking quality of soft red winter wheat. Plant Dis. 85:423–429.

Eyal, Z. 1999. The septoria tritici and stagonospora nodorum blotch diseases of wheat. Eur. J. Plant Pathol. 105:629–641.

Eyal, Z., A.L. Scharen, M.D. Huffman, and J.M. Prescott. 1985. Global insights into virulence frequencies of *Mycosphaerella graminicola*. Phytopathology 75:1456–1462.

Eyal, Z., A.L. Scharen, J.M. Prescott, and M. van Ginkel. 1987. The septoria diseases of wheat: Concepts and methods of disease management. CIMMYT, D.F., Mexico.

Faris, J.D., and T.L. Friesen. 2005. Identification of quantitative trait loci for race-nonspecific resistance to tan spot in wheat. Theor. Appl. Genet. 111:386–392.

Fernandez, M.R., R.M. DePauw, and J.M. Clarke. 2001. Reaction of common and durum wheat cultivars to infection of kernels by *Pyrenophora tritici-repentis*. Can. J. Plant Pathol. 23:158–162.

Fernandez, M.R., R.P. Zentner, B.G. McConkey, and C.A. Campbell. 1998. Effects of crop rotations and fertilizer management on leaf spotting diseases of spring wheat in south-western Saskatchewan. Can. J. Plant. Sci. 78:489–496.

Fried, P.M., D.R. MacKenzie, and R.R. Nelson. 1981. Yield loss caused by *Erysiphe graminis* f. sp. *tritici* on single culms of Chancellor wheat and four multilines. J. Plant Dis. Prot. 88:256–264.

Friesen, T.L., S. Ali, S. Kianian, L.J. Francl, and J.B. Rasmussen. 2003. Role of host sensitivity to Ptr ToxA in development of tan spot of wheat. Phytopathology 93:397–401.

Friesen, T.L., S. Ali, K.K. Kleim, and J.B. Rasmussen. 2005. Population genetic analysis of a global collection of *Pyrenophora tritici-repentis*, causal agent of tan spot of wheat. Phytopathology 95:1144–1150.

Friesen, T.L., E.D. De Wolf, and L.J. Francl. 2001a. Source strength of wheat pathogens during combine harvest. Aerobiology 17:293–299.

Friesen, T.L., J.B. Rasmussen, C.Y. Kwon, S. Ali, L.J. Francl, and S.W. Meinhardt. 2001b. Reaction to Ptr ToxA-insensitive wheat mutants to *Pyrenophora tritici-repentis* race 1. Phytopathology 92:38–42.

Friesen, T.L., Z. Zhang, P.S. Solomon, R.P. Oliver, and J.D. Faris. 2008. Characterization of the interaction of a novel *Stagonospora nodorum* host-selective toxin with a wheat susceptibility gene. Plant Physiol. 146:682–693.

Gale, L.R., T.J. Ward, V. Balmas, and H.C. Kistler. 2007. Population subdivision of *Fusarium graminearum* sensu stricto in the upper midwestern United States. Phytopathology 97:1434–1439.

Garcia, C., and D. Marshall. 1992. Observations on the ascogenous stage of *Septoria tritici* in Texas. Mycol. Res. 96:65–70.

Gartner, B.H., M. Munich, G. Kleijer, and F. Mascher. 2008. Characterisation of kernel resistance against

Fusarium infection in spring wheat by baking quality and mycotoxin assessments. Eur. J. Plant Pathol. 120:61–68.

Gilbert, J., and W.G.D. Fernando. 2004. Epidemiology and biological control of *Gibberella zeae / Fusarium graminearum*. Can. J. Plant Pathol. 26:464–472.

Gilbert, J., and S.M. Woods. 2001. Leaf spot diseases of spring wheat in southern Manitoba farm fields under conventional and conservation tillage. Can. J. Plant Sci. 81:551–559.

Gilbert, J., S.M. Woods, and U. Kromer. 2008. Germination of ascospores of *Gibberella zeae* after exposure to various levels of relative humidity and temperature. Phytopathology 98:504–508.

Goodwin, S.B. 2007. Back to basics and beyond: Increasing the level of resistance to Septoria tritici blotch in wheat. Aust. Plant Pathol. 36:532–538.

Goswami, R.S., and H.C. Kistler. 2004. Heading for disaster: *Fusarium graminearum* on cereal crops. Mol. Plant Pathol. 5:515–525.

Griffey, C., and M. Das. 1994. Inheritance of adult plant resistance to powdery mildew in Knox 62 and Massey winter wheats. Crop Sci. 34:641–646.

Griffey, C.A., M.K. Das, and E.L. Stromberg. 1993. Effectiveness of adult plant resistance in reducing loss to powdery mildew in winter wheat. Plant Dis. 77:618–622.

Guenther, J.C., and F. Trail. 2005. The development and differentiation of *Gibberella zeae* (anamorph: *Fusarium graminearum*) during colonization of wheat. Mycologia 97:229–237.

Guo, X.W., W.G.D. Fernando, and H.Y. Seow-Brock. 2008. Population structure, chemotype diversity, and potential chemotype shifting of *Fusarium graminearum* in wheat fields of Manitoba. Plant Dis. 92:756–762.

Gustafson, G.D., and G. Shaner. 1982. Influence of plant age on the expression of slow-mildewing resistance in wheat. Phytopathology 72:746–749.

Hanson, J.D., M.A. Liebig, S.D. Merrill, D.L. Tanaka, J.M. Krupinsky, and D.E. Stott. 2007. Dynamic cropping systems: Increasing adaptability amid an uncertain future. Agron. J. 99:939–943.

Hautea, R.A., W.R. Coffman, M.E. Sorrells, and G.C. Bergstrom. 1987. Inheritance of partial resistance to powdery mildew in spring wheat. Theor. Appl. Genet. 73:609–615.

Henze, M., M. Beyer, H. Klink, and J.A. Verreet. 2007. Characterizing meteorological scenarios favorable for *Septoria tritici* infections in wheat and estimation of latent periods. Plant Dis. 91:1445–1449.

Hess, D.E., and G. Shaner. 1987. Effect of moisture and temperature on development of Septoria tritici blotch in wheat. Phytopathology 77:215–219.

Holtzer, T.O., R.L. Anderson, M.P. McMullen, and F.B. Peairs. 1996. Integrated pest management of insects, plant pathogens, and weeds in dryland cropping systems of the Great Plains. J. Prod. Agric. 9:200–208.

Hoorne, C., L. Lamari, J. Gilbert, and G.M. Ballance. 2002. First report of *Mycosphaerella graminicola*, the sexual state of *Septoria tritici*, in Manitoba, Canada. Can. J. Plant Pathol. 24:445–449.

Hsam, S.L.K., and F.J. Zeller. 2002. Breeding for powdery mildew resistance in common wheat (*Triticum aestivum* L.). p. 219–238. *In* R.R. Belanger, W.R. Bushnell, A.J. Dik, and T.L.W. Carver (ed.) The powdery mildews: A comprehensive treatise. APS Press, St. Paul, MN.

Hunter, T., R.R. Coker, and D.J. Royle. 1999. The teleomorph stage, *Mycosphaerella graminicola*, in epidemics of septoria tritici blotch on winter wheat in the UK. Plant Pathol. 48:51–57.

Hunter, H., and H.M. Leake. 1933. Recent advances in agricultural plant breeding. Blakiston's Son & Co., Philadelphia, PA.

Imani, Y., A. Ouassou, and C.A. Griffey. 2002. Virulence of *Blumeria graminis* f. sp. *tritici* populations in Morocco. Plant Dis. 86:383–388.

Inch, S., and J. Gilbert. 2003. The incidence of *Fusarium* species recovered from inflorescences of wild grasses in southern Manitoba. Can. J. Plant Pathol. 25:379–383.

Ireta, M.J., and L.S. Gilchrist. 1994. Fusarium head scab of wheat (*Fusarium graminearum* Schwabe). Wheat Special Rep. No. 21b. CIMMYT, D.F., Mexico.

Jakobson, I., H. Peusha, L. Timofejeva, and K. Jarve. 2006. Adult plant and seedling resistance to powdery mildew in a *Triticum aestivum* × *Triticum militinae* hybrid line. Theor. Appl. Genet. 112:760–769.

Johnson, J.W., P.S. Baenziger, W.T. Yamazaki, and R.T. Smith. 1979. Effects of powdery mildew on yield and quality of isogenic lines of Chancellor wheat. Crop Sci. 19:349–352.

Keller, M., B. Keller, G. Schachermayr, M. Winzeler, J.E. Schmid, P. Stamp, and M.M. Messmer. 1999. Quantitative trait loci for resistance against powdery mildew in a segregating wheat × spelt population. Theor. Appl. Genet. 98:903–912.

Kema, G.H.J., J.G. Annone, R. Sayoud, C.H. Van Silfhout, M. Van Ginkel, and J. de Bree. 1996a. Genetic variation for virulence and resistance in the wheat–*Mycosphaerella graminicola* pathosystem: I. Interactions between pathogen isolates and host cultivars. Phytopathology 86:200–212.

Kema, G.H.J., E.C.P. Verstappen, M. Todorova, and C. Waalwijk. 1996b. Successful crosses and molecular tetrad and progeny analysis demonstrate heterothallism in *Mycosphaerella graminicola*. Curr. Genet. 30:251–258.

Kim, Y.K., G. Brown-Guedira, T.S. Cox, and W.W. Bockus. 2004. Inheritance of resistance to Stagonospora nodorum leaf blotch in Kansas winter wheat cultivars. Plant Dis. 88:530–536.

King, J.E., R.J. Cook, and S.C. Melville. 1983. A review of *Septoria* diseases of wheat and barley. Ann. Appl. Biol. 103:345–373.

Krupinsky, J.M. 1992. Grass hosts of *Pyrenophora tritici-repentis*. Plant Dis. 76:92–95.

Krupinsky, J.M., and D.L. Tanaka. 2001. Leaf spot diseases on winter wheat influenced by nitrogen, tillage, and haying after a grass–alfalfa mixture in the Conservation Reserve Program. Plant Dis. 85:785–789.

Krupinsky, J.M., D.L. Tanaka, S.D. Merrill, M.A. Liebig, M.T. Lares, and J.D. Hanson. 2007. Crop sequence effects on leaf spot diseases of no-till spring wheat. Agron. J. 99:912–920.

Lamari, L., and C.C. Bernier. 1989a. Evaluation of wheat lines and cultivars to tan spot (*Pyrenophora tritici-repentis*) based on lesion type. Can. J. Plant Path. 11:49–56.

Lamari, L., and C.C. Bernier. 1989b. Virulence of isolates of *Pyrenophora tritici-repentis* on 11 wheat cultivars and cytology of the differential host reactions. Can. J. Plant Pathol. 11:284–290.

Lamari, L., R. Sayoud, M. Boulif, and C.C. Bernier. 1995. Identification of a new race in *Pyrenophora tritici-repentis*: Implications for the current pathotype classification system. Can. J. Plant Pathol. 17:312–318.

Lamari, L., S.E. Strelkov, A. Yahyaoui, M. Amedov, M. Saidov, M. Djunusova, and M. Koichibayev. 2005. Virulence of *Pyrenophora tritici-repentis* in the countries of the Silk Road. Can. J. Plant Pathol. 27:383–388.

Leath, S., and K.L. Bowen. 1989. Effects of powdery mildew, triadimenol seed treatment, and triadimefon foliar sprays on yield of winter wheat in North Carolina. Phytopathology 79:152–155.

Leath, S., and M. Heun. 1990. Identification of powdery mildew resistance genes in cultivars of soft red winter wheat. Plant Dis. 74:747–752.

Leslie, J.F., L.L. Anderson, R.L. Bowden, and Y.W. Lee. 2007. Inter- and intra-specific genetic variation in *Fusarium*. Int. J. Food Microbiol. 119:25–32.

Limpert, E., F. Godet, K. Muller, and K. Muller. 1999. Dispersal of cereal mildews across Europe. Agric. For. Meteorol. 97:293–308.

Liu, Z., J.D. Faris, S.W. Meinhardt, S. Ali, J.B. Rasmussen, and T.L. Friesen. 2004. Genetic and physical mapping of a gene conditioning sensitivity in wheat to a partially purified host-selective toxin produced by *Stagonospora nodorum*. Phytopathology 94:1056–1060.

Ma, H., and G.R. Hughes. 1995. Genetic control and chromosomal location of *Triticum timopheevi* derived resistance to Septoria nodorum blotch in durum wheat. Genome 38:332–338.

Mains, E.B. 1934. Inheritance of resistance to powdery mildew, *Erysiphe graminis tritici* in wheat. Phytopathology 24:1257–1261.

Markell, S.G., and L.J. Francl. 2003. Fusarium head blight inoculum: Species prevalence and *Gibberella zeae* spore type. Plant Dis. 87:814–820.

McCartney, C.A., A.L. Brule-Babel, L. Lamari, and D.J. Somers. 2003. Chromosomal location of a race-specific resistance gene to *Mycosphaerella graminicola* in the spring wheat ST6. Theor. Appl. Genet. 107:1181–1186.

McKendry, A.L., G.E. Henke, and P.L. Finney. 1995. Effects of Septoria leaf blotch on soft red winter wheat milling and baking quality. Cereal Chem. 72:142–146.

McMullen, M., R. Jones, and D. Gallenberg. 1997. Scab of wheat and barley: A re-emerging disease of devastating impact. Plant Dis. 81:1340–1348.

Mergoum, M., P.K. Singh, S. Ali, E.M. Elias, J.A. Anderson, K.D. Glover, and T.B. Adhikari. 2007. Reaction of elite wheat genotypes from the northern Great Plains of North America to Septoria diseases. Plant Dis. 91:1310–1315.

Mesterhazy, A. 1995. Types and components of resistance to Fusarium head blight of wheat. Plant Breed. 114:377–386.

Miedaner, T., C.J.R. Cumagun, and S. Chakraborty. 2008. Population genetics of three important head blight pathogens *Fusarium graminearum*, *F. pseudograminearum* and *F. culmorum*. J. Phytopathol. 156:129–139.

Mingeot, D., N. Chantret, P.V. Baret, A. Dekeyser, N. Boukhatem, P. Sourdille, G. Doussinault, and J.M. Jacquemin. 2002. Mapping QTL involved in adult plant resistance to powdery mildew in the winter wheat line RE714 in two susceptible genetic backgrounds. Plant Breed. 121:133–140.

Minoshima, H., L.E. Jackson, T.R. Cavagnaro, S. Sanchez-Moreno, H. Ferris, S.R. Temple, S. Goyal, and J.P. Mitchell. 2007. Soil food webs and carbon dynamics in response to conservation tillage in California. Soil Sci. Am. J. 71:952–963.

Miranda, L.M., J.P. Murphy, D. Marshall, and S. Leath. 2006. *Pm34*: A new powdery mildew resistance gene transferred from *Aegilops tauschii* Coss. to common wheat (*Triticum aestivum* L.). Theor. Appl. Genet. 113:1497–1504.

Nass, H.A., W.L. Pedersen, D.R. MacKenzie, and R.R. Nelson. 1981. The residual effects of some defeated powdery mildew resistance genes in isolines of winter wheat. Phytopathology 71:1315–1318.

Nelson, L.R., and D. Marshall. 1990. Breeding wheat for resistance to *Septoria nodorum* and *Septoria tritici*. Adv. Agron. 44:257–277.

Niewoehner, A.S., and S. Leath. 1998. Virulence of *Blumeria graminis* f. sp. *tritici* on winter wheat in the eastern United States. Plant Dis. 82:64–68.

Nyvall, R.F. 1970. Chlamydospores of *Fusarium roseum* 'Graminearum' as survival structures. Phytopathology 60:1175–1177.

Odvody, G.N., M.G. Boosalis, and J.E. Watkins. 1981. Development of pseudothecia during progressive saprophytic colonization of wheat straw by *Pyrenophora trichostoma*. p. 33–35. *In* R.M. Hosford, Jr. (ed.) Tan spot of wheat and related diseases Workshop, Fargo, ND. July 1981. North Dakota Agric. Exp. Stn., North Dakota State Univ., Fargo, ND.

Olesen, J.E., J. Petersen, J.V. Mortensen, and L.N. Jorgensen. 2003. Effects of rates and timing of nitrogen fertilizer on disease control by fungicides in winter wheat: 2. Crop growth and disease development. J. Agric. Sci. 140:15–29.

Oliver, R.E., X. Cai, R.C. Wang, S.S. Xu, and T.L. Friesen. 2008. Resistance to tan spot and Stagonospora nodorum blotch in wheat–alien species derivatives. Plant Dis. 92:150–157.

Paillard, S., I. Goldringer, J. Enjalbert, G. Doussinault, C. de Vallavieille-Pope, and P. Brabant. 2000. Evolution of resistance against powdery mildew in winter wheat populations conducted under dynamic management: I. Is specific seedling resistance selected? Theor. Appl. Genet. 101: 449–456.

Parks, R., I. Carbone, J.P. Murphy, D. Marshall, and C. Cowger. 2008. Virulence structure of the eastern U.S. wheat powdery mildew population. Plant Dis. 92:1074–1082.

Parry, D.W., P. Jenkinson, and L. McLeod. 1995. Fusarium ear blight (scab) in small grain cerealsV—a review. Plant Pathol. 44:207–238.

Peairs, F.B., B. Bean, and B.D. Gossen. 2005. Pest management implications of reduced fallow periods in dryland cropping systems in the Great Plains. Agron. J. 97:373–377.

Pugh, G.W., H. Johann, and J.G. Dickson. 1933. Factors affecting infection of wheat heads by *Gibberella saubinetii*. J. Agric. Res. 46:771–797.

Sanderson, F.R. 1972. A *Mycosphaerella* species as the ascogenous state of *Septoria tritici* Rob. and Desm. N. Z. J. Bot. 10:707–710.

Scharen, A.L. 1999. Biology of the *Septoria/Stagonospora* pathogens: An overview. p. 19–22. *In* M. van Ginkel, A. McNab, and J. Krupinsky (ed.) *Septoria* and *Stagonospora* diseases of cereals: A compilation of global research. CIMMYT, D.F., Mexico.

Schilder, A.M.C., and G.C. Bergstrom. 1992. The dispersal of conidia and ascospores of *Pyrenophora tritici-repentis*. p. 96–99. *In* L.J. Francl, J.M. Krupinsky, and M.P. McMullen (ed.) Advances in tan spot research. Proc. Int. Tan Spot Workshop, Fargo, ND. 25–26 June 1992. North Dakota Agric. Exp. Stn., North Dakota State Univ., Fargo, ND.

Schnieder, F., G. Koch, C. Jung, and J.A. Verreet. 2001. Genotypic diversity of the wheat leaf blotch pathogen *Mycosphaerella graminicola* (anamorph) *Septoria tritici* in Germany. Eur. J. Plant Pathol. 107:285–290.

Schroeder, H.W., and J.J. Christensen. 1963. Factors affecting resistance of wheat to scab caused by *Gibberella zeae*. Phytopathology 53:831–838.

Scott, P.R., F.R. Sanderson, and P.W. Benedikz. 1988. Occurrence of *Mycosphaerella graminicola*, teleomorph of *Septoria tritici*, on wheat debris in the UK. Plant Pathol. 37:285–290.

Shah, D.A., G.C. Bergstrom, and P.P. Ueng. 2001. Foci of Stagonospora nodorum blotch in winter wheat before canopy development. Phytopathology 91:642–647.

Shaner, G. 1973. Evaluation of slow-mildewing resistance of Knox wheat in the field. Phytopathology 63:867–872.

Shaw, M.W., and D.J. Royle. 1989. Airborne inoculum as a major source of *Septoria tritici* (*Mycosphaerella graminicola*) infections in winter wheat crops in the UK. Plant Pathol. 38:35–43.

Shoemaker, R.A., and C.E. Babcock. 1989. *Phaeosphaeria*. Can. J. Bot. 67:1500–1599.

Simon, M.R., F.M. Ayala, C.A. Cordo, M.S. Roder, and A. Borner. 2007. The use of wheat/goatgrass introgression lines for the detection of gene(s) determining resistance to septoria tritici blotch (*Mycosphaerella graminicola*). Euphytica 154:249–254.

Simon, M.R., and C.A. Cordo. 1997. Inheritance of partial resistance to *Septoria tritici* in wheat (*Triticum aestivum*): Limitation of pycnidia and spore production. Agronomie 17:343–347.

Singh, P.K., M. Mergoum, S. Ali, T.B. Adhikari, E.M. Elias, and G.R. Hughes. 2006. Identification of new sources of resistance to tan spot, Stagonospora nodorum blotch, and Septoria tritici blotch of wheat. Crop Sci. 46:2047–2053.

Singh, P.K., M. Mergoum, and G.R. Hughes. 2007. Variation in virulence to wheat in *Pyrenophora tritici-repentis* population from Saskatchewan, Canada, from 2000 to 2002. Can. J. Plant Pathol. 29:166–171.

Somasco, O.A., C.O. Qualset, and D.G. Gilchrist. 1996. Single gene resistance to *Septoria tritici* blotch in the spring wheat cultivar 'Tadinia'. Plant Breed. 115:261–267.

Somers, D.J., G. Fedak, and M. Savard. 2003. Molecular mapping of novel genes controlling Fusarium head blight resistance and deoxynivalenol accumulation in spring wheat. Genome 46:555–564.

Starkey, D.E., T.J. Ward, T. Aoki, L.R. Gale, H.C. Kistler, D.M. Geiser, H. Suga, B. Toth, J. Varga, and K. O'Donnell. 2007. Global molecular surveillance reveals novel Fusarium head blight species and trichothecene toxin diversity. Fungal Genet. Biol. 44:1191–1204.

Strelkov, S.E., and L. Lamari. 2003. Host–parasite interactions in tan spot [*Pyrenophora tritici-repentis*] of wheat. Can. J. Plant. Pathol. 25:339–349.

Strelkov, S.E., L. Lamari, R. Sayoud, and R.B. Smith. 2002. Comparative virulence of chlorosis-inducing races of *Pyrenophora tritici-repentis*. Can. J. Plant Pathol. 24:29–35.

Summerell, B.A., and L.W. Burgess. 1989. Factors influencing survival of *Pyrenophora tritici-repentis*: Stubble management. Mycol. Res. 93:38–40.

Sutton, J.C. 1982. Epidemiology of wheat head blight and maize ear rot caused by *Fusarium graminearum*. Can. J. Plant Pathol. 4:195–209.

Svec, M., and M. Miklovicova. 1998. Structure of populations of wheat powdery mildew (*Erysiphe graminis* DC f. sp. *tritici* Marchal) in central Europe in 1993–1996: 1. Dynamics of virulence. Eur. J. Plant Pathol. 104:537–544.

Szunics, L., L. Szunics, G. Vida, Z. Bedo, and M. Svec. 2001. Dynamics of changes in the races and virulence of wheat powdery mildew in Hungary between 1971 and 1999. Euphytica 119:143–147.

Thomas, M.R., R.J. Cook, and J.E. King. 1989. Factors affecting the development of *Septoria tritici* in winter wheat and its effect on yield. Plant Pathol. 38:246–257.

Tucker, D.M., C.A. Griffey, S. Liu, G. Brown-Guedira, D.S. Marshall, and M.A. Saghai Maroof. 2007. Confirmation of three quantitative trait loci conferring adult plant resistance to powdery mildew in two winter wheat populations. Euphytica 155:1–13.

van Ginkel, M., and A.L. Scharen. 1988. Host pathogen relationships of wheat and *Septoria tritici*. Phytopathology 78:762–766.

Ward, T.J., R.M. Clear, A.P. Rooney, K. O'Donnell, D. Gaba, S. Patrick, D.E. Starkey, J. Gilbert, D.M. Geiser, and T.W. Nowicki. 2008. An adaptive evolutionary shift in Fusarium head blight pathogen populations is driving the rapid spread of more toxigenic *Fusarium graminearum* in North America. Fungal Genet. Biol. 45:473–484.

Wilkinson, C.A., J.P. Murphy, and R.C. Rufty. 1990. Diallel analysis of components of partial resistance to *Septoria nodorum* in wheat. Plant Dis. 74:47–50.

Wright, K.H., and J.C. Sutton. 1990. Inoculation of *Pyrenophora tritici-repentis* in relation to epidemics of tan spot of winter wheat in Ontario. Can. J. Plant Pathol. 12:149–157.

Xu, S.S., T.L. Friesen, and A. Mujeeb-Kazi. 2004. Seedling resistance to tan spot and Stagonospora nodorum blotch in synthetic hexaploid wheats. Crop Sci. 44:2238–2245.

Yu, J.B., G.H. Bai, W.C. Zhou, Y.H. Dong, and F.L. Kolb. 2008. Quantitative trait loci for Fusarium head blight resistance in a recombinant inbred population of Wangshuibai/Wheaton. Phytopathology 98:87–94.

Zhan, J., R.E. Pettway, and B.A. McDonald. 2003. The global genetic structure of the wheat pathogen *Mycosphaerella graminicola* is characterized by high nuclear diversity, low mitochondrial diversity, regular recombination, and gene flow. Fungal Genet. Biol. 38:286–297.

Chapter 8
Nematodes Which Challenge Global Wheat Production

Richard W. Smiley and Julie M. Nicol

SUMMARY

(1) Effects of cereal cyst (*Heterodera*) and root-lesion (*Pratylenchus*) nematodes on wheat are difficult to identify and control. Symptoms are nonspecific and easily confused with stress from nutrient deficiency, drought, or disease.
(2) Multiple species of *Heterodera* and *Pratylenchus* are capable of damaging wheat. Identification of species is difficult and procedures based on comparative morphology can be unreliable. Identification of species is now assisted by molecular tools.
(3) *Heterodera* species discussed in this chapter reproduce only on hosts within the Poaceae, and individual species are highly heterogeneous for virulence to specific host genotypes. Many pathotypes occur within *H. avenae*, and the same is anticipated for *H. filipjevi* and *H. latipons*. No pathotypic variation within either *P. neglectus* or *P. thornei* has been reported on wheat, but both species multiply in a wide range of monocot and dicot hosts. Mixtures of *Heterodera* species or pathotypes, and *Pratylenchus* species, may occur within individual fields.
(4) Field sanitation is important because these nematodes multiply on many weed species and volunteer cereals. Cereal cyst nematode can be controlled by rotating wheat with a noncereal, a resistant cultivar, or weed-free fallow. Root-lesion nematode is best managed by rotating resistant and tolerant wheat cultivars with other poor hosts.
(5) Resistance and tolerance are genetically independent, and cultivars resistant or tolerant to one species are not necessarily resistant or tolerant to another species. Root-lesion nematode resistance is quantitative and cereal cyst nematode is controlled by single-gene resistance. Molecular markers have been developed to identify genes and quantitative trait loci for resistance in seedlings.
(6) Molecular tests to identify and quantify nematodes in commercial soil testing laboratories will allow more effective surveys of populations. Greater collaboration is needed between research institutions, organizations, and countries.

INTRODUCTION

Plant-parasitic nematodes are tiny but complex animals (unsegmented roundworms) anatomically differentiated for feeding, digestion, locomotion, and reproduction (Barker et al., 1998). Most species are transparent, vermiform (eel-shaped), and 0.5–2 mm long. They puncture cells and damage plants mechanically and chemically, reducing plant vigor, inducing lesions, rots, deformations, galls, or root knots, and predisposing plants to infection by root-infecting fungi. World crop production is thought to be reduced 10% by damage from plant-parasitic nematodes (Whitehead 1997).

Plant-parasitic nematode species that live in the soil represent one of the most difficult pest problems to identify, demonstrate, and control. A particular challenge is to clearly identify damage by cereal nematodes, as the symptoms are nonspecific and easily confused with other ailments such as nitrogen deficiency, water availability, and other disease. Farmers, pest management advisors, and scientists routinely underestimate or fail to recognize their impact on wheat.

Procedures to sample, extract, identify, and quantify plant-parasitic nematodes are both technically challenging and time-consuming. Extensive training is necessary to distinguish genera and species (Varma 1995; Mai and Mullin 1996; Siddiqi 2000), which is essential for implementation of appropriate management strategies.

The damage threshold, defined as the number of nematodes to give a specific yield loss, is determined by both environmental and genotypic factors. The threshold generally is decreased when plant growth is stressed by drought, poor soil nutrition, impediments to root penetration, or adverse temperature. The threshold is increased by partial or full resistance reactions by a given cultivar. Damage caused by cereal nematode is likely to be greater where limited rotation or cultivar options exist, especially in rainfed cereal monoculture, including "rotations" of winter wheat with summer fallow. Unlike the visually obvious and more vastly studied cereal rusts, the global knowledge of economic importance is less well known and understood due to difficulties working with soil-inhabiting nematodes.

The most important plant-parasitic species affecting wheat are in the genera *Heterodera* (cyst), *Pratylenchus* (root-lesion), *Meloidogyne* (root knot), *Ditylenchus* (stem), *Tylenchorhynchus* and *Merlinius* (stunt), *Paratrichodorus* (stubby-root), and *Anguina* (seed-gall) (Rivoal and Cook 1993; McDonald and Nicol 2005; Nicol and Rivoal 2007; Bockus et al., 2009).

This chapter will focus on two nematodes of primary global importance to wheat. The global distribution of cereal cyst nematode species and pathotypes is clearly a major economic constraint to rainfed wheat production systems, especially where monocultures are dominant. Root-lesion nematode species are also important but appear to have a more restricted distribution.

CEREAL CYST NEMATODE

Cyst nematodes are the most studied plant-parasitic nematodes on wheat (Cook and Noel 2002; Nicol 2002; Nicol et al., 2003). Although the "*Heterodera avenae* group" (Handoo 2002) is a complex of 12 species and intraspecific pathotypes that invade roots of cereals and grasses, three main species are the most economically important: *Heterodera avenae*, *H. filipjevi*, and *H. latipons* (Rivoal and Cook 1993, McDonald and Nicol 2005).

Heterodera avenae is economically important in temperate wheat-producing regions throughout the world, including North and South Africa, East and West Asia, Australia, Europe, the Indian Subcontinent, the Middle East, and North America. *Heterodera latipons* occurs mostly throughout the Mediterranean region but also in Asia and Europe. This species was recently described as widespread and economically important in key wheat growing provinces of China (Peng et al., 2007). *Heterodera filipjevi* was recently detected in North America (Smiley et al., 2008) and also has an increasingly recognized wide distribution across northern Europe and continental climates of Central and West Asia, as well as the Middle East and Indian Subcontinent. Only one species is generally identified in most regions but mixtures of species may also occur in individual fields (Abidou et al., 2005).

Less prevalent species of cyst nematode associated with wheat include *H. arenaria, H. bifenestra, H. hordecalis, H. mani, H. pakistanensis, H. pratensis, H. zeae*, and *Punctodera punctata*.

Symptoms and epidemiology

Plants with roots heavily damaged by *H. avenae* appear initially as pale green seedlings that lack vigor (Color Plate 16). Mature plants are often severely stunted (Color Plate 17). Plants with visual damage often occur in patches but may also occur over entire fields, particularly under monocultures of susceptible cereals and when combined with inadequate plant nutrition or other stress.

Symptoms on roots are specific to host species. Wheat (*Triticum* species) and barley (*Hordeum vulgare* L.) roots invaded by *Heterodera avenae* branch excessively at locations where juveniles invade, resulting in a bushy or knotted appearance (Color Plate 18a). Root symptoms often do not become recognizable until 1–3 months after planting, depending upon climatic conditions and spring or winter wheat growth habit. Oat (*Avena sativa* L.) roots invaded by *H. avenae* are shortened and thickened but do not exhibit the knotted symptom.

Heterodera species complete only one generation per crop season. Juveniles penetrate epidermal and cortical cells of young root segments in the zone of elongation. They enter the stele, where they induce the formation of a specialized feeding cell called a syncytium. Females are fertilized by males and 100–600 eggs are retained in the female body.

Mature females become sedentary and embedded in the root. The presence of the white swollen female body (0.5–2 mm; about the size of a pin head) is diagnostic. It can be seen around the flowering time of wheat. One or more females are generally visible at the point of abnormal root proliferation. They protrude from the root surface, glisten when wet, and are white-gray. They are best viewed by washing a root sample and observing under low magnification (Color Plate 18b), because their presence among knotted roots is often obscured by adhering soil. The females are attached loosely and are easily dislodged when soil is washed from roots.

Upon death of host roots the female body wall dies and hardens into a resistant dark-brown cyst of a similar size as a soil particle. These cysts mostly dislodge into the soil as the wheat roots decompose. The cyst protects eggs and juveniles during periods between hosts. Eggs inside cysts may remain viable for several years. Emergence of juveniles from brown cysts requires a period of dormancy (diapause) that differs among species and climatic regions. Diapause characteristics must be understood before damage by these nematodes can be effectively managed. Emergence of juveniles is triggered by specific interactions between soil temperature and moisture, and these conditions may be overcome to some extent by exudates from host roots (Ismail et al., 2000; Scholz and Sikora 2004; McDonald and Nicol 2005). Well-established infestations exhibit ecotypic differences in which peak numbers of infective juveniles in soil generally coincide with the traditional wheat sowing and seedling growth stages in each geographic region (Rivoal and Cook 1993).

Cereal cyst nematode is not strongly restricted by soil type but damage is often greatest in light-textured, well-drained soils such as sands. The damage threshold varies with soil type, climate, and cultivar, and with nematode species, virulence, and ecotype characteristics. These variable influences on plant damage make it difficult to directly relate initial populations with reduction in grain yield (Bonfil et al., 2004).

Causal organisms

The most economically important cyst nematode species on wheat include *H. avenae* Wollenweber, *H. latipons* Franklin, and *H. filipjevi* (Madzhidov) Stone. Two important nematodes previously reported as *H. avenae* have been reclassified. The so-called Gotland strain of *H. avenae* is now accepted as *H. filipjevi* (Bekal et al., 1997; Ferris et al., 1999), making this species more reported than previously thought. Most recently, *H. avenae* pathotype Ha13 in Australia was redescribed as *H. australis* (Subbotin et al., 2002); that designation has not yet been widely accepted.

Identification of cereal cyst nematodes is complex and has traditionally been based on comparative morphology and diagnostic keys (Luc et al., 1988; Handoo 2002). Techniques based on protein or DNA differences using RFLP are now available to facilitate identification to a species level and to study phylogenetic relationships (Subbotin et al., 1996, 1999, 2000, 2001, 2003; Bekal et al., 1997; Andrés et al., 2001a,b; Mokabli et al., 2001; Rivoal et al., 2003). Further work is needed to convert these RFLP-based primers into the more cost- and time-efficient PCR-based probes.

One of the major challenges to controlling cereal cyst nematodes is occurrence of individuals within species and also among populations from different regions that is highly variable in virulence and in reproductive capacity (fitness) characteristics on the same host (Rivoal et al., 2001; Mokabli et al., 2002). Moreover, individual species within the *H. avenae* group are highly heterogeneous with respect to virulence to specific host genotypes (Cook and Rivoal 1998; Cook and Noel, 2002; McDonald and Nicol, 2005).

Virulence groups (pathotypes) are differentiated (Cook and Noel 2002; McDonald and Nicol 2005) by testing unknown populations against a matrix of cereals in "The International Cereal Test Assortment for Defining Cereal Cyst Nematode Pathotypes," which was developed by Andersen and Andersen (1982). The test distinguishes three primary groups based on host resistance reactions of three barley cultivars carrying the resistance genes *Rha1*, *Rha2*, and *Rha3*. Additional barley, oat, and wheat differentials are used to define pathotypes within each group. The most widely distributed *H. avenae* populations in Europe, North Africa, and Asia are in groups 1 and 2 (Al-Hazmi et al., 2001; Cook and Noel 2002; Mokabli et al., 2002; McDonald and Nicol 2005). Pathotypes in group 3 are prevalent in Australia, Europe, and North Africa (Rivoal and Cook 1993; Mokabli et al., 2002). Unfortunately, the pathotype concept is incomplete because it was established to differentiate northern European populations of *H. avenae* and is increasingly incapable of clearly defining resistance reactions achieved with populations in other regions. For instance, three undescribed pathotypes were recently reported from China (Nicol and Rivoal 2007; Peng et al., 2007), and the existing pathotype matrix does not define North American populations (R.W. Smiley, unpublished data). The Test Assortment therefore greatly underestimates polymorphism for *H. avenae* (Cook and Noel 2002; McDonald and Nicol 2005), *H. latipons*, and *H. filipjevi*. The Test Assortment needs to be revised to capture new sources of resistance and pathogen variation.

Management

To achieve effective control of cereal cyst nematodes it is necessary to reduce the population below the economic threshold for damage. This requires definitive studies on population dynamics and yield losses on representative local cultivars under natural field conditions. Cultural practices based on rotational combinations of nonhosts (noncereals), resistant cultivars, and clean fallow can effectively control these nematodes. Restricting hosts to 50% of the time in heavier soils and 25% in lighter soils can cause dramatic reductions in the population of *H. avenae*. However, these management strategies each require a full understanding of the virulence and diapause characteristics for the local nematode populations, and of the effectiveness and durability of the resistance gene(s) deployed against that nematode population.

The use of host-plant resistance is one of the most effective methods of controlling cereal nematodes. Resistance is defined as the ability of the host to inhibit nematode multiplication (Cook and Evans 1987). Ideally resistance should be combined with tolerance, which is the ability of the host plant to maintain yield potential in the presence of the nematode (Cook and Evans 1987). The use of cultivars that are both resistant and tolerant offers the best control option, in addition to being environmentally sustainable and requiring no additional equipment or cost. However, the use of resistance requires a sound knowledge of the virulence spectrum for the targeted species and pathotypes. Wheat cultivars resistant to *H. avenae* populations in one region may be fully susceptible to populations in other regions. This was shown for Australian cultivars evaluated in Israel (Bonfil et al., 2004) and for the cultivar Raj MR1 in India, which is effective in Rajasthan but not in the Punjab (A.K. Singh, pers. comm.). Also, although not frequently reported, repeated plantings of wheat, barley, and oat cultivars with a single gene for resistance to *H. avenae* have led to selection of new virulent pathotypes over prolonged time periods, overcoming host-plant resistance (Lasserre et al., 1996; Cook and Noel 2002), in addition to possibly increasing damage from root-lesion nematode (Lasserre et al., 1994).

It is also possible to manage damage by rotating resistant cereals with susceptible crop species. However, local knowledge of resistance reactions is essential for effective use of this practice. For instance, rye (*Secale cereale* L.) and certain cultivars of triticale (*Triticosecale rimpaui* Wittm.) are resistant. Oat is resistant to *H. avenae* in Australia and several Mediterranean countries but susceptible in northern Europe (McDonald and Nicol 2005). Moreover, resistant cultivars from one region may be exposed to mixtures of species in other regions, as exemplified in Israel by oat cultivars that are resistant to *H. avenae* and susceptible to *H. latipons* (Mor et al., 1992).

Host resistance will continue to be the most profitable and easily applied management procedure. However, resistance will only be used by farmers if the cultivars also contain a level of tolerance (yield performance) which is comparable to other commonly cultivated wheat cultivars. Sources of resistance to *H. avenae* populations worldwide have been collated and reviewed and, where possible, have had their genetic location and gene designation reported (Table 8.1) (Rivoal et al., 2001; Nicol 2002; Nicol et al., 2003; McDonald and Nicol 2005; Nicol and Rivoal 2007). All of the sources of resistance reported against cereal cyst nematode to date feature single-gene inheritance. Six *Cre* genes for *H. avenae* resistance in wheat (*Cre2* to *Cre7*) and the *Rkn2* gene for resistance to both *H. avenae* and *Meloidogyne naasi* (Jahier et al., 1998) were derived from *Aegilops* species. Other resistance genes were derived from *Triticum aestivum* (*Cre1* and *Cre8*) and *Secale cereale* (*CreR*). Several other sources of resistance (*CreX* and *CreY*) are also reported, but their genetic control and gene designation are still unknown. Most of these resistance genes have been introgressed into hexaploid wheat.

The *Cre1* gene is highly effective against populations of *H. avenae* from Europe, North Africa, and North America (Fig. 8.1) and moderately effective or ineffective against populations in Australia and Asia (Rivoal et al., 2001; Mokabli et al., 2002). Populations of *H. filipjevi* in India and *H. latipons* in Syria differ in virulence to the *Cre1* gene, compared with *H. avenae* (Mokabli et al., 2002). In Turkey, the *Cre1* gene appears effective against *H. filipjevi*, but *Cre3* is not. The *Cre3* gene is effective against Australian populations (Vanstone et al., 2008) but not European populations of *H. avenae* (de Majnik et al., 2003; Safari et al., 2005) or *H. filipjevi* in Turkey. The *Cre2* and *Cre4* resistance genes from *Aegilops* and an unidentified resistance gene from the wheat line AUS4930 offer promise against an array of *Heterodera* species and pathotypes (Nicol et al., 2001). An International Root Disease Resistance Nursery containing seven of the known *Cre* genes is coordinated by CIMMYT to establish the value of these genes in different regions of the world.

Molecular markers have been developed to identify genes for resistance to *H. avenae* in barley

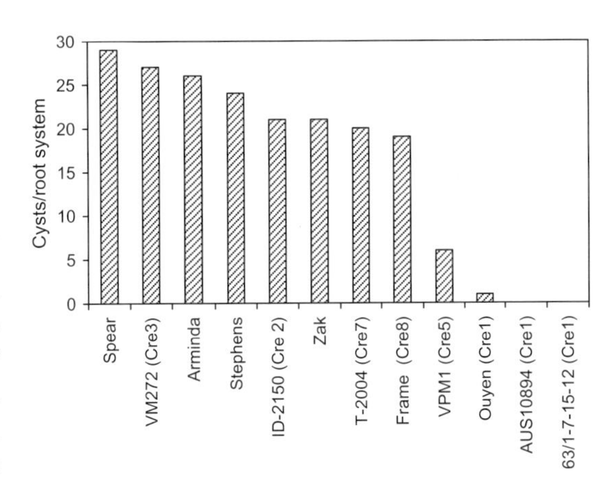

Fig. 8.1 Relative number of cysts for an Oregon population of *Heterodera avenae* developing on root systems of 12 wheat cultivars or lines; the identity of a *Cre* resistance gene is indicated if present.

Table 8.1 Principal sources of genes used to breed wheat for resistance to the cereal cyst nematode *Heterodera avenae* [*Ha*], unless stated otherwise.

Cereal Species	Cultivar or Line	Origin	Resistance Gene(s)[a,b]	Response to Pathotypes[b,c]	Use in Cultivars
Wheat					
Triticum aestivum	Loros, AUS10894	?[d]	*Cre1*[e] (formerly *Ccn1*), on chromosome 2BL	pR to several pathotypes	NW Europe, Australia; NW US under evaluation
	Katyil	Australia	*Ccn1*	S, India	Australia
	Festiguay	Australia	*Cre8* (formerly *CreF*) on chromosome 7L? Recent analysis suggests 6B	pR to Ha13	Australia
	AUS4930 = Iraq 48	Iraq	Possibly identical genetic location as *Cre1*; also resistant to *P. thornei*	R to several pathotypes and *Heterodera* species and *P. thornei*	Australia, France, CIMMYT; under evaluation
	Molineux	Australia	Chromosome 1B (14% resistance)	R to Ha13	Australia
	Raj MR1 (Raj Molya Rodhak1)	Landrace from Nidge, Turkey AUS 15854 × J-24	One dominant gene	R only to some populations of *H. avenae*, appears S to Indian *H. filipjevi*	Released cultivar in northern India in 2002
Triticum durum	Psathias 7654, 7655, Sansome, Khapli	?	?	S to some pathotypes, pR to others	
Triticale and rye					
Triticosecale	T701-4-6	Australia	*CreR* on chromosome 6RL	R to Ha13	Australia
	Drira (= Ningadhu)	Australia	?	R to Ha13	Australia
	Tahara	Australia	?	R to Ha13	
	Salvo	Poland	?		UK
Secale cereale	R173 Family		*CreR* on chromosome 6RL	R to Ha13	Australia
Wild grass relatives of wheat					
Aegilops tauschii	CPI 110813	Central Asia	*Cre4* on chromosome 2DL	R to Ha13	Australian synthetic hexaploid lines
Aegilops tauschii	AUS18913	?	*Cre3* on chromosome 2DL	R to Ha13	Australian advanced breeding lines
Aegilops peregrina (= *Ae. variabilis*)	1		*Cre(3S) with Rkn2* on chromosome 3S; *CreX*, not yet located		
Aegilops longissima	18	?	?	R to four French pathotypes and *Meloidogyne naasi*	France
Aegilops geniculata	79; MZ1, MZ61, MZ77, MZ124	?		R and pR to several pathotypes	France—under evaluation
Aegilops triuncialis	TR-353	?	*Cre7 (formerly CreAet)*	R and pR to several pathotypes	France—under evaluation
Aegilops ventricosa	VPM 1		*Cre5* (formerly *CreX*), on chromosome 2AS	R to several pathotypes	Spain—under evaluation
	11; AP-1, H-93-8		*Cre2* (formerly *CreX*) on genome N^v		
	11; AP-1, H-93-8, H-93-35		*Cre6*, on chromosome $5N^v$		

[a]*Sources:* Reviews and references in Rivoal and Cook (1993), Cook and Rivoal (1998), McDonald and Nicol (2005), and Nicol and Rivoal (2007).
[b]Characterized single-gene resistance to cereal cyst nematode.
[c]R = resistant, pR = partially resistant, S = susceptible.
[d]? = no published scientific studies conducted.
[e]Marker implemented in commercial breeding program; refer to Ogbonnaya et al. (2001b).

and wheat (Eastwood et al., 1994; Williams et al., 1994, 2006; Kretschmer et al., 1997; Barr et al., 1998; Paull et al., 1998; Eagles et al., 2001; Ogbonnaya et al., 2001a,b; Martin et al., 2004; Barloy et al., 2007). Some of these markers have been used in marker-assisted selection and for pyramiding genes for resistance.

Practices other than crop rotation and planting resistant cultivars are less efficient; however, components of these strategies could still form part of an integrated pest management approach to nematode control. Other cultural methods (Nicol and Rivoal 2007) include manipulating the sowing time to minimize the impact of the major hatching period, as when winter wheat is planted during autumn in cool, temperate regions where the major hatch occurs during spring. *Heterodera avenae* populations can also be reduced by planting a susceptible host as a trap crop prior to the major hatching period, thereby encouraging a maximum hatching efficiency by a plant stand that is then killed before new cysts are developed. The greatest crop loss occurs when nutrients or water become limiting for maximum plant growth potential at any point during the growing season. Crop damage is therefore minimized by supplying optimal plant nutrition (Color Plate 16) and, where possible, supplemental water during intervals of drought.

Biological control products are not commercially available, but *H. avenae* populations in some locations are maintained below an economic threshold by fungal and bacterial parasites of eggs and juveniles (Kerry 1987; Kerry and Crump 1998; Ismail et al., 2001).

A low rate of nematicide application can provide effective and economical control of cereal cyst nematode in wheat (Brown 1987). However, current environmental concerns associated with these chemicals eliminate them as a viable alternative for use by farmers. They will, however, continue to be an important research tool for studying yield loss and population dynamics.

Once introduced into a region or country it is very difficult to minimize the spread of cyst nematodes. They are efficiently disseminated by all means of soil movement, including minute amounts of soil that contaminate equipment, by animals and plant products, and by soil that is moved by water and wind. Rapid dissemination together with increased reporting of *H. avenae* is especially well illustrated in China, where this nematode was first reported in 1987 and is now reported in at least eight provinces (Nicol and Rivoal 2007; Peng et al., 2007). Likewise, *H. avenae* was first reported in the western US in 1974 and is now reported in at least seven states (Smiley et al., 1994, 2005c).

ROOT-LESION NEMATODE

At least eight species in the genus *Pratylenchus* are parasitic to wheat (De Waele and Elsen 2002; Nicol 2002; Nicol et al., 2003; McDonald and Nicol 2005; Castillo and Vovlas 2007). Four species (*P. crenatus*, *P. neglectus*, *P. penetrans*, and *P. thornei*) occur throughout the world in temperate cereal-producing regions.

Pratylenchus neglectus and *P. thornei* are the species most often associated with yield loss in wheat and are emphasized in this chapter. One or both species occur in Australia, Europe, the Indian Subcontinent, the Mediterranean Basin, the Middle East, West Asia, North Africa, and North America (Nicol and Rivoal 2007). *Pratylenchus thornei* is considered the most economically important species on wheat and has reduced yields as much as 85% in Australia, 37% in Mexico, 70% in Israel, and 50% in the US (Armstrong et al., 1993; Nicol and Ortiz-Monasterio 2004; Smiley et al., 2005a). *Pratylenchus neglectus* also causes losses up to 37% in the US (Smiley et al., 2005b).

Symptoms and epidemiology

Pratylenchus species are migratory root endoparasites capable of multiplying in a wide range of monocot and dicot host species (Loof 1978; Vanstone and Russ 2001a,b; Vanstone et al., 2008). They live freely in soil and may become entirely embedded in root tissue but never lose the ability to migrate within the root or back into soil.

Root-lesion nematodes puncture and migrate through root epidermal and cortical cells (Color Plate 19a). Tissue degradation results in lesions that favor greater colonization by root-rotting fungi. These activities reduce the ability of roots to produce branches and absorb water and nutrients. Cortical degradation and reduced branching (Color Plate 19b) often are not visible until plants are 6 or more weeks old, and these symptoms are often confused with root rots caused by *Pythium* and *Rhizoctonia*. Interactions of root-lesion nematodes, fungal pathogens, other plant-parasitic nematodes, and insect pests have been reported (Lasserre et al., 1994; Taheri et al., 1994; Smiley et al., 2004a,b).

Foliar symptoms are nonspecific (Van Gundy et al., 1974; Orion et al., 1984; Doyle et al., 1987; Thompson et al., 1995; Smiley et al., 2005a,b). Intolerant plants with roots heavily damaged by root-lesion nematodes may exhibit poor vigor, yellowing and premature death of lower leaves, stunting, reduced tillering, and reduced grain yield and grain quality. Damaged wheat plants are less capable of extracting soil water and exhibit stress and wilting earlier than undamaged plants as soil moisture becomes limiting for plant growth. Plants that become infested while growing under drought stress are more likely to suffer yield loss (Nicol and Ortiz-Monasterio 2004).

Pratylenchus species associated with wheat are not strongly restricted by soil type and may attain damaging population levels even in the very driest (250 mm annual precipitation) rainfed wheat-producing regions. Large populations have been detected throughout the depth of root growth in deep soils (Taylor and Evans 1998; Thompson et al., 1999; Ophel-Keller et al., 2008). *Pratylenchus* species can survive in an inactive, dehydrated state (anhydrobiosis) in roots and soil during dry conditions (Glazer and Orion 1983; Talavera and Vanstone 2001). Individuals entering host roots after emerging from anhydrobiosis multiply more rapidly than individuals that have not been subjected to dormancy. Populations of *Pratylenchus* often decline during long fallow periods between crops but high rates of survival have also been reported (Orion et al., 1984; Talavera and Vanstone 2001).

Causal organisms

Pratylenchus neglectus (Rensch) Filipjev Schuurmanns & Stekhoven and *P. thornei* Sher & Allen often occur as mixtures in the same soil. Both species are parthenogenic, with males generally being rare or absent. In contrast, species such as *P. penetrans* (Cobb) Filipjev and Schuurmans Stekhoven are amphimictic, with populations having both males and females.

All species of *Pratylenchus* retain a vermiform body shape (Color Plate 19a) with many being about 0.5 mm long and 0.02 mm in diameter. Life cycles range from 45 to 65 days depending on species and environmental variables. Females deposit about one egg per day in root tissue or in soil. First-stage juveniles molt to second-stage juveniles within the egg. One second-stage juvenile emerges from each egg about 1 week after the egg was deposited. Two additional molts within 35–40 days result in the adult stage. All juvenile and adult stages are parasitic. The number of nematodes in root tissue increases exponentially through the growing season.

Identification of *Pratylenchus* to the species level is an essential prerequisite for most control strategies. However, identification is difficult because the few morphological characteristics of taxonomic value for differentiating *Pratylenchus* species are, without exception, replete with large ranges of intraspecific variation, often including overlapping ranges and shapes. Therefore, procedures to differentiate species based on comparative morphology (Loof 1978; Filho and Huang 1989; Handoo and Golden 1989) are always difficult and can be unreliable. Modern techniques are based on detection of differences of proteins or DNA (Ibrahim et al., 1995; Ouri and Mizukubo 1999; Uehara et al., 1999; Andrés et al., 2000; Waeyenberge et al., 2000; Carta et al., 2001; Al-Banna et al., 2004; Carrasco-Ballesteros et al., 2007; Castillo and Vovlas 2007). The PCR or

RFLP procedures are particularly useful for identifying species. Species-targeted real-time PCR procedures have been developed to differentiate and quantify *P. neglectus* and *P. thornei* in a single DNA extract from soil (G.P. Yan and R.W. Smiley, unpublished data).

Biological diversity among populations has been reported for six *Pratylenchus* species (De Waele and Elsen 2002), including *P. neglectus* on potato (Hafez et al., 1999) but not for *P. thornei*. The potential impact of *P. neglectus* heterogeneity or pathotypes on wheat has not been reported, but results from screening specific wheat and barley genotypes against populations of both *P. neglectus* and *P. thornei* in Australia, Mexico, Turkey, and the US have thus far been uniform for both species across all countries, as have been observations from crop rotation experiments and commercial practices.

Management

Management of root-lesion nematodes is best approached by integrating crop rotations (Fig. 8.2) and planting wheat cultivars that are both resistant (Fig. 8.3) and tolerant (Fig. 8.4). Rotations alone are somewhat limited due to the polyphagous nature of *P. neglectus* and *P. thornei*. The greatest long-term production efficiency will therefore be achieved with wheat cultivars that are both resistant and tolerant to the most economically important *Pratylenchus* species in a region or to both species where mixtures occur (Thompson et al., 2008; Vanstone et al., 2008). However, cultivars with resistance to *P. neglectus* are not necessarily resistant to *P. thornei* (Fig. 8.3), and vice versa (Farsi et al., 1995).

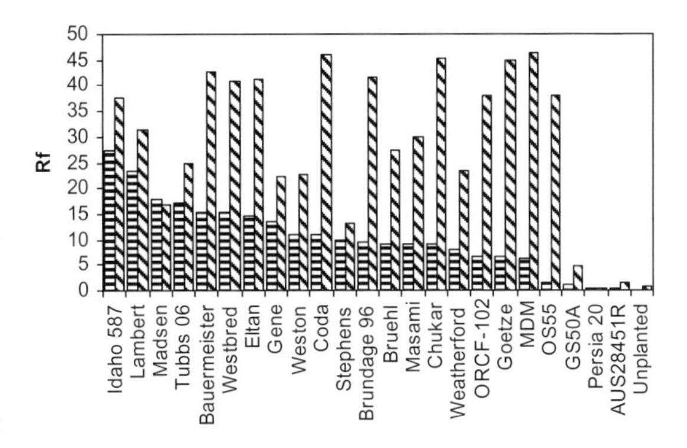

Fig. 8.3 Reproductive factor (Rf) for *Pratylenchus thornei* (left) and *P. neglectus* (right) in 20 Pacific Northwest (US) winter wheat cultivars, in 3 lines carrying genes for resistance, and in unplanted soil; Rf = P_f P_i^{-1} × 100, where P_f = final number after 16 weeks' growth and P_i = initial population in soil (750 nematodes per kilogram).

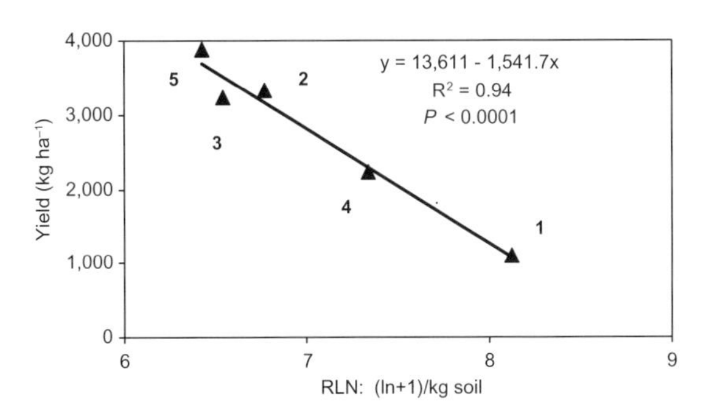

Fig. 8.2 Relationship between numbers of root-lesion nematodes (RLN) and grain yield for winter wheat in five crop-rotation and tillage-management treatments in a 300-mm precipitation zone of Oregon (US); means across 3 years for no-till annual winter wheat (1), rotations of winter wheat with cultivated fallow (2) or with chemical fallow (3), rotation of no-till winter wheat and no-till winter pea (4), and rotation of no-till winter wheat, no-till spring barley, and chemical fallow (5).

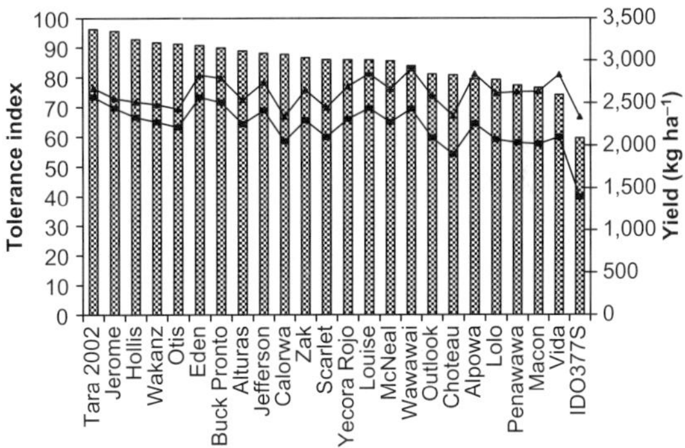

Fig. 8.4 Tolerance index (bars) and grain yield for Pacific Northwest (US) spring wheat cultivars produced in untreated soil (■) or in soil treated with nematicide (▲) to reduce numbers of *Pratylenchus neglectus*; grain yield in untreated soil correlated with tolerance index (R^2 = 0.80, $P < 0.0001$).

Cultivars tolerant to *P. neglectus* are not necessarily tolerant to *P. thornei*, and vice versa. Furthermore, resistance and tolerance to each of these root-lesion nematode species are genetically independent; a cultivar can be resistant and intolerant, susceptible and tolerant, or another combination.

Successive or frequent crops of susceptible wheat cultivars elevate populations of *P. neglectus* and *P. thornei* and increase the level of risk to subsequent intolerant crops. Many cultivars of mustard [*Brassica juncea* (L.) Czern., *Sinapsis alba* L.], canola (*Brassica napus* L.), lentil (*Lens culinaris* Medik.), and chickpea (*Cicer arietinum* L.) also increase the population of *P. neglectus* or *P. thornei*, or both, with multiplication capacities differing greatly for each combination of *Pratylenchus* species and host cultivar (Bernard and Montgomery-Dee 1993; Castillo et al., 1998; Potter et al., 1999; Fatemy et al., 2006).

Crops that restrict multiplication of *P. neglectus* and/or *P. thornei*, such as some cultivars of barley, safflower (*Carthamus tinctorius* L.), triticale, flax (*Linum usitatissimum* L.) and field pea (*Pisum sativum* L.), reduce the nematode population and improve the yield potential for subsequent intolerant wheat crops (Van Gundy et al., 1974; Heide 1975; Esmenjaud et al., 1990; Lasserre et al., 1994; Farsi et al., 1995; Thompson et al., 1995; Hollaway et al., 2000; Taylor et al., 2000; Riley and Kelley 2002; Smiley et al., 2004b). However, results from these studies indicate hosting ability is species- and cultivar-specific, within both legumes and cereals. Therefore, hosting-ability studies must be conducted with local cultivars.

Wheat cultivars exhibiting tolerance to *P. neglectus* and *P. thornei* have been deployed effectively in Australia (Vanstone et al., 1998, 2008; Thompson et al., 1999, 2008; Nicol et al., 2001), and studies are underway in the US to identify tolerant germplasm (Fig. 8.4). However, tolerance alone is not considered an effective long-term management strategy.

Resistance is the most important and economical strategy for reducing populations of root-lesion nematodes. Unlike the single-gene inheritance of cereal cyst nematode resistance, root-lesion nematode resistance is quantitative and controlled by several genes, making the prospect of developing effective resistance more challenging. However, many sources of resistance to *P. thornei* have been identified in commercial wheat cultivars, in Middle East landrace lines, and in wheat relatives such as *Aegilops* species (Table 8.2) (Thompson and Haak 1997; Nicol et al., 1999, 2001, 2003; Nombela and Romero 1999; Thompson et al., 1999; Hollaway et al., 2000; Zwart et al., 2004, 2005; Tokay et al., 2006; Sheedy et al., 2008). Several lines are especially interesting in that they exhibit resistance to both *P. neglectus* and *P. thornei* (Zwart et al., 2005; Nicol et al., 2007; Sheedy et al., 2007, 2008). Introgression of dual-resistance sources (Fig. 8.3) into commercial cultivars would eliminate the need for farmers to identify *Pratylenchus* to the species level before selecting a resistant cultivar. Also of particular interest are lines that convey high levels of both tolerance and resistance to *P. thornei* or *P. neglectus*.

Phenotypic identification of resistance, coupled with molecular biology, has been used to investigate the genetic control and location of resistance genes and the identification of resistance markers. Quantitative trait loci (QTLs) for resistance to *P. thornei* have been identified on chromosomes 1B, 2B, 3B, 4D, 6D, and 7A (Schmidt et al., 2005; Zwart et al., 2005, 2006; Tokay et al., 2006). The *P. neglectus* resistance gene *Rlnn1* occurs on chromosome 7A, and a molecular marker can identify its presence in seedlings (Williams et al., 2002). A resistance gene for *P. neglectus* was also identified on chromosome 4D (Zwart et al., 2005).

Research on resistance to *P. neglectus* and *P. thornei* has included development of simple sequence repeat (SSR) markers for tracking QTLs in breeding programs. Associations between markers and resistance reactions have been sufficiently consistent to demonstrate the potential for applying marker-assisted selection to the improvement of *Pratylenchus* resistance in wheat. This process is actively practiced by Australian and CIMMYT international wheat breeding programs, using the *Rlnn1* marker (Williams et al., 2002).

Table 8.2 Principal sources of genes used to breed wheat for resistance to root-lesion nematodes *Pratylenchus neglectus* (*Pn*) and *P. thornei* (*Pt*).

Cereal Species	Cultivar or Line	Origin	Resistance Gene(s)[a,b]	Response to Pathotypes[b,c]	Use in Cultivars
Triticum aestivum	GS50a	Australia—reselection from cv. Gatcher	Major QTL mapped to 6D		Australia
	AUS4930 = Iraq 48	Iraq	QTLs mapped to 1B, 2B, and 6D	R to *Pt* but also portrays R to *Ha*	Australia, CIMMYT—under investigation
	Reselection of Excalibur	Australian cv. Excalibur	QTL mapped to 7AL	R to *Pn* (*Rlnn1*), on chromosome 7AL	Australia, CIMMYT
	Croc_1/*Ae. tausch.* (224)//Opata	Primary synthetic	QTLs mapped to 1B and 3B	R to *Pt*	CIMMYT
	CPI133872	Primary synthetic	QTLS mapped to 2B, 4D, 6A, 6D	R to *Pt* and *Pn*	Australia
	W-7984 × Opata 85		QTLs mapped to 2B and 6D	R to *Pt*	Australia
	AUS4926	Middle eastern landrace	QTLS mapped to 1B, 2B, 3B, and 6D	R to *Pt*	Australia
	AUS13124	Middle eastern landrace	QTLs mapped to 2B, 3B, 6D, and 7A	R to *Pt*	Australia
Aegilops tauschii	CPI 110872			R to *Pt* and *Pn*	
Aegilops geniculata	MZ10, MZ61, MZ96, MZ144	Middle East and West Asia		pR to *Pt*, several also portray R to *Ha*	

[a]*Sources:* Reviews and references in Rivoal and Cook (1993), Cook and Rivoal (1998), McDonald and Nicol (2005), and Nicol and Rivoal (2007).
[b]Characterized QTLs associated with multigene resistance to root-lesion nematodes.
[c]R = resistant, pR = partially resistant, *Ha* = *Heterodera avenae*.

Other management practices are less effective in managing *Pratylenchus* populations. Field sanitation during the fallow phase is as important as during the in-crop phase, because *Pratylenchus* species multiply on many weed species in the genera *Avena*, *Brassica*, *Bromus*, *Carrichtera*, *Emex*, *Heliotropium*, *Hordeum*, *Malva*, *Raphanus*, *Rumex*, and *Tribulis* (Vanstone and Russ 2001a,b) and on volunteer cereals including oat, wheat, and triticale (Forge et al., 2000). The presence of susceptible weeds or crop species between planted crops allows *Pratylenchus* to increase population density over a greater interval of the cropping system (Smiley et al., 2004b).

Tillage has reportedly increased, decreased, or had no effect on populations of *Pratylenchus* in wheat (Smiley et al., 2004b). In North America it appears that the greatest impact of conservation cropping systems will be associated with the frequency of host crops or weeds rather than the presence, absence, or intensity of tillage. While cultivation and fallow have appeared to reduce populations of *P. neglectus* and *P. thornei* (Orion et al., 1984; Thompson 1992; Nombela et al., 1998; Smiley et al., 2004b; Strausbaugh et al., 2004), these interpretations were based mostly upon soil samples collected at shallow depth. Shallow samples accurately characterize *Pratylenchus* populations in shallow soils but not in many deep soils where these nematodes may be found as deep as 150 cm and where peak populations may vary from shallow to as deep as 60–

90 cm depending upon variables such as intensity of surface cultivation or seasonal rainfall (Taylor and Evans 1998; Thompson et al., 1999; Ophel-Keller et al., 2008).

The reproductive rate for *P. thornei* increases as soil temperature increases from 15 to 20 °C (Castillo et al., 1996). Irrigated wheat yield in *P. thornei*–infested fields was improved by delaying planting by one month, presumably because seedlings overwintered with lower populations than for early planting (Van Gundy et al., 1974). In a contrasting use of temperature treatment, mulching a field with polyethylene film for 6–8 weeks over the hot summer period suppressed *P. thornei* populations by 50% (Di Vito et al., 1991).

As with cereal cyst nematode, *Pratylenchus neglectus* and *P. thornei* are often more damaging to crops in drier than wetter regions. The economic threshold for damage is expected to be lower for low-rainfall environments than for crops produced with supplemental irrigation or in areas of greater precipitation especially during the growing season (Orion et al., 1984; Castillo et al., 1995; Nicol and Ortiz-Monasterio 2004).

Chemical nematicides are effective and are widely used in research (Taylor et al., 1999; Thompson et al., 1995; Smiley 2005a,b) but are not economically feasible, registered, or environmentally appropriate for managing these parasites on wheat. Biological control agents are not commercially available for *Pratylenchus* species on wheat. Bacterial parasites of *Pratylenchus* have been reported (Ornat et al., 1999) but are unlikely to be successfully adapted for managing these migratory species that are well adapted to highly diverse soils and climates.

Green-manure crops can be used to sanitize soil through biofumigation in regions where water is not a limiting factor for wheat growth. Several *Pratylenchus* species are capable of multiplying in roots of sudangrass [*Sorghum drummondii* (Nees ex Steud.) Millsp. & Chase] and many *Brassica* species. Populations may remain high where these crops are grown to maturity for seed or forage harvest. However, when green tissue from some of these crops is incorporated into soil it is, in some cases, capable of reducing the elevated population to preplant levels or below (Bernard and Montgomery-Dee 1993; Potter et al., 1998, 1999; Forge et al., 2000).

FUTURE PERSPECTIVES

In most regions, and especially in the developing world, the impact of cereal cyst and root-lesion nematodes on wheat yield has not been documented, because extraction of nematodes is not a normal practice for the husbandry of wheat. Even where nematodes are known to be damaging the identity of the species and pathotype (for cyst nematodes) complex is often poorly defined.

More intensive surveys are required to more clearly identify the following: (i) areas where cereal cyst and root-lesion nematodes are present; (ii) species and pathotypes; (iii) occurrences of mixed populations; and (iv) the magnitude and frequency of yield loss. Further development of molecular identification and quantification techniques and adoption of these procedures by commercial soil diagnostic laboratories will allow more rapid and effective surveys of populations in areas where nematodes are currently not monitored. A commercial testing program in South Australia is of particular interest (Ophel-Keller et al., 2008). A DNA extract from soil is used to quantitatively estimate inoculum levels of multiple fungal pathogens and populations of several nematode species, including *H. avenae*, *P. neglectus*, and *P. thornei*. Levels of disease risk, established from many years of collected field population and yield loss data, are communicated back to farmers through a network of agronomic advisors.

Many wheat breeding programs are not breeding for resistance to cereal cyst and root-lesion nematodes. Reasons may include a lack of understanding of the importance of the issue, limited financial, technical or institutional support for this disciplinary research, or lack of field test sites with adequate uniformity of infestation and absence of significant impacts by other parasitic nematode species and soilborne fungal pathogens. Additional wheat production efficiency and profitability will be attained by improving the level of nematode resistance and tolerance in wheat cultivars produced on highly infested fields.

Development of additional and improved molecular markers will allow more rapid integration of resistance into commercial cultivars. Genetic transformations may also aid in the development of nematode-resistant germplasm. For cyst nematodes, molecular techniques will also facilitate greater precision in studies of resistance mechanisms (Seah et al., 2000; Andrés et al., 2001a; Montes et al., 2003, 2004).

Global complementation among regional or national research programs has proven to be highly beneficial for identifying and deploying germplasm with higher levels of resistance and tolerance to cereal cyst and root-lesion nematodes. However, these global efforts currently lack effective funding and coordination, limiting the ability to realize the benefits already known to exist. Greater collaboration is therefore needed between advanced research institutions, international organizations such as CIMMYT and ICARDA, and scientists in countries where these nematodes are known to be a problem. These collaborative efforts will provide greater understanding of the complexity, economic importance, and control of *Heterodera* and *Pratylenchus* populations, and of pathotype evolution or selection for *Heterodera* species.

REFERENCES

Abidou, H., A. El-Ahmed, J.M. Nicol, N. Bolat, R. Rivoal, and A. Yahyaoui. 2005. Occurrence and distribution of species of the *Heterodera avenae* group in Syria and Turkey. Nematol. Mediterr. 33:195–201.

Al-Banna, L., A.T. Ploeg, V.M. Williamson, and I. Kaloshian. 2004. Discrimination of six *Pratylenchus* species using PCR and species-specific primers. J. Nematol. 36:142–146.

Al-Hazmi, A.S., R. Cook, and A.A.M. Ibrahim. 2001. Pathotype characterisation of the cereal cyst nematode, *Heterodera avenae*, in Saudi Arabia. Nematology 3:379–382.

Andersen, S., and K. Andersen. 1982. Suggestions for determination and terminology of pathotypes and genes for resistance in cyst-forming nematodes, especially *Heterodera avenae*. EPPO Bull. 12:379–386.

Andrés, M., T. Melillo, A. Delibes, M.D. Romero, and T. Bleve-Zacheo. 2001a. Changes in wheat root enzymes correlated with resistance to cereal cyst nematodes. New Phytol. 152:343–354.

Andrés, M.F., J. Pinochet, A. Hernández-Dorrego, and A. Delibes. 2000. Detection and analysis of inter- and intraspecific diversity of *Pratylenchus* spp. using isozyme markers. Plant Pathol. 49:640–649.

Andrés, M.F., M.D. Romero, M.J. Montes, and A. Delibes. 2001b. Genetic relationships and isozyme variability in the *Heterodera avenae* complex determined by isoelectric focusing. Plant Pathol. 50:270–279.

Armstrong, J.S., F.B. Peairs, S.D. Pilcher, and C.C. Russell. 1993. The effect of planting time, insecticides, and liquid fertilizer on the Russian wheat aphid (Homoptera: Aphididae) and the lesion nematode (*Pratylenchus thornei*) on winter wheat. J. Kans. Entomol. Soc. 66:69–73.

Barker, K.R., G.A. Pederson, and G.L. Windham (ed.) 1998. Plant and nematode interactions. ASA, Madison, WI.

Barloy, D., J. Lemoine, P. Abelard, A.M. Tanguy, R. Rivoal, and J. Jahier. 2007. Marker-assisted pyramiding of two cereal cyst nematode resistance genes from *Aegilops variabilis* in wheat. Mol. Breed. 20:31–40.

Barr, A.R., K.J. Chalmers, A. Karakousis, J.M. Kretschmer, S. Manning, R.C.M. Lance, J. Lewis, S.P. Jefferies, and P. Langridge. 1998. RFLP mapping of a new cereal cyst nematode resistance locus in barley. Plant Breed. 117:185–187.

Bekal, S., J.P. Gauthier, and R. Rivoal. 1997. Genetic diversity among a complex of cereal cyst nematodes inferred from RFLP analysis of the ribosomal internal transcribed spacer region. Genome 40:479–486.

Bernard, E.C., and M.E. Montgomery-Dee. 1993. Reproduction of plant parasitic nematodes on winter rapeseed (*Brassica napus* spp. *oleifera*). J. Nematol. 25(4S):863–868.

Bockus, W.W., R.L. Bowden, R.M. Hunger, W.L. Morrill, T.D. Murray, and R.W. Smiley (ed.) 2009. Compendium of wheat diseases and insects. 3rd ed. APS Press, St. Paul, MN.

Bonfil, D.J., B. Dolgin, I. Mufradi, and S. Asido. 2004. Bioassay to forecast cereal cyst nematode damage to wheat in fields. Precision Agric. 5:329–344.

Brown, R.H. 1987. Control strategies in low-value crops. p. 351–387. *In* R.H. Brown and B.R. Kerry (ed.) Principals and practice of nematode control in crops. Academic Press, Sydney, Australia.

Carrasco-Ballesteros, S., P. Castillo, B.J. Adams, and E. Pérez-Artés. 2007. Identification of *Pratylenchus thornei*, the cereal and legume root-lesion nematode, based on SCAR-PCR and satellite DNA. Eur. J. Plant Pathol. 118:115–125.

Carta, L.K., A.M. Skantar, and Z.A. Handoo. 2001. Molecular, morphological and thermal characters of 19 *Pratylenchus* spp. and relatives using the D3 segment of the nuclear LSU rRNA gene. Nematropica 31:195–209.

Castillo, P., A. Gomez-Barcina, and R.M. Jiménez-Diaz. 1996. Plant parasitic nematodes associated with chickpea in southern Spain and effect of soil temperature on reproduction of *Pratylenchus thornei.* Nematologica 42:211–219.

Castillo, P., R.M. Jiménez-Diaz, A. Gomez-Barcina, and N. Vovlas. 1995. Parasitism of the root-lesion nematode *Pratylenchus thornei* on chickpea. Plant Pathol. 44:728–733.

Castillo, P., and N. Vovlas. 2007. *Pratylenchus*, Nematoda, Pratylenchidae: Diagnosis, biology, pathogenicity and management. Nematol. Monogr. Perspect. 6:1–530.

Castillo, P., N. Vovlas, and R.M. Jiménez-Diaz. 1998. Pathogenicity and histopathology of *Pratylenchus thornei* populations on selected chickpea genotypes. Plant Pathol. 47:370–376.

Cook, R., and K. Evans. 1987. Resistance and tolerance. p. 179–231. *In* R.H. Brown and B.R. Kerry (ed.) Principles and practice of nematode control in crops. Academic Press, Sydney, Australia.

Cook, R., and G.R. Noel. 2002. Cyst nematodes: *Globodera* and *Heterodera* species. p. 71–105. *In* J.L. Starr, R. Cook, and J. Bridge (ed.) Plant resistance to parasitic nematodes. CAB Int., Wallingford, UK.

Cook, R., and R. Rivoal. 1998. Genetics of resistance and parasitism. p. 322–352. *In* S.B. Sharma (ed.) The cyst nematodes. Chapman and Hall, London, UK.

de Majnik, J., F.C. Ogbonnaya, O. Moullet, and E.S. Lagudah. 2003. The *Cre1* and *Cre3* nematode resistance genes are located at homeologous loci in the wheat genome. Mol. Plant Microbe Interact. 16:1129–1134.

De Waele, D., and A. Elsen. 2002. Migratory endoparasites: *Pratylenchus* and *Radopholus* species. p. 175–206. *In* J.L. Starr, R. Cook, and J. Bridge (ed.) Plant resistance to parasitic nematodes. CAB Int., Wallingford, UK.

Di Vito, M., N. Greco, and M.C. Saxena. 1991. Effectiveness of soil solarization for control of *Heterodera ciceri* and *Pratylenchus thornei* on chickpeas in Syria. Nematol. Mediterr. 19:109–111.

Doyle, A.D., R.W. McLeod, P.T.W. Wong, S.E. Hetherington, and R.J. Southwell. 1987. Evidence for the involvement of the root-lesion nematode *Pratylenchus thornei* in wheat yield decline in northern New South Wales. Aust. J. Exp. Agric. 27:563–570.

Eagles, H.A., H.S. Bariana, F.C. Ogbonnaya, G.J. Rebetzke, G.L. Hollamby, R.J. Henry, P.H. Henschke, and M. Carter. 2001. Implementation of markers in Australian wheat breeding. Aust. J. Agric. Res. 52:1349–1356.

Eastwood, R.F., E.S. Lagudah, and R. Appels. 1994. A directed search for DNA sequences tightly linked to cereal cyst nematode resistance genes in *Triticum tauschii.* Genome 37:311–319.

Esmenjaud, D., R. Rivoal, and H. Marzin. 1990. Numbers of *Pratylenchus* spp., (Nematoda) in the field on winter wheat in different cereal rotations. Nematologica 36:317–226.

Farsi, M., V.A. Vanstone, J.M. Fisher, and A.J. Rathjen. 1995. Genetic variation in resistance to *Pratylenchus neglectus* in wheat and triticales. Aust. J. Exp. Agric. 35:597–602.

Fatemy, S., E. Abootorabi, N. Ebrahimi, and F. Aghabeigi. 2006. First report of *Pratylenchus neglectus* and *P. thornei* infecting canola and weeds in Iran. Plant Dis. 90:1555.

Ferris, V.R., S.A. Subbotin, A. Ireholm, Y. Spiegel, J. Faghini, and J.M. Ferris. 1999. Ribosomal DNA sequence analysis of *Heterodera filipjevi* and *H. latipons* isolates from Russia and comparisons with other nematode isolates. Russ. J. Nematol. 7:121–125.

Filho, A.C.C., and C.S. Huang. 1989. Description of *Pratylenchus pseudofallax* n. sp. with a key to species of the genus *Pratylenchus* Filipjev, 1936 (Nematoda: Pratylenchidae). Rev. Nématol. 12:7–15.

Forge, T.A., R.E. Ingham, D. Kaufman, and J.N. Pinkerton. 2000. Population growth of *Pratylenchus penetrans* on winter cover crops grown in the Pacific Northwest. J. Nematol. 32:42–51.

Glazer, I., and D. Orion. 1983. Studies on anhydrobiosis of *Pratylenchus thornei.* J. Nematol. 15:333–338.

Hafez, S.L., A. Al-Rehiayani, M. Thornton, and P. Sundararaj. 1999. Differentiation of two geographically isolated populations of *Pratylenchus neglectus* based on their parasitism of potato and interaction with *Verticillium dahliae.* Nematotropica 29:25–36.

Handoo, Z.A. 2002. A key and compendium to species of the *Heterodera avenae* Group (Nematoda: Heteroderidae). J. Nematol. 34:250–262.

Handoo, Z.A., and A.M. Golden. 1989. A key and diagnostic compendium to the species of the genus *Pratylenchus* Filpjev, 1936 (lesion nematodes). J. Nematol. 21:202–218.

Heide, A. 1975. Studies on the population dynamics of migratory root nematodes in cereal monocultures as well as in alternating cereal growing. Arch. Phytopathol. Pflanzenschutz 11:225–232.

Hollaway, G.J., S.P. Taylor, R.F. Eastwood, and C.H. Hunt. 2000. Effect of field crops on density of *Pratylenchus neglectus* and *P. thornei* in southeastern Australia: Part 2. *P. thornei.* J. Nematol. 32(4S):600–608.

Ibrahim, S.K., R.N. Perry, and R.M. Webb. 1995. Use of isoenzyme and protein phenotypes to discriminate between six *Pratylenchus* species from Great Britain. Ann. Appl. Biol. 126:317–327.

Ismail, S., R.P. Schuster, and R.A. Sikora. 2000. Factors affecting dormancy of the Mediterranean cereal cyst nematode *Heterodera latipons* on barley. Meded.-Fac. Landb. Toeg. Biolog. Wetensch., Univ. Ghent. 65(2b):529–535.

Ismail, S., R.A. Sikora, and R.P. Schuster. 2001. Occurrence and diversity of egg pathogenic fungi of the Mediterranean cereal cyst nematode *Heterodera latipons.* Meded.-Fac. Landb. Toeg. Biolog. Wetensch., Univ. Ghent. 66(2b): 645–653.

Jahier, J., R. Rivoal, M.Q. Yu, P. Abélard, A.M. Tanguy, and D. Barloy. 1998. Transfer of genes for resistance to cereal cyst nematode from *Aegilops variabilis* Eig to wheat. J. Genet. Breed. 52:253–257.

Kerry, B.R. 1987. Biological control. p. 233–263. *In* R.H. Brown and B.R. Kerry (ed.) Principles and practice of nematode control in crops. Academic Press, Sydney, Australia.

Kerry, B.R., and D.H. Crump. 1998. The dynamics of the decline of the cereal cyst nematode, *Heterodera avenae*, in four soils under intensive cereal production. Fundam. Appl. Nematol. 21:617–625.

Kretschmer, J.M., K.J. Chalmers, S. Manning, A. Karakousis, A.R. Barr, A.K.M.R. Islam, S.J. Logue, Y.W. Choe, S.J. Barker, R.C.M. Lance, and P. Langridge. 1997. RFLP mapping of the Ha2 cereal cyst nematode resistance gene in barley. Theor. Appl. Genet. 94:1060–1064.

Lasserre, F., F. Gigault, J.P. Gauthier, J.P. Henry, M. Sandmeier, and R. Rivoal. 1996. Genetic variation in natural populations of the cereal cyst nematode (*Heterodera avenae* Woll.) submitted to resistant and susceptible cultivar of cereals. Theor. Appl. Genet. 93:1–8.

Lasserre, F., R. Rivoal, and R. Cook. 1994. Interactions between *Heterodera avenae* and *Pratylenchus neglectus* on wheat. J. Nematol. 26:336–344.

Loof, P.A.A. 1978. The genus *Pratylenchus* Filipjev, 1936 (Nematoda: Pratylenchidae): A review of its anatomy, morphology, distribution, systematics and identification. Swed. Univ. Agric. Sci., Res. Infor. Centre, Uppsala.

Luc, M., A.R. Maggenti, and R. Fortuner. 1988. A reappraisal of Tylenchina (Nemata): 9. The family Heteroderidae Filip'ev & Schuurmans Stekhoven, 1941. Rev. Nématol. 11:159–176.

Mai, W.F., and P.G. Mullin. 1996. Plant parasitic nematodes: A pictorial key to genera. Cornell University Press, Ithaca, NY.

Martin, E.M., R.F. Eastwood, and F.C. Ogbonnaya. 2004. Identification of microsatellite markers associated with the cereal cyst nematode resistance gene *Cre3* in wheat. Aust. J. Agric. Res. 55:1205–1211.

McDonald, A.H., and J.M. Nicol. 2005. Nematode parasites of cereals. p. 131–191. *In* M. Luc, R.A. Sikora, and J. Bridge (ed.) Plant parasitic nematodes in subtropical and tropical agriculture. CAB Int., Wallingford, UK.

Mokabli, A., S. Valette, J.-P. Gauthier, and R. Rivoal. 2002. Variation in virulence of cereal cyst nematode populations from North Africa and Asia. Nematology 4:521–525.

Mokabli, A., S. Valette, and R. Rivoal. 2001. Différenciation de quelques espèces de nématodes à kystes des céréales et des graminées par électrophorèse sur gel d'acétate de cellulose. Nematol. Mediterr. 29:103–108.

Montes, M.J., I. López-Braña, and A. Delibes. 2004. Root enzyme activities associated with resistance to *Heterodera avenae* conferred by gene *Cre7* in a wheat/*Aegilops triuncalis* introgression line. J. Plant Physiol. 161:493–495.

Montes, M.J., I. López-Braña, M.D. Romero, E. Sin, M.F. Andrés, J.A. Martín-Sánchez, and A. Delibes. 2003. Biochemical and genetic studies of two *Heterodera avenae* resistance genes transferred from *Aegilops ventricosa* to wheat. Theor. Appl. Genet. 107:611–618.

Mor, M., E. Cohn, and Y. Spiegel. 1992. Phenology, pathogenicity and pathotypes of cereal cyst nematodes, *Heterodera avenae* Woll. and *H. latipons* (Nematoda: Heteroderidae) in Israel. Nematologica 38:444–501.

Nicol, J.M. 2002. Important nematode pests of cereals. p. 345–366. *In* B.C. Curtis, S. Rajaram, and G. Macpherson (ed.) Bread wheat: Improvement and production. FAO Plant Production and Protection Series, No. 30. FAO, Rome, Italy.

Nicol, J.M., N. Bolat, A. Bagci, R.T. Trethowan, M. William, H. Hekimhan, A.F. Yildirim, E. Şahin, H. Elekçioğlu, H. Toktay, B. Tunali, A. Hede, S. Taner, H.J. Braun, M. van Ginkel, M. Keser, Z. Arisoy, A. Yorgancılar, A. Tulek, D. Erdurmus, O. Buyuk, and M. Aydogdu. 2007. The international breeding strategy for the incorporation of resistance in bread wheat against the soil borne pathogens (dryland root rot and cyst and lesion nematodes) using conventional and molecular tools. p. 125–137. *In* H.T. Buck, J.E. Nisi, and N. Salomón (ed.) Wheat production in stressed environments. Springer Publishing, Dordrecht, The Netherlands.

Nicol, J.M., K.A. Davies, T.W. Hancock, and J.M. Fisher. 1999. Yield loss caused by *Pratylenchus thornei* on wheat in South Australia. J. Nematol. 31: 367–376.

Nicol, J.M., and I. Ortiz-Monasterio. 2004. Effects of the root-lesion nematode, *Pratylenchus thornei*, on wheat yields in Mexico. Nematology 6:485–493.

Nicol, J.M., and R. Rivoal. 2007. Global knowledge and its application for the integrated control and management of nematodes on wheat. p. 243–287. *In* A. Ciancio and K.G. Mukerji (ed.) Integrated management and biocontrol of vegetable and grain crops nematodes. Springer Publishing, Dordrecht, The Netherlands.

Nicol, J., R. Rivoal, S. Taylor, and M. Zaharieva. 2003. Global importance of cyst (*Heterodera* spp.) and lesion nematodes (*Pratylenchus* spp.) on cereals: Distribution, yield loss, use of host resistance and integration of molecular tools. Nematol. Monogr. Perspect. 2:1–19.

Nicol, J.M., R. Rivoal, R.M. Trethowan, M. van Ginkel, M. Mergoum, and R.P. Singh. 2001. CIMMYT's approach to identify and use resistance to nematodes and soil-borne fungi, in developing superior wheat germplasm. p. 381–389. *In* Z. Bedö and L. Láng (ed.) Wheat in a global environment. Kluwer Academic Publishers, Dordrecht, The Netherlands.

Nombela, G., A. Navas, and A. Bello. 1998. Effects of crop rotations of cereals with vetch and fallow on soil nematofauna in central Spain. Nematologica 44:63–80.

Nombela, G., and M.D. Romero. 1999. Host response to *Pratylenchus thornei* of a wheat line carrying the *Cre2* gene for resistance to *Heterodera avenae*. Nematology 1:381–388.

Ogbonnaya, F.C., S. Seah, A. Delibes, J. Jahier, I. López-Braña, R.F. Eastwood, and E.S. Lagudah. 2001a. Molecular-genetic characterisation of a new nematode resistance gene in wheat. Theor. Appl. Genet. 102:623–629.

Ogbonnaya, F.C., N.C. Subrahmanyam, O. Moullet, J. de Majnik, H.A. Eagles, J.S. Brown, R.F. Eastwood, J. Kollmorgen, R. Appels, and E.S. Lagudah. 2001b. Diagnostic DNA markers for cereal cyst nematode resistance in bread wheat. Aust. J. Agric. Res. 52:1367–1374.

Ophel-Keller, K., A. McKay, D. Hartley, Herdina, and J. Curran. 2008. Development of a routine DNA-based testing service for soilborne diseases in Australia. Australas. Plant Pathol. 37:243–253.

Orion, D., J. Amir, and J. Krikun. 1984. Field observations on *Pratylenchus thornei* and its effects on wheat under arid conditions. Rev. Nématol. 7:341–345.

Ornat, C., S. Verdejo-Lucas, F.J. Sorribas, and E.A. Tzortzakakis. 1999. Effect of fallow and root destruction on survival of root-knot and root-lesion nematodes in intensive vegetable cropping systems. Nematotropica 29:5–16.

Ouri, Y., and T. Mizukubo. 1999. Discrimination of seven *Pratylenchus* species (Nematoda: Pratylenchidae) in Japan by PCR-RFLP analysis. Appl. Entomol. Zool. 34:205–211.

Paull, J.G., K.J. Chalmers, A. Karakousis, J.M. Kretschmer, S. Manning, and P. Langridge. 1998. Genetic diversity in Australian wheat varieties and breeding material based on RFLP data. Theor. Appl. Genet. 96:435–446.

Peng, D., D. Zhang, J.M. Nicol, S. Chen, L. Waeyenberge, M. Moens, H. Li, W. Tang, and I.T. Riley. 2007. Occurrence, distribution and research situation of cereal cyst nematode in China. p. 350–351. *In* Proc. Int. Plant Protection Conf., 16th, Glasgow, Scotland, UK. 15–18 Oct. 2007. Br. Crop Prod. Counc., Alton, Hampshire, UK.

Potter, M.J., K. Davies, and A.J. Rathjen. 1998. Suppressive impact of glucosinolates in *Brassica* vegetative tissues on root lesion nematode *Pratylenchus neglectus*. J. Chem. Ecol. 24:67–80.

Potter, M.J., V.A. Vanstone, K.A. Davies, J.A. Kirkegaard, and A.J. Rathjen. 1999. Reduced susceptibility of *Brassica napus* to *Pratylenchus neglectus* in plants with elevated root levels of 2-phenylethyl glucosinolate. J. Nematol. 31:291–298.

Riley, I.T., and S.J. Kelley. 2002. Endoparasitic nematodes in cropping soils of Western Australia. Aust. J. Exp. Agric. 42:49–56.

Rivoal, R., S. Bekal, S. Valette, J.-P. Gauthier, M.B.H. Fradj, A. Mokabli, J. Jahier, J. Nicol, and A. Yahyaoui. 2001. Variation in reproductive capacity and virulence on different genotypes and resistance genes of Triticeae, in the cereal cyst nematode species complex. Nematology 3:581–592.

Rivoal, R., and R. Cook. 1993. Nematode pests of cereals. p. 259–303. *In* K. Evans, D.L. Trudgill, and J.M. Webster (ed.) Plant parasitic nematodes in temperate agriculture. CAB Int., Wallingford, UK.

Rivoal, R., S. Valette, S. Bekal, J.-P. Gauthier, and A. Yahyaoui. 2003. Genetic and phenotypic diversity in the graminaceous cyst nematode complex, inferred from PCR-RFLP of ribosomal DNA and morphometric analysis. Eur. J. Plant Pathol. 109:227–241.

Safari, E., N.N. Gororo, R.F. Eastwood, J. Lewis, H.A. Eagles, and F.C. Ogbonnaya. 2005. Impact of *Cre1*, *Cre8* and *Cre3* genes on cereal cyst nematode resistance in wheat. Theor. Appl. Genet. 110:567–572.

Schmidt, A.L., C.L. McIntyre, J. Thompson, N.P. Seymour, and C.J. Liu. 2005. Quantitative trait loci for root lesion nematode (*Pratylenchus thornei*) resistance in Middle-Eastern landraces and their potential for introgression into Australian bread wheat. Aust. J. Agric. Res. 56:1059–1068.

Scholz, U., and R.A. Sikora. 2004. Hatching behaviour and life cycle of *Heterodera latipons* Franklin under Syrian agro-ecological conditions. Nematology 6:245–256.

Seah, S., C. Miller, K. Sivasithamparam, and E.S. Lagudah. 2000. Root responses to cereal cyst nematode (*Heterodera avenae*) in hosts with different resistance genes. New Phytol. 146:527–533.

Sheedy, J.G., R.W. Smiley, S.A. Easley, and A.L. Thompson. 2007. Resistance reaction of Pacific Northwest spring wheat and barley cultivars to root-lesion nematode; *Pratylenchus neglectus*. p. CF022. *In* Plant disease management reports. Vol. 1. APS Press, St. Paul, MN.

Sheedy, J.G., R.W. Smiley, S.A. Easley, and A.L. Thompson. 2008. Resistance of Pacific Northwest spring wheat and barley cultivars to root-lesion nematode; *Pratylenchus thornei*. p. N007. *In* Plant disease management reports. Vol 2. APS Press, St. Paul, MN.

Siddiqi, M.R. 2000. Tylenchida: Parasites of plants and insects. 2nd ed. CAB Int., Wallingford, UK.

Smiley, R.W., J.A. Gourlie, R.G. Whittaker, S.A. Easley, and K.K. Kidwell. 2004a. Economic impact of Hessian fly (Diptera: Cecidomyiidae) on spring wheat in Oregon and additive yield losses with Fusarium crown rot and lesion nematode. J. Econ. Entomol. 97:397–408.

Smiley, R.W., R.E. Ingham, W. Uddin, and G.H. Cook. 1994. Crop sequences for winter wheat in soil infested with cereal cyst nematode and fungal pathogens. Plant Dis. 78:1142–1149.

Smiley, R.W., K. Merrifield, L.-M. Patterson, R.G. Whittaker, J.A. Gourlie, and S.A. Easley. 2004b. Nematodes in dryland field crops in the semiarid Pacific Northwest United States. J. Nematol. 36:54–68.

Smiley, R.W., R.G. Whittaker, J.A. Gourlie, and S.A. Easley. 2005a. *Pratylenchus thornei* associated with reduced wheat yield in Oregon. J. Nematol. 37:45–54.

Smiley, R.W., R.G. Whittaker, J.A. Gourlie, and S.A. Easley. 2005b. Suppression of wheat growth and yield by *Pratylenchus neglectus* in the Pacific Northwest. Plant Dis. 89:958–968.

Smiley, R.W., R.G. Whittaker, J.A. Gourlie, S.A. Easley, and R.E. Ingham. 2005c. Plant-parasitic nematodes associated with reduced wheat yield in Oregon: *Heterodera avenae*. J. Nematol. 37:297–307.

Smiley, R.W., G.P. Yan, and Z.A. Handoo. 2008. First record of the cereal cyst nematode *Heterodera filipjevi* in Oregon. Plant Dis. 92:1136.

Strausbaugh, C.A., C.A. Bradley, A.C. Koehn, and R.L. Forster. 2004. Survey of root diseases of wheat and barley in southeastern Idaho. Can. J. Plant Pathol. 26:167–176.

Subbotin, S.A., H.J. Rumpenhorst, and D. Sturhan. 1996. Morphological and electrophoretic studies on populations of the *Heterodera avenae* complex from the former USSR. Russ. J. Nematol. 4:29–38.

Subbotin, S.A., D. Sturhan, H.J. Rumpenhorst, and M. Moens. 2002. Description of the Australian cereal cyst nematode *Heterodera australis* sp. n. (Tylenchida: Heteroderidae). Russ. J. Nematol. 10:139–148.

Subbotin, S.A., D. Sturhan, H.J. Rumpenhorst, and M. Moens. 2003. Molecular and morphological characterization of the *Heterodera avenae* species complex (Tylenchida: Heteroderidae). Nematology 5:515–538.

Subbotin, S.A., A. Vierstraete, P. De Ley, J. Rowe, L. Waeyenberge, M. Moens, and J.R. Vanfleteren. 2001. Phylogenetic relationships within the cyst-forming nematodes (Nematoda, Heteroderidae) based on analysis of sequences from the ITS regions of ribosomal DNA. Mol. Phylogenet. Evol. 21:1–16.

Subbotin, S.A., L. Waeyenberge, and M. Moens. 2000. Identification of cyst forming nematodes of the genus *Heterodera* (Nematoda: Heteroderidae) based on the ribosomal DNA-RFLP. Nematology 2:153–164.

Subbotin, S.A., L. Waeyenberge, I.A. Molokanova, and M. Moens. 1999. Identification of *Heterodera avenae* group species by morphometrics and rDNA-RFLPs. Nematology 1:195–207.

Taheri, A., G.J. Hollamby, and V.A. Vanstone. 1994. Interaction between root lesion nematode, *Pratylenchus neglectus* (Rensch 1924) Chitwood and Oteifa 1952, and root rotting fungi of wheat. N. Z. J. Crop Hortic. Sci. 22:181–185.

Talavera, M., and V.A. Vanstone. 2001. Monitoring *Pratylenchus thornei* densities in soil and roots under resistant (*Triticum turgidum durum*) and susceptible (*Triticum aestivum*) wheat cultivars. Phytoparasitica 29:29–35.

Taylor, S.P., and M.L. Evans. 1998. Vertical and horizontal distribution of and soil sampling for root lesion nematodes (*Pratylenchus neglectus* and *P. thornei*) in South Australia. Australas. Plant Pathol. 27:90–96.

Taylor, S.P., G.J. Hollaway, and C.H. Hunt. 2000. Effect of field crops on population densities of *Pratylenchus neglectus* and *P. thornei* in southeastern Australia: Part 1. *P. neglectus*. J. Nematol. 32(4S):591–599.

Taylor, S.P., V.A. Vanstone, A.H. Ware, A.C. McKay, D. Szot, and M.H. Russ. 1999. Measuring yield loss in cereals caused by root lesion nematodes (*Pratylenchus neglectus* and *P. thornei*) with and without nematicide. Aust. J. Agric. Res. 50:617–622.

Thompson, J.P. 1992. Soil biotic and biochemical factors in a long-term tillage and stubble management experiment on a vertisol: 2. Nitrogen deficiency with zero tillage and stubble retention. Soil Tillage Res. 22:339–361.

Thompson, J.P., P.S. Brennan, T.G. Clewett, J.G. Sheedy, and N.P. Seymour. 1999. Progress in breeding wheat for tolerance and resistance to root-lesion nematode (*Pratylenchus thornei*). Australas. Plant Pathol. 28:45–52.

Thompson, J.P., and M.I. Haak. 1997. Resistance to root-lesion nematode (*Pratylenchus thornei*) in *Aegilops tauschii* Coss., the D-genome donor to wheat. Aust. J. Agric. Res. 48:553–559.

Thompson, J.P., J. Mackenzie, and R. Amos. 1995. Root-lesion nematode (*Pratylenchus thornei*) limits response of wheat but not barley to stored soil moisture in the Hermitage long-term tillage experiment. Aust. J. Exp. Agric. 35:1049–1055.

Thompson, J.P., K.J. Owen, G.R. Stirling, and M.J. Bell. 2008. Root-lesion nematodes (*Pratylenchus thornei* and *P. neglectus*): A review of recent progress in managing a significant pest of grain crops in northern Australia. Australas. Plant Pathol. 37:235–242.

Tokay, H., C.L. McIntyre, J.M. Nicol, H. Ozkan, and H.İ. Elekçioğlu. 2006. Identification of common root-lesion nematode (*Pratylenchus thornei* Sher et Allen) loci in bread wheat. Genome 49:1319–1323.

Uehara, T., A. Kushida, and Y. Momota. 1999. Rapid and sensitive identification of *Pratylenchus* spp. using reverse dot blot hybridization. Nematology 1:549–555.

Van Gundy, S.D., J.G. Perez, L.H. Stolzy, and I.J. Thomason. 1974. A pest management approach to the control of *Pratylenchus thornei* on wheat in Mexico. J. Nematol. 6:107–116.

Vanstone, V.A., G.J. Hollaway, and G.R. Stirling. 2008. Managing nematode pests in the southern and western regions of the Australian cereal industry: Continuing progress in a challenging environment. Australas. Plant Pathol. 37:220–234.

Vanstone, V.A., A.J. Rathjen, A.H. Ware, and R.D. Wheeler. 1998. Relationship between root lesion nematodes (*Pratylenchus neglectus* and *P. thornei*) and performance of wheat varieties. Aust. J. Exp. Agric. 38:181–188.

Vanstone, V.A., and M.H. Russ. 2001a. Ability of weeds to host the root lesion nematodes *Pratylenchus neglectus* and *P. thornei*: I. Grass weeds. Australas. Plant Pathol. 30:245–250.

Vanstone, V.A., and M.H. Russ. 2001b. Ability of weeds to host the root lesion nematodes *Pratylenchus neglectus* and *P. thornei*: II. Broad-leaf weeds. Australas. Plant Pathol. 30:251–258.

Varma, M.K. 1995. Dictionary of plant nematology. Vedams Books Int., New Delhi.

Waeyenberge, L., A. Ryss, M. Moens, J. Pinochet, and T.C. Vrain. 2000. Molecular characterization of 18 *Pratylenchus* species using rDNA restriction fragment length polymorphism. Nematology 2:135–142.

Whitehead, A.G. 1997. Plant nematode control. CAB Int., Wallingford, UK.

Williams, K.J., J.M. Fisher, and P. Langridge. 1994. Identification of RFLP markers linked to the cereal cyst nematode resistance gene (*Cre*) in wheat. Theor. Appl. Genet. 89:927–930.

Williams, K.J., S.P. Taylor, P. Bogacki, M. Pallotta, H.S. Bariana, and H. Wallwork. 2002. Mapping of the root lesion nematode (*Pratylenchus neglectus*) resistance gene *Rlnn1* in wheat. Theor. Appl. Genet. 104:874–879.

Williams, K.J., K.L. Willsmore, S. Olson, M. Matic, and H. Kuchel. 2006. Mapping a novel QTL for resistance to cereal cyst nematode in wheat. Theor. Appl. Genet. 112:1480–1486.

Zwart, R.S., J.P. Thompson, and I.D. Godwin. 2004. Genetic analysis of resistance to root-lesion nematode (*Pratylenchus thornei*) in wheat. Plant Breed. 123:209–212.

Zwart, R.S., J.P. Thompson, and I.D. Godwin. 2005. Identification of quantitative trait loci for resistance to two species of root-lesion nematode (*Pratylenchus thornei* and *P. neglectus*) in wheat. Aust. J. Agric. Res. 56:345–352.

Zwart, R.S., J.P. Thompson, J.G. Sheedy, and J.C. Nelson. 2006. Mapping quantitative trait loci for resistance to *Pratylenchus thornei* from synthetic hexaploid wheat in the International Triticeae Mapping Initiative (ITMI) population. Aust. J. Agric. Res. 57:525–530.

Chapter 9
Insects Which Challenge Global Wheat Production

David R. Porter, Marion O. Harris, Louis S. Hesler, and Gary J. Puterka

SUMMARY

(1) Insect pests of wheat are dynamic, adaptable, universal, genetically variable, and persistent.
(2) Damage from insect infestations can vary from minimal yield and end-use-quality loss to death of the plant, depending on the growth stage of the plant and level of infestation.
(3) In addition to direct feeding damage, many insects vector important plant diseases, thus compounding the level of plant injury and subsequent loss to wheat production.
(4) Recommended control for insect infestations is via integrated pest management (IPM) technologies that may include cultural control, chemical control, biological control, and host-plant resistance.
(5) Cultural control of insect pests depends primarily on avoiding exposure of wheat seedlings to pests by manipulating planting date. The effectiveness of this tactic is highly variable and dependent on characteristics of the environment and the pest life cycle.
(6) A variety of chemical insecticides are available for insect control; however, the cost of application and the environmental concerns associated with their use reduce their appeal in wheat production.
(7) Natural enemies of wheat insect pests provide a level of biological control on pest populations and can supplement other components of the IPM approach. Several parasitoids and predators of important wheat pests are active in wheat fields worldwide.
(8) Host-plant resistance serves as the foundation of a successful IPM approach to controlling insect pests. Resistance genes, found within wheat and from related species, have been successfully utilized for many years to dramatically reduce the incidence of insect pest damage.

HESSIAN FLY

Economic impact and distribution

The Hessian fly (*Mayetiola destructor* Say) (Color Plate 20) is an important pest of wheat (Barnes 1956; Berzonsky et al., 2003; Harris et al., 2003). Throughout a single growing season, several generations of Hessian fly can be produced that can infest winter and spring wheat. During outbreaks of the Hessian fly, crop failure and losses of millions of dollars in production are common (Pedigo 2002).

Evidence points to the Hessian fly having a center of origin in the Fertile Crescent (Barnes 1956). From there, Hessian fly populations spread west throughout the Mediterranean region, north and west into Europe, and east into central Asia and Siberia. On islands in the Mediterranean, the Hessian fly has been recognized as a pest of wheat since "time immemorial" (Barnes 1956). The Hessian fly has been a well-known pest in Ukraine since 1847 and in Morocco, Algeria, and Tunisia since the early 1900s.

The Hessian fly was first reported in North America in 1777 and in New Zealand in 1888 and, in both cases, probably traveled via ships that carried soil or infested wheat stems. In North America, the Hessian fly is now found in all major wheat-growing regions, from east coast to west coast and from Georgia to the northern limits of wheat production in Canada. Populations have not been found in the wheat-growing regions of southern Asia, such as India, or in Southeast Asia, such as China. In the southern hemisphere, New Zealand is the only country reported to have the Hessian fly.

Biology, plant damage, and control methods

Relative to many other insect pests, the population dynamics of the Hessian fly are poorly understood for a number of reasons. The first is that the Hessian fly lives on many other grasses besides wheat, with many of these grasses occurring outside of managed agricultural fields (Barnes 1956; Harris et al., 2003). Larvae have been found on grass species in seventeen genera of the tribe Triticeae and one genus of the tribe Bromeae. Some of these grasses are crop or forage plants, including barley (*Hordeum vulgare* L.), rye (*Secale cereale* L.), triticale (× *Triticosecale rimpaui* Wittm.), and brome grass (*Bromus* spp.), while many others are common wild grass species of *Agropyron*, *Elymus*, *Hordeum*, and *Aegilops*, including the wild progenitors of domesticated wheat such as *Aegilops tauschii*. It is not known whether the Hessian fly populations on wild grasses contribute to periodic outbreaks in commercial wheat fields.

The second reason for our poor understanding of Hessian fly population dynamics is that, unlike many other insect and microbial pests of wheat, the Hessian fly is rarely part of annual surveys of wheat damage. This is because it is very difficult to quickly score a wheat plant for the presence of a Hessian fly larva. Unlike many aphids and pathogens, the Hessian fly larva does not elicit a visually obvious symptom, for example, yellow or white lesions or twisted leaf growth. This means that the plant must be carefully dissected to find the larva in its concealed feeding location, either between two leaf sheaths at the base of a juvenile plant or between a leaf sheath and stem in an older plant.

Natural enemies also make significant contributions to population dynamics of the Hessian fly, especially during the one to two generations that develop during late summer and autumn (Barnes 1956). A large number of parasitoids and predators are known both in areas where the Hessian fly originated (present-day Syria) and in areas where it was introduced. Predators and parasitoids commonly attack Hessian fly eggs, which remain on leaf surfaces for 3 to 5 days. Parasitoids also are common but do not kill their host until the third larval instar, when the Hessian fly larva's damage to the plant is complete. Thus, future crops have a reduced risk of attack, but the current crop probably does not benefit from the presence of the parasitoid.

A single Hessian fly larva is sufficient for inflicting damage to the wheat plant (Berzonsky et al., 2003). However, the extent and type of damage depends on the growth stage of the plant. During the seedling stage, the larva (the developmental stage that feeds on the plant) feeds at the base of the plant between the leaf sheaths. Impact is unusually severe because the larva manipulates resource allocation within the plant, with resources normally allocated to plant growth redirected to insect growth (Anderson and Harris 2006; Harris et al., 2006). This attack is associated with death of the seedling or, at the very least, death of the main stem. Yield is negatively impacted, with 60%–100% fewer seeds per plant produced. During stem elongation, the larva feeds on the stem just above the first or second node. This weakens the stem and produces the most easily recognized symptom of Hessian fly attack, that is, broken culms. In most cases, seeds are produced but they are smaller in number and shriveled. In this case, both yield and milling quality are negatively impacted.

The Hessian fly is somewhat unusual as an insect pest in that insecticides are rarely used (Berzonsky et al., 2003). Reasons for this include the availability of highly effective noninsecticidal methods, the difficulty of monitoring Hessian fly populations, and the unpredictable, localized, and/or sporadic nature of outbreaks. Noninsecticidal methods include host-plant resistance, planting after the "fly-free date," and removal of hosts that provide a "green bridge" between the harvest of one crop and the planting of the next. Planting after the fly-free date relies on temperature models that predict the last possible date during autumn when adult females are present in fields. Winter wheat crops planted after this date escape egg laying and subsequent larval attack during the vulnerable seedling stage. In areas with mild winters, this method is not used because adult emergence, female egg laying, and larval attack can occur during short periods of warmer weather. The effectiveness of removing bridging hosts such as volunteer wheat depends on whether there are other grasses in the area that can serve as hosts.

Utilization of host-plant resistance

In North America, the use of host-plant resistance has a long history of success (Berzonsky et al., 2003; Harris et al., 2003). Indeed, within 10 years of the discovery of Hessian fly populations on Long Island in 1777, a highly effective plant resistance trait was discovered and used to suppress the outbreak that threatened wheat, which was one of the new Republic's most important exports (Hunter 2001). Since that time, over 30 genes effective against the Hessian fly have been designated *H1* through *H32*. Table 9.1 shows the reactions of major sources of resistance in wheat to several biotypes (Ratcliffe and Hatchett 1997).

These genes have been found in both wild and domesticated grasses. Most commonly, this resistance is conditioned by dominant alleles at major resistance loci. Resistance has a dramatic effect (Berzonsky et al., 2003; Harris et al., 2003). Host acceptance of the adult female and neonate larva remains the same, but the larva dies soon after attack. The resistant seedling shows only minor growth disturbances (Anderson and Harris 2006).

Resistance to the Hessian fly can be compromised by the development of parasite virulence via modifications in matching avirulence (*Avr*) genes (Stuart et al., 2007). For example, three resistance (*R*) genes, *H3*, *H5*, and *H6*, were overcome by virulent biotypes after 15, 9, and 22 years, respectively (Foster et al., 1991). Starting with Gallun (1977), classical Mendelian genetic principles have been used to investigate the relationship between *H* genes and corresponding Hessian fly *Avr* genes (Stuart et al., 2007). The genetics of the interaction fit the gene-for-gene model initially developed by Flor (1946). This type of gene-for-gene interaction appears to be unique to Hessian fly–wheat and is considered

Table 9.1 Hessian fly resistance genes and biotype interactions in wheat.

		Hessian Fly Biotype															
Wheat Genotype & *H* gene	Origin of *H* gene	GP	A	B	C	D	E	F	G	H	I	J	K	L	M	N	O
		Reaction to biotype[a]															
Caldwel *H3*	*T. aestivum*	R	R	S	R	S	S	R	S	R	R	S	R	S	S	R	S
Magnum *H5*	*T. aestivum*	R	R	R	R	R	R	R	R	S	S	S	S	S	S	S	S
Monon *H6*	*T. turgidum*	R	R	R	S	S	R	S	S	R	R	R	S	S	R	S	S
Seneca *H7H8*	*T. aestivum*	R	S	S	S	S	R	R	R	R	S	S	S	S	R	R	R

[a]R and S indicate resistant and susceptible reactions, respectively.

rare among other plant–insect interactions (Harris et al., 2003).

Despite the development of virulence in Hessian fly populations, the strategy of deploying a single *H* gene, followed by deployment of a different *H* gene when the first is overcome by virulent biotypes, is considered cost-effective, especially for cereal crops such as wheat (Cox and Hatchett 1986; Berzonsky et al., 2003; Harris et al., 2003). Indeed, single *H* genes transferred to elite wheat genotypes and deployed in agriculture have eliminated the need for other control tactics in North America and North Africa, where the Hessian fly is a serious pest (Pedigo 2002). It has been suggested that *H* genes overcome by virulent biotypes can be reintroduced at a later time (Foster et al., 1991), because the modified *Avr* gene that conferred Hessian fly virulence to the *H* gene will be eliminated from the Hessian fly population when that *H* gene is removed as a source of selection pressure.

BIRD CHERRY–OAT APHID

The bird cherry–oat aphid (BCOA), *Rhopalosiphum padi* L. (Color Plate 21), is a common aphid pest of wheat worldwide (Vickerman and Wratten 1979). Bird cherry (*Prunus padus* L.) and closely related tree species are primary hosts at higher latitudes, while oat (*Avena sativa* L.), wheat, other cereals, and many grasses are secondary hosts from spring through fall.

Biology, plant damage, and control methods

The BCOA has multiple, asexual generations on secondary hosts (e.g., wheat). Migrants are produced in autumn and fly to primary hosts where mating occurs and overwintering eggs are deposited. In the northern hemisphere, parthenogenic BCOA routinely overwinter on cereals and grasses at latitudes below 40°N.

The BCOA was first identified as a wheat pest because it vectors viruses that cause barley yellow dwarf (BYD), one of the most important viral diseases of cereals (Lister and Ranieri 1995). However, since 1980, several studies have shown direct impact to wheat by BCOA without confounding effects of BYD. Seedling infestation has produced maximum grain yield loss (24%–65%), whereas yield loss from boot-stage infestations was lower and less frequent, and no yield loss resulted from infestation at later stages (Kieckhefer et al., 1995; Voss et al., 1997). Despite the lack of visible symptoms associated with its feeding, BCOA causes equal or greater yield loss as do cereal aphids that cause visible symptoms (i.e., greenbug and Russian wheat aphid) (Pike and Schaffner 1985; Kieckhefer et al., 1995). Yield loss from seedling infection by *Barley yellow dwarf virus* (BYDV) is greater than from seedling infestation by BCOA, and combined effects of BCOA and BYDV are greater than either treatment alone (Riedell et al., 1999, 2007). The BCOA also reduces baking quality of wheat (Basky et al., 2006).

Insecticides have been widely used to manage BCOA and reduce incidence of BYD (Burnett and Plumb 1998). However, problems are associated with insecticide use. It may not always provide economic return (Royer et al., 2005), BCOA has developed insecticide resistance (Cheng et al., 2004), and environmental costs are associated with insecticide use. Alternative use of natural enemies and host-plant resistance may be advantageous.

Natural enemies to BCOA occur as fungal and viral pathogens, polyphagous predators, aphidophagous predators, and hymenopterous parasitoids (Carter et al., 1980). However, the different kinds of natural enemies may vary in their efficacy for managing BCOA and limiting BYD (Sanderson et al., 1992).

Fungal epizootics affect cereal aphid populations, but viral epizootics have not been reported (Latgé and Papierok 1988). However, despite reaching epizootic levels at times, fungal entomopathogens depend on relatively high humidity and high aphid densities (Basky and Hopper 2000). Consequently, mycoses are typically low in BCOA populations, with incidence greatest at or near times of peak aphid abundance. Typical levels of natural infection therefore may not limit BCOA populations and BYD incidence below economic levels (Sanderson et al., 1992).

Polyphagous predators (e.g., spiders, ground beetles) often colonize cereal fields before aphids and have been associated with low aphid densities (Sanderson et al., 1992; Schmidt et al., 2004). By stemming initial BCOA populations, polyphagous predators have potential to prevent damage by BCOA and limit BYD (Östman et al., 2003).

Aphidophagous predators, for example, lady beetles [*Coleomegilla maculata* (DeGeer) and *Hippodamia convergens* Guérin-Méneville] and their larvae, hover fly [*Allograpta obliqua* (Say)] larvae, and lacewing [*Chrysoperla plorabunda* (Fitch)] larvae, consume large numbers of BCOA and contribute to biological control of cereal aphids. However, these predators typically peak during surges in cereal aphid populations and generally do not preempt sizeable aphid infestations (Sanderson et al., 1992; Elliott and Kieckhefer 2000). Moreover, predator activity causes BCOA to move among cereal plants and may contribute to increased BYD transmission and incidence (Bailey et al., 1995; Smyrnioudis et al., 2001).

Parasitoids of BCOA typically occur at low levels in cereal fields, and parasitism tends to peak at or around the time of maximum aphid abundance (Sanderson et al., 1992; Basky and Hopper 2000). Nonetheless, parasitism levels may be adequate to preclude some treatments for aphids and thus have been included in aphid treatment thresholds (Giles et al., 2003). Moreover, parasitoid attack may make BCOA more sedentary and help reduce spread of BYDV (Smyrnioudis et al., 2001).

Utilization of host-plant resistance

Host-plant resistance to BCOA has been identified in wheat and other monocotyledonous plants, and it has been derived from dicotyledonous transgenes (Migui and Lamb 2003; Liang et al., 2004; Hesler et al., 2005; Dunn et al., 2007). Although several promising sources of resistance to BCOA are available, none have been incorporated into contemporary wheat cultivars.

Host-plant resistance may differ in its applicability to managing BCOA and limit BYD. Models predict that BCOA is most sensitive to resistance mechanisms that produce high nymphal mortality, prolong early growth stages, and lower birth rate near the time of head emergence (Wiktelius and Pettersson 1985). The BCOA is less sensitive to mechanisms that increase the proportion of winged forms and decrease landing rates on plants. Plant resistance traits that limit winged forms may suppress BYD by reducing BCOA dispersal. Also, traits that inhibit BCOA from reaching the phloem may prevent inoculation and acquisition of BYDV (Gibson and Plumb 1977).

Tolerance is a form of host-plant resistance that allows plants to recover or limit the effects of infestation. It may limit direct yield loss from BCOA (Dunn et al., 2007) but could allow BCOA population buildup and promote secondary spread of BYDV (Gibson and Plumb 1977). Thus, tolerance may need to be coupled with traits conferring BYD resistance.

Although promising sources of resistance to BCOA have been identified in wheat, resistance genes have not been characterized or routinely utilized to develop new resistant wheat cultivars. While some of these sources of resistance to BCOA may indirectly limit BYD, conversely some forms of BYDV resistance may indirectly suppress BCOA population growth. Transgenic wheat plants with coat-protein-mediated resistance to BYDV maintained a lower virus titer than susceptible, conventional wheat and conventional, BYD-tolerant wheat lines. They also lowered fecundity, shortened the reproductive period, and reduced the number of progeny of BCOA compared with noninfected transgenic plants (Jiménez-Martínez et al., 2004) The authors suggested that BYDV-resistant transgenic plants could further limit BYD under field conditions by reducing BCOA population growth.

Finally, host-plant characteristics such as resistance traits and plant morphology can impact the effectiveness of natural enemies. Theoretically, plant resistance based on tolerance should not affect aphid fitness, and thus no indirect negative effects on natural enemies are predicted (Brewer and Elliott 2004). Empirical studies have generally supported this prediction. In contrast, a resistance trait that causes greater mortality or reduced population growth is hypothesized to negatively impact life histories of natural enemies. Plant morphology also may impact natural enemy effectiveness against aphids, with natural enemies favored generally by morphological traits that enhance their access to cereal aphids, for example, reducing the number of leaf hairs to facilitate foraging activity by natural enemies. Most studies on host plant–natural enemy interactions have focused on other cereal aphids, so future research on this topic with BCOA is needed.

GREENBUG

Economic impact and distribution

The greenbug (*Schizaphis graminum* Rondani) is a serious, perennial aphid pest of wheat with a worldwide distribution. A 1993 survey revealed that over 3 million hectares (41%) of dryland and 500,000 ha (93%) of irrigated wheat in the western US were infested with greenbug (Webster and Amosson 1995). Damage from greenbug during such periodic outbreaks can be substantial, such as 1976, when losses exceeded $80 million in Oklahoma alone (Starks and Burton 1977). In central Oklahoma, greenbug outbreaks generally occurred following a year of above-normal precipitation during the spring and summer; above-normal temperature during the winter, spring, and fall; and below-normal temperature during the summer (Rogers et al., 1972).

Biology, plant damage, and control methods

The greenbug is a small (approximately the size of a sesame seed) light green sap-sucking arthropod pest of wheat (Color Plate 22). It feeds on sieve tube sap through a proboscis, shaped like a soda straw and called a stylet. This feeding activity results in distinguishable leaf damage symptoms (yellow-red necrotic lesions at the feeding site surrounded by a larger area of chlorosis) that eventually lead to general necrosis of the leaf. Greenbug feeding on seedling winter wheat during the fall causes irreversible damage to plants. Population densities of 30 aphids per culm, with a feeding period of 7 days, caused reductions of at least 40% in grain weight (Kieckhefer and Kantack 1988). In addition to leaf damage, greenbug feeding reduces root length and dry weight (Burton 1986; Riedell and Kieckhefer 1995). Thus, greenbug infestations of winter wheat at the seedling stage may adversely impact the plant's ability to withstand environmental stresses such as drought and heat during later developmental stages, which is particularly important during grain filling.

In addition to direct plant damage caused by feeding, the greenbug is also a vector of viruses, notably *Barley yellow dwarf virus*. The first confirmed widespread barley yellow dwarf epidemic transmitted by greenbug into winter wheat occurred in southern Idaho during 1977 to 1978 (Forster 1990).

Although the greenbug has been known to exist since the 1880s, it was not until the 1950s, when greenbug-resistant wheat began to be developed, that greenbug populations or biotypes were identified that differed in their ability to damage resistant plants. Biotypes are genetically distinct populations, and each biotype is a phenotypic expression of an indefinite number of genotypes (Puterka and Peters 1990). Greenbugs are classified to a particular biotype primarily based on the plant virulence gene(s) they have in common, as manifested by their ability to damage plants with known resistance genes. Eleven greenbug biotypes (named A through K) have been identified using multiple wheat and sorghum (*Sorghum bicolor* L.) genotypes containing greenbug resistance genes to differentiate among these greenbug populations.

The most common method of controlling greenbug in wheat has been the use of chemical insecticides. Treatment thresholds vary by plant growth stage (e.g., 2 to 4 aphids per tiller at the seedling stage, 2 to 8 aphids per tiller when plants are 7 to 15 cm in height, and 8 to 20 aphids per tiller when the plants are 15 to 40 cm in height) and by crop condition (Royer et al., 1997b). Chemicals listed for use against greenbug in wheat include chlorpyrifos, dimethoate, disulfoton, imidacloprid, malathion, methyl parathion, parathion, and a mix of parathion and methyl parathion (Royer et al., 1997a).

In addition to the extra costs associated with the use of insecticides and their potential for contributing to environmental contamination, the greenbug has demonstrated the ability to develop resistance to insecticides. Resistance to several organophosphorous insecticides, including disulfoton and dimethoate, has been reported in Texas and Oklahoma (Peters et al., 1975; Teetes et al., 1975).

The greenbug has a number of natural enemies (predators and parasitoids) that may suppress populations during the growing season. Small parasitic wasps lay eggs inside the body of the aphid causing the aphid to swell and turn a tan color as the immature wasp feeds inside. This swollen, tan-colored aphid is called a mummy. As with BCOA, some important greenbug predators include lady beetles, lacewing larvae, and hover fly larvae (Royer et al., 1997b).

While insecticides and natural enemies are important control measures, plant resistance plays a pivotal role in the development of any sustainable approach to controlling greenbug damage in wheat. Insecticides and natural enemies can be successfully included in an integrated pest management (IPM) approach, with plant resistance forming the foundation upon which to build (Quisenberry and Schotzko 1994).

Utilization of host-plant resistance

Efforts to develop greenbug-resistant wheat began in the 1950s with the identification of greenbug-resistant DS 28A selected from durum wheat (Dahms et al., 1955). DS 28A was resistant to greenbug in the field at the time, and its single recessive gene was assigned the gene symbol *gb* (Curtis et al., 1960). From these initial efforts, a series of wheat germplasm and cultivars was developed and released in response to newly identified greenbug biotypes. Porter et al. (1997) present details of the reports of biotype identification and wheat resistance gene development and deployment.

Tyler et al. (1987) reviewed the status of greenbug biotypes and sources of resistance in wheat. They assigned designations for five genes (*gb1*, *Gb2*, *Gb3*, *Gb4*, and *Gb5*, in which *g* indicates recessive inheritance and *G* indicates dominant inheritance) for the sources of resistance in the respective wheat germplasm lines DS 28A, Amigo, Largo, CI17959, and CI17882. Since that time, one additional source of resistance has been reported in wheat. Porter et al. (1991) reported identification of greenbug biotype G resistance in a wheat–rye translocation germplasm (GRS-1201). GRS-1201 has a single, dominant gene (*Gb6*) located on the 1RS arm of the wheat–rye translocation chromosome T1AL·1RS (or 1RS·1BL) and is resistant to biotypes B, C, E, G, and I (Porter et al., 1994). Table 9.2 summarizes

Table 9.2 Greenbug resistance genes and biotype interactions in wheat.

Germplasm	Resistance Gene	Origin	Greenbug Biotype B	C	E	F	G	H	I	K
			Reaction to biotype[a]							
DS 28A	*gb1*	*T. turgidum durum*	S	S	S	R	S	S	S	S
Amigo	*Gb2*	*S. cereale*	R	R	S	S	S	S	S	S
Largo	*Gb3*	*T. tauschii*	S	R	R	S	S	R	R	R
CI 17959	*Gb4*	*T. tauschii*	S	R	R	S	S	S	R	R
CI 17882	*Gb5*	*T. speltoides*	S	R	R	S	S	S	R	R
GRS 1201	*Gb6*	*S. cereale*	R	R	R	S	R	S	R	R

[a]R and S indicate resistant and susceptible reactions, respectively.

the reactions of all sources of resistance in wheat to greenbug biotypes reported by Harvey et al. (1991), Porter et al. (1994), and Harvey et al. (1997).

Molecular markers for greenbug resistance in wheat have seen limited use to date. The *Gb5* resistance gene on chromosome 7S of *T. speltoides* (Tausch) Gren. was transferred to an interstitial chromosome segment of 7AL in 'Pavon' wheat via the *ph1b* mutation (Dubcovsky et al., 1998). Gene *Gb5*, located on the resultant 7AS·7AL–7S#1L·7AL chromosome, provides resistance to greenbug biotypes C, E, I, and K (Table 9.2). In the presence of the wild-type *Ph1* locus, chromosome segment 7S#1 does not recombine with wheat chromosome 7A. Therefore, only one of several RFLP markers associated with 7S#1 is needed to track *Gb5* (Dubcovsky et al., 1998). A wheat germplasm line (designated UCRBW98-2) carrying 7AS·7AL–7S#1L·7AL with *Gb5* was released for breeding purposes (Lukaszewski et al., 2000).

Markers have also been identified for resistance genes *Gb2* and *Gb6* located on the 1RS arm of the wheat–rye translocation chromosome T1AL·1RS (Graybosch et al., 1999). As in the previous example of a translocated chromosome segment, in the presence of the wild-type *Ph1* locus, 1RS is inherited as a non-recombined block of genes. Graybosch et al. (1999) identified secalin proteins and rye-specific PCR markers on 1RS that could effectively track *Gb2* and *Gb6*.

RUSSIAN WHEAT APHID

Economic impact and distribution

The Russian wheat aphid (RWA), *Diuraphis noxia* (Mordvilko) (Color Plate 23), is a significant pest of wheat and barley. This aphid occurs throughout the major wheat producing areas of the world except Australia. Its origin is thought to be in the wheat producing region of central Asia. The RWA was not considered a serious pest of cereal crops until 1978, when it was discovered causing extensive damage to wheat in South Africa (Walters 1984). The RWA gained international pest status after being discovered in Mexico infesting wheat in 1980. From there it moved into the US via Texas in 1986 and rapidly spread throughout the primary wheat producing states in the western half of the US by 1987 to 1988 (Morrison 1988). The RWA continues to be one of the most important pests of dryland wheat and barley in the US and South Africa, although it is of minor importance elsewhere.

Nearly 1 million hectares of the total 27 million hectares of wheat planted in the western US was treated for RWA at a cost of $17 million in 1987. Insecticide costs in combination with wheat yield losses caused by RWA damage exceeded $53 million, with about one-half of this total incurred by the state of Colorado alone (Webster et al., 1994). Losses to cereal crops totaled $893 million from 1987 to 1993. The impact of the RWA on wheat production became negligible in the central US Great Plains after 1994, following the release of 'Halt', a RWA-resistant wheat cultivar which carried the *Dn4* resistance gene (Quick et al., 1996).

Wheat and barley are the main cultivated hosts for RWA. Triticale and oat also can serve as hosts. However, the RWA must utilize volunteer cereal growth and other Graminaceous hosts to survive between summer grain harvest and the next planting event in the fall. Just as important to RWA ecology are the wild grass hosts in the US that include *Agropyron* spp., *Elymus* spp., *Pascopyrum* spp., *Bromus* spp., and *Aegilops* spp. (Armstrong et al., 1991; Hammon and Bishop 1997). With its broad host range, RWA can exist in the US without the presence of wheat or barley. However, cultivated hosts such as wheat and barley provide RWA with the opportunity to exploit a host crop monoculture and thus become economically significant.

Biology, plant damage, and control methods

The life cycle of the RWA is typical of most aphids, whereby it reproduces parthenogenically as female viviparae from spring to fall under warm temperatures. Generation times range from 8 to

42 days (Aalbersberg et al., 1987), and females can produce 13 to 46 nymphs per generation. Reproductive rates increase and generation time is reduced as air temperature increases. The RWA is holocyclic in its native range. Parthenogenic females give rise to males and females in the fall to produce eggs that overwinter and hatch in the spring. However, RWA is only known to overwinter in the US as parthenogenic females.

The RWA preferentially feeds on new plant growth and damages the leaves by causing characteristic longitudinal white streaking, purple discoloration, and rolling of the leaves. Prolonged infestations lead to prostrate plant growth, stunting, and heads being trapped by the rolled leaves. Ultimately, the damage leads to reduced seed weight, end-use quality, and yield (Walters, 1984). Infestations that occur from the boot growth stage onward are the most damaging to yield (Kriel et al., 1986). Rolling of the plant leaves plays a critical role in the aphid's survival by providing a tubular refuge from predators, parasites, and insecticides (Webster et al., 1987).

Management strategies for the control of RWA include cultural, biological, chemical, and host-plant resistance. Host-plant resistance has been a cornerstone in managing RWA in the US and South Africa. Cultural control options in the US consist of delaying the planting of winter wheat or barley until November to avoid early fall infestations, and planting spring grains in early March to establish strong seedlings prior to spring RWA infestations (Hammon et al., 1996). The control of volunteer wheat and barley after harvest has eliminated sources for fall infestations (Walters 1984). These methods are very useful and are typically employed whenever they fit a grower's production system.

Biological control agents that have been evaluated for RWA management in the US include predators, parasitoids, and pathogens. Life-table studies on RWA determined that key predators, *Hippodamia* spp. and *Chrysoperla* spp., had no effect on aphid densities or plant biomass (Randolph et al., 2002). The only endemic parasitoid, *Diaeretiella rapae* (McIntosh), parasitized RWA at very low levels (<5%). Similarly, three species of common aphid fungal pathogens were ineffective control agents (<2.5% aphid control) for RWA in dryland wheat (Wriaght et al., 1993). Biological control agents are, however, important components of IPM programs for cereals.

Chemical control by insecticides has been the most widely used method for RWA management in the absence of resistant cereal cultivars. Systemic or mixtures of contact and systemic insecticides (e.g., chlorpyrifos, dimethoate, disulfoton, imidacloprid, malathion, methyl parathion, parathion) are used when aphid levels reach an economic infestation level of 5% to 20% infested tillers per plant, depending on time of season and plant growth stage. Plants can withstand higher levels of infestation as they mature.

Utilization of host-plant resistance

Host-plant resistance has been the most effective and economical means of managing RWA in the US and South Africa. The first breeding effort to develop RWA-resistant wheat was initiated in South Africa shortly after the introduction of the RWA in 1978. Genes conferring resistance to RWA were identified in *T. aestivum* L. and designated as *Dn1*, *Dn2*, and *Dn5* (Dutoit 1989), which led to the first RWA-resistant cultivar (TugelaDn). By 2006, 27 resistant cultivars had been released in South Africa (Tolmay et al., 2007). South African and US researchers have now characterized nine resistance genes for wheat. Dominant genes *Dn1*, *Dn2*, *Dn4*, *Dn5*, *Dn6*, *Dn8*, and *Dn9* originated from *T. aestivum*. The recessive *dn3* gene originated from *T. tauschii*. Gene *Dn7* is a dominant gene resulting from the intergeneric transfer from rye to common wheat (Marais et al., 1994). The *Dn4* gene (Quick et al., 1996) was incorporated into seven RWA-resistant cultivars, which successfully managed RWA in the US from 1995 to 2002.

A RWA population can contain biotypes that have the ability to damage previously resistant cultivars. Their occurrence can have a serious impact on managing RWA with host-plant resistance. Biotypes were first reported in RWA populations from the former Soviet Union, Europe, and the Middle East in 1989. The RWA from Syria and Kirghiz regions were shown to severely damage wheat with the *Dn4* gene (Puterka et al., 1992). In 2003, a new biotype of RWA appeared in Colorado which seriously damaged wheat that carried the *Dn4* gene. This biotype, designated RWA2, could acutely damage wheat with any one of eight of the nine *Dn* resistance genes, with the exception of *Dn7* (Haley et al., 2004). Biotypes RWA3, RWA4, and RWA5 were soon discovered which differentially damaged *Dn1* to *Dn9* resistance in wheat (Burd et al., 2006). By 2005, RWA2 had already dominated the biotype complex in the western US (Puterka et al., 2007). The extensive distribution and predominance of RWA2 indicated that wheat cultivars containing the *Dn4* gene would have little value in managing RWA. Fortunately, the primary sources of RWA1 resistance in barley, STARS 9301B and STARS 9577B, have remained resistant to all known RWA biotypes (Puterka et al., 2006). Table 9.3 summarizes the reactions of all sources of resistance in wheat to Russian wheat aphid biotypes.

The origin of these new biotypes critically remains undetermined. Wheat breeders in the US are currently focusing efforts to move *Dn7* resistance into wheat and find new sources of resistance to these biotypes. The threat of new RWA biotypes to wheat production is not limited to the US but also has become a problem in South Africa (Tolmay et al., 2007). Effective deployment of RWA resistance in cereals will rest on a thorough characterization of biotypic diversity and testing of candidate resistance genes in the field, as well as vigilantly monitoring RWA biotype frequency after resistance gene deployment.

Table 9.3 Russian wheat aphid resistance genes and biotype interactions in wheat.

Resistance	Russian Wheat Aphid Biotype				
Gene	RWA1	RWA2	RWA3	RWA4	RWA5
	Reaction to biotype[a]				
Dn1	S	S	S	S	S
Dn2	R	S	S	S	S
dn3	R	S	S	S	S
Dn4	R	S	S	R	R
Dn5	R	S	S	S	R
Dn6	R	S	S	R	R
Dn7	R	R	S	S	R
Dn8	S	S	S	S	S
Dn9	S	S	S	S	S

[a]R and S indicate resistant and susceptible reactions, respectively.

FUTURE PERSPECTIVES

Advances in controlling insect pests of wheat will be dependent on research, development, and deployment of new and better components of the integrated pest management tactic. Better monitoring, understanding, and forecasting of insect pest population buildup and movement is needed to design and proactively implement IPM measures. New natural enemies for biological control and new genetic sources of resistance in wheat are needed that are effective, economical, and durable.

The recent discovery of the chemical structure of the sex pheromone produced by the adult female Hessian fly may lead to inexpensive and convenient methods for monitoring Hessian fly populations. Artificial sex pheromone lures deployed in sticky traps would catch adult males and should provide much-needed information on population dynamics, phenology, and geographic distribution.

Research is needed to improve understanding of the dynamics of fungal entomopathogens and ways of manipulating field conditions to favor epizootics, including perhaps the use of mycoinsecticides. Also, different natural enemy guilds interact, and their interactions may affect net suppression of insect pests. Evaluation of these interactions is largely unexplored, and future studies involving specific pests and particular combinations of natural enemies are needed.

Potentially, the area where additional research and development would have the most dramatic positive impact on insect pest control is through the introduction of new resistance genes to wheat and the combination of multiple resistance genes for deployment in a single cultivar. Resistance genes currently used in available cultivars have come from within the wheat genome or from closely related species. Each of these resistance genes has provided protection for a very small subset of specific biotypes of aphids or Hessian fly. Consequently, multiple resistance genes are needed to provide broad-spectrum protection across multiple biotypes of pests that typically constitute field populations. The advent of molecular markers associated with these resistance genes will now allow breeders to more effectively combine genes to provide protection against pest populations with known biotypic composition.

Success in wheat resistance to insect pests has typically been short-lived due to the tremendous inherent genetic diversity for virulence within pest populations. Sources of durable resistance may have to come from unrelated species of wheat via genetic transformation.

REFERENCES

Aalbersberg, Y.K., F. Dutoit, M.C. van der Westhuizen, and P.H. Hewitt. 1987. Development rate, fecundity and lifespan of apterae of the Russian wheat aphid, *Diuraphis noxia* (Mordvilko) (Homoptera: Aphididae), under controlled conditions. Bull. Entomol. Res. 77:629–635.

Anderson, K.G., and M.O. Harris. 2006. Does R gene resistance allow wheat to prevent plant growth effects associated with Hessian fly (Diptera: Cecidomyiidae) attack? J. Econ. Entomol. 99:1842–1853.

Armstrong, J.S., M.R. Porter, and F.B. Peairs. 1991. Alternate hosts of the Russian wheat aphid (Homoptera: Aphididae). J. Econ. Entomol. 84:1691–1694.

Bailey, S.M., M.E. Irwin, G.E. Kampmeier, C.E. Eastman, and A.D. Hewings. 1995. Physical and biological perturbations: Their effect on the movement of apterous *Rhopalosiphum padi* (Homoptera: Aphididae) and localized spread of barley yellow dwarf virus. Environ. Entomol. 24:24–33.

Barnes, H.F. 1956. Gall midges of economic importance: Vol. VII. Gall midges of cereal crops. Crosby Lockwood & Son, London.

Basky, Z., A. Fonagy, and B. Kiss. 2006. Baking quality of wheat flour affected by cereal aphids. Cereal Res. Commun. 34:1161–1168.

Basky, Z., and K.R. Hopper. 2000. Impact of plant density and natural enemy exclosure on abundance of *Diuraphis noxia* (Kurdjumov) and *Rhopalosiphum padi* (L.) (Hom., Aphididae) in Hungary. J. Appl. Entomol. 124:99–103.

Berzonsky, W.A., H. Ding, S.D. Haley, M.O. Harris, R.J. Lamb, R.I.H. McKenzie, H.W. Ohm, F.L. Patterson, F. Peairs, D.R. Porter, R.H. Ratcliffe, and T.G. Shanower. 2003. Breeding wheat for resistance to insects. Plant Breed. Rev. 22:221–296.

Brewer, M.J., and N.C. Elliott. 2004. Biological control of cereal aphids in North America and mediating effects of host plant and habitat manipulations. Annu. Rev. Entomol. 49:219–242.

Burd, J.D., D.R. Porter, G.J. Puterka, S.D. Haley, and F.B. Peairs. 2006. Biotypic variation among North American Russian wheat aphid (Homoptera: Aphididae) populations. J. Econ. Entomol. 99:1862–1866.

Burnett, P.A., and R.T. Plumb. 1998. Present status of controlling Barley yellow dwarf virus. p. 448–458. *In* A. Hadidi, R.K. Khetarpal, and H. Koganezawa (ed.) Plant virus disease control. APS Press, St. Paul, MN.

Burton, R.L. 1986. Effect of greenbug (Homoptera: Aphididae) damage on root and shoot biomass of wheat seedlings. J. Econ. Entomol. 79:633–636.

Carter, N.I., F.G. McLean, A.D. Watt, and A.F.G. Dixon. 1980. Cereal aphids: A case study and review. p. 271–348. *In* H. Coaker (ed.) Advances in applied biology. Vol. 5. Academic Press, New York.

Cheng, H.Z., S.Y. Zheng, Y.Y. Guo, S.Z. Zhao, and P. Wang. 2004. The resistance of wheat aphids to several insecticides. Henan Agric. Sci. 6:50–53.

Cox, T.S., and J.H. Hatchett. 1986. Genetic model for wheat/Hessian fly interactions: Strategies for deployment of resistance genes in wheat cultivars. Environ. Entomol. 15:24–31.

Curtis, B.C., A.M. Schlehuber, and E.A. Wood, Jr. 1960. Genetics of greenbug, *Toxoptera graminum* (Rond.), resistance in two strains of common wheat. Agron. J. 52:599–602.

Dahms, R.G., T.H. Johnston, A.M. Schlehuber, and E.A. Wood, Jr. 1955. Reaction of small-grain varieties and hybrids to greenbug attack. Okla. Agric. Exp. Stn. Tech. Bull. T-55, Stillwater.

Dubcovsky, J., A.J. Lukaszewski, M. Echaide, E.F. Antonelli, and D.R. Porter. 1998. Molecular characterization of two *Triticum speltoides* interstitial translocations carrying leaf rust and greenbug resistance genes. Crop Sci. 38:1655–1660.

Dunn, B.L., B.F. Carver, C.A. Baker, and D.R. Porter. 2007. Rapid phenotypic assessment of bird cherry–oat aphid resistance in winter wheat. Plant Breed. 126:240–243.

Dutoit, F. 1989. Components of resistance in three bread wheat lines to Russian wheat aphid (Homoptera: Aphididae). J. Econ. Entomol. 82:1251–1253.

Elliott, N.C., and R.W. Kieckhefer. 2000. Response by coccinellids to spatial variation in cereal aphid density. Population Ecol. 42:81–90.

Flor, H.H. 1946. Genetics of pathogenicity in *Melampsora lini*. J. Agric. Res. 73:335–357.

Forster, R.L. 1990. The 1985 barley yellow dwarf epidemic in winter wheat involving barley yellow dwarf virus transmitted by *Schizaphis graminum* and wheat streak mosaic. p. 266–274. *In* P.A. Burnett (ed.) World perspectives on barley yellow dwarf. CIMMYT, Mexico, D.F.

Foster, J.E., H. Ohm, F. Patterson, and P. Taylor. 1991. Effectiveness of deploying single gene resistances in wheat for controlling damage by the Hessian fly (Diptera: Cecidomyiidae). Environ. Entomol. 20:964–969.

Gallun, R.L. 1977. Genetic basis of Hessian fly epidemics. Ann. N.Y. Acad. Sci. 287:223–229.

Gibson, R.W., and R.T. Plumb. 1977. Breeding plants for resistance to aphid infestation. p. 473–500. *In* K.F. Harris and K. Maramorosch (ed.) Aphids and virus vectors. Academic Press, New York, NY.

Giles, K.L., D.B. Jones, T.A. Royer, N.C. Elliott, and S.D. Kindler. 2003. Development of a sampling plan in winter wheat that estimates cereal aphid parasitism levels and predicts population suppression. J. Econ. Entomol. 96:975–982.

Graybosch, R.A., J.H. Lee, C.J. Peterson, D.R. Porter, and O.K. Chung. 1999. Genetic, agronomic and quality comparisons of two 1AL.1RS–wheat–rye chromosomal translocations. Plant Breed. 118:125–130.

Haley, S.D., F.B. Peairs, C.B. Walker, J.B. Rudolph, and T.L. Randolph. 2004. Occurrence of a new Russian wheat aphid biotype in Colorado. Crop Sci. 44:1589–1592.

Hammon, R.W., and J. Bishop. 1997. Alternate host plants of Russian wheat aphid in Colorado. Colo. Agric. Exp. Stn. Tech. Rep. TR97–2, Ft. Collins.

Hammon, R.W, C.H. Pearson, and F.B. Peairs. 1996. Winter wheat planting date effect on Russian wheat aphid (Homoptera: Aphididae) and a plant virus complex. J. Kans. Entomol. Soc. 69:302–309.

Harris, M.O., T.P. Freeman, O. Rohfritsch, K.G. Anderson, S.A. Payne, and J.A. Moore. 2006. Virulent Hessian fly (Diptera: Cecidomyiidae) larvae induce a nutritive tissue during compatible interactions with wheat. Ann. Entomol. Soc. Am. 99:305–316.

Harris, M.O., J.J. Stuart, M. Mohan, S. Nair, R.J. Lamb, and O. Rohfritsch. 2003. Grasses and gall midges: Plant defense and insect adaptation. Annu. Rev. Entomol. 48:549–577.

Harvey, T.L., K.D. Kofoid, T.J. Martin, and P.E. Sloderbeck. 1991. A new greenbug virulent to E-biotype resistant sorghum. Crop Sci. 31:1689–1691.

Harvey, T.L., G.E. Wilde, and K.D. Kofoid. 1997. Designation of a new greenbug, biotype K, injurious to resistant sorghum. Crop Sci. 37:989–991.

Hesler, L.S., Z. Li, T.M. Cheesbrough, and W.E. Riedell. 2005. Population growth of *Rhopalosiphum padi* on conventional and transgenic wheat. J. Entomol. Sci. 40:186–196.

Hunter, B. 2001. Rage for grain: Flour milling in the Mid-Atlantic, 1750–1815. PhD diss. Univ. of Delaware, Newark.

Jiménez-Martínez, E.S., N.A. Bosque-Pérez, P.H. Berger, and R.S. Zemetra. 2004. Life history of the bird cherry–oat aphid, *Rhopalosiphum padi* (Homoptera: Aphididae), on transgenic and untransformed wheat challenged with barley yellow dwarf virus. J. Econ. Entomol. 97:203–212.

Kieckhefer, R.W., J.L. Gellner, and W.E. Riedell. 1995. Evaluation of the aphid-day standard as a predictor of yield loss caused by cereal aphids. Agron. J. 87:785–788.

Kieckhefer, R.W., and B.H. Kantack. 1988. Yield losses in winter grains caused by cereal aphids (Homoptera: Aphididae) in South Dakota. J. Econ. Entomol. 81:317–321.

Kriel, C.F., P.H. Hewitt, M.C. van der Westhuizen, and M.C. Walters. 1986. Russian wheat aphid *Diuraphis noxia* (Mordvilko): Population dynamics and effect on grain yield in the Western Orange Free State. J. Entomol. Soc. S. Afr. 49:317–335.

Latgé, J.P., and B. Papierok. 1988. Aphid pathogens. p. 323–335. *In* A.K. Minks and P. Harrewihn (ed.) Aphids: Their biology, natural enemies and control (World crop pests Vol. 2B). Elsevier, New York.

Liang, H., Y-F. Zhu, Z. Zhu, D-F. Sun, and X. Jia. 2004. Obtainment of transgenic wheat with the insecticidal lectin from snowdrop (*Galanthus nivalis* agglutinin; GNA) gene and analysis of resistance to aphid. Acta Genetica Sinica 31:189–194.

Lister, R.M., and R. Ranieri. 1995. Distribution and economic importance of barley yellow dwarf. p. 29–53. *In* C.J. D'Arcy and P.A. Burnett (ed.) Barley yellow dwarf: 40 years of progress. APS Press, St. Paul, MN.

Lukaszewski, A.J., D.R. Porter, E.F. Antonelli, and J. Dubcovsky. 2000. Registration of UCRBW98–1 and UCRBW98–2 wheat germplasms with leaf rust and greenbug resistance genes. Crop Sci. 40:590.

Marais, G.F., M. Horn, and F. Dutoit. 1994. Intergeneric transfer (rye to wheat) of gene(s) for Russian wheat aphid resistance. Plant Breed. 113:265–271.

Migui, S.M., and R.J. Lamb. 2003. Patterns of resistance to three aphid species among wheats in the genus *Triticum* (Poaceae). Bull. Entomol. Res. 93:323–333.

Morrison, W.P. 1988. Current distribution and economic impact. p. 4–8. *In* F.B. Peairs and S.D. Pilcher (ed.) Proc. Russian Wheat Aphid Workshop, 2nd, Denver, CO. 11–12 Oct. 1988. Colorado State Univ., Ft. Collins.

Östman, O., B. Ekbom, and J. Bengtsson. 2003. Yield increase attributable to aphid predation by ground-living polyphagous natural enemies in spring barley in Sweden. Ecol. Econ. 45:149–158.

Pedigo, L.P. 2002. Entomology and pest management. 3rd ed. Prentice Hall, Englewood Cliffs, NJ.

Peters, D.C., E.A. Wood, Jr., and K.J. Starks. 1975. Insecticide resistance in selections of the greenbug. J. Econ. Entomol. 68:339–340.

Pike, K.S., and R.L. Schaffner. 1985. Development of autumn populations of cereal aphids, *Rhopalosiphum padi* (L.) and *Schizaphis graminum* (Rondani) (Homoptera: Aphididae), and their effects on winter wheat in Washington state. J. Econ. Entomol. 78:676–680.

Porter, D.R., J.D. Burd, K.A. Shufran, J.A. Webster, and G.L. Teetes. 1997. Greenbug (Homoptera: Aphididae) biotypes: Selected by resistant cultivars or preadapted opportunists? J. Econ. Entomol. 90:1055–1065.

Porter, D.R., B. Friebe, and J.A. Webster. 1994. Inheritance of greenbug biotype G resistance in wheat. Crop Sci. 34:625–628.

Porter, D.R., J.A. Webster, R.L. Burton, G.J. Puterka, and E.L. Smith. 1991. New sources of resistance to greenbug in wheat. Crop Sci. 31:1502–1504.

Puterka, G.J., J.D. Burd, and R.L. Burton. 1992. Biotypic variation in a worldwide collection of Russian wheat aphid (Homoptera: Aphididae). J. Econ. Entomol. 85:1497–1506.

Puterka, G.J., J.D. Burd, D.W. Mornhinweg, S.D. Haley, and F.B. Peairs. 2006. Response of resistant and susceptible barley to infestations of five *Diuraphis noxia* (Kurdjumov), (Homoptera: Aphididae) biotypes. J. Econ. Entomol. 99:2151–2163.

Puterka, G.J., J.D. Burd, D. Porter, K. Shufran, C. Baker, B. Bowling, and C. Patrick. 2007. Distribution and diversity of Russian wheat aphid (Homoptera: Hemiptera) biotypes in North America. J. Econ. Entomol. 100:1679–1684.

Puterka, G.J., and D.C. Peters. 1990. Sexual reproduction and inheritance of virulence in the greenbug, *Schizaphis graminum* (Rondani). p. 289–318. *In* R.K. Campbell and R.D. Eikenbary (ed.) Aphid-plant genotype interactions. Elsevier, Amsterdam, The Netherlands.

Quick, J.S., G.E. Ellis, R.M. Normann, J.A. Stromberger, J.F. Shannahan, F.B. Peairs, J.B. Rudolf, and K. Lorenz. 1996. Registration of 'Halt' wheat. Crop Sci. 36:210.

Quisenberry, S.S., and D.J. Schotzko. 1994. Integration of plant resistance with pest management methods in crop production systems. J. Agric. Entomol. 11:279–290.

Randolph, T.L., M.K. Kroening, J.B. Rudolph, F.B. Peairs, and R. F. Jepson. 2002. Augmentative releases of commercial biological control agents for Russian wheat aphid management in winter wheat. Southwest. Entomol. 27:37–44.

Ratcliffe, R.H., and J.H. Hatchett. 1997. Biology and genetics of the Hessian fly and resistance in wheat. p.47–56. *In* K. Bondari (ed.) New developments in entomology. Research Signpost, Scientific Information Guild, Trivandrum, India.

Riedell, W.E., and R.W. Kieckhefer. 1995. Feeding damage effects of three aphid species on wheat root growth. J. Plant Nutr. 18:1881–1891.

Riedell, W.E., R.W. Kieckhefer, S.D. Haley, M.A.C. Langham, and P.D. Evenson. 1999. Winter wheat responses to bird cherry–oat aphid and barley yellow dwarf virus infection. Crop Sci. 39:158–163.

Riedell, W.E., S.L. Osborne, and A.A. Jaradat. 2007. Crop mineral nutrient and yield responses to aphids or barley yellow dwarf virus in spring wheat and oat. Crop Sci. 47:1553–1560.

Rogers, C.E., R.D. Eikenbary, and K.J. Starks. 1972. A review of greenbug outbreaks and climatological deviations in Oklahoma. Environ. Entomol. 1:664–668.

Royer, T.A., K.L. Giles, and N.C. Elliott. 1997a. Insect pests of small grains and their control. FS 7176. Oklahoma State Univ. Coop. Ext. Serv., Stillwater, OK.

Royer, T.A., K.L. Giles, and N.C. Elliott. 1997b. Small grain aphids in Oklahoma. FS 7183. Oklahoma State Univ. Coop. Ext. Serv., Stillwater, OK.

Royer, T.A., K.L. Giles, T. Nyamanzi, R.M. Hunger, E.G. Krenzer, N.C. Elliott, S.D. Kindler, and M. Payton. 2005. Economic evaluation of the effects of planting date and application rate of imidacloprid for management of cereal aphids and barley yellow dwarf in winter wheat. J. Econ. Entomol. 98:95–102.

Sanderson, T.A., M.J. Maudsley, and A.F.G. Dixon. 1992. The relative role of natural enemies and weather in determining cereal aphid abundance. Asp. Appl. Biol. 31:1–9.

Schmidt, M.H., U. Thewes, C. Thies, and T. Tscharntke. 2004. Aphid suppression by natural enemies in mulched cereals. Entomol. Exp. Appl. 113:87–93.

Smyrnioudis, I.N., R. Harrington, S.J. Clark, and N. Katis. 2001. The effect of natural enemies on the spread of barley yellow dwarf virus (BYDV) by *Rhopalosiphum padi* (Hemiptera: Aphididae). Bull. Entomol. Res. 91:301–306.

Starks, K.J., and R.L. Burton. 1977. Greenbugs: Determining biotypes, culturing, and screening for plant resistance with notes on rearing parasitoids. USDA Tech. Bull. 1556, U.S. Gov. Print. Office, Washington, DC.

Stuart, J.J., M.S. Chen, and M.O. Harris. 2007. Hessian fly. p. 93–102. *In* C. Kole and W. Hunter (ed.) Genome mapping and genomics in animals. Vol. 4. Insects. Springer, Berlin, Germany.

Teetes, G.L., C.A. Schaefer, J.R. Gipson, R.C. McIntyre, and E.E. Latham. 1975. Greenbug resistance to organophosphorous insecticides on the Texas High Plains. J. Econ. Entomol. 68:214–216.

Tolmay, V.L., R.C. Lindeque, and G.J. Prinsloo. 2007. Preliminary evidence of a resistance-breaking biotype of the Russian wheat aphid, *Diuraphis noxia* (Kurd.) (Homoptera: Aphididae), in S. Africa. Afr. Entomol. 15:228–230.

Tyler, J.M., J.A. Webster, and O.G. Merkle. 1987. Designations for genes in wheat germplasm conferring greenbug resistance. Crop Sci. 27:526–527.

Vickerman, G.P., and S.D. Wratten. 1979. The biology and pest status of cereal aphids (Hemiptera: Aphididae) in Europe: A review. Bull. Entomol. Res. 69:1–32.

Voss, T.S., R.W. Kieckhefer, B.F. Fuller, M.J. McLeod, and D.A. Beck. 1997. Yield losses in maturing spring wheat caused by cereal aphids (Homoptera: Aphididae) under laboratory conditions. J. Econ. Entomol. 90:1346–1350.

Walters, M.C. (ed.) 1984. Progress in Russian wheat aphid (*Diuraphis noxia* Mord.) research in the Republic of South Africa. Tech. Commun. No. 191, Dep. of Agric., Div. of Agric. Information, Pretoria, South Africa.

Webster, J.A., and S. Amosson. 1995. Economic impact of the greenbug in the western United States: 1992–1993. Great Plains Agric. Counc. Publ. 155, Stillwater, OK.

Webster, J.A., S. Amosson, L. Brooks, G. Hein, G. Johnson, D. Legg, W. Massey, P. Morrison, F. Peairs, and M. Weiss. 1994. Economic impact of the Russian wheat aphid in the western United States: 1992–1993. Great Plains Agric. Counc. Publ. 152, Stillwater, OK.

Webster, J.A., K.J. Starks, and R.L. Burton. 1987. Plant resistance studies with *Diuraphis noxia* (Homoptera: Aphididae), a new United States wheat pest. J. Econ. Entomol. 80:944–949.

Wiktelius, S., and J. Pettersson. 1985. Simulations of bird cherry–oat aphid population dynamics: A tool for developing strategies for breeding aphid-resistant plants. Agric. Ecosyst. Environ. 14:159–170.

Wriaght, S.P., T.J. Poprawski, W.L. Meyer, and F.B. Peairs. 1993. Natural enemies of Russian wheat aphid (Homoptera: Aphididae) and associated cereal aphid species in spring-planted wheat and barley in Colorado. Environ. Entomol. 22:1383–1391.

Chapter 10

Temporally and Spatially Dependent Nitrogen Management for Diverse Environments

William R. Raun, Ivan Ortiz-Monasterio, and John B. Solie

SUMMARY

(1) As annual world fertilizer nitrogen (N) consumption approaches 100 million tonnes, agriculture in particular must manage fertilizer N more efficiently for grain production and to minimize adverse environmental impact. Current practices where almost all nitrogen is applied preplant for wheat production are not environmentally sensitive nor do they optimize fertilizer use. Methods to increase nitrogen-use efficiency are sorely needed.

(2) Preplant soil testing that can determine the soil-available ammonium and nitrate forms of nitrogen is valuable, but nitrogen response varies considerably from year to year and from field to field.

(3) Nitrogen-rich strips applied preplant in wheat production fields can assist farmers in determining accurate midseason fertilizer N rates.

(4) A variant of this methodology, ramp calibration strips, offers midseason visual interpretation of N demand and an applied method for optimizing fertilizer N.

INTRODUCTION

Nitrogen (N) fertilizer is the most expensive input for cereal production worldwide. What is probably most important about nitrogen requirements in cereal crop production is that the demand changes drastically from field to field and from one year to the next. Of all the information that should be communicated to farmers in any locale is that this temporal and spatial dependency influences optimum nitrogen fertilizer rates. Long-term winter wheat (*T. aestivum* L.) research in the southern Great Plains of the US has shown that the average fertilizer-N application rates would have been correct in the ensuing year only 20% of the time; yet, using the same rate from one year to the next is common practice for wheat, maize (*Zea mays* L.), and rice (*Oryza sativa* L.) farmers worldwide. In addition, extensive research has shown that indigenous soil N across the landscape can vary several-fold, resulting in very different N recommendations depending on the location within the field. In this chapter, we describe two alternatives which improve upon current methods for determining fertilizer N rates and have the flexibility to be used in environments ranging from 260-ha fields in the High Plains of the US to 1-ha fields in Sub-Saharan Africa (SSA).

A direct measure of N requirement can be made with the ramp calibration strip (RCS), which is superimposed on farmer practices at or near planting. A range of N rates (zero to more than sufficient) is applied mechanically or by hand. The RCS is urgently needed in wheat production because N response varies considerably from year to year and from field to field. Soil testing procedures for NH_4-N and NO_3-N are valuable, but when taken at or near planting they cannot compensate for subsequent effects of the environment, especially in winter wheat that usually encumbers over 240 days in its growth cycle. The RCS allows farmers to make a visual or optical-sensor interpretation midway through the growing cycle and an ensuing N rate adjustment based on expected N response. Similarly, the N-rich-strip methodology that envelopes yield potential prediction has proven to be a reliable method for improving fertilizer N application in cereal production (www.nue.okstate.edu). Applying all N preplant for wheat, well known to be an inefficient method of applying fertilizer N, is no longer advisable, since field-tested alternatives are available.

NITROGEN-USE EFFICIENCY AS A DRIVER OF NEW TECHNOLOGY

Creating quantifiable methodologies to determine optimal rates of soil nutrient inputs that can be used both in intensive, large-scale, high-technology production systems and in developing-world single-hectare farms is a daunting task. Nonetheless, we must make the effort to package scientific results in forms applicable to all agricultural environments encountered in wheat production regions around the world. None of the existing methods for determining plant nutrient needs are easily scalable or readily adoptable by farmers in these diverse environments. The approaches described herein are scalable, and they have the potential of being adopted in myriad situations in developed versus developing regions, in stress versus limited-stress environments, or by small-scale versus large-scale farms.

In both wheat and maize production systems, preplant application has been documented as being the most inefficient method of applying fertilizer N (Mahler et al., 1994; Randall et al., 2003). Split-N application increases nitrogen-use efficiency (NUE) in cereal production systems by taking advantage of the improved efficiency of applying N fertilizer midway through the growing season. Nitrogen-use efficiency is defined as grain N uptake in fertilized plots minus grain N uptake in nonfertilized plots, divided by the N rate applied. With this in mind, methods that improve upon present approaches for midseason fertilizer N applications will ultimately deliver increased NUE, will improve farmer profitability, and will minimize adverse environmental impact.

Recent research to develop optical sensors and the requisite agronomic science has now provided tools to allow optimization of N use, which is the most significant input for wheat production. Of the more than 56 million tonnes of fertilizer N applied to all cereal crops each year, 66% is not recovered by the crop, some of which is immobilized in the soil and the remainder is lost via denitrification, volatilization, gaseous plant N loss, leaching, and surface runoff (Raun and Johnson 1999). If NUE could be improved from the current level of 33% (Raun and Johnson 1999) to more than 50%, the annual world savings would exceed $10.8 billion. Since 1992, scientists at Oklahoma State University have been working on the development of an optical-sensor-based approach to improve fertilizer NUE in wheat and other cereal crops. In 1998, the first active lighting sensor was successfully used to accurately predict cereal grain yield potential from midseason vegetative readings. Output from this sensor was later adapted to accurately predict midseason fertilizer N needs and to improve NUE (Raun et al., 2002). Farmers using the sensor on wheat and maize have realized average increases in revenue exceeding $20/ha. While this technology has been successfully extended in Mexico, India, Argentina, Pakistan, Australia, Canada, and the US, it must be modified to produce an inexpensive optical pocket sensor that can be extended even further. Science-based decision-making algorithms must be modified for the crops and needs of developing-world farmers.

Judicious and prudent use of N fertilizer has never been more important, especially considering the rapid escalation of natural gas price and the adverse environmental effects of irresponsible N use. This importance is amplified by the yield and nutritional penalties from not having applied sufficient N. The key to successfully introducing these technologies and scientific developments to farmers in the developed and developing world is the creation and extension of a comprehensive package of technology, science, education, and financing. In the absence of an entire package or system these technologies and scientific breakthroughs will not be adopted. The optical-sensor-based systems further described can provide the agronomic science and sensing technology components of this package.

CASE STUDY: WHAT DEFINES DIVERSE ENVIRONMENTS

Improved N management has never been more important than it is today, whether or not the environment is diverse and challenged in some way. Because fertilizer N used in cereals accounts for more than 60% of the total N (Alexandratos 1995, p. 190) used worldwide (90 million tonnes), maize, wheat, and rice farmers are the obvious target when accusations of mismanagement arise. An examination of fertilizer consumption and production data from Sub-Saharan Africa (SSA) and the US illustrate the importance of addressing the problem of imprecise use of N fertilizer (Table 10.1).

Today, SSA has a population exceeding 699 million persons. In 2005, SSA produced 97,317,420 t of cereal grain on 88,435,068 ha, or a mean of 1.10 t ha^{-1}. A total of 26,801,040 ha of maize was harvested in SSA, with a total production of 40,473,062 t, or a mean maize yield of 1.51 t ha^{-1} (Food and Agriculture Organization 2007). Wheat, sorghum (*Sorghum bicolor* L.), rice, and millet (*Pennisetum glaucum* L.) comprise the majority of the remaining cereal production (Table 10.1). Alternatively, the US produced 364,019,526 t of cereal grain on 56,404,000 ha, resulting in 6.50 t ha^{-1}. Fertilizer N consumption for SSA in 2005 was 1,307,443 t, of which 60% was estimated to be consumed for cereal production (Alexandratos 1995, p. 190). This translates into an anemic average N rate of 4 kg ha^{-1} for more than 88 million hectares of cereals produced in SSA. In the US, 6,526,998 t of fertilizer N was consumed for cereal production, and the average annual N rate was 52 kg ha^{-1} for all cereals. While SSA represents 10% of the world population, it

Table 10.1 Production and nitrogen use statistics for cereal production in Sub Saharan Africa (SSA), the US, and worldwide.

	SSA	US	World
Population	**699,813,000**	**300,000,000**	**6,600,000,000**
Cereal production (ha)	**88,435,068**	**56,404,000**	**657,085,620**
Maize (ha)	26,801,040	30,081,820	138,163,504
Wheat (ha)	2,631,932	20,226,410	210,247,188
Sorghum (ha)	25,829,881	2,301,470	41,689,272
Rice (ha)	8,477,895	1,352,880	147,455,159
Millet (ha)	20,480,119	200,000	34,242,897
Cereal production (t)	**97,317,420**	**364,019,526**	**693,427,825**
Maize production (t)	40,473,062	280,228,384	601,815,839
Cereal yields (t ha^{-1})	1.10	6.45	1.06
Maize yields (t ha^{-1})	1.51	9.32	4.36
Fertilizer N (t)	1,307,443	10,878,330	84,746,304
Fertilizer N, cereals (t)	784,466	6,526,998	50,847,782
N rate, cereals (kg ha^{-1})	**3.99**	**52.07**	**34.82**
N fertilizer costs ($)	706,019,220	5,874,298,200	45,763,004,160

Source: Food and Agriculture Organization (2007).

consumes less than 1.5% of the world fertilizer N. The US consumed 10,878,330 t of total fertilizer N in 2005, or 13% of the world total, with less than 5% of the world population. However, in either case, farmers likely are inaccurately and imprecisely estimating the actual nitrogen demand of the crop.

Malakoff (1998) estimated that excess N flowing down the Mississippi River was valued at over $750 million. With increased N prices, that value now exceeds $1.0 billion per year. This becomes increasingly important considering that SSA spent only $706 million on fertilizer N in 2005 for 88,435,068 ha of cereal production, while the US spent $5.8 billion on N fertilizer for over 56,404,000 ha of cereal production. While cereal farmers are making some effort to improve upon their fertilizer use efficiency, it is disturbing to note that the excesses from fertilizer N loss that end up in the Mississippi River each year exceed the total amount of N fertilizer applied for cereal production in SSA.

The underlying message is that existing technologies for optimizing the use of nitrogen fertilizers can be increased in all agricultural environments. The consequence has been either gross overapplication of N in regions with high levels of mechanization and abundant supplies of N fertilizer or gross underapplication in regions of limited N supply and small farms. The failure of existing technologies to determine optimal N application rates has been a compelling incentive to develop robust, simple, and flexible methods for determining N application rates to correct N deficiencies.

IS NITROGEN NEEDED

The fundamental question that must be answered for all nutrients is, "Is it needed?" The appropriate application rate is consequential to determining whether or not added amounts of the nutrient in question are needed. While simple and seemingly straightforward, this question is not always properly addressed, regardless of productivity level of a particular operation. With current approaches to soil testing, the decision to apply N is determined using chemical analysis of surface- and subsurface-soils for NH_4-N and NO_3-N levels. If these surface and subsurface levels of inorganic N are high, the demand for N can be small (Ferguson et al., 1991, 2002). However, increased amounts of inorganic N are normally associated with an increased risk of NO_3-N leaching (Andraski et al., 2000). Despite the ability to detect excesses in inorganic N in soil profiles, soil testing in US wheat production is not routine. As a result, future broad adoption of soil testing for N in the developed or developing world is not expected, nor is it expected to be a future solution of N management in wheat. In this regard, the identification of any technology that improves upon poor N-use efficiency present today will help in the long term.

Two fundamental reasons for the shortfall in using soil chemical analysis to determine N levels is the high mobility of nitrate N and the variation in available nitrate N as a consequence of changes in the soil environment from year to year. This problem is further compounded by variable availability of N throughout a given field. The former is termed *temporal variability* and the latter *spatial variability*.

Importance of spatial variability on N requirement

The applied question of whether or not N should be applied can be answered by applying a N-rich strip (NRS) in each field and each year. The NRS is essentially a rate of preplant fertilizer that will ensure no N deficiency is encountered in the crop throughout the growth cycle. Preplant fertilizer is by far the most inefficient method of applying N and that should be discouraged. Any methodology that works to partition the total amount of N to be applied will provide improved NUE versus preplant N applications. The NRS, as it is applied in a portion or strip of the field as the name implies, serves simply as a guide for midseason N determination. Having the NRS in no way replaces the need to apply a modest rate of preplant N.

As was reported by Ortiz-Monasterio and Raun (2007), many farmers worldwide have historically applied excessive amounts of N, and as a result, soil-profile inorganic N levels can be quite high. The demand for midseason N is then gauged via observable differences in plant growth at the time top-dress or side-dress N is usually applied. This can be further refined using a GreenSeeker normalized difference vegetation index (NDVI) sensor (Ukiah, CA; www.ntechindustries.com), combined with online algorithms for wheat and other crops (www.nue.okstate.edu) that determine precise N fertilizer needs as discussed later in this chapter. The NDVI measurement essentially provides an accurate estimate of the total amount of biomass present by accounting for the amount of red and near-infrared light absorbed (or reflected) by the plant canopy (Moges et al., 2004). By having an estimate of total biomass, yield potential can then be estimated by dividing biomass by the total number of days from planting to sensing in which growing degree day values [GDD = $(T_{min} + T_{max})/2 - 4.4°C$] were positive. This estimate, which is termed the *in-season estimate of yield*, essentially provides a value for growth rate, or biomass produced per day (Raun et al., 2005).

In other studies examining spatial variability, researchers have found that optimal nitrogen fertilizer rates vary widely from field to field (Cerrato and Blackmer 1991; Schmitt and Randall 1994; Bundy and Andraski 1995). In regions where spatial variability is present, using the average optimum N rate across a given region would be inadvisable in many of the fields. Other studies have evaluated both temporal and spatial variability and quantified their relative importance. It is interesting that two of these studies, one with rainfed maize in the midwestern US and the other with irrigated wheat in the Yaqui Valley of Mexico, concluded that spatial variability was three times more important than temporal variability (Babcock 1992; Lobell et al., 2004).

Importance of temporal and spatial variability combined

Recent work by Raun et al. (2008) highlights the importance of placing a NRS or RCS in each field, every year. The RCS approach is a variant of the NRS: instead of having one high N rate and the farmer-practice N application, an automated system is developed to apply a range of N rates at fixed intervals. When evaluated during the crop season, the RCS provides a visual response curve (Raun et al., 2008). This is recommended because the demand for fertilizer N differs each year as a function of the environment, and in each field as a function of the soil type and previous management practices. Therefore, the only way to decipher N need is to index temporal and spatial variability. Similar research by Miao et al. (2006) showed that the economically optimum N rate for maize production averaged 125 kg ha^{-1}, but that it varied from 93 to 195 kg ha^{-1}.

Long-term winter wheat grain yield data from check (no N applied) and fertilized (112 kg N ha^{-1}) plots are reported in Fig. 10.1.

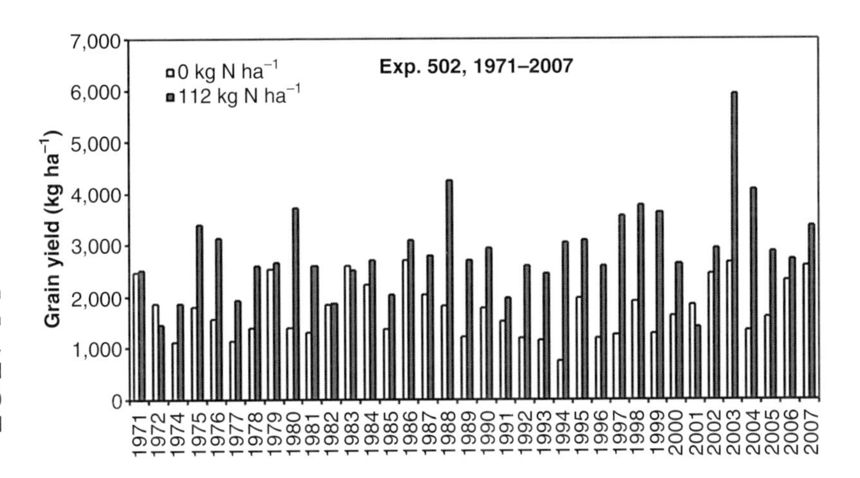

Fig. 10.1 Winter wheat grain yield means from long-term Experiment 502 at Lahoma, Oklahoma. Plots received no N fertilizer (0 kg N ha^{-1}) or were annually fertilized (112 kg N ha^{-1}).

These long-term trials were conducted in an area that received the same management practices from year to year. In this case, winter wheat was grown under conventional tillage, planted in October, and harvested between June and July. Evident from this work is that the demand for fertilizer N changed dramatically each year and this demand was unpredictable, similar to maize results reported by Miao et al. (2006). The variable demand for fertilizer N represents the influence of the environment (rainfall and temperature), soil type, and previous crop management on yield potential, which in turn can affect N responsiveness. Factors which impact yield potential are numerous, unpredictable, and change each year. This could include variable plant stand due to moisture stress, delayed planting due to excessive autumn rainfall, surface soil crusting, in-season moisture stress, diseases (various), lodging, soil type, and previous history of fertilizer use.

From the same data presented in Fig. 10.1, the optimum mean fertilizer N rate that resulted in maximum wheat grain yields from 1971 to 2006 was 56 kg N ha^{-1} (±42 kg N ha^{-1}), and the optimum rate varied from 1 to 156 kg N ha^{-1} (Fig. 10.2). Thus, fertilizing based on the mean optimum rate was only appropriate in 6 of 35 years, or 17% of the time. Oklahoma State University has numerous long-term experiments which document fertilizer N, P, and K response as a function of time in winter wheat (Raun 2008a). All of the long-term winter wheat trials scattered across the state of Oklahoma confirmed this observation that optimum N rates change drastically from year to year.

If the optimum N rates are known to change significantly both spatially and temporally, it is intuitive that yield levels, and yield potential, will change accordingly. Hence Raun et al. (2002, 2005) focused on the development of algorithms that could predict yield potential and determine N rates based on projected removal. The central component of this approach is recognizing the need to predict yield potential from sensor readings based on midseason NDVI. While straightforward, it is complicated by the fact that N responsiveness, or the response index (RI), must be determined separately. In each field, N is applied in one strip at a rate where N will not be limiting through the season. While it is recognized that this method of application is highly inefficient, it is only being used in one portion of the field. In this regard, it is important to recognize that N responsiveness is independent of yield potential. Furthermore, although N responsiveness like yield potential is dependent upon spatial and temporal variability, the same variables (total rainfall) can impact RI and YP_0 differently, where YP_0 is the yield potential that can be achieved with no added fertilizer applied.

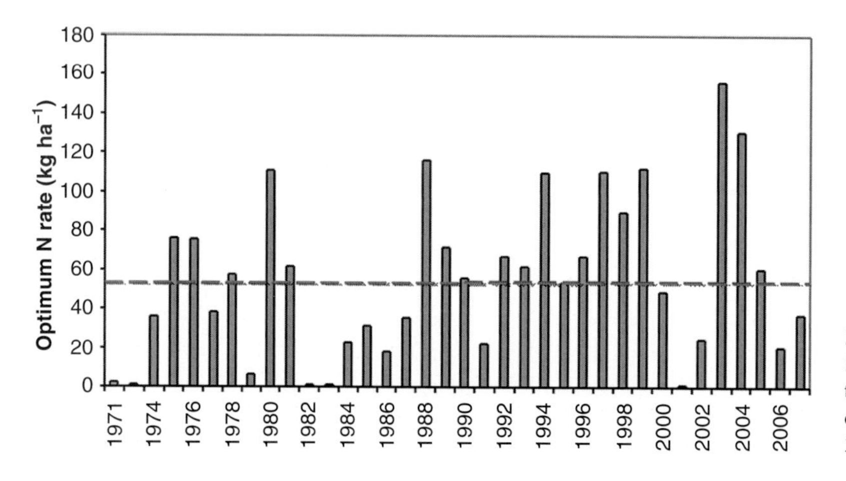

Fig. 10.2 Optimum fertilizer N rate, by year, from long-term Experiment 502 and the overall mean from 1971 to 2006, at Lahoma, Oklahoma.

NUTRIENT DEFICIENCIES OTHER THAN N

Where N is available, a costly mistake that can be made is applying this precious resource when it is not needed, or when yield response is limited by something else. On-farm trials that clearly target yield response as a result of the application of one factor while holding all others in nonlimiting amounts is an especially useful research and extension tool. Elimination of other confounding effects, including soil pH, soil test P, K, S, and/or micronutrients can be easily detected from comprehensive laboratory analyses. Submitting a surface (0–15 or 0–30 cm) soil sample for routine and/or comprehensive analysis is always recommended to eliminate the possible needs and/or correction of a deficiency or toxicity of another element. Soil testing to a certain extent is very much a scalable technology that historically has been underutilized, and as such, underappreciated.

The experimental method requires that we do our best to remove the influence of all possible factors, excluding the variable of interest. Thus, a N-rate experiment on P-deficient soils requires that we apply a sufficient rate of P so that P is not limiting. In this same light, we cannot recommend a N application in a low-vigor area of the field that may actually be P-deficient. The low-vigor area could also be low in soil organic matter, or it may have a shallow A-horizon, or low pH, or one of a host of other yield constraints. Similarly, in portions of the field that exhibit high vigor without some kind of reference, it would be difficult to discern the maximum level of vigor if N was moderately limiting at that end of the spectrum. The only way to know is to have a reference strip for N response.

PREDICTION OF YIELD POTENTIAL

Nitrogen fertilizer is critical for plant growth and grain yield of cereal crops, of which maize, wheat, and rice are the principal crops grown in the world. It has been very difficult to optimize the application of nitrogen fertilizer because it is rapidly transformed in soils and, when in the NO_3-N form, is highly mobile. Nitrogen fertilizer is also relatively expensive because of the high energy costs associated with its manufacture. Because N is critical for the production of cereal grains, farmers may resort to one of two extreme N application strategies: (i) apply N fertilizer in excess when sufficient fertilizer is available and cost is not excessive or (ii) apply little or none if only limited amounts of N are available and costs are high. In either case, there is no inexpensive scientifically based device or management system available for use by farmers to determine nitrogen fertilizer rates. Examining the evolving science and technology for managing N fertilizer and how it can be delivered to a range of environments and production systems reveals that midseason sensing technologies will likely play a role. This is especially evident in Fig. 10.1, where the mean yields from the fertilized plots (112 kg N ha^{-1}) varied from 1,500 to 6,000 kg ha^{-1}. The mean yield of the fertilized plots and unfertilized plots was 2,862 ± 860 and 1,720 ± 517 kg ha^{-1}, respectively, over this 35-year period. Using these numbers, the range in grain produced per kilogram of N applied was 13.4–53.5 kg. The common rule for winter wheat (2 lb N per bushel of wheat you hope to grow) translates to 30.0 kg grain per kilogram of N. Clearly, the demand for N is a function of yield potential, and that is known to change considerably by year and by field.

Focusing on the ability to predict grain yield midway through the growing season will become increasingly more important as we strive to match N fertilizer needs with final removal (Raun et al., 2001). Before proceeding, is it possible to predict yield potential in the middle of the growing season? Early work by Raun et al. (2001) showed that winter wheat grain yield potential could be predicted using multiple NDVI sensor readings collected midseason. This same approach was later modified to use one NDVI sensor reading, adjusted for the number of days from planting to sensing (Raun et al., 2005). The in-season estimated yield (INSEY) was successful in predicting yield potential when evaluated across 30 locations and 6 years. Other researchers have attempted to predict wheat yield potential, but the procedures require complicated stochastic methods using data collected from previous harvests (Brooks et al., 2001).

PREDICTION OF N RESPONSIVENESS INDEPENDENT OF YIELD POTENTIAL

Yield prediction is but part of the process when arriving at an optimal fertilizer N rate. As noted earlier the crop's response to N fertilizer changes radically from year to year as a function of the rate of conversion of organic N to inorganic N (Johnson and Raun 2003), as well as from field to field as a function of soil type and previous management. Furthermore, the same NDVI sensors used to predict wheat grain yield potential can also be used on the NRS and the area surrounding the NRS (hereafter termed *farmer practice*) to predict N responsiveness, or what is termed the response index (RI, estimated as NDVI collected from the NRS divided by NDVI from the farmer practice) (Mullen et al., 2003). Mullen et al. (2003) showed that the RI from NDVI readings (RI_{NDVI}) was highly correlated with final grain yield in the NRS divided by final grain yield in the farmer practice, or the harvest response index ($RI_{harvest}$). This positive and strong relationship confirmed that N responsiveness could be predicted across a wide range of yield levels from midseason indirect measurements, such as NDVI.

Once it was recognized that crop response to N fertilizer could be predicted from early-season sensor readings, this agronomic component was combined with yield potential at no additional N (YP_0) (Raun et al., 2001) to estimate crop yield with sufficient N, YP_N (Raun et al., 2005). The fundamental relationship was $YP_0 \times RI = YP_N$. By estimating the amount of N taken up by the crop (yield potential if fertilized, YP_N, and yield potential without added fertilizer, YP_0), the deficit or added amount needed to achieve maximum or near-maximum yields was that amount of N required to produce the predicted difference in grain yield (YP_N-YP_0). This calculation was further refined by accounting for the expected efficiency of the midseason N applied (usually between 0.5 and 0.7 based on agronomic data collected from a wide range of environments) and the maximum possible grain yield for each specific environment.

The improved midseason N management approaches have resulted in documented increases in NUE exceeding 15% (Raun et al., 2002). Extensive on-farm evaluation of the Sensor Based Nitrogen Rate Calculator for wheat and maize has shown a minimum increase in farmer revenue of \$22.00 ha^{-1} when using the recommended N rate in wheat (Ortiz-Monasterio and Raun 2007), and over \$40.00 ha^{-1} in maize.

The majority of this work has focused on delivering improved N rates at the field scale where only temporal variability is addressed. However, spatial variability, as well as temporal variability, can be addressed with this methodology and is more profitable (Biermacher at al., 2006), but that requires increased investment in sensing equipment that can be adapted to virtually any fertilizer applicator for midseason fertilizer application in cereals (NTech Industries, www.ntechindustries.com). Their systems employ the algorithm $YP_0 \times RI = YP_N$ as discussed earlier in this section, and they too have delivered increased farmer profit using this approach for winter wheat, spring wheat, and maize.

The combined knowledge of fluctuating yield levels spatially within a field and changes in additional N availability from one year to the next demands that a N application rate be determined midway through the crop season. This, coupled with knowledge of plant stands and vigor obtained from early season growth, provides an accurate method of applying a judicious fertilizer rate.

MIDSEASON N APPLICATIONS CAN RESULT IN MAXIMUM YIELDS

Provided N is applied later, early-season N stress seldom results in decreased grain yield in winter wheat or spring wheat. Thus, midseason N applications are much more efficient, and as a result of this improved efficiency, lower amounts of N are required relative to a preplant N application to produce the same level of yield (Morris et al.,

2006). In winter wheat, the life cycle from planting to harvest can exceed 240 days. Unlike spring wheat and hybrid maize production cycles that seldom exceed 110 days (planting to physiological maturity), the growth cycle of winter wheat is long enough that corrections can be made in-season, if a deficiency exists. Raun et al. (2002) showed that N applications delayed until Feekes growth stage 5 (Large 1954) could still result in maximum or near-maximum grain yields over four sites and two years in winter wheat. It is important to note differences between wheat and maize, since maize will remove almost double the amount of total N in the grain in one-half the time. Because of the shorter growth cycle and increased total N uptake, getting behind early in the growth cycle of maize can lead to decreased grain yields even if additional N is applied (Varvel et al., 1997). In addition, wheat can better recover than maize from midseason N applications because wheat can produce later tillers to compensate, whereas maize cannot set additional ears.

DETERMINATION OF MIDSEASON N RATE

The response to fertilizer N is dependent on the supply of nonfertilizer N (e.g., mineralized from soil organic matter, deposited in rainfall) in any given year. This is attributed to the extensive differences in annual rainfall and temperature and associated change in crop need (temporal variability), which directly influences how much nonfertilizer N is used by the crop.

The NRS, where N is not limiting, defines the sensor NDVI value at which N fertilizer is no longer limited. The highest NDVI measurement along the NRS can be used to calculate the maximum potential yield. Thus when N is not limiting for the year of measurement, YP_{max} is the maximum yield that can be expected within the most productive area in a field.

Three equations are used to calculate N fertilizer rates. Yield potential or YP_0 (Mg ha^{-1}) in winter wheat can be calculated directly by equation 10.1, where INSEY equals NDVI divided by the number of days from planting to sensing in which growing degree day values are positive [$GDD = (T_{min} + T_{max})/2 - 4.4°C$]:

$$YP_0 = 0.590e^{258.2 INSEY} \tag{10.1}$$

In spring wheat, INSEY is equal to NDVI divided by the number of days from planting to sensing, since with few exceptions all days in the spring wheat cycle will have a positive value for GDD. For winter wheat, many days may occur in which the mean temperature does not exceed 4.4°C, and as such there is no plant growth. Parameters for equation 10.1 are crop-specific (e.g., winter wheat, spring wheat) and are published at http://www.nue.okstate. The following equations predict the potential yield with additional N fertilizer, or YP_N, for two sets of conditions:

$$YP_N = YP_0 \times RI_{NDVI}, \quad \text{if } NDVI_{FieldRate} \geq 0.25 \text{ and } YP_N < YP_{max} \tag{10.2}$$

or

$$YP_N = YP_{max}, \qquad \text{if } YP_0 \times RI_{NDVI} \leq YP_{max}$$

Our observations over several years indicate that values of NDVI < 0.25 occur on bare soil or on soil with wheat stands so poor at Feekes growth stage 5 that they will not produce appreciable yields. The second set of conditions for YP_N simply state that YP_N cannot exceed YP_{MAX}, where YP_{MAX} is an agronomic optimum identified by farmers and scientists in specific regions.

For wheat, the top-dress N requirement can then be calculated by:

$$\boldsymbol{R} = 2.39\,(YP_N - YP_0)/\eta \tag{10.3}$$

in which $\boldsymbol{R}$ is the N application rate in kg ha^{-1}, 2.39 is the percentage N contained in wheat grain, and η is the expected efficiency from top-dress N application, generally between 0.5 and 0.7. This expected use-efficiency can vary widely, but should fall within this range for midseason N applications based on our observations.

"RAMP" METHOD OF DETERMINING MIDSEASON N RATE

The GreenSeeker NDVI sensor system has shown promise to farmers in the developed world for predicting yield and N responsiveness, and ultimately for providing a refined midseason fertilizer N rate. However, simplified technology may be more affordable in the developing world or for smaller-scale developed world farmers. As a result, we developed the ramp calibration strip approach that can be used to decipher midseason fertilizer N rates, in much the same fashion, but which does not require the more sophisticated sensor currently used. Similar to the approach described in the previous section, this methodology assumes that waiting until midseason to apply up to one-half of the required N fertilizer can result in maximum or near-maximum yields. The RCS applicator that we developed applies 16 incremental N rates at 3- to 6-m intervals over a total distance of 45–90 m (number of rates, intervals, and distances can be adjusted depending on the crop and other conditions) (Raun et al., 2008). Because the RCS is superimposed on the farmer preplant N practice, producers can observe plant responsiveness over the range of rates to determine the optimum top-dress N rate. Whether determined visually or with simplified active-reflectance-based sensors, the point where midseason visual growth differences no longer exist is the appropriate top-dress N rate. Where adequate but not excessive preplant N is available, the ramp interpolated rate provides a direct method to determine how much midseason N should be applied to achieve the maximum yields based on growth response evidenced within the RCS.

This approach has now been extensively adopted, and several farmers have ingeniously developed their own versions of the RCS with differing ranges in N rates, widths, and methods of application, depending on their production system, form of tillage, and crop being grown. For each of these variants on the RCS, methodology descriptions on how to build them are reported by Raun (2008b).

For the winter wheat example in Color Plate 24 (taken near Feekes growth stage 5), the RCS top-dress rate would have been near 80 kg N ha^{-1}, the point where total biomass was maximized. Where adequate but not excessive preplant N is available, the RCS interpolated rate (Fig. 10.3) provides an applied method to determine how much midseason N should be applied to achieve the maximum yields based on growth response evidenced within the RCS. This type of response is typical and one that can be deciphered either visually or using a hand-held Greenseeker NDVI sensor. As visualized in Color Plate 24 and plotted in Fig. 10.3, it is important to note that the response curves will take on a wide range of forms, and all are highly dependent on temporal vari-

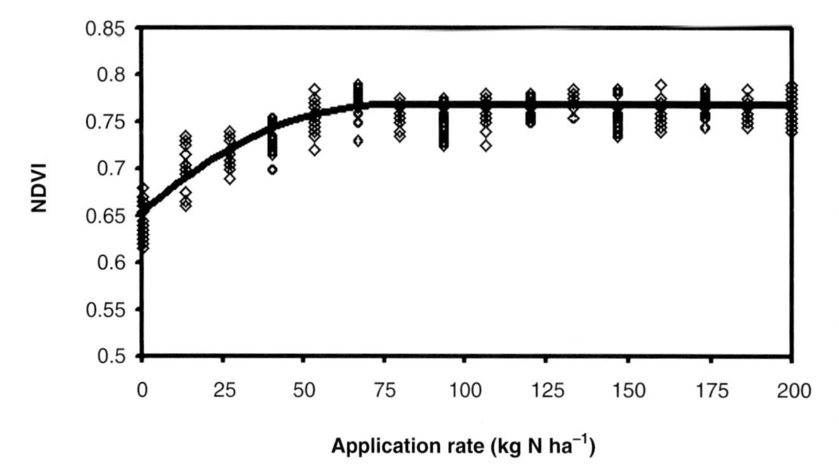

Fig. 10.3 Typical optical sensor measurements from a N-calibration ramp as a function of N application rate. Curves from this graph, adjusted for improvements in N uptake efficiencies realized by mid-growing-season application of N, can be used to determine optimum N application.

ability. Likewise, the optimum N rate determined from visual or sensor interpretation of the RCS varies accordingly, thus requiring that N rates be adjusted each year accordingly.

FUTURE PERSPECTIVES

There are still many unanswered needs relating to improved N management in cereals. The most important is generating improved yield-prediction equations for midseason N applications. Current work is tailored at using soil-profile moisture in addition to NDVI values to refine yield estimates, and that is certainly appropriate in rainfed environments where moisture is often limiting. If yield potential can be predicted, nutrient removal can be easily estimated by multiplication with known grain and straw N concentrations. Basing ensuing fertilizer N rates on projected removal is logical. Also, added work is needed to tailor N applications to the growth stage where N assimilation is the greatest. Applying N at the exact time when N demand is the greatest will likely provide a significant boost in resultant NUE.

REFERENCES

Alexandratos, N. (ed.) 1995. World agriculture: Towards 2010, an FAO study. FAO and John Wiley & Sons, West Sussex, England.

Andraski, T.W., L.G. Bundy, and K.R. Brye. 2000. Crop management and corn nitrogen rate effects on nitrate leaching. J. Environ. Qual. 29:1095–1103.

Babcock, B.A. 1992. The effects of uncertainty on optimal nitrogen applications. Rev. Agric. Econ. 14:271–280.

Biermacher, J.T., F.M. Epplin, B.W. Brorsen, J.B. Solie, and W.R. Raun. 2006. Maximum benefit of a precise nitrogen application system for wheat. Precision Agric. 7:193–204.

Brooks, R.J., M.A. Semenov, and P.D. Jamieson. 2001. Simplifying Sirius: Sensitivity analysis and development of a meta-model for wheat yield prediction. Eur. J. Agron. 14:43–60.

Bundy, L.G., and T.W., Andraski. 1995. Soil yield potential effects on performance of soil nitrogen tests. J. Prod. Agric. 8:561–568.

Cerrato, M.E., and A.M. Blackmer. 1991. Relationship between leaf nitrogen concentrations and the nitrogen status of corn. J. Prod. Agric. 4:525–531.

Ferguson, R.B., G.W. Hergert, J.S. Schepers, C.A. Gotway, J.E. Cahoon, and T.A. Peterson. 2002. Site-specific nitrogen management of irrigated maize: Yield and soil residual nitrate effects. Soil Sci. Soc. Am. J. 66:544–553.

Ferguson, R.B., J.S. Schepers, G.W. Hergert, and R.D. Lohry. 1991. Corn uptake and soil accumulation of nitrogen: Management and hybrid effects. Soil Sci. Soc. Am. J. 55:875–880.

Food and Agriculture Organization. 2007. FAOSTAT: Agricultural production [Online]. Available at http://faostat.fao.org (verified 19 Feb. 2008).

Johnson, G.V., and W.R. Raun. 2003. Nitrogen response index as a guide to fertilizer management. J. Plant Nutr. 26:249–262.

Large, E.C. 1954. Growth stages in cereals: Illustration of the Feekes Scale. Plant Pathol. 3:128–129.

Lobell, D.B., J.I. Ortiz-Monasterio, and G.P. Asner. 2004. Relative importance of soil and climate variability for nitrogen management in irrigated wheat. Field Crops Res. 87:155–165.

Mahler, R.L., F.E. Koehler, and L.K. Lutcher. 1994. Nitrogen source, timing of application, and placement: Effects on winter production. Agron. J. 86:637–642.

Malakoff, D. 1998. Death by suffocation in the Gulf of Mexico. Science 281:190–192.

Miao, Y., D.J. Mulla, P.C. Robert, and J.A. Hernandez. 2006. Within-field variation in corn yield and grain quality responses to nitrogen fertilization and hybrid selection. Agron. J. 98:129–140.

Moges, S.M., W.R. Raun, R.W. Mullen, K.W. Freeman, G.V. Johnson, and J.B. Solie. 2004. Evaluation of green, red and near infrared bands for predicting winter wheat biomass, nitrogen uptake, and final grain yield. J. Plant Nutr. 27:1431–1441.

Morris, K.B., K.L. Martin, K.W. Freeman, R.K. Teal, D.B. Arnall, K. Desta, W.R. Raun, and J.B. Solie. 2006. Mid-season recovery to nitrogen stress in winter wheat. J. Plant Nutr. 29:727–745.

Mullen, R.W., K.W. Freeman, W.R. Raun, G.V. Johnson, M.L. Stone, and J.B. Solie. 2003. Identifying an in-season response index and the potential to increase wheat yield with nitrogen. Agron. J. 95:347–351.

Ortiz-Monasterio, J.I., and W. Raun. 2007. Reduced nitrogen for improved farm income for irrigated spring wheat in the Yaqui Valley, Mexico, using sensor based nitrogen management. J. Agric. Sci. 145:1–8.

Randall, G.W., J.A. Vetsch, and J.R. Huffman. 2003. Corn production on a subsurface-drained Mollisol as affected by time of nitrogen application and nitrapyrin. Agron. J. 95:1213–1219.

Raun, W.R. 2008a. Long-term soil fertility experiments at Oklahoma State University [Online]. Available at http://www.nue.okstate.edu/Long_Term_Experiments.htm (verified 18 Feb. 2008).

Raun, W.R. 2008b. Predicting the potential response to applied N [Online]. Available at http://www.nue.okstate.edu/Index_RI.htm (verified 19 Feb. 2008).

Raun, W.R., and G.V. Johnson. 1999. Improving nitrogen use efficiency for cereal production. Agron. J. 91:357–363.

Raun, W.R., G.V. Johnson, M.L. Stone, J.B. Solie, E.V. Lukina, W.E. Thomason, and J.S. Schepers. 2001. In-season prediction of potential grain yield in winter wheat using canopy reflectance. Agron. J. 93:131–138.

Raun, W.R., J.B. Solie, G.V. Johnson, M.L. Stone, R.W. Mullen, K.W. Freeman, W.E. Thomason, and E.V. Lukina. 2002. Improving nitrogen use efficiency in cereal grain production with optical sensing and variable rate application. Agron. J. 94:815–820.

Raun, W.R., J.B. Solie, M.L. Stone, K.L. Martin, K.W. Freeman, R.W. Mullen, H. Zhang, J.S. Schepers, and G.V. Johnson. 2005. Optical sensor based algorithm for crop nitrogen fertilization. Commun. Soil Sci. Plant Anal. 36:2759–2781.

Raun, W.R., J.B. Solie, R.K. Taylor, D.B. Arnall, C.J. Mack, and D.E. Edmonds. 2008. Ramp calibration strip technology for determining mid-season N rates in corn and wheat. Agron. J. 100:1088–1093.

Schmitt, M.A., and G.W. Randall. 1994. Developing a soil nitrogen test for improved recommendations for corn. J. Prod. Agric. 7:328–334.

Varvel, G.E., J.S. Schepers, and D.D. Francis. 1997. Ability for in-season correction of nitrogen in corn using chlorophyll meters. Soil Sci. Soc. Am. J. 61:1233–1239.

Chapter 11

Grain Yield Improvement in Water-Limited Environments

Greg J. Rebetzke, Scott C. Chapman, C. Lynne McIntyre, Richard A. Richards, Anthony G. Condon, Michelle Watt, and Anthony F. van Herwaarden

SUMMARY

(1) Climate change threatens to reduce rainfall and increase rainfall variability in many of the world's rainfed and irrigated wheat-growing regions.

(2) Breeding for improved productivity in water-limited environments has been successful but at rates of genetic gain well below that observed in favorable environments. Breeders should not target selection for severe drought, as performance under severe stress does not correlate well in intermediate- to higher-yielding environments where growers derive most of their income.

(3) Genetic gain under drought will rely on multidisciplinary skills that focus on breeding but with support from the following: simulation modeling; improved physiological understanding and rapid, cost-efficient phenotyping; and molecular techniques aimed at more efficient screening.

(4) Greatest benefits arising from modeling and physiology reflect an understanding of the frequency and nature of drought stress, and traits with potential to improve water-use efficiency.

(5) The physiological complexity of adaptation to drought together with the known polygenic control of productivity traits will likely limit the implementation of transgenes in breeding programs. Transgenic approaches largely focus on assessment of traits affecting cell and plant survival when these are unlikely to increase productivity in managed cropping systems.

(6) Tremendous potential lies in the use of molecular markers for major genes such as resistance to disease and soil constraints, thereby enabling breeders to enrich populations for important alleles conferring adaptation to drought.

INTRODUCTION

The distribution of wheat production environments is broad, encompassing a range of temperature and radiation regimes, and both irrigated and nonirrigated, rainfed environments. Rosegrant et al. (1997) predicted that global demand for wheat for human and animal consumption will increase by as much as 40% by 2020, highlighting the need for improved productivity. Despite this growing demand, many factors have potential to impact crop productivity globally, including climate change, scarcity of water resources, and land-use change and degradation. Climate change will impact through direct and indirect effects of increasing atmospheric CO_2 levels, and warming air and ocean temperatures (IPCC 2007). For example, average air temperatures have risen globally approximately 0.8 °C since 1850. Rising global temperatures will likely affect cropping through higher day and night temperatures, inducing faster crop development, increasing the rate of crop water loss via increases in evapotranspiration (ET), and increasing respiration to increase the rate of carbon loss (Food and Agriculture Organization 2002).

Drought affects an estimated 65 million hectares of wheat worldwide. In these water-limited environments, wheat yields are commonly reduced to 50% or less of the irrigated yield potential (Byerlee and Morris 1993). Changing precipitation patterns with global warming are predicted to reduce rainfall in much of the middle latitudes, increase rainfall variability, and reduce total water available for dryland cropping. In Southern Australia, for example, wheat yields are predicted to decline an estimated 13% to 32% with global warming (Luo et al., 2005). In addition to effects of increased temperature and ET, irrigated cropping may suffer reductions in runoff and groundwater recharge, and diversions toward maintenance of river flows. Unsustainable groundwater use is decreasing groundwater reserves in the Indo-Gangetic plains of northern India to threaten India's grain harvest (Postel 1999). Excessive removal of groundwater estimated globally at 2,000 km^3 $year^{-1}$ also threatens irrigation in the US, Mexico, many Mediterranean countries, and China. Equally as important, expansion of urban areas and a declining quality of existing pastoral and cropping lands is decreasing the area of productive soils for use in agriculture. For example, an estimated 7 million hectares of arable land in India are currently affected by increasing salinity with potential to substantially reduce crop yields (Martinez-Beltran and Manzur 2005).

The effect of climate change on productivity and ability to meet anticipated demand for wheat will likely have its greatest impact in the developing world, where approximately 60% of food crops are rainfed. Of the remaining 40%, demand for limited water resources will likely cause conflict between the requirements for domestic and industry use, and water for sustaining crop production (Food and Agriculture Organization 2002). Agriculture currently accounts for 70% to 85% of all water withdrawals (Gleick 2003). With the need for increased food production, it is likely that water for agricultural use will increase as a percentage of total renewable water resources, the extent of increase varying on existing water requirements. An estimated 1.2 billion people already live in parts of the world lacking adequate water (Maris 2008). It is predicted that one in five developing countries, and particularly those of northern and eastern Africa and parts of Asia, will be confronted with water scarcity by 2030. Further, increasing industrialization in countries like China will see increased competition for water from manufacturing, let alone from population growth (Kijne 2003).

CLIMATE AND CROP GROWTH

Wheat production worldwide is undertaken across a range of environments. CIMMYT categorized the world's wheat-growing areas into broad, "megaenvironments," reflecting regional differences in environmental and biotic factors (Reynolds et al., 2006). Breeding may then be undertaken with selection for targeted adaptation to one or more agro-ecological regions. These regions may then be further split in recognition of environmental patterns peculiar to regions within the broader megaenvironment. For example, the drought-prone megaenvironment (ME) 4 is partitioned according to rainfall distribution as follows: ME 4A (postanthesis water stress), ME 4B (preanthesis water stress), and ME 4C (continuous water stress).

Figure 11.1 shows average annual rainfall distribution and air temperatures for locations across the Australian wheatbelt. These conditions are typical of environments where wheat is grown across the world. In many parts of the world, including Australia, wheat is sown when moisture is adequate for germination and emergence. Air temperatures are cool and vapor-pressure deficits (VPD) commonly low at this time (Nix 1975). The wheat seedling grows slowly, generating leaf area

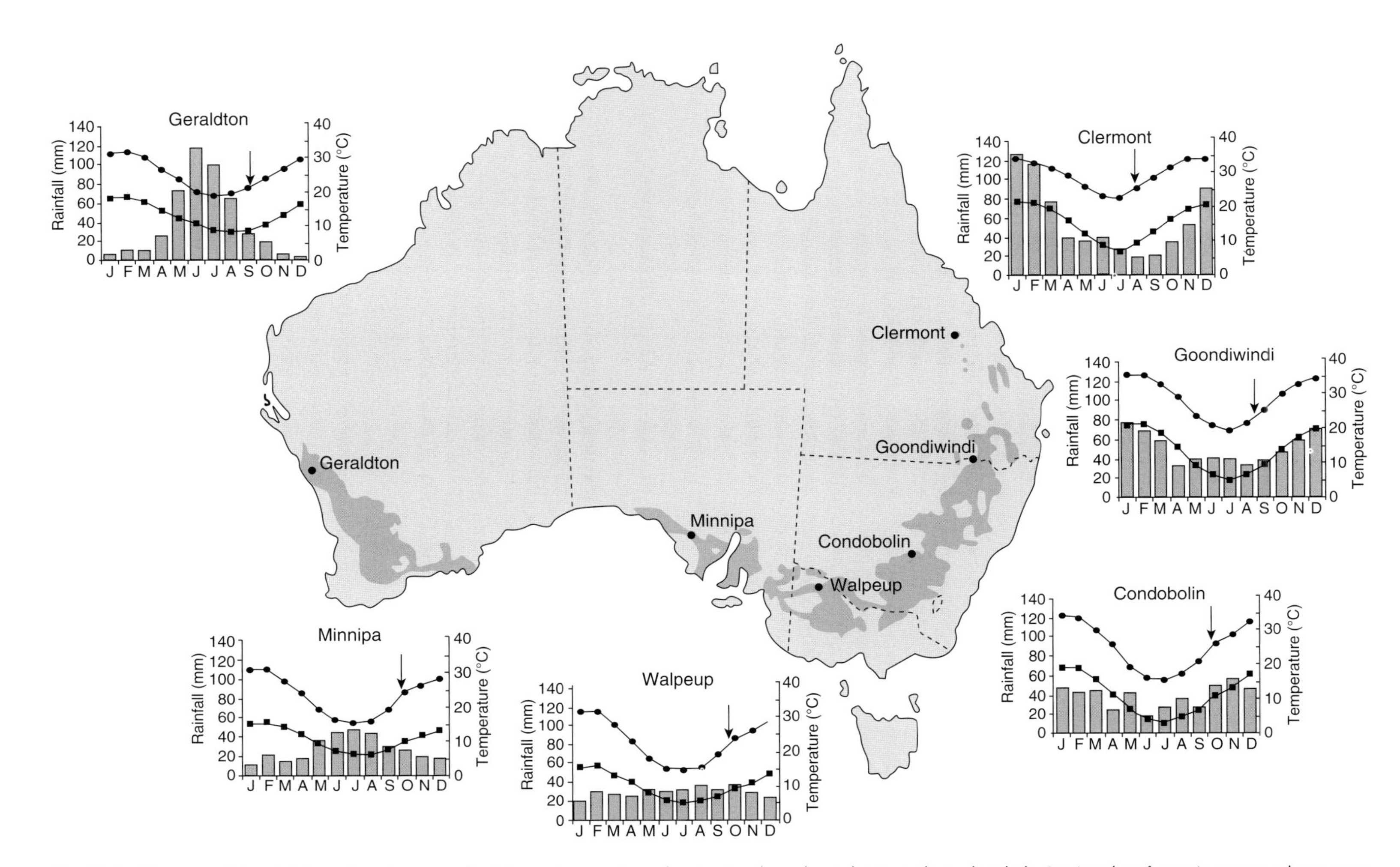

Fig. 11.1 Mean monthly rainfall, and maximum and minimum temperature, for six sites throughout the Australian wheatbelt. Sowing date for main-season wheat crops is commonly late May to early June. Arrows indicate approximate anthesis date for main-season sowings at each site.

and accumulating biomass under cool conditions of low radiation load. Moisture necessary for growth up to anthesis is provided either through irrigation or rainfall, including that stored in the soil prior to sowing (Fig. 11.1). Vapor pressure deficits are typically low during the vegetative period, although transient water deficits may arise if available soil water becomes depleted. After flowering, rising temperature and radiation increase VPD and evaporative demand (Fig. 11.1). In many parts of the world and particularly in Mediterranean environments (e.g., North Africa–southern Europe, Southern Australia, and the US Central Plains), the period from anthesis to crop maturity coincides with a reduction in rainfall. The wheat crop is increasingly reliant on irrigation or moisture stored and available in the soil profile (e.g., locations Geraldton and Minnipa in Fig. 11.1). If transpiration demand of the crop and evaporation from the soil surface (i.e., ET) together exceed potential soil water supply, then the crop experiences drought stress (Nix 1975).

Even when water is freely available across the season, crops typically experience short periods of transient water deficit. These small-scale events lower leaf conductance and photosynthesis, and slow leaf expansion. But these perturbations may contribute little to change in final performance if adequate water is available around key periods for growth and yield formation (e.g., around pollen meiosis and grain set) (Passioura 2002). Perhaps the greatest effect on grain number and yield arises for water deficit around flowering. For example, Fischer (1973) initiated water deficit at different times commencing 20 days before flowering. He found a 60% to 80% reduction in kernel number per spike with water stress at pollen meiosis (approx. 10 days before anthesis) compared with a reduction of only 20% to 40% with stress at flowering. Similarly, Saini and Aspinall (1981) demonstrated greater sensitivity to water stress in the male gametophyte (especially pollen mother cell meiosis), with male sterility in about 40% of florets but with little effect on female fertility.

The influence of postanthesis water deficit on grain yield reflects direct and indirect changes on kernel size. Drought influences cell division in the endosperm to reduce cell and starch granule number and decrease kernel size (Nicolas et al., 1985). A lack of water can also reduce grain weight through reductions in current photosynthesis and reduced leaf area (via canopy desiccation and senescence), which in turn reduce transpiration and accelerate the rate of grain filling, partly via increased crop temperature (Brooks et al., 1982). Collectively these effects decrease kernel size unless kernel number has already been greatly reduced by earlier stress. Reductions in kernel size contribute to an increase in the proportion of shriveled kernels, lower harvest index and grain yield, and decreased grain quality (van Herwaarden and Richards 2002; Ruuska et al., 2006). Agronomic factors such as excessive nitrogen can contribute to premature crop senescence where excessive leaf area can exhaust available soil water, leaving little water for grain filling (van Herwaarden et al., 1998).

WATER-LIMITED YIELD POTENTIAL

Factors during the season can impact effective water use by the growing wheat crop. In water-limited environments, final biomass and yield are often related to total crop ET (Taylor et al., 1983 and references therein). French and Schultz (1984) first identified that few wheat crops in southern Australia ever achieved potential grain yield for a given growing season ET. This observation has recently been extended to a meta-analysis of almost 700 experimental and farmer's wheat crops worldwide (Sadras and Angus 2006) (Fig. 11.2). Among all these crops, the yields of approximately 97% of them lay below the water-limited grain yield potential of 22 kg ha^{-1} mm^{-1} or aboveground biomass of 55 kg ha^{-1} mm^{-1}, assuming a water-limited harvest index of 0.40. The value of 22 kg ha^{-1} mm^{-1} represents a "boundary" or "upper limit" to grain yield under rainfed conditions.

Sadras and Rodriguez (2007) have since extended the model of French and Schultz (1984) to demonstrate how other climate factors, such as increased VPD and decreased fraction of diffuse

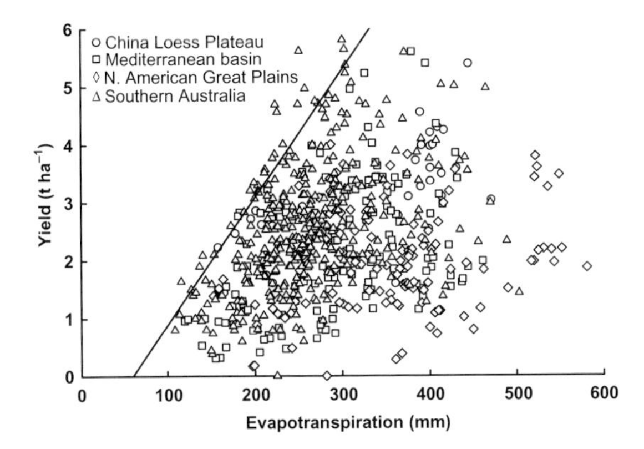

Fig. 11.2 Plot of wheat yield and seasonal evapotranspiration for 691 published studies representing four main global dryland cropping regions. The solid line represents the French and Schultz (1984) frontier and depicts water-limited grain yield (*ca.* 22 kg ha^{-1} mm^{-1}). Deviations from the slope reflect agronomic, phenology (matching demand for water with supply), and other factors limiting productivity (Sadras and Angus 2006).

radiation, act to decrease the realized yield water-use efficiency (WUE). The intercept of approximately 60 mm on the ET axis of Fig. 11.2 is considered to represent water lost through soil evaporation or runoff (or drainage), and increases for locations where numerous small rainfall events occur within-season (e.g., Mediterranean environments) rather than prior to sowing. An increased frequency of such events can contribute to large water losses through soil evaporation (e.g., estimated to be in excess of 110 mm by French and Schultz 1984).

Agronomic factors can explain why the water-limited yield potential is rarely achieved (French and Schultz 1984; Sadras and Angus 2006; Sadras and Rodriguez 2007). These include restricted water capture (e.g., factors influencing root growth such as late sowing, poor crop establishment, root disease, soil salinity, and acidification), competition with weeds, nonoptimal sowing density or arrangement, and leaf disease. In some cases, the rainfall distribution over the season is simply out-of-phase with periods of high water demand in the crop, and there is no opportunity for the crop to use this water. However, for many of these water-limiting factors, there are genetic opportunities available to overcome these constraints. Examples of these will be provided throughout this chapter.

As described earlier, the water demand–supply balance during the season impacts both biomass productivity and the partitioning of biomass to set and fill grains. Hence, it is helpful to attempt to independently characterize this water balance and use the information during the evaluation of plant breeding trials. Saulescu and Kronstad (1995) proposed that a "simulated entry" (i.e., yield estimated using weather and soil data in a dynamic crop simulation model) be calculated for each wheat trial and used as an unbiased check to account for environmental effects. Voltas et al. (1999b) used environmental indices to explain genotype × environment interaction in barley (*Hordeum vulgare* L.) evaluation trials, but these indices were based on average values of climate variables during different crop stages and were not integrative of the water supply–demand balance. For sorghum (*Sorghum bicolor*), Chapman et al. (2000) showed that a water demand–supply index estimated from a crop simulation model could explain consistent differences in genetic correlations of line performance between areas of the sorghum cropping region of northern Australia.

CHARACTERIZING TARGET ENVIRONMENTS

Selection for improved performance under drought is challenging, owing to changes in line ranking (i.e., genotype × environment interaction) across environments (Ceccarelli 1994). To enhance efforts in genetic improvement of wheat yield, it has been argued that it is useful to broadly define rainfed environment types around timing and average rainfall amount (Chapman et al., 2000). An example of such environmental characterization is described by Chapman (2008) for wheat grown in the northern wheat-growing region of Australia. There spring wheat is sown in autumn and grows largely on stored soil water from monsoonal rains. Mean farm yields are 1.5 to 2.5 t ha^{-1} and cropping practices are aimed at conserving subsoil moisture for the major yield-determining phases of crop growth toward the end of the season. Yet dryland farmers also desire wheat cultivars that perform well in wetter years (yields of 3–5 t ha^{-1}), because this is where profits are maximized. However, due to interannual variation in rainfall, breeders cannot reliably sample the range of on-farm environments in each year. Hence, it is of value to describe the nature of environments experienced over the long term, and consider how well these are sampled.

Long-term weather records (116 years, to 2006) for five locations representing the northern Australian wheat region were analyzed by Chapman (2008) to calculate a weekly water demand—supply (or stress) index, whereby a lower score equates to greater stress. The simulations encompassed several management regimes and planting dates for each part of the region. To focus on the key growth periods for sensitivity to stress, the index was centered at flowering and extended 450 degree-days (base temperature of 0 °C) before and after flowering. The stress indices were then summarized by cluster analysis to produce five drought stress patterns (Fig. 11.3). The major patterns identified were mild or no stress (no. 5, 22% of location–season combinations), preflowering stress (no. 2, 25%), and terminal stress [53%; beginning either around flowering (no. 4, 29%), booting (no. 3, 10%), or prior to booting (no. 1, 14%)]. The occurrence of preflowering drought was greater in the more northern locations, while postflowering drought was greatest in the most southerly locations of the region.

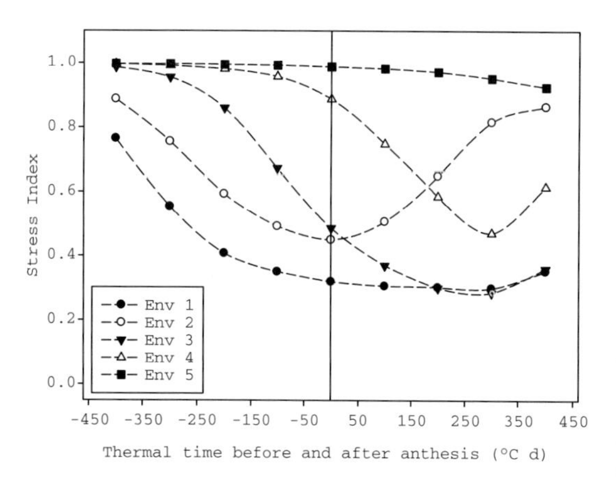

Fig. 11.3 Seasonal drought stress patterns derived from the APSIM-wheat cropping simulation model in 6 locations and 116 years for the northern wheat region of Australia. See text for description and frequency of occurrence of the drought stress patterns.

There have been some efforts to extend the concept of environmental characterization to breeding so as to understand change in genotype ranking with changing water supply. For example, Cooper et al. (1997) used performance for a common set of wheat genotypes to characterize environments, some of which included full or supplementary irrigation. Chapman (2008) extended the concept of a stress index for a data set of 18 genotypes evaluated in 12 years at 22 locations. Fifty-eight percent of the trials experienced a mild degree of water stress from just before flowering to mid–grain fill (ET1). A moderate to severe water stress in the period prior to flowering and through grain filling (ET2) occurred in the remaining 42% of trials. Without classification for environment type, the genotype × environment interaction variance was substantially larger than the genotypic variance (Table 11.1) to reduce heritability for yield and slow potential genetic advance. When classification of environment type was included in the analysis, the genotypic variance in the mild-stress environments (ET1) was estimated at three times the size of the genotypic variance within the moderate-stress environments (ET2). Examination of the data showed that predominantly later-flowering genotypes accounted for the genotype × environment

Table 11.1 Genotype and genotype × environment analyses of yield in 76 wheat trials and for trials classified within two drought environment types (after Chapman 2008).

Source	Variance Component	Standard Error
Genotype	0.083	0.027
Genotype × environment	0.120	0.007
Genotypes within ET1	0.142	0.047
Genotypes within ET2	0.038	0.014
Genotype × ET group	0.115	0.006
Residual variance	0.096	

interaction due to poor performance in the more-droughted (ET2) environments (Chapman 2008). This example illustrates the need for breeders to manage (e.g., CIMMYT simulated drought; Trethowan et al., 2005) or at least interpret trials in such a way as to sample the different types of drought via timing and location of trials. However, frequently the interannual variation is such that breeding evaluation trials should be characterized individually if drought stress pattern is to be used in the characterization of genotype performance (Chapman 2008).

BREEDING FOR IMPROVED PERFORMANCE UNDER DROUGHT

The identification and release of cultivars with the capacity to perform well across a range of environments represents a major challenge to wheat breeders. Genetic gain for grain yield has been greatest in managed, high-yielding environments, particularly with the development of improved disease resistance and introduction of dwarfing genes to increase harvest index and reduce lodging (Richards et al., 2002). In contrast, genetic progress has been smallest in water-limited environments (Araus et al., 2002) with long-term yield gains of less than 0.5% $year^{-1}$ reported for many water-limited environments (Byerlee and Morris 1993).

Development of broadly adapted cultivars is difficult when the cropping region encounters both well-watered and water-limited environments (Calhoun et al., 1994). To date, essentially two strategies have guided selection for yield under drought. The first reflects selection for improved performance under favorable conditions and an expectation that line ranking will be substantially maintained under less favorable conditions. The second strategy focuses on selection for improved productivity in targeted, water-limited environments. In this case, selection may take the form of empirical selection for yield *per se*, or take an analytical selection approach targeting physiological and/or developmental traits underpinning the known biology of adaptation to drought.

Yield potential and genetic gain in water-limited environments

Heritability and genetic gain are typically higher for wheat grown in favorable environments (Ud-Din et al., 1992; Cooper et al., 1997). Access to favorable testing environments may therefore provide an opportunity to select among lines for high yield potential. However, there are few instances where selection in well-watered environments has translated to broad adaptation in very low-yielding, water-limited environments. This partly reflects large genotype × environment interaction and a subsequently low genetic correlation for grain yield across contrasting stress levels (Ud-Din et al., 1992; Cooper et al., 1997). Genetic variance and heritability are also sometimes lower in water-limited environments (Ud-Din et al., 1992).

Using a theoretical framework, Rosielle and Hamblin (1981) showed tolerance to stress and mean productivity were negatively correlated when the genetic variance under stress was reduced compared to nonstress conditions. In turn, selection for tolerance to stress commonly results in reduced yield in nonstress environments and a decrease in mean productivity. However, selection for improved mean productivity generally increases yields in both stress and nonstress environments. Simmonds (1991) indicated that indirect selection for low-yielding environments via productivity in intermediate- to high-yielding environments was ineffective. This was particularly true when genotype × environment interactions reflect crossover changes in line ranking common to performance under lower-yielding, water-limited conditions (Ceccarelli 1994). This assertion has been confirmed empirically by Ud-Din et al. (1992) and Cooper et al. (1997). Both showed that selection for improved productivity in stress and other low-yielding conditions must be undertaken in appropriate target environments.

Selection for improved yield under drought has produced yield gains of 4.4% yr^{-1} in the CIMMYT semiarid wheat breeding program (Trethowan et al., 2002). Selection of semiarid, adapted lines with improved productivity under drought showed greater realized genetic gain than lines selected only for high yield potential (cf. 4.4% and 0.3% yr^{-1}) when evaluated together in the same yielding environments (yields of 4 t ha^{-1} or less). Progress of lines from the semiarid program was also modest (0.9% yr^{-1}) under favorable conditions (above 4 t ha^{-1}). The semiarid wheat program undertakes a "shuttle-breeding" approach, selecting under alternating drought and nondrought conditions. This is thought to allow selection for genes conditioning broader adaptation including improved disease resistance and healthier root systems (Trethowan et al., 2002, 2005). Despite their improved disease resistance, the slower gain under drought of lines selected for high yield potential is consistent with expectation (Rosielle and Hamblin 1981; Ceccarelli 1994).

The success of indirect selection for yield in well-watered versus water-limited environments may reflect the yield levels in the target population of environments (TPE). For example, drought in the UK results in a 15% reduction in the national average yield of 8 t ha^{-1} (Foulkes et al., 2002), whereas Trethowan et al. (2002) defined drought at yield levels of 4 t ha^{-1}. In some cropping regions (e.g., parts of Australia, the Mediterranean Basin, and China), water limitation may reduce yields to 1 t ha^{-1} or less. Breeders may be less interested in selection in these very low-yielding environments, as returns to growers are small and growers may graze or cut these crops for hay. Cooper et al. (1997) found that an irrigated, low water-stress nursery (mean yield ~ 4.5 t ha^{-1}) best predicted yields in similar low-stress target environments (3.9 t ha^{-1}; r = 0.89) but that this correlation decreased to 0.59 (3.4 t ha^{-1}) and 0.38 (3.1 t ha^{-1}) for moderate water stress, and to −0.08 (2.5 t ha^{-1}) for severe water-stress environments. Similarly, in barley, the rank correlation with the most favorable environments (yield = 5.2 t ha^{-1}) was poorest (r = −0.61) with the lowest-yielding environments (2.4 t ha^{-1}) sampled (Voltas et al., 1999a). Despite this, Cooper et al. (1997) emphasized the need to include one or more "high-input" trials during evaluation to ensure that lines with potential in wetter years were not accidentally overlooked through poor sampling of the variable on-farm environments. Similarly, CIMMYT's shuttling between drought and well-watered environments in their semiarid wheat breeding program seems to have favored improved disease resistance and retention of genes for broader adaptation and response to favorable conditions.

Selection for yield under stress and nonstress conditions has long been a contentious issue among breeders (Ceccarelli 1994). Strategies have been developed to remove the influence of high absolute yield and enable selection on relative yield in dry and well-watered environments. As mentioned previously, CIMMYT uses selection across alternating cycles of drought and well-watered conditions for their semiarid breeding program. Other approaches range in principle from regression-based, yield-stability parameter estimation (Finlay and Wilkinson 1963) to selection indices that capture the reduction in yield caused by stress when compared with performance in favorable environments [e.g., the drought susceptibility index described by Fischer and Maurer (1978), and the superiority measure of Lin and Binns (1988)]. In the latter, relative performance is measured for lines grown side-by-side, enabling assessment under the same conditions of timing and severity of water limitation. Lines with an ability to maintain yield in drought compared to well-watered conditions are seemingly less affected by drought. Studies have indicated that measures such as the drought susceptibility index tend to favor genotypes with low yield potential (in the absence of drought) and relatively high yield under drought, whereas superiority measures favor performance in higher-yielding environments (Clarke et al., 1992). Despite this, there is potential for use of relative performance under stress and nonstress conditions to reduce flowering date effects in genetic studies targeting yield under drought.

Evaluation of line performance under drought can be difficult if it is reliant on sampling of random environments (Cooper et al., 1997). Development of managed environments can allow repeatable screening of lines for known severity, timing, and frequency of water limitation reflective of the type of drought experienced in the

TPE. Ideally, managed environments should be uniform, repeatable, and allow differentiation of genotypes as well as the assessment of mechanisms or traits and putative quantitative trait loci (QTLs) or candidate genes associated with performance under drought. Site uniformity and control of water availability should aid in reducing the error variance and reduce potential for genotype × season interaction to increase heritability and genetic variance. Also, the site should be managed appropriately to reduce the potential for root disease. Use of combinations of drip and gravity-fed irrigation has assisted in the development of managed droughts for screening wheat lines at CIMMYT (Trethowan et al., 2005). The potential also exists to develop large rain-out shelters with the capacity for controlling water availability. Such a system has been established using drip irrigation for large-scale screening of rice in China (Liu et al., 2006).

Physiological breeding

The literature on eco-physiology of natural plant communities commonly refers to the ability of a plant to survive under water limitation as drought resistance or tolerance. Indeed, natural communities have evolved many mechanisms that allow plants to survive under conditions of severe water deficit (Passioura 2002). Under managed cropping systems, "fitness" is rarely described in terms of plant survival. Rather, agricultural fitness reflects productivity across a range of water-availability patterns. In many production systems growers make much of their income in favorable years when yields are greatest. Similarly, in many cropping regions, breeding companies receive royalties on the quantity of grain delivered. Hence, when there is strong incentive to boost productivity in favorable environments or conditions, there may be little incentive to focus specifically on selection for performance in droughted environments. As indicated, selection for improved drought tolerance is challenging. Productivity under drought reflects a culmination of many complex processes (Tuberosa et al., 2002), themselves under polygenic and epistatic control of QTLs with small effects (Maccaferri et al., 2008; Rebetzke et al., 2008a,b). Recognition of this complexity by some breeding companies is reflected in a change from selection for yield under drought to selection for greater water-use efficiency and subsequently greater yield for each additional unit of water used (Passioura 2006a).

Why should a genetic approach to improving performance under drought be adopted? Growers currently implement a range of agronomic practices, such as improved tillage systems or use of break crops to increase WUE and yield. However, these practices may bring extra risk and potential cost to the grower, particularly as growers target the maximal WUE (Fig. 11.4). This is because many of these practices incur financial costs that may not be recovered if seasonal conditions deteriorate in dryland environments. Despite the potential for greater return, these costs expose growers to greater risks and the likelihood of reduced return. Genetic opportunities exist to increase WUE (Richards et al., 2002) with little additional cost to the grower.

It is intuitively appealing to consider yield in a physiological framework. An accumulation of research knowledge underpins the integration of biological processes contributing to fitness or yield in droughted environments. Further, many features of physiological traits make them

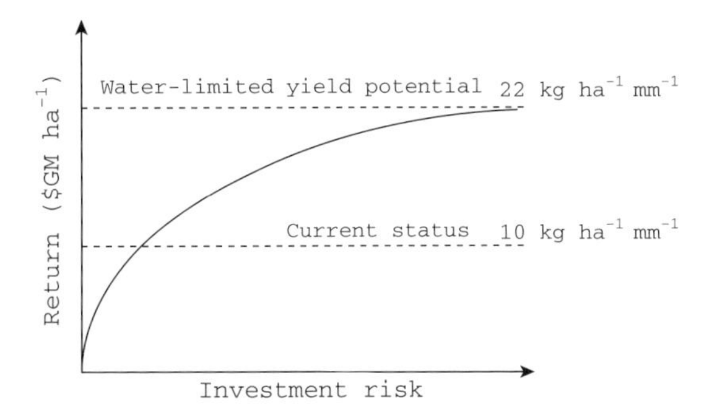

Fig. 11.4 Hypothesized increase in grower return (gross margin per hectare) with increasing system input and subsequent investment risk. The relationship demonstrates that increasing agronomic inputs (e.g., fertilizer, break crops to reduce root disease) produce potential increases in water-use efficiency (WUE) and yield but at increasing financial risk to the grower. Genetic increases in WUE add minimal increase to the cost to growers (after P. Carberry, pers. comm.).

potentially useful for implementation in a breeding program: (i) increased genetic variability; (ii) high heritability; (iii) opportunity for out-of-season selection; (iv) simplicity and lower cost relative to yield selection; (v) applicability to marker-assisted selection; (vi) an allowing for targeted selection of traits and therefore genes from unadapted donors; and (vii) the capacity to be assessed singly or with other traits in simulation modeling to determine their value to breeding (after Richards 2006). Yet despite this potential there are few examples where physiological traits have been successfully implemented in breeding programs (Jackson et al., 1996; Richards 2006). Lack of adoption may reflect the following: (i) trait type (survival and not productivity based); (ii) low heritability and/or low genetic correlation with yield; (iii) environment-specific expression; (iv) repulsion-phase linkages for the target gene and other important genes (especially in wild donors); and (v) high cost of phenotyping. Finally, implementation requires good communication for integration from the physiologist and molecular biologist to the breeder.

BREEDING TOOLS

Indirect selection via correlated traits

Breeding traditionally employs direct selection for genetic gain of target traits. However, the success of direct selection is contingent on populations containing adequate additive genetic variance and high narrow-sense heritability. The trait should also be simple and inexpensive to measure, particularly in mass selection where many families may be evaluated. Narrow-sense heritability (h^2) can be calculated on a line-mean basis as follows:

$$h^2_{\text{line-mean}} = \sigma^2_A \,/\, (\sigma^2_A + \sigma^2_{AE} \,/n_e + \sigma^2_{\text{residual}} \,/n_r n_e),$$

in which σ^2_A, σ^2_{AE}, and $\sigma^2_{\text{residual}}$ are estimates of the additive, additive × environment, and residual variances, respectively, and n_e and n_r are the number of environments and replications per environment, respectively. The additive genetic variance provides a measure of the effect of substituting one allele at a locus for another and can be estimated through measures of the level of inbreeding and the genetic relationship among sibs via the covariance among relatives (Falconer and Mackay 1996). Where heritability is low, alternate populations containing greater additive genetic variance (and high mean) should be considered, or sampling should be modified to reduce nongenetic variation.

Alternatively, surrogate traits may occur for characteristics that are costly or difficult to measure such as aerial biomass or grain yield. If the additive genetic correlation (r_A) between two traits (X and Y), and their narrow-sense heritabilities h^2_X and h^2_Y are known, the correlated response of trait Y to selection on trait X ($\Delta_{GY.X}$) can be predicted by:

$$\Delta_{GY.X} = k\ \sigma_{pY}\ h_Y\ h_X\ r_A, \qquad (11.1)$$

in which k is the standardized selection differential, and σ_{pY} is the phenotypic standard deviation for trait Y (Falconer and Mackay 1996). If $h_Y < h_X r_A$, then selection for trait X will result in greater change in trait Y than direct selection for Y. As a proxy for r_A, the genetic correlation for two variables can be readily estimated from analysis of covariance. Linkage disequilibrium (population type), chromosomal linkage, and pleiotropy can all contribute toward two traits being genetically correlated. However, owing to recombination, only pleiotropic effects are likely to maintain a genetic association over cycles of crossing and selection in a breeding program. Most studies report only phenotypic correlations, which differ from genetic correlations as phenotypic correlations also contain environmental and sampling covariance components (Searle 1961). Importantly, only the genetic covariance component of this correlation (more specifically, the additive correlation) is responsive to selection.

Several traits have been reported as under pleiotropic control with other traits at one or more loci. For example, carbon isotope discrimination (CID) measured prior to anthesis has shown a strong additive genetic correlation with grain yield and biomass (Rebetzke et al., 2002). Indirect selection for yield and biomass is more effective using CID when assessment is made on single plots. This is because large genotype × environment interaction and residual variances reduce narrow-sense heritability for yield and biomass in nonreplicated plots. The relative benefit of correlated genetic gain is reduced with replicated testing over blocks and environments, which increases heritability for yield but less so for biomass. Notwithstanding, CID offers potential for screening nonreplicated families in the early stages of yield testing, such as occurs when little seed is available or while early generations are still genetically heterogeneous. By selecting families with low CID, the breeder is restricting the costly stages of replicated yield testing across environments to those families with a greater likelihood of high yield in water-limited environments (e.g., in the release of the high water-use efficiency wheat cultivars from Australia, 'Drysdale' and 'Rees'). Further, by fixing favorable alleles for CID, greater emphasis can be placed on selection for other genes important for adaptation to water-limited environments.

Finally, it must be well established that indirect selection produces a correlated change in the desired trait in the TPE. Thus the value of traits conferring specific adaptation (e.g., CID vs. seedling vigor and their effects on water use; Condon et al., 2004) must be recognized with selection for the target environment type and their frequency in the TPE.

High-throughput phenotyping

The dynamic nature of rainfed, water-limited environments makes selection for improved grain yield a challenge to wheat breeders. Improved understanding of genotype × environment interaction and subsequent characterization of environment types, coupled with improved experimental designs (Singh et al., 2003) and statistical methodology (Smith et al., 2001), have contributed to reduced error variances to increase heritability and response to selection for yield. Similarly, improved experimental techniques are being employed to reduce experimental error in measuring yield. For example, removal of border rows at maturity will reduce the effects of intergenotypic competition between adjacent plots, while breeders are careful in selecting uniform sites and/or mapping field variability for known soil constraints (e.g., micronutrients or salinity).

There is substantial genetic diversity in wheat and its relatives, but adoption may be limited by repulsion-phase linkages with detrimental genes (e.g., *Cre1* and pinched grain), thereby slowing adoption and making assessment of new diversity a potentially expensive undertaking. Furthermore, genetically and physiologically complex traits such as water-soluble carbohydrate (WSC) concentration, or traits expensive to measure such as CID, may limit their adoption and subsequent selection in breeding programs despite their value in breeding. New tools are becoming increasingly available for phenotyping and selection of complex traits by breeders in early and later stages of family evaluation. Spectral information obtained through near-infrared (NIR) spectroscopy is inexpensive and reliable with potential as a surrogate for leaf nitrogen, CID, and WSC concentration (Ruuska et al., 2006). For example, Fig. 11.5 shows predicted versus actual WSC concentrations for wheat genotypes grown at different plant densities using a general NIR

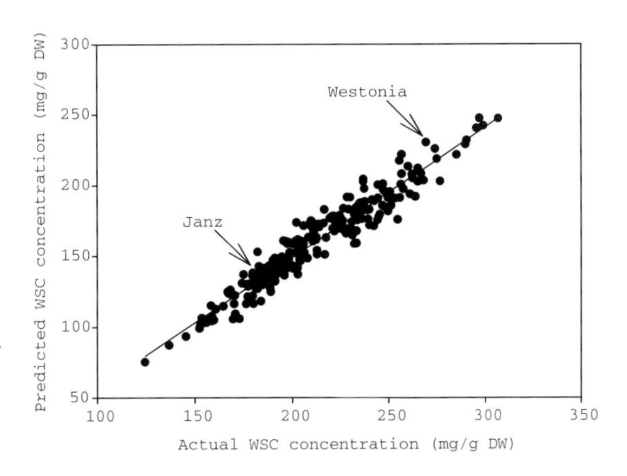

Fig. 11.5 Water-soluble carbohydrate (WSC) concentration measured for 22 wheat genotypes sown at 5 densities plotted against predicted WSC concentration using a NIR calibration equation developed for a wheat mapping population evaluated in southern Australia. High ('Westonia') and intermediate ('Janz') WSC concentration genotypes are shown for comparison ($Y = -35 + 0.92X$, $r^2 = 0.93$).

calibration equation developed for an unrelated wheat population. A similar prediction using NIR is given for CID measured using a mass spectrometer for wheat lines evaluated across multiple years and sites (Fig. 11.6). In both cases the rankings of low and high predicted extremes are consistent with rankings based on actual values. Remote sensing of multispectral reflectance enables rapid assessment of biomass and green leaf area, canopy architecture (as photosynthetically active radiation absorption), plant nitrogen, and water status of large breeding populations assessed as plots (Babar et al., 2006). Preliminary indications are that spectral reflectance indices are repeatable with high heritability and are correlated with grain yield across contrasting irrigation regimes (Babar et al., 2006).

Other seemingly less complex tools have application for high-throughput phenotyping in breeding programs. Use of digital cameras and availability of free or inexpensive digitizing software can be used for rapid assessment of vegetation indices, ground cover, or plant establishment counts for large numbers of lines (Casadesús et al., 2007). Similarly, Fig. 11.7 illustrates how image analysis of scanned harvest grain can be used to provide indirect estimates of kernel size for a population of 191 genotypes evaluated under drought. Three-hundred seed were scanned at low resolution (300 dpi) for each genotype, and area per kernel was nondestructively determined to provide an estimate of the mean, range, and variance for kernel size of each genotype sample.

Plant water status can be readily assessed from canopy temperatures using handheld, infrared thermometers (Olivares-Villegas et al., 2007) or through stomatal conductance values from viscous-flow porometers (Rebetzke et al., 2001b; Kirkegaard et al., 2007). Thermal imaging and the availability of infrared cameras provide good resolution around leaf temperatures (Chaerle et al., 2007). Indirect estimates of photosynthetic capacity may be obtained through leaf nitrogen or chlorophyll content from a soil–plant analytical development (SPAD) chlorophyll meter and together with measures of stomatal conductance may provide indirect measures of transpiration efficiency.

Despite the success in trait identification for implementation in breeding, there are few documented successes where physiological traits have been integrated into wheat breeding programs. A number of factors may be limiting adoption, with

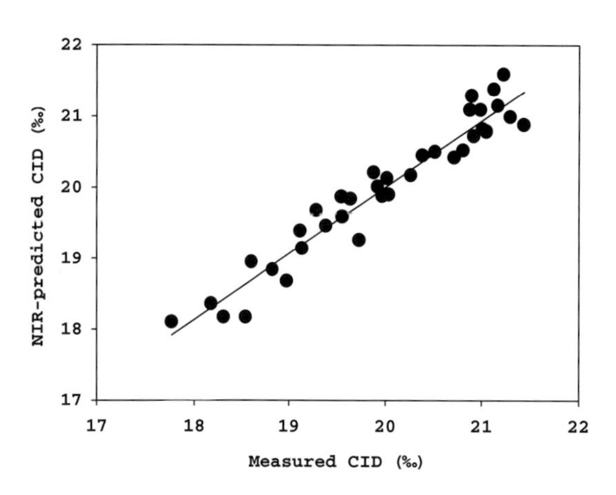

Fig. 11.6 Plot of carbon isotope discrimination (CID) measured on wheat leaves for three cultivars assessed across a range of nitrogen treatments, sites, and plant densities against CID predicted using a NIR calibration (Y = 1.3 + 0.93X, r^2 = 0.93).

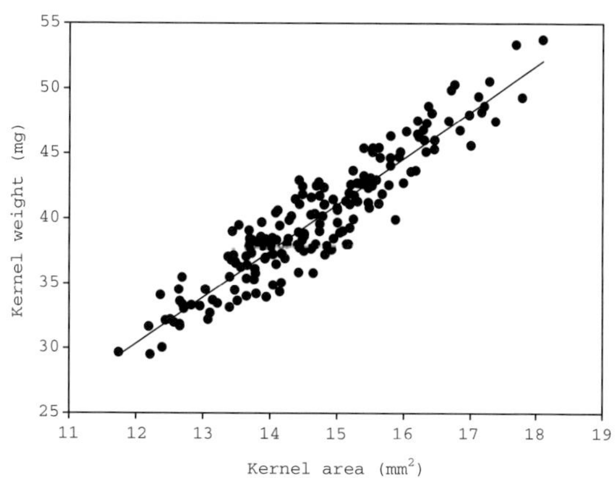

Fig. 11.7 Relationship of kernel area and kernel weight for 191 doubled-haploid progeny from the Sunco/Tasman mapping population. Kernel area was determined from approximately 300 harvested seed (per line) scanned into a single image using a flatbed scanner. The area was determined using AnalySis image analysis software (Y = −12 + 3.6X, r^2 = 0.84).

some of the reasons being considered earlier and in Richards (2006). However, of equal importance is the need for breeding teams that integrate skills from physiology to genetics and breeding. Such complementation currently exists for pathologists and breeders but should also be extended to physiologists if meaningful selection for physiological traits is to be implemented, as in the case of the commercial release of the high transpiration efficiency wheat cultivar Drysdale (Fig. 11.8). Finally, the value of accurate phenotyping is not restricted to breeding programs and is considered a critical requisite to good QTL analysis.

Quantitative trait loci

Little is known or understood of the genetic basis of wheat performance under drought. Substantial progress has been made on the development and availability of molecular markers and their subsequent integration into genetic maps across a number of genotyped populations (Gupta et al., 2008). The use of molecular techniques is complementary to the use of physiological approaches in breeding wheat with improved performance under drought. Further, a range of different

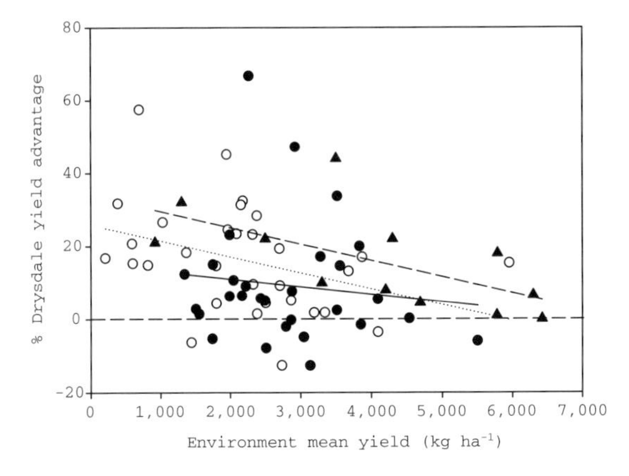

Fig. 11.8 Yield advantage of low carbon isotope discrimination–selected cultivar Drysdale grown side-by-side with its recurrent parent 'Hartog' at 31 sites in 2004 (○), 29 sites in 2005 (●), and 12 CSIRO sites (▲) from 1995 to 1998. Mean grain yield advantage was 16%, 9%, and 17% for Australian Grains Technology Breeding Company (AGT) 2004 and 2005, and CSIRO trials, respectively. (AGT data provided courtesy of Steven Jeffries.)

molecular markers provides greater opportunity for trait dissection into component QTLs, and insight into the underlying genetic basis for variation and covariation in a trait. For example, numbers of genes and their interactions, and gene action, can be determined even for low-heritability traits. Information around breeding complexity including linkage, linkage disequilibrium, and pleiotropy may also be gleaned through mapping studies, particularly if extended to multiple populations (Rebetzke et al., 2007a). Key processes involved in plant growth and development can then be inferred from what is already known of the physiology.

For example, development of sequence-based perfect markers for the *Rht-B1b* and *Rht-D1b* dwarfing genes has confirmed their effects on tissue insensitivity to endogenous gibberellins (Ellis et al., 2002), and subsequent reductions in both coleoptile and early leaf area development (Rebetzke et al., 2001a). Currently, PCR-based markers are being developed for gibberellin-sensitive dwarfing genes *Rht4*, *Rht5*, *Rht12*, and *Rht13* for breeding of long coleoptile, high early-vigor wheat with short stature (Ellis et al., 2004).

Perhaps the greatest opportunity for the use of linked markers in breeding for improved performance under drought will apply to selection for traits with low heritability, recessive gene expression, or high cost of measurement, such as root growth (Tuberosa et al., 2002) or WSC concentration (Rebetzke et al., 2008b). In the absence of recombination, markers linked to a target QTL have a heritability of 100%, thereby allowing rapid generation advance. From the indirect selection formula (equation 11.1), the benefit of using a 100%-heritable molecular marker over direct selection for grain yield (i.e., $\Delta_{GY.X} / \Delta_{GY}$) is r_A / h_{yield}. This can be extended to multiple QTLs so that the sum of absolute genetic correlations with yield must exceed h_{yield}. Assuming the marker is diagnostic (robust marker and gene association), and evaluation and generation costs are similar to that of yield, marker-aided selection in variable environments should produce greater genetic gain for yield. However, this is likely to be limited to traits controlled by fewer QTLs and to the likelihood of identifying genotypes containing all or most QTLs in smaller populations commonly used by breeders (Bonnett et al., 2005). Enrichment strategies may aid in overcoming large numbers of QTLs in biparental- and backcross-based breeding programs.

Marker-assisted selection is likely to provide the greatest benefit to breeders targeting novel traits from distantly related germplasm. For example, markers are extremely useful in identifying recombinants when targeting positive genes linked in repulsion to genes of negative effect. Several QTLs have been identified with potential for selection of improved performance in water-limited environments (Table 11.2). Some of these reflect major genes with known effects on improved root growth (e.g., saline tolerance or root disease resistance) to increase water use, while the functional basis of others is not well understood (e.g., Kirigwi et al., 2007).

Critical to successful implementation of linked markers in breeding are well-defined QTLs for improved productivity. Experimental populations under evaluation should be representative of the target populations under selection. Genotyping should be adequate in development toward a good molecular map with even marker coverage and quality mapping (good markers and map development); phenotyping should be sound and

Table 11.2 Traits with potential for improving wheat performance in water-limited environments. Details are also provided on genetic control for each trait.

Trait	Ease of Screening	Heritability	Chromosomal Location of Genes	Source
Root health				
Aluminum tolerance	Simple	High	4D	Delhaize et al. (2004)
Boron tolerance	Difficult	High	7B, 7D	Jefferies et al. (2000)
Cereal cyst nematode	Difficult	Low	2B, 2D	Lagudah et al. (1997)
Salt stress				
Sodium exclusion	Simple	High	2A, 4D	Huang et al. (2006)
Sodium tolerance	Difficult	High	5A	Byrt et al. (2007)
Drought stress				
Phenology	Simple	High	2A, 2B, 2D, 3A, 3B, 5A, 5B, 5D, 6A, 6B, 7A, 7B	Snape et al. (2001)
Osmotic adjustment	Difficult	Moderate	7A	Morgan and Tan (1996)
Carbon isotope disc. (leaf)	Difficult	High	1B, 1D, 2D, 3B, 4A, 4B, 4D, 5A, 7A, 7B	Rebetzke et al. (2008b)
Carbon isotope disc. (grain)	Difficult	High	1D, 2A, 2D, 4B, 4D, 6D, 7B	Rebetzke et al. (2008b)
Canopy temperature	Simple	Moderate	1B, 2B, 3B, 4A	Pinto et al. (2008)
Stem carbohydrates	Difficult	Moderate	1A, 2B, 2D, 3B, 4B, 5B, 6B, 7A, 7B	Rebetzke et al. (2008a)
Stomatal conductance	Difficult	Low	1B, 2A, 2B, 2D, 4A, 4B, 4D, 7A, 7B	G.J. Rebetzke, unpublished data
Glaucousness	Simple	Moderate	2B, 2D	Tsunewaki and Ebana (1999)
Staygreen	Difficult	Moderate	2B, 2D	Verma et al. (2004)
Leaf rolling	Simple	High	Unknown	Sirault et al. (2008)
Early vigor	Simple	High	2D, 4B, 4D, 5A	Rebetzke et al. (2001a)
Coleoptile length	Simple	Moderate	2B, 2D, 4A, 4B, 4D, 5D, 6B	Rebetzke et al. (2007b)
Harvest index	Difficult	High	2B, 2D, 4B, 4D	Ellis et al. (2002)
Photosynthetic capacity[a]	Simple	Moderate	1B, 1D, 2D, 3B, 4A, 4B, 4D, 5B, 6B, 7A, 7B	G.J. Rebetzke, unpublished data
Restricted-tillering	Simple	High	1A	Spielmeyer and Richards (2004)
Rate-of-grain-filling	Difficult	Uncertain	Unknown	Whan et al. (1996)
Root biomass	Difficult	Low	1B	Waines and Ehdaie (2007)

[a]Surrogates for photosynthetic capacity (SPAD, SLW, SLN, leaf N content).

repeatable, and be performed in environments representative of the target environment. Among the various marker types available, Diversity Array Technology markers (DArTs) show good potential as an inexpensive high-throughput marker tool and provide opportunity toward whole-genome mapping (Gupta et al., 2008). Finally, methods are being developed aimed at efficient marker implementation strategies in breeding programs. These strategies may vary depending on breeding program structure and goals, genetic complexity of traits, and cost and type of markers (Wang et al., 2007).

Functional genomics and beyond

Water stress triggers a wide variety of plant responses, including biochemical, physiological, and cellular structure changes in plant cells. High-throughput analysis of messenger RNA expression—made possible through advances in microarray technology—enables temporal, spatial, and genotypic variation in levels of gene expression to be studied for a very large number of genes in response to water stress. Knowledge of the genes involved in the drought response is important to understanding the underlying biochemical and physiological basis of traits and of QTLs involved in traits which will indirectly influence breeding strategies. Candidate genes can be used as perfect markers for the traits they influence, and allelic variants of these candidate genes may be readily converted to high-throughput, single-nucleotide polymorphism (SNP) markers (Cogan et al., 2006; Tuberosa and Salvi 2006). Finally, candidate genes may have potential as transgenes; manipulation of expression of these candidate genes may provide enhanced drought tolerance and increased yields of wheat in stress environments (see Umezawa et al., 2006 and references therein).

High-throughput analysis of messenger RNA (mRNA) expression has revealed that a large number of genes are differentially expressed in plants in response to water stress. Two broad categories of genes are involved in response to water stress: functional genes that protect against the stress and regulatory genes that control signal transduction and gene expression (Umezawa et al., 2006). The first group of genes includes osmoprotectants, chaperones, reactive oxygen species (ROS) scavengers, turnover transport proteins, membrane modifiers, and detoxification proteins. The second group of genes includes signaling molecules and transcription factors.

Transcriptomics enables high-throughput investigation of changes in mRNA expression levels, for example in response to drought. Proteomic and metabolomic profiling enable investigation of the effects of post-transcriptional and post-translational regulation, which complements information generated by mRNA profiling and provides further insight into the response of plants to water stress. For example, proteomic studies in wild watermelon (*Citrullus lanatus*) (Yoshimura et al., 2008) suggest that proteins produced early in drought stress are involved in root morphogenesis and carbon–nitrogen metabolism, while proteins involved in lignin and cell wall synthesis are switched on later. Proteomic and metabolomic studies of maize xylem sap also found changes in levels of phenylpropanoid compounds under drought stress (Alvarez et al., 2008).

Genetical genomics, the integration of genetic recombination and the raw power of genomics (Jansen and Nap 2001), assists in the elucidation of the genic basis of drought response traits, in the identification of candidate genes suitable for use as markers, and in the identification of regulatory genes controlling traits of interest. By using structured germplasm in genomic studies, gene expression variation is more likely to have a genetic basis that can be subsequently exploited in cultivar development programs (Xue et al., 2006, 2008). Genes that are differentially expressed in response to water stress that collocate with QTLs for drought-related traits are more likely to contribute to the expression of that trait. Furthermore, levels of expression of genes can be mapped; collocation with QTLs for drought-relevant traits can indicate a genetic role of a functional or regulatory gene in the expression of the trait. A recent study by Jordan et al. (2007) identified both *cis*-acting (functional gene) and *trans*-acting (regulatory gene) expression QTLs (eQTL) for traits such as grain weight, maturity, and quality traits in wheat. Furthermore, genetical genomics studies using the phenotypic tails of a segregating population resulted in a suite of potential candidate genes for two important drought-related traits, transpiration efficiency (Xue et al., 2006) and WSC concentration (Xue et al., 2008), that were enriched for likely candidate genes. These genes included those involved in growth, photosynthesis, drought-response, and carbohydrate metabolism, some of which collocated with QTLs for these traits.

There have been numerous successful attempts to manipulate expression of functional genes and genes involved in signal transduction, as a first step in evaluating their role in plant response to water stress. Manipulation of functional genes related to osmolyte metabolism, stress-responsive proteins, ROS-scavenging proteins—and of other genes in a range of plant species including Arabidopsis (*Arabidopsis thaliana*), tobacco (*Nicotiana tabacum*), rice (*Oryza sativa* L.), and wheat (see Umezawa et al., 2006 and references therein)—have resulted in enhanced dehydration or desiccation tolerance, "survivability," or "drought-tolerance" in laboratory-based assays. Transcription factors (TFs) are also postulated in the drought response through regulation of expression of downstream target genes, especially members of large TF families, such as AP2/ERF (including DREB/CBF), bZIP (including AREB/ABF), NAC, MYB, MYC, Cys2His2 zinc-finger, and WRKY (see Umezawa et al., 2006, and references therein). Overexpression of TFs of different TF families has resulted in enhanced levels of survivability in model plant species (e.g., Arabidopsis), as well as a few studies in wheat and tomato (*Solanum lycopersicum*). These TFs can act to enhance or repress transcription. Several genes encoding signaling factors that function in drought response have also been identified (see Umezawa et al., 2006, and references therein). Manipulation of these genes has resulted in enhanced survivability and frequently tolerance to more than one stress in laboratory-based assays.

A key issue that arises from these functional genomics studies is the nature of the phenotypic evaluation of transgenic plants or the drought screening protocol used to generate the material for RNA isolation. Most of the previously mentioned studies report evaluation of survival in model species Arabidopsis, tobacco, and rice using laboratory-scale drought screening under fast-developing water stress. It is likely that water-stress conditions in the field will generate very different gene expression responses versus those in laboratory-grown material (Passioura 2006b), and thus the relevance of these gene expression changes to actual productivity in the field is unknown. Furthermore, survivability and/or tolerance of severe tissue desiccation have not yet been associated with increased productivity in the field. Therefore considerable care must be taken with the conclusions drawn from laboratory-scale experiments prior to validation in field-grown material. More recently, functional genomics studies have increasingly sourced field-grown material for laboratory studies, or undertaken field evaluation of transgenic lines. For example, in Xue et al. (2006, 2008), RNA was isolated from tissue collected from field-grown material for functional genomics assays to identify candidate genes for drought-related traits, WSC and CID.

More recently, field performance of transgenic crop plants was evaluated; transgenic maize (*Zea mays*) plants overexpressing a NF-Y TF or an osmolyte maintaining gene, vacuolar H^+-translocating inorganic pyrophosphatase (H^+PPase), were shown to confer a grain yield advantage in water-limited field environments (Li et al., 2007; Nelson et al., 2007), while transgenic wheat (Bahieldin et al., 2005) and rice (Xiao et al., 2007) plants overexpressing a stress-responsive, stress-protective, late embryogenesis abundant (LEA) protein, were found to increase yield in water-stressed field environments. While these few studies are promising, the variable nature of rainfed, water-limited environments and the polygenic control of traits with a proven role in increasing production in water-limited environments suggests that manipulation of single genes is unlikely to be broadly applicable to increasing crop productivity under drought.

DEFINING THE BREEDING TARGET

The multifaceted nature of drought and the requirement for improved efficiency of water use dictate a reliance on multiple traits that account for the nature, timing, and variability of water limitation in the target environment. Empirical selection for yield in dry environments has led wheat breeders in many parts of the world to develop well-adapted wheat cultivars primarily through selection for earlier flowering and reduced height (Richards et al., 2002; Slafer et al., 2005). As alleles for appropriate phenology and height become fixed, the opportunity then arises for identification and selection of alleles for new traits such as greater early vigor, increased transpiration efficiency, or greater remobilization of stored stem carbohydrates, which extend adaptation to water-limited environments.

In its simplest form, grain yield (GY) reflects the partitioning of dry matter to grain and can be expressed mathematically as:

$$GY = TDM \times HI,$$

in which TDM is the total aboveground dry matter or biomass at harvest maturity, and HI is harvest index, a measure of the total dry matter allocated to grain. Genetic advances in grain yield have largely been achieved through increases in harvest index (Austin et al., 1980; Perry and D'Antuono 1989; Shearman et al., 2005; Zhou et al., 2007). Increases in harvest index have been attributable to changes in phenology (anthesis date) and incorporation of height-reducing genes, including the major *Rht-B1b* and *Rht-D1b* dwarfing genes promoted widely during the Green Revolution. The yield component most affected through this change is kernel number (Perry and D'Antuono 1989; Shearman et al., 2005; Zhou et al., 2007). There appears to be a limit to the proportion of biomass partitioned to the grain, and this value is estimated to be 50% to 60% (Austin et al., 1980). Thus future genetic gain in yield through empirical breeding is likely to arise through subtle changes in harvest index, coupled to an increase in total biomass.

There is some evidence that genotypic variation for total biomass exists in cultivated (Austin et al., 1980; Shearman et al., 2005) and novel genetic sources (e.g., chromosomal segments 1BL.1RS or 7DL.7Ag—Foulkes et al., 2007). However, their effects are commonly small, making selection difficult. Phenotypic selection for biomass is challenging as it is difficult to measure accurately and has low heritability, especially in early, segregating generations of a breeding program (Rebetzke et al., 2002).

Processes of yield determination can be further partitioned to reflect use of available soil water, as done in the widely used model first enunciated by Passioura (1977). The model describes grain yield determination under water-limited conditions in terms of water use or evapotranspiration (ET), efficiency of water use (WUE), and harvest index:

$$GY = ET \times WUE \times HI$$

Crop WUE can be further partitioned into two components reflecting the portion of total water transpired by the crop (T/ET) and the transpiration efficiency of biomass production (W). The resulting framework is given as:

$$GY = ET \times T/ET \times W \times HI$$

This is not the only possible partition of grain yield, as others have been developed to reflect grain yield components (e.g., kernel number and size), and others around radiation capture and conversion to dry matter. The benefit in the Passioura framework reflects the broad processes by which crops actually achieve yield in water-limited environments (Condon et al., 2004). A number of characteristics have been identified affecting ET, T/ET, W, and HI, and thereby yield improvement under drought. Many of these traits are physiologically independent, allowing genetic effects to be accumulated through selection (e.g., early vigor to increase T/ET and CID to increase W; Condon et al., 2004). However, as indicated in Condon et al. (2004), traits affecting W have the potential to affect other components in the framework. Some of these traits will now be discussed in greater detail.

Increasing water uptake

Increases in crop ET are strongly associated with increased grain yield in wheat. For example, across approximately 200 irrigated and dryland wheat experiments grown in the southern US, Musick et al. (1994) reported a correlation of ET and grain yield of 0.87. In turn, increased water uptake is an obvious means for increasing productivity in droughted environments. A larger root system has the potential to access greater soil moisture, while allowing increased access to soil nutrients and reducing root lodging through increased plant anchorage. Smaller root systems, particularly on coarse-textured soils early in the growing season, may intercept less moisture and soluble nutrients, which in turn increases deep drainage and nutrient leaching, affecting water-table quality and long-term environmental sustainability (Passioura 2002).

Despite the potential benefits of increased root growth, few wheat studies have demonstrated a yield benefit with increased water uptake in water-limited environments. Crop rotations, particularly with the use of *Brassica* species such as canola, reduce the incidence of soilborne diseases, increasing water use and yield of the following wheat crop (Kirkegaard et al., 1994). Using a simulation model, Manschadi et al. (2006) showed that delaying water uptake from deep in the soil profile increased water availability for postanthesis grain growth and improved grain yield. In carefully managed field studies, Kirkegaard et al. (2007) showed that extraction of an additional 11 mm of subsoil moisture from soil depths exceeding 1.3 m contributed an extra 0.6 tonne of grain per hectare. Marginal WUE of this additional yield was 59 kg ha^{-1} mm^{-1}, approximately three times greater than for water used earlier in the season. The value of deep rooting in extracting subsoil moisture was then assessed with simulation modeling of wheat grown at multiple sites and using over 100 years of historical rainfall data (Lilley and Kirkegaard 2007). Extraction of moisture from below 1.2 m provided a yield benefit of 0.1 to 0.6 t ha^{-1} and marginal WUE of 30 to 36 kg ha^{-1} mm^{-1}. The high marginal WUE reported in Kirkegaard et al. (2007) was rarely observed and restricted only to high-rainfall seasons. The authors conclude there is real value in development of deep roots in increasing post-anthesis water use and yield but only in deeper soils and in seasons where adequate rainfall will provide subsoil moisture.

Repeatable genotypic variation has been reported for factors contributing to greater root length and biomass in wheat (Hurd 1968). In many cases this variation reflects adaptation to hostile subsoils. For example, genes have been identified and undergone selection for traits controlling resistance to root disease (Lagudah et al., 1997) and other soil constraints, including soil salinity (Byrt et al., 2007) and soil acidity (Delhaize et al., 2004). Genotypic variation has also been reported for the ability of wheat roots to penetrate hard soil pans (Botwright et al., 2007), to produce secondary branching for increased nitrogen uptake (Palta et al., 2007), and to access soil moisture at greater depth (Hurd 1968). Selection for narrower xylem vessel diameter has the potential to increase hydraulic resistance and delay water use to later in the season. Richards and Passioura (1989) backcrossed genes from a Turkish landrace into two commercial wheat backgrounds to reduce seminal root xylem diameter from about 65 to 55 µm and increase grain yield between 3% and 11% in water-limited environments.

Surveys across a range of field studies indicated wheat roots grow downward an average of 0.8 to 1.1 cm per day, with final rooting depth depending on the duration of the vegetative phase, soil wetness, and soil type (Lilley and Kirkegaard 2007). Conceivably then, a 2-week delay in sowing date has the potential to reduce rooting depth by 10 to 15 cm. Selection for the different flowering genes (e.g., *VRN_*, *PPD_*, and *EPS*) will allow development of cultivars that can be sown earlier to extend the duration of the vegetative phase without changing anthesis date (Richards et al.,

2007). Similarly, genetic factors influencing leaf area duration (e.g., stay-green and leaf disease resistance) may also aid in extending the period for assimilation and continued root growth and nitrogen uptake (Gregory et al., 2005).

Other genetic factors affecting aboveground growth have been reported to strongly influence shoot–root partitioning and root growth. For example, selection for greater early growth (seedling vigor) is associated with greater root growth early in the season (Palta et al., 2007). Figure 11.9 shows relatively reduced root-length density and root number for low-vigor selection F25 compared to its high-vigor, full-sib F20 and parent 'Vigour 18'. This difference was maintained with increasing soil depth. Similarly, Fig. 11.10 illustrates increased seminal root number and greater root length via selection for greater early vigor in a recurrent selection program targeting greater shoot vigor. Genetically vigorous wheats

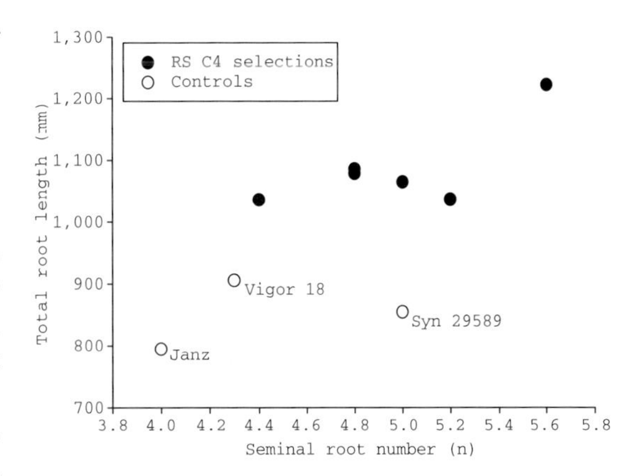

Fig. 11.10 Total root length and seminal root number measured at the one-leaf stage on six high-shoot-vigor cycle 4 recurrent selection lines (●) and controls (○): high shoot vigor, 'Vigour 18', low shoot vigor, 'Janz', and high root vigor, synthetic selection 29589.

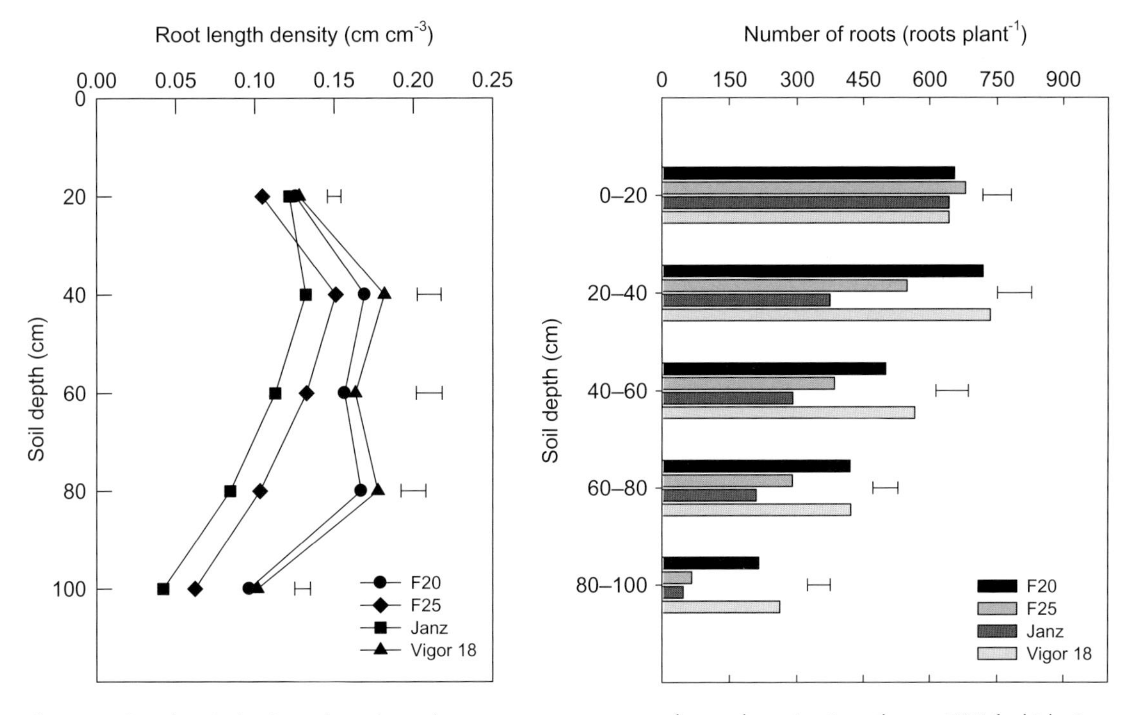

Fig. 11.9 Root-length density and numbers of roots at commencement of stem elongation (growth stage Z31) for high-vigor wheat parent 'Vigour 18' and progeny (high- and low-vigor selections F20 and F25, respectively), and low-vigor cultivar Janz. Leaf areas at Z31 were 267, 224, 179, and 170 cm^2 per plant for Vigour 18, F20, F25, and Janz, respectively. The least significant difference is given for testing genotypic differences at each root depth (Palta et al., 2007).

have also been demonstrated to produce greater root biomass when grown on unploughed soils of high bulk density typical of direct-drilled tillage systems (Watt et al., 2005).

Another interesting aboveground effect is associated with the restricted-tillering (*tin*) gene in wheat. Studies undertaken using near-isogenic lines varying for the *tin* gene show increased partitioning of assimilate to roots for lines containing the restricted-tillering gene (Fig. 11.11). The pleiotropic effect of the *tin* gene and change in root growth are consistent across genetic backgrounds and repeatable across a range of environments. However, not all genetic factors influencing tillering affect root growth. For example, Palta et al. (2007) contrasted lines lacking the *tin* gene but varying for tiller number and found no relationship for tillering ability and root growth.

Substantial diversity exists for root vigor and components in wheat (Manschadi et al., 2006; Palta et al., 2007). However, root traits are difficult and costly to measure, and they commonly show low repeatability (O'Toole and Bland 1987). There are few reports of inheritance for root growth, and with the exception of Richards and Passioura (1989), there is little demonstration of selection aimed at modifying root parameters not associated with overcoming subsoil constraints. Looking forward, perhaps the three greatest challenges confronting breeders and physiologists alike will be (i) identification of appropriate root-system architecture for the target environment(s), (ii) development of appropriate and cost-effective phenotyping methodologies, and (iii) development of robust molecular markers for use in breeding. Nevertheless, there are encouraging signs for selection of modified root systems in other species. For example, thicker and longer roots are considered important in rainfed upland rice systems. A marker-based selection program aimed at development of upland rice with greater root length and thickness produced lines with increased root length in irrigated and rainfed environments (Steele et al., 2007). However, no yield advantage was observed with any of the derived lines.

Stem carbohydrate production

Grain filling in cereals relies on carbon fixed during postanthesis assimilation and transported directly to grain, and carbon remobilized from assimilates stored in vegetative tissues (Schnyder 1993). Under conditions favorable for grain filling, up to 90% of grain dry weight is derived from

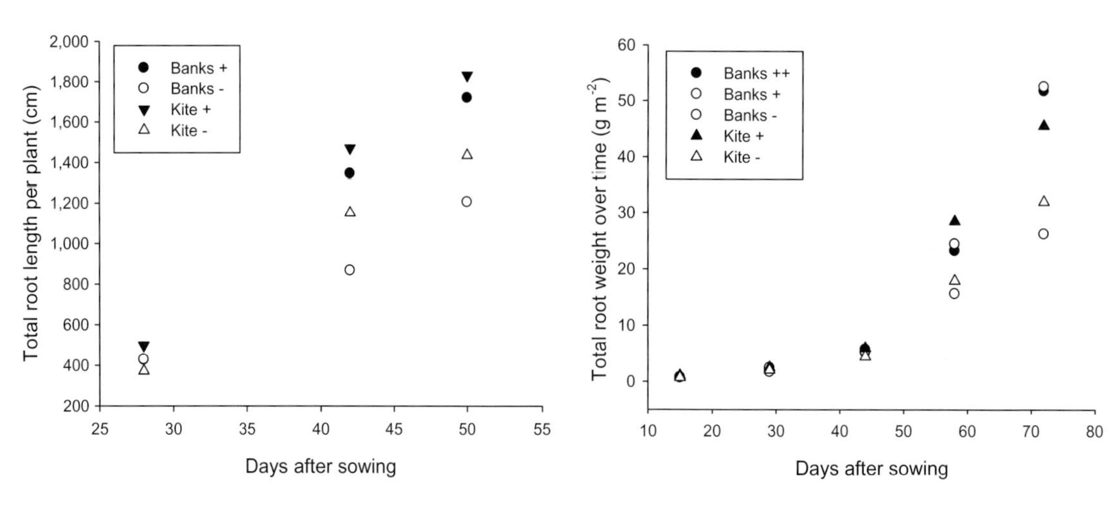

Fig. 11.11 Genotypic differences in root parameters for reduced-tillering (+ and ++) and free-tillering (–) 'Banks' and 'Kite' near-isogenic wheat lines evaluated in a cool glasshouse (left) and with field sowing (right). Presence of the reduced-tillering *tin* gene was associated with significant differences in root growth 43+ days after sowing in the glasshouse, and 58+ days after sowing in the field (Hendriks 2004).

postanthesis assimilation (Bidinger et al., 1977). However, conditions in rainfed environments are rarely favorable as drought (and disease) can inhibit photosynthesis and/or decrease leaf area to reduce carbon supply to developing grain. This reduced carbon supply commonly results in lower grain yields and the production of small and shrivelled kernels that devalue the crop and reduce returns to growers (Ruuska et al., 2006).

Assimilated carbon can accumulate temporarily in the stem and leaf sheath during the later growth stages of cereals. This nonstructural carbohydrate pool accumulates prior to, during, and after anthesis, and can be later remobilized and transported to developing grain (Schnyder 1993). In wheat, this stem reserve carbohydrate is stored in the form of WSC, which principally consists of fructosyl-oligosaccaharides (fructan) (Ruuska et al., 2006). Total WSC may attain levels of more than 40% of total stem dry weight in wheat, and its remobilization can make a significant contribution to final grain yield and kernel size (Schnyder 1993 and references therein). This contribution varies from season to season and can be as high as 30% to 50% of total grain weight when conditions are unfavorable for photosynthesis and between 10% and 20% when conditions are favorable (Bidinger et al., 1977; Schnyder 1993; van Herwaarden et al., 1998).

The importance of stem WSC reserves under drought-stress conditions has been demonstrated in the field (van Herwaarden et al., 1998; Foulkes et al., 2002; Yang et al., 2007; Rebetzke et al., 2008b) and predicted in crop modeling studies (Asseng and van Herwaarden 2003). Genotypic variation for WSC is large and repeatable across diverse environments in wheat (Fig. 11.12) (Ruuska et al., 2006). Breeding-era studies have demonstrated increases in WSC concentration with release of new wheat cultivars for some Australian wheat breeding programs (e.g., Western Australia in Fig. 11.13) and at CIMMYT in Mexico (Fig. 11.14) (van Herwaarden and Richards 2002). However, no relationship was observed for release of wheat cultivars in eastern Australian breeding programs (Fig. 11.13). Release of higher-yielding, UK wheat cultivars

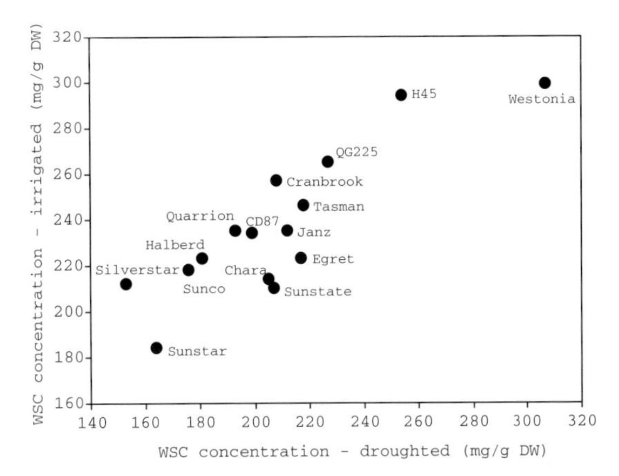

Fig. 11.12 Relationship of mean water-soluble carbohydrate (WSC) concentration at early grain-fill for different Australian wheat genotypes sampled at water-limited and irrigated sites ($Y = 86 + 0.73X$, $r^2 = 0.74$) (Rebetzke et al., 2008b).

from 1972 to 1995 was associated with increasing WSC concentration (Shearman et al., 2005). The environment in Western Australia is typically Mediterranean, so that the crop grows on current rainfall. It would be advantageous in this environment to develop large anthesis biomass with the potential to remobilize a large portion of this dry matter to grain (Fischer 1979).

In studies undertaken across multiple environments, Rebetzke et al. (2008b) observed positive associations (r_g = 0.27–0.43) between WSC concentration and grain yield, and between WSC concentration and kernel size (r_g = 0.60–0.62), in three unrelated wheat populations. Lines high for WSC concentration (milligrams WSC per gram dry weight) commonly were earlier for flowering, were reduced in plant stature, and produced fewer tillers per plant, leading to reduced anthesis and final plant biomass (Rebetzke et al., 2008b). Nitrogen content was also lower for high-WSC lines, consistent with high WSC measured in wheat grown with low fertilizer nitrogen (van Herwaarden et al., 1998). Lines actually selected for high WSC content (i.e., WSC per ground area, or g m^{-2}) produced larger kernel size and higher grain yields but maintained higher tiller numbers, leading to increased

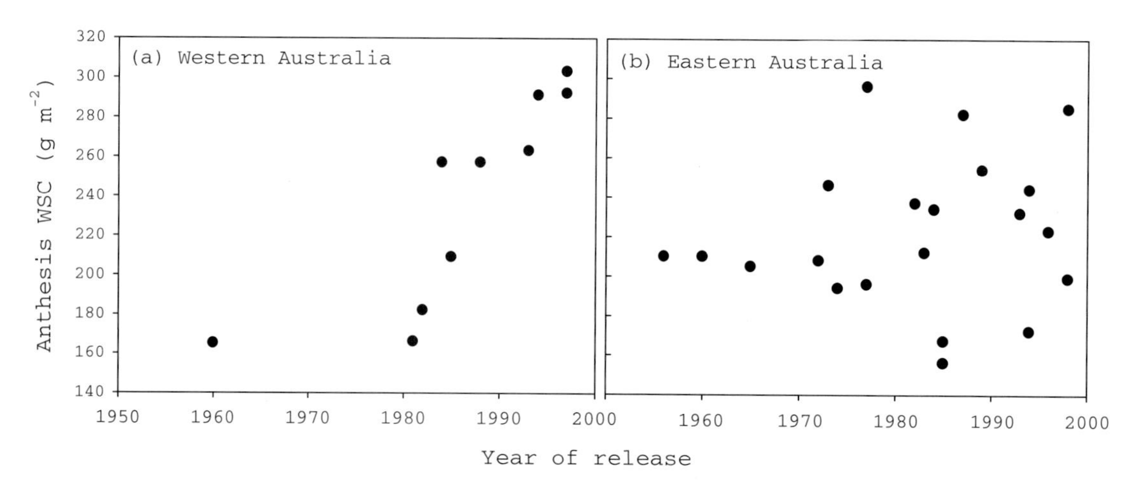

Fig. 11.13 Water-soluble carbohydrate content measured at anthesis for (a) Western Australian and (b) Eastern Australian wheat cultivars released over 40 years and evaluated together in the field.

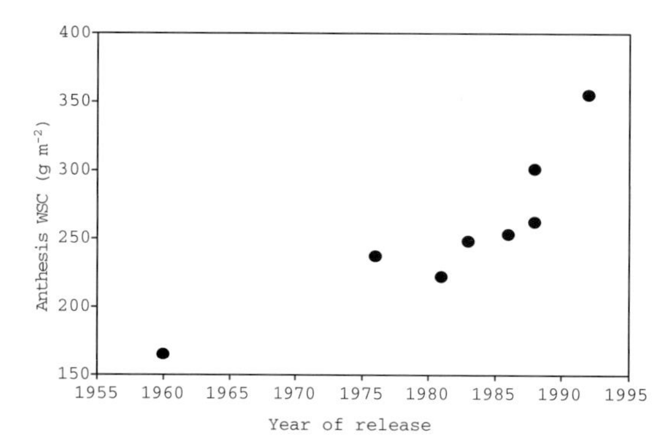

Fig. 11.14 Water-soluble carbohydrate content measured at anthesis for CIMMYT wheat cultivars released over 40 years and evaluated together in the field.

anthesis and final biomass. Days to heading, plant height, and nitrogen contents were similar for low- and high-selected WSC content lines (Rebetzke et al., 2008b). Therefore, it is important that, when selecting, breeders distinguish between genotypic differences in WSC concentration and WSC content.

Genetic studies have characterized WSC concentration with low to moderate genotype × environment interaction, which may lead to higher heritability and expected selection response (Rebetzke et al., 2008b). Extension to WSC content showed lower heritability, owing to larger errors when sampling anthesis biomass. The few reported studies have differed markedly in the number of QTLs for WSC. For example, Yang et al. (2007) reported a single QTL for WSC measured 14 days after flowering. This contrasted with Rebetzke et al. (2008b), who identified between 8 and 16 QTLs across 3 populations for WSC concentration sampled at the same developmental stage. Table 11.2 shows QTLs common to two or more populations in Rebetzke et al. (2008b). Physiologically these QTLs appeared robust, as many of them collocated with QTLs for WSC content, WSC per tiller, and plant nitrogen content (Fig. 11.15).

Other opportunities exist for selection of stem carbohydrates in wheat populations. The use of NIR provides a robust assessment of WSC without the need for expensive chemical analysis (Fig. 11.5). Other workers (Blum 1998 and references therein) have demonstrated the potential for large-scale population screening of remobilized WSC using chemical desiccants to remove green leaf area.

Tiller production

Current wheat cultivars initiate more tillers than they can sustain. Cool, moist conditions and high nutrient levels soon after emergence are very favorable for tiller initiation. Typically each plant produces between 5 and 20 tillers at stem elonga-

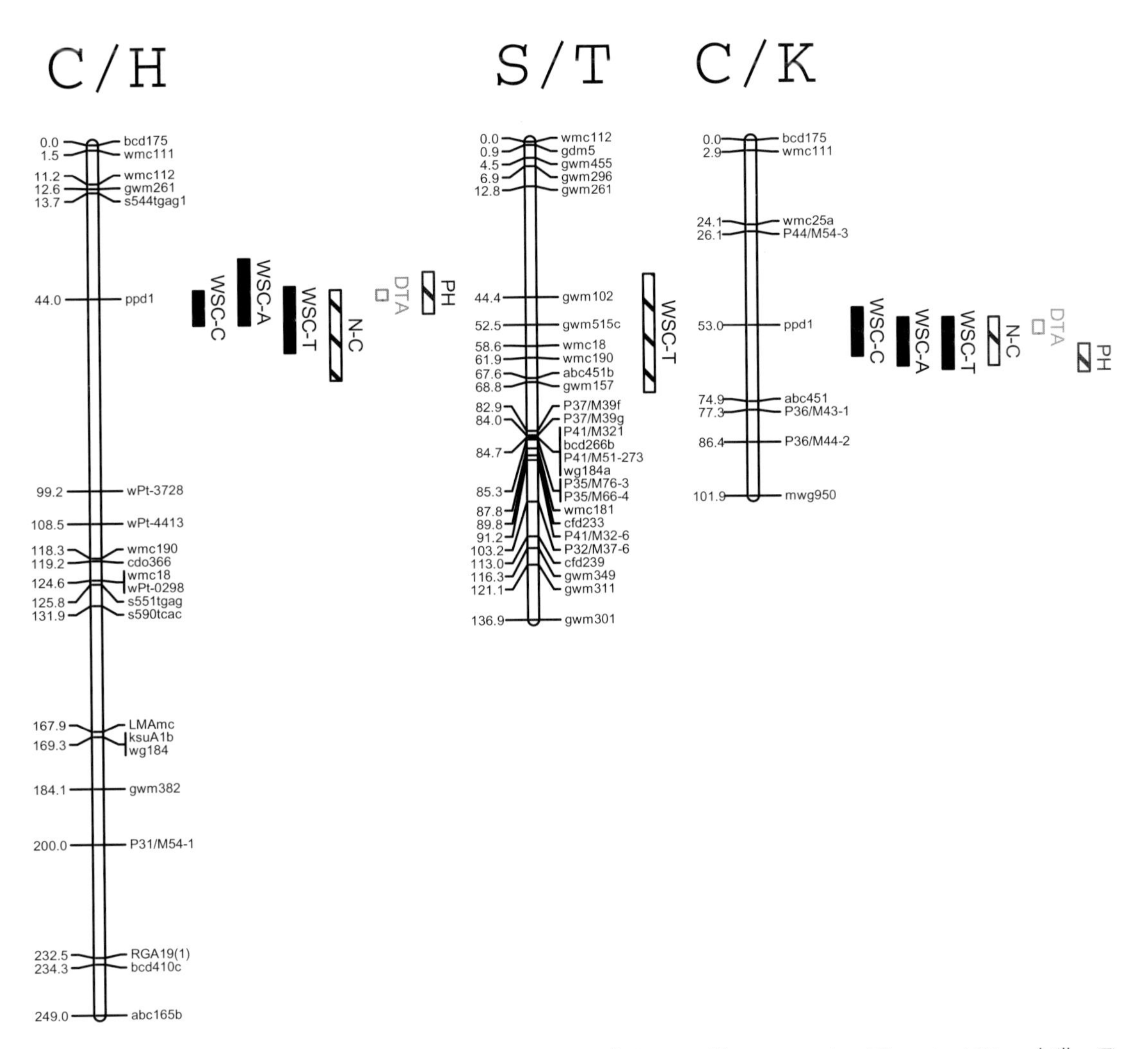

Fig. 11.15 Collocation of water-soluble carbohydrate (WSC) and nitrogen (N) concentration (C), content (A), and tiller (T) QTL measured 180 ºC d postanthesis with the chromosome 2DS *Ppd1* varying in the Cranbrook/Halberd and CD87/Katepwa populations but not the *Ppd1*-absent Sunco/Tasman population (DTA = days to anthesis and PH = plant height). The QTLs were derived from measurements in three to seven environments.

tion with a maximum of 2 to 8 surviving to produce fertile spikes at maturity (Fig. 11.16). The numbers of tillers surviving to maturity is dependent on genotype and environmental conditions (including agronomic considerations such as nutrition, sowing density, and arrangement) (Duggan et al., 2005). Development of drought prior to, and after, stem elongation commonly lowers tiller survival, leading to reduced spike number (Hendriks 2004). Tiller senescence represents a waste of water, critical under water-limited conditions while dying tillers remobilize little carbon to surviving spikes (Berry et al., 2003). Further, in reliably water-limited environments, retention of large tiller numbers can contribute to excessive water use (Islam and Sedgely 1981). Together, these indicate the need to control excessive tiller production when water is limiting, though in some regions tillering enhancement may be a desirable target to support the dual

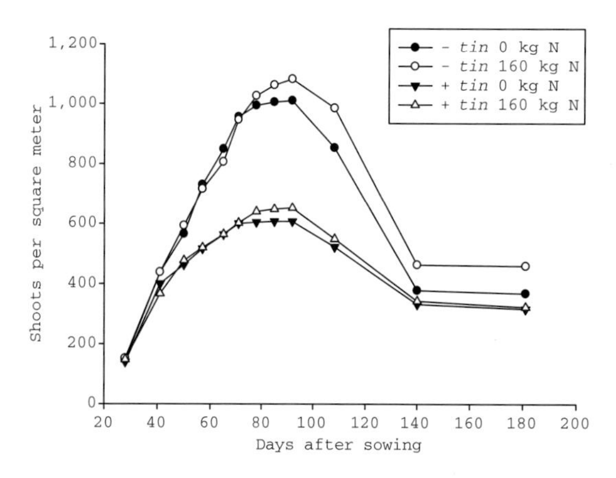

Fig. 11.16 Change in shoot number over time for sister lines with (+) and without (–) the tiller inhibition (*tin*) gene when supplied with starting fertilizer and then either 0 or 160 kg ha^{-1} additional fertilizer (Duggan et al., 2005).

production of wheat pasture and grain in a single cropping season.

De-tillering studies by Jones and Kirby (1977) and Islam and Sedgley (1981) indicate the potential for increasing grain yield through reductions in the number of nonsurviving shoots. The ability to genetically control tiller number has major potential in managing target tiller and spike number for a given environment or management regime. Genotypic variation exists to modify tiller number. For example, simply reducing the period from sowing to terminal spikelet to hasten flowering can reduce tiller number (Rebetzke et al., 2008b). A major tiller inhibition gene (*tin*) has been located linked to glume pubescence (*Hg*) and the microsatellite marker *Xgmw136* on chromosome 1AS (Spielmeyer and Richards 2004). This gene reduces potential tiller number from as many as 15 to between 1 and 5 tillers per plant depending on genetic background (Fig. 11.16) (Hendriks 2004; Duggan et al., 2005). Under water-limited conditions, near-isogenic lines containing the *tin* gene produced greater WSC, increasing kernel size and reducing the proportion of shriveled kernels (Mitchell et al., 2006). Further, lines containing the *tin* gene increased partitioning of carbon to roots, increasing root biomass (Hendriks 2004; Fig. 11.11).

Early leaf area development

In Mediterranean-type environments, crops are typically sown on the first rains, and water for growth is supplied as current rainfall. Here, faster leaf area development should reduce soil evaporation to increase crop water-use efficiency, yield (López-Castañeda and Richards 1994), and competitiveness with weeds (Coleman et al., 2001). Greater early vigor may also benefit crops through increased root growth early in the season (Fig. 11.9 and 11.10; Palta et al., 2007), while improved light interception should increase crop growth rate, biomass, and grain yield for late-sown wheat crops or in environments where crop duration is shorter (Regan et al., 1997). Movement toward environmental sustainability and concerns over increasing diesel prices have increased interest in conservation farming and especially reduced tillage. However, hard soils and retained stubble common to conservation farming slow early growth, reducing crop biomass. Greater intrinsic early vigor has potential to overcome constraints imposed by reduced tillage (Watt et al., 2005).

Genotypic increases in early vigor have been shown to be associated with greater biomass and grain yield for wheat grown in Mediterranean environments (Whan et al., 1991; Botwright et al., 2002). Despite the potential benefits for greater early vigor in cereals, there has been little evidence of targeted breeding for this trait, outside of efforts to improve stand establishment and promote vegetative canopy closure in early-planted, dual-purpose management systems (Carver et al., 2001). This may partly reflect low heritability and reduced selection response for seedling leaf area and biomass (Rebetzke and Richards 1999). Numerous morphological factors with moderate to high heritability contribute to increased early vigor in wheat. Variation in early vigor is associated with differences in rate of seedling emergence, kernel and embryo size, coleoptile tiller size, and specific leaf area of seedling leaves (López-Castañeda et al., 1996; Rebetzke and Richards 1999; Rebetzke et al., 2004).

Development of wheat cultivars with the capacity to emerge from deep sowing (110 mm) would benefit growers in arid regions. Late rains commonly delay sowing, leading to reduced aerial biomass and grain yield (Mahdi et al., 1998). Often sufficient moisture for germination is available deeper in the soil profile, but the shorter coleoptile of current semidwarf wheat cultivars prevents successful establishment (Schillinger et al., 1998). Deep sowing commonly results in few, typically later-emerging seedlings, producing small relative growth rates and leaf area, reducing seedling biomass (Hadjichristodoulou et al., 1977; Rebetzke et al., 2007b). In turn, later emerging plants have lower biomass at anthesis, fewer spikes, and lower final biomass and yield (Mahdi et al., 1998; Rebetzke et al., 2007b) (Fig. 11.17).

Numerous studies (e.g., Schillinger et al., 1998; Rebetzke et al., 2007b) have demonstrated a positive association between coleoptile length and plant number with deep sowing (Fig. 11.17). Shorter coleoptiles and poor emergence have commonly been associated with the presence of *Rht-B1b* and *Rht-D1b* dwarfing genes (Schillinger et al., 1998; Rebetzke et al., 2007a,b). Many of the alternative, gibberellin-sensitive dwarfing genes (e.g., *Rht4*, *8*, *12*, and *13*) reduce plant height with little or no effect on coleoptile length (Ellis et al., 2004). Indeed, studies have demonstrated the potential of *Rht8* in the development of semi-dwarf, long-coleoptile wheat targeted at sowing depths exceeding 110 mm (Schillinger et al., 1998; Rebetzke et al., 2007b). In addition to replacement of the *Rht-B1b* and *Rht-D1b* dwarfing genes with alternative dwarfing genes, genomic regions independent of height have been identified to increase coleoptile length in wheat (Table 11.2) (Rebetzke et al., 2001a, 2007b).

Extensive evaluation across irrigated and water-limited environments has been undertaken for semidwarf lines containing alternative dwarfing genes and representing a range of genetic backgrounds (Fig. 11.18). This evaluation has demonstrated a generally similar yield performance of alternative dwarfing genes compared with existing *Rht-B1b* and *Rht-D1b* genes. One exception was the extreme height-reducing *Rht12*, which despite its high harvest index, showed unstable performance owing to reduced biomass (data not shown). The potential also exists in development of reduced-height wheat that combines *Rht-B1b* or *Rht-D1b* with alternative dwarfing genes such as *Rht8* to increase partitioning to grain and reduce crop lodging without compromising establishment. Table 11.3 shows the increased harvest

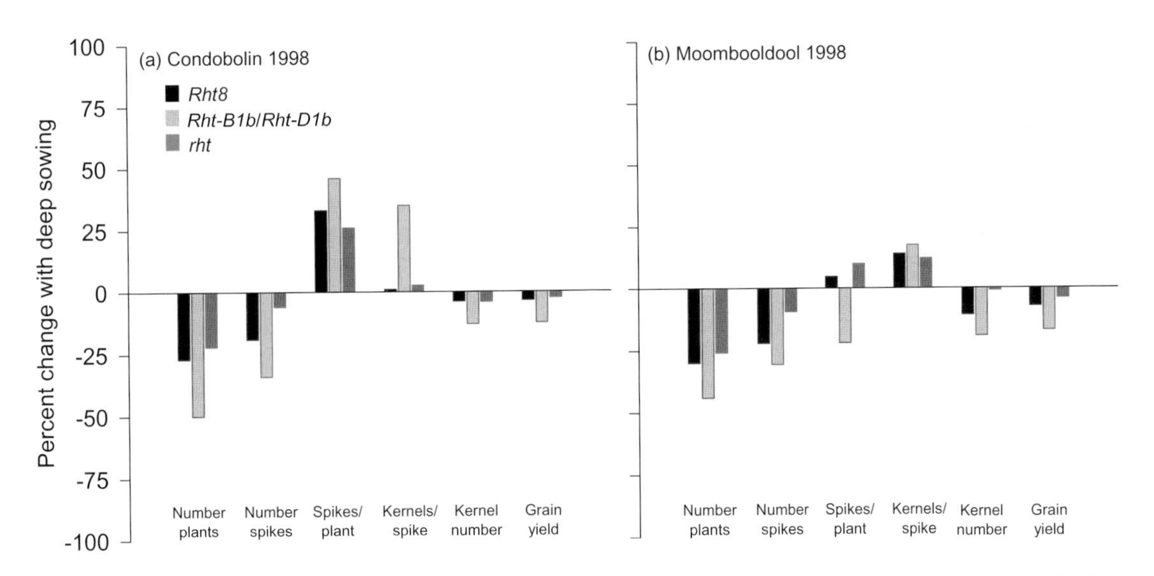

Fig. 11.17 Change (%) in different agronomic characteristics with deep-sowing at 110 mm expressed relative to the shallow 50-mm sowing depth at (a) Condobolin and (b) Moombooldool, Australia, in 1998. Bars represent (left to right) *Rht8*, *Rht-B1b/D1b*, *rht* (after Rebetzke et al., 2007b).

index and reduced lodging of *Rht-B1b*/*Rht-D1b*+*Rht8* "double-dwarfs" in two wheat backcross populations evaluated in multiple environments. Despite their reduced heights, coleoptile lengths of the doubled-dwarfs were not different from the *Rht-B1b* or *Rht-D1b* semidwarfs evaluated in the same study.

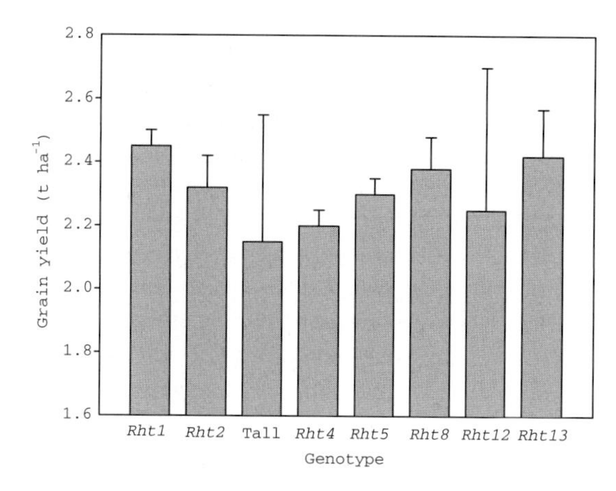

Fig. 11.18 Mean grain yield for wheat sister lines containing various dwarfing genes and representing a range of spring wheat backgrounds. Yields were obtained from experiments conducted in 2 years at up to 10 locations.

Transpiration efficiency

Transpiration efficiency (TE), or the ratio of net photosynthesis to water transpired, is an important component of crop water-use efficiency (biomass divided by water used during growth) in environments where stored soil water accounts for a major portion of crop water use (Condon et al., 2004). Variation in TE at the leaf level is negatively related to leaf intercellular CO_2 concentration (c_i), but both TE and c_i are difficult to measure. Carbon isotope discrimination is positively associated with c_i and therefore negatively correlated with TE (Condon et al., 1990). Use of CID has potential in breeding programs, as it integrates TE over the period in which dry matter is assimilated and is simple to measure on large numbers of families (Condon et al., 2004). In a breeding program targeting adaptation to water-limited environments, indirect selection for high biomass and yield via low CID can be more

Table 11.3 Mean values for a range of traits measured on lines genotyped for presence or absence of different height-reducing genes in two wheat populations.

Population/Dwarfing Genotype	Plant Height (cm)	Grain Yield (t ha^{-1})	Total Biomass (t ha^{-1})	Harvest Index	Kernel Number (no.m^{-2})	Lodging Score (1–9)	Coleoptile Length (mm)
CM18/Trident							
Rht-D1b,Xgwm261$_{192}$[a]	73	4.49	11.09	0.405	12,315	2.4	91
Rht-D1b,Xgwm261$_{208}$	78	4.48	11.85	0.378	11,947	2.4	90
Rht-D1a,Xgwm261$_{192}$	86	4.19	11.36	0.369	10,940	3.8	109
Rht-D1a,Xgwm261$_{208}$	97	3.84	11.71	0.328	10,159	3.9	112
cv. Trident	79	4.76	12.14	0.392	12,081	1.5	92
Chuan-mai 18	90	3.88	11.05	0.351	9,581	1.8	100
LSD	3	0.28	1.41	0.021	883	0.6	9
2*Chara/HM10S							
Rht-B1b,Xgwm261$_{192}$[a]	73	4.24	11.16	0.380	11,910	1.9	93
Rht-B1b,Xgwm261$_{165}$	79	4.17	11.30	0.369	12,157	2.2	91
Rht-B1a,Xgwm261$_{192}$	84	4.21	11.47	0.367	11,760	2.9	105
Rht-B1a,Xgwm261$_{165}$	90	3.97	11.71	0.339	11,474	3.0	107
cv. Chara	72	4.23	11.34	0.373	11,882	1.3	84
HM10S	69	3.97	10.61	0.374	11,507	2.0	125
LSD	5	0.37	1.81	0.022	1,389	0.8	11

LSD ($p < 0.05$) for statistical testing of differences among genotypic classes.

[a]*Rht-D1b* (syn. *Rht2*); *Rht-D1a* (*rht2*); *Rht-B1b* (*Rht1*); *Rht-B1a* (*rht1*); *Xgwm261*$_{192}$ (*Rht8*); *Xgwm261*$_{208}$ (*rht8*); *Xgwm261*$_{165}$ (*rht8*).

efficient than direct selection of either production trait in early generations (Rebetzke et al., 2002).

The wheat cultivars Drysdale and Rees were selected from a BC_2-derived population developed from the high-CID recurrent parent 'Hartog'. In side-by-side studies of Drysdale and Hartog undertaken in about 30 sites in each of two years, AGT breeding company established an average yield advantage of 16% and 9% for 2004 and 2005, respectively (Fig. 11.8). The greatest yield advantage occurred in the drier of the two years and in the driest of sites.

Evaluation in water-limited environments of BC_2-derived, sister lines showed CID to be negatively correlated with aerial biomass and yield (Rebetzke et al., 2002). In comparisons with the recurrent parent Hartog, low-CID selections achieved greater yield through increases in harvest index and biomass (Fig. 11.19). Kernel size was the primary yield component affected by selection for low CID. In contrast, high CID has often been associated with higher leaf conductance, increased water use, and growth. Hence positive relationships for CID, biomass, and yield are commonly obtained in irrigated environments where water supply was not a major constraint to yield (Fischer et al., 1998), or in Mediterranean environments where soil water availability preceding anthesis was plentiful (Araus et al., 2002). Thus the opportunity exists to select for high CID to increase potential yield where water for crop growth is abundant.

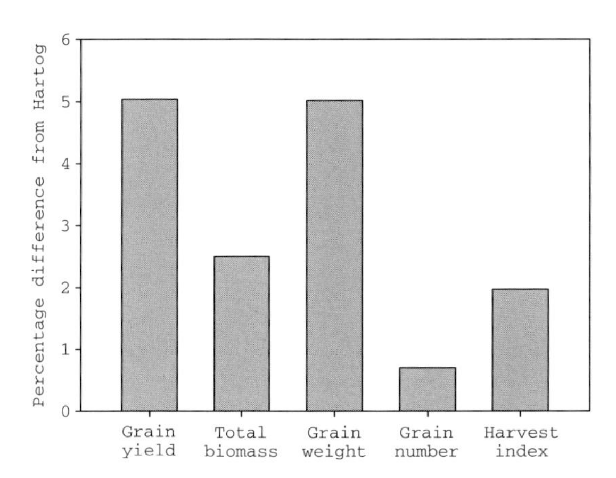

Fig. 11.19 Change in grain yield and yield components with selection for reduced carbon isotope discrimination in BC_2-derived lines developed from recurrent parent 'Hartog' evaluated in nine environments.

Repeatable genotypic variation has been reported in wheat for TE (Solomon and Labuschagne 2004) and CID (Rebetzke et al., 2002, 2006). These reports emphasize that broad- and narrow-sense heritability of CID is high when expressed on a single-plot or line-mean basis. Further, analysis of mating designs employing progeny from either F_1 or segregating generations has shown TE and CID to be under strong additive genetic control with little evidence for nonadditive gene action (Solomon and Labuschagne 2004; Rebetzke et al., 2006). Together, these results indicate that family selection for altered TE and/or CID in early generations will maintain altered CID with inbreeding.

Few QTL analyses have been reported for CID in wheat. Genotypic distributions for CID are commonly gaussian with evidence for transgressive segregation (Rebetzke et al., 2008a). QTL studies have been undertaken for wheat at CSIRO for multiple populations across different environments (Table 11.2). These indicate a large number of genomic regions of small effect, many of which are common across populations (Rebetzke et al., 2008a). Some of these regions collocate with regions for plant development and plant height, as well regions for the stomatal-related traits, canopy temperature, stomatal conductance, and leaf chlorophyll content (Fig. 11.20; Table 11.2). Reduced CID is commonly associated with later flowering, highlighting the need for caution when relating CID to yield in a terminal drought study of populations in which flowering is not carefully monitored (Sayre et al., 1995; Condon et al., 2002). Similarly, lower CID is associated with reduced plant height and, particularly, the presence of *Rht-B1b* and *Rht-D1b* dwarfing genes (Fig. 11.20).

A number of the CID QTLs identified for leaf tissue collocated with QTLs for CID measured on mature grain (Fig 11.20). Yet despite this commonality, alleles for increased CID are sometimes not consistent for leaf and grain CID QTLs. For example, parents contributing low leaf-tissue CID alleles at the *Ppd*, *Rht-B1*, and *Rht-D1* loci

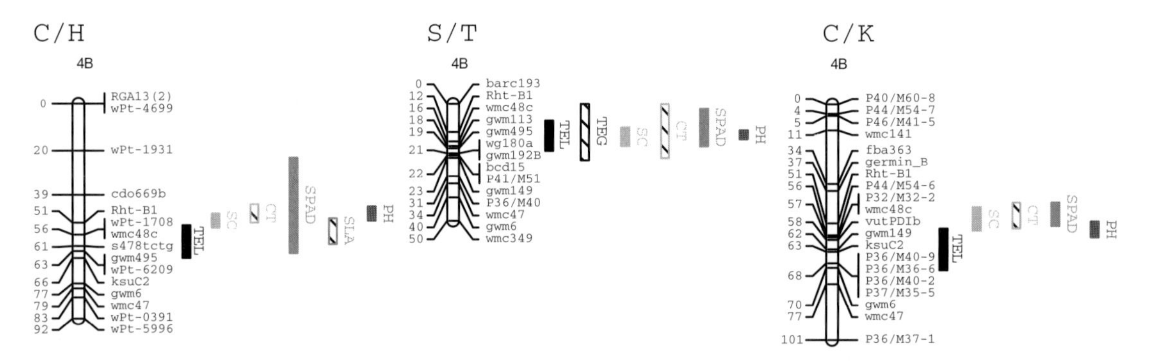

Fig. 11.20 Collocation of leaf (TEL) and grain (TEG) transpiration efficiency, and stomatal aperture trait QTL (SC = stomatal conductance, CT = canopy temperature, SPAD = chlorophyll content, SLA = specific leaf area) and plant height (PH) QTL for chromosome 4B measured in the Cranbrook/Halberd (C/H), Sunco/Tasman (S/T), and CD87/Katepwa (C/K) populations; TEG was determined for the S/T population only.

contributed high-CID alleles for CID measured on mature grain (Fig. 11.20). This inconsistency may account for the low genetic correlation for leaf and grain CID ($r_g = 0.24$, $n = 159$). This change in allelic state highlights the need to consider variation in developmental traits and the potential for confounding environmental factors with yield when measuring CID on grain. Further, it highlights the importance of controlling variation in development when indirectly selecting for yield via leaf or grain CID (*cf.* Ferrio et al., 2007).

The expense of measuring CID with a mass spectrometer (approximately US$40 per sample) underscores the need to identify surrogate traits for screening large populations. Colocation of CID QTLs with QTLs for stomatal-related traits gives some indication of the underlying physiological basis for variation in TE in wheat. For example, the 4BS low-CID QTL was associated with increased chlorophyll content, greater stomatal conductance, and cooler canopy temperatures (Fig. 11.20). This suggests the potential for indirect selection for CID using an infrared thermometer or SPAD chlorophyll meter, perhaps as early-generation selection on head rows. Use of NIR also offers the potential for inexpensive assessment of CID in large segregating wheat populations (Fig. 11.6).

Maintenance of leaf area

A growing wheat crop must accumulate and then maintain leaf area toward development of final biomass and yield. Maintenance of excessive leaf area when water is limiting can contribute to unnecessary water loss and sometimes lead to haying-off and reduced grain yields (Fischer 1979; van Herwaarden et al., 1998). On the other hand, reductions in leaf area, particularly around anthesis and throughout grain filling, can reduce assimilation and accumulation of carbon necessary for grain set and grain filling. The importance of extending the duration of green leaf area as a means of maintaining postflowering photosynthesis and supply of assimilated carbon has been widely recognized (Richards 2000; Foulkes et al., 2004). Between 50% and 90% of final grain weight and yield is derived from postanthesis assimilation (Bidinger et al., 1977; Schynder 1993). Loss of leaf area and subsequent reductions in grain yield have been reported in wheat (Gelang et al., 2000), while daily delays in flag leaf senescence were associated with an estimated 0.17 t ha^{-1} increase in grain yield and 2.7 kg ha^{-1} increase in nitrogen content (Gooding 2007).

A common response of water-stressed plants is accelerated and premature leaf senescence, leading to a reduction in the amount of photosynthetic leaf tissue (Sirault et al., 2004). Genotypic increases in leaf area duration can be achieved via selection for genes that delay the onset and rate of leaf senescence (termed stay-green) (Thomas and Howarth 2000) or for genes that protect the leaf from stresses affecting leaf senescence (e.g., disease, heat, and water stress). Genotypic variation in the ability to maintain leaf area under drought has been reported in other crops including sorghum (Harris et al., 2007). Genotypic variation has also been described in wheat for leaf-glaucousness (Kulwal et al., 2003), leaf rolling (Color Plate 25) (Sirault et al., 2008), and photosynthetic stay-green (Spano et al., 2003; Verma et al., 2004; Christopher et al., 2008). Reduced leaf glaucousness (greater waxiness) reduced leaf temperature and increased photosynthetic rate, increasing transpiration efficiency in wheat (Richards et al., 1986). Durum stay-green mutants showed a 10% increase in grain yield when water-stressed in pots (Spano et al., 2003), while the yield advantage of cultivar Seri was related to its stay-green phenotype (Christopher et al., 2008). Verma et al. (2004) also reported an association between maintenance of green leaf area and grain yield for wheat sister lines evaluated under optimal and water-limited conditions.

Plants experiencing severe water stress commonly show symptoms of leaf rolling (Sirault 2007). Similarly, genotypes with a greater capacity to roll leaves reduce leaf area to intercept less sunlight. Reduction in radiation load should reduce leaf warming and transpiration, slowing further leaf dehydration. Leaf area is subsequently maintained to allow photosynthesis to continue after drought has abated (G.J. Rebetzke, unpublished data; Sirault 2007). The effect of leaf-rolling has been demonstrated in wheat by Clarke (1986), who showed leaf rolling reduced effective leaf area by up to 48% and water loss by up to 84%. Similarly, excised leaves of leaf-rolling genotypes showed reductions in leaf area of 50% in 16 minutes compared with nonrolling genotypes, where reduction in leaf area was much slower at 42 minutes (Sirault 2007). Genotypic variation in leaf rolling has been positively associated with kernel size (Sirault 2007) and yield under drought (Clarke et al., 1991; Nachit et al., 1992). Despite evidence for genotypic differences in leaf rolling, not all variation is of adaptive value, with the potential for rolling in some genotypes to be associated with poor root growth or other hydraulic resistance (Sirault 2007).

Little is known of genetic control of leaf rolling in wheat. Rebetzke et al. (2001c) demonstrated that selection for leaf rolling in wheat can be readily undertaken under well-watered conditions. Sirault et al. (2008) subsequently demonstrated that leaf rolling under well-watered and water-limited conditions was under strong additive genetic control and showed high narrow-sense heritability ($h^2 = 0.80$ to 0.97).

FUTURE PERSPECTIVES

Drought reflects a dynamic process that is not always consistent in the timing or amount of rainfall for a target region. Hence empirical selection for yield in improving adaptation is a challenge for breeders. A better understanding of the frequency and nature of target environments through modeling, coupled with a strong physiological understanding of adaptation, has potential to identify useful traits for selection in breeding programs. The challenge now is the identification of genetic sources of variation for such traits, development of inexpensive high-throughput screening tools, and appropriate evaluation of selected lines in target or related-constructed environments. Molecular markers will assist in indirect selection of improved water uptake through development of better root systems. However, the polygenic control of many yield-related, physiological traits will reduce effectiveness of marker-assisted selection. Similarly, the value of transgenes needs to be assessed in field environments. Further, evaluation of transgenes should focus less on survival as this is unlikely to be of value in wheat production systems, and efforts directed at assessment for productivity or yield with water limitation. Future efforts should focus on efficient marker-implementation strategies for alleles of small genetic effect, and development of inexpensive phenotyping methods for use in early-generation selection where screening may be among thousands of lines. Finally, efforts should be directed at further assessment of candidate traits singly or in combination for use in breeding.

REFERENCES

Alvarez, S., E.L. Marsh, S.G. Schroeder, and D.P. Schachtman. 2008. Metabolomic and proteomic changes

in the xylem sap of maize under drought. Plant Cell Environ. 31:325–340.
Araus, J.L., G.A. Slafer, M.P. Reynolds, and C. Royo. 2002. Plant breeding and drought in C3 cereals: What should we breed for? Ann. Bot. 89:925–940.
Asseng, S., and A.F. van Herwaarden. 2003. Analysis of the benefits to wheat yield from assimilates stored prior to grain filling in a range of environments. Plant Soil 256:217–229.
Austin, R.B., J. Bingham, R.D. Blackwell, L.T. Evans, M.A. Ford, C.L. Morgan, and M. Taylor. 1980. Genetic improvements in winter wheat yields since 1900 and associated physiological changes. J. Agric. Sci. (Cambridge) 94:675–689.
Babar, M.A., M. van Ginkel, A.R. Klatt, B. Prasad, and M.P. Reynolds. 2006. The potential of using spectral reflectance indices to estimate yield in wheat grown under reduced irrigation. Euphytica 150:155–172.
Bahieldin, A., H.T. Mahfouz, H.G. Eissa, O.M. Saleh, A.M. Ramadan, and I.A. Ahmed. 2005. Field evaluation of transgenic wheat plants stably expressing the *HVA1* gene for drought tolerance. Physiol. Plant. 123:421–427.
Berry, P.M., J.H. Spink, M.J. Foulkes, and A. Wade. 2003. Quantifying the contributions and losses of dry matter from non-surviving shoots in four cultivars of winter wheat. Field Crops Res. 80:111–121.
Bidinger, F., R.B. Musgrave, and R.A. Fischer. 1977. Contribution of stored pre-anthesis assimilation to grain yield in wheat and barley. Nature 270:431–433.
Blum, A. 1998. Improving wheat grain filling under stress by stem reserve mobilisation. Euphytica 100:77–83.
Bonnett, D.G., G.J. Rebetzke, and W. Spielmeyer. 2005. Strategies for efficient implementation of molecular markers in wheat breeding. Mol. Breed. 15:75–85.
Botwright, T.L., A.G. Condon, G.J. Rebetzke, and R.A. Richards. 2002. Field evaluation of early vigour for genetic improvement of grain yield in wheat. Aust. J. Agric. Res. 53:1137–1145.
Botwright, T.L., E. Pasquin, and L.J. Wade. 2007. Genotypic differences in root penetration ability of wheat through thin wax layers in contrasting water regimes and in the field. Plant Soil 301:135–149.
Brooks, A., C.F. Fenner, and D. Aspinall. 1982. Effects of water deficit on endosperm starch granules and on grain physiology of wheat and barley. Aust. J. Plant Physiol. 9:423–436.
Byerlee, D., and M. Morris. 1993. Research for marginal environments: Are we underinvested? Food Policy 18:381–393.
Byrt, C.S., J.D. Platten, W. Spielmeyer, R.A. James, E.S. Lagudah, E.S. Dennis, M. Tester, and R. Munns. 2007. HKT1;5-like cation transporters linked to Na^+ exclusion loci in wheat, *Nax2* and *KNa1*. Plant Physiol. 143:1918–1928.
Calhoun, D.S., G. Gebeyehu, A. Miranda, S. Rajaram, and M. van Ginkel. 1994. Choosing evaluation environments to increase wheat grain yield under drought conditions. Crop Sci. 34:673–678.
Carver, B.F., I. Khalil, E.G. Krenzer, and C.T. MacKown. 2001. Breeding winter wheat for a dual-purpose management system. Euphytica 119:231–234.
Casadesús J., Y. Kaya, J. Bort, M.M. Nachit, J.L. Araus, S. Amor, G. Ferrazzano, F. Maalouf, M. Maccaferri, V. Martos, H. Ouabbou, and D. Villegas. 2007. Using vegetation indices derived from conventional digital cameras as selection criteria for wheat breeding in water-limited environments. Ann. Appl. Biol. 150:227–236.
Ceccarelli, S. 1994. Specific adaptation and breeding for marginal conditions. Euphytica 77:205–219.
Chaerle, L., I. Leinonen, H.G. Jones, and D. Van der Straeten. 2007. Monitoring and screening plant populations with combined thermal and chlorophyll fluorescence imaging. J. Exp. Bot. 58:773–784.
Chapman, S.C. 2008. Use of crop models to understand genotype by environment interactions for drought in real-world and simulated plant breeding trials. Euphytica 161:195–208.
Chapman, S.C., M. Cooper, G.L. Hammer, and D.G. Butler. 2000. Genotype by environment interactions affect-ing grain sorghum: II. Frequencies of different seasonal patterns of drought stress are related to location effects on hybrid yields. Aust. J. Agric. Res. 51:209–221.
Christopher, J.T., A.M. Manschadi, G.L. Hammer, and A.K. Borrell. 2008. Developmental and physiological traits associated with high yield and stay-green phenotype in wheat. Aust. J. Agric. Res. 59:354–364.
Clarke, J.M. 1986. Effect of leaf rolling on water loss in *Triticum* spp. Can. J. Plant Sci. 66:885–891.
Clarke, J.M., R.M. DePauw, and T.F. Townley-Smith. 1992. Evaluation of methods for quantification of drought tolerance in wheat. Crop Sci. 32:723–728.
Clarke, J.M., I. Romagosa, and R.M. DePauw. 1991. Screening durum wheat germplasm for dry growing conditions: Morphological and physiological criteria. Crop Sci. 31:770–775.
Cogan, N.O., R.C. Ponting, A.C. Vecchies, M.C. Drayton, J. George, P.M. Dracatos, M.P. Dobrowolski, T.I. Sawbridge, K.F. Smith, G.C. Spangenberg, and J.W. Forster. 2006. Gene-associated single nucleotide polymorphism discovery in perennial ryegrass (*Lolium perenne* L.). Mol. Gen. Genet. 276:101–112.
Coleman, R.K., G.S. Gill, and G.J. Rebetzke. 2001. Identification of quantitative trait loci for traits conferring weed competitiveness in wheat (*Triticum aestivum* L.). Aust. J. Agric. Res. 52:1235–1246.
Condon, A.G., G.D. Farquhar, and R.A. Richards. 1990. Genotypic variation in carbon isotope discrimination and transpiration efficiency in wheat: Leaf gas exchange and whole plant studies. Aust. J. Plant. Physiol. 17:9–22.
Condon, A.G., R.A. Richards, G.J. Rebetzke, and G.D. Farquhar. 2002. Improving intrinsic water-use efficiency and crop yield. Crop Sci. 42:122–131.
Condon, A.G., R.A. Richards, G.J. Rebetzke, and G.D. Farquhar. 2004. Breeding for high water use efficiency. J. Exp. Bot. 55:2447–2460.
Cooper, M., R.E. Stucker, I.H. Delacy, and B.D. Harch. 1997. Wheat breeding nurseries, target environments, and indirect selection for grain yield. Crop Sci. 37:1168–1176.

Delhaize, E., P.R. Ryan, D.M. Hebb, Y. Yamamoto, Y. Sasak, and H. Matsumoto. 2004. Engineering high level aluminium tolerance in barley with the *ALTM1* gene. Proc. Natl. Acad. Sci. USA 10:15249–15254.

Duggan, B.L., R.A. Richards, A.F. van Herwaarden, and N.A. Fettell. 2005. Agronomic evaluation of a tiller inhibition gene (*tin*) in wheat: I. Growth and partitioning of assimilate. Aust. J. Agric. Res. 56:179–186.

Ellis, M.H., G.J. Rebetzke, P. Chandler, D.G. Bonnett, W. Spielmeyer, and R.A. Richards. 2004. The effect of different height reducing genes on the early growth of wheat. Funct. Plant Biol. 31:583–589.

Ellis, M.H., W. Spielmeyer, K. Gale, G.J. Rebetzke, and R. A. Richards. 2002. Perfect markers for the *Rht-B1b* and *Rht-D1b* dwarfing mutations in wheat (*Triticum aestivum* L.). Theor. Appl. Genet. 105:1038–1042.

Falconer, D.S., and T.F.C. Mackay. 1996. Introduction to quantitative genetics. 4th ed. Longman Group Ltd., Essex, UK.

Ferrio, J.P., M.A. Mateo, J. Bort, O. Abdalla, J. Voltas, and J.L. Araus. 2007. Relationships of grains $\delta^{13}C$ and $\delta^{13}O$ with wheat phenology and yield under water-limited conditions. Ann. Appl. Biol. 150:207–215.

Finlay, K.W., and G.N. Wilkinson. 1963. The analysis of adaptation in a plant breeding programme. Aust. J. Agric. Res. 14:742–754.

Fischer, R.A. 1973. The effect of water stress at various stages of development on yield processes in wheat. p. 233–241. *In* R.D. Slaytner (ed.) Plant response to climatic factors. Proc. Uppsala Symp., Uppsala, Sweden. 1970. UNESCO, Paris, France.

Fischer, R.A. 1979. Growth and water limitation to dryland wheat yield in Australia: A physiological framework. J. Aust. Inst. Agric. Sci. 45:83–94.

Fischer, R.A., and R. Maurer. 1978. Drought resistance in spring wheat cultivars: I. Grain yield responses. Aust. J. Agric. Res. 29:897–912.

Fischer, R.A., D. Rees, K.D. Sayre, Z.M. Lu, A.G. Condon, and A.L. Saavedra. 1998. Wheat yield progress associated with higher stomatal conductance and photosynthetic rate, and cooler canopies. Crop Sci. 38:1467–1475.

Food and Agriculture Organization. 2002. World agriculture: Towards 2015/2030. An FAO perspective [Online]. Available at http://www.fao.org/docrep/005/y4252e/y4252e00.htm (verified 11 Aug. 2008).

Foulkes, M.J., J.W. Snape, V.J. Shearman, M.P. Reynolds, O. Gaju, and R. Sylvester-Bradley. 2007. Genetic progress in yield potential in wheat: Recent advances and future prospects. J. Agric. Sci. (Cambridge) 145:17–29.

Foulkes, M.J., R. Sylvester-Bradley, and R.K. Scott. 2002. The ability of wheat cultivars to withstand drought in UK conditions: Formation of grain yield. J. Agric. Sci. (Cambridge) 138:153–169.

Foulkes, M.J., R. Sylvester-Bradley, A.J. Worland, and J.W. Snape. 2004. Effects of a photoperiod-response gene *Ppd-D1* on yield potential and drought resistance in UK winter wheat. Euphytica 135:63–73.

French, R.J., and J.E. Schultz. 1984. Water use efficiency of wheat in a Mediterranean-type environment: I. The relation between yield, water use and climate. Aust. J. Agric. Res. 35:743–764.

Gelang, J., H. Pleijel, E. Sild, H. Danielsson, S. Younis, and G. Sellden. 2000. Rate and duration of grain filling in relation to flag leaf senescence and grain yield in spring wheat (*Triticum aestivum*) exposed to different concentrations of ozone. Physiol. Plant. 110:366–375.

Gleick, P.H. 2003. Water use. Annu. Rev. Environ. Res. 28:275–314.

Gooding, M.J. 2007. Influence of foliar diseases and their control by fungicides on grain yield and quality in wheat. p. 567–582. *In* H.T. Buck, J.E. Nisi, and N. Salomón (ed.) Wheat production in stressed environments. Springer, Dordrecht, The Netherlands.

Gregory, P.J., M.J. Gooding, K.E. Ford, P.W. Hendriks, J.A. Kirkegaard, and G.J. Rebetzke. 2005. Genotypic and environmental influences on the performance of crop root systems. Aspects Appl. Biol. 73:1–10.

Gupta, P.K., S. Rustgi, and R.R. Mir. 2008. Array-based high-throughput DNA markers for crop improvement. Heredity 101:5–18.

Hadjichristodoulou, A., A. Della, and J. Photiades. 1977. Effect of sowing depth on plant establishment, tillering capacity and other agronomic characters of cereals. J. Agric. Sci. (Cambridge) 89:161–167.

Harris, K., P.K. Subudhi, A. Borrell, D. Jordan, D. Rosenow, H. Nguyen, P. Klein, R. Klein, and J. Mullet. 2007. Sorghum stay-green QTL individually reduce post-flowering drought-induced leaf senescence. J. Exp. Bot. 58:327–338.

Hendriks, P.W. 2004. Root and shoot growth in wheats varying for presence of the *tin*, reduced-tillering gene. Honors thesis. Établissement National d'Enseignement Supérieur Agronomique de Dijon, France.

Huang, S., W. Spielmeyer, E.S. Lagudah, R.A. James, J.D. Platten, E.S. Dennis, and R. Munns. 2006. A sodium transporter (HKT7) is a candidate for *Nax1*, a gene for salt tolerance in durum wheat. Plant Physiol. 142:1718–1727.

Hurd, E.A. 1968. Growth of roots of seven varieties of spring wheat at high and low moisture levels. Agron. J. 60:201–205.

IPCC. 2007. Climate change 2007: The physical science basis. Summary for policy makers. Contribution of working group I to the fourth assessment report of the intergovernmental panel on climate change [Online]. Available at http://www.ipcc.ch/ (verified 11 Aug. 2008).

Islam, T.M.T., and R.H. Sedgley. 1981. Evidence for a 'uniculm effect' in spring wheat (*Triticum aestivum* L.) in a Mediterranean environment. Euphytica 30:277–282.

Jackson, P., M. Robertson, M. Cooper, and G. Hammer. 1996. The role of physiological understanding in plant breeding from a breeding perspective. Field Crops Res. 49:11–37.

Jansen, R.C., and J.P. Nap. 2001. Genetical genomics: The added value from segregation. Trends Genet. 17:388–391.

Jefferies, S.P., M.A. Pallotta, J.G. Paull, A. Karakousis, J.M. Kretschmer, S. Manning, A.K.M.R. Islam, P. Langridge, and K.J. Chalmers. 2000. Mapping and validation of chromosome regions conferring boron toxicity tolerance in wheat (*Triticum aestivum*). Theor. Appl. Genet. 101:767–777.

Jones, H.G., and E.J.M. Kirby. 1977. Effects of manipulation of number of tillers and water supply on grain yield in barley. J. Agric. Sci. (Cambridge) 88:391–397.

Jordan, M.C., D.J. Somers, and T.W. Banks. 2007. Identifying regions of the wheat genome controlling seed development by mapping expression quantitative trait loci. Plant Biotechnol. J. 5:442–453.

Kijne, W. 2003. Unlocking the water potential of agriculture. FAO, Rome, Italy.

Kirigwi, F.M., M. Van Ginkel, G. Brown-Guedira, B.S. Gill, G.M. Paulsen, and A.K. Fritz. 2007. Markers associated with a QTL for grain yield in wheat under drought. Mol. Breed. 20:401–413.

Kirkegaard, J.A., P.A. Gardner, J.F. Angus, and E. Koetz. 1994. Effects of *Brassica* crops on the growth and yield of wheat. Aust. J. Agric. Res. 45:529–545.

Kirkegaard, J.A., J.M. Lilley, G.N. Howe, and J.M. Graham. 2007. Impact of subsoil water use on wheat yield. Aust. J. Agric. Res. 58:303–315.

Kulwal, P.L., J.K. Roy, H.S. Balyan, and P.K. Gupta. 2003. QTL mapping for growth and leaf characters in bread wheat. Plant Sci. 164:267–277.

Lagudah, E.S., O. Moullet, and R. Appels. 1997. Map-based cloning of a gene sequence encoding a nucleotide binding domain and a leucine-rich region at the *Cre3* nematode resistance locus of wheat. Genome 40:659–665.

Li, B., A. Wei, C. Song, N. Li, and J. Zhang. 2007. Heterologous expression of the *TsVP* gene improves the drought resistance of maize. Plant Biotechnol. J. 6:146–159.

Lilley, J.M., and J.A. Kirkegaard. 2007. Seasonal variation in the value of subsoil water to wheat: Simulation studies in southern New South Wales. Aust. J. Agric. Res. 58:1115–1128.

Lin, C.S., and M.R. Binns. 1988. A superiority measure of cultivar performance for cultivar × location data. Can. J. Plant Sci. 68:193–198.

Liu, H., H. Mei, X. Yu, G. Zhou, G. Liu, and L. Luo. 2006. Towards improving the drought tolerance of rice in China. Plant Genet. Res. 4:47–53.

López-Castañeda, C., and R.A. Richards. 1994. Variation in temperate cereals in rainfed environments: III. Water use and water-use efficiency. Field Crops Res. 39:85–98.

López-Castañeda, C., R.A. Richards, G.D. Farquhar, and R.E. Williamson. 1996. Seed and seedling characteristics contributing to variation in early vigor among temperate cereals. Crop Sci. 36:1257–1266.

Luo, Q.Y., W. Bellotti, M. Williams, and B. Bryan. 2005. Potential impact of climate change on wheat yields in South Australia. Agric. For. Meteorol. 132:273–285.

Maccaferri, M., M.C. Sanguineti, S. Corneti, et al. 2008. Quantitative trait loci for grain yield and adaptation of durum wheat (*Triticum durum* Desf.) across a wide range of water availability. Genetics 178:489–511.

Mahdi, L., C.J. Bell, and J. Ryan. 1998. Establishment and yield of wheat (*Triticum turgidum* L.) after early sowing at various depths in a semi-arid Mediterranean environment. Field Crops Res. 58:187–196.

Manschadi, M., J. Christopher, P. deVoil, and G.L. Hammer. 2006. The role of root architectural traits in adaptation of wheat to water-limited environments. Funct. Plant Biol. 33:823–837.

Maris, E. 2008. More crop per drop. Nature 452:273–277.

Martinez-Beltran, J., and C.L. Manzur. 2005. Overview of salinity problems in the world and FAO strategies to address the problem. p. 311–313. *In* Proc. Int. Salinity Forum, Riverside, CA. 25–28 April 2005. U.S. Salinity Laboratory, USDA-ARS, Riverside, CA.

Mitchell, J.H., S.C. Chapman, G.J. Rebetzke, and S. Fukai. 2006. Reduced tillering wheat lines maintain kernel weight in dry environments. *In* N. Turner, T. Acuna, and R.C. Johnson (ed.) Ground-breaking stuff. Proc. Australian Agronomy Conf., 13th, Perth, Western Australia. 10–14 September 2006. The Regional Institute Ltd., Perth, Western Australia. Available at http://www.regional.org.au/au/asa/2006/ (verified 8 Sept. 2008).

Morgan, J.M., and M.K. Tan. 1996. Chromosomal location of a wheat osmoregulation gene using RFLP analysis. Aust. J. Plant Physiol. 23:803–806.

Musick, J.T., O.R. Jones, B. Stewart, and D.A. Dusek. 1994. Water–yield relationship for irrigated and dryland wheat in the US Southern plains. Agron. J. 86:980–986.

Nachit, M.M., M.E. Sorrells, R.W. Zobel, H.G. Gauch, W.R. Coffman, and R.A. Fischer, R.A. 1992. Association of morpho-physiological traits with grain yield and components of genotype-environment interaction in durum wheat. J. Genet. Breed. 46:363–368.

Nelson, D.E., P.P. Repetti, T.R. Adams, R.A. Creelman, J. Wu, D.C. Warner, D.C. Anstrom, R.J. Bensen, P.P. Castiglioni, M.G. Donnarummo, B.S. Hinchey, R.W. Kumimoto, D.R. Maszle, R.D. Canales, K.A. Krolikowski, S.B. Dotson, N. Gutterson, O.J. Ratcliffe, and J.E. Heard. 2007. Plant nuclear factor Y (NF-Y) B subunits confer drought tolerance and lead to improved corn yields on water-limited acres. Proc. Natl. Acad. Sci. USA 104:16450–16455.

Nicolas, M.E., R.M. Gleadow, and M.J. Dalling. 1985. Effect of post-anthesis drought on cell division and starch accumulation in developing wheat grains. *Ann. Bot.* 55:433–444.

Nix, H.A. 1975. The Australian climate and its effects on grain yield and quality. p. 184–226. *In* A. Lazenby and E.M. Matheson (ed.) Australian field crops: Vol. 1. Wheat and other temperate cereals. Angus and Robertson, Sydney, Australia.

Olivares-Villegas, J.J., M.P. Reynolds, and G.K. McDonald. 2007. Drought-adaptive attributes in the Seri/Babax hexaploid wheat population. Funct. Plant Biol. 34:189–203.

O'Toole, J.C., and W.L. Bland. 1987. Genotypic variation in crop plant root systems. Adv. Agron. 41:91–145.

Palta, J.A., I.R.P. Fillery, and G.J. Rebetzke. 2007. Restricted-tillering wheat does not lead to greater investment in roots and early N uptake. Field Crops Res. 104:52–59.

Passioura, J.B. 1977. Grain yield, harvest index, and water use of wheat. J. Aust. Inst. Agric. Sci. 43:117–121.

Passioura, J.B. 2002. Environmental biology and crop improvement. Funct. Plant Biol. 29:537–546.
Passioura, J.B. 2006a. Increasing crop productivity when water is scarce—from breeding to field management. Agric. Water Manage. 80:176–196.
Passioura, J.B. 2006b. The perils of pot experiments. Funct. Plant Biol. 33:1075–1079.
Perry, M.W., and M.F.D. D'Antuono. 1989. Yield improvement and associated characteristics of some Australian spring wheat cultivars introduced between 1860 and 1982. Aust. J. Agric. Res. 40:457–472.
Pinto, S., S.C. Chapman, C.L. McIntyre, R. Shorter, and M.P. Reynolds. 2008. QTL for canopy temperature response related to yield in both heat and drought environments. *In* R. Appels, R. Eastwood, E. Lagudah, P. Langridge, M. Mackay, L. McIntyre, and P. Sharp (ed.) Proc. Int. Wheat Genet. Symp., 11th, Brisbane, Australia. 24–28 August 2008. Sydney University Press, Sydney, Australia. (Available at http://ses.library.usyd.edu.au/bitstream/2123/3351/1/P172.pdf) (Verified 20 Dec. 2008.)
Postel, S.L. 1999. Pillar of sand: Can the irrigation miracle last? W.W. Norton & Co., New York.
Rebetzke, G.J., R. Appels, A.D. Morrison, R.A. Richards, G.K. McDonald, M.H. Ellis, W. Spielmeyer, and D.G. Bonnett. 2001a. Quantitative trait loci on chromosome 4B for coleoptile length and early vigour in wheat (*Triticum aestivum* L.). Aust. J. Agric. Res. 52:1221–1234.
Rebetzke, G.J., T.L. Botwright, C.S. Moore, R.A. Richards, and A.G. Condon. 2004. Genotypic variation in specific leaf area for genetic improvement of early vigour in wheat. Field Crops Res. 88:179–189.
Rebetzke, G.J., A.G. Condon, R.A. Richards, R. Appels, and G.D. Farquhar. 2008a. Quantitative trait loci for carbon isotope discrimination are repeatable across environments and wheat mapping populations. Theor. Appl. Genet. 118:123–137.
Rebetzke, G.J., A.G. Condon, R.A. Richards, and G.D. Farquhar. 2002. Selection for reduced carbon-isotope discrimination increases aerial biomass and grain yield of rainfed bread wheat. Crop Sci. 42:739–745.
Rebetzke, G.J., A.G. Condon, R.A. Richards, and J.J. Read. 2001b. Phenotypic variation and sampling for leaf conductance in wheat (*Triticum aestivum* L.) breeding populations. Euphytica 121:335–341.
Rebetzke, G.J., M.H. Ellis, D.G. Bonnett, and R.A. Richards. 2007a. Molecular mapping of genes for coleoptile growth in bread wheat (*Triticum aestivum* L.). Theor. Appl. Genet. 114:1173–1183.
Rebetzke, G.J., A.D. Morrison, R.A. Richards, D.G. Bonnett, and C.S. Moore. 2001c. Genotypic variation for leaf rolling ability in wheat. p. 172–175. *In* R. Eastwood (ed.) Proc. Wheat Breeding Assembly, 10th, Mildura, Australia. 16–21 September 2001. Wheat Breeding Soc. Australia, Horsham Victoria, Australia.
Rebetzke, G.J., and R.A. Richards. 1999. Genetic improvement of early vigour in wheat. Aust. J. Agric. Res. 50:291–301.
Rebetzke, G.J., R.A. Richards, A.G. Condon, and G.D. Farquhar. 2006. Inheritance of reduced carbon isotope discrimination in bread wheat (*Triticum aestivum* L.). Euphytica 150:97–106.
Rebetzke, G.J., R.A. Richards, N.A. Fettell, M. Long, A.G. Condon, and T.L. Acuna. 2007b. Genotypic increases in coleoptile length improves wheat establishment, early vigour and grain yield with deep sowing. Field Crops Res. 100:10–23.
Rebetzke, G.J., A.F. van Herwaarden, C. Jenkins, S. Ruuska, L. Tabe, D. Lewis, M. Weiss, N. Fettell, and R.A. Richards. 2008b. Quantitative trait loci for water-soluble carbohydrates and associations with agronomic traits in wheat. Aust. J. Agric. Res. 59:891–905.
Regan, K.L., K.H.M. Siddique, D. Tennant, and D.G. Abrecht. 1997. Grain yield and water use efficiency of early maturing wheat in low rainfall Mediterranean environments. Aust. J. Agric. Res. 48:595–603.
Reynolds, M.P., G.J. Rebetzke, A. Pellegrineschi, and R. Trethowan. 2006. Drought adaptation in wheat. p. 401–436. *In* J.M. Ribaut (ed.) Drought adaptation in cereals. Haworth Press, New York.
Richards, R.A. 2000. Selectable traits to increase crop photosynthesis and yield of grain crops. J. Exp. Bot. 51:447–458.
Richards, R.A. 2006. Physiological traits used in the breeding of new cultivars for water-scarce environments. Agric. Water Manage. 80:197–211.
Richards, R.A., and J.B. Passioura. 1989. A breeding program to reduce the diameter of the major xylem vessel in the seminal roots of wheat and its effect on grain yield in rain-fed environments. Aust. J. Agric. Res. 40:943–950.
Richards, R.A., H.M. Rawson, and D.A. Johnson. 1986. Glaucousness in wheat: Its development and effect on water-use efficiency, gas exchange and photosynthetic tissue temperatures. Aust. J. Plant Physiol. 13:465–473.
Richards, R.A., G.J. Rebetzke, A.G. Condon, and A.F. van Herwaarden. 2002. Breeding opportunities for increasing the efficiency of water use and crop yield in temperate cereals. Crop Sci. 42:111–121.
Richards, R.A., M. Watt, and G.J. Rebetzke. 2007. Physiological traits and cereal germplasm for sustainable agricultural systems. Euphytica 154:409–425.
Rosegrant, M.W., M.A. Sombilla, R.V. Gerpacio, and C. Ringler. 1997. Global food markets and US exports in the twenty-first century. *In* Illinois World Food and Sustainable Agriculture Program Conf., Urbana-Champaign, IL. 27 May 1997. Univ. Illinois, Urbana-Champaign, IL. (Available at http://66.102.1.104/scholar?hl=en&lr=&q=cache:TkppHyK7FOQJ:www.aces.uiuc.edu/~ILwfood/papers/demsup.PDF+author:%22Rosegrant%22+intitle:%22Global+food+markets+and+US+exports+in+the+twenty-first+ . . . %22+) (verified 20 Dec. 2008.)
Rosielle, A.A., and J. Hamblin. 1981. Theoretical aspects of selection for yield in stress and non-stress environments. Crop Sci. 21:943–946.
Ruuska, S., G.J. Rebetzke, A.F. van Herwaarden, R.A. Richards, N. Fettell, L. Tabe, and C. Jenkins. 2006. Genotypic variation for water soluble carbohydrate accumulation in wheat. Funct. Plant Biol. 33:799–809.

Sadras, V.O., and J.F. Angus. 2006. Benchmarking water-use efficiency of rainfed wheat in dry environments. Aust. J. Agric. Res. 57:847–856.

Sadras, V.O., and D. Rodriguez. 2007. The limit to wheat water-use efficiency in eastern Australia: II. Influence of rainfall patterns. Aust. J. Agric. Res. 58:657–669.

Saini, H.S., and D. Aspinall. 1981. Effect of water deficit on sporogenesis in wheat (*Triticum aestivum* L.). Ann. Bot. 48:623–633.

Saulescu, N.N., and W.E. Kronstad. 1995. Growth simulation outputs for detection of differential cultivar response to environmental factors. Crop Sci. 35:773–778.

Sayre, K.D., E. Acevedo, and R.B. Austin. 1995. Carbon isotope discrimination and grain yield for three bread wheat germplasm groups grown at different levels of water stress. Field Crops Res. 41:45–54.

Schillinger, W.F., E. Donaldson, R.E. Allan, and S.S. Jones. 1998. Winter wheat seedling emergence from deep sowing depths. Agron. J. 90:582–586.

Schnyder, H. 1993. The role of carbohydrate storage and redistribution in the source–sink relations of wheat and barley during grain filling—a review. New Phytol. 123:233–245.

Searle, S.R. 1961. Phenotypic, genetic and environmental correlations. Biometrics 17:474–480.

Shearman, V.J., R. Sylvester-Bradley, R.K. Scott, and M.J. Foulkes. 2005. Physiological processes associated with wheat yield progress in the UK. Crop Sci. 45:175–185.

Simmonds, N.W. 1991. Selection for local adaptation in a plant breeding programme. Theor. Appl. Genet. 82:363–367.

Singh, M., R.S. Malhotra, S. Ceccarelli, A. Sarker, S. Grando, and W. Erskine. 2003. Spatial variability models to improve dryland field trials. Exp. Agric. 39:151–160.

Sirault, X.R.R. 2007. Leaf rolling in wheat. PhD diss. Australian National Univ.

Sirault, X.R.R., A.G. Condon, G.J. Rebetzke, and G.D. Farquhar. 2008. Genetic analysis of leaf rolling in wheat. *In* R. Appels, R. Eastwood, E. Lagudah, P. Langridge, M. Mackay, L. McIntyre, and P. Sharp et al. (ed.) Proc. Int. Wheat Genetics Symp., 11th, Brisbane, Australia. 24–28 Aug. 2008. Sydney University Press, Sydney, Australia. (Available at http://ses.library.usyd.edu.au/bitstream/2123/3213/1/P213.pdf) (Verified 20 Dec. 2008.)

Sirault, X.R.R., N. Fettell, A.G. Condon, and G.J. Rebetzke. 2004. Does leaf rolling slow water use to maintain leaf area in a terminal drought? p. 52–55. *In* C.K. Black, J.F. Panozza, and G.J. Rebetzke (ed.) Cereals 2004—Proc. 54th Australian Cereal Chem. Conf. and 11th Wheat Breeders Assembly, Canberra, Australia. 21–24 Sept. 2004. Aust. Cereal Chem. Soc., Melbourne, Australia.

Slafer, G.A., J.L. Araus, C. Royo, and L.F.G. Del Moral. 2005. Promising eco-physiological traits for genetic improvement of cereal yields in Mediterranean environments. Ann. Appl. Biol. 146:61–70.

Smith, A.B., B.R. Cullis, and R. Thompson. 2001. Analyzing variety by environment data using multiplicative mixed models and adjustments for spatial field trends. Biometrics 57:1138–1147.

Snape, J.W., K. Butterworth, E. Whitechurch, and A.J. Worland. 2001. Waiting for fine times: Genetics of flowering time in wheat. Euphytica 119:185–190.

Solomon, K.F., and M.T. Labuschagne. 2004. Inheritance of evapotranspiration and transpiration efficiencies in diallel F_1 hybrids of durum wheat (*Triticum turgidum* L. var. *durum*). Euphytica 136:69–79.

Spano, G., N. Di Fonzo, C. Perrotta, C. Platani, G. Ronga, D.W. Lawlor, J.A. Napier, and P.R. Shewry. 2003. Physiological characterization of 'stay green' mutants in durum wheat. J. Exp. Bot. 54:1415–1420.

Spielmeyer, W., and R.A. Richards. 2004. Comparative mapping of wheat chromosome 1AS which contains the tiller inhibition gene (*tin*) with rice chromosome 5S. Theor. Appl. Genet. 109:1303–1310.

Steele, K.A., D.S. Virk, R. Kumar, S.C. Prasad, and J.R. Witcombe. 2007. Field evaluation of upland rice lines selected for QTLs controlling root traits. Field Crops Res. 101:180–186.

Taylor, H.M., W.R. Jordan, and T.R. Sinclair. 1983. Limitations to efficient water use in crop production. ASA, CSSA, SSSA, Madison, WI.

Thomas, H., and C.J. Howarth. 2000. Five ways to stay green. J. Exp. Bot. 51:329–337.

Trethowan, R.M., M. Reynolds, K. Sayre, and I. Ortiz-Monasterio. 2005. Adapting wheat cultivars to resource conserving farming practices and human nutritional needs. Ann. Appl. Biol. 146:405–413.

Trethowan, R.M., M. van Ginkel, and S. Rajaram. 2002. Progress in breeding wheat for yield and adaptation in global drought affected environments. Crop Sci. 42:1441–1446.

Tsunewaki K., and K. Ebana. 1999. Production of near-isogenic lines of common wheat for glaucousness and genetic basis of this trait clarified by their use. Genes Genet. Syst. 74:33–41.

Tuberosa, R., and S. Salvi. 2006. Genomics-based approaches to improve drought tolerance of crops. Trends Plant Sci. 11:405–412.

Tuberosa, R., S. Salvi, M. Corinna, C. Sanguineti, P. Landi, M. Maccaferri, and S. Conti. 2002. Mapping QTLs regulating morpho-physiological traits and yield: Case studies, shortcomings and perspectives in drought-stressed maize. Ann. Bot. 89:941–963.

Ud-Din, N., B.F. Carver, and A.C. Clutter. 1992. Genetic analysis and selection for wheat yield in drought-stressed and irrigated environments. Euphytica 62:89–96.

Umezawa, T., M. Fujita, Y. Fujita, K. Yamaguchi-Shinozaki, and Y. Shinozaki. 2006. Engineering drought tolerance in plants: Discovering and tailoring genes to unlock the future. Curr. Opin. Biotechnol. 17:113–122.

van Herwaarden, A.F., J.F. Angus, R.A. Richards, and G.D. Farquhar. 1998. 'Haying-off', the negative grain yield response of dryland wheat to nitrogen fertilizer: II. Carbohydrate and protein dynamics. Aust. J. Agric. Res. 49:1083–1093.

van Herwaarden, A.F., and R.A. Richards. 2002. Water soluble carbohydrate accumulation in stems is related to breeding progress in Australia wheats. p. 878–882 *In* J.A. McComb (ed.) Plant breeding for the 11th millennium.

Proc. Australasian Plant Breeding Conf., 12th, Perth, Western Australia. 12–15 Sept. 2002. Aust. Plant Breeding Soc., Perth, Western Australia.

Verma, V., M.J. Foulkes, A.J. Worland, R. Sylvester-Bradley, P.D.S. Caligari, and J.W. Snape. 2004. Mapping quantitative trait loci for flag leaf senescence as a yield determinant in winter wheat under optimal and drought-stressed environments. Euphytica 135:255–263.

Voltas, J., I. Romagosa, A. Lafarga, A.P. Armesto, A. Sombrero, and J.L. Araus. 1999a. Genotype by environment interaction for grain yield and carbon isotope discrimination of barley in Mediterranean Spain. Aust. J. Agric. Res. 50:1263–1271.

Voltas, J., F.A. van Eeuwijk, J.L. Araus, and I. Romagosa. 1999b. Integrating statistical and ecophysiological analyses of genotype by environment interaction for grain filling of barley: II. Grain growth. Field Crops Res. 62:75–84.

Waines, J.G., and B. Ehdaie. 2007. Domestication and crop physiology: Roots of green-revolution wheat. Ann. Bot. 100:991–998.

Wang, J., D.G. Bonnett, S.C. Chapman, G.J. Rebetzke, and J. Crouch. 2007. Efficient use of marker-based selection in plant breeding. Crop Sci. 47:580–588.

Watt, M., J.A. Kirkegaard, and G.J. Rebetzke. 2005. A wheat genotype developed for rapid leaf growth copes well with the physical and biological constraints of unploughed soil. Funct. Plant Biol. 32:695–706.

Whan, B.R., G.P. Carlton, and W.K. Anderson. 1991. Potential for increasing early vigour and total biomass in spring wheat: I. Identification of genetic improvements. Aust. J. Agric. Res. 42:347–361.

Whan, B.R., G.P. Carlton, and W.K. Anderson. 1996. Potential for increasing rate of grain growth in spring wheat: I. Identification of genetic improvements. Aust. J. Agric. Sci. 47:17–31.

Xiao, B., Y. Huang, N. Tang, and L. Xioing. 2007. Over-expression of a *LEA* gene in rice improves drought resistance under the field conditions. Theor. Appl. Genet. 115:35–46.

Xue, G.P., C.L. McIntyre, S.C. Chapman, N.I. Bower, H.M. Way, A. Reverter, B. Clarke, and R. Shorter. 2006. Differential gene expression of wheat progeny with contrasting levels of TE. Plant Mol. Biol. 61:863–881.

Xue, G.P., C.L. McIntyre, C.L. Jenkins, D. Glassop, A.F. van Herwaarden, and R. Shorter. 2008. Molecular dissection of variation in carbohydrate metabolism related to water soluble carbohydrate accumulation in stems of wheat (*Triticum aestivum* L.). Plant Physiol. 146:441–454.

Yang, D., R. Jing, X. Chang, and W. Li. 2007. Identification of quantitative trait loci and environmental interactions for accumulation and remobilization of water-soluble carbohydrates in wheat (*Triticum aestivum* L.) stems. Genetics 176:571–584.

Yoshimura, K., A. Masuda, M. Kuwano, A. Yokota, and K. Akashi. 2008. Programmed proteome response for drought avoidance/tolerance in the root of a C3 xerophyte (wild watermelon) under water deficits. Plant Cell Physiol. 49:226–241.

Zhou, Y., Z.H. He, X.C. Xia, and X.K. Zhang. 2007. Genetic improvement of grain yield and associated traits in the northern China winter wheat region from 1960 to 2000. Crop Sci. 47:245–253.

Chapter 12
Cutting Down on Weeds to Cut a Cleaner Wheat Crop

Drew J. Lyon, Robert E. Blackshaw, and Gurjeet S. Gill

SUMMARY

(1) Weeds compete with wheat for various growth factors, including nutrients, light, and water, resulting in reduced wheat yield and diminished profitability.

(2) Weeds can compromise wheat grain quality and marketability by reducing grain protein content, contributing to dockage and/or foreign material levels, and by increasing grain deterioration in storage.

(3) Herbicides have been hailed as one of the most important advances in agriculture and they now typically comprise 20%–30% of input costs in North American cropping systems. High input costs, crop injury and herbicide carryover concerns, the increasing incidence of herbicide-resistant weeds, and public concerns about the environmental and human health effects of pesticides are forcing reassessment of the heavy reliance on herbicides for weed management.

(4) More than 300 unique herbicide-resistant weed biotypes belong to more than 180 species around the world. The majority of herbicides used in wheat belong to just two site-of-action groups: Group 2, inhibitors of acetolactate synthase (ALS), and Group 4, synthetic auxins. Herbicide resistance in weeds is a major concern for wheat production worldwide.

(5) Integrated weed management (IWM) systems rely on the use of multiple tactics for weed control. Tactics include prevention, cultural practices such as crop rotation, mechanical methods such as tillage, and herbicide application.

(6) The patchy nature of many weed infestations provides an opportunity to reduce herbicide use through precision farming technologies, thereby reducing production costs and environmental impacts.

(7) In North America, intensification and diversification of the traditional wheat–fallow rotation, combined with the use of conservation or zero tillage (no-till), has reduced the impact of weeds in wheat. The benefits include increased water storage and water-use efficiency, higher annualized grain and forage yields, and greater net profitability.

(8) In Australia, widespread development of herbicide resistance in several important weed species has forced farmers to better manage weed seed production. Some of these management tactics are used in the rotation prior to the wheat phase (e.g., spray-topping pastures or crop-topping pulse crops), while others such as optimizing seeding date, seeding rate or crop density, and cultivar and herbicide choice are implemented during the wheat growing season.

IMPACT OF WEEDS ON WHEAT

Competition

Competition arises when two or more organisms seek a common resource whose supply falls below their combined demand (Donald 1963). The central proposition is that each individual in a population affects, and is affected by, other individuals in the population. Due to their presence in crop communities, weeds consume essential growth factors that would have been otherwise available to the crop. Consequently, crops grow more slowly in the presence of weeds, and they produce lower biomass and grain yield, reducing profitability for the farmer. The balance between supply and demand of various growth factors will vary during the growing season; consequently, the factor most actively competed for could change during the growing season.

Nutrients

Considerable evidence in the literature shows that neighboring plants compete with each other for the supply of essential mineral nutrients. Species or individuals with larger and more effective root systems are likely to absorb a greater share of the nutrient pool, thereby affecting the growth of other members of the plant community. However, even when roots occupy the same depths, competition may be less intense for nutrients with low mobility, for example, phosphate. In contrast, depletion shells of nitrate ions around active roots tend to be large (20–60 mm), and intermingling roots of crops and weeds are likely to differentially affect acquisition of this nutrient (Nye and Tinker 1977).

Manipulation of nitrogen fertilization is a promising cultural practice to reduce weed interference in crops. Nitrogen fertilizer placed in narrow bands below the crop rows compared with surface broadcast has been found to reduce the competitive ability of wild oat (*Avena fatua* L.) (Kirkland and Beckie 1998), foxtail barley (*Hordeum jubatum* L.) (Blackshaw et al., 2000a), and jointed goatgrass (*Aegilops cylindrica* Host) (Mesbah and Miller 1999) with wheat. The reduction in weed interference due to N placement (banding) is likely to be the result of improved access to the nutrient by the crop.

Increasing nitrogen inputs under weedy situations can exacerbate crop–weed competition. Carlson and Hill (1986) reported an increase in wheat yield loss from wild oat in treatments where preplant fertilizer nitrogen was applied (Table 12.1). It appears that the differences in responsiveness to nitrogen between the two species intensified competition for other growth factors such as light and water.

Light (shading)

In mixed plant communities, individuals of a more competitive species can reduce the amount of light reaching the leaves of neighboring plants of other competing species as a result of having greater height or leaf area. Due to interception by plant leaves, light quantity and quality (ratio of red to far-red wavelength) change as light passes through the canopy. Cudney et al. (1991) conducted a field study on wild oat–wheat competition under nonlimiting nitrogen and moisture conditions. They found that wild oat grew taller than wheat and had a greater proportion of its canopy above 60 cm at anthesis and maturity. The mathematical model developed by Cudney et al. (1991) predicted that competition from wild oat was due to reduced leaf area of wheat at early growth stages (possibly due to competition for soil factors) and reduced light penetration to wheat leaves at later growth stages (competition for light).

Table 12.1 Effect of wild oat density and the rate of preplant nitrogen on yield loss of spring wheat.

	Preplant N Rate (kg ha^{-1})		
	0	67	134
Wild Oat Density (plants m^{-2})	% loss		
8	8	19	20
16	15	31	30
32	23	45	55

Source: Adapted from Carlson and Hill (1986).

Water

Intermingled root systems at least partly exploit the same pool of water. The main evidence for this comes from calculations which show that the water potential at the midpoint between neighboring roots is similar to that at the root surfaces, until most of the available water has been removed from the soil (Lawlor 1972). Since plants suffer from water deficits in many parts of the world, at least in some periods on some soils, competition for water is likely to be widespread.

Convincing evidence for crop–weed competition comes from a study on soybean [*Glycine max* (L.) Merr.] in the US by Jones et al. (1997), in which they measured the impact of weeds on the sap flow through the crop. They found that weed-induced yield and sap flow reductions were very similar in magnitude, thereby indicating that water deprivation was the primary cause of yield reduction in soybean (Table 12.2). Although no such information is available for wheat, it is plausible that competition for water with weeds would have some significance, particularly during the reproductive phase of the crop.

As stated earlier, the balance between supply and demand of different growth factors will tend to vary during the growing season; as a result, the factor that is most actively competed for at the start of the growing season may not be the same factor most actively competed for later in the season. In the example of Jones et al. (1997), competition for water was the primary cause of weed effects on soybean. In other studies, competition for nitrogen has been shown to be the key factor involved in competition. In most situations, however, it is likely that competition is taking place for several growth factors simultaneously. Furthermore, the level of supply of one factor can influence the severity of competition for another factor. The example of Carlson and Hill (1986) illustrates this principle quite well, where increasing the supply of nitrogen aggravated competitive effects of wild oat on wheat, possibly due to greater competition for light and water.

Wheat grain yield

Weed species present and their density are the major determinants of the impact of weeds on wheat yield and quality. All available evidence indicates that the relationship between weed density and crop yield loss is hyperbolic or exponential in shape. Absolute yield or the relative yield (yield in the presence of weeds divided by weed-free yield) decreases asymptotically with increasing density of weeds. Therefore, all management programs endeavor to achieve weed-free crops, but this is becoming increasingly difficult because of the evolution of herbicide resistance in many major weed species around the world.

At the same weed density, yield loss caused by different weed species varies considerably. This could be due to several factors intrinsic to the weed species, including initial seedling size (related to seed size), relative growth rate, leaf area, canopy architecture, root growth, and distribution. A comparison of the impact of different weed species on wheat yield loss in Australia showed more than a 100-fold difference in percentage yield loss per weed plant (Gill and Davidson 2000). Most agricultural weeds have a

Table 12.2 Sap flow and grain yield of soybean grown either in monoculture, with common cocklebur (*Xanthium strumarium* L.), or with sicklepod (*Cassia obtusitolia* L.) at a fixed density.

Measured Species	Competitor Species	Sap Flow (kg ha^{-1} d^{-1})	Sap Flow Reduction (%)	Soybean Yield[a] (kg ha^{-1})	Yield Loss (%)
Soybean	None	48,000	—	1,553	—
Soybean	Common cocklebur	33,100*[a]	31	962*	38
Soybean	Sicklepod	22,700*	53	792*	51

Source: Adapted from Jones et al. (1997).
[a]Means followed by an asterisk are significantly different ($p < 0.05$) from the corresponding value in the weed-free soybean.

capacity to cause large reductions in wheat grain yield. For example, Anderson (1993) showed that 18 plants m^{-2} of jointed goatgrass reduced wheat grain yield by 27% in eastern Colorado. Similarly, Stougaard and Xue (2004) showed that grain yield of wheat declined by 54% as wild oat density increased to >400 plants m^{-2}.

In some weed species the expression of crop yield loss is heavily season-dependent. In a study over 3 years, Peterson and Nalewaja (1992) showed that wheat grain yield loss from competition with green foxtail [*Setaria viridis* (L.) Beauv.] at a similar density over seasons ranged from 0% to 47%. Environmental factors (early season temperature, precipitation, and soil texture) that influence seedling emergence pattern in green foxtail were found to be extremely important in determining the competitive impact of this weed species on wheat. Earlier, Blackshaw et al. (1981) also reported high year-to-year variability in the competitive effects of green foxtail on wheat yield.

Crop seeding rate, which impacts crop density, can also influence the competitive ability of wheat against weeds and can therefore be a useful weed management tool. In a recent study from Denmark, Olsen et al. (2005) showed that increasing wheat seeding rate from 200 to 700 seeds m^{-2} reduced weed biomass in all three spatial patterns investigated. They also found that weed biomass was lower and crop biomass was higher in planting patterns that represent a lower level of spatial aggregation. However, such results need to be interpreted with caution because there is evidence that at least in some situations, spatially aggregated systems (e.g., wide crop rows sown with discs) can result in much lower and delayed establishment of weed seedlings that tend to be less competitive with the crop (Chauhan et al., 2006a).

Stougaard and Xue (2004) showed that larger wheat seed size and higher density contributed to increased competitive ability of wheat with wild oat in Montana. The combined use of larger seed size and higher seeding rate resulted in a more competitive wheat crop, improving grain yields by 30%. Benefits from higher wheat density can also be related to suppression in weed reproductive output and lower grain contamination. In some experiments in the US Great Plains, Kappler et al. (2002) found that grain contamination (dockage) from jointed goatgrass was reduced at the rate of 6% for every 10 additional wheat plants above the threshold wheat density of 70 plants m^{-2}. Based on this research they concluded that increased seeding rates may be a good long-term investment as part of an integrated jointed goatgrass control program in winter wheat.

Martin et al. (1987) in New South Wales (Australia) showed that the optimum seeding rate of wheat increased in the presence of wild oat. However, they concluded that the gains from increasing wheat density beyond the weed-free optimum (100 plants m^{-2}) were too small to make it a viable alternative to herbicide use or crop rotation for wild oat control. More recently, Lemerle et al. (2004) reported that doubling wheat plant density from 100 to 200 plants m^{-2} halved rigid ryegrass (*Lolium rigidum* Gaudin) shoot dry matter.

Wheat grain quality and marketability

The preponderance of research reporting the effects of weeds on wheat have dealt with the impact of weed competition on grain yield. There are few reports in the literature on the effects of weeds on grain quality or nutritional value (Zimdahl 1990). The few reports that do exist paint a mixed picture of the importance of weeds to wheat grain quality.

Grain protein is an important quality factor in wheat. Several studies, including field and greenhouse studies, found no effect of weed competition on wheat grain protein or nitrogen content (Bell and Nalewaja 1968; Rooney 1991; Das and Yaduraju 1999). Other field studies found weeds reduced wheat grain protein in some years or locations (Young et al., 1994; Mason and Madin 1996). Still other studies noted significant increases in spring wheat protein content in some years or locations (Friesen et al., 1960; Nakoneshny and Friesen 1961). In a greenhouse study conducted at ambient and elevated carbon dioxide levels, suppression of grain nitrogen content increased as the size of the competing weed species increased (Thompson and Woodward 1994). Mason and Madin (1996) proposed that the effect of weeds on wheat grain protein is complex and depends on the interspecific competition for nitrogen and water at the end of the season and the relative importance of the competition for nitrogen and water in a particular situation.

Test weight is a critical wheat trade factor and serves as an indicator of milling quality. Wheat grain from plots infested with wild mustard [*Brassica kaber* (DC.) L.C. Wheeler] had higher test weight than from plots where wild mustard was controlled with herbicides or by hand weeding. The wheat kernels from weedy plots were also more uniform in size than those from plots where weeds were controlled (Burrows and Olson 1955). These unexpected findings likely reflected reduced tiller density of wheat in the presence of wild mustard, which resulted in fewer kernels per unit area but more uniform kernel size.

Weed seeds frequently contribute to grain dockage levels (Zimdahl 1990; Donald and Ogg 1991; Justice et al., 1994; Koscelny and Peeper 1997). Dockage consists of nonmillable material such as weed seeds, chaff, stems, and stones that can be removed from grain because it differs from grain in weight, size, or both (Webb et al., 1995). Dockage increases the chance of heating and deterioration of stored grain by supporting infestations of insects, microbes, or both (Jian et al., 2005). Populations of *Liposcelis bostrychophilus* Badonel, a cosmopolitan insect pest of stored grain, increased the most (10- to 16-fold) on cracked wheat grain, screenings that included various weed seeds, and wild buckwheat (*Polygonum convolvulus* L.) seed (Mills et al., 1992). Insects damage wheat grain by direct feeding, which reduces grain quality through reduction in grain weight, nutritional value, or germination. They also spread and encourage mold germination and increase grain rancidity. The presence of live insects or insect-damaged kernels can lower grain grade, which reduces the value of the grain in the marketplace.

Some weeds and volunteer or feral small grains contribute to foreign material levels in wheat (Dexter et al., 1984; Wray 1993; White et al., 2006). Foreign material is more difficult and costly to remove from grain than dockage material because it is similar to wheat in weight, size, and shape (Webb et al., 1995). The milling and/or baking performance of spring wheat was compromised by the presence of barley (*Hordeum vulgare* L.), wild oat, and domestic oat (*Avena sativa* L.) (Dexter et al., 1984). Foreign material in wheat quickly affects grain grade and consequently the value of the grain in the marketplace.

Weed seed contamination in wheat products can result in loss of export market access (P. Berglund, pers. comm.). For example, wild buckwheat seed can result in discolored specks in durum wheat products and has resulted in loss of market access to Japan for US growers. Wild oat in spring wheat flour can resemble insects in puffed wheat products and has caused loss of market access to South Korea.

Rainfall that occurs as wheat is ripening in the field often encourages weed growth that can challenge harvest operations. Weeds can increase the moisture content of harvested grain, and they reduce harvest speed and grain cleaning efficiency that may result in increased dockage and foreign material. Herbicides may be used prior to harvest to desiccate weeds and facilitate harvest; however, some herbicides used for this purpose can affect grain, flour, and breadmaking quality of hard red spring wheat (Manthey et al., 2004). Glyphosate applied at the hard dough stage can affect rheological properties and breadmaking quality of hard red spring wheat.

Another example of weeds indirectly causing a loss in wheat grain quality occurred in 1991, when teliospores of *Sporisorium neglecta* from infected heads of yellow foxtail [*Setaria glauca* (L.) Beauv.] growing in a Manitoba wheat field resulted in grain discoloration and a lower grain grade because more than 5% of the kernels were "naturally stained" (Thomas et al., 1998). Smut spores from weeds may also be a problem if they are confused with quarantined species such as karnal bunt (*Tilletia indica* Mitra).

CONTROLLING WEEDS WITH INTEGRATED WEED MANAGEMENT SYSTEMS

Preventative control

Prevention has been a cornerstone of weed control throughout history and remains a pillar of integrated weed management (IWM) systems. Preventative approaches are recognized as being highly cost-effective methods of weed management, but they are routinely overlooked (Jordan 1996). This occurs because preventative weed management involves complex integration of many practices with the goal of preventing the introduction, establishment, and dispersal of weed species. Weed prevention is often thought of as something that occurs at a regional or national level, where preventing introduction of new weed species is the main focus. However, weed prevention can be very effective at the individual farm level.

Preventative weed management should be implemented year-round and at all stages of crop production. Planting weed-free crop seed is a good starting point to prevent introduction of new weed species and to reduce overall weed densities. Additionally, crop seed should have high germination and vigor to ensure rapid crop establishment. Early crop emergence and rapid canopy closure will inhibit weed emergence and growth and thus contribute to reduced weed infestations (O'Donovan et al., 2007). Wheat farmers often plant farm-saved seed to reduce input costs, but this is not always a cost-effective practice. Quality of farm-saved wheat seed can be highly variable (Edwards and Krenzer 2006) and weed seed content can be high (Dastgheib 1989). Thus, a more economical long-term approach may be the purchase of certified seed each year.

Agricultural machinery disperses weeds from farm to farm and from field to field within a farm (Thill and Mallory-Smith 1997). Cultivation disperses weed species that propagate vegetatively, by seed, or by both methods. Cultivation will normally transport weed seed, rhizomes, or roots only a few meters within a field, but if propagules are embedded in soil or debris adhering to tractors or tillage implements, then weed migration easily occurs from field to field. This may result in the introduction of a new weed species in a particular field with serious consequences. Careful cleaning of tractor tires, seeders, and tillage equipment should be a high priority for all farmers but is routinely ignored in the rush of getting farm work completed in a timely manner.

Harvesting equipment has been shown to be a significant contributor to weed dispersal (Thill and Mallory-Smith 1997; Blanco-Moreno et al., 2004). Seed dispersal with grain harvesters is dependent on the number of weed seeds remaining on the plant at harvest and varies considerably with seed size and shape. Seed dispersal by combine harvesters has been found to range from 18 to 50 m (Ghersa et al., 1993; Blanco-Moreno et al., 2004). Shirtliffe and Entz (2005) documented that 74% of wild oat seed dispersed from a combine was distributed in the chaff spread on the field. Modifications to harvesters, such as chaff collectors, can markedly reduce in-field weed seed distribution (Matthews et al., 1996; Shirtliffe and Entz 2005). Combine harvesters should be thoroughly cleaned before moving to the next field or farm to prevent long-distance weed transport. Custom combining has been implicated in the spread of jointed goatgrass throughout the wheat-producing regions of the US (Donald and Ogg 1991).

Harvest operations that remove the entire crop plant greatly increase the likelihood that weed seed will also be harvested and potentially dispersed. Transport of hay and straw has been found to be a major source of long-distance movement of weed species (Zimdahl 2007). Thus, care is required when purchasing livestock feed from outside the local area.

Weed seed present in forage and grain fed to livestock can sometimes remain viable after animal ingestion. Weed seed viability in manure will depend on the weed species and on the animal consuming it. Weeds with hard seed coats have the greatest chance of surviving animal digestion. Blackshaw and Rode (1991) found that weed seed viability in beef cattle manure was near zero for grass species such as downy brome (*Bromus tectorum* L.) and foxtail barley but could be as high as 56% to 68% with hard-seeded species such as wild buckwheat, round-leaved mallow (*Malva pusilla* Sm.), and field pennycress (*Thalpsi arvense* L.). The general ranking of the ability of animals to kill weed seeds through ingestion is poultry > goats = sheep > pigs > horses = cattle (Harmon and Keim 1934; Neto et al., 1987). However, a small percentage of viable weed seed will likely remain after passage through all animals and, in many situations, would be adequate to start new weed infestations.

Studies have shown that at least 4 days are required to eliminate weed seeds from the digestive tract of many animals (Neto et al., 1987; Willms et al., 1995). Thus, keeping livestock in a confined area for a few days before moving them to a new field may limit weed seed dispersal through animal feces. This practice may be most useful when feeding forage acquired from outside the local region, thus limiting the spread of introduced species on a farm. The fermentation process of producing forage silage can markedly reduce weed seed viability (Blackshaw and Rode 1991) and thus is one means of reducing weed spread through animal feeds. Composting manure before spreading on agricultural land is another effective means of reducing weed seed dispersal. Compost temperatures of 55 to 60°C are required for several days to kill weed seeds, and compost piles or windrows must be turned a couple of times to ensure that viable weed seed does not persist on the outer edges (Grundy et al., 1998; Eghball and Lesoing 2000; Larney and Blackshaw 2003).

Irrigation water often contains weed seed that originates from species growing along the water corridor or in nearby fields. Kelley and Bruns (1975) documented seed of 137 species in irrigation water and calculated that 10,000 to 94,000 seeds ha^{-1} could be distributed during one season of irrigation. Wilson (1980) found seed of 34 weed species in irrigation water and determined that an irrigated field could receive as many as 48,000 seeds ha^{-1} in one year. Individual farmers may consider installing filters or decanting tanks in their irrigation systems to minimize weed seed dispersal. Alternatively, farmers within an irrigation district could collectively maintain a weed-free zone adjacent to irrigation canals that would greatly reduce weed seed contamination of irrigation water.

Controlling weeds on field margins and along fence rows and roadsides is another simple but highly effective weed preventative practice (Zimdahl 2007). Weed control may be accomplished by using herbicides or by mowing weeds before they produce viable seed. Tarping of grain trucks greatly reduces introduction of weeds on roadsides that may subsequently spread to neighboring fields.

Weed prevention is multifaceted and thus sometimes difficult to implement. It begins with awareness of how weeds spread in agricultural systems and then progresses to individual actions that restrict weed reproduction and spread. Wheat farmers need to be cognizant of the fact that weed preventative measures are among the most cost-effective means of weed control and thus should be given a high priority in IWM systems.

Cultural control

Diverse crop rotations are the cornerstone of all sustainable pest management and crop production systems (Karlen et al., 1994). Monoculture cropping facilitates an increase in weed species that are able to effectively compete with that crop or that are able to overcome competition through some avoidance mechanism (Liebman and Staver 2001). Weed species with similar life cycles to that of the crop (crop mimics) tend to be the greatest problem. Winter annual weeds proliferate in winter crops, and summer annual weeds dominate in spring-planted crops (Moyer et al., 1994).

Long-term rotation studies have demonstrated that the winter annual grass downy brome quickly becomes the dominant weed in continuous winter wheat production in Canada (Blackshaw et al., 2001a). However, by simply including spring canola (*Brassica rapa* L. or *B. napus* L.) in the rotation, downy brome was maintained at sufficiently low densities that crop yield was unaffected (Table 12.3) (Blackshaw 1994b). Loeppky and Derksen (1994) similarly reported that quackgrass [*Elytrigia repens* (L.) Nevski] populations

Table 12.3 Effect of two cropping sequences on downy brome plant density.

Year	Continuous Winter Wheat (plants m^{-2})	Winter Wheat–Canola (plants m^{-2})
1988	30	28
1989	54	25
1990	190	35
1991	400	70
1992	920	38
1993	740	40

Source: Adapted from Blackshaw (1994a).

Table 12.4 Effect of underseeded sweetclover in 1993 on weed growth during fallow in 1994 and spring wheat yield in 1995.

Crop Treatment	Weed Biomass (September 1994) (g m^{-2})	Wheat Yield (1995) (kg ha^{-1})
Field peas (1993)	446	2,970
Mustard (1993)	156	2,160
Field peas + sweetclover (1993–1994)	14	3,750
Mustard + sweetclover (1993–1994)	3	3,520

Source: Adapted from Blackshaw et al. (2001b).

could effectively be controlled through use of diverse crop rotations.

Fallow is often included in rotation with wheat in the semiarid Great Plains of North America. Fallow can effectively reduce weed populations (Blackshaw et al., 2001a; Anderson 2003), but it can negatively affect soil quality and expose the soil to erosion. Research has examined the usefulness of cover crops and green manure crops as partial fallow replacements. Moyer et al. (2000) documented that a winter rye (*Secale cereale* L.) cover crop planted after harvest of summer crops suppressed weed growth in the fall and early spring. Winter rye residue, after terminating the crop at heading in June, continued to suppress weeds for the remainder of the fallow period, likely due to combined physical and allelopathic effects (Teasdale 1996; Weston 1996). Another study found that underseeded biennial sweetclover [*Melilotus officinalis* (L.) Lam] reduced weed establishment after harvest and in the following spring before being terminated at the 90% bloom stage in late June (Blackshaw et al., 2001b). Sweetclover residue provided excellent weed suppression throughout the remaining portion of the fallow year. Wheat yield in the subsequent production year was higher due to fewer weeds and greater nitrogen availability from sweetclover nitrogen fixation (Table 12.4).

Diverse crops grown in rotation with wheat allow for greater herbicide choice over years and may avoid continuous use of the same herbicide with inadvertent selection for weed resistance. Additionally, crop diversity encourages operational diversity that, in turn, can facilitate improved weed management. Different crops are naturally planted and harvested at different times of the year. If sufficient differences exist in germination requirements between the rotational crop and potential weed species, then seeding date can be manipulated to benefit the crop. Early sown spring crops may out-compete weeds that require warmer soil temperatures for germination. For example, densities of the C_4 species green foxtail have declined in early planted spring crops such as canola or field pea (*Pisum sativum* L.) in zero-tillage systems that often have lower soil temperatures (Blackshaw 2005). Conversely, delayed seeding can be used to manage early spring germinating weeds such as kochia [*Kochia scoparia* (L.) Schrad.]. Alternating seeding dates over years is a desired weed management practice and one that farmers should try to implement.

Wheat cultivars can vary considerably in their competitiveness with weeds (Hucl 1998; Lemerle et al., 2001). Winter wheat cultivars have been identified that differ in their competitive ability with downy brome (Blackshaw 1994a). Wheat yield reductions caused by downy brome varied by as much as 30% depending on the cultivar grown (Table 12.5). Tall (non-semidwarf) cultivars (90–110 cm) had a height advantage over downy brome (70 cm) and cultivars with a spreading growth habit provided better interrow shading of downy brome. Increased competitive ability of wheat, or crops in general, has been associated

Table 12.5 Yield reduction of four winter wheat cultivars differing in growth habit when competing with 150 downy brome plants per square meter.

Cultivar	Height (cm)	Growth Habit	Yield Reduction (%) 1989	1990	1991
Norstar	110	Spreading	36	24	51
Redwin	90	Erect	39	33	52
Archer	70	Erect	42	40	63
Norwin	60	Spreading	53	62	81

Source: Adapted from Blackshaw (1994b).

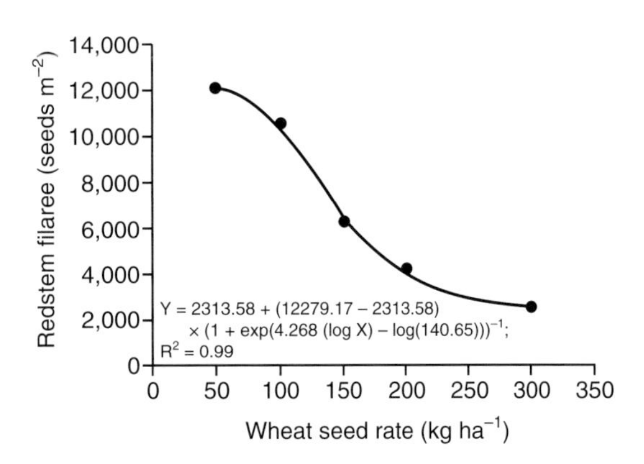

Fig. 12.1. Redstem filaree seed bank as affected by wheat seed rates applied in four consecutive years. [Source: Blackshaw et al. (2000b). Used with permission from the Weed Science Society of America and Allen Press Publishing.]

with early emergence, rapid leaf expansion forming a dense canopy, increased plant height, early vigorous root growth, and increased root size (Seavers and Wright 1999; Lemerle et al., 2001). Thus, wheat competitive ability can also be enhanced by seeding into a firm seedbed at an optimum depth of 3 to 5 cm. Packing the seed row will improve soil-to-seed contact and promote rapid wheat emergence.

The establishment of a wheat crop with a more uniform and dense plant distribution can increase its ability to suppress weeds (Mohler 2001). This is due to more rapid canopy closure that better shades weeds and to better root distribution that improves access to soil nutrients and water. Increasing seeding rate of wheat is one means of increasing its competitive ability with weeds (Walker et al., 2002; Lemerle et al., 2004; O'Donovan et al., 2005). A 4-year study found that an increase in wheat seeding rate from 50 to 300 kg ha^{-1} reduced redstem filaree [*Erodium cicutarium* (L.) L'Her. ex Ait.] biomass by 53% to 95% and increased wheat yield by 56% to 498% (Blackshaw et al., 2000b). Additionally, redstem filaree in the soil seed bank for future weed infestations was reduced by 79% (Fig. 12.1).

There is less potential to manipulate row spacing in wheat than in traditional row crops such as maize (*Zea mays* L.) or soybean, which are commonly grown in wider rows than the 15- to 30-cm row spacing common with wheat. Blackshaw et al. (1999) found that a decrease in wheat row spacing from 30 to 20 cm had little effect on foxtail barley biomass or crop yield in western Canada. However, Mertens and Jansen (2002) reported that reducing wheat row spacing from 30 to 10 cm in The Netherlands consistently reduced weed biomass but wheat yield remained constant.

Many agricultural weeds are high consumers of nutrients and therefore are capable of reducing available nutrients for crop growth (Di Tomaso 1995). Additionally, growth and competitive ability of many weed species is enhanced by higher soil nutrient levels. Research has determined that fertilizer timing and application method can markedly affect wheat–weed competition. Spring-applied versus fall-applied fertilizer often reduced weed biomass and increased spring wheat yield (Blackshaw et al., 2004, 2005a). Nitrogen fertilizer placed as subsurface bands, rather than surface broadcast, reduced the competitive ability of several weed species in wheat (Kirkland and Beckie 1998; Blackshaw et al., 2004). A field study utilizing ^{15}N-enriched liquid nitrogen fertilizer clearly documented greater nitrogen uptake by wheat, and often lower nitrogen uptake by weeds, when nitrogen was placed 10 cm below the soil surface (away from surface-germinating weeds) compared with surface broadcast (Blackshaw et al., 2002). Weed seed-bank data in a multiyear study indicated that the nitrogen fertilizer application method not only affects wheat–weed competition in any given year but is a critical component of long-term weed management (Fig. 12.2).

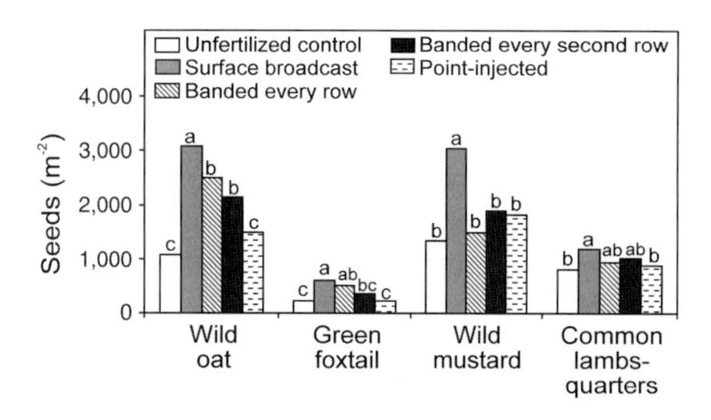

Fig. 12.2. Effect of nitrogen fertilizer application method in four consecutive years on the seed bank of wild oat, green foxtail, wild mustard (*Sinapis arvensis* L.), and common lambsquarters. Data within a weed species with the same letter are not significantly different according to Fisher's protected LSD test at the 5% probability level. [Source: Blackshaw et al. (2004). Used with permission from the Weed Science Society of America and Allen Press Publishing.]

Stubble burning after harvest can provide some measure of weed control (Ramussen 1995; Walsh and Powles 2007). Weed seed death with stubble burning increases with higher temperatures and longer burning duration. Studies have found that short-duration burns of 10 seconds require temperatures of 400 to 500 °C to kill weed seed. However, lower burn temperatures can be effective if the burn period is 50 seconds or longer. Large amounts of crop residue will result in longer burns at higher temperatures. Thus, one means of increasing burning efficiency is to concentrate harvest residues into narrow windrows during the harvest operation. Concentrated windrow burning can kill 99% and 80% of rigid ryegrass and wild radish (*Raphanus raphanistrum* L.) seeds, respectively (Walsh et al., 2005). An added benefit of concentrated windrow burning is that only 10% of the field area is burned, thus minimizing the risk of soil erosion and allowing remaining crop residues to improve or maintain soil quality.

Mechanical control by tillage

Use of tillage prior to planting crops is one of the oldest methods of agricultural weed management. However, the impact of tillage on a weed species depends on the interaction between the nature of soil disturbance and growth stage of the weed. The level of soil disturbance from presowing tillage and sowing operation is a major determinant of the vertical distribution of weed seeds in the soil (Yenish et al., 1992; Chauhan et al., 2006a). This change in the vertical distribution of weed seeds by tillage has recently been shown to result in lower seedling recruitment and greater decay of the seed bank of rigid ryegrass under no-till than the conventional high-disturbance tillage system (Chauhan et al., 2006a). However, in an earlier study, Buhler and Daniel (1988) showed much greater infestation of giant foxtail (*Setaria faberi* Herrm.) under no-till compared with high soil disturbance systems. Such radically different responses of these two grass species to no-till management reflects the complex interactions of weed life-history traits with biotic, abiotic, and management variables experienced at the site.

Concerns have been expressed regarding potential shifts in weed communities as a result of changing tillage practices. A recent study from Canada provided some evidence for weed species preference for different tillage systems. For example, perennial species such as Canada thistle [*Cirsium arvense* (L.) Scop.] and perennial sowthistle (*Sonchus arvensis* L.) were associated with reduced- and zero-tillage systems, but annual species were associated with a range of tillage systems. Some showed distinct preference for a particular type of tillage practice while others such as wild buckwheat and common lambsquarters (*Chenopodium album* L.) were equally abundant in all tillage systems (Thomas et al., 2004).

In Australia, farmers sometimes use a shallow cultivation in autumn to improve weed seed–soil contact, which can stimulate presowing weed germination and improve presowing weed kill (Gill and Holmes 1997). However, the success of this technique tends to be erratic and largely dependent on rainfall received after the initial cultivation. Reduced tillage systems can have a negative impact on weed control activity of herbicides such as trifluralin that are sensitive to photodegradation and volatilization. In studies by Chauhan et al. (2006b), trifluralin was shown to dissipate faster and provide less effective weed control under low disturbance tillage systems.

Chemical control

From the beginning of agriculture, farmers have been controlling weeds by different methods with varying levels of success. The advent of herbicides has been hailed as one of the most important advances in agriculture, and herbicides now typically comprise 20% to 30% of input costs in North American cropping systems (Derksen et al., 2002). Herbicides also hold top position among crop protection chemicals in Australia, where farmers in all cropping regions spend typically AUS$30–80 ha^{-1} on herbicides for weed control (Pratley 1996). In Australia the combination of large cropping areas, expensive labor, short growing seasons, fragile soils, and the imperative to minimize production costs have all contributed toward herbicides remaining a vital component of cropping systems (Walsh and Powles 2004). In most grain cropping systems in the world, herbicides are the dominant method of weed control.

Despite widespread adoption of herbicides, the reduced use of herbicides is attracting ever-increasing interest. Farmers cite crop injury and herbicide carryover, increased incidence of herbicide-resistant weeds, and public concerns about the environmental and human health effects of pesticides as issues forcing reassessment of the practice of weed management (Blackshaw 2006). According to Heap (2007), 314 unique herbicide-resistant weed biotypes belong to 183 species around the world. For a long time, prospects of evolution of resistance in weed populations to glyphosate were considered remote. However, resistance to glyphosate has now been confirmed in 11 weed species worldwide (Heap 2007). With no doubt the problem of herbicide resistance in weeds is truly global, is relevant to all cropping systems, including wheat-based systems, and is still expanding rapidly.

The widespread evolution of herbicide-resistant weed populations within intensive crop production systems would not be a major threat to the sustainability and profitability of cropping systems if new herbicide modes of action were introduced to replace those failing herbicides. However, the rate of introduction of new herbicides for world agriculture has slowed dramatically due to (i) the difficulty in discovering new herbicide modes of action with the necessary environmental properties and (ii) the strategic and substantial reduction in herbicide discovery programs by international agrichemical corporations (Ruegg et al., 2007). In 2005, only 11 companies invested significantly in crop protection research and development, compared with 35 companies in 1985.

A substantial increase in the cost of herbicide development, which is partly related to tightening regulatory systems, has negatively influenced "risk-taking" behavior of the industry. For example, preapproval costs related to toxicology and studies on pesticide fate in the environment have increased from US$28M during 1985–1990 to US$100M during 1990–1996 (Ruegg et al., 2007). Both these factors have contributed to the slowdown in discovery and release of new chemistries for agriculture. As it is very likely that the pace of introduction of new herbicide modes of action capable of controlling herbicide-resistant weeds will remain slow, there is a strong imperative to use the currently available herbicide resources in more sustainable ways. Recognition is growing worldwide that weed management must be built on a solid foundation of good crop husbandry. An integrated approach that incorporates a wide range of weed control methods is required (Walsh and Powles 2004).

A factor critical to long-term weed management is the principle of avoiding heavy reliance on single control methods such as one highly effective herbicide. To enable farmers to use diverse modes of actions and herbicide mixtures, the pesticide industry adopted a uniform system of herbicide classification based on the mechanism of herbicide action in the plant (Table 12.6). Recent simulation modeling research has shown that herbicide rotation in alternate years is markedly less effective as a strategy to delay resistance than the use of herbicide mixtures where both components are used at full rates (Diggle et al., 2003). However, herbicide combination or

Table 12.6 Classification of herbicides commonly used in wheat production according to primary site of action.

Group	Site of Action	Chemical Family	Common Name
2 (B)[a]	Inhibitors of acetolactate synthase (ALS), also called acetohydroxyacid synthase (AHAS)	Imidazolinone	Imazamox[b]
		Sulfonylamino-carbonyltriazolinone	Flucarbazone-sodium
			Propoxycarbazone
		Sulfonylurea	Chlorsulfuron
			Mesosulfuron
			Metsulfuron
			Prosulfuron
			Sulfosulfuron
			Thifensulfuron
			Triasulfuron
			Tribenuron
		Triazolopyrimidine	Florasulam
			Pyroxsulam
3 (K1)	Inhibitors of microtubule assembly	Dinitroaniline	Pendimethalin
			Trifluralin
4 (O)	Synthetic auxins	Phenoxy	2,4-D
			MCPA
		Benzoic acid	Dicamba
		Carboxylic acid	Aminopyralid
			Clopyralid
			Fluroxypyr
		Quinoline carboxylic acid	Quinclorac
6 (C3)	Inhibitors of photosynthesis at photosystem II site A	Nitrile	Bromoxynil
9 (G)	Inhibitor of 5-enolpyruvyl-shikimate-3-phosphate synthase (ESPS)	None	Glyphosate
14 (E)	Inhibitors of protoporphyrinogen oxidase (Protox)	Triazinone	Carfentrazone-ethyl
27 (F2)	Inhibitors of 4-hydroxyphenyl-pyruvate-dioxygenase (4-HPPD)	Pyrazole	Pyrosulfotole

Source: Adapted from Mallory-Smith and Retzinger (2003).
[a]Letter in parentheses is the classification according to the international Herbicide Resistance Action Committee (Schmidt 1998).
[b]Used in imazamox-resistant wheat cultivars only.

mixture strategy has the potential to increase herbicide load in the environment. Furthermore, the cost of using more herbicide is immediate, while the returns from delayed occurrence of herbicide resistance will be realized in the future (Diggle et al., 2003). A similar approach to population genetics modeling has produced recommendations of a judicious sequence of glyphosate followed by paraquat (double knockdown) to decrease the likelihood of glyphosate-resistance evolution (Neve et al., 2003).

The development of imidazolinone-tolerant wheat has provided producers with an effective tool for the management of winter annual grass weeds such as feral rye and jointed goatgrass, which have been difficult to control in winter wheat in the US and Canada (Geier et al., 2004). The adoption of imidazolinone-tolerant wheat is also increasing rapidly in southern Australia, primarily for the control of rigid brome (*Bromus rigidus* Roth). However, concerns arise about imidazolinone residues affecting performance of subsequent crops grown in the rotation. Furthermore, Group 2 herbicides have been repeatedly shown to be prone to resistance development in weeds and thus need to be integrated with other herbicide groups and nonchemical methods of weed control. It is possible that genetically modified wheat cultivars resistant to herbicides such as glyphosate or glufosinate could become available for production in the future. Considering the high value of glyphosate to the wheat industry, it is imperative that these cultivars be used in a manner that ensures long-term viability of this herbicide.

Biological control

Examples of commercially successful biological control are rare in annual cropping systems. Rhizobacteria have been screened for use as possible biological control agents of downy brome and jointed goatgrass in winter wheat (Kennedy et al., 1991; Kennedy and Stubbs 2007). Chemical analysis of the active fraction complex from strain D7 of *Pseudomonas fluorescens*, which showed promise for biocontrol of downy brome in agar studies, identified chromopeptides and other peptides, fatty acid esters, and a lipopolysaccharide matrix (Gurusiddaiah et al., 1994). Separation of any of the components of the matrix resulted in nearly total loss of activity. Despite some limited experimental success with rhizobacteria for biocontrol of weeds in wheat, no commercial application of this technology is currently available.

Cromar et al. (1999) contend that while most biological control efforts for weeds have centered on predators with specific feeding habits, invertebrates with opportunistic feeding strategies may provide the most effective broad-spectrum weed control. They urged use of management practices that conserve and encourage beneficial arthropod populations.

WEED SPATIAL VARIATION AND PRECISION FARMING

Spatial heterogeneity is common in crop production fields. Heterogeneity is observed as nonuniform plant vigor, crop yield, soil color, and pest infestation (Mortensen et al., 1998). Weed density and composition are not uniformly distributed across fields. Many weed populations are aggregated in patches of varying size and shape (Gerhards and Oebel 2006). Managing weed populations on less than a whole-field basis, also known as site-specific weed management, could result in a significant reduction in herbicide use with concomitant economic and environmental benefits (Mortensen et al., 1998; Timmermann et al., 2003).

The potential for herbicide use reduction with site-specific weed management varies between crops. Larger reductions are possible in competitive crops like wheat than in less competitive row crops like sugar beet (*Beta vulgaris* L.) or maize. In a 4-year experiment conducted on five fields, grass-herbicide use reductions using a GPS-guided sprayer were 90% in winter cereals (barley and wheat), 78% in maize, and 36% in sugar beet (Timmermann et al., 2003). Broadleaf herbicide use was reduced 60% in winter cereals, 11% in maize, and 41% in sugar beet. The cost of wild hemp (*Cannabis sativa* L. ssp. *spontanea*) control in Hungary was reduced 9%, and herbicide application was reduced 34%, with the use of site-specific weed control methods compared with broadcast spraying of the entire field (Reisinger et al., 2005). Wild hemp is very competitive against winter wheat and frequently forms large and stable patches in agricultural fields.

Site-specific weed management systems require accurate detection and location of weeds in fields. Two general approaches to weed detection in cultivated crops are available. One approach uses morphological differences such as leaf shape, hairiness, shininess, or plant structure to differentiate between the crop and weeds, while the other approach uses differences in spectral reflectance for plant differentiation (Girma et al., 2005). The detection of multiple weed species within a crop is challenging, particularly when attempting to detect grass weeds in a grass crop such as wheat.

Weed distribution is often associated with agronomic practices and field characteristics (Mortensen et al., 1998). Weed presence in Danish cereal grain fields was most dependent on cereal grain species and clay content of the soil (Andreasen et al., 1991). Large-seeded broadleaf species were associated with higher organic matter and lower elevation in Nebraska maize fields, while annual grasses were found in well-drained, higher areas (Mortensen et al., 1998). The correlation between soil properties and weed distribution may be used to improve the estimation of weed distribution by use of co-kriging methods. The prediction variance of *Lamium* spp. was improved 11% when silt content was included in co-kriging compared to kriging weed sample data alone (Heisel et al., 1999), indicating a preference of different weed species for certain soil characteristics.

Site-specific weed control has great potential to reduce herbicide use, but camera and application technologies, along with image analysis algorithms, must be improved before it is commercially viable (Gerhards and Oebel 2006). Herbicide savings with site-specific management must compensate for the costs of weed sampling, weed mapping, data processing, decision making, and site-specific application technology if it is to replace whole-field spraying (Timmermann et al., 2003).

PUTTING IT ALL TOGETHER: EXAMPLES OF EFFECTIVE SYSTEMS

Winter wheat in North America: Winter wheat–summer crop–fallow

In the US Great Plains dryland agriculture has developed around wheat production. Summer fallow, the practice of controlling all plant growth during the noncrop season, was quickly adopted to stabilize winter wheat production in the region. Winter wheat–fallow was the predominant crop rotation in the central Great Plains during most of the 20th century (Baumhardt and Anderson 2006).

Downy brome, jointed goatgrass, and feral rye cause significant economic loss in winter wheat–fallow production regions of the western US, particularly where conservation tillage is used (Lyon and Baltensperger 1995). In addition to wheat yield loss, jointed goatgrass and feral rye seed frequently contaminate winter wheat grain, resulting in economic loss from dockage and grade reduction. These three winter annual grass weeds have a similar life cycle and physiology to winter wheat, which limits effective control methods.

Plowing with a moldboard plow can effectively control these weeds (Donald and Ogg 1991; Kettler et al., 2000; Stump and Westra 2000), but it buries nearly all surface crop residues. Maintaining crop residues on the soil surface protects soil from wind and water erosion and increases the storage of soil water (Unger et al., 2006).

Several herbicides, including sulfosulfuron, propoxycarbazone, pyroxsulam, and imazamox, can provide selective control of one or more of these grass weeds. Imazamox can only be used with imazamox-tolerant wheat cultivars or serious crop injury will occur. Concerns with these products include high cost, long soil residual that restricts rotation flexibility, and development of weed resistance with frequent use.

Crop rotation with late spring-planted crops effectively controls these winter annual grass weeds in winter wheat (Daugovish et al., 1999). Table 12.7 shows the impact of adding a late spring-planted crop to the winter wheat–fallow rotation on jointed goatgrass seedling density, wheat grain contamination, and the soil seed bank. Similar results were reported for feral rye. Growing a winter wheat crop every 3 or 4 years rather than every other year promotes depletion of the soil seed bank as long as no plants are allowed to produce seed during the nonwheat portion of the rotation. This is easily accomplished if the rotational crops are not growing during the early spring when the use of nonselective herbicides or tillage can be used to kill emerged plants.

The intensification and diversification of the winter wheat–fallow rotation has not only helped to control winter annual grass weeds in wheat, but it has also increased water storage and water-use efficiency by reducing the length of the 16-month fallow period to 10 or 11 months and by replacing soil evaporation by crop transpiration (Baumhardt and Anderson 2006). Compared to winter wheat–fallow, winter wheat–summer crop–fallow has increased annualized grain and forage yields, increased net profitability, increased potentially active surface-soil organic C and N, and reduced yield loss in wheat due to soilborne disease.

Table 12.7 Jointed goatgrass seedling density in the spring and spikelet density after harvest, partitioned into harvested grain and the soil seed bank, after 6 years in a crop rotation study conducted at Sidney, Nebraska, from 1990 through 1997.

Crop Rotation[a]	Seedling Density	Harvested Grain	Soil Seed Bank
	Plants m^{-2}	------Spikelets m^{-2}------	
WW-Ft	9.6	86	548
WW-Fh	17	95	1,010
WW-F-F	0.01	2.0	0
WW-SF-F	0.15	1.3	0
WW-PM-F	0.07	2.6	0
Significance of contrast[b]			
2-yr vs. 3-yr	*	*	*
WW-Ft vs. WW-Fh	*	NS	*
Within 3-yr	NS	NS	NS

Source: Adapted from Daugovish et al. (1999).
[a]WW-Ft, winter wheat–fallow with fall tillage; WW-Fh, winter wheat with fall herbicides; WW-F-F, winter wheat–fallow–fallow; WW-SF-F, winter wheat–sunflower (*Helianthus annuus* L.)–fallow; WW-PM-F, winter wheat–proso millet (*Panicum miliaceum* L.)–fallow.
[b]NS and * indicate not significant and significant, respectively, at the 0.05 probability level.

Seed production of winter annual grass weeds can be reduced by combining cultural practices. Feral rye and jointed goatgrass seed production was reduced by applying nitrogen fertilizer 5 months before wheat seeding, increasing the wheat seeding rate, and planting a standard height cultivar (Anderson 1997). Standard height cultivars frequently yield less than many semidwarf cultivars. If a semidwarf cultivar is used, row spacing can be reduced to help compensate for the loss in weed competitiveness.

By combining crop rotation, cultural practices, and herbicides, winter annual grass weeds are effectively controlled in winter wheat and the entire cropping system is made more sustainable. The winter wheat–summer crop–fallow system is an example of an effective IWM system with implications beyond weed control.

Table 12.8 Crop production changes in Canada from 1995 to 2005.

Crop or System	2005 (million ha)	Change from 1995 (%)
Wheat	10	−20
Canola	5.5	+55
Barley	5	−5
Oat	2	−3
Field pea	1.4	+250
Corn	1.3	+3
Soybean	1.1	+35
Lentil	1	+190
Flax	0.8	+40
Canaryseed	0.4	+45
Mustard	0.25	−10
Dry bean	0.2	+110
Sunflower	0.1	+155
Forages	7.5	+20
Zero tillage	20	+100
Fallow	4	−70

Source: Statistics Canada (2006).

Spring wheat in North America

Zero tillage (no-till) has become a widely adopted agronomic practice in the spring wheat production areas of Canada and the northern US (Table 12.8). Research has shown that some weed species may become more prevalent with zero tillage, but overall weed densities decline with time (Derksen et al., 2002; Anderson 2003; Blackshaw 2005). Weed seed mortality tends to be greater when weed seeds are left on the soil surface compared with when they are buried in the soil with tillage. Additionally, crop residues on the soil surface may inhibit weed germination and growth through physical suppression and/or allelopathic interactions. Thus, zero tillage has contributed greatly to improved weed management as well as higher spring wheat yields.

Improved soil moisture conservation with zero tillage has allowed a greater variety of crops to be grown in recent years in the semiarid Canadian

Prairies (Table 12.8). Wheat-based rotations now include more oilseeds [e.g., canola, flax (*Linum usitatissimum* L.)] and pulses [e.g., field pea and lentil (*Lens culinaris* Medik.)]. Inclusion of forages such as alfalfa (*Medicago sativa* L.) or red clover (*Trifolium pretense* L.) in rotation with spring wheat, with the main goal of managing weeds, is gaining acceptance in areas where forage demand is high. Survey results indicate that 83% of farmers had lower weed densities after 2 to 4 years of forage production (Entz et al., 1995). Diverse crop rotations have resulted in lower weed populations in spring wheat.

Weed management in spring wheat can be improved by including fall-seeded crops in rotation (O'Donovan et al., 2007). Many spring-germinating weeds emerge after canopy closure of fall-seeded crops, which makes them noncompetitive. Winter wheat, winter rye, and winter triticale (*Triticosecale* spp.) are being more widely grown on the Canadian Prairies. Systematically changing planting dates and crop species prevents any one weed species from developing into a major problem (Derksen et al., 2002).

Spring wheat farmers in Canada are slowly but surely adopting IWM systems. Foxtail barley is an example of a weed species that became a greater problem with zero tillage (Blackshaw 2005). However, Blackshaw et al. (1999) determined that good control of this weed could be attained by combining crop rotation, higher wheat seeding rates, banded nitrogen fertilizer, and timely herbicide use in a multiyear approach. Farmer adoption of an IWM system for foxtail barley was one of the first success stories, and it occurred in part because the need was so great. Farmers will readily adopt new practices when they perceive a need for change and when those practices are effective and affordable.

Another multiyear study conducted at three locations assessed the merits of combining several crop production practices to manage weeds in the context of full or reduced herbicide rates in spring wheat and other major field crops of the Canadian Prairies (Blackshaw et al., 2005a,b). Factors included in the study were crop rotation, seeding date, seeding rate, fertilizer timing, and herbicide rate. The combination of earlier seeding date (3 weeks earlier), higher crop seeding rate (50% higher), and spring-applied subsurface-banded fertilizer resulted in the most competitive cropping system. Weeds were controlled with this IWM approach and it is notable that the weed seed bank was not greater after four continuous years of using 50% herbicide rates in a competitive cropping system at two of three sites. Farmers were impressed with the level and consistency of weed control in this study but were only truly convinced of the merits of these IWM systems when they were shown to be economically viable (Smith et al., 2006).

Spring wheat in Australia

Widespread development of herbicide resistance in several important weed species has forced Australian farmers to reassess their approach to weed management. A 2003 survey across the Western Australian wheatbelt showed that only 8% of rigid ryegrass and 17% of wild radish populations were susceptible to all the most commonly used in-crop selective herbicides (Owen et al., 2007; Walsh et al., 2007). Even with such widespread resistance, herbicides have remained an important component of weed management; but much greater thought is given to integration of herbicides with nonchemical weed control tactics.

Ley farming systems, in which pastures were rotated with wheat, was the most widely used farming system in Australia for many years. Although the system has lost some of its popularity due to changes in commodity prices, pastures are still rotated with wheat production over large areas of farm land. In this system, pastures offer an excellent opportunity to control herbicide-resistant rigid ryegrass with grazing by sheep and with nonselective herbicide application soon after rigid ryegrass flowers (Gill and Holmes 1997). This practice, referred to as *spray-topping*, has played an important role in the management of herbicide-resistant weed populations in pasture–wheat rotations in southern Australia. Gill and Holmes (1997) reported that the use of spray-topping in the year prior to cropping wheat, followed by shallow autumn cultivation before wheat sowing the next year, can provide rigid ryegrass control similar to that by selective herbicides. However, this system has greater complexity and requires effective integration of grazing pressure to prevent weeds that escape nonselective herbicides from setting seed (Table 12.9).

Opportunities exist to integrate different weed control tactics in a continuous cropping system to complement use of selective herbicides. A delay in seeding of wheat can achieve greater recruitment from the seed bank of rigid ryegrass (Gill and Holmes 1997). These presowing weed cohorts can then be killed by the application of glyphosate or paraquat as resistance to these nonselective herbicides is still rare in southern Australia. However, the effectiveness of this tactic is dependent on rainfall, which can be erratic, and wheat yield penalties up to 20% can occur when seeding is delayed 3 to 4 weeks past the optimum date (Anderson and Sawkins 1997; Hocking and Stapper 2001).

Table 12.9 Effectiveness of different weed control tactics on the management of annual ryegrass in southern Australia.

Weed Control Tactic	Effectiveness in Annual Ryegrass Control (%)
Autumn	
Stubble grazing	<20
Autumn burning of stubble	70 (20–95)[a]
Autumn cultivation to stimulate weed germination	25 (10–40)
Delayed sowing to kill weeds	40 (10–70)
Winter options	
Crop species, cultivar, seed rate (seed set reduction)	10–50
Weed kill with selective herbicides (herbicide selection based on resistance status)	90 (70–99)
Spring	
Spray-topping/crop-topping	85 (50–95)
Green manure crops (sprayed with glyphosate)	90 (70–95)
Hay cutting	80 (65–90)
Weed seed capture at harvest by chaff carts	60 (40–80)

Source: Adapted from Gill and Holmes (1997).
[a]Values in parentheses represent the range of control achieved.

Other weed management tactics that are being adopted by growers include higher crop seeding rates for improved crop competition with weeds (Lemerle et al., 2004). Crop-topping is another practice that has become important for weed population management for growers engaged in continuous cropping. In this technique, paraquat or glyphosate is sprayed to kill weed seeds when weeds are at the flowering to soft dough stage. Crop-topping is an expansion of the technique which was originally developed for use in pastures (spray-topping) and is now registered for use in pulse crops. If used at optimal timing, seed production of rigid ryegrass can be reduced by around 90% when using glyphosate or paraquat (Gill and Holmes 1997). However, weed seed-set reductions are often between 70% and 80% because of later-than-optimum timing of crop-topping treatments to avoid crop damage. Consequently, the weed seed bank in the subsequent wheat crop in the rotation is considerably reduced.

Other control tactics that have found some on-farm adoption include capture of weed seeds during the harvest operation as well as autumn burning of crop and weed stubble to kill weed seeds. Walsh and Parker (2002) showed that 75% to 85% of the seed entering the harvester can be captured in the chaff cart. However, the effectiveness of this technique can be strongly influenced by the level of weed seeds shed prior to harvest due to strong wind events and delays in the timing of harvest.

Stubble burning is an old method of weed management that can provide variable levels of weed-seed kill depending on the level of fuel present (i.e., residue load) and the level of seed burial. Recently there has been farmer adoption of a technique in which stubble and weed seeds are concentrated in windrows during harvest. Burning of these windrows not only results in greater effectiveness of the burn due to high fuel load, but it also reduces the erosion risk associated with whole-field burning (Walsh and Powles 2007).

There is a wide range of weed management tactics available to wheat growers in Australia. Some of these are used in the rotation prior to the wheat phase, e.g., spray-topping pastures or crop-topping pulse crops, while others such as time of seeding, seed rate (or crop density), cultivar selection, and herbicide choice are implemented during the wheat growing season. Integration of these tactics has provided effective management of weed populations in wheat.

FUTURE PERSPECTIVES

Preventative weed management is highly effective but routinely overlooked because it is complex and involves the integration of many practices. Greater understanding of the complex interactions involved in the introduction, establishment, and dispersal of weed species at the national, regional, and farm levels is necessary for greater adoption of these effective weed control strategies.

Weed populations adapt to changes in farm management. Farmers must strive to keep weeds off balance by frequently altering management practices and maximizing the competitive advantage of the crop at the expense of weeds. A good stand of healthy wheat can be very competitive with weeds and more effort needs to be put into developing integrated systems that maximize this advantage.

Overreliance on herbicides for weed control has led to weed resistance and weed shifts that complicate weed management. With a limited array of available herbicide classes, weed resistance will remain a challenge to wheat production in the future.

The ubiquitous use of glyphosate jeopardizes the sustainability of no-till systems that have proven so effective for diversifying and intensifying wheat production systems in semiarid environments. Proper glyphosate stewardship is critical to maintaining these effective production systems. Sustainable weed management in wheat will best be achieved through continued development and adoption of integrated weed management and crop production practices.

REFERENCES

Anderson, R.L. 1993. Jointed goatgrass (*Aegilops cylindrica*) ecology and interference in winter wheat. Weed Sci. 41:388–393.

Anderson, R.L. 1997. Cultural systems can reduce reproductive potential of winter annual grasses. Weed Technol. 11:608–613.

Anderson, R.L. 2003. An ecological approach to strengthen weed management in the semi-arid Great Plains. Adv. Agron. 80:33–62.

Anderson, W.K., and D. Sawkins. 1997. Production practices for improved grain yield and quality of soft wheats in Western Australia. Aust. J. Exp. Agric. 37:173–180.

Andreasen, C., J.C. Streibig, and H. Haas. 1991. Soil properties affecting the distribution of 37 weed species in Danish fields. Weed Res. 31:181–187.

Baumhardt, R.L., and R.L. Anderson. 2006. Crop choices and rotation principles. p. 113–139. *In* G.A. Peterson, P. W. Unger, and W.A. Payne (ed.) Dryland agriculture. ASA, CSSA, and SSSA, Madison, WI.

Bell, A.R., and J.D. Nalewaja. 1968. Competition of wild oats in wheat and barley. Weed Sci. 16:505–508.

Blackshaw, R.E. 1994a. Differential competitive ability of winter wheat cultivars against downy brome. Agron. J. 86:649–654.

Blackshaw, R.E. 1994b. Rotation affects downy brome (*Bromus tectorum*) in winter wheat (*Triticum aestivum*). Weed Technol. 8:728–732.

Blackshaw, R.E. 2005. Tillage intensity affects weed communities in agroecosystems. p. 209–221. *In* S. Injerjit (ed.) Invasive plants: Ecological and agricultural aspects. Birkhäuser Verlag, Basel, Switzerland.

Blackshaw, R.E. 2006. Evolving on-farm management systems: The Canadian experience. p. 17–21. *In* C. Preston, J.H. Watts, and N.D. Crossman (ed.) Proc. Australian Weeds Conf., 15th, Adelaide, South Australia. 24–28 Sept. 2006. Weed Management Society of South Australia, Adelaide, Australia.

Blackshaw, R.E., H.J. Beckie, L.J. Molnar, T. Entz, and J.R. Moyer. 2005a. Combining agronomic practices and herbicides improves weed management in wheat–canola rotations within zero-tillage production systems. Weed Sci. 53:528–535.

Blackshaw, R.E., F.J. Larney, C.W. Lindwall, P.R. Watson, and D.A. Derksen. 2001a. Tillage intensity and crop rotation affect weed community dynamics in a winter wheat cropping system. Can. J. Plant Sci. 81:805–813.

Blackshaw, R.E., L.J. Molnar, and H.H. Janzen. 2004. Nitrogen fertilizer timing and application method affect weed growth and competition with spring wheat. Weed Sci. 52:614–622.

Blackshaw, R.E., J.R. Moyer, R.C. Doram, and A.L. Boswell. 2001b. Yellow sweetclover, green manure, and its residues effectively suppress weeds during fallow. Weed Sci. 49:406–413.

Blackshaw, R.E., J.R. Moyer, K.N. Harker, and G.W. Clayton. 2005b. Integration of agronomic practices and herbicides for sustainable weed management in zero-till barley field pea rotation. Weed Technol. 19:190–196.

Blackshaw, R.E., and L.M. Rode. 1991. Effect of ensiling and rumen digestion by cattle on weed seed viability. Weed Sci. 39:104–108.

Blackshaw, R.E., G. Semach, and H.H. Janzen. 2002. Fertilizer application method affects nitrogen uptake in weeds and wheat. Weed Sci. 50:634–641.

Blackshaw, R.E., G. Semach, X. Li, J.T. O'Donovan, and K.N. Harker. 1999. An integrated weed management approach to managing foxtail barley (*Hordeum jubatum*) in conservation tillage systems. Weed Technol. 13:347–353.

Blackshaw, R.E., G. Semach, X. Li, J.T. O'Donovan, and K.N. Harker. 2000a. Tillage, fertilizer and glyphosate timing effects on foxtail barley (*Hordeum jubatum*) management in wheat. Can. J. Plant Sci. 80:655–660.

Blackshaw, R.E., G.P. Semach, and J.T. O'Donovan. 2000b. Utilization of wheat seed rate to manage redstem filaree (*Erodium cicutarium*) in a zero-till cropping system. Weed Technol. 14:389–396.

Blackshaw, R.E., E.H. Stobbe, and A.R.W. Sturko. 1981. Effects of seeding dates and densities of green foxtail (*Setaria viridis*) on the growth and productivity of wheat (*Triticum aestivum*). Weed Sci. 29:212–217.

Blanco-Moreno, J.M., L. Chamorro, R.M. Masalles, J. Recasens, and F.X. Sans. 2004. Spatial distribution of *Lolium rigidum* seedlings following seed dispersal by combine harvesters. Weed Res. 44:375–387.

Buhler, D.D., and T.C. Daniel. 1988. Influence of tillage systems on giant foxtail (*Setaria faberi*) and velvetleaf (*Abutilon theophrastii*) density and control in corn (*Zea mays*). Weed Sci. 36:642–647.

Burrows, V.D., and P.J. Olson. 1955. Reactions of small grains to various densities of wild mustard and the results obtained after their removal with 2,4-D or by hand. I. Experiments with wheat. Can. J. Agric. Sci. 35:68–75.

Carlson, H.L., and J.E. Hill. 1986. Wild oat (*Avena fatua*) competition with spring wheat: Effects of nitrogen fertilization. Weed Sci. 34:29–33.

Chauhan, B.S., G. Gill, and C. Preston. 2006a. Influence of tillage systems on vertical distribution, seedling recruitment and persistence of rigid ryegrass (*Lolium rigidum*) seed bank. Weed Sci. 54:669–676.

Chauhan, B.S., G. Gill, and C. Preston. 2006b. Tillage systems affect trifluralin bioavailability in soil. Weed Sci. 54:941–947.

Cromar, H.E., S.D. Murphy, and C.J. Swanton. 1999. Influence of tillage and crop residue on postdispersal predation of weed seeds. Weed Sci. 47:184–194.

Cudney, D.W., L.S. Jordan, and A.E. Hall. 1991. Effect of wild oat (*Avena fatua*) infestations on light interception and growth rate of wheat (*Triticum aestivum*). Weed Sci. 39:175–179.

Das, T.K., and N.T. Yaduraju. 1999. Effect of weed competition on growth, nutrient uptake and yield of wheat as affected by irrigation and fertilizers. J. Agric. Sci. 133:45–51.

Dastgheib, F. 1989. Relative importance of crop seed, manure, and irrigation water as sources of weed infestation. Weed Res. 29:113–116.

Daugovish, O., D.J. Lyon, and D.D. Baltensperger. 1999. Cropping systems to control winter annual grasses in winter wheat (*Triticum aestivum*). Weed Technol. 13:120–126.

Derksen, D.A., R.L. Anderson, R.E. Blackshaw, and B. Maxwell. 2002. Weed dynamics and management strategies for cropping systems in the northern Great Plains. Agron. J. 94:174–185.

Dexter, J.E., K.R. Preston, and K.H. Tipples. 1984. The effect of various levels of barley, wild oats and domestic oats on the milling and baking performance of hard red spring wheat. Can. J. Plant Sci. 64:275–283.

Diggle, A.J., P.B. Neve, and F.P. Smith. 2003. Herbicides used in combination can reduce the probability of herbicide resistance in finite weed populations. Weed Res. 43:371–382.

Di Tomaso, J.M. 1995. Approaches for improving crop competitiveness through the manipulation of fertilization strategies. Weed Sci. 43:491–497.

Donald, C.M. 1963. Competition between crop and pasture plants. Adv. Agron. 15:1–118.

Donald, W.W., and A.G. Ogg, Jr. 1991. Biology and control of jointed goatgrass (*Aegilops cylindrica*), a review. Weed Technol. 5:3–17.

Edwards, J.T., and E.G. Krenzer. 2006. Quality of farmer-saved wheat seed is variable in the southern Great Plains [Online]. Crop Manage. Available at www.plantmanagementnetwork.org/cm/ (verified 20 Dec. 2008). doi:10.1094/CM-2006-0531-01-RS.

Eghball, B., and G.W. Lesoing. 2000. Viability of weed seeds following manure windrow composting. Compost Sci. Util. 8:46–53.

Entz, M.H., W.J. Bullied, and F. Katepa. 1995. Rotational benefits of forage crops in Canadian prairie cropping systems. J. Prod. Agric. 8:521–529.

Friesen, G., L.H. Shebeski, and A.D. Robinson. 1960. Economic losses caused by weed competition in Manitoba grain fields. II. Effect of weed competition on the protein content of cereal crops. Can. J. Plant Sci. 40:652–658.

Geier, P.W., P.W. Stahlman, A.D. White, S.D. Miller, C.M. Alford, and D.J. Lyon. 2004. Imazamox for winter annual grass control in imidazolinone-tolerant winter wheat. Weed Technol. 18:924–930.

Gerhards, R., and H. Oebel. 2006. Practical experiences with a system for site-specific weed control in arable crops using real-time image analysis and GPS-controlled patch spraying. Weed Res. 46:185–193.

Ghersa, C.M., M.A. Martinez-Ghersa, E.H. Satorre, M.L. Van Esso, and G. Chichotky. 1993. Seed dispersal, distribution and recruitment of seedlings of *Sorghum halepense* (L.) Pers. Weed Res. 33:79–88.

Gill, G.S., and R.M. Davidson. 2000. Weed interference. p. 61–80. *In* B.M. Sindel (ed.) Australian weed management systems. R.G. and F.J. Richardson, Melbourne, Victoria, Australia.

Gill, G.S., and J.E. Holmes. 1997. Efficacy of cultural control methods for combating herbicide-resistant *Lolium rigidum*. Pestic. Sci. 51:352–358.

Girma, K., J. Mosali, W.R. Raun, K.W. Freeman, K.L. Martin, J.B. Solie, and M.L. Stone. 2005. Identification of optical spectral signatures for detecting cheat and ryegrass in winter wheat. Crop Sci. 45:477–485.

Grundy, A.C., J.M. Green, and M. Lennartsson. 1998. The effect of temperature on the viability of weed seeds in compost. Compost Sci. Util. 6:26–33.

Gurusiddaiah, S., D.R. Gealy, A.C. Kennedy, and A.G. Ogg, Jr. 1994. Isolation and characterization of metabolites from *Pseudomonas fluorescens*-D7 for control of downy brome (*Bromus tectorum*). Weed Sci. 42:492–501.

Harmon, G.W., and F.D. Keim. 1934. The percentage and viability of weed seeds recovered in the feces of farm animals and their longevity when buried in manure. J. Am. Soc. Agron. 26:762–767.

Heap, I. 2007. International survey of herbicide resistant weeds [Online]. Available at http://www.weedscience.org (verified 12 Oct. 2007).

Heisel, T., A.K. Ersbøll, and C. Andreasen. 1999. Weed mapping with co-kriging using soil properties. Precision Agric. 1:39–52.

Hocking, P.J., and M. Stapper. 2001. Effect of sowing time and nitrogen fertiliser on canola and wheat and nitrogen on Indian mustard. I: Dry matter production, grain yield, and yield components. Aust. J. Agric. Res. 52:623–634.

Hucl, P. 1998. Response to weed control by four spring genotypes differing in competitive ability. Can. J. Plant Sci. 78:171–173.

Jian, F., D.S. Jayas, and N.D.G. White. 2005. Movement and distribution of adult *Cryptolestes ferrugineus* (Coleoptera: Laemophloeidae) in stored wheat in response to temperature gradients, dockage, and moisture differences. J. Stored Prod. Res. 41:410–422.

Jones, R.E., Jr., R.H. Walker, and G. Wehtje. 1997. Soybean (*Glycine max*), common cocklebur (*Xanthium strumarium*) and sicklepod (*Senna obtusifolia*) sap flow in interspecific competition. Weed Sci. 45:409–413.

Jordan, N. 1996. Weed prevention: Priority research for alternative weed management. J. Prod. Agric. 9:485–490.

Justice, G.G., T.F. Peeper, J.B. Solie, and F.M. Epplin. 1994. Net return from Italian ryegrass (*Lolium multiflorum*) control in winter wheat (*Triticum aestivum*). Weed Technol. 8:317–323.

Kappler, B.F., D.J. Lyon, P.W. Stahlman, S.D. Miller, and K.M. Eskridge. 2002. Wheat plant density influences jointed goatgrass (*Aegilops cylindrica*) competitiveness. Weed Technol. 16:102–108.

Karlen, D.L., G.E. Varvel, D.G. Bullock, and R.M. Cruse. 1994. Crop rotations for the 21st century. Adv. Agron. 53:1–45.

Kelley, A.D., and V.F. Bruns. 1975. Dissemination of weed seeds by irrigation water. Weed Sci. 23:486–493.

Kennedy, A.C, L.F. Elliott, F.L. Young, and D.C. Douglas. 1991. Rhizobacteria suppressive to the weed downy brome. Soil Sci. Soc. Am. J. 55:722–727.

Kennedy, A.C., and T.L. Stubbs. 2007. Management effects on the incidence of jointed goatgrass inhibitory rhizobacteria. Biol. Control 40:213–221.

Kettler, T.A., D.J. Lyon, J.W. Doran, W.L. Powers, and W.W. Stroup. 2000. Soil quality assessment after weed-control tillage in a no-till wheat–fallow cropping system. Soil Sci. Soc. Am. J. 64:339–346.

Kirkland, K.J., and H.J. Beckie. 1998. Contribution of nitrogen fertilizer placement to weed management in spring wheat (*Triticum aestivum*). Weed Technol. 12:507–514.

Koscelny, J.A., and T.F. Peeper. 1997. Evaluation of registered herbicides for cheat (*Bromus secalinus*) control in winter wheat (*Triticum aestivum*). Weed Technol. 11:30–34.

Larney, F.J., and R.E. Blackshaw. 2003. Weed seed viability in composted beef cattle feedlot manure. J. Environ. Qual. 32:1105–1113.

Lawlor, D.W. 1972. Growth and water use of *Lolium perenne*. I. Water transport. J. Appl. Ecol. 9:79–98.

Lemerle, D., R.D. Cousens, G.S. Gill, S.J. Pelzer, M. Moerkerk, C.E. Murphy, D. Collins, and B.R. Cullis. 2004. Reliability of higher seeding rates of wheat for increased competitiveness with weeds in low rainfall environments. J. Agric. Sci. 142:395–409.

Lemerle, D., G.S. Gill, C.E. Murphy, S.R. Walker, R.D. Cousens, S. Mokhtari, S.J. Peltzer, R. Coleman, and D.J. Luckett. 2001. Genetic improvement and agronomy for enhanced wheat competitiveness with weeds. Aust. J. Agric. Res. 52:527–548.

Liebman, M., and C.P. Staver. 2001. Crop diversification for weed management. p. 322–374. *In* M. Liebman, C.L. Mohler, and C.P. Staver (ed.) Ecological management of agricultural weeds. Cambridge University Press, Cambridge, UK.

Loeppky, H.A., and D.A. Derksen. 1994. Quackgrass suppression with crop rotation. Can. J. Plant Sci. 74:193–197.

Lyon, D.J., and D.D. Baltensperger. 1995. Cropping systems control winter annual grass weeds in winter wheat. J. Prod. Agric. 8:535–539.

Mallory-Smith, C.A., and E.J. Retzinger, Jr. 2003. Revised classification of herbicides by site of action for weed resistance management strategies. Weed Technol. 17:605–619.

Manthey, F.A., M. Chakraborty, M.D. Peel, and J.D. Pederson. 2004. Effect of preharvest applied herbicide on breadmaking quality of hard red spring wheat. J. Sci. Food Agric. 84:441–446.

Martin, R.J., B.R. Cullis, and D.W. McNamara. 1987. Prediction of wheat yield loss due to competition by wild oats (*Avena* spp). Aust. J. Agric. Res. 38:487–499.

Mason, M.G., and R.W. Madin. 1996. Effect of weeds and nitrogen fertilizer on yield and grain protein concentration of wheat. Aust. J. Exp. Agric. 36:443–450.

Matthews, J.M., R. Llewellyn, T.G. Geeves, R. Jaechke and S.B. Powles. 1996. Catching weed seeds at harvest: A method to reduce annual weed populations. p. 684–685. *In* M. Asghar (ed.) Proc. Australian Agron. Conf., 8th, Toowoomba, Queensland. 30 Jan.–2 Feb. 1996. Australian Society of Agronomy.

Mertens, S.K., and J. Jansen. 2002. Weed seed production, crop planting pattern, and mechanical weeding in wheat. Weed Sci. 50:748–756.

Mesbah, A.O., and S.D. Miller. 1999. Fertilizer placement affects jointed goatgrass (*Aegilops cylindrica*) competition in winter wheat (*Triticum aestivum*). Weed Technol. 13:374–377.

Mills, J.T., R.N. Sinha, and C.J. Demianyk. 1992. Feeding and multiplication of a psocid, *Liposcelis bostrychophilus* Badonnel (Psocoptera: Liposcelidae), on wheat, grain screenings, and fungi. J. Econ. Entomol. 85:1453–1462.

Mohler, C.L. 2001. Enhancing the competitive ability of crops. p. 269–322. *In* M. Liebman, C.L. Mohler, and C.P. Staver (ed.) Ecological management of agricultural weeds. Cambridge University Press, Cambridge, UK.

Mortensen, D.A., J.A. Dieleman, and G.A. Johnson. 1998. Weed spatial variation and weed management. p. 293–309. *In* J.L. Hatfield, D.D. Buhler, and B.A. Stewart (ed.) Integrated weed and soil management. Ann Arbor Press, Chelsea, MI.

Moyer, J.R., R.E. Blackshaw, E.G. Smith, and S.M. McGinn. 2000. Cereal cover crops for weed suppression in a summer fallow–wheat cropping sequence. Can. J. Plant Sci. 80:441–449.

Moyer, J.R., E.S. Roman, C.W. Lindwall, and R.E. Blackshaw. 1994. Weed management in conservation tillage systems for wheat production in North and South America. Crop Prot. 13:243–258.

Nakoneshny, W., and G. Friesen. 1961. The influence of a commercial fertilizer treatment on weed competition in spring sown wheat. Can. J. Plant Sci. 41:231–238.

Neto, M.S., R.M. Jones, and D. Ratcliff. 1987. Recovery of pasture seed ingested by ruminants. I. Seed of six tropical pasture species fed to cattle, sheep and goats. Aust. J. Exp. Agric. 27:239–246.

Neve, P., A.J. Diggle, F.P. Smith, and S.B. Powles. 2003. Simulating evolution of glyphosate resistance in *Lolium rigidum* II: Past, present and future glyphosate use in Australian cropping. Weed Res. 43:418–427.

Nye, P.H., and P.B. Tinker. 1977. Solute movement in the soil–root system. Blackwell, Oxford, UK.

O'Donovan, J.T., R.E. Blackshaw, K.N. Harker, G.W. Clayton, and R. McKenzie. 2005. Variable plant establishment contributes to differences in competitiveness with wild oat among wheat and barley varieties. Can. J. Plant Sci. 85:771–776.

O'Donovan, J.T., R.E. Blackshaw, K.N. Harker, G.W. Clayton, J.R. Moyer, L.M. Dosdall, D.C. Maurice, and T.K. Turkington. 2007. Integrated approaches to managing weeds in spring-sown crops in western Canada. Crop Prot. 26:390–398.

Olsen, J., L. Kristensen, J. Weiner, and H.W. Griepentrog. 2005. Increased density and spatial uniformity increase weed suppression by spring wheat. Weed Res. 45:316–321.

Owen, M.J., M.J. Walsh, and S.B. Powles. 2007. Widespread occurrence of multiple herbicide resistance in Western Australian annual ryegrass (*Lolium rigidum*) populations. Aust. J. Agric. Res. 58:711–718.

Peterson, D.E., and J.D. Nalewaja. 1992. Green foxtail (*Setaria viridis*) competition with spring wheat (*Triticum aestivum*). Weed Technol. 6:291–296.

Pratley, J.E. 1996. Weed management research in crop production—a review. p. 17–22. *In* R.C.H. Shepherd (ed.) Proc. Australian Weeds Conf., 11th, Melbourne, Victoria. 30 Sept.–3 Oct. 1996. Weed Sci. Soc. of Victoria, Melbourne, Victoria, Australia.

Rasmussen, P.E. 1995. Effects of fertilizer and stubble burning on downy brome competition in winter wheat. Commun. Soil Sci. Plant Anal. 26:951–960.

Reisinger, P., E. Lehoczky, and T. Komives. 2005. Competitiveness and precision management of the noxious weed *Cannabis sativa* L. in winter wheat. Commun. Soil Sci. Plant Anal. 36:629–634.

Rooney, J.M. 1991. Influence of growth form of *Avena fatua* L. on the growth and yield of *Triticum aestivum* L. Ann. Appl. Biol. 118:411–416.

Ruegg, W.T., M. Quadranti, and A. Zoschke. 2007. Herbicide research and development: Challenges and opportunities. Weed Res. 47:271–275.

Schmidt, R.R. 1998. Classification of herbicides according to mode of action. Bayer Ag., Leverkusen, Germany.

Seavers, G.P., and K.J. Wright. 1999. Crop canopy development and structure influence weed suppression. Weed Res. 39:319–328.

Shirtliffe, S.J., and M.H. Entz. 2005. Chaff collection reduces seed dispersal of wild oat (*Avena fatua*) by combine harvester. Weed Sci. 53:465–470.

Smith, E.G., B.M. Upadhyay, R.E. Blackshaw, H.J. Beckie, K.N. Harker, and G.W. Clayton. 2006. Economic benefits of integrated weed management systems in field crops of western Canada. Can. J. Plant Sci. 86:1273–1279.

Statistics Canada. 2006. Field crop reporting series, Catalogue 22–002, 2006. Statistics Canada, Ottawa, Canada.

Stougaard, R.N., and Q. Xue. 2004. Spring wheat seed size and seeding rate effects on yield loss due to wild oat (*Avena fatua*) interference. Weed Sci. 52:133–141.

Stump, W.L., and P. Westra. 2000. The seedbank dynamics of feral rye (*Secale cereale*). Weed Technol. 14:7–14.

Teasdale, J.R. 1996. Contribution of cover crops to weed management in sustainable agricultural systems. J. Prod. Agric. 9:475–479.

Thill, D.C., and C.A. Mallory-Smith. 1997. The nature and consequence of weed spread in cropping systems. Weed Sci. 45:337–342.

Thomas, P.L., L.A. Cooke, and R.M. Clear. 1998. Quality loss in wheat caused by smut spores from yellow foxtail. Can. J. Plant Pathol. 20:111–114.

Thomas, A.G., D.A. Derksen, R.E. Blackshaw, R.C. Van Acker, A. Legere, P.R. Watson, G.C. Turnbull. 2004. A multistudy approach to understanding weed population shifts in medium- to long-term tillage systems. Weed Sci. 52:874–880.

Thompson, G.B., and F.I. Woodward. 1994. Some influences of CO_2 enrichment, nitrogen nutrition and competition on grain yield and quality in spring wheat and barley. J. Exp. Bot. 45:937–942.

Timmermann, C., R. Gerhards, and W. Kühbauch. 2003. The economic impact of site-specific weed control. Prec. Agric. 4:249–260.

Unger, P.W., W.A. Payne, and G.A. Peterson. 2006. Water conservation and efficient use. p. 39–85. *In* G.A. Peterson, P.W. Unger, and W.A. Payne (ed.) Dryland agriculture. ASA, CSSA, and SSSA, Madison, WI.

Walker, S.R., R.W. Medd, G.R. Robinson, and B.R. Cullis. 2002. Improved management of *Avena Ludoviciana* and

Phalaris paradoxa with more densely sown wheat and less herbicide. Weed Res. 42:257–270.
Walsh, M.J., P. Newman, and D. Chitty. 2005. Destroy wild radish and annual ryegrass seeds by burning narrow windrows. p. 159–163. *In* Agribusiness crop updates. Grains Research and Development Corp., Adelaide, Australia.
Walsh, M.J., M.J. Owen, and S.B. Powles. 2007. Frequency and distribution of herbicide resistance in *Raphanus raphanistrum* populations randomly collected across the Western Australian wheatbelt. Weed Res. 47:542–550.
Walsh, M.J., and W. Parker. 2002. Wild radish and ryegrass seed collection at harvest: Chaff carts and other devices. p. 37–38. *In* Agribusiness crop updates. Dep. Agriculture Western Australia, Perth, Australia.
Walsh, M.J., and S.B. Powles. 2004. Herbicide resistance: An imperative for smarter crop weed management. p. 76–77. *In* T. Fischer et al. (ed.) Proc. Int. Crop Sci. Congress, 4th, Brisbane, Australia. 26 Sept.–1 Oct. 2004. The Regional Institute, Gosford, Australia.
Walsh, M.J., and S.B. Powles. 2007. Management strategies for herbicide-resistant weed populations in Australian dryland crop production systems. Weed Technol. 21:332–338.
Webb, A., S.L. Haley, and S. Leetmaa. 1995. Enhancing US wheat export performance: The implications of wheat cleaning. Agribusiness 11:317–332.
Weston, L.A. 1996. Utilization of allelopathy for weed management in agroecosystems. Agron. J. 88:860–866.
White, A.D., D.J. Lyon, C.A. Mallory-Smith, C.R. Medlin, and J.P. Yenish. 2006. Feral rye (*Secale cereale*) in agricultural production systems. Weed Technol. 20:815–823.
Willms, W.D., S.N. Acharya, and L.M. Rode. 1995. Feasibility of using cattle to disperse cicer milkvetch (*Astragalus cicer* L.) seed in pastures. Can. J. Anim. Sci. 75:173–175.
Wilson, R. 1980. Dissemination of weed seeds by surface irrigation water in western Nebraska. Weed Sci. 28:87–92.
Wray, M.W. 1993. A survey of cereal admixture in wheat and barley grain 1988 to 1990. Asp. Appl. Bio. 35:195–206.
Yenish, J.P., J.D. Doll, and D.D. Buhler. 1992. Effects of tillage on vertical distribution and viability of weed seed in soil. Weed Sci. 40:429–433.
Young, F.L., A.G. Ogg, Jr., R.I. Papendick, D.C. Thill, and J.R. Alldredge. 1994. Tillage and weed management affects winter wheat yield in an integrated pest management system. Agron. J. 86:147–154.
Zimdahl, R.L. 1990. The effect of weeds on wheat. p. 11–32. *In* W.W. Donald (ed.) Systems of weed control in wheat in North America. Weed Science Society of America, Champaign, IL.
Zimdahl, R.L. 2007. Fundamentals of weed science. 3rd ed. Academic Press, San Diego, CA.

Plate 1 Individual plants of wild emmer wheat, *Triticum dicoccoides*, including several black-spike morphs, near Nahf, Upper Galilee. (Courtesy E. Nevo.)

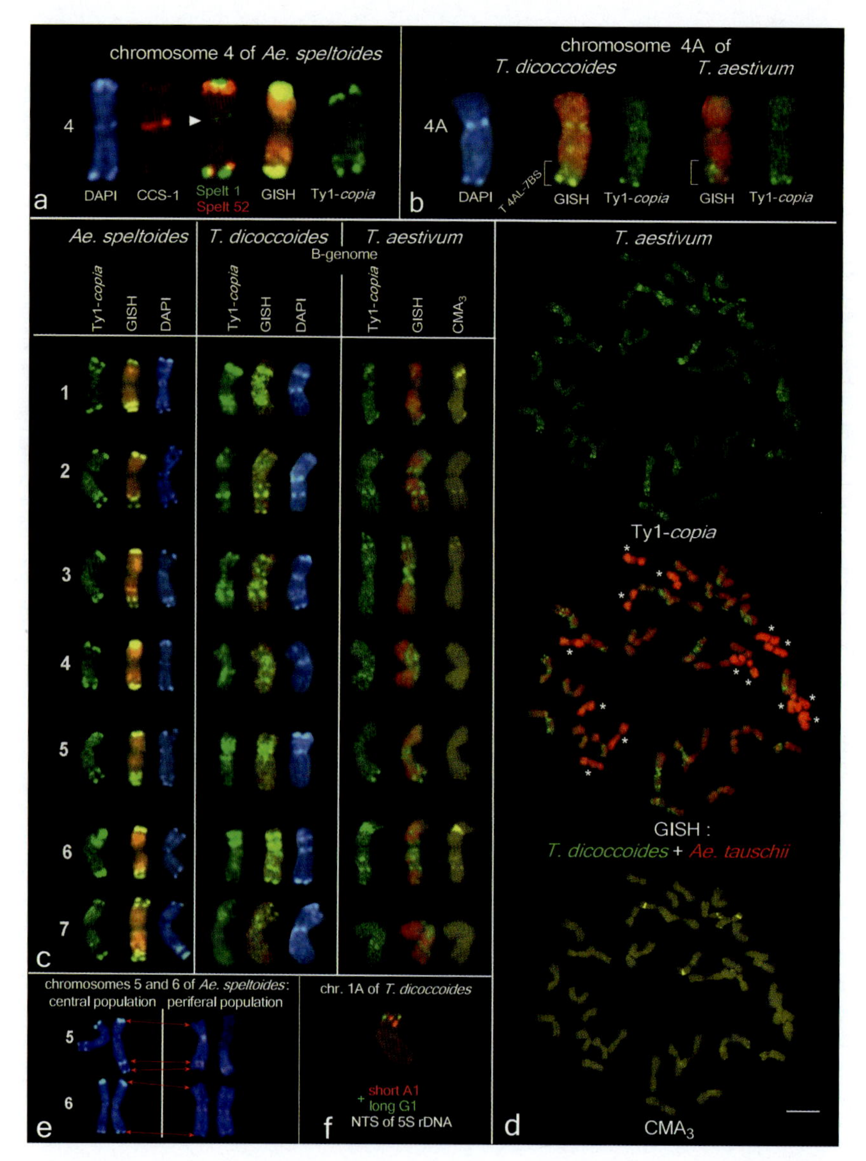

Plate 2 Chromosomal distribution of the repetitive DNA fraction in the wheat genome, and differential staining and in situ hybridization on the chromosomes of *Aegilops speltoides*, *Triticum dicoccoides*, and *T. aestivum*. (a) Chromosome 4 of *Ae. speltoides* (TS-84): from left to right—differential staining with DAPI revealing AT-enriched heterochromatic blocks; FISH with cereal-specific centromeric repeat CCS-1; FISH with species-specific *Spelt* 1 (green) and tribe-specific *Spelt* 52 (red) tandem repeats, chromosome-specific near-centromeric *Spelt 1* clusters are shown by an arrow; GISH with mix of genomic DNA of *Ae. speltoides* (green) + *Ae. bicornis* (red) revealing distal chromosomal clusters of faster evolving repetitive DNA; FISH with Ty1-*copia* retrotransposons. (b) Chromosome 4A of *T. dicoccoides* and *T. aestivum* ('Chinese Spring') after GISH with genomic DNA of *Ae. speltoides* (green) + *T. urartu* (red) for *T. dicoccoides*, and *T. dicoccoides* (green) + *Ae. tauschii* (red) for Chinese Spring, and FISH with Ty1-*copia* retrotransposons. Species-specific translocation 4AL–7BS revealed by GISH is shown. (c, d) Chromosomes of *Ae. speltoides*, B-genome chromosomes of *T. dicoccoides*, and *T. aestivum* (Chinese Spring) after FISH with Ty1-*copia* retrotransposons, GISH, and differential staining with AT-specific fluorochrome DAPI (*Ae. speltoides* and *T. dicoccoides*) and GC-specific chromomycin A3 (*T. aestivum*). GISH on metaphase chromosomes of *T. dicoccoides* with genomic probes of *Ae. speltoides* (green) + *T. urartu* (red); GISH on chromosomes of *T. aestivum* with genomic probe of *T. dicoccoides* (green) + *Ae. tauschii* (red). Intergenomic invasion—a substitution of part of the A-genome (reddish chromosomes) heterochromatin clusters by satellite DNA of the B-genome (green clusters) is revealed in *T. dicoccoides*. In *T. aestivum*, substitution of part of the "youngest" D-genome (14 red chromosomes marked by asterisk) heterochromatin by repetitive DNA of the AB-genome revealed in a far less degree. (e) Chromosomes 5 and 6 of *Ae. speltoides* from contrast populations as an example of significant reduction of the heterochromatin pattern in a marginal population. (f) Chromosome 1A of *T. dicoccoides* after FISH with short A1 (red) and long G1 (green) nontranscribed spacers (NTS) of 5S rRNA genes. Clusters of A1 and G1 show slightly different positions inside the rDNA cluster.

Plate 3 Tramlines allow precise application of pesticides and fertilizers while restricting compaction to traffic lanes. (Courtesy Phil Needham, Needham Ag, Calhoun, KY.)

Plate 4 Implements that increase harvest efficiency and speed, such as the stripper header, allow more rapid planting of a second or double crop behind wheat. (Courtesy Phil Needham, Needham Ag, Calhoun, KY.)

Plate 5 The first-hollow-stem stage of wheat is characterized by 1.5 cm (about the same diameter as a US dime) of hollow stem below the developing spike. This occurs before any hollow stem is visible above the soil surface.

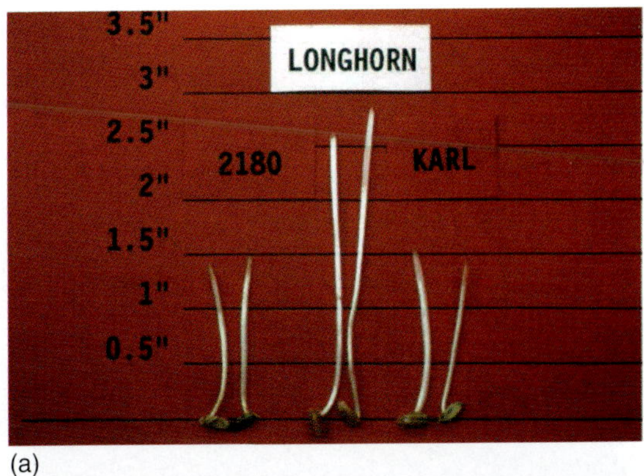

(a)

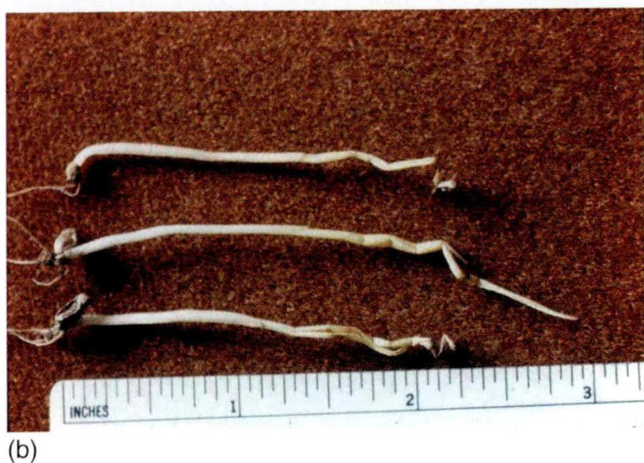

(b)

Plate 6 Coleoptile length varies among commercially released cultivars. (a) Coleoptile length is reduced as soil temperature increases. If wheat is sown deeper than the final coleoptile length, the first true leaf will emerge below the soil surface, take on an accordion-like appearance, and fail to produce a viable plant (b).

Plate 7 Compaction and plant damage due to cattle hooves reduce wheat yield in dual-purpose systems.

Plate 8 Previous crop residue can impact drill performance. Some crop residues, such as canola (a), are easier to penetrate with disc openers and allow optimal seed-to-soil contact. In contrast, hairpinning of residue can be a serious impediment when no-till planting wheat after wheat (b).

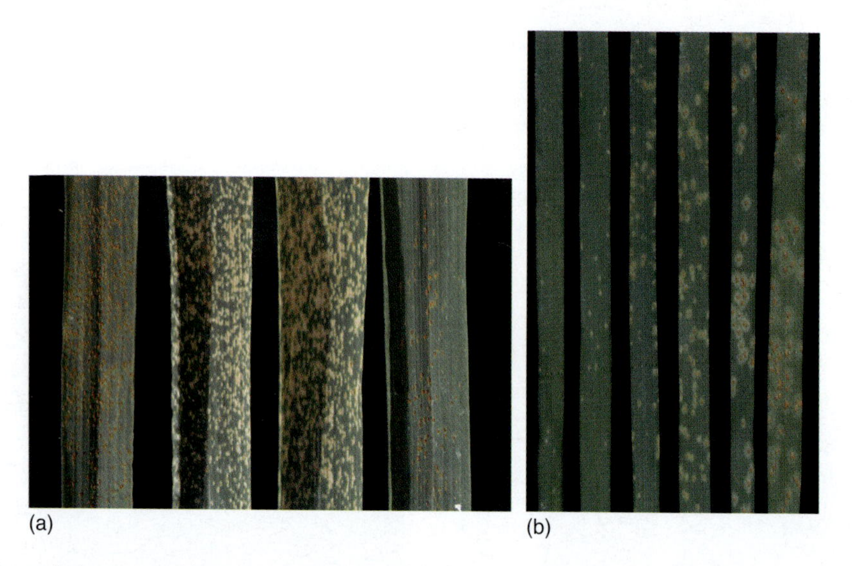

Plate 9 Leaf rust of wheat, caused by *Puccinia triticina*: (a) (from left to right) flag leaves of the susceptible wheat "Thatcher"; Thatcher with leaf rust gene *Lr12*; Thatcher with leaf rust gene *Lr13*; and Thatcher with leaf rust gene *Lr34*. The Thatcher lines with *Lr12* and *Lr13* have hypersensitive race-specific resistance and the line with *Lr34* has nonspecific resistance with fewer and smaller uredinia than Thatcher. (b) Seedling infection types on isogenic lines of wheat with single genes for leaf rust resistance. From left to right infection types increase from highly resistant to susceptible.

Plate 10 (a) Stripe rust of wheat, caused by *P. striiformis* f. sp. *tritici*. (b) High-temperature adult-plant resistance to stripe rust of wheat cultivar Alpowa.

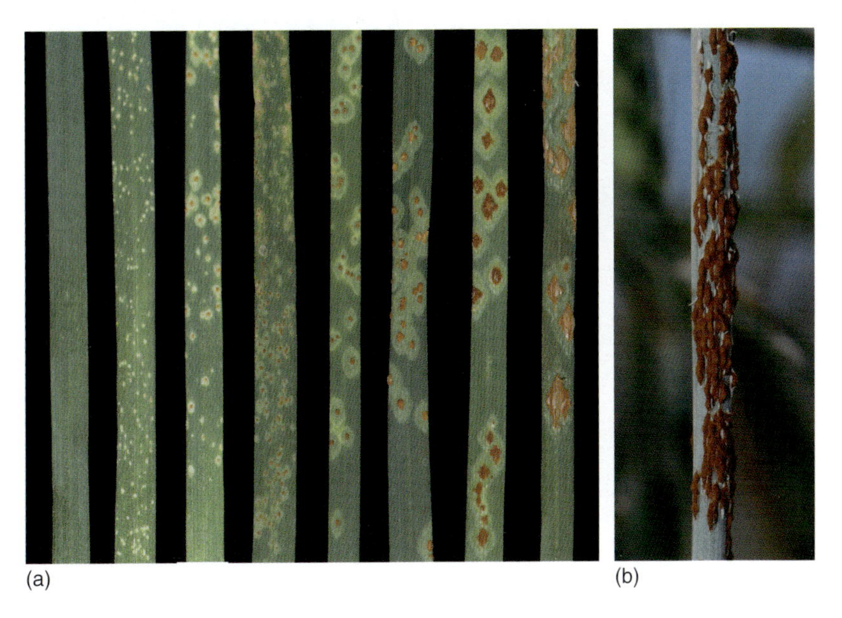

Plate 11 (a) Infection types of *Puccinia graminis* f. sp. *tritici* on wheat. From left to right the infection types increase from highly resistant to highly susceptible. (b) Uredinia of *P. graminis* f. sp. *tritici* on wheat stem tissue.

Plate 12. Lesions on subcrown internodes caused by (a) *Bipolaris sorokiniana* and (b) *Fusarium pseudograminearum*, (c) "spear tipping" of roots by *Rhizoctonia solani* AG-8, and (d) blackening of roots and basal stem by *Gaeumannomyces graminis* var. *tritici*. [(a & d) courtesy B.B. Bockus; (b & c) courtesy R.W. Smiley.]

Plate 14. (a) Improvement of wheat seedling stand by treating seed with metalaxyl (right) to reduce Pythium root-rot and damping-off (left), (b) Rhizoctonia bare patch symptoms in winter wheat during early spring, (c) patches of whiteheads caused by take-all, and (d) lodging of a winter wheat cultivar susceptible to eyespot (left) compared to a cultivar with a gene for resistance (right) [(a, b, and d) courtesy R.W. Smiley; (c) courtesy W.W. Willis.]

Plate 13. (a) Whiteheads caused by multiple root, crown, and culm rotting fungi and insect pests, (b) crown rot and (c, left) uniform browning of basal stem internodes by *Fusarium pseudograminearum*, and lesions of the culm main-stem (c, right) and leaf sheath (d) by *Oculimacula yallundae* ((a–c) courtesy R.W. Smiley; (d) courtesy T.D. Murray.)

(a)

(b)

Plate 15. Winter wheat with leaf stripes (a) and browning of vascular bundles (b) caused by *Cephalosporium gramineum*. (Courtesy R.W. Smiley.)

Plate 16 Visual symptoms of cereal cyst nematode (*Heterodera avenae*) damage on spring wheat are masked by a "doubled" fertilizer rate applied along the field border, in foreground; roots from this field are shown in Plate 18a. (Courtesy R.W. Smiley.)

Plate 17 Cereal cyst nematode (*Heterodera avenae*) causing patchy growth of winter wheat on a flat field. (Courtesy R.W. Smiley.)

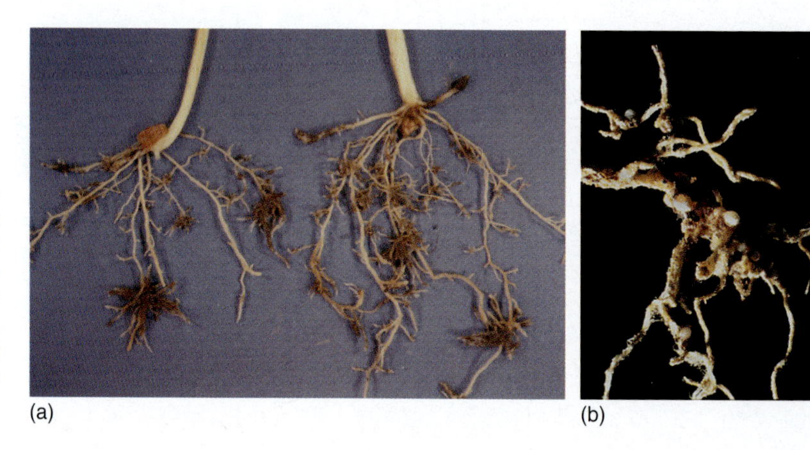

(a) (b)

Plate 18 Cereal cyst nematode (*Heterodera avenae*) on wheat roots, showing (a) abnormal branching of seedling roots at sites of invasion by juveniles (courtesy R.W. Smiley), and (b) cysts of adult females attached to mature roots (courtesy R.H. Brown).

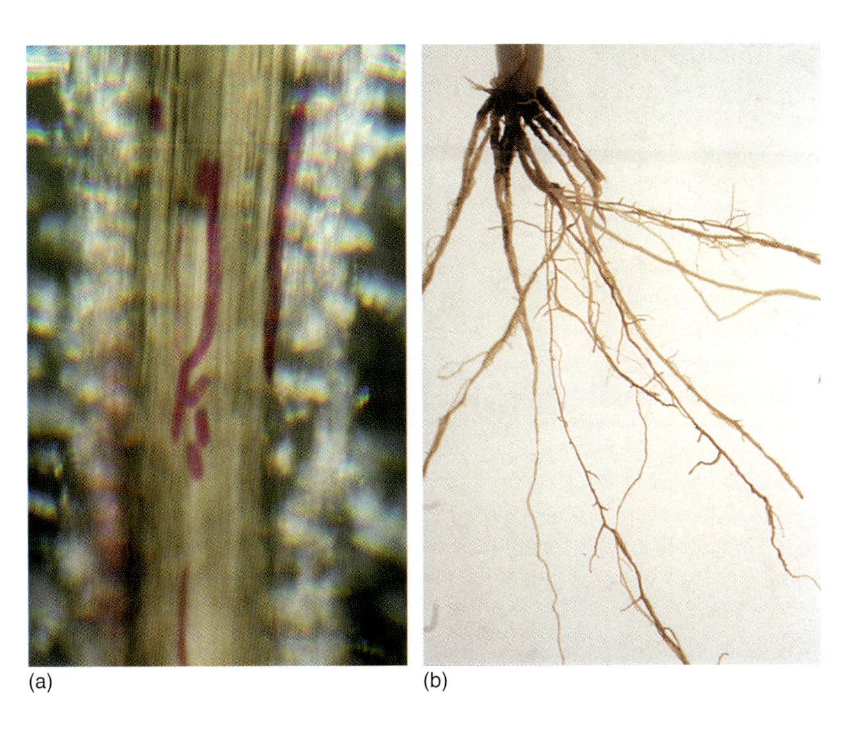
(a) (b)

Plate 19 Root-lesion nematode on wheat, showing (a) root cortex tissue stained to reveal an adult *Pratylenchus thornei* female (0.5 mm long) and eggs (courtesy R.W. Smiley), and (b) reduced root branching and cortical degradation by *P. neglectus* (courtesy V.A. Vanstone).

Plate 20 Hessian fly (*Mayetiola destructor* Say) adult female ovipositing at the base of a wheat leaf. Larvae from egg hatch crawl to the base of the leaf sheath to begin feeding. (Courtesy Marion Harris, North Dakota State University.)

Plate 21 Bird cherry–oat aphid (*Rhopalosiphum padi* L.) adults and nymphs feeding on wheat leaf. Damage is caused directly by feeding and also by vectoring viruses that cause barley yellow dwarf, one of the most important viral diseases of wheat worldwide. (Courtesy Tom Royer and Rick Grantham, Oklahoma State University.)

Plate 22 Greenbug (*Schizaphis graminum* Rondani) adult and nymphs feeding on wheat leaf. Resultant interveinal necrotic lesions are characteristic of susceptible wheat. (Courtesy Jack Dykinga, USDA-ARS.)

Plate 23 Russian wheat aphid [*Diuraphis noxia* (Mordvilko)] adult female feeding on wheat leaf. Feeding damage results in characteristic longitudinal white streaking, purple discoloration, and rolling of the leaves. (Courtesy Gary Puterka, USDA-ARS.)

Plate 24 From the bottom to upper right, the maximum N rate (150 kg N ha^{-1}), followed by the check (0 kg N ha^{-1}), and ensuing ramp calibration strip (RCS) in incremental rates up to the maximum (distance of 50 m), at Stillwater, Oklahoma. Maximum rates and increments may vary by environment.

Plate 25 Expression of the leaf-rolling phenotype early in the morning for the leaf-rolling B403D (L) and nonrolling "Diamondbird" (R) genotypes at Zadoks stage 59 (inflorescence completely emerged) when evaluated under irrigated conditions (courtesy of X.R.R. Sirault).

Plate 26. An F_2 winter wheat breeding nursery photographed in the spring at Mead, Nebraska, showing plots that survived (those with plants) or did not survive (bare soil areas) the preceding winter. This location is an example of a selection nursery. It has the most severe winter of the wheat growing areas in Nebraska and is used to identify plants that will survive the winter wherever wheat is grown in Nebraska. (Courtesy P.S. Baenziger.)

Plate 27. Aerial view of an off-season nursery in New Zealand used to advance generations and increase seed of lines for testing in multilocation trials. It also represents part of the scope and scale of modern wheat breeding programs. (Courtesy Southern Seed Technology, Ltd; photograph taken by Robert Lamberts.)

Plate 28. Early generation ($F_{2:4}$) multilocation trial used to measure productivity traits and end-use suitability traits. (Courtesy SPARC, AAFC; photograph taken by D. Schott.)

Plate 29. Five small-plot combines used to harvest spring wheat breeding trials near Swift Current, Saskatchewan. The photo also highlights the extensive field testing and the capital investment in technology to support modern wheat breeding research. (Courtesy SPARC, AAFC; photograph taken by D. Schott.)

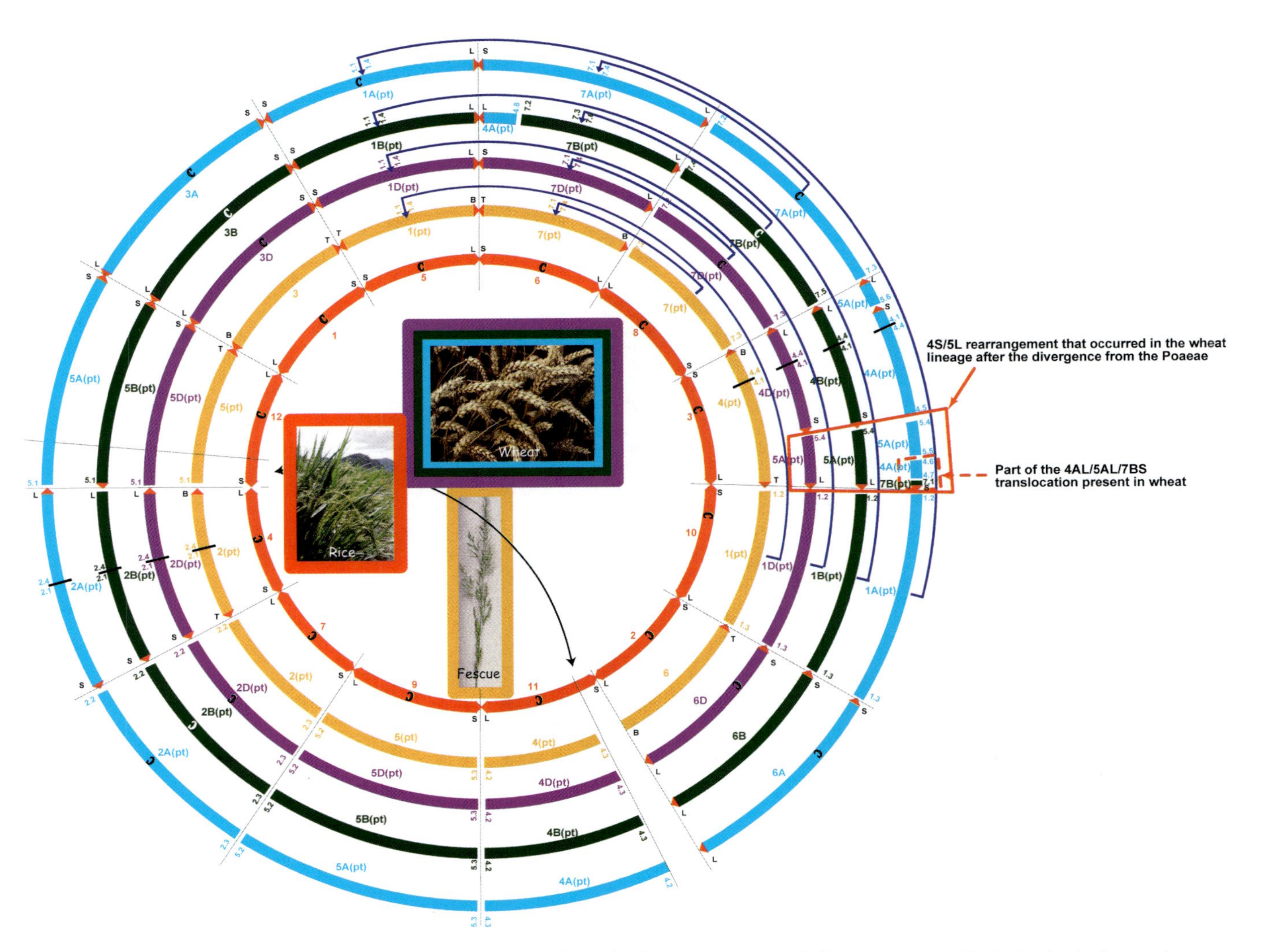

Plate 30 Relationship between the A, B, and D genomes of wheat, the fescue genome, and the rice genome. Red triangles indicate telomeres; C shows, where known, the location of the centromeres; (pt) after a chromosome number means part; S and L indicate the short and long arms of a chromosome, respectively. Where the orientation of a chromosome is unknown, one side of the chromosome is arbitrarily indicated with T (top) and the other side with B (bottom). Arrows indicate chromosomal rearrangements. In some cases, rearrangements are indicated by numbered chromosome segments.

Plate 31 Hexaploid wheat plants (cultivar Bobwhite) displaying photobleaching resulting from virus-induced gene silencing of phytoene desaturase (PDS) expression. Seedlings were infected with *Barley stripe mosaic virus* constructs carrying a fragment of the PDS gene 7 days after germination. The photograph was taken 14 days after viral inoculation. (Courtesy Steve Scofield and Amanda Brandt, USDA-ARS, West Lafayette, IN).

Plate 32 Representative stages in wheat transformation experiments. (a) Selection of donor material for embryos at the appropriate developmental stage (left two pictures); surface sterilization of immature caryopses (top right); and isolation of immature embryos (bottom right). (b) *Agrobacterium* cocultivation: A suspension of *Agrobacteria* is added to cultured scutella from immature embryos (top), which are then incubated for 2–3 days (bottom) to allow the transfer of T-DNA from the bacteria to the plant cells. (c) Starting components and results of biolistic transformation: The BioRad (Richmond, CA) PDS 1000/He particle-delivery device (left); immature embryos clustered in the center of a plate before bombardment (top right); and expression of Green Fluorescence Protein under blue light 2 days after bombardment with the *gfp* gene (bottom right). (d) Regeneration of shoots and roots from embryogenic calli in the presence of a selective agent. Note that only some of the plantlets thrive. (Photos courtesy of Caroline Sparks and Angela Doherty.)

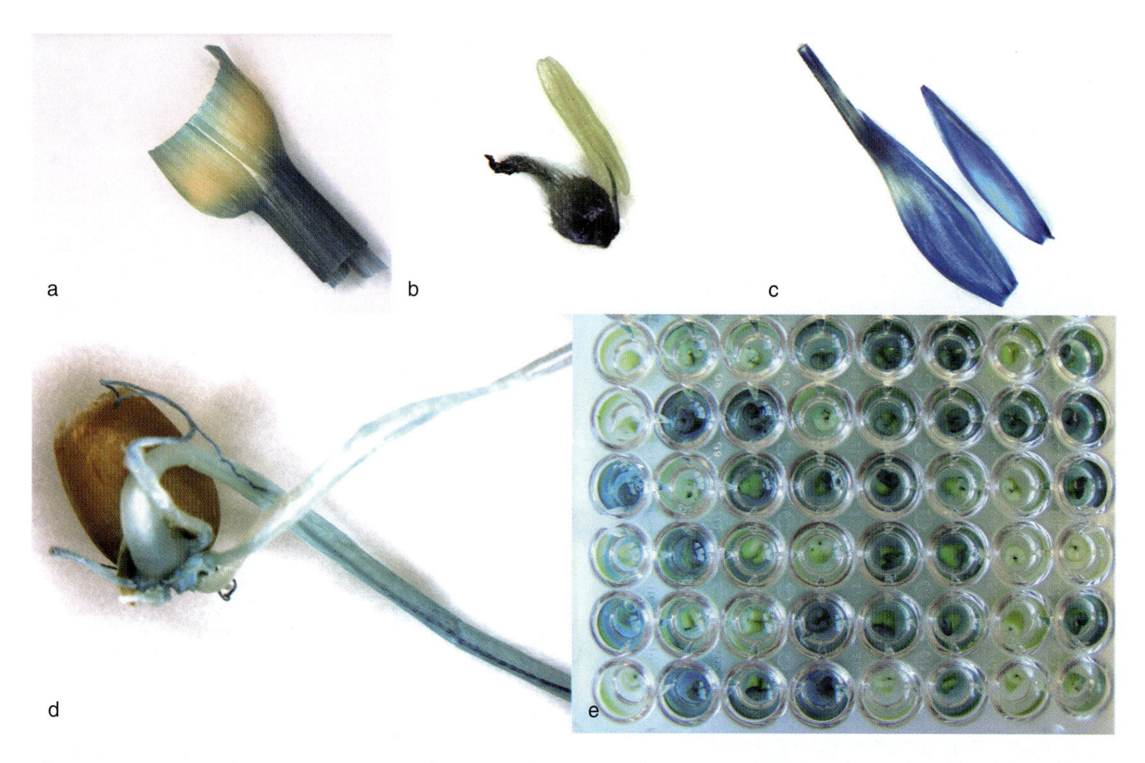

Plate 33 Detection of promoter activity in wheat using histochemical assay for the GUS reporter gene. Blue color indicates the presence of β-glucuronidase encoded by the GUS transgene(s). (a–d) The maize *Ubi1* promoter drives GUS expression in many different wheat tissues. Histochemical staining for GUS activity in a wheat plant homozygous for a Ubi::GUS transgene. Chlorophyll was removed from the green tissues by incubation in 70% ethanol. Tissues include (a) sheath at base of flag leaf, (b) ovary and anther before pollination, (c) glume, lemma, and palea from floret before anthesis, and (d) germinated seedling. (e) Endosperm half-seeds of progeny from a transgenic wheat plant heterozygous for a GUS transgene driven by a HMW-GS promoter. (Photos courtesy of Mara Guttman and Jeanie Lin.)

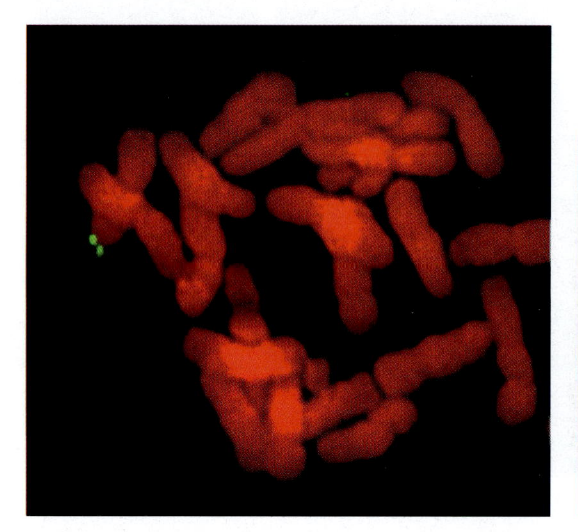

Plate 34 Fluorescent in situ hybridization (FISH) image of a transgene insertion at a distal location on a wheat chromosome. Note that the transgene locus has been replicated during the formation of paired chromatids. Root-tip mitosis metaphase spreads were probed with digoxygenin-labeled DNA, detected using FITC-avidin (flouroscein isothiocyanate), and counterstained in propidium iodide. (Photo courtesy of Jean Jacquet.)

Plate 35 Some of the great diversity of wheat-based food products.

Section III
Making of a Wheat Cultivar

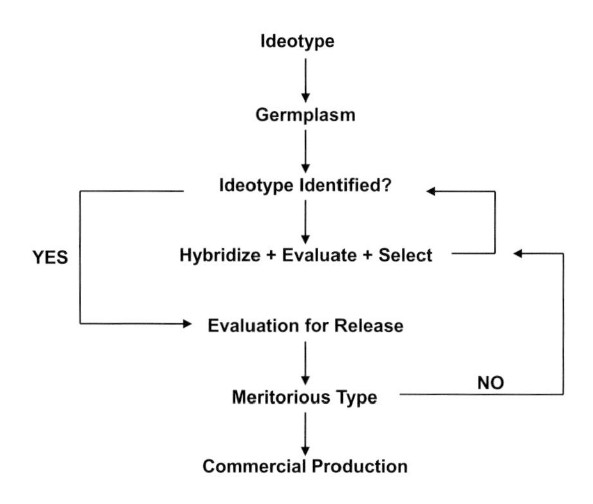

"Agriculture is the bridge between the natural ecosystem and the human social system."

Norman Borlaug

Chapter 13
Wheat Breeding: Procedures and Strategies

P. Stephen Baenziger and Ronald M. DePauw

SUMMARY

(1) Plant breeding, in general, and wheat breeding, in particular, have been practiced for centuries since wheat was domesticated. The predominant outcome of wheat breeding is a pureline cultivar that takes advantage of wheat being a self-pollinated plant.

(2) Principles underlying wheat improvement include biology of sexual recombination, Mendelian laws of inheritance, and selection. Efficient improvement in a trait requires knowledge of its heritability, the level of genetic variation, the mode of inheritance, and the number of genes controlling the trait.

(3) The commonly used breeding methods are pedigree, bulk, single-seed descent, doubled haploid, and backcross. Each method has its advantages and disadvantages. Wheat breeding is remarkably flexible, and these methods are often combined in practice to take advantage of their strengths and the selection environments that occur during cultivar development.

(4) Heritable genetic variation may be increased in a population by hybridization, mutation, tissue culture, or transformation (i.e., transgenic wheat). Though there is currently no commercial transgenic wheat, transformation and transgenic wheat cultivars will be necessary to ensure wheat as a crop remains competitive.

(5) Wheat breeding is a global enterprise and the exchange of germplasm for wheat improvement (hybridization) is covered by numerous international treaties and agreements.

(6) Some of the decisions all wheat breeders must make are which generations to select and when to end within-line selection. The earlier the generation at which within-line selection ends, the more heterogeneous the resultant line will be. The level of within-line heterogeneity will determine whether or not the line can be released as a cultivar, due to governmental regulations. Wheat breeders also must decide how to efficiently use their resources to select and evaluate lines in the diverse environments where wheat is grown.

(7) Molecular markers and genomics are changing breeding methodology by supplementing phenotypic selection with genotypic selection. Marker-assisted selection is very useful for selecting traits that have low heritability or that are expressed in the adult plant. Phenotypic selection on an adult plant requires considerably more time, resources, and space than selecting on seedling plants. The use of molecular markers will increase as they become less expensive and more readily available to wheat breeders.

(8) Molecular markers and the quantitative trait loci that they tag will be incorporated into growth and development models to help breeders select for new ideotypes with superior performance.

(9) Wheat breeding remains a public-sector and private-sector activity, often

depending upon regional opportunities, support, and laws. However, all plant breeding is becoming progressively more a private-sector activity. Regardless in which sector wheat breeding is done, it will remain a profession that is filled with the joy and satisfaction of creating new cultivars that can change human lives for the better.

BRIEF HISTORY OF WHEAT BREEDING

People express their preferences for food, feed, and milieu by deliberately choosing plants that meet these preferences. They have observed and retained propagules of the new variants that met their preferences. These new strains are perpetuated if they have a heritable advantage. Examples include mutations that did not disarticulate the spikelet from the spike and provided the free-threshing characteristic in which the lemma, palea, and glumes separated from the kernel to leave a naked wheat grain. The gatherers of grain from wild wheat relatives were most likely women, and therefore they were most likely the ones who observed these extremely valuable mutations for the production of food. These mutations contributed to a major reduction in the work required to gather and prepare wild wheat for food.

Principles underlying wheat improvement include biology of sexual recombination, Mendelian laws of inheritance, and selection. Those who first practiced plant improvement may never be known to us. We only know those who left a record that has been found. A summary of the history of principles of plant breeding and genetics has been presented by other authors (Acquaah 2007), which will be further condensed to salient features related to wheat improvement. Archeological records indicate that Assyrians and Babylonians artificially pollinated date palm by 700 BC. In 1694, R.J. Camerarius demonstrated sex in plants and suggested that hybridizing between different individuals might produce new plant types. Between 1771 and 1776 J. Köelreuter of Germany demonstrated that hybrid offspring of tobacco expressed traits from both parents. A century later in 1866, Gregor Mendel published his famous work with peas, formulating the laws of inheritance and postulating unit factors later called genes. Mendel formulated the concepts of assortment and independent segregation of unit factors that control traits. Mendel's work went unnoticed until rediscovered independently by Correns of Germany, de Vries of Holland, and von Tschermak of Austria. Selection within landraces and subsequent progeny testing was used to identify strains that performed better than the original population from which they were selected. This technique was practiced simultaneously in various countries: Knight, Le Couteur, and Sheriff in the UK; Vilmorin in France; and Rimpau, Heine, Beseler, and Strube in Germany (Bonjeau and Angus 2001). Two principles of selection evolved: (i) selection does not create genetic variation but only acts upon genetic variation already in the population and (ii) selection acts effectively only on heritable differences.

The realization that selection was effective only upon heritable variation led early plant breeders to use the nascent understanding of sexuality of plants to hybridize different wheat parents. The practice of hybridization followed by selection as a wheat improvement strategy was initiated in the latter part of the 19th century by Vilmorin (cultivar Dattel) in France about 1874 and by Wilhelm Rimpau ('Rimpaus früher Bastard') in Germany in 1875 (Bonjeau and Angus 2001). Almost simultaneously, crossbreeding and selection was practiced by Saunders ('Marquis') in Canada, by Pringle ('Pringle's Defiance') in the US, and by Robin ('Robin's Rust Proof') in Australia.

Initial wheat cultivar improvement and release was undertaken as private enterprise primarily in European countries. In the New World, federal and state governments supported cultivar improvement as a means to support the development of commercial agriculture (e.g., Australia, Canada, and the US). With implementation of intellectual property protection and means to track cultivars by seed pedigree programs and

DNA fingerprinting, private enterprise has also evolved to contribute to wheat improvement in these same countries. More recently the concept of participatory involvement in cultivar improvement has been promoted in some countries which lack strong national agricultural research service (Ceccarelli et al., 2000). Segregating populations from either simple or complex crosses are made available to farmers who then select within the population for strains that best fit their environment and combination of biotic and abiotic stresses.

Wheat cultivars released in the later half of the 20th century and in the 21st century are primarily purelines, with relatively few hybrids and multiline cultivars. At this time there are neither apomictic-derived cultivars nor clonally derived cultivars. The importance of apomictic or clonally derived cultivars will increase if hybrid wheat becomes important, as both methods are used to preserve or perpetuate genetically unstable lines (e.g., lines that cannot be maintained by selfing, which is the normal form of reproduction in wheat; Flament et al., 2001). Apomictic and clonal propagation of a pureline results in the same progeny as would selfing, so apomictic or clonal propagation of a pureline cultivar offers no advantage. However, if apomictic propagation were successful, it would certainly spur research in hybrid wheat.

THE CONTEXT OF APPLIED WHEAT BREEDING

In the following sections, we will describe various currently used and projected breeding methods to develop improved wheat cultivars. In comparing strengths and weaknesses of each method, it should be understood that the first consideration is whether the breeding method efficiently meets the breeding objective. In fact it is impossible to compare breeding methods without first describing the breeding objective. The second consideration is that every breeder is looking for the rare exception that becomes a cultivar. If developing cultivars were a common occurrence, there would be far more of them.

The two most important aspects of a segregating population are its mean and standard deviation—in this case the standard deviation among lines derived from the population. By knowing these two aspects, the breeder can use statistical estimations to predict which populations have highest probability of producing the best lines. For example, if a breeder had to choose between two populations, both with a mean of 4,000 kg ha^{-1} for grain yield and a standard deviation among the lines of 200 kg ha^{-1} (population I) or 400 kg ha^{-1} (population II), the breeder would be more interested in population II because the chance of developing a higher yielding line (e.g., 4,400 kg ha^{-1}) would be greater. Of course, most breeders are confronted with populations having different mean grain yield and standard deviation (e.g., 4,000 kg ha^{-1} with a standard deviation of 200 kg ha^{-1} vs. a population with a mean grain yield of 3,850 kg ha^{-1} and a standard deviation of 425 kg ha^{-1}) or with many breeding methods having very poor estimates of grain yield and no estimate of variation.

As virtually every breeding program has improving grain yield as its goal, this trait will be used as an example throughout the rest of the chapter. The other common goal is maintaining or improving end-use quality, though feed wheat and wheat forage systems also exist. Other examples of breeding objectives (e.g., disease or pest resistance, tolerance to abiotic stresses) will be used to highlight different aspects of wheat breeding as appropriate.

ACCESSING GENETIC RESOURCES

Food production and security depend on the wise use and conservation of agricultural biodiversity and genetic resources (Esquinas-Alcázar 2005). Crops and their wild relatives comprise the plant genetic resources for food and agriculture (PGRFA), or the genetic variability that provides the raw material for breeding new crop cultivars through classical breeding and biotechnological techniques. The center of origin of wheat is the Middle East. Today wheat is grown from low latitude to high latitude and from low altitude to high altitude. This incredible range of adaptation reflects the diversity of genes for adaptation that has been concentrated by breeders over the past 150 years, but improvement work has been ongoing for millennia. From the earliest times people worked with the germplasm available to them, usually only in their local areas. However, when farmers traveled to new locations and countries, they carried seeds with them to plant in their new environment. This led to new selection pressures on plant populations and subsequent development of new landraces or to new locally adapted cultivars. Having a keen eye for new types was as important then as now. Many of the now famous explorers doubled as plant collectors, gathering and exchanging seed during their visits. As mobility increased, so did exchanges. Embassies had agricultural attachés that sent seeds back to their own country. Percival (1921) lists genetic resources from every continent, including the now known sources of the dwarfing gene *Rht-B1b* (formerly *Rht1*) in landrace 'Daruma' from Japan.

As true today as it was throughout history, having access to genetic materials is a key to having a successful breeding program. The recognized importance of genetic resources led to the establishment of gene banks in many countries. The conservation, use, and exchange of genetic resources have long been in international discussions. Several international efforts have been made to identify, protect, and use PGRFA. In 1983 the Food and Agriculture Organization (FAO) established an intergovernmental Commission on Genetic Resources for Food and Agriculture (CGRFA). The CGRFA was the first international forum for the negotiation, development, and monitoring of international PGRFA agreements. The CGRFA has developed a comprehensive global system on PGRFA, which aims to ensure international cooperation and avoid duplication of efforts. Also in 1983, a nonbinding International Undertaking on Plant Genetic Resources was adopted by the FAO Conference as an element of the FAO Global System on Plant Genetic Resources. This was the forerunner of the International Treaty for Plant Genetic Resources for Food and Agriculture (ITPGRFA) (http://www.planttreaty.org/index_en.htm). Also, the international agricultural research centers (IARCs), in particular the International Plant Genetic Resources Institute (established in 1974), have promoted and facilitated international technical cooperation. Today the IARCs conserve more than 600,000 accessions *ex situ* in 11 gene banks, which according to some estimates might represent as much as 40% of the diversity that is maintained *ex situ* for the main crops (for wheat see http://www.cimmyt.org).

In 1992 the UN Environment Program adopted the Convention on Biological Diversity (CBD). The CBD provided a framework for the conservation and sustainable use of all biological diversity. However, it did not provide specific solutions for the unique features and problems that are related to agricultural biodiversity. Consequently, countries decided that the CGRFA should negotiate a legally binding international agreement that is specific to PGRFA and in harmony with the CBD. In November 2001 the FAO Conference adopted the ITPGRFA. The treaty provides a bridge between agriculture, commerce, and preservation of the environment. The treaty came into force on June 29, 2004. Its objectives are the conservation and sustainable use of plant genetic resources for food and agriculture and the fair and equitable sharing of benefits that arise from their use. The core of the treaty is its innovative Multilateral System of Access and Benefit-Sharing (MLS), which helps to ensure the availability of important genetic resources for research and plant breeding, as well as provide equitable sharing of benefits, including monetary benefits derived from commercialization. A key element of the treaty and the MLS is the Standard Material and Transfer Agreement (SMTA) that facilities the international movement of PGRFA and sharing of benefits.

A breeding program is a dynamic *in situ* gene pool. Because not all genotypes can be grown every year, the most valuable lines for a multiplicity of traits, or for a unique trait, are stored for future use. The storage of these materials is generally for a short period before placing them in a gene bank for long-term storage. Breeders conserve genetic resources by way of a working storage, short-term storage, and long-term storage.

When a new breeding program is initiated the breeder seeks genetic resources to be evaluated relative to the breeding objective or ideotype (specific trait). Informal networks are formed among breeders that are not necessarily confined locally, resulting in exchanges of materials of mutual interest among breeders in new or established breeding programs who understand the importance genetic variation.

With the advent of intellectual property protection it became possible to derive commercial benefit from unique genes or combinations of genes that control specific traits. These property rights have led both private companies and public institutions to seek commercial advantage of these gene combinations (discussed further later). The exchange of germplasm tends to be covered by material transfer and other legal agreements. The insertion of a patented gene into a previously unprotected cultivar can render the unprotected cultivar with the inserted gene unavailable for subsequent breeding, unless an agreement is made with the owner of the patented gene. Similarly, investors in cultivar development are seeking a return on their investment by not sharing new genetic materials for a specified period after its release. Though these concerns certainly affect plant breeding and plant breeders, most plant breeding companies and plant breeders recognize the critical need for germplasm exchange, and hence create mechanisms for germplasm exchange. They may not be as simple as before, but germplasm exchanges maintain an essential presence in contemporary wheat breeding programs.

METHODS TO GENERATE GENETIC VARIATION

Hybridization

We will consider mechanisms to generate genetic variation and subsequently examine methods for its assessment. Genetic variability originates via genetic recombination (through hybridization, also described as crossing), modification in chromosome composition or number, and mutations. With advances in biological knowledge and technology (e.g., recombinant DNA, gene transfer, and somaclonal variation), new mechanisms have been developed to augment and generate new genetic variability, that is, variation that does not exist within the species and its crossable relatives. However, it should be remembered that common wheat (*T. aestivum* L., $2n = 6x = 42$; genomes BBAADD, cytoplasm donor listed first) and durum wheat (*T. turgidum* ssp. *durum* Desf., $2n = 4x = 28$; genomes BBAA) are both polyploids with extensive interspecific and intergeneric germplasm resources (see Chapter 1).

A primary step in wheat cultivar improvement is to generate heritable genetic variation. By far, the most common way of generating genetic variation is to mate (cross) two or more wheat parents that have contrasting genotypes. Once the cross is made, it is critical to estimate the population size required for the chance occurrence of the desired recombinant (a line that has all of the desired genes). The number of different homozygous individuals that is possible from a single hybridization is 2^n, where n is the number of gene-pair differences between the two parents or the number of loci expected to segregate in the F_2 generation. The F_2 generation is the first generation of segregation and the preferred alleles can be in either a homozygous or heterozygous condition. Each gene-pair difference will be reduced in heterozygosity by 50% for each generation of inbreeding or selfing.

The consequences of number of gene-pair differences and inbreeding have a geometric impact on the frequency of the desirable alleles. Thomas and DePauw (2003) elaborated on a calculation put forward by Shebeski (1967) to estimate the proportion of plants within a population having all the desirable alleles. They defined the desirable gene quotient (DGQ) of a parental mating as that proportion of plants (P) in which a segregating locus is either heterozygous or homozygous for the preferred allele. For segregating alleles at locus 1 through locus j, the desirable gene quotient would be $\text{DGQ} = (P_1 \times P_2 \times P_3 \times \cdots \times P_j)$. In the absence of selection at these loci, DGQ reduces to an exponential function (P^n), where n is the number of loci segregating for the desired allele and P varies by generation of inbreeding (Table 13.1). Obviously, the reciprocal of DGQ is the average size of a population that contains one such desirable genotype. The minimum population size in which one desirable genotype occurs also changes exponentially (Table 13.2). Both Tables 13.1 and 13.2 assume independent segregation of genes (i.e., the loci are more than 50 map units apart) and gene frequency of one-half. If the genes are linked (i.e., less that 50 map units apart), the needed population sizes will be less than predicted (DGQ is higher) for coupling-phase linkage or greater than predicted (DGQ is lower) for repulsion-phase linkage.

The geometric changes which occur with increasing number of loci and decreasing frequency of genotypes with the desirable alleles upon inbreeding provide a rationale for early-generation selection, because the proportion of plants having all of the desirable alleles in either the homozygous or heterozygous condition decline upon inbreeding (Shebeski 1967). Historically the majority of breeders tend to select for qualitative traits first and quantitative traits after a few generations of inbreeding and selection. Because the frequency of F_2-derived lines with the complete complement of desirable alleles decreases with inbreeding, there are advantages to selecting for qualitative and quantitative traits simultaneously in the earliest possible generation. Based on the increased number of traits and therefore gene differences that breeders are working with in the 21st century and on considerations of Tables 13.1 and 13.2, breeding strategies should emphasize selecting for both quantitative and qualitative traits simultaneously in the early generations. Marker-assisted selection may be used to enrich gene frequencies in early generations. These concepts will be elaborated further in the section on early-generation selection and marker-assisted selection.

Mutations

Mutations, or changes in DNA sequence, can be caused by copying errors during cell division or by exposure to chemical mutagens, ultraviolet or ionizing radiation, or viruses. Variation caused by mutation is nontargeted and may generate novel

Table 13.2 The effect of inbreeding generation and number of gene differences (or segregating loci) on minimum population size.

Gene Differences	Minimum Population Size (1/DGQ[a])			
	F_2	F_4	F_6	DHL[b] or RIL
1	1	2	2	2
2	2	3	4	4
5	4	18	27	32
10	18	315	753	1,024
15	75	5,600	20,653	32,768
20	315	99,437	566,658	1,048,576

Source: Adapted from DePauw et al. (2007) with kind permission of Springer Science and Business Media.
[a]DGQ, desirable genotype quotient, from Table 13.1.
[b]DHL, doubled haploid line; RIL, random inbred line.

Table 13.1 The effect of inbreeding generation and number of gene differences (or segregating loci) on the desirable genotype quotient (DGQ), in which a desirable genotype has the preferred allele in either the homozygous or heterozygous condition at a locus.

Gene Differences	Desirable Genotype Quotient (DGQ)			
	F_2	F_4	F_6	DHL or RIL[a]
1	0.7500	0.5625	0.5156	0.5000
2	0.5625	0.3164	0.2659	0.2500
5	0.2373	0.0563	0.0364	0.0313
10	0.0563	0.0032	0.0013	0.0010
15	0.0134	0.0002	4.84E – 05	3.05E – 05
20	0.0032	1.01E – 05	1.76E – 06	9.54E – 07

Source: Adapted from DePauw et al. (2007) with kind permission of Springer Science and Business Media.
[a]DHL, doubled haploid line; RIL, random inbred line.

traits or novel substances. Mutations are the original source of variation for all genes. Mutagenesis is the process of generating mutations. If a mutation cannot be passed on through the germ line, hence is not heritable, it has little value in cultivar improvement in seed-propagated crops like wheat. A mutation may be used directly for cultivar release or used as new trait for incorporation into another background for release as a cultivar. As of 2007, 200 cultivars of *T. aestivum* L. and 30 cultivars of *T. turgidum* ssp. *durum* Desf. feature a mutation (http://www-infocris.iaea.org/MVD/).

Japanese wheat breeders were the first to domesticate spontaneous mutations for reduced height (Kihara 1984; Nonaka 1984). Two spontaneous mutations, *Rht8* (on chromosome 2D) and *Rht9* (7BS), occurred in Japanese landrace 'Akakomugi' (Gale and Youssefian 1985). In 1913 Strampelli recombined the short straw attributes and early maturity of Akakomugi with the productivity of 'Wilhelmina Tarwe' and the local adaptation of 'Rieti'. The cultivars Villa Gloria, Ardito, Mentana, and Damiano were derived from this combination of three parents, resulting in widespread use of *Rht8* in southern and central Europe (Borojevic and Borojevic 2005). Two other spontaneous reduced-height mutations in landrace Daruma were the source of *Rht-B1b* (formerly *Rht1*) and *Rht-D1b* (*Rht2*) in cultivar Norin 10, which was brought to North America by S.C. Salmon, deployed in the cultivar Gaines by O. Vogel, and introduced into the Centro Internacional de Mejoramiento de Maíz y Trigo (CIMMYT) gene pool by N. Borlaug (Gale et al., 1981). These spontaneous mutations contributed to the Green Revolution by expressing shorter and stronger straw that would not lodge under high inputs of nutrients and water.

A current example of novel trait introduction via mutation breeding is resistance to the imidazolinone (IMI) class of herbicides that inhibits acetohydroxyacid synthase (ALS), based on the induced semidominant nuclear gene coding for IMI-resistant ALS. Imidazolinone-resistant wheat was originally developed through seed mutagenesis of 'Fidel' winter wheat followed by screening with the herbicide (Newhouse et al., 1992). In further studies, Pozniak and Hucl (2004) mutated wheat and identified two additional IMI-tolerance genes and showed that having multiple genes for IMI tolerance was superior to having a single gene.

Their work illustrates a number of important concepts. The first is that common wheat is a hexaploid, often displaying gene redundancy (one copy of the homoeologous gene in each genome); so the three mutations for IMI-tolerance reside at homoeologous loci, in this case on chromosome group 6, but in different genomes (Pozniak et al., 2004). The second important point is that common wheat shares many of its genomes with its wild or weedy relatives. The original IMI-mutation was in the D genome, which caused concern because one of the weeds that the IMI-herbicides control is jointed goatgrass (*Aegilops cylindrica* Host, $2n = 4x = 28$, genomes CCDD). Hence the IMI tolerance on the D-genome of wheat could potentially be transferred to the D genome of jointed goatgrass via pollen flow, thus making jointed goatgrass tolerant to the herbicide (Zemetra et al., 1998; Tan et al., 2004). Imidazolinone-tolerant mutations on chromosomes 6A or 6B would have far less potential to be transferred to jointed goatgrass, because the A and B genomes are not present in the weed. Third, if a successful mutation can be induced in wheat, most likely it can and will occur naturally in nature. Spontaneous mutations for IMI tolerance in weeds are common. Hence, while gene flow is a concern, equally so should be spontaneous mutations in weedy species, whether or not they are related to the herbicide-tolerant crop. Imidazolinone-tolerant wheat cultivars have been released in Australia, Canada, and the US (Tan et al., 2004; Baenziger et al., 2006a).

Limitations of mutagenesis as a plant breeding technique include mutation drag (potentially deleterious or undesirable genes linked to the mutation that when carried with the mutation during crossing may hinder genetic improvement), low rate of production of viable mutants, higher frequency of recessive mutations than dominant mutations, and the random genetic site at which a mutation occurs. As one advantage, induced mutations can be used to understand gene function through the application of targeting induced local lesions in genomes (TILLING) (McCallum et al., 2000).

Variation from *in vitro* tissue culture

Wheat tissue culture is used for many purposes: (i) to rescue embryos from wide crosses made to transfer genes from wild relatives (Sharma and Ohm 1990; Raghavan 2003) or to create haploid plants (Laurie and Reymondie 1991); (ii) to use as source material for wheat transformation (Vasil 2007); (iii) to create doubled haploid plants (Shariatpanahi et al., 2006); (iv) to create somaclonal variation (Larkin et al., 1984). From a breeding perspective, the embryo rescue procedures are most important when they assist germplasm introgression from wild relatives into cultivated types or are used to create doubled haploid lines. The importance of genes from wild relatives for wheat breeding has been reviewed by Friebe et al. (1996), and it would be difficult to overestimate their importance. Many important cultivars have genes from wild relatives. In both cases, embryo rescue is needed because the endosperm does not form, or forms poorly, and cannot support the developing embryo, thus leading to embryo death without rescue.

Similarly, immature embryo culture is generally used for and most important when it is used as the recipient cells for wheat transformation. Immature embryo culture can also be used as donor cells for other tissue cultures such as suspension cell cultures (Fellers et al., 1995), which can then be used for transformation or for biochemical selection. Doubled haploid breeding and wheat transformation are discussed later in separate sections, and the use of wheat synthetics (one method of introgressing germplasm from wild relatives) is discussed in Chapter 16. Hence further discussion will be brief.

Wheat breeders should realize that almost all tissue culture techniques, especially those involving immature embryo culture and anther or microspore culture, are genotype-dependent. Basically some genotypes work better in culture than others (Shariatpanahi et al., 2006), so that one cannot assume any elite line will be transformed or that its microspores can be used to develop doubled haploids. The other aspect is that tissue culture techniques are known to create somaclonal and gametoclonal variation—that is, variation induced by the tissue culture protocol when somaclonal variation is found in somatic-derived cell cultures and regenerants and when gametoclonal variation is found in gamete-derived cell cultures and regenerants. Because all wheat lines have some level of heterogeneity, experiments must be carefully designed to separate somaclonal or gametoclonal variation from cultivar heterogeneity (Baenziger et al., 1989). Finally, when somaclonal or gametoclonal variation is found, they tend to be deleterious like most mutations. However, somaclonal or gametoclonal variation is not uniform in the regenerants, so by increasing the number of regenerants, those without deleterious effects can be identified.

Transgenic wheat and its impact on wheat breeding

In the early 1990s, wheat became the last major cereal grain species to be transformed (Vasil et al., 1992, 1993; Weeks et al., 1993). The current state of transgenic wheat research was recently reviewed (Vasil 2007) and will be discussed further in Chapter 18. As such, we will discuss transgenic wheat from a breeder's point of view, and specifically how it will affect future wheat breeding and its potential. It should also be understood that transgenic wheat has been called genetically modified (GM) wheat or genetically engineered wheat. In a chapter that discusses the genetic improvement of wheat, genetically modified or genetically engineered wheat could be used to describe what all wheat breeders do through traditional and evolving plant breeding methods; hence we will use the term transgenic wheat in this chapter.

As mentioned in our first section, the principles underlying wheat improvement include biology of sexual recombination, Mendelian laws of inheritance, and selection. Clearly, transformation can greatly change the amount and kind of genetic variability available to plant breeders. Current successful examples of transgenic traits include herbicide tolerance, pathogen resistance, insect resistance, abiotic stress tolerance (though this work may need further validation), and improved nutritional quality (reviewed by Vasil 2007). Though grain yield has been reported to have been increased by transgenes in greenhouses or growth chambers, field evaluations are much less promising (Meyer et al., 2007). However, this result should not be surprising as only with recent molecular marker technology have quantitative trait loci (QTLs) affecting grain yield been identified (Campbell et al., 2003; Cuthbert et al., 2008). Hence with our limited knowledge of grain yield (the exception being Ashikari et al. 2005), it is not surprising that a beneficial transgene for grain yield will be a challenge to identify initially and be used by a plant breeder.

Aside from the difficulty of identification, a related matter is whether the insertion of transgenes affect wheat grain yield. A possible reduction in grain yield could be due to the transgene insertion; it could be due to somaclonal variation from tissue culture during the recovery of the transformed cell or plant; or, as a result of cultivar heterogeneity, a single transformant could by chance represent a lower yielding biotype within the original cultivar. However, most transformations are done in highly transformable lines, and the resultant transgene is backcrossed into elite lines. Hence somaclonal variation or a lower yielding selection would be lost in the backcross. Most of the available data indicate that transgenic wheat performs very similarly to the parental line that does not carry the transgene (Bregitzer et al., 2006; Vasil 2007).

Though often overlooked when considering the importance of transgenic wheat, one area of transformation that can greatly assist wheat breeders is the ability to chemically or insertionally link useful genes so that they segregate as a unit or as a multiple-gene "locus" (Halpin 2005). As described in a subsequent section ("Methods of selecting while inbreeding to develop a cultivar"), the ultimate goal of plant breeding is to accumulate favorable alleles into one cultivar. As more desirable independently segregating genes are needed in a cultivar, the population size needs to be proportionately larger, because the chance of finding the line with the most desirable genes is lower (Tables 13.1 and 13.2). However, the DGQ is greater, and hence the predicted population size is lower, if the genes are linked in coupling phase.

In transgenic crops, the need to stack genes, or pyramid desirable genes, has become important as more beneficial transgenic traits are identified (Halpin 2005). While various methods are used to stack genes (e.g., iterative transformations, crossing independently transformed lines, cotransformation with different transgenes, or chemically linking the transgenes for a single insertion), methods which lead to co-insertion of the beneficial transgenes are preferred by plant breeders for their ease of manipulation. An example of chemically linked multiple genes would be a herbicide-selectable marker linked to another trait of interest (e.g., virus resistance). The resultant transgenic lines would be herbicide tolerant and virus resistant. Whenever the breeder selected for herbicide tolerance (often with a very easy selection assay), the selected lines would be virus resistant according to a simple F_2 segregation ratio of 3 : 1, herbicide-tolerant and virus-resistant–to–herbicide-and virus-susceptible.

A major limitation with using linked transgenes is that the needed stacked traits will differ by adaptation or geographic region. For example, while herbicide tolerance could be widely used, resistance to a specific pathogen may only be needed in a localized region; hence a different gene stack may be needed for different areas of adaptation. A second limitation is that while it might be best for wheat breeders to have a series of different inserted stacked traits, the practicality and expense of transgene regulation preclude this possibility. Currently every commercial transgene is heavily regulated and must undergo extensive testing. This testing presents a major economic impediment for having multiple and similar transgenic clusters undergo regulation. The high cost of regulatory approval, though it may not hinder initial transgenic discoveries by the public sector, requires cooperation with the private sector for commercialization.

One of the concerns with transgenic wheat is pollen flow to non-transgenic wheat cultivars and to wild or weedy relatives (Waines and Hegde 2003). Until there is consumer and regulatory acceptance of transgenic wheat, this will remain a problem for adequately testing or commercializing transgenic experimental lines. Though wheat is a self-pollinated crop, outcrossing rates are usually considered to be 1% or less for fields in proximity, though higher rates may be found with some cultivars (Martin 1990; Hucl and Matus-Cádiz 2001). These rates may be somewhat misleading in that most outcrossing which occurs in a commercial field represents sib matings (hence indistinguishable from selfing) within the cultivar. The only distinguishable outcrossing will occur when a plant of one cultivar is pollinated by a plant of a different cultivar. Using greenhouse studies, outcrossing rates varied from 0 to 3.5% depending upon the cultivar and year, with a few exceptions that were much higher (Lawrie et al., 2006). Studying commercial fields, Matus-Cádiz et al. (2007) found gene flow to occur at trace levels (≤0.01%) at distances up to 2.75 km. This level of outcrossing is acceptable for maintaining cultivar purity but may not meet the current regulatory requirements for the absence of transgenes in commercial fields.

METHODS TO ASSESS GENETIC VARIATION

Selection is based on methods to assess genetic variation. Selection is the process of determining the relative worth of individuals (genotypic or phenotypic) and propagating chosen individuals of a population from generation to generation. Traditionally selection has been based on the phenotype. Molecular markers linked with genes known to control traits [marker-assisted selection (MAS)] are being used to practice something akin to genotypic selection. DNA fragments that constitute the desired gene are the ultimate target in genotypic selection.

The phenotype is a combination of the genetic makeup or genotype of an individual, the environmental influence, and the interaction of genotype with the environment ($P = G + E + GE$). The variance is expressed as $V_{(P)} = V_{(G)} + V_{(E)} + V_{(GE)}$. The genetic component of variance may be further partitioned into components associated with additive or dominance (complete, partial, or overdominance) gene action, and the interaction among alleles at different loci:

$$V_{(G)} = V_{(A)} + V_{(D)} + V_{(I)}.$$

Variation for a trait may be discontinuous (qualitative) or continuous (quantitative). Response to diseases such as wheat rusts is an example of a discontinuous trait. Qualitative or discontinuous traits tend to be controlled by relatively few genes and environmental influences are limited. Such traits typically exhibit high response to selection. Conversely, quantitative traits exhibit continuous variation, tend to be controlled by a number of genes each with a relatively small influence, and are subject to modification by environment. Grain yield is an example of a quantitative trait. Response to selection for a quantitative trait is typically less than for a qualitative trait.

The goal of wheat genetic enhancement is an outcome or product, for example, an improved genotype for cultivation and use that is distinct, uniform, and stable. Efficient improvement in a trait requires knowledge of its heritability, the level of genetic variation, the mode of inheritance, and the number of genes controlling the trait.

Heritability is a measure of the degree to which a phenotype is genetically controlled and can be modified by selection. Broad-sense heritability estimates (H_B) are based on the total genetic variance ($V_{(G)}$) and are expressed mathematically as a ratio to the phenotype variance ($V_{(P)}$):

$$H_B = V_{(G)}/V_{(P)}.$$

Narrow-sense heritability (H_N) is the ratio of the additive variance (fixable component, $V_{(A)}$) to the phenotypic variance, and is expressed as:

$$H_N = V_{(A)}/V_{(P)}.$$

Excluding the error associated with their estimates, narrow-sense heritability should be no greater in magnitude than broad-sense heritability for a given trait. Holland et al. (2003) and Iqbal et al. (2007a,b) provide calculations of narrow- and broad-sense heritability for some agronomic traits and for grain protein content.

Selection results in discrimination among genetically variable individuals, whereby some are chosen to establish the next generation. Response to selection (R), also called genetic advance or genetic gain, is the difference between the mean phenotypic value of the progeny of selected individuals (a generation subsequent to selection) and the mean of the entire population before selection. A significant response to selection is represented by a significant change in the population mean between generations. Response to selection depends on three factors: (i) magnitude of variability, $V_{(P)}$; (ii) confounding effects of environmental and interaction components of variability on the genetic variation which determines heritability; and (iii) the proportion of the population selected, also known as the selection intensity. Response to selection (R) may be described mathematically as:

$$R = k\sqrt{V_{(p)}}H_N$$

In this equation k is the standardized selection differential, which takes into consideration the mean phenotypic value of the selected portion of the population, the mean phenotypic value of the entire population, the phenotypic standard deviation, and the proportion of the population selected. The equation to measure response to selection is fundamental to plant breeding and deserves repeated reflection by aspiring and rejuvenating breeders.

This equation for R estimates the response due to direct selection for a trait. For some traits, due to the difficulty or the time to measure a trait, indirect selection (selection for a trait using a related trait) is preferred (Falconer 1952; Ortiz et al., 2007). The equation, modified for indirect selection response, becomes:

$$R = k\sqrt{V_{(p)}}H_{N2}r_G$$

in which H_{N2} is the narrow-sense heritability of the related trait and r_G is the genetic correlation between the primary trait and related trait. Assuming the same selection intensity, and if $H_{N2}r_G > H_N$, then indirect selection is more effective in changing the primary trait mean than direct selection. A simple and contemporary example of indirect selection is selection based on molecular markers (for further reading, see Allard 1960; Falconer 1981; Acquaah 2007).

METHODS OF SELECTING WHILE INBREEDING TO DEVELOP A CULTIVAR

Two very early methods of plant breeding which do not necessarily begin with hybridization are mass selection and pureline selection. Both selection methods were used when landraces (a mixture of genetically different inbred lines with a low level of natural outcrossing) were brought to new areas. In this case, new phenotypes may appear due to the lines no longer being in the previous environment(s). As mentioned previously, selection does not create genetic variation but only acts upon genetic variation already in the population, and selection acts effectively only on heritable differences.

In mass selection, deleterious plants or off-types (plants which do not phenotypically appear as if they belong in the line) are removed from the population and the remaining plants are harvested in bulk (or *en masse*). It is a very good way to quickly develop more uniform cultivars but rarely improves the yield of the cultivar as most of its individual plants are retained. Mass selection is rarely used today to create new lines but is commonly used to purify advanced lines. Roguing to remove off-types (contaminants) or variants (inherent variation) in a breeder or foundation seed field is a form of mass selection.

Pureline selection differs from mass selection in that individual plants are selected, their progeny are evaluated, and the best plant-derived progeny is used to become the cultivar. Pureline selection attempts to identify and propagate the best individual in a line, whereas mass selection intends to keep the important aspects of the population. As landraces or populations are heterogeneous, pureline selections are often considerably different than the original parental source.

The designation between mass selection and pureline selection is often blurred in practice. An example was provided by the release of 'Redland' wheat (Schmidt et al., 1989). The parent cultivar, Brule (Schmidt et al., 1983), was heterogeneous for its reaction to stem rust (caused by *Puccinia graminis* Pers.: Pers. f. sp. *tritici* Eriks. & E. Henn.). Hence 100 lines were sampled from Brule and their progeny grown in the field for 2 years. After selection for phenotypic and stem rust resistance uniformity, 24 lines were composited to form Redland. Hence, in this example, pureline selection was used to identify individual lines, but to regain a portion of the heterogeneity of Brule, 24 lines (as opposed to only one line in traditional pureline breeding methods) were combined to form Redland. As might be expected, Redland was slightly different from Brule (4 cm shorter and 0.5 day earlier, Schmidt et al., 1989). As with mass selection, pureline selection is most commonly used today to purify or select a better line from an advanced line.

However, most breeding methods begin with hybridizing two or more parent lines. Following the hybridization, each heterozygous gene pair will be reduced in heterozygosity by 50% for each generation of inbreeding (selfing). It should be remembered that wheat is a naturally self-pollinated crop and that, with the rare exception of natural outcrossing, the progeny are created by selfing. With y heterozygous gene pairs, the proportion of homozygous individuals after w generations of self-fertilization is expressed mathematically as $[(2^w - 1)/2^w]^y$. By the F_6 generation of a biparental mating with 10 gene-pair differences, 73% of the lines will be homozygous for some combination of the 10 gene-pairs. Allard (1960) illustrated graphically the rates of fixation of loci as y and w change.

The plant breeder practices some form of selection during each generation of inbreeding until satisfied that the selected materials will constitute a distinct, uniform, stable improved cultivar (required by many seed laws and certification standards). Various methods of selection during inbreeding will be discussed. A key feature of breeding is that different methods of selecting (breeder selection or natural selection) may be applied at various generations to meet objectives and resources available. Breeder selection is an active process in which the breeder selects the type of plants or families that he or she wants. Natural selection is done by nature, though careful selection of environments by the breeder can optimize natural selection.

Pedigree selection

Pedigree selection is a method of breeding in which individual plants are selected from a segregating population of known parents (Love 1927). Hence breeder selection (sometimes referred to as artificial selection) and natural selection occur in every generation. An identity is assigned to individual selections in each generation. The progeny of an individual will be assigned an identity so its progenitor is known. Documentation enables the breeder to trace progeny–parent relationships to the original hybridization of parents. The criteria of selection may be phenotypic, genotypic, or a combination of both. Family relationships may also be used in the evaluation of data to make selections. The pedigree method is labor-intensive but provides the most genetic information about a selection, as it is based upon parent–progeny genetic analyses. For example in a segregating population, if the progeny of a selected plant Z is uniform for the trait, plant Z must have been homozygous for the trait.

Pedigree selection also allows the breeder to create a knowledge base using episodic traits during the progeny evaluation to know the characteristics of the selected new line. For example, if an important disease occurs once every 3 years during the progeny evaluation, the breeder will know if the progeny family was uniform or segregating for the trait (e.g., the parent plant was homozygous or heterozygous for the genes controlling the trait). Thus, once the selection history has been created, the breeder can know the response of the selected new line to various abiotic and biotic stresses that occurred during its development.

Pedigree breeding is used to create new progeny lines that are transgressive (better than) either parent for the breeding objective. Because there are numerous progenies from every population, the breeder can have an intuitive understanding of both the mean and standard deviation for each population. At this stage selection may occur among populations, as some populations are discarded entirely and other populations may be given varying priority. However, the unit of selection within a population is usually the plant or a plant within a progeny-row, and the harvested seed can come from a plant or from a single spike on the plant. If the whole plant is harvested, a small yield plot can be planted the subsequent generation to estimate grain yield. The grain yield from these plots generally provides a poor estimate of line performance, because the plots are usually not replicated within or across environments. They may still provide critical information on yield-related factors such as plant architecture, kernel size, or pest resistance. However, if a single spike is harvested, seed supply is insufficient to estimate yield, so selection is usually based upon highly heritable and often qualitative traits.

Only in the later generations, when the within-progeny-row variation is small and the entire family is harvested, is seed supply sufficient for replicated yield trials. In most breeding programs, there is a natural progression or hierarchy of selection, where the easily measured and highly heritable traits are selected first in the early generations. The poorer lines or those missing key traits are quickly eliminated, often termed negative selection. After the better lines from a population have been selected and there are fewer lines, more expensive tests are done, such as extensive milling and baking testing or other large-scale assays.

In summary, the pedigree breeding method is not only the most labor-intensive method but also the method that provides the most genetic information; hence in areas where labor is inexpensive, it is widely used. It is generally used to create new lines and cultivars that combine the best traits from parent lines.

Bulk selection

Bulk selection (developed by the Swedish wheat breeder Nilsson-Ehle as described by Newman 1912; compared with the pedigree method by Love 1927) is similar to mass selection in that plants are chosen which express individual advantages, and a sample of the aggregate of the seed is propagated in the next cycle of inbreeding. In this case, the breeder often relies extensively on natural selection or relatively simple selection techniques within the bulk population (syn. bulk) for removing unwanted types or retaining desirable types, as the population is harvested *en masse* with no progeny testing. Usually, bulk breeding is much less labor-intensive than the pedigree system because the populations are harvested in bulk, often with a small-plot combine that may harvest a plot in less than 1 minute. Bulk breeding also uses less space than pedigree breeding per population. A bulk plot may not be much larger than a progeny plot in the pedigree breeding system.

The specific strategy of choosing individuals to be propagated may constitute either negative selection or positive selection. Removal of undesirable types or those that do not meet the breeding objective would exemplify negative selection. Negative selection reduces the gene frequency of undesirable types, and often, the majority of the population is retained. An example would be removal of a moderately low frequency of excessively tall and/or late-maturing plants in a topcross F_1 population or succeeding generations, with the remaining majority of the F_1 or later-generation plants being harvested in bulk. Likewise removal of off-types or variants during multiplication of a cultivar is negative mass selection.

Two other common examples of selection in bulk populations are (i) application of an imidazolinone-class herbicide to populations that segregate for tolerance and (ii) growing populations that segregate for winter and spring growth habit in exclusively winter or spring growth habit environments (Color Plate 26). In the first case, the single-gene IMI tolerance is largely a recessive trait, so that after spraying, only plants that are homozygous for herbicide tolerance remain, thus ridding the population of susceptible homozygous or heterozygous plants. In the latter case, spring growth habit is typically dominant to winter growth habit (elaborated in Chapter 3), so that if the population segregating for winter and spring growth habit is planted in a spring environment, the winter types do not vernalize and never flower, and thus are removed from the population. However, the heterozygous plants do remain. If the population segregating for winter and spring growth habit is planted in a winter environment and the winter is sufficiently cold, the spring types are killed by the winter conditions and thus are removed from the population.

Retention of desirable types that meet the breeding objective would be typical of positive mass selection. Positive mass selection increases the gene frequency of desirable types, and generally a small fraction of the population is retained and advanced to the next generation with an aggregate seed sample. An example would be retention of F_2 plants that express a high degree of resistance to a complex of diseases, such as rusts (*Puccinia* spp.), bunts (*Tilletia* spp. and *Ustilago* spp.), and *Fusarium* spp., and that meet the target expression of plant stature and architecture. Only 5% of an F_2 population might be retained.

Molecular markers may be applied at some stage to aid in either positive or negative selection. Molecular markers can have a major impact on shifting gene frequency, especially in a topcross F_1 or backcross F_1 population (Knox and Clarke 2007).

One challenge with bulk breeding is to optimize selection for desirable types while minimizing unwanted plant-to-plant competition in the bulk (Harlan and Martini 1938). For example, with the advent of semidwarf wheat cultivars, numerous studies showed that tall wheat plants shaded the desirable and shorter semidwarf plants (Khalifa and Qualset 1974, 1975). The loss of semidwarf lines in both the mechanical mixtures and bulk populations occurred despite the semidwarf line in the mixture or the semidwarf segregants in the bulk being higher yielding. Of course, competition in a bulk could be reduced by changing the bulk planting system to one that was less competitive, such as using thinly spaced bulks (Khalifa and Qualset 1975) or by keeping the lines in the bulk for fewer generations so that the impact of competitive effects would be less.

As mentioned previously, breeders would ideally like to compare populations on the basis of their line means and standard deviations. Busch et al. (1974) showed that the grain yield of F_4 and F_5 bulks was correlated with the yield of $F_{2:5}$ lines when the bulks and lines were grown in the same environment, thus reducing the G × E effect. Interestingly, the highest yielding line came from a cross between high-yielding and low-yielding parents, but the highest frequency of high-yielding lines came from crosses between strictly high-yielding parents. Hence they concluded that bulks from crosses between high-yielding (elite) parents could be evaluated for grain yield and would be indicative of progeny lines developed from the bulk populations. Later Cregan and Busch (1977) found that by using bulks derived from adapted material, the yield of the F_2 bulks was positively correlated with the yield of their derived F_5 lines and that the amount of variation among the F_5 lines did not seem to affect the yield of the highest yielding derived lines. This research is also noteworthy because the process of handling the bulks to develop the lines included an initial screen for many disease resistance traits, illustrating how early-generation breeder selection and bulk breeding can be integrated.

From a practical standpoint, G × E interactions would affect the applicability of the results from both of these studies because (i) early-generation bulk yield trials normally occur in years preceding progeny-line evaluation, and (ii) often the bulks are evaluated in one or a few locations whereas the progeny lines are tested for adaptation in many more locations. Also, bulk populations are rarely grown in replicated trials to precisely estimate their grain yield. However, the ease of planting and harvesting bulks often leads to bulks being planted at more than one location. This allows the breeder to assess their performance in different environments and to avoid generation loss in areas where breeding nurseries can be lost to hail, winterkilling, drought, or other unforeseen disasters. Even with these concerns, information on early-generation bulks can be valuable to wheat breeders as they decide on which populations to focus their efforts (Busch et al., 1974).

Single-seed descent

Single-seed descent is a method to achieve homozygosity while often practicing minimal selection (Goulden 1939). Brim (1966) referred to the procedure as a modified pedigree selection method, most likely because he suggested including selection in the early generations. In its current usage, the objective is usually to develop a random sample of the 2^n possible homozygous individuals from a parental mating. Though often done without selection, single-seed descent does not preclude selection during generation advance (Brim 1966). Selection and evaluation often are delayed until inbred lines have been produced, when selection can be based on data from replicated field trials for agronomic performance, biotic and abiotic stress nurseries, and end-use quality testing. The method consists of selfing a random sample of F_2-derived plants in each generation and advancing only one seed per plant. This process attempts to represent each genotype present in the F_2 regardless of merit. Techniques to accelerate generation advancement may be used to grow two to three generations per year. This makes single-seed descent ideally suited for spring wheat, which does not require vernalization, but single-seed descent is also used in winter wheat breeding.

The rationale for practicing single-seed descent is to reduce the time to achieve near-homozygosity, to reduce operational costs and record keeping to achieve near-homozygosity, and to maintain maximum genetic variation while inbreeding. Natural selection does not influence gene frequency in this procedure, unless genotypes differ in their ability to produce viable seed under the growing conditions used. Genotypes which do not produce seed will be lost, and genotypes which are very or less prolific would be represented equally by a single seed in the following generation. If any phenotypic selection is practiced, it is based on the individual plant (hence must be for highly heritable traits) without regard for progeny performance.

Knott and Kumar (1975) compared single-seed descent with early-generation yield testing and found that, in general, early-generation yield testing identified higher yielding lines on average. The lower yielding lines were removed from the population, indicating that in early generations it is often easier to cull poor-performing lines than to select high-performing lines. However, there were very few differences among the higher yielding lines; hence the authors felt the early-generation testing was not worth the extra effort. Because plant breeders are most interested in the high-performing lines, they concluded that single-seed descent would have considerable value. However, the authors did not consider the lost opportunities from single-seed descent. From Table 13.3 it is noted that the highest frequency of all 20 desirable alleles, from a cross segregating for 20 target loci, will be present primarily in a heterozygous condition. Because single-seed descent only samples one seed per individual, the probability of selecting individuals with the maximum number of desirable alleles is greatly reduced.

Probably the greatest concern with single-seed descent is that it should not be used, if without selection, in crosses expected to produce greater genetic variation. As can be seen in Table 13.1, with as few as five segregating loci, only 3 of 100 single-seed descent lines (or random inbred lines) would be expected to have the desired allele at all 5 targeted loci. With 10 segregating loci, only 1 plant of 1,000 would have the desired allele at all 10 loci. Hence most plant breeders restrict single-seed descent to their elite × elite crosses, in which many of the favorable alleles are already fixed; hence segregation for important traits is less. Also, in these crosses, poor segregants occur at much lower frequency, making the need to cull in early generations much less.

In addition to being used in applied breeding programs, single-seed descent is often used to create mapping populations (Marza et al., 2006), where a random sample of lines needs to be developed which best represents the genetic

Table 13.3 Classes of genotypes, their distributions, and percentage frequency in that portion of an F_2 population which has all the more desirable alleles segregating at 20 loci.

Classes of Desirable Genotypes Heterozygous or Homozygous at the Specified Number of Loci		Number of Kinds of Genotypes in Each Class	Relative Frequency of Each Kind of Genotype	Relative Frequency of Each Class of Genotype	Percentage of Each Class of Genotype	Percentage of Each Class of Genotype in the Total F_2 Population
A Heterozygous	Homozygous	**B**[a] $[1 + (2^m - 1)]^n$	**C** 2^n	**D** **B** × **C**	**E** **D** × $100/3^n$	**F** **E**/4^n
0	20	1	1	1	2.87E – 08[b]	9.1E – 11
1	19	20	2	40	1.14E – 06	3.6E – 09
2	18	190	4	760	2.18E – 05	6.9E – 08
3	17	1,140	8	9,120	0.0003	8.3E – 07
4	16	4,845	16	77,520	0.0022	7.1E – 06
5	15	15,504	32	496,128	0.0142	4.5E – 05
6	14	38,760	64	2,480,640	0.0711	0.0002
7	13	77,520	128	9,922,560	0.2846	0.0009
8	12	125,970	256	32,248,320	0.9249	0.0029
9	11	167,960	512	85,995,520	2.4663	0.0078
10	10	184,756	1,024	189,190,144	5.4259	0.0172
11	9	167,960	2,048	343,982,080	9.8653	0.0313
12	8	125,970	4,096	515,973,120	14.7980	0.0469
13	7	77,520	8,192	635,043,840	18.2129	0.0578
14	6	38,760	16,384	635,043,840	18.2129	0.0578
15	5	15,504	32,768	508,035,072	14.5703	0.0462
16	4	4,845	65,536	317,521,920	9.1064	0.0289
17	3	1,140	131,072	149,422,080	4.2854	0.0136
18	2	190	262,144	49,807,360	1.4285	0.0045
19	1	20	524,288	10,485,760	0.3007	0.0010
20	0	1	1,048,576	1,048,576	0.0301	9.5E – 05
Totals		1,048,576		3.49E + 09	100	

Minimum population for chance presence of each genotype = 4^{20} = 1.09951 E + 12

Note: Calculations inspired by Allard (1960) and Shebeski (1967).

[a]Expansion of the bionomial in which n is the number of gene differences and m is the number of generations of selfing beyond the F_2.

[b]E is 10 raised to the power which follows E.

distribution in the F_2. In this case, single-seed descent is done strictly without selection.

As mentioned earlier, most wheat breeders would like to know the mean and standard deviation of the lines derived by their breeding populations, so they can concentrate on working in the best populations and quickly discard those with less potential. Breeders also want to know which segregating lines within a population have the greater potential to develop superior inbred lines. Thus the mean and standard deviation both among populations and among segregating lines within a population are important. In single-seed descent breeding, as well as the doubled haploid breeding method described next, wheat breeders can easily estimate both the population mean and standard deviation among derived lines. However, in practice, this is rarely done because once the homozygous lines are developed, new crosses have been made, and it is difficult to return to the better cross to generate additional homozygous lines. Rather plant breeders view their lines as a finished product of the cross or population and move on to other crosses and populations.

Doubled haploid breeding

Doubled haploid technology generates homozygous lines from haploid tissue. The method involves production of plants from haploid tissue and doubling the chromosomes (Guzy-Wróbelska and Szarejko 2003). The resultant plant will be completely homozygous and homogeneous. Haploid tissue may be produced by chromosomal elimination in wide crosses or by the direct use of pollen grains, microspores, or ovules as haploid tissue (Schaeffer et al., 1979; Laurie and Bennett 1988; Laurie and Reymondie 1991). The chromosome numbers of the haploid plant are doubled by application of the alkaloid colchicine, which interferes with microtubular activity. Embryo rescue methods may be used to propagate haploid tissue, especially in the case of chromosome elimination in wide crosses where the endosperm does not form. As two predominant methods are available to create doubled haploids—the wheat-by-maize system (Laurie and Bennett 1988) and the anther culture or microspore system (Guzy-Wróbelska and Szarejko 2003)—it is important to know if the lines developed by both doubled haploid breeding systems are equivalent. In their work, Guzy-Wróbelska and Szarejko (2003) compared in single rows the yields of anther-culture-derived doubled haploids, maize-pollination-derived doubled haploids, and single-seed descent lines and found the lines developed by each method to be similar or equivalent. In larger plots they compared the anther-culture-derived doubled haploids to maize-pollination-derived doubled haploids and again found them to be equivalent. Additional testing in more environments may identify small differences, but at this time, both methods create lines with similar agronomic value. Considering the top 10% of the created single-seed descent and doubled haploid lines (a method to remove any lines that might be poor performers due to gametoclonal variation), again no differences were detected, indicating all three methods develop lines useful for plant breeders.

However, in a recent study, Guzy-Wróbelska et al. (2007) found that the recombination frequency in anther-culture-derived doubled haploids was much higher than in the maize-derived doubled haploids, as determined by the genetic map developed from the anther-culture doubled haploids being 41% larger than the map developed from the maize-derived doubled haploids. They did not compare the recombination frequency in the two doubled haploid methods to the single-seed descent lines, so it is unknown which recombination frequency is closer to the single-seed descent process used to create mapping populations. While the recombination frequency was different, the gene order was identical with one exception.

Double haploidy is an expensive method but requires the least amount of time to develop inbred lines, especially when breeding winter wheat, where the vernalization requirement slows single-seed descent breeding. Doubled haploidy and single-seed descent have the same end point: production of a random set of inbred lines for subsequent assessment. Both doubled haploid and single-seed descent methods have a lower desirable genotype quotient compared with lines derived from earlier inbreeding generations, especially the F_2 (Table 13.1); hence they tend to be produced from narrow crosses where there are fewer important genes segregating.

The major difference between single-seed descent and doubled haploid breeding is in cumulative recombination frequency, as the single-seed descent process involves more segregating generations of selfing where effective recombination can occur (Riggs and Snape 1977). Because most authors have found few or small differences between the three methods, one can conclude for the populations in question that the additional recombination was not beneficial. This result may be to due to the relatively narrow crosses commonly used in single-seed descent and doubled haploid plant breeding. If past history repeats itself, the methods to create doubled haploids will become less expensive and will feature fewer culture-induced variants.

Two additional aspects of doubled haploid methods are important. First, recovery of mutants is very efficient at the haploid level because each locus is hemizygous and expressed (especially important for recessive alleles). Second, while most programs create haploid plants and then double all of them, selection with molecular markers at the haploid level is possible for many traits, thus allowing only selected plants to undergo the chromosome doubling process. If it is not possible to select at the haploid state, selection using only homozygous diploid lines is very efficient for studying mutants, because both target and nontarget mutants will be expressed and not hidden by heterozygosity. Similarly, using molecular markers in selection becomes more efficient because the heterozygous lines have been removed.

Finally, with the rapid speed that doubled haploid lines theoretically can be made, considerable theory has developed on how to use doubled haploid lines to estimate population means and among-line standard deviations (Choo et al., 1979; Choo and Reinbergs 1979; Snape et al., 1984; Caligari et al., 1985; Choo 1988; Baenziger 1996; Baenziger et al., 2001). While much of the research attempts to answer theoretical concerns with plant breeding methodology and understanding a given crop, a very practical use of doubled haploid breeding can be to estimate the value of a cross. Reinbergs et al. (1976) estimated that as few as 20 doubled haploid lines in barley (*Hordeum vulgare* L.) will accurately estimate among-progeny line mean and standard deviation. Hence even if doubled haploid technology is expensive and few lines can be made, it could allow breeders to identify the most important crosses with which to work.

Doubled haploid cultivars have been released in a number of countries and some have become dominant cultivars. For example in Canada in 2007, three of the five most widely grown cultivars in the market class Canada Western Red Spring (CWRS) were doubled haploid cultivars. 'Lillian' (DePauw et al., 2005) accounted for about 15% of the CWRS acreage, 'Superb' accounted for 12% (T.F. Townley-Smith, pers. comm.), and 'McKenzie' (Graf et al., 2003) accounted for 7%. 'Snowbird' (Humphreys et al., 2007b) and 'Kanata' (Humphreys et al., 2007a) accounted for all of the Canada Western Hard White acreage. 'Andrew' (Sadasivaiah et al., 2004) accounted for 99% of the Canada Western Soft White Spring acreage.

Backcrossing

Backcrossing is a method of recurrent hybridization by which a desirable allele for a specific trait is substituted for the alternative or undesirable allele, as initially proposed by Briggs (1938). The parental source of the desirable allele is designated the donor parent, and the parent used repeatedly in hybridization is the recurrent parent. The recurrent parent is reconstituted at the rate of $1 - (1/2)^n$, in which n is the number of recurrent crosses or backcrosses. The effectiveness of the backcross method depends upon (i) the heritability of the trait, (ii) the degree to which the expression of the trait is independent of background genes for its expressivity, (iii) expression of the trait or its detection phenotypically (or with markers) in the F_1 for subsequent backcrossing, (iv) linkage of the desirable gene with undesirable genes from the donor, and (v) the number of backcrosses necessary to recover a desirable level of the recurrent parent phenotype. Also important is to use a recurrent parent that will retain its value during the backcrossing procedure, as the backcross-derived line is equivalent to the recurrent parent plus the added trait. Examples of successful backcross-developed lines include 'Prowers' (Quick et al., 2001a) and 'Prairie Red' (Quick et al., 2001b).

The backcross method has been used effectively as a short-term breeding strategy to incorporate dominant genes for the control of devastating pathogens, such as that causing stem rust, in otherwise highly productive and adapted cultivars (Campbell et al., 1967; Campbell 1970; Green and Campbell 1979). The emergent stem rust race Ug99 (race TTKSK) in East Africa has virulence to gene *Sr31* derived from the 1RS·1BL translocation, and another variant of Ug99 (TTKST) subsequently identified has virulence to genes *Sr24* and *Sr31*. Both genes are deployed in numerous cultivars on a global scale. The Global Rust Initiative (http://www.globalrust.org/index.cfm?m=1) is a concerted effort to develop a global response to the emergence of these devastating virulent races of stem rust. Backcross breeding will be one of the strategies used to incorporate resistance genes into adapted cultivars in various countries. This illustrates a good example of opportunities to use DNA molecular markers linked to known genes that confer some resistance—for example, *Sr2* (Hayden et al., 2004) and *Sr26* (Mago et al., 2005)—without actually having to introduce the pathogen into regions where it does not exist, thus putting local wheat production at great risk should the introduced pathogen escape from the testing site. Final verification may be obtained in the field in countries where the pathogen already exists, after the backcrossing has been completed. Of course, many other genes and traits have been incorporated using backcross and molecular marker breeding (Dubcovsky 2005).

In addition to adding a gene to a recurrent parent, backcrossing is used to make alloplasmic lines (Kofoid and Maan 1982). Alloplasmic lines are used to study the effects of different cytoplasms on various traits. However, the most common use of alloplasmic lines was in developing cytoplasmic male sterility (CMS) for hybrid wheat production (reviewed by Edwards 2001). The three components of CMS hybrid wheat are a restorer line (R-line), the maintainer line (B-line), and the male-sterile line (A-line). Once a good B-line × R-line combination has been identified, the B-line needs to be converted into an A-line via backcrossing. Usually at least six backcrosses are needed for this conversion, but it should be understood that many of these backcrosses can occur while the A-line seed is being increased to commercial seed quantities. Every time an A-line is crossed by the same B-line to increase seed, it is a backcross. Hence a desirable B-line is converted by making two or three backcrosses by hand in the greenhouse or field to a previously existing A-line, followed by a variable number of crosses (i.e., backcrosses) that are needed to obtain commercial volumes of A-line seed for hybrid wheat production.

MAJOR ISSUES ALL WHEAT BREEDERS FACE

Early- vs late-generation selection

Reflections on desirable gene quotient (Table 13.1) and minimum population size (Table 13.2) reinforce the theory that the earliest inbreeding generations have the highest frequency of genotypes with the desirable alleles, albeit predominately in the heterozygous condition. The challenge then is to identify those genotypes in the earliest possible generation and to select them for further inbreeding.

In the 20-gene example a minimum of 315 F_2 plants would have to be retained for the chance occurrence of one containing all the desirable alleles (Table 13.2). From Table 13.3 it can be determined that two classes of genotypes occur with the largest frequency (18% each): one class homozygous for the desirable allele at 7 loci and heterozygous at the remaining 13 loci, and another class homozygous and heterozygous at 6 and 14 loci, respectively. The challenge is to identify these unique F_2 genotypes and concentrate subsequent selection among F_2-derived lines while continuing to inbreed. Alternatively it is preferable to discard genotypes which are low yielding or fail to have the desirable alleles (e.g., they are susceptible to diseases), which allows the breeder to allocate resources to select within more promising early generation lines.

As tools are developed to enable whole-genome scans, it may become possible to know which F_2 plants have the highest proportion of desirable alleles. For example, with codominant markers for all 20 desirable genes, one could potentially multiplex all 20 markers to identify the genotype with all desirable alleles, followed by subsequent inbreeding and continued selection among the remaining heterozygous loci to identify the genotype which has all 20 desirable alleles in a homozygous condition. While this example is for progeny from a biparental cross, whole-genome scans can similarly identify the best F_1 plants for selection in more complex crosses (three-way and double crosses).

Early-generation yield testing of individual lines or families has been assessed by various researchers with diverse results, but generally the consensus is that high-yielding materials can be distinguished from low-yielding materials. Lupton and Whitehouse (1957) reported a significant correlation coefficient of 0.56 for grain yields of F_4 versus F_5 lines of winter wheat grown in successive years. Shebeski (1967), Briggs and Shebeski (1971), and DePauw and Shebeski (1973) reported significant correlations for grain yield between generations of F_3, F_4, and F_5. These studies used frequently repeated systematic controls to address spatial variability. To overcome the cost of using repeated controls, statistical techniques such as moving-mean (Townley-Smith and Hurd 1973) and nearest-neighbor (Wilkinson et al., 1983) analyses have been used to partition genetic variation from environmental variation.

In Australia, Fischer et al. (1989) and Gras and O'Brien (1992) reported significant early inter-generation correlations for end-use quality traits. Early-generation testing for quality has been the basis for improvement in spring bread wheat (Lukow 1991) and in durum wheat in Canada (Clarke 2005).

Some contemporary plant breeding programs exploit these early-generation relationships to select simultaneously for qualitative as well as quantitative traits (DePauw et al., 2007). The F_2 plants are selected for resistance to diseases in nurseries with high pressure from diseases such as rusts and for other simply inherited traits such as plant height. The F_3 generation is used to multiply seed for subsequent multilocation yield trials and disease nurseries (Color Plate 27). These F_3 nurseries can be used to screen for simply inherited traits such as diseases, time to maturity, and plant architecture. Nonreplicated F_4 yield-trial nurseries are grown at multiple locations, and specialized disease nurseries may be established if these diseases are not reliably expressed in the yield-trial nurseries (Color Plate 28).

The harvested grain may subsequently be used for end-use quality testing. Prior to harvest, heads are selected to establish families of those lines which express the highest concentration of desirable alleles for all quantitative and qualitative traits. This procedure is very expensive and is suitable for application to populations which have a very high midparent value and are expected to have complementary genes for all traits to make a "field-ready" cultivar (Color Plate 29).

Mechanization and computerization have enabled breeders to handle very large populations to enhance their opportunities to identify the rare recombinants with all the desired traits. Mechanized seeders have replaced human-powered seeders in the past 60 years (Fig. 13.1). Many wheat breeding programs assess fivefold more genetic materials today than 30 years ago. Most breeding programs are able to score more traits on more lines each generation, analyze the data, and interpret it than 30 years ago. Molecular markers are being integrated as another tool to enhance efficiency and achieve genetic gain through selection.

Impact of molecular markers on wheat breeding

From a breeder's perspective, identifying QTLs will be very useful in understanding how important agronomic traits are inherited (Campbell

(a)

(b)

Fig. 13.1 (a) Human-powered V-belt seeders planting multirow yield trials in southern Saskatchewan, 1930s (courtesy SPARC, AAFC) and (b) self-propelled, reduced-tillage, air-conditioned seeder for planting multirow yield trials (courtesy SPARC, AAFC; photograph taken by D. Schott).

et al., 2003), and in many cases QTL discovery may benefit from the extraordinary cytological resources available in wheat (Berke et al., 1992). However, identifying QTLs and their use may be retrospective. Considering the major yield QTL identified by Campbell et al. (2004), Mahmood et al. (2004) found that many high-yielding contemporary cultivars had the favorable QTL molecular variant from 'Wichita' for grain yield. They also found that many of the western Nebraska cultivars adapted to dryland production had the less favorable molecular variant from 'Cheyenne'. In western Nebraska, the QTLs from Wichita and Cheyenne had very small effects, and thus it was not surprising that the lower yielding Cheyenne variant was often found. However, many contemporary lines had neither the Wichita nor Cheyenne variant, but something different. Hence if selection had been for the favorable Wichita variant, many adapted western Nebraska lines would have been discarded, as well as a few high-yielding lines that surprisingly had the Cheyenne variant. Also, all the lines with neither the Wichita nor the Cheyenne variant would have been discarded. These lines may or may not have an important molecular variant at that locus, but many were high-yielding lines, regardless of where the QTLs affecting yield are derived.

Clearly QTL detection and marker-assisted selection will be widely used to add important QTLs to lines that lack them and to pyramid genes; however, wheat breeding with its constant input of new germplasm will evolve from marker-assisted selection to marker-assisted breeding, whereby phenotyping (if carefully done) will be coupled with inexpensive markers to allow populations with novel QTLs to be used immediately (Bernardo and Yu 2007).

Quantitative trait loci have played an important role in identifying chromosomal regions that control useful traits such as grain protein concentration. The next step is dissection of the QTL so that the desirable portion of a chromosomal region is clearly marked and ultimately only the gene controlling the trait is transferred. For example high grain protein content was detected in a wild population of *T. turgidum* L. ssp. *dicoccoides* (Avivi 1978). Subsequently, the chromosomal region controlling high grain protein content from the Israel accession FA15–3 was transferred to hexaploid wheat cultivar Glupro (ND643) by Dr. R. Frohberg, North Dakota State University. Glupro had elevated grain protein concentration (later determined to be controlled by *Gpc-B1*), but this trait was linked with low grain volume weight. Using restriction fragment length polymorphism (RFLP) analysis Mesfin et al. (1999) identified a 15-cM segment of the 6BS region of Glupro to be in common with the original source FA15–3. A PCR-based marker specific to the *XNor-B2* locus (Khan et al., 2000) was used to transfer *Gpc-B1* in the 6BS region from the breeding line 90B07-AU2B (parentage 'Pasqua'*2/Glupro) to BW621. A selected line was subsequently released as 'Lillian' (DePauw et al., 2005). Uauy et al. (2005) reported that the marker *Xucw71* was closely linked to *Gpc-B1* (0.3 cM). DePauw et al. (2007) provided evidence using the nucleolus organizer region (NOR) marker and *Xucw71_5utr* cleavage amplification polymorphic sequence (CAPS) marker that the chromosomal region with *Gpc-B1* in Lillian was smaller than in the parent 90B07-AU2B and grandparent Glupro.

Another novel use of molecular markers has been to develop a triple hemizygous stock to fix multiple Fusarium head blight (FHB, caused by *Fusarium graminearum* Schwabe) QTLs on 3BS, 5AS, and 6BS (Thomas et al., 2006). The rapid fixation was achieved by hemizygosity of the chromosome arms that carry the QTLs (3BS, 5AS, and 6BS) in the F_1, followed by reversion to euploidy upon inbreeding. Elite hemizygous parents were developed by combining nonreciprocal translocations on 3BL, 5AL, and 6BL. Even though rapid fixation of the FHB QTLs and an increase in the number of lines with all three QTLs are expected, the disadvantage remains that any undesirable genes on the same chromosome arm will be retained, as no recombination within the chromosome arm occurs.

THE PRACTICE OF WHEAT BREEDING

In the preceding sections, we have described much of the theory and research laying the foundation of present-day wheat breeding. In this section, we will attempt to illustrate how the theory and research is actually used to create new cultivars. At the international level, CIMMYT recently reviewed their strategy for cultivar development (Ortiz et al., 2007) and their review is an excellent outline of practical plant breeding methods.

Extension of the theory

It should be remembered that normally it takes at least 7 years to develop a spring wheat cultivar and 12 years to develop a winter wheat cultivar. For a 35-year career a wheat breeder's germplasm will completely turn over three to five times. This concept is important to understanding that program continuity is critical, and that when a new technology is said to not affect wheat breeding for 25 years, from a breeder's perspective that is not a long time. Also, plant breeding really does not have a start or stop point but is a continuum where new crosses are made every year, populations or lines are selected and advanced every year, and if the program is productive, cultivars are released every few years and more frequently than every 7 or 12 years.

A typical breeding program can be described by the flow diagram in Fig. 13.2. Every breeding program starts with an objective, such as improving grain yield. Once the objective is decided, the most common way to introduce new variation into a breeding program is to introduce new germplasm. However, if the objective is not achieved by introduced germplasm, the next step is to make a cross or sexual hybrid between parents that have all or most of the complementary traits. Every program must decide which parents to use based upon the objective and the appropriate type of cross: single (A/B), three-way (A/B//C), double or four-way cross (A/B//C/D), and others.

Once there was considerable debate on whether elite-by-elite crosses (containing the best alleles)

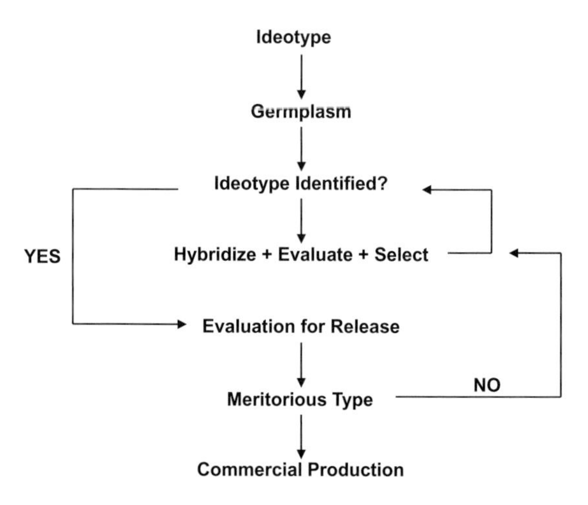

Fig. 13.2. Flow chart of steps to develop and commercialize a wheat cultivar.

were preferred to elite-by-nonelite crosses, with the idea that the nonelite lines might have rare or unused beneficial alleles (Busch et al., 1974; reviewed by Baenziger and Peterson 1992). Virtually every program now relies on elite-by-elite crosses for most of their breeding effort. Simply, these crosses have the highest likelihood of producing new cultivars. As such, most breeders heavily intermate their own material, followed by crossing to elite material in their region. In this effort, most breeders can use single or three-way crosses because the parents have equivalent or nearly equivalent value. Only when new diseases or pests or new traits are identified and where the resident breeding program is unlikely to have useful genetic diversity for the trait will the breeder use less-adapted germplasm. This could include elite germplasm from another region or even a wild or alien species.

Once the breeder decides to cross to nonelite material, rarely are single crosses used except as an interim cross to a three-way cross or backcross. The value of a three-way cross is that nonelite germplasm will contribute only 25% of the genes in the segregating population, including those which are truly targeted. If the germplasm source has very few useful genes, a backcross can be used to further reduce the amount of the nonelite germplasm genes in the subsequent progeny. Most progeny lines that are developed from a cross are discarded, a few are used as parents, and very rarely is a progeny line released as a cultivar. Hence germplasm that has few useful genes might be used as a parent to develop lines (without backcrossing), which in turn may become parents for future crosses (effectively an extended cycle and "backcross").

Despite the research on which crosses yield the best progeny (reviewed by Baenziger and Peterson 1992), crossing remains a relatively simple process in wheat, while evaluating parents and using specific mating systems to determine parental value are cumbersome. Thus most breeders make a large number of crosses knowing that most will fail to produce a cultivar. Often their progeny will become new parents for future cultivars or again parents. All breeding programs are fundamentally recurrent selection programs as previously alluded (Acquaah 2007), with differing periods of time for inbreeding, evaluation, and intermating.

Once the objective is decided and the cross is made, the breeder must decide which breeding method(s) to use. In our preceding discussion on breeding methods, each method was presented in a pure form. In practice, every breeder modifies the method to fit the available resources. Also, during the 7 to 12 years it takes to release a cultivar, different opportunities arise for selection, which require a highly flexible approach.

For example, many breeding programs may start as a bulk breeding program to allow natural selection (perhaps winter survival) or, in the case of IMI-tolerant wheat, artificial selection (herbicide application to remove the susceptible plants) to efficiently remove unwanted types. Often only one or two generations of selection are needed to remove the undesired types, so thereafter the breeding method might change. For example, in the later generations when it is more difficult to select lines phenotypically, perhaps single-seed descent will be used to rapidly advance generations.

Alternatively if key traits can be selected in the field during the first two generations of inbreeding, pedigree breeding may be used to identify the best F_2 or F_3 plants for creating progeny rows. This procedure is an example of allele enrichment, or moving the population to having more desired types genetically and phenotypically as further described later.

Most breeding programs are so large that their selection procedures have to be efficient and repeatable. In this area, great care needs to be taken as to where the early generations are grown, because they are often nonreplicated and if they are lost will the work and resources used to develop the plant material will be lost as well. Loss of a key nursery often impacts the genetic progress by more than a year by disrupting the acquisition of data upon which to make selections and by losing the genetic materials themselves. At the same time, natural selection can be advantageous for moving populations to the desired phenotype (Baenziger and Peterson 1992; Baenziger et al., 2006b). Ideally selection nurseries, or nurseries which magnify useful differences and hence make selection easier, are extremely valuable in the selection phase of wheat breeding. For example in Nebraska, early-generation winter wheat populations are grown at Mead, Nebraska, an area of relative unimportance to wheat production, because any plant that can survive there will survive anywhere in Nebraska. This location provides an effective environment to differentiate and select for an adequate level of winter survival throughout the northern Great Plains (Color Plate 26).

The value of a selection site is predicated upon the selected phenotype being representative of the desired genotype. In the previous example, winter survival is a good example, because spring growth habit genes at most loci are dominant to genes for winter growth habit; hence the desired phenotype is usually the genetically homozygous recessive plant. Similarly, optical kernel sorting for recessive traits can be extremely useful for enriching the segregating population for the desired phenotype or genotype. Examples include selection for white kernels in a population segregating for red and white kernels, or waxy wheat in a population segregating for waxy and nonwaxy kernels, or hard kernels in a population segregating for kernel hardness (Dowell et al., 2006). Similarly, perfect markers can be highly successful for allele enrichment in populations, as noted previously. The advantage of phenotypic selection, either in selection nurseries or by optical kernel sorting, is that it represents direct selection and can handle a large number of seeds or plants at little cost. The main disadvantage is that the traits must be highly heritable. The main advantage of MAS is that it can involve DNA from seed or leaf tissue for low-heritability traits, traits that are difficult to measure, or traits that can only be evaluated in the adult-plant stage. As an indirect selection procedure, the main disadvantage of MAS is the marker may not be diagnostic for the trait of interest depending on the population (Ellis et al., 2007), and though it is becoming less expensive, MAS remains expensive if used extensively.

Previously we mentioned allele enrichment strategies. By far the most common method of allele enrichment is using molecular markers where the markers identify early-generation progeny that have the QTL allele of interest. Those without it can be discarded. DePauw et al. (2007) reported how the *Xucw71_5utr* CAPS marker is used to enrich the frequency of *Gpc-B1* for elevated protein content in topcross F_1 populations. In crosses in which *Gpc-B1* occurs at 25% gene frequency (1*Gpc-B1gpc-B1*:1*gpc-B1gpc-B1*), the heterozygotes are retained and the homozygotes which lack the gene are discarded, producing a shift in gene frequency from 25% to 50%. In other topcross F_1 populations in which the frequency of *Gpc-B1* is 75% (1*Gpc-B1gpc-B1*:1*Gpc-B1Gpc-B1*), only the homozygotes with *Gpc-B1* are retained, producing a shift in gene frequency from 75% to 100%.

However, due to cost and population size requirements for using markers, reduction in population size is advised first by removing undesirable types for other traits before the application of marker selection. For example, consider the goal to develop a hard wheat cultivar that is resistant to Fusarium head blight. In the F_2 progeny of a single cross between a soft wheat cultivar with two QTLs for resistance to Fusarium head blight and a hard wheat cultivar susceptible to Fusarium head blight, it might be best to optically sort and remove the soft kernels first, so only the remaining hard kernels would be tested using markers for Fusarium head blight resistance genes. To compare the unsorted with the sorted populations, the frequency of hard (1/4) times the frequency of homozygous Fusarium head blight resistant lines (1/16) equals 0.0156, or 1.6% of the population. Using 96-well plates for MAS and 4 plates (i.e., 384 individuals), 6 individuals could be selected. However, if the soft kernels were removed first, then 1/16 of the remaining population would be homozyogous resistant, or potentially 24 of 384 individuals. Remember that while 24 individuals is a small population size to represent other segregating alleles in the population, it is fourfold greater than the 6 individuals available in the absence of selection for hardness; initial population size could be scaled up to generate a more desirable size of the enriched population.

Cultivar release

The last stage before releasing a line or dropping it from the program (the more common event) is extensive field evaluation (Color Plate 29). The generation when single-plant selection is terminated depends upon the level of homogeneity desired in the released cultivar. Heterogeneity in a released cultivar is a direct result of the genetic differences between the parents and the final generation of single-plant selection (Baenziger et al., 2006b). In the preceding example, with 10 segregating loci and selection in the F_6 generation, 73% of the lines were homozygous at all loci. However, 27% of the lines were heterozygous at one or more loci, and segregation of these heterozygous loci with continued selfing will lead to heterogeneity in later generations. If plant selection ends in the F_4 generation, only 26% of the lines will be homozygous at all 10 loci and 74% of the lines will be heterozygous at one or more loci, meaning that they will produce heterogeneous lines in future generations. Heterogeneity for traits not related to morphological appearance or value for cultivation and use may be tolerated in a finished cultivar, such as variation in some gliadin subunits. Where heterogeneity is tolerated (e.g., in many parts of the US), within-line selection often ends in the F_4 generation. However, in Europe with more stringent uniformity (homogeneity) standards, final single-plant selection is delayed until later generations (F_6 or later).

Understanding the phenotype

All plant breeders understand that the measured phenotype is a function of the genotype, the environment in which the genotype is grown, and the G × E interaction (Acquaah 2007). An inbred line developed or selected only at one location—a single environment—is likely to exhibit a narrow or specific adaptation to that environment. This approach is like developing a line in a greenhouse and then expecting it to be adapted across a broad range of environmental conditions. Hence a prerequisite for release of a cultivar is that its phenotype must be thoroughly understood. In evaluating an experimental line it needs to be tested in those environments where it may eventually be grown and also in those environments where it should not be grown, because growers should know which cultivars may or may not be grown on their farms. Regional performance trials are usually very expensive, because they must be located on and away from the main breeding nurseries to adequately represent the target environments (Ortiz et al., 2007).

Resource allocation is extremely important in this phase of wheat breeding, and numerous researchers have studied the optimal numbers of replications, locations, and years to estimate line performance (Comstock and Moll 1963; Bos 1983; Carter et al., 1983; Crossa 1990; Gauch and Zobel 1996; Cullis et al., 1996a,b, 2000; Yau 1997). The inference space from the dataset generated provides the basis for predicting a cultivar's future performance. The importance of having the most efficient regional performance assessment strategy is that the cost per plot is relatively similar, though it will change with more locations; hence available resources will largely determine the total number of plots that can be used. With a fixed number of plots, determining how to divide them among replications and locations within a year is critical, as is the determiniation of the most efficent experimental design and subsequent statistical analysis (Smith et al., 2005).

Though not unique to wheat, it should be understood that with the exception of wheat produced for irrigated production, rarely are rainfed nurseries consistently devoid of spatial variation. Hence methods that can remove spatial variation are necessary (Stroup et al., 1994). Removing spatial variation requires augmented or replicated designs. Single-replication trials which are popular in some crops are used in wheat only where seed is limited.

If the variation among locations is expected to be similar to the variation among years, then it is possible to substitute locations for years. Knowing the number of years that lines need to be evaluated is important to provide the grower with sufficient data over environments that they can make educated decisions on cultivar selection (Eskridge 1990). To quickly release a line, breeders usually try to substitute locations for years with the hope they have a sufficient number of environments (e.g., location-years) to adequately represent the target environment. In the authors' experience, at least 30 environments (location-years) over at least 2 years are needed to estimate line performance.

In reviewing line performance data across environments, the most troublesome statistical effect is G × E interaction caused by crossover interactions (e.g., Line A is better than Line B in some environments and Line B is better than Line A in other environments; Baker 1988; Crossa et al., 1993; Russell et al., 2003). At first glance having lines perform better in one set of environments than in other environments seems nothing more than making a recommendation for different target environment, which is routinely done. However, the difficulty arises when a location in different years (hence different environments) clusters with a different target environment. For example, a drought-stressed location might be clustered with low-yielding environments, whereas the same location with more rainfall will be clustered with medium- or higher-yielding environments. While it is important to understand the underlying environmental causes of the G × E interaction (Cullis et al., 1996b; Basford and Cooper 1998), it remains largely impossible to predict future environmental conditions.

Perhaps the importance of understanding the environment and developing lines for broad adaptation is best exemplified by the shuttle breeding program of CIMMYT (Ortiz et al., 2007), which has global responsibilities for maize and wheat improvement. As such, CIMMYT has partitioned the world into megaenvironments and, within those megaenvironments, has clustered similar environments that share common attributes such as high night temperatures. The megaenvironments and the subclusters within megaenvironments are useful for directing germplasm exchange and evaluation, and they reduce the problems of crossover interactions within subclusters.

Initially the shuttle breeding program for wheat improvement was designed for rapid generation advance, but by using two diverse locations in Mexico, the program also selected for lines which were photoperiod insensitive and were exposed to greatly different diseases and soil types. These lines could be widely grown and became the wheat cultivars of the Green Revolution. As the importance of the international effort grew, so did the need to understand their international target environments and G × E interactions. It also highlighted that the original shuttle breeding effort needed fine-tuning. Ortiz et al. (2007) proposed an updated and expanded shuttle breeding program, which includes a global effort to breed wheat cultivars that are evaluated and can be successfully grown in "hot spot" regions for various biotic and abiotic stresses. Coupled with the extensive international testing program are sophisticated statistical analyses and physiological assays to efficiently characterize their germplasm. Due to its broad adaptation and excellent disease resistance, the CIMMYT germplasm is viewed as a global genetic resource and treasure.

BREEDING HYBRID WHEAT

This section will provide a brief update on the review by Edwards (2001). While hybrid wheat research was largely conducted in the private sector but has been reduced in developed countries, hybrid crops and especially hybrid cereals that are normally self-pollinated continue to be researched in developing and emerging economies (e.g., hybrid rice, *Oryza sativa* L., in Asia, Food and Agriculture Organization 2004; and hybrid wheat in India, Matuschke et al., 2007). In these countries, the adoption has been spectacular for rice (15,000,000 ha in China and 200,000 ha in India in 2001–2002) and less so for wheat (25,000 ha in India in 2005). The economic benefits of hybrid wheat were found to be farm-size neutral, with small landholders benefiting as well as or better than larger landholders (Matuschke et al., 2007). Access to information and credit were critical to the adoption of hybrid wheat technology. The key for developing or emerging economies in developing hybrid wheat is that they seem to be more patient with the needed long-term investment that will successfully bring hybrid wheat to market.

As for the merits of hybrid wheat, progress continues to be made on the hybridization process using CMS (Chen 2003). In the US southern Great Plains for the period from 1975 to 1995, wheat hybrid performance was compared to contemporary pureline performance (Koemel et al., 2004). The wheat hybrids had higher grain yields than purelines, but were not more or less stable than purelines. In one way these results are remarkable in that throughout this period, the investment in human and other resources would most likely be equal or higher in pureline breeding programs. Hence the hybrid advantage was developed with less resources.

IMPORTANCE OF TECHNOLOGY

To this point, we have understated the importance of new technology. Clearly, advances in technology have drastically changed wheat breeding. The popularity of the bulk breeding method coincided with the development of the small plot combine, which can quickly harvest bulk populations. Breeding for minimum-tillage or no-till environments is only possible with better, small-plot drills. Sophisticated experimental designs and analyses are due to the revolution in computing power. Similarly, backcrossing for traits expressed in the adult plant has been greatly enhanced by molecular markers, which allows the breeder to select desirable plants for adult traits at the seedling stage. Wheat breeding will continue to evolve based upon emerging technologies.

FUTURE PERSPECTIVES

If the past is prologue to the future, wheat breeding will retain its long-successful methods and improve them using new technology. While sexual hybridization will remain the main way for creating new variation, germplasm developed through targeting induced local lesions in genomes (TILLING)(McCallum et al., 2000) will add precision and efficiency to the effective use of mutations. Though currently not available commercially, transgenic wheat will be available in the future, because its potential and competitive advantages manifested in other crops are too great to preclude its adoption. New or traditional (e.g., organic) markets will be developed for those not desiring transgenic foods.

As molecular marker technology continues to decrease in cost and increase in efficiency, marker-based breeding methods (Bernardo and Yu 2007) will become more prevalent. However, in order to truly use these methods, precise and reliable phenotypic evaluations will be needed to link with the marker data. In this case, the best statistical experimental and genetic designs will need to be used, which may drive programs to doubled haploid technologies where homozygous lines are evaluated phenotypically, at least to determine marker values and epistatic interactions. The advantage of using doubled haploid lines in this process is that elite doubled haploid lines can be used directly for release and, as all breeding programs are a form of recurrent selection, numerous advantages exist for doubled-haploid-based recurrent selection (Baenziger et al., 2001).

As markers and the underlying genes that affect major traits are identified, it may be possible to incorporate genes into simulation models for growth and development and productivity (Baenziger et al., 2004). There are two important reasons for this research. The first is that, despite the hundreds of environments that a candidate line will be tested in before release, a breeder can only estimate the thousands of farms (environments) where a successful cultivar may be grown. Though it may be unrealistic for a crop model to predict on the basis of genotype how a given cultivar will perform in environments that the cultivar has never been tested—especially for those extreme environments outside the range of tested environments—a gene-enhanced wheat model could greatly expand the available information to help a grower decide among cultivars. Past attempts to incorporate environmental information to explain QTL×environment interactions have been modestly successful at best (Campbell et al., 2004).

Gene-enhanced crop models may also help wheat breeders develop an improved understanding of how best to continue making breeding progress, which is always measured by increased productivity. Conceptually, this is like walking up a hill (achieving higher grain yields), always going higher, with the implicit assumption that one is walking up the highest hill or there may only be one hill (the target environments). However, consider the possibility of several hills and the current hill is not the highest (one could achieve greater yields with a completely different set of germplasm). How would a breeder know to walk down from a hill (give up grain yield) to walk to a higher hill (one with higher grain yield potential)? Crop models may provide the needed guidance and confidence to make these decisions. Agronomically important genes, such as photoperiod insensitivity or semidwarfing genes, were present in many lines before breeders understood their value in specific environments. For example, Strampelli used photoperiod insensitivity and dwarfing gene *Rht8* from the Japanese landrace Akakomugu long before the genetic and physiological basis were understood (Gale and Youssefian 1985). Plant breeding is always concept-driven, and models may help develop the plant phenotypes and genotypes needed for future success.

One final area that will need to be considered in any discussion on wheat breeding is who will employ the wheat breeders of the future—the public sector or the private sector (Morris et al., 2006). As both authors of this chapter are public-sector plant breeders, we fully recognize that our conclusions may be biased. In this chapter, we have described wheat breeding methods that are used by public and private breeders alike for developing cultivars for large and small landowners. However, in our careers we have witnessed the privatization of the formerly public breeding programs in England and the apparent transition from public to private-oriented programs in Australia. Also, within the global plant breeding enterprise, the trend is clearly toward the private sector (Morris et al., 2006). Hence, it is worthwhile to speculate on the appropriate role of public and private plant breeding (see also http://km.fao.org/gipb/index.php?option=com_frontpage&Itemid=1 and http://www.ag-innovation.usask.ca/final%20policy%20briefs/MallaGray_11.pdf).

In regions where the private sector cannot be financially successful, clearly the public sector is needed to create new cultivars. Internationally, this has been the role of CIMMYT and it partnership with the national agricultural research centers. In developing and developed countries, ineffective intellectual property rights (remedied by the International Union for the Protection of New Varieties of Plants; http://www.upov.int/index_en.html) and farmer-saved seed often prevented private breeding programs from being financially successful. However, with newer seed laws, private companies have succeeeded in most major wheat-producing regions (e.g., India, North America, Europe, South America, Australia). In some regions, private investment and activity has ebbed and flowed, as have the number of companies. In these regions, the public breeding sector will most likely remain strong until it is clear that the private sector will provide the necessary continuity for a vibrant agriculture.

For some products, such as hybrid wheat, the commercial sector is ideally suited for developing, increasing, and marketing the seed. For hybrid wheat, the public sector lacks the continuum of skills or resources to successfully develop hybrids. However, the private sector, especially in market economies, often lacks the patience or resources for the long-term research needed to develop hybrids. The need for private long-term investment may be driven by the public sector having made limited investment in this long-term research effort to support private-sector activities.

The public sector is also necessary to train new plant breeders (Baenziger 2006; Morris et al., 2006); however, the size and scope of modern plant breeding activities will increasingly place more demand on practical training in breeding through internships in the private sector. The public sector is also needed to develop and validate new plant breeding theory, as rarely will these activities have the financial rewards that private investors desire. In fact, some countries (including developed countries) with little public plant breeding capability can no longer effectively undertake research in plant breeding methodology. In reality, relatively few countries have universities that can train the next generation of plant breeders (Morris et al., 2006).

Returning to the appropriate roles for public and private wheat breeding, the public sector should be careful not to harm the private sector. A successful private-sector plant breeding effort should be viewed as a triumph in market-based economies, because it can free up public resources for other public needs and, through its taxes, support plant breeding research. However, because plant breeding research is validated by successful products, public plant breeders often compete with private breeders to the detriment of the latter. Many public plant breeding programs fund their research, in part, through commercial arrangements (research and development fees) which are similar to the royalties that private-sector companies charge. With this commercial orientation, the differences between the public and private sector have blurred to the point that public research priorities sometimes become skewed to the commercial release of cultivars to the detriment of the public sector's role in educating the next generation of plant breeders or developing new public knowledge on plant breeding methodology or developing novel genetic resources. In fact, the surest way to be privatized may be for the public sector to become so similar to the private sector in its goals that both sectors appear indistiguishable. With commercial success, public sector programs become a target for privatization.

Farmer organizations that wish to invest in the genetics of their crops are looking at quasi-non-governmental organizations. These organizations become a private–public partenership to deliver field-ready cultivars. Alternatively producer organizations might contract with a public institution to breed wheat cultivars. Gray and Malla (2007), citing studies on rates of return in agricultural research, reported a 4.6 : 1 benefit–costs ratio on producer levies to fund wheat cultivar development in Canada.

While discussion on the most appropriate role for the public and private sectors in wheat breeding will continue for years to come, what will remain is the joy and satisfaction of creating new cultivars that can change human lives for the better. That remains the challenge for the current and next generation of wheat breeders.

WEBLIOGRAPHY

http://www.ag-innovation.usask.ca/—Home page for CAIRN, the Canadian Agricultural Innovation Network, which has the objective to increase understanding of agricultural innovation and to aid in the development of public policy that promotes both product and process innovation in the Canadian agriculture and food sector.

http://www.cimmyt.org—International Center for Maize and Wheat Improvement. CIMMYT bred the lines that led to the Green Revolution.

http://km.fao.org/gipb/index.php?option=com_frontpage&Itemid=1—Global Partnership Initiative for Plant Breeding Capacity Building. Includes a global discussion on needs for plant breeding.

http://www.fao.org/rice2004/en/f-sheet/factsheet6.pdf—This website provides the most current information on hybrid rice production, which is difficult to find otherwise.

http://www.globalrust.org/index.cfm?m=1—Global Rust Initiative website that was developed due to a new strain of stem rust on wheat which threatens food security globally and is currently found in Africa and the Middle East.

http://grainscanada.gc.ca/information/fhb-e.htm—This is the Canadian website for Fusarium head blight (FHB) and contains information on research on this disease. Similar websites are available in other countries.

http://www-infocris.iaea.org/MVD/—This website includes a database on cultivars derived directly or developed by crossing with a mutant cultivar or line.

http://maswheat.ucdavis.edu/—This website reports on activities of a major grant to US wheat researchers who used marker-assisted selection to introgress or backcross numerous agronomically important genes. It provides a wealth of resources on marker-assisted selection.

http://www.planttreaty.org/index_en.htm—International treaty on plant genetic resources for food and agriculture. It contains a wealth of information on the ethical and legal exchange of germplasm.

http://www.upov.int/index_en.html—Main website for discussion of plant breeders' rights (intellectual property rights) and the protection of new varieties, and home page for the International Union for the Protection of New Varieties of Plants (UPOV).

REFERENCES

Acquaah, G. 2007. Principles of plant breeding and genetics. Blackwell Publishing Ltd., Malden, MA.

Allard, R.W. 1960. Principles of plant breeding. Wiley, New York, NY.

Ashikari, M., H. Sakakibara, S. Lin, T. Yamamoto, T. Takashi, A. Nishimura, E.R. Angeles, Q. Qian, H. Kitano, and M. Matsuoka. 2005. Cytokinin oxidase regulates rice grain production. Science 309:741–745.

Avivi, L. 1978. High protein concentration in wild tretraploid *Triticum dicoccoides* Korn. p. 372–380. *In* S. Ramanujam (ed.) Proc. Int. Wheat Genet. Symp., New Delhi, India. 23–28 Feb. 1978. Kapoor Art Press, New Delhi, India.

Baenziger, P.S. 1996. Reflections on doubled haploids in plant breeding. p.35–48. *In* S.M. Jain, S.K. Sopory, and R.E. Veileux (ed.) *In vitro* haploid production in higher plants: Vol. 1. Fundamental aspects and methods. Kluwer Academic Publishers, Norwell, MA.

Baenziger, P.S. 2006. Plant breeding training in the U.S. HortScience 41:40–44.

Baenziger, P.S., B. Beecher, R.A. Graybosch, D.D. Baltensperger, L.A. Nelson, J.M. Krall, Y. Jin, J.E. Watkins, D.J. Lyon, A.R. Martin, M–S. Chen, and G-H. Bai. 2006a. Registration of 'Infinity CL' wheat. Crop Sci. 46:975–977.

Baenziger, S., K.M. Kim, and K. Haliloglu. 2001. Wheat in vitro breeding. p. 979–1000. *In* A.P. Bonjeau and W.J. Angus (ed.) The world wheat book: A history of wheat breeding. Lavoisier Publishing, Paris, France.

Baenziger, P.S., G.S. McMaster, W.W. Wilhelm, A. Weiss, and C.J. Hays. 2004. Putting genes into genetic coefficients. Field Crops Res. 90:133–143.

Baenziger, P.S., and C.J. Peterson. 1992. Genetic variation: Its origin and use for breeding self-pollinated species. p. 69–92. *In* T.M. Stalker and J.P. Murphy (ed.) Plant breeding in the 1990s. CAB Int., Wallingford, UK.

Baenziger, P.S., W.K. Russell, G.L. Graef, and B.T. Campbell. 2006b. Improving lives: 50 years of crop breeding, genetics and cytology (C-1). Crop Sci. 46:2230–2244.

Baenziger, P.S., D.M. Wesenberg, V.M. Smail, W.L. Alexander, and G.W. Schaeffer. 1989. Agronomic performance of wheat doubled haploid lines derived from cultivars by anther culture. J. Plant Breed. 103:101–109.

Baker, R.J. 1988. Tests for crossover genotype–environmental interactions. Can. J. Plant Sci. 68:405–410.

Basford, K.E., and M. Cooper. 1998. Genotype × environment interactions and some considerations of their implications for wheat breeding in Australia. Aust. J. Agric. Res. 49:153–174.

Berke, T.G., P.S. Baenziger, and R. Morris. 1992. Location of wheat quantitative trait loci affecting agronomic performance of seven traits using reciprocal chromosome substitutions. Crop Sci. 32:621–627.

Bernardo, R., and J. Yu. 2007. Prospects for genomewide selection for quantitative traits in maize. Crop Sci. 47:1082–1090.

Bonjeau, A.P., and W.J. Angus (ed.) 2001. The world wheat book: A history of wheat breeding. Lavoisier Publishing, Paris, France.

Borojevic K., and K. Borojevic. 2005. Historic role of the wheat variety Akakomugi in Southern and Central European wheat breeding programs. Breed. Sci. 55:253–256.

Bos, I. 1983. The optimum number of replications when testing lines or families on a fixed number of plots. Euphytica 32:311–318.

Bregitzer, P., A.E. Blechl, D. Fiedler, J. Lin, P. Sebesta, J. Fernandez De Soto, O. Chicaiza, and J. Dubcovsky. 2006. Changes in high molecular weight glutenin subunit composition can be genetically engineered without affecting wheat agronomic performance. Crop Sci. 46:1553–1563.

Briggs, F.N. 1938. The use of the backcross in crop improvement. Am. Nat. 72:285–292.

Briggs, K.G., and L.H. Shebeski. 1971. Early generation selection for yield and breadmaking quality of hard red spring wheat. Euphytica 20:453–463.

Brim, C.A. 1966. A modified pedigree method of selection in soybeans. Crop Sci. 6:220.

Busch, R.H., J.C. Janke, and R.C. Frohberg. 1974. Evaluation of crosses among high and low yielding parents of spring wheat (Triticum aestivum L.) and bulk prediction of line performance. Crop Sci. 14:47–50.

Caligari, P.D., S.W. Powell, and J.L. Jinks. 1985. The use of doubled haploids in barley breeding: 2. An assessment of univariate cross prediction methods. Heredity 54:353–358.

Campbell, A.B. 1970. Neepawa hard red spring wheat. Can. J. Plant Sci. 50:752–753.

Campbell, A.B., R.G. Anderson, G.J. Green, D.J. Samborski, and K.R. Johnson. 1967. Note on Manitou hard red spring wheat. Can. J. Plant Sci. 47:330.

Campbell, B.T., P.S. Baenziger, K.M. Eskridge, H. Budak, N.A. Streck, A. Weiss, K.S. Gill, and M. Erayman. 2004. Using environmental covariates to explain genotype × environment and QTL × environment interactions for agronomic traits on chromosome 3A of wheat. Crop Sci. 44:20–627.

Campbell, B.T., P.S. Baenziger, K.S. Gill, K.M. Eskridge, H. Budak, M. Erayman, I. Dweikat, and Y. Yen. 2003.

Identification of QTLs and environmental interactions associated with agronomic traits on chromosome 3A of wheat. Crop Sci. 43:1493–1505.

Carter, T.E., Jr., J.W. Burton, J.J. Cappy, D.W. Israel, and H.R. Boerma. 1983. Coefficients of variation, error variances, and resource allocation in soybean growth analysis experiments. Agron. J. 75, 691–696.

Ceccarelli, S., S. Grando, R. Tutwiler, J. Baha, A.M. Martini, H. Salahieh, A. Goodchild, and M. Miachel. 2000. A methodological study on participatory barley breeding: I. Selection phase. Euphytica 111:91–104.

Chen, Q.F. 2003. Improving male fertility restoration of common wheat for *Triticum timopheevi* cytoplasm. Plant Breed. 122:401–404.

Choo, T.M. 1988. Cross prediction in barley using doubled-haploid lines. Genome 30:366–371.

Choo, T.M., B.R. Christie, and E. Reinbergs. 1979. Doubled haploids for estimating genetic variances and a scheme for population improvement in self-pollinated crops. Theor. Appl. Genet. 54:267–271.

Choo, T.M., and E. Reinbergs. 1979. Doubled haploids for estimating genetic variances in the presence of linkage and gene association. Theor. Appl. Genet. 55:129–132.

Clarke, J.M. 2005. Durum wheat improvement in Canada. p. 921–938. *In* C. Royo, M.M. Nachit, N. Di Fonzo, J.L. Araus, W. H. Pfeiffer, and G.A. Slafer (ed.) Durum wheat breeding: Current approaches and future strategies. Food Products Press, New York.

Comstock, R.E., and R.H. Moll. 1963. Genotype–environment interactions. p. 164–196. *In* W.D. Hansen and H.F. Robinson (ed.) Publ. 982. National Academy of Sciences, National Research Council, Washington, DC.

Cregan, P.B., and R.H. Busch. 1977. Early generation bulk hybrid yield testing of adapted hard red spring wheat crosses. Crop Sci. 17:887–891.

Crossa, J. 1990. Statistical analyses of multilocation trials. Adv. Agron. 44:55–85.

Crossa, J., P.L. Cornelius, M.S. Seyedsadr, and P. Byrne. 1993. A shifted multiplicative model cluster analysis for grouping environments without genotypic rank change. Theor. Appl. Genet. 85:577–586.

Cullis, B.R., A. Smith, C. Hunt, and A. Gilmour. 2000. An examination of the efficiency of Australian crop evaluation programmes. J. Agric. Sci. 135:213–222.

Cullis, B.R., F.M. Thomson, J.A. Fisher, A.R. Gilmour, and R. Thompson. 1996a. The analysis of the NSW wheat variety database: I. Modelling trial error variance. Theor. Appl. Genet. 92:21–27.

Cullis, B.R., F.M. Thomson, J.A. Fisher, A.R. Gilmour, and R. Thompson. 1996b. The analysis of the NSW wheat variety database: II. Variance component estimation. Theor. Appl. Genet. 92:28–39.

Cuthbert, J.L., D.J. Somers, A.L. Brûlé-Babel, P.D. Brown, and G.H. Crow. 2008. Molecular mapping of quantitative trait loci for yield and yield components in spring wheat (*Triticum aestivum* L.). Theor. Appl. Genet. 117: 595–608.

DePauw, R.M., R.E. Knox, F.R. Clarke, H. Wang, M.R. Fernandez, J.M. Clarke, and T.N. McCaig. 2007. Shifting undesirable correlations. Euphytica 157:409–415.

DePauw, R.M., and L.H. Shebeski. 1973. An evaluation of an early generation yield testing procedure in *T. aestivum* L. Can. J. Plant Sci. 53:465–470.

DePauw, R.M., T.F. Townley-Smith, G. Humphreys, R.E. Knox, F.R. Clarke, and J.M. Clarke. 2005. Lillian hard red spring wheat. Can. J. Plant Sci. 85:397–401.

Dowell, F.E., E.B. Maghirang, R.A. Graybosch, P.S. Baenziger, D.D. Baltensperger, and L.E. Hansen. 2006. An automated near-infrared system for selecting individual kernels based on specific quality characteristics. Cereal Chem. 83:537–543.

Dubcovsky, J. 2005. The Wheat CAP. Applied wheat genomics [Online]. Available at http://maswheat.ucdavis.edu/ (verified 24 Mar. 2008).

Edwards, I. 2001. Hybrid wheat. p. 1019–1045. *In* A.P. Bonjeau and W.J. Angus (ed.) The world wheat book. A history of wheat breeding. Lavoisier Publishing, Paris, France.

Ellis, M.H., D.G. Bonnett, and G.J. Rebetzke. 2007. A 192 bp allele at the *Xgwm261* locus is not always associated with the *Rht8* dwarfing gene in wheat (*Triticum aestivum* L.). Euphytica 157:209–214.

Eskridge, K.M., 1990. Selection of stable cultivars using a safety-first rule. Crop Sci. 30:369–374.

Esquinas-Alcázar, J. 2005. Protecting crop genetic diversity for food security: Political, ethical and technical challenges. Nat. Rev. Genet. 6:946–953.

Falconer, D.S. 1952. The problem of environment and selection. Am. Nat. 86:293–298.

Falconer, D.S. 1981. Introduction to quantitative genetics. Longman Group, New York.

Fellers, J.P., A.C. Guenzi, and C.M. Taliaferro. 1995. Factors affecting the establishment and maintenance of embryogenic callus and suspension cultures of wheat (*Triticum aestivum* L.). Plant Cell Rep. 15:232–237.

Fischer, R.A., L. O'Brien, and K.J. Quail. 1989. Early generation selection in wheat: 2. Grain quality. Aust. J. Agric. Res. 40:1135–1142.

Flament, P., G. Grimanelli, and O. Leblanc. 2001. Apomixis for wheat improvement. p. 1001–1015. *In* A.P. Bonjeau and W.J. Angus (ed.) The world wheat book: A history of wheat breeding. Lavoisier Publishing, Paris, France.

Food and Agriculture Organization. 2004. Hybrid rice for food security: FAO factsheets, 6 [Online]. Available at http://www.fao.org/rice2004/en/f-sheet/factsheet6.pdf (verified 23 Mar. 2008).

Friebe, B., J. Jiang, W.J. Raupp, R.A. McIntosh, and B.S. Gill. 1996. Characterization of wheat–alien translocation conferring resistance to diseases and pests: Current status. Euphytica 91:59–87.

Gale, M., G.A. Marshall, and M.V. Rao. 1981. A classification of the Norin 10 and Tom Thumb dwarfing genes in British, Mexican, Indian and other hexaploid bread wheat varieties. Euphytica 30:355–361.

Gale, M.D., and S. Youssefian. 1985. Dwarfing genes in wheat. p. 1–35. *In* G.E. Russell (ed.) Progress in plant breeding. Butterworths, London.

Gauch, H.G., Jr. and R.W. Zobel. 1996. Optimal replication in selection experiments. Crop Sci. 36:838–843.

Goulden, C.H. 1939. Problems in plant selection. p. 132–133. *In* R.C. Burnett (ed.) Proc. Int. Genetic Congress, Edinburgh. Cambridge University Press, Cambridge, UK.

Graf, R.J., P. Hucl, B.R. Orshinsky, and K.K. Kartha. 2003. McKenzie hard red spring wheat. Can. J. Plant Sci. 83:565–569.

Gras, P.W., and L. O'Brien. 1992. Application of a 2-gram mixograph to early generation selection for dough strength. Cereal Chem. 69:254–257.

Gray, R., and S. Malla. 2007. The rate of return to agricultural research in Canada. p. 1–11. Canadian Agricultural Innovation Research Network, No. 11, 2007 [Online]. Available at http://www.ag-innovation.usask.ca/final%20policy%20briefs/MallaGray_11.pdf (verified 23 Mar. 2008).

Green, G.J., and A.B. Campbell. 1979. Wheat cultivars resistant to *Puccinia graminis tritici* in Western Canada: Their development, performance and economic value. Can. J. Plant Sci. 1:3–11.

Guzy-Wróbelska, J., A. Labocha-Pawlowska, M. Kwasniewski, and I. Szarejko. 2007. Different recombination frequencies in wheat doubled haploid populations obtained through maize pollination and anther culture. Euphytica 156:173–183.

Guzy-Wróbelska, J., and I. Szarejko. 2003. Molecular and agronomic evaluation of wheat doubled haploid lines obtained through maize pollination and anther culture methods. Plant Breed. 122:305–313.

Halpin, C. 2005. Gene stacking in transgenic plants–the challenge for 21st century plant biotechnology. Plant Biotechnol. J. 3:141–155.

Harlan, H.V., and M.L. Martini. 1938. The effect of natural selection in a mixture of varieties. J. Agric. Res. 57:189–199.

Hayden, M., H. Kuchel, and K. Chalmers. 2004. Sequence tagged microsatellites for the *Xgwm533* locus provide new diagnostic markers to select for the presence of stem rust resistance gene *Sr2* in bread wheat (*Triticum aestivum* L.). Theor. Appl. Genet. 109:1641–1647.

Holland, J.B., W.E. Nyquist, and C.T. Cervantes-Martinez. 2003. Estimating and interpreting heritability for plant breeding: An update. Plant Breed. Rev. 22:9–111.

Hucl, P., and M. Matus Cádiz. 2001. Isolation distances for minimizing out-crossing in spring wheat. Crop Sci. 41:1348–1351.

Humphreys, D.G., T.F. Townley-Smith, E. Czarnecki, O.M. Lukow, B. Fofana, J.A. Gilbert, B.D. McCallum, T.G. Fetch, Jr., and J.G. Menzies. 2007a. Kanata hard white spring wheat. Can. J. Plant Sci. 87:879–882.

Humphreys, D.G., T.F. Townley-Smith, E. Czarnecki, O. Lukow, B. McCallum, T. Fetch, J. Gilbert, and J. Menzies. 2007b. Snowbird hard white spring wheat. Can. J. Plant Sci. 87:301–305.

Iqbal, M., A. Navabi, D.F. Salmon, R.C. Yang, B.M. Murdoch, S.S. Moore, and D. Spaner. 2007a. Genetic analysis of flowering and maturity time in high latitude spring wheat. Euphytica 154:207–218.

Iqbal, M., A. Navabi, D.F. Salmon, R.C. Yang, and D. Spaner. 2007b. Simultaneous selection for early maturity, increased grain yield and elevated protein content in spring wheat. Plant Breed. 126:244–250.

Khalifa, M.A., and C.O. Qualset. 1974. Intergenotypic competition between tall and dwarf wheats: I. In mechanical mixtures. Crop Sci. 14:795–799.

Khalifa, M.A., and C.O. Qualset. 1975. Intergenotypic competition between tall and dwarf wheats: II. In hybrid bulks. Crop Sci. 15:640–644.

Khan, I.A., J.D. Procunier, D.G. Humphreys, G. Tranquilli, A.R. Schlatter, S. Marcucci-Poltri, R. Frohberg, and J. Dubcovsky. 2000. Development of PCR-based markers for a high grain protein content gene from *Triticum turgidum* ssp. *dicoccoides* transferred to bread wheat. Crop Sci. 40:518–524.

Kihara, H., 1984. Origin and history of 'Daruma'—a parental variety of Norin 10. p. 13–19. *In* S. Sakamoto (ed.) Proc. Int. Wheat Genet., 6th, Kyoto, Japan. 28 Nov.–3 Dec. 1983. Plant Germ-Plasm Institute, Fac. Of Agric., Kyoto Univ.

Knott, D.R., and J. Kumar. 1975. Comparison of early generation yield testing and a single seed descent procedure in wheat breeding. Crop Sci. 15:295–299.

Knox, R.E., and F.R. Clarke. 2007. Molecular breeding approaches for enhanced resistance against fungal pathogens. p. 321–357. *In* Z.K. Punja, S. De Boer, and H. Sanfacon (ed.) Biotechnology and plant disease management. CAB Int., Oxfordshire, UK.

Koemel, J.E., Jr., A.C. Guenzi, B.F. Carver, M.E. Payton, G.H. Morgan, and E.L. Smith. 2004. Hybrid and pureline hard winter wheat yield and stability. Crop Sci. 44:107–113.

Kofoid, K.D., and S.S. Maan. 1982. Agronomic and breadmaking performance of fertile alloplasmic wheats. Crop Sci. 22:725–729.

Larkin, P.J., S.A. Ryan, R.I.S. Brettell, and W.R. Scowcroft. 1984. Heritable somaclonal variation in wheat. Theor. Appl. Genet. 67:443–455.

Laurie, D.A., and M.D. Bennett. 1988. The production of haploid wheat plants from wheat × maize crosses. Theor. Appl. Genet. 76:393–397.

Laurie, D.A., and S. Reymondie. 1991. High frequencies of fertilization and haploid seedling production in crosses between commercial hexaploid wheat varieties and maize. Plant Breed. 106:182–189.

Lawrie, R.G., M.A. Matus-Cádiz, and P. Hucl. 2006. Estimating out-crossing rates in spring wheat cultivars using the contact method. Crop Sci. 46:247–249.

Love, H.H. 1927. A program for selecting and testing small grains in successive generations following hybridization. Agron. J. 19:705–712.

Lukow, O.M. 1991. Screening of bread wheats for milling and baking quality—a Canadian perspective. Cereal Foods World 36:497–501.

Lupton, F.G.H., and R.N.H. Whitehouse. 1957. Studies on the breeding of self-pollinating cereals: I. Selection methods in breeding for yield. Euphytica 6:169–184.

Mago, R., H.S. Bariana, I.S. Dundas, W. Spielmeyer, G.J. Lawrence, A.J. Pryor, and J.G. Ellis. 2005. Development of PCR markers for the selection of wheat stem rust resistance genes *Sr24* and *Sr26* in diverse wheat germplasm. Theor. Appl. Genet. 111:496–504.

Mahmood, A., P.S. Baenziger, H. Budak, K.S. Gill, and I. Dweikat. 2004. The use of microsatellite markers for the detection of genetic similarity among winter bread wheat lines for chromosome 3A. Theor. Appl. Genet. 109:1494–1503.

Martin, T.J. 1990. Outcrossing in twelve hard red winter wheat cultivars. Crop Sci. 30:59–62.

Marza, F., G.H. Bai, B.F. Carver, and W.C. Zhou. 2006. Quantitative trait loci for yield and related traits in the wheat population Ning 7840 × Clark. Theor. Appl. Genet. 112:688–698.

Matus-Cádiz, M.A., P. Hucl, and B. Dupuis. 2007. Pollen-mediated gene flow in wheat at the commercial scale. Crop Sci. 47:573–579.

Matuschke, I., R.R. Mishra, and M. Qaim. 2007. Adoption and impact of hybrid wheat in India. World Dev. 35:1422–1435.

McCallum, C.M., L. Comai, E.A. Greene, and S. Henikoff. 2000. Targeting induced local lesions in genomes (TILLING). Plant Physiol. 123:439–442.

Mesfin A., R.C. Frohberg, and J.A. Anderson. 1999. RFLP markers associated with high grain protein from *Triticum turgidum* L. var. *dicoccoides* introgressed into hard red spring wheat. Crop Sci. 39:508–513

Meyer, F.D., L.E. Talbert, J.M. Martin, S.P. Lanning, T.W. Greene, and M.J. Giroux. 2007. Field evaluation of transgenic wheat expressing a modified ADP-glucose pyrophosphorylase large subunit. Crop Sci. 47:336–342.

Morris, M., G. Edmeades, and E. Pehu. 2006. The global need for plant breeding capacity: What roles for the public and private sectors? HortScience 41:30–39.

Newhouse, K.A., W.A. Smith, M.A. Starrett, T.J. Schaefer, and B.K. Singh. 1992. Tolerance to imidazolinone herbicides in wheat. Plant. Physiol. 100:882–886.

Newman, L.H. 1912. Plant breeding in Scandinavia. Canadian Seed Growers' Association, Ottawa, Canada.

Nonaka, S. 1984. History of wheat breeding in Japan. p. 593–599. *In* S. Sakamoto (ed.) Proc. Int. Wheat Genet., 6th, Kyoto, Japan. 28 Nov.–3 Dec. 1983. Plant Germ-Plasm Institute, Fac. Agric., Kyoto Univ.

Ortiz, R., R. Trethowan, G.O. Ferrara, M. Iwanaga, J.H. Dodds, J.H. Crouch, J. Crossa, and H.J. Braun. 2007. High yield potential, shuttle breeding, genetic diversity, and a new international wheat improvement strategy. Euphytica 157:365–384.

Percival, J. 1921. The wheat plant, a monograph. Duckworth and Co., London.

Pozniak, C.J., I.T. Birk, L.S. O'Donoughue, C. Ménard, P.J. Hucl, and B.K. Singh. 2004. Physiological and molecular characterization of mutation-derived imidazolinone resistance in spring wheat. Crop Sci. 44:1434–1443.

Pozniak, C.J., and P.J. Hucl. 2004. Genetic anaylsis of imidazolinone resistance in mutation-derived lines of common wheat. Crop Sci. 44:23–30.

Quick, J.S., J.A. Stromberger, S. Clayshulte, B. Clifford, J. J. Johnson, F.B. Peairs, J.B. Rudolph, and K. Lorenz. 2001a. Registration of 'Prowers' wheat. Crop Sci. 41:928–929.

Quick, J.S., J.A. Stromberger, S. Clayshulte, B. Clifford, J. J. Johnson, F.B. Peairs, J.B. Rudolph, and K. Lorenz. 2001b. Registration of 'Prairie Red' wheat. Crop Sci. 41:1362–1363.

Raghavan, V. 2003. One hundred years of zygotic embryo culture investigations. *In Vitro* Cell Dev. Biol. Plant 39:437–442.

Reinbergs, E., S.J. Park, and L.S.P. Song. 1976. Early identification of superior crosses by the doubled haploid technique. Z. Pflanzenzuecht. 76:215–224.

Riggs, T.J., and J.W. Snape. 1977. Effects of linkage and interaction in a comparison of theoretical populations derived by diploidized haploid and single seed descent methods. Theor. Appl. Genet. 49:111–115.

Russell, W.K., K.M. Eskridge, D.A. Travnicek, and F.R. Guillen-Portal. 2003. Clustering environments to minimize change in rank of cultivars. Crop Sci. 43:858–864.

Sadasivaiah, R.S., S.M. Perkovic, D.C. Pearson, B. Postman, and B.L. Beres. 2004. Registration of 'AC Andrew' wheat. Crop Sci. 44:696–697.

Schaeffer, G.W., P.S. Baenziger, and J.W. Worley. 1979. Haploid plant development from anthers and *in vitro* embryo culture of wheat. Crop Sci. 19:697–702.

Schmidt, J.W., V.A. Johnson, P.J. Mattern, A.F. Drier, D.V. McVey, and J.H. Hatchet. 1983. Registration of 'Brule' wheat. Crop Sci. 23:1223.

Schmidt, J.W., V.A. Johnson, P.J. Mattern, A.F. Drier, D.V. McVey, and J.H. Hatchet. 1989. Registration of 'Redland' wheat. Crop Sci. 29:491.

Shariatpanahi, M.E., D. Belogradova, L. Hessamvazriri, E. Heberle-Bors, A. Touraev. 2006. Efficient embryogenesis in freshly isolated and cultured wheat (*Triticum aestivum* L.) microspores without stress pre-treatment. Plant Cell Rep. 25:1294–1299.

Sharma, H.C., and H.W. Ohm. 1990. Crossability and embryo rescue enhancement in wide crosses between wheat and three Agropyron species. Euphytica 49:209–214.

Shebeski, L. 1967. Wheat and breeding. p. 252–271. *In* K.F. Nielsen (ed.) Proc. Canadian Centennial Wheat Symp. Modern Press, Saskatoon, Saskatchewan, Canada.

Smith, A.B., B.R. Cullis, and R. Thompson. 2005. The analysis of crop cultivar breeding and evaluation trials: An overview of current mixed model approaches. J. Agric. Sci. 143:449–462.

Snape, J.W., A.J. Wright, and E. Simpson. 1984. Methods for estimating gene numbers for quantitative characters using doubled haploid lines. Theor. Appl. Genet. 67:143–147.

Stroup, W.W., P.S. Baenziger, and D.K. Mulitze. 1994. A comparison of methods for removing spatial variation from wheat yield trials. Crop Sci. 34:62–66.

Tan, S., R.R. Evans, M.L. Dahmer, B.K. Singh, and D.L. Shaner. 2004. Imidazolinone-tolerant crops: History, current status and future. Pest Manag. Sci. 61:246–257.

Thomas, J.B., and R.M. DePauw. 2003. Dodging the exponential challenge of Fusarium head blight resistant cultivars. *In* R. Clear (ed.) Proc. Canadian Workshop on Fusarium Head Blight, 3rd, Winnipeg, MB. 9–12 Dec. 2003. Canadian Grain Commission, Winnipeg. Available at http://grainscanada.gc.ca/information/fhb-e.htm (verified 24 Mar. 2008).

Thomas, J.B., C. Hiebert, D. Somers, R.M. DePauw, S. Fox, and C. McCartney. 2006. Accelerating the transfer of resistance to fusarium head blight in wheat (*Triticum aestivum* L.). p. 701–706. *In* H.T. Buck, J.E. Nisi, and N. Salomon (ed.) Wheat production in stressed environments. Proc. Int. Wheat Genet. Conf., 7th, Mar del Plata, Argentina. 27 Nov–2 Dec. 2006. Springer, Dordrecht, The Netherlands.

Townley-Smith, T.F., and E.A. Hurd. 1973. Use of moving means in wheat yield trials. Can. J. Plant Sci. 53:447–450.

Uauy, C., J.C. Brevis, X. Chen, I. Khan, L. Jackson, O. Chicaiza, A. Distelfeld, T. Fahima , and J. Dubcovsky. 2005. High-temperature adult-plant (HTAP) stripe rust resistance gene *Yr36* from *Triticum turgidum* ssp. *dicoccoides* is closely linked to the grain protein content locus *Gpc-B1*. Theor. Appl. Genet. 122:97–105.

Vasil, I. 2007. Molecular genetic improvement of cereals: Transgenic wheat (*Triticum aestivum* L.). Plant Cell Rep. 26:1133–1154.

Vasil, V., A.M. Castillo, M.E. Fromm, and I.K. Vasil. 1992. Herbicide resistant fertile transgenic wheat plants obtained by microprojectile bombardment of regenerable embryogenic callus. Biotechnology 10:667–674.

Vasil, V., A.M. Castillo, M.E. Fromm, and I.K. Vasil. 1993. Rapid production of transgenic wheat plants by direct bombardment of cultured immature embryos. Biotechnology 11:1553–1558

Waines, J.G., and S.G. Hegde. 2003. Intraspecific gene flow in bread wheat as affected by reproductive biology and pollination ecology of wheat flowers. Crop Sci. 43:451–463.

Weeks, T.J., O.D. Anderson, and A.E. Blechl. 1993. Rapid production of multiple independent lines of fertile transgenic wheat (*Triticum aestivum*). Plant Physiol. 102: 1077–1084.

Wilkinson, G.N., S.R. Eckert, T.W. Hanock, and O. Mayo. 1983. Nearest neighbour (NN) analysis of field experiments. J. R. Stat. Soc. B. 45:151–200.

Yau, S.K. 1997. Efficiency of alpha-lattice designs in international variety trials of barley and wheat. J. Agric. Sci. 128:5–9.

Zemetra, R.S., J. Hansen, C.A. Mallory-Smith. 1998. Potential for gene transfer between wheat (*Triticum aestivum*) and jointed goatgrass (*Aegilops cylindrica*). Weed Sci. 46:313–317.

Chapter 14

State of QTL Detection and Marker-Assisted Selection in Wheat Improvement

Daryl J. Somers and Gavin Humphreys

SUMMARY

(1) Quantitative trait locus (QTL) analysis is the merger of genotyping and phenotyping data to draw associations of traits with specific chromosome regions; QTL analysis hinges on having a good genetic map of the wheat population.

(2) Wheat genetic maps have improved for over 20 years in both number and quality of markers, beginning with restriction fragment length polymorphism (RFLP) and advancing to single nucleotide polymorphisms (SNPs).

(3) Hundreds of QTLs have been published for disease, agronomic, and quality traits in wheat, including simply inherited and complex, polygenic traits.

(4) Association mapping is emerging as a new approach for mapping quantitative traits. Wheat and the associated technical resources appear to be amenable for association mapping.

(5) In the future, microarray-based gene expression data will be a new source of phenotypes to map gene expression profiles in relation to measurement of the physical properties of wheat.

(6) Wheat is generally bred and improved by selection for visual traits such as height, lodging, and grain yield. This is labor-intensive and marker-assisted selection (MAS) can assist with selection for these traits.

(7) Wheat breeding programs can combine MAS and doubled-haploid technology, first to screen parents with markers and then to screen haploids prior to chromosome doubling, to reduce costs and increase efficiencies of breeding.

(8) Perfect markers, or markers within genes controlling traits, are ideal, with no recombination expected between the trait and gene.

(9) Field performance and end-use quality testing of wheat breeding materials will continue to be the ultimate evaluation methods in the development of new wheat cultivars.

INTRODUCTION

Wheat is a crop of great historical significance. Its domestication approximately 10,000 years ago marks a turning point for humanity. The cultivation of crops including wheat allowed early humans to move from hunter–gatherers to agriculturalists. Wheat continues to be of great importance in our age. It is the most widely cultivated and is third to maize and rice in global production. In 2000, global production was approximately 571 million tonnes from 211 million hectares. Wheat supplies humanity with 20% of its caloric intake (Western Organization of Resource Councils 2002). Thus, it is not surprising that wheat breeding continues to develop the crop and that the most modern plant breeding techniques are being applied to its improvement.

This chapter will examine the use of quantitative trait locus (QTL) analysis; QTL analysis is a merger of both genotype data, which is largely DNA-based, and phenotype data. The statistical analysis of these datasets attempts to find associations between genotype and phenotype and thus specific genetic loci are able to explain significant amounts of phenotypic variation in a trait (Doerge 2002). The application of QTL analysis in wheat breeding is widespread and is clearly having a positive effect on both the precision and speed at which wheat is improved in wheat breeding programs.

BREEDING BY VISUAL SELECTION

During the development of a wheat cultivar, a wheat breeder strives to combine multiple traits into breeding lines, which are subsequently evaluated for their potential as new cultivars. Important wheat breeding traits can be divided into three general areas: agronomic, pests, and end-use quality.

Agronomic traits are generally those of most direct importance to producers, including maturity, height, lodging, grain yield, and test weight. Breeding for these characteristics has traditionally been done through visual evaluation in early generations (i.e., F_2–F_4), followed by replicated field trials in multiple environments. Field observations on days to heading or days to maturity are usually recorded in number of days. In cases where many lines (>50) must be rated per trial, maturities may be estimated using a relative scale [i.e., 1 (early) to 5 (late)]. Plant height may be measured, or as with maturities a relative scale rating may be given to breeding lines. Lodging is usually scored using a scale rating (i.e., 1–9). Grain yield is determined by weighing the grain harvested from a yield plot of a standardized size and seeding rate. Test weight is an estimate of seed packing density, which indicates grain soundness and can be related to end-use quality. It is normally measured as weight of a standardized volume (i.e., 0.5 L).

There are numerous pests that can pose important wheat breeding objectives depending on the agro-ecological zone for which the cultivar is destined. Pests can include insects, nematodes, fungi, bacteria, and viruses, and a pest in one area of the world may be absent in another. Traditionally, breeding for resistance has been conducted through *in vivo* screening for resistance to the pest of segregating populations. Disease incidence or insect feeding is habitually rated using a scoring system based on observations of wheat breeding material in the presence of the pest. Genes for resistance to pests are derived from other wheat cultivars, and novel sources of resistance are often sourced from wheat relatives. Recombinant DNA technologies now permit the transfer of genes from nonrelated species. For example, resistance to glyphosate has been transferred to wheat from bacteria using particle bombardment of plant tissue cultured *in vitro* and regeneration of plants from resistant plant tissue.

Wheat can be used to make many products, including bread, cookies, cakes, noodles, and pasta. In some parts of the world, it is also an important livestock feed and, more recently, wheat is being used to produce ethanol destined for industrial purposes such as automobile fuel. Breeding objectives for end-use quality will usually focus on a particular trait such as bread-making quality for hard wheat or cookie quality for soft wheat. Basic end-use quality tests include protein content, milling quality, dough rheological testing, and baking tests, as elaborated further in Chapter 20. A list of commonly used wheat quality tests is given in Table 14.1. While some of these end-use traits can be measured indirectly using near-infrared technology, much of the milling and rheological testing requires laborious, time-consuming, small-scale milling or wet chemistry tests to emulate industrial-scale milling or baking processes. The rigorous testing is necessary because milling and baking companies demand consistency in wheat that is targeted to ongoing industrial end-uses. Large wheat exporting nations may have wheat end-use quality standards that are rather rigid and new cultivars must adhere to these standards. Quantitative trait locus analysis and deployment of molecular breeding has the potential to reduce the time and cost of some of these laboratory tests (Koebner 2004). The potential exists to select for good end-use quality in early generations by genotypic selection

Table 14.1 Wheat quality traits typically evaluated as part of the Canada Western Red Spring variety registration testing.

Wheat Quality Trait	Method
Grain protein content	Near-infared reflectance
Flour protein content	Modified Dumas, AACC Method 46–30
Milling yield	Buhler, AACC Method 26–21
Hagberg falling number	Perten, AACC method 56–81
Flour ash content	Combustion, AACC Method 08–01
Flour color	Agtron, AACC Method 14–30
Grain hardness	Rotap, AACC Method 55–30
Absorption and dough mixing	Farinograph, AACC Method 54–21
Baking test	Can. short process (Preston et al., 1982)

[a]Approved methods of the American Association of Cereal Chemists (AACC 2000).

for favored alleles in specific chromosome intervals.

COMPLEX TRAITS AND GENE PYRAMIDING

Many of the traits with which wheat breeders are concerned are complex multigenic traits, such as grain yield, maturity, and resistance to Fusarium head blight (FHB, caused by *Fusarium graminearum* Schwabe). Environmental effects are often important in these traits, and both additive gene action and pleiotropic effects can be important. Wheat breeders are continually attempting to increase grain yield, which can often be accomplished through the use of new germplasm sources. However, quality traits, which are also usually controlled by multiple genes, are often static or change rather slowly. The need for novel germplasm to create new genetic combinations that would increase grain yield is generally at odds with the need to retain gene complexes selected over several generations for desirable end-use quality.

In the case of disease resistance, wheat breeders and geneticists have worked for decades to develop cultivars with improved disease resistance through the introduction of novel resistance genes from other germplasm, including wild relatives (i.e., *Lr19*). Further, durable resistance has been generated by combining several resistance genes into a single genotype (gene pyramid) and by the deployment of genes that give horizontal resistance (i.e., *Lr34*). Gene pyramids are desirable because defeated resistance genes can be combined with effective genes to build new resistance combinations that, in some instances, can be exceedingly difficult for pathogens to circumvent. For example, the leaf rust resistance in the Canadian spring wheat cultivar Pasqua is believed to contain five leaf rust resistance genes (Dyck 1993). No report exists of the Pasqua leaf rust resistance combination being overcome by a virulent race of *Puccinia triticina* Eriks. (B. McCallum, pers. comm.). This scenario provides another application of QTL analysis and molecular breeding. First, QTL analysis can assist in identifying the genomic location of multigenic disease resistance traits, and this can be followed by molecular breeding strategies (Somers et al., 2005) to pyramid multiple genes effective against a common pathogen, such as *Lr34* + *Lr19* + *Lr21*.

GENETIC MAPPING

In order to perform QTL analysis and marker-assisted selection (MAS) for wheat improvement, we must first consider genotyping and the development of genetic maps. Over the past 20 years, several types of DNA markers have been developed, which were often relevant to certain genetic applications, crop species, and research budgets. Briefly, the primary marker types include random amplified polymorphic DNA (RAPD), restriction fragment length polymorphism (RFLP), amplified fragment length polymorphism (AFLP), microsatellites or simple sequence repeats (SSRs), and single nucleotide polymorphisms (SNPs). All of these types of markers and derived variations have been used in wheat genetic mapping. Today, SSRs prevail as the dominant marker system (Bryan et al., 1997; Roder et al., 1998; Somers et al., 2004), though SNPs are quickly maturing as a robust DNA marker system. In addition, a novel type of marker platform that merges RFLP and microarray technology (diversity array technology, DArT) was developed in Australia that is particularly useful in biparental crosses and quick assembly of a genetic map (http://www.diversityarrays.com/; Semagn et al., 2006). The method is microarray-based (hybridization-based), fast, and relatively inexpensive. Generally, the primary attributes of a good DNA marker are that it (i) is PCR-based, (ii) has low cost per data point, (iii) has codominant allele detection, and (iv) has high throughput. The SSR and SNP markers satisfy all of these requirements (Table 14.2).

Genetic maps are routinely developed from biparental crosses. Typically, the parents of the cross are inbred to limit heterogeneity of alleles of progeny, and it is common to develop either F_2, BC_1, doubled haploid (DH), or recombinant inbred line (RIL) populations, depending on the final application, time, and budget. The RIL population passes through several meiotic events in development, thus creating more breakpoints along the chromosomes. The other three population types feature only one meiotic event in the F_1 generation, which is used to develop the final mapping population. Table 14.3 summarizes a RIL versus DH population to exemplify this idea. Consistent with theoretical expectation, the RIL population has very near twice the number of breakpoints per chromosome as the DH population. The progeny in the population segregate at many points in the genome, and this can be detected by DNA markers such as SSRs. Genetic maps are then constructed by collecting the genotype data over many progeny and calculating recombination distances between the segregating loci (Perretant et al., 2000; Somers et al., 2004).

Finally, QTL analysis is the merger between the genetic map information and phenotypic data collected on progeny from the same biparental cross that was mapped. Quantitative trait locus analysis is essentially a regression analysis to determine associations between loci along

Table 14.3 Comparison of recombinant inbred line (RIL) versus doubled haploid (DH) populations for chromosome breakpoints on chromosome 1A.

Cross	Population Type	Chromosome 1A Breakpoints		
		Total	Range	Mean
Synthetic × Opata	RIL	59	0–7	3.0
SC2180V2 × AC Karma	DH	30	0–4	1.5

Note: Based on random selection of 20 individuals per population.

Table 14.2 A comparison of molecular marker technology and application to marker-assisted selection.

Marker Type[a]	Inheritance	Abundance	Polymorphism	Development Cost	Application Cost[b]
RFLP	Codominant	Low	Medium	Medium	High
RAPD	Present/absent	Medium	Low	Low	Low
AFLP	Present/absent	Medium	Medium	Low	Medium
SSR	Codominant	Medium	High	High	Medium
SNP	Codominant	High	Low	High	Low
DArT	Present/absent	Medium	Medium	Medium	Low

[a]RFLP, restriction fragment length polymorphism; RAPD, random amplified polymorphic DNA; AFLP, amplified fragment length polymorphism; SSR, simple sequence repeat (microsatellites); SNP, single nucleotide polymorphism; DArT, diversity array technology.

[b]Includes time and labor costs and is based on current detection platforms.

the chromosome and variation in the trait phenotype (Doerge 2002; Wang et al., 2004). Further, the additive effects of each allele can be determined which associates the parental alleles with a favorable and a nonfavorable phenotype.

The three main factors affecting QTL analysis include (i) size of the population, (ii) density of markers and genetic spacing on the genetic map, and (iii) quality of the phenotypic data. The size of the population determines the distribution of crossover points in the genetic map and thus directly affects the resolution of the QTL analysis. Likewise, the marker density will directly affect the resolution of the QTL analysis. Both of the former attributes are easily controlled and the genotyping data is often discrete or absolute. Hence the success of QTL analysis is most affected by the quality of the phenotypic data. Poor data can lead to detection of spurious QTLs or improper estimation of the effect of a real QTL. It follows that complex traits such as yield, preharvest sprouting tolerance, or FHB resistance require a high degree of replication and careful measurement of traits within each replication. More simply inherited traits such as leaf rust resistance, or presence and absence of morphological traits, can be tested with fewer replications, but care must still be taken with trait measurement.

EARLY PROGRESS AND DEVELOPMENTS

Genetic maps

Wheat genetic mapping has always been fortunate to have a stable research investment and an evergrowing abundance of genetic markers. This began with RFLP markers, initially developed at the John Innes Centre, Norwich, UK, and the research laboratory headed by Dr. Mike Gale (Chao et al., 1989; Devos et al., 1993). This was followed by an international effort to map the wheat genome, with RFLP markers developed from many laboratories around the world. This effort was coordinated by the International Triticae Mapping Initiative (ITMI) that was initiated in 1989. The international wheat community decided to create a single, robust genetic map from the biparental cross of W7984 (synthetic) × Opata. The progeny consisted of 115 RILs and the cross was genetically wide, revealing >50% polymorphism across the genome. The cross was particularly useful for mapping the D genome, since the cross used a synthetic × bread wheat design. This cross is still in use today and is distributed around the world. Recently, the same ITMI population was re-created and is composed of >200 DH lines and >1,500 RILs (J.P. Gustafson, M.E. Sorrells, pers. comm., with D.J. Somers).

Wheat is a polyploid species carrying three different genomes designated A, B, and D. Geneticists can employ aneuploid lines, such as the wheat nullisomic–tetrasomic (NT) lines, to show from which chromosome each marker DNA fragment is derived, either through PCR or hybridization. This technique is still used heavily today with the development of SNP markers (Somers et al., 2003; http://wheat.pw.usda.gov/SNP/project.html) and other genome-specific markers. Therefore wheat genetic maps are composed of correctly labeled chromosomes, since the NT lines were developed through classical cytological studies and techniques that could identify the chromosomes based on mitotic staining patterns.

A critical milestone reached in wheat research was establishing early genetic maps of wheat that identified the chromosomes and provided a set of DNA markers that could be used to anchor future genetic maps. The ITMI map was a RFLP-based map and served to anchor future maps and also to facilitate the transition between RFLP- and PCR-based maps that employed either AFLP or SSR markers. In 1998, the first robust genetic map of wheat developed with SSRs (GWM) was released by Dr. Marion Roder and colleagues (Roder et al., 1998). This was followed by a second ITMI map with markers developed for the D-genome (GDM) (Pestsova et al., 2000). These maps consisted of >220 SSRs and 55 SSRs, respectively, distributed across the genome of the ITMI population and anchored to chromosomes using the available RFLP data on the same cross.

This was a milestone, since these maps provided inexpensive, PCR-based markers to then construct or anchor new maps to perform QTL analysis. Another milestone reached was the development of the first SSR-based consensus map of wheat (Somers et al., 2004), which positioned >1,200 then–publicly available SSRs on a single genetic map and facilitated new map construction, chromosome anchoring, and comparative mapping with SSRs.

Consensus map

Since publication of the first SSR-based genetic map of wheat (Roder et al., 1998), other genetic maps have been produced and the number of SSRs has increased severalfold. Comparative mapping between wheat genetic maps facilitated more marker identification in smaller regions of the genome. This was done by lining up drawings of maps and identifying markers in syntenic regions on different maps (Fig. 14.1). In 2004, a microsatellite consensus map was developed from a union of four wheat genetic maps (Somers et al., 2004), and this placed >1,200 loci onto a single map of wheat. This single map is very useful for anchoring new wheat genetic maps based on SSRs and identifying multiple markers on chromosome arms that may be deployed for MAS. For example, the gene *Lr16* was mapped to the terminal end of 2BS (McCartney et al., 2005b), along with the gene for Orange blossom wheat midge (*Sm1*) (Thomas et al., 2005); SSRs in the region are useful today for selection of lines for these pest resistance traits.

Progress in marker technology

The largest advance in technology related to genotyping and genetic mapping came with PCR-based markers, and in particular SSRs. Development of SSRs was proceeding in many crop species through the use of genomic DNA libraries enriched for short tandem repeats of DNA (microsatellites). The polyploid nature of wheat made this particularly difficult, since genome-specific markers are much more preferable for genetic mapping. Aside from the work at Institut für Pflanzengenetik und Kulturpflanzenforschung (IPK) by M. Roder and colleagues, the Wheat Microsatellite Consortium (WMC), a public and private effort with >30 members, developed >800 SSRs between 1997 and 2004. This set of WMC microsatellites has been used worldwide along with GWM and GDM SSRs to generate many genetic maps of wheat (Chalmers et al., 2001; Somers et al., 2004, 2006; McCartney et al., 2006). More recently, the USDA developed and mapped 540 SSR markers labeled BARC (Beltsville Agricultural Research Center) (Song et al., 2005), and a joint venture between Genoplante and Institut Scientifique de Recherche Agronomique (INRA) research center in Clermont Ferrand, France, developed 252 SSRs labelled CFA, CFD, and CFT (http://wheat.pw.usda.gov/GG2/index.shtml).

In addition to SSR development, capillary electrophoresis became available and affordable for many institutes, which facilitated high-throughput genotyping at low cost. The equipment from Applied Biosystems Inc. (Foster City, California), in particular, was well suited for SSR detection and facilitates automated allele calling. When SSRs are abundant and the use of a high-throughput detection platform is available, QTL analysis becomes much more feasible and is only limited by the time to create populations and collect phenotypic data over several environments.

Aside from searching for SSRs within genomic DNA libraries and sequences, a bioinformatics approach was developed to find SSR sequences within the growing public wheat database of expressed sequence tags (ESTs) (Kantety et al., 2002). This approach made sense since the cost of sequencing had already been absorbed by the EST sequencing efforts earlier, and only the cost of primer synthesis and PCR would be required to demonstrate the utility of these SSRs. This idea was also attractive since it would provide markers from within expressed genes, and gene sequence conservation could facilitate cross-species application or comparative mapping in other cereal species (Sorrells et al., 2003; La Rota and Sorrells 2004).

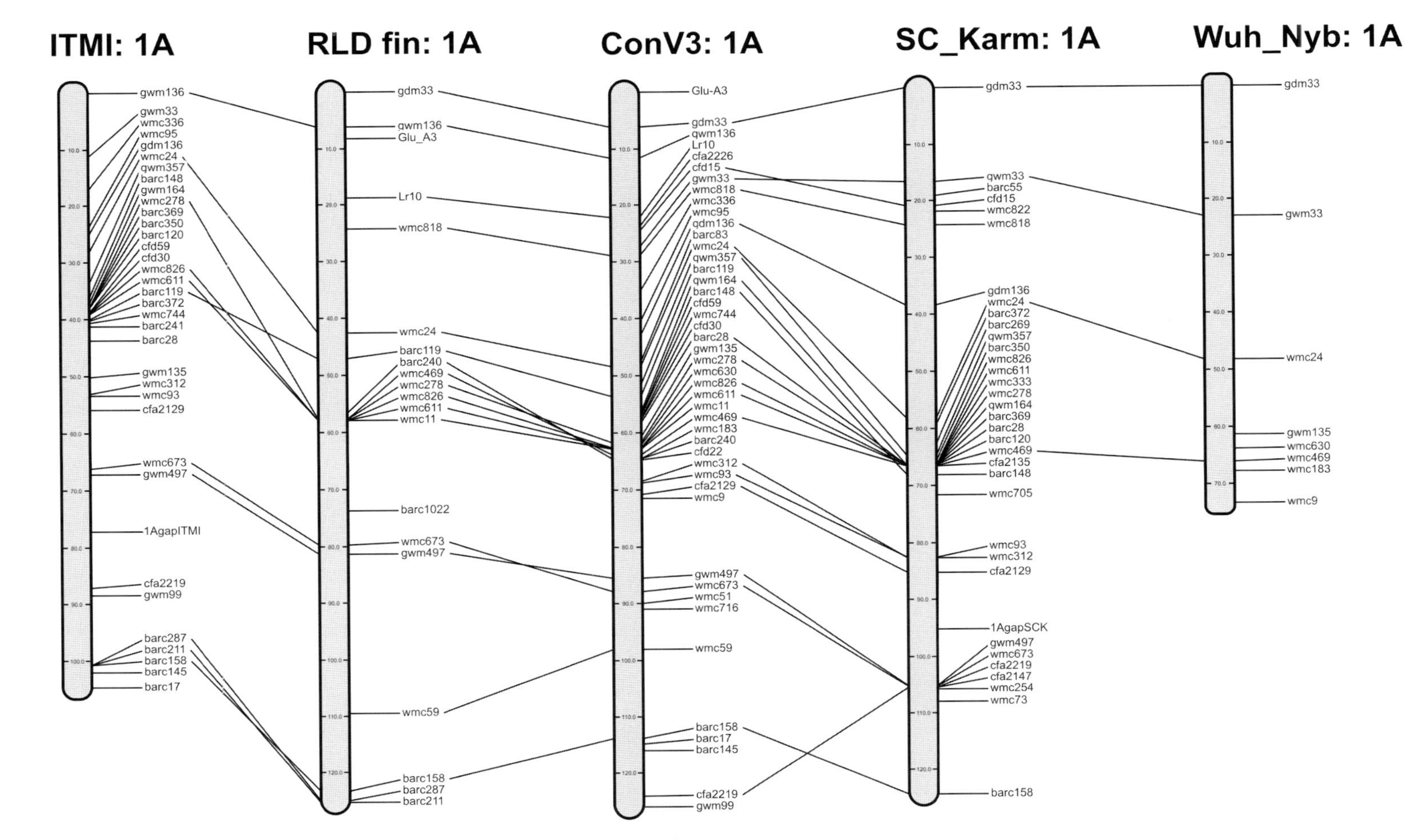

Fig. 14.1 Comparative map of chromosome 1A between four individual, biparental maps (ITMI, RLD fin, SC_Karm, Wuh_Nyb) and the wheat microsatellite consensus map (ConV3); SSR marker names are indicated on the right.

CURRENT PROGRESS IN QTL ANALYSIS AND DEPLOYMENT OF MAS

It is sometimes stated that wheat breeding is as much art as science. Through controlled hybridization followed by selfing, wheat breeders generate new genetic variability. Selection of inbred lines derived from crosses can lead to novel germplasm and new cultivars. However, traditional breeding methods are laborious, time-consuming, and costly. In Canada, approximately 10 generations are required to develop wheat lines suitable for cultivar evaluation. Selection for desirable traits is required during pureline development to ensure that cultivars possess the traits necessary for adoption by producers. The modern era of molecular linkage mapping began in the 1980s (Botstein et al., 1980). Twenty years later, high-density molecular linkage maps have been developed for most major cereal crops, including wheat (Somers et al., 2004). Molecular mapping and marker development for agronomic, disease, and quality traits (i.e., McCartney et al., 2005a, 2006) now provide robust tools that can be used to effectively and efficiently select for traits of interest.

Single-gene traits and complex traits

Robust markers have been developed for many important traits such as resistance to leaf rust, stem rust (caused by *Puccinia graminis* Pers.:Pers. f. sp. *tritici* Eriks. & E. Henn.), or common bunt (caused by *Tilletia tritici* [syn. *T. caries*]), and insect resistance, protein content, and glutenin alleles. A list of genes and traits is provided in Table 14.4, which is not intended to be exhaustive. The use of MAS for important single traits is desirable, because phenotypic selection for certain traits requires destructive tests. Other traits are expensive to measure such as grain cadmium content (Penner et al., 1995). The evaluation of quality traits requires a minimum grain sample that impedes quality analyses in early generations. Molecular marker assays are largely free of such constraints (Radovanovic and Cloutier 2003; Kuchel et al., 2007a). Selection can be conducted on single seeds to ensure traits are present before planting. The actual plants used in crossing or backcrossing can be directly assayed for desirable traits (Kuchel et al., 2005, 2007a). In some programs, breeders have combined MAS and DH techniques to improve wheat. Parents are screened before DH production, and the haploid plants are screened prior to chromosome doubling to generate populations of DH plants that are fixed for the gene(s) of interest. Parents can be assayed for a single-gene trait (e.g., *Lr39*, Gold et al., 1999; Glu-1Bx7, Radovanovic and Cloutier 2003) to ensure the DH population will include the trait(s) of interest.

The effectiveness of early-generation selection with markers could at times be compromised due to poor marker reliability (i.e., RAPDs) or excessive linkage distances (Kuchel et al., 2007a). At present, microsatellite markers are most commonly used in molecular mapping studies. Microsatellite markers are highly polymorphic and are reliable and repeatable. Microsatellite markers are amenable to high-throughput data collection, including 96-well-plate DNA extraction, 384-well PCR assays, as well as capillary electrophoresis systems. Microsatellite linkage maps provide a reliable framework that can be used in an ongoing fashion to map traits of interest, including both Mendelian and quantitatively inherited traits (McCartney et al., 2005b, 2006; Quarrie et al., 2005; Fofana et al., 2007).

Recurrent selection

Since SSRs are abundant on the wheat genetic map and high-throughput genotyping technology is available, the prospect of performing genome-wide recurrent genotype selection is made possible. For example, in a backcross breeding strategy, the restoration of the recurrent parent could be accelerated by selection of the desired alleles at a modest number of loci across the genome. Somers et al. (2005) demonstrated this approach by selecting at 40 to 70 loci in different backcross populations and restoring 95% of the recurrent genetic background in two cycles of backcrossing. Without marker assistance, the same level of restoration would require at the minimum three to four backcrossing cycles.

Table 14.4 Examples of DNA markers for major genes controlling important agronomic, disease, and cereal quality traits in wheat.

	Marker		
Trait	Gene	Type[a]	Reference
Agronomic traits			
Growth habit	*VRN-A1*	CAPS	Sherman et al. (2004)
Preharvest sprouting	*Vp-1B*	STS	Yang et al. (2007)
Plant height	*Rht-B1b, -D1b*	STS	Ellis et al. (2002)
Plant height	*Rht8*	SSR	Korzun et al. (1998)
Aluminum tolerance	*Alt-BH*	RFLP	Riede and Anderson (1996)
Boron tolerance	—	RFLP	Jefferies et al. (2000)
Pest resistance			
Common bunt	*Bt10*	STS	Laroche et al. (2000)
Cereal cyst nematode	*Cre1, Cre3*	STS	Ogbonnaya et al. (2001a,b)
Hessian fly	*H6*	STS	Dwiekat et al. (2002)
Hessian fly	*H9*	SCAR	Kong et al. (2005)
Leaf rust	*Lr28*	STS	Naik et al. (1998)
Durable leaf rust	*Lr34*	STS	Spielmeyer et al. (2005)
Loose smut	—	SCAR	Knox et al. (2002)
Powdery mildew	*Pm24*	AFLP	Huang et al. (2000)
Stem rust	*Sr39*	SCAR	Gold et al. (1999)
Stripe rust	*Yr26*	STS	Wang et al. (2008)
Wheat midge	*Sm1*	SCAR	Thomas et al. (2005)
Wheat streak mosaic virus	*Wsm1*	STS	Talbert et al. (1996)
Cereal quality			
Flour color	—	STS	Parker and Langridge (2000)
High protein content	*Gpc-B1*	STS	Distelfeld et al. (2006)
Storage proteins	*Glu-A1,B1,D1*	STS	Radovanovic and Cloutier (2003)
Polyphenol oxidase	—	STS	He et al. (2007)
Yellow pigment color	*Psy-A1*	STS	He et al. (2008)

[a]CAPS, cleavage amplification polymorphic sequence; STS, sequence tagged site; RFLP, restriction fragment length polymorphism; SSR, simple sequence repeat (microsatellites); SCAR, sequence characterized amplified region; AFLP, amplified fragment length polymorphism.

Another utility of whole-genome genotyping is construction of chromosome-substitution lines or advanced-backcross substitution lines containing single chromosome segments introgressed into a common genetic background. This process was well described by Huang et al. (2004) and can be useful for validation and characterization of QTLs.

Replicated field analysis

The importance of replicated field trials in many environments for QTL analysis cannot be underestimated. One of the most useful predictors of effective, robust QTLs is the detection of a QTL within a defined chromosome interval in multiple environments. This is the essential evidence a wheat breeder will insist on to commit to using MAS for a trait. Similar to breeding practices, it is more useful and relevant to test and measure traits in multiple environments rather than replicate extensively within few environments. Examples of highly replicated QTL analysis include testing for grain yield (Kuchel et al., 2007b,c). These experiments included 18 environments and could test for the persistence of QTLs over environments.

Ultimately, wheat breeders can consider this information and determine how robust a particular QTL is and whether it is worth committing resources toward MAS and reliance on the genotyping data for selection. In some cases, such as yield or a complex disease reaction such as FHB resistance, breeders are more likely to rely on well established, robust QTLs that were detected over multiple environments for enrichment of favorable alleles in populations. The alternative includes testing populations over multiple environments, which can take years.

Haplotype analysis

Haplotype analysis includes genotyping a collection of lines over a defined genetic or physical DNA interval and representing each line as a multilocus genotype. This can be at the level of a gene sequence (Jung et al., 2004), depicting SNPs within a few hundred nucleotides, or at the genetic level depicting SSR alleles present over centimorgan distances (McCartney et al., 2004). It is now becoming routine to perform a haplotype analysis in multiple regions of the genome where desirable QTLs have been discovered and mapped.

In wheat, haplotype analysis is best performed with multiallelic markers flanking and within the significance interval of the QTL. Wheat lines carrying the desired phenotype, and thus alleles, are compared to lines carrying undesirable phenotypes by collecting genotypic data across the interval on a set of germplasm. In some cases, the desired phenotype can be associated with a specific haplotype consisting of a few genetic loci. This strategy may also help to position the contributing gene to a smaller interval.

There are good examples of successful haplotype analysis for leaf rust (*Lr16*) resistance (McCartney et al., 2005b) and FHB resistance (Liu and Anderson, 2003; McCartney et al., 2004). For example, the haplotype surrounding *Lr16* on 2BS is unique to the donor source line. No other susceptible line tested carried this haplotype and thus deployment of markers is facilitated. Likewise, the haplotype at FHB resistance loci derived from Sumai 3 is quite unique at *FHB1* on 3BS and *FHB2* on 6BS (Cuthbert et al., 2006a,b). The haplotype approach is easily extended to include any mapped QTL.

Gene cloning and perfect markers

Once genetic mapping or QTL analysis is complete, the possibility of cloning a gene or mapping a candidate gene that controls a specific trait is very real. Gene cloning can be done via positional cloning and the use of large insert libraries of wheat, of which several are available. The gene sequence is used to develop a perfect marker in the sense that there is no recombination between the trait measured and the gene-based marker.

It has been demonstrated that the success of MAS is related to linkage distance between the marker and the actual trait (Kuchel et al., 2005). Thus, MAS applications would ideally be conducted using perfect markers which are those that have been developed from the sequence of the cloned genes. Genes have been cloned for numerous high-molecular-weight (HMW) and low-molecular-weight (LMW) glutenin genes as well as genes coding for gliadin proteins (Cloutier et al., 2001; de Bustos et al., 2001; Radovanovic and Cloutier 2003). Robust molecular markers have been developed for specific HMW glutenin alleles. For example, a PCR-based marker for the Bx7 allele from the cultivar Glenlea was developed, because it was shown that the Glenlea Bx7 allele was required for the "extra-strong" dough strength of this wheat cultivar (Radovanovic et al., 2002). Grain hardness genes, or the puroindoline genes, have been cloned and perfect markers for *Pina-D1* and *Pinb-D1* are available (Giroux et al., 2000; Limello and Morris 2000).

Ellis et al. (2002) reported perfect markers for the *Rht-B1b* and *Rht-D1b* dwarfing genes. These markers were found to be quite robust since they correctly categorized 19 wheat cultivars from 5 countries of known *Rht* type. The *Rht* markers were deemed perfect because they are designed to detect the base-pair change responsible for the semidwarf phenotype. Recent cloning of *Lr* genes (Feuillet et al., 2003 [*Lr10*]; Huang et al., 2003 [*Lr21*]; Cloutier et al., 2007 [*Lr1*]) will permit the development of robust markers for important wheat diseases such as leaf rust. Sherman et al. (2004) reported the development of *VRN-A1*-specific markers developed with the aid of the cloned gene sequence. PCR-based primers were developed that amplified a 810-bp segment containing diagnostic nucleotide changes, including an *Aci*I restriction enzyme site which permitted the development of a cleavage amplification polymorphic sequence (CAPS) based marker. The robust *VRN-A1* CAPS marker can be used in molecular breeding applications and to determine the allelic state of germplasm and breeding material.

COMPLEX TRAITS

Many agriculturally important traits, such as grain yield, are quantitatively inherited and phenotypes are distributed over a range. The distribution tends to indicate that the trait is controlled by multiple genes, each giving a small effect, and/or the trait is influenced by environmental conditions such that trait expression is controlled by a combination of genetic and environmental components. The mapping of quantitative trait loci is an important step toward the identification of the single genes that control a significant proportion of the phenotypic variation of quantitatively inherited traits (Lander and Botstein 1989). The continued development of high-density molecular maps (Somers et al., 2004) will permit the fine-mapping that is necessary to move from genomic regions to individual genes responsible for quantitatively inherited traits.

This progression is evidenced in the isolation of QTLs associated with Fusarium head blight, a destructive disease which reduces grain yield and negatively impacts end-use quality of wheat. Pathogenesis can include toxin production which has serious health implications even when present in low levels, for both livestock and humans (Gilbert and Tekauz 2000). Various QTLs for FHB resistance have been identified on several wheat chromosomes, including 2D, 3A, 3B, 4B, 5A, 5B, 6B, 7A, and 7B (McCartney et al., 2007), and these impact different components of host-plant resistance. Arguably, the most important QTL associated with FHB resistance is from the Chinese wheat cultivar Sumai-3 and is located on chromosome 3BS (Liu and Anderson 2003; Zhou et al., 2003). Recently, the locus on 3BS was mapped as a Mendelian trait and named *Fhb1* (Cuthbert et al., 2006b). Similarly, the FHB-resistance QTL identified on chromosome 6B has been mapped and named *Fhb2* (Cuthbert et al., 2006a). The move from large chromosomal segments to sectors closely linked to the gene(s) controlling the trait of interest will facilitate marker-assisted selection for quantitative traits such as FHB resistance.

Precocious sprouting of mature wheat grain can occur when wet, humid conditions exist prior to harvest. Mature wheat grain can germinate while plants are still in the field. The kernels are considered to be sprout-damaged, when there are signs of growth such as swollen, exposed germs and the appearance of root and shoot tips. Preharvest sprouting (PHS) can have a serious impact on grain yield and on end-use quality (Derera 1989; Kruger 1989), including noodle quality (Nagao 1995). Preharvest sprouting resistance is a complex trait affected by genotype, environment, plant diseases, and spike morphology. Thus, molecular markers should be effective tools to select for PHS resistance, because markers are unaffected by the growing environment.

Because of the importance of this trait to wheat production, several QTLs for PHS resistance have been identified in wheat (Anderson et al., 1993; Bailey et al., 1999; Kato et al., 2001; Osa et al., 2003; Mares et al., 2005; Mori et al., 2005). These QTLs for PHS resistance have been mapped to at least 10 of the 21 wheat chromosomes including 1A, 3A, 3B, 3D, 4A, 4B, 5A, 5D, 6B, and 7D. The location(s) of QTLs varied considerably depending on the specific mapping population, but PHS-resistance QTLs on group 3 and 4 chromosomes have been reported in multiple studies (Anderson et al., 1993; Kato et al., 2001; Groos et al., 2002; Kulwal et al., 2005; Mares et al., 2005; Mori et al., 2005). The importance of QTLs on the group 3 chromosomes appears to be related to the pleiotropic effects of the seed coat color alleles and the *TaVp1* genes. In maize, the *Vp1* gene reportedly codes for a dormancy-related transcription factor (Osa et al., 2003). Introgression of major QTLs for PHS resistance can increase grain dormancy (Kottearachchi et al., 2006). Marker-assisted selection for PHS resistance is desirable since it can be applied to early segregating generations, it does not require elaborate sprouting-test facilities, and is independent of the environment.

Wheat end-use quality consists of a group of complex traits which are largely quantitatively inherited. Further, growing environments usually have an important influence on the end-use quality phenotype. McCartney et al. (2006) mapped QTLs for 41 end-use quality traits in the population RL4452 × 'AC Domain'. In this study, 99 QTLs were reported on 18 chromosomes, although 44 QTLs mapped to 3 QTL clusters on chromosomes 1B, 4D, and 7D. On chromosome 1B, 14 QTLs mapped close to the *Glu-B1* locus. The important role that glutenin storage proteins play in end-use quality has been long established (see review by Bushuk 1998) and validated by others (Kuchel et al., 2006; Groos et al. 2007). Twenty QTLs mapped close to a plant-height QTL, likely the *Rht-D1* gene, and 10 QTLs mapped close to a QTL for maturity on 7D, thus demonstrating the important interactions between genes controlling agronomic traits and end-use quality. Major QTLs for protein content, water absorption, and dough strength parameters were also mapped. Noodle quality traits were mapped to 5B and 5D, and the largest QTL for b* value (yellow pigment color) mapped to 7AL, close to the location of the phytoene synthase (*Psy-1*) gene as reported by He et al. (2008). Groos et al. (2007) reported seven QTLs for breadmaking quality scores on eight different chromosomes and that breadmaking QTLS were generally coincident QTLs for dough rheology, protein content, or flour viscosity, indicating the interrelationship between quality traits.

Selection for end-use quality is normally conducted in later generations in a wheat breeding program because of limited grain supply and the cost of rheological and baking tests. Thus, MAS for end-use quality at earlier stages of the breeding process would be highly desirable. Marker-assisted selection of specific glutenin subunits has been used for some time with the intent to select for dough strength (Radovanovic and Cloutier 2003). However, there has been limited use of markers for more complex quality traits, possibly because the relationship between QTLs and end-use quality has been unclear. Recent mapping efforts (Kuchel et al., 2006; McCartney et al., 2006; Groos et al., 2007) should provide valuable candidate QTLs for use in molecular breeding of end-use quality.

FUTURE DEVELOPMENTS AND USES OF QTL ANALYSIS AND MAPPING

Many researchers would agree that SSR markers will continue to provide useful tools and insight into the organization of the wheat genome and be useful for resolving QTLs. It is reasonable to assume that SNP markers covering the whole genome is a technology that will evolve and, when coupled with an inexpensive, highly multiplexed detection platform, could overtake SSRs as the most applied marker system. The advantage of SNPs is the relative abundance compared to SSRs, but this is balanced against the level of polymorphism, which is lower than SSRs (Somers et al., 2003). Currently SSR genotyping costs range from $0.25 to $0.50 per datapoint (labor cost excluded), depending on the level of multiplexing, source of consumables, and detection platform. Single nucleotide polymorphism genotyping via the Illumina bead array system can reduce this cost to approximately $0.05 per datapoint and can dramatically increase the throughput with multiplex levels of 1,536 datapoints per run.

Association mapping

Many research groups in wheat and other crop species are turning their attention toward association mapping (Flint-Garcia et al., 2003; Rafalski and Morgante 2004). This is a natural extension of chromosome interval haplotype analysis, the difference being association analysis may be performed as a full genome analysis. To accomplish association analysis in wheat, the degree of linkage disequilibrium (LD) is first estimated. A measurement of LD will predict the marker density required across the genome to accommodate association mapping. It is estimated that LD decay in wheat occurs over 2–5 cM (Chao et al., 2007;

Somers et al., 2007). Given a wheat genome map length of 2,500 cM (Somers et al., 2004), then 2,500 markers (1 marker cM^{-1}) would be required to exceed the level of LD decay. This is the primary driver behind SNP development, as it could provide the needed marker density for efficient association mapping, although the concerns with low polymorphism levels with SNPs are still untested.

The advantage of association mapping versus QTL analysis in biparental crosses lies largely in population development. Populations for association mapping are simply a collection of existing wheat germplasm that can be characterized by genotyping. The collection makeup is easily controlled both for size and genetic content. The population can be diverse, such as a global collection of wheat accessions (Maccaferri et al., 2004), or it can be restricted to accessions adapted to a certain country or region (Chao et al., 2007; Somers et al., 2007). Since the population undergoes genotyping, it can be tailored to contain a diverse collection of alleles across the genome or have limited variation at certain loci. In contrast, biparental crosses and derived populations require time (1–4 years) and effort to produce; selection of the two parents determines the makeup of the mapping population. The QTL analysis is then restricted to the pairs of alleles segregating at each locus in the population. Hence another advantage of association mapping is the ability to analyze associations of traits with multiple alleles, including rare alleles.

The final advantage of association mapping populations concerns the number of meiotic events represented in the collection of wheat. In most cases, the accessions within an association mapping population will each have been derived through many meiotic events. Thus a high-resolution genotypic comparison of wheat accessions within the association mapping population will show multiple crossover events within a defined marker interval. In contrast, a typical doubled-haploid mapping population derived from a biparental cross will have undergone a single meiotic event, and the number of crossovers is limited. As a result, if a high-density marker system is available for both populations, the association mapping population has more potential to provide precise marker–trait associations with fewer individual lines in the population, compared with biparental crosses.

Gene expression analysis

One of the newest applications of QTL analysis is mapping gene expression patterns. Microarray-based analysis of gene expression, such as the Affymetrix platform and the Affymetrix wheat gene chip (http://www.affymetrix.com),

has facilitated highly parallel quantification of individual genes under controlled biological experiments. The Affymetrix wheat gene chip has in excess of 55,000 genes represented. Expression of individual genes can be regarded as quantitative phenotypic data on this platform. This is referred to as *genetical genomics* (Jansen and Nap 2001). Thus if gene expression data is collected from a biparental mapping population, then an SSR map of this population can be used to perform QTL analysis on individual gene expression patterns. Further, if the mapping population is characterized for agronomically important traits or seed quality attributes, the gene expression QTL (eQTL) can also be aligned with these traits.

This approach was taken recently with a doubled haploid population that was fully mapped with SSRs and fully phenotyped for many agronomic and seed quality traits (McCartney et al., 2005a, 2006). The microarray gene expression analysis of this same population used mRNA extracted from developing seeds (5 days postanthesis). The result was mapping >500 eQTLs across the genome. The eQTL results, when aligned with existing phenotypic traits, should succeed at identifying areas of the genome and candidate genes that control seed quality traits (Jordan et al., 2007).

FUTURE PERSPECTIVES

More than 20 years has passed since the development of the first, crude molecular genetic maps of wheat, beginning with RFLP markers and advancing to SNP and DArT markers, with increasing marker density and lower genotyping costs. To improve QTL analysis, map densities must be adequate across the genome and the markers must be rich in polymorphism, such as SSRs. Further, the technical aspects of increased genotyping throughput along with accurate, reliable allele sizing for robust loci would greatly improve our ability to make genetic maps and perform QTL analysis. Capillary electrophoresis and microarray-based genotyping are having a significant influence on genotyping and this will continue in the future, enabling laboratories to share technical information developed from a common genotyping platform.

Quantitative trait locus analysis itself is largely a statistical exercise combining genotype and phenotype data and has seen less dramatic change over time, but software applications continue to be improved for wheat researchers. One can speculate that QTL analysis will play an important role in future wheat genome analysis, since the genome is large and whole-genome sequencing for gene identification is still immature. The trend will continue for using QTLs as a starting point to fine-map traits or genes of interest in wheat, leading to gene cloning and perfect markers for wheat breeders.

The development of association mapping strategies is just beginning. Again, marker density in specific chromosome regions is sometimes sufficient, but whole-genome analysis will likely require 2,500–3,000 robust markers across the genome. This is where SNPs or single-feature polymorphism (microarray-based polymorphisms) can improve marker densities and facilitate association mapping.

Field performance and end-use quality testing of wheat breeding materials will continue to be the ultimate evaluation methods in the development of new wheat cultivars. Nevertheless, the use of molecular markers to test and select for qualitative traits is an accepted practice among wheat breeders. Thus, there is an ongoing need to develop markers for important breeding traits, both qualitative and quantitative. Marker development and deployment will likely become increasingly "a breeder responsibility," moving from the genetics laboratory to the screening lab which will be part of the breeding program. It will be imperative for future wheat breeders to be well-versed in molecular approaches to wheat cultivar development.

Fortunately, new advances in DNA technology are facilitating marker development and deployment. Robotics-based liquid handling systems and capillary electrophoresis technologies permit high-throughput DNA extraction and marker screening so that entire breeding populations can be screened for the trait(s) of interest. As more and more DNA markers are developed using the same marker technologies (i.e., microsatellites), multiplexing of markers for different genes of the same trait or for multiple traits in the same screen will be facilitated.

Cloning of candidate genes for the traits of interest will continue to permit the development of the "perfect markers," which is required to maximize the effectiveness and efficiency of marker-assisted breeding. The fine-mapping of genes that control a significant proportion of the phenotypic variability of quantitatively inherited traits has begun, and this effort will need to continue so that marker-assisted breeding can be effectively applied to quantitatively inherited traits.

REFERENCES

AACC. 2000. Approved methods of the American Association of Cereal Chemists. 10th ed. AACC Press, St. Paul, MN.

Anderson, J.A., M.E. Sorrells, and S.D. Tanksley. 1993. RFLP analysis of genomic regions associated with resistance to preharvest sprouting in wheat. Crop Sci. 33:453–459.

Bailey, P.C., R.S. McKibbin, J.R. Lenton, M.J. Holdsworth, J.E. Flintham, and M.D. Gale. 1999. Genetic map locations for orthologous *Vp1* genes in wheat and rice. Theor. Appl. Genet. 98:281–284.

Botstein, D., R.L. White, M. Skolnick, and R.W. Davis. 1980. Construction of a genetic linkage map in man using restriction fragment length polymorphisms. Am. J. Hum. Genet. 32:314–31.

Bryan, G.J., A.J. Collins, P. Stephanson, A. Orry, J.B. Smith, and M.D. Gale. 1997. Isolation and characterization of microsatellites from hexaploid bread wheat. Theor. Appl. Genet. 94:557–563.

Bushuk, W. 1998. Wheat breeding for end-product use. Euphytica 100:137–145.

Chalmers, K.J., A.W. Campbell, J. Kretschmer, A. Karakousis, P.H. Henschke, S. Pierens, N. Harker, M. Pallotta, G.B. Cornish, M.R. Shariflou, L.R. Rampling, A. McLauchlan, G. Daggard, P.J. Sharp, T.A. Holton, M.W. Sutherland, R. Appels, and P. Langridge. 2001. Construction of three linkage maps in bread wheat, *Triticum aestivum*. Aust. J. Agric. Res. 52:1089–1119

Chao, S., P.J. Sharp, A.J. Worland, R.M.D. Koebner, and M.D. Gale. 1989. RFLP-based genetic maps of homoeologous group 7 chromosomes. Theor. Appl. Genet. 78:495–504.

Chao, S., W. Zhang, J. Dubcovsky, and M. Sorrells. 2007. Evaluation of genetic diversity and genome-wide linkage disequilibrium among U.S. wheat (*Triticum aestivum* L.) germplasm representing different market classes. Crop Sci. 47:1018–1030.

Cloutier, S., B.D. McCallum, C. Loutre, T.W. Banks, T. Wicker, C. Feuillet, B. Keller, and M.C. Jordan. 2007. Leaf rust resistance gene *Lr1*, isolated from bread wheat (*Triticum aestivum* L.) is a member of the large psr567 gene family. J. Plant Mol. Biol. 65:93–106.

Cloutier, S., C. Rampitsch, G.A. Penner, and O.M. Lukow. 2001. Cloning and expression of a LMW-i glutenin gene. J. Cereal Sci. 33: 143–154.

Cuthbert, P.A., D.J. Somers, and A. Brule-Babel. 2006a. Mapping *Fhb2* on chromosome 6BS: A gene controlling Fusarium head blight resistance in bread wheat (*Triticum aestivum* L.). Theor. Appl. Genet. 114:429–437.

Cuthbert, P.A, D.J. Somers, J. Thomas, S. Cloutier, and A. Brule-Babel. 2006b. Fine mapping *Fhb1*, a major gene controlling Fusarium head blight resistance in bread wheat (*Triticum aestivum* L.) Theor. Appl. Genet. 112:1462–1472.

de Bustos, A., P. Rubio, C. Soler, P. Garcia, and N. Jouve. 2001. Marker-assisted selection to improve HMW-glutenins in wheat. Euphytica 119:69–73.

Derera, N.F. 1989. The effects of preharvest rain. p. 2–14. *In* N.F. Derera (ed.) Preharvest sprouting in cereals. CRC Press Inc., Boca Raton, FL.

Devos, K.M., T. Millan, and M.D. Gale. 1993. Comparative RFLP maps of homoeologous group 2 chromosomes of wheat, rye, and barley. Theor. Appl. Genet. 85:784–792.

Distelfeld, A., C. Uauy, T. Fahima, and J. Dubcovsky. 2006. Physical map of the wheat high-grain protein content gene *Gpc-B1* and development of a high-throughput molecular marker. New Phytol. 169:753–763.

Doerge, R.W. 2002. Mapping and analysis of quantitative trait loci in experimental populations. Nat. Rev. Genet. 3:43–52.

Dwiekat, I., W. Zang, and H. Ohm. 2002. Development of STS markers linked to Hessian fly resistance gene *H6* in wheat. 2002. Theor. Appl. Genet. 105:766–770.

Dyck, P.L. 1993. The inheritance of leaf rust resistance in the wheat cultivar Pasqua. Can. J. Plant Sci. 73:903–906.

Ellis, M.H., W. Speilmeyer, K.R. Gale, G.J. Rebetzke, and R.A. Richards. 2002. "Perfect" markers for the *Rht*-B1b and *Rht*-D1b dwarfing genes in wheat. Theor. Appl. Genet. 105:1038–1042.

Feuillet, C., S. Travella, N. Stein, L. Albar, A. Nublat, and B. Keller. 2003. Map-based isolation of the leaf rust disease resistance gene *Lr10* from the hexaploid wheat (*Triticum aestivum* L.) genome. Proc. Natl. Acad. Sci. USA 100:15253–15258.

Flint-Garcia, S.A., J.M. Thornsbery, and E.S. Buckler. 2003. Structure of linkage disequilibrium in plants. Annu. Rev. Plant Biol. 54:357–374.

Fofana, B., D.G. Humphreys, S. Cloutier, C.A. McCartney, and D.J. Somers. 2007. Mapping quantitative trait loci controlling common bunt resistance in a doubled haploid population derived from the spring wheat cross RL4452 × AC Domain. Mol. Breed. 21:317–325 [Online]. Available at http://springerlink.com/content/p62g875422q7q1jt/ (verified 12 Mar. 2008).

Gilbert, J., and A. Tekauz. 2000. Review: Recent developments in research on Fusarium head blight of wheat in Canada. Can. J. Plant Pathol. 22:1–8.

Giroux, M.J., L. Talbert, D.K. Habernicht, S. Lanning, A. Hemphill, and J.M. Martin. 2000. Association of puroindoline sequence type and grain hardness in hard red spring wheat. Crop Sci. 40:370–374.

Gold, J., D. Harder, T.F. Townley-Smith, T. Aung, and J. Procunier. 1999. Development of a molecular marker for rust resistance genes *Sr39* and *Lr35* in wheat breeding lines. Electron. J. Biotechnol. 2:1–6.

Groos, C., E. Bervas, E. Chanliaud, and G. Charmet. 2007. Genetic analysis of bread-making quality scores in bread wheat using a recombinant inbred line population. Theor. Appl. Genet. 115:313–323.

Groos, C., G. Gay, M.-R. Perretant, L. Gervais, M. Bernard, F. Dedryver, and G. Charmet. 2002. Study of the relationship between pre-harvest sprouting and grain color by quantitative trait loci analysis in a white × red grain bread-wheat cross. Theor. Appl. Genet. 104:39–47.

He, X.Y., Z.H. He, L.P. Zhang, D.J. Sun, C.F. Morris, E.P. Feurst, and X.C. Xia. 2007. Allelic variation of polyphenol oxidase (PPO) genes located on chromosomes 2A and 2D and the development of functional markers for the PPO genes in common wheat. Theor. Appl. Genet. 115:47–58.

He, X.Y., L.P. Zhang, Z.H. He, Y.P. Wu, Y.G. Xiao, C.X. Ma, and X.C. Xia. 2008. Characterization of phytoene synthase 1 gene (*Psy1*) located on common wheat chromosome 7A and development of a functional marker. Mol. Breed. [Online]. Available at http://www.springer.com/life+sci/plant+sciences/journal/11032 (verified 10 Mar. 2008).

Huang, L., S.A. Brooks, W. Li, J.P. Fellers, H.N. Trick, and B.S. Gill. 2003. Map-based cloning of leaf rust resistance gene *Lr21* from the large and polyploid genome of bread wheat. Genetics 164:655–664.

Huang, X.Q., S.L.K. Hsam, F.J. Zeller, G. Wenzel, and V. Mohler. 2000. Molecular mapping of wheat powdery mildew resistance gene *Pm24* and marker validation for molecular breeding. Theor. Appl. Genet. 101:407–414.

Huang, X.Q., H. Kempf, M.W. Ganal, and M.S. Röder. 2004. Advanced backcross QTL analysis in progenies derived from a cross between a German elite winter wheat variety and a synthetic wheat (*Triticum aestivum* L.). Theor. Appl. Genet. 109:933–943.

Jansen, R.C., and J-P. Nap. 2001. Genetical genomics: The added value from segregation. Trends Genet. 17:388–391.

Jefferies S.P., M.A. Pallotta, J.G. Paull, A. Karakousis, J.M. Kretschmer, S. Manning, A.K. Islam, P. Langridge, and K.J. Chalmers. 2000. Mapping and validation of chromosome regions conferring boron toxicity tolerance in wheat (*Triticum aestivum*). Theor. Appl. Genet. 101:767–77.

Jordan, M.C., D.J. Somers, and T.W. Banks. 2007. Identifying regions of the wheat genome controlling seed development by mapping expression quantitative trait loci. Plant Biotechnol. J. 5:442–453.

Jung, M., A. Ching, D. Bhattramakki, M. Dolan, S. Tingey, M. Morgante, and A. Rafalski. 2004. Linkage disequilibrium and sequence diversity in a 500-kbp region around the *adh1* locus in elite maize germplasm. Theor. Appl. Genet. 109:681–689.

Kantety, R.V., M. La Rota, D.E. Matthews, and M.E. Sorrells. 2002. Data mining for simple sequence repeats in expressed sequence tags from barley, maize, rice, sorghum and wheat. Plant Mol. Biol. 48:501–510.

Kato, K., W. Nakamura, T. Tabiki, H. Miura, and S. Sawada. 2001. Detection of loci controlling seed dormancy on group 4 chromosomes of wheat and comparative mapping with rice and barley genomes. Theor. Appl. Genet. 102:980–985.

Knox, R.E., J.G. Menzies, N.K. Howes, J.M. Clarke, T. Aung, and G.A. Penner. 2002. Genetic analysis of resistance to loose smut and an associated DNA marker in durum wheat doubled haploids. Can. J. Plant Pathol. 24:316–322.

Koebner, R.M.D. 2004. Marker assisted selection in the cereals: The dream and the reality. p. 317–329. *In* P.K. Gupta and R.K. Varshney (ed.) Cereal genomics. Kluwer Academic Publishers, Dordrecht, The Netherlands.

Kong, L., H.W. Ohm, S.E. Cambron, and C.E. Williams. 2005. Molecular mapping determines that Hessian fly resistance gene H9 is located on chromosome 1A of wheat. Plant Breed. 124:525–531.

Korzun, V., M.S. Roder, M.W. Ganal, A.J. Worland, and C.N. Law. 1998. Genetic analysis of the dwarfing gene (*Rht8*) in wheat: Part I. Molecular mapping of *Rht8* on the short arm of chromosome 2D of bread wheat (*Triticum aestivum* L.). Theor. Appl. Genet. 96:1104–1109.

Kottearachchi, N.S., N. Uchino, K. Kato, and H. Miura. 2006. Increased grain dormancy in white-grained wheat by introgression of preharvest sprouting tolerance QTLs. Euphytica 152:421–428.

Kruger, J.E. 1989. Biochemistry of preharvest sprouting in cereals. p. 61–84. *In* N.F. Derera (ed.) Preharvest sprouting in cereals. CRC Press Inc., Boca Raton, FL.

Kuchel, H., R. Fox, J. Reinheimer, L. Mosionek, N. Willey, H. Bariana, and S.P. Jefferies. 2007a. The successful application of a marker-assisted wheat breeding strategy. Mol. Breed. 20:295–308.

Kuchel, H., P. Langridge, L. Mosionek, K. Williams, and S.P. Jefferies. 2006. The genetic control of milling yield, dough rheology and baking quality of wheat. Theor. Appl. Genet. 112:1487–1495.

Kuchel, H., K. Williams, P. Langridge, H.A. Eagles, and S.P. Jefferies. 2007b. Genetic dissection of grain yield in bread wheat: I. QTL analysis. Theor. Appl. Genet. 115:1029–1041.

Kuchel, H., K. Williams, P. Langridge, H.A. Eagles, and S.P. Jefferies. 2007c. Genetic dissection of grain yield in bread wheat: II. QTL-by-environment interaction. Theor. Appl. Genet. 115:1015–1027.

Kuchel, H., G. Ye, R. Fox, and S.P. Jefferies. 2005. Genetic and economic analysis of a targeted marker-assisted wheat breeding strategy. Mol. Breed. 16:67–78.

Kulwal, P.L., N. Kumar, A. Gaur, P. Khurana, J.P. Khurana, A.K. Tyagi, H.S. Balyan, and P.K. Gupta. 2005. Mapping of a major QTL for pre-harvest sprouting tolerance on chromosome 3A in bread wheat. Theor. Appl. Genet. 111:1052–1059.

Lander, E.S., and D. Botstein. 1989. Mapping mendelian factors underlying quantitative traits using RFLP linkage maps. Genetics 121:185–199.

Laroche, A., T. Demenke, D.A. Gaudet, B. Puchalski, M. Frick, and R. McKenzie. 2000. Development of PCR marker for rapid identification of the *Bt-10* gene for common bunt resistance in wheat. Genome 43:217–223.

La Rota, M., and M.E. Sorrells. 2004. Comparative DNA sequence analysis of mapped wheat ESTs reveals the complexity of genome relationships between rice and wheat. Funct. Integr. Genomics 4:34–46.

Limello, M., and C.F. Morris. 2000. A leucine to proline mutation in puroindoline b is frequently present in hard wheats from Northern Europe. Theor. Appl. Genet. 100:1100–1107.

Liu, S., and J.A. Anderson. 2003. Marker assisted evaluation of Fusarium head blight resistant wheat germplasm. Crop Sci. 43:760–766.

Maccaferri, M., M.C. Sanguineti, E. Noli, and R. Tuberosa. 2004. Population structure and long-range linkage disequilibrium in a durum wheat elite collection. Mol. Breed. 15:271–290.

Mares, D., K. Mrva, J. Cheong, K. Williams, B. Watson, E. Storlie, M. Sutherland, and Y. Zou. 2005. A QTL located on chromosome 4A associated with dormancy in white- and red- grained wheats of diverse origin. Theor. Appl. Genet. 111:1357–1364.

McCartney, C.A., D.J. Somers, G. Fedak, and W. Cao. 2004. Haplotype diversity at Fusarium head blight QTLs in wheat. Theor. Appl. Genet. 109:262–271.

McCartney, C.A., D.J. Somers, G. Fedak, R.M. DePauw, J. Thomas, S.L. Fox, D.G. Humphreys, O. Lukow, M.E. Savard, B.D. McCallum, J. Gilbert, and W. Cao. 2007. The evaluation of FHB resistance QTLs introgressed into elite Canadian spring wheat germplasm. Theor. Appl. Genet. 20:209–221.

McCartney, C.A., D.J. Somers, D.G. Humphreys, O. Lukow, N. Ames, J. Noll, S. Cloutier, and B.D. McCallum. 2005a. Mapping quantitative trait loci controlling agronomic traits in the spring wheat cross RL4452 × 'AC Domain'. Genome 48:870–883.

McCartney, C.A., D.J. Somers, O. Lukow, N. Ames, J. Noll, S. Cloutier, D.G. Humphreys, and B.D. McCallum. 2006.

QTL analysis of quality traits in the spring wheat cross RL4452 × 'AC Domain'. Plant Breed. 125:565–575.

McCartney, C.A., D.J. Somers, B.D. McCallum, J. Thomas, D.G. Humphreys, J.G. Menzies, and P.D. Brown. 2005b. Microsatellite tagging of the leaf rust resistance gene *Lr16* on wheat chromosome 2BS. Mol. Breed. 15:329–337.

Mori, M., N. Uchino, M. Chono, K. Kato, and H. Miura. 2005. Mapping QTLs for grain dormancy on wheat chromosome 3A and the group 4 chromosomes, and their combined effect. Theor. Appl. Genet. 110:1315–1323.

Nagao, S. 1995. Detrimental effect of sprout damage on wheat flour products. p. 3–8. *In* K. Noda and D.J. Mares (ed.) Proc. Symp. on Pre-harvest Sprouting in Cereals, 7th, Osaka, Japan. July 1995, Center for Academic Societies, Osaka, Japan.

Naik, S., K.S. Gill, V.S. Prakasa Rao, V.S. Gupta, S.A. Tamhankar, S. Pujar, B.S. Gill, and P.K. Ranjekar. 1998. Identification of a STS marker linked to *Aegilops speltoides*–derived leaf rust resistance gene *Lr28* in wheat. Theor. Appl. Genet. 97:535–540.

Ogbonnaya, F.C., S. Seah, A. Delibes, J. Hahier, I. Lopez-Brana, R.F. Eastwood, and E.S. Lagudah. 2001a. Molecular-genetic characterization of a new nematode resistance gene in wheat. Theor. Appl. Genet. 102:623–629.

Ogbonnaya, F.C., N.C. Subrahmanyam, I. Moullet, J. de Majnik, H.A. Eagles, J.S. Brown, R.F. Eastwood, J. Kollmorgen, R. Appels, and E.S. Lagudah. 2001b. Diagnostic DNA markers for cereal cyst nematode in bread wheat. Aust. J. Agric. Res. 52:1367–1374.

Osa, M., K. Kato, M. Mori, C. Shindo, A. Torada, and H. Miura. 2003. Mapping QTLs for seed dormancy and the Vp1 homologue on chromosome 3A in wheat. Theor. Appl. Genet. 106:1491–1496.

Parker, G.D., and P. Langridge. 2000. Development of a STS marker linked to a major locus controlling flour colour in wheat (*Triticum aestivum* L.). Mol. Breed. 6:169–174.

Penner, G.A., J. Clarke, L.J. Bezte, and D. Leisle. 1995. Identification of RAPD markers linked to a gene governing cadmium uptake in durum wheat. Genome 38:543–547.

Perretant, M.R., T. Cadalen, G. Charmet, P. Sourdille, P. Nicolas, C. Boeuf, M.H. Tixier, G. Branlard, S. Bernard, and M. Bernard. 2000. QTL analysis of bread-making quality in wheat using a doubled haploid population. Theor. Appl. Genet. 100:1167–1175.

Pestsova, E., M.W. Ganal, and M.S. Roder. 2000. Isolation and mapping of microsatellite markers specific for the D genome of bread wheat. Genome 43:689–697.

Preston, K.R., R.H. Kilborn, and H.C. Black. 1982. The GRL pilot mill: II. Physical and baking properties of flour milled from Canadian red spring wheat. Can. Inst. Food Sci. Technol. J. 15:29–36.

Quarrie, S.A., A. Steed, C. Calestani, et al. 2005. A high-density genetic map of hexaploid wheat (*Triticum aestivum* L.) from the cross Chinese Spring × SQ1 and its use to compare QTLs for grain yield across a range of environments. Theor. Appl. Genet. 110:865–880.

Radovanovic, N., and S. Cloutier. 2003. Gene-assisted selection for high molecular weight glutenin subunits in wheat doubled haploid breeding programs. Mol. Breed. 12:51–59.

Radovanovic, N., S. Cloutier, D. Brown, D.G. Humphreys, and O.M. Lukow. 2002. Genetic variance for gluten strength contributed by high molecular weight glutenin protein. Cereal Chem. 79:843–849.

Rafalski, A., and M. Morgante. 2004. Corn and humans: Recombination and linkage disequilibrium in two genomes of similar size. Trends Genet. 20:103–111.

Riede, C.R., and J.A. Anderson. 1996. Linkage of RFLP markers to an aluminum tolerance gene in wheat. Crop Sci. 36:905–909.

Roder, M.S., V. Korzun, K. Wandehake, J. Planschke, M.H. Tixier, P. Leroy, and M.W. Ganal, 1998. A microsatellite map of wheat. Genetics 149:2007–2023.

Semagn, K., A. Bjornstad, H. Skinnes, A.G. Maroy, Y. Tarkegne, and M. Williams. 2006. Distribution of DArT, AFLP, and SSR markers in a genetic linkage map of a doubled haploid hexaploid wheat population. Genome 49:545–555.

Sherman, J.D., L. Yan, L. Talbert, and J. Dubcovsky. 2004. A PCR marker for growth habit in common wheat based on allelic variation at the *VRN-A1* gene. Crop Sci. 44:1832–1838.

Somers, D.J., T. Banks, R. DePauw, S. Fox, J. Clarke, C. Pozniak, and C. McCartney. 2007. Genome-wide linkage disequilibrium analysis in bread wheat and durum wheat. Genome 50:557–567.

Somers, D.J., G. Fedak, J. Clarke, and W. Cao. 2006. Mapping of FHB resistance QTLs in tetraploid wheat. Genome 49:1586–1593.

Somers, D.J., P. Isaac, and K. Edwards. 2004. A high density microsatellite consensus map for bread wheat (*Triticum aestivum* L.). Theor. Appl. Genet. 109:1105–1114.

Somers, D.J., R. Kirkpatrick, M. Moniwa, and A. Walsh. 2003. Mining single nucleotide polymorphisms from hexaploid wheat ESTs. Genome 49:431–437.

Somers, D.J., J. Thomas, R. DePauw, S. Fox, G. Humphreys, and G. Fedak. 2005. Assembling complex genotypes to resist *Fusarium* in wheat (*Triticum aestivum* L.). Theor. Appl. Genet. 111:1623–1631.

Song, Q.J., J.R. Shi, S. Singh, E.W. Fickus, J.M. Costa, J. Lewis, B.S. Gill, R. Ward, and P.B. Cregan. 2005. Development and mapping of microsatellite (SSR) markers in wheat. Theor. Appl. Genet. 110:550–560.

Sorrells, M.E., M. La Rota, C.E. Bermudez-Kandianis, et al. 2003. Comparative DNA sequence analysis of wheat and rice genomes. Genome Res. 13:1818–1827.

Spielmeyer, W., R.A. McIntosh, J. Kolmer, and E.S. Lagudah. 2005. Powdery mildew resistance and *Lr34/Yr18* genes for durable resistance to leaf and stripe rust cosegregate at a locus on the short arm of chromosome 7D of wheat. Theor. Appl. Genet. 111:731–735.

Talbert, L.E., P.L. Bruckner, L.Y. Smith, R. Sears, and T.J. Martin. 1996. Development of PCR markers linked to resistance to wheat streak mosaic virus in wheat. Theor. Appl. Genet. 93:463–467.

Thomas, J., N. Fineberg, G. Penner, C. McCartney, T. Aung, I. Wise, and B. McCallum. 2005. Chromosome location and markers of *Sm1*: A gene of wheat that conditions antibiotic resistance to orange wheat blossom midge. Theor. Appl. Genet. 15:183–192.

Wang, S., C.J. Basten, P. Gaffney, and Z-B. Zeng. 2004. Windows QTL cartographer 2.0. User manual. Bioinformatics Research Center, North Carolina State University, Raleigh, NC.

Wang, C., Y. Zhang, D. Han, Z. Kang, G. Li, A. Cao, and P. Chen. 2008. SSR and STS markers for stripe rust resistance gene *Yr26*. Euphytica 159:359–366.

Western Organization of Resource Councils. 2002. World wheat facts [Online]. Available at http://www.worc.org/pdfs/WorldWheatFacts.pdf (verified 12 Mar. 2008).

Yang, Y., X.L. Zhao, L.Q. Xia, X.M. Chen, X.C. Xiz, Z.H. He, and M. Roder. 2007. Development and validation of a *Viviparous-1* STS marker for pre-harvest sprouting tolerance in Chinese wheats. Theor. Appl. Genet. 115:971–980.

Zhou, W-C., F.L. Kolb, G-H. Bai, L.L. Domier, L.K. Boze, and N.J. Smith. 2003. Validation of a major QTL for scab resistance with SSR markers and use of marker-assisted selection in wheat. Plant Breed. 122:40–46.

Chapter 15
Genome Organization and Comparative Genomics

Katrien M. Devos, Jaroslav Doležel, and Catherine Feuillet

SUMMARY

(1) Detailed genetic maps have been generated for wheat using restriction fragment length polymorphism (RFLP), simple sequence repeat (SSR), single nucleotide polymorphism (SNP), and diversity array technology (DArT) markers. In addition, markers have been located to chromosome bins using overlapping sets of deletion lines. All maps are available on GrainGenes.

(2) Common sets of markers can be used to establish the relationship between the wheat genomes and those of other species within the Triticeae tribe, the Pooideae subfamily, and even the *Poaceae* family. Knowledge about the relationship between different genomes has allowed exploitation of information and resources across species.

(3) Analysis at the DNA sequence level of regions that are orthologous at the map level has shown that marker orders are highly conserved between species that diverged as much as 60 million years ago. Small rearrangements such as inversions, translocations, duplication, and deletion of one or a few genes are, however, common evolutionary events.

(4) Map-based gene isolation in wheat has become a reality. In addition to the wheat resources, comparative information, in particular between wheat and sequenced grass genomes, is an integral part of map-based cloning strategies in wheat. The challenges of gene isolation in wheat are demonstrated in nine success stories.

(5) A global physical map of the polyploid wheat genome is needed to facilitate genetic and genomic analyses in wheat. The complexity of this task can be reduced by constructing physical maps in diploid wheat species. An alternative approach is to isolate individual chromosomes or chromosome arms using flow-sorting and build physical maps for single chromosomes. The physical map of hexaploid wheat chromosome 3B has been described as a case study.

(6) The component of the wheat genome that is of most interest is the gene space. Current projections for the number of genes present in the hexaploid wheat genome vary from 108,000 to 300,000. Genes are not randomly distributed in the wheat genome, but are organized in gene islands that increase in number and size from the centromere to the telomere. The distal chromosome regions have higher evolutionary rates which, in turn, lead to lower levels of intergenomic colinearity.

(7) The next big target is to obtain the sequence of the wheat genome. Questions that have been raised include whether the entire wheat genome or only the gene space should be sequenced, and whether Sanger sequencing technology and/or new generation sequencing technologies should be used. Information content and cost will have to be balanced in the final decision.

(8) We should already start thinking beyond the genomic sequence. Functional tool development has been initiated and includes the generation of populations by TILLING (Targeted Induced Local Lesions IN Genomes), virus-induced gene silencing (VIGS), and improved transformation protocols. Knowledge of all genes and their function will allow highly efficient and targeted improvement of wheat.

MAPPING

Genetic mapping in wheat started in earnest in the mid-1980s following the development of restriction fragment length polymorphisms (RFLPs) as a marker system (Botstein et al., 1980). Since then, detailed maps have been developed for diploid, tetraploid, and hexaploid wheat (Chao et al., 1989; Gill et al., 1991b; Liu and Tsunewaki 1991; Devos et al., 1992; Hart et al., 1993; Xie et al., 1993; Gale et al., 1995; Nelson et al., 1995a,b,c; Jia et al., 1996; Marino et al., 1996; Blanco et al., 1998; Röder et al., 1998; Messmer et al., 1999; Peng et al., 2000; Zhang et al., 2004b). In addition to genetic maps, chromosome bin maps have been generated using sets of overlapping deletion lines (Kota et al., 1993; Delaney et al., 1995a,b; Mickelson-Young et al., 1995; Gill et al., 1996a; Conley et al., 2004; Hossain et al., 2004; Linkiewicz et al., 2004; Miftahudin et al., 2004; Munkvold et al., 2004; Peng et al., 2004; Randhawa et al., 2004). Deletion mapping does not require intervarietal polymorphism and thus provides a mechanism for generating high-density maps in crops such as wheat that have a narrow genetic base. The physical maps provided a first insight into the distribution patterns of genes in the large wheat genome and are a valuable tool for studying genome organization. Genetic maps have important agronomic applications. They form the framework for the mapping of traits using quantitative trait analyses as further elaborated in Chapter 14, for gene-tagging and, more recently, for the isolation of genes underlying these traits (see subsequent section on map-based cloning).

Genetic mapping

The first RFLP mapping in wheat was accomplished using cDNAs and low copy *Pst*I genomic clones as markers. In hexaploid wheat, genes are generally triplicated. While polyploidy complicates genetic mapping to a degree, it also provides advantages. The buffering capacity of the hexaploid background has allowed the development of series of nullisomic–tetrasomic (NT) lines. In each NT line, a chromosome pair has been replaced with a homoeologous pair (Sears 1954). These lines have been extremely useful for determining the chromosomal location of markers. Hybridization of RFLP markers to DNA extracted from the NT lines has revealed that nearly all cDNAs and about 50% of *Pst*I genomic clones detect copies on all three genomes in hexaploid wheat (Devos et al., 1992). When these markers are used for mapping, however, polymorphisms are rarely obtained between the parents of even wide mapping populations for all three copies. The main problem in the construction of detailed wheat genetic maps, therefore, was not the hexaploid nature of the crop, but the low level of variation.

To enhance polymorphism levels, and thus the number of markers that can be incorporated in a genetic map, segregating populations were often generated from wide crosses involving synthetic wheat (Gale et al., 1995; Nelson et al., 1995c). Alternatively, mapping was done in a diploid wheat progenitor or relative such as *Aegilops tauschii* and *Triticum monococcum*, which showed higher levels of variation (Dubcovsky et al., 1996; Boyko et al., 1999). Loci detected with the same probe in different genomes mapped to homoeologous locations. This showed that marker orders were highly conserved among the A, B, and D genomes of wheat.

To determine the location of centromeres on the genetic maps, markers were hybridized to a set of ditelosomic lines. Ditelosomic lines are disomic for the absence of a chromosome arm, and they exist for the majority of the chromosomes (Sears and Sears 1979). The centromeres typically colocalized with large clusters of markers. Although marker orders were generally highly conserved among the A, B, and D genomes, the maps confirmed that the wheat genome had undergone several chromosomal translocations. The best characterized rearrangement involved chromosome arms 4AL, 5AL, and 7BS (Liu et al., 1992; Devos et al., 1995; Nelson et al., 1995a). The reciprocal translocation between chromosome arms 4AL and 5AL took place in the A-genome diploid progenitor of wheat. This was followed by a pericentric inversion in chromosome 4A which resulted in the long arm becoming the physically shorter arm. A second reciprocal translocation took place between the 5AL region on 4AL and 7BS. In addition to the 4AL/5AL/7BS translocation, a 2BS/6BS translocation has been mapped (Devos et al., 1993b; Qi et al., 2006), and pericentric inversions have been identified in 'Chinese Spring' chromosomes 2B, 3B, 4A, 4B, 5A, and 6B (Qi et al., 2006). No rearrangements have been observed in the D genome, and the D genome is therefore considered to have retained the most ancestral chromosome organization.

As new polymerase chain reaction (PCR)–based marker systems were developed, these markers were used to populate the wheat genetic maps (Röder et al., 1998; Stephenson et al., 1998; Somers et al., 2004). Microsatellite markers tend to detect relatively high polymorphism levels and thus are a great tool for marker-assisted selection (Prasad et al., 1999; Qiu et al., 2006; Hiebert et al., 2007; Houshmand et al., 2007; Kumar et al., 2007; Sharma et al., 2007; Tsilo et al., 2007, 2008). However, microsatellites also tend to be chromosome-specific and hence provide little information on the relationship between the three wheat genomes. This is particularly true for microsatellites isolated from genomic sequences, but it is also often the case for microsatellites derived from 5′ or 3′ untranslated regions of expressed sequence tags (ESTs). Other marker systems that are currently in use in wheat are single nucleotide polymorphisms (Somers et al., 2003; http://probes.pw.usda.gov:8080/snpworld/disclaimer.jsp) and Diversity Array Technology (DArT) markers (Akbari et al., 2006). The latter technology is hybridization-based and allows the genotyping in parallel of several thousand loci (Jaccoud et al., 2001).

Typical features of the wheat genetic maps include the high conservation of colinearity between the homoeologous genomes and the clustering of markers in the centromeric regions. The latter is due to a reduction in recombination around the centromere. While mapping the centromeres on the genetic maps was easy due to the existence of ditelosomic lines, capping the genetic maps with telomeres proved to be a greater challenge. The ends of seven chromosome arms have been defined by mapping telomere-associated sequences beyond the most distal marker on a linkage group (Mao et al., 1997). The maps that have been generated over the years can be viewed on GrainGenes (http://wheat.pw.usda.gov/GG2/index.shtml). In addition, a composite wheat map resulting from the integration of 12 maps and 2,700 loci has been compiled (http://rye.pw.usda.gov/cmap/).

Deletion mapping

A number of alien chromosomes, including chromosomes 2C of *Ae. cylindrica*, 3C of *Ae. triuncialis*, and $2S^l$ and $4S^l$ of *Ae. sharonensis*, carry gametocidal genes (*Gc*) (McIntosh et al., 1998). It has been proposed that the primary function of a *Gc* factor is to induce chromosome breaks in gametes that lack the *Gc* factor (Endo 1990). The broken chromosome ends form telomeres and can thus be stably transmitted to the progeny. This characteristic of *Gc* genes has been exploited to generate a large collection of terminal deletions of varying size (Endo and Gill 1996). By using overlapping sets of deletions, a chromosome arm is *de facto* divided into small chromosome regions or bins. For mapping, each deletion line is scored for the presence or absence of a marker and this process allocates the marker to a particular bin. The disadvantage of deletion mapping is that markers within a bin cannot be ordered. The advantage is that no polymorphism is required for mapping. The first deletion maps were produced with RFLP markers (Werner et al., 1992; Delaney et al., 1995a,b; Mickelson-Young et al., 1995; Gill et al., 1996a,b). More recently, some 16,000 EST loci have been mapped across the seven wheat homoeologous groups (Qi et al., 2004).

Comparisons between the genetic maps and deletion maps have shown that genes are located mainly toward the ends of chromosomes. In fact, the decrease in marker density from the centromeric to the telomeric regions that can be seen in the genetic maps is inversely correlated with the gene density gradient on the physical maps. The distal chromosome regions tend to be gene-rich and have high recombination rates. These are also the regions that harbor most of the agronomic traits (Qi et al., 2004). Centromeric regions, on the other hand are gene-poor and largely devoid of recombination leading to marker clustering on the genetic map. These trends are clear even when only a small number of deletion lines is used per chromosome arm and bin sizes are relatively large.

A few studies have tried to quantify the size of the gene-rich regions in wheat. On the short arm of the group 1 chromosomes, it was estimated that 70% of the genes were contained within two gene-rich regions that occupied only 14% of the arm (Sandhu et al., 2001). Global projections for the entire wheat genome were that 94% of the genes were located within gene-rich regions comprising 29% of the wheat genome (Erayman et al., 2004). Knowing the precise organizational patterns of genes is very important when considering strategies for sequencing the wheat genome (see "Toward sequencing the wheat genome").

COMPARATIVE GENETICS

Cross-species genetic mapping was initiated in the early 1990s. The use of common sets of DNA probes, mostly RFLP markers, in the construction of genetic maps of related species provided insight into genome relationships. Following the development of genomic resources such as large EST collections, bacterial artificial chromosome (BAC) libraries, and whole-genome sequence information, the early map-based comparative analyses could be refined and complemented with detailed studies of colinearity at the DNA sequence level. The comparative information enhanced the use of related germplasm as a source of novel alleles in breeding, helped to elucidate mechanisms of genome evolution, and allowed exploitation of knowledge and resources across species. The latter was particularly important for large-genome species such as wheat. The 17,000-Mb wheat genome had long been considered intractable for map-based cloning, and the potential to use small-genome models such as rice (*Oryza sativa* L.) as intermediates to conduct chromosome walking opened the door to gene isolation in wheat. Although technical advances made chromosome walking in wheat a reality at the start of the 21st century, knowledge about comparative relationships remained important. It allowed whole-genome sequence data from rice and other grass species to be exploited for targeted marker development in wheat. In the current era of new sequencing technologies, comparative information of model species may aid with the assembly of shotgun sequence of large-genome relatives. In subsequent sections, we will review the knowledge about wheat—grass comparative relationships.

Comparative mapping

Triticeae tribe

The Triticeae tribe contains wheat and its relatives, with *Triticum* and *Aegilops* species being the most closely related to wheat at an estimated divergence age of less than 5 million years ago (MYA) (Huang et al., 2002b; Devos et al., 2005a), followed by rye (*Secale cereale*) at approximately 7 MYA (Huang et al., 2002a), and barley (*Hordeum vulgare*) at approximately 10–13 MYA (Wolfe et al., 1989; Gaut 2002) (Fig. 15.1). Wheat relatives such as rye and various *Aegilops* species are often used as donors of novel genes or alleles for wheat improvement (McIntosh et al., 2003). The first comparative genetic maps constructed between species within the Triticeae tribe had as a goal to identify the extent of synteny between

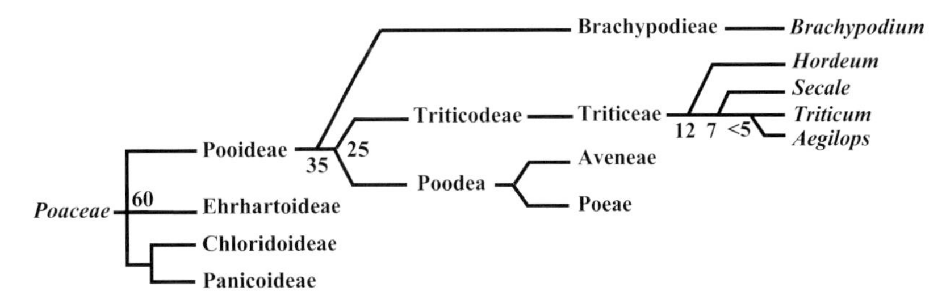

Fig. 15.1 Schematic representation of the phylogenetic relationship of the genus *Triticum* to other grass species. The approximate age of some of the branching points is indicated in millions of years ago (MYA).

Table 15.1 Chromosomes involved in rearrangements in different Triticeae species.

Species	Rearranged Chromosome Arm[a]	Type of Rearrangement
Wheat—A genome	4L, 5L	4AL–5AL and 4AL–7BS, translocations
Wheat—A genome	4A	Pericentric inversion
Wheat—B genome	7S	4AL–7BS, translocation
Wheat—D genome	—	
Ae. speltoides	—	
Ae. longissima	$4S^aL$, $7S^aL$	$4S^aL$–$7S^aL$, translocation
Ae. umbellulata	1UL, 2U, 3US, 4US, 4UL, 5UL, 6US, 6UL, 7U	4UL–5UL, translocation; 2U and 7U, pericentric inversions; others, translocations and inversions[c]
Rye	2RS, 3RS, 4RL, 5RL, 6RL, 7RS, 7RL	4RL–5RL, 3RL–6RL, 4RS–7RL, 2RS–6RS, 6RS–7RL and 2RS–4RL, translocations; 6R, pericentric inversion[b]
Barley	1HL	Paracentric inversion
H. bulbosum	1HL	Paracentric inversion

[a]Rearrangements are described relative to the D genome of wheat.
[b]Based on the model proposed by Devos et al. (1993).
[c]Due to the complex structure of some of the *Ae. umbellulata* chromosomes, it was not possible to determine the nature of all rearrangements that had taken place in the shaping of the *Ae. umbellulata* chromosomes.

the genomes of wheat and those of potential alien gene donors, with the rationale that the level of colinearity between donor and recipient genomes would affect recombination and thus introgression of the alien fragment into the wheat genome.

Homoeology between Triticeae chromosomes had previously been established on the basis of their ability to compensate for the loss of different wheat chromosomes. In 1987, Naranjo and colleagues identified several chromosome regions of rye that paired with nonhomoeologous wheat chromosomes (Naranjo et al., 1987). Genetic mapping of the rye genome using wheat RFLP markers confirmed that six of the seven rye chromosomes had undergone reciprocal translocations and/or inversions relative to wheat (Devos et al., 1993a). Similar comparative analyses of wheat with barley (Dubcovsky et al., 1996), *Ae. umbellulata* (Zhang et al., 1998), *Ae. longissima* (Zhang et al., 2001), *Hordeum bulbosum* (Salvo-Garrido et al., 2001), and *Ae. speltoides* (Luo et al., 2005) showed that large numbers of rearrangements were the exception rather than the norm. Chromosomes of all but one species, *Ae. umbellulata*, were either completely syntenic (*Ae. speltoides*) or carried a single rearrangement (*Ae. longissima*, barley, and *H. bulbosum*) relative to the D genome of wheat. In *Ae. umbellulata*, all seven chromosomes have been involved in at least one translocation, resulting in single *Ae. umbellulata* chromosomes having homoeology to segments of four or more wheat chromosomes (Zhang et al., 1998). A list of the chromosomes involved in rearrangements in different Triticeae species is given in Table 15.1. Most Triticeae genomes, including the A, B, and D genomes, have largely retained their ancestral configuration over the past 10–13 million years, which is the estimated age of the Triticeae tribe (Wolfe et al., 1989; Gaut 2002).

Pooideae subfamily

The Pooideae subfamily includes the supertribes Triticodeae and Poodea. The former contains the Triticeae tribe, and the latter the Aveneae and Poeae tribes. The two supertribes diverged approximately 25 MYA (Gaut 2002) (Fig. 15.1). Oat (*Avena sativa*) is the major cereal belonging to the Aveneae tribe, and the Poeae tribe includes the forage crops ryegrass (*Lolium perenne*) and fescue (*Festuca pratensis*). Comparative relationships have been established between wheat and oat (Van Deynze et al., 1995a), and between wheat, ryegrass, and fescue (Jones et al., 2002; Alm et al., 2003). The *Lolium* and fescue genomes, each of which has a basic chromosome number of seven, are highly colinear with the wheat genome. The major difference between the two Poeae species and wheat is the presence of a segment with homology to the long arm of wheat chromosome 5 in the distal region of the short arm of a chromosome that is otherwise completely syntenic with wheat chromosome 4. This 4S/5L rearrangement is different from the 4L/5L translocation that is present in some Triticeae genomes (Table 15.1).

The wheat–oat relationship is more complex. Oat, as wheat, has a basic chromosome number of seven, and at least five of the seven oat chromosomes have undergone rearrangements relative to wheat (Van Deynze et al., 1995a). Most of these rearrangements are specific to oat or, possibly, the Aveneae tribe. The 4S/5L rearrangement that differentiates the Poeae and Triticeae species is present in oat, indicating that it took place either in the lineage leading to the supertribe Poodeae or in the Triticodeae lineage.

In recent years, understanding the relationship between the genomes of wheat and *Brachypodium distachyon* has become a priority. All major crops within the Pooideae subfamily have large genomes with genome sizes between 2,000 and 5,500 Mb. The closest small-genome wheat relatives are species belonging to the genus *Brachypodium*, which diverged from wheat some 35 MYA (Bossolini et al., 2007) (Fig. 15.1). Two *Brachypodium* species have received research attention as potential model systems for wheat. *Brachypodium sylvaticum* is a tetraploid species with a basic chromosome number of 7 and a 1C DNA content of 460 Mb (Bennett and Smith 1991). A BAC library of *B. sylvaticum* has been constructed and used in the comparative analysis of wheat–*Brachypodium* relationships at the sequence level as discussed further in the section, "Colinearity at the DNA sequence level." *Brachypodium distachyon* is a diploid ($2n = 2x = 10$) with a 320-Mb genome, and this species has been selected for whole-genome shotgun (WGS) sequencing. Currently, a 4X shotgun sequence is available (http://www.brachybase.org/) and an 8X sequence is expected to be released in spring 2009. A comparative map of the *B. distachyon* genome is under development and will provide information on the relationship at the map level between the five *B. distachyon* chromosomes and the seven wheat homoeology groups. Preliminary analysis of the 4X *B. distachyon* sequence with bin-mapped wheat ESTs has indicated that four of the five *B. distachyon* chromosomes may be colinear with a single wheat homoeology group each. The fifth and largest *B. distachyon* chromosome appears to show homoeology with three wheat chromosome groups (M. Bevan, pers. comm.).

Poaceae family

Pairwise comparisons at the map level between genetic maps of wheat and those of non-Pooideae grasses include an analysis of the wheat homoeologous group 7 chromosomes with maize (*Zea mays*) chromosome 9 (Devos et al., 1994) and whole-genome studies of wheat–rice (Ahn et al., 1993; Kurata et al., 1994) and wheat–maize relationships (Van Deynze et al., 1995b). Despite 50–70 million years of evolutionary divergence between grass species (Wolfe et al., 1989) (Fig. 15.1), variation in chromosome numbers ($n = 12$ for rice, $n = 10$ for maize, and $n = 7$ for wheat), and large differences in genome size (1C = 400 Mb for rice, 1C = 2,500 Mb for maize, and 1C =

17,000 Mb for hexaploid wheat), the wheat, rice, and maize genomes appear to be highly colinear. Loci that map to non-colinear locations in different genomes are usually detected by multicopy probes and most likely represent paralogues rather than orthologues. Furthermore, considering that the grass ancestor has undergone a whole-genome duplication some 70 million years ago, apparent orthologues detected by single-copy probes may, in fact, be paralogues if different gene copies were deleted in different grass lineages, thereby complicating comparative analyses (Paterson et al., 2004). Breakdown of synteny tended to occur mostly toward the distal chromosome regions (Kurata et al., 1994).

While the early comparative maps relied on the cross-mapping of RFLP markers, which meant that only polymorphic loci could be mapped, advances in genomics in both wheat and rice allowed much more detailed comparisons to be conducted. By 2004, wheat deletion maps containing more than 16,000 loci had been constructed (Qi et al., 2004). The sequence of the rice genome had also been completed (International Rice Genome Sequencing Project 2005). Rice orthologues for the mapped wheat ESTs and their physical location could be identified *in silico* from the rice genomic sequence. This showed that the homoeologous group 1 chromosomes of wheat (W1S–cent–W1L) corresponded to rice 5S (R5S)–cent–R10–R5L, with 'cent' the centromere position in wheat. Wheat group 2 chromosomes (W2S–cent–W2L) corresponded to R4S–R7L–cent–R7S–R4L, W3S–cent–W3L to R1S–cent–R1L, W4S–cent–W4L to the proximal part of R3L–R11–cent–R3S, W5S–cent–W5L to R12L–cent–R9–distal part of 3RL, W6S–cen–W6L to R2S–cent–R2L, and W7S–cent–W7L to R6S–R8L–cent–R8S–R6L (La Rota and Sorrells 2004). These relationships do not take into account the complex translocation involving chromosomes 4A, 5A, and 7B. The effect of this rearrangement on the comparative relationship with rice can be seen in Color Plate 30.

Despite the clear evidence of syntenic blocks, some 35% of wheat markers that appeared to be single-copy in wheat mapped to nonsyntenic positions in rice, suggesting that interchromosomal rearrangements involving small regions or potentially single markers are commonplace. Non-colinear markers were found across the entire genome, but were particularly prevalent in the satellite region on 1BS, the most distal bin on chromosome 4AL, and the bin adjacent to the 6BS satellite (La Rota and Sorrells 2004). Wheat–rice colinearity was also difficult to establish for the centromeric regions. No breakdown in synteny was apparent in the distal chromosome regions, as had previously been noted by Kurata et al. (1994).

While the D genome of wheat is generally considered to be the most ancestral configuration within the Triticeae tribe, comparative analyses demonstrate that the entire Triticeae lineage underwent at least one gross chromosomal rearrangement since its divergence from the Aveneae–Poeae lineage. The structure of *Lolium* and fescue chromosome 4, which differs from wheat and other Triticeae species by a 4S/5L rearrangement represents, in fact, the ancestral chromosome configuration. Chromosomes for which the region orthologous to the long arm of the group 5 chromosomes in wheat (this breakpoint is proximal to the 4L/5L translocation that characterizes some Triticeae species) is fused to the distal region of 4S show orthology to rice chromosome 3 over their entire length (Color Plate 30). Translocation of the distal region of the short arm of the ancestral Pooid chromosome 4 to the long arm of chromosome 5 led to a break in synteny with the long arm of rice chromosome 3 (Color Plate 30). The 4S/5L breakpoint is also absent in *B. distachyon*, providing further evidence that the absence of this rearrangement represents the more ancestral chromosome configuration.

Colinearity at the DNA sequence level

Large-insert genomic library development in wheat started in the late 1990s with BAC libraries, first of the diploid wheats, *T. monoccoccum* (Lijavetzky et al., 1999) and *Ae. tauschii* (Moullet et al., 1999), and later of the tetraploid *T. turgidum* ssp. *durum* (Cenci et al., 2003) and the hexaploid *T. aestivum* (Allouis et al., 2003). These large-insert libraries provided the opportunity to conduct comparative analyses at the DNA sequence level between wheat and other grass species, mainly the small-genome model rice. Examples of such studies are comparisons between orthologous rice, sorghum (*Sorghum bicolor*), barley, and *T. monococcum* BACs selected to contain WG644, a marker linked to vernalization response (Ramakrishna et al., 2002); between a *T. monococcum* BAC clone carrying the grain hardness loci *Gsp-1*, *Pina-A*, and *Pinb-A* and the orthologous region on rice chromosome 12 (Chantret et al., 2004); between five regions on the short arm of chromosome $1A^mS$ totaling some 1.5 Mb and identified by the markers BCD1434, low-molecular-weight glutenins, and disease resistance genes *SRLK*, *Lrk10*, and *Lr10* and the orthologous genomic sequence on the short arm of rice chromosome 5 (Guyot et al., 2004); between a 250-kb region of hexaploid wheat and the orthologous regions on rice and *B. sylvaticum* (Griffiths et al., 2006); and between the wheat *Lr34* region and the orthologous regions in rice and *B. sylvaticum* (Bossolini et al., 2007). The main observations from these studies are that (i) gene content and order are relatively well conserved, but that rearrangements such as small deletions, duplications, inversions, and/or translocations are common evolutionary events (Fig. 15.2) and (ii) intergenic regions are expanded in wheat relative to the smaller rice and *Brachypodium* genomes. The expansion is caused mainly by the amplification and insertion of retrotransposons in the wheat genome.

The extent to which synteny is disrupted varies with chromosome region. One factor that has been shown to affect levels of colinearity is the rate of recombination. Synteny between the homoeologous wheat chromosomes is inversely correlated with recombination rates (Akhunov et al., 2003a). The higher recombinogenic distal chromosome regions carry more nonsyntenic markers than the more recombinational-inert centromeric regions. This has also been observed in wheat–rice colinearity studies (Akhunov et al., 2003b; See et al., 2006). Another factor that may influence the extent to which colinearity between orthologous regions is conserved is the type of genes that are present in the region under investigation. For example, Guyot et al. (2004) found that only 4 of 20 genes predicted to be present in a 638-kb wheat $1A^mS$ sequence were found in colinear positions on rice 5S. However, the genes that were present in non-colinear positions were mostly storage protein and disease resistance genes. Both types of genes are highly specialized and often appear in clusters of locally duplicated genes that provide templates for rearrangements through homologous recombination. Non-colinearity of storage protein genes, due to

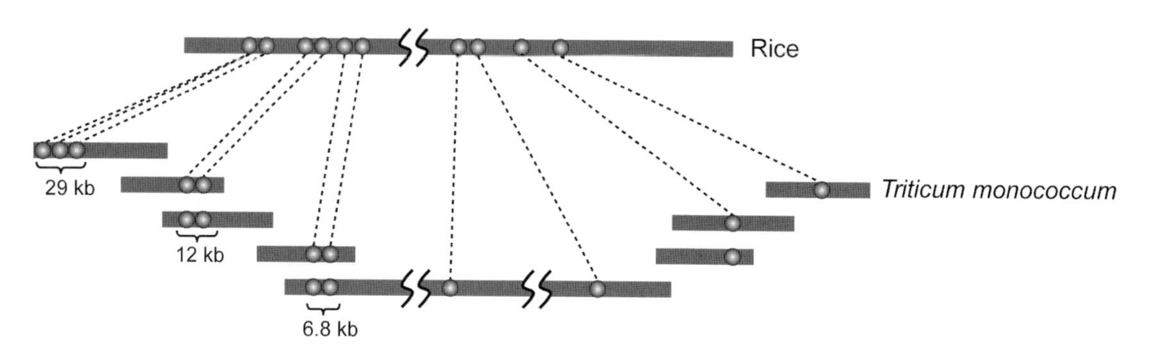

Fig. 15.2 Gene distribution and microcolinearity with rice of the *VRN-1* region in *T. monococcum*. Circles indicate genes. The size of gene islands in *T. monococcum* is indicated. Microcolinearity between *T. monococcum* and rice is highly conserved. The only break in colinearity is caused by a gene duplication present in *T. monococcum*.

differential copy numbers or complete absence of the genes, has been shown for maize zein genes (Song et al., 2002) and the wheat prolamins (Gao et al., 2007). Disease resistance genes have also been shown to undergo rapid rearrangements, leading to non-colinearity of these genes in cross-species comparisons (Gallego et al., 1998; Leister et al., 1998; Brueggeman et al., 2002).

Disruption of colinearity is evident in both wheat—rice and wheat—*Brachypodium* comparisons. The efficacy of a model organism tends to decline with increasing phylogenetic distance. *Brachypodium* is therefore expected to be a better model for wheat than rice. However, this is not a given as, for yet unknown reasons, some genomes are much more prone to rearrangements. The rice genome is generally considered to be a very stable genome (Ilic et al., 2003). Others, such as the maize and pearl millet (*Pennisetum glaucum*) genomes, seem to undergo rearrangements much more frequently (Devos et al., 2000; Ilic et al., 2003). The stability of the *Brachypodium* genome has, so far, been investigated in the tetraploid *B. sylvaticum* and is underway in the diploid *B. distachyon*. Griffiths et al. (2006) analyzed gene colinearity across the *Ph1* region of wheat, and orthologous regions in *B. sylvaticum* and rice. Of the 34 genes identified in this region in wheat, five were missing in both rice and *B. sylvaticum*, seven were missing in rice but not in *B. sylvaticum*, and one was missing in *B. sylvaticum* but was present in rice. In a similar study across the *Lr34* region by Bossolini et al. (2007), 11 and 10 of 15 wheat genes were present in conserved locations in *B. sylvaticum* and rice, respectively. An inversion was identified that differentiated the Pooid genomes from rice. Sequence similarity was greater between *B. sylvaticum* and wheat than between rice and wheat, dating the phylogenetic distance between *Brachypodium* and wheat at 35 MY. Preliminary comparative analyses at the DNA sequence level between wheat and *B. distachyon* also indicate that colinearity is often disrupted (K.M. Devos, unpublished data). Interestingly, there is also some indication that colinearity between rice and *B. distachyon* might be more extensive, at least in some regions, than colinearity of either with wheat. Based on this limited data, it would appear that *B. distachyon* will be a useful, but by no means a perfect, model for wheat genomics.

MAP-BASED CLONING

Isolation of the first plant gene using a map-based cloning approach that involved chromosome walking was achieved in *Arabidopsis thaliana* in 1992 (Arondel et al., 1992; Giraudat et al., 1992). Chromosome walking, however, is fraud with difficulties when applied to large-genome species. Because of the high content of repetitive DNA—more than 80% in wheat (Flavell et al., 1974)—most probes isolated from the ends of large-insert clones are repetitive and cannot be used to generate contigs that span a gene of interest. The actual chromosome walking step can be omitted if flanking markers for the gene of interest are identified that span a physical distance smaller than the average insert size of a large-insert clone. This approach has been termed *chromosome landing* (Tanksley et al., 1995) and was first applied in tomato (Martin et al., 1993). The chromosome landing approach shifted the challenge in map-based gene isolation from the walking step to finding markers that are tightly linked to the trait. Typically, this requires mapping populations of a few thousand gametes to achieve the necessary resolution and the ability to develop markers for the region of interest. Targeted marker development is facilitated by comparative information as it allows orthologous regions in other species to be tapped for markers. In the current era of whole-genome sequencing, including the completed genomes of rice, sorghum, maize, and *B. distachyon*, sequence information is the most important source of markers for fine-mapping in wheat.

The first successful map-based cloning experiments in wheat were concluded in 2003, some 10 years behind Arabidopsis. The lag was due to the large size of the wheat genome, which made construction of large-insert libraries and chromosome walking difficult. After all, constructing a

6X BAC library of a 17,000 Mb genome requires 1 million clones with an average insert size of 100 kb, and is not a small undertaking. Early gene-isolation strategies in wheat therefore used the smaller rice genome as a vehicle for map-based cloning. Markers flanking the gene that controlled a trait of interest in wheat were used to initiate a chromosome walk in rice and, in turn, markers identified in rice were mapped in wheat until two rice markers were found that closely flanked the target gene in wheat. The underlying assumption is that the rice sequence spanning the two flanking markers contains the rice homologue of the trait that is being analyzed in wheat. Depending on the physical size of the region, the rice sequence should therefore yield one or multiple candidate genes that can be functionally characterized. This strategy was used by Moore and colleagues in their early efforts to isolate the wheat *Ph1* gene, which controls homologous pairing (Foote et al., 1997). The pitfall of this approach is that, since colinearity between rice and wheat is hardly ever perfect, the target gene may have been subjected to a rearrangement in either the wheat or the rice lineage. This was the case for the *cdc2* genes, which are hypothesized to underlie the *Ph1* phenotype in wheat but are absent in the orthologous rice region (Griffiths et al., 2006). A walk in rice to isolate the barley gene *Rpg1*, a resistance gene for stem rust caused by *P. graminis* f. sp. *tritici*, also came to a halt with the absence of a *Rpg1* orthologue in rice (Han et al., 1999).

Chromosome walking in rice to isolate genes in wheat became less laborious as the rice genomic sequence became available, but remained a hit-or-miss process. This was particularly true for the isolation of genes that undergo rapid reorganization such as disease resistance genes. This prompted wheat researchers to start the enormous endeavor of generating BAC libraries for wheat. The first libraries were generated for the diploid wheats, *T. monococcum* and *Ae. tauschii*. The *T. monococcum* library, developed from the genotype DV92, consisted of 276,000 clones with an average insert size of 115 kb and covered the genome at 5.6X (Lijavetzky et al., 1999). The first *Ae. tauschii* library, developed from genotype Aus 18913, consisted of 144,000 clones with an average insert size of 120 kb, representing 4.2X genome coverage (Moullet et al., 1999). Several other *Ae. tauschii* BAC libraries, totaling 300,000 clones and representing 12.8X coverage, have since been constructed of the accession AL8/78 (http://www.plantsciences.ucdavis.edu/Dubcovsky/BAC-library/ITMIbac/ITMIBAC.htm). In polyploid wheat, a 5X BAC library was generated for the tetraploid *T. turgidum* ssp. *durum* cv. Langdon (516,000 clones) (Cenci et al., 2003) and a 9.3X library for the hexaploid *T. aestivum* cv. Chinese Spring (1,200,000 clones) (Allouis et al., 2003). A number of other diploid, tetraploid, and hexaploid wheat libraries representing fewer genome-equivalents are also available (Liu et al., 2000; Nilmalgoda et al., 2003; also see http://www.plantsciences.ucdavis.edu/Dubcovsky/BAC-library/ITMIbac/ITMIBAC.htm).

Positional cloning has focused on genes that are located in the more gene-rich high-recombinant regions of the wheat genome. To date, four disease resistance genes, three vernalization response genes, a domestication-related gene, and a homologous pairing-control gene have been isolated and validated (Faris et al., 2003; Feuillet et al., 2003; Huang et al., 2003; Yan et al., 2003, 2004, 2006; Yahiaoui et al., 2004; Griffiths et al., 2006; Simons et al., 2006; Cloutier et al., 2007).

Disease resistance genes

The four disease resistance genes that have been isolated to date include three conferring resistance to leaf rust (caused by *P. triticina* Eriks.), *Lr1*, *Lr10*, and *Lr21*, and one conferring resistance to powdery mildew [caused by *Blumeria graminis* (DC) E.O. Speer f. sp. *tritici*], *Pm3b*. Isolation of *Lr21*, the first disease resistance gene cloned in wheat, was facilitated by the high genetic-to-physical distance ratio in the region surrounding *Lr21*. Furthermore, a RFLP marker was available at the start of the cloning experiment that cosegregated with the resistance phenotype and which was, in fact, the resistance gene. The strategy used in the isolation of *Lr10* was much more typical and involved screening a large mapping population for recombination events in the region of interest, using a combination of marker development, contiging, and genetic mapping to narrow down the target region, sequencing of the target region and, finally, confirming the identity of the candidate genes by mutation analysis and/or transformation. The isolation procedures used for *Lr21*, *Lr10*, *Lr1*, and *Pm3b* are described later and provide examples of how pitfalls, such as the lack of polymorphism for a marker expected to be closely linked to the target gene and the absence of a BAC library of the resistant cultivar, can be managed. The four studies also demonstrate the different techniques that can be used to confirm the identity of the resistance gene candidates. Validation methods include resistance expressed in susceptible cultivars after stable or transient transformation with the candidate resistance gene, breakdown of resistance in lines that have undergone virus-induced gene silencing (VIGS) of the resistance gene after regions of the candidate gene were introduced as double-stranded RNA, and comparative analysis of the sequence of the candidate gene in several independent mutants that have lost the resistance following irradiation or treatment with a mutagen such as ethylmethylsulfonate (EMS).

Lr21

Gene *Lr21* originated from *Ae. tauschii* and was first incorporated in the wheat cultivar Thatcher in the 1970s (Rowland and Kerber 1974). It was later shown to be allelic to *Lr40*, another *Ae. tauschii* gene employed in enhancing the resistance of bread wheat to leaf rust (Huang and Gill 2001). Gene *Lr21* was an ideal candidate for map-based cloning. The gene mapped to the distal region of chromosome arm 1DS, which was gene-rich and highly recombinogenic. Furthermore, the introgressed *Ae. tauschii* fragment that carried *Lr21* provided increased polymorphism levels which facilitated mapping. Linkage analysis had already shown that the RFLP marker *XksuD14* was closely linked with *Lr21*, and fine mapping in a population of 520 F_2 plants placed *XksuD14* 0.1 cM distal to *Lr21*. The 3′ and 5′ ends of the KSUD14 sequence cosegregated with *Lr21*. The KSUD14 sequence was used to isolate a cosmid clone from an *Lr21*-containing *Ae. tauschii* accession. The 43-kb insert contained seven predicted genes, one of which was a nucleotide-binding site (NBS)–leucine-rich repeat (LRR) type gene with homology to KSUD14. Transformation experiments confirmed that KSU14 was, in fact, the *Lr21* gene (Huang et al., 2003).

Lr10

The leaf rust resistance gene *Lr10* originated from common wheat and is located in the distal region of chromosome arm 1AS. A screen of near-isogenic lines carrying different leaf rust resistance genes in a Thatcher background with a probe that encoded a serine–threonine protein kinase was the first step that ultimately led to the isolation of the *Lr10* gene. The serine–threonine protein kinase was a member of a multigene family, but a unique fragment was identified in the Thatcher line carrying *Lr10*. This fragment, designated *Lrk10*, cosegregated with *Lr10* in a population of 128 F_2 plants developed from the cross between the resistant near-isogenic line Thatcher*Lr10* and the susceptible cultivar Frisal (Feuillet et al., 1997). The *Lrk10* gene was subsequently shown in a larger population to be located 0.8 cM distal to *Lr10* and was used, together with the marker cMWG645 which is located 1.2 cM proximal to *Lr21*, to identify recombination events in the 2-cM region spanning *Lr10* (Stein et al., 2000). Ninety-six recombination events were found in 3,120 F_2 plants. The use of closely linked flanking markers, identified in a population of a few hundred gametes, to screen a population of a few thousand gametes for individuals that carry a recombination event in the region of interest is standard procedure in positional cloning and limits fine-mapping efforts to informative plants only.

A marker that cosegregated with *Lr10* in the fine-mapping population was used to screen the *T. monococcum* DV92 BAC library, and two overlapping BAC clones totaling 270 kb were identified. One end of the contig mapped distally of *Lr10*, but no proximal marker was identified. To extend the contig proximally, the most proximal of the two BAC clones was sequenced at low redundancy. Low-pass sequencing is considered standard for identifying low-copy sequences that can be mapped onto the genetic map and used for contig extension. In wheat, BAC-end sequences, the markers traditionally used in a chromosome walk, have limited use due to the high-repeat content of the genome. A low-copy marker obtained by Stein et al. (2000) from the sequence information identified a new BAC clone that extended the contig by 80 kb to form a 350-kb contig that spanned *Lr10*. Some 211 kb of *T. monococcum* DV92 was shotgun-sequenced to 8X redundancy and shown to carry five putative genes, including two resistance gene analogues, *rga1* and *rga2* (Wicker et al., 2001).

A potential problem with map-based cloning in a line that is not known to have the phenotype of interest is that the gene may be absent from that line. Haplotype analysis showed that this was not the case for DV92. The identity of *rga1* as *Lr10* was demonstrated by the identification of three independent mutations in *rga1* but not *rga2* in EMS mutants that had lost resistance to an avirulent race of leaf rust carrying *AvrLr10*. Transformation of the susceptible wheat cultivar Bobwhite with the *rga1* gene conferred resistance and confirmed that *rga1* encoded *Lr10*.

Lr1

The leaf rust resistance gene *Lr1* present in common wheat germplasm had been mapped to the distal region of chromosome arm 5DL (Feuillet et al., 1995). Saturation of this region with SSR and RFLP markers from 5DL and from orthologous barley and oat regions in two fine-mapping populations, consisting of 2,826 and 832 F_2 plants, delimited the location of *Lr1* to a 0.16-cM region. *Lr1* was flanked proximally by *Xpsr567* at 0.04 cM and distally by *Xabc718* at 0.12 cM (Ling et al., 2003). Screening of *Ae. tauschii* BAC libraries with PSR567 led to the isolation of the partial resistance gene analogues 567A, 567B, and 567C (Ling et al., 2003) and three additional low-copy markers (Cloutier et al., 2007). These markers were subsequently used to screen a BAC library of the hexaploid wheat cultivar Glenlea (Nilmalgoda et al., 2003), which contains the *Lr1* resistance gene. A second round of screening was carried out with markers developed from the positive BAC clones. All BAC clones identified during the two rounds of screening were fingerprinted and hybridized with PSR567. Additional sequence information was obtained from Glenlea for PSR567A, B, and D, which were originally cloned from *Ae. tauschii* and were non-polymorphic in the mapping population. The markers used in the BAC screens, together with the newly developed PSR567 markers, were used for genetic mapping in 400 F_1-derived doubled haploid lines and in a recombinant inbred population, also of some 400 lines. The *Lr1* gene was shown to cosegregate with one of the *Xpsr567* fragments, designated RGA567–5. Two overlapping BAC clones containing RGA567–5 were shotgun sequenced. Four open reading frames were identified, three of which were separated from *Lr1* by recombination events. Transformation of the susceptible cultivar Fielder, which lacks *Lr1*, with RGA567–5 confirmed its identity as *Lr1*. In a second confirmation experiment, *Lr1* activity in the resistant cultivar Thatcher*Lr1* was down-regulated using VIGS, resulting in virulence of the leaf rust *AvrLr1* gene.

Pm3

At least 10 alleles have been identified for the powdery mildew resistance gene *Pm3*. The gene had been mapped to the distal region of wheat chromosome arm 1AS, flanked by the markers WHS179 and BCD1434 (Hartl et al., 1993; Ma et al., 1994). In an F_2 population of 1,340 plants, WHS179 mapped 0.9 cM proximal to *Pm3b*. The end clone of a BAC, identified with BCD1434, mapped 3.9 cM distal to *Pm3b* (Yahiaoui et al., 2004). Existing wheat maps yielded a SSR marker, PSP2999, that mapped 0.07 cM distal to *Pm3b* and was used as the starting point for a chromosome walk. Walking was conducted simultaneously in *T. monococcum* accession DV92 and in *T. durum* cultivar Langdon. Two cycles of low-pass sequencing of selected BAC clones, mapping of low-copy sequences identified within the BACs, and rescreening of the BAC libraries led to a contig in *T. durum* that spanned the *Pm3b* locus. The physical-to-genetic distance for that region was shown to be 900 kb cM^{-1}. The contig contained three resistance-gene-like (RGL) sequences, designated *TdRGL-1*, *TdRGL-2*, and *TdRGL-3*, which were members of a large gene family on the wheat group 1 chromosomes (Yahiaoui et al., 2004). Using primers to conserved regions, five RGL genes were amplified from chromosome 1A from a near-isogenic line containing *Pm3b*. Four of these were mapped, but none cosegregated with *Pm3b*. No specific markers could be developed for the fifth RGL gene. Differences in gene content in the *Pm3* region between *T. monococcum*, *T. durum*, and *Pm3*-resistant and *Pm3*-susceptible *T. aestivum* lines, identified through a combination of sequencing and hybridization experiments, led to the isolation of a DNA fragment that cosegregated with *Pm3b* and was 100% identical to the RGL gene that had not been mapped previously. The identity of this gene was confirmed both by analysis of a γ-irradiation mutant that had become susceptible to powdery mildew due to a single base-pair deletion in the *Pm3* gene, and a transient single-cell assay showing reduced fungal growth in cells bombarded with the *Pm3* candidate gene.

Genes involved in adaptation

Five genes involved in adaptation have been isolated using a map-based cloning approach in wheat. This includes the three vernalization response genes *VRN-1*, *VRN-2*, and *VRN-3* (as elaborated in Chapter 3), the gene that confers the spelt phenotype to wheat, *Q*, and the gene that controls homologous pairing in polyploid wheat, *Ph1*. In contrast to the isolation of the disease resistance genes, which were rarely conserved between wheat and rice, the map-based cloning strategies employed for the isolation of *VRN-1*, *VRN-2*, *VRN-3*, *Q*, and *Ph1* relied heavily on the use of the rice genomic sequence as a source of markers.

VRN-1, VRN-2, and VRN-3

Vernalization is the requirement for a period of cold to accelerate flowering initiation. It is a mechanism to protect the cold-sensitive meristems from damage during the winter. In the diploid *T. monococcum*, vernalization response is regulated mainly by two genes, *VRN-1* and *VRN-2*: *VRN-2* is located distally on chromosome arm 5AL in the region translocated from 4AL (Dubcovsky et al., 1998); *VRN-1* is located more proximally on 5AL. In hexaploid wheat, the main vernalization gene is *VRN-1* with homoeoloci on chromosome arms 5AL, 5BL, and 5DL. Spring habit is dominant at the *VRN-1* loci, but recessive at the *VRN-2* loci. Because winter habit is ancestral, the *VRN-2* genes can confer spring habit in hexaploid wheat only if homozygous recessive at the *VRN-A2*, *VRN-B2*, and *VRN-D2* loci. This explains why phenotypic variation at the *VRN-2* locus has not been observed in hexaploid wheat. A third vernalization response locus, *VRN-4*, was first reported in the 1960s and located on wheat chromosome arm 7BS (Law 1966). This gene was identified in a Chinese Spring ('Hope 7B') substitution line. Because the presence of *VRN-4* in the cultivar Hope could not be confirmed in a subsequent study by Goncharov and Gaidalenok (1994), it was suggested that *VRN-4* might, in fact, be *VRN-1* that had been translocated from the group 5 chromosomes to 7BS during the development of the Chinese Spring (Hope 7B) substitution line. Isolation of the *VRN-4* gene showed that this gene is indeed different from *VRN-1*. Based on its orthology to the barley *VRN-H3* gene, the wheat *VRN-4* gene was renamed *VRN-3* (McIntosh et al., 2007).

Low-density genetic mapping had shown the *VRN-1* gene to be flanked distally by WG644 and proximally by CDO708. These markers were used to identify the orthologous region in rice, which was subsequently used as a source of markers for the *VRN-1* region. Two markers were used to screen an F_2 population of 3,095 *T. monococcum* F_2 plants for recombination events surrounding *VRN-1* (Yan et al., 2003). All rice genes present in the region spanned by those two markers were mapped in the population of informative recombinants. The recombinant progeny were also test-crossed to fully classify their genotype at the *VRN-1* locus. Two genes, *AP1* and *AGLG1*, were completely linked to *VRN-1* and were used to screen a *T. monococcum* DV92 BAC library. Additional screens were carried out with markers that closely flanked *VRN-1*, resulting in three *T. monococcum* contigs. One of the contigs contained *AP1* and two other markers that flanked *VRN-1* proximally at 0.02 cM; a second contig contained *AGLG1*; and the third contig contained markers that flanked *VRN-1* distally, with the closest marker located 0.02 cM from *VRN-1*. The two gaps in the *T. monococcum* contig were bridged by contigs in rice and sorghum. Comparative sequence analysis of the orthologous wheat, rice, and sorghum regions showed, with the exception of a single gene duplication in sorghum and wheat compared to rice, perfect colinearity. No new genes were identified and *AP1* and *AGLG1* remained the best candidates for *VRN-1*. Expression profiling and sequence variation identified in different *VRN-A^{m}1* alleles identified *AP1* as the most likely candidate for *VRN-1* (Yan et al., 2003).

A similar methodology was used for cloning the *VRN-A^{m}2* gene in the distal region of 5A^{m}L. Two markers that flanked the *VRN-2* gene in a low-resolution map were used to identify recombinant plants from a large (5,698 gametes) mapping population (Yan et al., 2004). Physical mapping was conducted simultaneously in *T. monoccoccum*, barley, and rice. Markers developed from the BAC clones were mapped in the recombinant population, which was also phenotyped for vernalization response. This resulted in *VRN-2* being delimited to a 0.04-cM interval, spanned in *T. monococcum* by four BAC clones that totaled some 440 kb. Sequencing and annotation of these BAC clones identified one pseudogene and eight genes, three of which were completely linked to *VRN-2*. Two of the genes, designated *ZCCT1* and *ZCCT2*, were the result of a duplication that occurred some 14 MYA and showed homology to the *constans* (*CO*) and *CO*-like proteins in Arabidopsis; *CO* is the key gene in the Arabidopsis photoperiod pathway, making these two genes good candidates for *VRN-2*. No function could be assigned to the third gene. A phylogenetic analysis of the *ZCCT* genes and their homologues in the A genome of tetraploid wheat, in barley, rice, and Arabidopsis suggested that the lineage that gave rise to the *ZCCT* genes originated in the grasses and that the *ZCCT* genes most likely arose as a cold-adaptation response in temperate cereals. Messenger RNA of *ZCCT1* was detected in the leaves and apices of unvernalized plants and gradually decreased during vernalization; *ZCCT2* had a similar expression pattern in the leaves but could not be detected in the apices, which are the critical meristems for transition from vegetative to reproductive phase. This made *ZCCT1* the preferred candidate for *VRN-2* (Yan et al., 2004).

Sequence analysis of the *ZCCT1* gene in winter and spring *T. monococcum* accessions showed that some of the spring *ZCCT1* alleles carried a single base-pair mutation that resulted in the substitution of an amino acid that was conserved in all ZCCT and CO-like proteins analyzed. In other spring accessions, the entire *ZCCT1* gene had been deleted. The deletion of the *ZCCT1* gene also appeared to be the cause of differentiation between winter and spring habit at the barley *VRN-H2* locus. The identity of *ZCCT1* as *VRN-2* was validated using an RNAi transgenic approach. Down-regulation of *ZCCT1* in hexaploid wheat by transformation with an RNAi construct that contained part of the *T. monococcum* *ZCCT1* gene accelerated flowering. Transgenic plants had reduced *ZCCT1* and enhanced *AP1* levels, consistent with the hypothesis that *ZCCT1* is a repressor that is derepressed by vernalization (Yan et al., 2004).

The cloning of *VRN-3* represents an example of how comparative information can be used to identify candidate genes for the trait of interest (Yan et al., 2006). The *VRN-B3* locus was mapped to a 6-cM interval in a population of 82 recombinant substitution lines. The most closely linked marker was ABC158, which mapped 1 cM proximal to *VRN-B3*. The ABC158 orthologue in rice was located 50 kb proximal to *Hd3a*; *Hd3a* is orthologous to the Arabidopsis gene *FLOWER-*

ING LOCUS T (*FT*), and both genes are an important part of the photoperiod response pathway (Kojima and Ogihara 1998). To test *FT* as a candidate for *VRN-B3*, a PCR-based marker was developed based on the available sequence information for the barley *FT* orthologue. The wheat *FT* orthologue cosegregated with *VRN-B3*. Cosegregation between *FT* and *VRN-H3* was also observed in two low-resolution barley mapping populations and also in a high-density mapping population of 800 F_2 plants. Physical mapping using *FT* and closely linked markers was conducted in barley and resulted in three contigs separated by two gaps. One representative BAC clone for each of the contigs was sequenced. The genetic map location of two markers developed from the BAC sequence narrowed the location of *VRN-H3* down to a 0.2-cM interval that contained only the *FT* gene. Spring habit in the Chinese Spring (Hope 7B) substitution line was caused by the presence of a long terminal repeat (LTR) retrotransposon that had inserted into the *FT* promotor. The *FT* transcript levels in winter plants were low and were upregulated by vernalization. In spring plants, *FT* transcript levels increased with development. The *FT* gene's identity as *VRN-3* was confirmed by transgenic analysis (Yan et al., 2006).

Q

The *Q* gene confers the square-headed and free-threshing phenotype of domesticated wheat and also affects some other domestication-related characters. Phenotypic analysis of CS deletion lines showed that *Q* was located in a submicroscopic region on 5AL defined by the terminal deletion lines 5AL-7 and 5AL-23 (Faris and Gill 2002). Differential cDNA and AFLP display on the 5AL-7 and 5AL-23 lines identified a number of markers located in that region. Mapping of these markers in a low-resolution mapping population of 190 F_2 plants delineated the *Q* region to 2 cM. The marker most closely linked to *Q*, at 0.7 cM, was used to initiate a chromosome walk, consisting of multiple cycles of screening the *T. monococcum* BAC library, sequencing of selected BAC clones, marker development, and genetic mapping of these markers. The result was a 263-kb contig (Faris et al., 2003). The closest distal marker to *Q* was located at 0.2 cM, but no proximal markers were identified. Three markers spanning approximately 70 kb cosegregated with *Q*. One of the markers was the orthologue of the *Arabidopsis* floral homoeotic gene *APETALA2* (*AP2*).

Since *Q* had been hypothesized to be a major regulatory gene for flower development, *AP2* was considered a good candidate for *Q*. Comparison of a Chinese Spring (*Q*) EST for *AP2* present in Genbank with the sequence of the *T. monococcum* (*q*) *AP2* gene identified at least two amino acid differences, which could be critical to *AP2* activity. Twenty-six fast neutron mutants that had lost the square-headed phenotype were homozygous or hemizygous for deletions in the *AP2* gene. The *Q* gene is ineffective in the hemizygous condition and plants with a single functional *AP2* gene have the speltoid phenotype. The region spanned by one of the deletions confirmed that the 260 kb contig indeed spanned the *Q* region (Faris et al., 2003). Validation of *AP2* as the gene that underlies *Q* was accomplished by isolating the full-length *AP2* gene from Chinese Spring wheat and comparing the Chinese Spring sequence with homologues isolated from Chinese Spring EMS mutants. All three EMS mutants with speltoid heads carried single base-pair mutations resulting in either an amino acid substitution or affecting intron–exon splice sites. The amino-acid substitution greatly reduced the ability of the *AP2* protein to form a homodimer (Simons et al., 2006). The identity of *AP2* as *Q* was also confirmed by transgenic analysis.

Ph1

The pairing control gene, *Ph1*, limits pairing of chromosomes in tetraploid and hexaploid wheats to homologous chromosomes (Riley and Chapman 1958). It is thought that pairing control originated in wheat following polyploidization, because *Ph1* activity is absent in the diploid progenitors (Chapman and Riley 1970). The isolation of *Ph1* was complicated by the fact that no phenotypic variation existed at this locus. Attempts to induce variation using EMS mutagenesis were also unsuccessful (Wall et al., 1971).

In 1977, the *Ph1* gene was mapped to the long arm of chromosome 5B using an X-ray mutant, designated *Ph1b* (Sears 1977). The mutant carried a deletion that included *Ph1* and was later shown to encompass some 70 Mb of DNA (Gill et al., 1993b). Using additional mutant lines, the *Ph1* region was subsequently reduced to less than 3 Mb in wheat (Gill et al., 1993b) and to an orthologous 400-kb region in rice, spanned by the markers *Xrgc846* and *Xpsr150A* (Roberts et al., 1999). Synteny between wheat and rice was generally conserved in the *Ph1* region (Foote et al., 1997), but mapping was hampered by the technical difficulty of hybridizing genes from rice, which diverged some 50 MYA from wheat, to the large wheat genome. Therefore, *B. sylvaticum*, a Pooid species 35 MY distant from wheat, was used as an intermediate (Foote et al., 2004). *Brachypodium sylvaticum* BAC clones were identified with rice probes, sequenced at low redundancy, and used as a source for the development of new markers. The *Brachypodium* markers were used for mapping, and also for screening a Chinese Spring bread wheat BAC library (Griffiths et al., 2006). This resulted in five BAC contigs of 500 kb, 300 kb, 1.2 Mb, 1 Mb, and 200 kb. Thirty-three wheat BACs spanning the 5 contigs were sequenced. This region contained 34 genes. Expression profiling of all 34 genes failed to show variation in lines with and without *Ph1*. Homology with genes of known function suggested that the *cdc2* genes, which affect chromosome condensation, were the best candidates for *Ph1*. The *cdc2* genes form a gene cluster on the wheat group 5 chromosomes with at least four members on chromosome 5B, one of which is chromosome-specific. The *cdc2* gene cluster on 5B was interrupted by an insertion from the subtelomeric region from chromosome arm 3AL. The authors concluded that a tandem array of a 2.3-kb repeat unit, present within this 3AL segment, together with the *cdc2* genes fulfilled the functional criteria for *Ph1* activity. This was supported by the fact that the 5B-specific *cdc2* gene and 2.3-kb array were present in all tetraploid wheats tested, but not in *Ae. speltoides*, the diploid species most closely related to the presumably extant B-genome progenitor. Validation of the *Ph1* candidates is in progress.

PHYSICAL MAPPING IN HEXAPLOID WHEAT

Hexaploid wheat is grown on over 95% of the worldwide wheat-growing area. Breeders are interested in utilizing its sequence to accelerate genetic improvements for the growing demands for high-quality food produced in an environmentally sensitive, sustainable, and profitable manner. A first step in accessing the genome sequence lies in the establishment of a physical map. However, the construction of a physical map of the hexaploid wheat genome is a daunting task due to its size, prevalence of repetitive DNA, and presence of three homoeologous genomes. To establish a physical map with 15-fold genome coverage, about 2.1 million BAC clones with an average size of 120 kb would need to be fingerprinted, assembled into contigs, and anchored to the genetic maps. While fingerprinting millions of BAC clones is feasible using high-information-content fingerprinting techniques (Luo et al., 2003; Meyers et al., 2004), it is not clear if the existing technology permits their specific assembly into contigs that faithfully represent the individual chromosomes. Moreover, anchoring the homoeologous BAC contigs to the genetic maps would be extremely time-consuming, would increase costs significantly, and would require the development of a large set of genome-specific markers that are not yet available in wheat. Thus, to handle the super-sized hexaploid wheat genome, some kind of complexity reduction would be very useful. Until now, two main strategies have been considered. The first one involves the construction of physical maps from diploid progenitors of hexaploid wheat such as *Ae. tauschii*, and the second one is to construct BAC libraries from flow-sorted chromosomes or chromosome arms and generate physical maps for each chromosome arm individually.

Constructing physical maps in the diploid progenitors of hexaploid bread wheat avoids problems due to polyploidy, but this approach suffers from some limitations. While *Ae. tauschii* represents a good model for the D genome of hexaploid wheat, as it was involved in hybridization with tetraploid emmer wheat 8,000 to 10,000 years ago, and while the A genome donor species, *T. urartu*, is also known, the diploid progenitor(s) of the B genome of hexaploid wheat has not been identified yet and is possibly extinct (Feldman 2001). Furthermore, the hexaploid wheat genome does not correspond structurally and functionally to the sum of its three ancestral diploid genomes because of genome rearrangements triggered by polyploidization (Feldman and Levy 2005) and translocations that have occurred since the polyploidization events. Therefore, physical maps of the diploid species related to the A, B, and D genomes of hexaploid wheat are not perfect substitutes for the physical map of the 21 chromosomes of hexaploid wheat. Finally, the genomes of diploid progenitors are still quite large (1C = ~ 5,000 Mb; i.e., twice the size of the maize genome), making the construction of their physical maps a huge exercise. Thus, even if physical maps of diploid wild ancestors can be used as a framework to assist in assembling physical maps and support positional cloning in hexaploid wheat, they will not provide the full breadth of information that breeders need from hexaploid wheat to accelerate its improvement.

The complexity of constructing a physical map of the hexaploid wheat genome can be reduced by using a chromosome-based strategy (Doležel et al., 2007). In this approach, BAC libraries are constructed from isolated hexaploid wheat chromosomes or chromosome arms and are used for the construction of individual physical maps. As the relative size of wheat chromosomes and their arms ranges from 3.6% to 5.8% and from 1.3% to 3.4% of the genome, respectively (Gill et al., 1991a), with the largest chromosome representing 1,000 Mb, this approach offers a significant reduction in complexity. The development of physical maps in a stepwise manner, one chromosome (arm) at a time, also provides the opportunity to divide the task between laboratories in an internationally coordinated effort.

Constructing subgenomic BAC resources

The critical first step toward the development of subgenomic BAC resources is the dissection of the wheat genome into individual chromosomes and/or chromosome arms. The construction of BAC libraries requires micrograms of high-molecular-weight DNA that must be isolated from millions of chromosomes. Flow cytometry is the only method capable of isolating such large amounts of chromosomes with high purity and intact DNA. Flow cytometric chromosome analysis and sorting (flow cytogenetics) involves the measurement of fluorescence of mitotic chromosomes constrained to flow in a single file within a fluid stream. Chromosomes stained with a DNA-binding fluorochrome are classified according to their relative DNA content (the emitted fluorescence is approximately proportional to chromosome size) and then displayed as a histogram of relative DNA content, or flow karyotype. Ideally, each chromosome is represented by a single peak on the flow karyotype. To sort desired particles, the fluid stream is broken into droplets. A droplet carrying the particle of interest is electrically charged, deflected during a passage through an electrostatic field, and collected in a suitable vessel (Doležel et al., 2004).

There are three basic prerequisites for flow cytogenetics to be suitable for generating subgenomic BAC libraries: (i) the availability of samples (liquid suspensions of intact chromosomes) suitable for flow cytometry; (ii) the possibility of discerning specific chromosomes on a flow karyotype; and (iii) the preservation of high-molecular-weight DNA during sample preparation and sorting. In the first report on chromosome sorting in wheat (Wang et al., 1992), the samples were prepared by disrupting protoplasts prepared from suspension of cultured cells. However, this approach suffered from genetic instability of cultured cells and low chromosome yield (Schwarzacher et al., 1997). In the same year, Doležel et al. (1992) published a high-yielding protocol for the preparation of suspensions of intact chromosomes by mechanical homogenization of formaldehyde-fixed root tips. Major advantages of the protocol include the ease with which seedlings can be handled, the karyological stability and high degree of cell cycle synchrony in root tips, and the mechanical stability of isolated chromosomes. The protocol, including its modification in which the fixation step was omitted (Lee et al., 1997; Gill et al., 1999), stimulated the recent progress in flow cytogenetics of wheat (Vrána et al., 2000; Kubaláková et al., 2002).

The availability of suspensions of intact chromosomes facilitated the analysis of wheat chromosomes using flow cytometry (Lee et al., 1997; Gill et al., 1999; Vrána et al., 2000). The chromosome content of individual peaks on the wheat flow karyotype was determined by Vrána et al. (2000), who found that only chromosome 3B formed a discrete peak and could be sorted as a single chromosome. The other chromosomes formed composite peaks I, II, and III (Fig. 15.3a). Thus, with the exception of chromosome 3B, flow-sorting techniques at that time could not be used to dissect the wheat genome into individual chromosomes. To overcome this obstacle, Kubaláková et al. (2002) screened a large set of wheat cultivars and landraces to detect structural chromosome changes and chromosome polymorphisms that would allow additional chromosomes to be separated, but the observed alterations in flow karyotypes were small and did not provide a reliable basis for sorting other chromosomes.

Kubaláková et al. (2002) then suggested that most wheat chromosome arms could be sorted from lines carrying telocentric chromosomes, or telosomes. Telosomes originate by centric fission of a chromosome and, after synthesis of new telomeres, they can be maintained as stable cytogenetic stocks. A complete set of telosome lines has been developed in Chinese Spring (Sears 1954). Chinese Spring has been chosen as a model genotype for wheat cytogenetics and genomics (Gill et al., 2004) and will be used by the International Wheat Genome Sequencing Consortium (IWGSC, http://www.wheatgenome.org) for the construction of the hexaploid wheat physical map. Due to their relatively small size (DNA content),

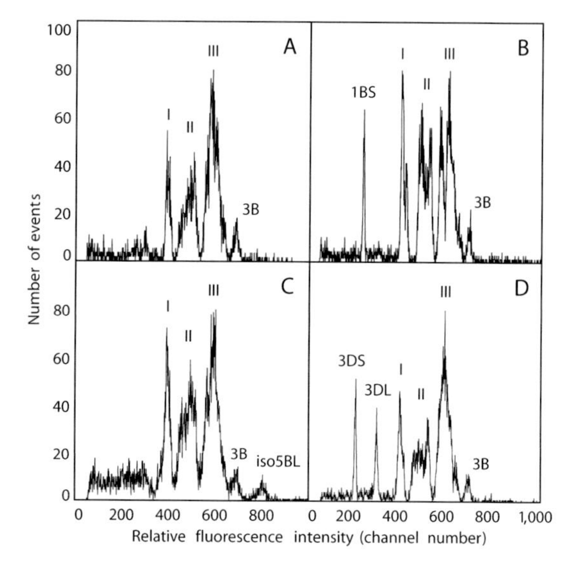

Fig. 15.3 Histograms of relative fluorescence intensities (flow karyotypes) obtained after flow cytometric analysis of DAPI-stained suspensions of mitotic chromosomes of hexaploid wheat. (a) The flow karyotype—Chinese Spring—consists of the chromosome 3B peak, the small composite peak I containing chromosomes 1D, 4D, and 6D, and the two large composite peaks II and III containing the remaining 17 chromosomes. (b) Flow karyotype obtained after the analysis of ditelosomic line Dt1BS of cultivar Pavon. The peak representing the short arm of chromosome 1B is well discriminated, which facilitates its sorting. (c) The analysis of a Chinese Spring line carrying an isochromosome for the long arm of chromosome 5B (iso5BL) results in a flow karyotype with an additional peak to the right of the peak of chromosome 3B. (d) The flow karyotype of the double ditelosomic 3D line (dDt3D) of Chinese Spring is characterized by additional peaks representing the short arm telosome 3DS and the long arm telosome 3DL, which facilitates their simultaneous sorting.

telosomes form extra peaks on the left side of flow karyotypes and do not overlap with peaks of other chromosomes (Fig. 15.3b). The only exceptions are the long arms of chromosomes 3B and 5B that overlap with peak I, and the long arms of 4A, 1B, and 7B, whose peaks are too close to peak I to be separated reliably. These arms, however, can be sorted from stocks carrying them as isochromosomes, that is, chromosomes with two identical arms that originate by centric fusion of two identical telosomes (Fig. 15.3c).

These advances made it possible to sort chromosomes at high speeds, typically 5 to 30 chromosomes per second, and to collect several hundred thousand chromosomes per day. Moreover, the use of double ditelosomic (dDt) lines (i.e., lines carrying both arms of a chromosome as pairs of telosomes) enabled the simultaneous sorting of the long and short chromosome arms (Fig. 15.3d). Depending on the size of the arm, the typical daily yield ranges from 100 to 200 ng DNA and, within several weeks, sufficient micrograms of DNA are available for BAC library construction. However, it was not clear that the DNA of flow-sorted chromosomes would be intact, a critical aspect, until Vrána et al. (2000) demonstrated that DNA of flow-sorted wheat chromosomes was of high molecular weight. Subsequently, Šimková et al. (2003) developed a protocol for preparing high-molecular-weight DNA from flow-sorted chromosomes, and this study opened a way to the construction of high-quality, chromosome (arm)–specific BAC libraries.

Construction of the first-ever chromosome-specific BAC library from a eukaryote genome, a BAC library specific for wheat chromosome 3B, was reported by Šafář et al. (2004). The library was constructed from 1.8 million 3B chromosomes from Chinese Spring (approximately 4 μg DNA) that were sorted in 18 working days. In total, 4,000 seeds were used to prepare 200 samples of chromosome suspensions for flow sorting. The library was constructed using a pIndigoBAC vector and the 67,968 BAC clones were ordered in 177 × 384-well plates. With an average insert size of about 103 kb, as estimated by pulse field electrophoresis, the library represents 6.2-fold coverage of chromosome 3B (995 Mb) (Šafář et al., 2004).

The reproducibility of BAC library construction from sorted chromosomes was confirmed by Janda et al. (2004), who constructed a library from chromosomes 1D, 4D, and 6D, that were sorted as a group from the composite peak I of the Chinese Spring flow karyotype (Table 15.2, Fig. 15.3a). While these experiments demonstrated the feasibility of developing genomic resources from sorted chromosomes, the inability to sort other chromosomes individually or in small groups necessitated the construction of libraries from chromosome arms. The first library from a wheat chromosome arm (1BS) was reported by Janda et al. (2006). With only 65,280 BAC clones, the library provided 14.5-fold coverage of the short arm. Similar libraries have

Table 15.2 BAC libraries that have been constructed from flow-sorted chromosomes of hexaploid wheat.

	BAC library (*Hin*dIII cloning site)						
Cultivar	Chromosome	Molecular Size (Mb)	Genome Fraction (%)	Number of Clones	Mean Insert Size (kb)	Chromosome Coverage	Reference
Chinese Spring	1D, 4D, 6D	1,964	11.6	87,168	85	3.4x	Janda et al. (2004)
Chinese Spring	1D, 4D, 6D	1,964	11.6	181,632	102	8.4x	Šafář et al. (pers. comm.)
Chinese Spring	3B	993	5.9	67,968	103	6.2x	Šafář et al. (2004)
Hope	3B	993	5.9	92,160	78	6.0x	Šimková et al. (pers. comm.)
Pavon	1BS	314	1.9	65,280	82	14.5x	Janda et al. (2006)
Chinese Spring	3AS	360	2.1	55,296	80	10.9x	Šimková et al. (2007)
Chinese Spring	3AL	468	2.8	55,296	106	10.2x	Šafář et al. (pers. comm.)
Chinese Spring	3DS	321	1.9	36,864	110	11.0x	Šafář et al. (pers. comm.)
Chinese Spring	3DL	449	2.7	64,512	105	12.2x	Šafář et al. (pers. comm.)
Chinese Spring	7DS	381	2.0	49,152	114	12.2x	Šimková et al. (pers. comm.)

been created from the short arms of chromosomes 3A, 3D, and 7D and from the long arms of 3A and 3D. Table 15.2 provides information about chromosome BAC libraries constructed in wheat so far.

Advantages of subgenomic BAC resources

The most obvious advantage of chromosome-based BAC libraries is their specificity. By applying these libraries for physical mapping and positional cloning, most of the difficulties due to homoeology are avoided (see below). The level of contamination of the libraries with other chromosomes depends on the accuracy of sorting. Although it is possible to sort wheat chromosomes that are 97% pure (Kubaláková et al., 2002), typically, purities in large-scale chromosome sorting experiments have varied from 88% to 91% (Janda et al., 2004, 2006; Šafář et al., 2004). The identity of the sorted particles is confirmed by fluorescence *in situ* hybridization (FISH) or primed *in situ* DNA labeling (PRINS) using probes and primers that provide chromosome-specific fluorescent labeling (Vrána et al., 2000; Šafář et al., 2004), as well as by screening of the BAC libraries with chromosome-specific markers (Janda et al., 2004, 2006; Šafář et al., 2004). The results show that chromosome-specific BAC libraries contain about 10% contaminating clones. However, the first physical mapping and positional cloning experiments conducted with the chromosome-specific BAC libraries (see below) indicated that this low level of contamination has no effect on the accuracy of the results and therefore does not compromise the value of the chromosome-specific BAC libraries.

In addition to simplifying the physical mapping by targeting single chromosomes, the use of chromosome BAC libraries offers important logistical advantages. With the number of clones ranging from 4×10^4 to 1×10^5 (ordered in 104 to 260 384-well plates), the libraries occupy limited freezer space, and maintenance (replication, pooling) and screening are easier and cheaper than for genomic BAC libraries that comprise 10 to 30 times more clones (1,000 to >2,600 384-well plates) (Allouis et al., 2003; Nilmalgoda et al., 2003; Ling and Chen 2005; Ratnayaka et al., 2005). Finally, the possibility of approaching the hexaploid wheat genome one chromosome at a time allows the establishment of a strategy for physical mapping that is based on international collaborations in which individual laboratories develop physical maps of specific chromosomes and chromosome arms for a reasonable cost and labor investment, with progress in each laboratory independent of the others.

An apparent limitation of the chromosome BAC libraries is the insert size, which is at the lower range of published genomic BAC libraries (Allouis et al., 2003; Nilmalgoda et al., 2003; Ling and Chen, 2005; Ratnayaka et al., 2005). Although the chromosome BAC libraries contain clones with inserts up to 200 kb, the presence of clones with inserts shorter than 50 kb compromises the average insert size (J. Šafář and H. Šimková, pers. comm.). Due to the small amount of starting DNA, the original protocol of chromosome BAC library construction involved only one DNA size-selection step (Šafář et al., 2004). Recently, Šafář and Šimková (J. Šafář and H. Šimková, pers. comm.) improved the protocol so that it is possible to make two size-selection steps even with small amounts of starting DNA. BAC libraries made using this protocol have average insert sizes of about 110 kb (Table 15.2).

While the use of the aneuploid lines from Chinese Spring is appropriate for the construction of a hexaploid wheat reference physical map, the development of physical maps at target loci from genotypes that carry specific traits of interest such as disease resistance will be needed in some cases. One option is to generate nongridded genomic BAC libraries from desired genotypes that are suitable for PCR screening (Ma et al., 2000; Isidore et al., 2005). These libraries do not need to have a high genome coverage and will generally be used only for a few screening steps to identify clones of interest. Another option is to generate chromosome-specific BAC libraries. If a physical contig spanning a region of interest is available from the Chinese Spring reference map, the second library can have a lower number of clones with shorter inserts. This makes BAC library construction less demanding and faster. Recently, Šimková and colleagues (H. Šimková, pers. comm.) created an ordered BAC library from 1×10^6 3B chromosomes of cultivar Hope in only 4 weeks, including chromosome sorting. This specific BAC resource is being used in map-based cloning of the *Sr2* gene (W. Spielmeyer, pers. comm.) after a contig of 1 Mb had been established on the Chinese Spring 3B reference physical map (C. Feuillet, unpublished data). However, this strategy is only applicable to chromosome 3B, and a group of 1D, 4D, and 6D chromosomes that can be sorted from any hexaploid wheat. To sort other chromosomes, telosomic stocks will have to be developed from specific genotypes. This may require less than 2 years (A. Lukaszewski, pers. comm.).

In addition to facilitating the construction of physical maps and map-based cloning, chromosome (arm)–specific BAC libraries aid in the development of genome-specific markers for marker-assisted selection. This is highly relevant for polyploid species such as wheat in which the presence of homoeologous sequences hampers the efficient development of genome-specific markers. It has been shown recently that BAC-end sequences from chromosome-specific BAC libraries can be used very efficiently to develop a large number of genome-specific markers for genetic and physical mapping in wheat (Paux et al., 2006) by using the junctions between repetitive elements (Devos et al., 2005b). The main advantage of these insertion site based polymorphism (ISBP) markers or repeat junction markers (RJMs) is that they are very specific, polymorphic, and distributed homogeneously along the chromosomes in contrast to EST or SSR markers that can show some bias in genomic distribution.

Chromosome-based approach offers more than subgenomic BAC libraries

Development of subgenomic BAC libraries has been the most attractive use of flow cytogenetics in wheat. However, other applications of sorted chromosomes deserve attention. The first one is the development of DNA markers from particular chromosomal regions for which BAC libraries are not yet available. One option is to create a short-insert DNA library from a small number of flow-sorted chromosomes, enrich the library for particular microsatellite motifs, and use microsatellite-containing DNA clones to develop SSR markers (Požárková et al., 2002). Although not yet explored in wheat, the results that were obtained in rye demonstrate the potential of this approach in species with complex genomes (Kofler et al., 2007). Another unexplored option is to isolate markers from chromosomal DNA using the DArT technology (Wenzl et al., 2004). Chromosome specificity of candidate markers can be ascertained prior to genetic mapping by PCR or hybridization on DNA from NT lines. An alternative is to perform PCR on a small number (100–500) of chromosomes that can be sorted in a short time (Požárková et al., 2002). A more powerful approach involves the use of DNA amplified from flow-sorted chromosomes as probes on DNA arrays for mass parallel mapping of DNA sequences (Šimková et al., 2008).

A marriage of flow and molecular cytogenetics has resulted in a powerful tool for cytogenetic mapping. Traditionally this has been done by FISH on mitotic metaphase spreads from root-tip meristems (Jiang and Gill 2006). The number of metaphases that can be analyzed limits the throughput, and the sensitivity and specificity of FISH are often compromised by the presence of cell wall and cytoplasmic debris. As flow-sorted chromosomes are free of cellular remnants and can be sorted onto a microscopic slide in large numbers, they are ideal targets for FISH and PRINS. This facilitates physical mapping of DNA sequences (Kubaláková et al., 2002, 2005) and detection of rare structural variants (Kubaláková et al., 2003). A further advantage of using flow-sorted chromosomes is that they can be stretched longitudinally to improve the spatial resolution of cytogenetic mapping up to 100-fold compared with mitotic metaphase chromosomes (Valárik et al., 2004).

Until now, FISH has been used in wheat mainly to determine the genomic distribution of DNA sequences, to identify individual chromosomes, and to study structural changes in chromosomes (Zhang et al., 2007). However, the development of physical contig maps will need cytogenetic mapping to ascertain the position of some BAC contigs, to determine their orientation, and to estimate contig gaps. In other crops, this has been done by FISH with BAC clones (Cheng et al., 2002; Budiman et al., 2004). Unfortunately, BAC FISH has been hampered in wheat by the prevalence of repetitive DNA, which has resulted in dispersed FISH signals (Zhang et al., 2004b; Janda et al., 2006). Thus, although FISH on flow-sorted chromosomes offers a powerful approach for cytogenetic mapping, its use to support the development of physical contig maps in wheat is hampered by the lack of suitable probes that localize to single loci.

Physical map of chromosome 3B—a case study

The availability of a BAC library specific for chromosome 3B (Šafář et al., 2004) provided an opportunity to test the feasibility of constructing a physical map of the hexaploid wheat genome using a chromosome-specific-based approach (Gill et al., 2004). To establish a physical map of chromosome 3B, high-information-content (HIFC) fingerprints were generated from the 67,968 BAC clones of the 3B BAC library using a modified SNaPshot protocol (Luo et al., 2003) and assembled into contigs using the FPC software (Soderlund et al., 2000). Using a single ABI 3730 XL capillary sequencer (Applied Biosystems, Foster City, California), it was possible to fingerprint all BAC clones of the 3B BAC library within 10 weeks. To date, the physical map consists of about 1,000 contigs that cover nearly 80% of the chromosome (Paux et al., 2008).

The most laborious task in the construction of a physical map is the anchoring of the fingerprint contigs to the genetic map. In most of the plant genome physical mapping projects published so far, anchoring has been performed through hybridization with RFLP and overgo probes that correspond to gene sequences and/or by PCR using microsatellites, cleaved amplified polymorphic site (CAPS), or sequence tagged site (STS) markers. To anchor the 3B physical contigs to the genetic map, PCR screening was performed with more than 2,000 molecular markers following two approaches. The first strategy was to use markers (SSR, EST) that have been located either on genetic or cytogenetic maps of chromosome 3B (Munkvold et al., 2004) to screen the BAC library and anchor the corresponding FPC contigs. The second approach comprised marker development from the BAC contigs and placement of the markers onto the genetic maps. To do that, 19,400 BAC-end sequences were generated representing a cumulative length of nearly 11 Mb (1.1% of the chromosome length) distributed among the contigs of chromosome 3B (Paux et al., 2006). The systematic identification of junctions between transposable elements allowed the development of retrotransposon-based insertion polymorphism (RBIP) markers (Flavell et al., 1998) that consist of PCR amplicons spanning the junctions. This type of marker has several advantages: (i) it is highly abundant since more than 80% of the wheat genome consists of transposable elements that are frequently nested within one another; (ii) it is mostly unique in the genome as there is a low chance that the same insertion event occurred at another locus; (iii) it is genome-specific since it originated from a specific chromosome sequence (Paux et al., 2006). More than 700 ISBP markers have been developed across chromosome 3B and have been used for anchoring the contigs on the genetic and cytogenetic maps (Paux et al., 2008). Finally, a minimum tiling path (MTP) has been established for chromosome 3B by selecting clones with minimal overlap. The MTP can now be used for structural and functional studies on chromosome 3B.

It is very cost-effective to use PCR-based methods to perform the anchoring of physical maps to genetic maps. However, with large genomes such as wheat, the number of PCRs required to identify all BAC clones bearing a single target marker sequence can become very high. To reduce significantly the number of reactions, several pooling strategies have been proposed, all of which rely on concentrating the BAC library into pools that represent overlapping groups of clones. The most powerful is six-dimensional pooling (Klein et al., 2000) that reduces by a factor of 384 the number of PCRs needed to unambiguously obtain the address of a BAC clone in a single experiment compared with nonpooling strategies (Yim et al., 2007). However, six-dimensional pools are expensive to produce and require automated platforms that are not available in every laboratory. Because of these limitations, lower dimensional pools, such as three-dimensional (3D) pools that consist of pools of 384-well plates, rows, and columns, are used often. One of the drawbacks of 3D-pools is an increased number of ambiguous BAC addresses after screening. Additional PCR reactions must then be performed to eliminate the ambiguity, which reduces the efficiency of the 3D pooling strategy. To increase efficiency and reduce the cost of the anchoring during construction of the 3B chromosome physical map, new software called *Elephant* (*Ele*ctronic *ph*ysical map *an*choring *t*ool) was developed (Paux et al., 2007). This freely available Perl script combines BAC contig information generated by FPC with results of BAC library pool screening to identify BAC addresses with a minimal number of PCR reactions. Using *Elephant* during the construction of the 3B physical map, we have shown that a one-dimensional pool screening can be sufficient to anchor a BAC contig while reducing the number of PCRs by 384-fold.

We can summarize several important features from these preliminary data on the construction of the first chromosome-based physical map of a hexaploid wheat chromosome:

1. The chromosome-based strategy is feasible and even with a single BAC library constructed at 8X redundancy, it is possible to generate contigs ranging from 300 kb to 4 Mb that are suitable for map-based cloning and that cover about 80% of the chromosome.
2. Systematic BAC-end sequencing of the MTP is a powerful approach to developing genome-specific markers from each of the wheat chromosomes and chromosome arms for contig anchoring and for breeding.
3. Clone contamination (10%) is inherent to the construction of BAC libraries from sorted chromosomes but does not interfere with the establishment of the physical contigs.
4. A single laboratory equipped with a capillary sequencer and basic robotic equipment can establish a physical map of a wheat chromosome, making it possible to complete the physical map of the 20 remaining chromosomes of hexaploid wheat as an international collaborative effort—one "stone" at a time.

ORGANIZATION AND EVOLUTION OF THE WHEAT GENOME

The C-value paradox was formulated in 1971 by C.A. Thomas to convey the rather perplexing absence of a correlation between genome size and organismal complexity (Thomas 1971). This mystery was resolved with the discovery of noncoding DNA although, even to date, many questions still remain regarding the role of repeated DNA in genome function and evolution. Sample sequencing in *Ae. tauschii* has indicated that the repeat fraction consists mainly of LTR retrotransposons (56%) and DNA transposons (14.5%). Tandem repeats (2.8%), non-LTR retrotransposons (2.6%), and miniature inverted-repeat transposable elements (MITES) (1.8%) make up a small proportion of the repeats. A further 22.8% of the repeat fraction consists of as yet unclassified repeats. Low-copy DNA comprises 8.4% of the genomic sequences and, of these, 2.5% has homology to known genes (Li et al., 2004).

A first indication that genes and repetitive DNA were not evenly distributed in the wheat genome came from cytological staining techniques such as C- and N-banding that stained the constitutive heterochromatic regions of a chromosome. These regions were shown to correlate with the presence of long arrays of tandem repeats (Flavell et al., 1987). Studies with overlapping sets of terminal deletion lines showed that gene density generally declined along the telomere–centromere axis (Werner et al., 1992; Gill et al., 1993a, 1996a,b; Delaney et al., 1995a,b; Mickelson-Young et al., 1995; Qi et al., 2004). More detailed information on the organization of genes and repeats within gene-rich regions was provided by the sequence analysis and annotation of BAC clones (Wicker et al., 2001, 2003b; San-Miguel et al., 2002; Yan et al., 2003; Gu et al., 2004; Kong et al., 2004). In most cases, these BACs were preselected to contain a gene of interest. A similar study using randomly selected BAC clones assessed how gene densities differ in gene-rich and gene-poor regions (Devos et al., 2005b; K.M. Devos, J.L. Bennetzen, and P. San Miguel, unpublished data). These studies confirmed that gene-rich regions generally contain higher gene numbers but also showed that considerable variation exists in gene densities within gene-rich regions, ranging from 0 genes per 100-kb sequence to 1 gene per 5 to 8 kb.

Organization of genes and repeats

Most of the information currently available on the organization of genes and repeats in the wheat genome is derived from regions that were sequenced as part of a map-based cloning effort. They thus represent either distal or interstitial chromosome regions where recombination is relatively high. The only data available on the organization of genes and repeats in proximal regions of the wheat genome come from a project that involved sequencing of randomly selected BAC clones that were subsequently mapped to wheat deletion bins (K.M. Devos, J.L. Bennetzen, and P. San Miguel, unpublished data).

To evaluate to what extent patterns of gene density and organization vary along the telomere–centromere axis, we can compare the composition of sequenced regions derived from telomeric, interstitial, and proximal chromosome bins. The relative gene density (RGD) for each of these bins is known from large-scale EST mapping and reflects whether the number of ESTs that map to a particular bin is proportional to the size of the bin (Conley et al., 2004; Hossain et al., 2004; Linkiewicz et al., 2004; Miftahudin et al., 2004; Munkvold et al., 2004; Peng et al., 2004; Randhawa et al., 2004). If genes were evenly distributed along a chromosome, all regions would have a relative gene density of 1. Therefore, values between 0.8 and 1.2, >1.5, and <0.5 are considered intermediate, high, and low gene densities, respectively. Of course, it should be taken into account that most of these bins are relatively large and that a bin of medium gene density most likely consists of a combination of high-, medium-, and low-density regions.

This is exemplified by the region defined by breakpoints at fraction lengths 0.55 and 0.76 on chromosome arm 5BL for which the RGD is 1.08 (Linkiewicz et al., 2004). In a smaller scale mapping study using a larger set of deletion lines, Gill et al. (1996a) showed that *Ph1* and three other markers were located in a submicroscopic bin at fraction length 0.55, five markers were mapped to the 0.55–0.59 region, three to the 0.59–0.75 region, and four to the 0.75–0.76 region. Translating these marker densities to RGDs led to values varying from 0.26 for the 0.59–0.75 region to >5 for the (sub)microscopic bins at fraction lengths 0.55 and 0.75. The organizational patterns of genes and repeats present in a few regions for which several hundred kilobases of precisely annotated sequence is available and are considered representative for distal, interstitial, and proximal chromosome locations are discussed below.

The 211-kb contig that spans the *Lr10* gene physically maps to the distal 14% of chromosome arm 1AS (Schachermayr et al., 1997). This region has a RGD of 3.7 (Peng et al., 2004). A total of five genes has been annotated in the 211-kb sequence (GenBank accession number AF326781). Three of the genes, one of which is a resistance gene analogue, are located within a 31-kb segment. A second cluster of two genes, a resistance gene analogue and a nodulin-like gene, encompass some 40 kb. The two gene islands are separated by more than 140 kb. Overall gene density in this 211-kb region was 1 gene per 42 kb, with gene densities reaching 1 gene per 10–20 kb in the gene islands.

The *VRN-2* gene maps to the distal 13% of chromosome arm 5AL, which has a RGD of 1.97 (Linkiewicz et al., 2004). Approximately 440 kb of contiguous sequence for that region contains eight genes and one pseudogene, indicating an overall gene density of 1 gene per 55 kb. The genes are organized in three islands, one with three genes and the other two islands each containing two genes. The three-gene island spans 15 kb, and the two-gene islands span 22 kb and 28 kb. The gene density within the islands is thus in the range of 1 gene per 5–24 kb. The distance between islands varies from 52 kb to 224 kb. The highest gene density found in this region was 1 gene per 5 kb, and the lowest gene density was 1 gene per 300 kb.

The *VRN-1* gene is located interstitially on chromosome arm 5AL in a segment defined by fraction lengths 0.68 and 0.78 (Sarma et al., 1998). The RGD in the larger 0.35–0.78 region is 1.18, which is considered average (Linkiewicz et al., 2004). Two overlapping BAC clones and three individual BAC clones, comprising a total of 550 kb, have been sequenced for that region. Nine genes and one pseudogene have been identified, resulting in an average density of 1 gene per 61 kb. Seven of the genes were organized in three islands with local gene densities of 1 gene per 3.4 kb, 6.2 kb, and 9.6 kb (SanMiguel et al., 2002; Yan et al., 2003; Fig. 15.2). The other two genes were found on separate BAC clones. Thus, within the 0.2-cM interval that spanned *VRN-1*, gene densities varied from being comparable to those in rice to what is expected to be an average gene density in wheat.

While the *VRN-1*, *VRN-2*, and *Lr10* regions were selected because of their agronomic importance, this was not the case for the 4DL13 region. The 4DL13 region is located on a segment of chromosome arm 4DL defined by fraction lengths 0.56 and 0.71 and has a RGD of 1.57. The sequenced region consists of nine BAC clones totaling 1 Mb. Of the five annotated genes, two clustered in a 14-kb region. The three remaining genes were separated by intergenic distances of 140 and 190 kb (A. Massa, J. Dvorak, P. Rabinowicz, and K.M. Devos, unpublished data). The overall gene density for the region is 1 gene per 200 kb.

To obtain representative examples of the gene organization in proximal chromosome regions, six BAC clones that mapped to centromeric bins on different chromosomes and had been annotated for gene content were randomly selected (K.M. Devos, P. San Miguel, and J.L. Bennetzen, unpublished data). The size of the bins varied from 27% to 45% of a chromosome arm. Relative gene densities were <0.8 in five of the bins and 1.16 in the 7BS centromeric bin. Three of the BAC clones contained no genes, and the other three contained one gene each.

These studies confirmed the overall trend seen in RGD values obtained from deletion mapping, that is, gene numbers decrease when moving from the telomere toward the centromere. Lower gene numbers are translated into fewer gene islands and larger intergenic distances. In the distal chromosome regions, most genes are clustered with two or three genes occupying 40 kb or less. Gene densities in these islands approach those observed in the 400-Mb rice genome (International Rice Genome Sequencing Project 2005). Relatively fewer of these islands are seen in the interstitial chromosome regions, and more genes are present as singletons. When gene islands are present, gene densities, however, remain in the order of 1 gene per 5 to 20 kb, irrespective of the location in the genome. In the proximal regions, gene islands are largely lacking and genes are separated by large intergenic distances.

Intergenic regions consist mainly of retrotransposons that have inserted into one another in a nested fashion, similar to the patterns found in maize (Wicker et al., 2001; SanMiguel et al., 2002; Gu et al., 2004; Kong et al., 2004; Fig. 15.4). While retrotransposons have occasionally been found in genes (Harberd et al., 1987; Martienssen and Baulcombe 1989), most retrotransposons insert into other repetitive sequences. This could be an active selection mechanism, devised by the plant to both preserve gene function and control retrotransposon activity through the insertion of new elements (Bennetzen 2000). On the other

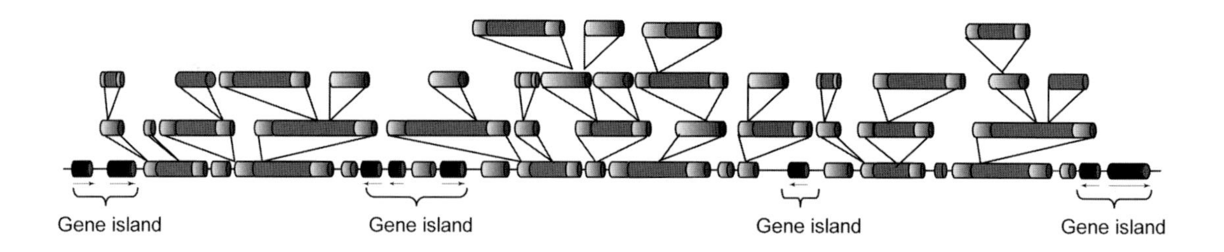

Fig. 15.4 Schematic representation of the organization of genes and repeats in the wheat genome. Genes are shown in black with an arrow, repeats in grey. Repeats consist mainly of LTR retrotransposons that are inserted in one another in a nested fashion.

hand, it is also possible that insertions into genes are generally selected against because of reduced fitness caused by the insertion. In wheat, as in other plant species, most retrotransposons are less than 6 million years old. Insertion times can be dated by analyzing the amount of nucleotide variation that differentiates the two LTRs of a retrotransposon, which are identical at the time of insertion (SanMiguel et al., 1998). Because most retrotransposons identified in the wheat genome have amplified in recent times, their particular location in the genome is rarely conserved even between closely related *Triticum* species that are estimated to have diverged less than 3 MYA (SanMiguel et al., 2002; Wicker et al., 2003b; Gu et al., 2004). The most likely reason for the absence of or lack of recognition of older elements is that these tend to erode over time, mostly through a process called illegitimate recombination (Devos et al., 2002; Wicker et al., 2003b).

Other elements that are found in the intergenic regions are non-LTR retrotransposons and CACTA elements (Wicker et al., 2003b; Kong et al., 2004), which are DNA transposons that contain terminal inverted repeats that terminate in the motif CACTA. Many CACTA elements, a number of which have been found to be associated with genes, have undergone deletions and do not carry a functional transposase, which makes them difficult to identify (Wicker et al., 2003a). Nevertheless, CACTA elements have been shown to be high-copy elements and are the second largest group of transposable elements in the wheat genome (Li et al., 2004).

Evolution of the wheat genome

The increase in gene density from centromere to telomere invokes the question of whether this gradient is caused by an effective increase in the number of genes in the distal versus proximal chromosome regions or a differential insertion and/or removal of retrotransposons in different chromosome regions. Akhunov and colleagues addressed the first part of this question by examining genes that were detected at paralogous loci on different chromosomes or in different bins of the same chromosome (Akhunov et al., 2003b). The likely ancestral locus was identified by examining which of the wheat paralogues had an orthologue in rice that fits the pattern of known wheat–rice syntenic relationships. This study revealed that the majority of the duplicated loci were located in the distal chromosome regions and that their distribution was positively correlated with recombination rates (Akhunov et al., 2003b). Interestingly, gene-deletion rates were also significantly higher in the high-recombinant regions than in the low-recombinant regions (Dvorak and Akhunov 2005), suggesting that recombination has a mechanistic role in both duplication and deletion. Recombination will also affect the rate with which these new gene arrangements are fixed in the population (Gaut et al., 2007). Advantageous genes will be fixed faster and deleterious genes will be purged faster in the high- than in the low-recombinant regions. So, through interplay of rearrangements and selection, recombination leads to an increase in the density of mostly duplicated gene copies in the distal chromosome regions. This is true for both tandem and dispersed gene duplicates.

A corollary of the enhanced evolutionary rates at the ends of chromosomes is that the levels of colinearity between wheat and other grass genomes will be lower in the distal than in the proximal regions (Akhunov et al., 2003a; See et al., 2006). If the duplication occurred in the wheat lineage, then the duplicated gene copy will not have a true orthologue in rice. If, on the other hand, the duplication pre-dated the wheat–rice divergence, distal paralogues will have a higher chance than proximal copies to be differentially deleted in the two species, again leading to a disturbance of syntenic relationships that will affect mostly distal regions. Distal genes that are nonessential may also accumulate mutations faster, simply because recombination is inherently mutagenic. For essential genes, this will be counteracted by natural selection, but duplicated gene copies will be largely free from such a constraint.

Retrotransposons, on the other hand, are present at higher frequencies in the proximal compared with the distal chromosome regions. The uneven distribution of retrotransposons in the wheat genome may be caused by the preferential insertion of LTR-retrotransposons into other repeats. As mentioned earlier, this may serve to preserve gene function as well as to potentially inactivate existing elements in the genome through the insertion of new copies. Accumulation of retrotransposons in low-recombinant regions may also be facilitated by the fact that selection is inefficient in these regions. However, a study in rice found that retrotransposon removal occurred equally efficient in centromeric versus euchromatic regions (Ma et al., 2007). At least in rice, preferential insertion of retrotransposons thus seems to be the main cause of the inverse relationship between gene density and number of retrotransposons. Wheat has a steeper recombination gradient than rice, and we cannot exclude that, in wheat, removal of retrotransposons in wheat is affected by recombination.

TOWARD SEQUENCING THE WHEAT GENOME

Great strides have been made in wheat genomics over the past 10 years. Nevertheless, making a quantum leap forward in our understanding of wheat genome organization, gene function, and evolutionary mechanisms will require sequencing of the wheat genome. While the international wheat community is unified in its goal of obtaining a whole-genome sequence of wheat, the discussion is still ongoing on the most appropriate sequencing strategies. Early discussions focused on whether to sequence the hexaploid bread wheat genome or a diploid relative such as *Ae. tauschii*. Bread wheat is the economically most important Triticeae species. A key question was whether the gene content of the A, B, and D genomes was sufficiently different to justify the financial resources needed for sequencing all three genomes. The development of cheap sequencing technologies has shifted the challenge from generating the sequence to assembling the reads into a high-quality draft sequence. The debate now focuses largely on which technology or combination of technologies will give the highest information content for the lowest cost.

The most complete genome sequence is obtained when sequencing is done BAC by BAC, following the construction of a physical map. This approach has been used to sequence the Arabidopsis genome (The Arabidopsis Genome Initiative 2000), has been used by the International Rice Genome Sequencing Project to sequence the rice genome (International Rice Genome Sequencing Project 2005), and is also being used to sequence the maize genome (http://www.maizesequence.org/). Physical maps for both the diploid *Ae. tauschii* (J. Dvorak and M.-C. Luo, pers. comm.) and for hexaploid wheat (see section

"Physical mapping in hexaploid wheat") are being constructed, and sequencing the wheat genome BAC by BAC thus remains an option. Other potential strategies are to limit sequencing to the gene space using gene-enrichment techniques (Rabinowicz et al., 1999; Yuan et al., 2003) or by judiciously choosing BAC clones that are likely to contain genes as has been done for *Lotus japonicus* and *Medicago truncatula* (Young et al., 2005). The new sequencing technologies also provided new opportunities as well as challenges for sequencing the wheat genome. The advantages and disadvantages of different sequencing strategies, and the costs and benefits of applying these strategies to the wheat genome, are discussed in subsequent sections.

Sanger sequencing

Hierarchical genome sequencing

Sequencing a genome BAC by BAC requires the availability of an anchored physical map. After establishment of a MTP, each BAC in the MTP is sequenced and assembled separately. The weakest link in this approach is the physical map, and the level of genome coverage provided by the physical map is the major factor that will determine the completeness of the sequence. The critical limiting factor in applying this approach to the large wheat genome is the cost. Fingerprinting hexaploid wheat BAC libraries at 15X coverage will cost roughly $5 million. Sequencing a MTP of some 210,000 BAC clones at 8X redundancy without finishing and using Sanger sequencing will cost around $125 million (calculated at $0.50/read). Construction of chromosome-specific BAC libraries allows the burden of physical mapping and sequencing of the hexaploid wheat genome to be spread internationally over different laboratories. While the BAC-by-BAC approach would undoubtedly yield the most complete sequence, a price tag of $6 million for sequencing an individual chromosome may be too high for many laboratories to participate.

Whole-genome shotgun sequencing

Whole-genome shotgun sequencing, which involves the sequencing of random genomic clones, has been successfully used for sequencing small genomes. It is more cost-effective for sequencing small genomes than hierarchical sequencing, because no physical map is required. For large genomes, however, the physical map construction represents less than 5% of the sequencing cost and so the financial gain is not high. Furthermore, assembling an entire genome, some 16,800 Mb in the case of hexaploid wheat, compared with assembling a few hundred kilobases at the most when assembly is done BAC by BAC is, to say the least, problematic. In rice, several WGS sequences were produced in addition to the hierarchical sequence of the rice genome (Goff et al., 2002; Yu et al., 2002, 2005). The latest assemblies of both the *indica* and *japonica* genomes, sequenced at 6X, have an N50 value of around 23 kb for contigs and 30 kb for scaffolds. The N50 value is the size at which 50% of all base pairs are incorporated in contigs with this minimal length. By combining information from both the 6X *indica* and *japonica* WGS assemblies, these contigs were further assembled into superscaffolds of 8 to 10 Mb (Yu et al., 2005). Whole-genome shotgun assemblies can also be improved by incorporating information from BAC-end sequences and BAC lengths obtained from fingerprint data (Warren et al., 2006). The largest genome that has been sequenced to date by the WGS approach is around 3,500 Mb (http://www.ncbi.nlm.nih.gov/genomes/leuks.cgi). Applying WGS sequencing to wheat and assembling the sequence, at least for the low-copy regions of the genome, would be feasible. The cost, although cheaper than BAC-by-BAC sequencing, would still be considerable. A 6X shotgun Sanger sequence of the hexaploid wheat genome would require some 120 million reads and carry a price tag of around $60 million.

Sequencing of gene-rich BAC clones

Sequencing of gene-rich BAC clones relies on the assumption that genes are clustered in gene-rich regions and that these regions can be selected effectively. We can use our current knowledge of the organization of the wheat genome, obtained through a combination of physical (deletion) mapping and BAC sequencing, to estimate the size of the gene space in wheat. To do this, we first need to know what the gene number is in wheat. Paux et al. (2006) analyzed about 11 Mb of BAC-end sequences derived from chromosome 3B for the presence of genes. Considering that about 1.2% of the BAC-end sequences consisted of coding regions, and assuming an average gene size (coding region only) of 2 kb, the 1-Gb-large 3B chromosome was estimated to contain 6,000 genes (Paux et al., 2006). Extrapolating to the entire B genome gave an estimate of 36,000 genes. Extrapolation from 800 kb of WGS sequence, on the other hand, led to an estimate of approximately 98,000 genes per genome (Rabinowicz et al., 2005). The large gene number is thought to be due to a recent amplification of pseudogenes in the hexaploid wheat genome. Nevertheless, since neither study discriminated between functional genes and pseudogenes, this does not explain the large discrepancy in the number of genes estimated from BAC-end and WGS sequences. This difference is inherent to the data sets and is not due to the annotation methodology used.

Reannotation of the Rabinowicz data set by Paux and colleagues led to a similarly high gene number (107,000 per wheat genome). A preliminary annotation of genes in 6.7 Mb of sequence obtained from randomly selected BAC clones suggested the presence of 190,000 genes in the hexaploid wheat genome, or some 63,000 per genome (X. Xu, K.M. Devos, P. San Miguel, J.L. Bennetzen, unpublished data), which is close to the average of the Paux et al. (2006) and Rabinowicz et al. (2005) studies.

Assuming that the gene number in wheat (including both functional genes and pseudogenes) is 195,000 and that the average size of a wheat coding region is 2 kb, the wheat gene space would occupy approximately 390 Mb. Of course, if the gene space is isolated from BAC clones, then we also have to take intergenic distances into account. From the *Lr10* and *VRN-2* studies, we have learned that average gene densities, considering not only gene islands but also interisland distances, are in the range of 1 gene per 40–55 kb in the distal chromosome regions. Assuming that this gene density is maintained along the telomere–centromere axis—although we know that this assumption is incorrect—we would have to sequence 8,800 Mb or about 50% of the wheat genome to obtain a minimum of 95% of the genes. Preliminary annotation of 66 randomly selected BAC clones has indicated that less than 50% of the genes are organized in gene islands and that the remainder is present in the genome as singletons (X. Xu, K.M. Devos, P. San Miguel, and J.L. Bennetzen, unpublished data). This study also suggests that obtaining 94% of the wheat genes by selecting gene-containing BAC clones would require sequencing around 50% of the hexaploid wheat genome. The cost of this approach would be approximately $65 million.

Sequencing the gene space using gene-enrichment methodologies

The rationale behind gene-enriched sequencing strategies is that they provide a large proportion of the genes without having to sequence the repeats. Since the gene space is expected to be of similar size in large and small genomes, sequencing the gene-space is a cost-effective way of obtaining the genes, particularly in large genomes. The traditional way of obtaining gene sequences is by end-sequencing cDNA libraries to produce ESTs. In wheat, approximately 1 million ESTs are currently available in GenBank (Ogihara et al., 2003; Zhang et al., 2004a; Houde et al., 2006; Mochida et al., 2006). The only plant species for which higher numbers of ESTs are available are rice (1.2 million), maize (1.4 million), and Arabidopsis (1.5 million). In Arabidopsis, it has been estimated that the ESTs represent only around 60% of the annotated genes (The Arabidopsis Genome Initiative 2000). Similar, or lower, figures are available for other species, suggesting that EST sequencing alone does not capture all genes present in a genome (Barbazuk et al., 2005). Lowly expressed genes and highly specialized genes that are transcribed only during specific developmental stages, in specific tissues, or under specific environmental conditions will generally not be represented in EST databases. These genes might be obtained using gene-enrichment techniques that target gene-containing genomic DNA fragments. These methods are based either on the different kinetics with which low-copy sequences and repeats undergo reassociation following denaturation (high C_0t; Yuan et al., 2003) or the differential methylation of genes and repeats. Transcribed genes tend to be hypomethylated while repeats generally carry 5-methyl groups at CG and CNG locations. Preferential digestion of genic regions with methylation-sensitive restriction enzymes (Emberton et al., 2005) and selective propagation of hypomethylated fragments in methylation-restrictive bacterial hosts (Rabinowicz et al., 1999) are two methods that can be employed to generate libraries of cloned genomic fragments enriched for genes. Pilot studies of these methodologies in maize showed that high C_0t and methyl-filtered reads tagged >95% of genes (Springer et al., 2004) and reduced the effective size of the genome to be sequenced at least fivefold (Barbazuk et al., 2005).

Application of the high C_0t technology to hexaploid wheat genomic DNA resulted in a 13.7-fold enrichment in gene sequences, a 5.8-fold enrichment in unknown low-copy sequences, and a 3-fold reduction in repetitive DNA (Lamoureux et al., 2005). The gene enrichment obtained in wheat was slightly higher than in maize (4.2%), which is probably a reflection of the lower relative gene content and higher repeat content in wheat compared with the maize genome. When methyl filtration was applied to wheat, the outcome was rather surprising. In maize, the gene-enrichment factor (GEF), which is the percentage of genes present in a methyl-filtered library relative to the percentage of genes present in a WGS library, was 13.15. One would expect the GEF to increase with increasing genome size (Rabinowicz et al., 2005), but in hexaploid wheat, the GEF was only 4.7. A low GEF was also observed in diploid wheat. Barley, on the other hand, had an expected GEF of 18.7. In diploid wheat, the low GEF appeared to be due mainly to the high percentage of unmethylated repeats. Methyl filtration is therefore unlikely to be useful as an enrichment strategy in diploid wheat (Li et al., 2004; Rabinowicz et al., 2005). In hexaploid wheat, the proportion of repeats in the methyl-filtered library was similar to that in other grasses, but the low GEF was due to the high number of genes identified in the WGS library. Many of these genes appeared to be methylated and are likely pseudogenes (Rabinowicz et al., 2005). It should be noted that the gene number of 98,000 identified by Rabinowicz et al. (2005) in the WGS library was considerably higher than the 36,000 genes estimated from 3B BAC-end sequences (Paux et al., 2006). If the GEF in wheat is indeed artificially deflated due to the presence of large numbers of methylated pseudogenes in the WGS (control) libraries, then methyl filtration does provide actual enrichment of active genes in hexaploid wheat.

One problem with gene-enrichment techniques is that the gene sequences are not placed in a genomic context. To obtain information on the location of the genes in the genome, the cDNA or genomic fragments need to be superimposed on a low-redundancy BAC draft sequence. Another technology that could aid in connecting the sequences generated by gene-enrichment techniques is methylation spanning linker libraries. These are large-insert clones generated with methylation-sensitive restriction enzymes. These clones have low-copy sequences at their ends, and the remaining sequence consists of methylated repeats. Hence they span the distance between neighboring genes (Yuan et al., 2002). This technique would work for connecting genes present in the more distal chromosome regions of wheat where intergenic distances are likely to fall in the range of a BAC cloning vector, but it might not be technically feasible to span the large intergenic distances that are likely to exist in the proximal chromosome regions.

New-generation sequencing technologies

A series of different sequencing technologies has been developed in recent years, and more are under development. Currently, the three leading systems are the Roche Diagnostics Corporation (Indianapolis, Indiana) GS FLX system (previously 454), the Illumina (San Diego, California) Genome Analyzer (previously Solexa), and the Applied Biosystems (Foster City, California) SOLiD DNA sequencer. The sequencing technologies used by the three companies differ, but for all platforms the massive numbers of sequencing reactions can be conducted in parallel, producing several hundred megabases or even a few gigabases of sequence data per run for a few thousand dollars.

Before discussing if and how these technologies can be applied for sequencing the wheat genome, it is important to compare advantages and disadvantages of the different instruments. All three instruments have the capability to run paired-end reads, which will aid in the assembly process. Roche GS FLX Titanium produces approximately 1.2 million reads of 400 bp (500 Mb total data). The Illumina Genome Analyzer produces some 40 million reads of 35 bp, or a total of 1.4 Gb per run. For paired-end reads on the ABI SOLiD, read lengths are 25 bp and the total output is in the range 1.5–2 Gb of sequence data. Sequence accuracy is similar for the three instruments. While these technologies allow a genome to be sequenced inexpensively, new strategies will need to be developed that will allow these small reads to be assembled correctly. Wicker et al. (2006) demonstrated that 100-bp reads, generated from four barley BAC clones at 10X redundancy using the 454 pyrosequencing technology, could be assembled quite easily for regions that were single-copy in the BAC clones. The efficiency with which pooled BAC clones can be assembled can be further enhanced by bar-coding individual BACs or BAC subpools. Currently, the Illumina Genome Analyzer and ABI SOLiD instruments are used mainly to resequence species for which a reference genome is already available. However, software is being development for the *de novo* assembly of short reads (Pop and Salzberg 2008), and the use of these technologies in wheat should be explored.

It seems most likely that the wheat genome ultimately will be sequenced using a combination of these new-generation sequencing technologies. One possible cost-efficient scenario would be to produce a low-redundancy draft of the wheat genome using GS FLX Titanium sequencing of pooled BAC clones originating from the physical map-based MTPs. This could be achieved in about 20 runs per diploid genome. Alternatively, DNA isolated from flow-sorted chromosomes or chromosome arms could be shotgun-sequenced by GS FLX Titanium. The resulting 2X draft, which would allow the assembly of most of the genic regions, could then be used as template for the assembly of short Illumina or SOLiD reads generated at 20X. Sequencing each genome at 20X using the Illumina or SOLiD technology would require 50 to 60 runs per genome. The total cost for sequencing the hexaploid wheat genome in this fashion would be, at most, $2 million. Add a few million dollars for the bioinformatics, and the entire wheat genome, or at least the genic portion of it, could be completed with these new-generation sequencing technologies for a fraction of the cost of sequencing the entire genome or even the gene space using Sanger technology. As sequence technologies continue to develop at a fast pace—one of the latest developments is real-time sequencing from single molecules that can produce read lengths of a few kilobases (Eid et al., 2009)—it is expected that it will be possible to produce a high-quality draft sequence of the wheat genome for this price tag or less in the coming years.

FUTURE PERSPECTIVES

Forward genetics—starting from a trait and isolating the gene(s) that control that trait—is now possible in wheat, although map-based cloning remains highly labor-intensive. This process will be greatly facilitated in the coming years by the availability of a physical map of the wheat genome to which the sequence of the gene space or even the entire wheat genome has been anchored.

Reverse genetics, whereby the gene is the starting point and elucidation of its function is the aim, is in its infancy. Technologies and resources that have been developed in wheat to aid this endeavor are targeted induced local lesions in genomes (TILLING) populations and VIGS. In TILLING, chemical mutagenesis is used to create a population of mutants. Primers, made against the gene that is the target of the functional analysis, are used to screen the DNA of the M2 mutant population using a pooled approach (McCallum et al., 2000). Key to the success of TILLING is the high density of mutations that can be achieved. This density is much higher in polyploids compared with diploids because most genes are redundant and the organism is thus buffered against mutations that would be lethal in a diploid. In tetraploid and hexaploid wheat, respectively, mutation rates of 1 per 40 bp and 24 bp have been obtained by Slade et al. (2005) and of 1 per 45 and 60 kb by Dubcovsky and colleagues (J. Dubcovsky, pers. comm.). Considering the high mutation rate, it is feasible to find mutations in the orthologous genes on the A, B, and D genomes. These genes then have to be combined into a single plant to assess the phenotype. This has been successfully demonstrated in the combination of EMS mutations in waxy alleles, which has led to the creation of a waxy wheat in which the synthesis of amylose was greatly reduced (Slade et al., 2005).

A second reverse-genetics tool available in wheat is VIGS using *Barley stripe mosaic virus* (BSMV) (Scofield et al., 2005), which is a positive-sense single-strand RNA virus. Double-stranded RNA, which occurs as an intermediate in viral replication, triggers sequence-specific degradation. When transcribed fragments corresponding to a wheat gene are cloned into the viral genome, the mRNA produced by the endogenous target gene will be degraded, leading to gene silencing. The potential of VIGS in wheat was first demonstrated by silencing phytoene desaturase (PDS) expression. Suppression of PDS, an enzyme in the carotenoid pigment biosynthetic pathway, results in photolysis of chlorophyll in the affected tissues. Photobleaching was observed 10 days after infection with the BSMV:PDS construct (Scofield et al., 2005; Color Plate 31).

Virus-induced gene silencing was then used to assess the functional identity of the *Lr21* gene candidate. A 174-bp fragment from the 3′ UTR of the *Lr21* candidate was cloned into the BSMV genome. Inoculation of a wheat line carrying the *Lr21* resistance gene first with the BSMV:*Lr21* construct and then with a leaf rust isolate that is avirulent to *Lr21* led to virulent interactions in some areas of the leaf. Similar results were obtained with three other genes, *RAR1*, *SGT1*, and cytosolic *HSP90*, which have been shown to function in many of the NBS-LRR resistance pathways (Scofield et al., 2005). The products of all four genes are thus required for *Lr21*-mediated resistance. This study demonstrated the use of VIGS for functional analysis of genes that are expressed in the leaves of seedlings. Since then, VIGS protocols have also been developed for silencing genes in the spikes of wheat, making VIGS a tool that can be employed for the rapid screening of gene candidates (S. Scofield, pers. comm.).

Research is also being conducted to facilitate wheat transformation technologies. Wheat transformation using biolistics is robust and can reach efficiencies of over 60% with select genotypes (Pellegrineschi et al., 2002). The drawback of biolistic DNA delivery is that it often results in complex integration patterns. This has led to the development of *Agrobacterium*-mediated gene transfer, which generally results in the insertion of single or few copies of the transgene in a cost-efficient manner. Transformation efficiencies obtained in wheat with *Agrobacterium* are generally lower than the best published values for biolistic transfers, but are likely to increase as protocols are further developed and optimized (Jones 2005). The main drawback of both systems is that fertile plants need to be regenerated from transformed somatic cells via somatic embryogenesis, a process that is time-consuming, is highly genotype-dependent, and can create somaclonal variants. These disadvantages can be overcome by using an *in planta* transformation system, modeled on the "dipping" protocol that exists in Arabidopsis (Bechtold et al., 1993; Clough and Bent 1998). Dipping wheat inflorescences into *Agrobacterium* infiltration medium has yielded a small number of successful transformation events

(Agarwal et al., 2008). Integration events were confirmed by both Southern and segregation analysis, and were shown to be mainly single copy. In an alternative approach, transformed plants were obtained after piercing a region of the embryonic apical meristem in imbibed seeds with a needle that had been dipped in *Agrobacterium* inoculum (Supartana et al., 2006). Transformation was confirmed in the T_1 generation by PCR, Southern analysis, and by isolation and sequence analysis of wheat DNA flanking the T-DNA insertion sites. If the efficiency of the *in planta* transformation technologies can be enhanced, they will provide a powerful tool for developing functional genomics tools.

There is thus real potential for being able to conduct large-scale structural and functional analyses in wheat in the coming years. In addition to having the sequence of all wheat genes available, the wheat research community will have access to the necessary functional genomics tools to elucidate the function of the genes that are key to determining the agricultural importance of wheat. Detailed knowledge about the action mechanisms in wheat of genes identified by QTL and association studies as underlying specific traits, of genes shown in model species to underlie particular pathways, and of previously "unknown" genes annotated from sequence data will provide a means for highly efficient and targeted improvement of wheat.

REFERENCES

Agarwal, S., S. Loar, C. Steber, and J. Zale. 2008. Floral transformation of wheat. *In* H. Jones and P. Shewry (ed.) Methods in biotechnology. Humana Press, Totawa, NJ.

Ahn, S., J.A. Anderson, M.E. Sorrells, and S.D. Tanksley. 1993. Homoeologous relationships of rice, wheat and maize chromosomes. Mol. Gen. Genet. 241:483–490.

Akbari, M., P. Wenzl, V. Caig, J. Carling, L. Xia, S.Y. Yang, G. Uszynski, V. Mohler, A. Lehmensiek, H. Kuchel, M. J. Hayden, N. Howes, P. Sharp, P. Vaughan, B. Rathmell, E. Huttner, and A. Kilian. 2006. Diversity arrays technology (DArT) for high-throughput profiling of the hexaploid wheat genome. Theor. Appl. Genet. 113:1409–1420.

Akhunov, E.D., A.R. Akhunova, A.M. Linkiewicz, et al. 2003a. Synteny perturbations between wheat homoeologous chromosomes caused by locus duplications and deletions correlate with recombination rates. Proc. Natl. Acad. Sci. USA 100:10836–10841.

Akhunov, E.D., A.W. Goodyear, S. Geng, et al. 2003b. The organization and rate of evolution of wheat genomes are correlated with recombination rates along chromosome arms. Genome Res. 13:753–763.

Allouis, S., G. Moore, A. Bellec, R. Sharp, P. Faivre-Rampant, K. Mortimer, S. Pateyron, T.N. Foote, S. Griffiths, M. Caboche, and B. Chalhoub. 2003. Construction and characterisation of a hexaploid wheat (*Triticum aestivum* L.) BAC library from the reference germplasm 'Chinese Spring'. Cereal Res. Commun. 31:331–338.

Alm, V., C. Fang, C.S. Busso, K.M. Devos, K. Vollan, Z. Grieg, and O.A. Rognli. 2003. A linkage map of meadow fescue (*Festuca pratensis* Huds.) and comparative mapping with other *Poaceae* species. Theor. Appl. Genet. 108:25–40.

Arondel, V., B. Lemieux, I. Hwang, S. Gibson, H.M. Goodman, and C.R. Somerville. 1992. Map-based cloning of a gene controlling omega-3-fatty-acid desaturation in Arabidopsis. Science 258:1353–1355.

Barbazuk, W.B., J.A. Bedell, and P.D. Rabinowicz. 2005. Reduced representation sequencing: A success in maize and a promise for other plant genomes. BioEssays 27:839–848.

Bechtold, N., J. Ellis, and G. Pelletier. 1993. *In-planta Agrobacterium*-mediated gene-transfer by infiltration of adult *Arabidopsis thaliana* plants. C. R. Acad. Sci. 316:1194–1199.

Bennett, M.D., and J.B. Smith. 1991. Nuclear DNA amounts in angiosperms. Philos. Trans. R. Soc. Lond., Ser. B 334:309–345.

Bennetzen, J.L. 2000. Transposable element contributions to plant gene and genome evolution. Plant Mol. Biol. 42:251–269.

Blanco, A., M.P. Bellomo, A. Cenci, C. De Giovanni, R. D'Ovidio, E. Iacono, B. Laddomada, M.A. Pagnotta, E. Porceddu, A. Sciancalepore, R. Simeone, and O.A. Tanzarella. 1998. A genetic linkage map of durum wheat. Theor. Appl. Genet. 97:721–728.

Bossolini, E., T. Wicker, P.A. Knobel, and B. Keller. 2007. Comparison of orthologous loci from small grass genomes *Brachypodium* and rice: Implications for wheat genomics and grass genome annotation. Plant J. 49:704–717.

Botstein, D., R.L. White, M. Skolnick, and R.W. Davis. 1980. Construction of a genetic linkage map in man using restriction fragment length polymorphisms. Am. J. Hum. Genet. 32:314–331.

Boyko, E.V., K.S. Gill, L. Mickelson-Young, S. Nasuda, W.J. Raupp, J.N. Ziegle, S. Singh, D.S. Hassawi, A.K. Fritz, D. Namuth, N.L.V. Lapitan, and B.S. Gill. 1999. A high-density genetic linkage map of *Aegilops tauschii*, the D-genome progenitor of bread wheat. Theor. Appl. Genet. 99:16–26.

Brueggeman, R., N. Rostoks, D. Kudrna, A. Kilian, F. Han, J. Chen, A. Druka, B. Steffenson, and A. Kleinhofs. 2002. The barley stem rust-resistance gene *Rpg1* is a novel disease-resistance gene with homology to receptor kinases. Proc. Natl. Acad. Sci. USA 99:9328–9333.

Budiman, M.A., S.B. Chang, S. Lee, T.J. Yang, H.B. Zhang, H. de Jong, and R.A. Wing. 2004. Localization of *jointless-2* gene in the centromeric region of tomato chromosome 12 based on high resolution genetic and physical mapping. Theor. Appl. Genet. 108:190–196.

Cenci, A., N. Chantret, X. Kong, Y. Gu, O.D. Anderson, T. Fahima, A. Distelfeld, and J. Dubcovsky. 2003. Construction and characterization of a half million clone BAC library of durum wheat (*Triticum turgidum* ssp. *durum*). Theor. Appl. Genet. 107:931–939.

Chantret, N., A. Cenci, F. Sabot, O. Anderson, and J. Dubcovsky. 2004. Sequencing of the *Triticum monococcum Hardness* locus reveals good microcolinearity with rice. Mol. Genet. Genomics 271:377–386.

Chao, S., P.J. Sharp, A.J. Worland, E.J. Warham, R.M.D. Koebner, and M.D. Gale. 1989. RFLP based genetic maps of wheat homoeologous group 7 chromosomes. Theor. Appl. Genet. 78:495–504

Chapman, V., and R. Riley. 1970. Homoeologous meiotic chromosome pairing in *Triticum aestivum* in which chromosome 5B is replaced by an alien homoeologue. Nature 226:376–377.

Cheng, Z.K., C.R. Buell, R.A. Wing, and J.M. Jiang. 2002. Resolution of fluorescence *in-situ* hybridization mapping on rice mitotic prometaphase chromosomes, meiotic pachytene chromosomes and extended DNA fibers. Chromosome Res. 10:379–387.

Clough, S.J., and A.F. Bent. 1998. Floral dip: A simplified method for *Agrobacterium*-mediated transformation of *Arabidopsis thaliana*. Plant J. 16:735–743.

Cloutier, S., B.D. McCallum, C. Loutre, T.W. Banks, T. Wicker, C. Feuillet, B. Keller, and M.C. Jordan. 2007. Leaf rust resistance gene *Lr1*, isolated from bread wheat (*Triticum aestivum* L.) is a member of the large psr567 gene family. Plant Mol. Biol. 65:93–106.

Conley, E.J., V. Nduati, J.L. Gonzalez-Hernandez, et al. 2004. A 2600-locus chromosome bin map of wheat homoeologous group 2 reveals interstitial gene-rich islands and colinearity with rice. Genetics 168:625–637.

Delaney, D.E., S. Nasuda, T.R. Endo, B.S. Gill, and S.H. Hulbert. 1995a. Cytologically based physical maps of the group-2 chromosomes of wheat. Theor. Appl. Genet. 91:568–573.

Delaney, D.E., S. Nasuda, T.R. Endo, B.S. Gill, and S.H. Hulbert. 1995b. Cytologically based physical maps of the group 3 chromosomes of wheat. Theor. Appl. Genet. 91:780–782.

Devos, K.M., M.D. Atkinson, C.N. Chinoy, R.L. Harcourt, R.M.D. Koebner, C.J. Liu, P. Masojc, D.X. Xie, and M. D. Gale. 1993a. Chromosomal rearrangements in the rye genome relative to that of wheat. Theor. Appl. Genet. 85:673–680.

Devos, K.M., M.D. Atkinson, C.N. Chinoy, C. Liu, and M.D. Gale. 1992. RFLP based genetic map of the homoeologous group 3 chromosomes of wheat and rye. Theor. Appl. Genet. 83:931–939.

Devos, K.M., J. Beales, Y. Ogihara, and A.N. Doust. 2005a. Comparative sequence analysis of the *Phytochrome C* gene and its upstream region in allohexaploid wheat reveals new data on the evolution of its three constituent genomes. Plant Mol. Biol. 58:625–641.

Devos, K.M., J.K.M. Brown, and J.L. Bennetzen. 2002. Genome size reduction through illegitimate recombination counteracts genome expansion in *Arabidopsis*. Genome Res. 12:1075–1079.

Devos, K.M., S. Chao, Q.Y. Li, M.C. Simonetti, and M.D. Gale. 1994. Relationship between chromosome *9* of maize and wheat homoeologous group *7* chromosomes. Genetics 138:1287–1292.

Devos, K.M., J. Dubcovsky, J. Dvořák, C.N. Chinoy, and M.D. Gale. 1995. Structural evolution of wheat chromosomes 4A, 5A, and 7B and its impact on recombination. Theor. Appl. Genet. 91:282–288.

Devos, K.M., J. Ma, A.C. Pontaroli, L.H. Pratt, and J.L. Bennetzen. 2005b. Analysis and mapping of randomly chosen BAC clones from hexaploid bread wheat. Proc. Natl. Acad. Sci. USA 102:19243–19248.

Devos, K.M., T. Millan, and M.D. Gale. 1993b. Comparative RFLP maps of the homoeologous group-2 chromosomes of wheat, rye, and barley. Theor. Appl. Genet. 85:784–792.

Devos, K.M., T.S. Pittaway, A. Reynolds, and M.D. Gale. 2000. Comparative mapping reveals a complex relationship between the pearl millet genome and those of foxtail millet and rice. Theor. Appl. Genet. 100:190–198.

Doležel, J., J. Číhalíková, and S. Lucretti. 1992. A high-yield procedure for isolation of metaphase chromosomes from root-tips of *Vicia faba* L. Planta 188:93–98.

Doležel, J., M. Kubaláková, J. Bartoš, and J. Macas. 2004. Flow cytogenetics and plant genome mapping. Chromosome Res. 12:77–91.

Doležel, J., M. Kubaláková, E. Paux, J. Bartoš, and C. Feuillet. 2007. Chromosome-based genomics in the cereals. Chromosome Res. 15:51–66.

Dubcovsky, J., D. Lijavetzky, L. Appendino, and G. Tranquilli. 1998. Comparative RFLP mapping of *Triticum monococcum* genes controlling vernalization requirement. Theor. Appl. Genet. 97:968–975.

Dubcovsky, J., M-C. Luo, G-Y. Zhong, R. Bransteitter, A. Desai, A. Kilian, A. Kleinhofs, and J. Dvořák. 1996. Genetic map of diploid wheat, *Triticum monococcum* L. and its comparison with maps of *Hordeum vulgare* L. Genetics 143:983–999.

Dvorak, J., and E.D. Akhunov. 2005. Tempos of gene locus deletions and duplications and their relationship to recombination rate during diploid and polyploid evolution in the *aegilops–triticum* alliance. Genetics 171:323–332.

Eid, J., A. Fehr, J. Gray, et al. 2009. Real-time DNA sequencing from single polymerase molecules. Science 323:133–138.

Emberton, J., J.X. Ma, Y.N. Yuan, P. SanMiguel, and J.L. Bennetzen. 2005. Gene enrichment in maize with hypomethylated partial restriction (HMPR) libraries. Genome Res. 15:1441–1446.

Endo, T.R. 1990. Gametocidal chromosomes and their induction of chromosome mutations in wheat. Jpn. J. Genet. 65:132–152.

Endo, T.R., and B.S. Gill. 1996. The deletion stocks of common wheat. J. Hered. 87:295–307.

Erayman, M., D. Sandhu, D. Sidhu, M. Dilbirligi, P.S. Baenziger, and K.S. Gill. 2004. Demarcating the gene-rich regions of the wheat genome. Nucleic Acids Res. 32:3546–3565.

Faris, J.D., J.P. Fellers, S.A. Brooks, and B.S. Gill. 2003. A bacterial artificial chromosome contig spanning the major domestication locus *Q* in wheat and identification of a candidate gene. Genetics 164:311–321.

Faris, J.D., and B.S. Gill. 2002. Genomic targeting and high-resolution mapping of the domestication gene *Q* in wheat. Genome 45:706–718.

Feldman, M. 2001. Origin of cultivated wheat. p. 3–56. *In* A.P. Bonjean and W.J. Angus (ed.) The world wheat book: A history of wheat breeding. Lavoisier Publishing, Paris, France.

Feldman, M., and A.A. Levy. 2005. Allopolyploidy—a shaping force in the evolution of wheat genomes. Cytogenet. Genome Res. 109:250–258.

Feuillet, C., M. Messmer, G. Schachermayr, and B. Keller. 1995. Genetic and physical characterization of the *LR1* leaf rust resistance locus in wheat (*Triticum aestivum* L.). Mol. Gen. Genet. 248:553–562.

Feuillet, C., G.M. Schachermayr, and B. Keller. 1997. Molecular cloning of a new receptor-like kinase gene encoded at the *Lr10* disease resistance locus of wheat. Plant J. 11:45–52.

Feuillet, C., S. Travella, N. Stein, L. Albar, A. Nublat, and B. Keller. 2003. Map-based isolation of the leaf rust disease resistance gene *Lr10* from the hexaploid wheat (*Triticum aestivum* L.) genome. Proc. Natl. Acad. Sci. USA 100:15253–15258.

Flavell, R.B., M.D. Bennett, A.G. Seal, and J. Hutchinson. 1987. Chromosome structure and organisation. p. 211–268. *In* F.G.H. Lupton (ed.) Wheat breeding, its scientific basis. Chapman and Hall Ltd., London.

Flavell, R.B., M.D. Bennett, J.B. Smith, and D.B. Smith. 1974. Genome size and proportion of repeated nucleotide sequence DNA in plants. Biochem. Genet. 12:257–269.

Flavell, A.J., M.R. Knox, S.R. Pearce, and T.H.N. Ellis. 1998. Retrotransposon-based insertion polymorphisms (RBIP) for high throughput marker analysis. Plant J. 16:643–650.

Foote, T.N., S. Griffiths, S. Allouis, and G. Moore. 2004. Construction and analysis of a BAC library in the grass *Brachypodium sylvaticum*: Its use as a tool to bridge the gap between rice and wheat in elucidating gene content. Funct. Integr. Genomics 4:26–33.

Foote, T., M. Roberts, N. Kurata, T. Sasaki, and G. Moore. 1997. Detailed comparative mapping of cereal chromosome regions corresponding to the *Ph1* locus in wheat. Genetics 147:801–807.

Gale, M.D., M.D. Atkinson, C.N. Chinoy, R.L. Harcourt, J. Jia, Q.Y. Li, and K.M. Devos. 1995. Genetic maps of hexaploid wheat. p. 29–40. *In* Z.S. Li and Z.Y. Xin (ed.) Proc. Int. Wheat Genet. Symp., 8th, Beijing, China. 20–25 July 1993. China Agricultural Scientech Press, Beijing, China.

Gallego, F., C. Feuillet, M. Messmer, A. Penger, A. Graner, M. Yano, T. Sasaki, and B. Keller. 1998. Comparative mapping of the two wheat leaf rust resistance loci *Lr1* and *Lr10* in rice and barley. Genome 41:328–336.

Gao, S.C., Y.Q. Gu, J.J. Wu, D. Coleman-Derr, N.X. Huo, C. Crossman, J.Z. Jia, Q. Zuo, Z.L. Ren, O.D. Anderson, and X.Y. Kong. 2007. Rapid evolution and complex structural organization in genomic regions harboring multiple prolamin genes in the polyploid wheat genome. Plant Mol. Biol. 65:189–203.

Gaut, B.S. 2002. Evolutionary dynamics of grass genomes. New Phytol. 154:15–28.

Gaut, B.S., S.I. Wright, C. Rizzon, J. Dvorak, and L.K. Anderson. 2007. Opinion—Recombination: An underappreciated factor in the evolution of plant genomes. Nat. Rev. Genet. 8:77–84.

Gill, B.S., R. Appels, A-M. Botha-Oberholster, C.R. Buell, J.L. Bennetzen, B. Chalhoub, F.G. Chumley, J. Dvořák, M. Iwanaga, B. Keller, W. Li, W.R. McCombie, Y. Ogihara, F. Quetier, and T. Sasaki. 2004. A workshop report on wheat genome sequencing: International Genome Research on Wheat Consortium. Genetics 168:1087–1096.

Gill, K.S., K. Arumuganathan, and J.H. Lee. 1999. Isolating individual wheat (*Triticum aestivum*) chromosome arms by flow cytometric analysis of ditelosomic lines. Theor. Appl. Genet. 98:1248–1252.

Gill, B.S., B. Friebe, and T.R. Endo. 1991a. Standard karyotype and nomenclature system for description of chromosome bands and structural aberrations in wheat (*Triticum aestivum*). Genome 34:830–839.

Gill, K.S., B.S. Gill, and T.R. Endo. 1993a. A chromosome region-specific mapping strategy reveals gene-rich telomeric ends in wheat. Chromosoma 102:374–381.

Gill, K.S., B.S. Gill, T.R. Endo, and E.V. Boyko. 1996a. Identification and high density mapping of gene rich regions in chromosome group 5 of wheat. Genetics 143:1001–1012.

Gill, K.S., B.S. Gill, T.R. Endo, and Y. Mukai. 1993b. Fine physical mapping of *Ph1*, a chromosome-pairing regulator gene in polyploid wheat. Genetics 134:1231–1236.

Gill, K.S., B.S. Gill, T.R. Endo, and T. Taylor. 1996b. Identification and high-density mapping of gene-rich regions in chromosome Group *1* of wheat. Genetics 144:1883–1891.

Gill, K.S., E.L. Lubbers, B.S. Gill, W.J. Raupp, and T.S. Cox. 1991b. A genetic linkage map of *Triticum tauschii* (DD) and its relationship to the D genome of bread wheat (AABBDD). Genome 34:362–374.

Giraudat, J., B.M. Hauge, C. Valon, J. Smalle, and F. Parcy. 1992. Isolation of the *Arabidopsis Abi3* gene by positional cloning. Plant Cell 4:1251–1262.

Goff, S.A., D. Ricke, T.H. Lan, et al. 2002. A draft sequence of the rice genome (*Oryza sativa* L. ssp. *japonica*). Science 296:92–100.

Goncharov, N.P., and R.F. Gaidalenok. 1994. Role of chro mosomes of wheat homeological group-7 in growth habit control. Genetika 30:1234–1237.

Griffiths, S., R. Sharp, T.N. Foote, I. Bertin, M. Wanous, S. Reader, I. Colas, and G. Moore. 2006. Molecular characterization of *Ph1* as a major chromosome pairing locus in polyploid wheat. Nature 439:749–752.

Gu, Y.Q., D. Coleman-Derr, X-Y. Kong, and O.D. Anderson. 2004. Rapid genome evolution revealed by comparative sequence analysis of orthologous regions from four triticeae genomes. Plant Physiol. 135:459–470.

Guyot, R., N. Yahiaoui, C. Feuillet, and B. Keller. 2004. *In silico* comparative analysis reveals a mosaic conservation of genes within a novel colinear region in wheat chromosomes 1AS and rice chromosome 5S. Funct. Integr. Genomics 4:47–58.

Han, F., A. Kilian, J.P. Chen, D. Kudrna, B. Steffenson, K. Yamamoto, T. Matsumoto, T. Sasaki, and A. Kleinhofs. 1999. Sequence analysis of a rice BAC covering the syntenous barley *Rpg1* region. Genome 42:1071–1076.

Harberd, N.P., R.B. Flavell, and R.D. Thompson. 1987. Identification of a transposon-like insertion in a *Glu-1* allele of wheat. Mol. Gen. Genet. 209:326–332.

Hart, G.E., R.A. McIntosh, and M.D. Gale. 1993. Linkage maps of *Triticum aestivum* (hexaploid wheat, $2n = 42$, Genomes A, B & D) and *T.tauschii* ($2n = 14$, Genome D). p. 6.204–6.219. *In* S.J. O'Brien (ed.) Genetic maps: Locus maps of complex genomes. Cold Spring Harbor Laboratory Press, Woodbury, NY.

Hartl, L., H. Weiss, F.J. Zeller, and A. Jahoor. 1993. Use of RFLP markers for the identification of alleles of the *Pm3* locus conferring powdery mildew resistance in wheat (*Triticum aestivum* L). Theor. Appl. Genet. 86:959–963.

Hiebert, C.W., J.B. Thomas, D.J. Somers, B.D. McCallum, and S.L. Fox. 2007. Microsatellite mapping of adult-plant leaf rust resistance gene *Lr22a* in wheat. Theor. Appl. Genet. 115:877–884.

Hossain, K.G., V. Kalavacharla, G.R. Lazo, et al. 2004. A chromosome bin map of 2148 expressed sequence tag loci of wheat homoeologous group 7. Genetics 168:687–699.

Houde, M., M. Belcaid, F. Ouellet, J. Danyluk, A.F. Monroy, A. Dryanova, P. Gulick, A. Bergeron, A. Laroche, M.G. Links, L. MacCarthy, W.L. Crosby, and F. Sarhan. 2006. Wheat EST resources for functional genomics of abiotic stress. BMC Genomics 7:149 [Online]. Available at http://www.pubmedcentral.nih.gov/articlerender.fcgi?artid=1539019 (verified 8 May 2008).

Houshmand, S., R.E. Knox, F.R. Clarke, and J.M. Clarke. 2007. Microsatellite markers flanking a stem solidness gene on chromosome 3BL in durum wheat. Mol. Breed. 20:261–270.

Huang, L., S.A. Brooks, W. Li, J.P. Fellers, H. Trick, and B.S. Gill. 2003. Map-based cloning of a leaf rust resistance gene *Lr21* from the large polyploid genome of bread wheat. Genetics 164:655–664.

Huang, L., and B.S. Gill. 2001. An RGA-like marker detects all known *Lr21* leaf rust resistance gene family members in *Aegilops tauschii* and wheat. Theor. Appl. Genet. 103:1007–1013.

Huang, S., A. Sirikhachornkit, J.D. Faris, X. Su, B.S. Gill, R. Haselkorn, and P. Gornicki. 2002a. Phylogenetic analysis of the acetyl-CoA carboxylase and 3-phosphoglycerate kinase loci in wheat and other grasses. Plant Mol. Biol. 48:805–820.

Huang, S., A. Sirikhachornkit, X. Su, J. Faris, B.S. Gill, R. Haselkorn, and P. Gornicki. 2002b. Genes encoding plastid acetyl-CoA carboxylase and 3-phosphoglycerate kinase of the *Triticum/Aegilops* complex and the evolutionary history of polyploid wheat. Proc. Natl. Acad. Sci. USA 99:8133–8138.

Ilic, K., P.J. SanMiguel, and J.L. Bennetzen. 2003. A complex history of rearrangement in an orthologous region of the maize, sorghum, and rice genomes. Proc. Natl. Acad. Sci. USA 100:12265–12270.

International Rice Genome Sequencing Project. 2005. The map-based sequence of the rice genome. Nature 436:793–800.

Isidore, E., B. Scherrer, A. Bellec, K. Budin, P. Faivre-Rampant, R. Waugh, B. Keller, M. Caboche, C. Feuillet, and B. Chalhoub. 2005. Direct targeting and rapid isolation of BAC clones spanning a defined chromosome region. Funct. Integr. Genomics 5:97–103.

Jaccoud, D., K. Peng, D. Feinstein, and A. Kilian. 2001. Diversity arrays: A solid state technology for sequence information independent genotypin. Nucleic Acids Res. 29:e25–31.

Janda, J., J. Bartoš, J. Šafář, M. Kubaláková, M. Valárik, J. Číhalíková, H. Šimková, M. Caboche, P. Sourdille, M. Bernard, B. Chalhoub, and J. Doležel. 2004. Construction of a subgenomic BAC library specific for chromosomes 1D, 4D and 6D of hexaploid wheat. Theor. Appl. Genet. 109:1337–1345.

Janda, J., J. Šafář, M. Kubaláková, J. Bartoš, P. Kovářová, P. Suchánková, S. Pateyron, J. Číhalíková, P. Sourdille, H. Šimková, P. Faivre-Rampant, E. Hřibová, M. Bernard, A. Lukaszewski, J. Doležel, and B. Chalhoub. 2006. Advanced resources for plant genomics: A BAC library specific for the short arm of wheat chromosome 1B. Plant J. 47:977–986.

Jia, J., T.E. Miller, S.M. Reader, K.M. Devos, and M.D. Gale. 1996. RFLP-based maps of the homoeologous group 6 chromosomes of wheat and their application in the tagging of *Pm12*, a powdery mildew resistance gene transferred from *Aegilops speltoides* to wheat. Theor. Appl. Genet. 92:559–565.

Jiang, J.M., and B.S. Gill. 2006. Current status and the future of fluorescence *in situ* hybridization (FISH) in plant genome research. Genome 49:1057–1068.

Jones, H.D. 2005. Wheat transformation: Current technology and applications to grain development and composition. J. Cereal Sci. 41:137–147.

Jones, E.S., N.L. Mahoney, M.D. Hayward, I.P. Armstead, J.G. Jones, M.O. Humphreys, I.P. King, T. Kishida, T. Yamada, F. Balfourier, G. Charmet, and J.W. Forster. 2002. An enhanced molecular marker based genetic map of perennial ryegrass (*Lolium perenne*) reveals comparative relationships with other *Poaceae* genomes. Genome 45:282–295.

Klein, P.E., R.R. Klein, S.W. Cartinhour, P.E. Ulanch, J. Dong, J.A. Obert, D.T. Morishige, S.D. Schlueter, K.L. Childs, M. Ale, and J.E. Mullet. 2000. A high-throughput AFLP-based method for constructing integrated genetic and physical maps: Progress towards a sorghum genome map. Genome Res. 10:789–807.

Kofler, R., C. Schlotterer, and T. Lelley. 2007. SciRoKo: A new tool for whole genome microsatellite search and investigation. Bioinformatics 23:1683–1685.

Kojima, T., and Y. Ogihara. 1998. High-resolution RFLP map of the long arm of chromosome 5A in wheats and its synteny among cereals. Genes Genet. Syst. 73:51–58.

Kong, X-Y., Y.Q. Gu, F.M. You, J. Dubcovsky, and O.D. Anderson. 2004. Dynamics of the evolution of orthologous and paralogous portions of a complex locus region in two genomes of allopolyploid wheat. Plant Mol. Biol. 54:55–69.

Kota, R.S., K.S. Gill, B.S. Gill, and T.R. Endo. 1993. A cytogenetically based physical map of chromosome 1B in common wheat. Genome 36:548–554.

Kubaláková, M., P. Kovářová, P. Suchánková, J. Číhalíková, J. Bartoš, S. Lucretti, N. Watanabe, S.F. Kianian, and J. Doležel. 2005. Chromosome sorting in tetraploid wheat and its potential for genome analysis. Genetics 170:823–829.

Kubaláková, M., M. Valárik, J. Bartoš, J. Vrána, J. Číhalíková, M. Molnár-Láng, and J. Doležel. 2003. Analysis and sorting of rye (*Secale cereale* L.) chromosomes using flow cytometry. Genome 46:893–905.

Kubaláková, M., J. Vrána, J. Číhalíková, H. Šimková, and J. Doležel. 2002. Flow karyotyping and chromosome sorting in bread wheat (*Triticum aestivum* L.). Theor. Appl. Genet. 104:1362–1372.

Kumar, M., O.P. Luthra, N.R. Yadav, L. Chaudhary, N. Saini, R. Kumar, I. Sharma, and V. Chawla. 2007. Identification of microsatellite markers on chromosomes of bread wheat showing an association with karnal bunt resistance. Afr. J. Biotechnol. 6:1617–1622.

Kurata, N., G. Moore, Y. Nagamura, T.N. Foote, M. Yano, Y. Minobe, and M.D. Gale. 1994. Conservation of genome structure between rice and wheat. Bio/Technology 12:276–278.

Lamoureux, D., D.G. Peterson, W.L. Li, J.P. Fellers, and B.S. Gill. 2005. The efficacy of Cot-based gene enrichment in wheat (*Triticum aestivum* L.). Genome 48:1120–1126.

La Rota, M., and M.E. Sorrells. 2004. Comparative DNA sequence analysis of mapped wheat ESTs reveals the complexity of genome relationships between rice and wheat. Funct. Integr. Genomics 4:34–46.

Law, C.N. 1966. The location of genetic factors affecting a quantitative character in wheat. Genetics 53:487–493.

Lee, J.H., K. Arumuganathan, Y. Yen, S. Kaeppler, H. Kaeppler, and P.S. Baenziger. 1997. Root tip cell cycle synchronization and metaphase-chromosome isolation suitable for flow sorting in common wheat (*Triticum aestivum* L.). Genome 40:633–638.

Leister, D., J. Kurth, D.A. Laurie, M. Yano, T. Sasaki, K.M. Devos, A. Graner, and P. Schulze-Lefert. 1998. Rapid reorganization of resistance gene homologues in cereal genomes. Proc. Natl. Acad. Sci. USA 95:370–375.

Li, W., P. Zhang, J.P. Fellers, B. Friebe, and B.S. Gill. 2004. Sequence composition, organization and evolution of the core Triticeae genomes. Plant J. 40:500–511.

Lijavetzky, D., G. Muzzi, T. Wicker, B. Keller, R. Wing, and J. Dubcovsky. 1999. Construction and characterization of a bacterial artificial chromosome (BAC) library for the A genome of wheat. Genome 42:1176–1182.

Ling, P., and X.M. Chen. 2005. Construction of a hexaploid wheat (*Triticum aestivum* L.) bacterial artificial chromosome library for cloning genes for stripe rust resistance. Genome 48:1028–1036.

Ling, H.Q., Y. Zhu, and B. Keller. 2003. High-resolution mapping of the leaf rust disease resistance gene *Lr1* in wheat and characterization of BAC clones from the *Lr1* locus. Theor. Appl. Genet. 106:875–882.

Linkiewicz, A.M., L.L. Qi, B.S. Gill, et al. 2004. A 2500-locus bin map of wheat homoeologous group 5 provides insights on gene distribution and colinearity with rice. Genetics 168:665–676.

Liu, C.J., K.M. Devos, C.N. Chinoy, M.D. Atkinson, and M.D. Gale. 1992. Non-homoeologous translocations between group 4, 5 and 7 chromosomes in wheat and rye. Theor. Appl. Genet. 83:305–312.

Liu, Y.G., K. Nagaki, M. Fujita, K. Kawaura, M. Uozumi, and Y. Ogihara. 2000. Development of an efficient maintenance and screening system for large-insert genomic DNA libraries of hexaploid wheat in a transformation-competent artificial chromosome (TAC) vector. Plant J. 23:687–695.

Liu, Y.G., and K. Tsunewaki. 1991. Restriction fragment length polymorphism (RFLP) analysis in wheat: II. Linkage maps of the RFLP sites in common wheat. Jpn. J. Genet. 66:617–633.

Luo, M.C., K. Deal, Z.L. Yang, and J. Dvorak. 2005. Comparative genetic maps reveal extreme crossover localization in the *Aegilops speltoides* chromosomes. Theor. Appl. Genet. 111:1098–1106.

Luo, M.C., C. Thomas, F.M. You, J. Hsiao, O.Y. Shu, C.R. Buell, M. Malandro, P.E. McGuire, O.D. Anderson, and J. Dvorak. 2003. High-throughput fingerprinting of bacterial artificial chromosomes using the SNaPshot labeling kit and sizing of restriction fragments by capillary electrophoresis. Genomics 82:378–389.

Ma, Z-Q., M.E. Sorrells, and S.D. Tanksley. 1994. RFLP markers linked to powdery mildew resistance genes *Pm1*, *Pm2*, *Pm3*, and *Pm4* in wheat. Genome 37:871–875.

Ma, Z., S. Weining, P.J. Sharp, and C.J. Liu. 2000. Non-gridded library: A new approach for BAC (bacterial artificial chromosome) exploitation in hexaploid wheat (*Triticum aestivum*). Nucleic Acids Res. 28:e106.

Ma, J.X., R.A. Wing, J.L. Bennetzen, and S.A. Jackson. 2007. Plant centromere organization: A dynamic structure with conserved functions. Trends Genet. 23:134–139.

Mao, L., K.M. Devos, L. Zhu, and M.D. Gale. 1997. Cloning and genetic mapping of wheat telomere-associated sequences. Mol. Gen. Genet. 254:584–591.

Marino, C.L., J.C. Nelson, Y-H. Lu, M.E. Sorrells, P. Leroy, N.A. Tuleen, C.R. Lopes, and G.E. Hart. 1996. Molecular genetic maps of the group 6 chromosomes of hexaploid wheat (*Triticum aestivum* L. em. Thell). Genome 39:359–366.

Martienssen, R.A., and D.C. Baulcombe. 1989. An unusual wheat insertion sequence (WIS*1*) lies upstream of an α-amylase gene in hexaploid wheat, and carries a minisatellite array. Mol. Gen. Genet. 217:401–410.

Martin, G.B., S.H. Brommonschenkel, J. Chunwongse, A. Frary, M.W. Ganal, R. Spivey, T.Y. Wu, E.D. Earle, and S.D. Tanksley. 1993. Map-based cloning of a protein-

kinase gene conferring disease resistance in tomato. Science 262:1432–1436.
McCallum, C.M., L. Comai, E.A. Greene, and S. Henikoff. 2000. Targeted screening for induced mutations. Nat. Biotechnol. 18:455–457.
McIntosh, R.A., K.M. Devos, J. Dubcovsky, W.J. Rogers, C.F. Morris, R. Appels, D.J. Somers, and O.A. Anderson. 2007. Catalogue of gene symbols for wheat: 2007 supplement. Annu. Wheat Newsl. 53:159–180.
McIntosh, R.A., G.E. Hart, K.M. Devos, M.D. Gale, and W.J. Rogers. 1998. Catalogue of gene symbols for wheat. p. 1–325. *In* A.E. Slinkard (ed.) Proc. Int. Wheat Genetics Symp., 9th, Vol. 5, Saskatoon, Saskatchewan. 2–7 Aug. 1998. Univ. Ext. Press, Saskatoon, Saskatchewan, Canada.
McIntosh, R.A., Y. Yamazaki, K.M. Devos, J. Dubcovsky, W.J. Rogers, and R. Appels. 2003. Catalogue of gene symbols for wheat. p. 1–34. Proc Int. Wheat Genet. Symp., 10th, Vol. 4, Paestum, Italy. 1–6 Sept. 2003. Istituto Sperimentale per la Cerealcoltura, Paestum, Italy.
Messmer, M.M., M. Keller, S. Zanetti, and B. Keller. 1999. Genetic linkage map of a wheat × spelt cross. Theor. Appl. Genet. 98:1163–1170.
Meyers, B.C., S. Scalabrin, and M. Morgante. 2004. Mapping and sequencing complex genomes: Let's get physical. Nat. Rev. Genet. 5:578–588.
Mickelson-Young, L., T.R. Endo, and B.S. Gill. 1995. A cytogenetic ladder-map of the wheat homoeologous group-4 chromosomes. Theor. Appl. Genet. 90:1007–1011.
Miftahudin, K. Ross, X.F. Ma, et al. 2004. Analysis of expressed sequence tag loci on wheat chromosome group 4. Genetics 168:651–663.
Mochida, K., K. Kawaura, E. Shimosaka, N. Kawakami, I. Shin, Y. Kohara, Y. Yamazaki, and Y. Ogihara. 2006. Tissue expression map of a large number of expressed sequence tags and its application to in silico screening of stress response genes in common wheat. Mol. Gen. Genomics 276:304–312.
Moullet, O., H.B. Zhang, and E.S. Lagudah. 1999. Construction and characterisation of a large DNA insert library from the D genome of wheat. Theor. Appl. Genet. 99:305–313.
Munkvold, J.D., R.A. Greene, C.E. Bertmudez-Kandianis, et al. 2004. Group 3 chromosome bin maps of wheat and their relationship to rice chromosome 1. Genetics 168:639–650.
Naranjo, T., A. Roca, P.G. Goicoechea, and R. Giraldez. 1987. Arm homoeology of wheat and rye chromosomes. Genome 29:873–882.
Nelson, J.C., M.E. Sorrells, A.E. Van Deynze, Y-H. Lu, M.D. Atkinson, M. Bernard, P. Leroy, J.D. Faris, and J.A. Anderson. 1995a. Molecular mapping of wheat: Genes and rearrangements in homoeologous groups 4, 5 and 7. Genetics 141:721–731.
Nelson, J.C., A.E. Van Deynze, E. Autrique, M.E. Sorrells, Y-H. Lu, M. Bernard, and P. Leroy. 1995b. Molecular mapping of wheat: Homoeologous group 3. Genome 38:525–533.
Nelson, J.C., A.E. Van Deynze, E. Autrique, M.E. Sorrells, Y-H. Lu, M. Merlino, M.D. Atkinson, and P. Leroy. 1995c. Molecular mapping of wheat: Homoeologous group 2. Genome 38:516–524.
Nilmalgoda, S.D., S. Cloutier, and A.Z. Walichnowski. 2003. Construction and characterization of a bacterial artificial chromosome (BAC) library of hexaploid wheat (*Triticum aestivum* L.) and validation of genome coverage using locus-specific primers. Genome 46:870–878.
Ogihara, Y., K. Mochida, Y. Nemoto, K. Murai, Y. Yamazaki, I. Shin, and Y. Kohara. 2003. Correlated clustering and virtual display of gene expression patterns in the wheat life cycle by large-scale statistical analyses of expressed sequence tags. Plant J. 33:1001–1011.
Paterson, A.H., J.E. Bowers, and B.A. Chapman. 2004. Ancient polyploidization predating divergence of the cereals, and its consequences for comparative genomics. Proc. Natl. Acad. Sci. USA 101:9903–9908.
Paux, E., F. Legeai, N. Guilhot, A.F. Adam-Blondon, M. Alaux, J. Salse, P. Sourdille, P. Leroy, and C. Feuillet. 2007. Physical mapping in large genomes: Accelerating anchoring of BAC contigs to genetic maps through *in silico* analysis. Funct. Integr. Genomics 8:29–32.
Paux, E., D. Roger, E. Badaeva, G. Gay, M. Bernard, P. Sourdille, and C. Feuillet. 2006. Characterizing the composition and evolution of homoeologous genomes in hexaploid wheat through BAC-end sequencing on chromosome 3B. Plant J. 48:463–474.
Paux, E., P. Sourdille, J. Salse, et al. 2008. A physical map of the 1-gigabase bread wheat chromosome 3B. Science 322:101–104.
Pellegrineschi, A., L.M. Noguera, B. Skovmand, R.M. Brito, L. Velazquez, M.M. Salgado, R. Hernandez, M. Warburton, and D. Hoisington. 2002. Identification of highly transformable wheat genotypes for mass production of fertile transgenic plants. Genome 45:421–430.
Peng, J., A.B. Korol, T. Fahima, M.S. Röder, Y.I. Ronin, Y.C. Li, and E. Nevo. 2000. Molecular genetic maps in wild emmer wheat, *Triticum dicoccoides*: Genome-wide coverage, massive negative interference, and putative quasi-linkage. Genome Res. 10:1509–1531.
Peng, J.H., H. Zadeh, G.R. Lazo, et al. 2004. Chromosome bin map of expressed sequence tags in homoeologous group 1 of hexaploid wheat and homoeology with rice and Arabidopsis. Genetics 168:609–623.
Pop, M., and S.L. Salzberg. 2008. Bioinformatics challenges of new sequencing technology. Trends Genet. 24:142–149.
Požárková, D., A. Koblìžková, B. Román, A.M. Torres, S. Lucretti, M. Lysák, J. Doležel, and J. Macas. 2002. Development and characterization of microsatellite markers from chromosome 1-specific DNA libraries of Vicia faba. Biol. Plant. 45:337–345.
Prasad, M., R.K. Varshney, A. Kumar, H.S. Balyan, P.C. Sharma, K.J. Edwards, H. Singh, H.S. Dhaliwal, J.K. Roy, and P.K. Gupta. 1999. A microsatellite marker associated with a QTL for grain protein content on chromosome arm 2DL of bread wheat. Theor. Appl. Genet. 99:341–345.
Qi, L.L., B. Echalier, S. Chao, et al. 2004. A chromosome bin map of 16,000 expressed sequence tag loci and distri-

bution of genes among the three genomes of polyploid wheat. Genetics 168:701–712.

Qi, L.L., B. Friebe, and B.S. Gill. 2006. Complex genome rearrangements reveal evolutionary dynamics of pericentromeric regions in the Triticeae. Genome 49:1628–1639.

Qiu, Y.C., X.L. Sun, R.H. Zhou, X.Y. Kong, S.S. Zhang, and J.Z. Jia. 2006. Identification of microsatellite markers linked to powdery mildew resistance gene *Pm2* in wheat. Cereal Res. Commun. 34:1267–1273.

Rabinowicz, P.D., R. Citek, M.A. Budiman, A. Nunberg, J.A. Bedell, N. Lakey, A.L. O'Shaughnessy, L.U. Nascimento, W.R. McCombie, and R.A. Martienssen. 2005. Differential methylation of genes and repeats in land plants. Genome Res. 15:1431–1440.

Rabinowicz, P.D., K. Schutz, N. Dedhia, C. Yordan, L.D. Parnell, L.D. Stein, W.R. McCombie, and R. Martienssen. 1999. Differential methylation of genes and retrotransposons facilitates shotgun sequencing of the maize genome. Nat. Genet. 23:305–308.

Ramakrishna, W., J. Dubcovsky, Y-J. Park, C. Busso, J. Emberton, P.J. SanMiguel, and J.L. Bennetzen. 2002. Different types and rates of genome evolution detected by comparative sequence analysis of orthologous segments from four cereal genomes. Genetics 162:1389–1400.

Randhawa, H.S., M. Dilbirligi, D. Sidhu, et al. 2004. Deletion mapping of homoeologous group 6–specific wheat expressed sequence tags. Genetics 168:677–686.

Ratnayaka, I., M. Baga, D.B. Fowler, and R.N. Chibbar. 2005. Construction and characterization of a BAC library of a cold-tolerant hexaploid wheat cultivar. Crop Sci. 45:1571–1577.

Riley, R., and V. Chapman. 1958. Genetic control of the cytologically diploid behaviour of hexaploid wheat. Nature 182:713–715.

Roberts, M.A., S.M. Reader, C. Dalgliesh, T.E. Miller, T.N. Foote, L.J. Fish, J.W. Snape, and G. Moore. 1999. Induction and characterization of *Ph1* wheat mutants. Genetics 153:1909–1918.

Röder, M.S., V. Korzun, K. Wendehake, J. Plaschke, M-H. Tixier, P. Leroy, and M.W. Ganal. 1998. A microsatellite map of wheat. Genetics 149:2007–2023.

Rowland, G.G., and E.R. Kerber. 1974. Telocentric mapping in hexaploid wheat of genes for leaf rust resistance and other characters derived from *Aegilops-squarrosa*. Can. J. Genet. Cytol. 16:137–144.

Šafář, J., J. Bartoš, J. Janda, A. Bellec, M. Kubaláková, M. Valárik, S. Pateyron, J. Weiserová, R. Tušková, J. Číhalíková, J. Vrána, H. Šimková, P. Faivre-Rampant, P. Sourdille, M. Caboche, M. Bernard, J. Doležel, and B. Chalhoub. 2004. Dissecting large and complex genomes: Flow sorting and BAC cloning of individual chromosomes from bread wheat. Plant J. 39:960–968.

Salvo-Garrido, H., D.A. Laurie, B. Jaffe, and J.W. Snape. 2001. An RFLP map of diploid *Hordeum bulbosum* L. and comparison with maps of barley (*H. vulgare* L.) and wheat (*Triticum aestivum* L.). Theor. Appl. Genet. 103:869–880.

Sandhu, D., J.A. Champoux, S.N. Bondareva, and K.S. Gill. 2001. Identification and physical localization of useful genes and markers to a major gene-rich region on wheat group 1S chromosomes. Genetics 157:1735–1747.

SanMiguel, P.J., B.S. Gaut, A.P. Tikchonov, Y. Nakajima, and J.L. Bennetzen. 1998. The paleontology of intergene retrotransposons of maize. Nat. Genet. 20:43–45.

SanMiguel, P.J., W. Ramakrishna, J.L. Bennetzen, C.S. Busso, and J. Dubcovsky. 2002. Transposable elements, genes and recombination in a 215-kb contig from wheat chromosome $5A^{m}$. Funct. Integr. Genomics 2:70–80.

Sarma, R.N., B.S. Gill, T. Sasaki, G. Galiba, J. Sutka, D.A. Laurie, and J.W. Snape. 1998. Comparative mapping of the wheat chromosome 5A *Vrn-A1* region with rice and its relationship to QTL for flowering time. Theor. Appl. Genet. 97:103–109.

Schachermayr, G.M., C. Feuillet, and B. Keller. 1997. Molecular markers for the detection of the wheat leaf rust resistance gene *Lr10* in diverse genetic backgrounds. Mol. Breed. 3:65–74.

Schwarzacher, T., M.L. Wang, A.R. Leitch, N. Miller, G. Moore, and J.S. Heslop-Harrison. 1997. Flow cytometric analysis of the chromosomes and stability of a wheat cell-culture line. Theor. Appl. Genet. 94:91–97.

Scofield, S.R., L. Huang, A.S. Brandt, and B.S. Gill. 2005. Development of a virus-induced gene-silencing system for hexaploid wheat and its use in functional analysis of the *Lr21*-mediated leaf rust resistance pathway. Plant Physiol. 138:2165–2173.

Sears, E.R. 1954. The aneuploids of common wheat. Mo. Agric. Exp. Stn. Res. Bull. 572:1–59.

Sears, E.R. 1977. An induced mutant with homoeologous pairing in common wheat. Can. J. Genet. Cytol. 19:585–593.

Sears, E.R., and L.M.S. Sears. 1979. The telocentric chromosomes of common wheat. p. 389–407. *In* S. Ramanujam (ed.) Proc. Int. Wheat Genet. Symp., 5th, New Delhi, India. 23–28 Feb. 1978. Indian Soc. Genet. and Plant Breeding, New Delhi, India.

See, D.R., S. Brooks, J.C. Nelson, G. Brown-Guedira, B. Friebe, and B.S. Gill. 2006. Gene evolution at the ends of wheat chromosomes. Proc. Natl. Acad. Sci. USA 103:4162–4167.

Sharma, R.C., E. Duveiller, and J.M. Jacquemin. 2007. Microsatellite markers associated with spot blotch resistance in spring wheat. J. Phytopathol. 155:316–319.

Šimková, H., J. Číhalíková, J. Vrána, M.A. Lysák, and J. Doležel. 2003. Preparation of HMW DNA from plant nuclei and chromosomes isolated from root tips. Biol. Plant. 46:369–373.

Šimková, H., J.T. Svensson, P. Condamine, E. Hřibová, P. Suchánková, P.R. Bhat, J. Bartoš, J. Šafář, T.J. Close, and J. Doležel. 2008. Coupling amplified DNA from flow-sorted chromosomes to high-density SNP mapping in barley. BMC Genomics 9:294.

Simons, K.J., J.P. Fellers, H.N. Trick, Z.C. Zhang, Y.S. Tai, B.S. Gill, and J.D. Faris. 2006. Molecular characterization of the major wheat domestication gene *Q*. Genetics 172:547–555.

Slade, A.J., S.I. Fuerstenberg, D. Loeffler, M.N. Steine, and D. Facciotti. 2005. A reverse genetic, nontransgenic approach to wheat crop improvement by TILLING. Nat. Biotechnol. 23:75–81.

Soderlund, C., S. Humphray, A. Dunham, and L. French. 2000. Contigs built with fingerprints, markers, and FPCV4.7. Genome Res. 10:1772–1787.

Somers, D.J., P. Isaac, and K. Edwards. 2004. A high-density microsatellite consensus map for bread wheat (*Triticum aestivum* L.). Theor. Appl. Genet. 109:1105–1114.

Somers, D.J., R. Kirkpatrick, M. Moniwa, and A. Walsh. 2003. Mining single-nucleotide polymorphisms from hexaploid wheat ESTs. Genome 46:431–437.

Song, R., V. Llaca, and J. Messing. 2002. Mosaic organization of orthologous sequences in grass genomes. Genome Res. 12:1549–1555.

Springer, N.M., X. Xu, and W.B. Barbazuk. 2004. Utility of different gene enrichment approaches toward identifying and sequencing the maize gene space. Plant Physiol. 136:3023–3033.

Stein, N., C. Feuillet, T. Wicker, E. Schlagenhauf, and B. Keller. 2000. Subgenome chromosome walking in wheat: A 450-kb physical contig in *Triticum monococcum* L. spans the *Lr10* resistance locus in hexaploid wheat (*Triticum aestivum* L.). Proc. Natl. Acad. Sci. USA 97:13436–13441.

Stephenson, P., G.J. Bryan, J. Kirby, A.J. Collins, K.M. Devos, C.S. Busso, and M.D. Gale. 1998. Fifty new microsatellite loci for the wheat genetic map. Theor. Appl. Genet. 97:946–949.

Supartana, P., T. Shimizu, M. Nogawa, H. Shioiri, T. Nakajima, N. Haramoto, M. Nozue, and M. Kojima. 2006. Development of simple and efficient *in planta* transformation method for wheat (*Triticum aestivum* L.) using *Agrobacterium tumefaciens*. J. Biosci. Bioeng. 102:162–170.

Tanksley, S.D., M.W. Ganal, and G.B. Martin. 1995. Chromosome landing—a paradigm for map-based gene cloning in plants with large genomes. Trends Genet. 11:63–68.

The Arabidopsis Genome Initiative. 2000. Analysis of the genome sequence of the flowering plant *Arabidopsis thaliana*. Nature 408:796–815.

Thomas, C.A. 1971. The genetic organization of chromosomes. Annu. Rev. Genet. 5:237–256.

Tsilo, T.J., Y. Jin, and J.A. Anderson. 2007. Microsatellite markers linked to stem rust resistance allele *Sr9a* in wheat. Crop Sci. 47:2013–2020.

Tsilo, T.J., Y. Jin, and J.A. Anderson. 2008. Diagnostic microsatellite markers for the detection of stem rust resistance gene *Sr36* in diverse genetic backgrounds of wheat. Crop Sci. 48:253–261.

Valárik, M., J. Bartoš, P. Kovářová, M. Kubaláková, J.H. De Jong, and J. Doležel. 2004. High-resolution FISH on super-stretched flow-sorted plant chromosomes. Plant J. 37:940–950.

Van Deynze, A.E., J.C. Nelson, L.S. O'Donoughue, S.N. Ahn, W. Siripoonwiwat, S.E. Harrington, E.S. Yglesias, D.P. Braga, S.R. McCouch, and M.E. Sorrells. 1995a. Comparative mapping in grasses: Oat relationships. Mol. Gen. Genet. 249:349–356.

Van Deynze, A.E., J.C. Nelson, E.S. Yglesias, S.E. Harrington, D.P. Braga, S.R. McCouch, and M.E. Sorrells. 1995b. Comparative mapping in grasses: Wheat relationships. Mol. Gen. Genet. 248:744–754.

Vrána, J., M. Kubaláková, H. Šimková, J. Číhalíková, M.A. Lysák, and J. Doležel. 2000. Flow sorting of mitotic chromosomes in common wheat (*Triticum aestivum* L.). Genetics 156:2033–2041.

Wall, A.M., R. Riley, and M.D. Gale. 1971. The position of a locus of chromosome 5B in *Triticum aestivum* affecting homoeologous meiotic pairing. Genet. Res. 18:329–339.

Wang, M.L., A.R. Leitch, T. Schwarzacher, J.S. Heslop-Harrison, and G. Moore. 1992. Construction of a chromosome-enriched HpaII library from flow-sorted wheat chromosomes. Nucleic Acids Res. 20:1897–1901.

Warren, R.L., D. Varabei, D. Platt, X. Huang, D. Messina, S.P. Yang, J.W. Kronstad, M. Krzywinski, W.C. Warren, J.W. Wallis, L.W. Hillier, A.T. Chinwalla, J.E. Schein, A.S. Siddiqui, M.A. Marra, R.K. Wilson, and S.J.M. Jones. 2006. Physical map-assisted whole-genome shotgun sequence assemblies. Genome Res. 16:768–775.

Wenzl, P., J. Carling, D. Kudrna, D. Jaccoud, E. Huttner, A. Kleinhofs, and A. Kilian. 2004. Diversity Arrays Technology (DArT) for whole-genome profiling of barley. Proc. Natl. Acad. Sci. USA 101:9915–9920.

Werner, J.E., T.R. Endo, and B.S. Gill. 1992. Toward a cytologically based physical map of the wheat genome. Proc. Nat. Acad. Sci. USA 89:11307–11311.

Wicker, T., R. Guyot, N. Yahiaoui, and B. Keller. 2003a. CACTA transposons in Triticeae: A diverse family of high-copy repetitive elements. Plant Physiol. 132:52–63.

Wicker, T., E. Schlagenhauf, A. Graner, T.J. Close, B. Keller, and N. Stein. 2006. 454 sequencing put to the test using the complex genome of barley. BMC Genomics 7:275–285.

Wicker, T., N. Stein, L. Albar, C. Feuillet, E. Schlagenhauf, and B. Keller. 2001. Analysis of a contiguous 211 kb sequence in diploid wheat (*Triticum monococcum* L.) reveals multiple mechanisms of genome evolution. Plant J. 26:307–316.

Wicker, T., N. Yahiaoui, R. Guyot, E. Schlagenhauf, Z-D. Liu, J. Dubcovsky, and B. Keller. 2003b. Rapid genome divergence at orthologous low molecular weight glutenin loci of the A and A^m genomes of wheat. Plant Cell 15:1186–1197.

Wolfe, K.H., M. Gouy, Y-W. Yang, P.M. Sharp, and W-H. Li. 1989. Date of the monocot–dicot divergence estimated from chloroplast DNA sequence data. Proc. Natl. Acad. Sci. USA 86:6201–6205.

Xie, D.X., K.M. Devos, G. Moore, and M.D. Gale. 1993. RFLP-based genetic maps of the homoeologous group 5 chromosomes of bread wheat (*Triticum aestivum* L.). Theor. Appl. Genet. 87:70–74.

Yahiaoui, N., P. Srichumpa, R. Dudler, and B. Keller. 2004. Genome analysis at different ploidy levels allows cloning of the powdery mildew resistance gene *Pm3b* from hexaploid wheat. Plant J. 37:528–538.

Yan, L., D. Fu, C. Li, A. Blechl, G. Tranquilli, M. Bonafede, A. Sanchez, M. Valarik, S. Yasuda, and J. Dubcovsky. 2006. The wheat and barley vernalization gene *VRN3* is an orthologue of *FT*. Proc. Natl. Acad. Sci. USA 103:19581–19586.

Yan, L.L., A. Loukoianov, A.E. Blechl, G. Tranquilli, W. Ramakrishna, P. SanMiguel, J.L. Bennetzen, V. Echenique, and J. Dubcovsky. 2004. The wheat VRN2 gene is a flowering repressor down-regulated by vernalization. Science 303:1640–1644.

Yan, L., A. Loukoianov, G. Tranquilli, M. Helguera, T. Fahima, and J. Dubcovsky. 2003. Positional cloning of the wheat vernalization gene *VRN1*. Proc. Natl. Acad. Sci. USA 100:6263–6268.

Yim, Y.S., P. Moak, H. Sanchez-Villeda, T.A. Musket, P. Close, P.E. Klein, J.E. Mullet, M.D. McMullen, Z. Fang, M.L. Schaeffer, J.M. Gardiner, E.H. Coe, and G.L. Davis. 2007. A BAC pooling strategy combined with PCR-based screenings in a large, highly repetitive genome enables integration of the maize genetic and physical maps. BMC Genomics 8:47–59.

Young, N.D., S.B. Cannon, S. Sato, D. Kim, D.R. Cook, C.D. Town, B.A. Roe, and S. Tabata. 2005. Sequencing the genespaces of *Medicago truncatula* and *Lotus japonicus*. Plant Physiol. 137:1174–1181.

Yu, J., S.N. Hu, J. Wang, et al. 2002. A draft sequence of the rice genome (*Oryza sativa* L. ssp. *indica*). Science 296:79–92.

Yu, J., J. Wang, W. Lin, et al. 2005. The genomes of *Oryza sativa*: A history of duplications. PLoS Biol. 3:266–281.

Yuan, Y.N., P.J. SanMiguel, and J.L. Bennetzen. 2002. Methylation-spanning linker libraries link gene-rich regions and identify epigenetic boundaries in *Zea mays*. Genome Res. 12:1345–1349.

Yuan, Y.N., P.J. SanMiguel, and J.L. Bennetzen. 2003. High-Cot sequence analysis of the maize genome. Plant J. 34:249–255.

Zhang, D., D.W. Choi, S. Wanamaker, et al. 2004a. Construction and evaluation of cDNA libraries for large-scale expressed sequence tag sequencing in wheat (*Triticum aestivum* L.). Genetics 168:595–608.

Zhang, P., B. Friebe, B. Gill, and R.F. Park. 2007. Cytogenetics in the age of molecular genetics. Aust. J. Agric. Res. 58:498–506.

Zhang, H., J. Jia, M.D. Gale, and K.M. Devos. 1998. Relationship between the chromosomes of *Aegilops umbellulata* and wheat. Theor. Appl. Genet. 96:69–75.

Zhang, P., W.L. Li, J. Fellers, B. Friebe, and B.S. Gill. 2004b. BAC-FISH in wheat identifies chromosome landmarks consisting of different types of transposable elements. Chromosoma 112:288–299.

Zhang, H., S.M. Reader, X. Liu, J.Z.Jia, M.D. Gale, and K.M. Devos. 2001. Comparative genetic analysis of the *Aegilops longissima* and *Ae. sharonensis* genomes with common wheat. Theor. Appl. Genet. 103:518–525.

Chapter 16
Synthetic Wheat—An Emerging Genetic Resource

Richard M. Trethowan and Maarten van Ginkel

SUMMARY

(1) Hexaploid wheat can be reconstituted by natural intercrossing, induced chromosome doubling, and embryo rescue to produce primary "synthetic" wheat.
(2) Combining variability from both modern durum wheat and ancestral tetraploids with *Aegilops tauschii* has produced new genetic variation for a range of biotic, abiotic, and quality-related traits.
(3) Direct and indirect evidence indicates that much of the newly observed genetic diversity in synthetic wheat is novel. Synthetic derivatives, developed by crossing primary synthetics with adapted cultivars, have been developed with enhanced resistance to biotic and abiotic stresses.
(4) The exploitation of synthetic wheat is still in its infancy. In the future, combining novel genetic diversity in synthetic wheat with that existing in the wheat gene pool can be expected to significantly enhance the adaptation and marketability of wheat.

INTRODUCTION

Wheat is grown on more than 220 million hectares globally, producing more than 600 million tonnes of grain annually (Rajaram and van Ginkel 2001; Pfeiffer et al., 2005). It is the most traded cereal grain, with 10% of total production sold in international markets (Marathee and Gomez-MacPherson 2001), and is the most widely grown of all the cereals; production spans the equator to latitudes of more than 60°N and 50°S, and altitudes of up to 3,000 m above sea level. Wheat is grown in dry, rainfed environments and under irrigation in highly productive river valleys, primarily in the developing world, and is the primary source of calories for millions of people. Average global wheat consumption is 73 kg per capita per annum and can be as high as 166 kg per capita in some North African and central Asian countries (Marathee and Gomez-MacPherson 2001). Hexaploid bread wheat (*Triticum aestivum* L.) is the primary species grown and consumed, although tetraploid durum wheat (*T. turgidum* ssp. *durum*), comprising approximately 6% to 8% of total wheat production, is an important commodity in some areas, such as the Mediterranean region (Elias and Manthey 2005).

A range of different wheat products are made and consumed, as further discussed in Chapters 20 and 21. Flat bread (such as chapattis), noodles, and steamed bread tend to dominate in developing countries, reflecting consumer preferences in India and China, the world's largest wheat producers. In contrast, leavened bread made from hard-grained wheat flour, and biscuits, cakes, and confectionary products produced from soft-grained wheat, are prevalent in western countries, though increasingly these are being consumed in developing countries. Pastas, couscous, bulgur, and local breads are processed from durum wheat.

Tetraploid wheat was domesticated some 10,000–12,000 years ago, either in the Middle East, eastern Turkey, or western Iran, where the habitats of *Aegilops speltoides*, the probable donor of the B genome, and *T. urartu*, the donor of the A genome, overlap (Feldman 2001). This amphiploid, called wild emmer or *T. turgidum* ssp. *dicoccoides*, had a brittle rachis, was difficult to thresh, and was therefore unsuitable for cultivation. Mutation later gave rise to *T. turgidum* ssp. *dicoccum*, which was free-threshing with a nonbrittle rachis and was more suitable for cultivation. Some 8,000 years ago, these cultivated tetraploids spread from the Fertile Crescent region into western Iran, where they likely hybridized with *Aegilops tauschii*, the donor of the D genome (Kihara 1944; McFadden and Sears 1946) (syn. *Ae. squarrosa* and *T. tauschii*, see Chapter 1), giving rise to hexaploid wheat. Feldman (2001) believed that this hybridization involved the cultivated tetraploid, *T. dicoccum*, rather than wild emmer. As cultivation spread from the Fertile Crescent, so did the geographic range of wheat. Evidence suggests that wheat cultivation began in 6500 BC in the Fertile Crescent, spreading as far as modern-day France and Egypt by 4000 BC, and east to India and eastern China by 3000 and 1500 BC, respectively (Feldman 2001).

In the process of evolution and domestication, both tetraploid and hexaploid wheat passed through significant genetic bottlenecks, greatly reducing genetic variability in their cultivated forms. Mutation gave rise to free-threshing, nonbrittle domesticated forms of einkorn and emmer wheat which were selected and grown by farmers, thereby greatly truncating existing variation. The formation of hexaploid wheat was subject to even greater truncation. Evidence suggests that the spontaneous hybridization of *T. dicoccum* and *Ae. tauschii* likely involved a few independent events, creating a significant founder effect (Appels and Lagudah 1990).

Farmers have continually improved the adaptation and productivity of wheat over the centuries through continuous selection of better plants as seed stocks for the following year. Large-scale wheat breeding based on planned hybridizations only began in the early 20th century, following the rediscovery of Mendel's laws of heredity (Sneep 1966). The success of modern wheat breeding has been built upon exploitation of useful genetic variation in the wheat gene pool. Much of the improvement in wheat productivity can be attributed first to improved disease resistance and second to the maintenance of this resistance over time. As pathogens mutate and overcome host resistance, so have plant breeders sought new sources of resistance from an ever-dwindling reservoir of still-effective genes. Significant but reduced genetic progress has been made in the improvement of wheat for tolerance to abiotic stresses, largely because genetic variation is less plentiful, the inheritance of these traits is complex, and heterogeneity in the environment limits response to selection.

In their search for new genetic variation, wheat breeders have used adapted cultivars, landraces, and translocated chromosome segments from wild relatives to extend the pool of useful genes. Nevertheless, variability in existing cultivars and landraces is still subject to the founder effect and alien translocations are difficult to produce, are often unstable, and can bring with them unfavorable gene linkages. The *de novo* synthesis of new primary synthetics is an effective way to overcome the founder effect in wheat. Hundreds of accessions of *Ae. tauschii* are held in gene banks around the world, with still more *in situ* in native habitats, and extensive collections of cultivated and wild tetraploids are readily available. Variability among these wild diploid and tetraploid progenitors of hexaploid wheat has likely changed through natural selection over thousands of years. Many of these species have been collected in some of the harshest environments on earth, and it is likely that useful genes for adaptation to these conditions have accumulated with time. This chapter explores the genetic variability available in primary synthetic hexaploid wheat and the application of this variation in wheat breeding.

PRIMARY SYNTHETIC HEXAPLOID WHEAT

The first published amphiploid was produced in the late 19th century from a cross between wheat and rye (*Secale cereale* L.) (Wilson 1876). With the advent of colchicine in the 1930s it became possible to develop hybrids between wheat and *Aegilops* spp. These successful hybridizations demonstrated that species could be artificially formed via allopolyploidy (Feldman 2001). The first artificial hybridization between tetraploid wheat and *Ae. tauschii*, and hence the first primary synthetic, was reported by McFadden and Sears (1946). Large-scale development of synthetic wheat began at the International Maize and Wheat Improvement Center (CIMMYT) in the mid-1980s (van Ginkel and Ogbonnaya 2007). Many of these primary synthetics have been screened for various traits and crossed in breeding programs around the world.

Figure 16.1 describes the three types of crosses generally made to produce new primary synthetic hexaploid wheat. This genetic resource can have wide genetic diversity and can vary considerably in agronomic type. Synthetics formed from modern durum wheat tend to carry less diversity than synthetics formed from either of the tetraploid emmer types, but they have better agronomic type and are therefore more easily used in applied wheat breeding.

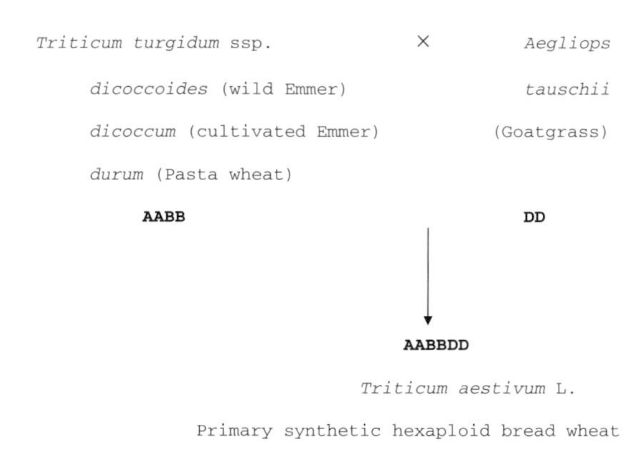

Fig. 16.1 Typical crosses that give rise to primary synthetic hexaploid wheat.

New genetic variability for tolerance to biotic stress

Rust diseases

The primary synthetics have proven to be a valuable source of genetic variability for disease resistance. Rust is the most important disease of wheat worldwide and is controlled by both race-specific and race-nonspecific genes. Much of the variability in race-specific or major gene resistances has eroded with time as pathogen mutation has overcome the effectiveness of these genes, particularly when deployed singularly under disease pressure. Major gene resistance for stem rust (caused by *Puccinia graminis* f. sp. *tritici*) and for leaf rust (caused by *P. triticina* Ericks.) has been reported in synthetic wheat (Kerber and Dyck 1969, 1978; Villareal et al., 1992; Innes and Kerber 1994). Aguilar-Rincón et al. (2000) also reported resistance to leaf rust among a set of five primary synthetics and determined that the genes conferring resistance were different, indicating they could be pyramided in cultivars to improve resistance. However, it is not clear whether or not these genes differ from those previously reported in the wheat gene pool. Assefa and Fehrmann (2004) found that 8% of 169 *Ae. tauschii* accessions exposed to stem rust were resistant. When primary synthetics were developed from these resistant sources, however, expression varied from resistant to susceptible due to poorly understood suppression. Among the materials they tested, several primary synthetics were found to carry different resistance genes.

Interestingly, a number of authors also reported both suppression and overexpression of stripe rust (caused by *P. striiformis* Westend. f. sp. *tritici*) resistance in primary synthetics. Kema et al. (1995) reported that stripe rust resistance expressed in both *Ae. tauschii* and *T. dicoccum* was not expressed in the resultant primary synthetic. Similarly, Ma et al. (1995a) examined 74 primary synthetics and concluded that stripe rust resistance present in the tetraploid and diploid progenitors was not expressed or was partially expressed in some combinations. They found both major gene and adult-plant (or race-nonspecific) resistance expressed among these materials. In a further evaluation of stripe rust resistance the same authors (Ma et al., 1995b) evaluated 34 durum wheat genotypes, 278 *Ae. tauschii* accessions, and 267 primary synthetics in the field. Among the respective groups, 46%, 23%, and 12% were found to have seedling resistance to stripe rust, indicating the presence of major genes. Nevertheless, some seedling-susceptible primary synthetics also displayed adult-plant resistance, indicating the presence of potentially novel race-nonspecific genes, which have yet to be exploited.

Septoria diseases and tan spot

Septoria spp. are also important causes of diseases of wheat, affecting almost 50 million hectares worldwide (Gilchrist and Dubin 2002). The most widely spread is *Septoria tritici* Desm. (speckled leaf blotch), which is prevalent in higher rainfall years in many environments. May and Lagudah (1992) evaluated a wide range of *Ae. tauschii* accessions and a number of primary synthetics and found that 90% of the *Ae. tauschii* accessions and more than 60% of the primary synthetics were resistant to *S. tritici*. In crosses of a few resistant primary synthetics with cultivated wheat, they concluded that a single dominant gene conferred resistance in the synthetic materials and that this resistance was different from that found in cultivated types. However, they did not determine the genetic basis of resistance in the remaining primary synthetics.

Stagonospora nodorum (Berk.) Castellani and Germano (syn. *Septoria nodorum*) is less widely spread but still an important disease. Nicholson et al. (1993) determined the location of resistance in a primary synthetic developed from *T. dicoccum* and concluded that resistance was associated with chromosomes 5D and, to a lesser extent, 3D and 7D. Loughman et al. (2001) tested a set of 433 *Ae. tauschii* accessions and found resistance to be widespread. However, this resistance was either not expressed or only partially expressed in primary synthetics. They concluded that expression was dependent on the specific tetraploid × *Ae. tauschii* combination and that resistance expressed in the best synthetic wheat was equivalent to that in moderately resistant bread wheat cultivars. Xu et al. (2004) tested 120 primary synthetics and their respective durum parents for reaction to *Septoria nodorum* and *Pyrenophora tritici-repentis* (Died.) Drechs, the causal organism for tan spot disease, and found that 47% and 30% of the primary synthetics were resistant to these respective diseases. These diseases are expected to increase in importance as reduced- or zero-tillage management systems become even more widespread, as the fungus overseasons on stubble and crop debris retained on the soil surface.

Karnal bunt

Although Karnal bunt (caused by *Tilletia indica* Mitra) is not an important disease globally, its incidence can limit global wheat trade. For this reason many countries have imposed zero-tolerance restrictions on the importation of Karnal bunt–infected grain, making immunity a goal for breeders. Partial resistance to Karnal bunt can be found in wheat cultivars originating from India and China. However, higher levels of resistance, either through the discovery of immunity or new genes with additive effect, are sought.

Immunity to Karnal bunt was first reported in synthetic amphiploids by Multani et al. (1988). Villareal et al. (1994b) later reported that 49% of the primary synthetics they evaluated were immune to Karnal bunt. The 3-year mean infection of the synthetic materials was <1%, compared to 56% for the susceptible check cultivar. They subsequently determined that resistance was either dominant or partially dominant among a subset of the resistant primary synthetics (Villareal et al., 1994a). However, it has not been possible to transfer the observed complete immunity to cultivated wheat, indicating that resistance may be in part due to the tenacious glumes of most primary synthetics (Valenzuela-Herrera et al., 2006).

Fusarium and powdery mildew diseases

Fusarium graminearum Schwabe, commonly known as head scab or Fusarium head blight, reduces yield and produces toxin-infected grain in high-rainfall environments. Resistance to scab is incomplete and a number of quantitative trait loci (QTL) for resistance has been identified (Liu and Anderson 2003; Pumphrey et al., 2007). Nevertheless, the search for new additive variation or immunity has been extended to synthetic wheat. Although complete resistance has not been found, significant variation for resistance has been reported in primary synthetic wheat (Oliver et al., 2005). In some cases, the resistance is not associated with *Ae. tauschii* but with the tetraploid donor. Hartel et al. (2004) found scab resistance in *T. dicoccoides* associated with chromosome 3A that was expressed in the resultant primary synthetic. Four of these primary synthetics have since been registered as sources of scab resistance (Berzonsky et al., 2004). There are a number of mechanisms of scab resistance (Mesterhazy 1995), and evidence suggests that the synthetics confer type II resistance, which limits fungal spread within the spike (Hartel et al., 2004). Resistance to powdery mildew [*Blumeria graminis* (DC) E.O. Speer f. sp. *tritici*] has also been reported, in association with chromosomes 5D and 7D (Lutz et al., 1995).

Insect pests

Insect damage can significantly reduce wheat yield and quality. Resistance to a range of insect pests has been reported in primary synthetic wheat. The Russian wheat aphid (RWA), *Diuraphis noxia* (Mordvilko), is widespread across Asia and North America, causing significant crop losses when infestations are severe. Resistance in hexaploid bread wheat has been generally associated with the D genome or with rye introgressions (Lage et al., 2004). Resistant primary synthetic wheat accessions, based on crosses to resistant *Ae. tauschii*, have been reported (Nkongolo et al., 1991; Mujeeb-Kazi et al., 2000b). Lage et al. (2004) evaluated 58 primary synthetics developed by crossing *T. dicoccum* with *Ae. tauschii*. The *T. dicoccum* parents were resistant and the *Ae. tauschii* parents were susceptible to RWA. Although the level of resistance in the resultant synthetics was marginally less than the tetraploid parent, effective levels of resistance were observed. Clearly, this variability in combination with D genome or rye-segment resistance would improve the effectiveness of the resistance of cultivated wheat to RWA.

The same set of 58 primary synthetics was also evaluated for greenbug (*Schizaphis graminum* Rondani) resistance (Lage et al., 2003). The greenbug absorbs plant sap and injects a toxin that causes tissue death; subsequent yield losses can be significant. Interestingly, while the expression of resistance was suppressed in some combinations, epistatic interaction gave rise to some synthetics with higher levels of resistance than either parent. Similarly, Smith and Starky (2003) found more than 33% of 149 synthetic materials based on resistant *Ae. tauschii* sources were highly resistant to greenbug.

Hessian fly (*Mayetiola destructor* Say) is widespread across Asia, Europe, and North America. Larvae feed on stem sap, weakening the plant and reducing yield. Resistance has been found and described in *Ae. tauschii* and primary synthetic wheat (Hatchett et al., 1981; Hatchett and Gill 1981). A gene for resistance, *H32* located on chromosome 3D, was recently identified in the primary synthetic parent of the International Triticae Mapping Initiative (ITMI) population (Sardesai et al., 2005). In addition, two primary synthetics resistant to Hessian fly were recently registered for breeding applications (Xu et al., 2006), although their respective resistance genes may be allelic to the previously described *H26* and *H13* genes (Wang et al., 2006).

Soilborne nematodes

Soilborne nematodes have been shown to limit wheat production in some areas of the world and remain undiagnosed in many more. Cereal cyst nematode (*Heterodera avenae* Wollenweber) can be an important constraint in southern Australia, western Asia, and parts of India. A resistance gene designated *Cre3* was found in a primary synthetic and subsequently transferred to adapted wheat (Eastwood et al., 1991). Molecular markers are now available for *Cre3* and routinely used in wheat breeding (Martin et al., 2004). Root lesion nematode (*Pratylenchus thornei* Sher and Allen and *P. neglectus* Rensch) limits wheat yields in Australia and parts of North Africa and western Asia. Zwart et al. (2005) found QTL associated with resistance to both *P. thornei* and *P. neglectus* in a cross with a primary synthetic.

In perhaps the most comprehensive evaluation of primary synthetic wheat, Ogbonnaya et al. (2008) considered 253 synthetics representing 192 unique *Ae. tauschii* accessions and 39 durum wheat genotypes for resistance to cereal cyst nematode, root lesion nematode (both *P. thornei* and *neglectus*), *S. tritici* and *nodorum*, tan spot, leaf rust, stem rust, and stripe rust. Table 16.1 summarizes their findings. Sources of resistance for all of these diseases were found, ranging from only 1% of tested materials for *P. neglectus* up to 73% for *S. tritici*. Although the uniqueness of genes conferring resistance in the primary synthetics has in many instances yet to be established, significant variation is clearly present for reaction to many of the important pathogens and pests affecting wheat in the synthetic wheat gene pool.

Table 16.1 Percentage of lines resistant to various diseases among 253 primary synthetic wheat accessions.

Disease	Percentage of Resistant Lines
Cereal cyst nematode	10
Root lesion nematode (*P. neglectus*)	1
Root lesion nematode (*P. thornei*)	21
Septoria nodorum	10
Septoria tritici	73
Tan spot	10
Leaf rust	15
Stem rust	40
Stripe rust	24

Source: Summarized from Ogbonnaya et al. (2008).

New genetic variability for tolerance to abiotic stress

Abiotic stresses severely limit wheat productivity and product quality in many environments worldwide. Limited available moisture is the most significant of all stresses. It affects crops in most wheat-growing regions and is expected to increase in intensity with climate change. Even the high-yielding river valleys of China and India are projected to suffer moisture deficit as water tables fall and water for irrigation becomes scarce (Reeves et al., 2001). Populations of *Ae. tauschii* and the wild and cultivated emmers have evolved over thousands of years in some of the harshest environments on earth across North Africa and western Asia. Natural selection has in all probability skewed gene frequency in favor of abiotic stress-related gene complexes that, when combined to reconstitute synthetic wheat, produce genetic variation previously unseen in the hexaploid wheat gene pool.

Drought

Although the database of derived synthetic performance under drought is expanding, along with anecdotal evidence of the superior performance of primary synthetic wheat, there remains little published evidence. This probably reflects the difficulty of assessing grain yield of agronomically poor, primary materials that are difficult to thresh. Nevertheless, Villareal et al. (1998) and Villareal and Mujeeb-Kazi (1999) assessed primary synthetics under managed postanthesis drought stress in the Sonoran Desert in northwestern Mexico. They concluded that many synthetics were superior in drought adaptation to their durum progenitors and the adapted hexaploid wheat check cultivar, Seri 82.

Salinity and waterlogging

There is more evidence of salt tolerance in primary synthetic wheat, largely related to the ease with which controlled-environment assays can be applied. Salinity reduces crop yields in irrigated and, to a lesser extent, dryland cropping areas worldwide. According to Szabolzs (1994) 7% of the world's soils are salt-affected, and Mujeeb-Kazi and Diaz de Leon (2002) estimated that up to 10% of the wheatland in south Asia, Iran, Libya, Egypt, and Mexico is salt-affected.

Shah et al. (1987) evaluated primary synthetic wheat accessions and concluded that the D genome from *Ae. tauschii* conferred salt tolerance via Na^+ exclusion. The ability to maintain low Na^+ and high K^+ in leaves is associated with salt tolerance. Gorham (1990) used hydroponic culture to test a range of synthetic hexaploids based on different tetraploid progenitors. He reported good salinity tolerance based on the $K^+:Na^+$ ratio among the hexaploid synthetics and their diploid donors, but concluded that this character has been lost in the evolution of modern durum wheat. Schachtman et al. (1992) tested the salt tolerance of five primary synthetics developed from *Ae. tauschii* materials varying in their degree of salt tolerance. They found that salt tolerance of the *Ae. tauschii* accessions was expressed in the hexaploid primary synthetics and that this tolerance could be attributed to maintenance of seed weight under salt stress. Pritchard et al. (2002) and Dreccer et al. (2004) reported similar results using a much wider range of primary synthetics.

Waterlogging and salinity often limit wheat yield in combination (Konukcu et al., 2006), and waterlogging-tolerant materials such as the CIMMYT line 'Ducula' have been identified (Boru et al., 2001). To augment this variation, primary synthetic wheat has been screened for tolerance to waterlogging and useful variation was identified. Four primary synthetics with good performance in flooded irrigation basins were registered as useful sources of tolerance to waterlogging (Villareal et al., 2001).

Micronutrient imbalance

Of the micronutrient imbalances that limit wheat yields worldwide, the most widespread is Zn deficiency (Cakmak and Tanksley 2000). Primary synthetic hexaploids have been shown to grow well in Zn-deficient soils. Cakmak et al. (1999) tested two synthetic hexaploids, one based on *Ae. tauschii*, the other based on a cross between tetraploid wheat and *T. monococcum*. Zinc efficiency, or ability to grow in Zn-deficient soil, of the two diploid species was expressed in the resultant hexaploids. Similarly, Genc and McDonald (2004) studied 30 primary synthetic hexaploids and compared these to Zn-efficient modern cultivars. They found primary synthetics that were significantly more Zn-efficient (100% efficiency) compared with the best modern materials (85%).

Acidic or low pH soils limit wheat yield in many areas, and genetic variation in synthetic wheat has been found. Zhou et al. (2007) tested 239 primary synthetic accessions for tolerance to aluminum toxicity, a major limitation in acidic soils, and found six tolerant genotypes. They conclude that this tolerance was likely to be different from that previously observed on the basis of lineage. Synthetic hexploid wheat has also been reported with enhanced tolerance to other micronutrient imbalances (Dreccer et al., 2003).

Temperature stress

High-temperature stress limits wheat yield, particularly in lower-latitude environments, affecting up to 70 million hectares (Pfeiffer et al., 2005). As with drought, high-temperature stress is expected to increase with climate change. Significant variation for tolerance to high-temperature stress among wild *Triticum* and *Aegilops* species, including *Ae. tauschii*, has been reported (Ehdaie and Waines 1992; Zaharieva et al., 2001). This tolerance is expressed when combined in hexaploid synthetic backgrounds, although the effectiveness of this germplasm may be limited to specific environments. Yang et al. (2002) compared primary synthetics at 30/25 °C (day/night temperature regime) with adapted cultivars and reported that chlorophyll content, grain-filling duration, grain yield, and kernel weight were negatively correlated with a heat-susceptibility index, but that grain yield was positively correlated at the 20/15°C regime. They concluded that heat tolerance was associated with low yield in the absence of high temperature and that the synthetics may be useful for improving wheat in regions where stress from high temperature occurs frequently. The synthetics could, therefore, provide useful variation for breeders targeting the 7 million hectares of wheat limited by consistently high temperatures (Reynolds et al., 1994).

Freezing temperatures can also limit wheat yield. However, it appears that the potential of synthetic wheat as a source of variation is limited. Although the synthetics do not appear to offer significant new variation for cold tolerance in the vegetative stage (Limin and Fowler 1982, 1993), they may contain useful variation for tolerance to frost at flowering (Maes et al., 2001). Maes et al. (2001) found that floret death under freezing temperatures was delayed in primary synthetics with pubescent glumes by up to 4 minutes. Although statistically significant, the biological significance of these results is not clear.

Preharvest sprouting

Preharvest sprouting is a significant problem in areas where rainfall occurs during harvest. The grain sprouts in the head, reducing both grain yield and economic value of the grain. Seed dormancy is the primary mechanism of tolerance in wheat, and red-grained wheat is generally more tolerant than white-grained types (Gale 1989). Gatford et al. (2002) screened *Ae. tauschii* accessions for preharvest sprouting tolerance and the synthetic hexaploids derived from them. They concluded that tolerance from the D-genome donor was expressed, primarily as seed dormancy, in the resultant synthetic. It was later shown that this source of preharvest sprouting tolerance is different from that previously identified in the wheat gene pool (Ogbonnaya et al., 2007b). The contribution of this dormancy source to the improvement of preharvest sprouting in white-grained wheat is yet to be confirmed.

Grain quality attributes

Wheat processing and product quality are important breeding objectives in many parts of the world, and the definition of quality varies among regions. Three general determinants of wheat quality are grain hardness or endosperm texture, protein content, and protein quality. Hard-grained wheat is required for processing leavened and flat breads and some noodle products, as milling induces starch damage which, in turn, increases water absorption, an important criterion for the development of these products. The *Ae. tauschii* accessions are generally soft-grained, as are the primary synthetics produced from them. The hard endosperm trait arose from a mutation at the *Ha* or hardness locus on chromosome 5D (Lillemo et al., 2006). This locus confers production of puroindoline a (gene *Pina*) and puroindoline b (gene *Pinb*). The two linked genes confer soft endosperm when in their wild-type allelic state (*Pina-D1a*/*Pinb-D1a*). Synthetic wheat introduces seven new alleles of *Pina* and six alleles of *Pinb* from *Ae. tauschii*, all of which confer soft grain (Lillemo et al., 2006). Hence it is necessary to cross primary synthetics with hard-grained sources if the associated array of hard-grained products is expected.

Wheat protein content is largely influenced by the environment and can be inversely related to yield (Peña et al., 2002), although a gene from *T. dicoccoides* has been reported to improve protein content without adversely affecting yield (Davies et al., 2006). Protein quality is important in determining dough extensibility and is largely controlled by high- and low-molecular-weight glutenins and the gliadins. The *Glu-D1* locus of *Ae. tauschii* contributes alleles not found in cultivated bread wheat (William et al., 1993; Peña et al., 1995; Pfluger et al., 2001). Peña et al. (1995) found that synthetics derived from a common durum wheat had better overall quality and bread loaf volume when they possessed the allelic variants 5 + 12 or 1·5 + 10 than when they had any other *Glu-D1* encoded glutenin subunit. Similarly Nelson et al. (2006), in an analysis of quality characteristics in the ITMI population, found recombinant inbred lines with quality superior to both parents. They concluded that the tetraploid and diploid parents of the synthetic materials contributed favorable alleles for quality. Primary synthetics with higher grain protein content, longer kernels, greater kernel weight, and improved SDS-sedimentation compared to Seri 82, an adapted cultivar, have also been observed (Lage et al., 2006).

Grain quality is not only defined in terms of product processing, but also encompasses nutritional quality. Zinc, Fe and vitamin A deficiencies, particularly in women and children, occur in many developing countries where wheat is the primary source of calories (Ortiz-Monasterio et al., 2007). No variation for vitamin A or beta-carotene content has been discovered in the wheat gene pool. However, significant variation in Zn and Fe grain concentration has been identified in a range of materials, ranging from adapted cultivars to landraces and the tetraploid and diploid progenitors of wheat (Ortiz-Monasterio et al., 2007). Calderini and Ortiz-Monasterio (2003) tested a range of primary synthetics and concluded that their higher grain micronutrient concentration was a result of higher nutrient uptake efficiency.

STRATEGIES FOR USING PRIMARY SYNTHETICS IN APPLIED WHEAT BREEDING

Primary synthetic hexaploids are a valuable source of variability for a host of traits conferring resistance to biotic and abiotic stresses. However, while they can be directly crossed with cultivated wheat and do not require embryo rescue, breeders must be aware of potential problems when introducing this variation into their cultivar development program.

Much of the new variation in primary synthetics is not useful variation. When crossing with primary synthetic wheat, it is advisable to make at least one backcross and to grow significantly larger BC_1F_1 and BC_1F_2 generations, thereby allowing greater selection pressure for desired agronomic type in the segregating generations. If the tetraploid used to make the synthetic is less adapted, such as *T. dicoccum* or *T. dicoccoides*, then even larger populations will be required. Characteristics of the primary synthetic which must be removed are tenacious glumes and rachis, speltoide head shape, generally poor agronomic type, and depending on market requirements, the soft red-grained seed and generally poor processing quality.

An additional limitation is the relatively high frequency of F_1 hybrid necrosis when crossing primary synthetics with adapted wheat. Hybrid necrosis is controlled by genes at two loci named *Ne1* and *Ne2* (Pukhalskiy et al., 2000). In the heterozygous state for both loci (*Ne1/ne1* and *Ne2/ne2*), necrosis occurs and plants die in early development. Bread wheat and durum wheat differ in the frequency of these alleles; hence crosses between primary synthetic wheat carrying the durum complement of necrosis genes and bread wheat will often produce hybrid necrosis. If a specific cross combination is sought by the breeder and experience indicates that hybrid necrosis is a problem, it will be necessary to conduct a bridging cross with a compatible parent before top-crossing the resultant F_1 with the desired cultivar.

The breeder should be prepared to conduct two cycles of breeding before materials eligible for commercialization are identified. Experience at CIMMYT indicated that when derivatives expressing the desired trait from the first round of backcrossing with primary synthetics were crossed with adapted wheat, the resultant progeny were significantly better adapted (Lage and Trethowan 2008). Alleles controlling grain quality are, to a large extent, randomly distributed in primary synthetic wheat. Favorable linkage blocks built up over time and conserved in many breeding programs are disrupted when adapted cultivars are crossed with primary synthetics. Many breeders, particularly those from developed countries where specific market classes are targeted because of associated price premiums, tend to be reluctant to use primary synthetics. In these instances a concerted parent building approach is required before this new variation can be used extensively.

Segregating generations can be handled using strategies favored by local programs, although at CIMMYT the larger population sizes and larger numbers of selected plants led to the adoption of a selected bulk methodology (Singh et al., 1998) to maintain gene frequency and reduce costs. Under this strategy, F_2 plants are individually selected on the basis of rust reaction, plant height, flowering date and spike size, color, shape, and threshability. Seed of the selected plants are bulked to form the F_3. This process continues until near-homozygous lines with the desired suite of characters are identified from among F_6-derived head rows. The best derived materials enter CIMMYT international yield trials following yield and quality evaluation, and the most elite enter the crossing nursery for continued improvement. A modified pedigree method is sometimes employed for specific crosses when molecular markers for multiple, simply inherited traits are used (William et al., 2007).

PERFORMANCE OF DERIVED SYNTHETICS

Derived synthetics are hexaploid bread wheat lines that contain a primary synthetic wheat in their pedigree. Primary synthetics are clearly a source of potentially new variation for tolerance to some of the key biotic and abiotic constraints affecting wheat. Wheat breeders have begun using this variation to improve the adaptation and product quality of wheat. This section explores some of the progress made in applied breeding.

Resistance to biotic stress

Variation for disease resistance in primary synthetic wheat has been used extensively in wheat breeding programs. Several synthetic-derived wheat cultivars with enhanced stripe rust resistance have been released in China. Resistance in 'Chuanmai 38' (Zhang et al., 2006a), 'Chuanmai 42' (Zhang et al., 2006b) and 'Chuanmai 47' (Li et al., 2007) was reported to be controlled by different single dominant genes. Singh and Huerta-Espino (2000) estimated that synthetic wheat has contributed up to four new seedling resistance genes for stripe rust in the CIMMYT advanced wheat germplasm. The synthetic-derived wheat cultivar Carmona, registered in Spain, is reported to be resistant to foliar diseases and to be well adapted to zero-tillage management (van Ginkel and Ogbonnaya 2007). Mujeeb-Kazi et al. (2001) transferred Karnal bunt resistance from primary synthetic wheat into adapted wheat and reported an equivalent level of resistance. Newly derived synthetics have also been registered as trait-specific germplasm with enhanced resistance to *S. tritici* (Mujeeb-Kazi et al., 2000a) and spot blotch (caused by *Cochliobolus sativus* Ito and Kurib Drechs. Ex Daster).

Tolerance to abiotic stress

Perhaps the most exciting application of new genetic variation from synthetic wheat is in the improvement of wheat for tolerance to abiotic stresses. Significant variation for simply inherited resistance to many diseases exists within the wheat gene pool, but less heritable variation has been available for abiotic stress tolerance with more complex inheritance. While there is a dearth of published evidence on the performance of primary synthetics under drought, much more evidence is documented on the performance of derived materials. Trethowan et al. (2000, 2003, 2007) evaluated derived synthetics under terminal drought stress in Mexico using limited irrigation and found derivatives with up to 23% higher yield than their recurrent parents and 33% higher yield than the check cultivar.

While it is possible to see such responses in one environment under one set of environmental conditions, the real value of this tolerance depends on its transferability to other environments and years. Synthetic wheat has been used extensively in CIMMYT's bread wheat breeding program since the early 1990s (van Ginkel and Ogbonnaya 2007). Currently, approximately 50% of the pedigrees in CIMMYT's rainfed breeding program contain at least one synthetic progenitor (Trethowan et al., 2003). Derived synthetics were distributed worldwide in CIMMYT's international yield trial network, and collaborators in many countries have returned performance data. One such trial is the Semi-Arid Wheat Yield Trial (SAWYT), which targets rainfed environments. Lage and Trethowan (2008) analyzed the yield performance of those materials from the deployment of the first synthetic derivatives in yield trials in 1996, when they comprised 8% of the total entries, to 2006 when 46% of the entries were derived from synthetic wheat. Between the years 1996 and 2006, the average contribution of primary synthetics to pedigrees in the SAWYT decreased from 75% to 19% based on coefficients of parentage. The extent of this reduction was supported by subsequent microsatellite (SSR) analyses of recent derivatives, which indicated that only a small portion of the primary synthetic genome remained (Zhang et al., 2005). The average yield rank of the synthetic-derived materials in the 50-entry SAWYT improved from 30th to 25th position over the period. Synthetic derivatives were identified that were superior to the local check cultivars across a wide range of global environments and yield levels. Those derivatives possessing a smaller portion of the primary synthetic genome tended to perform better, indicating that at least one backcross is required to improve agronomic type sufficiently to exploit underlying variation for yield and adaptation.

CIMMYT screens the derived materials in the Sonoran Desert in northwestern Mexico using limited irrigation to simulate specific stress regimes. There was interest in Australia, where wheat is frequently grown in moisture-stressed conditions, in testing the value of the materials bred and selected in Mexico. When the synthetic derivatives developed in Mexico were tested under Australian conditions in multienvironment trials, a yield advantage between 8% and 30% compared with the best local check cultivar was observed (Ogbonnaya et al., 2007a). In a separate analysis of a different set of 156 derivatives, 56% of the materials were found to be higher yielding than the best local check cultivar across multiple locations (Dreccer et al., 2007). Both authors noted strong genotype × environment interactions and concluded that northern Australian wheat-growing environments were similar to those in Mexico, while southern environments ranked materials differently. The most likely driver of this north–south division was phenology; later-maturing materials tended to perform better in cooler southern areas of Australia (Mathews et al., 2007). Azizinya et al. (2005) evaluated synthetic derivatives developed in Mexico in Iran and found lines that outperformed the local check cultivars, producing heavier, longer spikes with better spikelet fertility.

The preceding results show that the performance of synthetic derivatives developed under managed stress in one country can be transferred to dry environments in other locations around the world. In a different comparison, Gororo et al. (2002) produced synthetic hexaploid derivatives from primary synthetics developed in Australia and tested those materials in both Australia and Mexico. The synthetic derivatives produced in Australia were up to 32% higher yielding than their recurrent parent and exceeded the check in 38 of 42 comparisons in multienvironment trials in both countries.

The evaluation of derivatives in multienvironment trials clearly indicates that useful variation for adaptation, particularly to drought stress, can be obtained from synthetic wheat. Reynolds et al. (2007) and Reynolds and Trethowan (2007) evaluated synthetic backcross derivatives under drought stress and found that the 15%-to-33% yield advantage of the derivatives compared with their recurrent parents could be attributed to a greater investment in root biomass deeper in the soil profile. Interestingly, this investment in deep roots at the expense of surface roots led to a 46% lower root–shoot ratio under drought stress. Synthetic derivatives were also able to better maintain greater seed weight under drought and high-temperature stress (Trethowan et al., 2005). Larger seed size may also have favorable implications for other characters, such as improved milling yield (Marshall et al., 1986).

Many synthetic derived materials show improved stay-green or green-leaf duration, whereby leaves and stems maintain their photosynthetic capability longer in the crop cycle than plants with less stay-green capability. This character has been implicated in conferring heat tolerance in wheat (Reynolds et al., 1994). Figure 16.2 shows the performance of several synthetic wheat derivatives tested in the field in northwestern Mexico. These materials were sown in mid-January, 8 weeks later than the optimal planting date for wheat, thereby subjecting the crop to temperatures in excess of 35 °C during anthesis and subsequent grain filling. The plots were fully irrigated to ensure drought stress did not confound the results. Synthetic derivatives with 30% higher yield compared to the commonly grown check cultivar were identified. However, the transferability of this response to other environments is yet to be determined. While significant variation among primary synthetics for tolerance to salinity, low soil Zn, B toxicity, and waterlogging has been found (see previous section, "Primary synthetic hexaploid wheat") and used by plant breeders in their crossing programs (Villareal and Mujeeb-Kazi 1999; Reynolds et al., 2005), no published evidence exists to date of the performance of the derived materials.

Ortiz-Monasterio et al. (2007) reported an attempt to improve the nutritional value of adapted wheat cultivars by transferring high Zn and Fe concentration from a *T. dicoccum*–based primary synthetic wheat. However, as most micronutrients are found in the aleurone layer, it is difficult to simultaneously improve yield and micronutrient concentration; increasing yield decreases the aleurone–endosperm ratio, thereby diluting grain Fe and Zn concentration. The resultant synthetic derivatives were developed by selecting for Zn and Fe concentration in Mexico. They were subsequently tested in multienvironment yield trials across Pakistan, and derivatives higher yielding than the check cultivars and recurrent parents were observed. However, the Zn concentration of these lines did not change and the improvement in Fe concentration, while significant, was not high enough to have an impact on human nutrition. To date, little variation in endosperm Fe and Zn concentration has been found both within and external to the synthetic wheat gene pool.

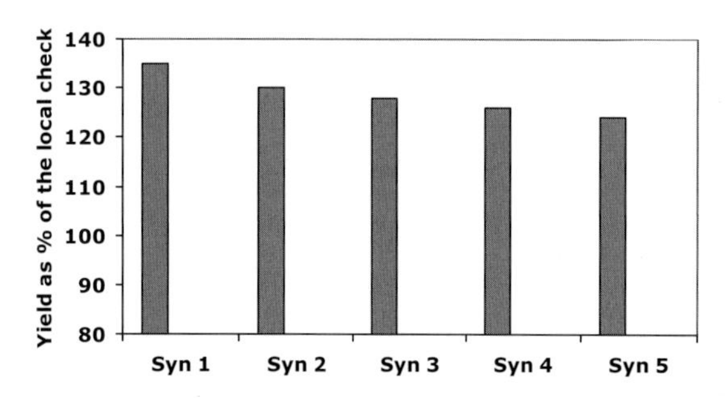

Fig. 16.2 Grain yield of five synthetic derivatives in late-sown trials compared with the locally adapted check cultivar in Sonora, Mexico, 2004. All lines were higher yielding than the check at $P < 0.01$.

FUTURE PERSPECTIVES

The wealth of synthetic hexaploid germplasm produced has been only partially characterized, and variation for hitherto unreported traits, such as resistance to take-all (caused by *Gaeumannomyces graminis* var. *tritici* Sacc.) or variation for beta-carotein, may still yet be found. The exploration of wild tetraploid germplasm, particularly *T. dicoccum* and *T. dicoccoides*, as potential sources of new variation in synthetic hexaploid wheat backgrounds is very much in its infancy.

Attempts have been made to synthesize tetraploid wheat from crosses of *T. monococcum* × *T. urartu*, *T. sinskajae* × *T. urartu*, *T. boeoticum* × *T. urartu*, *Ae. tauschii* × *Agropyron cristatum* among others (Gandilyan et al., 1986; Martin et al., 1999). However, these new tetraploids have had little impact on wheat breeding, as exploitation of the wealth of variation in existing wild and cultivated tetraploid forms has proven to be more successful. Huang et al. (1999) estimated that there is more variability among tetraploid wheats than hexaploid forms, underscoring the value of this approach. A more successful alternative to developing synthetic tetraploids has been to directly cross durum wheat with *T. boeticum* and *T. urartu* for the improvement of stripe rust resistance (Valkoun 2001).

As much of the variation in synthetic wheat is not agronomically useful, the challenge for plant breeders is to identify and introduce useful genes into their elite materials. Even though Zhang et al. (2005) found that little of the primary synthetic parent remained after crossing and selection, the introduction of synthetic materials into the CIMMYT wheat breeding program has significantly increased the latent diversity among more recently developed advanced lines (Warburton et al., 2006).

The constraints facing the global wheat improvement research community are becoming ever more difficult. Exhaustion of genetic variability for disease resistance, increasing human population pressures, reduced availability of water for agriculture, and a changing climate present significant challenges. Synthetic hexaploid wheat is likely to be an important source of new genetic variation for the mitigation of these constraints.

REFERENCES

Aguilar-Rincón, V.H., P.R. Singh, and J. Huerta-Espino. 2000. Inheritance of resistance to leaf rust in four synthetic hexaploid wheats. Agrociencia 34:235–246.

Appels, R., and E.S. Lagudah. 1990. Manipulation of chromosomal segments from wild wheat for the improvement of bread wheat. Aust. J. Plant Physiol. 17:253–366.

Assefa, S., and H. Fehrmann. 2004. Evaluation of *Aegilops tauschii* Coss. for resistance to wheat stem rust and inheritance of resistance genes in hexaploid wheat. Genet. Resour. Crop Evol. 51:663–669.

Azizinya, S., M.R. Ghanadha, A.A. Zali, B.Y. Samadi, and A. Ahmadi. 2005. An evaluation of quantitative traits related to drought resistance in synthetic wheat genotypes in stress and non-stress conditions. Iran. J. Agric. Sci. 36:281–293.

Berzonsky, W.A., K.D. Hartel, S.F. Kianian, and G.D. Leach. 2004. Registration of four synthetic hexaploid wheat germplasm lines with resistance to Fusarium head blight. Crop Sci. 44:1500–1501.

Boru, G., M. van Ginkel, W.E. Kronstad, and L. Boersma. 2001. Expression and inheritance of tolerance to waterlogging stress in wheat. Euphytica 117:91–98.

Cakmak, O., S. Eker, A. Ozdemir, N. Watanabe, and H.J. Braun. 1999. Expression of high zinc efficiency of *Aegilops tauschii* and *Triticum monococcum* in synthetic hexaploid wheats. Plant Soil 215:203–209.

Cakmak, O., and M. Tanksley. 2000. Possible roles of zinc in protecting plant cells from damage by reactive oxygen species. New Phytol. 146:185–205.

Calderini, D.F., and I. Ortiz-Monasterio. 2003. Are synthetic hexaploids a means of increasing grain element concentrations in wheat? Euphytica 134:169–178.

Davies, J., W.A. Berzonsky, and G.D. Leach. 2006. A comparison of marker-assisted and phenotypic selection for high grain protein content in spring wheat. Euphytica 152:117–134.

Dreccer, M.F., M.G. Borgognone, F.C. Ogbonnaya, R.M. Trethowan, and B. Winter. 2007. CIMMYT-selected synthetic bread wheats for rainfed environments: Yield evaluation in Mexico and Australia. Field Crops Res. 100:218–228.

Dreccer, M.F., F.C. Ogbonnaya, and M.G. Borgognone. 2004. Sodium exclusion in primary synthetic wheats. p. 118–121. *In* C.K. Black, J.F. Panozzo, and G.J. Rebetzke (ed.) Proc. Aust. Cereal Chem. Conf., 54th & Wheat Breeders Assembly, 11th, Canberra. 21–24 Sept. 2004. Canberra Cereal Chemistry Division, RACI.

Dreccer, M.F., F. Ogbonnaya, G. Borgognone, and J. Wilson. 2003. Boron tolerance is present in primary synthetic wheats. p. 1130–1132. *In* N.E. Pogna, M. Romano, E.A. Pogna, and G. Galterio (ed.) Proc. Int. Wheat Genet. Symp., 10th, Paestum, Italy. 1–6 Sept. 2003. Istituto Sperimentale per la Cerealcoltura, Paestum, Italy.

Eastwood, R.F., E.S. Lagudah, R. Appels, M. Hannah, and J.F. Kollmorgen. 1991. *Triticum tauschii*: A novel source of resistance to cereal cyst nematode (*Heterodera avenae*). Aust. J. Agric. Res. 42:69–77.

Ehdaie, B., and J.G. Waines. 1992. Heat resistance in wild *Triticum* and *Aegilops*. J. Genet. Breed. 46:221–227.

Elias, E.M., and F.A. Manthey. 2005. End products: Present and future uses. p. 63–86. *In* C. Royo, M. Nachit, N. Di Fonzo, J.L. Araus, W.H. Pfeiffer, and G.A. Slafer (ed.) Durum wheat breeding. Haworth Press, Binghamton, NY.

Feldman, M. 2001. Origin of cultivated wheat. p. 3–53. *In* A.P. Bonjean and W.J. Angus (ed.) The world wheat book: A history of wheat breeding. Lavoisier Publishing, Paris, France.

Gale, M.D. 1989. The genetics of preharvest sprouting in cereals, particularly in wheat. p. 85–110. *In* N.F. Derera (ed.) Preharvest field sprouting in cereals. CRC Press, Boca Raton, FL.

Gandilyan, P.A., Zh.O. Shakaryan, and E.A. Petrosyan. 1986. Synthesis of new emmers and tetraploid speltoids, and problems of phylogeny in the wheat genus. Biolog. Zh. Armenii 39:5–15.

Gatford, K.T., P. Hearnden, F. Ogbonnaya, R.F. Eastwood, and G.M. Halloran. 2002. Novel resistance to pre-harvest sprouting in Australian wheat from the wild relative *Triticum tauschii*. Euphytica 126:67–76.

Genc, Y., and G.K. McDonald. 2004. The potential of synthetic hexaploid wheats to improve zinc efficiency in modern bread wheat. Plant Soil 262:23–32.

Gilchrist, L., and H.J. Dubin. 2002. Septoria diseases of wheat. p. 273–277. *In* B.C. Curtis, S. Rajaram, and H. Gomez-Macpherson (ed.) Bread wheat: Improvement and production. FAO Plant Production and Protection Series, No. 30. FAO, Rome, Italy.

Gorham, J. 1990. Salt tolerance in the *Triticeae*: Ion discrimination in synthetic hexaploid wheats. J. Exp. Bot. 41:623–627.

Gororo, N.N., H.A. Eagles, R.F. Eastwood, M.E. Nicolas, and R.G. Flood. 2002. Use of *Triticum tauschii* to improve yield of wheat in low-yielding environments. Euphytica 123:241–254.

Hartel, K.D., W.A. Berzonsky, S.F. Kianian, and S. Ali. 2004. Expression of a *Triticum turgidum* var. *dicoccoides* source of Fusarium head blight resistance transferred to synthetic hexaploid wheat. Plant Breed. 123:516–519.

Hatchett, J.H., and B.S. Gill. 1981. D-genome sources of resistance in *Triticum tauschii* to Hessian fly. J. Hered. 72:126–127.

Hatchett, J.H., T.J. Martin, and R.W. Livers. 1981. Expression and inheritance of resistance to Hessian fly in synthetic hexaploid wheats derived from *Triticum tauschii* (Coss) Schmal. Crop Sci. 21:731–734.

Huang, L., E. Millet, J. Rong, J.F. Wendel, Y. Anikster, and M. Feldman. 1999. Restriction fragment length polymorphism in wild and cultivated tetraploid wheat. Isr. J. Plant Sci. 47:213–224.

Innes, R.L., and E.R. Kerber. 1994. Resistance to wheat leaf rust and stem rust in *Triticum tauschii* and inheritance in hexaploid wheat of resistance transferred from *T. tauschii*. Genome 37:813–822.

Kema, G.H.J., W. Lange, and C.H. Silfhout. 1995. Differential suppression of stripe rust resistance in synthetic wheat hexaploids derived from *Triticum turgidum* ssp. *dicoccoides* and *Aegilops squarrosa*. Phytopathology 85:425–429.

Kerber, E.R., and P.L. Dyck. 1969. Inheritance in hexaploid wheat of leaf rust resistance and other characters derived from *Aegilops squarrosa*. Can. J. Genet. Cytol. 11:639–647.

Kerber, E.R., and P.L. Dyck. 1978. Resistance to stem and leaf rust on wheat in *Aegilops squarrosa* and transfer of resistance of a gene for stem rust resistance to hexaploid wheat. p. 358–364. *In* S. Ramanujam (ed.) Proc. Int. Wheat Genet. Symp., 5th, New Delhi, India. 23–28 Feb. 1978. Indian Society of Genetics and Plant Breeding, New Delhi India.

Kihara, H., 1944. Discovery of the DD analyser, one of the ancestors of *Triticum vulgare*. Agric. Hortic. 19:889–890.

Konukcu, F., J.W. Gowing, and D.A. Rose. 2006. Dry drainage: A sustainable solution to waterlogging and salinity problems in irrigation areas? Agric. Water Manage. 83:1–12.

Lage, J., B. Skovmand, and S.B. Andersen. 2003. Expression and suppression of resistance to greenbug (Homoptera: Aphididae) in synthetic hexaploid wheats derived from *Triticum dicoccum* × *Aegilops tauschii* crosses. J. Econ. Entomol. 96:202–206.

Lage, J., B. Skovmand, and S.B. Andersen. 2004. Field evaluation of emmer wheat derived synthetic hexaploid wheats for resistance to Russian wheat aphid (Homoptera: Aphididae). J. Econ. Entomol. 97:1065–1070.

Lage, J., B. Skovmand, R.J. Peña, and S.B. Andersen. 2006. Grain quality of *Triticum dicoccum* × *Aegilops tauschii* derived synthetic hexaploid wheats. Genet. Resour. Crop Evol. 53:955–962.

Lage, J., and R.M. Trethowan. 2008. CIMMYT's use of synthetic hexaploid wheat in breeding for adaptation to rainfed environments globally. Aust. J. Agric. Res. 59:461–469.

Li, J., H.T. Wei, X.R. Hu, Z.S. Peng, and W.Y. Yang. 2007. Molecular mapping of stripe rust resistance gene in synthetic derivative Chuanmai 47. J. Agric. Biotechnol. 15:318–322.

Lillemo, M., F. Chen, X.C. Xia, M. William, R.J. Pena, R.M. Trethowan, and Z.H. He. 2006. Puroindoline grain hardness alleles in CIMMYT bread wheat germplasm. J. Cereal Sci. 44:86–92.

Limin, A.E., and D.B. Fowler. 1982. The expression of cold hardiness in *Triticum* species amphiploids. Can. J. Genet. Cytol. 24:51–56.

Limin, A.E., and D.B. Fowler. 1993. Inheritance of cold hardiness in *Triticum aestivum* × synthetic hexaploid wheat crosses. Plant Breed. 110:103–108.

Liu, S.X., and J.A. Anderson. 2003. Marker assisted evaluation of Fusarium head blight resistant wheat germplasm. Crop Sci. 43:760–766.

Loughman, R., E.S. Lagudah, M. Trottet, R.E. Wilson, and A. Mathews. 2001. Septoria nodorum blotch resistance in *Aegilops tauschii* and its expression in synthetic amphiploids. Aust. J. Agric. Res. 52:1393–1402.

Lutz, J., S.L.K. Hsam, E. Limpert, and F.J. Zeller. 1995. Chromosomal location of powdery mildew resistance genes in *Triticum aestivum* L. (common wheat). 2. Genes *Pm2* and *Pm19* from *Aegilops squarrosa* L. Heredity 74:152–156.

Ma, H., R.P. Singh, and A. Mujeeb-Kazi. 1995a. Resistance to stripe rust in *Triticum turgidum*, *T. tauschii* and their synthetic hexaploids. Euphytica 82:117–124.

Ma, H., R.P. Singh, and A. Mujeeb-Kazi A. 1995b. Suppression/expression of resistance to stripe rust in synthetic hexaploid wheat (*Triticum turgidum* × *T. tauschii*). Euphytica 83:87–93.

Maes, B., R.M. Trethowan, M.P. Reynolds, M. van Ginkel, and B. Skovmand. 2001. The influence of glume pubescence on spikelet temperature of wheat under freezing conditions. Aust. J. Plant Physiol. 28:141–148.

Marathee, J., and H. Gomez-MacPherson. 2001. Future world supply and demand. p. 1107–1116. *In* A.P. Bonjean and W.J. Angus (ed.) The world wheat book: A history of wheat breeding. Lavoisier Publishing, Paris, France.

Marshall, D.R., D.J. Mares, H.J. Moss, and F.W. Ellison. 1986. Effects of grain shape and size on milling yields in wheat. II. Experimental studies. Aust. J. Agric. Res. 37:331–342.

Martin, A., A. Cabrera, E. Esteban, P. Hernandez, M.C. Ramirez, and D. Rubiales. 1999. A fertile amphiploid between diploid wheat (*Triticum tauschii*) and crested wheatgrass (*Agropyron cristatum*). Genome 42:519–524.

Martin, E.M., R.F. Eastwood, and F.C. Ogbonnaya. 2004. Identification of microsatellite markers associated with the cereal cyst nematode resistance gene *Cre3* in wheat. Aust. J. Agric. Res. 55:1205–1211.

Mathews, K.L., S.C. Chapman, R.M. Trethowan, W. Pfeiffer, M. van Ginkel, J. Crossa, T. Payne, I. DeLacy, P.N. Fox, and M. Cooper. 2007. Global adaptation patterns of Australian and CIMMYT spring bread wheats. Theoret. Appl. Genet. 115:819–835.

May, C.E., and E.S. Lagudah. 1992. Inheritance in hexaploid wheat of Septoria tritici blotch resistance and other characteristics derived from *Triticum tauschii*. Aust. J. Agric. Res. 43:433–442.

McFadden, E.S., and E.R. Sears. 1946. The origin of *Triticum spelta* and its free-threshing hexaploid relatives. J. Hered. 37:81–89.

Mesterhazy, A. 1995. Types and components of resistance to Fusarium head blight of wheat. Plant Breed. 114:377–386.

Mujeeb-Kazi, A., and J.L. Diaz de Leon. 2002. Conventional and alien genetic diversity for salt tolerant wheats: Focus on current status and new germplasm development p. 69–82. *In* R. Ahmad and K.A. Malik (ed.) Prospects for saline agriculture. Kluwer Academic Publishers, Dordrecht, The Netherlands.

Mujeeb-Kazi, A., G. Fuentes-Davila, R.L. Villareal, A. Cortes, V. Roasas, and R. Delgado. 2001. Registration of 10 synthetic hexaploid wheat and six bread wheat germplasms resistant to Karnal bunt. Crop Sci. 41:1652–1653.

Mujeeb-Kazi, A., L.I. Gilchrist, R.L. Villareal, and R. Delgado. 2000a. Registration of 10 wheat germplasms resistant to *Septoria tritici* leaf blotch. Crop Sci. 40:590–591.

Mujeeb-Kazi, A., B. Skovmand, M. Henry, R. Delgado, and S. Cano. 2000b. New synthetic hexaploids (*Triticum dicoccum/Aegilops tauschii*): Their production, cytology, and utilization as a source for Russian wheat aphid resistance. Annu. Wheat Newsl. 46:79–81.

Multani, D.S., H.S. Dhaliwal, P. Singh, and K.S. Gill. 1988. Synthetic amphiploids of wheat as a source of resistance to Karnal bunt (*Neovossi indica*). Plant Breed. 101:122–125.

Nelson, J.C., C. Andreescu, F. Breseghello, P.L. Finney, D.G. Gualberto, C.J. Bergman, R.J. Pena, M.R. Perretant, P. Leroy, C.O. Qualset, and M.E. Sorrells. 2006. Quantitative trait locus analysis of wheat quality traits. Euphytica 149:145–159.

Nicholson, P., H.N. Rezannor, and A.J. Worland. 1993. Chromosomal location of resistance to *Septoria nodorum* in a synthetic hexaploid wheat determined by the study of chromosomal substitution lines in 'Chinese Spring' wheat. Plant Breed. 110:177–184.

Nkongolo, K.K., J.S. Quick, A.E. Limin, and D.B. Fowler. 1991. Sources and inheritance of resistance to Russian wheat aphid in *Triticum* species amphiploids and *Triticum tauschii*. Can. J. Plant Sci. 71:703–708.

Ogbonnaya, F., F. Dreccer, G. Ye, R.M. Trethowan, D. Lush, J. Shepperd, and M. van Ginkel. 2007a. Yield of synthetic backcross-derived lines in rainfed environments of Australia. Euphytica 157:321–336.

Ogbonnaya, F.C., M. Imtiaz, and R.M. DePauw. 2007b. Haplotype diversity of preharvest sprouting QTLs in wheat. Genome 50:107–118.

Ogbonnaya, F.C., M. Muhammad, H. Bariana, M. Shankar, G. Hollaway, R. Trethowan, E. Lagudah, and M. van Ginkel. 2008. Mining synthetic hexaploids for multiple disease resistance to improve bread wheat. Aust. J. Agric. Sci. 59:421–431.

Oliver, R.E., X. Cai, S.S. Xu, X. Chen, and R.W. Stack. 2005. Wheat–alien species derivatives: A novel source of resistance to Fusarium head blight in wheat. Crop Sci. 45:1353–1360.

Ortiz-Monasterio, J.I., N. Palacios, E. Meng, K. Pixley, R. Trethowan, and R.J. Peña. 2007. Enhancing the mineral and vitamin content of wheat and maize through plant breeding. J. Cereal Sci. 46:293–307.

Peña, R.J., R.M. Trethowan, W.H. Pfeiffer, and M. van Ginkel. 2002. Quality (end-use) improvement in wheat: Compositional, genetic, and environmental factors. J. Crop Prod. 5:1–37.

Peña, R.J., J. Zarco-Hernandez, and A. Mujeeb-Kazi. 1995. Glutenin subunit compositions and bread-making quality characteristics of synthetic hexaploid wheats derived from *Triticum turgidum* × *Triticum tauschii* (Coss.) Schmal crosses. J. Cereal Sci. 21:15–23.

Pfeiffer, W.H., R.M. Trethowan, M. van Ginkel, I. Ortiz-Monasterio, and S. Rajaram. 2005. Breeding for abiotic stress tolerance in wheat. p. 401–489. *In* M. Ashraf and P.J.C. Harris (ed.) Abiotic stresses: Plant resistance through breeding and molecular approaches. Haworth Press, Binghamton, NY.

Pfluger, L.A., R. D'Ovidio, B. Margiotta, R. Pena, A. Mujeeb-Kazi, and D. Lafiandra. 2001. Characterisation of high- and low-molecular weight glutenin subunits associated to the D genome of *Aegilops tauschii* in a collection of synthetic hexaploid wheats. Theor. Appl. Genet. 103:1293–1301.

Pritchard, D.J., P.A. Hollington, W.P. Davies, J. Gorham, J.L. Diaz de Leon, and A. Mujeeb-Kazi. 2002. K+/Na+ discrimination in synthetic hexaploid wheat lines: Transfer of the trait for K+/Na+ discrimination from *Aegilops tauschii* into a *Triticum turgidum* background. Cereal Res. Commun. 30:261–267.

Pukhalskiy, V.A., S.P. Martynov, and T.V. Dobrotvorskaya. 2000. Analysis of geographical and breeding-related distribution of hybrid necrosis genes in bread wheat (*Triticum aestivum* L.). Euphytica 114:233–240.

Pumphrey, M.O., R. Bernardo, and J.A. Anderson. 2007. Validating the *Fhb1* QTL for Fusarium head blight resistance in near-isogenic wheat lines developed from breeding populations. Crop Sci. 47:200–206.

Rajaram, S., and M. van Ginkel. 2001. Mexico: 50 years of international wheat breeding. p. 579–608. *In* A.P. Bonjean and W.J. Angus (ed.) The world wheat book: A history of wheat breeding. Lavoisier Publishing, Paris, France.

Reeves, T.G., P.L. Pingali, S. Rajaram, and K. Cassaday. 2001. Crop and natural resource management strategies to foster sustainable wheat production in developing countries. p. 23–36. *In* Z. Bedo and L. Lang (ed.) Wheat in a global environment. Kluwer Academic Publishers, Dordrecht, The Netherlands.

Reynolds, M.P., M. Balota, M.I.B. Delgado, I. Amani, and R.A. Fischer. 1994. Physiological and morphological traits associated with spring wheat yield under hot, irrigated conditions. Aust. J. Plant Physiol. 21:717–730.

Reynolds, M., F. Dreccer, and R. Trethowan. 2007. Drought-adaptive traits derived from wheat wild relatives and landraces. J. Exp. Bot. 58:177–186.

Reynolds, M.P., A. Mujeeb-Kazi, and M. Sawkins. 2005. Prospects for utilising plant-adaptive mechanisms to improve wheat and other crops in drought- and salinity-prone environments. Ann. Appl. Biol. 146:239–259.

Reynolds, M.P., and R.M. Trethowan. 2007. Physiological interventions in breeding for adaptation to abiotic stress. p. 129–146. *In* J.H.J. Spiertz, P.C. Struik, and H.H. van Laar (ed.) Scale and complexity in plant systems research, gene–plant–crop relations. Springer, Dordrecht, The Netherlands.

Sardesai, N., J.A. Nemacheck, S. Subramanyam, and C.E. Williams. 2005. Identification and mapping of *H32*, a new wheat gene conferring resistance to Hessian fly. Theor. Appl. Genet. 111:1167–1173.

Schachtman, D.P., E.S. Lagudah, and R. Munns. 1992. The expression of salt tolerance from *Triticum tauschii* in hexaploid wheat. Theor. Appl. Genet. 84:714–719.

Shah, S.H., J. Gorham, B.P. Forster, and R.G.W. Jones. 1987. Salt tolerance in the Triticeae: The contribution of the D-genome to cation selectivity in hexaploid wheat. J. Exp. Bot. 38:254–269.

Singh, R.P., and J. Huerta-Espino. 2000. Sources and genetic basis of variability of major and minor genes for yellow rust resistance in CIMMYT wheats. p. 144–151. *In* Proc. Regional Wheat Workshop for Eastern, Central and Southern Africa, 11th, Addis Ababa, Ethiopia. 18–22 Sept. 2000. CIMMYT, D.F., Mexico.

Singh, R.P., S. Rajaram, A. Miranda, J. Huerta-Espino, and E. Autrique. 1998. Comparison of two crossing and four selection schemes for yield, yield traits, and slow rusting resistance to leaf rust in wheat. Euphytica 100:35–43.

Smith, C.M., and S. Starky. 2003. Resistance to greenbug (Heteroptera: Aphididae) biotype I in *Aegilops tauschii* synthetic wheats. J. Econ. Entomol. 96:1571–1576.

Sneep, J. 1966. Some facts about plant breeding before the discovery of Mendelism. Euphytica 15:135–140.

Szabolzs, I. 1994. Soils and salinization. p. 3–11. *In* M. Pessarakli (ed.) Handbook of plant crops stress. Marcel Dekker, New York.

Trethowan, R.M., J. Borja, and A. Mujeeb-Kazi. 2003. The impact of synthetic wheat on breeding for stress tolerance at CIMMYT. Annu. Wheat Newsl. 49:67–69.

Trethowan, R.M., and M.P. Reynolds. 2007. Drought resistance: Genetic approaches for improving productivity under stress. p. 289–300. *In* H.T. Buck, J.E. Nisi, and N. Salomon (ed.) Proc. Int. Wheat Conf., 7th, Mar del Plata, Argentina. 27 Nov.–2 Dec. 2005. Springer, New York.

Trethowan, R.M., M.P. Reynolds, K.D. Sayre, and I. Ortiz-Monasterio. 2005. Adapting wheat cultivars to resource conserving farming practices and human nutritional needs. Ann. Appl. Biol. 146:404–413.

Trethowan, R.M., M. van Ginkel, and A. Mujeeb-Kazi. 2000. Performance of advanced bread wheat × synthetic hexaploid derivatives under reduced irrigation. Annu. Wheat Newsl. 46:87–88.

Valenzuela-Herrera, V., V.H. Aguilar-Rincón, G. García-De los Santos, R. Trethowan, and G. Fuentes-Dávila. 2006. Herencia de la resistencia a *Tilletia indica* (Mitra) del sintetico hexaploide Ruff (*Triticum turgidum* L.)/*T. tauschii* (Coss.) Schmalh. × *Triticum aestivum* L. Rev. Mexicana Fitopatol. 24:153–159.

Valkoun, J.J. 2001. Wheat pre-breeding using wild progenitors. Euphytica 119:17–23.

van Ginkel, M., and F. Ogbonnaya. 2007. Using synthetic wheats to breed cultivars better adapted to changing production conditions. Field Crops Res. 104:86–94.

Villareal, R.L., O. Banuelos, J. Borja, and A. Mujeeb-Kazi. 1998. Drought tolerance of synthetic wheats (*Triticum turgidum* × *Aegilops tauschii*). Annu. Wheat Newsl. 44:142–144.

Villareal, R.L., and A. Mujeeb-Kazi. 1999. Exploiting synthetic hexaploids for abiotic stress tolerance in wheat. p. 542–552. *In* (ed.) Proc. Regional Wheat Workshop for Eastern, Central and Southern Africa, 10th, Univ. Stellenbosch, South Africa. 14–18 Sept. 1998. CIMMYT, D.F., Mexico.

Villareal, R.L., A. Mujeeb-Kazi, E. del Toro, J. Crossa, and S. Rajaram. 1994a. Agronomic variability in selected *Triticum turgidum* × *T. tauschii* synthetic hexaploid wheats. J. Agron. Crop Sci. 173:307–317.

Villareal, R.L., A. Mujeeb-Kazi, G. Fuentes-Davila, S. Rajaram, and E. del Toro. 1994b. Resistance to karnal bunt (*Tilletia indica* Mitra) in synthetic hexaploid wheats derived from *Triticum turgidum* × *T. tauschii*. Plant Breed. 112:63–69.

Villareal, R.L., K. Sayre, O. Banuelos, and A. Mujeeb-Kazi. 2001. Registration of four synthetic hexaploid wheat (*Triticum turgidum*/*Aegilops tauschii*) germplasm lines tolerant to waterlogging. Crop Sci. 41:274.

Villareal, R.L., R.P. Singh, and A. Mujeeb-Kazi. 1992. Expression of resistance to *Puccinia recondita* f. sp. *tritici* in synthetic hexaploid wheats. Vorträge Pflanzenzuecht. 24:253–255.

Wang, T., S.S. Xu, M.O. Harris, J.G. Hu, L.W. Liu, and X.W. Cai. 2006. Genetic characterization and molecular mapping of Hessian fly resistance genes derived from *Aegilops tauschii* in synthetic wheat. Theor. Appl. Genet. 113:611–618.

Warburton, M.L., J. Crossa, J. Franco, M. Kazi, R. Trethowan, S. Rajaram, W. Pfeiffer, P. Zhang, S. Dreisigacker, and M. van Ginkel. 2006. Bringing wild relatives back into the family: Recovering genetic diversity of CIMMYT bread wheat germplasm. Euphytica 149:289–301.

William, M., R.J. Peña, and A. Mujeeb-Kazi. 1993. Seed protein and isozyme variations in *Triticum tauschii* (*Aegilops squarrosa*). Theor. Appl. Genet. 87:257–263.

William, H.M., R.M. Trethowan, and E.M. Crosby-Galvan. 2007. Wheat breeding assisted by markers: CIMMYT's experience. Euphytica 157:307–319.

Wilson, A.S. 1876. On wheat and rye hybrids. Trans. Proc. Bot. Soc. Edinb. 12:286–288.

Xu, S.S., X. Cai, T. Wang, M.O. Harris, and T.L. Friesen. 2006. Registration of two synthetic hexaploid wheat germplasms resistant to Hessian fly. Crop Sci. 46:1401–1402.

Xu, S.S., T.L. Friesen, and A. Mujeeb-Kazi. 2004. Seedling resistance to tan spot and Stagonospora nodorum blotch in synthetic hexaploid wheats. Crop Sci. 44:2238–2245.

Yang, J., R.G. Sears, B.S. Gill, and G.M. Paulsen. 2002. Growth and senescence characteristics associated with tolerance of wheat–alien amphiploids to high temperature under controlled conditions. Euphytica 126:185–193.

Zaharieva, M., E. Gaulin, M. Havaux, E. Acevedo, and P. Monneveux. 2001. Drought and heat responses in the wild wheat relative *Aegilops geniculata* Roth: Potential interest for wheat improvement. Crop Sci. 41:1321–1329.

Zhang, P., S. Dreisigacker, A.E. Melchinger, J.C. Reif, A. Mujeeb-Kazi, M. van Ginkel, D. Hoisington, and M.L. Warburton. 2005. Quantifying novel sequence variation and selective advantage in synthetic hexaploid wheats and their backcross-derived lines using SSR markers. Mol. Breed. 15:1–10.

Zhang, Y., W.Y. Yang, Y.L. Peng, J. Li, and Y.L. Zheng. 2006a. Inheritance of resistance for Chinese wheat stripe rust races in a new common wheat variety Chuanmai 42 derived from synthetics between *Triticum durum* × *Aegilops tauschii*. Acta Phytophylacica Sinica 33:287–290.

Zhang, Y., W.Y. Yang, Y.L. Zheng, Y. Yu, X.R. Hu, Y.L. Peng, and J.X. Yang. 2006b. Inheritance of resistance for Chinese wheat stripe rust races in a new common wheat variety Chuanmai 38 derived from synthetics (*Triticum durum* × *Aegilops tauschii*). Southwest China J. Agric. Sci. 19:14–18.

Zhou, L., G. Bai, B. Carver, and D. Zhang. 2007. Identification of new sources of aluminium resistance in wheat. Plant Soil 297:105–118.

Zwart, R.S., J.P. Thompson, and I.D. Godwin. 2005. Identification of quantitative trait loci for resistance to two species of root-lesion nematode (*Pratylenchus thornei* and *P. neglectus*) in wheat. Aust. J. Agric. Res. 56:345–352.

Chapter 17
Success in Wheat Improvement

Jackie C. Rudd

SUMMARY

(1) Wheat grain yield throughout the world has increased by two- or threefold since 1950. The UK currently has the highest yield in the world at 7,700 kg ha^{-1}. China has experienced the greatest percentage gain, increasing yield by 450% since the 1960s.

(2) Historically, the genetic component of grain yield improvement in wheat has been estimated at 50%. Evidence now indicates this fraction may have increased in recent years.

(3) Annual genetic gain for grain yield has averaged about 40 kg ha^{-1}, or about 1%. Gains have been reported even at very low production levels, but they have been highest in the most productive environments.

(4) Reduced height, increased harvest index, and greater kernel number per unit area are primary contributing factors to improved yield, even among semidwarf cultivars. New studies have shown that simultaneous improvement in harvest index and biomass occurred in recent years.

(5) The mean yields of wheat and soybean (*Glycine max* L.) in the US increased from about 1,200 kg ha^{-1} in 1950 to near 3,000 kg ha^{-1} in recent years. During this same period, maize (*Zea mays* L.) yield increased from 2,000 kg ha^{-1} to over 9,000 kg ha^{-1}. These statistics, and the fivefold (or greater) difference in program support for maize versus wheat across public and private domains, clearly show the value of investing in plant breeding.

WORLD YIELD GAINS

World wheat grain yield rose slowly or not at all during the first half of the 20th century but has increased by two- or threefold during the past 50 years (Calderini and Slafer 1998). Table 17.1 shows the yield progression by decade since 1961 for the 15 leading wheat producing countries, ranked according to 2001-to-2007 production data. The yield across all countries averaged 1,300 kg ha^{-1} in the 1960s but is close to 3,000 kg ha^{-1} currently, an increase of 40 kg ha^{-1} yr^{-1}. As expected, these numbers differ by country. Grain yield in the UK has averaged 7,700 kg ha^{-1} this decade, while the yield in Kazakhstan averaged just over 1,000 kg ha^{-1}.

The rate of gain also has varied widely among countries. France, Germany, and the UK all had gains over 100 kg ha^{-1} yr^{-1}. China's annual gain was 88 kg ha^{-1} yr^{-1} and clearly showed the greatest percentage change. Wheat yield in China during the 1960s was 940 kg ha^{-1}; it has increased by a factor of 4.5 to 4,190 kg ha^{-1} so far this decade (Table 17.1 and Fig. 17.1). India and Pakistan increased wheat yields by 40–50 kg ha^{-1} yr^{-1}, while yields in the US, Russia, Canada, Turkey, Argentina, and Iran have shown yield gains between 20 and 35 kg ha^{-1} yr^{-1} (Table 17.1). The annual yield increase in Australia has been 12.9 kg ha^{-1} yr^{-1} since 1961. Australia experienced severe droughts in 2002, 2006, and 2007, and as a result, the yield for the current decade is lower than for the 1990s. Australia did have substantial

Table 17.1 Wheat grain yields by decade of the 15 highest wheat producing countries according to 2001–2007 production.

	1961–1970	1971–1980	1981–1990	1991–2000	2001–2007	Annual Gain 1961–2007	Annual Production 2001–2007
	Grain Yield (kg ha^{-1})						1,000,000 Mg
China	940	1,620	2,850	3,610	4,190	88.2	96.3
India	940	1,380	1,920	2,500	2,670	47.8	70.5
US	1,810	2,130	2,450	2,620	2,710	24.3	54.2
Russia[a]				1,610	1,960	28.1	45.6
France	3,170	4,440	5,780	6,850	6,840	104.3	35.2
Canada	1,510	1,790	1,890	2,290	2,340	23.0	22.3
Germany	3,540	4,450	5,710	6,960	7,320	105.3	22.3
Pakistan	910	1,330	1,680	2,070	2,460	39.7	20.3
Turkey	1,140	1,610	1,940	2,040	2,170	26.7	19.7
Australia	1,210	1,260	1,440	1,790	1,510	12.9	18.8
Ukraine[a]				2,780	2,610	–52.0	15.6
UK	3,990	4,740	6,540	7,560	7,710	107.8	14.3
Argentina	1,370	1,560	1,850	2,280	2,450	29.3	14.2
Iran	780	930	1,080	1,610	2,110	33.0	13.4
Kazakhstan[a]				870	1,070	17.5	12.6
World	1,300	1,720	2,220	2,600	2,780	40.4	598.3

Decade means, annual gain (from linear regression analysis of yield against year), and annual production were calculated from FAOSTAT online data (http://faostat.fao.org/).
[a]Data not available before 1992.

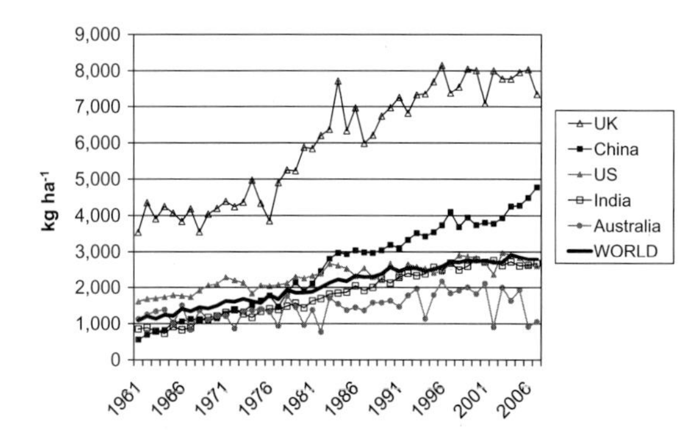

Fig. 17.1 Wheat grain yields from 1961 to 2007. The UK currently has the highest grain yields in the world. China, India, and the US represent the top three countries for wheat production. Australia is one of the top 10 countries for production, yet with low yields. Data assembled from FAOSTAT online resources (http://faostat.fao.org/).

yield gains in the 1980s and 1990s (Table 17.1 and Fig. 17.1). Yield in the Ukraine is currently less for this decade than it was in the 1990s. Figure 17.1 illustrates the annual yields of China, India, the US, Germany, Australia, and the entire world. Although strong annual variation has occurred within countries, the trend of increasing wheat grain yields is clear.

GENETIC COMPONENT OF GRAIN YIELD IMPROVEMENT

The two most frequently cited factors contributing to the increase in grain yield are improved cultivars and N fertilizer application (Bell et al., 1995; Austin 1999; Chloupek et al., 2004). Chloupek et al. (2004) showed the yield increase of wheat in Germany was 114 kg ha^{-1} yr^{-1} over the 40-year period from 1960 to 2000, and that N fertilizer use (and other inputs) accounted for 27% of that gain.

Bell et al. (1995) estimated that wheat yield gains in the Yaqui Valley of northwest Mexico from 1968 to 1990 were mainly attributed to improved cultivars (28%) and N fertilizer (48%). Recent studies from the International Maize and Wheat Improvement Center (CIMMYT) indicated that fertilizer use has leveled off in the Yaqui Valley while yields are still increasing, but perhaps at a reduced rate (Reynolds et al., 1999; Rajaram and Braun 2008). Reynolds et al. (1999) attributed this to an increase in N-use efficiency by new cultivars, indicating that the genetic component of grain yield improvement has increased.

Schmidt (1984) used yield data obtained from uniform regional nurseries in the US to determine genetic gain from 1958 to 1980. In comparison to the long-term check cultivars used in the nurseries, the yield advantage of the top experimental lines increased from 25% in 1959 to 46% in 1979, an annual rate of gain of 0.74%. He then compared this genetic gain with the on-farm wheat yields in the US across the same time period and determined that approximately one-half of on-farm yield gains could be attributed to breeding for yield. Brancourt-Hulmel et al. (2003) reported that between 33% and 63% of the yield gains in France were genetically driven.

Wheat grain yield in the UK increased linearly by 110 kg ha^{-1} yr^{-1} from 1950 to 1995, but N use peaked around 1985 (Austin 1999). The continued increase in grain yield after 1985, without an increase in N, implies that genetic improvement played a vital role in later years. Figure 17.1 illustrates that yield increases in the UK have not always been constant. It appears that growth was relatively slow from 1961 through 1976, followed by a period of rapid growth from 1976 to 1996, and no growth from 1996 to 2007. It is encouraging that Shearman et al. (2005) showed that genetic gain in the UK is still occurring. Cultivar replacement and management changes take place over several years, and new cultivars may not yet be fully adopted. Calderini and Slafer (1998) analyzed yield trends in 21 countries and found that gains were often bilinear or trilinear, having different slopes over different periods of years. These qualitative changes were likely related to major shifts in cultivars or management practices.

Based on the type of evidence just cited, it is generally projected that 50% of the increase in wheat grain yield is due to the adoption of new cultivars (genetic), while the other 50% is due to improved management practices. In reality, these are hard to separate empirically or theoretically. Borlaug (2007) stressed that the Green Revolution was not the result of introducing high-yielding semidwarf wheat cultivars or N fertilizers alone but the product of plant breeding and agronomic practices working in combination.

EMPIRICAL ESTIMATION OF GENETIC GAIN

Numerous authors have reviewed genetic gains in wheat, including Austin et al. (1989), Calderini and Slafer (1998), Austin (1999), Reynolds et al. (1999), Brancourt-Hulmel et al. (2003), Foulkes et al. (2007), and Rajaram and Braun (2008). These authors point to difficulties in quantifying genetic gain independently of crop management changes, but all agree that genetic gain estimates are needed to detect any degree of breeding progress over time. There are two methods commonly used to estimate genetic gains. One is to examine wheat cultivar performance trials over a period of time and compare new cultivar performance with a common control cultivar; examples are Schmidt (1984) and Donmez et al. (2001). Another method is to compare old and new cultivars in common yield trials (Cox et al., 1988; Brancourt-Hulmel et al., 2003; Shearman et al., 2005). Cox et al. (1988) describe this method as "a direct comparison of old and new cultivars, but (taking) place on the new cultivars' turf."

Table 17.2 summarizes 18 representative studies from 11 countries. Not surprisingly, newer wheat cultivars yielded more than older cultivars, but the rates of gain and the associated yield components varied greatly.

Grain yield

The data summarized in Table 17.2 reveal a wide range in estimates of genetic gain for grain yield among environments. Most data were collected in multiple natural or modified environments. The natural environments were used to simulate on-farm results. The input levels in these experiments were similar to what was commonly used

Table 17.2 Genetic gain experiments in wheat.

Reference	Country	Period	Grain Yield Gain (kg ha^{-1} yr^{-1})	Biomass Yield	HI	K m^{-2}	K spk^{-1}	Spk m^{-2}	K wt.	Vol. wt.	Head	Ht	Environment
Acreche et al. (2008)	Spain	1940–2004	0–55 0–1.5%	0	+	+			0			–	Protected
Brancourt-Hulmel et al. (2003)	France	1946–1992	36–63 0.6–0.9%	0	+	+	0	0	0			–	Varied N and fungicide levels
Calderini et al. (1995)	Argentina	1920–1990	50 1.0%	0	+	+	+	0	– (old) + (new)			–	High input, protected
Cox et al. (1988)	US	1919–1987	0–31 0.7–1.4%	0	+				+	+		–	Great Plains winter, natural
De Vita et al. (2007)	Italy, durum	1900–1990	16–25	0	+	+	+	+	0		–	–	Varied N level
Donmez et al. (2001)	US	1873–1995	12 0.44%	+ (new)	+	+	+	0	0		0	–	Great Plains winter, natural
Giunta et al. (2007)	Italy, durum	1900–1999	22–29	0	+	+	+	+	0		–	–	Varied N level and planting date
Guarda et al. (2004)	Italy	1900–1994	19–43		+	+			–		–	–	Varied N level
Rodrigues et al. (2007)	Brazil	1940–1992	45 1.5%	+ (new)	0	+	+	+	0			–	High input, protected
Sankaran et al. (2000)	India	1965–1990	0–35 0–1.3%			+					+		Planting dates
Sayre et al. (1997)	Mexico	1962–1988	67 0.9%	0	+	+	0	0	0	+	0	0	High input, protected
Shearman et al. (2005)	UK	1972–1995	119 1.2%	+ (new)	+	+		+	0			0	High input, protected
Slafer and Andrade (1993)	Argentina	1920–1980		0		+	+	+	–		0		High input, protected
Underdahl et al. (2008)	US	1968–2006	15–60 0.6–2.5%			0	0	0	+	+		0	Great Plains spring,
Waddington et al. (1986)	Mexico	1950–1982	59 1.1%	+ (new)	0	+	+	0	–		0		High input, protected
White and Wilson (2006)	Ireland	1977–1991	37 0.6%	+ (new)	0								High input, protected
Zhou et al. (2007a)	China	1960–2000	32–72 0.5–1.2%	0	+	0	0	0	0		–	–	Northern winter, protected
Zhou et al. (2007b)	China	1949–2000	14–41 0.3–0.7%	0	+	0	0	0	+		0	–	Southern spring, protected

Grain yield gains are given in kilograms per hectare per year (kg ha^{-1} yr^{-1}) and percent per year. Aboveground biomass yield, harvest index (HI), kernels per square meter (K m^{-2}), K per spike (K spk^{-1}), K weight (K wt.), volume weight (Vol. wt.), days to heading (Head), and plant height (Ht) either increased (+), decreased (–), or was not associated (0) with year of cultivar release for a given study. Blanks indicate that the parameter was not reported.

in the region being studied. Some investigators used different N fertilizer levels, fungicide treatments, and planting dates to create different yield levels.

Generally, the highest rates of gain were measured at the highest level of productivity. Shearman et al. (2005) reported gains of 119 kg ha^{-1} yr^{-1} in the UK during the period from 1972 through 1995. This increase was found at a very high yield level, in which the 2-year mean of the three most recently released cultivars exceeded 11,000 kg ha^{-1}. This contrasted sharply with data from the US Great Plains winter wheat production areas, where yields of the newest cultivars varied from 2,000 to 6,000 kg^{-1} ha^{-1} depending on the environment (Cox et al., 1988; Donmez et al., 2001). Where N rates were varied, yield gains were higher at the higher N rates, but significant gains were also detected at the low N rates. Although the absolute gains (in kilograms per hectare per year) were higher under the highest production environments, the rate of gain on a percentage basis was similar across environments, averaging about 1% per year. Rajaram and Braun (2008) reviewed yield gains within the CIMMYT program and demonstrated that absolute genetic yield gains were higher under irrigated environments, but the percentage gains were similar.

Some genetic gain estimations employed fungicides and insecticides as necessary to control diseases, while others involved natural conditions (see Environment in Table 17.2). In the unprotected environments, yield was reduced by leaf rust (caused by *Puccinia triticina* Eriks.) (Cox et al., 1988; Donmez et al., 2001; Underdahl et al., 2008), stem rust (*P. graminis* Pers.:Pers. f. sp. *tritici* Eriks. & E. Henn.) (Cox et al., 1988), Fusarium head blight (*Fusarium graminearum* Schwabe) (Underdahl et al., 2008), and tan spot (*Pyrenophora tritici-repentis* Died.) (Cox et al., 1988). Genetic improvement in grain yield was positively related with resistance to diseases targeted by the breeding programs (Fusarium head blight, leaf rust, and stem rust). The highest percentage gain reported in Table 17.2 was 2.5% per year in the presence of a Fusarium head blight epidemic in North Dakota, where the most recently released cultivars offered improved resistance to Fusarium head blight (Underdahl et al., 2008). On the other hand, no genetic gain for grain yield was realized under a heavy tan spot epidemic in Kansas (Cox et al., 1988).

Yield components

Parallel with these changes in wheat yield has been a consistent reduction in height, increase in harvest index (HI), and increase in kernel number per unit area. These are all documented to be phenotypical expressions of the semidwarf (*Rht*) genes that were introduced into wheat cultivars beginning around 1960 (Borlaug 2007; Fischer 2007; Ortiz et al., 2007). The increases in HI and kernel number are not entirely related to the introduction of semidwarf genes. Several studies have shown that these two yield components have continued to increase among semidwarf cultivars (Sayre et al., 1997; Abbate et al., 1998; Sankaran et al., 2000; Shearman et al., 2005; White and Wilson 2006).

Greater kernel number per unit area can be the result of more spikes per unit area and/or more kernels per spike. There was more often a trend for newer cultivars to produce more kernels per spike, but evidence was found for an increase in the number of spikes. De Vita et al. (2007) and Giunta et al. (2007) showed that newer Italian durum wheat cultivars produced more spikes per unit area and more kernels per spike. Some studies showed no increase with year of release for kernels per spike or the number of spikes. Hence genetic variability may be present, where some cultivars have relatively more spikes and some have relatively more kernels per spike, but there was no definitive trend over time of cultivar release.

Kernel weight generally has not changed over time, although Calderini et al. (1995) observed an increase among newer cultivars in Argentina. Zhou et al. (2007b) found that kernel weight increased in southern China, where large seed was a common selection criterion among breeding programs (He et al., 2001). Kernel weight also increased among North Dakota spring wheat cultivars, where a consistent and larger kernel size is desired for milling yield (Underdahl et al., 2008). They attributed some of the increase in kernel size to increased resistance to Fusarium head blight. Likewise Cox et al. (1988) attributed the increase in kernel weight in their study to increased resistance to stem rust and leaf rust. In three cases (Waddington et al., 1986; Slafer and Andrade 1993; Guarda et al., 2004), kernel weight had been reduced, indicating that breeders must actively select for kernel size. Three studies revealed an increase in grain volume weight, a trait that had been selected for by the respective breeding program (Cox et al., 1988; Sayre et al., 1997; Underdahl et al., 2008).

Although genetic variability for aboveground biomass may exist, often changes in grain yield are not accompanied by changes in biomass when comparing old cultivars with new cultivars. However, when comparing among only contemporary cultivars, parallel increases in grain yield and biomass may be found (Table 17.2). This is important since several authors have suggested that future yield advances will likely derive from increases in biomass (Austin 1999; Reynolds et al., 1999; Fischer 2007; Foulkes et al., 2007). They speculated that since so much progress to date has been made from increasing HI, breeders should turn their attention to increasing biomass. Fischer (2007) pointed out that further gains in HI are possible since most wheat cultivars have a HI of about 45% and the purported upper limit to HI is 60% (Austin 1980).

The physiological basis of genetic improvement in wheat has been reviewed extensively, such as Fischer (2007), Reynolds et al. (1999), and Foulkes et al. (2007). Increased HI is associated with altered partitioning to the spike that in turn leads to more kernels through more fertile florets and reduced kernel abortion. Shearman et al. (2005) showed genetic gain in preanthesis radiation-use efficiency and water-soluble carbohydrate content at anthesis. They concluded that genetic gain for grain yield in UK wheat cultivars was both sink- and source-related. An improved growth rate before anthesis led to an increase in the number of grains per meter squared (sink), and the increase in stem soluble carbohydrate reserves provided a larger source for grain filling. Slafer and Andrade (1993) found that in some new cultivars spike development occurred earlier and thus were more competitive for photosynthate. Many researchers agree that even with the increased kernel number in new cultivars, wheat is still sink-limited (Fischer 2007; Foulkes et al., 2007; Reynolds 2007).

WHEAT YIELD GAINS IN LIGHT OF OTHER CROPS

Wheat yield in the UK has reached an unprecedented level of 7,700 kg ha^{-1}, with Germany and France not far behind (Table 17.1). This level is comparable to maize yields in the US only 10 years ago (Fig. 17.2). Wheat production in the UK is intensively managed. Standard management practices include 200 kg ha^{-1} nitrogen fertilizer, two applications of foliar fungicides, and the application of a plant growth regulator. The breeding programs have obviously been successful developing cultivars for this system. Shearman et al. (2005) found that contemporary cultivars in the UK had higher HI and produced more biomass and more spikes per unit area than the cultivars they replaced. Most cultivars in their study were considered feed-grade wheat; thus the observed genetic gain could have exceeded the level which might be expected with the added complexity of multiple selection criteria for improved breadmaking performance.

China produces more wheat than any other country in the world (Table 17.1). The gain in

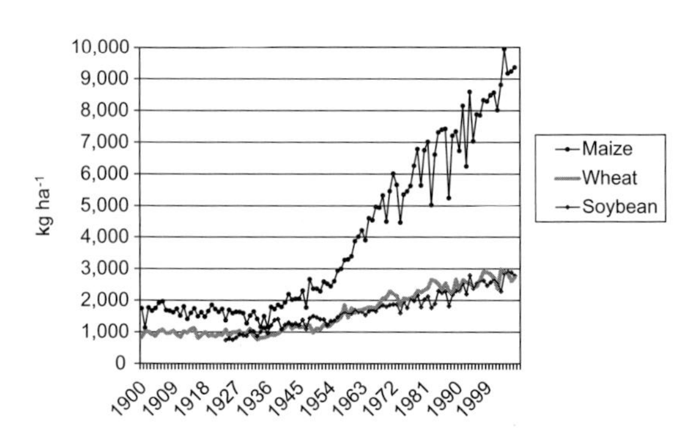

Fig. 17.2 Maize, wheat, and soybean yields in the US from 1900 to 2007. Data assembled from USDA Agricultural Statistical Service online resources (http://www.nass.usda.gov/).

production that started in the early 1960s has continued at a steady pace (Fig. 17.1). He et al. (2001) described wheat management in China as intensive with high inputs. They listed 36 national and provincial breeding programs, each employing multiple scientists. Each of these programs is highly focused and breeds for a relatively small target area. Cultivar replacement has been rapid, or about four to six times in the past 50 years (He et al., 2001). Over 90% of the grain in China is used to make steamed bread and noodles; breadmaking quality has not been part of the selection criteria until recently (He et al., 2001). The genetic gain for grain yield was well documented by Zhou et al. (2007a,b). Improved cultivars in southern China have shown higher grain yield, shorter stature, stronger straw, and increased kernel weight. Characteristics of newer cultivars in northern China are higher grain yield, shorter stature, stronger straw, and earlier heading date.

Total production and genetic gain for grain yield have been lower in areas that are subject to erratic abiotic stress such as Australia, Canada, and the US Great Plains. Cox et al. (1988) compared historic and contemporary cultivars in six environments and found a wide range in mean yield and genetic gain among environments. No relationship existed between genetic gain and grain yield at a particular location, but genetic gain was dependent more on the particular yield-limiting stress factor at that location. Genetic progress was greater under a certain stress condition targeted by the local breeding program. This is supported by the work of Carver et al. (2001) that showed selection in a grain-only management system did not necessarily result in genetic gain in a system that included grazing. Based on this information, the breeding program integrated grazing into their selection process.

Figure 17.2 charts the grain yield increase of three major crops in the US throughout this past century: maize, wheat, and soybean. Grain yield of all three crops did not change substantially from 1900 to about 1940 and then began to increase. In a review of breeding advances in maize, Duvick (2005) described the trends in maize yield. Nitrogen fertilizer use started around 1960 and increased until leveling off in 1985. Hybrid maize production started in the mid-1930s and by 1945, maize hybrids were grown on nearly 100% of the maize acres. Duvick (2005) also reported that genetic gain for maize yield in the US from 1930 to 2001 was 77 kg ha^{-1} yr^{-1}, a value not much different than those reported for wheat at higher production levels (Table 17.2).

In 1982, 36 PhD-level scientists worked for private industry in the US in soybean breeding, 23 in wheat, and 155 in maize (reviewed by Heisey et al., 2001). By 1994, there were 101 PhD-level soybean breeders in the private sector, 54 wheat breeders, and 510 maize breeders (Frey 1996). Also in 1994 there were 55, 77, and 35 public soybean, wheat, and maize breeders (Frey 1996). Perhaps the largest contribution to maize yields in the US has been the added investment in research made possible by hybrid seed sales and the elimination of farmer-saved seed. In that light, Fig. 17.2 could look very different if expressed on a per-scientist basis.

FUTURE PERSPECTIVES

The research encapsulated in this chapter helps to appraise genetic progress and establish benchmarks of where wheat breeding is today. It can also provide direction for the future. As has been said, "if you keep doing what you have been doing, you keep getting what you have been getting." Plant breeding is an amazing science. Identify superior progeny, recombine, and select again. Small incremental changes have moved wheat yields from 1,000 kg ha^{-1} to 3,000 kg ha^{-1} in only the past 50 years. Where will yields be 50 years from now? Past experience tells us that if the current path continues, then world wheat yields will increase by 40 kg ha^{-1} yr^{-1} as they have the past 50 years. But the science which surrounds wheat breeding is much different than 50 years ago. As expounded elsewhere in this book, our knowledge of a wheat plant continues to expand,

and we have new technologies to strengthen mainstream breeding efforts. For example, Reynolds (2007) found that increasing biomass before anthesis would increase grain yield, and Babar et al. (2006) and Prasad et al. (2007) recently proposed to do precisely that using spectral reflectance. New technologies to accelerate and improve selection efficiency are becoming a routine part of breeding programs. Marker-assisted selection and doubled haploid production are components of many breeding programs today (Rajaram 2001; Brennan and Martin 2007). Transgenic wheat and perhaps hybrid wheat will likely be commercialized in the not too distant future.

Brennan and Martin (2007) evaluated the rate of return on new technologies in plant breeding and suggested that savings from new technologies can be reinvested in the breeding program to increase genetic gain. Investments in plant breeding have paid excellent rates of return (Frisvold et al., 2003; Pardey et al., 2006). Fernandez-Cornejo (1999) reported that rates of return on plant breeding can vary from 30% to as much as 90%. Wheat breeding is no exception. With continuous investment in new technologies and mainstream breeding, genetic gain in wheat will accelerate and sustained yield increases of 100 kg ha^{-1} yr^{-1} are not unrealistic.

REFERENCES

Abbate, P.E., F.H. Andrade, L. Lazaro, J.H. Bariffi, H.G. Berardocco, V.H. Inza, and F. Marturano. 1998. Grain yield increase in recent Argentine wheat cultivars. Crop Sci. 38:1203–1209.

Acreche, M.M., G. Briceno-Felix, J.A.M. Sanchez, and G. A. Slafer. 2008. Physiological bases of genetic gains in Mediterranean bread wheat yield in Spain. Eur. J. Agron. 28:162–170.

Austin, R.B. 1980. Physiological limitations to cereal yields and ways of reducing them by breeding. p. 3–19. *In* R.G. Hurd, P.V. Bisco, and C. Dennit (ed.) Opportunities for increasing crop yields. Pitman, London.

Austin, R.B. 1999. Yield of wheat in the United Kingdom: Recent advances and prospects. Crop Sci. 39:1604–1610.

Austin, R.B., M.A. Ford, and C.L. Morgan. 1989. Genetic improvement in the yield of winter wheat: A further evaluation. J. Agric. Sci. 112:295–301.

Babar, M.A., M.P. Reynolds, M. Van Ginkel, A.R. Klatt, W.R. Raun, and M.L. Stone. 2006. Spectral reflectance to estimate genetic variation for in-season biomass, leaf chlorophyll and canopy temperature in wheat. Crop Sci. 46:1046–1057.

Bell, M.A., R.A. Fischer, D. Byerlee, and K. Sayre. 1995. Genetic and agronomic contributions to yield gains: A case study for wheat. Field Crops Res. 44:55–65.

Borlaug, N.E. 2007. Sixty-two years of fighting hunger: Personal recollections. Euphytica 157:287–297.

Brancourt-Hulmel, M., G. Doussinault, C. Lecomte, P. Berard, B. Le Buanec, and M. Trottet. 2003. Genetic improvement of agronomic traits of winter wheat cultivars released in France from 1946 to 1992. Crop Sci. 43:37–45.

Brennan, J.P., and P.J. Martin. 2007. Returns to investment in new breeding technologies. Euphytica 157:337–349.

Calderini, D.F., M.F. Dreccer, and G.A. Slafer. 1995. Genetic-improvement in wheat yield and associated traits—a reexamination of previous results and the latest trends. Plant Breed. 114:108–112.

Calderini, D.F., and G.A. Slafer. 1998. Changes in yield and yield stability in wheat during the 20th century. Field Crops Res. 57:335–347.

Carver, B.F., I. Khalil, E.G. Krenzer, and C.T. MacKown. 2001. Breeding winter wheat for a dual-purpose management system. Euphytica 119:231–234.

Chloupek, O., P. Hrstkova, and P. Schweigert. 2004. Yield and its stability, crop diversity, adaptability and response to climate change, weather and fertilisation over 75 years in the Czech Republic in comparison to some European countries. Field Crops Res. 85:167–190.

Cox, T.S., J.P. Shroyer, L. Ben-Hui, R.G. Sears, and T.J. Martin. 1988. Genetic improvement in agronomic traits of hard red winter wheat cultivars from 1919 to 1987. Crop Sci. 28:756–760.

De Vita, P., O.L. Nicosia, F. Nigro, C. Platani, C. Riefolo, N. Di Fonzo, and L. Cattivelli. 2007. Breeding progress in morpho-physiological, agronomical and qualitative traits of durum wheat cultivars released in Italy during the 20th century. Eur. J. Agron. 26:39–53.

Donmez, E., R.G. Sears, J.P. Shroyer, and G.M. Paulsen. 2001. Genetic gain in yield attributes of winter wheat in the Great Plains. Crop Sci. 41:1412–1419.

Duvick, D.N. 2005. The contribution of breeding to yield advances in maize (*Zea mays* L.). Adv. Agron. 86:83–145.

Fernandez-Cornejo, J. 1999. An exploration of data and information on crop seed markets, regulation, industry structure, and research and development. Bull. 786. USDA-ERS.

Fischer, R.A. 2007. Understanding the physiological basis of yield potential in wheat. J. Agric. Sci. (Cambridge) 145:99–113.

Foulkes, M.J., J.W. Snape, V.J. Shearman, M.P. Reynolds, O. Gaju, and R. Sylvester-Bradley. 2007. Genetic progress in yield potential in wheat: Recent advances and future prospects. J. Agric. Sci. (Cambridge). 145:17–29.

Frey, K.J. 1996. National plant breeding study—1: Human and financial resources devoted to plant breeding research and development in the United States in 1994. Spec. Rep. 98. Iowa Agric. and Home Econ. Exp. Stn., Ames, IA.

Frisvold, G.B., J. Sullivan, and A. Raneses. 2003. Genetic improvements in major US crops: The size and distribution of benefits. Agric. Econ. 28:109–119.

Giunta, F., R. Motzo, and G. Pruneddu. 2007. Trends since 1900 in the yield potential of Italian-bred durum wheat cultivars. Eur. J. Agron. 27:12–24.

Guarda, G., S. Padovan, and G. Delogu. 2004. Grain yield, nitrogen-use efficiency and baking quality of old and modern Italian bread-wheat cultivars grown at different nitrogen levels. Eur. J. Agron. 21:181–192.

He, Z.H., S. Rajaram, Z.Y. Xin, and G.Z. Huang (ed.) 2001. A history of wheat breeding in China. CIMMYT, D.F., Mexico.

Heisey, P.W., C.S. Srinivasan, and C. Thirtle. 2001. Public sector breeding in a privatizing world. Bull. 772. USDA-ERS.

Ortiz, R., R. Trethowan, G.O. Ferrara, M. Iwanaga, J.H. Dodds, J.H. Crouch, J. Crossa, and H.J. Braun. 2007. High yield potential, shuttle breeding, genetic diversity, and a new international wheat improvement strategy. Euphytica 157:365–384.

Pardey, P.G., N. Beintema, S. Dehmer, and S. Wood. 2006. Agricultural research: A growing global divide? Int. Food Policy Res. Inst., Washington, DC.

Prasad, B., B.F. Carver, M.L. Stone, M.A. Babar, W.R. Raun, and A.R. Klatt. 2007. Potential use of spectral reflectance indices as a selection tool for grain yield in winter wheat under Great Plains conditions. Crop Sci. 47:1426–1440.

Rajaram, S. 2001. Prospects and promise of wheat breeding in the 21(st) century. Euphytica 119:3–15.

Rajaram, S. and H-J. Braun. 2008. Wheat yield potential. p. 103–107. *In* Reynolds M.P., J. Pietragalla, and H-J. Braun, (ed.) Proc. Int. Symp. on Wheat Yield Potential: Challenges to International Wheat Breeding, Ciudad Obregon, Sonora, Mexico. 20–24 March 2006. CIMMYT, D.F., Mexico.

Reynolds, M. 2007. Association of source/sink traits with yield, biomass and radiation use efficiency among random sister lines from three wheat crosses in a high-yield environment. J. Agric. Sci. (Cambridge) 145:3–16.

Reynolds, M.P., S. Rajaram, and K.D. Sayre. 1999. Physiological and genetic changes of irrigated wheat in the post-green revolution period and approaches for meeting projected global demand. Crop Sci. 39:1611–1621.

Rodrigues, O., J.C.B. Lhamby, A.D. Didonet, and J.A. Marchese. 2007. Fifty years of wheat breeding in Southern Brazil: Yield improvement and associated changes. Pesqui. Agropecu. Bras. 42:817–825.

Sankaran, V.M., P.K. Aggarwal, and S.K. Sinha. 2000. Improvement in wheat yields in northern India since 1965: Measured and simulated trends. Field Crops Res. 66:141–149.

Sayre, K.D., S. Rajaram, and R.A. Fischer. 1997. Yield potential progress in short bread wheats in northwest Mexico. Crop Sci. 37:36–42.

Schmidt, J.W. 1984. Genetic contributions to yield gains in wheat. *In* W.R. Fehr (ed.) Genetic contributions to yield gains of five major crop plants. CSSA Spec. Publ. 7. CSSA, Madison, WI.

Shearman, V.J., R. Sylvester-Bradley, R.K. Scott, and M.J. Foulkes. 2005. Physiological processes associated with wheat yield progress in the UK. Crop Sci. 45:175–185.

Slafer, G.A., and F.H. Andrade. 1993. Physiological attributes related to the generation of grain yield in bread wheat cultivars released at different eras. Field Crops Res. 31:351–367.

Underdahl, J.L., M. Mergoum, J.K. Ransom, and B.G. Schatz. 2008. Agronomic traits improvement and associations in hard red spring wheat cultivars released in North Dakota from 1968 to 2006. Crop Sci. 48:158–166.

Waddington, S.R., J.K. Ransom, M. Osmanzai, and D.A. Saunders. 1986. Improvement in the yield potential of bread wheat adapted to northwest Mexico. Crop Sci. 26:698–703.

White, E.M., and F.E.A. Wilson. 2006. Responses of grain yield, biomass and harvest index and their rates of genetic progress to nitrogen availability in ten winter wheat varieties. Ir. J. Agric. Food Res. 45:85–101.

Zhou, Y., Z.H. He, X.X. Sui, X.C. Xia, X.K. Zhang, and G.S. Zhang. 2007a. Genetic improvement of grain yield and associated traits in the Northern China winter wheat region from 1960 to 2000. Crop Sci. 47:245–253.

Zhou, Y., H.Z. Zhu, S.B. Cai, Z.H. He, X.K. Zhang, X.C. Xia, and G.S. Zhang. 2007b. Genetic improvement of grain yield and associated traits in the southern China winter wheat region: 1949 to 2000. Euphytica 157:465–473.

Chapter 18
Transgenic Applications in Wheat Improvement

Ann E. Blechl and Huw D. Jones

SUMMARY

(1) The most reliable DNA introduction methods that lead to fertile integrative transformants of wheat are *Agrobacterium* and biolistics. These two methods result in random integration of one or more copies of the transgenes at one to a few locations in the genome.

(2) Both durum (*Triticum durum* L.) and bread wheat (*T. aestivum* L.) have been genetically transformed by biolistics and *Agrobacterium* methods. Reported efficiencies range from less than 0.2% to 70%, with typical frequencies less than 5%.

(3) Although many different cultivars of wheat have been transformed, not all are amenable to current methods, and transformation efficiencies are genotype-dependent.

(4) Genetic transformation has been used to confirm the identities of several agronomically important genes of wheat previously isolated by map-based cloning. These include genes that determine grain hardness, three major vernalization genes, the Q domestication gene, three leaf rust resistance genes, and the *GPC-B1* gene that influences grain protein, zinc, and iron contents. In each case, the functions of candidate gene sequences were verified by using transformation to test the effects of their introduction or of increases or decreases in their expression on traits of transgenic wheat plants.

(5) Thus far, traits most frequently targeted for change include herbicide resistance; resistance to *Fusarium*, powdery mildew (caused by *Blumeria graminis* f. sp. *tritici*), and other fungi; drought and/or salt tolerance; seed protein and starch compositions for enhanced end-use properties; and mineral availability in seeds by accumulation of phytase.

(6) The majority of transgenic wheat plants containing new genes for high-molecular-weight glutenin subunits (HMW-GS) and bialaphos resistance show no effects on nontargeted traits at either the molecular level of transcription and metabolome or at the whole-plant level as assessed by agronomic performance in field trials. Flours made from seeds of these plants exhibit a wide range of potentially useful mixing properties.

(7) Current limitations to the application of wheat transformation for commercial crop improvement include an inability to control the location, copy numbers, or linkage relationships of transgene integration, lack of predictability of the effects of genome context on expression of newly integrated genes, and societal opposition to the presence of genetically engineered wheat in the food supply.

INTRODUCTION

Wheat transformation is the introduction of DNA into wheat cells via means other than sexual hybridization. For the purposes of this chapter, transformation is a process that results in integration and inheritance of new DNA. By these criteria, the first published report of successful wheat transformation was by Vasil et al. (1992). DNA can also be introduced and expressed transiently in a nonintegrative form, but this does not result in inheritance of new genetic material.

Wheat transformation allows experimentalists to introduce DNA from any source. The genes for new or improved traits no longer need come from sexually compatible relatives; even novel synthetic genes made *in vitro* can be utilized. However, engineering completely novel traits into plants is challenging and in practice, considerations of mRNA and protein stability and how introduced proteins may interact with existing pathways can limit the choice and donor species of candidate genes. In a genetic transformation experiment, only a single or a few genes are introduced, allowing, at least in theory, changes to be made without modifying the background genotype. This differs from traditional breeding by which two entire genomes are combined in an initial cross. To recover the original genotype with a new trait after such a cross, several rounds of backcrossing need to be undertaken with selection for the new DNA of interest in each round. Even in extensively backcrossed material, genes closely linked to the gene of interest from the source parent will persist in the progeny, so-called linkage drag. In contrast, genetic transformation allows the experimenter to introduce only known genes. An interesting special case exists when the introduced DNA has been isolated solely from a sexually compatible species (so-called intra- or cis-genics) (Rommens 2007). Although the methods used are different, nothing other than gene location, and possibly expression levels, may distinguish the plants produced in this way from the products of conventional crossing or mutation breeding.

In this chapter, we provide a short description of transformation methods and recount the results of published transformation experiments aimed at understanding and/or improving a number of wheat agronomic and end-use properties. We also discuss some of the current limitations of wheat transformation technology and prospects for its impact on wheat improvement. We confine our coverage to wheat transformation studies published in English, with occasional reference to results from transformation studies in rice (*Oryza sativa* L.) or model species.

There are several excellent published reviews of wheat transformation, some of which cover in-depth topics only mentioned in passing in this chapter. Patnaik and Khurana (2001) present a detailed history of transformation methods. Vasil (2007) brings both historical and futuristic perspectives to the subject, emphasizing how this technology will help solve problems facing agriculture worldwide. Other good overviews of this subject are by Sahrawat et al. (2003) and Jones (2005). *Agrobacterium*-based transformation is treated in detail by Jones et al. (2005) and Shrawat and Lörz (2006).

WHEAT TRANSFORMATION: METHODS AND RESULTS

Wheat transformation protocols consist of several steps: culturing of explants that contain cells capable of multiplying and organizing into somatic embryos; introduction of DNA into those totipotent cells; identification of cell clusters or embryos that have integrated the incoming DNA into chromosomes; and regeneration of fertile plants.

Targets for wheat transformation

In wheat, there are few cells that are known to be capable of division, organization into somatic embryos (differentiation), and regeneration of shoots and roots into fertile plants. To serve as transformation targets, such cells must be located in regions (usually surfaces) that are accessible to and viable after DNA introduction. In wheat, the most commonly used targets for transformation are scutellar cells from immature embryos that are typically 10 to 15 days after anthesis (early soft-dough stage, Feekes scale 11.2; Large 1954). These cells can be transformed immediately upon culture or, more commonly, after 2 to 6 days of precultivation to form embryogenic calli (Weeks et al., 1993, and many subsequent reports). Color Plate 32a illustrates the processes for selection and culture of immature embryos: searching wheat plants for heads at the appropriate developmental stage, surface sterilization of immature caryopses, and aseptic excision of embryos. For some genotypes, it is important to prevent germination of the zygotic embryo by removing the embryo axis before culturing the scutellar cells (Nehra et al., 1994; Li et al., 2003a,b). For other genotypes, the embryo is placed in contact with the medium. In all cases, the media contain 2,4-dichlorophenoxyacetic acid, an auxin that inhibits germination and promotes dedifferentiation and division of scutellar cells into embryogenic calli. The culture media also contain carbon and nitrogen sources for heterotrophic growth in the dark.

Another explant shown to be useful for transformation of many cultivars of both durum and bread wheat is the immature inflorescence, from which meristematic regions can be triggered to proliferate in culture (Rasco-Gaunt and Barcelo 1999; Lamacchia et al., 2001). Other wheat explants used for transformation include 5- to 7-month-old embryogenic callus cultures (Vasil et al., 1992); protoplasts from embryogenic suspension cultures (He et al., 1994); germinating pollen tubes (Chong et al., 1998; Sawahel and Hassan 2002); anther cultures (Brisibe et al., 2000); shoot apical meristems (Ahmad et al., 2002; Sticklen and Oraby 2005; Zhao et al., 2006); leaf basal segments (Chugh and Khurana 2003b); mature embryos (Supartana et al., 2006) and calli derived from them (Patnaik et al., 2006). The usefulness of these tissues and their amenability to *in vitro* culture depends strongly on genotype, primarily because of varietal differences in wound responses to methods of DNA introduction (Parrott et al., 2002), and in competence for regeneration into fertile plants via a defined tissue culture regimen (Takumi and Shimada 1997; Iser et al., 1999; Pellegrineschi et al., 2002b).

DNA delivery methods and integration

There are many ways by which DNA has been introduced into wheat cells, including bombardment with DNA-coated metal particles termed biolistics (Vasil et al., 1992, and many others subsequently); cocultivation with *Agrobacterium*, a bacterium with a natural ability to conjugate DNA to plant cells (Cheng et al., 1997, and many others subsequently); permeabilization by electric current (electroporation)(He et al., 1994; He and Lazzeri 1998; Sorokin et al., 2000); vortexing with silicon carbide fibers or whiskers (Sawahel and Saker 1997; Brisibe et al., 2000); combination of *Agrobacterium* and silicon fiber wounding (Singh and Chawla 1999); laser micropuncture (Badr et al., 2005); uptake by germinating pollen tubes (Chong et al., 1998, and others subsequently); and microinjection (Simmonds et al., 1992; Ponya et al., 1999). Among these, *Agrobacterium* and biolistics have reproducibly yielded transformed wheat plants in several laboratories.

For *Agrobacterium*-mediated transformation, bacteria cocultivated with immature embryos (Color Plate 32b) transfer DNA into plant cells in discrete, protein-coated units called T-DNAs, which are delimited by left and right border sequences (McCullen and Binns 2006). The biolistics method consists of coating plasmid DNA onto tungsten or gold particles and propelling the coated particles through the cell wall with the force of a pressure pulse of helium gas (Klein et al., 1987), exploding gun powder (Sanford 1988), or electric discharge (McCabe and Christou 1993). Color Plate 32c shows some of the components of a biolistics transformation experiment: the helium-driven commercially available BioRad PDS 1000/He particle-delivery device (Kikkert 1993), immature embryos on agar medium before bombardment, and expression of a gene encoding green fluorescent protein (GFP) in scutellar cells after bombardment. Green fluorescent protein is encoded by the *gfp* gene from jellyfish (*Aequorea victoria*). The protein fluoresces green in living cells when blue or ultraviolet (460–490 nm) light is shone on them (Color Plate 32c) (Chalfie et al., 1994; Pang et al., 1996; Jordan 2000).

The transforming DNA consists of three components needed for expression in wheat cells: a promoter, a coding sequence, and a transcription terminator (Fig. 18.1). Promoters are DNA sequences that control gene expression, usually found in front of coding sequences. Although they are not themselves translated into proteins, promoter sequences have informational content: they specify when and where a gene will be transcribed. They can be thought of as switches that are triggered in response to intracellular and external environmental cues. The coding sequence is transcribed into RNA that is usually translated into protein. The transcription terminator is placed after the coding sequence and specifies the end of the RNA transcript and the site of polyA addition for messenger RNAs.

A typical transformation experiment includes at least two expression units, one incorporating a gene intended only to allow selection of transformed cells (selection gene), and one or two experimental genes that are expected to affect the trait being modified or studied, referred to as the gene(s) of interest. When the targeted trait is herbicide resistance, the gene of interest is also a selection gene, so only one gene needs to be introduced. For bombardment, the two expression units are usually on separate plasmid DNAs (Fig. 18.1a). Joining is not necessary because the separate plasmids are usually integrated together into

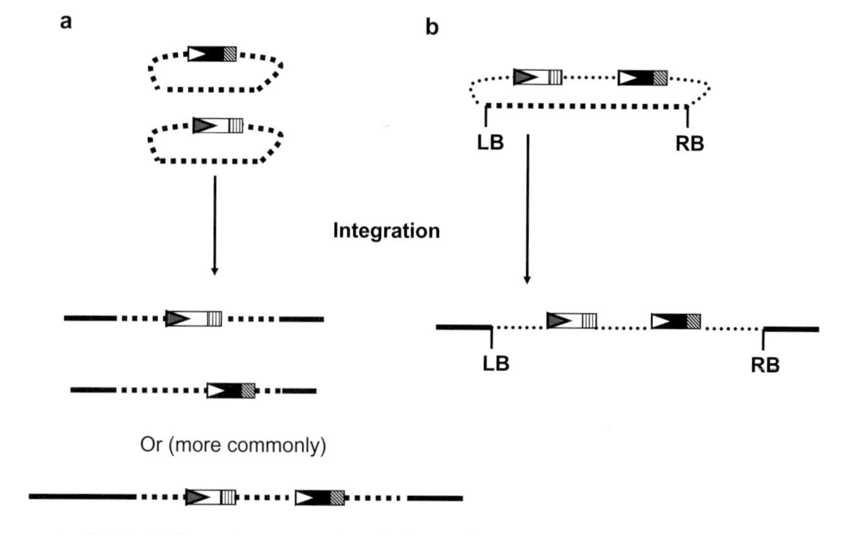

Fig. 18.1 Structures of DNAs before (above arrows) and after (below arrows) biolistic (a) and *Agrobacterium* (b) mediated transformation. The coding regions of selection genes and genes of interest are shown as open and filled rectangles, respectively. Promoters and their direction of transcription are depicted by open or gray arrowheads. Noncoding regions that include transcription terminators and located 3′ to the coding regions are shown as striped rectangles. The promoters and 3′ noncoding regions may be the same or different for the selection gene and gene of interest. Bold dotted lines are plasmid backbone sequences that include origins of replication for propagation in bacteria and antibiotic-resistance genes for bacterial transformation. Dotted lines are nongenic regions of T-DNA between the left (LB) and right borders (RB) for *Agrobacterium* transfer. The simplest single-copy structures that can result from integration are shown below the vertical arrows. Solid lines depict wheat genomic DNA of any length. For biolistic transformations, the selection gene and gene of interest can integrate at either unlinked or linked locations. Drawings are not to scale.

the same site(s) in the genome (Pellegrineschi et al., 2001; Rasco-Gaunt et al., 2001; Wright et al., 2001; Permingeat et al., 2003). For *Agrobacterium* transformation, both gene cassettes are placed on a single plasmid between the left and right border sequences of the transferred (T-) DNA (Fig. 18.1b). Unless there are rearrangements after the transfer to plant cells, all the genes on a single T-DNA integrate together.

Integration into the host cell chromosomes occurs sometime after entry of the DNA into the cell. An informative experiment was reported by Lonsdale et al. (1998), who followed expression patterns of DNA introduced into wheat callus cells by biolistics. The experiment employed a luciferase gene, which encodes an enzyme that emits light during conversion of its substrate to its product (Ow et al., 1986). Transient luciferase gene expression was visible in the wheat calli as light emissions in many different cells shortly after bombardment (similar to the GFP expression shown in Color Plate 32c). The emissions peaked at 48 hours, after which they declined, becoming undetectable 10 to 20 days after bombardment. Around 30 days after bombardment, a few new foci of gene expression could be seen; presumably these were the descendents of single cells that had stably incorporated the luciferase gene. This experiment suggests that integration can occur days or even weeks after the DNA is introduced.

The process by which DNA integrates into the host chromosomes is not well understood. Figure 18.1 shows the simplest structures that result from biolistic (Fig. 18.1a) and *Agrobacterium*-mediated (Fig. 18.1b) transformations. For biolistics, using linear DNA to coat the particles yields at least as many transformants as using circular plasmid DNA (Uze et al., 1999; Yao et al., 2006). Genes that originally are on different circular plasmids usually end up co-integrated at the same site (Fig. 18.1a bottom). For example, Campbell et al. (2000) found that three separate plasmids were cotransformed in 9 (36%) of the 25 regenerated plants they assessed. Cotransformation frequency for two separate plasmids was 66% in 88 transgenic events studied by Rasco-Gaunt et al. (2001) and 85% of 32 plants studied by Stoger et al. (1998). These results suggest that, prior to integration, either recombination occurs between circular plasmid DNAs or the circles are linearized and then ligated to one another by their ends. Kohli et al. (2003) review the data on the integration structures that result from biolistic transformation and discuss molecular mechanisms that could account for them.

For *Agrobacterium*-mediated transformation, the structure of the integrating DNA is better understood (reviewed in McCullen and Binns 2006). In the presence of an inducer, *de novo* synthesis of the T-strand is initiated within the bacterium, usually at single-stranded nicks, using the T-DNA as template. There is evidence that the T-strand synthesis is initiated at the right border, proceeds in the 5' to 3' direction along the T-DNA and usually terminates at the left border. The displaced T-strand is coated with VirE2, a single-stranded nucleic acid binding protein, and the whole complex is transferred to plant cells, where it integrates at one or two loci in the plant genome (Fig. 18.1b).

Identification of transformants

As can be seen by the transient expression of GFP in Color Plate 32c, many target cells receive and express DNA after transformation. However, integration of the incoming DNA typically occurs in only a few percent of these cells, necessitating the use of selection methods to identify cells in which the DNA persists and ultimately integrates. Several selection systems have been successful in favoring the preferential survival of transformed cells during division and regeneration (Table 18.1). Each system consists of a selective agent added to the media, used in conjunction with a selection gene, usually from bacteria, that confers an advantage to cells containing it on that media. Most selection systems rely on a cytotoxic agent, typically a herbicide or antibiotic. Cells with the appropriate resistance genes either make versions of the target that are insensitive to the agent or detoxify the agent. Cells lacking the resistance genes stop growing and eventually die.

Table 18.1 Selection systems used to identify wheat transformants.

Selection Agent(s)	Target in Plant Cell	Selection Gene	Encoded Enzyme	Selection Gene Source	Reference for First Use in Wheat Transformation
Geneticin, paramomycin, kanamycin	Inhibits protein synthesis	*nptII*	Neomycin phosphotransferease II	*E. coli*	Nehra et al. (1994)
Phosphinothricin, bialaphos, glufosinate	Inhibits glutamine synthetase	*bar*	Phosphinothricin acetyl transferase	*Streptomyces hygroscopicus*	Vasil et al. (1992)
Hygromycin	Inhibits protein synthesis	*hph*	Hygromycin phosphotransferase	*E. coli*	Ortiz et al. (1996)
Glyphosate	Aromatic amino acid synthesis, specifically the EPSPS[a] enzyme	*aroA*:CP4 *GOX*	Resistant EPSPS Glyphosate oxidoreductase	*Agrobacterium* Another bacterium	Zhou et al. (1995)
Cyanamide	Unknown	*cah*	Cyanamide hydratase	Soil fungus *Myrothecium verrucaria*	Weeks et al. (2000)
Mannose	None[b]	*pmi*	Phosphomannose isomerase	*E. coli*	Wright et al. (2001)

[a]5-enopyruvylshikimate-3-phosphate synthase.
[b]Selection gene provides a metabolic advantage that allows nonphotosynthetic wheat cells in culture to utilize mannose as a carbon source.

The common procedures for wheat transformation employ herbicides based on phosphinothricin (Thompson et al., 1987), glyphosate (Zhou et al., 1995; Hu et al., 2003), or the aminoglycoside antibiotics G418, paramomycin, kanamycin, or hygromycin (Table 18.1). Partly driven by a desire to avoid the use of antibiotic resistance genes, nutrient conversion selection methods have been developed. In these systems, cells that acquire the selection gene have a metabolic advantage and outgrow the cells that are not transformed. An example of a metabolic selection scheme that has been used for identifying wheat transformants is the sugar mannose in conjunction with the phosphomannose isomerase (*pmi*) gene. Cells containing the *pmi* gene can use mannose as a carbon source, while cells lacking it cannot grow without some other sugar in the medium (Wright et al., 2001; Gadaleta et al., 2006). The cyanamide selection system combines elements of both cytotoxicity and metabolic advantage. Cyanamide is toxic to plant cells, but the cyanamide hydratase (*cah*) gene enables them to convert the herbicide to urea, simultaneously detoxifying it and making it into a usable nitrogen source (Weeks et al., 2000).

An alternative to using selection systems is to employ a nondestructive screening method to identify cells that have received new genes and then to hand-pick those expressing cells. This strategy often uses reporter genes that visually mark the presence and/or location of gene activity, as pictured in Color Plate 32c and described previously for *gfp* and the luciferase gene (Table 18.2). Direct detection of the introduced DNA by polymerase chain reaction (PCR) can also be employed as a screening method (Permingeat et al., 2003; Zhang et al., 2006). For wheat transformation experiments, in which only a few percent of cells are ultimately stably transformed, such procedures are labor intensive. In addition, DNA need not be integrated to be detected by PCR or reporter gene activity, so that transiently transformed tissues are also identified. Some researchers have employed a combination of screening and

Table 18.2 Reporter genes used in wheat transformation.

Common Name	Reporter Gene (Source)/ Encoded Protein	Assays	Main Uses in Wheat Transformation	Advantages	Disadvantages	References[a]
GUS	*uidA (E. coli)*/ β-glucuronidase enzyme	Histochemical staining for *in situ* localization; fluorescence for quantification	Patterns of promoter activity (Plate 33); marker for transformed cells and tissues	Product of histochemical reaction, insoluble, remains at site of reaction	Need to get substrate into cells; cells damaged and killed by assay	Jefferson et al. (1987); Stoger et al. (1999b)
GFP	*Gfp* (jellyfish)/green fluorescence protein	Fluoresces in response to blue light	Marker for transformed cells and tissues (Plate 32c)	Nondestructive assay, only light needed to visualize	Photosynthetic and many other wheat tissues autofluoresce at similar wavelengths	Chalfie et al. (1994); Jordan (2000)
Luciferase	*Luc* (firefly)/ luciferase	Converts substrate to luminescence in presence of ATP	Indicator of gene expression	Short-lived, so gives picture of current, not cumulative, expression	Need to get substrate into cells	Ow et al. (1986); Lonsdale et al. (1998)
Anthocyanin	*C1* + *B-Peru* or *Lc* (Maize)/myb and basic Helix-loop-helix transcription factors	Visual color change	Marker for transformed cells and tissues	Nondestructive visual assay, no substrate needed	Two genes usually required	Goff et al. (1992); Doshi et al. (2007)

[a]First reference is for early description of reporter gene(s); second reference is an example of its use in transformed wheat.

selection methods (Huber et al., 2002) or two screening methods (Doshi et al., 2007).

Regeneration of fertile plants

The fourth process in plant transformation is the regeneration of cells containing newly integrated DNA into a fertile plant that can pass the transgene to its progeny as part of its chromosomes.

Shortly after *in vitro* culture, some of the cells in the original explant begin to divide and form callus tissue. It is at this stage that the DNA is introduced. The callus cells continue dividing and during 2 to 6 weeks of culture in the dark, clusters of cells begin to differentiate into somatic embryos. During this period after *Agrobacterium* transformation, the media include antibiotics to inhibit bacterial growth. Transfer of the embryogenic calli to media with reduced auxin levels and incubation under light/dark conditions induces the somatic embryos to differentiate shoots (Color Plate 32d). The regeneration media includes the selection agent to favor transformed shoots and, for *Agrobacterium* transformation, antibiotics that inhibit bacterial growth. Shoots are subsequently transferred to selection media with compositions that favor formation of roots (Color Plate 32d). The rooted plantlet can then be transferred to a potting medium, and after 7 to 10 days of acclimation to lower humidity, to a greenhouse.

Efficiency of wheat transformation

The typical efficiencies with which stable wheat transformants are recovered vary from 1% to 5%. Efficiencies are usually reported as the percentage of individual starting explants (e.g., cultured wheat embryos) that give rise to transformed plants (T_0), although some authors report efficiencies as the percentage of explants that give rise to stably transformed callus lines. The number of integrative transformants is most accurately assessed by Southern blot analysis of DNA from regenerated T_0 plants, or by inheritance of DNA or transgene-encoded phenotypes. Use of PCR on DNA from T_0 plants or assays for transgene-encoded phenotypes in such plants may give an overestimate of transformation efficiency, because these methods can detect nonintegrated genes or transient gene expression, respectively. This is particularly important when plant tissues have been incubated with *Agrobacterium*, which can persist within regenerating plants and give false positives in PCR screening.

For both bread and durum wheat, biolistic transformation efficiency varies greatly among cultivars (Takumi and Shimada 1997; Iser et al., 1999; Rasco-Gaunt et al., 2001; Varshney and Altpeter 2001; Pellegrineschi et al., 2002a,b). Other factors implicated in transformation efficiency are donor plant health (Harvey et al., 1999; Pellegrineschi et al., 2002a), donor plant age (Pastori et al., 2001), the type of explant (Cheng et al., 2003), and the selection system used (Ortiz et al., 1996; Witrzens et al., 1998; Cheng et al., 2003; Przetakiewicz et al., 2004). DNA delivery is generally not a limiting factor for stable transformation, and the conditions for optimal transient expression are not necessarily the same as those for the highest efficiencies of stable transformation, both for the biolistics (Altpeter et al., 1996a) and *Agrobacterium* (Wu et al., 2003) approaches. The highest reported efficiencies for bombardment were 20% when the *pmi* selection method was used (Reed et al., 2001; Wright et al., 2001) and 70% when a super-transformable line of 'Bobwhite' was used (Pellegrineschi et al., 2002b). The efficiency of the first *Agrobacterium* transformation was 1.1% to 1.6% (Cheng et al., 1997), but recent improvements in the method have increased it to as high as 10% with 3% to 5% being typical (Huber et al., 2002; Cheng et al., 2003; Chugh and Khurana 2003a; Hu et al., 2003; Khanna and Daggard 2003; Wu et al., 2003).

APPLICATIONS OF WHEAT TRANSFORMATION

Although the ultimate goal of wheat transformation is to improve the crop's agronomic and utilization properties, the objectives of individual experiments depend on current genetic understanding of the targeted trait. At the basic end of the research spectrum are reverse genetic experiments to confirm the identity and characterize the DNA sequences of wheat genes that control or modulate specific traits. These approaches rely on data provided by classical genetics and genomics approaches. At the applied end of the spectrum, biotechnology uses genes already well-characterized and/or expected to modify a trait, and aims to either change existing wheat gene expression levels or add genes that are new to wheat. One of the great strengths of the biotechnology approach is that one experiment can yield insights into gene-controlled processes and produce plants with new and useful properties that can be propagated. In the following sections, we give several examples of each of these applications, but first we discuss some general experimental design considerations.

Promoters

As described previously and depicted in Fig. 18.1, a typical transformation experiment includes at least two expression units that need to function in plant cells: the selection gene and one or two experimental or candidate genes. An important consideration in designing transformation experiments is the choice of promoter to drive expression of the transgene(s). With recombinant DNA methods, the experimenter can fuse any promoter to any coding sequence *in vitro*, thereby specifying when and where a protein will be made in the plant. Novel fusions of gene sequences made *in vitro* are designated with double colons separating their sections, that is, promoter::coding sequence:: transcription termination signal sequence. The different segments can be from natural genes or made synthetically.

Selection genes must be active during the initial stages of cell proliferation and/or differentiation in tissue culture and, if they are to be used to identify transformants after regeneration, they must be active in at least some tissues of regenerated plants. In nearly all cases, selection genes used in wheat transformation experiments thus far have been controlled by constitutive promoters, which drive high levels of transcription in all or nearly all tissue types at nearly all stages of development under a wide variety of environmental conditions. The most commonly used promoters for selection genes in wheat are those from the maize (*Zea mays* L.) ubiquitin (*Ubi1*) (Christensen et al., 1992; Stoger et al., 1999b; Rooke et al., 2000) and rice actin (*Act1*) genes (Zhang et al., 1991), and the 35*S* gene promoter from *Cauliflower mosaic virus* (Becker et al., 1994). The *Ubi1* and *Act1* promoters include the first introns of their respective genes. In these and many other genes, the first intron increases the levels of gene expression relative to those of the promoter alone (Rasco-Gaunt et al., 2003; Bourdon et al., 2004). The native 35*S* promoter does not have an intron, so a first intron from a cereal gene, such as maize *ADH1*, is often included in transformation constructs to ensure high expression levels in wheat (Rasco-Gaunt et al., 2003; Bourdon et al., 2004). The use of constitutive promoters for selection genes also gives the experimenter an easily assayed phenotype—herbicide or antibiotic resistance—with which to follow transgene inheritance. The activity of the *Ubi1* promoter in several different transgenic wheat tissues is visualized using the GUS reporter gene in Color Plate 33a–d.

For the genes of interest or candidate genes, promoter choice can vary widely, depending on the aims of the experimenter. By judicious selection of promoters, the experimenter can limit transgene expression to only tissues or conditions where it is needed. This conserves resources for the transformed plant and ensures that the transgene-encoded mRNAs and proteins do not have unintended effects in nontarget tissues. For example, applications aimed at changing wheat seed properties can employ endosperm-specific promoters, such as those for genes that encode a high-molecular-weight glutenin subunit (HMW-GS) (Lamacchia et al., 2001) or the puroindoline proteins (Wiley et al., 2007). The activity of a HMW-GS gene promoter in the endosperm of transformed wheat plants segregating for a HMW-GS::GUS transgene is shown in Color Plate 33e. Transgenes aimed at controlling pathogens can employ promoters induced by infection.

For example, in their experiments aimed at controlling powdery mildew infection, Altpeter et al. (2005) used a *GSTA1* gene promoter that is pathogen-induced and primarily expressed in the leaf epidermal cells. Transgenes aimed at providing drought resistance can use promoters induced by stress, such as that of the barley (*Hordeum vulgare* L.) *HVA1* gene (Hong et al., 1992). Unfortunately, there are few promoter sequences known in wheat or other cereals whose expression levels, inducibility, and tissue specificity are well-defined. Furthermore, an experimenter may not know when and where a gene is best expressed to have the desired impact. For these reasons, and because high levels of expression are usually desired, most wheat transformation experiments to date have utilized one of the three constitutive promoters described earlier or the gene's own native promoter to express genes of interest.

Applications for functional genomics

Many transformation experiments in wheat have the goal of identifying the gene sequences and molecular bases underlying trait expression. Broad strategies for this kind of functional genomics approach are increasing gene expression, decreasing gene expression (suppression), and complementation, whereby a functional gene is added to a genotype that lacks it. The latter can only be used in conjunction with transformation when the genotype that lacks a gene function can be transformed at a reasonable efficiency. The gene overexpression and suppression strategies have been more often employed because these experiments can be done with genotypes chosen for ease of transformation.

Overexpression of a gene sequence can easily be achieved by transformation to introduce more copies of a native gene or to introduce constructions consisting of the gene's coding region under control of a nonnative (heterologous) but highly expressed promoter. To suppress native gene expression, early experiments utilized DNA constructions which were transcribed into antisense RNAs that read in the direction opposite the native mRNAs that were translated into protein. Such transcripts would hybridize to their complementary sense transcripts, forming double stranded (ds)RNAs that could not be translated into proteins (Fig. 18.2). Thus, antisense RNAs decreased gene expression by disabling homologous coding messenger RNAs.

Some transformants containing constructs designed to increase gene expression of native genes were found instead to reduce the expression of those genes. This paradoxical phenomenon, easily visualized in experiments to change petunia flower color (Jorgensen 1995), was called sense suppression or transgene-mediated cosuppression (Matzke and Matzke 1995). It could be explained by the unintended formation of antisense RNA when newly integrated sequences were transcribed from promoters brought into proximity either by integration near

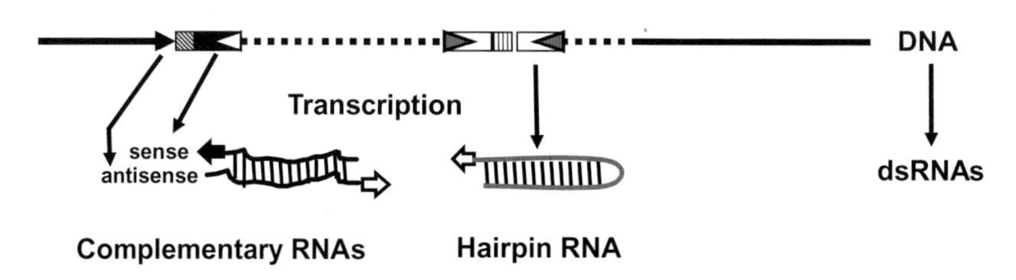

Fig. 18.2 Arrangement of transgenes that can result in promoter read-through from adjacent wheat genomic DNA, or adjacent transgenes that integrate in opposite orientations and lose one or both functional transcription terminators. Such transgenes can be transcribed to form antisense (ending in open arrow) or hairpin RNAs. A promoter in wheat genomic DNA is depicted as a solid horizontal arrowhead. The sense RNA transcribed from the gene-of-interest promoter ends in a filled arrow. Complementary regions between strands that form double-stranded RNA are indicated by vertical lines. Other symbols are defined in Fig. 18.1. Drawings are not to scale.

native gene promoters or by rearrangements in the integrating DNA that result in reversal of the coding sequence or promoter orientations (Fig. 18.2). In effect, the sense construct is converted into an antisense construct. Overexpression sense constructs can also suppress homologous native genes at the DNA level in some transgenic events, precluding their transcription, but the molecular mechanisms underlying these interactions are not understood (Matzke and Matzke 1995).

In recent years it has become clear that in eukaryotic cells, dsRNAs have much wider impacts than prevention of translation. It is now known that dsRNAs are perceived by plant cells as viral replicons, triggering a multicomponent response called RNA interference (RNAi), which results in the degradation of the dsRNAs as well as all closely related single-stranded mRNAs (Ossowski et al., 2008). The dsRNAs are cut into small fragments by the dicer RNAse, and these fragments are incorporated as targeting guides into complexes that degrade all homologous RNAs. As understanding of this phenomenon has grown from studies in model systems such as tobacco (*Nicotiana tabacum* L.), it has become clear that more effective than antisense transcripts for triggering the RNAi response are hairpin RNAs. These single RNA molecules include complementary sequences in reverse orientation so that they can fold back on themselves to form regions of dsRNA (Fig. 18.2). Such constructions reliably suppress the expression of related sequences at the RNA level. For a detailed review of the application of RNAi in wheat functional genomics, see Fu et al. (2007).

RNA interference is a powerful tool for functional genomics in wheat, because the suppression effects of the interfering construct are dominant, making it reliable for generating losses of gene function even though the polyploid genome contains multiple alleles of each gene. Moreover, experimenters can target either individual genes or entire gene families, simply by including unique or shared gene sequences, respectively, in the region of the construct that becomes double-stranded. Travella et al. (2006) used RNAi to obtain phenocopies in hexaploid wheat of mutants that had been previously characterized in diploid model plants. Their RNAi construct for the phytoene desaturase gene down-regulated all three homoeologous genome copies, resulting in seedlings with photobleached leaves. Introduction of an RNAi construct for the *EIN2* gene, which encodes an ethylene signal transduction factor, resulted in wheat insensitive to ethylene. In both cases, different transgenic lines showed different levels of reductions of the targeted transcripts, which was reflected in the range of severity of the phenotypes in the different transformed lines. The RNAi-induced phenotypes were stable for at least two selfing generations.

Overexpression, antisense RNA, RNAi, and complementation constructs have all been successfully employed to confirm the identity of wheat gene sequences isolated either by map-based cloning or by homology to genes with proven functions in model plant systems with sequenced genomes such as Arabidopsis (*Arabidopsis thaliana*), rice, or Brachypodium (*Brachypodium distachyon* L.) (www.brachypodium.org).

An interesting wheat gene whose identity and pleiotropic effects were confirmed by genetic transformation is the *Q* gene, a key gene in the domestication of wheat that had been linked to the square spike and free-threshing (fragile glume combined with tough rachis) characteristics, with minor pleiotropic effects on glume shape and tenacity, rachis fragility, spike length, plant height, and spike-emergence time. Simons et al. (2006) used map-based cloning to isolate a candidate sequence for the Q gene that encoded a transcription factor. To verify its identity, they transformed wheat with an overexpression construct and obtained some plants with increased expression and some showing cosuppression. The range of phenotypes of the transgenic plants was correlated with expression levels of the candidate transgene, confirming that a single gene controls both spike compactness and the free-threshing character, the major traits that had been attributed to the *Q* locus.

To understand the molecular basis for leaf rust resistance, candidate sequences for the *Lr1*, *Lr10*, and *Lr21* genes were identified by map-based cloning (Cloutier et al., 2007; Feuillet et al., 2003; Huang et al., 2003, respectively). Each sequence encoded recognizable gene-for-gene resistance proteins with typical coiled coil, leucine-rich repeat, and nucleotide-binding-site domains. Genetic transformation was used to prove that the sequences did indeed control the resistance phenotypes. In the case of *Lr1*, transformation of a susceptible cultivar with the candidate gene conferred resistance to a *Puccinia triticina* race carrying avirulence gene *Avr*1. The extent of resistance was correlated with transgene dosage and expression levels (Cloutier et al., 2007). In the case of *Lr10*, overexpression of the candidate sequence under control of the *Ubi1* promoter conferred enhanced resistance to the expected races of leaf rust (Feuillet et al., 2003). In the case of *Lr21*, transformation with the candidate sequence complemented the absence of the equivalent allele in 'Fielder', a cultivar that had been susceptible to the tester leaf rust isolate PRTUS6 (Huang et al., 2003).

Transformation has been instrumental in identifying some of the key components in the complex network of wheat genes that control the transition from vegetative to reproductive states. In early experiments, Chong et al. (1998) reported that an antisense construct to the wheat *ver203* gene delayed heading in the first generation of wheat plants transformed by the pollen-tube method. In later experiments an antisense version of *VER2*, a related gene expressed in young leaves surrounding the shoot apex, resulted in a 44-day delay in heading after vernalization of winter wheat (Yong et al., 2003). In wheat plants carrying a sense-suppressed *WAP1* transgene, a wheat gene with similarities to the Arabidopsis gene *AP1* that induces flowering, heading times were delayed regardless of vernalization or photoperiod (Murai et al., 2003). The authors concluded that the *WAP1* gene product accelerated the autonomous phase transition from vegetative to flowering states (Murai et al., 2003). In the first application of RNAi strategy to confirm the identification of a gene sequence isolated by map-based cloning, Yan et al. (2004) showed that decreased expression of the vernalization gene *VRN2* accelerated flowering time in winter wheat by more than four weeks, proving their hypothesis that *VRN2* encodes a repressor of flowering. Members of the same group also used RNAi to show that the *VRN1* gene encoded an inducer of flowering; reductions in *VRN1* transcripts in spring wheat transformed with an RNAi construct delayed the transition of the shoot apex from vegetative to reproduction growth, increasing the time to heading by 2 to 3 weeks (Loukoianov et al., 2005). Continuing this approach, members of the same group used map-based cloning to isolate a candidate gene encoding a transcription factor that they hypothesized was the product of a third vernalization gene, *VRN3*. Transformation of winter wheat with a sense construct of this gene resulted in its overexpression and accelerated flowering (Yan et al., 2006).

Wheat transformation also was instrumental in verifying the sequence and pleiotropic effects of the Grain Protein Content (*GPC-B1*) gene, which modulates protein, zinc, and iron contents of wheat grain. A quantitative trait locus from *Triticum turgidum* ssp. *diccocoides* was found to increase the levels of these three nutrients in domesticated durum wheat carrying a segment of the 6B chromosome from *T. dicoccoides*. The same segment accelerated senescence of the flowering plant. To show that a single gene was responsible for all of these phenotypes, Uauy et al. (2006) designed an RNAi construct to decrease expression of all genes homologous to *GPC-B1* in a domesticated hexaploid wheat cultivar. The resultant transgenic plants senesced more than 3 weeks later than non-transgenic siblings and had more than 30% reductions in seed protein, zinc, and iron contents. Having the complete sequence of the *T. dicoccoides GPC-B1* allele is allowing breeders to use molecular markers to introgress it into domesticated wheat.

Applications to understand or modify seed properties

Wheat supplies about 20% of the food calories consumed by the world's population. Thus it is not surprising that understanding the genetic foundations of wheat end-use quality was one of the earliest and remains a frequent objective of transformation experiments. Most studies have employed genes encoding the seed storage proteins which comprise gluten, the protein network that underlies wheat dough properties and allows the making of leavened wheat products (Shewry et al., 2003). Among the various protein fractions and subunits associated with dough elasticity and extensibility, the strongest correlation occurs between HMW-GS composition and dough strength. Indeed HMW-GS composition explained 45% to 70% of the variation in dough strength among European cultivars (Shewry et al., 2003).

Once the genes encoding HMW-GS had been cloned and wheat transformation methods became routine and efficient, several laboratory groups applied biotechnology to change HMW-GS composition. Genes encoding subunits Ax1 (Altpeter et al., 1996b; Barro et al., 1997), Dx5 (Barro et al., 1997), Dy10 (Blechl et al., 2007), and a hybrid between Dx5 and Dy10 (Blechl and Anderson 1996), each under control of its native HMW-GS promoter, were added to various wheat backgrounds (referenced in Shewry et al., 2003). In a typical example, the transgene-encoded Ax1 subunit accumulated to as high as 2.3% of the seed protein of normal fertile plants (Altpeter et al., 1996b). The expression of Ax1 in bread wheat with five original subunits or in durum wheat with two original subunits resulted in doughs with increased strength and improved rheological properties in all lines except one with very high expression levels (He et al., 1999; Alvarez et al., 2001; Popineau et al., 2001; Vasil et al., 2001; Barro et al., 2003; Darlington et al., 2003; Rakszegi et al., 2008). The transgenic flours with improved mixing characteristics produced bread loaves with slightly larger volumes and similar crumb grain, compared with the nontransformed parent (Vasil et al., 2001; Darlington et al., 2003). The overexpression of the *1Dx5* gene in bread wheat with two or five original subunits or in durum wheat with two original subunits resulted in doughs that were difficult to hydrate and, at expression levels greater than 2.3-fold the native *1Dx5* gene, could not be mixed in a 2-gram mixograph without blending (Rooke et al., 1999; Alvarez et al., 2001; Popineau et al., 2001; Rakszegi et al., 2005; Blechl et al., 2007). It has been hypothesized that the different effects of the Ax1 and Dx5 subunits relate to the presence of a fifth cysteine residue in Dx5 that is available for intermolecular disulfide bond formation, versus four cysteine residues in all other x-type subunits (Shewry et al., 2003). Bread loaves baked from the transgenics with high levels of Dx5 had very low volumes and poor crumb grain (Darlington et al., 2003). Transgenic bread wheat with elevated levels of HMW-GS Dy10 produced doughs with longer development times and greater mixing tolerances than the non-transgenic parent (Blechl et al., 2007). Even doughs with 5.4 times the native levels of Dy10 could be mixed, showing that the Dy10 subunit's contribution to the glutenin polymer differed from that of the Dx5 subunit (Blechl et al., 2007; Gadaleta et al., 2008a). The contrasting effects of similar levels of *1Dx5* and *1Dy10* overexpression on mixing properties can be seen in Fig. 18.3.

The first examples of transgene-mediated cosuppression in wheat were observed with sense constructs of HMW-GS genes (Blechl et al., 1998; Alvarez et al., 2000). Such transgenic wheat flours have reduced levels of all HMW-GS, poor dough development (Alvarez et al., 2001; Uthayakumaran et al., 2003; He et al., 2005), no resistance to extension (Uthayakumaran et al., 2003), and are unsuitable for making either bread or tortillas (Uthayakumaran et al., 2003). Recently, Yue et al. (2008) introduced an RNAi construct into bread wheat that decreased expression of *1Dx5* and *1Bx7*, without affecting the expression levels of the other three native HMW-GS genes (*1Ax2**, *1By9*, and *1Dy10*). The resulting flour had greatly reduced dough development time and mixograph stability (Yue et al., 2008).

Starch composition is another major seed component important to wheat end-use properties. Starch composition was altered by transformation in two studies. Båga et al. (1999) used an antisense construction for the gene encoding Starch Branching Enzyme I (SBE I), under control of the rice *Act1* promoter, to

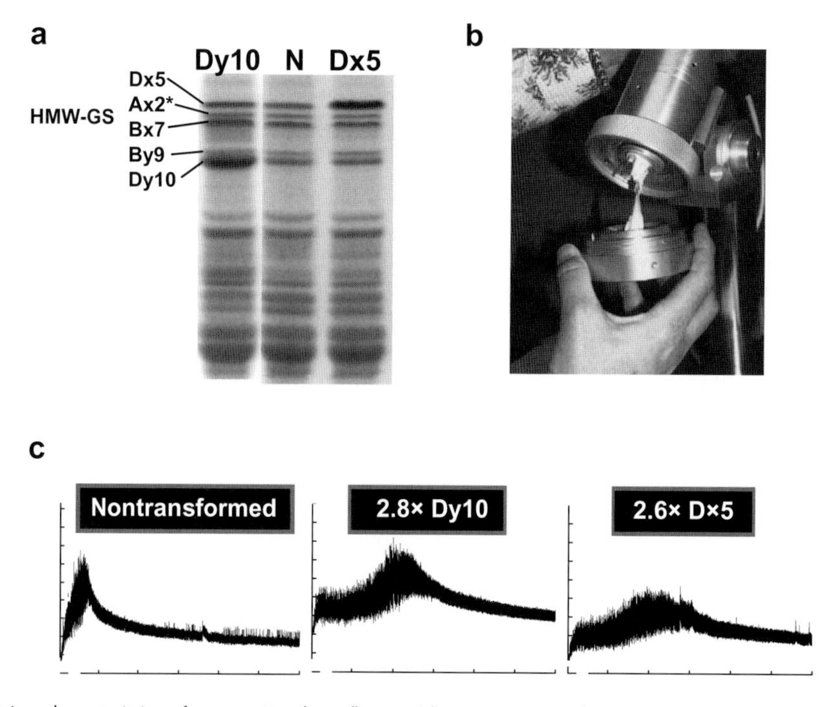

Fig. 18.3 Mixing characteristics of transgenic wheat flours with increases in either HMW-GS Dy10 or Dx5. (a) SDS-PAGE of seed proteins of transgenic wheat lines containing increased levels of Dy10 (left lane) or Dx5 (right lane) compared with their nontransformed parent (N). (b) Picture of 2-g mixograph instrument, showing the dough after development. The mixograph is used to measure changes in resistance over time. (c) Comparison of mixograph traces for the three flours whose seed proteins are shown in (a). Resistance in arbitrary units is plotted against time for 30 minutes. Increases in either Dy10 or Dx5 improve dough strength and mixing tolerance as shown by the thicker traces, longer times to peak resistance, and lower slopes after peak resistance, compared with the nontransformed parent. However, the peak resistance achieved by mixing dough with extra Dy10 is higher than the peak for the dough with extra Dx5, indicating that the effects of the two subunits on dough development are different. (Photo courtesy of Frances Dupont.)

obtain transgenic plants with lower starch-branching activities in their kernels. One transgenic plant had only 10% of the wild-type SBE I activity; its seed starch had less crystalline structure and gelatinized at a lower temperature than starch from the nontransformed parental seed. Regina et al. (2006) used RNAi constructs to suppress expression of the genes encoding SBEII a and/or b isoforms. Seeds from plants with suppression of both isoforms contained starch that was over 70% amylose, the linear unbranched form that is more resistant to mammalian digestion. Rats fed wholemeal prepared from transgenic seed exhibited improved indices of large bowel function but the same growth rates as rats fed meal from non-transgenic seed (Regina et al., 2006).

Transformation has played a role in elucidating the genetic basis for grain hardness, a characteristic that determines how much force is needed to mill grain to flour. The force requirement determines how much starch damage occurs during milling, which is associated with water absorption and bran contamination of white flour. Transformation experiments were used to prove that the puroindoline genes *Pina-D1* and *Pinb-D1* together control grain hardness. Addition of *Pinb-D1a* (wild-type allele) under control of a HMW-GS gene promoter to hard red spring wheat 'HiLine', which contains a partially functional mutant *Pinb-D1b* allele, resulted in soft grain texture (Beecher et al., 2002). Similarly, the addition of *Pina-D1a* complemented the null mutation (*Pina-D1b*) in the hard white spring wheat Bobwhite, resulting in soft grain texture (Martin et al., 2006). Transformants of HiLine with *Pina-D1a* and/or *Pinb-D1a* controlled by a HMW-GS promoter showed that both puroindoline proteins are needed to achieve maximum softness (Hogg et al., 2004).

Third-generation selfed progeny of these lines were grown in the field and their harvested grain was milled and baked into bread. The lines transformed with either the overexpressed *Pina* or *Pinb* genes had mean single-kernel characterization system (SKCS) hardness scores of 29 and 52, respectively, compared with the mean SKCS hardness score of 84 for the nontransformed parent HiLine. The transgenic lines also exhibited decreased total flour yield, but increased break-flour yield, compared with their nontransformed parent line. Dough made from the transgenic flour had similar mixing properties to that of the nontransformed parent flour, but had decreased water absorption. The transgenic flour produced loaves with smaller volumes and lower crumb grain scores (Hogg et al., 2005; Martin et al., 2007). However, the soft transgenic lines performed better than their nontransformed parent in cookie-baking tests (Martin et al., 2007). Differences between the transgenic lines and their nontransformed parent were similar to those between nontransformed soft and hard wheats.

Further confirmation that the *Pin* genes underlie grain hardness came from overexpression of the *Pina* coding sequence under control of the *Ubi1* promoter. In some lines, increases in puroindoline a from the added transgenes led to softer texture, while in other lines cosuppression was triggered, resulting in undetectable levels of transgene and endogenous *Pina* expression and thus hard grain texture (Xia et al., 2008).

Transformation experiments aimed at improving the nutritional value of wheat grain have employed genes from other species. In an attempt to raise iron levels in cereal seeds, Drakakaki et al. (2000) transformed wheat and rice with a soybean (*Glycine max* L.) gene that encodes ferritin under control of the maize *Ubi1* promoter. However, the anticipated increase in stored iron was only detected in vegetative tissues, not in seeds. To improve phosphate and mineral availability in wheat seeds, Brinch-Pedersen et al. (2000) added a gene that encodes phytase, an enzyme that degrades the antinutrient phytic acid. Their first attempt used an *Aspergillus niger phyA* coding sequence under the control of the maize *Ubi1* promoter and resulted in transgenic wheat seeds with four times the nontransformed parent's secreted phytase activity (Brinch-Pedersen et al., 2000). More recently, the same group employed a secreted synthetic version of the *phyA* gene under control of the endosperm-specific wheat HMW-GS promoter and obtained levels of phytase activity in seeds that were 6.5 times those of the nontransformed parent (Gregersen et al., 2005). In this case, the coding region was designed to contain the codons that are most optimal for translation in wheat endosperm.

Another important seed trait is sink strength, which determines the efficiency of carbon and nitrogen translocation from photosynthetic tissue to the developing seed. Smidansky et al. (2002) transformed wheat with a maize sequence that encoded a mutant version of the large subunit of ADP-glucose pyrophosphorylase that was insensitive to allosteric inhibition. The coding sequence was under control of either its native promoter (Smidansky et al., 2002) or one from a wheat HMW-GS gene (Meyer et al., 2004). Fifth-generation selfed descendents of one resultant line had increases of 31% in total biomass and 38% in seed weight per plant (Smidansky et al., 2002). The rates of photosynthesis in the transgenic plants were increased, compared with those of the nontransformed parent, but only in high levels of light. Concentrations of glucose, fructose, and sucrose peaked in the flag leaves 7 and 14 days after flowering, but increases in the transgenic seed ADP-glucose and UDP-glucose were not evident until maturity (Smidansky et al., 2007). The yield increases of the transgenic plants were also evident in four years of field trials at three locations, but only when conditions were not limiting, that is, in well-spaced plantings (8–9 plants per 3 m row) under irrigation. Transgenic and non-transgenic plants had the same yields in rainfed plots planted at standard densities (Meyer et al., 2007).

Preharvest sprouting compromises wheat end-use quality. Wilkinson et al. (2005) were able to reduce the extent of preharvest sprouting by transforming wheat with an oat *VP1* gene under control of the *Ubi1* promoter. The oat gene transcript was more efficiently spliced to coding mRNA than the endogenous wheat *VP1* transcripts, resulting in higher levels of the VP1 protein, which is both a positive regulator of maturation and a negative regulator of germination. Wheat transformants containing the oat *VP1* gene showed 35% reductions in the number of seeds that germinated 9 and 14 weeks after anthesis under cool moist conditions (Wilkinson et al., 2005).

Applications to improve pathogen and pest resistance

Another major target for wheat genetic transformation has been increased resistance to various plant pathogens and pests. Several strategies (Dahleen et al., 2001) have been aimed at improving resistance to Fusarium head blight caused by *Fusarium graminearum* Schwabe, a disease of wheat that decreases yield and contaminates the grain with mycotoxins. Several genes have shown promise for limiting the spread of *Fusarium* within the spike in greenhouse trials (Type II resistance): (i) one encoding deoxynivalenol acetyltransferase, targeted to reduce toxicity of the mycotoxin that facilitates spread of the fungus (Okubara et al., 2002); (ii) several encoding cereal-defense proteins, including thaumatin-like proteins (tlps) from barley (Mackintosh et al., 2007) and rice (Chen et al., 1999); (iii) class II β-1,3-glucanases from barley (Mackintosh et al., 2007) and resistant wheat cultivar Sumai 3 (Anand et al., 2003); (iv) Ribosomal Inhibitory Protein (RIP) b32 from maize (Balconi et al., 2007); (v) acidic chitinases from resistant wheat cultivar Sumai 3 (Anand et al., 2003) and rice (Chen et al., 1999); and (vi) α-purothionin from wheat (Mackintosh et al., 2007). To achieve widespread expression throughout the plant, all coding sequences were controlled by the maize *Ubi1* promoter, except for the maize *b32* gene, which was controlled by the 35*S* promoter.

Makandar et al. (2006) used a strategy designed to invoke a plantwide defense system to combat Fusarium head blight. They transformed wheat with the Arabidopsis *NPR1* gene under control of the maize *Ubi1* promoter. This gene encodes an inducer of systemic acquired resistance. By expressing this regulator throughout development, the investigators hoped to arm the plant with multiple defense compounds before it was challenged by the pathogen.

Greenhouse tests of transgenic plants carrying each of these constructs showed that Type II resistance improved 20 to 50% compared with their nontransformed parents. However, when several of these transgenic plants were challenged with *Fusarium* infection in field experiments (Okubara et al., 2002; Anand et al., 2003; Mackintosh et al., 2007), only those expressing the wheat α-thionin, barley β-1,3-glucanase, and barley tlp showed any improvement in resistance (Mackintosh et al., 2007). Among these, only a line carrying the glucanase transgene showed reduction of multiple disease indices in the field, including deoxynivalenol accumulation, the percentage of visually scabby kernels, and disease severity, compared with the nontransformed parent (Mackintosh et al., 2007).

The potential for control of some other fungal pathogens has been tested by genetic transformation with genes encoding plant defense proteins or compounds with known antifungal activity. Resveratrol is a member of the stilbene family of phytoalexins. It serves as an antifungal defense protein in grape (*Vitis vinifera* L.) and is also believed to have positive effects on human health. Fettig and Hess (1999) transformed wheat plants with the grapevine stilbene sythase gene under control of the maize *Ubi1* promoter. Four of seven plants expressed the transgene and were shown to accumulate resveratrol. Serazetdinova et al. (2005) transformed wheat with a resveratrol synthase gene, *vst1* or *vst2* from grapevine and a pinosylvin synthase gene from pine (*Pinus sylvestris* L.), each under control of the stress-inducible *vst1* or *vst2* promoters. The plants accumulated stilbene derivatives that were similar to, but more hydrophilic than, resveratrol or pinosylvin. The transgenic plants expressing the resveratrol synthase genes showed 19%–27% reductions in symptoms after infection with the biotrophic leaf rust pathogen *Puccinia triticina* Eriks. and 42%–71% reductions in symptoms after infection with the facultative biotroph *Septoria nodorum* Berk (Serazetdinova et al., 2005). In contrast, wheat plants containing a pinosylvin synthase gene controlled by a *vst* promoter were not protected from either pathogen.

Clausen et al. (2000) obtained resistance to stinking smut infection, caused by *Tilletia caries (*DC.) Tul. & C. Tul. 1847, by transforming two Swiss wheat cultivars with the coding region of a viral antifungal protein KP4, under control of the *Ubi1* promoter. One transformed line for each cultivar showed a 30% decrease in symptoms in the greenhouse (Clausen et al., 2000) and a 10% decrease in field trials (Schlaich et al., 2006).

Powdery mildew resistance has been the target of several wheat transformation experiments. A diagnostic test for resistance used in these reports is measurement of the number and/or size of colonies formed on detached leaves 5 to 7 days after inoculation with the pathogen. By this criterion, Bliffeld et al. (1999) found increased resistance to powdery mildew in transgenic wheat plants secreting a barley seed class II chitinase controlled by the *Ubi1* promoter. Investigators in the same laboratory group expressed a secreted barley RIP under control of the 35*S* promoter and an intron derived from the *Rice tungro bacilliform virus* (Bieri et al., 2000). Leaves of two of the expressing lines had decreased mildew colony formation of 40% and 80%. Expanding their strategy in later experiments, the same group expressed either the barnase RNAse under control of the 35*S* promoter and an intron, or a barley seed chitinase and barley 1,3-β-glucanase under control of the *Ubi1* and *Act1* promoters, respectively. None of the transgenic plants with the new constructs showed more resistance than the original barley 35*S*::RIP lines (Bieri et al., 2003). Combining the transgene loci for the three barley proteins by crossing did not increase protection.

Oldach et al. (2001) expressed either an *Aspergillus giganteus* antifungal protein or barley class II chitinase with the *Ubi1* promoter in transgenic wheat. Leaves expressing the antifungal protein had reductions of 40% to 50% in colony formation for both powdery mildew and leaf rust inoculated with 80 to 100 spores per square centimeter. The degree of protection was strongly dependent on the inoculation dose. The barley chitinase transgene conferred no protection (Oldach et al., 2001). Zhao et al. (2006) reported reduced symptoms in both greenhouse and field tests of second-generation transgenic wheat plants expressing a tobacco β-1,3-glucanase controlled by the 35*S* promoter. Roy-Barman et al. (2006) reported increases up to 50% in powdery mildew resistance scored in the detached leaf assay of second-generation transgenic plants expressing a lipid transfer protein from *Allium* under control of the *Ubi1* promoter. Altpeter et al. (2005) used the wheat pathogen-induced *GSTA1* promoter to express a wheat peroxidase and oxalate oxidase in leaf epidermis. (These proteins are normally found only in inner leaf cells.) Leaves from the transgenic plants carrying the peroxidase construct had less surface area colonized by powdery mildew after infection than those of the nontransformed parent. Expression of the oxalate oxidase was not protective. The plants containing the peroxidase in epidermal cells responded to powdery mildew attack more frequently than controls, with a hypersensitive response comprising localized cell death and production of hydrogen peroxide (Schweizer 2008).

Viral infections have been reduced in many different plants by triggering cosuppression of viral RNAs via overexpression of viral coat protein or replicase genes (Beachy 1997). In wheat, this strategy has had mixed success. Sivamani et al. (2000b) expressed the replicase NIb of *Wheat streak mosaic virus* (WSMV) and found that symptoms after WSMV infection of the transgenic plants were reduced or absent. The same group transformed wheat with the gene encoding the WSMV coat protein. Although no coat protein was detected, one of five transgenic lines had milder symptoms and a lower virus titer after WSMV infection (Sivamani et al., 2002). Neither of the lines that had shown greenhouse resistance was resistant in 2 years of field trials (Sharp et al., 2002). In later investigations, some of these same researchers found that the coat protein transgenes had been silenced in the second- and third-generation progeny of the transformants (Li et al., 2005b).

Zhang et al. (2001) expressed a bacterial ribonuclease III, which degrades double-stranded RNA, under control of the *Ubi1* promoter. When the transgenic wheat plants were infected with *Barley stripe mosaic virus* in a greenhouse, they exhibited no infection symptoms and had reduced virion accumulation compared to their nontransformed parent (Zhang et al., 2001). Jiménez-Martínez et al. (2004a) studied the interactions of the bird cherry–oat aphid (*Rhopalosiphum padi* L.) vector with transgenic wheat plants expressing *Barley yellow dwarf virus* (BYDV) coat protein. Compared with nontransformed plants, BYDV virions were less efficiently transferred by the aphid after feeding on transgenic plants (Jiménez-Martínez and Bosque-Pérez 2004), and BYDV-infected transgenic plants were less attractive to the aphids than BYDV-infected nontransformed plants (Jiménez-Martínez et al., 2004b). The virus titer was lower in transgenic plants than in non-transgenic plants (Jiménez-Martínez et al., 2004a).

The potential of various protease inhibitors to control insect infestation in wheat has been tested by genetic transformation. Altpeter et al. (1999) expressed a barley trypsin inhibitor under control of the *Ubi1* promoter. Seeds from some of their transgenic plants accumulated the inhibitor to levels of 1.1% of their protein. Early instar larvae of the grain moth *Sitotroga cerealella* had reduced survival and weights when fed transgenic seeds compared to larvae fed seeds from nontransformed wheat. Growth and survival of leaf-feeding grasshoppers (*Melanoplus sanguinipes*) were not different on transgenic and non-transgenic wheat lines (Altpeter et al., 1999). Bi et al. (2006) produced three transgenic wheat lines expressing a cowpea (*Vigna unguiculata* L.) trypsin inhibitor gene under control of the 35*S* promoter. Seeds from the transgenic plants had 44%–67% less damage when exposed to grain moths for 3 months (Bi et al., 2006). Hesler et al. (2005) found no effects on the bird cherry–oat aphid when they were fed transgenic wheat expressing a potato (*Solanum tuberosum* L.) proteinase inhibitor II gene under control of its own promoter.

Stoger et al. (1999a) used the maize *Ubi1* or a rice phloem-specific sucrose synthase gene promoter to express the coding region for a lectin from snowdrop, *Galanthus nivalis* agglutinin (GNA). The transgenic wheat accumulated GNA to levels as high as 0.2% of leaf soluble protein. Neonate nymphs of the grain aphid *Sitobion avenae* that were allowed to colonize the transgenic plants for 16 days produced fewer offspring than nymphs that colonized non-transgenic plants. The transgenic plants had no effect on the survival of either the grain aphid (Stoger et al., 1999a) or of the cereal aphid *Metopolophium dirhodum* or of a cereal aphid biocontrol fungus *Pandora neoaphidis* (Shah et al., 2005).

Vishnudasan et al. (2005) attempted to obtain nematode resistance in durum wheat by expressing a potato serine proteinase inhibitor under control of the rice *Act1* promoter. Ten-day-old seedling progeny of heterozygous transgenic plants inoculated with nematodes were taller and had higher seed numbers and weights compared with the non-transgenic controls. The increases were well-correlated with the different levels of proteinase inhibitor accumulated in the different transgenic plants.

Applications to improve tolerance of abiotic stress

Transgenic strategies to improve the drought and salt tolerance of wheat by expressing either stress-inducible gene regulators or increasing levels of putative osmoprotectants show some promise. Transgenic wheat expressing the barley *HVA1* gene under control of the *Ubi1* promoter had greater water-use efficiency and dry weight of both shoots and roots than the nontransformed parent under water-deficit conditions in the greenhouse (Sivamani et al., 2000a). In a series of field trials, these transgenic plants had greater total biomass and grain yield at two of four dryland locations and across all locations. The relative water content of the leaves was correlated with barley *HVA1* expression levels.

The Arabidopsis *DREB-1A* gene encodes a transcription factor that regulates genes involved in drought, cold, and salt tolerance. Transgenic wheat plants expressing the *DREB-1A* gene under control of the Arabidopsis desiccation-inducible *rd29a* gene promoter showed a 10-day difference in the onset of wilting compared with nontransformed plants, after water was withheld in greenhouse experiments (Pellegrineschi et al., 2004).

Xue et al. (2004) transformed wheat plants with the Arabidopsis Na^+/H^+ antiporter gene *ATNHX1* under control of the 35*S* promoter. The encoded vacuolar transport protein sequesters sodium ions in the vacuole. Plants that expressed the transgene exhibited more efficient germination (84% vs. 68%) and produced about 1.5 times greater biomass under severe saline conditions (100–150 mM NaCl) than nontransformed plants in the greenhouse. The transgenic plants also had 30% to 50% higher grain yield in the field under saline conditions than the nontransformed parent (Xue et al., 2004).

Wheat plants containing a *Vigna aconitifolia* δ-pyrroline-5-carboxylate synthetase gene under control of the 35*S* promoter overaccumulated proline in their leaves at 12 times the control level and could grow and set seed in 200 mM NaCl (Sawahel and Hassan 2002). Expression of the same gene under control of the stress-inducible ABA-responsive element from the barley *HVA22* gene promoter produced 2.5 times the amount of proline in control plants at the booting stage, 14 days after withholding water (Vendruscolo et al., 2007). Biochemical measurements in the stressed plants led the authors to conclude that the accumulated proline acts by protecting plant tissues from oxidative damage rather than by mediating osmotic adjustment.

Abebe et al. (2003) expressed the mannitol biosynthesis *mtlD* gene from *E. coli* under control of the *Ubi1* promoter in transgenic wheat and found that the transgenic plants could better tolerate 100–150 mM salt and water-withholding treatments. The increases in intracellular mannitol levels measured in calli and in the fifth leaf of stressed transgenic plants were too small to account for the protection by osmotic adjustment.

One published report documents an attempt to improve wheat frost tolerance by genetic transformation. Khanna and Daggard (2006) transformed wheat with a fusion of the rice *Act1* promoter to the coding region from a flounder (*Pseudopleuronectes americanus* L.) antifreeze protein gene that had been modified to contain wheat-favored codons. The secreted protein accumulated to levels of 1.6% of leaf protein. An electrolyte leakage assay showed that flag leaves were significantly more resistant to frost damage than those of the nontransformed parent at temperatures between −1 and −7°C.

Other applications

Transformation technology has been used to introduce completely new traits into wheat. When the goal is herbicide resistance, the selection gene and the gene of interest are identical. Glyphosate-resistant or Roundup Ready wheat plants were directly selected by scientists at Monsanto Corporation after *Agrobacterium* transfer of an *Agrobacterium*-derived 5-enolpyruvylshikimate-3-phosphate synthase (*EPSPS*) coding region under control of various promoter and intron combinations (Hu et al., 2003). Resistance to herbicides with the active ingredient chloroacetanilide (e.g., alachlor) was introduced into wheat by genetic transformation with a maize glutathione S-transferase gene under control of the *Ubi1* promoter (Milligan et al., 2001). Not surprisingly, many of the transformed plants from experiments that have used *Ubi1*::*bar* as a selection gene are resistant to glufosinate-based herbicides (Gopalakrishnan et al., 2000).

Male-sterile but otherwise normal wheat plants were produced by expressing the barnase RNAse under control of maize or rice promoters whose expression is confined to tapetal tissue that nourishes the male gametophyte (De Block et al., 1997). Single-chain antibodies that could be useful pharmaceuticals for diagnosis of cancer or eye diseases were produced by expression of their coding regions in wheat under the control of seed-specific promoters. Levels of the pharmacological proteins were reported to be as high as 30 $\mu g\ g^{-1}$ dry weight (Stöger et al., 2000), and the proteins were stable for at least one year in dry seeds (Brereton et al., 2007).

Impacts on production agriculture

As of this writing, no transgenic wheat exists in commercial production in the US or elsewhere. Applications of wheat transformation to date have mostly been in the areas of gene discovery and proof-of-concept experiments. For the latter, genes were added to wheat with the expectation of modifying a targeted trait. Some of these transgenes were well expressed and, in most of the published cases, made small but detectable differences in measurements or symptoms in small-scale lab or greenhouse tests. Relatively few of these transgenic lines has been submitted to field trials to test for expression levels and stability under standard wheat production conditions in multiple environments. Thus, in most of the experiments summarized earlier, it remains unknown whether such transgenic lines offer a recognizable benefit for production agriculture. There are, however, two types of transgenic wheat for which the published results of several field tests show that the modified traits have the stability that wheat breeders and producers would demand before commercial release. These are Roundup Ready wheat and transgenic wheat with altered HMW-GS content.

Altered HMW-GS transgenic wheat was among the first transformation products produced by several different groups. This material has since been subjected to field trials to test for stability of transgene expression and to ascertain how the presence of selection genes and HMW-GS transgenes affects yield or other agronomic properties. In these plants, the *bar* selection gene is under control of the *Ubi1* promoter, and the HMW-GS transgenes are under control of their native endosperm-specific promoters.

Vasil et al. (2001) observed that field performance of six independent lines carrying HMW-GS *1Ax1* and *Ubi1::bar* transgenes was equivalent to that of their nontransformed parent. Expression of the HMW-GS transgenes was stable through three generations of growth in the greenhouse and in fourth-generation plants grown in the field. Barro et al. (2002) grew four lines carrying *1Ax1* and/or *1Dx5* and selection transgenes, three lines carrying only the HMW-GS transgenes, two lines without any transgenes that had segregated from the transgenic plants (null segregants), and two nontransformed parental lines for two years in field experiments. For traits other than protein composition, consistent differences were observed only for heading date and number of spikelets per spike. All transgenic lines headed a few days later and produced more spikelets per spike than the non-transgenic lines (Barro et al., 2002). No consistent significant differences were observed for grain yield. The third through fifth generations of a subset of the same lines were grown in three separate years at two UK sites (Shewry et al., 2006). Entries in this trial included three lines carrying *1Ax1* or *1Dx5* transgenes, one null segregant line, and two nontransformed parental lines. Transgenic and non-transgenic lines exhibited the same levels of variability for grain nitrogen and kernel weight between years and locations. The expression levels of the transgene-encoded HMW-GS showed no greater variability between years and locations than expression of the native HMW-GS genes. The mixing properties of the HMW-GS transgenics showed the same differences from their parents, regardless of their growth environment (Shewry et al., 2006). The authors concluded that transgene-encoded changes had the same stability across environments and generations as conventional genetic traits. Rakszegi et al. (2005) grew one transgenic line from this set and its parent in an arid continental climate for 3 years. The transgenic and control lines had the same grain yields; expression of the *1Dx5* transgene remained at high levels, four times that of native HMW-GS genes.

Bregitzer et al. (2006) compared the field performance of 50 transgenic wheat lines with altered HMW-GS composition versus four lines with only the *Ubi1::bar* selection gene, 10 non-transgenic sister lines (null segregants), and the nontransformed parent across multiple years and locations. The null segregants and the lines containing only the selection gene did not differ in yield or heading date from the nontransformed control. The HMW-GS transgenic group had lower mean grain yield than the nontransformed control group, but most of the difference was attributable to decreased yields of seven individual lines at all locations. The majority of the transgenic lines were indistinguishable in agronomic performance from their non-transgenic sibs and parent. Expression of the HMW-GS transgenes was the same between years. The authors concluded that changes in HMW-GS composition can be made by genetic transformation without necessarily altering wheat growth, development, or seed characteristics.

Among the more than 3,000 wheat plants directly selected for transgene-encoded glyphosate resistance (Hu et al., 2003), one was selected by Monsanto Corporation as a candidate for commercial release as Roundup Ready wheat. This line contained a single insertion of a single copy of a construct containing two *EPSPS* genes, one under control of the rice *Act1* promoter and intron and the other under control of the 35*S* promoter and an intron from a maize heat shock protein gene. In addition to its simple integration structure, the criteria for selection of this line included stable expression of the EPSPS enzyme over nine generations, inheritance of resistance as a stable dominant trait, survival of 1.68 kg acid-equivalent per hectare spray application of glyphosate, more than 80% fertility after 3.36 kg acid-equivalent of glyphosate, and good agronomic performance in field trials at 13 or 14 locations over 3 years (Hu et al., 2003). In the field, this line showed no damage after application of up to 4 L glyphosate per hectare. Yield and agronomic traits of the transgenic line were indistinguishable from the nontransformed parent, whether or not glyphosate was applied to the transgenic plants (Zhou et al., 2003). Blackshaw and Harker (2002) found a 4%–16% yield advantage of the transgenic line when glyphosate was used to control weeds compared with conventional weed control measures. In addition, this line had reduced dockage content (Wilson et al., 2003).

Based on this line's performance and characteristics, Monsanto Corporation petitioned the US Animal and Plant Health Inspection Service and the Food and Drug Administration for their approvals for commercial release. As a result, glyphosate-resistant wheat became the subject of several other published investigations. It was found to be equivalent to nontransformed wheat in chicken feeding studies (Kan and Hartnell 2004). The results of a risk assessment for commercial release of the transgenic glyphosate-resistant wheat were published by Peterson and Shama (2005), who concluded that it posed no more risk to the environment, humans, livestock, and wildlife than release of imidazdolone-resistant wheat obtained by standard mutagenesis. Harvey et al. (2003) found no difference between transgenic glyphosate-resistant wheat and its non-transgenic parent for survival and growth of greenbug [*Schizaphis graminum* (Rondani)], Russian wheat aphid [*Diuraphis noxia* (Mordvilko)], or wheat curl mite [*Aceria tosichella* (Keifer)], nor did they find any difference in survival of the transgenic and non-transgenic lines after infestation with the insects or in transmission of *Wheat streak mosaic virus* from curl mite. Lupwayi et al. (2007) found no consistent or large effects on soil microorganism communities when glyphosate-resistant wheat and canola (*Brassica napus* L.) were rotated at six Canadian prairie sites. Unexpectedly, sexually derived advanced breeding lines carrying the resistance transgene were found to be transiently resistant to leaf rust and stem rust (latter caused by *Puccinia graminis* Pers.:Pers. f. sp. *tritici* Eriks.

& E. Henn.) for 21 and 35 days after application of glyphosate (Anderson and Kolmer 2005).

Although the candidate transgenic wheat performed as well as or better than its parent in these various tests, Monsanto elected to withdraw its petitions for US commercial production of Roundup Ready wheat in spring 2004. This decision was attributed to opposition by trading partners in Europe and Asia (Wilson et al., 2003) and to the relatively low demand for weed control in US spring wheat (Skostad 2004). Hard red spring wheat constitutes only 20% of US exports and plantings have decreased since 1997 due to the threat of losses to Fusarium head blight and the northward expansion of growing regions suitable for maize and soybean production (Skostad 2004).

An interesting aspect of the history of transgenic glyphosate-resistant wheat is that less than 1% of the approximately 3,000 lines produced by *Agrobacterium* transformation and direct selection for glyphosate resistance met all the criteria for commercial release imposed by Monsanto (Hu et al., 2003). Even fewer of the transgenic lines produced in parallel by bombardment met those criteria (Hu et al., 2003). In the next section, we describe the limitations of current wheat transformation methods that result in large numbers of transgenic plants that are neither informative to basic researchers nor useful for commercial release in production agriculture.

LIMITATIONS OF WHEAT TRANSFORMATION TECHNOLOGY

Ideally, a wheat transformation experiment could be applied to any genotype and would result in a plant that contains a predictable number of copies of a gene or genes of interest integrated at a single site. The transgene would be expressed at the required levels when and where its product was needed. It would not have any other effects on the plant except those directly affected by the presence of the gene product in appropriate tissues. The transformed plant would be identical to its parent in growth, development, and yield except for the properties controlled and/or influenced by the transgene protein product. In practice, there are several limitations of current transformation methodology that can result in less-than-ideal outcomes. None of the difficulties encountered in recovering useful transformants are unique to wheat and some problematic outcomes of transformation have been better studied in other transgenic crop and model species (Filipecki and Malepszy 2006).

Genotype

Not all wheat genotypes can be transformed by current methods. Only some genotypes respond to *in vitro* culture to produce embryogenic callus. The calli derived from some genotypes cannot recover from wounding by particle bombardment or co-cultivation with *Agrobacterium*. Furthermore, some genotypes of embryogenic calli are not able to differentiate into shoots and roots in response to exogenous hormone changes in culture. A completely genotype-independent method of wheat transformation is unlikely in the near future, since none are known even for model plants such as tobacco and Arabidopsis. As long as genotype limitations persist, conventional breeding will be needed to move transgene-encoded traits into locally adapted cultivars. For breeders, a useful property of transgenes is that they almost always include, by their very nature, new sequences that can serve as molecular markers for following inheritance.

Structures of integrated transgenes

The simplest structures for integrated transgenes are shown in Fig. 18.1, but transgene insertions often have higher copy numbers. For biolistic transformations, transgene copy number can vary from 1, in a third of transformants, to more than 100 (Rasco-Gaunt et al., 2001). The median copy number among 25 lines investigated by Rasco-Gaunt et al. (2001) was in the range of 3 to 5. In the first report of *Agrobacterium* transformation of wheat, Cheng et al. (1997) found that one-third of 26 transformants had a single transgene copy versus only 17% of 77 transformants made by the same laboratory using biolistics. One-half of the *Agrobacterium* transformants had 2 to 3 copies (Cheng et al., 1997). The consensus of several reports is that one-third to two-thirds of *Agrobacterium* transformants of wheat are single-copy insertions (Cheng et al., 1997, 2003; Hu et al., 2003; Khanna and Daggard 2003; Wu et al., 2006).

For *Agrobacterium*-mediated transformation, the DNA which resides between border sequences usually integrates as a predictable structure (Fig. 18.1b), but sometimes rearrangement or truncation can occur. Moreover, one large study found that the majority of *Agrobacterium* transformants contained plasmid sequences from beyond the left border (Fig. 18.4a) (Wu et al., 2006). This would result whenever the transfer DNA did not efficiently and precisely end at the left border sequence. Such DNA can include antibiotic resistance genes from the plasmid backbone.

Integration after biolistic transformation can result in complex multi-transgene structures. The most detailed studies of these integration sites use fluorescent *in situ* hydridization (FISH) (Color Plate 34) and/or Southern blot hybridizations to visualize the orientations and dispersal of transgenes (Abranches et al., 2000; Jackson et al., 2001; Rooke et al., 2003). Genes that originally were on separate plasmids usually become linked before or during integration, forming tandem (Fig. 18.4b top) or nontandem (Fig. 18.4b middle) arrays in the chromosome. In some cases, the transgenes are interrupted by stretches of unknown or presumably host genomic DNA (Fig. 18.4b bottom). A study of eight independent wheat lines transformed with a single plasmid showed that in three lines, integration sites in metaphase chromosomes were separated by more than 1 Mb of genomic DNA and in one line, the sites were on opposite chromosome arms (Abranches et al., 2000). The remainder of the lines had single integration sites. In studies of 13 independent

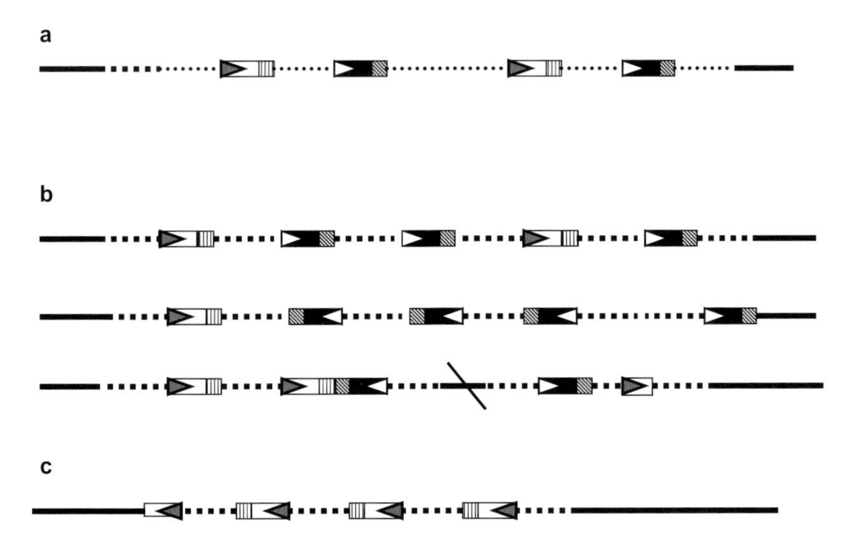

Fig. 18.4 Complex integration structures that can result from *Agrobacterium* (a) or biolistic (b,c) transformations. Symbols are as in Fig. 18.1, and drawings are not to scale. (a) In *Agrobacterium*-mediated transformations, plasmid DNA beyond the T-DNA left border (bold dotted line) and/or multiple T-DNA copies can be integrated. (b) In biolistic transformations, multiple copies can integrate in tandem (top drawing) or in inverse orientations (middle drawing). Gene copies can be separated by genomic DNAs (slashed solid line), and/or genes can become disrupted or rearranged (bottom drawing). (c) Possible promoter read-through from the integrated genes into adjacent wheat genomic DNA from the promoter of a truncated selection gene lacking its transcription terminator and part of its coding region.

transgenic wheat lines by Jackson et al. (2001), visualization of transgene loci by fiber-FISH showed that the most common structures (8 of 13) contained one to two transgene copies. The next most common structure (4 of 13) consisted of large tandem arrays of transgenes spanning 32 to 125 kb and interspersed with unknown DNA. A single transgenic event contained a 77-kb tandem array of plasmids with one 7-kb interruption (Jackson et al., 2001). Rooke et al. (2003) studied six transgenic wheat lines transformed with HMW-GS and selection genes. They found that the transgenes were present in 1 to about 15 copies at 1 to 3 separate loci. Some copies of genes in the same locus were clustered but separated by genomic DNA. Some were present as tandem repeats and others were rearranged or truncated. In two cases, the HMW-GS gene could be segregated from the selection gene (Rooke et al., 2003).

Complex integration structures in wheat transformants are difficult to fully characterize because of the size and repetitive nature of the wheat genome. Since commercialization requires extensive molecular characterization of the transgenic plant, plants destined for commercial applications must have simple insertion structures. A change in the structure of the DNA used for bombardment, originally implemented to reduce the occurrence of antibiotic resistance genes in transgenic plants, shows promise for reducing the complexity of integration sites in biolistic transformants. In this method, plasmid backbone sequences are physically removed from the DNA before transformation. The resulting linear DNAs comprise "minimum gene cassettes," consisting only of the sequences needed for expression of the genes in wheat: functionally linked promoter, coding, and transcription termination sequences. Bombardment with these so-called clean genes has been shown in rice, wheat, and other crops to result in lower transgene copy numbers and less-complex insertions (Fu et al., 2000; Agrawal et al., 2005; Yao et al., 2006; Gadaleta et al., 2008b). Transformation efficiencies are at least as high as those with intact plasmid DNA (Yao et al., 2006), resulting in integration of one to four copies of each transgene. These results for both rice and wheat suggest that, for biolistic transformation, the sequences in plasmid backbones of circular DNAs facilitate interactions among separate plasmids, either by homologous recombination or as hotspots for illegitimate recombination, and that such interactions can lead to formation of multigene arrays before integration of the whole unit into the chromosome (Kohli et al., 2003).

To the extent that co-integration of the selection gene and the gene(s) of interest occurs, it becomes more difficult to find progeny of the transformants that contain only the gene of interest. Linkage of the selection gene and gene of interest can sometimes be advantageous in that the phenotype of the selection gene can be used to easily follow inheritance. This can be of value when the gene of interest has a phenotype that is costly or difficult to assay, such as grain protein content or resistance to Fusarium head blight. However, transgenic plants that retain selection genes are becoming increasingly undesirable for several reasons (Natarajan and Turna 2007). For basic research, such plants cannot be retransformed using the same selection gene. For plants destined for commercial production, the presence of unnecessary new proteins in the food supply could make transgenic wheat less acceptable to the general public. For plants destined for testing or release into field environments, the absence of herbicide resistance genes (in cases where resistance is not the targeted trait) is desirable to prevent their widespread dispersal via pollen or seed mixing with cultivated nontransformed wheat and weedy relatives, to which they could confer a selective advantage (Matus-Cadiz et al., 2004; Brule-Babel et al., 2006). In the US, jointed goatgrass (*Aegilops cylindrica* Host, CCDD) is the only known wild wheat relative that can produce, albeit at low frequencies, semifertile hybrids with domesticated wheat (reviewed in Hegde and Waines 2004; Hanson et al., 2005; Schoenenberger et al., 2005). In parts of the world where wild wheat relatives flourish, the possibility of gene flow via pollen transmittal would be more problematic (Weissmann et al., 2005; Zaharieva and Monneveux 2006; Felber et al., 2007).

For all these reasons, it would be valuable to eliminate selection genes from the final transformed plant. Several strategies have been employed to achieve this goal. To increase the likelihood of separate integration sites, the selection genes and genes of interest can be transformed on separate DNAs in biolistic experiments or on separate T-DNAs housed on the same or different binary plasmids for *Agrobacterium* transformation. For the latter, experiments in rice and tobacco have shown that the two T-DNAs integrate at unlinked loci in up to about a quarter of *Agrobacterium* transformants (Komari et al., 1996). There are a few reports that suggest that use of "clean genes" without plasmid backbones for biolistic transformation increases the likelihood of obtaining transgenics with selection genes that can be segregated from other transgenes (Yao et al., 2006; Gadaleta et al., 2008b). Another method to eliminate selection genes is to flank them with target sites for site-specific recombinases. Srivastava et al. (1999) demonstrated that *Ubi1*::*bar* transgenes flanked by *lox* sites were excised from their integration loci when plants containing them were crossed to another transformant that expressed the *cre* recombinase. Segregation of the *cre* transgene in the following generation resulted in plants with neither *Ubi1*::*bar* nor *cre* transgenes. A third strategy is to use the mannose/*pmi* or another metabolic-advantage selection system. The *pmi* gene is only useful in tissue culture and confers no advantage for plant growth and fertility during field propagation. A fourth strategy is to use transformation methods and genotypes that are so efficient that transformants can be directly identified by DNA screening without the aid of selection (Zhang et al., 2006).

Integration location

Transgene loci produced by biolistic transformation of wheat have been detected in most parts of chromosomes, including telomeric (Color Plate 34), subtelomeric, intercalary, and centromeric regions (Abranches et al., 2000; Jackson et al., 2001). Little information is available for the chromosomal locations of *Agrobacterium* transformants of wheat, but much is known from experiments in the model plant Arabidopsis. In the absence of selection, Kim et al. (2007) recovered Arabidopsis transformants in all regions of the genome, implying that integration is completely random. The use of selection genes to identify Arabidopsis transformants results in recovery of transformants with integration sites mainly in gene-rich regions of the genome, where expression of the selection gene is favored (Alonso et al., 2003).

The lack of integration site predictability means that a given genome site cannot be targeted a second time using either *Agrobacterium* or biolistic transformation. Thus, when traits require expression of several different genes to be manifested or when multiple different traits are desired in a single background, all genes must be included in the initial transformation to have some chance of being linked. Otherwise, the different transgenes will have to be brought together by crossing. Assembling multiple desirable traits in a single genotype is a problem that breeders routinely face. Ow (2005) proposed a strategy for using site-specific recombination to stack multiple transgenes in a single location, but the efficiency of such a process relative to random integration is not yet known.

Inheritance anomalies

Many wheat transgenes exhibit normal inheritance upon selfing of regenerated plants. However, there have been several reports that some events resulting from biolistic transformation exhibit poor transmission of transgenes, as manifested in fewer than expected transgenic progeny (Cannell et al., 1999; Campbell et al., 2000; Rasco-Gaunt et al., 2001), and even complete loss of transgenes between generations in a minority of transformed lines (Srivastava et al., 1996; Stoger et al., 1998; Iser et al., 1999). Some investigators report occasional rearrangements within loci between generations (Srivastava et al., 1996), while others noted a lack of rearrangement over five generations (Demeke et al., 1999). The mechanism(s) by which some transgenes are lost or rearranged is not yet known. Poor or absent transmission could be a consequence of the complex nature of many biolistic integration structures in that the interspersion of transgenes and native DNA in large arrays may be inherently unstable due to either recombination or poor pairing between homologous chromosomes in heterozygotes. However, in a single report in which both detailed structures of loci and inheritance are reported for a population of transformed wheat plants, no correlation was found between locus complexity and transmission rates (Stoger et al., 1998).

Transgene expression levels and stability

Even if transgenes are stably integrated and predictably inherited, they may not be expressed in the first or subsequent generations. The content of the genomic DNA that flanks the integration locus can affect transgene expression via poorly understood processes known as position effects (Matzke and Matzke 1998). For example, some regions of chromosomes are tightly condensed (heterochromatic) and not conducive to efficient transcription (Fischer et al., 2006). Promoters of nearby native genes could produce read-through transcripts in either sense or antisense orientations, or otherwise influence the levels or tissue specificity of transgene expression (Fig. 18.2).

Several reports indicate progressive silencing of selection genes and other transgenes in wheat transformants during generation advance (Srivastava et al., 1996; Cannell et al., 1999; Mitchell et al., 2004; Przetakiewicz et al., 2004). Expression can be silenced at either the DNA (Fettig and Hess 1999; Howarth et al., 2005) or RNA (Wegel et al., 2005) level. Silencing is usually observed in early generations (Karunaratne et al., 1996; Iser et al., 1999) and can occur on some transgenes but not others within the same plant. For example, Li et al. (2003b) found that the *nptII* selection gene was silenced in seven of nine first-generation plants of one transformant, while the co-transformed *uidA* gene remained active in all nine of the same plants. Chen et al. (1998, 1999) proposed that some sequences are more prone to silencing than others, based on silencing of a rice chitinase driven by the 35*S* gene but expression of *bar* driven by *Ubi1* within the same plant.

Complexity of insertion sites may play a role in silencing. Cheng et al. (1997) found that 98% of *Agrobacterium* transformants coexpressed two transgenes, compared with 42%–62% of biolistics transformants. Bourdon et al. (2002) observed decreases rather than increases in luciferase expression in homozygous progeny of originally well-expressed transgenic lines, but the expression levels of transgenics carrying a similar construct that included an additional intron were not affected by homozygosity (Bourdon et al., 2004).

Transgene silencing can also be triggered by sexual hybridization. When a GUS::*nptII* fusion transgene, whose expression had been stable for four generations, was crossed to a nontransformed wheat, expression was lost and the transgene was methylated (Demeke et al., 1999).

The sporadic occurrence of transgene silencing is a source of frustration for many researchers using transgenic plants. For example, several wheat lines containing transgenes aimed at pest resistance could not be properly evaluated in field studies, because they had been silenced by the time sufficient seed were available for field testing. Following WSMV-coat protein expression and resistance manifested in first-generation plants in the greenhouse, both were lost in second- and third-generation plants tested in the field (Li et al., 2005b). Anand et al. (2003) found that 20 of 24 lines containing various wheat defense-protein-coding regions driven by *Ubi1* showed gene silencing between the first and second seed generations.

Even when transgenes are transcribed, their encoded nonnative mRNA or protein may not accumulate in wheat cells (Sparks et al., 2001). Messenger RNA processing signals can differ even between two cereals. For example, functional transcripts of the maize *Rp1-D* rust resistance gene did not accumulate in transgenic plants of either wheat or barley because of premature polyadenylation (Ayliffe et al., 2004).

Unintended effects of transformation, transgene insertion, or expression

A side effect of transformation protocols is the sporadic occurrence of mutations and chromosome rearrangements that result from tissue culture rather than the integration of any transgenes (somaclonal variation) (Qureshi et al., 1992; Philips et al., 1994; Ivanov et al., 1998). The phenotypic effect of such variation may be partially masked in wheat because of its polyploid nature. Somaclonal variants can be easily removed by crossing or backcrossing, since they are not likely to be linked to the transgenes.

A potentially serious consequence of the randomness of transgene integration is the possibility that the insertion results in changes in the expression of the native chromosomal DNA at the integrated locus (Filipecki and Malepszy 2006). Most simply, the transgene could physically interrupt a native gene by virtue of its integration, resulting in a null mutation. Alternatively, the transgene promoter could read into adjacent native genes, changing their expression patterns (Fig. 18.4c). Such changes could be manifested as reduced fitness of the transgenic line or as changes in expression of genes not expected to be affected by the transgene itself.

Constitutive transgene expression can result in inappropriate accumulation of the gene product in tissues where it has a detrimental effect on plant development or fertility. When the maize *Ubi1* promoter was used to drive expression of two different wheat peptidyl prolyl *cis-trans* isomerase isozymes, the transgenic plants and/or their seeds were abnormal (Kurek et al., 2002). Likewise, wheat plants transformed with a *Ubi1::Lr10* gene construct had significantly lower kernel weight than their nontransformed parent in the absence of leaf rust (Romeis et al., 2007). Even nonconstitutive promoters can result in transgene expression that interferes with normal plant growth or seed set. Transgenic wheat plants carrying a cell-cycle regulator gene under control of the wheat ADP-glucose pyrophosphorylase large subunit gene promoter had abnormal spikes, decreased seed set, and 50% less viable pollen than their nontransformed parents (Chrimes et al., 2005).

Only a few experiments have been reported thus far that detected subtle changes in nontarget gene expression in transgenic wheat. A survey of common seed metabolites was conducted in three transgenic wheat lines expressing added HMW-GS genes and in three non-transgenic controls grown in three separate years at two sites (Baker et al., 2006). While small differences occurred between one transgenic line and its parent for maltose or sucrose contents and for some free amino acids, none of the other differences exceeded environmental differences between control genotypes. In addition, no significant differences were detected among the same sets of HMW-GS transgenic, selection-gene transgenics, and nontransformed parental plants when leaf and endosperm transcripts were surveyed (Baudo et al., 2006). The authors concluded that, except for changes in the relative abundances of seed storage protein types predicted by the expression of the HMW-GS transgenes, composition of the transgenic seeds was substantially equivalent to that of the parent and showed no more variability than that among traditionally bred cultivars. Rakszegi et al. (2005), growing the same set of lines for three years in a different location, also observed that while the HMW-GS lines had the same yields as their nontransformed parents, the transgenics had slight increases in kernel hardness and decreases in kernel size.

Horváth-Szanics et al. (2006) used proteomics to examine the albumin and globulin families of seed proteins in transgenic wheat plants expressing *bar*-encoded herbicide resistance transgenes. When they compared transgenic and control plants under drought stress, they noted higher levels of several proteinase inhibitor proteins (15–27 kDa) in the transgenics (Horváth-Szanics et al., 2006). They hypothesized that the transgenics were more stressed than non-transgenics by the drought treatment.

Gregersen et al. (2005) used hybridization to measure expression of 9,000 genes at three time-points in developing seeds of transgenic wheat plants expressing an *Aspergillus* phytase gene under control of a HMW-GS promoter. They detected minor differences in the timing of accumulation of transcripts for some seed storage and other abundant proteins, compared with seeds from non-transgenic control plants. By the last time-point at 32 days after pollination, the transcript compositions of transgenic and control seeds were the same (Gregersen et al., 2005).

Practical considerations

In summary, so far no consistent or large differences have been detected between transgenic and non-transgenic wheat plants in transcripts, proteins, or metabolites, except for those expected to be directly affected by transgene expression. Despite the limitations of transformation technology discussed earlier, the majority of transgenic wheat lines best studied in field trials so far—those with HMW-GS and *Ubi1::bar* or glyphosate resistance transgenes—have the same yields and other agronomic characteristics as their non-transgenic parents.

The practical consequences of the random insertions and unpredictable structures of integrated transgenes are that multiple transformation events for each construct may need to be screened to find those that behave predictably in terms of inheritance, expression, and lack of non-target effects. For functional genomics research, 8 to 10 independent transformants are usually sufficient to determine the identities of genes with major effects. For applied research that introduces new genes into wheat, 20 or more transgenics may be needed to find those with the desired phenotypes. For transgenic plants destined for commercialization, hundreds of different transgenic events may be needed to find one line with stable inheritance, an easily defined integration structure, high and appropriate expression levels, desirable agronomic characteristics, and, in some cases, no selection genes.

FUTURE PERSPECTIVES

Indisputably, wheat transformation will have a major impact on wheat breeding, because it will continue to be a method for identifying the gene sequences that determine or affect important agronomic and utilization traits. The proven success of RNA interference to generate allele- and family-specific down-regulation of genes in the wheat polyploid genome has no peer in technology at this time. Determining the sequences of important genes is more than just an academic exercise: knowing exactly which sequence controls or influences a given trait provides wheat breeders with a perfect molecular marker, not only for detecting the presence or absence of a gene but also for identifying relevant individual alleles of that gene. Gene sequences are also needed by geneticists seeking to detect useful chemically induced mutations in known wheat genes by targeting induced local lesions in genomes (TILLING) (Slade et al., 2005). Such alleles can potentially provide novel variation for genes of interest.

Wheat transformation could benefit from technology improvements in efficiency, in widening of the range of genotypes to which it can be applied, in precision of insertion, in better control of insertion structures, and in better understanding of promoter activity. Basic research progress in model plants with smaller genomes, such as Arabidopsis and rice, is likely to yield better understanding of position effects, transgene silencing, and the complete set of features that allow predictable expression of transgenes in new genomic contexts. More promoters are needed whose activity is restricted to a limited number of plant tissues and/or developmental stages. For most types of pest resistances, it would be desirable to have a promoter with high levels of activity in green tissues or roots, but no activity in wheat seeds. Another type of promoter that would be valuable is one that can be activated by external cues that an experimenter or grower could control, for example, the ethanol-inducible promoter described by Li et al. (2005a). Such promoters need not come from wheat, but would have to be active in wheat. In addition, the inclusion of recognition sequences for site-specific recombinases could allow marker gene excision, site-specific integration, and stacking of multiple transgenes in a single plant genomic location (Ow 2005).

Over the last decade, efforts have been made to develop methods for the direct transformation of plant germline cells, such as microspores, ovules, or meristem cells, which develop directly into differentiated plant tissues and ultimately seeds. Limited success was achieved until it was demonstrated in Arabidopsis that transformants could be generated by germinating seeds in the presence of *Agrobacterium* (Feldmann and Marks 1987), and later that infiltration of floral tissues with *Agrobacterium* could also produce transformants (Bechtold et al., 1993). These are important developments because transformation methods via germ-line tissues are potentially more efficient, less labor-intensive, less dependent on tissue genotype for regeneration capability, and not prone to the somaclonal variation that can affect culture-based transgenics. Recently, successful applications of germ-line transformation methods were reported in wheat (Supartana et al., 2006; Zale and Steber 2006). Despite the advantages of *in planta* transformation methods, it is too early to say what efficiency levels will be routinely achieved by using them in wheat.

The new standard for transgenic crops is likely to include the absence of antibiotic- and herbicide-resistant selection genes that are remnants of the transformation process. Particularly for applications destined for release into the commercial marketplace (other than herbicide resistance), the preference clearly exists among consumers, food manufacturers, and environmentalists for the absence of selection genes and their products in the food supply (Natarajan and Turna 2007).

When transgenic wheat plants will find a way into US commercial production is difficult to predict. Although transgenic maize, soybean, and cotton (*Gossypium hirsutum* L.) have been planted and harvested for at least a decade without adverse effects on either the food supply or the environment, opposition to transgenic wheat is high, particularly in Japan and Europe. For many persons, "genetically modified" wheat and rice are especially problematic because these grains comprise a significant percentage of the human diet.

Assuming barriers to commercial production of transgenic wheat plants are eventually overcome, what traits are the most likely targets for improvement? For producers, resistance to Fusarium head blight, especially to initial infection (Type I resistance), would be very useful to combine with existing traditional sources of Type II resistance. Broad-spectrum resistance to wheat rusts would also be valuable. Drought and/or salt tolerance could expand the growing areas of wheat to more marginal land. Improving the suitability of wheat straw for biofuel production could expand marketplace demand for wheat. Improvements in grain nutritional composition—such as increases in the quantity and quality of protein and increased levels or bioavailability of iron—would be of direct benefit to consumers. Some have proposed using biotechnology to reduce the allergenicity of wheat for persons with celiac disease, but we consider this application unlikely to be successful, because many different classes of wheat seed proteins, including several that are important for wheat breadmaking quality, contain peptides that are toxic to celiac patients (Hamer 2005; Howdle 2006).

Bringing a genetically engineered crop plant into the marketplace requires considerable resource investment in terms of generating large numbers of plants, characterizing them extensively over several generations, securing intellectual property, and safety testing each transgenic line before release. The first two processes are similar to their counterparts in traditional breeding, but the latter two are much more costly for transgenic than for traditionally bred cultivars. At this time, relatively few traits are considered to have sufficient value to undertake commercial development of transgenic wheat, especially when the additional costs of seed segregation and identity preservation in the marketplace are considered (Wilson et al., 2003; Johnson et al., 2005). Nevertheless, the rapid increase in worldwide areas planted with genetically engineered maize, cotton, and soybean cultivars during the first decade of their availability is proof that such traits are of value to producers

(Brookes and Barfoot 2006). With adoption of the new refinements in wheat transformation technology listed in this section, which will allow integration and expression of only the genes of interest without unintended effects on other plant traits or on the food safety of the grain, such tangible benefits may eventually override perceived risks.

REFERENCES

Abebe, T., A.C. Guenzi, B. Martin, and J.C. Cushman. 2003. Tolerance of mannitol-accumulating transgenic wheat to water stress and salinity. Plant Physiol. 131:1748–1755.

Abranches, R., A.P. Santos, E. Wegel, S. Williams, A. Castilho, P. Christou, P. Shaw, and E. Stoger. 2000. Widely separated multiple transgene integration sites in wheat chromosomes are brought together at interphase. Plant J. 24:713–723.

Agrawal, P.K., A. Kohli, R.M. Twyman, and P. Christou. 2005. Transformation of plants with multiple cassettes generates simple transgene integration patterns and high expression levels. Mol. Breed. 16:247–260.

Ahmad, A., H. Zhong, W.L. Wang, and M.B. Sticklen. 2002. Shoot apical meristem: In vitro regeneration and morphogenesis in wheat (*Triticum aestivum* L.). In Vitro Cell. Dev. Biol. Plant 38:163–167.

Alonso, J.M., A.N. Stepanova, T.J. Leisse, et al. 2003. Genome-wide insertional mutagenesis of *Arabidopsis thaliana*. Science 301:653–657.

Altpeter, F., I. Diaz, H. McAuslane, K. Gaddour, P. Carbonero, and I.K. Vasil. 1999. Increased insect resistance in transgenic wheat stably expressing trypsin inhibitor CMe. Mol. Breed. 5:53–63.

Altpeter, F., A. Varshney, O. Abderhalden, D. Douchkov, C. Sautter, J. Kumlehn, R. Dudler, and P. Schweizer. 2005. Stable expression of a defense-related gene in wheat epidermis under transcriptional control of a novel promoter confers pathogen resistance. Plant Mol. Biol. 57:271–283.

Altpeter, F., V. Vasil, V. Srivastava, E. Stöger, and I.K. Vasil. 1996a. Accelerated production of transgenic wheat (*Triticum aestivum* L.) plants. Plant Cell Rep. 16:12–17.

Altpeter, F., V. Vasil, V. Srivastava, and I.K. Vasil. 1996b. Integration and expression of the high-molecular-weight glutenin subunit *1Ax1* gene into wheat. Nature Biotechnol. 14:1155–1159.

Alvarez, M.L., M. Gómez, J.M. Carrillo, and R.H. Vallejos. 2001. Analysis of dough functionality of flours from transgenic wheat. Mol. Breed. 8:103–108.

Alvarez, M.L., S. Guelman, N.G. Halford, S. Lustig, M.I. Reggiardo, N. Ryabushkina, P. Shewry, J. Stein, and R.H. Vallejos. 2000. Silencing of HMW glutenins in transgenic wheat expressing extra HMW subunits. Theor. Appl. Genet. 100:319–327.

Anand, A., T. Zhou, H.N. Trick, B.S. Gill, W.W. Bockus, and S. Muthukrishnan. 2003. Greenhouse and field testing of transgenic wheat plants stably expressing genes for thaumatin-like protein, chitinase and glucanase against *Fusarium graminearum*. J. Exp. Bot. 54:1101–1111.

Anderson, J.A., and J.A. Kolmer. 2005. Rust control in glyphosate tolerant wheat following application of the herbicide glyphosate. Plant Dis. 89:1136–1142.

Ayliffe, M.A., M. Steinau, R.F. Park, L. Rooke, M.G. Pacheco, S.H. Hulbert, H.N. Trick, and A.J. Pryor. 2004. Aberrant mRNA processing of the maize *Rp1-D* rust resistance gene in wheat and barley. Mol. Plant Microbe Interact. 17:853–864.

Badr, Y.A., M.A. Kereim, M.A. Yehia, O.O. Fouad, and A. Bahieldin. 2005. Production of fertile transgenic wheat plants by laser micropuncture. Photochem. Photobiol. Sci. 4:803–807.

Båga, M., A. Repellin, T. Demeke, K. Caswell, N. Leung, E.S. Abdel-Aal, P. Hucl, and R.N. Chibbar. 1999. Wheat starch modification through biotechnology. Stärke 51:111–116.

Baker, J.M., N.D. Hawkins, J.L. Ward, A. Lovegrove, J.A. Napier, P.R. Shewry, and M.H. Beale. 2006. A metabolomic study of substantial equivalence of field-grown genetically modified wheat. Plant Biotechnol. J. 4:381–392.

Balconi, C., C. Lanzanova, E. Conti, T. Triulzi, F. Forlani, M. Cattaneo, and E. Lupotto. 2007. Fusarium head blight evaluation in wheat transgenic plants expressing the maize *b-32* antifungal gene. Eur. J. Plant Pathol. 117:129–140.

Barro, F., P. Barceló, P.A. Lazzeri, P.R. Shewry, J. Ballesteros, and A. Martín. 2003. Functional properties of flours from field grown transgenic wheat lines expressing the HMW glutenin subunit 1Ax1 and 1Dx5 genes. Mol. Breed. 12:223–229.

Barro, F., P. Barceló, P.A. Lazzeri, P.R. Shewry, A. Martín, and J. Ballesteros. 2002. Field evaluation and agronomic performance of transgenic wheat. Theor. Appl. Genet. 105:980–984.

Barro, F., L. Rooke, F. Békés, P. Gras, A.S. Tatham, R. Fido, P.A. Lazzeri, P.R. Shewry, and P. Barceló. 1997. Transformation of wheat with high molecular weight subunit genes results in improved functional properties. Nature Biotechnol. 15:1295–1299.

Baudo, M.M., R. Lyons, S. Powers, G.M. Pastori, K.J. Edwards, M.J. Holdsworth, and P.R. Shewry. 2006. Transgenesis has less impact on the transcriptome of wheat grain than conventional breeding. Plant Biotechnol. J. 4:369–380.

Beachy, R.N. 1997. Mechanisms and application of pathogen-derived resistance in transgenic plants. Curr. Opin. Biotechnol. 8:215–220.

Bechtold, N., J. Ellis, and G. Pelletier. 1993. In-planta *Agrobacterium*-mediated gene-transfer by infiltration of adult *Arabidopsis-thaliana* plants. C. R. Acad. Sci. III Sci. Vie Life Sci. 316:1194–1199.

Becker, D., R. Brettschneider, and H. Lorz. 1994. Fertile transgenic wheat from microprojectile bombardment of scutellar tissue. Plant J. 5:299–307.

Beecher, B., A. Bettge, E. Smidansky, and M.J. Giroux. 2002. Expression of wild-type *pinB* sequence in transgenic wheat complements a hard phenotype. Theor. Appl. Genet. 105:870–877.

Bi, R.M., H.Y. Jia, D.S. Feng, and H.G. Wang. 2006. Production and analysis of transgenic wheat (*Triticum aestivum* L.) with improved insect resistance by the introduction of cowpea trypsin inhibitor gene. Euphytica 151:351–360.

Bieri, S., I. Potrykus, and J. Futterer. 2000. Expression of active barley seed ribosome-inactivating protein in transgenic wheat. Theor. Appl. Genet. 100:755–763.

Bieri, S., I. Potrykus, and J. Futterer. 2003. Effects of combined expression of antifungal barley seed proteins in transgenic wheat on powdery mildew infection. Mol. Breed. 11:37–48.

Blackshaw, R.E., and K.N. Harker. 2002. Selective weed control with glyphosate in glyphosate-resistant spring wheat (*Triticum aestivum*). Weed Technol. 16:885–892.

Blechl, A.E., and O.D. Anderson. 1996. Expression of a novel high-molecular-weight glutenin subunit gene in transgenic wheat. Nature Biotechnol. 14:875–879.

Blechl, A.E., H.Q. Le, and O.D. Anderson. 1998. Engineering changes in wheat flour by genetic transformation. J. Plant Physiol. 152:703–707.

Blechl, A., J. Lin, S. Nguyen, R. Chan, O.D. Anderson, and F.M. Dupont. 2007. Transgenic wheats with elevated levels of Dx5 and/or Dy10 high-molecular-weight glutenin subunits yield doughs with increased mixing strength and tolerance. J. Cereal Sci. 45:172–183.

Bliffeld, M., J. Mundy, I. Potrykus, and J. Futterer. 1999. Genetic engineering of wheat for increased resistance to powdery mildew disease. Theor. Appl. Genet. 98:1079–1086.

Bourdon, V., Z. Ladbrooke, A. Wickham, D. Lonsdale, and W. Harwood. 2002. Homozygous transgenic wheat plants with increased luciferase activity do not maintain their high level of expression in the next generation. Plant Sci. 163:297–305.

Bourdon, V., A. Wickham, D. Lonsdale, and W. Harwood. 2004. Additional introns inserted within the luciferase reporter gene stabilise transgene expression in wheat. Plant Sci. 167:1143–1149.

Bregitzer, P., A.E. Blechl, D. Fiedler, J. Lin, P. Sebesta, J.F. De Soto, O. Chicaiza, and J. Dubcovsky. 2006. Changes in high molecular weight glutenin subunit composition can be genetically engineered without affecting wheat agronomic performance. Crop Sci. 46:1553–1563.

Brereton, H.M., D. Chamberlain, R. Yang, M. Tea, S. McNeil, D.J. Coster, and K.A. Williams. 2007. Single chain antibody fragments for ocular use produced at high levels in a commercial wheat variety. J. Biotechnol. 129:539–546.

Brinch-Pedersen, H., A. Olesen, S.K. Rasmussen, and P.B. Holm. 2000. Generation of transgenic wheat (*Triticum aestivum* L.) for constitutive accumulation of an *Aspergillus* phytase. Mol. Breed. 6:195–206.

Brisibe, E.A., A. Gajdosova, A. Olesen, and S.B. Andersen. 2000. Cytodifferentiation and transformation of embryogenic callus lines derived from anther culture of wheat. J. Exp. Bot. 51:187–196.

Brookes, G., and P. Barfoot. 2006. Global impact of biotech crops: Socio-economic and environmental effects in the first ten years of commercial use [Online]. AgBioForum 9:139–151. Available at http://www.agbioforum.org/v9n3/index.htm (verified 5 Mar. 2008).

Brule-Babel, A.L., C.J. Willenborg, L.F. Friesen, and R.C. Van Acker. 2006. Modeling the influence of gene flow and selection pressure on the frequency of a GE herbicide-tolerant trait in non-GE wheat and wheat volunteers. Crop Sci. 46:1704–1710.

Campbell, B.T., P.S. Baenziger, A.M.S. Sato, and T. Clemente. 2000. Inheritance of multiple transgenes in wheat. Crop Sci. 40:1133–1141.

Cannell, M.E., A. Doherty, P.A. Lazzeri, and P. Barcelo. 1999. A population of wheat and tritordeum transformants showing a high degree of marker gene stability and heritability. Theor. Appl. Genet. 99:772–784.

Chalfie M., Y. Tu, G. Euskirchen, W.W. Ward, and D.C. Prasher. 1994. Green fluorescent protein as a marker for gene expression. Science 263:802–805.

Chen, W.P., P.D. Chen, D.J. Liu, R. Kynast, B. Friebe, R. Velazhahan, S. Muthukrishnan, and B.S. Gill. 1999. Development of wheat scab symptoms is delayed in transgenic wheat plants that constitutively express a rice thaumatin-like protein gene. Theor. Appl. Genet. 99:755–760.

Chen, W.P., X. Gu, G.H. Liang, S. Muthukrishnan, P.D. Chen, D.J. Liu, and B.S. Gill. 1998. Introduction and constitutive expression of a rice chitinase gene in bread wheat using biolistic bombardment and the *bar* gene as a selectable marker. Theor. Appl. Genet. 97:1296–1306.

Cheng, M., J.E. Fry, S.Z. Pang, H.P. Zhou, C.M. Hironaka, D.R. Duncan, T.W. Conner, and Y.C. Wan. 1997. Genetic transformation of wheat mediated by *Agrobacterium tumefaciens*. Plant Physiol. 115:971–980.

Cheng, M., T.C. Hu, J. Layton, C.N. Liu, and J.E. Fry. 2003. Desiccation of plant tissues post-*Agrobacterium* infection enhances T-DNA delivery and increases stable transformation efficiency in wheat. In Vitro Cell. Dev. Biol. Plant 39:595–604.

Chong, K., S.L. Bao, T. Xu, K.H. Tan, T.B. Liang, J.Z. Zeng, H.L. Huang, J. Xu, and Z.H. Xu. 1998. Functional analysis of the *ver* gene using antisense transgenic wheat. Physiol. Plant. 102:87–92.

Chrimes, D., H.J. Rogers, D. Francis, H.D. Jones, and C. Ainsworth. 2005. Expression of fission yeast *cdc25* driven by the wheat ADP-glucose pyrophosphorylase large subunit promoter reduces pollen viability and prevents transmission of the transgene in wheat. New Phytol. 166:185–192.

Christensen, A.H., R.A. Sharrock, and P.H. Quail. 1992. Maize polyubiquitin genes: Structure, thermal perturbation of expression and transcript splicing, and promoter activity following transfer to protoplasts by electroporation. Plant Mol. Biol. 18:675–689.

Chugh, A., and P. Khurana. 2003a. Herbicide-resistant transgenics of bread wheat (*T-aestivum*) and emmer wheat (*T-dicoccum*) by particle bombardment and *Agrobacterium*-mediated approaches. Curr. Sci. 84:78–83.

Chugh, A., and P. Khurana. 2003b. Regeneration via somatic embryogenesis from leaf basal segments and genetic transformation of bread and emmer wheat by particle bombardment. Plant Cell Tissue Organ Cult. 74:151–161.

Clausen, M., R. Krauter, G. Schachermayr, I. Potrykus, and C. Sautter. 2000. Antifungal activity of a virally encoded gene in transgenic wheat. Nature Biotechnol. 18:446–449.

Cloutier, S., B. McCallum, C. Loutre, T. Banks, T. Wicker, C. Feuillet, B. Keller, and M. Jordan. 2007. Leaf rust resistance gene *Lr1*, isolated from bread wheat (*Triticum aestivum* L.) is a member of the large *psr567* gene family. Plant Mol. Biol. 65:93–106.

Dahleen, L.S., P.A. Okubara, and A.E. Blechl. 2001. Transgenic approaches to combat Fusarium Head Blight in wheat and barley. Crop Sci. 41:628–637.

Darlington, H., R. Fido, A.S. Tatham, H. Jones, S.E. Salmon, and P.R. Shewry. 2003. Milling and baking properties of field grown wheat expressing HMW subunit transgenes. J. Cereal Sci. 38:301–306.

De Block, M., D. Debrouwer, and T. Moens. 1997. The development of a nuclear male sterility system in wheat: Expression of the barnase gene under the control of tapetum specific promoters. Theor. Appl. Genet. 95:125–131.

Demeke, T., P. Hucl, M. Båga, K. Caswell, N. Leung, and R.N. Chibbar. 1999. Transgene inheritance and silencing in hexaploid spring wheat. Theor. Appl. Genet. 99:947–953.

Doshi, K., F. Eudes, A. Laroche, and D. Gaudet. 2007. Anthocyanin expression in marker free transgenic wheat and triticale embryos. In Vitro Cell. Dev. Biol. Plant 43:429–435.

Drakakaki, G., P. Christou, and E. Stöger. 2000. Constitutive expression of soybean ferritin cDNA in transgenic wheat and rice results in increased iron levels in vegetative tissues but not in seeds. Transgenic Res. 9:445–452.

Felber, F., G. Kozlowski, N. Arrigo, and R. Guadagnuolo. 2007. Genetic and ecological consequences of transgene flow to the wild flora. Adv. Biochem. Eng. Biotechnol. 107:173–205.

Feldmann, K.A., and M.D. Marks. 1987. *Agrobacterium*-mediated transformation of germinating-seeds of *Arabidopsis-thaliana*: A non-tissue culture approach. Mol. Gen. Genet. 208:1–9.

Fettig, S., and D. Hess. 1999. Expression of a chimeric stilbene synthase gene in transgenic wheat lines. Transgenic Res. 8:179–189.

Feuillet, C., S. Travella, N. Stein, L. Albar, A. Nublat, and B. Keller. 2003. Map-based isolation of the leaf rust disease resistance gene *Lr10* from the hexaploid wheat (*Triticum aestivum* L.) genome. Proc. Natl. Acad. Sci. USA 100:15253–15258.

Filipecki, M., and S. Malepszy. 2006. Unintended consequences of plant transformation: A molecular insight. J. Appl. Genet. 47:277–286.

Fischer, A., I. Hofmann, K. Naumann, and G. Reuter. 2006. Heterochromatin proteins and the control of heterochromatic gene silencing in *Arabidopsis*. J. Plant Physiol. 163:358–368.

Fu, X.D., L.T. Duc, S. Fontana, B.B. Bong, P. Tinjuangjun, D. Sudhakar, R.M. Twyman, P. Christou, and A. Kohli. 2000. Linear transgene constructs lacking vector backbone sequences generate low-copy-number transgenic plants with simple integration patterns. Transgenic Res. 9:11–19.

Fu, D., C. Uauy, A. Blechl, and J. Dubcovsky. 2007. RNA interference for wheat functional gene analysis. Transgenic Res. 16:689–701.

Gadaleta, A., A.E. Blechl, S. Nguyen, M.F. Cardone, M. Ventura, J.S. Quick, and A. Blanco. 2008a. Stably expressed D genome-derived HMW glutenin subunit genes transformed into different durum wheat genotypes change dough mixing properties. Mol. Breed. 22:267–279.

Gadaleta, A., A. Giancaspro, A. Blechl, and A. Blanco. 2006. Phosphomannose isomerase, pmi, as a selectable marker gene for durum wheat transformation. J. Cereal Sci. 43:31–37.

Gadaleta, A., A. Giancaspro, A. Blechl, and A. Blanco. 2008b. A transgenic durum wheat line that is free of marker genes and expresses *1Dy10*. J. Cereal Sci. 48:439–445.

Goff, S.A., K.C. Cone, and V.L. Chandler. 1992. Functional analyses of the transcriptional activator encoded by the maize B gene: Evidence for a direct functional interaction between two classes of regulatory proteins. Genes Dev. 6:864–875.

Gopalakrishnan, S., G.K. Garg, D.T. Singh, and N.K. Singh. 2000. Herbicide-tolerant transgenic plants in high yielding commercial wheat cultivars obtained by microprojectile bombardment and selection on Basta. Curr. Sci. 79:1094–1100.

Gregersen, P.L., H. Brinch-Pedersen, and P.B. Holm. 2005. A microarray-based comparative analysis of gene expression profiles during grain development in transgenic and wild type wheat. Transgenic Res. 14:887–905.

Hamer, R.J. 2005. Coeliac disease: Background and biochemical aspects. Biotechnol. Adv. 23:401–408.

Hanson, B.D., C.A. Mallory-Smith, J. Price, B. Shafli, D.C. Thill, and R.S. Zemetra. 2005. Interspecific hybridization: potential for movement of herbicide resistance from wheat to jointed goatgrass (*Aegilops cylindrical*). Weed Technol. 19:674–682.

Harvey, T.L., T.J. Martin, and D. Seifers. 2003. Effect of Roundup Ready® wheat on greenbug, Russian wheat aphid, (Homoptera : Aphididae) and wheat curl mite, (Acari : Eriophyidae). J. Agric. Urban Entomol. 20:203–206.

Harvey, A., L. Moisan, S. Lindup, and D. Lonsdale. 1999. Wheat regenerated from scutellum callus as a source of material for transformation. Plant Cell Tissue Organ Cult. 57:153–156.

He, G.Y., H.D. Jones, R. D'Ovidio, S. Masci, M. Chen, J. West, B. Butow, O.D. Anderson, P. Lazzeri, R. Fido, and P.R. Shewry. 2005. Expression of an extended HMW subunit in transgenic wheat and the effect on dough mixing properties. J. Cereal Sci. 42:225–231.

He, G.Y., and P.A. Lazzeri. 1998. Analysis and optimisation of DNA delivery into wheat scutellum and tritordeum inflorescence explants by tissue electroporation. Plant Cell Rep. 18:64–70.

He, D.G., A. Mouradov, Y.M. Yang, E. Mouradova, and K.J. Scott. 1994. Transformation of wheat (*Triticum-aestivum* L.) through electroporation of protoplasts. Plant Cell Rep. 14:192–196.

He, G.Y., L. Rooke, S. Steele, F. Békés, P. Gras, A.S. Tatham, R. Fido, P. Barcelo, P.R. Shewry, and P.A. Lazzeri. 1999. Transformation of pasta wheat (*Triticum turgidum* L. var. *durum*) with high-molecular-weight glutenin subunit genes and modification of dough functionality. Mol. Breed. 5:377–386.

Hegde, S.G., and J.G. Waines. 2004. Hybridization and introgression between bread wheat and wild and weedy relatives in North America. Crop Sci. 44:1145–1155.

Hesler, L.S., Z. Li, T.M. Cheesbrough, and W.E. Riedell. 2005. Nymphiposition and population growth of *Rhoaplosiphum padi* L. (Homoptera : Aphididae) on conventional wheat cultivars and transgenic wheat isolines. J. Entomol. Sci. 40:186–196.

Hogg, A.C., B. Beecher, J.M. Martin, F. Meyer, L. Talbert, S. Lanning, and M.J. Giroux. 2005. Hard wheat milling and bread baking traits affected by the seed-specific overexpression of puroindolines. Crop Sci. 45:871–878.

Hogg, A.C., T. Sripo, B. Beecher, J.M. Martin, and M.J. Giroux. 2004. Wheat puroindolines interact to form friabilin and control wheat grain hardness. Theor. Appl. Genet. 108:1089–1097.

Hong, B., R. Barg, and T.-H.D. Ho. 1992. Developmental and organ-specific expression of an ABA- and stress-induced protein in barley. Plant Mol. Biol. 18:663–674.

Horváth-Szanics, E., Z. Szabó, T. Janáky, J. Pauk, and G. Hajás. 2006. Proteomics as an emergent tool for identification of stress-induced proteins in control and genetically modified wheat lines. Chromatographia 63:S143-S147.

Howarth, J.R., J.N. Jacquet, A. Doherty, H.D. Jones, and M.E. Cannell. 2005. Molecular genetic analysis of silencing in two lines of *Triticum aestivum* transformed with the reporter gene construct pAHC25. Ann. Appl. Biol. 146:311–320.

Howdle, P.D. 2006. Gliadin, glutenin or both? The search for the Holy Grail in coeliac disease. Eur. J. Gastroenterol. Hepatol. 18:703–706.

Hu, T., S. Metz, C. Chay, H.P. Zhou, N. Biest, G. Chen, M. Cheng, X. Feng, M. Radionenko, F. Lu, and J. Fry. 2003. *Agrobacterium*-mediated large-scale transformation of wheat (*Triticum aestivum* L.) using glyphosate selection. Plant Cell Rep. 21:1010–1019.

Huang, L., S.A. Brooks, W. Li, J.P. Fellers, H.N. Trick, and B.S. Gill. 2003. Map-based cloning of leaf rust resistance gene *Lr21* from the large and polyploid genome of bread wheat. Genetics 164:655–664.

Huber, M., R. Hahn, and D. Hess. 2002. High transformation frequencies obtained from a commercial wheat (*Triticum aestivum* L. cv. 'Combi') by microbombardment of immature embryos followed by GFP screening combined with PPT selection. Mol. Breed. 10:19–30.

Iser, M., S. Fettig, F. Scheyhing, K. Viertel, and D. Hess. 1999. Genotype-dependent stable genetic transformation in German spring wheat varieties selected for high regeneration potential. J. Plant Physiol. 154:509–516.

Ivanov, P., Z. Atanassov, V. Milkova, and L. Nikolova. 1998. Culture selected somaclonal variation in five *Triticum aestivum* L. genotypes. Euphytica 104:167–172.

Jackson, S.A., P. Zhang, W.P. Chen, R.L. Phillips, B. Friebe, S. Muthukrishnan, and B.S. Gill. 2001. High-resolution structural analysis of biolistic transgene integration into the genome of wheat. Theor. Appl. Genet. 103:56–62.

Jefferson, R., S. Burgess, and D. Hirsh. 1987. β-Glucuronidase from *Escherichia coli* as a gene-fusion marker. Proc. Natl. Acad. Sci. USA 83:8447–8451.

Jiménez-Martínez, E.S., and N.A. Bosque-Pérez. 2004. Variation in barley yellow dwarf virus transmission efficiency by *Rhopalosiphum padi* (Homoptera: Aphididae) after acquisition from transgenic and nontransformed wheat genotypes. J. Econom. Entomol. 97:1790–1796.

Jiménez-Martínez, E.S., N.A. Bosque-Pérez, P.H. Berger, and R.S. Zemetra. 2004a. Life history of the bird cherry–oat aphid, *Rhopalosiphum padi* (Homoptera: Aphididae), on transgenic and untransformed wheat challenged with barley yellow dwarf virus. J. Econ. Entomol. 97:203–212.

Jiménez-Martínez, E.S., N.A. Bosque-Pérez, P.H. Berger, R.S. Zemetra, H. Ding, and S.D. Eigenbrode. 2004b. Volatile cues influence the response of *Rhopalosiphum padi* (Homoptera: Aphididae) to barley yellow dwarf virus-infected transgenic and untransformed wheat. Environ. Entomol. 33:1207–1216.

Johnson, D.D., W. Lin, and G. Vocke. 2005. Economic and welfare impacts of commercializing a herbicide-tolerant, biotech wheat. Food Policy 30:162–184.

Jones, H.D. 2005. Wheat transformation: Current technology and applications to grain development and composition. J. Cereal Sci. 41:137–147.

Jones, H.D., A. Doherty, and H. Wu. 2005. Review of methodologies and a protocol for the *Agrobacterium*-mediated transformation of wheat. Plant Methods 1:5.

Jordan, M.C. 2000. Green fluorescent protein as a visual marker for wheat transformation. Plant Cell Rep. 19:1069–1075.

Jorgensen, R.A. 1995. Cosupression, flower color patterns, and metastable gene expression states. Science 268:686–691.

Kan, C.A., and G.F. Hartnell. 2004. Evaluation of broiler performance when fed Roundup-Ready wheat (Event MON 71800), control, and commercial wheat varieties. Poultry Sci. 83:1325–1334.

Karunaratne, S., A. Sohn, A. Mouradov, J. Scott, H.H. Steinbiss, and K.J. Scott. 1996. Transformation of wheat with the gene encoding the coat protein of barley yellow mosaic virus. Aust. J. Plant Physiol. 23:429–435.

Khanna, H.K., and G.E. Daggard. 2003. *Agrobacterium tumefaciens*-mediated transformation of wheat using a superbinary vector and a polyamine-supplemented regeneration medium. Plant Cell Rep. 21:429–436.

Khanna, H.K., and G.E. Daggard. 2006. Targeted expression of redesigned and codon optimised synthetic gene leads to recrystallisation inhibition and reduced electrolyte leakage in spring wheat at sub-zero temperatures. Plant Cell Rep. 25:1336–1346.

Kikkert, J.R. 1993. The biolistic(R) Pds-1000 He device. Plant Cell Tissue Organ Cult. 33:221–226.

Kim, S.-I., V. Gelvin, and S.B. Gelvin. 2007. Genome-wide analysis of *Agrobacterium* T-DNA integration sites in the Arabidopsis genome generated under non-selective conditions. Plant J. 51:779–791.

Klein, T.M., E.D. Wolf, R. Wu, and J.C. Sanford. 1987. High-velocity microprojectiles for delivering nucleic-acids into living cells. Nature (London) 327:70–73.

Kohli, A., R.M. Twyman, R. Abranches, E. Wegel, E. Stoger, and P. Christou. 2003. Transgene integration, organization and interaction in plants. Plant Mol. Biol. 52:247–258.

Komari, T., Y. Hiei, Y. Saito, N. Murai, and T. Kumashiro. 1996. Vectors carrying two separate T-DNAs for co-transformation of higher plants mediated by *Agrobacterium tumefaciens* and segregation of transformants free from selection markers. Plant J. 10:165–174.

Kurek, I., E. Stöger, R. Dulberger, P. Christou, and A. Breiman. 2002. Overexpression of the wheat FK506-binding protein 73 (FKBP73) and the heat-induced wheat FKBP77 in transgenic wheat reveals different functions of the two isoforms. Transgenic Res. 11:373–379.

Lamacchia, C., P.R. Shewry, N. Di Fonzo, J.L. Forsyth, N. Harris, P.A. Lazzeri, J.A. Napier, N.G. Halford, and P. Barcelo. 2001. Endosperm-specific activity of a storage protein gene promoter in transgenic wheat seed. J. Exp. Bot. 52:243–250.

Large, E.C. 1954. Growth stages in cereals—illustration of the Feekes scale. Plant Pathol. 3:128–129.

Li, B.C., K. Caswell, N. Leung, and R.N. Chibbar. 2003a. Wheat (*Triticum aestivum* L.) somatic embryogenesis from isolated scutellum: Days post anthesis, days of spike storage, and sucrose concentration affect efficiency. In Vitro Cell. Dev. Biol. Plant 39:20–23.

Li, R.Z., X.Y. Jia, and X. Mao. 2005a. Ethanol-inducible gene expression system and its applications in plant functional genomics. Plant Sci. 169:463–469.

Li, B.C., N. Leung, K. Caswell, and R.N. Chibbar. 2003b. Recovery and characterization of transgenic plants from two spring wheat cultivars with low embryogenesis efficiencies by the bombardment of isolated scutella. In Vitro Cell. Dev. Biol. Plant 39:12–19.

Li, Z.W., Y. Liu, and P.H. Berger. 2005b. Transgene silencing in wheat transformed with the WSMV-CP gene. Biotechnology 4:62–68.

Lonsdale, D.M., S. Lindup, L.J. Moisan, and A.J. Harvey. 1998. Using firefly luciferase to identify the transition from transient to stable expression in bombarded wheat scutellar tissue. Physiol. Plant. 102:447–453.

Loukoianov, A., L. Yan, A. Blechl, A. Sanchez, and J. Dubcovsky. 2005. Regulation of *VRN-1* vernalization genes in normal and transgenic polyploid wheat. Plant Physiol. 138:2364–2373.

Lupwayi, N.Z., K.G. Hanson, K.N. Harker, G.W. Clayton, R.E. Blackshaw, J.T. O'Donovan, E.N. Johnson, Y. Gan, R.B. Irvine, and M.A. Monreal. 2007. Soil microbial biomass, functional diversity and enzyme activity in glyphosate-resistant wheat–canola rotations under low-disturbance direct seeding and conventional tillage. Soil Biol. Biochem. 39:1418–1427.

Mackintosh, C.A., J. Lewis, L.E. Radmer, S. Shin, S.J. Heinen, L.A. Smith, M.N. Wyckoff, R. Dill-Macky, C.K. Evans, S. Kravchenko, G.D. Baldridge, R.J. Zeyen, and G.J. Muehlbauer. 2007. Overexpression of defense response genes in transgenic wheat enhances resistance to Fusarium head blight. Plant Cell Rep. 26:479–488.

Makandar, R., J.S. Essig, M.A. Schapaugh, H.N. Trick, and J. Shah. 2006. Genetically engineered resistance to Fusarium Head Blight in wheat by expression of Arabidopsis *NPR1*. Mol. Plant Microbe Interact. 19:123–129.

Martin, J.M., F.D. Meyer, C.F. Morris, and M.J. Giroux. 2007. Pilot scale milling characteristics of transgenic isolines of a hard wheat over-expressing puroindolines. Crop Sci. 47:497–506.

Martin, J.M., F.D. Meyer, E.D. Smidansky, H. Wanjugi, A.E. Blechl, and M.J. Giroux. 2006. Complementation of the *pina* (null) allele with the wild type *Pina* sequence restores a soft phenotype in transgenic wheat. Theor. Appl. Genet. 113:1563–1570.

Matus-Cadiz, M.A., P. Hucl, M.J. Horak, and L.K. Blomquist. 2004. Gene flow in wheat at the field scale. Crop Sci. 44:718–727.

Matzke, M.A., and A.J.M. Matzke. 1995. How and why do plants inactivate homologous (trans)genes? Plant Physiol. 107:679–685.

Matzke, A.J.M., and M.A. Matzke. 1998. Position effects and epigenetic silencing of plant transgenes. Curr. Opin. Plant Biol. 1:142–148.

McCabe, D., and P. Christou. 1993. Direct DNA transfer using electric-discharge particle-acceleration (Accell(Tm) Technology). Plant Cell Tissue Organ Cult. 33:227–236.

McCullen, C.A., and A.N. Binns. 2006. *Agrobacterium tumefaciens* and plant cell interactions and activities required for interkingdom macromolecular transfer. Annu. Rev. Cell Dev. Biol. 22:101–127.

Meyer, F.D., E.D. Smidansky, B. Beecher, T.W. Greene, and M.J. Giroux. 2004. The maize Sh2r6hs ADP-glucose pyrophosphorylase (AGP) large subunit confers enhanced AGP properties in transgenic wheat (*Triticum aestivum*). Plant Sci. 167:899–911.

Meyer, F.D., L.E. Talbert, J.M. Martin, S.P. Lanning, T.W. Greene, and M.J. Giroux. 2007. Field evaluation of transgenic wheat expressing a modified ADP-glucose pyrophosphorylase large subunit. Crop Sci. 47:336–342.

Milligan, A.S., A. Daly, M.A.J. Parry, P.A. Lazzeri, and I. Jepson. 2001. The expression of a maize glutathione S-transferase gene in transgenic wheat confers herbicide tolerance, both *in planta* and *in vitro*. Mol. Breed. 7:301–315.

Mitchell, R.A.C., P.A. Joyce, H. Rong, V.J. Evans, P.J. Madgwick, and M.A.J. Parry. 2004. Loss of decreased-rubisco phenotype between generations of wheat transformed with antisense and sense *rbc*S. Ann. Appl. Biol. 145:209–216.

Murai, K., M. Miyamae, H. Kato, S. Takumi, and Y. Ogihara. 2003. *WAP1*, a wheat *APETALA1* homolog, plays a central role in the phase transition from vegetative to reproductive growth. Plant Cell Physiol. 44:1255–1265.

Natarajan, S., and J. Turna. 2007. Excision of selectable marker genes from transgenic crops as a concern for environmental biosafety. J. Sci. Food Agric. 87:2547–2554.

Nehra, N.S., R.N. Chibbar, N. Leung, K. Caswell, C. Mallard, L. Steinhauer, M. Baga, and K.K. Kartha. 1994. Self-fertile transgenic wheat plants regenerated from isolated scutellar tissues following microprojectile bombardment with 2 distinct gene constructs. Plant J. 5:285–297.

Okubara, P.A., A.E. Blechl, S.P. McCormick, N.J. Alexander, R. Dill-Macky, and T.M. Hohn. 2002. Engineering deoxynivalenol metabolism in wheat through the expression of a fungal trichothecene acetyltransferase gene. Theor. Appl. Genet. 106:74–83.

Oldach, K.H., D. Becker, and H. Lorz. 2001. Heterologous expression of genes mediating enhanced fungal resistance in transgenic wheat. Mol. Plant Microbe Interact. 14:832–838.

Ortiz, J.P.A., M.I. Reggiardo, R.A. Ravizzini, S.G. Altabe, G.D.L. Cervigni, M.A. Spitteler, M.M. Morata, F.E. Elias, and R.H. Vallejos. 1996. Hygromycin resistance as an efficient selectable marker for wheat stable transformation. Plant Cell Rep. 15:877–881.

Ossowski, S., R. Schwab, and D. Weigel. 2008. Gene silencing in plants using artificial microRNAs and other small RNAs. Plant J. 53:674–690.

Ow, D.W. 2005. 2004 SIVB Congress Symposium Proceeding: Transgene management via multiple site-specific recombination systems. In Vitro Cell. Dev. Biol. Plant 41:213–219.

Ow, D.W., K.V. Wood, M. Deluca, J.R. De Wet, D.R. Helinksi, and S.H. Howell. 1986. Transient and stable expression of the firefly luciferase gene in plant cells and transgenic plants. Science 234:856–859.

Pang, S.Z., D.L. DeBoer, Y. Wan, G.B. Ye, J.G. Layton, M.K. Neher, C.L. Armstrong, J.E. Fry, M.A.W. Hinchee, and M.E. Fromm. 1996. An improved green fluorescent protein gene as a vital marker in plants. Plant Physiol. 112:893–900.

Parrott, D.L., A.J. Anderson, and J.G. Carman. 2002. *Agrobacterium* induces plant cell death in wheat (*Triticum aestivum* L.). Physiol. Mol. Plant Pathol. 60:59–69.

Pastori, G.M., M.D. Wilkinson, S.H. Steele, C.A. Sparks, H.D. Jones, and M.A.J. Parry. 2001. Age-dependent transformation frequency in elite wheat varieties. J. Exp. Bot. 52:857–863.

Patnaik, D., and P. Khurana. 2001. Wheat biotechnology: A minireview [Online]. Electron. J. Biotechnol. 4 (2): 74–102. Available at http://www.ejbiotechnology.info/content/vol4/issue2/index.html#review (verified 5 Mar. 2008).

Patnaik, D., D. Vishnudasan, and P. Khurana. 2006. *Agrobacterium*-mediated transformation of mature embryos of *Triticum aestivum* and *Triticum durum*. Curr. Sci. 91:307–317.

Pellegrineschi, A., R.M. Brito, L. Velazquez, L.M. Noguera, W. Pfeiffer, S. McLean, and D. Hoisington. 2002a. The effect of pretreatment with mild heat and drought stresses on the explant and biolistic transformation frequency of three durum wheat cultivars. Plant Cell Rep. 20:955–960.

Pellegrineschi, A., S. McLean, M. Salgado, L. Velazquez, R. Hernandez, R.M. Brito, M. Noguera, A. Medhurst, and D. Hoisington. 2001. Transgenic wheat plants: A powerful breeding source. Euphytica 119:133–136.

Pellegrineschi, A., L.M. Noguera, B. Skovmand, R.M. Brito, L. Velazquez, M.M. Salgado, R. Hernandez, M. Warburton, and D. Hoisington. 2002b. Identification of highly transformable wheat genotypes for mass production of fertile transgenic plants. Genome 45:421–430.

Pellegrineschi, A., M. Reynolds, M. Pacheco, R.M. Brito, R. Almeraya, K. Yamaguchi-Shinozaki, and D. Hoisington. 2004. Stress-induced expression in wheat of the *Arabidopsis thaliana DREB1A* gene delays water stress symptoms under greenhouse conditions. Genome 47:493–500.

Permingeat, H.R., M.L. Alvarez, G.D.L. Cervigni, R.A. Ravizzini, and R.H. Vallejos. 2003. Stable wheat transformation obtained without selectable markers. Plant Mol. Biol. 52:415–419.

Peterson, R.K.D., and L.M. Shama. 2005. A comparative risk assessment of genetically engineered, mutagenic, and conventional wheat production systems. Transgenic Res. 14:859–875.

Philips, R.L., S.M. Kaeppler, and P. Olhoft. 1994. Genetic instability of plant tissue cultures: Breakdown of normal controls. Proc. Natl. Acad. Sci. USA 91:5222–5226.

Ponya, Z., P. Finy, A. Feher, J. Mityko, D. Dudits, and B. Baranbas. 1999. Optimisation of introducing foreign genes into egg cells and zygotes of wheat (*Triticum aestivum* L.) via microinjection. Protoplasma 208:163–172.

Popineau, Y., G. Deshayes, J. Lefebvre, R. Fido, A.S. Tatham, and P.R. Shewry. 2001. Prolamin aggregation, gluten viscoelasticity, and mixing properties of transgenic wheat lines expressing 1Ax and 1Dx high molecular weight glutenin subunit transgenes. J. Agric. Food Chem. 49:395–401.

Przetakiewicz, A., A. Karas, W. Orczyk, and A. Nadolska-Orczyk. 2004. *Agrobacterium*-mediated transformation of polyploid cereals: The efficiency of selection and transgene expression in wheat. Cell. Mol. Biol. Lett. 9:903–917.

Qureshi, J.A., P. Hucl, and K.K. Kartha. 1992. Is somaclonal variation a reliable tool for spring wheat improvement? Euphytica 60:221–228.

Rakszegi, M., F. Békés, L. Láng, L. Tamas, P.R. Shewry, and Z. Bedö. 2005. Technological quality of transgenic wheat expressing an increased amount of a HMW glutenin subunit. J. Cereal Sci. 42:15–23.

Rakszegi, M., G. Pastori, H.D. Jones, F. Békés, B. Butow, L. Láng, Z. Bedö, and P.R. Shewry. 2008. Technological quality of field grown transgenic lines of commercial wheat cultivars expressing the 1Ax1 HMW glutenin subunit gene. J. Cereal Sci. 47:310–321.

Rasco-Gaunt, S., and P. Barcelo. 1999. Immature inflorescence culture of cereals: A highly responsive system for regeneration and transformation. p. 71–81. *In* H. Jones (ed.) Methods in molecular biology: Plant gene transfer and expression protocols. Humana Press, Totowa, NJ.

Rasco-Gaunt, S., D. Liu, C.P. Li, A. Doherty, K. Hagemann, A. Riley, T. Thompson, C. Brunkan, M. Mitchell, K. Lowe, E. Krebbers, P. Lazzeri, S. Jayne, and D. Rice. 2003. Characterisation of the expression of a novel constitutive maize promoter in transgenic wheat and maize. Plant Cell Rep. 21:569–576.

Rasco-Gaunt, S., A. Riley, M. Cannell, P. Barcelo, and P.A. Lazzeri. 2001. Procedures allowing the transformation of a range of European elite wheat (*Triticum aestivum* L.) varieties via particle bombardment. J. Exp. Bot. 52:865–874.

Reed, J., L. Privalle, M.L. Powell, M. Meghji, J. Dawson, E. Dunder, J. Suttie, A. Wenck, K. Launis, C. Kramer, Y.F. Chang, G. Hansen, and M. Wright. 2001. Phosphomannose isomerase: An efficient selectable marker for plant transformation. In Vitro Cell. Dev. Biol. Plant 37:127–132.

Regina, A., A. Bird, D. Topping, S. Bowden, J. Freeman, T. Barsby, B. Kosar-Hashemi, Z.Y. Li, S. Rahman, and M. Morell. 2006. High-amylose wheat generated by RNA interference improves indices of large-bowel health in rats. Proc. Natl. Acad. Sci. USA 103:3546–3551.

Romeis, J., M. Waldburger, P. Streckeisen, P.A.M. Hogervorst, B. Keller, M. Winzeler, and F. Bigler. 2007. Performance of transgenic spring wheat plants and effects on non-target organisms under glasshouse and semi-field conditions. J. Appl. Entomol. 131:593–602.

Rommens, C.M. 2007. Intragenic crop improvement: Combining the benefits of traditional breeding and genetic engineering. J. Agric. Food Chem. 55:4281–4288.

Rooke, L., F. Békés, R. Fido, F. Barro, P. Gras, A.S. Tatham, P. Barcelo, P. Lazzeri, and P.R. Shewry. 1999. Overexpression of a gluten protein in transgenic wheat results in greatly increased dough strength. J. Cereal Sci. 30:115–120.

Rooke, L., D. Byrne, and S. Salgueiro. 2000. Marker gene expression driven by the maize ubiquitin promoter in transgenic wheat. Ann. Appl. Biol. 136:167–172.

Rooke, L., S.H. Steele, P. Barcelo, P.R. Shewry, and P.A. Lazzeri. 2003. Transgene inheritance, segregation and expression in bread wheat. Euphytica 129:301–309.

Roy-Barman, S., C. Sautter, and B.B. Chattoo. 2006. Expression of the lipid transfer protein Ace-AMP1 in transgenic wheat enhances antifungal activity and defense responses. Transgenic Res. 15:435–446.

Sahrawat, A.K., D. Becker, S. Lutticke, and H. Lörz. 2003. Genetic improvement of wheat via alien gene transfer, an assessment. Plant Sci. 165:1147–1168.

Sanford, J.C. 1988. The biolistic process. Trends Biotechnol. 6:299–302.

Sawahel, W.A., and A.H. Hassan. 2002. Generation of transgenic wheat plants producing high levels of the osmoprotectant proline. Biotechnol. Lett. 24:721–725.

Sawahel, W.A., and M. Saker. 1997. Stable genetic transformation of mature wheat embryos using silicone carbide fibers and DNA imbibition. Cell. Mol. Biol. Lett. 3:421–429.

Schlaich, T., B.M. Urbaniak, N. Malgras, E. Ehler, C. Birrer, L. Meier, and C. Sautter. 2006. Increased field resistance to *Tilletia caries* provided by a specific antifungal virus gene in genetically engineered wheat. Plant Biotechnol. J. 4:63–75.

Schoenenberger, N., F. Felber, D. Savova-Bianchi, and R. Guadagnuolo. 2005. Introgression of wheat DNA markers from A, B and D genomes in early generation progeny of *Aegilops cylindrica* Host × *Triticum aestivum* L. hybrids. Theor. Appl. Genet. 111:1338–1346.

Schweizer, P. 2008. Tissue-specific expression of a defense-related peroxidase in transgenic wheat potentiates cell death in pathogen-attacked leaf epidermis. Mol. Plant Pathol. 9:45–57.

Serazetdinova, L., K.H. Oldach, and H. Lörz. 2005. Expression of transgenic stilbene synthases in wheat causes the accumulation of unknown stilbene derivatives with antifungal activity. J. Plant Physiol. 162:985–1002.

Shah, P.A., A.M.R. Gatehouse, S.J. Clark, and J.K. Pell. 2005. Wheat containing snowdrop lectin (GNA) does not affect infection of the cereal aphid *Metopolophium dirhodum* by the fungal natural enemy *Pandora neoaphidis*. Transgenic Res. 14:473–476.

Sharp, G.L., J.M. Martin, S.P. Lanning, N.K. Blake, C.W. Brey, E. Sivamani, R. Qu, and L.E. Talbert. 2002. Field evaluation of transgenic and classical sources of wheat streak mosaic virus resistance. Crop Sci. 42:105–110.

Shewry, P.R., N.G. Halford, A.S. Tatham, Y. Popineau, D. Lafiandra, and P.S. Belton. 2003. The high molecular weight subunits of wheat glutenin and their role in determining wheat processing properties. Adv. Food Nutr. Res. 45:219–302.

Shewry, P.R., S. Powers, J.M. Field, R.J. Fido, H.D. Jones, G.M. Arnold, J. West, P.A. Lazzeri, P. Barcelo, F. Barro, A.S. Tatham, F. Bekes, B. Butow, and H. Darlington. 2006. Comparative field performance over 3 years and two sites of transgenic wheat lines expressing HMW subunit transgenes. Theor. Appl. Genet. 113:128–136.

Shrawat, A.K., and H. Lörz. 2006. *Agrobacterium*-mediated transformation of cereals: A promising approach crossing barriers. Plant Biotechnol. J. 4:575–603.

Simmonds, J., P. Stewart, and D. Simmonds. 1992. Regeneration of *Triticum-aestivum* apical explants after microinjection of germ line progenitor cells with DNA. Physiol. Plant. 85:197–206.

Simons, K.J., J.P. Fellers, H.N. Trick, Z. Zhang, Y.-S. Tai, B.S. Gill, and J.D. Faris. 2006. Molecular characterization of the major wheat domestication gene *Q*. Genetics 172:547–555.

Singh, N., and H.S. Chawla. 1999. Use of silicon carbide fibers for *Agrobacterium*-mediated transformation in wheat. Curr. Sci. 76:1483–1485.

Sivamani, E., A. Bahieldin, J.M. Wraith, T. Al-Niemi, W.E. Dyer, T.H.D. Ho, and R. Qu. 2000a. Improved biomass productivity and water use efficiency under water deficit conditions in transgenic wheat constitutively expressing the barley *HVA1* gene. Plant Sci. 155:1–9.

Sivamani, E., C.W. Brey, W.E. Dyer, L.E. Talbert, and R. D. Qu. 2000b. Resistance to wheat streak mosaic virus in transgenic wheat expressing the viral replicase (*NIb*) gene. Mol. Breed. 6:469–477.

Sivamani, E., C.W. Brey, L.E. Talbert, M.A. Young, W.E. Dyer, W.K. Kaniewski, and R.D. Qu. 2002. Resistance to wheat streak mosaic virus in transgenic wheat engineered with the viral coat protein gene. Transgenic Res. 11:31–41.

Skostad, E. 2004. Monsanto pulls the plug on genetically modified wheat. Science 304:1088–1089.

Slade, A.J., S.I. Fuerstenberg, D. Loeffler, M.N. Steine, and D. Facciotti. 2005. A reverse genetic, nontransgenic approach to wheat crop improvement by TILLING. Nature Biotechnol. 23:75–81.

Smidansky, E.D., M. Clancy, F.D. Meyer, S.P. Lanning, N.K. Blake, L.E. Talbert, and M.J. Giroux. 2002. Enhanced ADP-glucose pyrophosphorylase activity in wheat endosperm increases seed yield. Proc. Natl. Acad. Sci. USA 99:1724–1729.

Smidansky, E.D., F.D. Meyer, B. Blakeslee, T.E. Weglarz, T.W. Greene, and M.J. Giroux. 2007. Expression of a modified ADP-glucose pyrophosphorylase large subunit in wheat seeds stimulates photosynthesis and carbon metabolism. Planta 225:965–976.

Sorokin, A.P., X.Y. Ke, D.F. Chen, and M.C. Elliott. 2000. Production of fertile transgenic wheat plants via tissue electroporation. Plant Sci. 156:227–233.

Sparks, C.A., C.K. Castleden, J. West, D.Z. Habash, P.J. Madgwick, M.J. Paul, G. Noctor, J. Harrison, R. Wu, J. Wilkinson, W.P. Quick, M.A.J. Parry, C.H. Foyer, and B.J. Miflin. 2001. Potential for manipulating carbon metabolism in wheat. Ann. Appl. Biol. 138:33–45.

Srivastava, V., O.D. Anderson, and D.W. Ow. 1999. Single-copy transgenic wheat generated through the resolution of complex integration patterns. Proc. Natl. Acad. Sci. USA 96:11117–11121.

Srivastava, V., V. Vasil, and I.K. Vasil. 1996. Molecular characterization of the fate of transgenes in transformed wheat (*Triticum aestivum* L). Theor. Appl. Genet. 92:1031–1037.

Sticklen, M.B., and H.F. Oraby. 2005. Shoot apical meristem: A sustainable explant for genetic transformation of cereal crops. In Vitro Cell. Dev. Biol. Plant 41:187–200.

Stöger, E., C. Vaquero, E. Torres, M. Sack, L. Nicholson, J. Drossard, S. Williams, D. Keen, Y. Perrin, P. Christou, and R. Fischer. 2000. Cereal crops as viable production and storage systems for pharmaceutical scFv antibodies. Plant Mol. Biol. 42:583–590.

Stoger, E., S. Williams, P. Christou, R.E. Down, and J.A. Gatehouse. 1999a. Expression of the insecticidal lectin from snowdrop (*Galanthus nivalis* agglutinin; GNA) in transgenic wheat plants: Effects on predation by the grain aphid *Sitobion avenae*. Mol. Breed. 5:65–73.

Stoger, E., S. Williams, D. Keen, and P. Christou. 1998. Molecular characteristics of transgenic wheat and the effect on transgene expression. Transgenic Res. 7:463–471.

Stoger, E., S. Williams, D. Keen, and P. Christou. 1999b. Constitutive versus seed specific expression in transgenic wheat: Temporal and spatial control. Transgenic Res. 8:73–82.

Supartana, P., T. Shimizu, M. Nogawa, H. Shioiri, T. Nakajima, N. Haramoto, M. Nozue, and M. Kojima. 2006. Development of simple and efficient *in planta* transformation method for wheat (*Triticum aestivum* L.) using *Agrobacterium tumefaciens*. J. Biosci. Bioeng. 102:162–170.

Takumi, S., and T. Shimada. 1997. Variation in transformation frequencies among six common wheat cultivars through particle bombardment of scutellar tissues. Genes Genet. Syst. 72:63–69.

Thompson, C.J., N.R. Movva, R. Tizard, R. Crameri, J.E. Davies, M. Lauwereys, and J. Botterman. 1987. Characterization of the herbicide-resistance gene *bar* from *Streptomyces hygroscopicus*. EMBO J. 6:2519–2523.

Travella, S., T.E. Klimm, and B. Keller. 2006. RNA interference-based gene silencing as an efficient tool for functional genomics in hexaploid bread wheat. Plant Physiol. 142:6–20.

Uauy, C., A. Distelfeld, T. Fahima, A. Blechl, and J. Dubcovsky. 2006. A *NAC* gene regulating senescence improves grain protein, zinc, and iron content in wheat. Science 314:1298–1301.

Uthayakumaran, S., O.M. Lukow, M.C. Jordan, and S. Cloutier. 2003. Development of genetically modified wheat to assess its dough functional properties. Mol. Breed. 11:249–258.

Uze, M., I. Potrykus, and C. Sautter. 1999. Single-stranded DNA in the genetic transformation of wheat (*Triticum aestivum* L.): Transformation frequency and integration pattern. Theor. Appl. Genet. 99:487–495.

Varshney, A., and F. Altpeter. 2001. Stable transformation and tissue culture response in current European winter wheats (*Triticum aestivum* L.). Mol. Breed. 8:295–309.

Vasil, I.K. 2007. Molecular genetic improvement of cereals: Transgenic wheat (*Triticum aestivum* L.). Plant Cell Rep. 26:1133–1154.

Vasil, I.K., S. Bean, J.M. Zhao, P. McCluskey, G. Lookhart, H.P. Zhao, F. Altpeter, and V. Vasil. 2001. Evaluation of baking properties and gluten protein composition of field grown transgenic wheat lines expressing high molecular weight glutenin gene 1Ax1. J. Plant Physiol. 158:521–528.

Vasil, V., A.M. Castillo, M.E. Fromm, and I.K. Vasil. 1992. Herbicide resistant fertile transgenic wheat plants obtained by microprojectile bombardment of regenerable embryogenic callus. Bio/Technology 10:667–674.

Vendruscolo, E.C.G., I. Schuster, M. Pileggi, C.A. Scapim, H.B.C. Molinari, C.J. Marur, and L.G.E. Vieira. 2007. Stress-induced synthesis of proline confers tolerance to water deficit in transgenic wheat. J. Plant Physiol. 164:1367–1376.

Vishnudasan, D., M.N. Tripathi, U. Rao, and P. Khurana. 2005. Assessment of nematode resistance in wheat transgenic plants expressing potato proteinase inhibitor (*PIN2*) gene. Transgenic Res. 14:665–675.

Weeks, J.T., O.D. Anderson, and A.E. Blechl. 1993. Rapid production of multiple independent lines of fertile transgenic wheat (*Triticum-aestivum*). Plant Physiol. 102:1077–1084.

Weeks, T., K.Y. Koshiyama, U. Maier-Greiner, T. Schäeffner, and O.D. Anderson. 2000. Wheat transformation using cyanamide as a new selective agent. Crop Sci. 40:1749–1754.

Wegel, E., R.H. Vallejos, P. Christou, E. Stöger, and P. Shaw, 2005. Large-scale chromatin decondensation induced in a developmentally activated transgene locus. J. Cell Sci. 118:1021–1031.

Weissmann, S., M. Feldman, and J. Gressel. 2005. Sequence evidence for sporadic intergeneric DNA introgression from wheat into a wild *Aegilops* species. Mol. Biol. Evol. 22:2055–2062.

Wiley, P., P. Tosi, A. Evrard, A. Lovegrove, H. Jones, and P. Shewry. 2007. Promoter analysis and immunolocalisation show that puroindoline genes are exclusively expressed in starchy endosperm cells of wheat grain. Plant Mol. Biol. 64:125–136.

Wilkinson, M., J. Lenton, and M. Holdsworth 2005. Transcripts of *Vp-1* homoeologues are alternatively spliced within the *Triticeae* tribe. Euphytica 143:243–246.

Wilson, W.W., E.L. Janzen, and B.L. Dahl. 2003. Issues in development and adoption of genetically modified (GM) wheats [Online]. AgBioForum 6:101–112. Available at http://www.agbioforum.org/v6n3/index.htm (verified 5 Mar. 2008).

Witrzens, B., R.I.S. Brettell, F.R. Murray, D. McElroy, Z.Y. Li, and E.S. Dennis. 1998. Comparison of three selectable marker genes for transformation of wheat by microprojectile bombardment. Aust. J. Plant Physiol. 25:39–44.

Wright, M., J. Dawson, E. Dunder, J. Suttie, J. Reed, C. Kramer, Y. Chang, R. Novitzky, H. Wang, and L. Artim-Moore. 2001. Efficient biolistic transformation of maize (*Zea mays* L.) and wheat (*Triticum aestivum* L.) using the phosphomannose isomerase gene, *pmi*, as the selectable marker. Plant Cell Rep. 20:429–436.

Wu, H., C. Sparks, B. Amoah, and H.D. Jones. 2003. Factors influencing successful *Agrobacterium*-mediated genetic transformation of wheat. Plant Cell Rep. 21:659–668.

Wu, H., C.A. Sparks, and H.D. Jones. 2006. Characterisation of T-DNA loci and vector backbone sequences in transgenic wheat produced by *Agrobacterium*-mediated transformation. Mol. Breed. 18:195–208.

Xia, L., H. Geng, X. Chen, Z. He, M. Lillemo, and C.F. Morris. 2008. Silencing of puroindoline a alters the kernel texture in transgenic bread wheat. J. Cereal Sci. 47:331–338.

Xue, Z.Y., D.Y. Zhi, G.P. Xue, H. Zhang, Y.X. Zhao, and G.M. Xia. 2004. Enhanced salt tolerance of transgenic wheat (*Tritivum aestivum* L.) expressing a vacuolar Na^+/H^+ antiporter gene with improved grain yields in saline soils in the field and a reduced level of leaf Na^+. Plant Sci. 167:849–859.

Yan, L., D. Fu, C. Li, A. Blechl, G. Tranquilli, M. Bonafede, A. Sanchez, M. Valarik, S. Yasuda, and J. Dubcovsky. 2006. The wheat and barley vernalization gene *VRN3* is an orthologue of *FT*. Proc. Natl. Acad. Sci. USA 103:19581–19586.

Yan, L., A. Loukoianov, A. Blechl, G. Tranquilli, W. Ramakrishna, P. SanMiguel, J.L. Bennetzen, V. Echenique, and J. Dubcovsky. 2004. The wheat *VRN2* gene is a flowering repressor down-regulated by vernalization. Science 303:1640–1644.

Yao, Q., L. Cong, J.L. Chang, K.X. Li, G.X. Yang, and G.Y. He. 2006. Low copy number gene transfer and stable expression in a commercial wheat cultivar via particle bombardment. J. Exp. Bot. 57:3737–3746.

Yong, W.-D., Y.-Y. Xu, W.-Z. Xu, X. Wang, N. Li, J.-S. Wu, T.-B. Liang, K. Chong, Z.-H. Xu, K.-H. Tan, and Z.-Q. Zhu. 2003. Vernalization-induced flowering in wheat is mediated by a lectin-like gene *VER2*. Planta 217:261–270.

Yue, S.J., H. Li, Y.W. Li, Y.F. Zhu, J.K. Guo, Y.J. Liu, Y. Chen, and X. Jia. 2008. Generation of transgenic wheat lines with altered expression levels of 1Dx5 high-molecular weight glutenin subunit by RNA interference. J. Cereal Sci. 47:453–161.

Zaharieva, M., and P. Monneveux. 2006. Spontaneous hybridization between bread wheat (*Triticum aestivum* L.) and its wild relatives in Europe. Crop Sci. 46:512–527.

Zale, J.M., and C.M. Steber. 2006. *In planta* transformation of wheat as a genomics tool. p. 275. *In* Abstracts, Plant and Animal Genome Conf. XIV, San Diego, CA. 14–18 Jan. 2006. Scherago International, New York.

Zhang, L.Y., R. French, W.G. Langenberg, and A. Mitra. 2001. Accumulation of barley stripe mosaic virus is significantly reduced in transgenic wheat plants expressing a bacterial ribonuclease. Transgenic Res. 10:13–19.

Zhang, W.G., D. McElroy, and R. Wu. 1991. Analysis of rice *Act1* 5' region activity in transgenic rice plants. Plant Cell 3:1155–1165.

Zhang, J.R., Q. Xiao, K.X. Li, M.J. Chen, J.L. Chang, L.T. Luo, Y Li, Y. Liu, P.R. Shewry, and G.Y. He. 2006. An optimal pooling strategy applied to high-throughput screening for rare marker-free transformants. Biotechnol. Lett. 28:1537–1544.

Zhao, T.J., S.Y. Zhao, H.M. Chen, Q.Z. Zhao, Z.M. Hu, B.K. Hou, and G.M. Xia. 2006. Transgenic wheat progeny resistant to powdery mildew generated by *Agrobacterium* inoculum to the basal portion of wheat seedling. Plant Cell Rep. 25:1199–1204.

Zhou, H., J.W. Arrowsmith, M.E. Fromm, C.M. Hironaka, M.L. Taylor, D. Rodriguez, M.E. Pajeau, S.M. Brown, C.G. Santino, and J.E. Fry. 1995. Glyphosate-tolerant *CP4* and *GOX* genes as a selectable marker in wheat transformation. Plant Cell Rep. 15:159–163.

Zhou, H., J.D. Berg, S.E. Blank, C.A. Chay, G. Chen, S.R. Eskelsen, J.E. Fry, S. Hoi, T. Hu, P.J. Isakson, M.B. Lawton, S.G. Metz, C.B. Rempel, D.K. Ryerson, A.P. Sansone, A.L. Shook, R.J. Starke, J.M. Tichota, and S.A. Valenti. 2003. Field efficacy assessment of transgenic Roundup Ready wheat. Crop Sci. 43:1072–1075.

Section IV
Making of a Wheat Industry

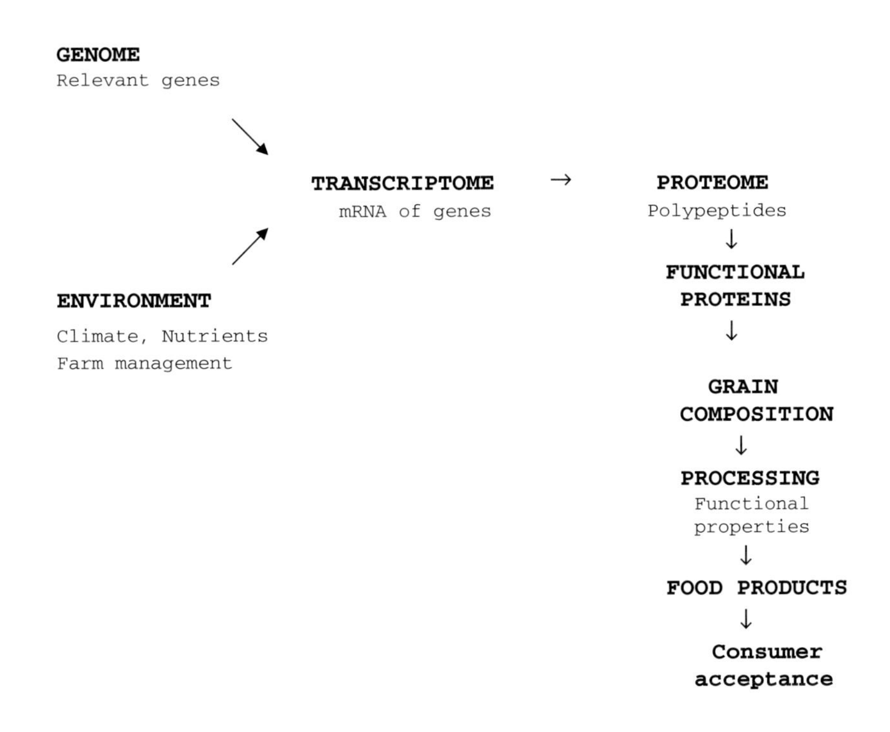

"Millers and bakers in France have found that bread is improved by putting into it a larger amount of gluten than is found in French or American wheats, and as consequence very hard wheats . . . are mixed with the others. These wheats cannot be raised in France, and must be imported, and they are the only kinds which are always sure to find a market in this country."

US Consul John Covert,
reporting to the US State Department
The New York Times, 1899

Chapter 19

Overview of Wheat Classification and Trade

Kendall L. McFall and Mark E. Fowler

SUMMARY

(1) Wheat generally is classified by its physical characteristics such as kernel color and texture. Physical evaluation is used to determine basic milling performance, and to some degree, storability.

(2) Wheat is purchased by millers based upon its value as determined by the baked or end products it can produce. Functional end-use quality attributes are more valuable than milling quality attributes.

(3) United States commodity exchanges currently function as the world's wheat price discovery instruments.

(4) The link from farmer to buyer can seem complex or distant, but in reality, buyers speak directly to suppliers through purchase choices. Buyers inform wheat producers and sellers which items are of most value, in the form of functional quality parameters or physical characteristics quoted in a tender, by revealing their willingness to pay.

(5) With nearly 600 million tonnes of wheat consumed globally each year and about 100 million tonnes of wheat traded annually by more than 30 countries, individual buyers do not impact supply or demand forces that control the market price of wheat.

INTRODUCTION

> Bread, symbol of life itself; in abundance the guarantee of well-being, in times of want the dream and the cry of the famine-ridden.
>
> "Flour for Man's Bread";
> Stork and Teague

Wheat, like rice (*Oryza sativa* L.) and maize or corn (*Zea mays* L.), is one of the world's most important staple grains. The economic and social significance of wheat is derived from its magnitude of production and matchless ability to generate diverse foods. Although also used to feed animals, wheat is the world's best grain for delivering gluten-containing flour that can be worked into bread. The gluten, made by mixing wheat protein and water, gives wheat flour dough the unique capability to hold gas bubbles, which help support the loaf until the heat of baking sets the structure firm. Just say the word bread, and people worldwide picture a product that is both sustaining and pleasurable. Consequently, people across the globe rise each morning to negotiate and trade information on this simple grain to determine its value and secure a constant supply for the world they help feed.

WORLD PRODUCTION

Worldwide, wheat producers annually harvest wheat crops totaling more than 600 million tonnes, a staggering quantity. Recently, however, average wheat consumption slightly exceeded production, leaving the world with less wheat to begin the next year. During the past 5 years, this imbalance was almost 10 million tonnes per year (Table 19.1). Though only a small percentage in any given year, the cumulative effect has greatly diminished world stocks. This highlights the need to understand both how wheat is described and how it is delivered from the producer to the miller.

Wheat production for all US classes of wheat (Table 19.2) averaged about 57.0 million tonnes from 2002 to 2007. This represents only 9.5% of the world's total production but cannot be completely consumed nationally; therefore, almost one-half of US wheat production is available for trade. This excess often represents nearly one-third of the world's traded wheat crop. Four other major wheat producing regions and countries contribute to global wheat trade annually: Argentina, Australia, Canada, and the EU. Other nations have moved in and out of the market, but none can match the consistent reliable supplies provided by the five major exporters.

The significance of the US wheat crop is magnified because the efficient transportation infrastructure across North America easily moves wheat from field to port. The US system of trade allows buyers from anywhere in the world to compete for wheat on a level playing field. These two factors—open market–price discovery and consistent supply—combine to make the US the foremost provider of market information for the world's wheat buyers and sellers.

GLOBAL WHEAT TRADE

International wheat trade is quantified annually and typically is about 100 million tones (World Agricultural Outlook Board 2008). Thus less than 20% of the wheat grown each year is traded or sold outside the country where it was produced. Recently, more wheat exporting countries have started to obtain a share of the increasingly competitive global wheat market (Fig. 19.1). United

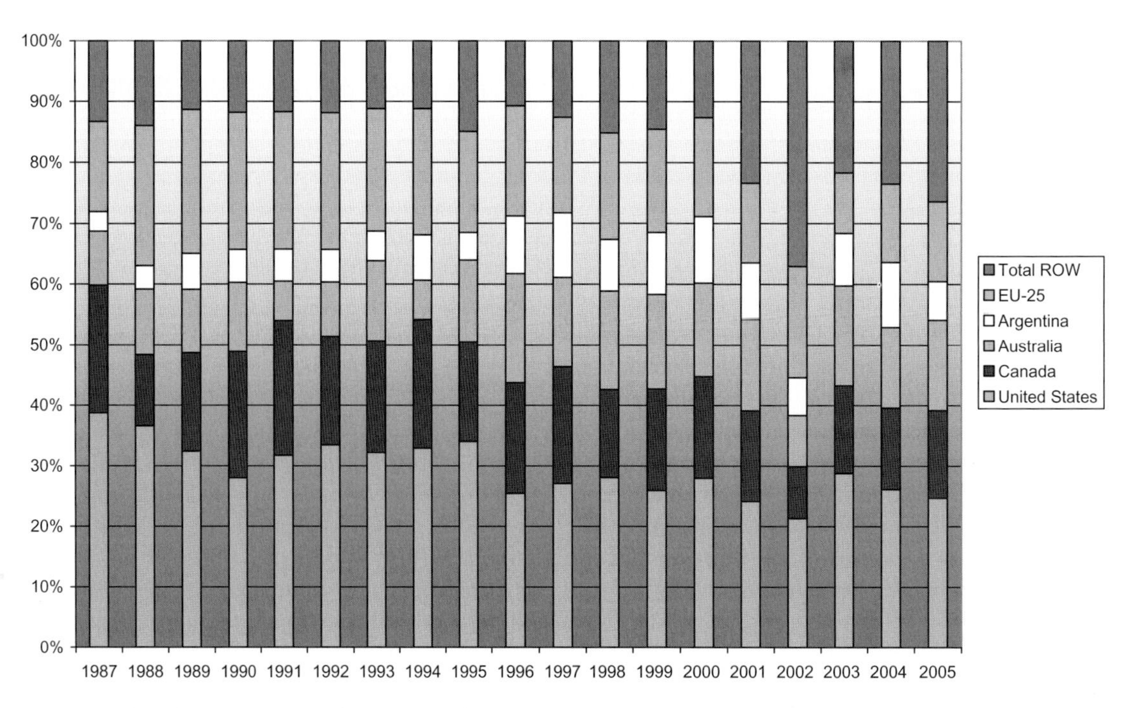

Fig. 19.1 The relative amount of wheat exported by the US and Canada has decreased as the rest of the world (ROW) has increased its share of the total market.

Table 19.1 Wheat and wheat flour production and use, from 1996–1997 to 2007–2008, in million tonnes.

Country or Region	1996/97	1997/98	1998/99	1999/00	2000/01	2001/02	2002/03	2003/04	2004/05	2005/06	2006/07	2007/08 (Proj)
Production												
Canada	29.8	24.3	24.1	26.9	26.5	20.6	16.2	23.6	25.9	26.8	25.3	20.1
Australia	22.9	19.2	21.5	24.8	22.1	24.3	10.1	26.1	22.6	24.5	10.6	13.1
Argentina	15.9	15.7	13.3	16.4	16.2	15.5	12.3	14.5	16.0	12.1	15.2	15.5
EU	118.0	115.0	125.0	114.7	124.2	113.6	124.8	106.9	136.8	122.9	124.8	119.7
Former USSR	63.0	80.5	56.0	64.8	63.1	90.1	96.9	60.9	86.5	91.7	86.0	93.7
Other Europe	8.9	16.0	14.5	11.9	12.1	14.7	12.7	7.5	15.4	13.2		
China	110.6	123.3	109.7	113.9	99.6	93.9	90.3	86.5	92.0	97.0	104.5	106.0
India	62.1	69.4	66.4	70.8	76.4	69.7	71.8	65.1	72.1	72.0	69.4	75.8
All other foreign	89.5	79.1	90.2	79.2	80.5	85.9	88.8	99.7	100.8	99.3	107.3	106.6
US	62.0	67.5	69.3	62.5	60.6	53.0	43.7	63.8	58.7	57.3	49.3	56.3
World total	582.6	610.0	590.0	585.8	581.5	581.1	567.7	554.6	626.7	616.8	593.0	606.7
Utilization												
US	35.4	34.2	37.6	35.4	36.2	32.4	30.4	32.5	31.9	32.3	31.4	29.8
Former USSR	68.9	72.0	63.9	65.0	64.0	70.2	75.4	65.9	72.7	75.5	73.6	76.6
China	107.6	109.1	108.3	109.3	110.3	108.7	105.2	104.5	102.0	101.0	101.0	100.5
All others	352.9	361.5	368.0	371.8	372.0	373.8	392.8	378.5	401.6	411.6	409.8	412.1
World total	564.9	576.8	577.8	581.5	582.5	585.2	603.8	581.4	608.2	620.5	615.8	619.1
Ending stocks	164.5	197.1	208.1	208.9	206.5	202.4	166.1	132.1	149.6	142.6	124.9	112.5

Note: Years 1996–1997 to 2004–2005 extracted from USDA (2006) and years 2006–2007 and 2007–2008 extracted from World Agricultural Outlook Board (2008).

Table 19.2 US wheat production and usage.

		2002/03	2003/04	2004/05	2005/06	2006/07	2007/08
Area							
Planted	Million hectares	24.40	25.13	24.16	23.15	23.19	24.44
Harvested		18.54	21.49	20.24	20.28	18.94	20.64
Supply							
Beginning stocks	Million tonnes	21.15	13.38	14.87	14.70	15.55	12.42
Production		43.71	63.82	58.74	57.29	49.32	56.25
Imports		2.11	1.71	1.92	2.22	3.32	2.45
Total supply		66.97	78.91	75.54	74.20	68.18	71.12
Usage							
Food	Million tonnes	25.00	24.82	24.76	24.90	25.40	25.72
Seed		2.30	2.16	2.11	2.11	2.22	2.34
Feed and residual		3.15	5.52	4.96	4.35	3.42	2.99
Total US usage		30.45	32.51	31.82	31.36	31.04	32.66
Exports		23.14	31.53	29.01	27.29	24.73	32.66
Total usage		53.59	64.04	60.84	58.66	55.77	63.72
Ending stocks	Million tonnes	13.38	14.87	14.70	15.55	12.42	7.40

States wheat exports have fallen steadily since peaking in the 1980s. According to the *USDA Wheat Situation Outlook Yearbook*, US wheat exports fell from a high of nearly 50% of the global wheat export market (48 million tonnes) in 1981 to about 25% (28 million tonnes) of world exports in 2005 (USDA 2006). The quantity of US wheat exports as a percentage of US wheat production also has decreased (Table 19.3). In the 1988–1989 marketing year, US wheat exports, as a percentage of total production, peaked at 78%. Since the 1999–2000 marketing year, the proportion of US wheat production that was exported has remained steady at approximately 50%.

During this same period, increased privatization of the global milling industry resulted in fewer public (government) tenders for large quantities of grain and more purchases by private flour milling companies, though in smaller quantities. Private flour milling companies, by nature of their proximity to the end user, place a greater emphasis on specific wheat quality characteristics. In contrast, public buyers tend to focus on price as the primary driver in a purchase decision.

New exporting countries (i.e., nontraditional competitors) represent the largest percentage increase in global wheat exports (USDA 2006). Similar to the US, other major wheat exporting countries—Argentina, Australia, Canada, and countries in the European Union—experienced little or no growth in wheat exports during the same period. The EU reported a small increase in recent exports, but this is mainly due to the addition of countries to the European Union. India, Russia, and other former Soviet Union Republics reversed their role in the global wheat market by becoming net exporters. More countries exporting wheat creates a more competitive global wheat market. Buyers have become more quality conscious and more sensitive to price fluctuations.

Of the major exporting countries, the US and Argentina are the only countries that do not include chemical or functional quality factors, such as protein content, as a part of their official classification and grading system. This information void must be filled in other ways. Commonly, nongrade quality measures are written into contracts as specifications for the exporter to meet. The montage of price versus quality

Table 19.3 Wheat exports by the five major exporters and other foreign exporters during the period 1987–2005.

	Australia		Canada		Argentina		EU-25[a]		Total Rest of World		US	
Year	%	Million Tonnes	%	Million Tonnes	%	Million Tonnes	%	Million Tonnes	%	Million Tonnes	%	Million Tonnes
1987	8.83	9.85	21.08	23.52	3.32	3.70	14.73	16.44	13.29	68.35	38.74	43.22
1988	10.74	11.30	11.82	12.44	3.84	4.03	23.03	24.22	13.96	66.66	36.62	38.51
1989	10.41	10.78	16.33	16.88	5.86	6.07	23.67	24.50	11.31	69.90	32.42	33.53
1990	11.32	11.76	20.93	21.72	5.39	5.58	22.55	23.41	11.79	74.74	28.03	29.10
1991	6.46	7.10	22.27	24.50	5.26	5.77	22.57	24.82	11.71	75.07	31.74	34.89
1992	8.95	9.85	17.91	19.71	5.32	5.85	22.49	24.74	11.86	73.22	33.48	36.85
1993	13.22	13.72	18.42	19.11	4.83	5.01	20.15	20.90	11.17	70.30	32.22	33.42
1994	6.47	6.34	21.23	20.85	7.45	7.32	20.73	20.36	11.18	65.87	32.93	32.34
1995	13.42	13.31	16.47	16.33	4.52	4.49	16.61	16.47	14.94	65.43	34.05	33.78
1996	17.98	19.22	18.24	19.52	9.54	10.21	18.03	19.27	10.71	79.67	25.50	27.27
1997	14.69	15.35	19.28	20.14	10.68	11.16	15.65	16.36	12.57	76.10	27.12	28.31
1998	16.26	16.47	14.52	14.70	8.45	8.57	17.49	17.72	15.18	72.84	28.10	28.47
1999	15.63	17.86	16.79	19.16	10.15	11.59	16.98	19.38	14.54	84.59	25.90	29.59
2000	15.42	15.92	16.77	17.31	10.92	11.27	16.27	16.79	12.63	74.33	28.00	28.91
2001	15.12	16.41	14.99	16.28	9.28	10.07	13.11	14.24	23.35	82.34	24.13	26.18
2002	8.42	9.15	8.66	9.42	6.22	6.75	18.36	19.95	37.04	85.49	21.30	23.14
2003	16.50	18.05	14.44	15.79	8.60	9.42	10.00	10.94	21.64	77.84	28.82	31.52
2004	13.31	14.75	13.50	14.97	10.68	11.84	12.96	14.37	23.46	81.93	26.10	28.93
2005	14.95	16.49	14.51	16.00	6.35	7.00	13.15	14.51	26.38	83.10	24.67	27.22

Note: Aggregate of differing local marketing years including Australia (Oct./Sept.), Canada (Aug./July), Argentina (Dec./Nov.), and EC-25 (July/June).
[a]Includes intra-EU trade.

discussions required to include nongrade specifications in a wheat contract are all part of price discovery negotiations in a buy-and-sell transaction.

Single-desk trading allows one entity to control the price, quantity, and quality of a commodity exported from a country. The single-desk organizations, the Canadian Wheat Board (CWB) and the Australian Wheat Board (AWB), meet market requirements by controlling wheat from the field to the customer. By adjusting their defined quality parameters within a class and the quantity of wheat available for export to a respective country, these organizations ensure a uniform supply with consistent quality. Some argue that a single-desk trading structure provides a competitive advantage in export markets. For example, the CWB includes protein content and cultivar in their grading and classification standards, and the AWB includes protein content, falling number, cultivar, and color specifications. Both the AWB and CWB can control the price, quantity, and cultivar of wheat available to the export market, and unlike the US marketing system, these boards include quality measures in their grading standards. Some argue that single-desk boards have an advantage when they protect certain customers by withholding supply from the open market. These types of transactions and their lack of transparency to the rest of the marketplace are often criticized.

In 1916, the US Congress passed the US Grain Standards Act (USGSA), which established quality standards for grain traded in the US. Wheat standards were first implemented in 1917 and have been revised numerous times. In May 2006, the US Federal Grain Inspection Service (FGIS) implemented a program to test wet gluten content in hard red winter and hard red spring wheat classes. This was an example of a change in the US marketing system that helped meet buyers' increased desire to include functional properties of wheat in contract specifications.

Developing test methods that allow quick, reliable evaluation of wheat functional properties helps wheat exporters to meet customer requirements without having to control the entire crop as a single-desk entity.

FUNDAMENTAL WHEAT CLASSIFICATION CRITERIA

All wheat classifications were originally derived from observable physical distinctions present in wheat. The most commonly used characteristic is the color of the outer layers of the kernel, often described as either red or white. The red description refers to the reddish brown color of the seed's protective layers, referred to by millers as the bran coat. The white description refers to the yellowish tan color of the bran coat found in some regions and classes; this color is genetically distinct from the more common color in red wheat cultivars.

Another common observable characteristic is the texture or hardness of seed, which often is described as hard or soft. At extremes, this texture difference is easily determined by bite. Though hardly objective, this texture determination method played a role in the early days of wheat identification.

Planting and growing cycles also are used to identify wheat. Wheat is described as a winter wheat if planted in late summer or fall and harvested in late spring or early summer, remaining in the ground through the entire winter. Spring wheat is planted in spring and harvested in late summer or early fall. This general classification is most common in areas frequently subjected to very harsh winters with extremely low temperatures. Hard red spring (HRS) wheat is an example of a US wheat class identified by endosperm texture (hard), color (red), and the planting and growing cycle (spring). Wheat identification and classification is approached slightly differently in each major exporting country. A general overview of some classification factors can help us understand what is available when buying wheat and why millers around the world value some wheat classes more than others.

US SYSTEM OF WHEAT CLASSIFICATION

The US FGIS became an agency of the USDA in 1976 under the US Grain Standards Act. In 1994, FGIS merged with the Packers and Stockyards Administration to form a new agency, Grain Inspection, Packers and Stockyards Administration (GIPSA). This agency administers a nationwide system for officially inspecting and weighing grain and other commodities. The Grain Standards Act requires, with few exceptions, official inspection and weighing of export grain sold by grade. The GIPSA does not change the standards each year to reflect the fair average of the crop; rather, standards remain fixed until specifically revised and are changed only after consultation with grain industry stakeholders.

The US system divides wheat into eight broad classifications for marketing efficiency. The classes represent several hundred wheat cultivars. The US system does not regulate or control release of new cultivars. Instead, the market at large determines, by supply and demand, what cultivars will be planted and harvested for sale. Therefore, standards describe a class as distinct and separate from other classes; but in the US, a class can encompass multiple cultivars. The highest value of a particular class at a given point in time assumes no mixture, intentional or unintentional, with another class. The six classes of economic significance are hard red winter (HRW), soft red winter (SRW), hard red spring (HRS), soft white (SW), hard white (HW), and durum. To allow trade of devalued wheat classes that have been mistakenly blended together or wheat of an undetermined class, the standards also classify wheat as mixed or unclassed.

Grade factors

Grading standards used in the US are shown in Table 19.4. In the US system, grade or grade requirements are determined based on grain when it is free from dockage, which is all matter other than wheat that has been removed with an approved device according to FGIS instructions (http://archive.gipsa.usda.gov/reference-library/handbooks/equipment/apprlist-a.pdf).

Table 19.4 Wheat grades and grade requirements.

Grading Factors	Grades US No. 1	2	3	4	5
	Minimum limits of				
Test weight (lb bu^{-1})					
Hard red spring or White club	58.0	57.0	55.0	53.0	50.0
All other classes and subclasses	60.0	58.0	56.0	54.0	51.0
Test weight (kg hL^{-1})					
Hard red spring or White club	76.4	75.1	72.5	69.9	66.0
Durum	78.2	75.6	73.0	70.4	66.5
All other classes and subclasses	78.9	76.4	73.8	71.2	67.3
	Maximum percentage limits of				
Defects					
Damaged kernels					
Heat (part of total)	0.2	0.2	0.5	1.0	3.0
Total	2.0	4.0	7.0	10.0	15.0
Foreign material	0.4	0.7	1.3	3.0	5.0
Shrunken and broken kernels	3.0	5.0	8.0	12.0	20.0
Total[a]	3.0	5.0	8.0	12.0	20.0
Wheat of other classes[b]					
Contrasting classes	1.0	2.0	3.0	10.0	10.0
Total[c]	3.0	5.0	10.0	10.0	10.0
Stones	0.1	0.1	0.1	0.1	0.1
	Maximum count limits of				
Other material					
Animal filth	1	1	1	1	1
Castor beans	1	1	1	1	1
Crotalaria seeds	2	2	2	2	2
Glass	0	0	0	0	0
Stones	3	3	3	3	3
Unknown foreign substance	3	3	3	3	3
Total[d]	4	4	4	4	4
Insect-damaged kernels in 100 g	31	31	31	31	31

[a]Includes damaged kernels (total), foreign material, shrunken and broken kernels.
[b]Unclassed wheat of any grade may contain not more than 10.0% of wheat of other classes.
[c]Includes contrasting classes.
[d]Includes any combination of animal filth, castor beans, crotalaria seeds, glass, stones, or unknown foreign substances.

Dockage, reported as a percentage, can also include underdeveloped, shriveled, and small pieces of wheat kernels removed during the approved cleaning process. Grades are assigned using numeric designations of US No. 1 for the highest grade through US No. 5 for the lowest. Any sample that fails to meet the criteria for US No. 5, or has a commercially objectionable odor, is designated as sample grade, or wheat of distinctly low quality.

Foreign material, shrunken and broken kernels, damaged kernels, and total defects are important in determining grade requirements for wheat designated as US No. 1 through US No. 5; these materials negatively affect flour yield and quality and often are removed during cleaning.

An additional quality measure is test weight, which is reported in pounds per Winchester bushel (approximately 1.25 ft^3) in the US and in kilograms per hectoliter in many other parts of the world:

$$\text{kg hL}^{-1} = (1.292 \times \text{lb bu}^{-1}) + 1.419, \text{ for most wheat, and}$$

$$\text{kg hL}^{-1} = (1.292 \times \text{lb bu}^{-1}) + 0.630, \text{ for durum.}$$

Higher test weight (i.e., greater wheat mass in the same volume) indicates better quality wheat and suggests easier processing and greater flour yield. However, higher test weight does not always translate into measurable milling yield improvement, because test weight as a single factor does not account for other important factors. These may include kernel uniformity, size, and shape, and kernel-to-kernel surface friction or other environmental factors that affect compaction of wheat into the test device. As a result, other measures have been developed to help millers understand the milling quality of a sample. These may include nongrade determinates such as thousand kernel weight discussed in a later section or other measures of wheat size and weight determined by the single-kernel characterization system (SKCS 4100, Perten Instruments, Huddinge Sweden).

Because US grain grading standards focus on physical characteristics related to milling quality and, to some extent, storability, these standards often fail to accurately assess the functional quality characteristics required to meet end-use performance standards. Understanding functional quality, defined as the wheat doing what the processor wants it to do for a given end product, requires information beyond FGIS grades and grade requirements.

Nongrade factors

Nongrade quality factors often influence value and expected outcomes for the buyer, but they are not part of the official US grade and grade requirements. They include moisture, protein, and ash contents, enzyme activity, physical dough characteristics, and baking characteristics. A more technical description of the methodology and instrumentation used to measure many of these quality factors is provided in Chapter 20. They are mentioned here in the context of understanding miller and baker preferences that often are expressed in purchase specifications.

Moisture content

Moisture content is the amount of water by weight present in a wheat sample. Too much moisture can lead to spoilage during storage, and in many cases, the difference between low wheat moisture content and flour moisture allowed by a baker is equivalent to the miller's profit margin. Although wheat is discounted at the first point of sale (farmer to elevator) when the moisture level exceeds 13.5%, the reverse is not the case. Farmers do not receive premiums for low-moisture wheat, which makes moisture valuable to millers. The US market treats excessive moisture as a discount to value but does not always take into consideration the value of low moisture.

Protein content

Protein content is measured by determining the amount of nitrogen present in the wheat sample, and with the conversion factor of 5.7, it is expressed as a percentage of the sample weight at an understood moisture level. For wheat sold in the US, protein is expressed as a percentage on a 12.0% moisture basis. In Canada, wheat is sold with protein content quoted as a percentage on a 13.5% moisture basis. Others buy and sell grain and grain products on a dry (0.0%) basis. Flour buyers often require a quote for protein content on a 14.0% moisture basis. When communicating about the level of an attribute in grain or a grain product sample, establishing the moisture level at which the attribute is expressed is critical.

Wheat ash content

Wheat ash content also is expressed as a percentage at a given moisture basis. Ash is the mineral residue present after incinerating a wheat or flour sample. The amount of minerals present in a wheat sample often varies among cultivars and growing environments. Millers and grain buyers must understand the amount of minerals present in a wheat sample, because bakers often specify the maximum allowable mineral content of flour.

Understanding the specification along with the inherent mineral content of the wheat to be ground allows the miller to make fine adjustments in mill settings to exclude the outermost layers of the endosperm, where minerals are more concentrated. By purchasing wheat with lower ash content, millers are sometimes able to extend the extracted percentage of flour from the wheat berry. This might be as small as a 0.5% unit extraction improvement but can add significantly to the miller's profit margin.

Kernel weight

Kernel weight of wheat is measured in commercial sectors as the weight (in grams) of 1,000 wheat kernels, or thousand-kernel weight (TKW). This value is influenced to some degree by wheat size and density but is less susceptible to some of the environmental factors that influence kernel compaction when measuring test weight. Thousand-kernel weight supplements information provided by the test weight measurement. Two samples of similar test weight could differ in TKW. Millers view samples with higher TKW as having potentially greater flour yield.

Grain hardness

Class descriptions specify wheat as hard or soft, but although this general view of hardness is understood in trade and has long been used for classifying wheat in the cultivar development process, more definitive hardness measures are necessary. Several methods exist for measuring and differentiating wheat hardness. One parameter commonly used to describe hardness within a class is the particle size index (PSI). The PSI is expressed as a percentage of flour produced by grinding a sample on a specific laboratory mill. A softer wheat produces more flour as a percentage of the original wheat sample. The SKCS measures the force required to crush a kernel of wheat. A softer wheat requires less crushing force than a harder wheat. The results are presented as a mean value for 300 kernels with a standard deviation.

Millers are interested in hardness not because this characteristic can help determine soundness of the crop but because knowledge of the hardness level may reveal possible processing changes that might be required. For example, the amount of time required for water to penetrate into the kernel increases with wheat hardness. This is not a concern when a miller purchases consistently within a class from the same area. However, if a miller purchased HRS wheat, a very hard class of wheat, as a replacement for HRW wheat, a medium hard class of wheat, the time required for moisture penetration would increase significantly.

Falling number

Falling number is an indication of the level of α-amylase enzyme activity in grain. Enzyme activity precedes the sprouting process in wheat and is undetectable under current grading systems, which recognize only visible sprout damage as undesirable. The falling number method uses starch as the medium for determining enzyme activity. The measurement device analyzes viscosity by the resistance of a flour and water paste to a falling stir rod. Results are expressed in seconds. A high falling number (e.g., more than 300 seconds) indicates minimal enzyme activity and sound wheat quality. As enzyme activity increases, the falling number value decreases and soundness of wheat is questionable. A falling number value less than 250 indicates that enzyme activity is substantial and the wheat sample is sprout-damaged regardless of any detectable visible damage.

Falling number is important to millers because enzyme activity affects baked product quality. Too much enzyme activity means too much sugar is present and too little starch is available to aid in product support structure. This can result in very sticky dough during processing and poor texture in finished products. There is no remedy for wheat with excessive enzyme activity. Low enzyme activity can be adjusted by adding additional enzymes to flour prior to baking. Once present, enzymes cannot be removed from wheat or flour, making low falling number a serious problem for millers and bakers.

Starch viscosity

Viscosity analysis is a measure of starch properties and enzyme activity resulting from sprouting wheat and typically is performed by measuring resistance of flour and water slurries to a stirring action. When the slurry is heated, starch granules absorb water, thickening the mixture into a paste-like consistency. A thicker slurry has greater resistance to the stirring action, which indicates low enzyme activity and indicates water-holding capacity of the starch. Viscosity analysis in the absence of high enzyme activity is important in products such as cakes, pancakes or batters, and coatings.

Wet gluten content

Gluten is the primary protein component of the structure formed when water and flour are mixed.

It provides the stretch and resistance to stretching (i.e., elasticity) in dough. Wet gluten content is determined by washing flour or a ground wheat sample with a saltwater solution to remove starch from the sample. Gluten content is expressed as a percentage on a 14% moisture basis. It also has been used to estimate gluten quality or strength by subjecting the wet gluten material to centrifugal forces through a mesh. The percentage remaining is called the gluten index and provides an indication of gluten strength. Wet gluten is used extensively as a product specification and is highly indicative of the protein content of wheat flour.

Dough performance

Several instruments are used throughout wheat commerce to assess developmental and extensional properties of the dough. The farinograph measures and records the resistance of dough to mixing with paddles. Results from the farinograph test include water absorption, arrival time, stability time, peak time, departure time, and mixing tolerance index. This is one of the most commonly used flour quality tests for bread wheat markets, and a farinograph curve reveals several clues about end-product quality (Fig. 19.2). Because water is required to make dough, the amount of water needed to develop a dough to its optimum consistency is an important indicator of value. A flour capable of holding more water as a dough can be advantageous for certain products but is an absolute necessity when formulating a bread product from a specific wheat or flour sample.

The farinograph measures water-holding ability by first defining optimum dough development and then measuring the amount of water required to reach that point. Hence farinograph absorption is the amount of water needed to center the farinograph curve on the 500-Brabender-unit line. Other measures such as the time it takes for optimal dough formation (peak time), the time a dough can be mixed before its consistency weakens (stability time), and the degree of dough softening during mixing (mixing tolerance index) also are helpful in understanding how a given grain sample will perform when mixed and indicate, to some degree, the quality of gluten formed (i.e., dough strength).

An extensograph determines dough extensibility and resistance to extension by measuring the force required to stretch the dough with a hook until it breaks. An extensograph curve indicates resistance to initial stretching (resistance) and resistance to breakage (extensibility) (Fig. 19.3). The extensograph test results help millers and bakers understand the gluten strength and breadmaking characteristics of a wheat or flour sample. It is also helpful in determining the effects of

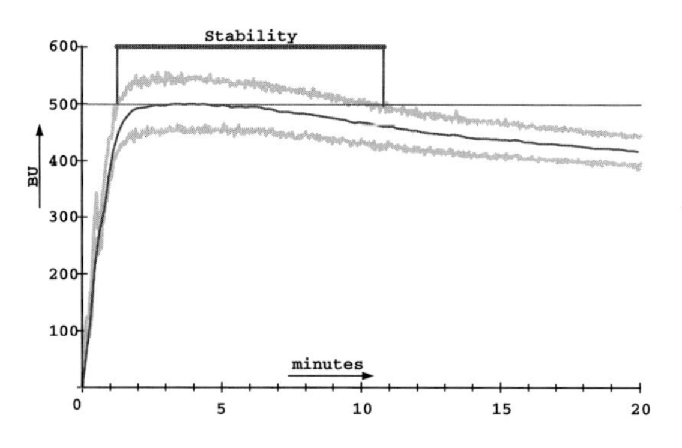

Fig. 19.2 The farinograph curve helps the end user predict mixing properties of a flour sample. Stability, estimated from the farinograph curve (Brabender units, BU, vs. minutes of mixing time), is considered first by bread makers as an indicator of the dough's ability to maintain optimum consistency while mixing. It is also another indicator of strength.

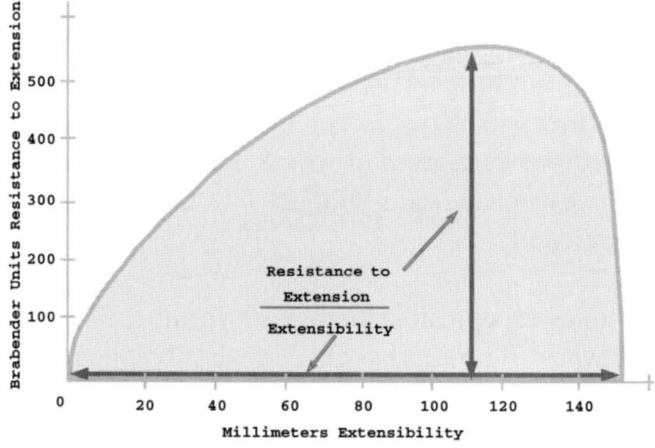

Fig. 19.3 From an extensograph curve, the ratio of resistance to extension and extensibility indicate the balance between strength and the limit to stretch before breaking of the dough.

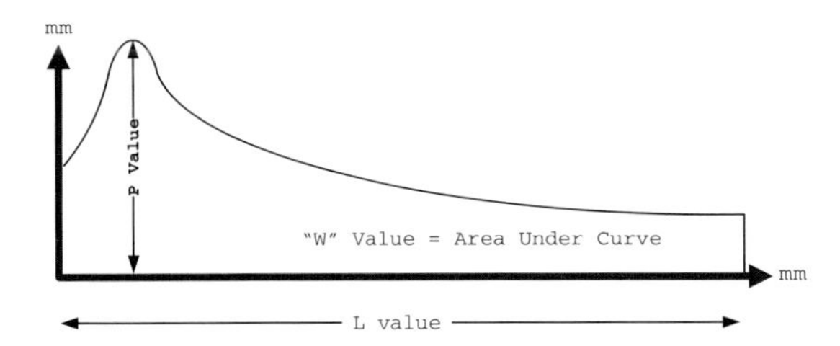

Fig. 19.4 The *P* value from an alveograph curve represents the force required to blow a bubble from the dough, whereas the *L* value is the extensibility of the dough before breaking. The *W* value incorporates both *P* and *L* and is considered an overall indicator of dough strength.

fermentation time or an additive on dough performance. Extensograph results also can be helpful in understanding how dough will perform in sheeted products such as pizza skins.

The alveograph is similar to the extensograph; however, it measures dough by mimicking the gas pressure created by yeast during fermentation. The force required to inflate and burst a bubble of dough is graphed, and doughs with varying elastic strengths may be compared (Fig. 19.4). Dough that can be inflated to a very large bubble without breaking has high elasticity and is suitable for yeast-levened bread products. Millers and bakers use results of the alevograph test to ensure consistency. The alveograph is especially well suited for evaluating wheat with medium to weak dough strength, such as that intended for use in cakes, cookies, or other confectionery products.

Product performance

Although many chemical and functionality tests are available to measure dough quality, finished product performance is the only true measure of the performance quality of wheat. Evaluating wheat quality using a standardized finished product methodology attempts to eliminate all variables except those that exist within the wheat and wheat flour, and such an evaluation requires a diversity of tests as varied as the finished products. Essentially, there are as many ways to evaluate wheat as there are types of bread. Some groups have attempted to create standardized bake tests to establish a common language for and understanding of finished product quality. The American Association of Cereal Chemists developed numerous standardized methods for evaluating flour in a food product (AACC 2000). Millers must understand the product for which a particular flour type will be used and evaluate wheat choices based on tests that best represent that product.

CANADIAN SYSTEM OF CLASSIFICATION AND MARKETING

The Canadian Grain Commission (CGC) sets Canadian wheat standards and classifications in consultation with industry; however, the wheat classification approach in Canada is much more restrictive than in the US. In particular, Western Canada wheat cultivars must meet a visual appearance standard referred to as kernel visual distinguishability (KVD). The KVD standard ensures that wheat classes are readily identifiable throughout the handling system, and it has maintained kernel size and shape relatively uniform within a class over time. Eastern Canada wheat is not under the control of the Canadian Wheat Board for marketing, so KVD is not required for most eastern Canadian classes; Canada eastern white winter is the exception. The KVD standard is well established and has demonstrated benefits, but it has been criticized for limiting cultivar development to meet niche markets and for using visual criteria that are somewhat subjective.

Two of the eight Western Canada classes constitute the majority of Canada's wheat exports. The largest class, often 15.0 million tonnes produced per year, is Canada western red spring (CWRS). The second largest class, at nearly 5.0 million tonnes of production per year, is Canada western amber durum. The remaining classes—Canada western extra strong (CWES), Canada western red winter (CWRW), Canada prairie spring red (CPSR), Canada prairie spring white (CPSW), and Canada western soft white spring (CWSWS)—are considered minor because of their small influence on export markets.

Through the CGC, Canada also takes a different approach to grading wheat within a class, using primary standards such as moisture and protein content for purchase from the farmer and export standards meant to describe in more detail the quality and uniformity of the crop to processors. Primary standards are in place to grade wheat upon delivery by the farmer. Because all Western Canada wheat becomes the property of the CWB upon delivery to the elevator, export standards are used to grade wheat destined for export. These export standards include characteristics such as cleanliness, uniformity, and consistency, and they promote a safe, reliable supply. Canada, through the CWB, exports about 15 to 20 million tonnes of wheat each year. As a single-desk supplier, the CWB decides who receives Western Canada wheat in times of short supply. It is the CWB's best interest to sell wheat for the highest possible price, but the pricing system is often criticized for its lack of transparency. In the US, the market-determined price becomes the arbiter of trade.

AUSTRALIAN SYSTEM OF CLASSIFICATION AND MARKETING

Wheat classification in Australia has similarities to the US and Canadian systems. Similar to Canada, wheat in Australia is handled by a single-desk marketing authority. When originally established by government statute, the Australian Wheat Board, now AWB, acted as the sole wheat pooling and exporting authority for Australian wheat. In recent years, the newly organized AWB has become the major manager of Australia's grain. It evolved from a government agency to a publicly listed company and is owned primarily by Australian growers. Like the CWB, AWB acts as the only bulk exporter of wheat for its country's farmers. This could change; proposals are under consideration to divide some of the export responsibilities among regions and have bulk-handling companies export the wheat.

Australia divides its wheat into six separate classes: Australia prime hard, Australia hard, Australia premium white, Australia standard white, Australia soft, and Australia durum. The Australian cultivar development system is similar to the US system and less restrictive than the Canadian system.

PURCHASING DECISION MAKING

When purchasing wheat, international buyers can choose from among many classes and associated sets of quality attributes from a variety of countries. To understand this decision-making process, we must understand how wheat is offered to the market at large and how a buyer arrives at an understanding of price.

The price a buyer is willing to pay for wheat is a function of several factors: market factors, such as quantities of global supply and demand; governmental policies; payment terms of the contract; delivery time; quantity of shipment and quality attributes. With nearly 600 million tonnes of wheat consumed globally each year and, on average, more than 100 million tonnes of wheat traded annually by more than 30 countries, individual buyers do not impact supply or demand forces that control the market price of wheat (forces which are discussed further in Chapter 23).

Governmental policies that affect the cost of wheat include import and export restrictions, tariffs or quotas, and phytosanitary requirements. Many countries require imported plants and plant materials to have a phytosanitary certificate which verifies that the material does not contain biological contaminants, such as various fungi and molds that can contaminate wheat or unwanted plant seeds. In some cases, such as direct governmental purchases, buyers might have the ability to influence governmental policies. In most cases, however, buyers cannot control governmental restrictions.

Terms of payment are financial agreements between buyers and sellers to arrange payment for the wheat. Terms of payment affect the contract price for wheat exports, depending on which party is extending credit to the buyer and how long the seller is financing the wheat shipment. The longer the seller must carry the cost of the sale, the more the seller will increase the contract price to cover finance charges. Cost of financing is straightforward, and although it affects the total cost paid for a shipment of wheat, it does not change the base value or price of the wheat. As a separate part of the buy–sell transaction, finance charges can be excluded from price discussions. Thus, parts of the purchasing decision over which international wheat buyers have control are the wheat quality attributes they specify, the quantity of the shipment, the delivery time, and the location from which they purchase.

Wheat is purchased not as food but as a food ingredient, which explains why a high value is placed on certain functional characteristics. For instance, wheat is evaluated less on its ability to meet dietary requirements than on its ability to create a certain volume of bread. Two broad areas evaluated for trade value are functional quality of flour and physical characteristics related to milling yield. Milling performance is an important economic driver, but it has secondary consideration in price negotiation. To be considered for purchase, wheat must supply flour capable of making a product according to a baker's expectations. If it cannot, and no matter the expected milling performance, the miller will look elsewhere.

Quality drives purchase, but value defines the trading point at which buying and selling occur. Understanding these concepts is helpful in understanding the wheat trade. Value is a combination of quality, price, and supply. Quality, as discussed previously, is based on how well a particular lot of wheat meets a set of functional requirements. Price is determined by a functioning marketplace in which price is the arbiter of all supply and demand factors present at the time of trade. Supply, arguably a portion of the price equation, is held separate as a way to highlight the demand on millers to deliver a quality product over time. Consistency of supply affects a buyer's decision; wheat of sufficient quality offered at a fair market price can be less attractive when the quantity available is insufficient to maintain an adequate percentage in the wheat mix over a given length of time. In other words, a baker (the buyer of wheat's functional quality) first demands quality but also requires that the miller deliver that quality regularly over time. Millers meet this consistency requirement by locating or sourcing wheat lots of sufficient quantity that can be blended together to deliver repeatable end results.

Quality characteristics selected by wheat buyers depend on functional properties for flour that bakers need to produce products demanded in their markets. For example, the quality of wheat required in the Egyptian market to produce pita bread is different than wheat quality required in the South Asian market, where wet noodles are the primary product. Those characteristics are more appropriately described in Chapters 20 and 21.

GRAIN EXCHANGES

Historically, people met to haggle over the purchase and sale of grain when the farmer delivered it to market. The need for a more organized method of buying and selling resulted in creation of formal exchanges, or meeting places, such as the Kansas City Board of Trade (KCBT, chartered in 1876). Early trading was primarily in cash grains, meaning actual grain commodities were bought and sold.

Today, buying and selling extends to futures and options contracts in exchanges operated in Minneapolis, Minnesota (MGEX); Chicago, Illinois (CBOT); and Kansas City, Missouri (KCBT). Futures trading allows traders to buy and sell the right to a quantity of grain of an understood quality delivered at a specified time in the future. This mechanism reduces risk for millers and producers. Producers can "lock in" a selling price for wheat prior to harvest, and millers can set the buying price for wheat not yet delivered to the mill. "Trading" a future contract of purchase and delivery provides an efficient way for buyers and sellers to determine current as well as future value of a commodity. This simple idea has contributed to a smooth flow of grain from farmers to the world market for more than 100 years.

Open outcry system

Trading at the exchange is conducted by open outcry, a public auction-style event during which members of the exchange shout out bids to buy and offers to sell. The event is held in a pit in full view of others so trades are transparent. The exchange house (e.g., KCBT) does not set prices; rather, it acts as host for the daily auction. Prices are discovered each day through actual bids and offers by wheat traders who base their decisions on their own assessment of supply and demand factors and relationships between similar commodities. In Kansas City, traders negotiate on a known class, quantity, and quality of wheat at a specific geographical location. The KCBT wheat futures contract represents 5,000 bushels of HRW wheat graded at a minimum quality of US No. 2 and delivered during specific times to Kansas City. Other delivery points include Wichita, Hutchinson, Salina, and Abilene, Kansas.

Because only KCBT members can trade in the pits and there must be a buyer for every sale and a seller for every purchase, speculative traders also are part of an exchange. These traders perform the role of a risk taker. They help producers and processors by taking the opposite of the transaction when needed. Speculative traders can be a buyer or seller for a future position in the commodity. These investors intend neither to take delivery on the grain nor to supply it to a processor. They intend to profit from price changes in the contracts they buy and sell. A further enhancement to the open outcry system is found in the recent move to allow electronic trading of futures and options. This service is offered to the grain exchanges by the Chicago Mercantile Exchange (CME). The CME through its Globex platform allows both overnight and side-by-side trading of the wheat classes available in open outcry trading. This additional step has helped to increase trading opportunities for floor traders as well as electronic traders worldwide.

This system is attractive for both US farmers and the world buyers because prices and transactions that contribute to those prices are recorded and updated as they occur; the relative value of wheat is known to everyone regardless of participation in the auction. For both buyers and sellers, the ability to estimate value based on the day's trading in Kansas City, for example, and the logistical difference between origins is a matter of understanding differences in quality (e.g., US No. 2 HRW vs. Australia hard).

Farmer to elevator

Flow of wheat from farms through the distribution system to millers, whether down the road or overseas, happens almost seamlessly. Farmers harvest wheat when ripe and either store it in on-farm bins or deliver it to a local elevator for storage. Selling grain is not necessarily linked with delivery to the elevator. United States farmers retain control of ownership of their wheat until they sell. If farmers deliver wheat to an off-farm storage site, they pay a storage fee unless they sell their grain at the time of delivery. This practice differs from farmers operating in a country with a single-desk system.

In the simplest transaction, a farmer delivers grain at the time of harvest to the scale of the elevator to be weighed. Grain is sampled and evaluated for quality related to soundness and storability. The elevator quotes a cash price for the grain, which is the price they are willing to pay that day. Inherent in the price is either an agreed-upon sale to another elevator or mill or a speculative view of the price at which the elevator might be able to sell the grain in the future. The quoted price includes discounts for inferior quality or premiums for superior quality and is negotiated orally at the time of delivery. Weight of the grain is used to determine the number of bushels, and an elevator receipt is offered as proof of delivery for settlement. A farmer who does not agree to the offered price can either store the grain and hope for a future improvement in the price or take the grain to a competing elevator.

Elevator to world

Wheat owned by a US elevator is available worldwide. Grain trading companies trade their knowledge of available wheat and their beliefs about price to profitably move wheat from its present location to the buyer's location. Wheat moves through the distribution system by bids from processors and traders. The decision to move wheat from the local elevator to a terminal elevator or direct it to a flour mill across the country often is made by traders—who might never see the grain—completing the ownership transactions. Large terminal elevators with rail and barge access to large cities and international ports compete with other terminals across the region and world to supply wheat to buyers worldwide.

This competition begins when a country issues a tender to buy wheat. This tender, or buying intention, describes the terms under which the buyer would be willing to transact business. Quantity, wheat quality parameters, shipment timing, delivery method (barge, ship, or rail), and payment terms are part of the process. Grain exporters and traders compete for buyers' business by submitting offers to the complete tender or a portion of it. Ultimately, this supply competition works its way back to the elevator that has wheat available to ship. If enough elevators or small terminals are reluctant to part with the wheat, the price the exporter has to pay to accumulate the wheat increases and, ultimately, influences the price quoted to the buyer.

The link from farmer to buyer can seem complex or distant, but in reality, buyers speak directly to suppliers through purchase choices. Buyers inform wheat producers and sellers which items are of most value, in the form of functional quality parameters or physical characteristics quoted in a tender, by revealing their willingness to pay. The market tallies individual choices of the multitude of buyers to signal changes to sellers. Buyers' collective eagerness or reluctance to return to a particular region or country for supply can be felt directly by wheat sellers. Quality attributes desired by buyers at harvest can signal farmers which cultivars to plant over time. An advantage of the single-desk system is found in the speed at which this market signal prompts farmers and invokes responsiveness.

FUTURE PERSPECTIVES

Predicting trends in world wheat trade is folly; too many factors are involved, and these change annually. However, analyzing broad contributors to world grain trade can provide insight into their potential effects throughout the wheat industry. Next, broad contributors to the future of grain trade are considered (numbered), followed by anticipated effects on wheat trade:

1. Grain will be used as a source of liquid fuel for combustion engines.
   Future effects:
   a. Corn does not return to historic values of $2.00 per bushel.
   b. Wheat does not return to historic values of $3.00 to $4.00 per bushel.
   c. Wheat loses ground to rice as the staple crop.
2. Without intervening forces, wheat will continue to fall behind corn and other crops in farmers' planting decisions.
   Future effects:
   a. Major wheat producing countries pursue transgenic wheat research, and world buyers accept transgenic wheat as a viable alternative to dramatically higher food prices.
   b. Governments temporarily restrict exports to protect their wheat supply for food use.
   c. Governments alter agricultural policies, allowing more land area to be planted with fewer restrictions.
3. Freight costs increase in real cost and as a percentage of commodity value.
   Future effects:
   a. Wheat is increasingly traded within regions to limit the effect of freight costs.
   b. Regional commodity exchanges develop to better discover the value of wheat from nontraditional suppliers.
   c. Wheat trade by nontraditional suppliers increases as a percentage of total world

wheat trade, at the expense of US, Australian, and Canadian exports.

Editor's Note: The wheat marketing system in Australia has undergone rapid evolution subsequent to acceptance of this manuscript. As of June 30, 2008, the single desk/export monopoly held by the Australian Wheat Board was ended. Through the end of 2008, 21 companies have met eligibility requirements through application to government regulator Wheat Exports Australia and have been accredited to conduct export wheat business from Australia. Accredited companies include both Australian and multinational grain exporters (John Oades, pers. comm.).

REFERENCES

AACC. 2000. Approved methods of the American Association of Cereal Chemists. 10th ed. AACC Press, St. Paul, MN.

USDA. 2006. Wheat situation and outlook yearbook, WHS-2006s. USDA-ERS, Market and Trade Economics Division [Online]. Available at http://usda.mannlib.cornell.edu/usda/current/WHS-yearbook/WHS-yearbook-05-03-2006.pdf (verified 14 April 2008).

World Agricultural Outlook Board. 2008. World agricultural supply and demand estimates. WASDE-457, 9 April 2008. USDA-ERS, FAS [Online]. Available at http://www.usda.gov/oce/commodity/wasde/latest.pdf (verified 14 April 2008).

Chapter 20

Passing the Test on Wheat End-Use Quality

Andrew S. Ross and Arthur D. Bettge

SUMMARY

(1) The end uses of wheat encompass a multitude of products. For optimum results each product requires wheat flour with a specific array of functional properties that are suitable for the product's processing and quality needs. The factors that impact suitability for a chosen end use include various attributes such as kernel texture, starch properties, and arguably most important, gluten composition, which controls subsequent dough properties.

(2) From observing the processing and quality attributes of the major product sectors it becomes clear that for each product type the specific array of attributes required from the wheat flour is somewhat different, either qualitatively (e.g., the larger emphasis on starch characteristics for noodles) or quantitatively (e.g., the various optimum flour protein concentrations for the different products).

(3) To optimally match a cultivar or a sample of wheat with an end use, we first need to know the abundance of key components and the milling potential of the grain. Tools are available to perform compositional analysis and grain testing (moisture and protein content, kernel texture, grain soundness, ash content, and other components).

(4) Knowing the composition of wheat and flour only gives a cereal technologist a first approximation of the processing potential. To get a better prediction additional tools are needed that directly measure, or that predict, the functional attributes of the whole material (flour or dough) or its components (gluten, starch, arabinoxylans, etc.). These tools fall under the broad categories of starch and flour paste properties, solvent retention capacity, and dough testing and prediction of dough properties.

(5) The final arbiter is the quality of the end product. We emphasize only the most general and all-encompassing principles of end-use testing, a subject that could fill an entire volume on its own.

(6) The newest technologies are highlighted along with their potentials, first to improve the process of getting the right wheat to the right customer and second to improve our scientific understanding of wheat-based foods and their process intermediates.

INTRODUCTION

"There are countless different bakery products and even more combinations of ingredients, and if you add to this the possible differences in the processing then you have an infinite combination . . . We therefore have some sympathy with the view that if you want to know whether a flour will make a pizza, you have to make a pizza" (Catterall and Cauvain 2007).

The end uses of wheat encompass a mouthwatering variety of products: pan breads, flat breads, steamed breads, noodles, pasta, muffins, and bagels, as well as cookies, cakes, pastries, and other sweet goods. For optimum results each product requires a wheat flour with functional properties that are *suitable* for the product's processing and quality needs. Fortunately processors can find a useable spectrum of functional properties in the many wheat cultivars (and their derived market classes) that are grown throughout the world. There are a number of important "intrinsic" compositional and functionality factors that can vary between cultivars or classes, and which impact end use suitability. These factors include kernel texture, seed coat color, milling characteristics, enzymatic activity, wheat and flour protein contents, gluten protein composition and dough characteristics, and the thermal processing behavior of starch.

Arguably among these attributes it is the abundance and composition of the gluten component that is most important. In particular the unique viscoelastic properties of the hydrated and mechanically developed gluten sets wheat apart from other grains, even grains from closely related species in the family Triticeae, and which allows wheat to be the supreme, if not almost the only, source of flour for its signature product: leavened breads. Nonetheless gluten's unique properties also make it useful for other products. For example note the ease with which wheat flour and water are compounded into a cohesive dough at room temperature during noodlemaking. No other requirements are needed except that the water is incorporated in a somewhat homogeneous fashion with the hydrated flour particles, then is simply squeezed together to form the dough. This is contrasted to, say, rice noodles, which require a precooking step to gelatinize sufficient starch to allow the rice-noodle sheets to become cohesive enough to allow further processing.

Overall wheat grain quality encompasses more than the intrinsic factors noted previously. Additional factors are largely end-use independent and are therefore applicable to all traded wheat. These so-called trade-quality factors include minimum test weights, absence of defects (e.g., shrunken or broken kernels, insect damage, or excess moisture), and specified limits for admixtures of other materials (e.g., seeds of other plant species, insects or insect fragments, stones, and other nonwheat materials). Buyers anticipate that grain is well filled and may specify that the grain is free from sources of unwanted enzymatic activity, such as excessive levels of either preharvest or late-maturity α-amylase (Chakraborty et al., 2003; Wrigley 2006) or high levels of polyphenol oxidase (Fuerst et al., 2006).

Suitability can be examined by evaluating individual components in analytical procedures or by performing tests that integrate all or some of the compositional factors. Integrated tests examine either the functional properties of wheat or flour (e.g., flour paste viscosity or dough properties), or the processing characteristics and quality of the end products themselves as the final arbiters of suitability. Cultivar identification also gives some guidance to end-use suitability as a result of the highly heritable nature and genetic control of certain functionally significant components or attributes, such as gluten composition, starch composition, and kernel texture. The tests discussed here are applicable at many points in the overall enterprise of delivering wheat to the end user from the earliest stages of wheat cultivar development to assessment of commercial flours. Under ideal circumstances the different sectors of the grains industry would have a common vocabulary that would encompass common tests, common test results, and common understandings of functionality. In this manner the whole enterprise, from cultivar development to consumer satisfaction, might be seamlessly coordinated.

As a consequence of the many places that wheat needs to "pass the test on end-use quality," the techniques discussed here are relevant to subjects covered elsewhere in this book such as cultivar development, identification of quantitative trait loci (QTLs) for quality factors, and wheat classification and trading. In turn these assessment procedures are predicated on the biochemical and molecular bases of wheat quality that are detailed in the following chapter. This chapter aims to briefly spell out some basic processing and quality characteristics of major wheat-based foods, and then to focus on the principles of current and emerging technologies that can be used by all segments of the industry to assess grain and flour functionality.

The procedural intricacies of the techniques have been reviewed elsewhere (e.g., Rasper and Walker 2000) and can also be examined in published standard methods (AACC 2000; ICC 2006). Here these techniques will be viewed as tools to ensure that wheat "passes the test" on functionality and, where possible, new analytical methods and predictive tools that are aimed at improving the speed, accuracy, and precision of grain functionality assessments will be highlighted. Where the outputs of specific functional tests are mentioned in the first section on the characteristics of wheat-based foods, readers are directed to the relevant later sections, for example, on "Starch and flour properties" and "Dough testing and prediction of dough properties," for explanations of the specific terms used.

CHARACTERISTICS OF MAJOR WHEAT-BASED FOODS

Most modern wheat-based foods are made from ground or milled wheat rather than the whole berries. Milled wheat can be used as is, as whole-wheat flour (wholemeal), or as flour of greater or lesser levels of refinement. Flours range from atta, basically a wholemeal with the coarse bran removed (Quail 1996), to the finest of white patent flours that consist largely of only the central endosperm (Posner 2000). In durum milling for pasta production the desired end point is semolina, which has a coarser particle-size distribution than flour. Color should be appropriate to the level of flour refinement (e.g., whole-wheat flour is expected to have a brown hue and to be less bright than refined white flour). Flours or meals should have moisture content of less than 14% to maintain storage stability (primarily to restrict mold growth) and to ensure that flour flows without caking or clumping during transfer in mills and bakeries (Atwell 2001). Flours should also have specified protein contents for different products. For example, most flours used for pan breads have higher optimum flour protein content than flours destined for cookies (biscuits). Arguably most important, flour should have appropriate gluten composition—or in colloquial terms, protein quality—and therefore appropriate dough characteristics for end-use processing.

Essential bread requirements

Leavened and unleavened breads share a common group of processing requirements and end-product traits, the precise nature of which varies between bread types. All breads require flour that makes dough of a dependable and relatively soft consistency that can be easily molded into the desired shape, at a level of water absorption that allows the dough to be cohesive and elastic without undue stickiness. However, there is much variation in optimum water requirement, from as high as 70%–80% (flour basis; fb) for ciabatta, to as low as 55% (fb) for Arabic breads (Qarooni 1996). A key requirement of bread dough is that its elastic and viscous rheological behaviors are balanced. The desired end point is a dough that is strong enough to withstand the rigors of processing (mixing, fermentation, dividing, molding), and for leavened breads, the dough must be strong enough to hold the fermentation gases during proofing and yet deformable or extensible enough to rise easily and form the familiar aerated internal structures (crumb) in the finished products.

Straight-dough processes

Straight-dough breadmaking processes are the basic template for bread production. In this process all of the flour, water, and other ingredients are mixed together into dough in one step. The dough is then fermented fully and can be optionally degassed (punched or folded) at different times during fermentation (Cauvain and Young 2007). Straight-dough processes are used to make a large variety of breads, from white sandwich pan bread made by rapid dough processes to crusty whole-wheat artisan hearth bread made by long bulk-fermentation processes (Moore 2004).

Flour used for most pan breads generally has optimum flour protein content between 10.5% and 13%. For bread made from bulk-fermented dough, optimum flour protein content increases to 12% or more, and optimum dough strength generally increases with the length of the bulk-fermentation stage (Cauvain and Young 2006). For rapid straight-dough processes like the Chorleywood or other no-time or mechanical dough development (MDD) process, a relatively strong dough with short mixing requirement is advantageous. For example a dough that requires higher work input for optimum dough development than the 10–14 Wh kg^{-1} commonly used will be undermixed with the attendant quality deficits in the finished bread product (Wooding et al., 2000). (Note that work input is roughly analogous to mix time at a constant mix speed.)

The MDD process commonly employs oxidants to speed dough development and may also have relatively short or no bulk-fermentation times, relying on the final proof stage for the development of the characteristic bread flavors that are supplied by the yeast. Pyler (1972) contends that there are no detectable flavor deficiencies in pan bread made from MDD dough as a result of higher levels of residual sugars, which can subsequently participate in browning reactions during baking. We contend otherwise. For MDD and other rapid straight-dough processes, flour of slightly lower protein content than that required for the equivalent bread made using a bulk-fermentation straight-dough process can often be specified (Cauvain and Young 2007).

Sponge and dough and other pre-ferment processes

Pre-ferment processes are two-step processes that allow a portion of the fermentation to occur before the final dough is mixed. In sponge-and-dough processes, highlighted here as the archetype for pre-fermented systems, the first-stage sponge uses around one-half of the total flour in the formulation (AACC method 10-11 uses 60% of the total flour; AACC 2000). Variable proportions of the total water are used depending on the desired consistency. After the sponge has fermented it is combined with the other ingredients and remixed to form the final dough. The process then continues in a roughly parallel fashion to that used for straight-dough processing.

The sponge has a profound effect on flavor and gluten development, with the benefit of creating a softer more extensible dough after the second mixing (Cauvain and Young 2007). Claimed advantages of sponge-and-dough processes are greater flexibility in processing, better product uniformity, increased product volume, more desirable internal structures, and softer bread textures (Pyler 1988; Kulp and Ponte 2000). Sponge-and-dough processes can put more stress on a dough, particularly during gluten development in the sponge stage and when the sponge may need to rise for up to four hours until it breaks or decreases in volume (Pyler 1988). Bakers specify flour of at least 12% protein with equal or greater dough strength than specified for straight-dough processes (Cauvain and Young 2007). Although sponge-and-dough breads now make up the majority of breads made in the United States (Kulp and Ponte, 2000), preferments with familiar names such as "biga" and "polish" are also used in artisan baking (Figoni, 2004; Moore, 2004).

High-volume bread types

High-volume breads encompass an astonishing variety of shapes, sizes, crust colors, and crumb structures and textures (Faridi and Faubion 1998; Kulp and Ponte 2000; Moore 2004). The most common high-volume breads are probably white pan breads. The word "pan" simply indicates that they are baked in pans, either with or without a lid. White sandwich breads conform to the basic pattern for baked leavened breads: a thin, dry brown outer layer, or crust, and a soft interior, or crumb, of proportionately large volume in comparison to the crust. The crumb is generally lighter in color (creamy-white to white) and is made up of a fine cellular structure (Cauvain and

Young 2006). Alternatively, hearth breads are not confined in a pan during baking and are traditionally baked on the hearth, or sole, of the oven. Hearth breads generally have thicker and/or crispier crusts than pan breads baked in lidded pans, and have a wider range of crumb textures from soft to very chewy. Crumb structures of high-volume breads range from very fine-grained for sponge-and-dough derived white sandwich breads to more open structures such as those associated with traditional French baguettes (Seibel 1998).

When assessing the general breadmaking suitability of experimental breeder lines or large commercial grain lots, dough rheological properties, loaf volume, and crumb structure are generally emphasized. However, loaf volume on its own provides a valuable index of overall breadmaking potential (Graybosch et al., 1999). Other important factors are the form ratio (height-to-width ratio) of hearth breads (Færgestad et al., 2000; Sahlstrom et al., 2006), visual crumb characteristics such as cell uniformity, size, shape, and cell-wall thickness, and textural characteristics such as moistness, elasticity, and smoothness (Collar et al., 2005). Attributes like aroma, taste, and staling (firming characteristics) can be assessed when required. These latter factors are important to consumers but are greatly affected by formulation and process and therefore not necessarily key to wheat "passing the test."

For high-volume breads, loaf volume generally increases linearly with increasing flour protein content. This relationship, and its interaction with genotype, has been known for decades; that is, for incremental increases in flour protein content, different wheat genotypes have greater or lesser incremental increases in loaf volume (Finney and Barmore 1948; Briggs et al., 1969). Form ratio can also be affected. In hearth breads at equivalent flour protein content, weaker and less elastic doughs from wheat genotypes with glutenin subunits 2 + 12 (encoded by *Glu-D1a*) were shown to produce hearth breads with significantly lower form ratios (and lower volumes) than breads made from genotypes with glutenin subunits 5 + 10 (*Glu-D1d*) (Tronsmo 2003). The biochemical basis of the relationships between gluten composition and baking quality has been reviewed elsewhere (Gianibelli et al., 2001; Veraverbeke and Delcour 2002; Bekes et al., 2004) and is addressed in the next chapter.

Steamed breads

A variant on the high-volume bread template is steamed bread, which is most common in China and other parts of eastern Asia. As with other regional specialties, however, its popularity has spread well beyond its place of origin. Steamed bread is markedly different in appearance to baked bread as a result of the absence of a brown crust, having instead a white skin. Steaming maintains the water activity of the bread surface well above the level that allows for Maillard browning. Steamed bread is traditionally made with a sponge-and-dough process, although no-time straight-dough systems are increasingly used (Huang 1999). Leavening can be based on a sourdough starter or on yeast, or the steamed breads can be chemically leavened.

The optimum quality characteristics of steamed bread vary depending on geographical origin; hence types of steamed bread are commonly categorized as northern, southern, and Guandong styles. Respectively these are hard and dense, softer and less dense, and very soft and somewhat friable. The Guandong style breads are noticeably different in formulation with up to 10% sugar (fb) and 25% fat (fb), in contrast to the lean formulae for the other styles (Huang 1999). Specific steamed bread quality attributes vary between styles, but common attributes are the white color of both skin and crumb, high volume and volume per unit weight (specific volume), and a spread or form ratio of rounded buns that show a bold aspect (i.e., width roughly equal to height). A slumped aspect, where the width is much greater than the height, is undesirable. All styles should also have a shiny, smooth, and attractive external appearance (Huang et al., 1993).

Flour quality for steamed bread varies with the style being made. For northern style steamed bread, hard wheat was preferred in one study

(Huang et al., 1996), and overall bread quality improved with increased flour protein content up to around 10% protein in either soft or hard wheats (Rubenthaler et al., 1990; Huang et al., 1996). Quality leveled off thereafter with very high protein flours showing decreased overall quality (Huang et al., 1996). He et al. (2003) concluded that flour and dough with medium protein content, medium-to-strong gluten strength, and good extensibility were best for northern style steamed bread made by a mechanical mixing process. Weaker doughs performed better for hand mixing. Whiter flour with higher flour cooked-paste viscosity could also enhance northern style steamed bread quality (Crosbie et al., 1998; He et al., 2003). In addition to being softer in texture, southern style steamed bread has a more open crumb structure. Accordingly, flour of slightly lower protein content (9.5%–11% compared with 10%–12% for northern style) and of medium dough strength is more suitable (Crosbie et al., 1998). For the very soft and friable Guandong style bread, very white flour of 7.5%–8.5% protein with weak to medium dough strength is required (Crosbie et al., 1998).

Low-volume bread types

Low-volume or flat breads are common throughout the continent of Asia and in North Africa and are also found in a multitude of types. A major contrast within flat breads is the distinction between the two-layered types generically known as Arabic breads (pita) and one-layered types (e.g., barbari, chapatti, tanoor, lavash) (Qarooni 1996; Quail 1996). The general category of flat breads can also include flour tortillas and bread types with intermediate ratios of crust-to-crumb such as foccacia and ciabatta (Qarooni 1996).

Two-layered bread

Two-layered Arabic bread is commonly made by straight-dough processes (Qarooni 1996; Quail 1996). However, Amr and Ajo (2005) showed some quality benefits from using a sponge-and-dough process. Arabic bread is made from a relatively stiff dough (650–850 BU consistency on a farinograph, C.W. Brabender Instruments, Inc. South Hackensack, New Jersey) consisting of water, yeast, salt, and a high-extraction (75%–80%) flour (Qarooni 1996). The dough may be sheeted or die-cut into its final shape (Quail 1996). The relatively dry and stiff dough easily forms a skin in final proofing and in the initial stages of the typically high temperature (350–600°C) baking. The skin then forms a barrier preventing escape of the steam and thus allows an increase in the gas pressure in the baking dough piece. It is this pressure that splits the single dough piece into two layers, preferably of even thickness (Qarooni 1996; Quail 1996).

Arabic bread should be soft and flexible with well-separated layers. Each layer should have only a thin adhering layer of crumb that has a fine, even structure. The bread should have smooth, brown, and blister-free surfaces. The intensity and desired hue of the crust's brown color varies depending on bread type and local preferences. Bread softness and flexibility should be accompanied by some strength to allow rolling and folding of the bread without cracking or tearing (Qarooni 1996; Quail 1996).

In contrast to the linear increase in loaf volume with increasing flour protein content described for pan bread, overall quality of Arabic bread has been shown to have a roughly parabolic response to increased flour protein (Qarooni et al., 1988; Quail et al., 1991). Optimum flour protein content ranged from 10.5% to 12% with distinct declines in Arabic bread quality above 12% or below 9%. Moderately strong and extensible dough was considered most suitable (Quail 1996). Higher dough strength or dough elasticity was found to produce bread with a leathery texture. It can be concluded that Arabic bread is sensitive to changes in dough strength and elasticity and that there appears to be a threshold of dough elasticity beyond which a rapid decline in bread quality takes place (Qarooni et al., 1988; Quail et al., 1991; Toufeili et al., 1999). In our experience (A.S. Ross, unpublished data), overstrong doughs resist sheeting.

This is manifested as nonround shapes, tearing of the dough, and stress ridges on the dough surface. The stress ridges are heated first in baking, forming blisters. Bread baked from an overstrong dough also has a tough and unpalatably chewy texture.

Quail (1996) suggested that flour milled from hard white wheat was desirable for Arabic bread production, probably as a result of color and flavor advantages as well as higher water absorption than soft wheat flour. However, there is no consensus that hard wheat is better than soft wheat for Arabic bread production at equivalent protein content and dough strength. Farvili et al. (1995) compared flours from hard and soft wheats and showed the only deficit in bread made from the soft wheat flour to be pale crust color. Lower starch damage may give reduced opportunity for the production of reducing sugars for Maillard browning when flour is milled from soft wheat.

Single-layered bread

Single-layered bread includes leavened types such as baladi and a variety of unleavened types such as chapatti. Some leavened types are restricted from pocketing by docking the dough pieces (e.g., tanoor bread) (Qarooni 1996). Such a wide variety of single-layered breads makes it difficult to define a common set of flour quality requirements and to succinctly describe their optimum quality attributes. However, as all types need to be sheeted or otherwise flattened, a flour of moderate flour protein content, with moderate strength, and good extensibility would seem to be a common need. For example a dough with excessive resistance to extension (too strong) contracts after sheeting (Ur-Rehman et al., 2007a), and this is a problem for all thinly sheeted doughs. For "naan" bread, a flour of 12% protein with a moderately short farinograph mix time (2.5 min), moderately long farinograph stability (17 min), and a high viscoamylograph (C.W. Brabender Instruments, Inc., South Hackensack, New Jersey) paste viscosity was found suitable (Farooq et al., 2001). Ur-Rehman et al. (2007b) indicated that the protein content of chapatti flour should be low and that the dough should be extensible enough to allow stretching without later contraction. These authors suggested that sheeting quality was essential to the quality and shelf life of chapatti bread.

Asian noodles

Asian noodles are made from both soft and hard hexaploid wheat (*Triticum aestivum* L.), and in large-scale industrial practice are most commonly made by a mix, sheet, and cut process (Hou 2001; Crosbie and Ross 2004; Ross and Hatcher 2005). They can be differentiated on formulation (salted vs. alkaline) and the type of postcutting process applied: none (i.e., fresh or raw noodles), freezing, drying, steaming and drying, steaming and frying (instant noodles), parboiling, boiling then freezing or vacuum packing, and others (Hou 2001; Crosbie and Ross 2004). Noodles can also be differentiated on cross-sectional dimensions and general eating qualities, such as whether the noodles are firm or soft on biting and chewing.

A common requirement of all Asian noodle products is an appealing appearance. This is determined by a number of factors such as color (hue), gloss (Solah et al., 2007), geometry, and freedom from discoloration. Good appearance is often associated with low flour ash content, but this acts only as a proxy for low levels of bran admixture, which can be measured in other ways as discussed in the flour color section. The appearance attributes most emphasized are color (hue) and discoloration.

Noodle color varies from nearly white to deep yellow (Crosbie and Ross 2004). The exact hue is dependent on endosperm pigmentation and noodle formulation (Mares and Campbell 2001). Alkaline noodles are generally more yellow as a result of the expression of the yellow color of flavonoid pigments at high pH (Mares et al., 1997; Asenstorfer et al., 2006). Clean, clear coloration (i.e., lack of discoloration) is also desired. Discoloration most commonly results from admixture of bran particles in the flour, and its impact increases as flour extraction rates increase (Kruger et al., 1994; Hatcher and Symons 2000a). Larger bran particles are visible as discrete spots or specks in the dough or cooked noodle and constitute a product deficit termed speckiness. Using machine vision Hatcher and Symons (2000a,b) showed, not surprisingly, that straight-grade flours gave noodle sheets with more specks than patent flours, while salted noodle sheets had fewer visible specks than alkaline noodle sheets.

The presence of high levels of polyphenol oxidase (PPO), which is largely located in bran (Hatcher and Kruger 1993; Every et al., 2006), leads to a progressive darkening of noodle dough. Parallel darkening may occur that is not related to PPO, but the mechanism(s) of this additional darkening is unknown (Fuerst et al., 2006). Darkening is a problem for noodle makers who rest the dough during processing to improve dough handling or boiled-noodle eating quality. Excessive darkening shortens the time available for resting and, for sellers of fresh noodles, the time during which the product retains its desired attractive appearance.

Soft-bite noodles

Soft-bite noodles are almost all salted noodles; that is, the major formulation is flour, water, and sodium chloride or salt (e.g., udon). The textural attributes of these noodles are primarily derived from flour of moderately low protein content (9.0%–9.5%) milled from soft wheat with medium-strong dough characteristics, and crucially, starch that is relatively low in amylose (i.e., high swelling; Crosbie et al., 1998). Wheat starch with low amylose content is related to the absence of one or two copies of the enzyme granule bound starch synthase, which is in turn related to one or two null alleles at the three *Wx-1* loci (partial waxy wheat). Amylose content of starch decreases with increasing dosage of *Wx-1* null alleles (Geera et al., 2006). Flour containing 100% amylose-free or waxy starch (all three null *Wx-1* alleles) appears unsuitable in some, if not all, noodle applications (Park and Baik 2004). However, it may be possible to use waxy-wheat flour in blends with other flours or as a replacement for potato starch, which is sometimes used to achieve the desired softness and elasticity when the starch characteristics of the flour are not ideal.

Hard-bite noodles

Hard-bite noodle formulations range from simple flour, water, and salt (e.g., Chinese raw noodles), and flour, water, and alkali compounds (e.g., Hokkien noodles), to more complex flour, water, table salt, alkali, gum, and polyphosphate formulations in some instant noodles (Crosbie and Ross 2004). Generally the harder textured styles of noodles are made from hard wheat flours of moderately high to high protein content (10.5% to >13.0%; Crosbie and Ross 2004). Moderately strong to strong dough characteristics are preferred, and both normal and low-amylose starches may be selected depending on the desired textural attributes of the final product (Crosbie et al., 1998; Crosbie and Ross 2004).

Feedback from Asian noodle makers suggests that preferences for starch characteristics may be changing, with more manufacturers of hard-bite noodles considering possible textural advantages of lower-amylose *Wx*-null starch with regard to elastic bite characteristics. Previous recommendations for hard-bite noodles had emphasized low hot-paste viscosity (in the absence of α-amylase) (Miskelly and Moss 1985; Ross et al., 1997) or low swelling power (Ross et al., 1997; Corke and Bhattacharya 1999), that is, flour with normal starch. The possible change in preference is supported by evidence from Akashi et al. (1999), who indicated an improvement in alkaline noodle elasticity when the noodles were made from flour with reduced amylose content and higher swelling power.

Soft wheat products

Cookies and crackers

Cookies (biscuits in Anglo-English) are made with formulations consisting of flour, water, salt, and relatively high amounts of fats and sugar in comparison with other baked products. Cookies and crackers can be made with a number of processes, differentiated by the amounts of fat and sugar used. Wire-cut and deposited cookies have high levels of both ingredients, sheeted and molded cookies have intermediate levels, and crackers have even lower levels (Manley 2001; Hazelton et al., 2004). Cookies are generally made from soft wheat flour with low protein content (7%–9%; Faridi et al., 2000), and with moderately weak to weak dough characteristics. Cookies share a need, along with other products that are baked to very low finished moisture contents, to have flour that can make a workable dough with a relatively low amount of added water. Low-protein soft wheat fulfills this need in the first instance by reducing to a minimum the water absorbed by protein and by starch. Low absorption by starch is achieved through low levels of starch damage attained by soft wheat upon milling compared with hard wheat (Martin et al., 2007). Even within single cultivars cookie quality has been shown to decline with increased starch damage deliberately induced through additional disc milling (Barrera et al., 2007). Furthermore lower amounts of other hygroscopic components such as arabinoxylans (pentosans) and proteins also reduce the overall water requirements of soft wheat flour.

Zhang et al. (2007) showed with 17 Chinese soft-wheat cultivars grown in six site × year environments that water-soluble pentosan content was negatively correlated with cookie diameter. They also applied a cluster analysis that divided the cultivars into good, fair, and poor clusters (good represented by largest cookie diameter and largest diameter-to-height ratio). This analysis showed that the good cluster had the lowest water-soluble pentosan level and lowest water absorption at statistically equivalent flour protein content across the three clusters. Finnie et al. (2006) evaluated 7 soft white spring and 20 soft white winter wheat cultivars across 10 or more site×year environments and concluded that arabinoxylan content of grain and flour was primarily under genetic control. The authors concluded that genetic control was more evident for the important water-soluble arabinoxylans. Slade and Levine (1994) further attributed low sugar-snap cookie diameter to the formation of elastic 3-dimensional polymer networks by one, two, or all three of the high-polymer flour components: high-molecular-weight glutenins, arabinoxylans, and damaged starch. The variety of styles of cookies precludes a blanket summary of desired quality characteristics. However, the development and control of cookie spread during baking is the key to cookie quality.

Like cookies, crackers are also baked to low-moisture content. Crackers typically have a flaky and hard but friable (crisp) texture. To achieve the required texture, cracker dough is commonly laminated. According to Faridi et al. (2000) crackers can be categorized as saltine, chemically leavened, and savory. Saltine crackers are leavened by fermentation in a sponge-and-dough process. Flour for crackers tends to be somewhat higher in optimum protein (8.5%–10.0%; Faridi et al., 2000) than for cookies and may also be somewhat higher in dough strength coupled with good extensibility if optimized solely for cracker production (Hazelton et al., 2004). Cracker production also benefits from flour with low water absorption capacity, ideally milled from soft wheat. Feedback to the Oregon State University (OSU) breeding program from industrial participants in the US Pacific Northwest Wheat Quality Council also reflects the desire for slightly higher dough strength in soft wheat cultivars that are specialized for cracker production.

Cakes and batters

Cakes, pancakes, coatings, waffles, cake donuts, and similar products share a common process and base ingredients. They are produced from high-water-content batters and made from soft wheat flour with varying amounts of other ingredients: sugar, fat, emulsifiers, and leavening agents. Batter-based products rely largely on viscosity to control baking and end-product characteristics (Morris and Rose 1996). Batter viscosity needs to be sufficiently high to allow for the retention of leavening gasses, to allow the batter to cling to products when used as coatings, and to prevent settling out of batter components. Simultaneously, viscosity must be low enough to allow sufficient flow for proper product performance (e.g., pancake spread). However, viscosity in batters is not based on gluten formation (Morris and Rose 1996). If gluten forms to any great extent in batter-based baking systems, the resulting products are tough, chewy, and unacceptable to consumers. Cakes, for example, should have crumb texture that is tender and soft with little resilience, a cell structure that is somewhat less defined than breads but still uniform with densely packed small cells, and for high-ratio (more sugar than flour) cakes a silky structure (AACC approved method 10-90, AACC 2000; Cauvain and Young 2006).

To minimize gluten formation, the soft wheat flour used in batter-based products should be of low extraction (for cakes, 60% extraction flour, equivalent to about 0.23% ash, is frequently used), and of low protein content (7.0%–8.5%). The low extraction rate minimizes the starch damage that leads to greater water absorption and problems with the starch gelatinization profile (Atwell et al., 1988). Further the reduced amounts of bran and aleurone layer fragments in low-extraction flour reduce the amount of water-unextractable arabinoxylans, which can bind 10 times their weight in water (Kulp 1968). Increased water-unextractable arabinoxylan content leads to competition for water between these endogenous components and sugar as an ingredient, leading to viscosity and textural problems in the product and complications with formulation, handling, and baking.

To achieve proper viscosity, batters rely partly on small bubbles created by air incorporated at mixing (chiefly nitrogen) and by chemical leaven ing agents, which produce carbon dioxide and/or ammonia through the reaction of acid–base leavening pairs (Heidolph 1996). Greater numbers of small bubbles increase batter viscosity both initially and during baking. As the batter heats during baking, the bubbles expand, creating a light, fluffy product. Without the viscosity provided by leavening bubbles, the batter would become more fluid due to heating, until starch gelatinization occurred, deleteriously affecting product quality. Batters therefore must be sufficiently viscous to hold leavening bubbles throughout the mixing and baking process. Since gluten formation is minimized in high-water, low-protein, short mixing time batters, formation of a large carbohydrate matrix by oxidative gelation among water-extractable arabinoxylans through esterification of ferulic acid moieties has been suggested as the mechanism of viscosity creation (Izydorczyk et al., 1991; Bettge and Morris 2007).

Cake flour to be used for batters is frequently modified by chlorination or heat treatment to enhance functionality. Chlorination reduces the flour pH by creation of hydrochloric acid, and also affects proteins and starch-granule surface lipids through oxidation. The starch becomes more hydrophobic and gelatinizes and swells more quickly during baking (Seguchi 1990; Greenwell and Brock 1996). Chlorination is decreasing in use as a flour treatment, especially in countries other than the US. Heat treatment, as an alternative treatment, also increases the oxidation of the flour, providing some of the same beneficial changes to flour quality (Ozawa and Seguchi 2006).

Durum pasta

In large-scale industrial settings pasta is made from durum semolina and water. The relatively dry dough (water addition about 18%–25% of dry solids; Sissons 2004) is first mixed to blend the semolina and water and hydrate the semolina, sometimes under vacuum. The dough is then extruded under vacuum through a shaping die, and dried at controlled high temperatures (>60 °C), often with detailed time–temperature profiles (Sissons 2004; Cubadda et al., 2007). Fresh-pasta products, although still commonly made from durum semolina, are often made with a similar sheeting process to Asian noodles. However, this product sector is smaller in volume than the extruded and dried pasta sector.

Durum wheat for pasta production should be sound, be well filled, and contain >75% vitreous kernels, which is related to higher semolina yields when milled. Nonvitreous kernels tend to mill to fines, or milled particles which pass a US number 100 sieve (150 μm). Durum flour, to distinguish it from semolina, is defined as the material passing a US number 70 (212 μm) sieve (Donnelly and Ponte 2000). Sissons et al. (2000), using a single-kernel characterization system (SKCS 4100, Perten Instruments, Huddinge Sweden) to predict vitreousness, showed a weak but positive correlation between vitreousness and semolina yield.

Grain protein content should be high enough to produce semolina of >11% protein (Sissons 2004). The resultant semolina should be bright, yellow or amber in color, and free, as far as possible, from visible specks of bran. Semolina should have a uniform particle-size distribution, which allows even hydration of all particles during mixing. The protein content of semolina is widely considered the primary factor influencing pasta cooking and eating qualities. However, the semolina should also make a relatively strong dough. Strong gluten semolina has been associated with better cooking and eating qualities and allows better flexibility with regard to drying regimes (Cubadda et al., 2007).

The key quality attributes of pasta are appearance, cooking quality, and cooked texture. Dried pasta products should have a bright, clear yellow or amber coloration (yellowness CIE b* value >56 in spaghetti dried at 70°C), and the dried product should be free of visible cracking or checking (Sissons 2004). Cooking and texture quality factors include firmness, lack of surface disintegration, and lack of stickiness (Troccoli et al., 2000). The extent of surface disintegration can be conveniently estimated by directly measuring the solids-loss in the cooking water or by colorimetric measurement of the amylose content in the cooking water (Matsuo et al., 1992). Cooking losses have been shown to decrease with increasing semolina protein content (Malcolmson et al., 1993). Firmness can be assessed by sensory analysis or by instrumentation. Attempts have been made to standardize the instrumental assessment of pasta firmness (Sissons et al., 2008).

COMPOSITIONAL ANALYSIS AND GRAIN TESTING

Moisture

Moisture is a fundamentally important constituent of grain, flour, and finished end products. Major needs are satisfied by knowing moisture content—for example, whether grain can be stored without being susceptible to mold growth or how much tempering water to add to grain before milling. The performance of subsequent analyses on an equivalent-solids basis, and the reporting of other compositional analyses at constant moisture, is predicated on accurate knowledge of moisture content. Indeed knowledge of the amount of water already present in the material is paramount, since so many aspects of flour and dough technology require measured but variable water additions.

Moisture can be measured in a variety of ways but derivative methods need to refer back to the basic air-oven method: drying the sample in an oven at a specified high temperature (e.g., 130°C) and measuring weight loss as a proportion of the initial weight. One-stage drying is used for materials <13% moisture and two-stage drying for samples >13% (AACC method 44-15A, AACC 2000). Moisture can also be estimated by dielectric meter (capacitance) or by near-infrared spectroscopy (NIRS), or by conductance in the Perten SKCS 4100. Water is a strong absorber of near-infrared energy and the higher the moisture content, the more near-infrared energy is absorbed (Caddick 2007). Hence NIRS is well suited to the measurement of moisture content.

Moisture content is obviously critical. However, moisture may not be distributed evenly in a kernel or end-product structure, and exciting advances in monitoring moisture distribution have arisen, particularly with the development of magnetic resonance imaging (MRI) techniques. For example, MRI has been used to monitor the movement of moisture in wheat grain during drying (Ghosh et al., 2007) and to visualize non-homogeneous moisture distribution in the endosperm of kernels at equilibrium moisture (Song et al., 1998). Other work has used MRI to monitor the hydration of Japanese salted noodles during boiling, resolving that water penetration was faster in a lower protein sample compared to a sample made with higher protein flour (Kojima et al., 2004). In a similar way Horigane et al. (2006) monitored water penetration in dried and fresh spaghetti and were able to resolve differences in moisture diffusion between the two pasta types.

Protein content

Protein has been long considered the most important functional constituent of wheat grain and flour. Protein content is a significant factor in price determination in wheat trading, once the base price has been set through supply-and-demand considerations. Higher protein hard wheat, and lower protein soft wheat, generally command moderately higher prices. Protein content alone is only an approximate indicator of overall wheat functionality, but it has profound modifying effects on other functional properties. For example, regardless of how appropriate the gluten composition of a cultivar or grain lot is for breadmaking, if grain and hence flour protein content is too low, then nothing will save it.

Two chemical methods of determining nitrogen are in common usage: the Kjeldahl digestion method (AACC approved methods 46-09 to 46-13, AACC 2000) and the Dumas combustion method (AACC approved method 46-30, AACC 2000). The Dumas method is becoming ascendant primarily because of its relative speed, ease, and safety compared with the Kjeldahl method. However, the nitrogen release by digestion in the Kjeldahl method is less efficient than the release of nitrogen by pyrolysis in the Dumas method. This leads to small, but significantly higher, nitrogen contents in samples tested by the combustion method. Once converted to protein (in the case of wheat and wheat flour, as N × 5.7; AACC approved method 46-19, AACC 2000) this translates to an apparent increase in protein of 0.15%–0.25% in samples tested by combustion (Williams et al., 1998).

Wet (and dry) gluten determinations can also be used as a rough guide to grain or flour protein content and quality. Wet gluten can be derived manually or by machine (AACC approved methods 38-10 and 38-12A, AACC 2000). These methods are particularly useful as indicators of relative protein or gluten abundance where more sophisticated methods are unavailable. Wet gluten has been highly correlated with protein content determined as N × 5.7 (A.S. Ross, unpublished data), but this relationship does not always hold true (e.g., Pasha et al., 2007).

Near-infrared spectroscopy, in either transmission or diffuse reflectance modes, is a mature, robust, and accurate method for estimating protein contents in grain, meal, or flour. It was first applied in a breeding context more than 30 years ago. As a secondary or derivative method, NIRS is based on the correlation of sample spectra with the concentration of the constituent of interest as determined by a standard reference method (AACC approved methods 39-00, 39-10, 39-11, 39-25; AACC 2000). The success of NIRS in determining grain and flour protein content, as a result of its accuracy and precision, ease-of-use, reliability, and relatively low operational cost, has seen it become entrenched as a routine method and unlikely to be supplanted in the foreseeable future by other spectroscopic methods such as midinfrared spectroscopy (Reeves and Delwiche 1997). One change that is occurring is in instrument design, such as the new generation of diode-array instruments, which can substantially reduce analysis times and increase sample throughput (Osborne 2006).

Kernel texture

Kernel texture is another characteristic of wheat that must be known before one can decide the suitability of wheat for a given product sector. Soft wheat mills with lower starch damage than hard wheat (Williams 2000), and this difference is fundamental to the profound divergence of their optimal end uses. Though the distinction between hard and soft wheat is controlled by a pair of linked *Pin* (puroindoline) genes on chromosome 5D in hexaploid wheat (Morris 2002), variable hardness levels occur within both genetically hard and genetically soft wheat. In a preliminary investigation of a soft × extrasoft population, QTLs were detected on chromosomes 2A and 3D that in combination accounted for 49% of the phenotypic variance in kernel texture (Wang et al., 2007). This result indicates that additional genetic controls for kernel texture occur within soft wheat that are independent of the *Pin* loci.

Williams (2000) documented the history of wheat kernel texture measurement, reporting the first "single-kernel characterization system," which appears to have been a sensory method of assessing the force to crush kernels between the teeth (Cobb 1896, from Williams 2000). Following that, granularity measurement (particle-size index; PSI) appeared in 1935, NIRS in 1978, and the SKCS 4100 in the 1990s. The PSI is a measure of sample granularity after it is milled by a standard protocol. The amount of material passing a 75-μm sieve expressed as a proportion of the initial sample size is the PSI. Low PSI values indicate more granular material and therefore harder kernels since less material mills to fines and passes the 75-μm sieve (see AACC approved method 55-30, AACC 2000).

The Perten SKCS 4100 instrument was developed in response to objective classification needs in the US wheat industry. The primary measurement utilized by most SKCS users is the hardness index, which is based on an algorithm related to the raw crush-force–time profiles of each kernel in a sample. One advantage of the SKCS 4100 lies not only in its ability to report a hardness value for a sample of kernels but also in its ability to provide additional measurements of kernel diameter, kernel weight, and kernel moisture. Because the SKCS 4100 processes samples one kernel at a time, further advantages can be realized by allowing operators to monitor the variability or uniformity of a bulk sample for any of the aforementioned parameters. For example, wheat breeders may need to make hard × soft crosses to achieve goals for adaptation, disease resistance, or quality attributes. Early-generation breeding materials, such as F_4 head rows, can then be screened for kernel texture using the SKCS. The SKCS 4100 thus provides a rapid and reliable method of assessing both mean hardness and hardness distributions of early-generation samples, enabling advancement of lines to the appropriate hard or soft wheat nurseries as well as rapidly identifying samples still segregating for hardness. Identification of segregating lines is based on high standard deviations for hardness, or observations of bimodal distributions in the SKCS histograms of hardness values. The use of SKCS has been reviewed in detail (Osborne and Anderssen 2003).

Although the standard SKCS parameters are calculated and reported by the proprietary software, some workers have advocated the use of the available lower level data (e.g., raw crush-force profiles) (Fig. 20.1) to extract extra information value from the SKCS 4100 (Gaines et al., 1996; Osborne and Anderssen 2003). For example Osborne et al. (2007) showed agreement between maximum crush force on isolated endosperm cylinders and endosperm strength (maximum stress) derived from the SKCS 4100 raw crush-force profiles on whole kernels. Pearson et al. (2007) indicated that discrimination between hardness classes could be improved and misclassifications decreased by a factor of 2 by using SKCS lower level data to predict cumulative percentage of particles >21 μm, in addition to hardness index. Useful predictions of visually vitreous durum wheat kernels have also been derived using an algorithm derived from the raw crush-force profiles (Sissons et al., 2000). The utility of the SKCS

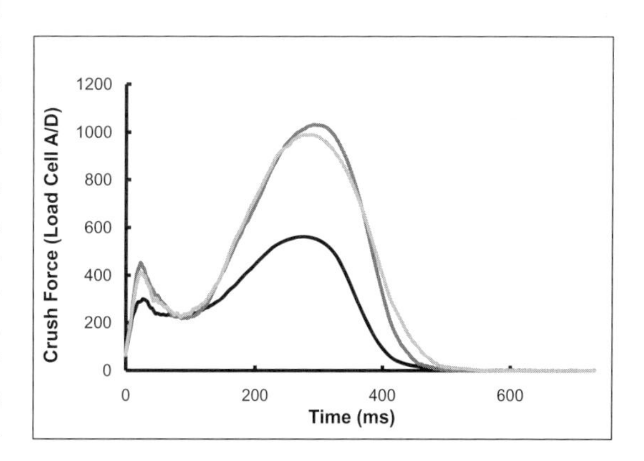

Fig. 20.1 Raw SKCS 4100 crush profiles for two hard wheat samples (upper curves) and one soft wheat sample (lower curve) harvested from Pendleton, Oregon, in 2005. [Adapted from Ohm et al. (2006).]

4100 for monitoring the conditioning and tempering of wheat prior to milling has also been established (Williams 2000; Edwards et al., 2007a).

The final of the three standard ways to assess kernel texture is NIRS (AACC approved method 39-70A, AACC 2000), as near-infrared reflectance is sensitive to differences in particle size of granular materials. As wheat grain becomes harder, grinding generates progressively larger particle-size distributions, and near-infrared absorption increases with particle size in granular materials. Accordingly NIRS in reflectance mode (using ground material) is a reliable method of assessing kernel texture. Vázquez et al. (2007) showed that standard errors of prediction <2% and coefficients of determination of >0.8 made reflectance NIRS a robust predictor of PSI when performed on ground grain. Applying NIRS *reflectance* analyses to whole grain gave unreliable predictions of PSI (standard errors of prediction >4% and coefficients of determination of <0.4). However, whole grain can be analyzed for hardness by NIRS *transmittance* (Osborne 2006), and this technique adequately differentiates wheat cultivars of different hardness (Williams 2000) or variations of hardness within hard × hard genetic populations (Giroux et al., 2000).

Grain soundness and α-amylase

The endosperm of wheat grain is a storehouse of polymeric materials that are ready to be mobilized to feed the developing embryo. Upon germination the large increase in the activity of the starch degrading enzyme α-amylase is the most striking event, potentially most detrimental to product quality, and easiest to track as a marker of the extent of germination. Although there is an optimum amount of α-amylase activity desired for best breadmaking performance (Goesaert et al., 2006), it is easily exceeded in preharvest sprouted wheat. Additionally bakers generally prefer to start with flour from sound grain and then to add controlled amounts of α-amylase from selected sources (fungal, cereal, or bacterial) to achieve the desired results (Goesaert et al., 2006). In breadmaking an excess of cereal α-amylase activity such as that which occurs with sprouting can cause sticky and slack dough, poor crumb structure and texture, and sticky crumb. However, α-amylase is by no means the only problem in sprouted wheat. Increases in endoproteases upon germination (Kruger and Reed 1988) are also associated with sticky and slack dough, and with increased darkening of noodles (Edwards et al., 1989). As a result of all of the above, sprouting of wheat grain in the head prior to harvest is a major cause of concern for growers since the loss of quality is generally reflected in large price discounts.

Damage from preharvest sprouting can be measured in a number of ways: visual examination (human or machine vision), autolytic tests, direct measurements of enzymatic activity, or increased presence of enzymatic proteins. Human visual examination is limited by the fact that detrimental levels of amylase may be already present by the time the germ first becomes visibly swollen, and the levels are almost certainly detrimental by the time the germ is split. In addition visual rankings of sprouted kernels are not always well correlated with instrumental determinations of sprouting such as Falling Number (FN; Barnard 2001). Neethirajan et al. (2007) investigated an interesting new machine-vision method to detect sprouted kernels using soft x-rays (λ = 1–100 nm). Sprouted kernels had regions that were softened by the amylase action, and these showed up as lighter shaded areas in the x-ray images. The authors were able to apply statistical treatments to the images that allowed distinctions between sound and sprouted kernels. This sidestepped one limitation of human visual assessments—the inability to quickly screen large numbers of seeds. The authors indicated that it will be necessary to adapt the method to bulk samples before it can be implemented in the grains industry.

Autolytic tests are those that employ the starch contained in the kernel as the substrate for amylase activity. These include the Falling Number (FN) (Perten Instruments, Huddinge, Sweden) test (AACC approved method 56-81B, AACC 2000), the stirring number test using the Rapid Visco-Analyzer (RVA) (Newport Scientific Pty. Ltd.,

Warriewood, NSW, Australia) (AACC approved method 22-08, AACC 2000), and an assessment using the viscoamylograph (VAG) (AACC approved method 22-10, AACC 2000). These tests can be applied to both flour and wholemeal, but the VAG is more commonly applied to flour alone. Operational principles of the latter two instruments are detailed in subsequent sections. Each of the three methods relies on the thermal and enzymatic breakdown of starch to thin the paste viscosity after the starch has been gelatinized *in situ*. Short FN times and low RVA stirring numbers or VAG viscosities are indicative of the presence of α-amylase.

Specifications for FN are widely quoted in grain classification schemes, as the FN test was the first one widely used for this application. Its entrenchment as a standard test and the use of FN specifications in buying and selling wheat speaks to both the usefulness of the test and the inertia of the wheat enterprise in taking up new methods. Wheat with FN greater than 350 s is generally considered to be sound. Wheat with FN below 250 s can be considered to contain enough α-amylase to have some detrimental effects in processing. Threshold values can change depending on the local environment and local product characteristics. In Italy, for example, the Synthetic Index of Quality (*Indice Sintetico di Qualità*) gives a FN threshold for the highest quality category of improver wheat (*frumento di forza*) of only 250 s (Foca et al., 2007).

The RVA has not been widely used for its original intent as a screening tool for preharvest sprouting (Ross et al., 1986, 1987). However, in recent times the Canadian Grain Commission has been investigating the RVA as a tool for assessing harvest receivals. Using over 4,400 hard red spring samples, Hatcher (2007) observed a polynomial relationship between RVA viscosity and FN with r^2 of 0.93 and a standard error of prediction of 14 s. Under the standard conditions delineated in the approved methods, the FN and the RVA-stirring number tests are not as sensitive to low levels of α-amylase as the viscoamylograph. This is because the stirring number and FN tests both use ballistic heating to high temperatures, and so have smaller windows of amylase activity than the viscoamylograph that heats at 1.5°C s^{-1} (see Chang et al., 1999, for discussion of FN). Temperature profiles are simple to customize and thereby alter the sensitivity of the RVA test, but this capability is very restricted in the FN test, where only small changes are available by changing the volume of material used (Chang et al., 1999).

More recently an immunoassay called the WheatRite test has been developed as a rapid test using antibodies specific to α-amylase molecules (Skerritt and Heywood 2000). The assay is directed at on-farm testing with the potential benefit that growers could avoid binning sprouted with sound grain, thus maximizing the grower's income potential in a rain-affected harvest. Alternatively one can use a laboratory based test to measure α-amylase directly using an enzymatic assay with an exogenous chromogenic substrate (e.g., AACC approved method 22-05, AACC 2000).

Polyphenol oxidase

Polyphenol oxidase is mostly regarded as a problem in noodlemaking, but Demeke and Morris (2002) and references therein also mentioned PPO-induced darkening in chapattis, Middle Eastern flat bread, and steamed bread. As noted earlier, PPO is associated with the bran layers in mature kernels (Kruger and Reed 1988), and total wheat PPO can be an order of magnitude greater than flour PPO depending on the level of flour refinement (Hatcher and Kruger 1993). Reducing the impact of PPO on wheat foods has been primarily addressed by selection of low-PPO lines in breeding programs; hence most effort has gone into developing whole-kernel screening methods to detect and reduce total kernel PPO, and then by inference flour PPO. Whole-kernel assays have used a variety of phenolic substrates, including catechol, phenol, tyrosine, and L-DOPA (Bettge 2004). AACC-approved method 22–85 (Bettge 2004) uses L-DOPA and is based on the work of Anderson and Morris (2001). In that study L-DOPA and catechol were equally good PPO substrates. However, L-DOPA has the advantage of not being toxic to the seeds, allowing seeds tested in L-DOPA to be planted if required. Problems with PPO-related darkening are not specifically related to preharvest sprouting. However, increases in PPO activity (Kruger and Reed 1988) and the putative increase in soluble phenolic substrates (Edwards et al., 1989) during germination are an undesirable combination.

Test milling

The milling properties of wheat are fundamental to the great majority of wheat uses in the world. Commercial millers have considerable control over the characteristics of the flours they produce. This begins with their choice of wheat classes. In some cases millers are able to choose both market class and region of production in order to get specific attributes such as restricted protein range, or regional selection might enable millers to limit or control the number and identities of cultivars in the grain lots they purchase. Millers also control the wheat blends (grists) they apply to the mill. Millers can also choose the type of mill flow, as well as roll speed, speed differentials, disposition of fluted rolls (e.g., sharp to sharp), and roll pressures in smooth reduction rolls. Further choices are available in the combinations of sieves and purifiers, and finally in the mill stream selections used to blend the final flour. All these strategies help millers to consistently produce flour with the desired qualities. Small-scale test mills have nowhere near the flexibility, but that may be a blessing when setting up standard test milling protocols.

Posner and Hibbs (1997) make a distinction between small laboratory-scale mills with preset roll gaps and roll speeds, and simple sifting schemes, versus larger experimental mills. The larger mills allow operators to have some control over the mill settings and the mills have more complex sifting schemes in order to optimize, as far as possible, the mill settings for the type of wheat being milled. Interested readers are encouraged to read Posner and Hibbs (1997) for a comprehensive review on experimental milling as well as AACC approved method 26-10A (AACC 2000) for further specific guidelines. According to Posner and Hibbs (1997) test milling needs to depict accurately the ease of separation of the endosperm, germ, and bran, the level of admixture of bran in ground endosperm, and the particle-size distribution of the flour. They also suggest that test milling should depict the flow of the mill stocks that would give some indication of sieving ease, although the latter is difficult to ascertain on small test mills.

There are a few important milling parameters that one can observe using a test mill. For example, overall flour yield, break flour yield (the amount of endosperm material of flour particle size released from the first break roll), flour ash (AACC approved method 08-01, AACC 2000; see following section), flour color, and particle-size distribution are elements of milling performance that can be compared across cultivars or blends. Breeding programs arguably focus on the first three elements, flour yield, break yield, and ash. The emphasis changes depending on whether soft wheat or hard wheat is being milled. For soft wheat break flour yield is emphasized. For hard wheat total flour yield is the first parameter taken into account. Flour ash is important in both instances. Candidate cultivars from several programs entered in the Pacific Northwest Wheat Quality Council are milled on a Miag pilot scale mill, where cumulative ash curves from 15 mill streams are generated. This type of analysis is a better indicator of overall milling quality, showing the ash contents of each mill stream. The sharpness of the corner of the plot where the flour streams transition to the shorts, "red-dog", and bran streams are an indicator of the relative ease of separation of bran and endosperm (Fig. 20.2). Sharper corners are more desired because they indicate retention of low ash levels as far as possible before the ash levels increase sharply at the offal streams.

Grain and flour ash

Ash is the mineral or inorganic residue left after burning wheat or flour at 500–600°C for as long as needed to completely oxidize all the combusti-

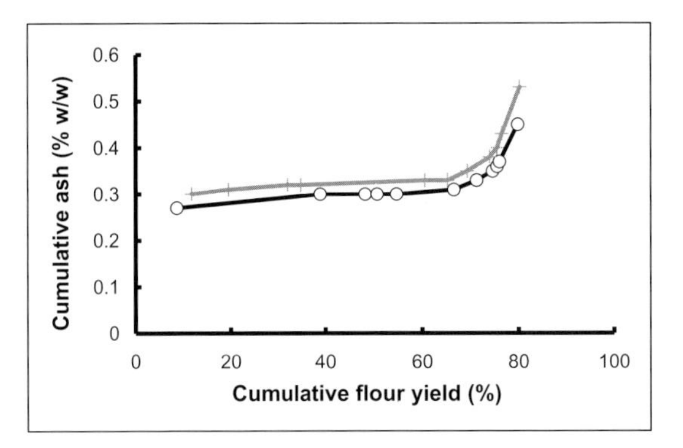

Fig. 20.2 Cumulative ash curves for two soft wheat genotypes, the cultivar Stephens (O) and the Oregon experimental line OR2050910 (+). Flour milled on a Miag Multomat flour mill at the USDA ARS Western Wheat Quality Laboratory, Pullman, Washington.

ble or organic material (AACC approved method 08-01, AACC 2000). Flour ash can also be predicted using NIRS (AACC approved method 08-21, AACC 2000), but Osborne (2007) cautions that NIRS calibrations are sensitive to matrix effects and may require recalibration at each grist change. As the bran layers of wheat are much higher in mineral content than the endosperm, the flour ash content is often considered the most reliable indicator of bran admixture and relative flour purity in flour milled from a common wheat sample. That view is challenged by other studies. One study (Peterson and Fulcher 2002) showed a nonsignificant correlation ($r = 0.20$) between flour ash (0.47%–0.58%, 14% moisture basis, mb) and flour bran measured by digital image analysis of fluorescent bran particles. Conversely Symons and Dexter (1992) showed highly significant relationships between pericarp fluorescence and flour ash, flour color grade, and dry flour brightness (CIE L*). Nonetheless they did caution that the relationships between flour color and ash established in one wheat class did not necessarily translate to other wheat classes.

The base level of whole-grain ash varies between environments (Dick and Matsuo 1988) and has been shown to be correlated with straight-grade flour ash. For example 2006 harvest data from the Pacific Northwest Wheat Quality Council showed a correlation ($r = 0.81$, $n = 19$, $p < 0.01$) between whole-grain ash and straight-grade flour ash, which is similar to the correlations between wheat and flour ash cited in Peterson and Fulcher (2002). Flour ash remains a valuable tool for monitoring flour refinement but it has evident limitations.

Flour color

Flour color is important, not only as an alternative way of monitoring flour refinement but also as a predictor of color in products, for example, skin hue of steamed breads, hue and discoloration of noodles, and crumb color in bread. Flour color is commonly measured using tristimulus color meters, and the CIE (Commission Internationale De l'Eclairage) L*a*b* color space is the most common of the available color spaces used in flour assessment. Other instruments, such as Agtron (green mode $\lambda = 546$ nm; AACC approved method 14-30, AACC 2000) and Kent–Jones flour color grader ($\lambda = 540$ nm), are also used. The 540-nm wavelength was apparently chosen for measuring color grade because it minimized the effect of yellowness and was also close to the maximum sensitivity of the human eye ($\lambda = 555$ nm) (Oliver et al., 1992).

Direct measurements of bran specks or total bran area also can be used as indicators of flour refinement and the potential for product discoloration. However, bran measurements, color grade, and Agtron do not indicate the chromaticity of the flour related to flour pigments. The latter can be measured directly (e.g., AACC approved method 14-50, AACC 2000) or viewed using the a* and b* axes in the CIE L*a*b* color space. Overall the ability of the tristimulus meters to define color within a three-dimensional coordinate system appears to be an advantage. Oliver et al. (1992) advocated a flour color index utilizing both the L* (brightness) and b* (yellowness) color coordinates. The Pekar slick test (AACC approved method 14-10, AACC 2000) is convenient as a quick method to compare flour color.

Measurements of flour color grade, Agtron value, and flour L* can have significant correlations with flour ash but this is not always so. Unconvincing associations were observed between flour ash and brightness (L*) of dry flour (maximum 25% of the variation in flour L* explained by flour ash) and between ash and L* of a flour slurry (maximum 16% of the variation in slurry L* explained by flour ash) (Zhang et al., 2005).

Speckiness

Excessive numbers of visible bran specks in flour and semolina can lead to quality deficits in noodles and pasta. Therefore flour (or semolina) speckiness is another element that can be important for wheat to pass the test on end-use quality. Specks can be assessed visually, or by machine vision in flour or semolina, or in dough or the finished product. Machine vision need not always be associated with complex image analysis systems. Hatcher et al. (2004) described moving from a CCD camera-based system to a system based on a flat-bed scanner for assessing speckiness of noodle sheets. The scanner system was able to count specks as well as monitor color shifts occurring as noodle sheets aged and darkened. A CCD-based laboratory-scale instrument was reported by Evers (1998). The speck value from this instrument has been used as part of a milling index (e. g., Osborne et al., 2007). In-line industrial scale versions are also available for monitoring flour streams in commercial mills. For particular applications, exploiting the fluorescence of some bran components may also be of value as mentioned previously.

STARCH AND FLOUR PROPERTIES

Total starch content

Starch can be analyzed simply as a component although it is more common to require analysis of its properties during thermal processing (see "Starch and flour paste viscosity and swelling power" below). Although starch content of the endosperm is negatively correlated with the abundance of the other major endosperm component protein, direct measurement of starch can be of value. Measurement of total starch essentially involves its enzymatic digestion using starch-specific enzymes (specific to glucose molecules linked α,1→4 or α,1→6) with subsequent measurement of the glucose using a linked glucose oxidase–peroxidase–chromogen assay (AACC approved methods 76-11 and 76-13, AACC 2000).

Starch damage

The water absorption capacity of damaged starch is a major factor in the overall absorption capacity of flour. Starch damage can be controlled by millers and therefore may require measurement. A number of methods are available. Enzymatic methods utilize the differential ability of damaged starch granules to swell and become susceptible to enzymatic hydrolysis at temperatures below those required for gelatinization (Farrand 1964; AACC approved methods 76-30A, 76-31, AACC 2000). An available amperometric method is based on the greater absorption of iodine by damaged starch and is effectively an extension of earlier dye-binding methods (Morgan and Williams 1995; Boyaci et al., 2004). Attempts have been made to increase the speed of existing tests. For example, Boyaci et al. (2004) used a refractometer to measure degrees-Brix of the solution resulting from an enzymatic digestion rather than using a reducing sugar determination. Successful NIRS calibrations have also been reported for starch damage. Miralbés (2004) reported an r^2 value of 0.94 and standard error of prediction of 1.63 on a validation sample set when calibrated against the amperometric SDMatic procedure (Chopin, Triplette et Renauld, Paris, France). Good agreement among the various starch-damage tests has been reported (Gibson et al., 1993; Morgan and Williams 1995; Lin and Czuchajowska 1996).

Starch and flour paste viscosity and swelling power

The viscoamylograph and its variants, and the Rapid Visco-Analyzer series of instruments, are the instruments most commonly used for paste viscosity analysis of starch and flour in wheat technology applications in the current era. A number of swelling power tests is available. The one in most common use in wheat applications is the flour swelling volume (FSV) test (Crosbie et al., 1992; AACC approved method 56-21, AACC 2000).

The principles of operation for both of the paste viscosity instruments are similar. A starch or flour slurry is cooked under controlled conditions, and either the resistance to the movement of a rotating sensor (RVA) or the pressure against a stationary sensor from a rotating sample and container (VAG) is detected mechanically (VAG only) or electronically (RVA, VAG-E, Micro-VAG). Outputs from both instruments are plots of apparent viscosity vs. time, and there are numerous variants of solids concentrations, rotational speeds, and heating and cooling rates that can be applied for specific purposes. Applications of the RVA (Crosbie and Ross 2007) and VAG (Rasper 1980) have been reviewed elsewhere.

The swelling power and FSV tests gelatinize starch or flour in excess water in a closed tube. The starch swells under minimal shear in comparison with the paste viscosity measurements, since only mild inversion is applied after the starch-flour suspension has swollen enough to prevent sedimentation. Volume, or height (as a proxy for volume), is recorded after centrifugation. Weight can also be recorded and the proportional increase in gel weight with respect to the dry sample weight has been reported as water-holding capacity (Wang and Seib 1996).

Operating under standard conditions (for RVA, AACC approved method 76-21, AACC 2000; for VAG, Rasper 1980), the pasting curves from the RVA and VAG have several common elements that are useful in determining the processing potential of a flour or starch sample (Fig. 20.3). In principle the curve shapes are similar. However, there are inherent differences between instruments and occasional apparent disagreements in results. These differences are often considered to be related to differences in heating rates. However, even when the RVA temperature profile was matched to the VAG heating and cooling rate of 1.5 °C min^{-1} and to the same hold times at 50 and 95 °C, differences in pasting curves were still evident and appear to be related to mixing geometry, apparent shear rates, and differences in solids concentrations (Deffenbaugh and Walker 1989). The apparent shear rate in the RVA has been calculated as 54 s^{-1} at 160 rpm (Booth and Bason 2007). The slower rotational speed (75 rpm) and the dissimilar mixing geometry of the VAG intuitively suggest lower shear rates (Deffenbaugh and Walker 1989), but the mixing geometry of the VAG is considered too complex to calculate shear rates in the complex flow field (Lagarrigue and Alvarez 2001). These types of differences make comparisons among the RVA, VAG, and any

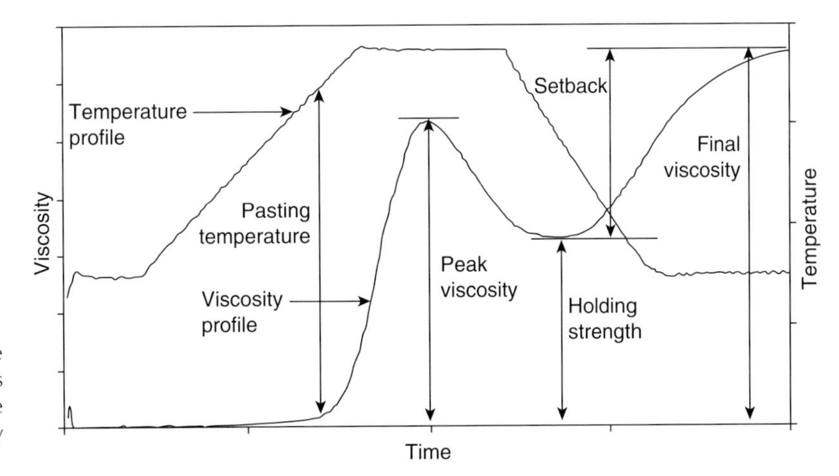

Fig. 20.3 Typical RVA curve showing the basic descriptors used to characterize the curve attributes. [*Source*: Batey (2007).]

other types of empirical rheometers quite difficult, and it is also no easy task to derive fundamental rheological data.

That said there are many disadvantages to standard rheometers for characterizing starch or starchy flour. Starch granules are dense (~1.5 g cm^{-3}) and thus sink like stones if not kept suspended in a turbulent flow. Hence clumping and settling out of granules before gelatinization can be problematic in concentric cylinder geometries. However, strategies such as gelatinizing part of the sample can be used to prevent sedimentation of granules (see Lagarrigue and Alvarez 2001).

Basic descriptors from a RVA pasting curve (Batey 2007; Batey and Bason 2007) are shown in Fig. 20.3. These are mostly consistent with descriptors for VAG curves described by Rasper (1980) and are largely valid for both flour and starch analyses. The first parameter that can be observed is pasting temperature, or the first detectable increase in apparent viscosity. This is not the gelatinization temperature. Many already irreversible structural changes have occurred in starch granules before swelling has progressed enough to be detected as a viscosity increase (Batey 2007). Further heating leads to a peak in the viscosity-vs.-time curve that is related in both height and timing to competing phenomena: granule swelling and the breakdown of the swollen granules under shear. Continued mixing of the paste at 95°C leads to a decrease in viscosity, which is commonly more evident under the shear conditions in the RVA. On cooling, the paste again increases in viscosity. The extent of this increase in viscosity is related often to starch amylose content. Flours containing starch higher in amylose have higher cooled paste viscosities as a result of the ability of amylose molecules to reassociate on cooling (Batey 2007).

Much valuable information has been created using paste viscosity analysis. The characteristic early pasting and high paste viscosity of reduced-amylose wheat were first detected with the VAG (Moss 1980). These characteristics are now known to be also associated with a relatively large decrease in paste viscosity with continued stirring. These diagnostic indicators from paste viscosity curves were subsequently exploited using both the RVA and VAG to identify wheat cultivars suitable for soft-bite noodles. High peak paste viscosity is associated with high FSV, and high FSV is also used to identify reduced-amylose wheats suitable for soft-bite noodles. The converse pasting characteristics—low peak viscosity in the absence of amylase and relatively low breakdown—have been identified as potentially beneficial for hard-bite noodles (Moss 1980). Crosbie et al. (2002) cautioned that the association between high RVA peak viscosity and cooked noodle softness could be masked if a method to inactivate the low levels of α-amylase present even in sound wheat was not applied, such as the use of a dilute solution of silver nitrate. In contrast the standard FSV test is relatively insensitive to amylase activity (Crosbie and Lambe 1993) and needs no such intervention. Higher paste peak viscosities have also been associated with increased cooking losses in durum pasta (Sissons and Batey 2003).

Given the multiple parameters that are available in a paste viscosity analysis of flour or starch, the rotating viscometers have an advantage over swelling power tests with regard to information value. However, the RVA, and in particular, the VAG, are limited by low throughput potential, even with the application of a rapid RVA profile (Crosbie et al., 2002). The FSV test may remain the method of choice in assessing breeding material for the presence of reduced-amylose phenotypes. However, for early-generation screening, NIRS calibrations for FSV (Crosbie et al., 2007) have been reported as well as an antibody test specific only for the *Wx-B1b* null allele ("Null-4A" gene) that is common in Australian wheat (Gale et al., 2004). This particular antibody test is blind to the *Wx-A1b* or *Wx-D1b* null alleles, which may limit its widespread use. There are reports of PCR-based assays that can identify all of the main *Wx* null alleles (*Wx-A1b*, *Wx-B1b*, *and Wx-D1b*) that could be used in breeding programs (e.g., Nakamura et al., 2002). However, this type of genetic screening does not indicate the complete response of the starch to thermal processing, which is also affected by variation in amylopectin structure (Konik-Rose et al., 2007) among other phenomena. Information on the amount of residual ungelatinized starch in finished products, or the degree-of-cook of extruded products and their process intermediates, can also be analyzed using paste viscosity testing (Whalen 2007).

SOLVENT RETENTION CAPACITY

As noted earlier, absorption capacity is a key indicator of flour utility. Water absorption can be predicted using the farinograph to measure the combined absorption of intrinsic flour components and gluten formation (see below) or by observing the ability of flour or wholemeal to absorb water in the presence of a large excess. This type of analysis can be further refined by adding selected solutes to the water to partition the absorption capacities of different flour components. The original test that measured absorption in an excess of solvent was the alkaline water retention capacity (AWRC) test. This test used a weak alkaline solution of 0.84% (w/v) sodium bicarbonate (AACC approved method 56-10, AACC 2000). High AWRC absorption is associated with poor cookie spread. The concept of this test was modified to the sugar water retention capacity test (Saunders et al., 1989 as cited in Slade and Levine 1994), which in turn was further refined as the solvent retention capacity (SRC) method (AACC approved method 56-11, AACC 2000).

In standard guise the SRC test uses four solvents for the flour: water (W-SRC) and three aqueous solutions, 50% w/w sucrose (Su-SRC), 2% w/w sodium carbonate (SC-SRC), and 5% w/w lactic acid (LA-SRC). The SRC is determined by measuring the weight increase of the flour pellet after suspension of the flour or meal in the solvent of interest and subsequent controlled centrifugation. The SRC is reported as a function of the original sample weight at 14% mb. This test was designed originally for evaluating soft wheat and has been adapted to smaller sample size and the use of wholemeal to make it more applicable in breeding programs (Bettge et al., 2002). Investigations of the utility of SRC have been extended to hard wheat, where it could show some promise (Xiao et al., 2006) such as prediction of water absorption capabilities in the absence of gluten development.

The principle of SRC testing is that solvent uptake is primarily associated with specific flour components: SC-SRC is an indicator of starch damage and therefore, indirectly, kernel texture; Su-SRC is an indicator of pentosan and gliadin content; and W-SRC is a general indicator of absorption capacity summed over all substances in the flour or meal (Bettge et al., 2002). The LA-SRC uses absorption characteristics of glutenins to predict dough or gluten strength and is correlated with sodium dodecyl sulfate–sedimentation volume (SDS-SV) in both hard (Xiao et al., 2006) and soft wheat (Guttieri et al., 2004; Gaines et al., 2006). Additionally in hard wheat LA-SRC has been shown to have better predictive value than SDS-SV over a narrow (1%) flour protein range than SDS-SV (Xiao et al., 2006), although confirmation of this result is required before it can be applied widely and with confidence. The correlation between LA-SRC and SDS-SV has led some researchers to adopt the practice of reporting either SDS-SV or LA-SRC but not both, as they tend to be redundant.

The absorptions determined by the four solvents are highly correlated with each other, and this is more evident for W-SRC, SC-SRC, and Su-SRC (unpublished data). Clearly there is overlap in what components are absorbing in each solvent. The originators of the technique did not suggest strict partitioning of the absorption to specific components, rather that the solvents emphasize absorption of the component of interest (e.g., starch damage for SC-SRC). An example is the anomalously high Su-SRC value of flour from the soft wheat cultivar Daws compared with its overall water absorption capacity (Bettge et al., 2002). This was considered to be a result of the known higher pentosan content of Daws. Interestingly wholemeal from this cultivar also had an anomalously high SC-SRC, indicating the possibility that the high pH of the sodium carbonate solution also partitions some water-unextractable pentosans into the aqueous phase. High pH is employed routinely to extract non-water-soluble pentosan or arabinoxylan materials from various plant sources.

Solvent retention capacity characteristics have been most often associated with cookie baking attributes of soft wheat, but individual SRC tests are not strongly predictive in all circumstances. In one study involving single cultivars across environments, individual SRC tests accounted for no more than 59% of the variation in sugar-snap cookie diameter (Guttieri et al., 2002). In the same study the narrow range of kernel texture values precluded a significant correlation between SC-SRC and SKCS hardness index. Since the SRC tests were designed as a group they might best be considered as a group in order to provide the most helpful profile of a flour sample. Guttieri et al. (2001) used multiple regression to account for 78% of the variation in cookie diameter using only Su-SRC and flour protein content. Using the same sample set these authors used cluster analysis to show that larger-diameter cookies could be made from flours with both low and high W-SRC, but larger diameter required low values for both SC-SRC and Su-SRC. Also moderately large cookie diameters were achieved by flours with low to moderately low W-SRC, SC-SRC, and Su-SRC but with relatively high LA-SRC. At the other end of the scale the cluster with the lowest cookie diameter was unambiguously differentiated from the others by high values for all four SRCs. Ram et al. (2005) used multiple regression analysis to account for 83% of the variation in farinograph water absorption using W-SRC, Su-SRC, SC-SRC, and grain protein content. The group of 192 genotypes used in that study varied in W-SRC from 53% to 71%, suggesting a mix of hard and soft kernel texture.

Additional solvents or variations on the theme of the SRC tests have already been suggested. These include a suggestion by Bettge and Morris (2001) that a sequential SRC test might enhance differentiation among cultivars. They suggested water followed by ethanol followed by lactic acid, sequentially exposing flour to each solvent in turn, without drying the pellet between SRC determinations. More developments are expected with the investigation of further additional solutions including those containing reducing agents and SDS (Miklus et al., 2004).

DOUGH TESTING AND PREDICTION OF DOUGH PROPERTIES

It is difficult to approach the subject of measuring dough properties without repeating what has been thoroughly reported and reviewed in the past 70 years. The older tests covered in this section have been the stalwarts of commerce and research in wheat for decades and show no signs of becoming any less utilized. This is despite some clear limitations of these older techniques and vigorous efforts to find better and faster ways of measuring or predicting dough characteristics.

SDS sedimentation volume

Determination of SDS sedimentation volume is a widely used method, particularly for coarse screening of dough strength potential in breeding programs. The test is based on the principle that gluten proteins swell in dilute lactic acid solution and that larger proteins—those related to higher dough strength—sediment more slowly and are associated with higher sedimentation volume. The test comes in a number of variants, and the newer variants are primarily adapted for stronger-gluten wheat or for microscale tests (AACC approved methods 56-60, 56-61A, and 56-63, AACC 2000). Actual SDS sedimentation volumes are highly dependent on flour protein content, so dividing the result by the flour protein content provides a corrected SDS sedimentation volume per unit protein, and a useful comparison of dough strength potential among samples. Indeed some breeders consider the raw SDS sedimentation volumes to be of little value, but use the protein-corrected volumes to identify lines with weaker dough attributes that, for example, may contain the 1RS·1BL translocation (B.F. Carver, pers. comm.).

Recording dough mixers

A recording dough mixer (RDM) is truly the workhorse of wheat science and technology and provides the basis of a variety of standard methods (e.g., AACC approved methods 54-21, 54-40A, and 54-50, AACC 2000). The importance of RDMs becomes more apparent when considering that subsequent testing of developed doughs requires a known and repeatable processing history. This is most easily achieved with a RDM. There are a number of RDMs in common use but importantly they share elements of data output and so the general features of RDMs can be addressed as a whole rather than mixer by mixer. This is despite the circumstance that the various RDMs possess different mixing geometries (e.g., z-arm versus planetary pin), different minutiae related to operation, differences in the precise nature of their outputs, and limitations that are unique to each type.

Using a mixograph (National Mfg. Co., Lincoln, Nebraska) trace as the archetype one can observe the common elements (Fig. 20.4). Data output can be analogue and read manually, but most RDMs have the capacity to give electronic outputs where the major descriptors of the mixing curves are available through machine analyses. An alternative method of assessing mixograms is by visually scoring the overall shape of the mixogram curve to provide a mixing tolerance score based on comparison with a set of reference mixogram curves (e.g., Baenziger et al., 2001).

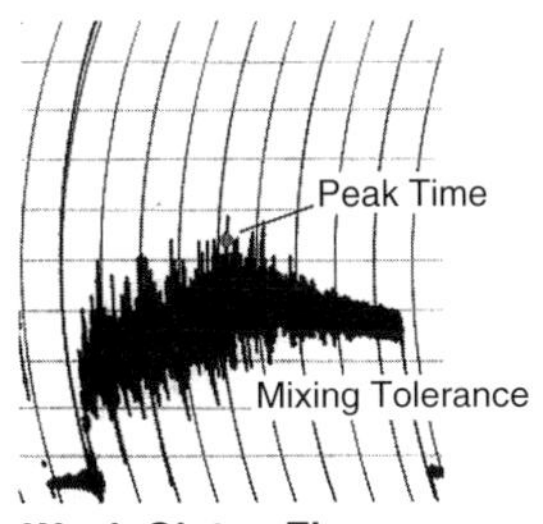

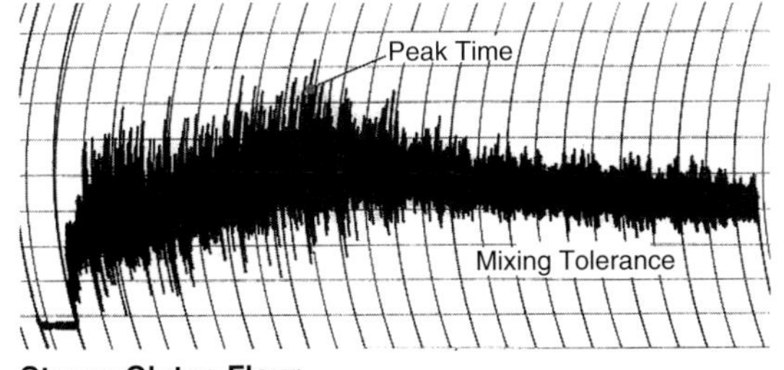

Fig. 20.4 Mixograms of typical weak (upper panel) and strong (lower panel) doughs. [*Source*: Wheat Marketing Center (2004).]

The RDM curve can be partitioned into two major portions: (i) a rising portion of the curve prior to the peak where the flour is hydrating and dough development is occurring; (ii) a descending portion where the developed dough breaks down to a greater or lesser extent as a result of continued mixing (overmixing). Differences in the time of arrival at peak development are indicative of differences in mixing times that in a broad sense are reflected in mix times in commercial production; elements of scale, geometry, and mixing intensity change from mixer to mixer and these aspects will dictate optimum mixing time in a specific mixer.

For a given mixer faster arrival at peak dough development can indicate faster hydration. Hydration speed can affect mix times in both z-arm and pin mixers. For example, Sapirstein et al. (2007) showed that reducing the particle size in a mix of durum semolina and flour (by inference, increasing surface area to volume ratio of the particles and therefore increasing hydration rate) shortened development time in both the farinograph and mixograph. Faster arrival at peak dough development in a general sense can also be related to weaker dough characteristics, as short mix times are often but not always associated with lower tolerances to overmixing. The mixers themselves vary in mixing intensity and a planetary pin-mixer like the mixograph has a higher rate of energy input (more intense mixing characteristics) than the available z-arm (e.g., farinograph) or blade mixers when these are run at standard speeds. For example, it is known that the amount of SDS-insoluble glutenin decreases as mixing continues to peak development (Don et al., 2003). Following from that Surel et al. (2006) compared three RDMs and showed that the SDS-insoluble fraction decreased to a minimum value of 15% of SDS-insoluble proteins extracted from flour after 4, 8, and 20 min in mixograph, farinograph, and alveograph (Chopin Technologies, Villeneuve-la-Garenne, France) mixers, respectively, indicating higher intensity in the mixograph pin-mixer. Availability of instruments such as the Brabender Docorder (Brabender Instruments, Inc., South Hackensack, New Jersey), the newer Brabender farinographs, and the Newport Scientific Doughlab (Newport Scientific Pty. Ltd., Warriewood, NSW, Australia), which can all be run at variable speeds, allows investigators to observe dough behaviors at variable mixing intensities within the same mixing geometry.

In the descending portion of mixing curves, rapid reduction in resistance to mixing (consistency), steeper angles of decline, and faster decreases in the bandwidth of the mixing curve are all indicative of lower tolerances to overmixing and generally weaker dough characteristics. The outputs from the RDMs are well able to rank material in breeding programs for mix times and indicators of mixing tolerances. The RDMs are also amenable in commercial applications to investigate the suitability of incoming grain or flour lots for a specific manufacturing facility through empirical relationships observed between RDM outputs and performance characteristics of specific mixers.

All of the RDMs have limitations and advantages. Notably the limitations of the mixograph are its inability to measure optimum water absorption, which needs to be calculated from flour protein content beforehand and adjusted by trial and error for the effects of other flour components. The z-arm mixers are able to measure water absorption of flour based on the amount of water needed to achieve a predetermined but arbitrary level of dough consistency. The commercially available z-arm mixers also have the advantage of jacketed bowls that allow temperature control, which is becoming more sophisticated as newer RDMs and newer models of existing RDMs are released. The mixograph has an advantage of somewhat shorter analysis times as a result of its higher mixing intensity. The RDMs are produced in various size ranges based on the amount of flour used: mixograph, 10 and 35 g of flour (a 2-g direct-drive prototype was also produced); farinograph, 10, 50, and 300 g; DoughLab, 50 and 300 g; Micro-DoughLab (prototype Z-arm mixer) (Newport Scientific Pty. Ltd., Warriewood, NSW, Australia) 4 g; Alveoconsistograph (Chopin Technologies, Villeneuve-la-Garenne, France), 250 g. Smaller mixer capacity and higher mixing intensity (shorter overall analysis times) are particularly applicable in breeding programs.

Recently attempts have been made to apply more rigorous mathematical analyses of RDM outputs, particularly to mixograms. Gras et al. (2000) analyzed mixograph outputs as a series of microextensions (also described as elongate-rupture-relax oscillations) and suggested that variation in curve bandwidth was a potentially useful parameter in assessing the extensional dough strength. Verbyla et al. (2007) used a Fourier transform analysis to determine the main cyclic patterns in high-resolution mixograph data. Patterns correlated with the movement of the mixing pins relative to fixed pins and three main frequencies were detected. Using only the major frequency the technique provided estimates of peak height and time-to-peak that correlated with standard computer-assisted estimates of mixing curve parameters. The Fourier analysis for the major frequency provided a repeatable estimate of bandwidth that was considered to give greater resolution among samples than computer-assisted bandwidth estimates.

Technology has further evolved to monitor dough development in commercial-scale mixers. Examples include NIRS, the use of mechanical probes to directly measure the physical state of the dough, and methods to measure mixer torque (Wilson et al., 1997; Wesley et al., 1998; Dempster et al., 2007). The output from a typical spiral mixer observed using the BRI-Australia Easymix system (BRI Research, North Ryde, NSW, Australia) to measure torque conformed to the basic template for mixing curves from the smaller RDMs (Rasper and Walker 2000). The use of NIRS to estimate dough mixing properties is discussed further below.

Measuring extensional properties of developed doughs

In commercial practice once a dough is developed in the mixer it is often subjected to a number of additional unit operations that extend the dough in length. In baking applications the most important is the biaxial extension encountered during fermentation, proofing, and oven rise. Noodle dough is not developed during mixing, but rather, dough development occurs during the uniaxial extension imposed during sheeting operations (Ross and Ohm 2006). Hence the outputs of RDMs do not necessarily describe dough attributes important to these extensional phases of processing. To do this dough attributes need to be measured while the dough is being stretched or otherwise extended and often after a specified period of rest that allows internal stresses in the dough to dissipate (relax). Dobraszczyk et al. (2000) reviewed the fundamentals of a number of extensional and dynamic instruments that are used to do this.

Uniaxial extension

The type of dough deformation most familiar to wheat technologists is the extension or stretching applied during uniaxial extension tests. Two basic scales of testing are applied: large (150 g flour) and small (10 g or less). The large-scale tests are done using the Brabender extensograph (Brabender Instruments, Inc., South Hackensack, New Jersey), and small-scale extension tests can be done using the Kieffer dough extensibility rig for the Stablemicrosystems TA texture meter series (Stable Micro Systems Ltd., Godalming, Surrey, UK). A custom-made prototype microextension tester was also reported by Australian CSIRO wheat researchers (e.g., Anderssen et al., 2004). The Simon research extensometer was also mentioned in other reviews (e.g., Rasper and Walker 2000). However, no studies using this instrument can be found in a search of the Food Science and Technology Abstracts between 1990 and November 2007.

At the most basic level the outputs of all the uniaxial extension tests can provide measurements of the maximum force required to extend the dough (maximum resistance, MXR, and often used as a descriptor of dough strength) and the distance the dough can be extended before rupture (extensibility, EXT) (Fig. 20.5). Further information can be gathered by measuring the work-to-rupture via the area under the force vs. time curve or other manipulation of the raw data, for example, the MXR:EXT ratio. Anderssen et al. (2004) advocated the use of EXT at MXR, and EXT between MXR and rupture as more useful measures of dough extensional rheology, as it relates to glutenin composition of wheat cultivars. This work was done with the CSIRO prototype microextension tester, but the conclusions may also be valid for the Brabender extensograph where MXR occurs at a variable point during the extension of the dough. However, for the Kieffer dough rig, MXR occurs very close to EXT at rupture and this type of analysis may lack utility for that variant of uniaxial dough extension. Mann et al. (2005) showed a figure specifically comparing Brabender extensograph output with output from a Kieffer extension rig that highlights this difference.

There has been interest in using uniaxial extension test results as the basis for examining fundamental dough rheology parameters. The complex geometry, reduction in cross-sectional area of the dough piece, and constant velocity extension (therefore, steadily decreasing strain rate) mean that the extensograph does not perform a genuine tensile strength test. However, MacRitchie and Lafiandra (1997) contend that elements of the molecular level phenomena can be elucidated using this type of uniaxial constant velocity testing. Nevertheless the utility of these empirical tests has been proven with regard to their abilities to categorize dough samples as, for example, strong and inextensible or weak and extensible in ways that are meaningful to plant breeders and commercial users of wheat flour.

Biaxial extension

It can be argued that the biaxial extension applied by the increasing pressure in gas cells during fermentation and initial heating in the oven is

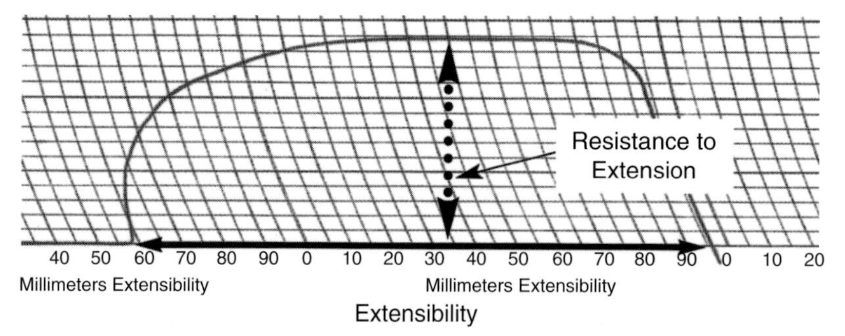

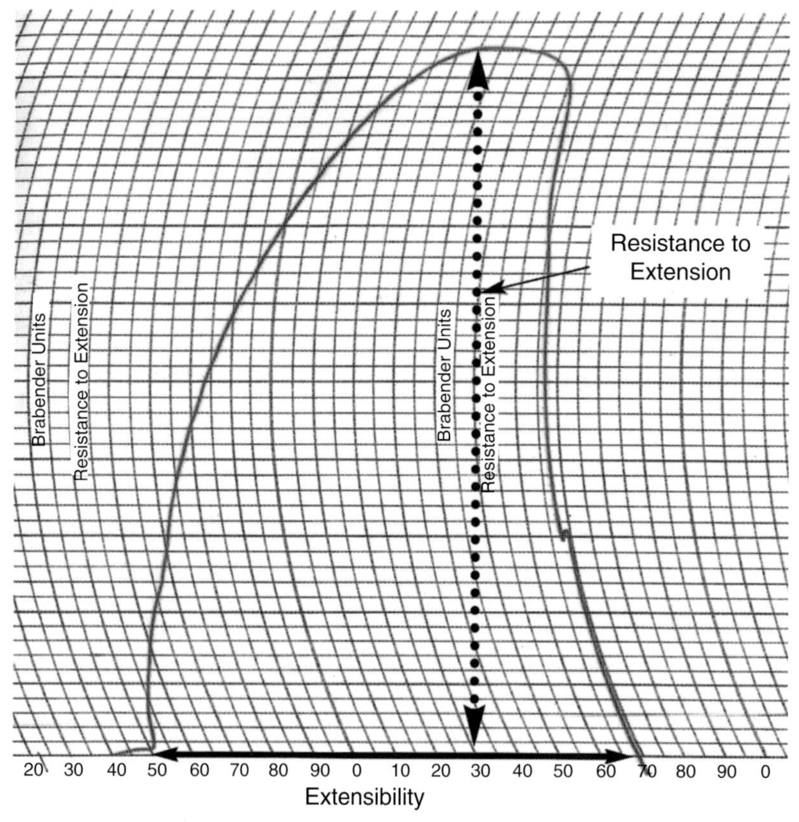

Fig. 20.5 Extensograms of typical weak (upper panel) and strong (lower panel) doughs. [*Source*: Wheat Marketing Center (2004).]

the most important phenomenon occurring in a dough after it has developed in mixing. Extensional properties are also important in doughs extended uniaxially such as extruded pasta dough (compressed biaxially) or sheeted dough (compressed uniaxially but restrained in one coordinate axis; Ross and Ohm 2006).

The most common instruments that apply biaxial extension by expanding a bubble of dough are the alveograph and the Dobraszczyk Roberts

(D/R) dough inflation system (Stable Micro Systems Ltd., Godalming, Surrey, UK). Alternatively biaxial extension can also be generated during uniaxial compression [e.g., lubricated squeezing flow (LSF); Bagley and Christianson 1986; Baltsavias et al., 1999; Ross and Ohm 2006; Liao et al., 2007].

The alveograph (AACC approved method 56-30A, AACC 2000) is primarily an empirical instrument and was originally designed for dough made from soft wheat flour. A dough sample is prepared under specified conditions, which originally dictated a fixed water addition of 50% (fb) designed for dough from soft wheat flours of low protein content. This has been shown to be a problem for assessing dough from hard wheat flour. Increased starch damage of hard wheat, and therefore absorption capacity of the flour, produced a stiffer dough that required more work to extend and was less extensible at the fixed water addition (Edwards and Dexter 1987). The newer alveoconsistograph overcomes this limitation by allowing users to adjust water to reach a predetermined consistency prior to further testing.

The alveograph records the bubble air pressure (overpressure) as a function of time until rupture of the bubble. Values of maximum overpressure (P or tenacity), time or distance or bubble volume at rupture (L or abscissa; Faridi and Rasper 1987), the total area under the curve (W or work to rupture), and the P/L (configuration) ratio are useful as indices of dough properties of relevance to end-product manufacture. For example, in a breadmaking application, Edwards et al. (2007b) showed alveograph L to be well correlated with loaf volume in bread made from durum wheat and similar in predictive capacity to work to extend the dough in a small-scale extension test. These researchers were also able to distinguish high-molecular-weight glutenin subgroups within their sample set based on alveograph P and W. Notably all the γ-gliadin 42 types had the expected weak dough characteristics as indicated by low P and W values. Maghirang et al. (2006) showed a general trend for a group of 100 US hard red spring wheats to have higher W than a comparable group of 100 US hard red winter wheats. Slade and Levine in a communication to the 2007 US Pacific Northwest Wheat Quality Council suggested that W at a fixed bubble volume was useful for comparisons of dough strength (e.g., higher W at constant bubble volume equals stronger dough in general).

The D/R dough inflation system is an attachment for the TA series of texture analyzers. The instrument crosshead displaces a piston that inflates the dough bubble. The texture-analyzer-based system is able to monitor bubble height and bubble pressure and to control the rate of bubble inflation. Later models of the TA series can also inflate the bubble at constant strain rates (exponentially increasing velocities in extension), which is valuable in experiments aimed at understanding fundamental dough rheology. In its simplest form, the D/R system can output a series of parameters that is similar to the alveograph outputs (P, L, W, and P/L). However, flexibility of the D/R system has been exploited to allow investigation of the fundamental properties of the inflating bubble. This has led to confirmation of the strain hardening characteristics of wheat flour dough and the association of greater strain hardening with increased bubble stability and larger loaf volume (Dobraszczyk et al., 2005). Biaxial extension as related to strain hardening was reviewed recently (van Vliet 2008).

Simpler in geometry and operation is the lubricated squeezing flow (LSF) technique (Fig. 20.6). This involves the squeezing of a laterally unconstrained dough sample between two parallel lubricated plates. The LSF technique has been applied to many types of dough, including bread dough (e.g., Bagley and Christianson 1986; Kokelaar et al., 1996; Sliwinski et al., 2004a; Rouillé et al., 2005; Stojceska et al., 2007), gluten "dough" (Kokelaar et al., 1996; Sliwinski et al., 2004b), fat-shortened dough (Baltsavias et al., 1999), and biscuit (cookie) dough (Sai-Manohar and Haridas-Rao 1999; Lee and Inglett 2006). There are also reports of LSF applied to noodle dough (Ross and Ohm 2006; Liao et al., 2007).

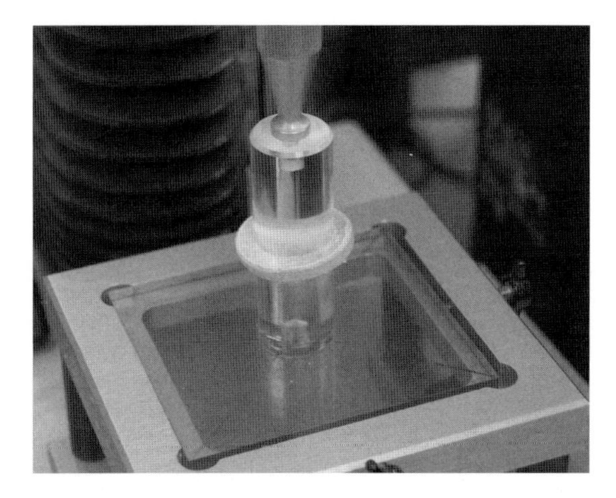

Fig. 20.6 Noodle dough undergoing lubricated squeezing flow rheometry.

A number of parameters can be measured or derived from the primary force-vs.-time curves, including the apparent biaxial extensional viscosity. Ross and Ohm (2006) held the probe at maximum strain for 30 s in order to observe the dissipation of forces within noodle dough, recording relaxation time (time for stress to decay to 1/e of maximum) and the residual force at 25 s after stopping the probe. These parameters were well able to differentiate alkaline and salted formulations from the same flour and differentiate strong from weak doughs. Liao et al. (2007) confirmed the utility of LSF for noodle dough, which by nature is not amenable to direct investigation with conventional cereal science rheology techniques. These applications of LSF to dough rheology highlight its future potential.

Other dough rheology tests

Many other tests have been applied or suggested as ways of monitoring dough rheology and/or subsequent end-product attributes such as loaf volume. Dynamic rheology is frequently used and reported in the literature. However, as the deformations are much smaller than those encountered in commercial processing, the utility of dynamic rheometry seems to be confined primarily to examination of fundamental structure–function relationships in dough. A lucid and recent discussion of the roles of the conventional empirical rheology tests and the fundamental, more universal rheology tests applied to investigate network structures in materials can be found in Mulvaney (2005a,b).

The use of ultrasound has also been investigated as a way of monitoring a number of aspects of dough behavior and the technique may be applicable for on-line or in-process doughs (Álava et al., 2007). Elmehdi et al. (2004) explained the principle: "As a low intensity ultrasonic pulse propagates through a dough sample, the characteristics of the pulse (velocity and attenuation coefficient) are affected by the properties of the dough." Ultrasonic velocity is affected by the stiffness or rigidity of the dough, and can be related to properties such as the dough's elastic or Young's modulus and potentially can be used to predict the quality of the derived products. Ultrasonics may also be able to monitor aspects of dough mixing by monitoring changes in the attenuation coefficient that are related to changes in dough density, which is known to change systematically during mixing (Elmehdi et al., 2004).

END-PRODUCT TESTING

End-product testing could indeed be a chapter or even a book in its own right. As a consequence we will discuss only some general principles about end-product testing that are broadly applicable. Even after the best predictions of end-use suitability of wheat from component or functionality tests, the wheat needs to be manufactured into a representative end product of the category being examined. Nonetheless laboratory-scale end-product manufacture is, like its dough mixing counterparts, only an approximation of the processes and formulations used in commercial-scale production. However, it still functions as the only fully integrated test that breeders, researchers, grain traders, or bakers or noodle makers can apply to measure the combined effects of all the functional components. To understand how wheat "passes the test" when applying laboratory-scale product tests and their attendant quality assessments, one only needs to scan through the desired characteristics of the wheat-based foods described at the beginning of the chapter.

There are myriad ways of assessing the characteristics of the finished products. For example, these include simple measures of loaf volume via displacement to complex image analyses of bread crumb structure. However, all end-product assessment techniques may be placed into two categories with regard to the experimental approach: sensory or instrumental. From another viewpoint there are also two further categories with regard to sensory assessments: intensity ratings and hedonic assessments such as consumer preference. Intensity ratings can be done either by instrument or with trained sensory panels (e. g., is it harder or softer, larger or smaller, or darker or lighter?). Hedonic assessments can be done only by humans and are the ultimate test. However, these are not often directly applicable during breeding or trading of wheat. However, wheat technologists through experience or direct comparisons can associate certain attribute intensities with consumer preferences in target markets (e.g., a liking for soft crumb texture in sponge-and-dough sandwich bread in North America).

Laboratory-scale production also necessitates some rationalization since no laboratory working with large sample numbers, or with only small amounts of flour (e.g., breeding programs), can cater to all the possible permutations of process and formulation that can be used even for a single product. For example, subdividing Asian noodle sample sets into only the simplest of categories with respect to formulation (alkaline and salted) doubles the required testing, not to mention the variations in the composition of the alkaline compounds used in commercial practice, or other formulation or process variants. Thus, reasoned approaches need to be applied to get the best information possible from the least amount of testing. For example, sugar-snap cookie diameter, straight-dough loaf volume, and cooked noodle or pasta hardness values are used respectively as indices of the overall quality of soft wheat, of the overall suitability of wheat for breadmaking, and of the eating qualities of noodles and pasta. One can refine the testing if there are known (and specific) end-use requirements. For example, wire-cut formulations should be used rather than sugar-snap formulations if the objective is to know the effect of flour composition on cookie tenderness.

Standard test methods and assessment procedures are available for many products and these may include the following: straight-dough and sponge-and-dough test baking of pan bread (AACC-approved methods 10-09 and 10-11, AACC 2000); baking quality of cakes (AACC-approved methods 10-15 and 10-90, AACC 2000); baking quality of cookie flour (AACC-approved methods 10-31B through 10-54, AACC 2000); and laboratory-scale pasta processing and assessment of firmness (AACC-approved methods 66-41, 66-42, and 66-50, AACC 2000). If specific approved standard methods are not available, then published methods or guidelines can be found, such as guidelines for laboratory-scale Asian noodle processing (Ross and Hatcher 2005) and texture assessment (Ross 2006). Guidelines for steamed bread processing and assessment can be found in the literature (Huang et al., 1996) or in the "Guidelines for Testing a Variety of Products" (AACC approved method 10-13A, supplement to AACC 2000). The latter also includes guidelines for frozen pizza dough, soft pretzels, flour tortillas, bagels, hamburger buns, French hearth breads, pita bread, cakes, and yeast-raised doughnuts. Furthermore specialists within product sectors are constantly refining the end-product test methods. An example of this is the work being done by the AACC-International Soft Wheat Technical Committee to update approved method 10–90 for cake baking to streamline the method so that only one bake at a fixed hydration level, rather than three bakes, would be required. The proposed modifications are also aimed at making the method more useful for estimating the ability of the flour to make a cake rather than observing the response of the flour to variations in formulation.

EMERGING OPPORTUNITIES

Spectroscopy

Of all the rapid tests applied to wheat grain from breeder to trader, one that seems most amenable to further exploitation is near-infrared spectroscopy. Some of the current and potential uses of NIRS have been discussed in earlier sections dealing with specific components or functional properties. A relatively recent review by Osborne (2006) on the use of NIRS in cereal breeding programs detailed the virtues of whole-grain NIRS testing, its speed, its nondestructive nature, and its ability to concurrently estimate multiple analytes. The review cites studies that deal specifically with the problem of small sample size in the early generations of inbreeding, speed and the use of newer and faster diode-array instruments, and other decisive factors related to the choice of NIRS instrumentation.

Apart from the basic and necessary analyses of protein, moisture, and hardness, NIRS calibrations have been developed for a number of other components. High-molecular-weight glutenin content, as a percentage of total protein, was predicted within an acceptable level of accuracy ($r^2 = 0.88$ and standard error of cross validation, 0.77; Bhandari et al., 2000). Elements of starch composition including amylose-to-amylopectin ratio (Wesley et al., 2003) and starch or flour swelling characteristics (Crosbie et al., 2007) have also been predicted with an accuracy claimed to be sufficient for screening early-generation breeding materials for their potential starch properties. In another study Delwiche et al. (2006) indicated that NIRS was *not* able to distinguish between wild-type and single null partial-waxy lines in durum wheat, although fully waxy genotypes were correctly categorized.

The accuracy of NIRS was also assessed for predicting numerous grain, milling, flour, dough, and breadmaking parameters from 198 US hard red winter and spring wheat and flour samples (Dowell et al., 2006b). For factors other than protein and moisture, NIRS only predicted flour yellowness with an accuracy claimed to be applicable in process control ($r^2 > 0.97$). Many other parameters were predicted with accuracies claimed to be suitable for coarse screening. However, of these parameters the only ones predicted with $r^2 > 0.70$ (once correlations with protein were removed) were test weight, flour color, free lipids, flour particle size, and the percentage of dark vitreous kernels. As many factors including loaf volume and mixograph, farinograph, and alveograph parameters were correlated with protein content, this relationship was considered to be the main influence on the NIRS predictions.

It is considered by some that the future possibilities for NIRS do not necessarily lie in the detection of an increasing number of component concentrations and/or functional traits but in exploitation of the technique for real-time testing (Scotter 2000), such as on loading conveyors, or sorting of single kernels for chosen traits (Dowell et al., 2006a). Delwiche (2005) summarized the emerging uses and new instrumentation available in NIRS. The issue of the correlation of many quality attributes with protein content was again highlighted as well as possible solutions to the challenge of finding robust methods of predicting "protein quality." This article also suggested that more physically robust NIRS instruments will allow their deployment in harsher environments such as in harvesters or flour mills. Other opportunities arise from the use of single-kernel NIRS instrumentation for high-speed sorting of seeds in many circumstances: from segregation of hard and soft kernels in breeding programs to commercial-scale separation and removal of fusarium-infected kernels from clean kernels to achieve acceptably low levels of deoxynivalenol.

Further possibilities lie in the use of other spectroscopic techniques, alone, or in tandem with NIRS. Raman spectroscopy has been used in conjunction with NIRS to improve NIRS-based predictions (Anderssen et al., 2005). Seabourn et al. (2005) reported the use of Fourier-transform (FT) midinfrared spectroscopy to monitor dough mixing in a mixograph. The ratio of α-helix to β-sheet structures as determined in the mid-IR range in the first minute of mixing was highly correlated with mixograph peak time. The authors considered that the FT midinfrared technique could predict mixograph peak time in a truncated mixing process (<1 min) based on changes in the secondary structure of the gluten proteins and that it could be employed to rapidly and accurately screen wheat lines in early generations of the breeding process.

Cultivar identification

Cultivar identification is an important part of "passing the test" for wheat. Several circumstances may allow cultivar identification to be of value: (i) identity preservation of one cultivar or a limited group of cultivars for a specific marketing purpose; (ii) as an enforcement backup for statutory or contractual requirements for cultivar declaration at delivery; (iii) as a method of establishing a priori knowledge of quality traits that are primarily under genetic control. Such traits include kernel texture (hard or soft), seed coat color (red or white), starch properties (with or without *Wx-1* null alleles), and dough attributes (via gluten composition).

Other than segregation by visually distinguishable market classes (where these exist) there are a number of techniques that have been applied in attempts to rapidly and accurately identify cultivars in the marketing stream where speed, coupled with reliability, is essential. These techniques include NIRS, microfluidic capillary electrophoresis (MFCE), and cultivar-identification methods based on molecular markers. Traditional methods such as identifying gliadin polymorphisms via electrophoresis are generally too slow and too complex for wide adoption as routine screening methods. Although each method has its virtues, limitations are evident where there are too many cultivars to identify, or where they are highly related, or where small admixtures of unwanted cultivars are allegedly present in a grain lot. In the latter case stringent adherence to correct sampling procedures is required as well as the application of relevant statistical procedures to determine confidence limits for percentages [see Clopper and Pearson (1934) for the exact binomial confidence interval and Agresti and Coull (1998) for further discussion]. For example, where 2 seeds are found to be from an incorrect cultivar in an individual subsample of 100 seeds, then the 95% exact binomial confidence interval indicates that there is a 0.2%–7% admixture of the incorrect cultivar in the whole sample. For 1 in 50, nominally the same proportion, the limits are 0.05%–10.7%. For 15 seeds in 50 (30%) of the incorrect cultivar, a proportion likely to provide functional changes in contrasting cultivars, the uncertainty at $p = 0.05$ is from 14% to 44%. In any technique that cannot handle mixed samples and is required to assess individual seeds to detect admixtures, these statistical stringencies need to be addressed in reporting of results.

The choice of the most appropriate technique is predicated on the purpose of cultivar identification. If for example one desires the separation of genotypes with fully waxy starch from partial waxy or wild-type starch, then NIRS may be sufficient (Delwiche et al., 2006), although simple screening of cut kernels with iodine to quickly identify seeds with endosperm that does not have the blue reaction of normal (or partial waxy) starch to iodine may be more cost-effective. If one needs to segregate genotypes on the basis of high-molecular-weight glutenin subunits, such as differentiating 2 + 12 genotypes from 5 + 10 types, then simple MFCE (Uthayakumaran et al., 2005) or PCR-based analyses (Xu et al., 2008) may suffice. Additionally the potential for MFCE to establish quantitative estimates of the presence of individual subunits may make it a candidate for predicting dough properties in wheat. For more complex distinctions between many cultivars, MFCE may be coupled with libraries of gliadin patterns from the cultivars expected to be found in a growing region so that pattern matching (Wrigley et al., 1992) is possible. Use of antibody-based tests is also possible (Skylas et al., 2001). The area of wheat cultivar identification was recently reviewed in brief (Wrigley et al., 2006).

FUTURE PERPECTIVES

For the cereal technologist there are many exciting advances on the horizon that promise better, faster, and more accurate predictions of the functional properties of wheat. Among these probably the most exciting areas of research are the microfluidic techniques for cultivar identification (and possibly predictions of dough characteristics), further advances in spectroscopy, automation of existing traditional tests, and the further refinement of dough testing procedures. In the area of dough testing, being arguably the most important functional attribute for breadmaking, there are some challenges to overcome if we are to advance this aspect of functionality testing to a new level. In the effort to improve the capabilities for measuring dough properties some would argue that there needs to be a new generation of physical dough-testing instruments. Ideally these would take measurements in units fundamental to physics and chemistry, in contrast to the undefined units that are the current *lingua franca* of dough rheology, and which are largely meaningless outside cereal science. Couching dough physical properties in fundamental terms could provide a way to help us better understand dough rheology at the molecular level, as has happened to some extent with measurements of the strain hardening properties of dough (Dobraszczyk et al., 2005). However, if a new generation of physical dough-testing instruments was to entirely supplant the older ones, any new instruments or techniques would also need to have the relative ease of use, and the predictive utility in commerce, of the older instruments. That said it may be simple enough to describe food or dough physical properties in fundamental units that are of value to researchers, but in a new generation of dough-testing equipment these measures would need to be of equal or better value to wheat processors than are the current measures with their empirical units.

REFERENCES

AACC. 2000. Approved methods of the American Association of Cereal Chemists. 10th ed. AACC Press, St. Paul, MN.

Agresti, A., and B. Coull. 1998. Approximate is better than 'exact' for interval estimation of binomial proportions. Am. Stat. 52:119–126.

Akashi, H., M. Takahashi, and S. Endo. 1999. Evaluation of starch properties of wheats used for Chinese yellow-alkaline noodles in Japan. Cereal Chem. 76:50–55.

Álava, J.M., S.S. Sahia, J. García-Álvarez, A. Turó, J.A. Chávez, M.J. García, and J. Salazar. 2007. Use of ultrasound for the determination of flour quality. Ultrasonics 46:270–276.

Amr, A., and R. Ajo. 2005. Production of two types of pocket-forming flat bread by the sponge and dough method. Cereal Chem. 82:499–503.

Anderson, J.V., and C.F. Morris. 2001. An improved whole-seed assay for screening wheat germplasm for polyphenol oxidase activity. Crop Sci. 41:1697–1705.

Anderssen, R.S., F. Bekes, P.W. Gras, A. Nikolov, and J.T. Wood. 2004. Wheat-flour dough extensibility as a discriminator for wheat varieties. J Cereal Sci. 39:195–203.

Anderssen, R.S., E. Carter, B.G. Osborne, and I.J. Wesley. 2005. Joint inversion of multi-modal spectroscopic data of wheat flours. Appl. Spectrosc. 59:920–925.

Asenstorfer, R.E., Y. Wang, and D.J. Mares. 2006. Chemical structure of flavonoid compounds in wheat (*Triticum aestivum* L.) flour that contribute to the yellow colour of Asian alkaline noodles. J. Cereal Sci. 43:108–119.

Atwell, W.A. 2001. Wheat flour. Eagan Press, St Paul, MN.

Atwell, W.A., L.F. Hood, D.R. Lineback, E. Varriano-Marston, and H.F. Zobel. 1988. The terminology and methodology associated with basic starch phenomena. Cereal Foods World 33:306–311.

Baenziger, P.S., D.R. Shelton, M.J. Shipman, and R.A. Graybosch. 2001. Breeding for end-use quality: Reflections on the Nebraska experience. Euphytica 119:95–100.

Bagley, E.B., and D.D. Christianson. 1986. Response of commercial chemically leavened doughs to uniaxial compression. p. 27–36. *In* H. Faridi and J.M. Faubion (ed.) Fundamentals of dough rheology. AACC Press, St Paul, MN.

Baltsavias, A., A. Jurgens, and T. van Vliet. 1999. Rheological properties of short doughs at large deformation. J. Cereal Sci. 29:33–42.

Barnard, A. 2001. Genetic diversity of South African winter wheat cultivars in relation to preharvest sprouting and falling number. Euphytica 119:107–110.

Barrera, G.N., G.T. Perez, P.D. Ribotta, and A.E. Leon. 2007. Influence of damaged starch on cookie and bread-making quality. Eur. Food Res. Technol. 225:1–7.

Batey, I.L. 2007. Interpretation of RVA curves. p. 19–30. *In* G.B. Crosbie and A.S. Ross (ed.) The RVA handbook. AACC International Press, St Paul, MN.

Batey, I.L., and M.L. Bason. 2007. Appendix 2: Definitions. p. 138–140. *In* G.B. Crosbie and A.S. Ross (ed.) The RVA handbook. AACC International Press, St Paul, MN.

Bekes, F., M.C. Gianibelli, and C.W. Wrigley. 2004. Grain proteins and flour quality. p. 416–423. *In* C.W. Wrigley, H. Corke, and C.E. Walker (ed.) The encyclopedia of grain science. Vol. 3. Elsevier Academic Press, Oxford, UK.

Bettge, A.D. 2004. Collaborative study on L-DOPA—wheat polyphenol oxidase assay (AACC Method 22–85). Cereal Foods World 49:338, 340–342.

Bettge, A.D., and C.F. Morris. 2001. Enhanced end-use quality prediction using sequential solvent retention capacity. *In* Abstracts, AACC Annu. Meet., Charlotte, NC. 14–18 Oct. 2001. AACC, St. Paul, MN. Available at www.aaccnet.org/meetings/2001/abstracts/a01ma432.htm (verified 10 Nov. 2007).

Bettge, A.D., and C.F. Morris. 2007. Oxidative gelation measurement and influence on soft wheat batter viscosity and end-use quality. Cereal Chem. 84:237–242.

Bettge, A.D., C.F. Morris, V.L. DeMacon, and K.K. Kidwell. 2002. Adaptation of AACC method 56–11, solvent retention capacity, for use as an early generation selection tool for cultivar development. Cereal Chem. 79:670–674.

Bhandari, D.G., S.J. Millar, and C.N.G. Scotter. 2000. Prediction of wheat protein and HMW-glutenin contents by near infrared (NIR) spectroscopy. Spec. Publ.—R. Soc. Chem. 261 (Wheat gluten):313–316.

Booth, R., and M.L. Bason. 2007. Principles of operation and experimental techniques. p. 1–17. *In* G.B. Crosbie and A.S. Ross (ed.) The RVA handbook. AACC International Press, St. Paul, MN.

Boyaci, I.H., P.C. Williams, and H. Köksel. 2004. A rapid method for the estimation of damaged starch in wheat flours. J. Cereal Sci. 39:139–145.

Briggs, K.G., W. Bushuk, and L.H. Shebeski. 1969. Protein quantity and quality as factors in the evaluation of bread wheats. Can. J. Plant Sci. 49:113–122.

Caddick, L. 2007. Grain moisture content and its measurement [Online]. Available at www.sgrl.csiro.au (verified 23 Oct. 2007).

Catterall, P., and S.P. Cauvain. 2007. Flour milling. p. 333–369. *In* S.P. Cauvain and L.S. Young (ed.) Technology of breadmaking. 2nd ed. Springer, New York.

Cauvain, S.P., and L.S. Young. 2006. Baked products: Science, technology and practice. Blackwell Publishing, Oxford, UK.

Cauvain, S.P., and L.S. Young (ed.) 2007. Technology of breadmaking. 2nd ed. Springer, New York.

Chakraborty, M., G.A. Hareland, F.A. Manthey, and L.R. Berglund. 2003. Evaluating quality of yellow alkaline noodles made from mechanically abraded sprouted wheat. J. Sci. Food Agric. 83:487–495.

Chang, S-Y., S.R. Delwiche, and N.S. Wang. 1999. Hydrolysis of wheat starch and its effect on the Falling Number procedure: Experimental observations. J. Sci. Food Agric. 79:19–24.

Clopper, C., and S. Pearson. 1934. The use of confidence or fiducial limits illustrated in the case of the binomial. Biometrika 26:404–413.

Cobb, N.A. 1896. The hardness of grain in the principal varieties of wheat. Agric. Gaz. N.S.W. 7:279–298.

Collar, C., C. Bollain, and A. Angioloni. 2005. Significance of microbial transglutaminase on the sensory, mechanical and crumb grain pattern of enzyme supplemented fresh pan breads. J. Food Eng. 70:479–488.

Corke, H., and M. Bhattacharya. 1999. Wheat products: 1. Noodles. p. 43–70. *In* C.Y.W. Ang, K. Liu, and Y-W. Huang (ed.) Asian food science and technology. Technomic Publishing, Lancaster, PA.

Crosbie, G.B., P.C. Chiu, and A.S. Ross. 2002. Shortened temperature program for application with a rapid visco analyser in prediction of noodle quality in wheat. Cereal Chem. 79:596–599.

Crosbie, G.B., S. Huang, and I.R. Barclay. 1998. Wheat quality requirements of Asian foods. Euphytica 100:155–156.

Crosbie, G.B., and W.J. Lambe. 1993. The application for the flour swelling volume test for potential noodle quality to wheat breeding lines affected by sprouting. J. Cereal Sci. 18:267–276.

Crosbie, G.B., W.J. Lambe, H. Tsutsui, and R.F. Gilmour. 1992. Further evaluation of the flour swelling volume test for identifying wheats potentially suitable for Japanese noodles. J. Cereal Sci. 15:271–280.

Crosbie, G.B., B.G. Osborne, I.J. Wesley, and T.D. Adriansz. 2007. Screening of wheat for flour swelling volume by near-infrared spectroscopy. Cereal Chem. 84:379–383.

Crosbie, G.B., and A.S. Ross. 2004. Asian wheat flour noodles. p. 304–312. *In* C.W. Wrigley, H. Corke, and C. E. Walker (ed.) The encyclopedia of grain science. Vol. 2. Elsevier Academic Press, Oxford, UK.

Crosbie, G.B., and A.S. Ross (ed.) 2007. The RVA handbook. AACC International Press, St. Paul, MN.

Cubadda, R.E., M. Carcea, E. Marconi, and M.C. Trivisonno. 2007. Influence of gluten proteins and drying temperature on the cooking quality of durum wheat pasta. Cereal Chem. 84:48–55.

Deffenbaugh, L.B., and C.E. Walker. 1989. Comparison of starch pasting properties in the Brabender viscoamylograph and the rapid visco-analyzer. Cereal Chem. 66:493–499.

Delwiche, S.R. 2005. Up and coming methods and instrumentation for end-use wheat quality measurements. p 283–290. *In* O.K. Chung and G.L. Lookhart (ed.) Proc. Int. Wheat Quality Conf., 3rd, Manhattan, KS. 22–26 May 2005. Grain Industry Alliance, Manhattan, KS.

Delwiche, S.R., R.A. Graybosch, L.E. Hansen, E. Souza, and F.E. Dowell. 2006. Single kernel near-infrared analysis of tetraploid (durum) wheat for classification of the waxy condition. Cereal Chem. 83:287–292.

Demeke, T., and C.F. Morris. 2002. Molecular characterization of wheat polyphenol oxidase (PPO). Theor. Appl. Genet. 104:813–818.

Dempster, R., M. Olewnik, and V.S. Smail. 2007. Determination of dough development using near infrared radiation. US Patent 7 312 451. Date filed: 22 June 2004. Date issued: 25 Dec. 2007.

Dick, J.W., and R.R. Matsuo. 1988. Durum wheat and pasta products. p. 507–547. *In* Y. Pomeranz (ed.) Wheat chemistry and technology. Vol. 1. 3rd ed. AACC Press, St. Paul, MN.

Dobraszczyk, B.J., G.M. Campbell, and Z. Ghan. 2000. Bread: A unique food. p. 183–232. *In* B.J. Dobraszczyk and D.A.V. Dendy (ed.) Cereals and cereal products:

Chemistry and technology. Aspen Publishers, Gaithersburg, MD.

Dobraszczyk, B.J., W. Li, E. Roberts, G.R. Mitchell, G.H. Mckinley, and T.S.K. Ng. 2005. Large deformation rheological testing of doughs: Rheological and polymer molecular structure–function relationships in breadmaking. p. 226–233. *In* O.K. Chung and G.L. Lookhart (ed.) Proc. Int. Wheat Quality Conf., 3rd, Manhattan, KS. 22–26 May 2005. Grain Industry Alliance, Manhattan, KS.

Don, C., W.J. Lichtendonk, J.J. Plijter, and R.J. Hamer. 2003. Understanding the link between GMP and dough: From glutenin particles in flour towards developed dough. J. Cereal Sci. 38:157–165.

Donnelly, B.J., and J.G. Ponte. 2000. Pasta: Raw materials and processing. p. 647–663. *In* K. Kulp and J.G. Ponte (ed.) Handbook of cereal science and technology. 2nd ed. Marcel Dekker, New York.

Dowell, F.E., E.B. Maghirang, R.A. Graybosch, P.S. Baenziger, D.D. Baltensperger, and L.E. Hansen. 2006a. An automated near-infrared system for selecting individual kernels based on specific quality characteristics. Cereal Chem. 83:537–543.

Dowell, F.E., E.B. Maghirang, F. Xie, G.L. Lookhart, R.O. Pierce, B.W. Seabourn, S.R. Bean, J.D. Wilson, and O.K. Chung. 2006b. Predicting wheat quality characteristics and functionality using near-infrared spectroscopy. Cereal Chem. 83:529–536.

Edwards, N., and J. Dexter. 1987. Alveograph—sources of problems in curve interpretation with hard common wheat flour. Can. Inst. Food Sci. Technol. J. 20:75–80.

Edwards, M.A., B.G. Osborne, and R.J. Henry. 2007a. Investigation of the effect of conditioning on the fracture of hard and soft wheat grain by the single-kernel characterization system: A comparison with roller milling. J. Cereal Sci. 46:64–74.

Edwards, N.M., K.R. Preston, F.G. Paulley, M.C. Gianibelli, T.N. McCaig, J.M. Clarke, N.P. Ames, and J.E. Dexter. 2007b. Hearth bread baking quality of durum wheat varying in protein composition and physical dough properties. J. Sci. Food Agric. 87:2000–2011.

Edwards, R.A., A.S. Ross, D.J. Mares, F.W. Ellison, and J.D. Tomlinson. 1989. Enzymes from rain damaged and laboratory germinated wheat: 1. Effects on product quality. J Cereal Sci. 10:157–167.

Elmehdi, H.M., J.H. Page, and M.G. Scanlon. 2004. Ultrasonic investigation of the effect of mixing under reduced pressure on the mechanical properties of bread dough. Cereal Chem. 81:504–510.

Evers, A.D. 1998. Branscan, a new concept in a traditional industry. p. 100–106. *In* L. O'Brien, A.B. Blakeney, A.S. Ross, and C.W. Wrigley (ed.) Proc. Aust. Cereal Chem. Conf., 48th, Cairns. 17–20 Aug. 1998. Royal Australian Chemical Institute, Melbourne, Australia.

Every, D., L.D. Simmons, and M.P. Ross. 2006. Distribution of redox enzymes in millstreams and relationships to chemical and baking properties of flour. Cereal Chem. 83:62–68.

Færgestad, E.M., E.L. Molteberg, and E.M Magnus. 2000. Interrelationships of protein composition, protein level, baking process and the characteristics of hearth bread and pan bread. J. Cereal Sci. 31:309–320.

Faridi, H., and J.M. Faubion. 1998. Wheat end uses around the world. AACC Press, St. Paul, MN.

Faridi, H., C.S. Gaines, and B.L. Strouts. 2000. Soft wheat products. p. 575–613. *In* K. Kulp and J.G. Ponte (ed.) Handbook of cereal science and technology. 2nd ed. Marcel Dekker, New York.

Faridi, H.A., and V.F. Rasper 1987. The alveograph handbook. AACC Press, St. Paul, MN.

Farooq, Z., S. Rehman, M.S. Butt, and M.Q. Bilal. 2001. Suitability of wheat varieties/lines for the production of leavened flat bread (naan). J. Res. Sci. 12:171–179.

Farrand, E.A. 1964. Flour properties in relation to the modern bread processes in the United Kingdom with special reference to α-amylase and starch damage. Cereal Chem. 41:98–111.

Farvili, N., C.E. Walker, and J. Qarooni. 1995. Effects of emulsifiers on pita bread quality. J. Cereal Sci. 21:301–308.

Figoni, P. 2004. How baking works: Exploring the fundamentals of baking science. John Wiley and Sons, Hoboken, NJ.

Finney, K.F., and M.A. Barmore. 1948. Loaf volume and protein content of hard winter and spring wheats. Cereal Chem. 25:291–312.

Finnie, S.M., A.D. Bettge, and C.F. Morris. 2006. Influence of cultivar and environment on water-soluble and water-insoluble arabinoxylans in soft wheat. Cereal Chem. 83:617–623.

Foca, G., A. Ulrici, M. Corbellini, and M.A. Pagani. 2007. Reproducibility of the Italian ISQ method for quality classification of bread wheats: An evaluation by expert assessors. J. Sci. Food Agric. 87:839–846.

Fuerst, E.P., J.V. Anderson, and C.F. Morris. 2006. Delineating the role of polyphenol oxidase in the darkening of alkaline wheat noodles. J. Agric. Food Chem. 54:2378–2384.

Gaines, C.S., P.F. Finney, L.M. Fleege, and L.C. Andrews. 1996. Predicting a hardness measurement using the single-kernel characterization system. Cereal Chem. 73:278–283.

Gaines, C.S., J. Frégeau-Reid, C. Vander Kant, and C.F. Morris. 2006. Comparison of methods for gluten strength assessment. Cereal Chem. 83:284–286.

Gale, K.R., M.J. Blundell, and A.S. Hill. 2004. Development of a simple, antibody-based test for granule-bound starch synthase Wx-B1b (Null-4A) wheat varieties. J. Cereal Sci. 40:85–92.

Geera, B.P., J.E. Nelson, E. Souza, and K.C. Huber. 2006. Granule bound starch synthase I (GBSSI) gene effects related to soft wheat flour/starch characteristics and properties. Cereal Chem. 83:544–550.

Ghosh, P.K., S. Digvir, D.S. Jayasa, M.L.H. Gruwel, and N.D.G. White. 2007. A magnetic resonance imaging study of wheat drying kinetics. Biosystems Eng. 97:189–199.

Gianibelli, M.C., O.R. Larroque, F. MacRitchie, and C.W. Wrigley. 2001. Biochemical, genetic, and molecular char-

acterization of wheat glutenin and its component subunits. Cereal Chem. 78:635–646.
Gibson, T.S., C.J. Kaldor, and B.V. McCleary. 1993. Collaborative evaluation of an enzymatic starch damage assay kit and comparison with other methods. Cereal Chem. 70:47–51.
Giroux, M.J., L. Talbert, D.K. Habernicht, S. Lanning, A. Hemphill, and J.M. Martin. 2000. Association of puroindoline sequence type and grain hardness in hard red spring wheat. Crop Sci. 40:370–374.
Goesaert, H., K. Gebruers, C.M. Courtin, K. Brijs, and J.A. Delcour. 2006. Enzymes in breadmaking. p.337–364. *In* Y.H. Hui, H. Corke, I. De Leyn, and N. Cross (ed.) Bakery products: Science and technology. Blackwell Publishing, Ames, IA.
Gras, P.W., H.C. Carpenter, and R.S. Anderssen. 2000. Modelling the developmental rheology of wheat-flour dough using extension tests. J. Cereal Sci. 31:1–13.
Graybosch, R.A., C.J. Peterson, G.A. Hareland, D.R. Shelton, M.C. Olewnik, H. He, and M.M. Stearns. 1999. Relationships between small-scale wheat quality assays and commercial test bakes. Cereal Chem. 76:428–433.
Greenwell, P., and C. Brock. 1996. Modified flour. US Patent 5 560 953. Date filed: 31 May 1995. Date issued: 1 October 1996.
Guttieri, M.J., C. Becker, and E.J. Souza. 2004. Application of wheat meal solvent retention capacity tests within soft wheat breeding populations. Cereal Chem. 81:261–266.
Guttieri, M.J., D. Bowen, D. Gannon, K. O'Brien, and E. Souza. 2001. Solvent retention capacities of irrigated soft white spring wheat flours. Crop Sci. 41:1054–1061.
Guttieri, M.J., R. McLean, S.P. Lanning, L.E. Talbert, and E.J. Souza. 2002. Assessing environmental influences on solvent retention capacities of two soft white spring wheat cultivars. Cereal Chem. 79:880–884.
Hatcher, D.W. 2007. The use of the RVA to predict Falling Number: The Canadian experience. Newport Sci. World 9:1–3.
Hatcher, D.W., and J.E. Kruger. 1993. Distribution of polyphenol oxidase in flour millstreams of Canadian common wheat classes milled to three extraction rates. Cereal Chem. 70:51–55.
Hatcher, D.W., and S.J. Symons. 2000a. Assessment of oriental noodle appearance as a function of flour refinement and noodle type by image analysis. Cereal Chem. 77:181–186.
Hatcher, D.W., and S.J. Symons. 2000b. Image analysis of Asian noodle appearance: Impact of hexaploid wheat with a red seed coat. Cereal Chem. 77:388–391.
Hatcher, D.W., S.J. Symons, and U. Manivannan. 2004. Developments in the use of image analysis for the assessment of oriental noodle appearance and colour. J. Food Eng. 61:109–117.
Hazelton, J.L., J.L. DesRochers, C.E. Walker, and C.W. Wrigley. 2004. Cookies, biscuits, and crackers: Chemistry of manufacture. p. 307–312. *In* C.W. Wrigley, H. Corke, and C.E. Walker (ed.) The encyclopedia of grain science. Vol. 1. Elsevier Academic Press, Oxford, UK.
He, Z.H., A.H. Liu, R.J. Peña, and S. Rajaram. 2003. Suitability of Chinese wheat cultivars for production of northern style Chinese steamed bread. Euphytica 131:155–163.
Heidolph, B.B. 1996. Designing chemical leavening systems. Cereal Foods World 41:118–126.
Horigane, A.K., S. Naito, M. Kurimoto, K. Irie, S. Yamada, H. Motoi, and M. Yoshida. 2006. Moisture distribution and diffusion in cooked spaghetti studied by NMR imaging and diffusion model. Cereal Chem. 83:235–242.
Hou, G. 2001. Oriental noodles. Adv. Food Nutr. Res. 43:141–193.
Huang, S. 1999. Wheat products: 2. Breads, cakes, cookies, pastries, and dumplings. p. 43–70. *In* C.Y.W. Ang, K. Liu, and Y-W. Huang (ed.) Asian food science and technology. Technomic Publishing, Lancaster, PA.
Huang, S., S. Betker, K. Quail, and R. Moss. 1993. An optimized processing procedure by response surface methodology (RSM) for Northern-style Chinese steamed bread. J. Cereal Sci. 18:89–102.
Huang, S., S-K. Yun, K. Quail, and R. Moss. 1996. Establishment of flour quality guidelines for Northern-style Chinese steamed bread. J. Cereal Sci. 24:179–185.
ICC. 2006. ICC standard methods. International Association for Cereal Science and Technology, Vienna, Austria.
Izydorczyk, M.S., C.G. Biliaderis, and W. Bushuk. 1991. Physical properties of water-soluble pentosans from different wheat varieties. Cereal Chem. 68:145–150.
Kojima, T.I., A.K. Horigane, H. Nakajima, M. Yoshida, and A. Nagasawa. 2004. T(2) map, moisture distribution, and texture of boiled Japanese noodles prepared from different types of flour. Cereal Chem. 81:746–751.
Kokelaar, J.J., T. van Vliet, and A. Prins. 1996. Strain hardening properties and extensibility of flour and gluten doughs in relation to breadmaking performance. J. Cereal Sci. 24:199–214.
Konik-Rose, C., J. Thistleton, H. Chanvrier, I. Tan, P. Halley, M. Gidley, B. Kosar-Hashemi, H. Wang, O. Larroque, J. Ikea, S. McMaugh, A. Regina, S. Rahman, M. Morell, and Z. Li. 2007. Effects of starch synthase IIa gene dosage on grain, protein and starch in endosperm of wheat. Theor. Appl. Genet. 115:1053–1065.
Kruger, J.E., M.H. Anderson, and J.E. Dexter. 1994. Effect of flour refinement on raw Cantonese noodle color and texture. Cereal Chem. 71:177–182.
Kruger, J.E., and G. Reed. 1988. Enzymes and color. p. 441–501. *In* Y. Pomeranz (ed.) Wheat chemistry and technology. Vol. 1. 3rd ed. AACC Press, St. Paul, MN.
Kulp, K. 1968. Pentosans of wheat endosperm. Cereal Sci. Today 13:414–417, 426.
Kulp, K., and J.G. Ponte. 2000. Breads and yeast leavened bakery foods. p. 539–573. *In* K. Kulp and J.G. Ponte (ed.) Handbook of cereal science and technology. 2nd ed. Marcel Dekker, New York.
Lagarrigue, S., and G. Alvarez. 2001. The rheology of starch dispersions at high temperatures and high shear rates: A review. J. Food Eng. 50:189–202.
Lee, S., and G.E. Inglett. 2006. Rheological and physical evaluation of jet-cooked oat bran in low calorie cookies. Int. J. Food Sci. Technol. 41:553–559.

Liao, H-J., Y-C. Chung, and J. Tattiyakul. 2007. Biaxial extensional viscosity of sheeted noodle dough. Cereal Chem. 84:506–511.

Lin, P.Y., and Z. Czuchajowska. 1996. Starch damage in soft wheats of the Pacific Northwest. Cereal Chem. 73: 551–555.

MacRitchie, F., and D. Lafiandra. 1997. Structure function relationships of wheat proteins. p. 293–324. *In* S. Damodaran and A. Paraf (ed.) Food proteins and their applications. Marcel Dekker, New York.

Maghirang, E.B., G.L. Lookhart, S.R. Bean, R.O. Pierce, F. Xie, M.S. Caley, J.D. Wilson, B.W. Seabourn, M.S. Ram, S.H. Park, O.K. Chung, and F.E. Dowell. 2006. Comparison of quality characteristics and breadmaking functionality of hard red winter and hard red spring wheat. Cereal Chem. 83:520–528.

Malcolmson, L.J., R.R. Matsuo, and R. Balshaw. 1993. Texture optimization of spaghetti using response surface methodology: Effects of drying temperature and durum protein level. Cereal Chem. 70:417–423.

Manley, D. 2001. Biscuit, cracker and cookie recipes for the food industry. Woodhead Publishing, Cambridge, UK.

Mann G., H. Allen, M.K. Morell, Z. Nath, P. Martin, J. Oliver, B. Cullis, and A. Smith. 2005. Comparison of small-scale and large-scale extensibility of dough produced from wheat flour. Aust. J. Agric. Res. 56:1387–1394.

Mares, D.J, and A.W. Campbell. 2001. Mapping components of flour and noodle colour in Australian wheat. Aust. J. Agric. Res. 52:1297–1309.

Mares, D.J., Y. Wang, and C.A. Cassidy. 1997. Separation, identification and tissue location of compounds responsible for the yellow colour of alkaline noodles. p. 114–117. *In* A.W. Tarr, A.S. Ross, and C.W. Wrigley (ed.) Proc. Cereal Chem. Conf., 47th, Perth. 14–19 Sept. 1997. Cereal Chemistry Division, Royal Australian Chemical Institute, Melbourne, Australia.

Martin, J.M., F.D. Meyer, C.F. Morris, and M.J. Giroux. 2007. Pilot scale milling characteristics of transgenic isolines of a hard wheat over-expressing puroindolines. Crop Sci. 47:497–506.

Matsuo, R.R., L.J. Malcolmson, N.M. Edwards, and J.E. Dexter. 1992. A colorimetric method for estimating spaghetti cooking losses. Cereal Chem. 69:27–29.

Miklus, M., M. Kweon, L. Slade, and H. Levine. 2004. Interpretation of solvent retention capacity profiles for wheat flour functionality and performance: Diagnostic responses to aqueous ethanol, reducing agents, and SDS as SRC solvents. *In* Abstracts, AACC TIA Joint Meeting, San Diego, CA. 29–22 Sept. 2004. AACC, St. Paul, MN. Available at www.aaccnet.org/meetings/2004/abstracts/a04ma218.htm (verified 10 Nov. 2007).

Miralbés, C. 2004. Quality control in the milling industry using near infrared transmittance spectroscopy. Food Chem. 88:621–628.

Miskelly, D.M., and H.J. Moss. 1985. Flour quality requirements for Chinese noodle manufacture. J. Cereal Sci. 3:379–387.

Moore, T.R. 2004. Breads. p. 109–119. *In* C.W. Wrigley, H. Corke, and C.E. Walker (ed.) The encyclopedia of grain science. Vol. 1. Elsevier Academic Press, Oxford, UK.

Morgan, J.E., and P.C. Williams. 1995. Starch damage in wheat flours: A comparison of enzymatic, iodometric, and near-infrared reflectance techniques. Cereal Chem. 72:209–212.

Morris, C.F. 2002. Puroindolines: The molecular genetic basis of wheat grain hardness. Plant Mol. Biol. 48: 633–647.

Morris, C.F., and S.P. Rose. 1996. Wheat. p. 3–54. *In* R.J. Henry and P.S. Kettlewell (ed.) Cereal grain quality. Chapman and Hall, London, UK.

Moss, H.J. 1980. The pasting properties of some wheat starches free of sprout damage. Cereal Res. Commun. 8:297–302.

Mulvaney, S.J. 2005a. Linear viscoelastic measurements and network structures: General comments and relevance to molecular structures. p. 219–224. *In* O.K. Chung and G.L. Lookhart (ed.) Proc. Int. Wheat Quality Conf., 3rd, Manhattan, KS. 22–26 May 2005. Grain Industry Alliance, Manhattan, KS.

Mulvaney, S.J. 2005b. Session facilitator discussion. p. 267–268. *In* O.K. Chung and G.L. Lookhart (ed.) Proc. Int. Wheat Quality Conf., 3rd, Manhattan, KS. 22–26 May 2005. Grain Industry Alliance, Manhattan, KS.

Nakamura, T., P. Vrinten, M. Saito, and M. Konda. 2002. Rapid classification of partial waxy wheats using PCR-based markers. Genome 45:1150–1156.

Neethirajan, S., D.S. Jayas, and N.D.G. White. 2007. Detection of sprouted wheat kernels using soft X-ray image analysis. J. Food Eng. 81:509–513.

Ohm, J-B., A.S. Ross, C.J. Peterson, and M. Beilstein. 2006. Relationships of Single Kernel Characterization System variables and milling quality in hard and soft white winter wheats. *In* Abstracts, World Grains Summit and 91st AACC-Int. Annu. Meet., San Francisco, CA. 17–20 Sept. 2006. AACC Int., St. Paul, MN. Available at http://www.aaccnet.org/meetings/2006/abstracts/p-312.htm (verified 7 March 2008).

Oliver, J.R., A.B. Blakeney, and H.M. Allen. 1992. Measurement of flour color in color space parameters. Cereal Chem. 69:546–551.

Osborne, B.G. 2006. Applications of near infrared spectroscopy in quality screening of early-generation material in cereal breeding programmes. J. Near Infrared Spectrosc. 14:93–101.

Osborne, B.G. 2007. Flours and breads. p. 281–296. *In* Y. Ozaki, W.F. McClure, and A.A. Christy (ed.) Near-infrared spectroscopy in food science and technology. John Wiley and Sons, Hoboken, NJ.

Osborne, B.G., and R.S. Anderssen. 2003. Single-kernel characterization principles and applications. Cereal Chem. 80:613–622.

Osborne, B.G., R.J. Henry, and M.D. Southan. 2007. Assessment of commercial milling potential of hard wheat by measurement of the rheological properties of whole grain. J. Cereal Sci. 45:122–127.

Ozawa, M., and M. Seguchi. 2006. Relationship between pancake springiness and interaction of wheat flour components caused by dry heating. Food Sci. Technol. Res. 12:167–172.

Park, C.S., and B.K. Baik. 2004. Significance of amylose content of wheat starch on processing and textural properties of instant noodles. Cereal Chem. 81:521–526.
Pasha, I., F.M. Anjum, M.S. Butt, and J.I. Sultan. 2007. Gluten quality prediction and correlation studies in spring wheats. J. Food Qual. 4:438–449.
Pearson, T., J. Wilson, J. Gwirtz, E. Maghirang, F. Dowell, P. McCluskey, and S. Bean. 2007. Relationship between single wheat kernel particle-size distribution and Perten SKCS 4100 hardness index. Cereal Chem. 84:567–575.
Peterson, D., and R.G. Fulcher. 2002. Variation in Minnesota HRS wheats: Bran content. J. Food Sci. 67:67–70.
Posner, E.S. 2000. Wheat. p. 1–29. *In* K. Kulp and J.G. Ponte (ed.) Handbook of cereal science and technology. 2nd ed. Marcel Dekker, New York.
Posner, E.S., and A.N. Hibbs. 1997. Chapter 2—experimental and laboratory milling. p. 31–62. *In* E.S. Posner and A.N. Hibbs (ed.) Wheat flour milling. AACC Press, St. Paul, MN.
Pyler, E.J. 1972. Baking science and technology. Vol. 2. Seibel Publishing Co., Chicago, IL.
Pyler, E.J. 1988. Baking science and technology. Vol. 2. 3rd ed. Sosland Publishing, Merriam, KS.
Qarooni, J. 1996. Flat bread technology. Chapman and Hall, New York.
Qarooni, J., H.J. Moss, R.A. Orth, and M. Wootton. 1988. The effect of flour properties on the quality of Arabic bread. J. Cereal Sci. 7:95–107.
Quail, K.J. 1996. Arabic bread production. AACC Press, St. Paul, MN.
Quail, K.J., G. McMaster, and M. Wootton. 1991. Flour quality tests for selected wheat cultivars and their relationship to Arabic bread quality. J. Cereal Sci. 14:131–139.
Ram, S., V. Dawar, R.P. Singh, and J. Shoran. 2005. Application of solvent retention capacity tests for the prediction of mixing properties of wheat flour. J. Cereal Sci. 42:261–266.
Rasper, V. 1980. Theoretical aspects of amylography. p. 1–6. *In* W.C. Shuey and K.H. Tipples (ed.) The amylograph handbook. AACC Press, St. Paul, MN.
Rasper, V., and C.E. Walker. 2000. Quality evaluation of cereals and cereal products. p. 505–537. *In* K. Kulp and J.G. Ponte (ed.) Handbook of cereal science and technology. 2nd ed. Marcel Dekker, New York.
Reeves, J.B., III, and S.R. Delwiche. 1997. Determination of protein in ground wheat by mid-infrared diffuse reflectance spectroscopy. Appl. Spectrosc. 51:1200–1204.
Ross, A.S. 2006. Review: Instrumental measurement of physical properties of cooked Asian wheat-flour noodles. Cereal Chem. 83:42–51.
Ross, A.S., G.B. Crosbie, and K.J. Quail. 1997. Physicochemical properties of Australian flours influencing the texture of yellow alkaline noodles. Cereal Chem. 74:814–820.
Ross, A.S., and D.W. Hatcher. 2005. Guidelines for the laboratory manufacture of Asian wheat-flour noodles. Cereal Foods World 50:296–304.
Ross, A.S., and J-B. Ohm. 2006. Sheeting characteristics of salted and alkaline Asian noodle doughs: Comparison with lubricated squeezing flow attributes. Cereal Foods World 50:191–196.
Ross, A.S., R.A. Orth, and C.W. Wrigley. 1986. Rapid screening for weather damage in wheat. p. 577–583. *In* D.J. Mares (ed.) Int. Symp. Pre-harvest Sprouting in Cereals, 4th, Port Macquarie, NSW. March 1986. Westview Press, Boulder, CO.
Ross, A.S., C.E. Walker, R.I. Booth, R.A. Orth, and C.W. Wrigley. 1987. The Rapid Visco-Analyzer: A new technique for the rapid estimation of sprout damage in cereals. Cereal Foods World 32:827–829.
Rouillé, J., G. Della Vallea, J. Lefebvre, E. Sliwinski, and T. van Vliet. 2005. Shear and extensional properties of bread doughs affected by their minor components. J. Cereal Sci. 42:45–57.
Rubenthaler, G.L., M.L. Huang, and Y. Pomeranz. 1990. Steamed bread: I. Chinese steamed bread formulation and interactions. Cereal Chem. 67:471–475.
Sahlstrom, S., A.B. Bævre, and R.A. Graybosch. 2006. Impact of waxy, partial waxy, and wildtype wheat starch fraction properties on hearth bread characteristics. Cereal Chem. 83:647–654.
Sai-Manohar, R, and P. Haridas-Rao. 1999. Effect of mixing method on the rheological characteristics of biscuit dough and the quality of biscuits. Eur. Food Res. Technol. 210:43–48.
Sapirstein, H.D., P. David, K.R. Preston, and J.E. Dexter. 2007. Durum wheat breadmaking quality: Effects of gluten strength, protein composition, semolina particle size and fermentation time. J. Cereal Sci. 45:150–161.
Scotter, C.N.G. 2000. NIR techniques in cereals analyses. p. 90–99. *In* B.J. Dobraszczyk and D.A.V. Dendy (ed.) Cereals and cereal products: Chemistry and technology. Aspen Publishers, Gaithersburg, MD.
Seabourn, B.W., F. Xie, and O.K. Chung. 2005. An objective and rapid method to determine dough optimum mixing time for early generation breeding lines using FT-HATR mid-infrared (IR) spectroscopy. p. 383. *In* O.K. Chung and G.L. Lookhart (ed.) Proc. Int. Wheat Quality Conf., 3rd, Manhattan, KS. 22–26 May 2005. Grain Industry Alliance, Manhattan, KS.
Seguchi, M. 1990. Study of wheat starch granule surface proteins from chlorinated wheat flours. Cereal Chem. 67:258–260.
Seibel, W. 1998. Wheat usage in western Europe. p. 93–125. *In* H. Faridi and J.M. Faubion (ed.) Wheat end uses around the world. AACC Press, St. Paul, MN.
Sissons, M. 2004. Pasta. p. 409–418. *In* C.W. Wrigley, H. Corke, and C.E. Walker (ed.) The encyclopedia of grain science. Vol. 2. Elsevier Academic Press, Oxford, UK.
Sissons, M.J., and I.L. Batey. 2003. Protein and starch properties of some tetraploid wheats. Cereal Chem. 80: 468–475.
Sissons, M.J., B.G. Osborne, R.A. Hare, S.A. Sissons, and R. Jackson. 2000. Application of the single-kernel characterization system to durum wheat testing and quality prediction. Cereal Chem. 2000. 77:4–10.
Sissons, M.J., L.M. Schlichting, N. Egan, W.A. Aarts, S. Harden, and B.A. Marchylo. 2008. Standardized method

for the instrumental determination of cooked spaghetti firmness. Cereal Chem. 85:440–444.

Skerritt, J.H., and R.H. Heywood. 2000. A five-minute field test for on-farm detection of pre-harvest sprouting in wheat. Crop Sci. 40:742–756.

Skylas, D.J., L. Copeland, W.G. Rathmell, and C.W. Wrigley. 2001. The wheat-grain proteome as a basis for more efficient cultivar identification. Proteomics 1:1542–1546.

Slade, L., and H. Levine. 1994. Structure function relationships of cookie and cracker ingredients. p. 23–142. *In* H. Faridi (ed.) The science of cookie and cracker production. Chapman and Hall, New York.

Sliwinski, E.L., P. Kolster, and T. van Vliet. 2004a. Large-deformation properties of wheat dough in uni- and biaxial extension: I. Flour dough. Rheologica Acta 43:306–320.

Sliwinski, E.L., P. Kolster, and T. van Vliet. 2004b. Large-deformation properties of wheat dough in uni- and biaxial extension: II. Gluten dough. Rheologica Acta 43:321–332.

Solah, V.A., G.B. Crosbie, S. Huang, K. Quail, N. Sy, and H.A. Limley. 2007. Measurement of color, gloss, and translucency of white salted noodles: Effects of water addition and vacuum mixing. Cereal Chem. 84:145–151.

Song, H.P., S.R. Delwiche, and M.J. Line. 1998. Moisture distribution in a mature soft wheat grain by three-dimensional magnetic resonance imaging. J. Cereal Sci. 27:191–197.

Stojceska, V., F. Butler, E. Gallagher, and D. Keehan. 2007. A comparison of the ability of several small and large deformation rheological measurements of wheat dough to predict baking behaviour. J. Food Eng. 83:475–482.

Surel, O., O. Darde, I. Tea, F. Violleau, and D. Kleiber. 2006. Suivi de la solubilisation des protéines par le sds au cours du pétrissage: Comparaison de trois pétrins. (In French, with English abstract). Sci. Aliments 26: 247–258.

Symons, S.J., and J.E. Dexter. 1992. Estimation of milling efficiency: Prediction of flour refinement by the measurement of pericarp fluorescence. Cereal Chem. 69:137–141.

Toufeili, I., B. Ismail, S. Shadarevian, R. Baalbaki, B.S. Khatkar, A.E. Bell, and J.D. Schofield. 1999. The role of gluten proteins in the baking of Arabic bread. J. Cereal Sci. 30:255–265.

Troccoli, A., G.M. Borrelli, P. De Vita, C. Fares, and N. Di Fonzo. 2000. Durum wheat quality: A multidisciplinary concept. J. Cereal Sci. 32:99–113.

Tronsmo, K.M., E.M. Færgestad, J.D. Schofield, and E.M. Magnus. 2003. Wheat protein quality in relation to baking performance evaluated by the Chorleywood bread process and a hearth bread baking test. J. Cereal Sci. 38:205–215.

Ur-Rehman, S., A. Paterson, and J.R. Piggott. 2007a. Chapatti quality from British wheat cultivar flours. LWT—Food Sci. Technol. 40:775–784.

Ur-Rehman, S., A. Paterson, and J.R. Piggott. 2007b. Optimisation of flours for chapatti preparation using a mixture design. J. Sci. Food Agric. 87:425–430.

Uthayakumaran, S., I.L. Batey, and C.W. Wrigley. 2005. On-the-spot identification of grain cultivar and wheat-quality type by Lab-on-a-chip capillary electrophoresis. J. Cereal Sci. 41:371–374.

van Vliet, T. 2008. Strain hardening as an indicator of bread-making performance: A review with discussion J. Cereal Sci. 48:1–9.

Vázquez, D., P.C. Williams, and B. Watts. 2007. NIR spectroscopy as a tool for quality screening. p. 527–533. *In* H.T. Buck, J.E. Nisi, and N. Salomón (ed.) Wheat production in stressed environments. Springer, New York.

Veraverbeke, W.S., and J.A. Delcour. 2002. Wheat protein composition and properties of wheat glutenin in relation to breadmaking functionality. Crit. Rev. Food Sci. Nutr. 42:179–208.

Verbyla, A.P., C. Saint-Pierre, C.J. Peterson, A.S. Ross, and R. Appels. 2007. Fourier modelling, analysis and interpretation of high-resolution mixograph data. J. Cereal Sci. 46:11–21.

Wang, G., J.M. Leonard, A.S. Ross, C.J. Peterson, and O. Riera-Lizarazu. 2007. Identifying QTLs for the extra-soft characteristic. *In* Abstracts, Plant and Animal Genome Conf. XV, San Diego, CA. 13–17 Jan. 2007. Scherago International, New York. Available at www.intl-pag.org/15/abstracts/PAG15_P05c_283.html (verified 24 Oct. 2007).

Wang, L., and P.A. Seib. 1996. Australian salt-noodle flours and their starches compared to U.S. wheat flours and their starches (1). Cereal Chem. 73:167–175.

Wesley, I.J., N. Larsen, B.G. Osborne, and J.H. Skerritt. 1998. Non-invasive monitoring of dough mixing by near infrared spectroscopy. J. Cereal Sci. 27:61–69.

Wesley, I.J., B.G. Osborne, R.S. Anderssen, S.R. Delwiche, and R.A. Graybosch. 2003. Chemometric localization approach to NIR measurement of apparent amylose content of ground wheat. Cereal Chem. 80:462–467.

Whalen, P.J. 2007. Extruded products and degree of cook. p. 75–84. *In* G.B. Crosbie and A.S. Ross (ed.) The RVA handbook. AACC International Press, St. Paul, MN.

Wheat Marketing Center. 2004. Wheat and flour testing methods: A guide to understanding wheat and flour quality. Wheat Marketing Center, Portland, OR.

Williams, P. 2000. Applications of the Perten SKCS 4100 in flour-milling. p. 7421–7424. *In* Association of Operative Millers Bulletin, March [Online]. Available at http://www.x-cd.com/opmillers/bulletins/65.pdf (verified 23 Dec. 2008). Williams, P., D. Sobering, and J. Antoniszyn. 1998. Protein testing methods at the Canadian Grain Commission. p. 37–47. *In* D.B. Fowler, W.E. Geddes, A.M. Johnston, and K.R. Preston (ed.) Wheat protein production and marketing. Proc. Wheat Protein Symp., Saskatoon, SK. 9–10 March 1998. Univ. Ext. Press, Univ. Saskatchewan, Canada.

Wilson, A.J., A.R. Wooding, and M.P. Morgenstern. 1997. Comparison of work input requirement on laboratory-scale and industrial-scale mechanical dough development mixers. Cereal Chem. 74:715–721.

Wooding, A.R., S. Kavale, A.J. Wilson, and F.L. Stoddard. 2000. Effects of nitrogen and sulfur fertilization on commercial-scale wheat quality and mixing requirements. Cereal Chem. 77:791–797.

Wrigley, C. 2006. Late-maturity alpha-amylase—apparent sprout damage without sprouting. Cereal Foods World 51:124–125.

Wrigley, C.W., I.L. Batey, F. Bekes, P.J. Gore, and J. Margolis. 1992. Rapid and automated characterisation of seed genotype using Micrograd electrophoresis and pattern-matching software. Appl. Theor. Electrophor. 3:69–72.

Wrigley, C.W., I.L. Batey, S. Uthayakumaran, and W.G. Rathmell. 2006. Modern approaches to food diagnostics for grain quality assurance. Food Aust. 58:538–542.

Xiao, Z.S., S.H. Park, O.K. Chung, M.S. Caley, and P.A. Seib. 2006. Solvent retention capacity values in relation to hard winter wheat and flour properties and straight-dough breadmaking quality. Cereal Chem. 83:465–471.

Xu, Q., J. Xu, C.L. Liu, C. Chang, C.P. Wang, M.S. You, B.Y. Li, and G.T Liu. 2008. PCR-based markers for identification of HMW-GS at *Glu-B1x* loci in common wheat. Cereal Sci. 47:394–398.

Zhang, Y., K. Quail, D.C. Mugford, and Z. He. 2005. Milling quality and white salt noodle color of Chinese winter wheat cultivars. Cereal Chem. 82:633–638.

Zhang, Q., Y. Zhang, Y. Zhang, Z. He, and R.J. Peña. 2007. Effects of solvent retention capacities, pentosan content, and dough rheological properties on sugar snap cookie quality in Chinese soft wheat genotypes. Crop Sci. 47:656–664.

Chapter 21

The Biochemical and Molecular Basis of Wheat Quality

Colin Wrigley, Robert Asenstorfer, Ian Batey, Geoffrey Cornish, Li Day, Daryl Mares, and Kolumbina Mrva

SUMMARY

(1) Wheat flour goes into pastries, cakes, cookies, breakfast cereals, confectionary, thickening agents, custards and sauces in western countries, and into a great diversity of foods in Asian and Middle Eastern countries—noodles, in their many forms, flat breads (pocket, Arabic, chapatti), and Chinese steamed breads. Wheat is also used in animal feeds. Starch and gluten from wheat provide raw materials for the manufacture of paper, adhesives, feed for aquaculture and pets, and biofuels.

(2) Distinct quality specifications apply for each of the diverse uses of wheat, involving the hardness and protein content of the grain, the water absorption of the milled flour, the viscosity of the starch, and the strength and extensibility of the resulting dough. Many of these specifications are directly attributable to specific proteins, which can be used to predict flour quality.

(3) The subunits of glutenin may be ranked according to their specific contributions to dough quality, and certain combinations are recommended as being suitable for quality targets in breeding or for quality-based segregation of grain at harvest. Specific puroindolines are suited to grain hardness requirements of flour milling.

(4) Starch pasting properties can be selected on the basis of alleles relating to isoforms of granule-bound starch synthase enzymes. Thus grain quality (suitability for processing) depends to a great extent on genotype or cultivar (the combination of genes), but also on the growth environment, especially on whether the developing grain has been subject to stress conditions.

(5) Lipid composition also affects processing quality. Nonpolar lipids, embedded within the gluten-protein matrix, may have a lubrication role in dough rheology. The polar glycolipids may be bound to the gliadin proteins hydrophilically and to the glutenin proteins hydrophobically, further affecting dough properties.

(6) The color of wheat-based food products depends on flour color and on enzymic reactions during mixing with water. For example, polyphenol oxidases (PPOs) cause darkening, especially of Asian noodles. Amylases, produced by sprouting or genetic defects, have adverse effects on the quality of most end products, particularly bread and noodles, causing stickiness, collapse of product structure, and poor color.

(7) New understanding of grain quality at the molecular level offers a basis to test for the causative compounds (or relevant genes), thereby simplifying the quality-screening process.

INTRODUCTION

Wheat flour is uniquely suited to the production of leavened bread, largely due to the viscoelastic properties of the gluten proteins. Thus, the protein content of wheat grain is a significant attribute in trade. Protein quality is also important, but it is not so readily measured as is protein quantity. Despite the importance of the protein component, it is said that wheat flour has just the "right" types of starch, lipid, and pentosan to suit it for breadmaking. It is thus appropriate to consider the full range of grain components when attempting to understand the biochemical basis of wheat quality.

We also need to study the range of uses of wheat; this range extends well beyond the narrow concept of "a loaf of bread," to include the great diversity of wheat-based foods around the world, as well as wheat's many industrial uses (Faridi and Faubion 1995). Each of these uses has its own unique set of grain- and flour-quality specifications, which are ultimately determined by the appropriate consumer, via the processor (Fig. 21.1). Provision of these quality specifications starts with the breeder, who is responsible for creating the genotype (cultivar), based ultimately on feedback from the customer. The quality potential "built in" by the breeder is, in turn, modified by growth and storage conditions (Fig. 21.1).

These quality specifications include phenotypic attributes such as dough quality, starch-pasting properties, possible grain defects, and enzymic activities. If these attributes can be reduced to the biochemical and molecular levels, the breeder's task is likely to be simplified, because inheritance is simpler for a specific target protein than for a quantitative trait such as dough quality.

DIVERSITY OF WHEAT UTILIZATION

The range of uses of wheat—western foods

In the western diet, pan bread is the most obvious food product made from wheat, generally "bread

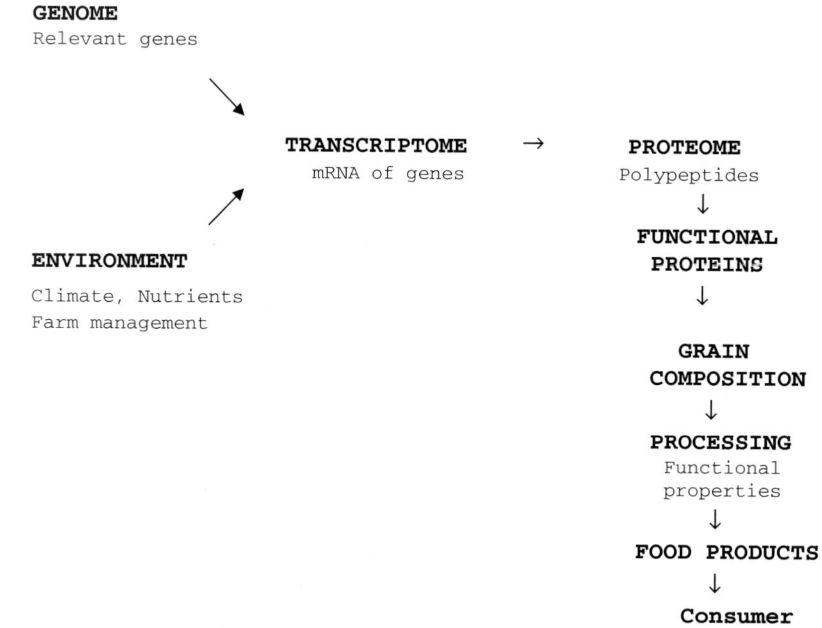

Fig. 21.1 The pathway of grain quality from breeder (genome) and grain-grower (environment) to the consumer, via molecular aspects of grain composition.

wheat"—the hexaploid species, *Triticum aestivum*. Even for this common product, there is a wide range of types, depending on whether wholemeal or white flour has been the raw material, depending on the shape and size of the loaf, and depending on the range of possible additives such as rye flour, milk solids, and whole or kibbled (coarsely ground) grains (Khan and Shewry 2009). In addition, different methods of production may have led to the final product presented to the customer. For example, the multiple sheeting method may have been used to develop the dough instead of conventional mixing, and even the latter may have involved the application of pressure to improve crumb structure. Specialty breads include products for special diets, such as gluten-free, yeast-free, and milk-free. The western diet also includes a wide range of pastries, cakes, and cookies (biscuits). Wheat flour is the main raw material for many food products such as soups, porridge, breakfast cereals, confectionary, sausages, thickening agents, condiments, custards, and sauces.

Animal feed constitutes a significant use of wheat (about 15% globally). In this case, wheat may make an important (low-cost) contribution to energy needs, while various other ingredients are included for other purposes, such as complementing the low lysine content of wheat. Grains may be fed as the whole grain, as partly milled, or in pellet form. This use of wheat is especially cost-sensitive, so the lower grades of wheat usually contribute to feed uses.

The pasta family of foods adds a further dimension of variety in its many forms—macaroni, spaghetti, and vermicelli (Fabriana and Lintas 1988). Durum wheat is the preferred raw material for pasta, and also for couscous and burghul (also known as bulgar, bulgur, or bulghur) (Bayram 2007). Pasta should not be confused with noodles, in their many forms, which are made from bread wheat. Durum is a species distinct from hexaploid bread wheat, being tetraploid of the species *T. durum*, also designated *T. turgidum* var. *durum*.

The range of uses of wheat—"exotic" foods

A much wider range of wheat-based foods exists beyond the traditional "western" range, but these, previously labeled as exotic, have in recent decades become familiar around the world (Faridi and Faubion 1995). These include foods traditional in Asian and Middle Eastern countries—noodles, in their many forms, flat breads (pocket, Arabic, chapatti), and Chinese steamed breads. Some of these are listed in Table 21.1 and illustrated in Color Plate 35.

Industrial uses of wheat

Significant amounts of wheat flour are used for separating starch from gluten (final product listed in Table 21.1). Much of the resulting "vital wheat gluten" is used to fortify the dough strength of flour for breadmaking and for special snack foods. Being a low-priced food protein, gluten has also found many food uses where protein supplementation is needed, for example, in meat and cheese products. This supplementation, in the form of vital dry gluten, may make use of the unique rheological properties of gluten, or in other cases denatured gluten is used when these properties are not appropriate.

Gluten also provides its unique functional properties to special animal feeds, especially for pet foods as a calf-milk substitute and for aquaculture (Day et al., 2006). Gluten-based packaging films and coatings are favored as edible and biodegradable air barriers. After modification by acidic or enzymatic hydrolysis, gluten may serve ingredient uses for emulsifying and foaming in foods. The many nonfood applications of gluten include grafting into resins and polymers, moisturizing in cosmetics, and foaming in hair-care products.

The starch from gluten washing finds many nonfood uses, especially for paper manufacture, and even for mineral refining. For home cooking use, wheat starch may find its way into the supermarket labeled as corn flour, although not derived from corn or maize (*Zea mays* L.). It is an

Table 21.1 Quality attributes preferred in wheat used for specific products.

Product	Grain Protein	Grain Hardness	Water Absorption	Dough Strength	Farinograph Dough Development Time	Peak Viscosity
Breads						
Pan bread						
Sponge-and-dough	>12%	Hard	High	Strong	High	High
Straight-dough	>12%	Hard	High	Medium	Medium	
Rapid-dough	>11%	Hard	High	Medium	Short	
Flat bread						
Middle Eastern	10.5–12%	Hard	High	Medium	Medium	Medium
Indian Subcontinent	10–12%	Hard	High	Medium	Short	
Tortillas	10.5–12%	Hard	Medium	Medium	Medium	
Steamed bread						
Northern style	11–13%	Hard	High	Medium	Medium	
Southern style	10–12%	Medium-hard	Low-medium	Medium	Medium	
Noodles						
Yellow alkaline	11–13%	Hard	High	High	Medium	Medium
White salted	10–12%	Soft and hard	Medium	Medium	Medium	High
Instant	10–12%	Hard	Medium	Medium	Medium	
Cookies and cakes	8–9%	Very soft	Low	Weak	Low	Medium
Pasta (from durum wheat)	>13%	Extremely hard	Not high (semolina)	Strong	High	Medium
Starch-gluten[a]	>13% preferred	Soft preferred	Low preferred	Medium	High	High preferred

Sources: Adapted from Wrigley and Bekes (2004) and Cornish (2007) and updated with advice from Di Miskelly and Bob Cracknell (pers. comm.).
Note: In all cases, good milling is required, giving a high yield of white flour.
[a]Genotypes preferred for starch-gluten manufacture should have soft grain, high protein content, and high proportions of large (A-type) starch granules. However, economic factors may outweigh these considerations.

ingredient of food products such as icing-sugar mix, baking powder, frosting, chewing gum, marshmallows, salad dressing, frozen desserts, and coffee whitener, in addition to a large number of nonfood uses (Lawton 2004). There is a further range of uses for chemically and physically modified wheat starch (BeMiller 2004).

The production of biofuels represents an increasing use of cereal grains, mainly via the fermentation of starch into ethanol. Corn (maize) is a primary raw material for this purpose, but wheat use for ethanol production has long been linked to the production of vital wheat gluten, with high-grade starch being used for special purposes, leaving poorer starch for fermentation (Maningat and Bassi 2004).

PROCESSING SPECIFICATIONS FOR WHEAT UTILIZATION

Wheat, once a commodity, is now traded to a significant extent with detailed specifications defined by processor requirements (see Chapter 19 and Khan and Shewry 2009). In addition, niche markets are arising for which even stricter specifications must be met. Basic specifications relate to soundness, lack of contaminants and defects, bulk weight (indicating grain plumpness), grain color (red or white), grain hardness (hard or soft), and protein content. Undesirable defects for all products include the presence of α-amylase activity, either from preharvest sprouting or from a genetic defect termed late-maturity α-amylase (LMA). For products with color specifications, low levels of polyphenol oxidase are desirable.

The more detailed range of specifications for individual products (listed in Table 21.1) provides a basis for breeders in targeting cultivars tailored to specific grades, as well as forming grade targets for postharvest segregation and marketing. The level of dough strength varies widely among these products, but good dough extensibility is generally desirable.

Pan bread

As the products in Table 21.1 are made from white flour, milling quality is a primary requirement (Posner and Hibbs 2005). The grain should provide a high yield of white flour. In addition, for bread baking, the miller prefers grain that will produce flour with high water absorption to increase economic returns. Hard-grained wheat is thus used to produce flour of moderate starch damage, ensuring high water absorption and adequate substrate to produce fermentable sugars.

High protein content and good dough strength are needed for long-fermentation baking procedures to ensure that the bubble structure developed during mixing does not collapse during fermentation. However, medium dough development time may be preferred to minimize the mixing energy required to develop the dough fully. A dough with long stability to overmixing offers flexibility in manufacture. Bread is also produced commercially by the rapid-dough and sponge-and-dough processes, and optimum ranges of protein content differ slightly from that for the straight-dough—bulk-fermentation process. Bright white flour (high L*) with low yellow color (low b*) is preferred, although lipoxygenases from added soy flour may bleach nonwhite color present in the original flour.

Flat bread

Flat bread, popular in Middle East and in the Indian Subcontinent, varies considerably in form and quality characteristics (Table 21.1). In general, its production involves short high-temperature baking. As various flat breads have a high ratio of surface area to volume, it is important that moisture is retained in the dough for as much as possible of the short baking time. The flour (or wheat meal) should thus have very high water absorption.

For pocket bread, a dough of high extensibility and medium dough strength is needed for the production of circular dough pieces of appropriate thickness. Poor dough strength results in an elliptical shape. Excessive dough strength causes the dough piece to shrink after sheeting, so that the dough is too thick, leaving a "doughy" loaf with blisters on the circumference.

Yellow alkaline noodles

Flour specifications for yellow alkaline noodles are similar to those for pan bread (Table 21.1), but regional differences occur across Asia in requirements for protein content. A high peak-paste viscosity for starch may not be as important for yellow alkaline noodles as for pan bread.

White salted noodles

However, high starch-paste viscosity is required to produce smooth white-salted Udon noodles to provide the required "al-dente" mouth feel. Color is important for this product, as Japanese consumers prefer cream rather than white Udon noodles. The target color registers as a high L* (flour brightness) and a medium b* (flour yellowness). Flour brightness is achieved by milling Udon flour to low extraction (typically 50%–60% flour yield) thus also achieving a low flour-ash content. Preferences in China and Korea lead to the milling of hard wheat to achieve white noodle color (thus a low b* value).

Cookies (biscuits) and cakes

Biscuit flour is milled from soft wheat to give low starch damage and to minimize water absorption, thereby reducing the baking time needed to dry out the product. An extensible dough of low strength ensures excellent sheeting characteristics, as is needed to produce biscuits with a consistent size and shape.

Cake flour is also milled from soft wheat to provide low water absorption. This type of flour is generally provided in the supermarket for household use, presumably on the assumption that much home use is for cake and cookie baking.

Chinese steamed bread

Heating dough by means of steam produces buns that are white, without the characteristic brown crust of oven-baked breads. Steamed breads are popular throughout China and in the wide range of Chinese diets around the world. Flour specifications for regional preferences of steamed bun are given in Table 21.1, indicating a range of protein content, with generally medium dough strength depending on the role of fat in the dough.

Starch-gluten manufacture

Desirable products of this manufacturing process are gluten of good strength and extensibility, and starch with minimal damage and high paste viscosity. To achieve these objectives, wheat milled should be strong and soft, but this combination has not been a common breeding target, so starch-gluten manufacturers must usually accept flour milled from strong hard wheat. Nevertheless, limited breeding efforts have recently favored the specifications of the starch-gluten industry. Furthermore, given the increasing importance of ethanol production for fuel, breeding targets for this application are being favored for suitable cereals.

Pasta

Durum wheat for pasta production is intrinsically very hard, with its dough-strength potential being selected for strong. Milling of the very hard durum grain produces semolina of coarse particle size in the range 130 to 550 μm. The gluten protein must be strong to form a continuous covering of the endosperm particles so that the pasta surface does not become sticky during extrusion and cooking.

PROTEIN COMPOSITION AND WHEAT QUALITY

Virtually all the flour specifications listed in Table 21.1 are related to protein composition, and many are directly produced by specific groups of proteins. Given the importance of proteins in wheat quality, it is now possible to specify marker proteins as a basis for predicting wheat or flour quality, reducing the need to perform the appropriate quality test, such as dough testing, baking, or starch pasting (Cornish et al., 2001, 2006).

However, in most cases the marker proteins are polypeptides (part of the proteome), not necessarily functional proteins. This distinction is shown in Fig. 21.1. Gene action results in the production of messenger RNA (the transcriptome), and thence the polypeptide, whose amino-acid sequence is specified by the respective DNA and RNA sequences. Thereafter, folding and disulfide bonding leads to the production of the functional protein then capable, for example, of enzymic activity. Proteome analyses have demonstrated the presence of about 1,700 polypeptides in developing wheat endosperm, but even this large number is only about 30% of the number of polypeptides that should be present, given the identification of over 6,000 mRNA species at this stage of endosperm development (Skylas et al., 2005).

Dough quality and functional proteins

The glutenin proteins exemplify the important distinction between polypeptide and functional protein. Traditionally, they have been examined as polypeptides (subunits), extracted after the rupture of all disulfide bonds. However, it is the native polymers of the glutenin subunits that make the contribution to dough strength; the subunits alone cannot contribute to dough strength (Table 21.2). In contrast to the glutenin proteins, the gliadin polypeptides are synthesized as single chains, with disulfide bonds formed within the polypeptide chain. Table 21.2 also lists the gene designations for the respective groups of gliadin and glutenin polypeptides.

The gliadin proteins have a molecular-weight range up to about 100,000 Da, contributing to the viscosity and extensibility of the dough. However, genetic assessments of the functional properties of the *Gli-1* controlled gliadins are confused by their tight linkage to the *Glu-3* genes controlling the low-molecular-weight (LMW) subunits of glutenin (Shewry et al., 2003).

The polymeric glutenin proteins have molecular sizes ranging up into the tens of millions of Daltons (Southan and MacRitchie 1999; Wieser et al., 2006). The great length of these chains is presumed to confer on wheat its unique dough-forming properties. Conversely, dough strength increases as the proportion of very large glutenin polymers increases. However, excessive dough strength is inappropriate even for products such as pan breads and pasta (Table 21.1). A balance of strength and extensibility is needed. The level of this balance depends on processing requirements, indicated in Table 21.1 as dough strength.

This balance is apparent in the molecular weight distribution of the polymeric glutenin proteins, and also in consideration of the balance of glutenin to gliadin, which can again be seen as a very broad molecular-weight distribution (Wrigley et al., 2006). Analysis of the molecular-weight distribution has generally been performed by size-exclusion

Table 21.2 Marker proteins (and thus genes) for specific aspects of grain quality.

Locus Designation	Polypeptide	Functional Protein	Quality Attribute
Gli-1	Gliadin proteins	Gliadin proteins	Modest contribution to dough viscosity and extensibility[a]
Gli-2	Gliadin proteins	Gliadin proteins	
Glu-1	HMW subunits of glutenin	Polymeric glutenin proteins	Major contribution to dough strength
Glu-3	LMW subunits of glutenin		Significant contribution to extensibility[a]
Pin-a	Puroindoline a	Puroindoline a	Grain hardness
Pin-b	Puroindoline b	Puroindoline b	Grain hardness
Wx-1	GBSS[b]	GBSS	Starch properties

[a]Contributions of *Gli-1*, *Gli-2*, and *Glu-3* require further investigation for more precise assessment.
[b]Granule-bound starch synthase.

HPLC, but this methodology does not provide fractionation up into the very large size range of the glutenin polymers. Field-flow fractionation (FFF) has no upper limit to its separation potential. In the examples of FFF profiles in Fig. 21.2, the upper limit of size distribution is not indicated precisely, because of difficulties in this set of experiments in obtaining sufficiently large proteins for calibration purposes.

There is still uncertainty about the structure of the polymeric glutenins. As Fig. 21.2 indicates, they cover a wide continuous molecular-weight range. There is also uncertainty about how the glutenin polypeptides are disulfide cross-linked together. A linear backbone of high-molecular-weight (HMW) subunits is likely, forming an elastic structure, possibly via the beta-spirals of their central domains (Wieser et al., 2006). No doubt, the very long polymers become entangled, thereby contributing to the combined effects of resistance to extension and viscous drag. These various attributes have earned the glutenin polymers the reputation of being "amongst the most complex aggregates in nature" (Wieser et al., 2006), as well as being among nature's largest proteins.

Dough quality and polypeptide composition

A few decades of intense study of the subunits of glutenin (reviewed by Shewry et al., 2003) have elucidated much of their contributions, as component polypeptides, to the dough properties of glutenin. The HMW subunits of glutenin (70,000 to 100,000 Da) have received more attention than the LMW subunits, historically, because the greater size range of the HMW subunits holds them at the top of SDS-gel electrophoresis patterns, without interference from other flour proteins. By contrast, initial extraction of gliadins is necessary for the LMW subunits (20,000–50,000 Da) to be seen clearly in SDS-electrophoretic patterns. The HMW and LMW subunits are coded by genes at distinct loci (Table 21.2) on the long and short arms, respectively, of the homoeologous group-1 chromosomes. The *Glu-1* locus is complex, involving pairs of paralogous x- and y-type genes. The frequencies of gluten-protein alleles in wheat worldwide have been reviewed by Bekes et al. (2007).

Allelic variation at the *Glu-1* loci (HMW subunits, Table 21.3) correlates with differences in the genetic potential for rheological properties (especially dough strength as R_{max}, the height of the extensigraph curve) and thus breadmaking quality (Payne et al., 1987; Cornish et al., 2001, 2006; Vawser et al. 2002; Eagles et al., 2006). Table 21.3 combines the results of these and other recent publications on the rankings of HMW subunits. This list goes beyond the maximum score of 4 allocated by Payne (1987) for HMW subunits 5 + 10, due to the great contribution of the over-expressed 7 subunit. The addition of a zero score should not imply that alleles on the bottom line contribute nothing to dough strength, but rather that they are the least contributors to R_{max} overall. Some alleles (Glu-A1p, Glu-B1b, and Glu-B1a) are not included in Table 21.3 as they are rare, and reliable estimates of their ranking is difficult because there is limited data about them. Furthermore, they are represented to a very small extent in current cultivars around the world. For example, normal subunit 7 was originally identified (with subunit 8) in the reference cultivar Chinese Spring. We now know that the band 7 of Chinese Spring's 7 + 8 pair differs slightly from what is more commonly found; thus the commonly encountered pair of subunits is designated

Table 21.3 Dough strength rankings and allele designations (as lowercase letters) for HMW subunits of glutenin (*Glu-1*).

Dough-Strength Score	*Glu-A1*	*Glu-B1*	*Glu-D1*
5		*al* (7̲ + 8)	
4			*d* (5 + 10)
3	a (1) *b* (2*)	*f* (13 + 16)	
2		*u* (7* + 8) i (17 + 18)	
1	c (null)	c (7 + 9)	a (2 + 12) c (4 + 12)
0		*d* (6 + 8) e (20x + 20y)	*b* (3 + 12) *f* (2.2 + 12)

Note: Subunit 7̲ is so indicated as it is overexpressed for this allele.

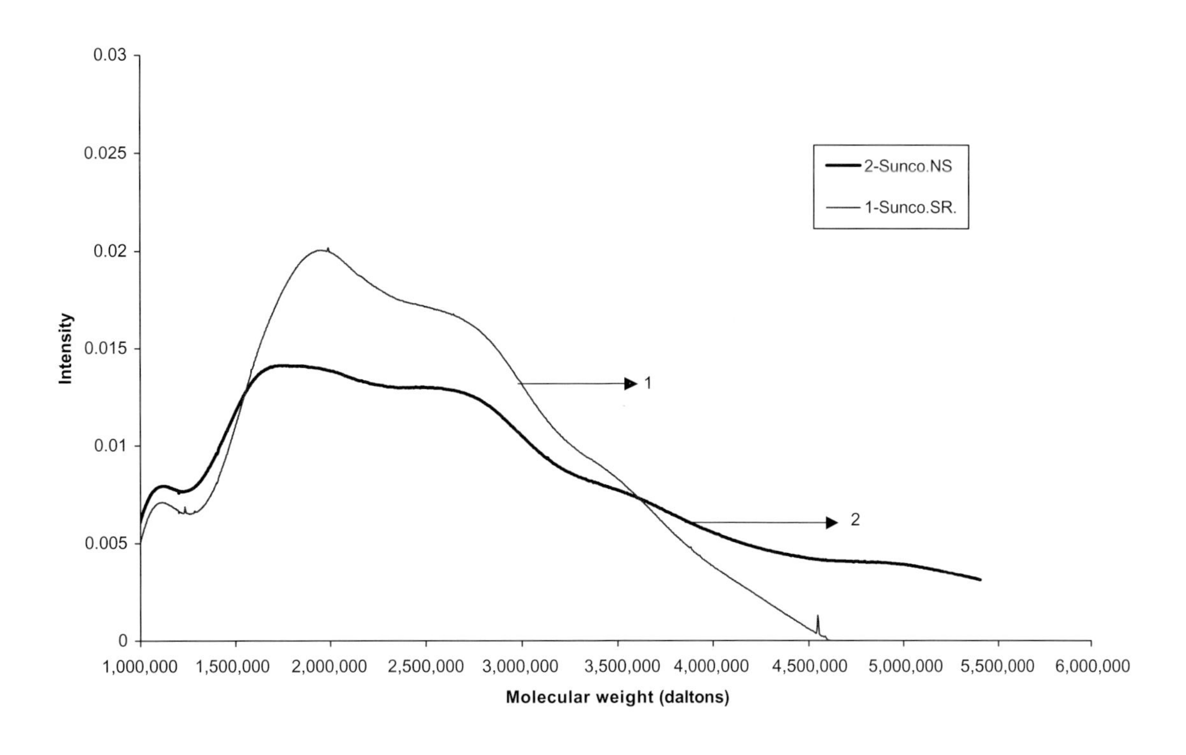

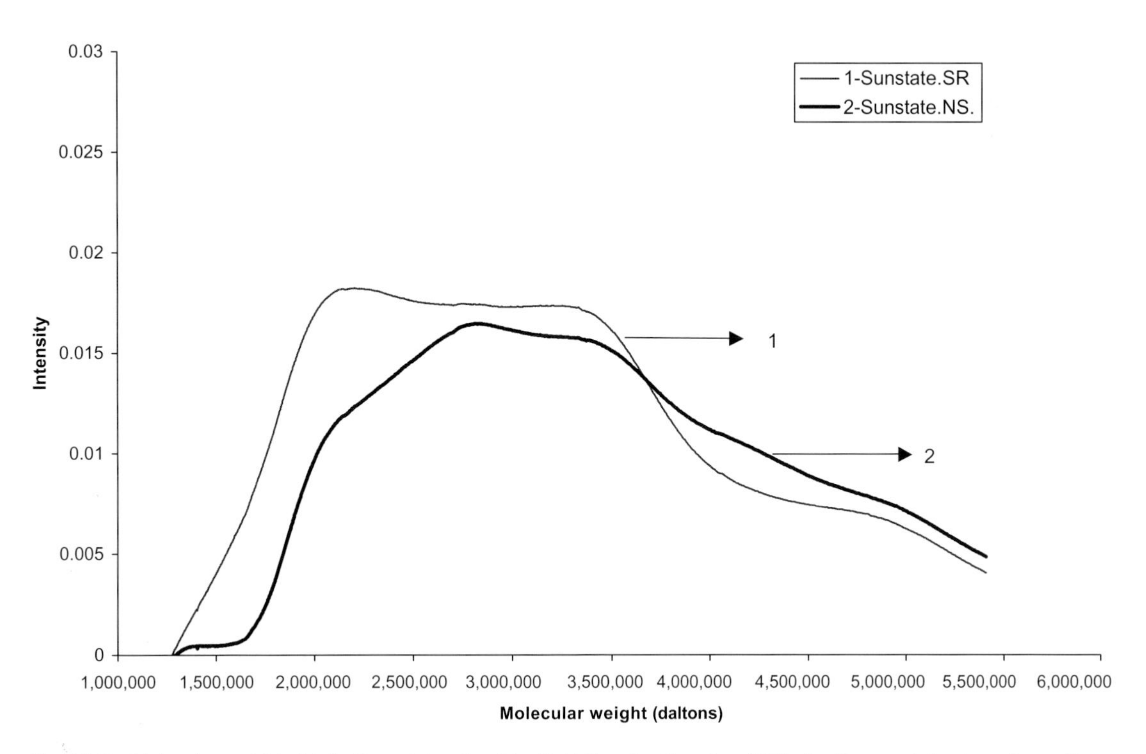

Fig. 21.2 Molecular-weight distribution, determined by field-flow fractionation (FFF), for flour proteins from the cultivars Sunco (top) and Sunstate (bottom), whose plants have undergone heat stress (SR) or have not been stressed (NS) (Daqiq 2002). Refer to Table 21.6 for dough properties of the respective flours.

7* + 8, which has been assigned the allele designation *Glu-B1u*, indicated for example in the cultivar Sunco. Normal 7 + 8 is rare, and there is insufficient rheological data for it to be included in Table 21.3.

There is much more (about threefold) of the LMW subunits in glutenin compared with the HMW subunits, but the contribution of the HMW subunits to dough resistance is more significant than would be expected by their smaller proportion in the glutenin fraction. The LMW subunits do contribute to dough strength, but their great importance relates to extensibility (Table 21.2). Therefore, selection against null alleles at *Glu-1* and *Glu-3* (e.g., *Glu-A1c* and *Glu-A3e*) has the important effect of increasing dough strength and extensibility (Table 21.4). Null alleles are thus seen as undesirable. At the other extreme, a few of the *Glu-1* alleles (e.g., *Glu-B1al*) are overexpressed (subunit 7 in Table 21.3), so that the resulting subunits appear in the endosperm in more than the normal amounts for other subunits; as a result, the presence of these alleles causes major changes (increases) in the genetic potential for both dough strength and extensibility. Overexpression of glutenin subunits has also been observed in transgenic wheat (Rooke et al., 1999).

The "*Glu-1*" score system of Payne et al. (1987) was designed to indicate the individual contributions of specific HMW subunits (and thus the associated genes) to baking quality. Despite the age of this scheme, and despite the lack of inclusion of the LMW subunits and overexpressed genes, the *Glu-1* score is an effective tool for breeders wanting to target the genetic potential of progeny for specific dough-quality attributes. Furthermore, predictions based on gluten alleles offer advantages at all stages of the "grain chain." More recently, fuller systems have been developed taking into account the results of subsequent research, such as is listed in Table 21.3. Given the complex nature of processing specifications (Table 21.1), it is appropriate to provide recommendations (Table 21.5) for the ideal combinations of glutenin alleles that would be expected to provide dough quality attributes appropriate to various processing needs.

A valuable application of this type of information is the selection of parent lines with sets of alleles suited to the targeted quality outcome. To assist in this respect, glutenin subunits have been catalogued for large numbers of cultivars and lines worldwide; they are available in McIntosh et al. (2003) or at the web site of AACC International http://www.aaccnet.org/grainbin/gluten_gliadin.asp.

Table 21.4 Dough-strength rankings, from highest (top) to lowest (bottom), for LMW glutenin subunits (*Glu-3*).

Glu-A3	*Glu-B3*	*Glu-D3*
d	b; d; g; m	d; f
b	h	e
c	a	a; c; b
f	c	
a		
e		

Table 21.5 Best quality alleles for various wheat products.

Genes	Pan Bread	Flat Bread	Yellow Alkaline Noodles	White Salted Noodles	Cookies and Cakes
Glu-A1	a, b	a, b	a, b	a, b	c
Glu-B1	c, f, i, u, al	c, f, i, u	c, f, i, u, al	c, f, i, u	d, e
Glu-D1	d	a	a, d	a	a, c
Glu-A3	b, d	b, c	b, d	b, c	b
Glu-B3	b, g	b, d, h	b, g	b, d, h	b, d
Glu-D3	a, b	a, b, c	a, b	a, b, c	a, b, c
Pina-D1	a	b	a	a	a
Pinb-D1	b	a	b	a	a
Wx-B1	b	a, b	a	b	a, b

Source: Adapted from Cornish (2007).

Grain hardness

The quality specifications for processing wheat-based foods (Table 21.1) include many factors in addition to dough quality, so allelic recommendations are also provided in Table 21.5 for appropriate puroindoline genes for grain hardness. The puroindolines are basic lipid-binding proteins, rich in cysteine, of about 13,000 Da (Gautier et al., 1994; Morris 2002; Jones et al., 2006). They belong to the 2S albumin superfamily of proteins.

The puroindolines are encoded by genes at two loci on the short arm of chromosome 5D: *Pina-D1* (two alleles, *a* and *b*) and *Pinb-D1* (many alleles, *a-g*, *l*, *p*, *q*) (Morris 2002; Jones et al., 2006; Wanjugi et al., 2008). Bread wheat generally contains two types of puroindolines differing slightly in size. Soft wheat possesses both puroindoline a and puroindoline b, due to the allele combination *Pina-D1a* and *Pinb-D1a* (Table 21.5). As their starch granules are loosely attached to the protein matrix, soft wheat crushes easily, producing largely intact starch granules and fine flour. On the other hand, the starch granules of hard wheat are tightly bound to the protein matrix, requiring greater milling energy and producing coarser flour with higher levels of starch damage.

The resulting distinctions between flours from hard and soft wheats suit them respectively for distinct products (Table 21.1). In hard wheat, a common combination of *Pin* alleles is the *Pina-D1b* null allele with *Pinb-D1a*. Other hard wheat genotypes have the combination of *Pina-D1a* with *Pinb-D1b* down to *Pinb-D1g*, and beyond alphabetically. Durum wheat lacks the *Pin* loci completely, with the result that durum grain is very hard (Jones et al., 2006).

The puroindoline alleles are not evenly distributed among germplasms across the world, so for any specific regional set of wheat genotypes, all alleles may not be represented. There is thus motivation for breeders to expand the range of puroindoline genes in use, thus to extend the range of endosperm textures available. The specific alleles recommended in Table 21.5 relate to hardness specifications for the products listed in Table 21.1.

Starch pasting properties

Table 21.5 also lists allele recommendations for the waxy locus (*Wx-B1*) for granule-bound starch synthase. White salted noodles (Table 21.1), in particular, benefit from the "Null-4A" gene (*Wx-B1b* null allele) as further discussed later in the section on starch granules and in Chapter 22.

Protein composition and genotype identification

Finally, protein composition has long served the valuable purpose of indicating genotype (cultivar) identity. In 1965, Zuckerkandl and Pauling (1965) classed proteins as "episemantic molecules," that is, their "meaning" (semantic significance) in relation to genotype lies a few steps away from the genome (Fig. 21.1). DNA was classed as a "primary semantide," and RNA as a "secondary semantide." This hierarchy draws attention to the possibility that environmental influences may corrupt the information about genetic origins when analyzing protein composition.

Nevertheless, grain protein composition has served well as a basis for determining the genetic identity of wheat samples for many years, based on the composition of gliadins and/or of glutenin subunits, using acidic or SDS-gel electrophoresis, reversed-phase HPLC, or capillary electrophoresis (Uthayakumaran et al., 2006). In this general application, the protein composition provides a "fingerprint," not necessarily providing any indication of relationship to grain quality. In the many years of routine use, the possible confusing factor of growth conditions has not posed problems, with the occasional exception of severe sulfur deficiency, which causes the proportion of ω-gliadins to become unusually high.

Application of principles: Defects explained

An understanding of composition–function relationships leads to the elucidation of ways in which grain defects reduce processing efficiencies. For example, knowledge of the reliance of dough strength on the very large glutenin polymers makes it clear that dough weakening is inevitable due to the action of proteases deposited in the developing grain by *Eurygaster* insects. As a result, grain that is severely damaged by *Eurygaster maura* is unsuitable for breadmaking purposes, and is thus downgraded at harvest (Sivri et al., 1999, 2004). If, in this case, dough quality has been lost due to a major reduction in the molecular-weight distribution of the gluten proteins, it should be possible to restore dough strength by re-forming cross-links between the glutenin polymers. This restoration has been demonstrated with the use of transglutaminase as a means of producing new cross-links between glutenin chains (Koksel et al., 2001).

Dough strength may also be reduced by the environmental influence of heat stress during grain filling (Table 21.6), due to interference with the process of disulfide-bond formation after polypeptide synthesis, thereby reducing the proportion of large glutenin polymers (Blumenthal et al., 1995). The mechanism of this defect is illustrated (Fig. 21.2) by the shift toward a lower molecular-weight distribution in the FFF profile of proteins from grain samples of plants that had been heat-stressed during grain filling (Daqiq 2002).

These examples illustrate the importance of disulfide bonds between the polymeric glutenins (Wrigley and Bekes 1999). The process of disulfide-bond formation is understood to continue slowly after the initial synthesis of the glutenin proteins, even continuing during grain storage if under warm conditions (e.g., at over 30°C for a few months) when flour milled from the grain is found to have increased dough strength (Wrigley and Bekes 1999). Lack of awareness of this phenomenon has caused difficulties in the past for grain handlers and millers.

LIPID COMPOSITION AND WHEAT QUALITY

In contrast to protein and starch, the lipid fraction is a minor component of the wheat grain, constituting about 3%–4% of whole grain weight, and even less (1%–2.5% by weight) in the endosperm (straight-run milled flour). When protein is removed from wheat flour, dough and breadmaking properties are completely lost. In contrast, dough properties and breadmaking capacity are largely retained when the lipids are removed from the flour. However, flour lipids do play an important role in the dough-mixing and baking processes. They interact and form complexes with gluten protein and contribute to the stabilization of gas-cell structure, thus having significant effects on loaf volume and on final texture.

Lipid composition and distribution

Wheat lipids are a complex mixture of components. They are unevenly distributed throughout the various parts of the wheat kernel. One-third of the total lipid fraction is located in the germ, which accounts for only about 4% of the total grain by weight; therefore, the germ has the highest lipid content (Fig. 21.3).

More than 20 classes of lipids exist in wheat, each containing numbers of separate entities (Morrison 1978; Hargin and Morrison 1980). They can be divided into two groups, the nonpolar and polar lipids. Triglycerides are the major nonpolar lipids, representing about 50% of total nonpolar lipids in wheat. They are deposited in spherosomes (oil droplets) bounded by a monolayer membrane, and this is the form in which plants usually store lipids. The remainder of the

Table 21.6 Loss of dough strength (as R_{max} in Brabender units) for grain harvested with or without three days of heat stress at two sites in northern NSW, Australia.

Variety (HMW Alleles)[a]	No Heat Stress	Heat Stressed	Change in Dough Strength Due to Heat
Sunco (*aua*)	550	205	63% loss
Sunstate (*aid*)	660	380	42% loss

[a]*Glu-1* alleles are indicated in order for the A, B, and D genomes.

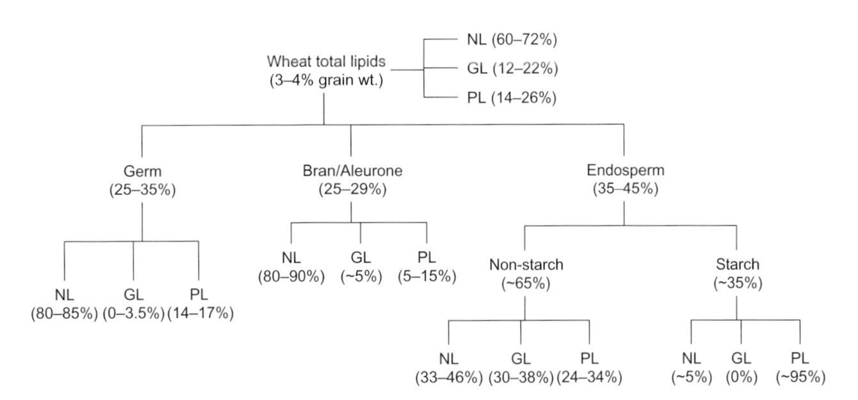

Fig. 21.3 Anatomical distribution of lipid classes in wheat grain. The distribution of total lipid is shown under the names of the three main anatomical parts of the grain, namely, the germ, bran–aleurone, and endosperm (NL, nonpolar lipids; GL, glycolipids; PL, polar lipids).

nonpolar lipids are mainly di- and monoglycerides, free fatty acids, and sterol esters.

The main components in the polar lipids are glycolipids and phospholipids found in all membranes, including the amyloplast membrane. They are organized in a reversed hexagonal liquid cystalline phase located in a protein matrix closely bound to starch granules in the endosperm of wheat kernels (Al Saleh et al., 1986). The principle glycolipids in whole kernels and starchy endosperm are monogalactosyldiglyceride (MGDG) and digalactosyldiglyceride (DGDG), with smaller amounts of the corresponding monoacyl lipids. The major phospholipids are phosphatidylcholine (PC), phosphatidylethanolamine (PE), and phosphatidylinositol (PI). Monoacylphosphoglycerides, or lysophospholipids, are usually regarded as degradation products of phospholipids.

In milled endosperm flour, the amount of glycolipids is often greater than the amount of phospholipids. Apart from those major acyl lipids, other lipids in wheat include sterols, and lipid-associated compounds such as carotenoids and tocopherols. Although carotenoids are very minor constituents, color contributed by carotenoids is an important factor in the use of cereal grains in food production, particularly in the use of durum wheat for pastamaking.

The nonpolar lipids are concentrated in the outer layers of the kernel, or the bran and aleurone layers (Hargin and Morrison 1980). The endosperm, the major fraction of wheat grain, has significantly lower lipid content than the other fractions (Fig. 21.3). Hence, the lipid composition and content in the flour can differ, depending on the milling process and the flour yield (Morrison and Hargin 1981). It is also important to note that lipid composition and content are highly dependent on combined genetic and agroenvironmental factors. Lipid hydrolysis can occur during storage of wheat grain and wheat flour due to the presence of enzymes, such as lipases and phospholipases. The consequent increase of free fatty acids can have deleterious effects on the quality of wheat flour (e.g., increased rancidity) and on the final product.

Lipids in flour (not wholemeal) can be subdivided into nonstarch and starch lipids. Starch lipids are predominantly of phospholipids (Fig. 21.3), and almost exclusively lysophospholipids, in particular lysophosphatidylcholine or lysolecithin. They are tightly bound as an inclusion compound with amylose. These amylose–lipid complexes may affect the properties of starch to some extent, but are not available for interacting with gluten protein during dough mixing and gluten formation. Consequently, starch lipids play little or no part in dough rheological properties during processing. The nonstarch lipids, which consist of all the endosperm lipids excluding those inside starch granules, comprise approximately two-thirds of the total wheat flour lipids, consisting of nonpolar lipids, predominantly triglycerides, and relatively high amounts of glycolipids and phospholipids (Fig. 21.3). It is the nonstarch flour lipids (in addition to lipid additives) that interact with flour proteins during dough mixing, with proven technical value in baking processing.

Interaction with gluten proteins

Lipids become bound or associated with gluten protein to form protein–lipid complexes during dough mixing (Chung 1986). Gluten that has been water-washed from flour contains much higher lipid content than flour. The lipid in gluten cannot be readily extracted using nonpolar solvents. However, the association of lipids with proteins in gluten often depends on the conditions used to isolate the gliadin and glutenin fractions. By dissolving gluten in dilute acetic acid, then fractionating it with 70% ethanol, more lipid, particularly polar lipids, are found in the gliadin protein fraction than in the glutenin protein fraction; but when gliadin and glutenin are fractionated in acid media by pH precipitation, more than 80% of gluten lipid is found to be associated with the glutenin protein and very little with the gliadin (Chung 1986).

Gluten is not composed of discrete molecules. It is a water-insoluble aggregated network in which protein–protein, lipid–lipid, and protein–lipid interactions can occur (Frazier et al., 1981). Therefore, physical techniques are more desirable to probe these interactions for the whole gluten system. These studies have shown that lipids are retained with the gluten network in a fairly nonspecific way through a combination of forces, involving the physical entrapment of lipids and also by polar and/or ionic bonding between protein and the surfaces of the lipid phases (Marion et al., 1987).

Our study (McCann et al., 2008) suggests that nonpolar lipids (primarily triglycerides), glycolipids, and phospholipids have different interactive mechanisms with gluten proteins. Nonpolar lipids are likely embedded within the gluten protein matrix as lipid vesicles, therefore providing a lubricating or plasticizing effect to gluten and dough rheology. Glycolipids may be involved in a gliadin–glycolipid–glutenin complex, in which glycolipids are bound to the gliadin proteins hydrophilically and to the glutenin proteins hydrophobically. Simultaneous binding of polar lipids to both protein groups may contribute structurally to the gas-retaining complex in gluten (Hoseney et al., 1970). Glycolipids may also likely be involved within the starch–glycolipid–gluten complex. It has been proposed that phospholipids interact with lipid-binding proteins to form biomolecular layers, with protein chains bound to the outer edges of a phospholipid leaflet array, probably via salt-type linkages between acidic groups of the phospholipids and the basic protein groups (Marion et al., 2003). This model is capable of providing gluten with the plasticity necessary for optimum baking characteristics (Marion et al., 2003).

The role of flour lipids in baking

One way of elucidating the role of lipids in baking is to remove the lipids from flour and to reconstitute the lipid fractions back to the defatted flour (MacRitchie and Gras 1973; MacRitchie 1981). Fractions of polar and nonpolar lipid affect loaf volume and crumb texture in opposite ways. Polar lipids, such as phospholipids and glycolipids, decrease the loaf volume until a threshold concentration is attained. Above this concentration, the loaf volume increases. These observations were interpreted as competition between the surface-active soluble proteins and polar lipids for the gas–water interface, followed by the progressive replacement of the interfaces by polar lipids (Gan et al., 1995). On the contrary, increasing the nonpolar lipid content of triglycerides and free fatty acids leads to a continuous decrease of bread loaf volume, because these lipids provide a new interface for surface-active components dispersed in the aqueous phase of dough and are also capable of destabilizing the interfacial protein films. Triglycerides present in the gluten matrix as vesicles may interfere with protein–protein interaction or complexation in the gluten matrix.

When the ratio of polar to nonpolar lipids is increased, the minimum of the loaf volume versus lipid content curve is shifted to lower lipid contents. If the percentage of polar to nonpolar lipids in a flour is varied from 0 to 100, at a constant lipid level, test-bake loaf volume increases approximately linearly (MacRitchie 1977). Higher loaf volume and better crumb texture in bread are favored by a high ratio of polar–nonpolar lipids and a higher content of native flour lipids (McCormack et al., 1991). However, the variation in loaf volume that can be attributed to lipids is relatively small, and by far the greatest variation is imposed by the effect of gluten protein quality.

Similar effects by lipids were also found for Arabic bread, steamed bread, cakes, and biscuits (Papantoniou et al., 2003). Despite differences in processing, the general features with respect to lipids have much in common. Crumb texture is influenced most and, like bread, differences in lipid have not been found to account for major variations in quality.

Dough structure and gas cell stabilization

A simple dough is made by mixing wheat flour with water. Doughs contain roughly 45%–50% water, in which approximately 35% water is taken up by binding to gluten protein and starch granules. The additional 15% water is in the formation of a separate liquid phase, "dough liquor," which can be separated from dough solids by ultracentrifugation. Air is also incorporated into the dough in the form of tiny bubbles during the later stages of dough mixing. No more air is introduced and no new bubbles are formed in the dough after mixing, although subdivision of existing bubbles can occur during subsequent molding steps. Cell rupture and coalescence may occur during proofing and baking. During fermentation, the gas produced by yeast dissolves in the liquid phase and diffuses into the gas cells. The liquid phase is essential for expansion and stability of the gas cells formed during dough mixing. Therefore the final loaf volume and texture is largely determined at the mixing stage.

The gas cells are surrounded by thin films with an aqueous interlamellae phase (Gan et al., 1995). Surface-active compounds (mainly the water-soluble proteins and polar lipids) are adsorbed to form monomolecular films at the gas–liquid interfaces of the dispersed gas bubbles. Competition for the interface occurs between polar lipids and soluble proteins. Since proteins form rigid films at the interfaces, the highly mobile polar lipids destabilize the protein films first. Thus increased polar-lipid content leads to the formation of condensed monolayers at air–water interfaces, providing more stable interfacial films between the gas and the liquid phases, and stabilizing the gas cells and the dough structure (MacRitchie 2003). This mechanism explains why, when adding polar lipids in a defatted flour, there is first a decrease of bread volume (caused by the instability of the gas cell structure). Then (above a threshold content) polar lipids improve the stability of the gas-cell structure during dough mixing and expansion, and thus the final loaf volume and texture.

Certain interfacially active proteins in wheat, such as nonspecific lipid-transfer proteins, puroindolines, and various α-amylase-trypsin inhibitors (CM3 proteins), may also be expected to form thin protein–protein or protein–lipid films lining the gas cells (Jones et al., 2006). Puroindolines are able to form very stable foams which have high resistance to destabilization by both neutral and polar lipids (Marion et al., 2003; Rouillé et al., 2005). Furthermore, puroindoline foams show a synergistic enhancement of the stability in the presence of polar lipids. They can act cooperatively with polar lipids in interfacial films resulting in increased foam stability. However, the involvement of these lipoproteins in gluten structure has not yet been investigated.

STARCH COMPOSITION AND WHEAT QUALITY

In contrast to lipid, a minor component of the grain, starch is the most important by amount, comprising almost two-thirds of the grain dry matter. However, for many years starch was considered to play only a minor role in wheat product quality. It has been only recently that the contribution of starch to flour processing quality has been recognized (Rahman et al., 2000). The main chemical and physical starch characteristics of direct importance to quality are the amylose content, the granule-size distribution, the temperature of gelatinization, the viscosity of starch gels, and swelling power. Some of these properties are linked to others; the two basic properties are amylose content and granule size.

Amylose content

The two chemical components of starch are amylose and amylopectin. Amylose is an essentially linear chain of glucose units linked by α-1,4-bonds between the reducing group of carbon-1 and the hydroxyl group of C-4 of the next glucose molecule. This reducing group in glucose itself normally exists in a cyclic structure with a hemi-acetal formed between the hydroxyl of C-5 and C-1. At the reducing end of the chain, C-1 is not bound to other glucose units. There are thousands of glucose units in an amylose chain, giving it a molecular weight of several hundred thousand Daltons. Amylopectin also consists of glucose units joined together in the same way as amylose. However, there are also occasional side branches produced when a bond is formed between the hydroxyl on C-6 and C-1 of another glucose molecule or chain. On these chains, additional branches may form, giving a highly branched structure with a molecular weight over one million Daltons, although most of the individual side chains contain less than thirty glucose units. Like amylose, amylopectin contains only a single reducing group in each molecule. The chain that contains the reducing group is known as the C-chain. Bound to it are B-chains and A-chains. A-chains have no further branches, while B-chains have one or more branches along their length.

Starch granules

In its native form, starch is deposited in granular form, ranging in size from <1 μm to about 40 μm (Briarty et al., 1979). These granules contain not only amylose and amylopectin, but also small amounts of protein, lipid, and nonstarch polysaccharides (Rahman et al., 2000). The protein may be bound within the granules (granule-bound proteins) or on the surface. The granule-bound proteins are enzymes that have been involved in starch synthesis, and have become trapped in the matrix while taking part in forming the granule. The surface-bound proteins often include artifacts, captured there during the isolation of starch from flour or wheat. One of the surface proteins has been linked with grain softness (Greenwell and Schofield 1986), although this "grain softness protein" or "friabilin" has since been shown to be a mixture of several closely related proteins known as puroindolines (Morris et al., 1994).

One of the granule-bound proteins, granule-bound starch synthase (GBSS), is associated with amylose synthesis in wheat. The enzyme is encoded by loci on chromosomes 7A, 7D, and 4A (Nakamura et al., 1995). The locus on chromosome 4A, known as *Wx-B1*, was translocated naturally from chromosome 7B during the evolution of primitive wheats (Table 21.2). The presence of the null allele at this locus has been linked to high noodle quality, especially for Japanese udon noodles (Zhao et al., 1998). The presence of all three GBSS enzymes encoded at their respective loci introduces a degree of redundancy in the synthesis of amylose, which usually comprises about 25%–30% of the starch. Removal of one of the enzymes encoded by any of the loci (to give what is known as a single null) results in starch containing less amylose, usually from 18% to 25%. Removal of a second enzyme reduces the amylose content further to about 12%–18%. Wheat with this combination is known as a double-null genotype, and both single- and double-null genotypes are also referred to as partial waxy. When all three forms of the enzyme are absent, the amylose content is reduced to almost zero in what is known as waxy wheat (Nakamura et al., 1995; Kiribuchi-Otobe et al., 1997). Other granule-bound proteins have been associated with branching and debranching of glucose chains during the biosynthesis of amylopectin (Rahman et al., 2000).

The size distribution of wheat starch granules is also important for processing quality. The distribution is bimodal, with a minimum in the vicinity of 10 μm. By weight, around 70%–80% of the starch is in the large granules, greater than 10 μm, but these account for only about 20% of the number of granules. In the washing isolation of starch and gluten from wheat commercially, separation of the large granules in a purified state requires much less energy and water than isolation of the small granules (Rahman et al., 2000). Thus, wheat with a reduced proportion of small granules would provide a raw material better suited to starch-and-gluten production, provided its milling and dough characteristics are satisfactory.

There is some confusion about the use of the terms "A- starch" and "B-starch" In a commercial setting, A-starch is the highly purified form comprising mostly large granules, with a very low protein content (<0.3%), while B-starch has a higher protein content. B-starch comprises predominantly small granules but with some large granules also present. In a biosynthetic view, A-granules comprise the large granules each formed in single amyloplasts (the starch synthetic cell) (Briarty et al., 1979). These are initiated early in the development of the wheat grain. B-granules and C-granules (both <10 μm) are initiated later in development and they are formed in an amyloplast that has budded from an existing amyloplast. The B-granules are initiated in a second phase of starch synthesis, while the C-granules are initiated at an even later phase (Bechtel et al., 1990). There is no clear morphological distinction between B- and C-granules, and both are found in commercial B-starch.

Small starch granules have a higher surface area to mass ratio than do larger granules. Binding of water by intact starch granules is proportional to the surface area, so small granules bind more water than the same mass of large granules (Larsson and Eliasson 1997). Water absorption is important in dough mixing and baking, and wheat with an increased proportion of small granules may show a higher processing quality for some purposes, to the extent that starch properties contribute to water absorption. There is also a size effect of the granules contributing to dough properties (Rasper and de Man 1980).

Gelatinization temperature

Gelatinization occurs when the internal crystalline structure of the starch granule is lost (it "melts") during heating in the presence of water, which is absorbed into the granule (Rahman et al., 2000). In its native state, the amylopectin exists in crystalline lamellae, and the ordered state of the granule results in a structure which exhibits a typical Maltese cross when viewed with transmitted polarized light under a microscope (birefringence). Gelatinized granules have lost this appearance. Watching the disappearance of this birefringence on a hot-stage microscope has been one of the methods used to determine the gelatinization temperature of starch. The other main method has been differential scanning calorimetry (DSC). Unlike pure organic crystalline materials, the degree of crystallinity of a collection of starch granules is not uniform. As a result, starch is observed to gelatinize over a range of several Celsius degrees. The point at which gelatinization begins is known as the onset temperature, the point at which most granules are melting is the peak temperature, and the point at which gelatinization is complete is known as the completion temperature, abbreviated T_o, T_p, and T_c, respectively. The range of gelatinization temperatures for starch from various grain species is much wider than that for different naturally occurring wheats. It is because wheat starch has a relatively narrow range of gelatinization temperatures that it is unique in its ability to interact with gluten proteins in the formation of bread. In reconstitution experiments, replacement of wheat starch with starch of other origins yields bread with reduced quality.

Viscosity of starch

Starch viscosity is usually measured with either the Visco-amylograph (C.W. Brabender Instruments, Inc., South Hackensack, New Jersey) or the Rapid Visco-Analyser (Newport Scientific Pty. Ltd., Warriewood, NSW, Australia). Although some authors refer to determining the gelatinization of starch with these instruments, it is not gelatinization as such that is being measured but rather "pasting." The important viscosity parameters measured during pasting are peak viscosity, trough viscosity or holding strength, and final viscosity. Both instruments utilize a controlled temperature profile that heats and then cools the starch in the presence of excess water, from an initial temperature (usually 50°C) to about 95°C and then back to 50°C. During the heating phase, the starch does gelatinize, but the observed viscosity change does not occur until the gelatinized starch granules rapidly absorb water and swell. Swelling of the granules requires their gelatinization, and it may occur immediately when the gelatinization temperature is reached or it may occur later at a higher temperature (Morrison et al., 1993).

Amylose content affects the granules' ability to absorb water; waxy starch absorbs water and swells at a lower temperature than starch containing amylose (Miura and Tanii 1993). After reaching a peak viscosity on heating, the viscosity reduces as the swollen granules are ruptured and degraded. The minimum viscosity reached (holding strength) is an indication of the stability of the starch gel at a high temperature. When the gel is cooled, its viscosity increases. This increase is almost entirely due to the amylose in the starch, as waxy starch shows very little viscosity increase in this stage of the process. The final viscosity reached is an indication of gel strength in products at lower temperatures, such as custards or even gravies.

The viscosity of starch is often related to quality, as it affects the processing and final gel structure of many products. For those products processed or utilized at high temperature (e.g., hot sizing in paper manufacture), the holding strength of the starch may be of more importance than the final viscosity.

Swelling power

The swelling power of starch is a measure that indicates its ability to absorb water and increase in size. As stated previously, amylose content is a major factor determining the ability of the granule to swell on heating in water (Crosbie 1991). Swelling power is particularly important in products where the granules are gelatinized but remain essentially intact. One such example is the quality of Japanese udon noodles. For these noodles, the softness of the cooked product is an important attribute, and this is imparted by the starch granule's ability to swell but maintain its integrity. In the selection of lines suitable for this end use, swelling power is used by the breeder as a screening tool (Crosbie 1991; Crosbie and Lambe 1993). In wheat, swelling power is determined by the GBSS alleles referred to above. A single-null allele confers a greater swelling power than the wild type with all three forms of GBSS. Double-null genotypes confer more swelling, while triple-null genotypes (waxy) have even greater swelling power. Peak viscosity in a viscogram is also a measure of swelling power, but it is less useful as a screening tool due to the requirements of sample size and time.

NONSTARCH POLYSACCHARIDE COMPOSITION AND WHEAT QUALITY

While starch predominates in the endosperm, the nonstarch polysaccharides are important components of the outer bran layers of the grain. They play an important role in the milling process, during which this material must be removed cleanly and efficiently from the endosperm (Posner and Hibbs 2005). The value of flour is greater than that of the nonendosperm parts of the grain. Thus, a high extraction rate for white flour is a top priority for the miller. As milling quality is a significant economic consideration, some pressure rests on the breeder to consider it as a quality objective, involving such aspects as grain hardness, bran thickness and germ size, as well as kernel size and shape (Fincher and Stone 1986). An obvious breeding goal is to aim for a high ratio of endosperm to whole-grain mass, but this goal conflicts with a yield-oriented goal of breeding for large germ size to provide early seedling vigor (Rebetzke et al., 2001). A molecular marker for milling yield was found on chromosome 5B in two distinct sets of doubled-haploid populations by Smith et al. (2001). They were unable to identify any associated aspect of grain chemistry. A quantitative trait locus has been found accounting for about one-third of the variation in arabino–xylan ratio on the long arm of chromosome 1B (Fincher and Stone 2004).

The nonstarch polysaccharides of the endosperm play a lesser but significant role in determining milling quality. Largely the remnants of the endosperm cell walls, these nonstarch polysaccharides may cause problems in milling by clogging the sieves. This fault may be another aspect of milling quality that can be resolved in the breeding process, given that there is considerable variation in the level of nonstarch polysaccharides in the endosperm. This variation covers the range from 0.5% to 2.3% for β-glucans in whole grain, according to Fincher and Stone (2004). However, there is much yet to be learned about the extent of heritability of this chemical trait.

Noncellulosic polysaccharides, especially pentosans (heteroxylans) and β-glucans, constitute a high proportion of the walls of the aleurone and endosperm cells (Fincher and Stone 2004). The beta-glucans are linear polysaccharides, unbranched, and polymerized through both (1→4) and (1→3) linkages. They range in molecular weight from about 50,000 to 3 million Daltons. They are poorly extractable into water at room temperature. Strong alkali (e.g., 4 M sodium hydroxide) is required for complete extraction.

Although the nonstarch polysaccharides are minor white-flour components (about 2%), their high water-binding capacity makes them significant in relation to dough properties. Despite the role of nonstarch polysaccharides in determining the extent of water absorption by a flour sample, they share this function with the protein and starch fractions, the latter's role depending greatly on the degree of starch damage. The extended conformation of water-soluble pentosans means that they contribute significantly to viscosity, possibly over 15 times more than the viscosity contribution of globular proteins of similar molecular weight (Fincher and Stone 1986). Their water-holding capacity has been reported to affect dough extensibility and loaf texture (D'Appolonia and Rayas-Duarte 1994). During mixing, the soluble pentosans become less extractable, presumably due to their interaction with proteins. Various oxidative bread "improvers" may accelerate this cross-linking reaction, thereby making a positive contribution to dough rheology.

The nonstarch polysaccharides are also significant in relation to dietary fiber and wellness for humans. This positive contribution to human nutrition is contrasted to the negative role of the nonstarch polysaccharides in the feed value of grain for nonruminants, for which they contribute antinutritive effects. Nonstarch polysaccharides create special feeding problems for poultry, giving the health difficulty of "sticky droppings" and lower available metabolizable-energy results than expected by the energy content of the grain. The soluble nonstarch polysaccharides (arabinoxylans, xylans, and β-glucans), in particular, have a negative effect as their content is inversely proportional to the energy availability of grain in poultry (Black 2004).

FLOUR COLOR AND WHEAT QUALITY

The color of flour (or of semolina in the case of durum wheat) and end-products represents important criteria of wheat quality, playing a significant role in determining the suitability of grain for particular products and markets. Color clearly has an immediate visual impact. Considerable effort has gone into the development of reflectance technologies and algorithms that can provide a reproducible, quantitative assessment of color that is related to color perception by the human eye. The result is a range of reflectance instruments that define the color of a sample in a three-dimensional color space using the coordinates L* (brightness), b* (yellow–blue), and a* (red–green).

Flour or semolina color is influenced by two major components: (i) bran speck size and number (speckiness), a function of the milling and sieving processes; (ii) the inherent color of the starchy endosperm, which constitutes the bulk of flour and semolina (Mares and Campbell 2001).

Brightness (L*) is determined by the overall light reflectance of the sample and is a function of particle size, particle-size distribution, and the absorption of light by the flour or semolina. The creaminess of bread wheat flour and the yellowness of durum semolina are measured as positive values of b* and are primarily related to the content of xanthophyll (Mares and Campbell 2001), mainly lutein and its mono- and di-fatty acid esters, and to the inherent color of the starchy endosperm, which constitutes the bulk of flour and semolina (Mares and Campbell 2001). During the development of modern bread-wheat cultivars, selection has been towards a white-creamy flour and relatively low levels of lutein. In contrast, the opposite is true of durum wheat, for which bright yellow pasta is preferred. Red-green pigments are not considered appropriate for wheat-based products, and materials with a* values substantially different from zero are generally discarded during the breeding process.

End-product color depends on the color of the flour or semolina, together with reactions and interaction with ingredients that occur during mixing with water, as well as subsequent processing and cooking. The process of mixing with water itself results in a lower brightness due to a reduction in the light reflecting capacity. In Asian alkaline noodles, the interaction between ingredients, alkaline salts and high pH, and flour constituents (flavone-C-diglycosides) results in the development of the yellow color that is considered desirable for this end-product (Asenstorfer et al., 2006). Variation in brightness between samples and change in brightness with time can be related partly to protein content and partly to the activity of enzymes.

ENZYMES AND WHEAT QUALITY

Enzymes are biological catalysts that facilitate complex biochemical reactions at temperatures that are relevant to living tissues. Wheat kernels contain a vast array of enzymes, not only in the living tissues of the embryo, scutellum, and aleurone, but in addition some enzyme activities are retained in the dead tissues of the seed coat and the starchy endosperm. Many of the enzymes in the living tissues are essential for general cellular metabolism and function, typical of all plant cells, and will not be discussed here. Other enzymes that may have had specific roles during grain development are retained in ripe grain. Some of these may interact with endogenous substrates during mixing and processing to produce compounds that affect product color or color stability.

Finally, there are enzymes that are normally at low levels (or not present) in dry, ripe grain but are activated during germination to remobilize the starch and protein stored in the starchy endosperm to feed the new seedling. If, for example, the grain becomes wet prior to harvest and starts to germinate (preharvest sprouting), or if the cultivar contains a genetic defect such as late-maturity α-amylase (LMA, otherwise referred to as prematurity α-amylase), these enzymes can be present at unacceptable levels in harvested grain. The result is a failure to meet receival standards, price dockages, and unsuitability for many end products.

Lipase and lipoxygenase

Lipases release free fatty acids from triglycerides and lead to the development of rancidity in wholemeal, germ (embryo), and products that contain germ that has not been stabilized or deactivated. Free fatty acids such as linoleic and linolenic acids can be oxidized by lipoxygenase (LOX) to produce hydroperoxides (Hessler et al., 2002). These in turn can initiate oxidative degradation of pigments such as lutein and β-carotene. As a consequence of this co-oxidation, lutein and lutein-ester content may decline during dough mixing, giving poorer colored end products, especially noodles and pasta. In addition to its effects on color, lipoxygenase has been associated with the formation of off-flavors and unpleasant odors. Substantial variation for LOX activity, including near-zero levels within bread and durum wheat cultivars, means that the effects of LOX may be addressed by breeding and selection (Mares and Mrva 2008a).

Polyphenol oxidase

As PPOs are located in the seed coat, most enzymatic activity is removed during the milling process (Hatcher and Kruger 1993). Unfortunately, as separation of endosperm from the germ and bran layers is not complete, sufficient activity can be retained in flour, where it causes darkening in products such as Asian noodles. Whereas bread wheat cultivars vary from low to high PPO, most modern durum cultivars have very low or near-zero levels of PPO. Enzymes in this group hydroxylate mono-phenols that are either present in the flour or released during mixing, forming o-dihydroxy phenols that are subsequently oxidized further to o-benzoquinones. o-Benzoquinones can polymerize nonenzymatically to form brown-black melanin pigments that are concentrated around bran particles, giving a distinct and unattractive specky appearance.

As with LOX, genetic sources of near-zero PPO have been identified to allow wheat breeders to develop cultivars with much improved color stability and reduced product darkening with time (Mares and Mrva 2008a). Darkening (poor color stability or loss of brightness) is a particular concern to manufacturers and consumers of Asian noodles and pasta products (Fuerst et al., 2006). In the absence of PPO, either through natural genetic variation or the inclusion of specific PPO inhibitors, there is still a substantial component of darkening. This non-PPO darkening is largely inhibited by heating products in boiling water for a short period, suggesting that it also involves enzyme activity. However, no specific enzyme inhibitors have been identified, and neither the mechanisms nor the flour constituents involved in non-PPO darkening are known.

Peroxidase

Peroxidases are also capable of oxidizing phenols to produce dark pigments, but in contrast to PPO, they require hydrogen peroxide as a cofactor. While peroxidases are prevalent in grain tissues, there is no convincing evidence that they contribute to product darkening. However, they have been implicated in the development of symptoms of black point, a defect associated with the characteristic deposition of black pigment in the seed coat overlying the embryo (Williamson 1997). This defect is associated with high humidity during grain development. The pigmented tissue breaks up during milling, leaving black specks in the flour or semolina. Grain with high levels of black point is not suitable for production of Asian noodles or pasta.

Enzymes in sprouted or LMA-affected grain

The major enzyme involved is α-amylase, which initiates the degradation of large starch molecules, particularly once the starch is gelatinized during the cooking process, producing smaller fragments that are then degraded to simple sugars by β-amylase and α-glucosidases. The latter appear to be constitutive enzymes that are already present in ripe grain. Excess α-amylase has an adverse effect on the quality of most end products, particularly bread and noodles, causing stickiness, collapse of product structure, and poor color (Kruger 1989).

High levels of α-amylase in ripe grain can arise via a number of mechanisms (Lunn et al., 2001). First, α-amylase produced in the seed coat during the early stages of grain development may be retained in ripe grain if the normal degradation process in inhibited, as for example sometimes occurs in frost-affected grain or in humid, persistently overcast conditions. Second, some genotypes have a genetic defect referred to as LMA, and high levels of α-amylase are synthesized during the mid-to-late stages of grain development, often in response to a cool-temperature shock (Mares and Mrva 2008b).

This defect is present in both durum and bread wheat and while an exhaustive survey has not been conducted, it appears to have been imported into the germplasm pools of most breeding programs around the world. It ranges in frequency from very high (>75%, synthetic hexaploid wheat), to medium-high (Mexico, UK, China,

Japan, South Africa) to relatively low (<10%–20%, in Australia where there has been a concerted effort to reduce the incidence of LMA). Pinpointing the source of the LMA genes is difficult; however, older LMA-prone genotypes include 'Lerma52', 'Mentana', 'Inia66', 'Bezostaya', and 'Professeur Marchal'.

Finally, rain on ripe wheat crops can trigger the grain to commence germination while still in the spike. This phenomenon of preharvest sprouting occurs in many wheat-growing areas of the world.

Proteases

The quality, shape, and appearance of many wheat-based products depend on the capacity of the storage proteins to form a gluten matrix. These proteins are broken down during germination. Endoproteases, capable of breaking large protein molecules into peptides that are then very susceptible to attack by a range of peptidases and exo-proteases, are generally at very low levels in ripe grain, unless it is badly sprouted.

SELECTION FOR WHEAT QUALITY IN BREEDING

A century ago, there was relatively little concern for grain quality in wheat breeding programs. To the extent that baking quality was considered, testing may have involved the "chewing test" to provide an estimate of grain hardness. An experienced breeder may have taken the next step of forming a dough ball in the mouth, followed by stretching it with the fingers, to gain some idea of gluten strength. In contrast, the modern breeder works in close collaboration with a cereal chemist to obtain information about the processing potential of advanced lines. This breeder-chemist combination may have first started in Australia, where William Farrer (breeder) requested estimates of milling and baking quality for his small grain samples (<100 g, or a few ounces) from Frederick Guthrie, the chemist of the Colony of New South Wales in about 1892 (Wrigley 1978).

Our knowledge of grain quality at the genetic level has recently progressed to the extent that many aspects of the complex traits associated with processing quality (Table 21.1) can be defined by the interactions between specific chemical components and between specific genes (Table 21.2). Obviously, however, the emerging picture of genetic control is complex, necessitating the breeder's attention to a number of genetic loci. There is the added complication that grain quality is modified by a diversity of growth conditions.

Accordingly, Table 21.5 suggests specific genes that could be expected to deliver suitable genetic potential for the manufacture of each of several food products. There is thus the suggestion to the breeder to target grain quality for a specific group of food products. Having done so, the strategy is to reduce the number of genes to be considered by "locking in" essential quality-related alleles in selecting the most suitable parent lines, that is, before making crosses. Consequently, only a limited range of marker proteins is likely to be present in the progeny, and this information facilitates the task of selecting lines with desirable quality on the basis of protein composition. The use of marker proteins as a basis for quality selection does not replace the need for actual milling and baking (or making other food products), but it offers good opportunity early in the selection process for the breeder to discard lines likely to be unsuitable, thus avoiding years of unnecessary propagation.

FUTURE PERSPECTIVES

Breeders are now proceeding to the next level of selecting for genetic potential, namely, to DNA analysis. This approach moves the selection process further to the left in Fig. 21.1, providing the advantage of simultaneously screening one sample of DNA for many nucleotide sequences that are characteristic of a diversity of agronomic and quality traits. This approach has so far proved more popular with traits such as disease resistance (more simply inherited than is grain quality), but good progress is being made for quality traits also (Shewry et al., 2003; Henry 2007).

Transformation with genes that govern known quality traits is also progressing. This approach offers the prospect of efficient introgression of attractive genes without the laborious need for repeated backcrossing (Shewry et al., 2003, 2006). Furthermore, such introgression may cross the normal boundaries of species and genus. The insertion of specific glutenin genes into triticale (Martinek et al., 2008) is an obvious approach to improving the breadmaking potential of this wheat–rye hybrid that has not lived up to hopes for its baking quality.

Transgenic technologies may also be used to manipulate the expression or overexpression of specific storage-protein genes, thereby creating novel dough properties (e.g., Rooke et al., 1999; Butow et al., 2003). Such approaches might provide "ingredient wheats" of extreme dough strength that might be blended with low-protein wheat with weak dough strength to provide adequate dough quality, thus avoiding the need to add vital dry gluten in baking. Alternatively, wheat that has been manipulated to exhibit dough of extreme extensibility might usefully blend with otherwise unsuitable wheats to suit the needs of biscuit (cookie) manufacture. On the other hand, conventional breeding methods may be used to incorporate the genes for storage proteins from ancestral relatives into modern wheat. Such manipulation has the potential to provide novel properties for dough quality (e.g., Hassani et al., 2008).

The contribution of the lipid component to flour quality should not be ignored when considering parameters to be used as a basis for selection in plant breeding programs, even though protein composition appears to be the major factor that determines flour quality. Wheat has a higher proportion of polar lipids than most other cereal grains; this appears to be an important factor contributing to its superior baking performance.

Even in a basic bread-dough system, multiple interactions occur among endogenous and/or added components. The protein–starch complexes, mediated by lipids, play important functional roles. More studies are needed to understand the association of lipids with dough proteins and the manner by which gas-cell structure is stabilized throughout dough mixing, proofing, and baking. This information might prove applicable in breeding programs which target intrinsic baking qualities, thus obviating the need for adding ingredients to wheat-based foods to produce the required functional properties.

Whereas the range of amylose–amylopectin ratio of wheat starch has been limited until recently, new wheat genotypes are being developed that offer the full range of starch composition in this respect. The development of waxy wheat (high amylopectin content) (Nakamura et al., 1995) is further advanced than wheat with high amylose content (Rahman et al., 2000). Both new types of grain offer opportunities for innovation in specialized and novel food products, especially in high-value snack foods.

REFERENCES

Al Saleh, A., D. Marion, and D.J. Gallant. 1986. Microstructure of mealy and vitreous wheat endosperms (*Triticum durum* L.) with special emphasis on location and polymorphic behaviour of lipids. Food Microstructure 5:131–140.

Asenstorfer, R.E., Y. Wang, and D.J. Mares. 2006. Chemical structure of flavonoid compounds in wheat (*Triticum aestivum* L.) flour that contribute to the yellow colour of Asian alkaline noodles. J. Cereal Sci. 43:108–119.

Bayram, M. 2007. Application of bulgur technology to food aid programs. Cereal Foods World 25:249–256.

Bechtel, D.B., I. Zayas, L. Kaleikau, and Y. Pomeranz. 1990. Size-distribution of wheat starch granules during endosperm development. Cereal Chem. 67:59–63.

Bekes, F., C.W. Wrigley, S. Uthayakumaran, C.R. Cavanagh, I.L. Batey, and W. Bushuk. 2007. Frequencies of gluten-protein alleles in a worldwide collection of over 5,600 wheat genotypes. p. 43–47. *In* G.L. Lookhart and P.K.W. Ng (ed.) Gluten proteins 2006. AACC Int., St. Paul, MN.

BeMiller, J.N. 2004. Starch: Modification. p. 219–223. *In* C. Wrigley, H. Corke, and C. Walker (ed.) Encyclopedia of grain science. Vol. 3. Elsevier, Oxford, UK.

Black, J.L. 2004. Animal feed. p. 11–20. *In* C. Wrigley, H. Corke, and C. Walker (ed.) Encyclopedia of grain science. Vol. 1. Elsevier, Oxford, UK.

Blumenthal, C., F. Bekes, P.W. Gras, E.W.R. Barlow, and C.W. Wrigley. 1995. Identification of wheat genotypes tolerant to the effects of heat stress on grain quality. Cereal Chem. 72:539–544.

Briarty, L.G., C.E. Hughes, and A.D. Evers. 1979. The developing endosperm of wheat—a stereological analysis. Ann. Bot. 44:641–658.

Butow, B.J., A.S. Tatham, A.W.J. Savage, S.M. Gilbert, P.R. Shewry, R.G. Solomon, and F. Bekes. 2003. Creating a balance—the incorporation of a HMW glutenin subunit into transgenic wheat lines. J. Cereal Sci. 38:181–187.

Chung, O.K. 1986. Lipid–protein interactions in wheat flour, dough, gluten, and protein fractions. Cereal Foods World 31:242–244, 246–247, 249–252, 254–256.

Cornish, G.B. 2007. Juggling quality genes. p. 93–97. *In* J.F. Panozzo and C.K. Black (ed.) Cereals 2007. Proc. Aust. Cereal Chem. Conf., 57th, Melbourne, VIC. 5–10 August 2007. Royal Aust. Chem. Inst., Melbourne, Australia.

Cornish, G.B., F. Bekes, H.M. Allen, and D.J. Martin. 2001. Flour proteins linked to quality traits in an Australian doubled haploid wheat population. Aust. J. Agric. Res. 52:1339–1348.

Cornish, G.B., F. Bekes, H.A. Eagles, and P.I. Payne. 2006. Prediction of dough properties for bread wheats. p. 243–280. *In* C.W. Wrigley, F. Bekes, and W. Bushuk (ed.) Gliadin and glutenin: The unique balance of wheat quality. AACC Int., St. Paul, MN.

Crosbie, G.B. 1991. The relationship between starch swelling properties, paste viscosity and boiled noodle quality in wheat flours. J. Cereal Sci. 13:145–150.

Crosbie, G.B., and W.J. Lambe. 1993. The application of the flour swelling volume test for potential noodle quality to wheat breeding lines affected by sprouting. J. Cereal Sci. 18:267–276.

D'Appolonia, B.L., and P. Rayas-Duarte. 1994. Wheat carbohydrates: Structure and functionality. p. 107–127. *In* W. Bushuk and V.F. Rasper (ed.) Wheat production, properties and quality. Chapman and Hall, London, UK.

Daqiq, L. 2002. Polymer size and shape in cereal processing. PhD diss. Univ. of Sydney, Australia.

Day, L., I.L. Batey, C.W. Wrigley, and M.A. Augustin. 2006. Gluten uses and food industry needs. Trends Food Sci. Technol. 17:82–90.

Eagles, H.A., K. Cane, R.F. Eastwood, G.J. Hollamby, H. Kuchel, P.J. Martin, and G.B. Cornish. 2006. Contributions of glutenin and puroindoline genes to grain quality traits in southern Australian wheat breeding programs. Aust. J. Agric Res. 57:179–186.

Fabriana, G., and C. Lintas (ed.) 1988. Durum wheat: Chemistry and technology. AACC Int., St. Paul, MN.

Faridi, H., and J. Faubion (ed.) 1995. Wheat end-uses around the world. AACC Int., St. Paul, MN.

Fincher, G.B., and B.A. Stone. 1986. Cell walls and their components in cereal grain technology. p. 207–295. *In* Y. Pomeranz (ed.) Advances in cereal science and technology. Vol. VII. AACC, St. Paul, MN.

Fincher, G.B., and B.A. Stone. 2004. Chemistry of nonstarch polysaccharides. p. 206–223. *In* C. Wrigley, H. Corke, and C. Walker (ed.) Encyclopedia of grain science. Vol. 1. Elsevier, Oxford, UK.

Frazier, P.J., N.W.R. Daniels, and P.W.R. Eggitt. 1981. Lipid–protein interactions during dough development. J. Sci. Food Agric. 32:877–897.

Fuerst, E.P., J.V. Anderson, and C.F. Morris. 2006. Delineating the role of polyphenol oxidase in the darkening of alkaline wheat noodles. J. Agric. Food Chem. 54:2378–2384.

Gan, Z., P.R. Ellis, and J.D. Schofield. 1995. Gas cell stabilisation and gas retention in wheat bread dough. J. Cereal Sci. 21:215–230.

Gautier, M.-F., M.-E. Aleman, A. Guirao, D. Marion, and P. Joudrier. 1994. *Triticum aestivum* puroindolines, two basic cystine-rich seed proteins: cDNA sequence analysis and developmental gene expression. Plant Mol. Biol. 25:43–57.

Greenwell, P., and J.D. Schofield. 1986. A starch granule protein associated with endosperm softness in wheat. Cereal Chem. 63:379–380.

Hargin, K.D., and W.R. Morrison. 1980. The distribution of acyl lipids in the germ, aleurone, starch and non-starch endosperm of four wheat varieties. J. Sci. Food Agric. 31:877–888.

Hassani, M.E., M.R. Shariflou, M.C. Gianibelli, and P.J. Sharp. 2008. Characterisation of ω-gliadin gene in *Triticum tauschii*. J. Cereal Sci. 47:50–67.

Hatcher, D.W., and J.E. Kruger. 1993. Distribution of polyphenol oxidase in flour millstreams of Canadian common wheat classes milled to three extraction rates. Cereal Chem. 70:51–55.

Henry, R.J. 2007. Genomics as a tool for cereal chemistry. Cereal Chem. 84:365–369.

Hessler, T.G., M.J. Thomson, D. Benscher, M.M. Nachit, and M.E. Sorrells. 2002. Association of a lipoxygenase locus, *Lpx-B1*, with variation in lipoxygenase activity in durum wheat seeds. Crop Sci. 42:1695–1700.

Hoseney, R.C., Y. Pomeranz, and K.F. Finney. 1970. Functional (breadmaking) and biochemical properties of wheat flour components: VI. Gliadin–lipid–glutenin interaction in wheat gluten. Cereal Chem. 47:135–140.

Jones, B., C. Morris, F. Bekes, and C.W. Wrigley. 2006. Proteins that complement the roles of gliadin and glutenin. p. 413–446. *In* C.W. Wrigley, F. Bekes, and W. Bushuk (ed.) Gliadin and glutenin: The unique balance of wheat quality. AACC Int., St. Paul, MN.

Khan, K., and P. Shewry (ed.) 2009. Wheat: Chemistry and technology. 4th ed. AACC Int., St. Paul, MN.

Kiribuchi-Otobe, C., T. Nagamine, T. Yanagisawa, M. Ohnishi, and I. Yamaguchi. 1997. Production of hexaploid wheats with waxy endosperm character. Cereal Chem. 74:72–74.

Koksel, H., D. Sivri, P.K.W. Ng, and J.F. Steffe. 2001. Effects of transglutaminase enzyme on fundamental rheological properties of sound and bug-damaged wheat flour doughs. Cereal Chem. 78:26–30.

Kruger, J.E. 1989. Biochemistry of preharvest sprouting in cereals. p. 61–84. *In* N.F. Derera (ed.) Preharvest field sprouting in cereals. CRC Press Inc., Boca Raton, FL.

Larsson, H., and A-C. Eliasson. 1997. Influence of the starch granule surface on the rheological behaviour of wheat flour dough. J. Texture Stud. 28:487–501.

Lawton, J.W. 2004. Starch: Uses of native starch. p. 195–202. *In* C. Wrigley, H. Corke, and C. Walker (ed.)

Encyclopedia of grain science. Vol. 3. Elsevier, Oxford, UK.

Lunn, G.D., B.J. Major, P.S. Kettlewell, and R.K. Scott. 2001. Mechanisms leading to excess *alpha*-amylase activity in wheat (*Triticum aestivum*, L.) grain in the U.K. J. Cereal Sci. 33:313–329.

MacRitchie, F. 1977. Flour lipids and their effects in baking. J. Sci. Food Agric. 28:53–58.

MacRitchie, F. 1981. Flour lipids: Theoretical aspects and functional properties. Cereal Chem. 58:156–158.

MacRitchie, F. 2003. Fundamentals of dough formation. Cereal Foods World 48:173–176.

MacRitchie, F., and P.W. Gras. 1973. The role of flour lipids in baking. Cereal Chem. 50:292–302.

Maningat, C.C., and S.D. Bassi. 2004. Fuel ethanol production. p. 406–415. *In* C. Wrigley, H. Corke, and C. Walker (ed.) Encyclopedia of grain science. Vol. 1. Elsevier, Oxford, UK.

Mares, D.J., and A.W. Campbell. 2001. Mapping components of flour and noodle colour in Australian wheat. Aust. J. Agric. Res. 52:1297–1309.

Mares, D.J., and K. Mrva. 2008a. Genetic variation for quality traits in synthetic wheat germplasm. Aust. J. Agric. Res. 59:406–412.

Mares, D.J., and K. Mrva. 2008b. Late-maturity α-amylase: Low falling number in wheat in the absence of preharvest sprouting. J. Cereal Sci. 47:6–17.

Marion, D., L. Dubreil, and J.P. Douliez. 2003. Functionality of lipids and lipid–protein interactions in cereal-derived food products. Oléagineux, Corps Gras, Lipides 10:47–56.

Marion, D., C. LeRoux, S. Akoka, C. Tellier, and D. Gallant. 1987. Lipid–protein interactions in wheat gluten: A phosphorus nuclear magnetic resonance spectroscopy and freeze-fracture electron microscopy study. J. Cereal Sci. 5:101–115.

Martinek, P., M. Vinterova, I. Buresova, and T. Vyhnanek. 2008. Agronomic and quality characteristics of triticale (× *Triticosecale* Wittmack) with HMW glutenin subunits *5* + *10*. J. Cereal Sci. 47:68–78.

McCann, T.H., D.M. Small, I.L. Batey, C.W. Wrigley, and L. Day. 2008. Protein–lipid interactions in gluten elucidated using acetic-acid fractionation. Food Chem. (in press).

McCormack, G., J. Panozzo, F. Bekes, and F. MacRitchie. 1991. Contributions to breadmaking of inherent variations in lipid content and composition of wheat cultivars: I. Results of survey. J. Cereal Sci. 13:255–261.

McIntosh, R.A., Y. Yamazaki, K.M. Devos, J. Dubcovsky, W.J. Rogers, and R. Appels. 2003. Catalogue of gene symbols for wheat. p. 34. *In* N.E. Pogna, M. Romanò, E. A. Pogna, and G. Galterio (ed.) Int. Wheat Genetics Symp., 10th, Paestum, Italy. Vol. 4. 1–6 September 2003. Istituto Sperimentale per la Cerealicoltura, Paestrum, Rome, Italy.

Miura, H., and S. Tanii. 1993. Endosperm starch properties in several wheat cultivars preferred for Japanese noodles. Euphytica 72:171–175.

Morris, C.F. 2002. Puroindolines: The molecular basis of wheat grain hardness. Plant Mol. Biol. 48:633–647.

Morris, C.F., G.A. Greenblatt, A.D. Bettge, and H.I. Malkawi. 1994. Isolation and characterization of multiple forms of friabilin. J. Cereal Sci. 20:167–174.

Morrison, W.R. 1978. Wheat lipid composition. Cereal Chem. 55:548–558.

Morrison, W.R., and K.D. Hargin. 1981. Distribution of soft wheat kernel lipids into flour milling fractions. J. Sci. Food Agric. 32:579–587.

Morrison, W.R., R.F. Tester, C.E. Snape, R. Law, and M.J. Gidley. 1993. Swelling and gelatinization of cereal starches: IV. Some effects of lipid-complexed amylose and free amylose in waxy and normal barley starches. Cereal Chem. 70:385–391.

Nakamura, T., M. Yamamori, H. Hriano, S. Hidaka, and T. Nagamine. 1995. Production of *waxy* (amylose-free) wheats. Mol. Gen. Genet. 248:253–259.

Papantoniou, E., E.W. Hammond, A.A. Tsiami, F. Scriven, M.H. Gordon, and J.D. Schofield. 2003. Effects of endogenous flour lipids on the quality of semisweet biscuits. J. Agric. Food Chem. 51:1057–1063.

Payne, P.I. 1987. Genetics of wheat storage proteins and the effect of allelic variation on bread-making quality. Annu. Rev. Plant Physiol. 38:141–153.

Payne, P.I., M.A. Nightingale, A.F. Krattiger, and L.M. Holt. 1987. The relationship between HMW glutenin subunit composition and the bread-making quality of British-grown varieties. J. Sci. Food Agric. 40:51–65.

Posner, E.S., and A.N. Hibbs (ed.) 2005. Wheat flour milling. 2nd ed. AACC Int., St. Paul, MN.

Rahman, S., Z. Li, I. Batey, M.P. Cochrane, R. Appels, and M. Morell. 2000. Genetic alteration of starch functionality in wheat. J. Cereal Sci. 31:91–110.

Rasper, V.F., and J.M. de Man. 1980. Effect of granule size of substituted starches on the rheological character of composite doughs. Cereal Chem. 57:331–340.

Rebetzke, G.J., R. Appels, A.D. Morrison, R.A. Richards, G. McDonald, M.H. Ellis, W. Spielmeyer, and D.G Bonnett. 2001. Quantitative trait loci on chromosome 4B for coleoptile length and early vigour in wheat (*Triticum aestivum* L.). Aust. J. Agric. Res. 52:1221–1234.

Rooke, L., F. Bekes, R. Fido, F. Barro, P. Gras, A.S. Tatham, P. Barcelo, P. Lazzeri, and P.R. Shewry. 1999. Overexpression of a gluten protein in transgenic wheat results in greatly increased dough strength. J. Cereal Sci. 30: 115–120.

Rouillé, J., G. Della Valle, M.F. Devaux, D. Marion, and L. Dubreil. 2005. French bread loaf volume variations and digital image analysis of crumb grain changes induced by the minor components of wheat flour. Cereal Chem. 82:20–27.

Shewry, P.R., N.G. Halford, and D. Lafiandra. 2003. Genetics of wheat gluten proteins. Adv. Genet. 49:111–184.

Shewry, P.R., D. Lafiandra, L. Tamas, and F. Bekes. 2006. Genetic manipulation of gluten structure and function. p. 363–385. *In* C.W. Wrigley, F. Bekes, and W. Bushuk (ed.) Gliadin and glutenin: The unique balance of wheat quality. AACC Int., St. Paul, MN.

Sivri, D., I.L. Batey, D.J. Skylas, L. Daqiq, and C.W. Wrigley. 2004. Changes in the composition and size distribution of endosperm protein from bug-damaged wheats. Aust. J. Agric. Res. 55:1–7.

Sivri, D., H. Koksel, H. Sapirstein, and W. Bushuk. 1999. Effects of bug protease (*Eurygaster maura*) on glutenin proteins. Cereal Chem. 76:816–820.

Skylas, D.J., D. Van Dyk, and C.W. Wrigley. 2005. Proteomics of wheat grain. J. Cereal Sci. 41:165–179.

Smith, A.B., B.R. Cullis, R. Appels, A.W. Campbell, G.B. Cornish, D. Martin, and H.M. Allen. 2001. The statistical analysis of quality traits in plant improvement programs with application to the mapping of milling yield in wheat. Aust. J. Agric. Res. 52:1207–1219.

Southan, M., and F. MacRitchie. 1999. Molecular weight distribution of wheat proteins. Cereal Chem. 76:827–836.

Uthayakumaran, S., C.W. Wrigley, I.L. Batey, W. Bushuk, and G. Lookhart. 2006. Genotype identification. p. 306–331 *In* C.W. Wrigley, F. Bekes, and W. Bushuk (ed.) Gliadin and glutenin: The unique balance of wheat quality. AACC Int., St. Paul, MN.

Vawser, M., G.B. Cornish, and K.W. Shepherd. 2002. Rheological dough properties of Aroona isolines differing in glutenin subunit composition. p. 53–58. *In* C.K. Black, J. F. Panozzo, C.W. Wrigley, I.L. Batey, and N. Larsen (ed.) Cereals 2002. Royal Aust. Chem. Inst., Melbourne, Australia.

Wanjugi, H.W., J.M. Martin, and M.J. Giroux. 2008. Influence of puroindolines A and B individually and in combination on wheat milling and bread traits. Cereal Chem. 84:540–547.

Wieser, H., W. Bushuk, and F. MacRitchie. 2006. The polymer structure of glutenin. p. 213–240. *In* C.W. Wrigley, F. Bekes, and W. Bushuk (ed.) Gliadin and glutenin: The unique balance of wheat quality. AACC Int., St. Paul, MN.

Williamson, P.M. 1997. Black point of wheat: In vitro production of symptoms, enzymes involved, and association with *Alternaria alternata*. Aust. J. Agric. Res. 48:13–19.

Wrigley, C.W. 1978. W.J. Farrer, and F.B. Guthrie: The unique breeder–chemist combination that pioneered quality wheats for Australia. Rec. Aust. Acad. Sci. 4:7–25.

Wrigley, C.W., and F. Bekes. 1999. Glutenin-protein formation during the continuum from anthesis to processing. Cereal Foods World 44:562–565.

Wrigley, C.W., and F. Bekes. 2004. Processing quality requirements for wheat and other cereal grains. p. 349–388. *In* R.L. Benech-Arnold and R.A. Sánchez (ed.) Handbook of seed physiology: Applications to agriculture. Haworth Press, Inc., New York.

Wrigley, C.W., F. Bekes, and W. Bushuk. 2006. Gluten: A balance of gliadin and glutenin. p. 3–32. *In* C.W. Wrigley, F. Bekes, and W. Bushuk (ed.) Gliadin and glutenin: The unique balance of wheat quality. AACC Int., St. Paul, MN.

Zhao, X.C., P.J. Sharp, G. Crosbie, I. Barclay, R. Wilson, I.L. Batey, and R. Appels. 1998. A single genetic locus associated with starch granules and noodle quality in wheat. J. Cereal Sci. 27:7–13.

Zuckerkandl, E., and L. Pauling. 1965. Molecules as documents of evolutionary history. J. Theor. Biol. 8: 357–366.

Chapter 22
New Uses for Wheat and Modified Wheat Products

Robert A. Graybosch, Rui Hai Liu, Ronald L. Madl, Yong-Cheng Shi, Donghai Wang, and Xiaorong Wu

SUMMARY

(1) Hard white wheat represents a major opportunity for increasing wheat use in Asian noodles and whole wheat products throughout the world in the future, as long as segregation can be achieved throughout the value chain.
(2) Alteration of the amylose–amylopectin ratio results in wheat flour and starch with modified characteristics, such as higher viscosity and reduced staling rate, longer storage time without development of rancidity, and potential for producing resistant starch ingredients to formulate foods with lower glycemic index.
(3) Reduction of enzymes, such as polyphenol oxidase, may offer value in food systems, such as Asian noodle products, by helping maintain a stable, light color.
(4) Reduction of phytic acid levels in wheat bran (or enhanced levels of phytase) may also offer additional value in animal nutrition applications.
(5) Whole-grain foods offer a wide range of phytochemicals with nutritional benefits that are only recently becoming known. Recognition of the health value for these micronutrients will support development of wheat with enhanced levels of target nutrients for new foods and food ingredients from wheat.
(6) Industrial wheat can serve as a feedstock for both ethanol production and animal feeding. Key modifications include waxy wheat starch, lower protein/higher starch content, and low phytate content, combined with high agronomic yield.

INTRODUCTION

In 2004, a summit was held at Kansas State University for the purpose of identifying key issues and opportunities that needed to be addressed by the wheat industry. All aspects of the industry were discussed, including genetic research, production and management, transportation and storage, and new or nontraditional uses to expand demand for wheat (*Triticum* spp.). Discussion of the last area relied substantially on a Sparks report prepared for the National Association of Wheat Growers in 2002 (NAWG 2002). Table 22.1 summarizes the results of that discussion.

Several topics identified in that report are still relevant today, although priorities may have changed. The topics reflect industry interest at the time, not necessarily feasible or economic options. Topics under "Kernel Pericarp Characteristics" included color, hardness, size, and white wheat. Hardness and size reflected millers' interest in reducing energy input or increasing yield during milling. The topic identified as most logical to pursue with highest priority was hard white wheat. It was already developed, it had

Table 22.1 Wheat value-added traits and new uses of wheat identified by the US wheat industry in 2004.

Trait Category	Trait	Volume Potential	Premium Potential	Technical Feasibility	Development Time	Development Cost	Priority
		Million t	USD t^{-1}		Year		
Kernel pericarp characteristics	Whiter pericarp & endosperm	1.36	7.35	Low	10	High	Medium
	Hardness (reduced milling energy requirement)	2.04	1.84	High	5	Medium	Medium
	Kernel size	1.36	1.10	Medium	5	Medium	Medium
	Hard white wheat	2.72	5.51	High	1	Low	High
Protein quality	Meat substitutes	0.27	0	High	2	Low	Low
	Gluten strength	2.72	7.35	Medium	6	High	High
	Speed of hydration	2.72	9.19	Low	7	High	High
	Tortilla quality	1.36	1.84	Medium	5	Medium	High
	Celiac friendly	2.72	11.0	Low	10	High	Medium
Starch characteristics	Low glycemic index	2.72	27.6	Low	10	High	Medium
	Waxy wheat	1.36	5.51	High	4	Medium	Low
Enzymatic and antioxidant activity	Low PPO	2.72	7.35	High	3	Medium	High
	High antioxidants	1.90	36.8	Medium	9	High	High
High yielding	Feed wheat	1.63	0	Low	7	Low	Low
	Wet milling	0.27	1.84	Low	7	Low	Low
By-product utilization	Cellulosic ethanol	2.72	14.0	High	9	Medium	Medium
	Strawboard	1.36	14.0	Medium	6	Low	Low
	Xylitol	0.14	14.0	Medium	3	Low	Low

clearly identified markets, and the markets had expressed growing demand.

Under "Protein Quality," the most important topics identified were gluten strength, hydration rate, and tortilla quality. While protein modification for the benefit of gluten-sensitive individuals was considered important, it was only given medium priority because of its perceived technical difficulty. Since sensitive individuals react to the gliadin proteins, it is unlikely that reactive sites on this major wheat protein can be modified without also affecting functionality.

Both topics under "Starch Characteristics" have received attention since 2004, even though they were not prioritized highly at that time. Low glycemic index was recognized as potentially beneficial in the weight control and diet foods markets. Since then, several foods were reformulated with a variety of components to reduce starch content or to modify digestibility of the starch so that it could be considered to contain "good carbs" with low glycemic index. The reduced-calorie content label for these foods provided an effective marketing tool.

Waxy wheat, with its altered starch content, has been under development without a clear market. However, as adapted cultivars approach limited release, food formulators are finding that it can replace the function of modified starch in certain baking applications. This could allow replacement of "modified starch" with preferred "starch" on food labels. Further modification may produce interesting new products. Fermentation of waxy starch is more efficient than wild-type starch with typical ratios of amylose to amylopectin. This may lead to potentially valuable uses for wheat starch in industrial applications.

Wheat with inherently low polyphenol oxidase (PPO) is considered very important for application in Asian food products, such as noodles and steamed bread. Thus hard white wheat with low PPO should have greater marketing value for

Asian foods. Antioxidants have recently been found to exist in some wheat cultivars at levels comparable to berries, fruits, and other foods. Improving the consistency of high antioxidant levels could offer another avenue for extending wheat-based foods to the nutraceutical market. The environmental effect on antioxidant activity is a key component overlaying the genetic potential. Understanding the environmental triggers could lead to a breakthrough in this area.

The potential to produce a high-yielding wheat cultivar for animal feeding, fermentation feedstock, and for other industrial uses was given a low priority in 2004, but this research area is receiving greater consideration as commodity grain prices are driven higher by ethanol production. For industrial uses, the approach now is to maximize agronomic yield, suppress protein content, convert the starch to a waxy type, and reduce phytate content. This wheat would need to be contract-grown or otherwise managed to assure its segregation from food-wheat cultivars.

Utilization of by-products from the wheat field include conversion of wheat straw to composite board products, hydrolysis and chemical conversion to xylitol for the sweetner market, and hydrolysis followed by fermentation to ethanol or other valuable chemicals. Priorities have rapidly changed since 2004, as the first cellulosic ethanol plant in Kansas is targeting wheat straw as its primary cellulosic feedstock source.

WHITE WHEAT

White wheat breeding

Winter wheat production in the Great Plains has been dominated by cultivation of hard red winter (HRW) types. The first successfully cultivated landraces, 'Turkey', 'Kharkof', and 'Crimea', were introduced from the Black Sea region of southwest Asia in the 1870s; all were red wheats (Cox et al., 1986). These lines, and various selections from them, formed the basis of all subsequent breeding efforts in the Great Plains. Red wheat came to dominate the region, which might very well be a historical accident, or in genetic terms, a "founder effect." Had the first widely adapted winter wheat cultivars in the region been white, the Great Plains might have developed into a region of white wheat cultivation.

During the past 25 years, however, it has become abundantly clear that many international markets favor white wheat. To diversify the product line of Great Plains wheat producers, hard winter and spring wheat breeders initiated programs to develop hard white wheat cultivars. This development began in earnest in the late 1960s when Kansas State University (KSU) wheat breeder, Dr. Elmer Heyne, was on sabbatical leave in Australia, where the crop was entirely white wheat. He recognized the potential of hard white wheat and realized that it may be even more adaptable to the growing conditions in the central Great Plains. Hard white wheat breeding slowly increased, mainly in Kansas, until 1990 when hard white (HW) was established as a new class of wheat by the USDA Federal Grain Inspection Service (later renamed the Grain Inspection, Packers and Stockyards Administration). This enhanced the efforts directed toward HW development in the KSU wheat breeding program and other breeding programs throughout the Great Plains.

Lower grain yields and greater occurrence of preharvest sprouting relative to HRW in early HW cultivars, including 'Arlin', 'Rio Blanco', and 'Oro Blanco', limited acceptance of HW wheat. 'Clark's Cream' (PI476305), a farmer-selected cultivar with excellent preharvest sprouting resistance, was released in 1972 (Anonymous 1972). Breeders at the Kansas Agricultural Experiment Station (KAES) transferred this sprouting resistance to improved experimental lines, which were released to other breeders in 1988. In 1998, the KAES released its first HW cultivars, Betty and Heyne, to foundation seed producers. These cultivars exhibited yield levels comparable to HRW cultivars in their areas of adaptation.

Following Kansas' lead (Morris and Paulsen 1992), nearly every remaining wheat breeding program in the Great Plains started to develop HW wheat by the early 1990s. Hard white wheat cultivars released in the region to date, and the respective developing party, include 'Intrada', 'Guymon' and 'OK Rising' (Oklahoma State University and USDA-ARS); Arlin, Betty, Heyne, 'Trego', 'Lakin', and 'Danby' (KSU); 'Nuplains', 'Antelope', 'Arrowsmith', and 'Anton' (USDA-ARS and University of Nebraska-Lincoln); 'Alice' (South Dakota State University); 'Avalanche' (Colorado State University); and Rio Blanco, Oro Blanco, 'Platte', 'NuFrontier', 'NuGrain', 'NuHorizon', 'NuHills', and 'NuDakota' (AgriPro-Coker).

Red grain color is inherited as a dominant trait in wheat and conferred by three genetic loci designated *R1*, *R2*, and *R3* (McIntosh et al., 2005). Presence of a dominant allele at any one of these loci will condition the red-grained phenotype. Intermatings of red wheats can yield white-grained progeny if the red wheat parents carry recessive alleles at complementary *R* loci. Many white-grained cultivars have been developed by intermatings of red-grained parents. Arlin was derived via selection of white-grained progeny from a bulk population of HRW and hard red spring parents that were intercrossed in 1981 (Sears et al., 1993). Alice descended from the cross of two red wheat parents, 'Abilene' and 'Karl'.

While the existing red wheat gene pool could provide genes for agronomic adaptation and disease resistance, introduction of genes for novel quality traits (see below) must be incorporated into HW wheat breeding efforts. Current efforts typically involve either intermatings of adapted white wheat breeding lines or matings between previously released HW wheats and adapted red wheats. For example, Antelope was derived from the mating of 'Pronghorn', a Nebraska-developed HRW wheat, and Arlin, a HW wheat (Graybosch et al., 2005). Because the Great Plains white wheat gene pool is small while the North American HRW gene pool is a large product of a 75-year breeding effort, matings with adapted red wheats are essential.

Red wheat will continue to provide parentage for development of new HW winter wheat cultivars. Segregation ratios in progeny of crosses (matings) of red-grained and white-grained parents will differ depending on the genotype of the red-grained parent. For example, matings of a red wheat parent possessing homozygous dominant alleles at *R1*, *R2*, and *R3* and any white wheat will yield white-grained progeny at a frequency of 1:64. The 1:64 ratio is the "worst-case" scenario. In a survey of 90 UK wheats, Flintham and Humphray (1993) found 41% carried a single dominant *R* allele, 41% carried two dominant *R* alleles, and only 18% had dominant alleles for all three loci. If results from the UK are representative of most wheat breeding pools, more favorable segregation ratios of 1/4 or 1/16 will be more often encountered. A red-grained parent of the genotype *R1R1r2r2r3r3* will yield white-grained progeny at a frequency of 1/4, after mating with a white wheat.

Still, wheat breeding progress requires the evaluation of large numbers of selected progeny, and the low frequency of white-grained progeny in matings derived from red wheat is an impediment to breeding progress. The process has, however, been greatly expedited by the development of automated seed sorting technology (Dowell et al., 2006). Such systems are capable of rapidly extracting white seed from early-generation segregating or mixed populations, allowing breeders to seed nearly pure white-grained populations, even after red wheat matings. Although color-sorting technology is available, breeding programs will rarely target only one wheat class. Great Plains wheat breeders likely will continue to develop and intermate both red and white wheat parents and use sorting technology to derive various gene pools to meet regional needs.

Late-season rainfall, especially if it occurs as wheat plants approach physiological maturity, can result in the initiation of germination *in situ* (i.e., preharvest sprouting), with a concomitant release of amylase enzymes. High amylase activity leads to starch digestion, resulting in a loss of test weight and diminished processing quality. In general, red wheat is more resistant to preharvest sprouting than white wheat (Morris and Paulsen 1992); but in typical years, preharvest sprouting rarely will be encountered with white wheat produced in the Great Plains, especially in sites west of the 100th meridian. However, genetic diversity exists among both red and white wheat for sprouting tolerance (Wu and Carver 1999).

Rio Blanco is reported to be among the most tolerant HW cultivars (Wu and Carver 1999); an older HW cultivar, Clark's Cream, also has served as a source of genes for preharvest sprouting tolerance (Morris and Paulsen 1992). Nuplains, a recently released HW wheat cultivar, has tolerance similar to Rio Blanco.

Tolerance to preharvest sprouting may be assessed by use of misting systems and determination of α-amylase activity; however, such approaches can encounter unacceptably high error frequencies with field-grown materials. Alternative approaches include germination studies of seed harvested at physiological maturity and rapidly dried (Wu and Carver 1999) or germination of such seed in the presence of the germination inhibitor, abscisic acid (ABA) (Morris et al., 1989). Germination in lines susceptible to preharvest sprouting will not be suppressed by ABA. Gibberellic acid (GA3) stimulates germination and acts as an antagonist of ABA (Gold and Duffus 1992).

Genetic studies on inheritance of preharvest sprouting tolerance generally indicate polygenic inheritance (Paterson and Sorrells 1990). However, the number of North American populations studied is quite limited, and studies on inheritance of the trait in more recent cultivars, such as Rio Blanco and Nuplains, are lacking. It is not known, for example, whether pyramiding of tolerance genes from different sources will result in genetic gains in improvement to sprouting tolerance. Also, while the pigments or their precursors present in red-grained cultivars confer resistance to preharvest sprouting, genetic diversity exists among red wheat genotypes for the trait. Red-grained cultivars might serve as a genetic reservoir for additional genes for tolerance to preharvest sprouting in white wheat. Such an approach was successfully employed by Hucl and Matus-Cadiz (2002) in the development of white spring wheat with improved seed dormancy.

Hard white wheat—consumer markets

The total supply of white wheat in the United States has remained steady at 9.2 million tonnes in 1989–1990 and 9.3 million tonnes in 2008–2009, with a peak of 12.9 million tonnes in 1996–1997 (Anonymous 2007). However, HW wheat production peaked in 2003, reflecting the 3-year incentive program for HW wheat in the 2002 Farm Act for harvest years 2003–2005. The HW harvest in those three seasons was 1.12 million tonnes, 1.03 million tonnes, and 0.96 million tonne, respectively. Kansas had more than one-half of the US HW acreage in 2003 based on incentive program enrollment data (Lin and Vocke 2004). This was followed by a 0.68 million tonne HW wheat harvest in 2006 (A. Harris, pers. comm.). This loss of production volume after the incentive program ended reinforced growers' concerns about additional risks of preharvest sprouting and handling costs associated with growing and delivering HW wheat, and offered no assurance of premiums over HRW wheat.

A major benefit of HW wheat is that it allows millers to achieve a higher milling yield, based on flour color—as much as 1%–3% more than HRW wheat. Though flour color is correlated with bran content of flour, higher milling yield and more nutrition might be achieved in HW flour without an unfavorable change in color. This was confirmed in a study by McFall et al. (2003), who conducted milling studies on a popular HRW cultivar and two HW cultivars. They plotted cumulative ash and Agtron whiteness values against milling yield. Ash values surpassed 0.5% ash at >75% milling yield for all three HW cultivars. The two HW cultivars produced 1.2%–2.3% higher flour yield, with Agtron whiteness values of 5.9%–10.7 units higher than the HRW cultivar. Agtron values showed that HW cultivars achieved the same whiteness at 85%–90% milling yield as the HRW cultivar showed at 40%. McFall et al. (2003) concluded that compared with HRW wheat, HW wheat could provide the same color of flour at higher milling yields, or that whole-wheat flour could be prepared from HW wheat nearly as white as standard-grade flour from HRW wheat.

Hard white wheat has utility in several domestic markets and is the preferred wheat of commerce in a number of international markets. The favorable end-use characteristics of HW wheat differ from HRW and are particularly well suited for whole-wheat products, as well as replacing HRW wheat in pan breads, tortillas, and oriental products (e.g., noodles and steamed bread). Whole wheat pan bread made from HW wheat is not only lighter colored, but also less bitter than bread made from HRW wheat at the same milling extraction, because white wheat bran contains fewer bitter components associated with a strong flavor. Tortillas are traditional Mexican flat breads made from either maize (*Zea mays* L.) or wheat flour. Tortillas are used for many traditional Mexican dishes and, increasingly, for a variety of non-Mexican foods such as "wraps." Wheat tortillas are twice as popular as corn tortillas in the US.

In Asia, wheat is consumed primarily in noodles and steamed bread. Wheat quality attributes (including functional protein characteristics), initial product color, color stability, and mellow gluten (Kruger 1996) are critical for these applications. Australian white wheat is known for its capability to produce noodles with a stable white or bright yellow color. This desirable color must be maintained for at least 24 hours to meet requirements of the Asian market. Enzymatic activity of polyphenol oxidase (PPO) is responsible for noodle discoloration (Corke and Bhattacharya 2001), and HW cultivars with very low PPO, such as Lakin and Platte, have been developed (Lin and Vocke 2004). Research at the Wheat Marketing Center in Portland, Oregon, has helped identify US wheat cultivars that perform best in noodle applications.

As learned from HW wheat production in the Great Plains, the grain-handling infrastructure needs to achieve and maintain segregation throughout the value chain. Capturing the value of any trait differentiated from commodity-based wheat will require segregation during storage and transportation, and ultimately, any higher value trait will face similar challenges.

LOW POLYPHENOL OXIDASE WHEAT

High levels of grain PPO have been associated with discoloration in a variety of fresh wheat products, including Asian noodles (Baik et al., 1995). Cultivated durum (*T. durum* L.) wheat has none to trace amounts of grain PPO activity. Among US HW bread wheat cultivars, Platte and Lakin have low levels of grain PPO but still produce more PPO than typical durum wheats (Sayaslan et al., 2005). The PPO activity of Platte was found to still be 3 times as high as that of the durum cultivars Ben and Renville (Bettge 2004). Characterization of the PPO levels of a given genotype is easily accomplished via a quantitative assay using L-dopa as a substrate (Anderson and Morris 2001). Such assays are, however, reliable in genetic studies only if conducted on samples from multiple locations or growing seasons; cultural environment will influence levels of PPO detected (Anderson and Morris 2001). DNA markers recently have been identified and linked to the low PPO trait in a number of genetic backgrounds (Raman et al., 2005; Sun et al., 2005). Either by DNA markers or by simple whole-seed assays of grain PPO using tyrosine or L-dopa substrates, selection for low PPO is readily accomplished in breeding programs.

Genetic studies in common hexaploid (*T. aestivum* L.) wheat (Demeke et al., 2001) have demonstrated PPO levels to be influenced by multiple genes arising from several chromosomes, including 2A, 2B, 3B, 3D, and 6B. The complex nature of the inheritance of PPO activity suggests that genetic complementation of low PPO mutants might provide a path toward development of more hexaploid wheat germplasm with nil grain PPO. Bernier and Howes (1994), using a tyrosine assay, found these hexaploid cultivars with grain PPO levels approaching that of durum wheat: 'Bihar 124' (from India), 'Little Club', 'Cadet', and 'Fielder' (US), 'Cook' and 'Tincurrin' (Australia), and 'Pitic 62' (Mexico). Intermatings of such low PPO lines might generate segregates with nil levels of the enzyme.

ALTERED STARCH

Altered starch breeding

Waxy (amylose-free) wheat

Until the 1990s, little was known of the inherent genetic variation in wheat starch composition, and essentially no breeding work to exploit such variation had been accomplished. Wild-type wheat starch is composed of approximately 75% amylopectin and 25% amylose. It long was thought that genetic variation did not exist for the ratio of these two polymers. However, in the early 1990s, Nakamura et al. (1992, 1993) devised strategies to separate gene products of the three loci encoding granule bound starch synthase (GBSS) in wheat. The enzyme GBSS is primarily responsible for amylose synthesis in endosperm.

Electrophoretic techniques demonstrated existence of nonfunctional (null) alleles in many wheat accessions and three distinct GBSS isoforms in wild-type wheat. Further, Nakamura et al. (1993) demonstrated that null alleles at one or more waxy loci led to production of wheat starch with reduced amylose content. Reduced amylose wheat is termed "partial waxy." Finally, by traditional intermatings of lines carrying the three null alleles, Nakamura et al. (1995) developed lines with no functional GBSS and no amylose. Amylose-free wheat, following the convention established in other cereal crops, is termed "waxy." Null alleles at two of the three waxy loci are common and are found in adapted US wheat cultivars (Graybosch et al., 1998). Null mutations at the third locus (*wx-D1*) are rare (Nakamura et al., 1995). Development of waxy wheat adapted to the US wheat belt has been slow, as the only available donors of the necessary *wx-D1* null allele were two poorly adapted lines, 'BaiHuo' and 'BaiHuoMai', both from China.

Eight possible GBSS genotypes exist in hexaploid wheat, and four possible combinations exist in durum wheat. It has yet to be established whether any one genotype would be preferred for production of modified food starch. In common wheat, genotypes do not differ in their gross starch granule morphology (Kim et al., 2003), but differences in functional properties do exist (Epstein et al., 2002; Kim et al., 2003). Waxy wheat also demonstrates different milling properties, specifically producing higher levels of damaged starch (Bettge et al., 2000; Kim et al., 2003). Little is known, however, of possible agronomic effects of various GBSS genotypes, especially with regard to grain yield.

Breeding waxy wheat cultivars presents the same challenge as breeding HW wheat. Populations derived from crosses between waxy and wild-type wheat will contain only 1/64 waxy progeny. This frequency is low, especially when one is attempting to introgress a trait from thoroughly unadapted sources. However, automated seed sorting technology (Dowell et al., 2006), coupled with near-infrared reflective spectroscopy (NIRS), can be used to produce populations enriched in waxy progeny. Delwiche and Graybosch (2002) demonstrated facility of NIRS to identify wheat grain based on amylose content. Subsequently, Delwiche et al. (2006) combined NIRS with automated seed sorting technology to extract waxy seed from mixed populations of waxy, partial waxy, and wild-type durum wheat. Application of NIRS in commercial settings will facilitate segregation of waxy wheat from wild-type wheat.

Waxy cultivars have been developed in Europe and used to produce a commercial flour (Westhove Wheat, Wisconsin) described as an "instant waxy wheat flour" (Anonymous 2006). A soft white waxy wheat cultivar, Waxy-Pen, was released in the Pacific Northwest region of the US (Morris and King 2007). Spring waxy wheat, with grain yield equal to that of check cultivars, has been produced (Graybosch et al., 2004). Waxy winter wheat grain yield continues to approach that of commercial wild-type cultivars (Graybosch 2005), and release of waxy winter wheat cultivars adapted to the Great Plains likely will occur by 2010.

High-amylose wheat

In maize, mutations at the *ae* (amylose extender) locus result in production of endosperm starch with elevated levels of amylose (Neuffer et al., 1997). The *ae* locus encodes a starch branching enzyme (SBEII). When this gene is nonfunctional, not only is starch amylose elevated, but the branching pattern of the remaining amylopectin is altered (Neuffer et al., 1997). Yamamori et al. (2000) were the first to report high-amylose mutants in wheat. Wheat grain lacking starch granule protein-1 (SGP-1) is reported to have an apparent amylose content of approximately 35% elevated above the typical 25% amylose content of wild-type wheat (Yamamori 2005). Li et al. (1999) reported SGP-1 to be a soluble starch synthase homologous to SSIIa of maize, which plays a role in synthesis of maize amylopectin (Zhang et al., 2004).

An alternate approach was used to develop wheat with an amylose content of approximately 70%. Regina et al. (2006) used RNA interference to down-regulate wheat starch-branching enzymes SBEIIa and SBEIIb. Suppression of only SBEIIb did not affect amylose content, but simultaneous suppression of both forms resulted in starch with 70% amylose.

Sweet wheat

Nakamura et al. (2006) combined the waxy mutant (lacking functional GBSS) with the *sgp-1* mutant (lacking soluble starch synthase SSIIa) to form a double-mutant type. Surprisingly, grains of this line had elevated levels of maltose and sucrose but also displayed shrunken kernels. Kernels were reported to have a sweet taste, similar to that of sweet corn. Flour from this line also had elevated sugar content, and incorporating it into baked goods resulted in sweeter-tasting products.

Altered starch characteristics

Unique waxy wheat flour properties

Waxy wheat starch consists mainly of amylopectin (a branched glucose polymer). In addition, waxy wheat kernels contain approximately 20% more lipids, 35% more arabinoxylan (pentosan), and 30% more β-glucan in their endosperm than wild-type wheat (Yasui et al., 1999; Sayaslan et al., 2006). Protein content of waxy flour can vary from 9.4% to 14.2% (Sayaslan et al., 2006). Sayaslan et al. (2006) reported that waxy wheat flour had about 3%–7% less total starch than normal or wild-type wheat flour, though in some cases waxy flour might be equivalent to normal wheat flour (Guan et al., 2007b).

Dry-milling of various waxy hard, soft, and spring wheats gave 3%–20% lower flour yields compared with the wild-type wheat (Yasui et al., 1999; Graybosch et al., 2003; Kim et al., 2003; Chibbar and Chakraborty 2005; Sayaslan et al., 2006). In some cases lower test weight accounted for lower flour yield; in others, the low flour yield could not be fully explained. Low flour yield does not seem to be related to gluten strength but has been attributed to elevated lipid and β-glucan content in waxy wheat (Yasui et al., 1999). These factors could cause poor bolting of flour (Neel and Hoseney 1984) or flaking of middling fractions during grinding on smooth rolls, especially for soft waxy wheat. Waxy wheat also produces higher levels of damaged starch during dry-milling (Bettge et al., 2000; Kim et al., 2003). Graybosch et al. (2003) suggested that waxy wheat endosperm might have higher crystallinity due to lack of amylose, and that could affect dry-milling of waxy wheat. The same flour yield can be obtained from waxy hard wheat as from normal hard wheat by roller-milling on a Buhler experimental mill, but the feed rate of waxy wheat kernels must be reduced or the waxy flour would clog the system during dry-milling (S. Garimella, pers. comm.). To realize the full potential of waxy wheat, flour yield needs to improve. Further research is needed to investigate the cause of reduced flour yield and how to improve dry-milling of waxy wheat.

The wet-milling process can produce two valuable products from waxy wheat flour: vital wheat gluten and waxy wheat starch. However, waxy hard wheat flour is not always suited for the wet-milling process, which fractionates the flour into vital wheat gluten, prime starch, tailings (B-grade starch), and water-soluble components. From 1999 to 2002, researchers at Kansas State University examined the wet-milling of 15 waxy wheat lines (Sayaslan 2002; Sayaslan et al., 2006). Wet-milling was accomplished by the dough-washing method. All 15 samples were mixed to a stiff dough, but only two dough samples remained cohesive during washing under a stream of water to remove starch. During washing, most of the dough balls lost their cohesiveness and became soft and runny. The soft, runny consistency made it impossible to remove the waxy wheat starch from the dough mass. In a similar experiment, flour from six waxy hard wheats, the wild-type HRW cultivar Karl 92, and the partial waxy HW cultivar Trego harvested in 2006 was examined by the dough-washing method (Guan et al., 2007b). Among the six waxy wheat samples, four provided starch recovery comparable to wild-type hard wheat flour. However, the waxy wheat flour dough was more sticky during the early dough-washing stages.

Seven waxy wheat flour samples and two wild-type flour samples also were processed by a flour-dispersion process (Sayaslan 2002; Sayaslan et al., 2006). Except for one soft waxy flour sample, recovery of gluten protein (79%–87%) and their purities (82%–88% protein) were comparable to wild-type flour in the dispersion process. However, purification of prime starch was somewhat more difficult and required more processing. Even with added washing the prime waxy wheat starch still contained 0.4%–0.5% protein compared with 0.3% protein in normal wheat starch.

Exact reasons for generally inferior wet-milling properties of waxy wheat are not known. To date, no published results report on protein structure in waxy hard wheat versus wild-type hard wheat. Some studies suggested elevated pentosan content in waxy flour might interfere with separation and purification of waxy wheat starch (Sayaslan 2002; Sayaslan et al., 2006). Others (Bettge et al., 2000) reported that waxy wheat starch granules are easily damaged by physical forces, which also might explain some of the difficulty in purifying waxy wheat starch granules. Guan et al. (2007b) reported that when a small amount of hemicellulase was added, waxy hard wheat flour dough became stronger, and during dough washing, the dough became less sticky. More research is needed to determine the factors that give good versus poor fractionation and to develop improved wet-milling processes to produce starch and vital wheat gluten from hard waxy wheat.

Waxy wheat starch structure and properties

Waxy wheat starch granules give an A-type x-ray diffraction pattern as do normal wheat starch granules (Hayakawa et al., 1997; Fujita et al., 1998). Waxy wheat starch has a higher degree of crystallinity (37%–44%) than normal wheat starch (29%–36%), presumably due to the higher percentage of amylopectin in waxy wheat starch (Fujita et al., 1998). The amylopectin fraction is responsible for the crystalline phase in wild-type and waxy starch (French 1984).

Thermal properties of waxy wheat starch, including gelatinization and retrogradation, have been determined by differential scanning calorimetry (DSC) and compared with those of normal wheat starch (Yasui et al., 1996; Hayakawa et al., 1997; Fujita et al., 1998; Sasaki et al., 2000; Yoo and Jane 2002). Gelatinization peak temperatures are 2–5 °C higher in waxy wheat starch (Yasui et al., 1996; Fujita et al., 1998; Hayakawa et al., 1997). Hayakawa et al. (1997) reported that the onset gelatinization temperature of waxy wheat starch was 1–3 °C higher than that of normal wheat starch, whereas Yasui et al. (1996) did not detect differences in onset and final temperatures. The endothermic enthalpy (ΔH) of gelatinization of waxy wheat starch was higher than for normal wheat starch (Yasui et al., 1996; Hayakawa et al., 1997; Sasaki et al., 2000). However, when ΔH was based on amylopectin content, little difference was observed between waxy wheat starch and normal wheat starch (Yasui et al., 1996; Hayakawa et al., 1997). Gelatinization temperatures of waxy wheat starch were about 9 °C lower than that of waxy maize starch, whereas the endothermic values (ΔH) of waxy wheat and waxy maize starches were identical (Hayakawa et al., 1997).

Because waxy wheat starch contains no or little amylose, no peak due to the dissociation of the lipid–amylose complex was observed by DSC when waxy wheat starch was heated in water. Starch lipid content of waxy starch, expressed as fatty acid methyl esters (FAME), were 0.07–0.17 g 100 g^{-1}, significantly lower than normal starch (Yasui et al., 1996). Interestingly, waxy wheat endosperm contains more fat than normal wheat starch (1.0% vs. 0.8%), but waxy starch granules contain much less fat (0.2% vs. 1.0%) (Yasui et al., 1996; Lumdubwong and Seib 2001). The lower lipid content in waxy wheat starch permits conversion into a family of high and low molecular weight maltodextrins that do not become rancid during drying and storage (Lumdubwong and Seib 2001).

After gelatinization, starch can retrograde during storage. Hayakawa et al. (1997) found that when stored at 4°C for one to three weeks, waxy wheat starch (20% solids) retrograded significantly less than normal wheat and waxy maize starch. In contrast, Yoo and Jane (2002) reported that after storage at 4°C for a week, the endothermic enthalpy value of melting for retrograded waxy wheat starch (33% solids) was higher (4.6 J g^{-1}) than that of one commercial wheat (3.8 J g^{-1}), and similar to that of another normal wheat ('Centura') starch. The percentage retrogradation ($\Delta H_{retrogradation}/\Delta H_{gelatinization} \times 100$) of waxy wheat, Centura wheat, and the commercial wheat starch was 33.7%, 45.1%, and 35.9%, respectively. In addition, Sasaki et al. (2000) studied the effects of amylose content on gelatinization, retrogradation, and pasting properties of wheat starch and found that the enthalpy value for retrograded starch correlated negatively with amylose content. Guan et al. (2007a) studied waxy wheat starch isolated from waxy hard wheat and found that the enthalpy values of the retrograded waxy starch (25% solids, after storage at 4°C for 7 days) was significantly lower (about 2.0 J g^{-1}) than that of waxy maize starch (5.9 J g^{-1}) and normal wheat starch (4.8 J g^{-1}) as measured by DSC (L. Guan, pers. comm.). More research should be conducted to clarify the cold storage stability of cooked waxy wheat starch and compare it with waxy maize starch and normal wheat starch.

The molecular structure of waxy wheat has been compared with that of normal wheat and other waxy starch types. Using high-performance anion exchange chromatography (HPAEC), Yasui et al. (1996) reported that the chain-length distribution profiles of the amylopectin fraction of waxy wheat lines were similar to their wild-type counterparts. Yoo and Jane (2002) studied the molecular structure of starch isolated from waxy wheat, amylose-reduced wheat, and wild-type HRW wheat, and showed that the peak chain-length among all starch types was at DP 12, as determined by HPAEC. However, average chain-lengths varied between DP 23.5 and 24.9. Waxy wheat amylopectin had no detectable extra-long branch-chains relative to amylopectin of normal wheat starch, but its molecular weight was greater.

Waxy wheat flour exhibits aberrant falling numbers independent of α-amylase activity (Graybosch et al., 2000; Abdel-Aal et al., 2002). Evidence indicates that starch granules in waxy wheat flour are more fragile and subject to breakdown under heat and mechanical shear, and they give a low-consistency flour slurry in the falling number test (Abdel-Aal et al., 2002; Chibbar and Chakraborty 2005). Waxy wheat starch generally displays a lower pasting temperature but higher peak viscosity than waxy maize starch and normal wheat starch as measured by a rapid viscoanalyzer (Kiribuchi-Otobe et al., 1997; Yasui et al., 1999; Grant et al., 2001; Yoo and Jane 2002; Abdel-Aal et al., 2002; Kim et al., 2003). Differences in pasting temperature and peak viscosity between waxy and normal wheat starches were significantly greater than the difference between their maize starch counterparts (Yoo and Jane 2002). Using a hot-stage microscope, Guan et al. (2007b) confirmed that waxy wheat starch granules began to swell at a lower temperature, exhibited higher swelling power, but eventually fragmented into smaller pieces. In contrast, normal wheat starch granules retained a rounded shape even after being heated to 90°C, presumably because the amylose–lipid complex restricted swelling of granules in normal wheat starch (Tester and Morrison 1990). Guan et al. (2007a) heated waxy wheat starch in water on a hot-stage microscope and recorded changes in morphology of waxy wheat granules continuously during heating.

Video-recorded images revealed that waxy wheat starch granules swelled greatly at 60–70 °C and disintegrated into many small fragments at 70–80 °C. No granular residues remained at 90 °C. In contrast, normal wheat starch granules started increasing in size at 57 °C, continued to swell above that temperature, and retained a rounded shape even at 90 °C. The easy fragmentation of waxy wheat granules explains the large breakdown in viscosity, greater susceptibility to α-amylase degradation, and low falling numbers of waxy wheat flour.

Limited work has been done on chemical modifications of waxy wheat starch (Reddy and Seib 1999; Bertolini et al., 2003; Hansen et al., 2007). Cross-linked waxy wheat starch showed higher thickening power compared with cross-linked waxy maize starch (Reddy and Seib 1999). Furthermore, the same modification produced better freeze–thaw stability in waxy wheat starch than in waxy maize starch. Bertolini et al. (2003) reported that waxy wheat starch generally exhibited higher reactivity with a cross-linking reagent (phosphorus oxychloride) than normal or partial waxy wheat starch, but with the substitution reagent propylene oxide, no differences in reactivity were observed among wheat starch types differing in amylose content. Higher levels of phosphoryl chloride were required to effect changes in pasting behavior of waxy maize starch compared with waxy wheat starch (Reddy and Seib 2000). Waxy wheat starch cross-linked at a low level gave a higher pasting consistency compared with cross-linked waxy maize starch. Cross-linked and hydroxypropylated or acetylated waxy wheat starch had similar consistency compared with similarly modified waxy maize starch, but the modified waxy wheat starch had lower gelatinization temperature and better freeze–thaw stability (Reddy and Seib 2000).

Partial waxy wheat has been shown to provide superior quality in certain Asian (wet) noodle applications (Epstein et al., 2002). Blends of waxy and wild-type wheat can give superior quality to Asian noodles and fresh flour tortillas (Guo et al., 2003a,b) and can be used to retard staling of baked products (Bhattacharya et al., 2002). Partial waxy wheat starch also may provide a better substrate than wild-type wheat in the production of modified food starch (Reddy and Seib 2000).

High-amylose wheat

High-amylose wheat and its potential applications are discussed in two reviews by Hung et al. (2006) and Regina et al. (2007). The first high-amylose wheat mutants had an apparent amylose content of 30.8 to 37.4% (proportion of total starch), as determined by colorimetric measurement, amperometric titration, and concanavalin A methods (Yamamori et al., 2000). Amylopectin structure in the amylose mutants was altered. Levels of chains with DP 6–10 increased, while chains with DP 11–25 decreased.

Starch granules from high-amylose wheat mutants have a number of unique features: (i) A-type starch granules are deformed, (ii) the x-ray diffraction pattern does not show any major peaks, and (iii) the DSC thermogram is broad and does not have a peak due to the melting of the amylose–lipid complex (Yamamori et al., 2000). Starch granule morphology was also altered in high-amylose wheat endosperm (>70% amylose) produced by RNA interference (Regina et al., 2006). Under polarized light more than 90% of the starch granules were not birefringent. The majority of high-amylose potato starch granules also show less birefringence under polarized light, and many have irregular surfaces with deep fissures in the center of the granules (Schwall et al., 2000). In contrast, high-amylose maize starch granules (>90%) typically show birefringence in the form of a typical Maltese cross under polarized light (Shi and Jeffcoat 2001). There are some elongated, tubular, or rodlike starch granules in high-amylose maize starch, and those granules do not give a Maltese cross under polarized light.

Dough properties and baking qualities of the high-amylose *sgp-1* wheat mutant were compared to that of both waxy and wild types (Morita et al., 2002; Hung et al., 2005). Bread baked from high-amylose flour had significantly lower loaf volume than bread from wild-type or waxy flour, and interior appearance of loaves of high-amylose and waxy forms were inferior to that of the wild type. However, high-amylose maize starch and flour have been the preferred starting material for producing ingredients with highly resistant starch content and/or total dietary fiber content (Thompson 2000; Shi and Liu 2002; Okoniewska et al., 2006). Large-bowel function was improved in rats fed a diet including high-amylose starch, suggesting this approach has potential to improve human health as well.

WHEAT PHYTOCHEMICALS

Wheat is one of the major grains in the human diet accounting for one-third of the total worldwide grain production. Whole wheat grains are composed of endosperm, germ, and bran. The endosperm makes up about 75%–80% of the grain weight, whereas the germ and bran weights may vary among different wheat cultivars. Whole grains are important components of the human diet as evidenced by their inclusion in the Food Guide Pyramid and US Dietary Guidelines (National Research Council 1989). However, emphasis on whole-grain consumption has been less than on fruits and vegetables. Previous nutritional guidelines placed grains and grain products at the base of the food guide pyramid to emphasize grains or grain product consumption as the foundation for a healthy diet (USDA 2000, 2005).

The health benefit of whole-grain consumption is partly derived from unique phytochemicals in whole grains which complement those in fruits and vegetables when consumed together. For example, various classes of phenolic compounds in grains include phenolic acids, anthocyanidins, quinones, flavonols, chalcones, flavones, flavanones, and amino phenolic compounds (Thompson 1994; Maillard and Berset 1995; Shahidi and Naczk 1995; Lloyd et al., 2000). Some of these phytochemicals, such as ferulic acid and diferulates, are predominantly found in grains but are not present in significant quantities in some fruits and vegetables (Shahidi and Naczk 1995; Bunzel et al., 2001). Grains also contain tocotrienols, tocopherols, and oryzanols (Thompson 1994; Lloyd et al., 2000). These phytochemicals play important structural and defensive roles in grains. The cereal species (or grain type) and cultivar influence the concentration of whole-grain phytochemicals (Adom et al., 2003). The most important groups of phytochemicals found in whole grains can be classified as phenolics, carotenoids, vitamin E compounds, lignans, β-glucan, inulin, and betaine.

Phenolics

Phenolics are compounds possessing one or more aromatic rings with one or more hydroxyl groups, and generally are categorized as phenolic acids, flavonoids, stilbenes, coumarins, and tannins (Liu 2004). Phenolics are the products of secondary metabolism in plants, providing essential functions in the reproduction and growth of the plants, acting as defense mechanisms against pathogens, parasites, and predators, as well as contributing to the color of plants. Phenolic compounds in our diet may provide health benefits associated with reduced risk of chronic diseases.

The concentration of phenolic compounds in whole-wheat grains is influenced by grain type, cultivar, and the part of the grain sampled (Adom and Liu 2002; Adom et al., 2003, 2005). These compounds usually exist as glycosides linked to various sugar moieties or as other complexes linked to organic acids, amines, lipids, carbohydrates, and other phenols. The most common phenolic compounds found in whole grains are phenolic acids and flavonoids.

Phenolic acids can be subdivided into two major groups, hydroxybenzoic acid and hydroxycinnamic acid derivatives. The former include p-hydroxybenzoic, protocatechuic, vannilic, syringic, and gallic acids. They are commonly present in the bound form and are typically components of a complex structure like lignins and hydrolyzable tannins. They can also be found in the form of sugar derivatives and organic acids in plant foods. Hydroxycinnamic acid derivatives include p-coumaric, caffeic, ferulic, and sinapic acids. They are mainly present in the bound form, linked to cell wall structural components such as cellulose, lignin, and proteins through ester bonds. Ferulic acid occurs primarily in the seeds and leaves of plants, mainly covalently conjugated to mono- and disaccharides, plant cell wall polysaccharides, glycoproteins, polyamines, lignin, and insoluble carbohydrate biopolymers. Wheat bran is a good source of ferulic acid, which is esterified to hemicellulose molecules in cell walls. Food processing, such as thermal processing, pasteurization, fermentation, and freezing, contributes to the release of this bound phenolic acid (Dewanto et al., 2002).

Ferulic, caffeic, p-coumaric, protocatechuic, and vannilic acids are present in almost all plants. Chlorogenic acids and curcumin are also major derivatives of hydroxycinnamic acids present in plants. Chlorogenic acids are the ester form of caffeic acids and are the substrate for enzymatic oxidation leading to browning. Curcumin is made of two ferulic acid molecules linked by a methylene in a diketone structure.

The common phenolic acids found in whole grains such as wheat include ferulic acid, vanillic acid, caffeic acid, syringic acid, and p-coumaric acid (Sosulski et al., 1982). Ferulic acid (*trans*-4-hydroxy-3-methoxycinnamic acid) is one of the most common phenolic acids found in wheat grains (Sosulski et al., 1982; Abdel-Aal et al., 2001; Yang et al., 2001; Adom et al., 2003). It is abundant in the aleurone, pericarp, and embryo cell walls of various grains, but occurs only in trace amounts in the endosperm (Smith and Hartley 1983). Ferulic acid and other phenolic acids protect whole grain kernels by providing both physical and chemical barriers through cross-linking carbohydrates, antioxidant activities to combat destructive radicals, and astringency that deters consumption by insects and animals (Hahn et al., 1983; Arnason et al., 1992). Higher concentrations of ferulic acid in grains increase dimerization, which affects the physical and chemical properties of grain structure. Caffeic acid and other ortho-phenolic acids have been linked to suppression of colon cancer in model animal systems (Drankhan et al., 2003; Carter et al., 2006).

Ferulic acid can exist in the free, soluble-conjugated, and bound forms in wheat grains. Bound ferulic acid was significantly higher (>93% of total) than free and soluble-conjugated ferulic acid in maize, wheat, oat (*Avena sativa* L.), and rice (*Oryza sativa* L.) (Adom and Liu 2002). The ratio of free, soluble-conjugated, and bound ferulic acid in maize and wheat was 0.1 : 1 : 100. The order of total ferulic acid content among the tested grains was maize > wheat > oat > rice (Adom and Liu 2002).

Grain ferulic acid content differs among cultivars. For example, Adom et al. (2003) observed significant differences (up to twofold) among 11 wheat cultivars for total ferulic acid, which existed mostly in the bound form (>97%) in all cultivars. Similarly, significant genetic variability in ferulic acid content was reported in durum wheat (threefold) and common wheat (twofold) (Lempereur et al., 1997). In another study, significant differences among wheat cultivars in ferulic acid content corresponded to levels of enzymes involved in phenolic acid metabolism (Régnier and Macheix 1996). Ferulic acid content was similar among cultivars during successive phases of grain development, but final concentrations in wheat were different among cultivars. Ferulic acid content also varied significantly for some wheat cultivars grown in different environments, with about a 13% difference in mean ferulic acid content (Abdel-Aal et al., 2001).

Carotenoids

Carotenoids are nature's most widespread pigments with yellow, orange, and red colors, and they have received substantial attention for their provitamin and antioxidant roles. Carotenoids are classified into hydrocarbons (carotenes) and their oxygenated derivatives (xanthophylls). More than 600 carotenoids have been identified in nature. They occur widely in plants, microorganisms, and animals. Carotenoids have a 40-carbon skeleton of isoprene units. The structure may be cyclized at one or both ends, have various hydrogenation levels, or possess oxygen-containing functional groups. Lycopene and β-carotene are examples of acyclized and cyclized carotenoids, respectively. Carotenoid compounds most commonly occur in nature in the all-trans form. The most characteristic feature of carotenoids is the long series of conjugated double bonds forming the central part of the molecule. This gives them their shape, chemical reactivity, and light-absorbing properties.

Carotenoids perform important functions in plants. They provide pigmentation essential for photosynthesis, reproduction, and protection. They provide yellow color in whole-grain flour. They may also act as antioxidants in lipid environments of many biological systems through their ability to react with free radicals and form less reactive products. Carotenoid radicals are stabilized by delocalization of unpaired electrons over the conjugated polyene chain of the molecule, allowing addition of other functional groups to many sites on the radicals (Britton 1995). Carotenoids are especially powerful against singlet oxygen generated from lipid peroxidation or radiation. β-Carotene, α-carotene, and β-cryptoxanthin have provitamin A activity. Zeaxanthin and lutein are the major carotenoids in the macular region (yellow spot) of the retina in humans.

Carotenoids commonly found in wheat grains are lutein, zeaxanthin, β-cryptoxanthin, β-carotene, and α-carotene (Britton 1995; Adom et al., 2003, 2005). Generally, lutein is the carotenoid present in the highest concentration in wheat, followed by zeaxanthin, and then β-cryptoxanthin. In the study mentioned previously (Adom et al., 2003) involving 11 wheat genotypes—which included a synthetic wheat and a combination of red or white, hard or soft, and winter or spring cultivars—lutein, zeaxanthin, and β-cryptoxanthin contents varied significantly among cultivars. Lutein content varied from 26.41 ± 1.40 to 143.46 ± 6.67 μg 100 g^{-1} grain, amounting to a 5.4-fold difference. Zeaxanthin content varied from 8.70 ± 0.75 to 27.08 ± 0.54 μg 100 g^{-1} grain, amounting to a threefold difference. The β-cryptoxanthin content varied from 1.12 ± 0.13 to 13.28 ± 0.43 μg 100 g^{-1} grain, amounting to a 12-fold difference. The synthetic wheat experimental line in that study, W7985, gave the lowest carotenoid concentrations of any of the genotypes. Such large genotypic differences in carotenoid content may open up new opportunities for breeding wheat cultivars with higher nutritional value (Adom et al., 2003).

Vitamin E

Vitamin E is the generic term used to describe a family of eight lipid-soluble antioxidants with two types of structures, the tocopherols (α-tocopherol, β-tocopherol, γ-tocopherol, δ-tocopherol) and tocotrienols (α-tocotrienol, β-tocotrienol, γ-tocotrienol, and δ-tocotrienol). Their basic structures comprise a 6-hydroxychroman group and a phytyl side chain made of isoprenoid units. The chroman group may be methylated at different positions to generate different compounds with vitamin activity. Both tocopherol and tocotrienol structures are similar except tocopherols contain saturated phytol side chains whereas the tocotrienols have three carbon–carbon double bonds in the phytol side chain.

Vitamin E compounds are found in many foods including whole grains, where they are mostly present in the germ fraction. The concentration of vitamin E compounds in whole grains are: 75 mg kg^{-1} dry weight (DW) total tocopherol in soft wheat and barley (*Hordeum vulgare* L.); 33 to 43 mg kg^{-1} DW β-tocotrienol in wheat; 45 mg kg^{-1} DW γ-tocopherol in maize; and 56 and 40 mg kg^{-1} DW α-tocotrienol in oat and barley, respectively (Panfili et al., 2003). The most important functions of vitamin E in the body are antioxidant activity and maintenance of membrane integrity. The free hydroxyl group on the aromatic ring is responsible for the antioxidant properties. The hydrogen atom from this group can be donated to free radicals, resulting in a resonance-stabilized vitamin E radical. Vitamin E has also been shown to play a role in immune function, in DNA repair, and other metabolic processes (Traber 1999).

Lignans

Lignans are a group of dietary phytoestrogen compounds that are made up of two coupled C6C3 units. The common plant lignans in the human diet include secoisolariciresinol, matairesinol, lariciresinol, pinoresinol, and syringaresinol. Plant lignans are found in a wide variety of plant foods including flax (*Linum usitatissimum* L.) seed, whole grains [maize, oat, wheat, and rye (*Secale cereale* L.)], legumes, fruits, and vegetables (Thompson et al., 1991). Qu et al. (2005) reported a range of secoisolariciresinol content from undetectable to 83 $\mu g\ g^{-1}$ in the wheat cultivars studied. When consumed, plant lignans such as secoisolariciresinol and matairesinol are converted to the mammalian lignans, enterodiol and enterolactone, by intestinal microflora in humans.

The mammalian lignans, enterodiol and enterolactone, have strong antioxidant activity and weak estrogenic activity that may account for their biological effects and health benefits (Thompson et al., 1991, 1996; Wang et al., 1994) and make them unique and very useful in promoting health and combating various chronic diseases. Enterodiol and enterolactone may protect against heart disease and hormone-related breast and prostate cancers (Adlercreutz and Mazur 1997; Johnsen et al., 2004). They may inhibit colon cancer cell growth and induce cell cycle arrest and apoptosis *in vitro* (Qu et al., 2005). Lower cancer rates have been associated with high intakes of dietary lignans (Adlercreutz and Mazur 1997). In a Danish study that followed 857 postmenopausal women, those eating the highest amounts of whole grains had significantly higher blood levels of enterolactone (Johnsen et al., 2004). Blood levels of enterolactone were inversely related to cardiovascular-related and all-cause death in Finnish men (Vanharanta et al., 2003), suggesting the protective effects of lignans against such conditions.

β-Glucan

β-Glucan is a group of linear polymers of glucose molecules connected by a 7:3 ratio of β-(1-4)- and β-(1-3)–linkages. Compared to cellulose with only β-(1-4)-linkages, the β-(1-3)–linkages interrupt β-(1-4)–linkages to make β-glucan more flexible, soluble, and viscous. β-Glucan is commonly found in cell walls of many grains, such as oat, barley, and wheat. The major biological effects of β-glucan include lowering blood cholesterol level, controlling blood sugar, and enhancing the immune system.

The regulating effects of β-glucan on blood cholesterol and sugar levels probably relate to its high viscosity property as a soluble fiber to bind cholesterol and bile acids and to facilitate their elimination from the body. β-Glucan is the main component responsible for the cholesterol-lowering effect of oat bran (Wood 1990; Davidson et al., 1991; Braaten et al., 1994b; Bell et al., 1999). Results from studies using either oat- or yeast-derived β-glucan show typical reductions of 10% for total cholesterol and 8% for LDL cholesterol after 4 weeks of use. This was accompanied by up to a 16% elevation in HDL cholesterol (Uusitupa et al., 1992; Behall et al., 1997; Bell et al., 1999). The FDA has approved the health claim that consumption of about 3 $g\ day^{-1}$ of β-glucan soluble fiber lowers blood cholesterol levels (FDA 1997). β-Glucan had an effect in controlling blood sugar in diabetic subjects, and was helpful in reducing the elevation in blood sugar levels after a meal, probably by delay of gastric emptying, allowing dietary sugar to be absorbed more gradually, or by possibly increasing the tissue sensitivity to insulin (Braaten et al., 1994a; Pick et al., 1996).

Phytosterols

Phytosterols are a collective term for plant sterols and stanols, which are similar in structure to cholesterol but differ in their side-chain groups. Plant sterols have a double bond in the sterol ring, while plant stanols lack a double bond in the sterol ring. The most common plant sterols are sitosterol, campesterol, and stigmaterol, and the most common plant stanols are sitostanol, campestanol, and stigmastanol.

Plant sterols and stanols are white crystalline powders with restricted lipid solubility. Esterification of plant sterols and stanols makes them more lipid-soluble with properties similar to those in normal edible fats and oils. The esterified forms can be easily incorporated into foods such as margarines and salad dressings. Upon intake, the ester is cleaved by lipases in the small intestine, and the plant sterol and stanol residues are released. Esterified plant stanols are the major form used in human clinical trials and in food fortification.

Plant sterols and stanols are found in oilseeds, unrefined vegetable oils, whole grains, nuts, and legumes. The Western diet provides 150 to 400 mg phytosterols per day. For vegetarians a normal diet could have an even higher dietary intake of plant sterols and stanols reaching 500 mg to 1 g (Ntanios 2001).

High intakes of plant sterols or stanols can lower serum total cholesterol and LDL cholesterol concentrations in humans (Miettinen et al., 1995; Hendriks et al., 1999). Phytosterols compete with cholesterol for micelle formation in the intestinal lumen and inhibit cholesterol absorption (Nissinen et al., 2002).

Dietary fiber, inulin, and resistant starch

Dietary fiber has been identified as an important component of a healthy diet. Whole-wheat grains are good sources of dietary fiber. Dietary fiber is defined as the components of plant cells that resist digestion by human digestion enzymes (Trowel and Burkitt 1986). For whole grains, such components include cellulose, hemicellulose, lignin, inulin, resistant starch, and other constituents distributed in the bran and endosperm parts of the grain. Consumption of whole-grain dietary fiber has been associated with reduced risk of chronic diseases.

Women with high cereal fiber intake showed a 34% lower risk of coronary heart disease events when compared to women with low cereal fiber intake (Wolk et al., 1999). Dietary fiber from fruits and vegetables did not exhibit the same effect in this study. Results from other studies have demonstrated the protective role of whole-grain dietary fiber against myocardial infarction (Rimm et al., 1996), coronary heart disease mortality (Pietinen et al., 1996), some cancers (Kasum et al., 2002), weight gain and diabetes (Meyer et al., 2000; Liu et al., 2003; Koh-Banerjee et al., 2004), and insulin resistance and metabolic syndromes (McKeown et al., 2004). Whole-grain dietary fiber may have these effects through multiple physiological mechanisms that include binding and eliminating cholesterol, binding bile acids, modulation of hormonal activity, stimulation of the immune system, facilitating toxicant transit through the digestive tract, production of short chain fatty acids in the colon, dilution of gut substances, lowering caloric content and glycemic index of foods, improving insulin response, providing bulk in foods, and scavenging free radicals. Inulin and resistant starch are the most studied dietary fiber components of whole grains.

Inulin is a mixture of fructose chains (2–60 fructose units) that vary in length and typically have a terminal glucose molecule (Niness 1999). Inulin is a natural storage carbohydrate found in several edible plants including chicory, artichoke, leek, onion, asparagus, wheat, barley, rye, garlic, and banana. The bond between fructose units in inulin is a β-(2-1) glycosidic linkage. Inulin is sometimes added to food products because of its sweet taste and texture. American diets typically provide about 2.6 g of inulin, with wheat (69%) and onions (23%) being primary sources (Moshfegh et al., 1999).

Inulin, when ingested, acts as a prebiotic to stimulate the growth of friendly and healthy intestinal bacteria (probiotics) that support good colon health. Inulin is a preferred food for the probiotics, lactobacilli and bifidobacteria, in the intestine and can stimulate the growth and improve the balance of these friendly bacteria in the colon (Roberfroid 1993; Bouhnik et al., 1994; Gibson et al., 1995; Gibson and Roberfroid 1995; Butel et al., 1997). Bifidobacteria have been shown to inhibit the growth of harmful bacteria, to stimulate the immune system, and to facilitate the absorption of minerals and synthesis of B vitamins. Gibson et al. (1995) reported that when subjects were given 15 g inulin per day for 15 days, the population of bifidobacteria increased by about 10% during that period. At the same time the population of pathogenic bacteria, such as *Clostridium perfringens* and diarrheogenic strains of *Escherichia coli*, decreased. Bifidobacteria digest inulin to produce short-chain fatty acids, such as acetic, propionic, and butyric acids. Acetic and propionic acids serve as energy sources for the liver, while butyric acid has cancer-preventing properties in the colon (Spiller 1994; Reddy et al., 1997). Butyric acid induced differentiation of normal colon cells, and induced the growth arrest and apoptosis in cancer cells (Archer et al., 1998; Avivi-Green et al., 2002; Ruemmele et al., 2003). These cellular activities may be responsible for butyrate's anticancer activity. Inulin may facilitate the absorption of calcium, magnesium, and iron in the colon due to the formation of short-chain fatty acids, such as acetic, propionic, and butyric acids, that affect pH in colon. Calcium and magnesium are important regulators of cellular activity and may help control the rate of cell turnover. High concentrations of calcium may aid formation of insoluble bile or salts of fatty acids and therefore may reduce the damaging effects of bile or fatty acids on colon cells (Topping and Clifton 2001).

Resistant starch resists upper intestinal digestion and passes into the lower intestine to be fermented by microflora in the colon. There are three types of resistant starch: physically trapped starch, resistant starch granules, and retrograded starch (Englyst et al., 1993; Muir et al., 1993). Physically trapped starch is trapped within food matrices that severely prevent or delay their interaction with digestive enzymes in the small intestine. They are commonly found in whole or partly ground grains, seeds, and legumes, and their concentration and distribution is affected by food processing techniques. Resistant starch granules have crystalline regions that are less susceptible to digestion by acid or α-amylase enzymes. Food processing techniques that gelatinize such starches can aid their digestion. Retrograded starch is formed from gelatinized starch that undergoes the process of retrogradation. High-amylose starch retrogrades faster than high-amylopectin starch. High-amylose starch can be retrograded to a form that resists dispersion in water and digestion by α-amylase. Retrograded starch may be generated during food processing (Englyst et al., 1993; Muir et al., 1993).

The physiological functions of resistant starch include improving glycemic response and colon health, providing lower calorie intake, and modulating fat metabolism. Food components that moderate blood sugar levels following food consumption provide health benefits such as reduced risk of developing diabetes, heart disease, and other chronic diseases. Resistant starch is fermented by microflora in the colon to produce short-chain fatty acids (acetate, propionate, and butyrate) to promote colon health (Avivi-Green et al., 2002; Ruemmele et al., 2003). Resistant starch used in products provides bulk and helps decrease the caloric content of foods. Resistant starch consumption has been shown to promote lipid oxidation and metabolism in human subjects (Higgins et al., 2004). In their study, addition of 5.4% resistant starch to the diet significantly increased lipid oxidation by 23% when compared to the control meal with 0% resistant starch. The results suggested replacement of 5.4% of total dietary carbohydrate with resistant starch significantly increased postprandial lipid oxidation and metabolism, and therefore could conceivably decrease fat accumulation in the long-term.

Betaine

Trimethyl glycine, also known as glycine betaine, or simply betaine, is found in a wide range of dietary sources (Fig. 22.1). Betaine was first discovered in sugar beet juice in the 19th century. Since then it has been found in many food sources, with wheat, spinach (*Spinacia oleracea* L.), and sugar beet (*Beta vulgaris* L.) being the richest plant dietary sources. Betaine functions as an osmolyte, protecting cells, proteins, and enzymes from environmental stress. It also functions as a methyl donor in the methionine cycle for conversion of homocysteine to methionine, primarily in the liver and kidneys. Betaine increases glutathione levels in the liver and improves antioxidant status. Low levels of betaine may contribute to various diseases, including coronary, cerebral, hepatic, and vascular diseases (Craig 2004).

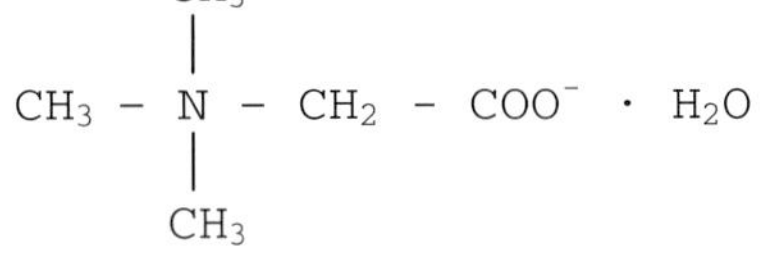

$$
\begin{array}{c}
\quad\quad CH_3 \\
\quad\quad | \\
CH_3 - N - CH_2 - COO^- \cdot H_2O \\
\quad\quad | \\
\quad\quad CH_3
\end{array}
$$

Fig. 22.1 Molecular structure of glycine betaine.

Betaine has been used as a dietary feed supplement in animal nutrition for more than 50 years.

Slow et al. (2005) reported that the primary source of betaine in the New Zealand human diet was from cereal grains and the primary form was glycine betaine. Earlier data from Waggle et al. (1967) shows that wheat betaine is concentrated in wheat bran and wheat germ. However, levels can vary depending on cultivar and possibly environmental factors.

Likes et al. (2007) tested several mill streams for betaine and choline content, and confirmed that the bran and germ fractions are richest in these components. Betaine levels in a mill stream generally reflect the bran content. This data reinforces the nutritional value of whole wheat and suggests that incorporation of whole-wheat products into the diet would reduce risk of many chronic diseases.

INDUSTRIAL WHEAT

Since wheat is so well recognized for its unique protein functional properties, its value as a human food often precludes its consideration for industrial applications. Such applications usually favor underutilized materials, such as co-products from other processes. Wheat co-products from the dry-milling process are more often used in animal feed rather than in industrial processes. Wet-milling of wheat for separation of gluten and starch fractions results in products used in industrial, feed, and food products. However, wheat cultivars are typically not bred for industrial uses.

The 2004 Wheat Summit held at Kansas State University reviewed options to invigorate and expand the wheat industry, including ideas to pursue new uses (Table 22.1). At that time, highest prioritized new uses of wheat were protein quality characteristics, low PPO activity, or high antioxidant content—all related to food applications. Industrial uses were not ranked highly at that time. However, in only 3 years, the landscape has changed and industrial potential is being reexamined. This has come about with the development of waxy wheat cultivars, new findings about the value of waxy starch in fermentation processes, and the rapid growth of ethanol production from cereal starch.

The concept of an industrial wheat "class" involves combining several genetic features that have not been combined before. First, industrial wheat would be targeted to nonfood uses (animal feed and fermentation), so the protein quantity and function would be reduced in importance. Indeed, protein content could be suppressed and quality ignored in this context. Starch content should be maximized, so low-protein cultivars would be favored. Second, high agronomic yield would be critical to achieve sufficient profit per unit area to compete with higher-value food wheat. Finally, the starch should be waxy, since there is growing evidence to suggest that waxy starch will convert to sugars and ferment to ethanol in approximately 70% of the time as normal cereal starch (see the following section: "Soft and waxy wheat for ethanol production"). This offers ethanol producers a significant productivity gain over normal starch, reflecting comparable value with other wheat types or even maize, which offer higher agronomic yield.

Breeding wheat for nonfood uses

Two types of wheat might have application in industrial settings, namely low-phytate and waxy types. Phytic acid or phytate (myo-inositol-1,2,3,4,5,6-hexakisphosphate, or Ins P6) is the most abundant storage form of phosphorus in seeds, but it is indigestible by humans and non-ruminant livestock. High phytate consumption can actually contribute to mineral depletion and deficiency (Raboy 2002). In humans and nonruminant animals, consumption of grain with reduced phytate levels can improve the absorption of essential micronutrients. Guttieri et al. (2004) identified a mutant designated Js-12-LPA that was characterized as having high levels of seed inorganic phosphate (HIP) and reduced phytate. The low-phytic-acid (LPA) trait had little effect on processing quality of hard wheat, but there was some detrimental effect on soft wheat quality attributes (Guttieri et al., 2006a). Grain yield was negatively affected in some genetic backgrounds; the effects, however, were inconsistent, suggesting that additional breeding effort could overcome this deficit (Guttieri et al., 2006b). Application of low-phytate mutants in wheat breeding promises the development of a more nutritious class of feed wheat, with diminished release of phosphorus into the environment.

Waxy wheat also has potential advantages as a feed source. Animals that have a low feed-conversion ratio are considered more efficient users of feed for increasing body mass. Broiler chickens fed a diet of waxy wheat had lower feed conversion ratios than birds fed wild-type samples (Pirgozliev et al., 2002). Kim et al. (2005) found improved starch digestibility in swine fed waxy wheat. Breeding efforts to combine the low phytate and waxy traits should result in a superior type of feed wheat.

Wheat conversion to ethanol

New market for wheat in ethanol industry

United States fuel ethanol production increased 300% from 2000 to 2006, with an annual output of 18.5 billion liters in 2006. Of this, 95% was produced from maize, approximately 4% from grain sorghum [*Sorghum bicolor* (L.) Moench], and less than 1% from wheat and other feedstocks. However, wheat has been used as a major feedstock for fuel ethanol production in Europe (70%) and Canada (15%), and wheat ethanol production will continue to increase in those countries (Smith et al., 2006).

Both wet-milling and dry-grind processes can be used for ethanol production from wheat. Because investment and operating costs for dry-grind plants are about one-half those for wet-milling plants, most of the recently built, small-to-medium size ethanol plants are dry-grind plants. Consequently, the percentage of fuel ethanol produced by the dry-grind process rose from 50% in 2001 to 82% in 2006 (Table 22.2), while ethanol production by the wet-milling process decreased (Renewable Fuels Association 2006, 2007). Because physical and chemical properties of wheat and functional characteristics of wheat components are different from those of

Table 22.2 Percentage of all grain based fuel ethanol produced in the US from two primary processes.

Year	Wet-Milling Process	Dry-Grind Process
	------------------ %	----------------
2001	50	50
2002	40	60
2003	33	67
2004	25	75
2005	21	79
2006	18	82

Source: (Renewable Fuels Association 2006, 2007).

maize, wheat ethanol plants may differ in many ways from maize ethanol plants (Warren et al., 1994).

Although both wheat and maize wet-milling processes have been used for ethanol production, they differ in how the protein is separated from starch. In maize wet-milling, the starting material (maize kernels) is first soaked in sulfur dioxide solution for 24 to 48 hours. The separation procedure following grinding is focused more on germ oil and starch for food markets than on maize protein which is used for animal feed. Products from ethanol production in maize wet-milling plants are typically ethanol, gluten meal, oil, gluten feed, and fiber (Fig. 22.2).

Wet-milling of wheat uses mostly wheat flour as starting material, and the separation process is focused more on wheat gluten. Separation of wheat protein and starch from wheat flour is based on their water insolubility, density, and particle size. Four processing technologies [Martin, Alfa-Laval/Raisio, hydrocyclone, and high-pressure disintegration (HD)] have been used industrially in the wet-milling of wheat flour. The first three processes are more common in North America, and the HD process is more popular in Europe (Cornell and Hoveling 1998; Sayaslan 2004). Therefore, the protein–starch separation section of a wheat wet-milling ethanol plant is very different from that in a maize wet-milling ethanol plant, and may vary between plants. Downstream processes from starch to fuel

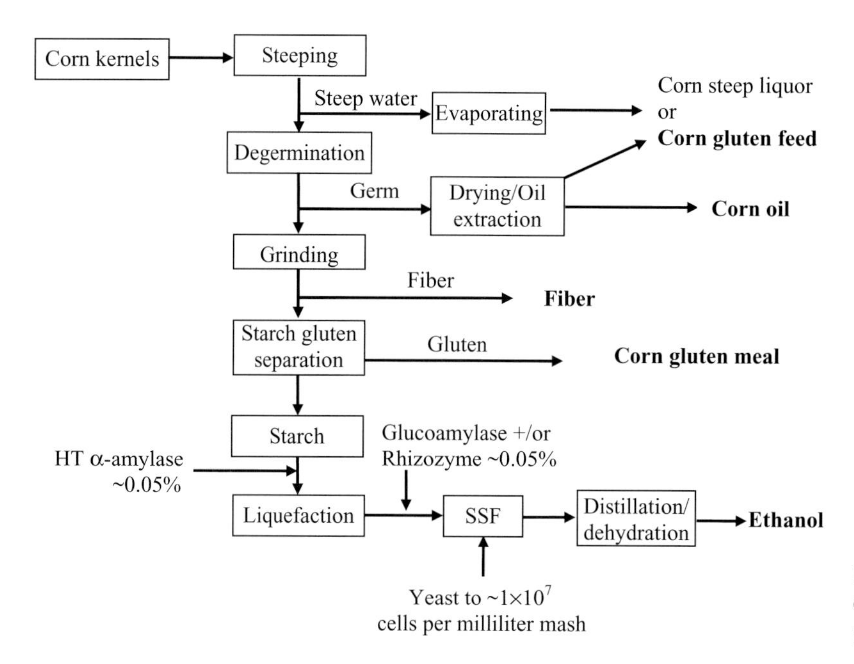

Fig. 22.2 A block-schematic diagram of maize wet-milling processes and products.

ethanol are the same for both maize and wheat wet-milling plants.

With wheat flour as a starting material, the major co-product of wheat wet-milling ethanol plants is vital wheat gluten. Current industrial processes can achieve 80%–85% recovery of wheat protein in the form of wheat gluten, which generally contains 80% protein, 3%–20% carbohydrates (mostly starch), 5%–8% total lipids, 1% pentosans, and 0.5%–1.5% ash on a dry-weight basis (Sayaslan 2004). Compared with the dry-grind process, ethanol yield from the wet-milling process is slightly lower because of the loss of some starch to the wheat gluten. However, if the drying process retains gluten vitality, it can be sold into the higher-value market of vital wheat gluten (\$4,000 t^{-1}), which exceeds the value for gluten meal from the maize wet-milling process. This compensates for the lower ethanol yield (Day et al., 2006).

The dry-grind process of wheat in conventional ethanol plants is similar to that of the maize dry-grind process, which includes grinding, liquefaction, simultaneous saccharification and fermentation (SSF), distillation/dehydration, and separation/concentration of spent grains.

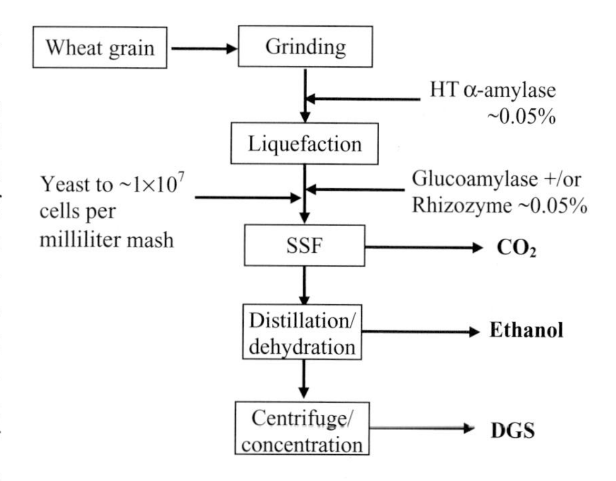

Fig. 22.3 A block-schematic diagram of wheat dry-grind processes and products.

The co-products are distiller's grain with solubles (DGS) and carbon dioxide. Maize DGS contains about 30% protein and is sold as a protein supplement for animal feed. Sales from DGS account for 15%–20% of the annual revenue of a maize dry-grind ethanol plant. Wheat DGS normally has a higher protein content than maize DGS (Wu et al., 1984; Rasco et al., 1987).

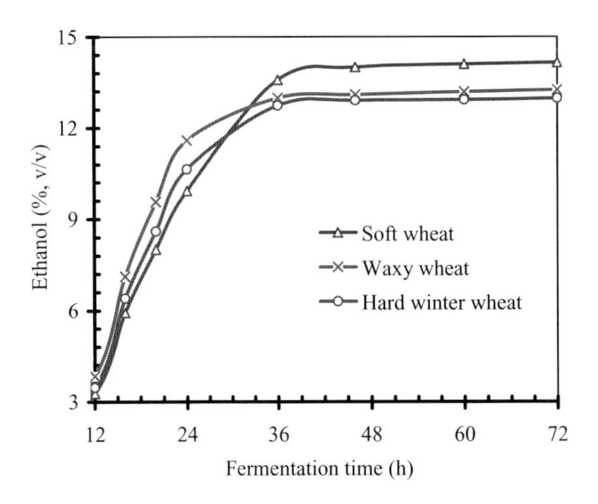

Fig. 22.4 Ethanol yield of different wheat classes.

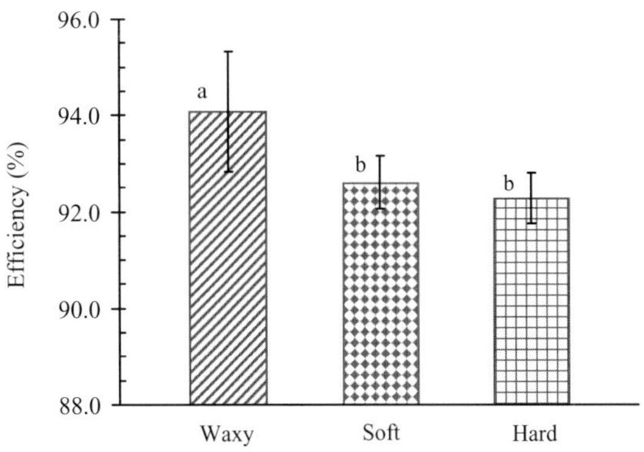

Fig. 22.5 Ethanol fermentation efficiency of different wheat classes.

Figure 22.3 is a simplified schematic diagram of the dry-grind process. The current dry-grind process is mostly designed for ethanol production from maize and may not be well tailored for wheat. Readers are referred to Kelsall and Lyons (2003) and the references listed in that citation.

Soft and waxy wheat for ethanol production

Potential ethanol yield is directly proportional to starch content or total fermentable carbohydrates, especially for dry-grind plants. Unpublished data (D. Wang) on three wheat classes—soft wheat, waxy wheat, and hard winter wheat (Fig. 22.4)—showed that soft wheat had a higher ethanol yield that was consistent with differences in respective starch content (68.7%, 60.4%, and 62.4%). Waxy wheat had higher ethanol yield than hard wheat, though its starch content was two percentage points lower. Waxy wheat starch may be more efficiently converted into fermentable sugars and ethanol than soft and hard wild-type wheat (Fig. 22.5), which is in agreement with fermentation results from other cereal grains (Wu et al., 2007).

Previous research has demonstrated waxy wheat to be a more efficient substrate for ethanol production. Wu et al. (2006) compared fermentation efficiency of waxy wheat to that of normal wheat, maize with varying amylose content, and waxy and nonwaxy sorghum. After prefermentation treatment by cooking at 95 °C, waxy wheat had the highest fermentation efficiency. Amylopectin might serve as a more efficient substrate for α-amylases used in the liquefaction step. The observed higher efficiency of ethanol production from waxy wheat requires commercial verification.

Lower temperatures are required to gelatinize waxy wheat (Graybosch et al., 2000), a necessary first step in the production of starch-derived ethanol. When evaluated in a rapid viscoanalyzer, waxy flour reached peak viscosity at 80 °C, while normal wheat flour attained this point at 95 °C. After cooking to 85 °C, waxy wheat starch essentially had lost all of its structure, making chains readily accessible for digestion by enzymes employed in the digestion process. Thus, lower energy input required for gelatinization of waxy starch represents another possible advantage in biofuel production. Waxy wheat intended for use as a source of biofuel should likely possess low grain protein content, as elevated levels of gluten protein might reduce efficiency of starch extraction, or render difficult the conversion from maize-based substrates to wheat-based.

Feedstock criteria for ethanol production

Feedstock selection may involve different criteria for the wet-milling process of wheat. Since the wet-milling co-product, vital wheat gluten, is a more valuable food ingredient, the protein starch separation process may vary between plants depending on feedstock. Wet-milling wheat ethanol plants should determine the choice of feedstock according to their technical processes for starch-protein separation, availability, protein and starch content, and market values of ethanol and co-products.

Basic standards for feedstock and all other raw materials, and strict adherence to these standards, are critical for smooth operation of the plant and quality assurance. The basic standards for feedstock should at least include limits of moisture content, starch and protein ranges, and tolerant levels for mycotoxins and other contaminants. Special attention should be given to mycotoxins as they tend to be concentrated severalfold in the co-products of vital wheat gluten, gluten meal, gluten feed, germ or germ oil, and DGS. The FDA has set limits for some of the common mycotoxins, such as aflatoxins, vomitoxin, fumonisins, and zearalenone in animal feeds. Wheat is well known for its susceptibility to Fusarium toxins (vomitoxin and zearalenone); therefore, quality standards must be enforced to prevent toxin-contaminated feedstock from entering the plant.

Evaluation techniques for feedstock and co-product quality

Currently, near infrared spectroscopy (NIRS) equipment is widely accepted in the fuel ethanol industry for instant measurement of moisture, protein, and starch content. It plays an important role in an ethanol plant, by allowing quality-assurance personnel to make on-site decisions as feedstock is delivered. Because accuracy of NIRS data largely relies on accuracy of the reference method and the completeness and size of the calibration database, NIRS readings for moisture and protein content are more reliable than starch estimates.

As reported by Wu et al. (2007), different starch sources differ in ethanol production potential. Ethanol yield from cereals with similar starch content can be significantly different. Only laboratory fermentation tests, which may take four to five days, can accurately evaluate ethanol yield potential of a feedstock. Most ethanol plants are equipped with an HPLC system, which can accurately determine the content of important components in the fermentation process, such as glucose and other sugars, ethanol, glycerol, acetic acid, and lactic acid. With these timely data in hand, engineers can closely monitor fermentation and other production processes and take immediate corrective or preventive measures when necessary.

Recent advances in technology

Technology development in the ethanol industry has lowered the production cost of fuel ethanol and made ethanol competitive on the energy market. The most significant technological advances include enhanced feedstock quality (grain yield and ethanol yield), improved characteristics of fermentation organisms, and reduction in energy and water consumption. Because 95% of the fuel ethanol in the US is produced from maize, innovations in ethanol production mostly impact the dry-grind process of maize (Rendleman and Shapouri 2007). Some innovations could be applicable directly, or after some modification, to the dry-grind process of wheat. However, most of the innovations described below are still in the developmental stage and have not been used in commercial production.

Fermentation technology

Most dry-grind plants operate on a batch basis. Some wet-milling plants run continually for several months to a year. Continuous operation provides cost-savings on equipment, maintenance, labor, and yeast; avoids peak utility demand; and achieves higher ethanol yield. Ingledew (2003) and Warren et al. (1994) provide more justification and features of continuous ethanol fermentation.

Development of new fermentation microorganisms with improved features, such as tolerance to higher ethanol concentrations (up to 18%–23% v/v) and higher temperature (up to 60°C), has greatly improved fermentation efficiency and speed in the past decade. Because fiber in DGS is considered a limitation for its application in nonruminant animal feed, many ethanol producers plan to convert the fiber in cereal grain or DGS into ethanol, which can increase ethanol yield and increase the value of DGS as a protein supplement. Genetically engineered yeast (Ho et al., 1998) and bacteria (Alterthum and Ingram 1989; Zhang et al., 1995) have been constructed to co-ferment both C5 and C6 sugars from lignocellulosic hydrolysates. Some pilot-scale and demonstration tests have been conducted using these engineered organisms, with promising results (Dien et al., 2003; Van Maris et al., 2006; Tsantili et al., 2007). Improvements are still needed before being used in commercial ethanol production.

Processing technology

New cereal cultivars with high ethanol yield, such as high total fermentables (HTF) from Pioneer Hi-Bred International, Inc. and high fermentable corn (HFC) from Monsanto Company, have been developed and introduced into the market. These new cultivars increase ethanol yield from 409 L t^{-1} to 432 L t^{-1} (Rendleman and Shapouri 2007). Fractionation of coarsely ground cereals into bran, germ, and endosperm parts not only expands the co-product profile and increases the value of co-products, but also increases the starch content of the feedstock used for ethanol production, thus indirectly enhancing production capacity of the ethanol plant (Wang et al., 1997; Karl et al., 2007).

Most energy consumption in an ethanol plant occurs in the cooking process during liquefaction and distillation, which accounts for up to 25% of the total production cost of ethanol. Raw starch saccharification not only greatly reduces the energy cost of ethanol production but also significantly lowers the content of residual sugar, organic acids, and glycerol in the spent grains. It also increases protein content and improves protein quality of DGS (Lewis et al., 2004; Lewis 2007). The claim was recently made for successful integration of fractionation (the BFrac process) with raw starch saccharification (the BPX process using an acid fungal amylase from Novozymes, Bagsværd, Denmark) in several newly built dry-grind ethanol plants (POET 2006). Integrated technologies have proven commercially viable (Bryan 2005; Lewis 2007) and should lead to higher ethanol yield, increased nutrient quality and flowability of dried DGS, lower plant emissions, and reduced energy costs by up to 15%.

FUTURE PERSPECTIVES

The future of wheat as a crop throughout the world depends on the value it can bring to the grower per unit of land on which wheat is grown. Hard white wheat is recognized as having higher demand in the world market. Wheat producing regions should expand HW production in the future and develop segregation mechanisms throughout the value chain.

Applications of altered amylose-to-amylopectin ratios in wheat flour offers a new area for food ingredients to replace many chemically modified products used today in the food industry. For example, use of high-amylose resistant starch may help formulate foods with lower glycemic index. Reduction of enzyme systems, such as polyphenol oxidase, may offer additional value in food systems required to maintain a stable, light color. Reduction of phytate levels in wheat bran may offer additional value in animal nutrition applications to enhance mineral absorption.

Whole-grain foods are becoming more popular as consumers realize the health value of micronutrients in these foods. More information on micronutrient content of wheat will lead to consumers associating nutrition with grains and will motivate breeders to enhance these components. Future wheat cultivars may have higher levels of target nutrients, and new ingredient concentrates from wheat will be available for food formulation.

Efforts to develop new sources of bioenergy will dominate much of the plant breeding activity in future decades. Wheat may play a role. Waxy wheat starch converts to ethanol faster than other starch. Wheat with lower protein and higher starch contents, in which the starch is waxy, would improve the potential of wheat as a grain feedstock for ethanol. High-starch cultivars have already shown value for animal feed. An industrial type of wheat that combines these features with low phytic acid content and high agronomic yield should be another opportunity for wheat production growth. In addition, conversion of wheat straw through fermentation or thermochemical routes should also contribute to biofuels and other products in the future.

The greatest challenges facing the wheat research community are (i) identifying commercial applications for specialty-wheat cultivars, (ii) convincing wheat growers to produce such cultivars, and (iii) developing a marketing system for the acquisition and sale, by grain handlers, of specialty wheat cultivars. Germplasm is presently available to incorporate traits of waxy endosperm, high-amylose starch, low PPO activity, and low phytate in hard white wheat backgrounds. However, well-differentiated commercial applications have only been developed for hard white wheat. Establishing well-defined, financially rewarding uses for additional types of novel wheat remains the greatest challenge.

REFERENCES

Abdel-Aal, E.S.M., P. Hucl, R.N. Chibbar, H.L. Han, and T. Demeke. 2002. Physicochemical and structural characteristics of flour and starches from waxy and nonwaxy wheats. Cereal Chem. 79:458–464.

Abdel-Aal, E.S.M., P. Hucl, F.W. Sosulski, R. Graf, C. Gillott, and L. Pietrzak. 2001. Screening spring wheat for midge resistance in relation to ferulic acid content. J. Agric. Food Chem. 49:3559–3566.

Adlercreutz, H., and W. Mazur. 1997. Phyto-estrogens and western diseases. Ann. Med. 29:95–120.

Adom, K.K., and R.H. Liu. 2002. Antioxidant activity of grains. J. Agric. Food Chem. 50:6182–6187.

Adom, K.K., M.E. Sorrells, and R.H. Liu. 2003. Phytochemical profiles and antioxidant activity of wheat varieties. J. Agric. Food Chem. 51:7825–7834.

Adom, K.K., M. Sorrells, and R.H. Liu. 2005. Phytochemicals and antioxidant activity of milled fractions of different wheat varieties. J. Agric. Food Chem. 53:2297–2306.

Alterthum, F., and L.O. Ingram. 1989. Efficient ethanol production from glucose, lactose, and xylose by recombinant *Escherichia coli*. Appl. Environ. Microbiol. 55:1943–1948.

Anderson, J.V., and C.F. Morris. 2001. An improved whole-seed assay for screening wheat germplasm for polyphenol oxidase activity. Crop Sci. 41:1697–1705.

Anonymous. 1972. 'Clark's Cream' (PI476305), a farmer-selected cultivar [Online]. Available at http://www.ars-grin.gov/cgi-bin/npgs/acc/search.pl?acid=Clark%27s+Cream (verified 7 Feb. 2008).

Anonymous. 2006. Bakery and snacks: Limagrain launches 'world's first' instant waxy wheat flour [Online]. Available at http://www.bakeryandsnacks.com/news-by-product/news.asp?id=71792&idCat=41&k=limagrain–wheat-flour (verified 29 Jan. 2008).

Anonymous. 2007. White wheat: Supply and disappearance [Online]. Available at http://www.ers.usda.gov/Data/Wheat/YBtable10.asp (verified 29 Jan. 2008).

Archer, S.Y., S. Meng, A. Shei, and R.A. Hodin. 1998. P21WAF1 is required for butyrate-mediated growth inhibition of human colon cancer cells. Proc. Natl. Acad. Sci. USA 95:6791–6796.

Arnason, J.T., J. Gale, B. Conilh de Beyssac, A. Sen, S.S. Miller, B.J.R. Philogene, J.D.H. Lambert, R.G. Fulcher, A. Serratos, and J. Mihm. 1992. Role of phenolics in resistance of maize grain to stored grain insects, *Prostphanus truncatus* (Horn) and *Sitophilus zeamais* (Motsch). J. Stored Prod. Res. 28:119–126.

Avivi-Green, C., S. Polak-Charcon, Z. Madar, and B. Schwartz. 2002. Different molecular events account for butyrate-induced apoptosis in two human colon cancer cell lines. J. Nutr. 132:1812–1818.

Baik, B.K., Z. Czuchajowska, and Y. Pomeranz. 1995. Discoloration of dough for oriental noodles. Cereal Chem. 72:198–205.

Behall, K.M., D.J. Scholfield, and J. Hallfrisch. 1997. Effect of beta-glucan level in oat fiber extracts on blood lipids in men and women. J. Am. Coll. Nutr. 16:46–51.

Bell, S., V.M. Goldman, and B.R. Bistrian. 1999. Effect of beta-glucan from oats and yeast on serum lipids. Crit. Rev. Food Sci. Nutr. 39:189–202.

Bernier, A.-M., and N.K. Howes. 1994. Quantification of variation in tyrosinase activity among durum and common wheat cultivars. J. Cereal Sci. 19:157–159.

Bertolini, A.C., E. Souza, J.E. Nelson, and K.C. Huber. 2003. Composition and reactivity of A- and B-type starch granules of normal, partial waxy, and waxy wheat. Cereal Chem. 80:544–549.

Bettge, A.D. 2004. Collaborative study on L-DOPA—wheat polyphenol oxidase assay (AACC Method 22–85). Cereal Foods World 49:338–342.

Bettge, A.D., M.J. Giroux, and C.F. Morris. 2000. Susceptibility of waxy starch granules to mechanical damage. Cereal Chem. 77:750–753.

Bhattacharya, M., S.V. Erazo-Castrejón, D.C. Doehlert, and M.C. McMullen. 2002. Staling of bread as affected by waxy wheat flour blends. Cereal Chem. 79:178–182.

Bouhnik, Y., B. Flourié, F. Ouarne, M. Riottot, N. Bisetti, F. Bornet, and J. Rambaud. 1994. Effects of prolonged ingestion of fructo-oligosaccharides on colonic bifidobacteria, fecal enzymes and bile acids in humans. Gastroenterology 106:A598-A604.

Braaten, J.T., F.W. Scott, and P.J. Wood. 1994a. High beta-glucan oat bran and oat gum reduce postprandial blood glucose and insulin in subjects with and without type 2 diabetes. Diabet. Med. 11:312–318.

Braaten, J.T., P.J. Wood, and F.W. Scott. 1994b. Oat beta-glucan reduces blood cholesterol concentration in hypercholesterolemic subjects. Eur. J. Clin. Nutr. 48:465–474.

Britton, G. 1995. Structure and properties of carotenoids in relation to function. FASEB J. 9:1551–1558.

Bryan, T. 2005. Changing the game. Ethanol Producer Mag. 8:58–63.

Bunzel, M., J. Ralph, J.M. Martia, R.D. Hatfield, and H. Steinhart. 2001. Diferulates as structural components in soluble and insoluble cereal dietary fibre. J. Sci. Food Agric. 81:653–660.

Butel, M.J., N. Roland, A. Hibert, F. Popot, A. Favre, A.C. Tessedre, M. Bensaada, A. Rimbault, and O. Szylit. 1997. Clostridial pathogenicity in experimental necrotising enterocolitis in gnotobiotic quails and protective role of bifidobacteria. J. Med. Microbiol. 47:391–399.

Carter, J.W., R.L. Madl, and F. Padula. 2006. Wheat antioxidants suppress intestinal tumor activity in min mice. Nutr. Res. 26:33–38.

Chibbar, R.N., and M. Chakraborty. 2005. Characteristics and uses of waxy wheat. Cereal Foods World 50:121–126.

Corke, H., and M. Bhattacharya. 2001. Quality of Asian noodles. p. 57–59. *In* O.K. Chung and J.L. Steele (ed.) Proc. Int. Wheat Quality Conf., 2nd, Manhattan, KS. 20–24 May 2001. Grain Industry Alliance, Manhattan, KS.

Cornell, H.J., and A.W. Hoveling. 1998. The wet-milling of wheat flour. p. 79–125. Wheat: Chemistry and utilization. CRC Press, Boca Raton, FL.

Cox, T.S., J.P. Murphy, and D.M. Rodgers. 1986. Changes in genetic diversity in the red winter wheat regions of the United States. Proc. Natl. Acad. Sci. USA 83:5583–5586.

Craig, S.A.S. 2004. Betaine in human nutrition. Am. J. Clin. Nutr. 80:539–49.

Davidson, M.H., L.D. Dugan, and J.H. Burns. 1991. Hypocholesterolemic effects of beta-glucan in oatmeal and oat bran: A dose-controlled study. JAMA 265:1833–1839.

Day, L., M.A. Augustin, I.L. Batey, and C.W. Wrigley. 2006. Wheat-gluten uses and industry needs. Trends Food Sci. Technol. 17:82–90.

Delwiche, S.R., and R.A. Graybosch. 2002. Identification of waxy wheat by near-infrared reflectance spectroscopy. J. Cereal Sci. 35:29–38.

Delwiche, S.R., R.A. Graybosch, L.E. Hansen, E. Souza, and F.E. Dowell. 2006. Single kernel near-infrared analysis of tetraploid (durum) wheat for classification of the waxy condition. Cereal Chem. 83:287–292.

Demeke, T., C.F. Morris, K.G. Campbell, G.E. King, J.A. Anderson, and H.G. Chang. 2001. Wheat polyphenol oxidase: Distribution and genetic mapping in three inbred line populations. Crop Sci. 41:1750–1757.

Dewanto, V., X.Z. Wu, and R.H. Liu. 2002. Processed sweet corn has higher antioxidant activity. J. Agric. Food Chem. 50:4959–4964.

Dien, B.S., M.A. Cotta, and T.W. Jeffries. 2003. Bacteria engineered for fuel ethanol production: Current status. Appl. Microbiol. Biotechnol. 63:258–266.

Dowell, F.E., E.B. Maghirang, R.A. Graybosch, P.S. Baenziger, D.D. Baltensperger, and L.E. Hansen. 2006. An automated single-kernel near-infrared trait selection system. Cereal Chem. 83:537–543.

Drankhan, K., J. Carter, R.L. Madl, C. Klopfenstein, F. Padula, Y. Lu, T. Warren, N. Schmitz, and D.J. Takemoto. 2003. Antitumor activity of wheats with high orthophenolic content. Nutr. Cancer 47 (2):188–194.

Englyst, H.N., S.M. Kingman, and J.H. Cummings. 1993. Resistant starch: Measurement in foods and physiological role in man. p. 137. *In* F. Meuser, D.J. Manners, and W. Seibel (ed.) Plant polymeric carbohydrates. Royal Soc. of Chem., Cambridge, UK.

Epstein, J., C.F. Morris, and K.C. Huber. 2002. Instrumental texture of white salted noodles prepared from recombinant inbred lines of wheat differing in the three granule bound starch synthase (waxy) genes. J. Cereal Sci. 35:39–50.

FDA. 1997. Food labeling: Health claims—oats and coronary heart disease—rules and regulations. Fed. Regist. 62:3584–3601.

Flintham, J.E., and S.J. Humphray. 1993. Red coat genes and wheat dormancy. Asp. Appl. Biol. 36:135–141.

French, D. 1984. Organization of starch granules. p. 183–212. *In* R.L. Whistler, J.N. BeMiller, and E.F. Paschal (ed.) Starch chemistry and technology. 2nd ed. Academic Press, New York.

Fujita, S., H. Yamamoto, Y. Sugimoto, N. Morita, and M. Yamamori. 1998. Thermal and crystalline properties of waxy wheat (*Triticum aestivum*) starch. J. Cereal Sci. 27:7–13.

Gibson, G.R., E.R. Beatty, X. Wang, and J.H. Cummings. 1995. Selective stimulation of bifidobacteria in the human colon by oligofructose and inulin. Gastroenterology 108:975–982.

Gibson, G.R., and M.B. Roberfroid. 1995. Dietary modulation of the human colonic microbiota: Introducing the concept of prebiotics. J. Nutr. 125:1401–1412.

Gold, C.M., and C.M. Duffus. 1992. The effect of gibberellic acid-insensitive dwarfing genes on pre-maturity α-amylase and sucrose relationships during grain development in wheat. p. 171–177. *In* M.K. Walker-Simmons and J.L. Reid (ed.) Pre-harvest sprouting in cereals. AACC Press, St. Paul, MN.

Grant, L.A., N. Vignaux, D.C. Doehlert, M.S. McMullen, E.M. Elias, and S. Kianian. 2001. Starch characteristics of waxy and nonwaxy tetraploid (*Triticum turgidum* L. var. *durum*) wheats. Cereal Chem. 78:590–595.

Graybosch, R.A. 2005. Development and characterization of waxy winter wheats. p. 113–122. *In* O.K. Chung and G.L. Lookhart (ed.) Proc. Int. Wheat Quality Conf., 3rd, Manhattan, KS. 22–26 May 2005. Grain Industry Alliance, Manhattan, KS.

Graybosch, R.A., G. Guo, and D.R. Shelton. 2000. Aberrant falling numbers of waxy wheats independent of α-amylase activity. Cereal Chem. 77:1–3.

Graybosch, R.A., C.J. Peterson, P.S. Baenziger, L.A. Nelson, B.B. Beecher, D.B. Baltensperger, and J.M. Krall. 2005. Registration of 'Antelope' hard white winter wheat. Crop Sci. 45:1661–1662.

Graybosch, R.A., C.J. Peterson, L.E. Hansen, S. Rahman, A. Hill, and J. Skerritt. 1998. Identification and characterization of U.S. wheats carrying null alleles at the *wx* loci. Cereal Chem. 75:162–165.

Graybosch, R.A., E. Souza, W. Berzonsky, P.S. Baenziger, and O.K. Chung. 2003. Functional properties of waxy wheat flours: Genotypic and environmental effects. J. Cereal Sci. 38:69–76.

Graybosch, R.A., E.J. Souza, W.A. Berzonsky, P.S. Baenziger, D.J. McVey, and O.K. Chung. 2004. Registration of nineteen waxy spring wheats. Crop Sci. 44:1491–1492.

Guan, L., P.A. Seib, and Y.C. Shi. 2007a. Morphology changes in waxy wheat, normal wheat and waxy maize starch granules in relation to their pasting properties. *In* Abstracts, Annu. Meet., AACC, Cereal Foods World 52 (4):A19.

Guan, L., P.A. Seib, and Y.C. Shi. 2007b. Wet-milling of starch from waxy wheat flours. *In* Abstracts, Annu. Meet., AACC, Cereal Foods World 52 (4):A42.

Guo, G., D.S. Jackson, R.A. Graybosch, and A.M. Parkhurst. 2003a. Wheat tortilla quality: Impact of amylose content adjustments using waxy wheat flour. Cereal Chem. 80:427–436.

Guo, G., D.S. Jackson, R.A. Graybosch, and A.M. Parkhurst. 2003b. Asian salted noodle quality: Impact of amylose content adjustments using waxy wheat flour. Cereal Chem. 80:437–445.

Guttieri, M.J., D. Bowen, J.A. Dorsch, V. Raboy, and E. Souza. 2004. Identification and characterization of a low phytic acid wheat. Crop Sci. 44:418–424.

Guttieri, M.J., K.M. Peterson, and E.J. Souza. 2006a. Milling and baking quality of low phytic acid wheat. Crop Sci. 46:2403–2408.

Guttieri, M.J., K.M. Peterson, and E.J. Souza. 2006b. Agronomic performance of low phytic acid wheat. Crop Sci. 46:2623–2629.

Hahn, D.H., J.M. Faubion, and L.W. Rooney. 1983. Sorghum phenolic acids, their high performance liquid chromatography separation and their relation to fungal resistance. Cereal Chem. 60:255–259.

Hansen, L.E., D.S. Jackson, R.A. Graybosch, J.D. Wilson, and R.L. Wehling. 2007. Characterization of chemically modified waxy, partially waxy, and wild-type tetraploid wheat starch. *In* Abstracts, Annu. Meet., AACC, Cereal Foods World 4:A44.

Hayakawa, K., K. Tanaka, T. Nakamura, S. Endo, and T. Hoshino. 1997. Quality characteristics of waxy hexaploid wheat (*Triticum aestivum* L.): Properties of starch gelatinization and retrogradation. Cereal Chem. 74:576–580.

Hendriks, H.F., J.A. Weststrate, T. Van Vliet, and G.W. Meijer. 1999. Spreads enriched with three different levels of vegetable oil sterols and the degree of cholesterol lowering in normocholesterolaemic and mildly hypercholesterolaemic subjects. Eur. J. Clin. Nutr. 53:319–327.

Higgins, J.A., D.R. Higbee, W.T. Donahoo, I.L. Brown, M.L. Bell, and D.H. Bessesen. 2004. Resistant starch consumption promotes lipid oxidation. Nutr. Metab. 1:8–19.

Ho, N.W.Y., Z. Chen, and A.P. Brainard. 1998. Genetically engineered *Saccharomyces* yeast capable of effective cofermentation of glucose and xylose. Appl. Environ. Microbiol. 64:1852–1859.

Hucl, P., and M. Matus-Cadiz. 2002. CDC EMDR-4, CDC EMDR-9 and CDC EMDR-14 spring wheats. Can. J. Plant Sci. 82:411–413.

Hung, P.V., T. Maeda, and N. Morita. 2006. Waxy and high-amylose wheat starches and flour: Characteristics, functionality and applications. Trends Food Sci. Technol. 17:448–456.

Hung, P.V., M. Yamamori, and N. Morita. 2005. Formation of enzyme-resistant starch in bread as affected by high-amylose wheat flour substitutions. Cereal Chem. 82:690–694.

Ingledew, W.M. 2003. Continuous fermentation in the fuel alcohol industry: How does the technology affect yeast? p. 135–143. *In* K.A. Jacques, T.P. Lyons, and D.R. Kelsall (ed.) The alcohol textbook: A reference for the beverage, fuel and industrial alcohol industries. 4th ed. Nottingham Univ. Press, Nottingham, UK.

Johnsen, N.F., H. Hausner, A. Olsen, I. Tetens, J. Christensen, K.E. Knudsen, K. Overvad, and A. Tjonneland. 2004. Intake of whole grains and vegetables determines the plasma enterolactone concentration of Danish women. J. Nutr. 134:2691–2697.

Karl, D.W., C.R. Anderson, A. Hart, and J. Owen. 2007. Corn fractionation method. U.S. Patent 20070184541. Date filed: 27 June 2005.

Kasum, C.M., D.R.J. Jacobs, K. Nicodemus, and A.R. Folsom. 2002. Dietary risk factors for upper aerodigestive tract cancers. Int. J. Cancer 99:267–272.

Kelsall, D.R., and T.P. Lyons. 2003. Grain dry milling and cooling procedures: Extracting sugars in preparation for fermentation. p. 10–21. *In* K.A. Jacques, T.P. Lyons, and D.R. Kelsall (ed.) The alcohol textbook: A reference for the beverage, fuel and industrial alcohol industries. 4th ed. Nottingham Univ. Press, Nottingham, UK.

Kim, W., J.W. Johnson, R.A. Graybosch, and C.S. Gaines. 2003. Physicochemical properties and end-use quality of wheat starch as a function of waxy protein alleles. J. Cereal Sci. 37:195–204.

Kim J., B. Mullan, and J. Pluske. 2005. A comparison of waxy versus non-waxy wheats in diets for weaner pigs: Effects of particle size, enzyme supplementation, and collection day on total tract apparent digestibility and pig performance. Anim. Feed Sci. Technol. 120:51–65.

Kiribuchi-Otobe, C.T. Nagamine, T. Yanagisawa, M. Ohnishi, and I. Yamaguchi. 1997. Production of hexaploid wheats with waxy endosperm character. Cereal Chem. 74:72–74.

Koh-Banerjee, P., M. Franz, L. Sampson, S. Liu, D.R. Jacobs, D. Spiegelman, W. Willett, and E. Rimm. 2004. Changes in whole-grain, bran, and cereal fiber consumption in relation to 8-y weight gain among men. Am. J. Clin. Nutr. 80:1237–1245.

Kruger, J.E. 1996. Noodle quality—what can we learn from the chemistry of breadmaking? p. 157–167. *In* J.E. Kruger, R.B. Matsuo, and J.W. Dick (ed.) Pasta and noodle technology. AACC Press, St. Paul, MN.

Lempereur, I., X. Rouau, and J. Abecassis. 1997. Genetic and agronomic variation in arabinoxylan and ferulic acid contents of durum wheat (*Triticum durum* L.) grain and its milling fractions. J. Cereal Sci. 25:103–110.

Lewis, S.M. 2007. Method for producing ethanol using raw starch. U.S. Patent 20070196907. Date filed: 20 Feb. 2007.

Lewis, S.M., S.E. Van Hulzen, J.M. Finck, and D.L. Roth. 2004. Method for producing ethanol using raw starch. U.S. Patent 20040234649. Date filed: 10 March 2004. Date published: 23 Sept. 2004.

Li, Z., X. Chu, G. Mouille, L. Yan, B. Kosar-Hashemi, S. Hey, J. Napier, P. Shewry, B. Clarke, R. Appels, M.K. Morell, and S. Rahman. 1999. The localization and expression of the class II starch synthases of wheat. Plant Physiol. 120:1147–1156.

Likes, R., R.L. Madl, S.H. Zeisel, and S.A.S. Craig. 2007. The betaine and choline content of a whole wheat flour compared to other mill streams. J. Cereal Sci. 46:93–95.

Lin, W., and G. Vocke. 2004. Hard white wheat at a crossroads. USDA-ERS Publ. WHS-04K-01. Available at http://www.ers.usda.gov/publications/whs/dec04/whs04K01/whs04K01.pdf (verified 29 Jan. 2008).

Liu, R.H. 2004. Potential synergy of phytochemicals in cancer prevention: Mechanism of action. J. Nutr. 134:3479S-3485S.

Liu, S., W.C. Willett, J.E. Manson, F.B. Hu, B. Rosner, and G. Colditz. 2003. Relation between changes in intakes of dietary fiber and grain products and changes in weight and development of obesity among middle-aged women. Am. J. Clin. Nutr. 78:920–927.

Lloyd, B.J., T.J. Siebenmorgen, and K.W. Beers. 2000. Effect of commercial processing on antioxidants in rice bran. Cereal Chem. 77:551–555.

Lumdubwong, N., and P.A. Seib. 2001. Low- and medium-DE maltodextrins from waxy wheat starch: Preparation and properties. Stärke 53:605–615.

Maillard, M.N., and C. Berset. 1995. Evolution of antioxidant activity during kilning: Role of insoluble bound phenolic acids of barley and malt. J. Agric. Food Chem. 43:1789–1793.

McFall, K.L., R.L. Madl, and J. Gilpin. 2003. Hard white wheat: Capitalizing on milling advantages. Kans. Agric. Exp. Stn., Kans. State Univ. Coop. Ext. Serv., Manhattan, KS.

McIntosh, R.A., K.M. Devos, J. Dubcovsky, W.J. Rogers, C.F. Morris, R. Appels, and O.D. Anderson. 2005. Catalogue of gene symbols for wheat. 2005 Supplement. Available at http://wheat.pw.usda.gov/ggpages/wgc/2005upd.html (verified 29 Jan. 2008).

McKeown, N.M., J.B. Meigs, S. Liu, E. Saltzman, P.W. Wilson, and P.F. Jacques. 2004. Carbohydrate nutrition, insulin resistance, and the prevalence of the metabolic syndrome in the Framingham Offspring Cohort. Diabetes Care 27:538–546.

Meyer, K.A., L.H. Kushi, D.R.J. Jacob, J. Slavin, T.A. Sellers, and A.R. Folsom. 2000. Carbohydrates, dietary fiber, incident type 2 diabetes mellitus in older women. Am. J. Clin. Nutr. 71:921–930.

Miettinen, T.A., P. Puska, H. Gylling, H. Vanhanen, and E. Vartiainen. 1995. Reduction of serum cholesterol with sitostanol-ester margarine in a mildly hypercholesterolemic population. N. Engl. J. Med. 333 (20):1308–1312.

Morita, N., T. Maeda, M. Miyazaki, M. Yamamori, H. Miura, and I. Ohtsuka. 2002. Dough and baking properties of high-amylose and waxy wheat flours. Cereal Chem. 79:491–495.

Morris, C.F., and G.E. King. 2007. Registration of 'Waxy-Pen' soft white spring waxy wheat. J. Plant Reg. 1:23–24.

Morris, C.F., J.M. Moffatt, R.G. Sears, and G.M. Paulsen. 1989. Seed dormancy and responses of caryopses, embryos and calli to abscisic acid in wheat. Plant Physiol. 90:643–647.

Morris, C.F., and G.L. Paulsen. 1992. Review: Research on pre-harvest sprouting resistance in hard red and white winter wheats at Kansas State University. p. 113–120. *In* M.K. Walker-Simmons and J.L. Reid (ed.) Pre-harvest sprouting in cereals. AACC Press, St. Paul, MN.

Moshfegh, A.J., J.E. Friday, J.P. Goldman, and J.K.C. Ahuja. 1999. Presence of inulin and oligofructose in the diets of Americans. J. Nutr. 129:14075–14115.

Muir, J.G., G.P. Young, K. O'Dea, D. Cameron-Smith, I.E. Brown, and G.R. Collier. 1993. Resistant starch—the neglected 'dietary fiber'? Implications for health. Diet. Fiber Bibliogr. Rev. 1:33–40.

Nakamura, T., T. Shimbata, P. Vrinten, M. Saito, J. Yonemaru, Y. Seto, H. Yasuda, and M. Takahama. 2006. "Sweet wheat." Genes Genet. Syst. 81:361–365.

Nakamura, T., M. Yamamori, S. Hidaka, and T. Hoshino. 1992. Expression of HMW wx protein in Japanese common wheat (*Triticum aestivum* L.) cultivars. Jpn. J. Breed. 42:681–685.

Nakamura, T., M. Yamamori, H. Hirano, and S. Hidaka. 1993. Decrease of waxy (Wx) protein in two common wheat cultivars with low amylose content. Plant Breed. 111:99–105.

Nakamura, T., M. Yamamori, H. Hirano, S. Hidaka, and T. Nagamine. 1995. Production of waxy (amylose-free) wheats. Mol. Gen. Genet. 248:253–259.

National Research Council. 1989. Food and Nutrition Board: Recommended dietary allowance. 10th ed. National Academy Press, Washington, DC.

NAWG. 2002. New and improved wheat uses audit. Natl. Assoc. Wheat Growers, Washington, DC.

Neel, D.V., and R.C. Hoseney. 1984. Factors affecting flowability of hard and soft wheat flours. Cereal Chem. 61:262–266.

Neuffer, M.G., E.H. Coe, and S.R. Wessler. 1997. Mutants of maize. Cold Spring Harbor Laboratory Press, Woodbury, NY.

Niness, K.R. 1999. Inulin and oligofructose: What are they? J. Nutr. 129:1402S-1406S.

Nissinen, M., H. Gylling, M. Vuoristo, and T.A. Miettinen. 2002. Micellar distribution of cholesterol and phytosterols after duodenal plant stanol ester infusion. Am. J. Physiol. Gastrointest. Liver Physiol. 282:G1009-G1015.

Ntanios, F. 2001. Plant sterol-ester-enriched spreads as an example of a new functional food. Eur. J. Lipid Sci. 103:102–106.

Okoniewska, M.K., W. Bindzus, I. Brown, R.A. Skorge, Y.C. Shi, and T.J. Shah. 2006. Flour composition with increased total dietary fiber, process of making, and uses thereof.

U.S. Patent 2 006 263 503. Date filed: 18 May 2006. Date published: 22 Nov. 2006.

Panfili, G., A. Fratianni, and M. Irano. 2003. Normal phase high-performance liquid chromatography method for the determination of tocopherols and tocotrienols in cereals. J. Agric. Food Chem. 51:3940–3944.

Paterson, A.H., and M.E. Sorrells. 1990. Inheritance of grain dormancy in white-kernelled wheat. Crop Sci. 30:25–30.

Pick, M.E., Z.J. Hawrysh, and M.I. Gee. 1996. Oat bran concentrate bread products improve long-term control of diabetes: A pilot study. J. Am. Dieticians Assoc. 96:1254–1261.

Pietinen, P., E.B. Rimm, P. Korhonen, A.M. Hartman, W.C. Willett, D. Albanes, and J. Virtamo. 1996. Intake of dietary fiber and risk of coronary heart disease in a cohort of Finnish men: The alpha-tocopherol, beta-carotene cancer prevention study. Circulation 94:2720–2727.

Pirgozliev, V.R., S.P. Rose, and R.A. Graybosch. 2002. Energy and amino acid availability to chickens of waxy wheat. Arch. Geflugelkd. 66:108–113.

POET. 2006. Broin companies reveals significant ethanol performance achievement of BPX™ Process, makes technology available outside its group [Online]. Available at http://www.poetenergy.com/news/showRelease.asp?id=47 (verified 23 Feb. 2008).

Qu, H., R.L. Madl, D.J. Takemoto, R.C. Baybutt, and W. Wang. 2005. Lignans are involved in the antitumor activity of wheat bran in colon cancer SW480 cells. J. Nutr. 135:598–602.

Raboy, V. 2002. Progress in breeding low phytate crops. J. Nutr. 132:503S-505S.

Raman, R., H. Raman, K. Johnstone, C. Lisle, A. Smith, P. Matin, and H. Allen. 2005. Genetic and in silico comparative mapping of the polyphenol oxidase gene in bread wheat (*Triticum aestivum* L.). Funct. Integr. Genomics 5:185–200.

Rasco, B.A., F.M. Dong, A.E. Hashisaka, S.S. Gazzaz, S.E. Downey, and M.L. San Buenaventura. 1987. Chemical composition of distillers' dried grains with solubles (DDGS) from soft white wheat, hard red wheat, and corn. J. Food Sci. 52:236–237.

Reddy, B.S., R. Hamid, and C.V. Rao. 1997. Effect of dietary oligofructose and inulin on colonic preneoplastic aberrant crypt foci inhibition. Carcinogenesis 18:1371–1374.

Reddy, I., and P.A. Seib. 1999. Paste properties of modified starches from partial waxy wheats. Cereal Chem. 76:341–349.

Reddy, I., and P.A. Seib. 2000. Modified waxy wheat starch compared to modified waxy corn starch. J. Cereal Sci. 31:25–39.

Regina, A., A.R. Bird, Z. Li, S. Rahman, G. Mann, E. Chanlaud, P. Berbezy, D. Topping, and M.K. Morell. 2007. Bioengineering cereal carbohydrates to improve human health. Cereal Foods World 52:182–187.

Regina A., A.R. Bird, D. Topping, S. Bowden, J. Freeman, T. Barsby, B. Kosar-Hashemi, Z. Li, S. Rahman, and M. Morell. 2006. High-amylose wheat generated by RNA interference improves indices of large-bowel health in rats. Proc. Natl. Acad. Sci. USA 103:3546–3551.

Régnier, T., and J.J. Macheix. 1996. Changes in wall-bound phenolic acids, phenylalanine and tyrosine ammonia-lyases, peroxidases in developing durum wheat grains (*Triticum turgidum* L. var. *durum*). J. Agric. Food Chem. 44:1727–1730.

Rendleman, C.M., and H. Shapouri. 2007. New technologies in ethanol production. USDA-AER Rep. 842. USDA, Office of Energy Policy and New Uses, Washington, DC.

Renewable Fuels Association. 2006. From niche to nation: Ethanol industry outlook 2006 [Online]. Available at http://www.ethanolrfa.org/objects/pdf/outlook/outlook_2006.pdf (verified 23 Dec. 2008).

Renewable Fuels Association. 2007. Building new horizon: Ethanol industry outlook [Online]. Available at http://www.ethanolrfa.org/objects/pdf/outlook/RFA_Outlook_2007.pdf (verified 29 Jan. 2008).

Rimm, E.B., A. Ascherio, E. Giovanucci, D. Spiegelman, M.J. Stampfer, and W.C. Willett. 1996. Vegetable, fruit and cereal fiber intake and risk of coronary heart disease among men. JAMA 275:447–541.

Roberfroid, M. 1993. Dietary fiber, inulin, and oligofructose: A review comparing their physiological effects. Crit. Rev. Food Sci. Nutr. 33:103–148.

Ruemmele, F.M., S. Schwartz, E.G. Seidman, S. Dionne, E. Levy, and M.J. Lentze. 2003. Butyrate induced Caco-2 cell apoptosis is mediated via the mitochondrial pathway. Gut 52 (1):94–100.

Sasaki, T., T. Yasui, and J. Matsuki. 2000. Effects of amylose content on gelatinization, retrogradation, and pasting properties of starches from waxy and nonwaxy wheat and their F_1 seeds. Cereal Chem. 77:58–63.

Sayaslan, A. 2002. Bench-style wet-milling of wheat flour: Development of a test to fractionate a highly sheared flour–water dispersion and its comparison with fractionation by the dough-washing test. PhD diss. Kansas State Univ., Manhattan, KS.

Sayaslan, A. 2004. Wet-milling of wheat flour: Industrial processes and small-scale test methods. Lebensm. Wiss. Technol. 37:499–515.

Sayaslan, A., P.A. Seib, and O.K. Chung. 2005. Wet-milling of flours from red, white and low-polyphenol oxidase white wheats. Food Sci. Technol. Int. 11:243–249.

Sayaslan, A., P.A. Seib, and O.K. Chung. 2006. Wet-milling properties of waxy wheat flours by two laboratory methods. J. Food Eng. 72:167–178.

Schwall, G.P., R. Safford, R.J. Westcott, R. Jeffcoat, A. Tayal, Y.C. Shi, M.J. Gidley, and S.A. Jobling. 2000. Production of very-high-amylose potato starch by inhibition of SBE A and B. Nat. Biotechnol. 18:551–554.

Sears, R.G., T.J. Martin, and J.P. Shroyer. 1993. Arlin hard white wheat. Kans. St. Univ. Coop. Ext. Serv. Publ. L-882, Manhattan. Available at http://www.oznet.ksu.edu/library/crpsl2/l882.pdf (verified 29 Jan. 2008).

Shahidi, F., and M. Naczk. 1995. Food phenolics: Sources, chemistry, effects, applications. Technomic Publishing Co., Inc., Lancaster, PA.

Shi, Y.C., and R. Jeffcoat. 2001. Structural features of resistant starch. p. 430–439. *In* B. McCleary and L. Prosky (ed.) Advanced dietary fibre technology. Blackwell Sci. Ltd., Oxford, UK.

Shi, Y.C., and Y. Liu. 2002. Cereal grains with high total dietary fiber and/or resistant starch content and their preparation thereof. U.S. Patent 20 020 197 373. Date filed: 14 March 2002. Date published: 11 Dec. 2002.

Slow, S., M. Donaggio, P.J. Cressey, M. Lever, P.M. George, and S.T. Chambers. 2005. The betaine content of New Zealand foods and estimated intake in the New Zealand diet. J. Food Compost. Anal. 18:473–485.

Smith, M.M., and R.D. Hartley. 1983. Occurrence and nature of ferulic acid substitution of cell wall polysaccharides in gramineous plants. Carbohydr. Res. 118:65–80.

Smith, T.C., D.R. Kindred, J.M. Brosnan, R.M. Weightman, M. Shepherd, and R. Sylvester-Bradley. 2006. Wheat as a feedstock for alcohol production. HGCA Res. Rev. 61. Available at http://www.ukwheat.com/document.aspx?fn=load&media_id=3194&publicationId=3588 (verified 29 Jan. 2008).

Sosulski, F., K. Krygier, and L. Hogge. 1982. Free, esterified, and insoluble-bound phenolic acids: 3. Composition of phenolic acids in cereal and potato flours. J. Agric. Food Chem. 30:337–340.

Spiller, G.A. 1994. Dietary fiber in health and nutrition. CRC Press, Boca Raton, FL.

Sun, D.J., Z.H. He, X.C. Xia, L.P. Zhang, C.F. Morris, R. Appels, W.J. Ma, and H. Wang. 2005. A novel STS marker for polyphenol oxidase activity in bread wheat. Mol. Breed. 16:209–218.

Tester, R.F., and W.R. Morrison. 1990. Swelling and gelatinization of cereal starches: I. Effects of amylopectin, amylose, and lipids. Cereal Chem. 67:551.

Thompson, L.U. 1994. Antioxidant and hormone-mediated health benefits of whole grains. Crit. Rev. Food Sci. Nutr. 34:473–497.

Thompson, D.B. 2000. Strategies for the manufacture of resistant starch. Trends Food Sci. Technol. 11:245–253.

Thompson, L.U., P. Robb, M. Serraino, and F. Cheung. 1991. Mammalian lignan production from various foods. Nutr. Cancer 16:43–52.

Thompson, L.U., M.M. Seidl, S.E. Rickard, L.J. Orcheson, and H.H. Fong. 1996. Antitumorigenic effect of a mammalian lignan precursor from flaxseed. Nutr. Cancer 26:159–165.

Topping, D.L., and P.M. Clifton. 2001. Short-chain fatty acids and human colonic function: Roles of resistant starch and nonstarch polysaccharides. Physiol. Rev. 81:1031–1064.

Traber, M.G. 1999. Vitamin E. p. 347–361. *In* M.E. Shils, J.A. Olson, M. Shike, and A.C. Ross (ed.) Modern nutrition in health and disease. 10th ed. Williams and Wilkins, Baltimore, MD.

Trowel, H., and D. Burkitt. 1986. Physiological role of dietary fiber: A ten year review. J. Dent. Child. 53:444–447.

Tsantili, I.C., M.N. Karim, and M.I. Klapa. 2007. Quantifying the metabolic capabilities of engineered *Zymomonas mobilis* using linear programming analysis. Microb. Cell Fact. 6:8.

USDA. 2000. Dietary guidelines for Americans. USDA, US Dep. Of Health and Human Services. Publ. 001-000-04681-1. Out of print. US Gov. Print. Office, Washington, DC.

USDA. 2005. Nutrition for your health: Dietary guidelines for Americans. USDA, US Dep. of Health and Human Services. Publ. 001-000-04719-1. US Gov. Print. Office, Washington, DC.

Uusitupa, M.I., E. Ruuskanen, and E. Makinen. 1992. A controlled study on the effect of beta-glucan-rich oat bran on serum lipids in hypercholesterolemic subjects: Relation to apolipoprotein E phenotype. J. Am. Coll. Nutr. 11:651–659.

Vanharanta, M., S. Voutilainen, T.H. Rissanen, H. Adlercreutz, and J.T. Salonen. 2003. Risk of cardiovascular disease-related and all-cause death according to serum concentrations of enterolactone: Kuopio ischaemic heart disease risk factor study. Arch. Intern. Med. 163:1099–1104.

Van Maris, A.J.A., D.A. Abbott, E. Bellissimi, J. van den Brink, M. Kuyper, M.A.H. Luttik, H.W. Wisselink, W.A. Scheffers, J.P. van Dijken, and J.T. Pronk. 2006. Alcoholic fermentation of carbon sources in biomass hydrolysates by *Saccharomyces cerevisiae*: Current status. Antonie Van Leeuwenhoek 90:391–418.

Waggle, D.H., M.A. Lambert, G.D. Miller, E.P. Farrell, and C.W. Deyoe. 1967. Extensive analyses of flours and millfeeds made from nine different wheat mixes: II. Amino acids, minerals, vitamins, and gross energy. Cereal Chem. 44:48–60.

Wang, C., T. Makela, T. Hase, H. Adlercreutz, and M.S. Kurzer. 1994. Lignans and flavonoids inhibit aromatase enzyme in human preadipocytes. J. Steroid Biochem. Mol. Biol. 50:205–212.

Wang, S., K. Sosulski, F. Sosulski, and M. Ingledew. 1997. Effect of sequential abrasion on starch composition of five cereals for ethanol fermentation. Food Res. Int. 30:603–609.

Warren, R.K., D.G. Macdonald, and G.A. Hill. 1994. The design and costing of a continuous ethanol process using wheat and cell recycle fermentation. Bioresour. Technol. 47:121–129.

Wolk, A., J.E. Manson, M.J. Stampfer, G.A. Colditz, F.B. Hu, F.E. Speizer, C.H. Hennekens, and W.C. Willett. 1999. Long-term intake of dietary fiber and decreased risk of coronary heart disease among women. JAMA 281:1998–2004.

Wood, P.J. 1990. Physicochemical properties and physiological effects of the (1–3)(1–4)-beta-d-glucan from oats. Adv. Exp. Med. Biol. 270:119–127.

Wu, J., and B.F. Carver. 1999. Sprout damage and preharvest sprout resistance in hard white winter wheat. Crop Sci. 39:441–447.

Wu, Y.V., K.R. Sexson, and A.A. Lagoda. 1984. Protein-rich residue from wheat alcohol distillation: Fractionation and characterization. Cereal Chem. 61:423–427.

Wu, X., R. Zhao, S.R. Bean, P.A. Seib, J.S. McLauren, R.L. Madl, M.R. Tuinstra, M.C. Lenz, and D. Wang. 2007. Factors impacting ethanol production from grain sorghum in the dry-grind process. Cereal Chem. 84:130–136.

Wu, X., R. Zhao, D. Wang, S.R. Bean, P.A. Seib, M.R. Tuinstra, and M. Campbell. 2006. Effects of amylase, corn protein, and corn fiber contents on production of ethanol from starch-rich media. Cereal Chem. 83:569–575.

Yamamori, M. 2005. High amylose wheat starch and wheat containing the same. US Patent 6 903 255. Date filed: 11 April 2004. Date issued: 7 June 2005.

Yamamori, M., S. Fujita, K. Hayakawa, J. Matsuke, and T. Yasui. 2000. Genetic elimination of starch granule protein, SGP-1, of wheat generates an altered starch with apparent high amylase. Theor. Appl. Genet. 101:21–29.

Yang, F., T.K. Basu, and B. Ooraikul. 2001. Studies on germination conditions and antioxidant content of wheat grain. Int. J. Food Sci. Nutr. 52:319–330.

Yasui, T., J. Matsuki, T. Sasaki, and M. Yamamori. 1996. Amylose and lipid contents, amylopectin structure, and gelatinization properties of waxy wheat (*Triticum aestivum* L.) starch. J. Cereal Sci. 24:131–137.

Yasui, T., T. Sasaki, and J. Matsuki. 1999. Milling and flour pasting properties of waxy endosperm mutant lines of bread wheat (*Triticum aestivum* L.). J. Sci. Food Agric. 79:687–692.

Yoo, S.J., and J.-L. Jane. 2002. Structural and physical characteristics of waxy and other wheat starches. Carbohydr. Polym. 49:297–305.

Zhang, X., C. Colleoni, V. Ratushna, M. Sirghie-Colleoni, M.G. James, and A. Meyers. 2004. Molecular characterization demonstrates that the *Zea mays* gene sugary codes for the starch synthase isoform SSIIa. Plant Mol. Biol. 54:865–879.

Zhang, M., C. Eddy, K. Deanda, M. Finkelstein, and S. Picataggio. 1995. Metabolic engineering of a pentose metabolism pathway in ethanologenic *Zymomonas mobilis*. Science 267:240–243.

Chapter 23
US Wheat Marketing System and Price Discovery

Kim B. Anderson and B. Wade Brorsen

SUMMARY

(1) The wheat marketing system and price discovery process is a complex system with many players. An efficient and mostly transparent system has evolved. A price change at any point in the system creates price changes at all other points. The system is like a spider web where each connecting point depends on all other connecting points.
(2) A futures contract price is used as the base price for negotiating a price between buyers and sellers for the sale and purchase of wheat. Supply and demand expectations are used by pit traders in the commodity exchanges to determine futures contract prices. Owners of wheat may use wheat futures contract prices to determine whether to sell wheat or hedge wheat for future delivery.
(3) Cash prices originate with the end users. Prices paid for wheat between end users and producers are determined by subtracting transportation and handling costs and profit margins for each handler.
(4) Competition between sellers and between buyers, and the transparency of the futures market, ensure that no market player has a significant advantage over another player.

INTRODUCTION

In the book *An Inquiry into the Nature and Cause of the Wealth of Nations*, Adam Smith (1776) introduces the economic principle called "the invisible hand." Smith's economic principle stipulates that entrepreneurs use resources under their control to produce the commodity that will maximize profit. Relating this to a free economy, entrepreneurs react only to their needs (profit) and are not concerned with the good of society as a whole. However, consumer demand alters commodity price relationships such that the good of society is met. The result is that, by reacting to prices and maximizing profit, entrepreneurs maximize consumer satisfaction. Price is the invisible hand that directs resource allocation for commodity production and distribution.

This chapter describes the US wheat marketing system. How prices determine the form, the place, the time, and the player who holds possession of a commodity will be addressed. Discussion will also focus on how the wheat marketing system facilitates the simultaneous discovery of price at each level, and how prices determine the allocation of resources to produce the commodity that is most desired by consumers. Adam Smith's invisible hand involves a complex system of price discovery and product flow from the point of initial production to the point of consumption.

MARKETING SYSTEM

As shown in Fig. 23.1, many players are required to make the wheat marketing system function.

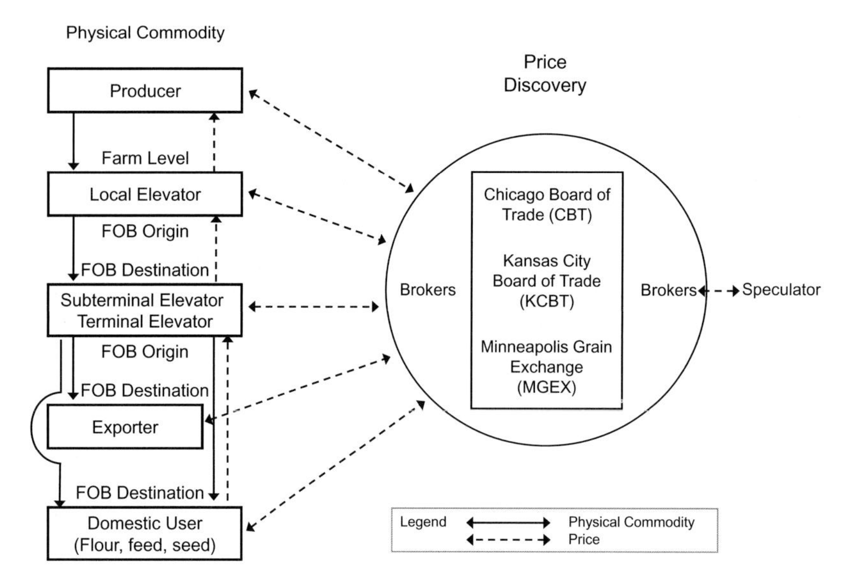

Fig. 23.1 Wheat marketing system participants, price discovery, and product flow.

Production is at one end of the marketing system, while domestic users are at the other end. Some players in the marketing system are not involved in the physical ownership or movement of wheat but are instrumental in the price discovery process. These players include commodity exchange pit traders, brokers, and speculators.

Other players provide the services necessary to assemble, store, and move wheat from the production level to the final user level. These players include producers, elevator managers, merchandisers, and end users. Elevator managers may reside at local elevators (first handlers), subterminals (assembling wheat for large shipments), and terminal elevators that sell to end users (such as exporters or flour millers). Merchandisers are normally employed by elevators to buy and sell wheat and to schedule transportation services, or they may be employed by flour mills to buy wheat and sell flour and wheat mids.

Producers use expected price to allocate land, labor, capital, and management practices to wheat production, thereby to produce the highest expected profit. Expected prices come from commodity exchange futures contract prices, which may be used to estimate harvest prices.

Price discovery and determination

Since prices determine what is produced (whether wheat or other cereal crops), what form (wheat class; hard red winter versus hard white winter) is produced, where it is delivered, and who owns the product throughout the marketing system, the first discussion will be about price discovery. Economic theory suggests that price is determined by supply and demand. This theory is illustrated in Fig. 23.2, where the vertical axis represents price per unit, and the horizontal axis represents quantity per unit of time. As price increases, the quantity demanded declines and the quantity produced increases. The demand curve (the straight line in this example) reflects how much wheat would be consumed (demanded) at each price level. The supply curve reflects how much farmers would produce (supply) at each price level. Price equilibrium is obtained where the demand and supply curves intersect.

The real world is not as simple as the theoretical world. In the real world, both supply and demand are unknown. At any point in time, no one knows precisely how much wheat is in storage, and no one knows how much wheat will be

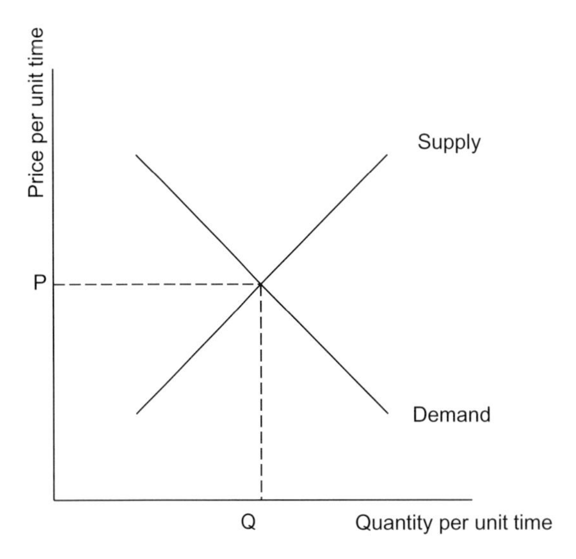

Fig. 23.2 Supply and demand curves determining equilibrium price, quantity demand, and quantity supplied.

consumed. Thus, the market players compile information from across the world and use this information to estimate the supply and demand curves. As new information is obtained and analyzed, the supply and demand estimates are adjusted. As supply and demand estimates change, the intersection point changes for the supply and demand curves, and price changes.

As wheat harvests are completed around the world, the wheat supply curve shifts to the left or to the right. A harvest that exceeds expectations will shift the supply curve to the right, which signals that more wheat will be delivered at a given price. A harvest that is below expectations will shift the supply curve to the left, which signals that less wheat will be delivered at a given price.

Wheat price discovery is a dynamic process. New information continuously causes expectations to change. As expectations change, price bids and offers change. To visualize this process, compare the marketing system and price discovery to a spider web. Any force that changes the tension at any connecting point on the web will have a ripple effect that changes the tension at every other connecting point. When one price relationship changes in the wheat marketing system, all other price relationships adjust, and new equilibrium prices result.

Commodity futures exchanges

In a dynamic marketing and price discovery system where the information flow requires continuous adjustment in prices, central locations are needed to facilitate price discovery. The futures commodity exchanges provide a central location where supply and demand information can be converted into a price. The price discovery process is transparent, and the resulting prices are available to all players in the market almost instantaneously. The success of futures exchanges is due to their convenience and extremely low transaction costs.

Three commodity exchanges are instrumental in US wheat price discovery. Chicago Board of Trade (CBT) wheat contracts are used mostly for soft red winter wheat. Kansas City Board of Trade (KCBT) wheat contracts are used for hard red winter wheat. Minneapolis Grain Exchange (MGEX) wheat contracts are used for spring wheat. Market players who buy and sell the physical commodity base their buy and sell price offers on quoted commodity exchange prices. Other market players are speculators, brokers, and pit traders who rarely own the physical commodity (Fig. 23.1).

Brokers receive sell and buy orders from speculators and owners or potential owners of the physical commodity. Once a broker receives a sell or buy order, the order is communicated to a pit trader to auction off the order, or the order may be sold or bought via the electronic trading system. Pit traders "fill" the sell order by selling to the highest bidder or fill the buy order by buying from the lowest bidder. The resulting price is communicated to the broker who then notifies the buyer or seller. The entire process of selling or buying a futures contract may be completed in less than one minute. The resulting price is recorded and released for public use.

Rights to publish commodity futures contract prices are sold to commercial distributers that sell subscriptions to public users. Commercially distributed commodity exchange prices may occur either in "real time" or by 10-minute delay. The exchanges also release price information on their websites. Exchange-released prices for the public use normally are 10-minute delayed prices.

Sometimes during the wheat marketing year (June 1 through May 31), more wheat may be sold than is demanded by end users. An example would be at harvest. At other times, end users may want to buy more wheat than producers are selling. When these conditions occur, the speculators become the sellers and buyers.

Speculators provide market liquidity, which allows wheat to be sold or bought anytime throughout the marketing year. Speculators buy when more wheat is being sold than is demanded, because when supply is greater than demand, prices are relatively low. Speculators sell when demand is greater than supply, as prices are relatively high. Speculators convert a static market into a dynamic market.

Market players who sell and buy wheat will use the appropriate commodity exchange to establish the cash price. The futures contract price is only part of the price determination process. The futures contract price is quoted for a specified class, quantity, and quality of wheat that must be delivered to a specific location during a designated time period. The futures contract price must be adjusted for quality, time, and location.

A central Oklahoma elevator buying wheat during harvest (June) may use the KCBT July wheat contract price as the basis for the cash purchase price. The KCBT July wheat contract price would be adjusted for location (Kansas City to central Oklahoma), time (July vs. June), and any quality differences between the futures contract specifications and the quality of the purchased wheat. The price difference between any cash price and the KCBT futures contract price is called the basis (basis = cash price − futures contract price). Wheat prices at each level in the marketing system may be established "basis" (based on) a futures contract price.

Hedges

Nearly all wheat that is bought or sold after the farm level (after producers sell the wheat) is hedged with a commodity exchange contract. A hedge is established by taking equal and opposite positions in two markets (the cash market and the futures market). A storage hedge is established by buying the physical commodity (wheat) and selling an equal amount of wheat using wheat futures contracts.

Local elevators normally "back-to-back" wheat purchased from producers to a subterminal elevator, a terminal elevator, or a flour miller. Wheat that is sold back-to-back by a local elevator is then sold by the purchaser using commodity futures contracts (storage hedge). If the local elevator does not back-to-back the wheat, the local elevator normally hedges the wheat by selling an equal amount of wheat futures contracts as a temporary substitute for selling the wheat.

Elevators that sell wheat to end users (flour mills or foreign buyers) may sell (forward contract) more wheat than they own. In this case they buy wheat futures contracts to cover the amount of wheat that was sold. As wheat is purchased, the wheat futures contracts are sold to offset the cash wheat purchases. Any change in the cash wheat price is mostly offset by changes in the futures contract price; thus the price at which the wheat was forward contracted to the end user is protected.

Wheat is rarely delivered against sold futures contracts, nor is delivery taken for bought futures contracts. Rather than delivering or taking delivery of the physical commodity, a hedge is reversed by buying the futures contracts back and selling the physical commodity on the cash market or conversely buying the physical commodity and selling futures contracts. Profit or loss from the futures contract sell or buy is offset by the loss or profit from the cash purchase and cash sale of the physical commodity. Cash and futures contract prices tend to follow the same price pattern. Without the futures market, elevators would have to accept more risk and would have to offer producers lower prices in order to be compensated for the cost of taking greater risk.

Cash price relationships

At each level in the wheat marketing system, two prices are normally found (Fig. 23.1). One is the price for received wheat, and the other is the price for shipped wheat. The delivery price may be referred to as the destination or "FOB destination" price, that is, "free on board" (delivered to) at destination. Producers delivering to the elevator receive the farm-level price, or in trade terms, the destination price.

Elevators selling wheat may sell FOB (origin) or destination. When buyers schedule and pay transportation costs, they offer a FOB origin price. If sellers schedule and pay transportation costs, they receive a destination price. The difference in the FOB origin and FOB destination prices is the cost of transportation.

PHYSICAL FLOW OF WHEAT

Some producers deliver wheat to subterminal or terminal elevators, or they may deliver directly to an end user. However, most producer-owned wheat is delivered to local elevators.

Local elevators mostly sell wheat to subterminal or terminal elevators, which in turn sell the wheat to domestic (flour or feed mills) or foreign (export) users. Larger local elevators may sell directly to domestic users but rarely will local elevators sell to foreign buyers.

There is an increase in producers selling directly to foreign buyers. Recently, a wheat producer in the Oklahoma Panhandle sold a unit train (10,614 tonnes) of hard white wheat to flour mills in Mexico. The producer worked directly with a multinational grain firm to coordinate handling, transportation, and payment.

CASH PRICES

The US wheat marketing system is competitive. Most sellers choose among several buyers, and most buyers choose among several sellers. Factors that determine to whom to sell or from whom to buy include price, availability of transportation equipment, quality specifications, timing, and the relationship between the seller and the buyer. Competition among buyers and competition among sellers keep profit margins competitive with other investments in the economy as a whole. If wheat is to be sold, it will be sold to the highest bidder. If wheat is to be bought, it will be bought from the elevator offering the lowest price.

Since wheat is a storable commodity, and the futures market offers contracts that may be used to protect a price while still owning the cash wheat, sellers may use futures contracts to price wheat and delay selling the wheat on the cash market. Futures market contracts provide an alternative to selling and delivering wheat immediately after buying it. When a merchandiser at a local elevator buys a producer's wheat, they compare the bids from several buyers with the hedge potential on the futures market to determine whether to sell or hedge the wheat. The same futures contract price is available to both sellers and buyers. If a seller or a buyer believes that cash prices are too high or too low relative to futures prices (basis is too strong or basis is too weak), futures contacts provide a temporary alternative to immediately participating in the cash market.

If a level can be identified in the wheat marketing system where the cash price is established, it would be when the end user buys wheat. Both terminal elevators and end users normally base bid and ask prices on a futures contract price. The final price is determined by negotiation. Competition among sellers and buyers impacts the negotiation process.

Terminal elevators subtract transportation and handling costs plus a profit margin to determine the price offered for wheat. Subterminal elevators subtract transportation and handling costs plus a profit margin to determine the price offered for wheat. Local elevators subtract transportation and handling costs plus a profit margin to determine the price offered for wheat. Producers evaluate transportation costs and elevators' price offers to determine where they will deliver and sell wheat.

To determine an offer price, each seller subtracts the transportation costs and all other costs associated with handling the wheat plus a profit margin from the cash price offered by the end user. This calculated offer price is then adjusted to match the elevator competitor's offer prices. The price paid for farmer-owned wheat is the end user price minus all transportation and handling costs plus the profit margin withheld by each handler.

QUALITY DISCOUNTS AND PREMIUMS

Most wheat prices are quoted for US No. 1 grade wheat (grades include US no. 1, 2, 3, 4, 5, and Sample Grade). Grades do not account for every quality factor that affects wheat price, because the end user may pay a premium for a factor such as baking quality, while a producer will not typically receive a premium for these characteristics. Except in years where substantial high-protein wheat is available, protein premiums may be available. Numerous other factors affect quality such as falling number but are not used to value producers' wheat because they are too costly to measure at the farm level. Wheat buyers do have some information about the baking or functional quality of wheat from a specific region of the country in a specific year. The value of other functional characteristics is at least partially reflected in prices. If a region of the country began growing low or high quality wheat, it would be reflected in prices. The functional quality of wheat is maintained partly by agricultural experiment stations and private companies choosing not to release wheat cultivars with inappropriate quality characteristics for the expected market class and region of production. Whereas the market works well in determining the overall price, it may not work as well in rewarding quality.

FUTURE PERSPECTIVES

Changes in the wheat marketing system will result from entrepreneurs finding more cost-effective methods to buy, sell, handle, and transport wheat or from changes in consumer demand. Recently, a grain merchandising cooperative was established by seven local grain cooperatives consisting of 23 locations in one state. The merchandising cooperative has grown to 13 grain cooperatives controlling 66 locations in three states. Merchandising efficiency was improved (costs reduced) by the ability to originate larger shipments meeting more specific quality characteristics. Transportation costs were reduced by shipping wheat from the location(s) storing wheat, meeting the contracted quality characteristics with the lowest transportation costs.

Supply and demand changes may also create changes in the marketing system. Research shows that flour millers may prefer hard white over hard red wheat but not enough to pay a premium. When hard white sprouting-resistant cultivars are developed, allowing hard white to compete more widely with hard red wheat in production, flour millers may buy the hard white wheat before buying hard red wheat, which would raise the price of hard white relative to hard red wheat.

The market searches for more efficient and lower-cost (higher profit) methods to originate and deliver wheat to meet consumer demand. Efficiency, costs, profit, and demand are reflected in prices. Adam Smith's invisible hand (price) will continue to ensure that societal needs are met.

REFERENCE

Smith, A. 1776. An inquiry into the nature and cause of the wealth of nations. Printed for A. Strahan and T. Cadell. Methuen and Co., Ltd., London. Available at http://www.econlib.org/library/Smith/smWN.html (verified 23 Dec. 2008).

Index

Page numbers followed by "f" indicate figures.
Page numbers followed by "t" indicate tables.
"Plate" refers to color illustrations grouped at end of book.